2022 24th European Conference on Power Electronics and Applications (EPE'22 ECCE Europe)

Hanover, Germany
5-9 September 2022

Pages 2689-3361

IEEE Catalog Number: CFP22850-POD
ISBN: 978-1-6654-8700-9

Copyright © 2022, The European Power Electronics and Drives Association
All Rights Reserved

*** This is a print representation of what appears in the IEEE Digital Library. Some format issues inherent in the e-media version may also appear in this print version.

IEEE Catalog Number: CFP22850-POD
ISBN (Print-On-Demand):   978-1-6654-8700-9
ISBN (Online):   978-9-0758-1539-9

Additional Copies of This Publication Are Available From:

Curran Associates, Inc
57 Morehouse Lane
Red Hook, NY 12571 USA
Phone: (845) 758-0400
Fax: (845) 758-2633
E-mail: curran@proceedings.com
Web: www.proceedings.com

2022 24th European Conference on Power Electronics and Applications (EPE'22 ECCE Europe)

Hanover, Germany
5-9 September 2022

Pages 2689-3361

IEEE Catalog Number: CFP22850-POD
ISBN: 978-1-6654-8700-9

TABLE OF CONTENTS

Dynamic Power Analysis of Inverter-Fed Drives Based on the Switching Period of the Power Electronics 1
    Alexander Stock

Stability Analysis in an Inverter-Dominant Microgrid Facing In-Rush Current of an Induction Machine 11
    Nastaran Fazli, David Hammes, Sidney Gierschner, Hans-Gunter Eckel

Self-Oscillating Capacitive Power Transfer with Multiple Receiver Capability and Coupling Path Adaption 22
    Norbert Seliger

An Electrically Driven Gas Compressor for Hydrogen Refueling Stations with Active Power Smoothing 30
    Alfred Rufer

Unsymmetrical Fault Behavior of PLL Based Grid-Connected Converters 39
    Philipp Hackl, Ziqian Zhang, Robert Schuerhuber

Stability Assessment and Optimization of MMC Energy Balancing for Drive Applications at Standstill using an Averaging Approach 49
    Qiuye Gui, Hendrik Fehr, Albrecht Gensior

Turn-On Losses Optimization for Medium Power SiC MOSFET Half-Bridge Module 59
    Pham Ha Trieu To, Felix Kayser, Hans-Günter Eckel

Oscillation Damping in a 500kW Hybrid Si/SiC Three-Level ANPC Inverter with Decoupling Capacitor 70
    Pham Ha Trieu To, Hans-Günter Eckel

Multi Busbar Sub-Module Modular Multilevel STATCOM with Partially Rated Energy Storage Configured in Sub-Stacks 80
    Chuantong Hao, Wenhao Ma, Michael Merlin, Paul Judge, Stephen Finney

Three-Phase ZVS Inverter with Variable and Fixed Frequency Operation Based on GaN Semiconductors 88
    Benedikt Kohlhepp, Michael Lutsch, Thomas Dürbaum

Influences of Conductor Positions and Fast Rising Impulse Voltages on the Line-End Coil Based on a Three-Phase High-Frequency Model 97
    Ting Helmholdt-Zhu, Volker Grabs

Simulation Tool for Optimization of Digital Active Gate Drive Sequence using Genetic Algorithm 108
    Hajime Takayama, Shuhei Fukunaga, Takashi Hikihara

Analysis of Balancing Algorithms for Quasi- Two/Three-Level Single Phase Operation of a Flying Capacitor Converter 115
    Stefan Mersche, Markus Bayer, Kai Rickert, Marc Hiller

Instability in Active Balancing Control of Dc Bus Voltages in VSC Converters Interconnected via Multi-Winding Transformers 125
    Duro Basic, Sami Siala

Online Learning-Based Islanding Detection Scheme for Grid-Connected Systems.. 135
Mohammed Ali Khan, V S Bharath Kurukuru, Rupam Singh

Difference in the Design Process of LCL Filters for Grid Connected VSI When using SiC/GaN
Instead of Si Semiconductors 145
Dennis Kampen, Lukas Fräger, Niklas Badenhop, Arthur Mambetow

Analysis and Design of a Resonant DC/DC Transformer in Modular Operation.. 152
Abraham López, Manuel Arias, Pablo F. Miaja, Arturo Fernández

Predictive Braking Algorithm for Soft Starter Driven Induction Motors.. 160
Hauke Nannen, Heiko Zatocil, Gerd Griepentrog

Ambient Electromagnetic Energy Harvesting Circuit using Rectennas Manufactured with
Stereolithography Resin 169
*Xuan Viet Linh Nguyen, Tony Gerges, Jacques Verdier, Philippe Lombard, Michel Cabrera,
Bruno Allard, Jean-Marc Duchamp, Philippe Benech*

Boost/Buck-Boost Based Grid Connected Solar PV Micro-Inverter with Reduced Number of
Switches and Having Power Decoupling Capability 178
Arup Ratan Paul, Arghyadip Bhattacharya, Kishore Chatterjee

Operation and Selection of Multilevel Power Converters for Doubly Fed Induction Generator-
Based Wind Turbines 187
Kapil Jha, Joseph Banda, Hridya I, Arvind Tiwari

A Detailed View on the Trapezoidal Operation for MMC Type Braking Chopper in Medium
Voltage Application... 195
Patrick Hofstetter, Viktor Hofmann, Dennis Karwatzki

Influence of Operating Frequency on High-Power Medium-Voltage Medium-Frequency
Transformers 203
Thomas B. Gradinger, Ralph M. Burkart, Marko Mogorovic

Output Power Characteristics of Isolated Secondary-Resonant SAB DC-DC Converter for Output
Voltage Variation 213
Shota Yamashita, Kohei Budo, Takaharu Takeshita

Hardware and Control Design of a High Precision Modular Power Converter Based on GaN
Technology for Particle Accelerator Magnets... 223
*Thomas Margreiter, Ivan De Cesaris, Maurizio Incurvati, Sebastien Pelletier, Martin
Schiestl, Ronald Stärz*

Battery Cycler to Generate Open Li-Ion Cell Aging Data and Models... 232
Matthias Luh, Thomas Blank

Function Blocks of a Highly-Integrated All-In-GaN Power IC for DC-DC Conversion 242
Michael Basler, Richard Reiner, Stefan Moench, Patrick Waltereit, Rüdiger Quay

Comparison of Redundancy Requirements for Modular Multilevel Converter Considering
Manufacturer Reliability Inputs and Mission Profile 251
Diego Velazco, Guy Clerc, Emmanuel Boutleux, Francois Wallart

Impact of Insulation and Cooling on Performance Due to Reliability-Oriented Design of Electrical
Machines 261
Lucas Vincent Hanisch, Jonas Franzki, Markus Henke

Long Switching Horizon Model Predictive Controller for High-Speed Integrated Modular Motor Drives 268
Martin Schiestl, Maurizio Incurvati, Ronald Starz, Markus Schmid

Standalone Power Management System for Flexible Piezo Electric Nano Generators (PENG) Based on the Co-Polymer P(VDF:TrFE) 279
Alexander Wölk, Mahmoud Shousha, Shashank Shekhawat Singh, Martin Haug, Lorandt Fölkel, Michael Brooks, Asier Alvarez, Andreas Petritz, Philipp Schäffner, Jonas Groten, Andreas Tschepp, Barbara Stadlober

Analysis and Estimation of Neutral-Point Voltage Balancing Ability of an Optimized Balancing Algorithm for Grid Connected Active-NPC Converter 289
Joseph Banda, Kapil Jha, Hridya Ittamveettil, Arvind Kumar Tiwari, Fernando Ramirez

A Direct Model Predictive Control Strategy of Back-To-Back Modular Multilevel Converters using Arm Energy Estimation 297
Akseli Hakkila, Antonios Antonopoulos, Petros Karamanakos

Study on Commutation Loop Inductance and Current Distribution to DC-Link Capacitors in a GaN Half-Bridge 307
Benedikt Kohlhepp, Samuel Faber, Jeremias Kaiser, Thomas Dürbaum

Cooperative Control of Online Impedance Spectroscopy Monitoring Method and Maximum Power Point Tracking Method for Photovoltaic Panels 315
Xin Wang, Zhixue Zheng, Michel Aillerie, Alexandre De Bernardinis, Jean–paul Sawicki, Marie-Cécile Péra, Daniel Hissel

Benefits of Switching from Si to SiC Modules with Further Converter Optimization 325
Antxon Arrizabalaga, Mikel Mazuela, Iosu Aizpuru, June Urkizu, Jon Aztiria

On the Reduction of Output Capacitance in Two-Level Three Phase PFC Boost Rectifier for Pulsating Loads 335
Tania C. Cano, Douglas Pedroso, Alberto Rodríguez, Ignacio Castro, Diego G. Lamar

Cognitive Insights into Metaheuristic Digital Twin Based Health Monitoring of DC-DC Converters 344
Abdul Basit Mirza, Kushan Choksi, Sama Salehi Vala, Krishna Moorthy Radha, Madhu Sudhan Chinthavali, Fang Luo

A Three-Phase Isolated Secondary-Resonant Single-Active-Bridge DC-DC Converter with a Delta-Star Connected Transformer 351
Atsushi Nishio, Kohei Budo, Mai Van Tuan, Takaharu Takeshita

A Novel Concept to Optimize Core Loss in Planar Magnetic Based on an Unbalanced-Flux-Approach 361
Sobhi Barg, Kent Bertilsson, Grover Torrico

Model Reduction using Singular Perturbation Methods for a Microgrid Application 370
Lasse Gnärig, Albrecht Gensior, Saioa Burutxaga Laza, Miguel Carrasco, Carsten Reincke-Collon

Drive Level Parameter Identification of an Induction Motor 380
Andreas Bünte, Alex Hald, Andreas Kirsch

Impedance Stability of Single-Phase LCL Grid-Connected Voltage Source Inverters with Wideband Gap Devices Under Different Control Approaches 390
Ramy Ali, Terence O'Donnell

Design and Modulation Optimization of an MMC Based Braking Chopper 400
  Viktor Hofmann, Patrick Hofstetter

Modeling the Arrangement of Drill Holes for Orthogonal Biasing in Controllable Inductors for Power Electronic Converters 411
  Jonas Pfeiffer, Christoph Drexler, Pierre Küster, Peter Zacharias, Michael Schmidhuber

A Sectorized FCS-MPC Transformerless SST for Power Transmission Application 421
  Gabriel Gaburro Bacheti, Renner Sartório Camargo, Emilio José Bueno, Marco Liserre, Lucas Frizera Encarnação

Inductance Estimation for Square-Shaped Multilayer Planar Windings 432
  Theofilos Papadopoulos, Antonios Antonopoulos

Cost and Efficiency Considerations in On-Board Chargers 442
  Marija Jankovic, Christian Felgemacher, Kevin Lenz, Aly Mashaly, Abdelmouneim Charkaoui

A Novel Combined Control of Ground Current and DC-Pole-To-Ground Voltage in Symmetrical Monopole Modular Multilevel Converters for HVDC Applications 451
  Pablo Briff, Amit Kumar

A PFC Boost Converter with Reduced Switching Losses Operating at a Fixed Switching Frequency 459
  Burkhard Ulrich

Predictive Control of Power Electronics Autotransformer for Mitigating Three-Phase Grid Current Unbalance in Railway Supply Systems 468
  Tabish Nazir Mir, Faysal Hardan, Masood Hajian, Tamer Kamel, Pietro Tricoli

Parameter Sensitivity of a MRAS-Based Sensorless Control for AFPMSM Considering Speed Accuracy and Dynamic Response at Multiple Parameter Variations 474
  Michael Brüns, Christian Rudolph, Tankred Müller

Synchronization Stability of a Grid Forming Converter Under the Effect of Current Limit in Voltage Dips with VI Based Current Limiting Method: Analysis and Solution 484
  Siam Hasan Khan, Markel Zubiaga Lazkano, Pedro Izurza, Alain Sanchez-Ruiz, Javier Cañas Aceña, Joseba Arza

Analytic Calculation of Touch and Leakage Currents of Non-Isolated EV Chargers using a Fast Common Mode Calculation Method and Non-Ideal Passive Component Models 493
  Christian Stutz, Sebastian Nielebock, Martin März

Triple-Phase-Shift Controlled Dual Active Bridge Converter with Variable Input Voltage in Auxiliary Railway Supply 504
  Martin Scohier, Olivier Deblecker, Carlos Valderrama

Loss Characterization Methodology for Soft Magnetic Nano-Crystalline Tape Materials in Coupled Inductors 514
  David Bohne, Valentin Wagner, Patrick Deck, Christian P. Dick

Substitution of Nanocrystalline Toroid by Laminated Ferrite Toroid in the Application of a Common-Mode Choke 525
  Lukas Reißenweber, Fritz Wohlrath, Alexander Stadler

Direct Active Stabilization of the DC-Link in Voltage-Source Converters 534
  Matthieu Bertin, Mohamad Koteich

Hardware-In-The-Loop Control of a Modular Induction Motor Drive in Power Electronics Education 544
Jens Peter Kaerst

Design and Efficiency Analysis of an LCL Capacitive Power Transfer System with Load-Independent ZPA 554
Francesco Musolino, Ahmed Abdullah, Mario Pavone, Fabio Ferreyra, Paolo Crovetti

A Pulse Generator Based on Transmission Line Transformer for Insulation Aging Test 562
Xiao Yu, Khanh-Hung Nguyen, Peter Zacharias

Design of a Single-Phase Common Mode and Differential Mode Inductor for Interleaved Converters 572
Jonathan Robinson, Gopal Mondal, Stefan Hänsel, Matthias Neumeister

Steady-State Analysis and Comparison of SSFB, SDFB and DSFB MMC-Based STATCOM 582
Mohamed Moez Belhaouane, Pierre Vermeersch, François Gruson, Pierre Rault, Sébastien Dennetiere, Xavier Guillaud

Current Distribution Control in Parallel Connected Power Converters with Continuous Output Voltage 593
Sabrina Ulmer, Andreas Brunner, Philipp Czerwenka, Gernot Schullerus, Ertugrul Sönmez

Optimized Pulse Pattern with Half-Wave Symmetry for 5-Level Converter 604
Jonas Weires, Pedro Leal Dos Santos, Steven Liu

Characterization of Si-IGBT Crosstalk with a Concentration on Power Circuit Parasitic Elements and the Device Operation Point 614
Amir Azam Rajabian, Sadegh Mohsenzade, Javad Naghibi, Kamyar Mehran

Impact of Higher Current Harmonics on Component Current Stress and Conduction Losses of Half-Bridge-Series-Resonant-Converters in Discontinuous Conduction Mode for High-Power Applications 624
Daniel Haake, Anton Grodnichev, Fabian Schnabel, Marco Jung

Control of a Zero-Voltage Switching Isolated Series-Resonant Power Circuit for Direct 3-Phase AC to DC Conversion 634
Yusuf Kosesoy, Remco Bonten, Henk Huisman, Jan Schellekens

Design of a Robust Voltage Control for Inverters with LC Filter Based on the Internal Model Control 641
Frederik Stallmann, Axel Mertens, Lukas Fräger

Influence of Power Semiconductor Device Variations on Pulse Shape of Nanosecond Pulses in a Solid-State Linear Transformer Driver 651
Raffael Risch, Anliang Hu, Jürgen Biela

Optimal Design of Integrated Motor Drives - Comparison of Topologies (2L/3L/Modular), PWM Variants, and Switch Technologies (Si/SiC/GaN) 662
Thilo Bringezu, Jürgen Biela

Distribution Transformer Voltage Control using a Single-Phase Matrix Converter 673
Rui Wang, Henk Huisman, Korneel Wijnands

Influence of Carrier-Based PWM Techniques on the Common-Mode Voltage and Common-Mode Current of Six-Phase Full-Bridge Inverters 681
Juris Arrozy, Esin Ilhan Caarls, Henk Huisman, Jorge L. Duarte, Lorenzo Ceccarelli

Mitigation of Dead-Time Effects on Transient DC Bias Elimination in Dual Active Bridge Link Current................... 689

    MK Kharabela Mohanta, Dipankar De, Silpashree Sahu, Alberto Castellazzi

Generalized Automated Tool for Analysis and Design of Multiphase Coupled Inductor Buck Converters 698

    Rana Asad Ali, Mahmoud Shousha, Martin Haug

Experimental Study of a Directly Oil-Cooled Electrical Machine for a Full-Electric Vehicle by using Low Viscosity Oil................... 709

    Huihui Xu, Georg Tobias Götz, Shimin Zhang, Rik W. De Doncker

Development of a Family of High Voltage Gain Step-Up Multi-Port DC-DC Converters for Fuel Cell-Based Hybrid Vehicular Power Systems 719

    Pouya Zolfi, Sina Vahid, Ayman El-Refaie

Bidirectional DC Circuit Breaker with Improved Performance During Commissioning and Reclosing................... 730

    Aditya Pogulaguntla, Venkata Raghavendra I, Satish Naik Banavath, Andrii Chub, T Sreekanth, Harish Sarma Krishnamoorthy

Modeling Method for Conducted Noise Flowing in Power Lines of DC/DC Converter 739

    Takato Hattori, Wataru Kitagawa, Takaharu Takeshita

High-Bandwidth Power Hardware-In-The-Loop for Motor and Battery Emulation at High Voltage Levels 749

    Manuel Fischer, Philipp Kemper, Johannes Herbold, Daniel Epping, Frank Puschmann

Analysis and Discussion of Different Three-Phase dv/dt Filter Topologies and the Influences of Their Filter Parameters on Losses and EMC 758

    Eric Fritze, Michael Meissner, Klaus F. Hoffmann, Kai-Uwe Rathjen, Stefan Dickmann, Oliver Woywode

State of Charge Prediction of Lithium-Ion Batteries Based on Artificial Neural Networks and Reduced Data 767

    Sebastian Pohlmann, Ali Mashayekh, Dominic Karnehm, Manuel Kuder, Antje Gieraths, Thomas Weyh

Investigation for Condensation Test Condition of HVIGBT Modules................... 777

    Kenji Hatori, Keiichi Nakamura, Wakana Noboru, Nils Soltau, Eugen Wiesner

Three Phase PV Inverter LCOE Optimization Considering Technological Choice 787

    Morteza Tadbiri Nooshabadi, Jean-Luc Schanen, Shahrokh Farhangi, Hossein Iman-Eini

Square Wave Operation to Reduce Pulsating Power in Isolated MMC-Based Ultrafast Chargers 798

    Ygor Pereira Marca, Maurice G. L. Roes, Jorge L. Duarte, Korneel Wijnands

Surge Current Protection for Railway Traction Applications................... 805

    Michael Gleissner, Mark-M. Bakran

Impedance-Based Analysis of HVDC Converter Control for Robust Stability in AC Power Systems................... 814

    André Schön, Andreas Lorenz, Rodrigo Alonso Alvarez Valenzuela

Class-E Push-Pull Resonance Converter with Load Variation Robustness for Industrial Induction Heating 825

    Janus Dybdahl Meinert, Benjamin Futtrup Kjærsgaard, Thore Stig Aunsborg, Asger Bjorn Jorgensen, Stig Munk-Nielsen, Sune Bro Duun

Review of Power Converter Topologies for Electrochemical Impedance Spectroscopy of Lithium-Ion Batteries 833
Hamzeh Beiranvand, Julius M. Placzek, Marco Liserre, Giorgia Zampardi, Doriano Constantino Brogioli, Fabio La Mantia

Design and Experimental Validation of a Voltage Sensing-Current Cancellation Common Mode Linear Active Filter 843
B. Mohamed Nassurdine, PE Lévy, D. Labrousse, JL Schanen, X. Maynard, S. Carcouet

Partial Discharges of Insulated Wires Under Impulses from Wide Bandgap Power Electronics 854
Ting Helmholdt-Zhu, Vivien Grau, Urs Obernolte

Analysis of a Droop-Based Power Controller for Three-Phase Microgrids 865
Andrea Lauri, Hossein Abedini, Davide Biadene, Tommaso Caldognetto, Paolo Mattavelli

Efficiently Paralleling GaN-Transistors for High Current and High Frequency Applications using a Butterfly Layout 873
Martin Wattenberg, Oscar Lorenz, Juan Sanchez

Data-Driven Decentralized Volt/Var Control for Smart PV Inverters in Distribution Systems 883
Yizhou Lu, Qianwen Xu, Lars Nordström

Study of Current Ripple Generators for Accelerated Ageing of Capacitors 891
Robert Keilmann, Hendrik Schefer, Regine Mallwitz

Intra-Arm Balancing Control of Cascaded Multi-Port Converter for Whole Power Unbalance Conditions 902
Takumi Yasuda, Jun-Ichi Itoh

Investigation of Creepage Distances on Printed Circuit Boards for Avionic Applications 912
Hendrik Schefer, Zhongqing Xu, Tobias Kopp, Regine Mallwitz, Michael Kurrat

A 20 kW, 3-Level Flying Capacitor 1500 V Inverter with Characterized GaN Devices for Grid-Tie Applications 922
Van Sang Nguyen, Anthony Bier, Hajar Es-Seghier, Ulrich Soupremanien, Gérard Delette, Stephane Catellani

New Analytical Model for Calculating HF-Losses in Litz Wire Regions Located Outside the E/U-CoreWindow of Transformers 933
Qingchao Meng, Jürgen Biela

Fast and Accurate Soft-Switching and Hard-Switching Losses Estimation for Power Converter, Application to the Dual Active Bridge (DAB) Converter 944
Francois Boige, Nicolas Videau, Adel Ziani, Bruno Guerrero, Julien Laclaverie

Influence of an Electrical Machine on the Dimension and Packaging of Multi-Machine Systems 952
Thomas Stöckl, Hans-Georg Herzog

Design of a Serial Impingement Cooling Heatsink for a 30 kW PV String Inverter 960
Paul Bruyere, Guillaume Piquet Boisson, Gaëtan Perez

Online Junction Temperature Measurement of SiC-MOSFETs via Gate Impedance using the Gate-Signal Injection Method 971
David Hirning, Luca Bauer, Johannes Ruthardt, Jörg Haarer, Philipp Ziegler, Jörg Roth-Stielow

Powercycling Test Bench with Realistic Loss Distribution and Temperature Ripples 980
Till-Mathis Plötz, Jan Fuhrmann, Hans-Günter Eckel

Design, Implementation and Characterization of an Integrated Current Sensing in GaN HEMT
Device by using the Current-Mirroring Technique 990
*Van-Sang Nguyen, René Escoffier, Stéphane Catellani, Murielle Fayolle-Lecocq, Jérémy
Martin*

GaN-Based Modular Multilevel Converter for Low-Voltage Grid Enables High Efficiency 999
Philip Kiehnle, Patrick Himmelmann, Marc Hiller

Energy Management of Smart Homes with Electric Vehicles using Deep Reinforcement Learning............ 1006
Xavier Weiss, Qianwen Xu, Lars Nordström

Simple and Low-Computational Losses Modeling for Efficiency Enhancement of Differential
Inverters with High Accuracy at Different Modulation Schemes.. 1015
Ahmed Shawky, Mokhtar Aly, Emad M. Ahmed, Samir Kouro, José Rodriguez

Estimation of Battery Parameters in Cascaded Half-Bridge Converters with Reduced Voltage
Sensors 1025
Nima Tashakor, Bita Arabsalmanabadi, Elham Hosseini, Kamal Al-Haddad, Stefan Goetz

Method to Analyze the Influence of Switching Behavior in Hard Switching Half Bridge Topologies
for Traction Application 1036
Dominik Nehmer, Michael Gleissner, Lukas Bergmann, Mark-M. Bakran

Impact of Aluminum Casing on High-Frequency Transformer Leakage Inductance and AC
Resistance.. 1046
*Reda Bakri, Xavier Margueron, Wendell Da Cunha Alves, Xavier Cimetiere, Frédéric Gillon,
Antoine Bruyere, Lucian Vatamanu*

Neural Networks-Generalized Predictive Control for MIMO Grid-Connected Z-Source Inverter
Model 1056
Navid Salehi, Herminio Martinez-Garcia, Guillermo Velasco-Quesada

Voltage Estimation for Diode-Clamped MMCs Based on a Simplified Neural Network 1064
Nima Tashakor, Davood Keshavarzi, Shady Banana, Stefan Goetz

A Non-Cooperative Game-Theoretic Distributed Control Approach for Power Quality
Compensators... 1074
*Claudio Burgos-Mellado, Victor Bucarey, Helmo K. Morales-Paredes, Diego Muñoz-
Carpintero*

A Comparative Analysis of Power Converter Topologies for Integration of Modular Batteries in
Electric Vehicles... 1083
*Alberto Cárcamo, Aitor Vázquez, Alberto Rodriguez, Diego G. Lamar, Marta M. Hernando,
Daniel Remón*

Design of a High-Dynamic Test Bench for Accelerated Dielectric Lifetime Testing with Adjustable
Voltage Slopes and Temperatures 1094
Hendrik Schefer, Lucas Hanisch, Tim-Hendrik Dietrich, Regine Mallwitz, Markus Henke

Novel Modulation Method for Common-Mode Noise Reduction in Solid-State Transformer Based
on ISOP Configuration.. 1104
Naoto Kikuchi, Hiroki Watanabe, Keisuke Kusaka, Jun-Ichi Itoh

Modular STATCOM for Compensation of Reactive Power and Voltage Asymmetry in Medium-Voltage Distribution Power Grids 1114
Josef Štengl, Tomáš Kormska, Jakub Talla, Zdenek Peroutka

Novel Method for Active Short Circuit (ASC) Tests of Power Module in Automotive Traction Application 1121
Tobias Appel, Arne Bieler

Short Circuit Performance and Current Limiting Mode of a Monolithically Integrated SiC Circuit Breaker for DC Applications Up to 800 V 1128
Norman Boettcher, Taro Takamori, Keiji Wada, Wataru Saito, Shin-Ichi Nishizawa, Tobias Erlbacher

Application of a HV Bipolar Square-Wave Voltage Generator for Qualification and Assessment of Energy Equipment 1137
Rico Fischer-Baeumer, Kai Gohrmann, Konrad Domes, Benjamin Sahan, Christian Staubach

A Decentralized and Communication-Free Control Algorithm of DC Microgrids for the Electrification of Rural Africa 1147
Lucas Richard, David Frey, Marie-Cecile Alvarez-Herault, Bertrand Raison

Universal Real-Time Model for Active Rectifiers in Versatile Totem-Pole PFC Configurations 1157
Axel Kiffe, Thorben Hoffstadt

Investigation of Core-Loss Mechanisms in Large-Scale Ferrite Cores for High-Frequency Applications 1167
Michael Baumann, Christoph Drexler, Jonas Pfeiffer, Jens Schueltzke, Erwin Lorenz, Michael Schmidhuber

Generation of Methodology for Making Benchmark Microgrids and Application in ESUSCON Microgrid 1177
Oscar Dorner, Patricio Mendoza-Araya

An Overview of Grid-Connection Requirements for Converters and Their Impact on Grid-Forming Control 1187
Paul Imgart, Mebtu Beza, Massimo Bongiorno, Jan R. Svensson

Modular Battery-Integrated Power Electronics-Modelling, Advantages, and Challenges 1197
Nima Tashakor, Jan Kacetl, Tomas Kacetl, Stefan Goetz

Design of Triple-Active Bridge Converter with Inherently Decoupled Power Flows 1207
Dong-Uk Kim, Byengjoo Byen, Byunghwang Jeong, Sungmin Kim

Application of a Multi-Winding Magnetic Component Characterization Method to Optimize Cross-Regulation Performances in DCM Flyback Converters 1216
Denis Motte-Michellon, Brahim Ramdane, Yves Lembeye, Bruno Cogitore

Application of an Electrostatic Machine in a Low-Voltage Microgrid 1226
Gabriel Ramos Huerta, Patricio Mendoza-Araya

Influences of Parasitic Capacitances in Wide Bandwidth Rogowski Coils for Commutation Current Measurement 1237
Philipp Ziegler, Tobias Festerling, Jorg Haarer, Philipp Marx, David Hirning, Jorg Roth-Stielow

Systematic Analysis of Oscillations in DC-Links of Fast Switching Power Electronics 1247
Tobias Fricke, Regine Mallwitz

EMI Mitigation Induced by an IGBT Driver Based on a Controlled Gate Current Profile 1256
  Daniel S. Martinez-Padron, Nicolas Patin, Eric Monmasson

An Accurate and Fast Model of Three-Level Three-Phase Dual-Active Bridge Converters in Real-Time Simulation 1266
  Ming Jia, Philipp Joebges, Rik W. De Doncker

A Calorimetric and Electrical Method for Measuring Loss Energies of Half-Bridges 1277
  Jörg Haarer, Mattea Eckstein, Philipp Ziegler, Philipp Marx, David Hirning, Jörg Roth-Stielow

Condition Monitoring Approach of a SiC Power Semiconductor using Turn-Off Delay with an Integration in a SiC Driver 1286
  Victor Golev, Ulf Schümann, Rando Raßmann, Jan Bockholt

Measurement Results of Multilevel Hysteresis Control for Paralleled Two-Level Converters 1294
  Magdalena Gierschner, Yves Hein, Hans-Günter Eckel, Christian Heien

Design and Development of a Short-Circuit Test Bench for Low-Voltage Direct Current Protection Devices 1300
  Simon Ravyts, Thomas Vandenbussche, Koen Stul, Jan Cappelle

A Novel Modified-TOGI Based PLL for the Three-Phase Unbalanced and Distorted Grid Conditions 1309
  Khanh-Hung Nguyen, Ahmad Ali Nazeri, Xiao Yu, Peter Zacharias

Comparison of Two and Three-Level AC-DC Rectifier Semiconductor Losses with SiC MOSFETs Considering Reverse Conduction 1319
  Guangyao Yu, Thiago Batista Soeiro, Jianning Dong, Pavol Bauer

Measurement Method for Simple Determination of Sinusoidal Large Signal Losses in Inductive Components 1328
  Peter Zacharias, Alejandro Aganza-Torres

A Novel Technique for the Suppression of the Displacement Current Through Power Module Base-Plate Capacitance 1336
  Mahmoud Saeidi, Ahmad Ali Nazeri, Rufad Zilic, Peter Zacharias

Analysis and Implementation of Effective Placement of EMC Capacitors for WBG Modules 1343
  Mahmoud Saeidi, Ahmad Ali Nazeri, Firas Jenhani, Peter Zacharias

Power Hardware-In-The-Loop Verification of a Cold Load Pickup Scenario for a Bottom-Up Black Start of an Inverter-Dominated Microgrid 1350
  Mina Mirzadeh, Robin Strunk, Tobias Erckrath, Axel Mertens

Detection of Incipient Inter-Turn Short-Circuit Faults by Artificial Intelligence Classifiers 1361
  Osman Örgüt, Ilker Sahin, Ece Olcay Günes

Modeling the Impact of Grid-Forming E-STATCOMs on Inter-Area System Oscillations 1371
  A. Bolzoni, N. Johansson, J. P. Hasler

Combining Schwarz-Christoffel Mappings and Biot-Savart Law to Calculate the High-Frequency Current Distribution Inside a Single Slot 1381
  Torben Fricke, Phil Leon Pickert, Babette Schwarz, Bernd Ponick

Standardised Switching Cell Building Block for Converter Design Optimisation with Detailed Electro-Thermal Model 1391
  Georgios Papadopoulos, Jürgen Biela

Design Procedure for Transformer-Based Solid-State Pulse Modulators with Damping Network 1402
  Spyridon Stathis, Juergen Biela

DC Bias Impact on Magnetic Core Losses at High Frequency 1413
  Bima Nugraha Sanusi, Ziwei Ouyang

Investigation of the Short-Circuit Type II Safe Operating Area of IGBTs....................................... 1424
  Madhu Lakshman Mysore, Mohamed Alaluss, Abhishek Maitra, Thomas Basler, Roman
  Baburske, Franz-Josef Niedernostheide, Hans-Joachim Schulze

Single Transformer, MMC Based MV Power Electronic Traction Transformer 1434
  Simon Fuchs, Simon Beck, Jürgen Biela

A New Power MOSFET Technology Achieves a Further Milestone in Efficiency 1445
  Ralf Siemieniec, Michael Hutzler, Cesar Braz, Tomasz Naeve, Elias Pree, Heimo Hofer,
  Ingmar Neumann, David Laforet

Experimental Evaluation of Battery Impedance and Submodule Loss Distribution for Battery
Integrated Modular Multilevel Converters 1456
  Arvind Balachandran, Tomas Jonsson, Lars Eriksson, Anders Larsson

Constant DC Power Infeed Grid Forming with Improved Ability to Ride-Through Unbalanced
Low-Voltage Faults... 1466
  Tayssir Hassan, Malte Eggers, Huoming Yang, Peter Teske, Sibylle Dieckerhoff

Constrained Long-Horizon Direct Model Predictive Control for Grid-Connected Converters with
LCL Filters 1476
  Mattia Rossi, Petros Karamanakos, Francesco Castelli-Dezza

Performance Evaluation of SiC-Based Isolated Bidirectional DC/DC Converters for Electric
Vehicle Charging... 1486
  Kaushik Naresh Kumar, Rafal Miskiewicz, Przemyslaw Trochimiuk, Jacek Rabkowski,
  Dimosthenis Peftitsis

Impact of Threshold Voltage Shifting on Junction Temperature Sensing in GaN HEMTs........................... 1497
  Burhan Etoz, Jose Ortiz Gonzalez, Arkadeep Deb, Saeed Jahdi, Olayiwola Alatise

Comparison of Power Cycling Results of Discrete GaN Cascodes for Automotive Power
Electronics with High Temperature Swings... 1506
  Florian Lippold, Philipp Hauenschild, Regine Mallwitz

Current Distortion Study for Hybrid Multi-Level Grid Inverter with Active Neutral-Point-Clamped
4-Leg Topology... 1515
  Jonas Steffen, Matthias Klee, Fabian Schnabel, Axel Seibel, Marco Jung

Dynamic Maximum Power Point Tracking Method Including Detection of Varying Partial Shading
Conditions for Photovoltaic Systems 1525
  Rosalie Rouphael, Nezha Maamri, Jean-Paul Gaubert

Novel Operation Mode of the Modular Multilevel Matrix Converter Based on a Dimensioning
Algorithm 1533
  Rebecca Dierks, Axel Mertens

On the Cosmic Ray Influence on the Electronics Design of a High Altitude Electric Aircraft 1543
Philippe Morey, Mauro Carpita

DC-Bus Control Considerations of Asymmetrical Multilevel Inverters with Embedded Buck-Boost
Converter 1551
Theodoros P. Mouselinos, Emmanuel C. Tatakis

A Seamless Modulation Strategy for Step-Up/Down Partial Power Processing Converter (SUD-
P3C).. 1561
*Chao Liu, Zhe Zhang, Ziwei Ouyang, Jiasheng Huang, Michael A. E. Andersen, Tiberiu
Gabriel Zsurzsan*

Performances Analysis of Non-Model-Based Speed Estimation Algorithms for Motor Drives 1569
*Gaetano Turrisi, Luigi Danilo Tornello, Giacomo Scelba, Giulio De Donato, Giuseppe
Scarcella*

A Method to Design Power Control System of Wayside Energy Storage System for Energy Saving
in DC-Electrified Railway 1580
Kota Sato, Keiichiro Kondo, Hiroyasu Kobayashi, Makoto Chida

A Reconfigurable Single-Stage Three-Phase Electric Vehicle DC Fast Charger Compatible with
Both 400V and 800V Automotive Battery Packs... 1590
Mojtaba Forouzesh, Yan-Fei Liu, Paresh C. Sen

Efficiency Improvement of Single-Stage AC-DC LLC Converter using a Line Cycle Synchronous
Rectifier (SR) Driving Strategy 1601
Mojtaba Forouzesh, Yan-Fei Liu, Paresh C. Sen

Influence of DC Supply Voltage Unbalances on the Performance of ARCP Inverters................................ 1611
Gholamreza Tabrizi, Sebastian Sprunck, Marco Jung

Grid-Forming Control for Enhanced Microgrid Interconnection 1620
Tobias Erckrath, Christian Bendfeld, Peter Unruh, Axel Seibel, Marco Jung

Low Phase Shift Filter for Current Sensing Based on the Difference Between AC Machine Models
with and Without Iron Losses.. 1631
Niklas Himker, Marcel Krümpelmann, Axel Mertens

Design and Analysis of a Voltage Clamping Active Delay Control Method for Series Connected
SiC MOSFETs... 1641
Rui Wang, Asger Bjørn Jørgensen, Hongbo Zhao, Stig Munk-Nielsen

Practical Implementation of a Concept for In-Situ Detection of Humidity-Related Degradation of
IGBT Modules... 1649
Benedikt Kostka, Axel Mertens

Design for Enhanced Noise Immunity of PCB Coils Used for Sensing Current Through Power
Devices.. 1658
Aamir Rafiq, Sumit Pramanick

Measurement Principle for Measuring High Frequency Bearing Currents in Electric Machines and
Drive Systems.. 1665
Benjamin Knebusch, Lennart Junemann, Pauline Holtje, Axel Mertens, Bernd Ponick

Climatically Induced Insulation Degradation in Power Semiconductor Modules of Wind Turbines............ 1674
Timo Lichtenstein, Sören Fröhling, Bernd Tegtmeier, Katharina Fischer

Comparison of Magnetic Noise Compensation Techniques for Dual Three-Phase Electrically Excited Synchronous Machines.. 1684
Jonas Henkenjohann, Jan Andresen, Axel Mertens

PCB Technology Comparison Enabling a 900V SiC MOSFET Half Bridge Design for Automotive Traction Inverters 1692
Matthias Spieler, Che-Wei Chang, Ayman El-Refaie, Muhammad H Alvi, Dong Dong, Rolando Burgos

Desaturated Turn-Off of Low-Saturation IGBTs with Clamping Method to Reduce Turn-Off Energy Losses... 1703
Vishwas Acharya Nayampalli, Hans-Günter Eckel

Impact of Bond Wire Configuration on the Power Cycling Capability of Discrete SiC-MOSFET Devices... 1713
Patrick Heimler, Nick Thönelt, Josef Lutz, Thomas Basler

A Low-Leakage, Low-Loss Magnetic Transformer Structure for High-Frequency Applications................. 1722
Allen Nguyen, Ajinkya Phanse, Michael Solomentsev, Alex J. Hanson

Temperature Distribution of an IGBT Chip During Repetitive Switching Events Under Consideration of Front-Side Ageing.. 1733
Christian Bäumler, Bo Zhang, Maximilian Goller, Xing Liu, Thomas Basler

Boosting Pilot-Diode Reverse-Conducting IGBTs Turn-ON and Reverse-Recovery Losses with a Simple Gate-Control Technique 1744
Daniel Lexow, Hans-Günter Eckel

Modeling of an Interleaved DC-DC Boost Converter for a Direct Model Predictive Control Strategy.. 1754
Thomas Effenberger, Hannes Böorngen, Eyke Liegmann, Michael Hoerner, Petros Karamanakos, Ralph Kennel

Static Analysis and Control Strategies of the Single Active Bridge Converter 1765
Alexis A. Gómez, Alberto Rodríguez, Marta M. Hernando, Diego G. Lamar, Javier Sebastián, Ibán Ayarzaguena, Jose Manuel Bermejo, Igor Larrazabal, David Ortega, Francisco Vázquez

Multi-Port Inductive Power Transfer System Considering Charging Auxiliary Battery in EVs.................... 1776
Zhuoqi Zhang, Ryosuke Ota, Ryohei Okada, Nobukazu Hoshi

Influence of IGBT and Diode Parameters on the Current Sharing and Switching-Waveform Characteristics of Parallel-Connected Power Modules.. 1785
Y. Ando, J. Sakai, K. Hatori, N. Soltau, E. Wiesner

Innovative Driving Scheme for Electrical Generators in More Electric Aircrafts Employing Series Active Filtering.. 1796
Nena Apostolidou, Nick Papanikolaou

Field-Measurement Based Hygrothermal Modelling of the Converter-Cabinet Climate in Wind Turbines.. 1804
Katharina Fischer, Katherina Gohler

A Multi-Mode Control Based Asymmetrical Dual-Active-Bridge Series-Resonant DC-DC Converter (DABSRC) 1815
M. Yaqoob, Grover Torrico, Wang Shuqin

Extended Balancing and Dimensioning of Capacitors in MMC Double Submodules 1824
    Ali Sharaf Addin, Christopher Dahmen, Thomas Brückner

Saliency Extraction and Torque Sharing Estimation of Dual Motor Drive using Special Current
Sensor Configuration 1834
    E. Rodriguez Montero, M. Vogelsberger, T. Wolbank

Soft-Switching Converter for Inductive Power Transfer System with Double-Sided LCC Resonant
Network 1844
    Ryohei Okada, Ryosuke Ota, Nobukazu Hoshi

Ultra Low Loss - MMC Submodules Favorable for SiC-FET Enabling High Functional Safety 1855
    Christopher Dahmen, Rainer Marquardt

Control of an Active Gate Driver for an Electric Vehicle Traction Inverter using Artificial Neural
Networks 1865
    Julius Wiesemann, Jacob Dumtzlaff, Axel Mertens

Cascaded H-Bridge Converter Designs for Future Short-Range All-Electric Aircraft Propulsion 1875
    Maximilian Hagedorn, Malte Lorenz, Axel Mertens

Overview and Evaluation of Energy Balancing Techniques for MMCs with Various Input and
Output Frequencies 1885
    Gyanendra Kumar Sah, Michael Schütt, Hans-Günter Eckel

Comparative Lifetime Estimations for IGBT Modules in Wind Turbine Converters 1895
    Christian Neumann, Hans-Gunter Eckel

Single-Phase, Five-Level Inverter with SPWM-Based Neutral Point Voltage Balancing Scheme 1906
    Dmytro Kondratenko, Arkadiusz Lewicki, Charles Odeh

Magnetic Core Evaluation Kit for the Comparison of Core Losses 1914
    Wilmar Martinez, Xiaobing Shen, Siqi Lin, Jens Friebe

Multi-Objective Optimization of Modular Multilevel Converter Systems 1923
    Nikolaus Patzelt, Christian Schlegel, Michail Vasiladiotis

Sizing of Hybrid Energy Storage System for Residential PV Applications 1933
    Xiangqiang Wu, Zhongting Tang, Tamas Kerekes

DC Bias Currents in Full-Bridge DC-DC Converters in Context of WBG Semiconductors and High
Switching Frequencies 1939
    Niklas Badenhop, Lukas Fräger, Dennis Kampen, Sascha Langfermann, Michael Owzareck

Parameter Tuning Method for Class Φ_2 Converters for High-Frequency Wireless Power Transfer
Applications 1947
    Yining Liu, Prasad Jayathurathnage, Jorma Kyyrä

Inductor Design Optimization using FEA Supervised Machine Learning 1955
    D. Cajander, I. Viarouge, P. Viarouge, D. Aguglia

Enabling Large-Scaled MMC EMT-RMS Co-Simulation by Data Exchange in the Loop (DXiL) 1966
    Xiong Xiao, Soham Choudhury, Martin Coumont, Jutta Hanson

Advanced Low-Voltage System-In-Package Half-Bridge MOSFET with Added Protection Features 1975
    S. Musumeci, V. Barba, F. Scrimizzi, C. Mistretta

Evaluation of Common-Mode Leakage Current of Aalborg-Type Transformerless PV Inverters 1985
  Georgios I. Orfanoudakis, Eftychios Koutroulis, Georgios Foteinopoulos, Weimin Wu

Multi-Frequency Traction-To-Auxiliary Integrated EV Drivetrain: Eliminating the Need for an
Auxiliary Power Module 1995
  Caniggia Viana, Mehanathan Pathmanathan, Peter W. Lehn

Potentials to Improve the Post-Fault Performance of a Fault-Tolerant Inverter System in Electrified
Aircraft Propulsion System 2003
  Yongtao Cao, Leon Fauth, Jens Friebe, Axel Mertens

Model Predictive Control-Enabled Fault Ride Through Operation Strategy for High Power Wind
Turbine 2011
  Pedro Catalán, Yanbo Wang, Zhe Chen, Joseba Arza

A Theoretical Comparison of Different Virtual Synchronous Generator Implementations on
Inverters 2021
  Patrick Körner, Andrea Reindl, Hans Meier, Michael Niemetz

Linear Flux-Switching Machine Design - A Multiobjective Optimization 2030
  Hendrik Marks, Henning Schillingmann, Sridhar Balasubramanian, Markus Henke

Single-Arm MMC-Based Converter for Transformerless Rail Interties 2038
  Simon Beck, Simon Fuchs, Jürgen Biela

Medium Voltage Diode Rectifier Design for High Step-Up DC-DC Converter 2049
  Pierre Le Métayer, Cyril Buttay, Drazen Dujic, Piotr Dworakowski

Fast Switching Planar Inductance Current Source ZETA Converter with Integrated Common Mode
Filter 2058
  Benjamin H. Zacher, Christian Schumann

System Level Simulation of Moisture Propagation and Effects in Wind Power Converters 2066
  Johannes C. Wenzel, Axel Mertens

PWM-Based Optimization-Free Active Voltage-Balancing Control of 7-Level Active Neutral-
Point-Clamped Flying-Capacitor Multicell Inverters 2073
  Vahid Dargahi

Model Predictive Power Sharing Algorithm for Fuel Cell Integration in a Dual Inverter Electric
Vehicle Drivetrain 2084
  Mehanathan Pathmanathan, Caniggia Viana, Sukhjit Singh, Peter W. Lehn

Comparative Evaluation of the 5-Phase Vienna and the 5-Phase PWM Rectifiers Under DC
Voltage Control 2092
  A. Dieng

Modelling and Control of a 50kW SiC-Based Isolated DAB Converter for Off-Board Chargers of
Electric Vehicles 2101
  Haaris Rasool, Manh Tuan Tran, Sajib Chakraborty, Joeri Van Mierlo, Thomas Geury,
  Mohamed El Baghdadi, Omar Hegazy

Impact of Cyber Attacks on Cost Oriented Power Routing Schemes in Microgrids 2110
  Kirti Gupta, Subham Sahoo, Bijaya Ketan Panigrahi, Frede Blaabjerg

Response of IGBT Chip Characteristics Due to Critical Stress 2119
  Kohei Yamauchi, Rik W. De Doncker

Mega-Hertz High-Power WPT System with Parallel-Connected Inverters using Current Balance Circuit..........2127

    Masamichi Yamaguchi, Keisuke Kusaka, Jun-Ichi Itoh

Investigation and Mitigation of Common-Mode Voltage in Four-Level NPC Converters Modulated by Redundant Level Modulation 2136

    Jun Wang, Wei Xu, Xibo Yuan, Lihong Xie

Ferrite Optimization for a Three-Phase Wireless Power Transfer System for Electric Vehicles 2145

    Shuang Nie, Mehanathan Pathmanathan, Peter W. Lehn

Frequency and Modulation Index Related Effects in Continuous and Discontinuous Modulated Y-Inverter for Motor-Drive Applications 2156

    Hamzeh J. Jaber, Alberto Castellazzi

Performance Evaluation of Sinusoidal-Flux Reluctance Machine for Improving Power Density with Reduced Torque and Input-Current Ripples 2164

    Kiwa Nagayasu, Masaki Iida, Kazuhiro Umetani, Mastaka Ishihara, Eiji Hiraki

Power Hardware-In-The-Loop Test of Low-Voltage Battery for a Plug-In Hybrid Electric Vehicle 2175

    Ronan German, Florian Tournez, Alain Bouscayrol, Aurelien Lievre, Betty Lemaire-Semail

Stability Analysis of DFIG System Connected with High-Frequency Capacitive Grid Based on Closed-Loop Current Control and Direct Power Control 2182

    Bin Hu, Heng Nian, Subham Sahoo, Frede Blaabjerg, Yaqian Zhang, Zixiao Xu

Full-Bridge Modular Multilevel Converter for the Four-Quadrant Supply of High Power Magnets in Particle Accelerators 2189

    Manuel Colmenero, Ricardo Vidal-Albalate, Francisco R. Blanquez, Ramon Blasco-Gimenez

Deep Neural Network for Magnetic Core Loss Estimation using the MagNet Experimental Database 2197

    Xiaobing Shen, Hans Wouters, Wilmar Martinez

Hybrid Circuit Board Structure for Power Electronics 2205

    Gerrit Braun, Deniz-Heinz Moldenhauer

Active Control of Gear Mesh Vibration using a Permanent-Magnet Synchronous Motor and Simultaneous Equation Method 2211

    Dominik Reitmeier

Research Laboratory for Testing Grid Connected Devices Under Grid Voltage / Grid Impedance Variations and Microgrid Conditions 2219

    Swen Bosch, Jochen Staiger, Heinrich Steinhart

Reducing the Impact of Skin Effect Induced Measurement Errors in M-Shunts by Deliberate Field Coupling 2230

    Hauke Lutzen, Jonas Müller, Vladimir Polezhaev, Till Huesgen, Nando Kaminski

Grid Forming Control for HVDC Systems: Opportunities and Challenges 2241

    Adil Abdalrahman, Ying-Jiang Häfner, Malaya Kumar Sahu, Khirod Kumar Nayak, Ashkan Nami

A Highly Integrated and Modular High Speed Electric Drive for Lightweight Electric Mountain Bikes 2251

    Matthias Hofer, Mario Nikowitz, Manfred Schrödl

Performance Enhancement of Power Conditioning Systems in More Electric Aircrafts 2257
Nick Rigogiannis, Nick Papanikolaou, Yongheng Yang

Steady State Simulations of a Hybrid HVAC/HVDC Network using OS Based ARM Devices 2266
Ioan Catalin Damian, Mircea Eremia

Experimental Comparison of FPGA-Implemented Model Predictive Voltage Control to Cascaded
Proportional Resonant Control for a Three-Phase Four-Wire Three-Level Grid-Forming Inverter of
250 kVA 2276
Jarren Lange, Dominik Schmies, Karl Stephan Stille, Joachim Böcker, Oliver Wallscheid

Experimental Study of Interleaved Y-Inverter Performance 2285
Yusuke Endo, Masataka Minami, Hamzeh J. Jaber, Alberto Castellazzi

Design of a GaN-Based Reconfigurable Resonant Converter for High Frequency On-Board
Charger of Battery Electric Vehicles 2293
*Manh Tuan Tran, Haaris Rasool, Dai Duong Tran, Mohamed El Baghdadi, Philippe Lataire,
Omar Hegazy*

Transient Liquid Phase Bond Reliability Evaluation of Die-Attach for Power Module Packaging 2301
Laxma R. Billa, Yangang Wang, Thomas Grant, Xiang Li, Harley Neal, Muhammad Morshed

Experimental Evaluation on Observer-Based Delay-Compensating Active Damping for LC-Filters 2308
Michael Schütt, Hans-Günter Eckel

Influence of Static Rotor Imbalance on the Roller Bearing Damage Due to Inverter-Induced
Bearing Currents 2316
Martin Weicker, Omid Safdarzadeh, Andreas Binder

Novel Current Balancing Method for HF Interleaved Converters with Reduced Control Effort 2327
Christian Beckemeier, Jens Friebe

dV/dt-Based Filter Design for Motor Inverters with Continuous Output Voltage 2334
Sabrina Ulmer, Stevan Bugarski, Gernot Schullerus, Ertugrul Sönmez

Evaluation of Core Losses in Transformers for Three-Phase Multi-Level DAB Converters 2344
Babak Khanzadeh, Yuriy Serdyuk, Torbjörn Thiringer

A Quasi-Offline Condition Monitoring Method of DC-Link Capacitor Banks in Accelerator Power
Converters 2355
*Timm Felix Baumann, Konstantinos Papastergiou, Raul Murillo Garcia, Dimosthenis
Peftitsis*

Minimizing Voltage Stress in Auxiliary Resonant Commutated Pole Inverters using Saturable
Inductors 2366
Markus Zocher, Norbert Grass, Ralph Kennel

Adaptive Dead-Time Control in a Resonant Wireless Power Transfer System 2375
Tim Krigar, Martin Pfost

Multilevel Battery Converter with Cascaded H-Bridges on Cell Level-Battery Management System
Or a Renewed Attempt for Power Electronic Building Blocks? 2383
*Max Rothenburger, Markus Horn, Xiao Yu, Gerold Schulze, Koenraad Muyllaert, Peter
Zacharias, Ludwig Brabetz, Hartmut Hillmer*

Design and Potential of EMI cm Chokes with Integrated DM Inductance 2392
Mohammad Ali, Rehnuma Bushra, Jens Friebe, Axel Mertens

Implementation Options of a Fully SiC Buck-CSI for Advanced Motor Drive Application.......................... 2402
    Yonghwa Lee, Alberto Castellazzi

Optimized Control Scheme to Achieve ZVS for the Complete Pre-Charging Phase of
Supercapacitors with a 500 kHz SiC- And GaN-Based Dual Active Bridge 2413
    Patrick Lenzen, Martin Pfost

Fault Blocking Capability in the DC-MMC with Reduced Number of Sub-Modules................................... 2422
    J. D. Páez, F. Morel, S. Bacha, P. Dworakowski

An Open-Source FEM Magnetic Toolbox for Calculating Electric and Thermal Behavior of Power
Electronic Magnetic Components 2432
    Nikolas Förster, Jonas Hölscher, Till Piepenbrock, Philipp Rehlaender, Oliver Wallscheid,
    Frank Schafmeister, Joachim Böcker

Comparison of Dual-Active-Bridge-Based Topologies for Single-Phase Single-Stage EV On-Board
Chargers 2441
    Daniel Gaona, Denis Pauls, Eduardo Facanha De Oliveira

Design Concepts for Medium Voltage DC Networks Supplying the Future Circular Collider (FCC)........... 2451
    Manuel Colmenero, Francisco R. Blanquez, Ramon Blasco-Gimenez

A Novel Dual CC-CV Output Wireless EV Charger with Minimal Dependency on Both Coil
Coupling and Load Variation 2462
    Subhranil Barman, Kishore Chatterjee

A High-Performance EMI Filter Based on Laminated Ferrite Ring Cores 2470
    Marcin Kacki, Marek S. Rylko, John G. Hayes, Charles R. Sullivan

Investigation of the Static Performance and Avalanche Reliability of High Voltage 4H-SiC
Merged-PiN-Schottky Diodes 2477
    Chengjun Shen, Saeed Jahdi, Phil Mellor, Juefei Yang, Erfan Bashar, Jose Ortiz-Gonzalez,
    Olayiwola Alatise

On Chain-Link Based Multi-Port Converters Able to Connect HVDC and MVDC to AC
Transmission Network... 2486
    Daniele Falchi, Oriol Gomis-Bellmunt, Eduardo Prieto-Araujo, Olivier Despouys

Voltage Control Scheme for Multilevel Interfacing PV Application: Real-Time MRAC-Based
Approach 2496
    Mohammad Sadegh Orfi Yeganeh, Mehdi Rahmani, Nenad Mijatovic, Tomislav Dragicevic,
    Frede Blaabjerg, Pooya Davari

Control Principles for Island Operation and Black Start by Offshore Wind Farms Integrating Grid-
Forming Converters... 2504
    Daniela Pagnani, Lukasz Kocewiak, Jesper Hjerrild, Frede Blaabjerg, Claus Leth Bak

Experimental Study of the Reduction and Removal of Turn-On Snubber for IGCT Based MMC
Submodule using Fast Silicon Diodes.. 2515
    Arthur Boutry, Cyril Buttay, Besar Asllani, Bruno Lefebvre, Eric Vagnon, Dong Dong

Characterisation of a Ferrite-Polymer Based Magnetic Material 2526
    Johan Le Leslé, Guillaume Lefevre, Julien Morand, Rémi Perrin, Pierre-Yves Pichon,
    Guillaume Regnat

Model Predictive-Based Control Technique for Fault Ride-Through Capability of VSG-Based Grid-Forming Converter... 2537
Mobina Pouresmaeil, Amir Sepehr, Basit Ali Khan, Jafar Adabi, Edris Pouresmaeil

Grounding Points in HV/MV Hybrid Transformer Auxiliary Converters....................................... 2544
Adrian Wiemer, Jürgen Biela

Non-Parasitic Induced Transient Overvoltage in ANPC Topology Due to Critical Switching Sequences 2554
Michael Geiss, Robert Kragl, Jürgen Thoma, Benjamin Volzer

Open-Delta SBC: A New Converter Topology with Low Number of Sub-Modules for MV Applications... 2564
D. Lanzarotto, P. B Steckler, K. Vershinin, F. Morel

Characterising the Effect of an Inverter on the Regulation of the AC Voltage using a Frequency Response Identification Technique 2574
Mohamed Aldarmon, Joan Marc Rodriguez, Adria Junyent-Ferre

Artificial-Intelligence Based DC-DC Converter Efficiency Modelling and Parameters Optimization 2581
Fanghao Tian, Diego Bernal Cobaleda, Wilmar Martinez

Analysis of the Loss Distribution of a 6 kW Two Stage Power Supply for 600 V DC Applications............. 2588
Lukas Fräger, Sascha Langfermann, Michael Owzareck, Dennis Kampen, Jens Friebe

Study on the Gate Loop Design and Its Impact on Switching Characteristics of GaN Transistors............... 2596
Xiaomeng Geng, Carsten Kuring, Oliver Hilt, Mihaela Wolf, Joachim Würfl, Sibylle Dieckerhoff

Analysis of Current Sharing in the Parallel Connection of GaN Transistors 2607
Frederik Stalleicken, Sibylle Dieckerhoff, Karsten Handt, Sebastian Nielebock

Verification of GaN-HEMT Spice Models using an S-Parameters Approach 2618
Alonso Gutierrez, Nasri Said, Emmanuel Marcault, Mathieu Gavelle

Power Loss Modelling of GaN HEMT-Based 3L-ANPC Three-Phase Inverter for Different PWM Techniques.. 2628
Salvatore Mita, Arjun Sujeeth, Giuseppe Aiello, Dario Patti, Francesco Gennaro, Giacomo Scelba, Mario Cacciato

Generalized Core and Winding Area Ratio - Trends for Inductors and Transformers in Power Electronics with High Switching Frequencies... 2638
Siqi Lin, Leon Fauth, Wilmar Martnez, Jens Friebe

Active Substrate Termination of Discrete and Monolithic Bidirectional GaN HEMTs in a T-Type Inverter 2644
Carsten Kuring, Yannic Lange, Xiaomeng Geng, Oliver Hilt, Mihaela Wolf, Joachim Würfl, Sibylle Dieckerhoff

Transformer Design Optimization and Comparison for a DC-DC Converter Used in PV Micro-Inverters... 2655
Tobias Manthey, Meriem Khader, Jens Friebe

Automated Gate Impedance Network Design for SiC MOSFETs using SPICE Solver Interfaced with MATLAB Environment 2661
Pawel Piotr Kubulus, Szymon Michal Beczkowski, Stig Munk-Nielsen, Asger Bjørn Jørgensen

An Improved Multi-Loop Resonant and Plug-In Repetitive Control Schemes for Three-Phase Stand-Alone PWM Inverter Supplying Non-Linear Loads 2670
Ahmad Ali Nazeri, Peter Zacharias

High Switching Frequency Operation of a Single-Phase Five-Level Hybrid Active Neutral Point Clamped Inverter with a Model Predictive Control Approach 2682
Mohammad Najjar, Mahdi Shahparasti, Rasool Heydari, Morten Nymand

Design of Planar Coupled Inductor Applied to Zero-Current Switching Clamped Current Converter 2689
Vinicius Freire Bezerra, Tobias Manthey, Montiê Alves Vitorino, Jens Friebe

Characterization of Online Junction Temperature of the SiC Power MOSFET by Combination of Four TSEPs using Neural Network 2698
Kanuj Sharma, Simon Kamm, Kevin Muñoz Barón, Ingmar Kallfass

Novel Extended Robust Disturbance Observer for Improved Cogging Force Compensation in Permanent Magnet Linear Motors 2706
Franz Luckert, Axel Mertens

Improvement of a Self-Powered Gate Driver Power Supply 2715
Mariana Raya, Oriol Aviñó, Sergio Busquets-Monge, Xavier Perpiñá, Miquel Vellvehi, Xavier Jordà

Optimization and Scaling of a Compact High-Power IGCT Capacitor Charger Based on Simulation and Measurements with a 300 kW/3.3 kV Demonstrator 2726
Felix Haag, Fabian Albrecht, Volker Brommer, Oliver Liebfried, Klaus F. Hoffmann

Multilayer Busbars for Medium Voltage ANPC Converter Dedicated to Battery Energy Storage Systems.. 2736
Mamadou Lamine Beye, Luc Bimmel, Anthony Bier, Jérémy Martin

A Simulation Model for SiC MOSFET Switching Transients Controlled by an Adaptive Gate Driver with the Capability of Reducing Switching Losses and EMI Across the Full Operating Range.. 2744
Zheming Li, Robert W. Maier, Mark-M. Bakran, Franz-J. Niedernostheide, Daniel Domes

Phase-Shift Modulation for Flying-Capacitor DC-DC Converters.. 2754
Philipp Rehlaender, Frank Schafmeister, Joachim Böcker

An EV Integrated Isolated DC Charger using a Six-Phase Synchronous Machine 2763
Sukhjit S Ghumman, Mehanathan Pathmanathan, Peter W Lehn

Configurable ISOP-IPOP DC-DC Converter for Universal Solid-State Transformer.................................. 2773
Pramod Apte, Jens Friebe, Lukas Fräger

Using System-On-Chip Boards for the Deployment of Controller for Verification and Prototyping 2780
Adeel Jamal, Gerd Griepentrog

Utilizing the Reactive Current Control Capability of an MMC-Fed AC/DC Converter for Volt-Second Balancing in Medium Frequency Transformers 2788
Kaveh Pouresmaeil, Maurice Roes, Jorge Duarte, Korneel Wijnands, Nico Baars, George Papafotiou

Cost Comparison for Different PV-Battery System Architectures Including Power Converter Reliability 2795
Martijn Deckers, Leander Van Cappellen, Glenn Emmers, Fereshteh Poormohammadi, Johan Driesen

Insulation Design and Analysis of a Medium Voltage Planar PCB-Based Power Bus Considering Interconnects and Ancillary Circuit Integration 2806
Joshua Stewart, Rolando Burgos, Dushan Boroyevich

Modular Multilevel Converter Control with using a General Space Vector PWM Method in Medium Voltage Hydro Power Application 2813
Chengjun Tang, Torbjörn Thiringer

A Technical Overview of Single-Stage Three-Port DC-DC-AC Converters 2824
Sebastian Neira, Zoe Blatsi, Michael M. C. Merlin, Javier Pereda

Common-Mode EMI Noise Modeling of Three-Level T-Type Inverter for Adjustable Speed Drive Systems.. 2835
Vefa Karakasli, Abdelmoumin Allioua, Gerd Griepentrog

A Condition Monitoring Scheme for Semiconductor Devices in Modular Multilevel Converters with Cascaded H-Bridge Submodules 2843
Mohsen Asoodar, Mehrdad Nahalparvari, Christer Danielsson, Hans-Peter Nee

Particular Requirements on Drive Inverters for Safe and Robust Operation on an Open Industrial DC Grid.. 2852
Simon Puls, Jan-Niklas Koch, Martin Ehlich, Holger Borcherding

Investigation About Operation and Performance of Gate Drivers for Power Electronics Converters for Cryogenic Temperatures... 2860
Mustafeez-Ul-Hassan, Yuxuan Wu, Vyacheslav Solovyov, Fang Luo

Synchronization Angle Determination in DVCSFO of DFIM Naval Propulsion.......................... 2869
Youssef Drimizi, Maria Pietrzak-David, Pascal Maussion

Power Control of LCR-DAB Converter with Phase Shift in Fixed Switching Frequency 2877
Seung-Hyuk Baek, Jaehong Lee, Seung-Hwan Lee, Sungmin Kim

A Simplified Braking Method for Direct Matrix Converter-Fed PMSM Drives with Consideration of Avoiding Regenerative Energy 2885
Jun Xie, Dustin Henneberg, Martin Suberski, Thomas Ellinger, Uwe Radel, Jürgen Petzoldt

Inverter-Machine Parametric Co-Design for Energy Efficient Electric Drives............................. 2893
Jaedon Kwak, Alberto Castellazzi

Bidirectional Cuk Converter in Partial-Power Architecture with Current Mode Control for Battery Energy Storage System in Electric Vehicles 2903
J. S. Artal-Sevil, J. Anzola, V. Ballestín-Bernad, I. Aizpuru

Design Space Exploration for a Capacitive 36V, 4A, 4:1 DCDC Converter with GaN Switches using a Performance-Cost-Matrix Including Uncommon Topologies.. 2912
Adrian Gehl, Malte Kempchen, Simon Disselkamp, Markus Olbrich, Bernhard Wicht

A Fast Control for a Three-Switch Multi-Input DC-DC Converter... 2919
Simone Cosso, Andrea Formentini, Mario Marchesoni, Massimiliano Passalacqua, Luis Vaccaro

Impact on the Torque and on the Copper Losses Under Fault-Tolerant Control of 5-Phase PMSG 2930
A. Dieng

Weighting Factor Design for FS-MPC in VSCs: A Brain Emotional Learning-Based Approach 2939
Mohammad Sadegh Orfi Yeganeh, Arman Oshnoei, Saeed Peyghami, Nenad Mijatovic, Tomislav Dragicevic, Frede Blaabjerg

A Strategy for Smooth Microgrid Transitions Without Phase Misalignment and Voltage Mismatch 2948
Gabriel Silva Rocha, Amiron Wolff Dos Santos Serra, Cesar Augusto Santana Castelo Branco, Hercules Araujo Oliveira, Jose Gomes De Matos, Luiz Antonio De Souza Ribeiro

Subtle Design and Performance Comparison of WF-FSM and DC-VRM for Large-Scale Direct-Drive Wind Power Generation 2958
Udochukwu B. Akuru, Maarten J. Kamper, Zi-Qiang Zhu

Analysis and Implementation of Different Non-Isolated Partial-Power Processing Architectures Based on the Cuk Converter 2967
J. S. Artal-Sevil, J. Anzola, V. Ballestín-Bernad, J. L. Bernal-Agustín

GaN HEMT and SiC Diode Commutation Cell Based Dual-Buck Single-Phase Inverter with Premagnetized Inductors and Negative Gate Driver Turn-Off Voltage 2977
Tobias Brinker, Hendrik Gräber, Jens Friebe

Determination of Optimal Associated Discrete Circuit Switch Model Parameters for Real-Time Simulation of Dual-Active Bridge Converters 2985
Marija Stevic, Ravinder Venugopal

Integrated Motor Drive: A Multidisciplinary Approach 2996
Betty Lemaire-Semail, Nadir Idir, Eric Semail, Souad Harmand

Hardware in the Loop Test of an Electric Aircraft Powertrain 3005
Sebastian Mönninghoff, Moritz Scholjegerdes, Kay Hameyer

A Multi-Port Smart Transformer for Green Airport Electrification 3014
Giampaolo Buticchi, Giovanni De Carne, Thiago Pereira, Kangan Wang, Xiang Gao, Jiajun Yang, Youngjong Ko, Zhixiang Zou, Marco Liserre

Improvement of EMI Filter Attenuation using Shielding 3022
Mohammad Ali, Rehnuma Bushra, Jens Friebe, Axel Mertens

Implementation of Onsite Junction Temperature Estimation for a SiC MOSFET Module for Condition Monitoring 3031
Farzad Hosseinabadi, Shahid Jaman, Sachin Kumar Bhoi, Md. Mahamudul Hasan, Sajib Chakraborty, Mohamed El Baghdadi, Omar Hegazy

Energy Storage Systems for Airborne Wind Generators 3037
Bakr Bagaber, Axel Mertens

Design Interactions of AC- And DC-Side Filters for Traction Drives with SiC Inverters 3048
Hedieh Movagharnejad, Benjamin Knebusch, Axel Mertens, Bernd Ponick

Investigation of an Interleaved Current-Fed Single Active Bridge DC-DC Converter for PV Applications 3059
Lucas Vinícius De Araújo Gomes, Tobias Manthey, Montiê Alves Vitorino, Jens Friebe

Real-Time Thermal Characterization of Power Semiconductors using a PSO-Based Digital Twin Approach 3067
Johannes Kuprat, Yoann Pascal, Marco Liserre

Self-Sensing Design and Control for an Induction Machine with an Additional Short-Circuited Rotor Coil.. 3075
Stefan Luecke, Axel Mertens

Calculating the Tractive Power and Power Conversion Efficiency of Battery Electric Vehicles using a Global Navigation Satellite System and a Road Elevation Database................................... 3084
Shinichi Domae, Alberto Castellazzi, Hamzeh J. Jaber, Tenghui Dong, Taketsune Nakamura

PCB Layer Optimization of Planar Medium Frequency Transformer for On-Board EV Chargers............... 3092
Fabian Groon, Hamzeh Beiranvand, Thiago Pereira, Görkem Can, Marco Liserre

Fault Current Capability Assessment of Low-Voltage Side Inverters in Smart-Transformers 3101
Thiago Pereira, Luis Camurca, Francisco Santos, Marco Liserre

Adaptive Resonant-Valley Switching for a GaN HEMT Direct AC-AC Auxiliary Resonant Commutated Pole Converter 3112
Kyle Steyn, Johan Beukes

The Variation of Core Loss in High-Frequency Transformers Under Different Load Conditions................. 3120
Navid Rasekh, Jun Wang, Xibo Yuan

A Complete PFC Inductor Design for Lighting Equipment Applications... 3130
Wai Keung Mo, Kasper M. Paasch, Thomas Ebel

Automatic Generation Control-Based Charging/Discharging Strategy for EV Fleets to Enhance the Stability of a Vehicle-To-Weak Grid System.. 3140
Majid Mehrasa, Mehrdad Gholami, Reza Razi, Khaled Hajar, Antoine Labonne, Ahmad Hably, Seddik Bacha

Model-Based Converter Control for the Emulation of a Wind Turbine Drive Train 3149
Alexander Ernst, Wilfried Holzke, Dawid Koczy, Nando Kaminski, Bernd Orlik

A Novel Grid-Demanded Power Point Tracking (GPPT) Control Method for Wind Turbines to Preserve Grid Stability with High Wind Energy Penetration 3159
David Matthies, Alexander Ernst, Henning Sauerland, René Reimann, Wilfried Holzke, Bernd Orlik

Extension and Implementation of a Model-Based Lifetime Monitoring System with Parallel Calculation of Multiple Power Semiconductors.. 3169
Steffen Menzel, Wilfried Holzke, Michael Hanf, Holger Groke, Bernd Orlik, Nando Kaminski

Smart Charging Strategy for Electric Vehicles using an Optimized Fuzzy Logic System............................ 3179
M. Gholami, M. Mehrasa, R. Razi, K. Hajar, A. Hably, S. Bacha, A. Labonne

Analysis and Discussion of a Concept for an Adjustable Inductance Based on an Impact of an Orthogonal Magnetic Field.. 3188
Guido Schierle, Michael Meissner, Klaus F. Hoffmann

A Field Programmable and Dynamic Configurable Power Electronic Converter Concept............................ 3198
Bjarte Hoff

DAB Converter Discrete ADRC Control into Real-Time CHIL Simulation of a MVDC/LVDC Power Grid 3206
Alessio Clerici, Riccardo Chiumeo, Diego Raggini, Alessandro Veroni

SNNFT: Sequential Neural Network-Fuzzy Thermal Early Warning System for Lithium-Ion Batteries... 3215
    Marui Li, Chaoyu Dong, Yunfei Mu, Qian Xiao, Jingming Cao, Hongjie Jia

Fine-Grained Dynamics Representation and Stability Analysis for MMC-Based Hybrid AC/DC Power Systems 3225
    Jingming Cao, Chaoyu Dong, Qian Xiao, Marui Li, Xiaodan Yu, Hongjie Jia

Adaptive Pontryagin's Minimum Principle-Inspired Supervised-Learning-Based Energy Management for Hybrid Trains Powered by Fuel Cells and Batteries 3235
    Hujun Peng, Feifei Li, Zhu Chen, Kai Deng, Sebina Jeschke, Kay Hameyer

A Case Study of Pole-Phase Changing Induction Machine Performance 3246
    Konstantina Bitsi, Sjoerd G. Bosga

New Topology of Superconducting Fault Current Limiter with Bypass Resistor 3254
    D. Baimel, Eli Barbi, S. Bronstein, N. Baimel, A. Kuperman

A Pre- And Discharge Unit for Capacitive DC-Links Based on a Dual-Switch Bidirectional Flyback Converter 3262
    Madlen Hoffmann, Martin März

Control and Integration of a Multiphase Brushless Wounded Synchronous Motor Drive 3272
    Remi Perrin, Guilherme Bueno-Mariani

A Way Forward to Achieve Interoperability in Multi-Vendor HVDC Systems 3282
    Adil Abdalrahman, Ying-Jiang Häfner, Philippe Maibach, Christoph Haederli

Model Predicitve Position Control of Electrical Drives on an Industrial PC 3292
    Fabian Karau, Michael Leuer

Bidirectional Active EMC Filter for Industrial Power Converters.. 3301
    Bernhard Wunsch, Stanislav Skibin, Ville Forsstrom

A General Method to Measure Parasitic Capacitance of Transformer using Guarding Technique 3309
    Shaokang Luan, Stig Munk-Nielsen, Bruce Wakelin, Magnus Hortans, Jan Schupp, Hongbo Zhao

Inductance Analysis of Electric Machines by Classical and Numerical Methods.......................... 3318
    J. J. Germishuizen, T. J. E. Miller

Dynamic Wireless Power Transfer DWPT Time Domain Model: Xyz Position and Speed Coupling Effect 3327
    Iosu Aizpuru, Eneko Agirrezabala, Mikel Mazuela, Unai Iraola, Estanis Oyarbide, Carlos Bernal

Dynamic Average Small Signal Model of the SAB Converter 3336
    Alexis A. Gómez, Alberto Rodríguez, Marta M. Hernando, Diego G. Lamar, Javier Sebastián, Ibán Ayarzaguena, Jose Manuel Bermejo, Igor Larrazabal, David Ortega, Francisco Vázquez

Algorithm for Optimal Selection of Drive Motor Transmission Combination.............................. 3344
    Santiago Ramos Garces, Dries Jacques, Stijn Derammelaere, Simon Houwen, Nick Van Oosterwyck, Bart Vanwalleghem

Evaluation of Drain-Source Voltage in Switch Transient Time Intervals as Gate Oxide Degradation Precursor of SiC Power MOSFETs... 3353
    Javad Naghibi, Sadegh Mohsenzade, Kamyar Mehran, Martin P. Foster

Active Output LLC Converter Topology 3362
  Hannes Börngen, Eyke Liegmann, Sriram Jagannath, Ralph Kennel

Short Circuit Type II and III Behavior of 1.2 kV Power SiC-MOSFETs 3373
  Xing Liu, Xupeng Li, Thomas Basler

Analog MPPT Comparison for Interplanetary Small Satellites Missions 3382
  C. Torres, A. Garrigós, J. M. Blanes, P. Casado, D. Marroquí, C. Orts

Feasibility Assessment of Variable-Speed Generator Set Concepts with Focus on Rating of Power
Electronic Equipment 3391
  Hendrik Fehr, Albrecht Gensior, Andreas Möckel, Frank Atzler, Tilo Roß, Carsten Reincke-
  Collon

Bus Voltage Regulation using Sequentially Switched ZVZCS Converters for Spacecraft Power
Systems 3401
  A. Garrigós, C. Orts, D. Marroquí, J. M. Blanes, C. Torres, P. Casado

A Standardized and Modular Power Electronics Platform for Academic Research on Advanced
Grid-Connected Converter Control and Microgrids 3411
  Frank S. R., Schulz D., Stefanski L., Schwendemann R., Hiller M.

Gate Input Capacitance Characterization for Power MOSFETs using Turn-On and Turn-Off
Switching Waveforms 3420
  Yota Nishitani, Michiko Inoue, Takashi Sato, Michihiro Shintani

AC Battery: Modular Layout with Cell-Level Degradation Control 3429
  Claudio Burgos-Mellado, Marcos Orchard, Diego Muñoz-Carpintero, Tomislav Dragicevic,
  Lorenzo Reyes-Chamorro, Jacqueline Llanos

Analysis of Test Methods for Measurement of Leakage and Magnetising Inductances in Integrated
Transformers 3440
  Sajad A. Ansari, Jonathan N. Davidson, Martin P. Foster, David A. Stone

A Topology-Morphing Series Resonant Converter for Photovoltaic Module Applications 3450
  Grigorios Sergentanis, Liliana De Lillo, Lee Empringham, C. Mark Johnson

A Novel Parameter for the Evaluation of Protective Circuits for IGBT Explosion Protection in
Submodules of MMC 3460
  Christoph Junghans, Hans-Guenter Eckel

Sub-Modules Switching Algorithms for Dual Active Bridge Modular Multilevel Converters to
Optimize Capacitor Voltage Deviation Versus Power Efficiency 3470
  Peizhou Xia, Chuantong Hao, Stephen Finney, Michael Merlin

Systematic Adaptive Robust State Feedback Control for Active Front-End Rectifiers 3480
  Aidar Zhetessov, Giri Venkataramanan

An Optimized Compensation Strategy of Direct Matrix Converter-Fed PMSM Drives with Field
Weakening Under Unbalanced Supply Conditions 3491
  Jun Xie, Dustin Henneberg, Martin Suberski, Manuel Kusebauch, Uwe Rädel, Jürgen
  Petzoldt

Double Inverter Concept for High-Speed Drives Without Motor Filters 3501
  Henning Kasten, Stephan Beineke, Matthias Bachmann

A Universal Single Stage Current-Fed Bidirectional Converter with Both AC and DC Input Power Source Compatibility 3511
 Manish Kumar, Sumit Pramanick, Bijaya Ketan Panigrahi

Optimization of Electric Vehicle Charge Scheduling with Consideration of Battery Degradation 3518
 Raka Jovanovic, Sertac Bayhan, Islam Safak Bayram

Onboard ESU Sizing and Dynamic IPT Charging Scenarios for a Tramway Application 3529
 Endika Bilbao Muruaga, Irma Villar, Florian Legay, Pierre Prenleloup, Jean-François Reynaud

Investigations on the Active Reduction of Common Mode Noise with Opposing Noise Sources 3536
 Philipp Marx, Felix Seybold, Philipp Ziegler, David Hirning, Jörg Roth-Stielow

Knowledge Based Grey Box Modeling of Inaccessible Circuits for System EMC-Simulation in Time Domain 3545
 Jan-Philipp Roche, Jens Friebe, Oliver Niggemann

Novel Quasi-Direct Rotor Position Estimator for Permanent Magnet Synchronous Machines Based on the Back-Electromotive Force using Current Oversampling 3555
 Georg Lindemann, Viktor Willich, Axel Mertens

Design Considerations for Fast On-State Voltage Measurement Circuits 3565
 Mathias C. J. Weiser, Manuel Rueß, Ingmar Kallfass

Analytical, FEM and Experimental Study of the Influence of the Airgap Size in Different Types of Ferrite Cores 3574
 Asier Arruti, Francisco Jose Perez-Cebolla, Jon Anzola, Iosu Aizpuru, Mikel Mazuela

Design Method of a High Frequency GaN-Based Half-Bridge with Bottom-Side Cooled Transistors using Multi-PCB Assembly 3582
 Loris Pace, Florian Chevalier, Thierry Duquesne, Nadir Idir

A 30 kW Dynamic Wireless Inductive Charging System for EVs 3590
 Zariff Meira Gomes, José Renes Pinheiro, Gilney Damm, Karim Kadem, Hassan Moussa

Dynamic Control of the Switching Behavior of SiC MOSFETs in Converter Operation 3599
 Jochen Henn, Laurids Schmitz, Rik W. De Doncker

A Series Resonant Balancing Converter for Bipolar DC Grids on Ships 3607
 Sachin Yadav, Zian Qin, Pavol Bauer

A V2G-Enabled Seven-Level Buck PFC Rectifier for EV Charging Application 3615
 Anekant Jain, Ritika Agarwal, Krishna Kumar Gupta, Sanjay K. Jain

Experimental Demonstration of a 2.2kW Active-Clamp Converter for High-Current Wide-Voltage-Transfer Ratio Applications 3625
 Philipp Rehlaender, Bastian Korthauer, Frank Schafmeister, Joachim Böcker

A Simplified Model for the Battery Ageing Potential Under Highly Rippled Load 3636
 Tomáš Kacetl, Jan Kacetl, Nima Tashakor, Stefan Goetz

System Modeling and Design of a Hybrid Renewable Energy System for a Cable Network Head-End Station in Rural Area 3646
 Tobias Schillinger, Thomas Schuhmann, Martin Eckart

Comparison of System-Level Availability in Industrial Grids 3655
    G. Emmers, J. Driesen

Ageing Mitigation and Loss Control in Reconfigurable Batteries in Series-Level Setups.......................... 3665
    Tomáš Kacetl, Jan Kacetl, Nima Tashakor, Stefan Goetz

Characterization of Conventional and Advanced Current Measurement Techniques Suitable for
WBG Semiconductor Devices.. 3676
    Severin Klever, André Thönnessen, Rik W. De Doncker

Zero-Sequence Voltage Reduces DC-Link Capacitor Demand in Cascaded H-Bridge Converters for
Large-Scale Electrolyzers by 40% 3686
    Roland Unruh, Frank Schafmeister, Joachim Böcker

Thermal Behavior Impact on the Electric Motor Shape Multi-Objective Optimization.......................... 3696
    Aissam Riad Meddour, Anthony Babin, Nassim Rizoug, Christopher Vagg, Richard Burke,
    Laid Degaa

Modelling Approaches of Power Systems Considering Grid-Connected Converters and Renewable
Generation Dynamics.. 3704
    Jaume Girona-Badia, Vinícius Albernaz Lacerda, Eduardo Prieto-Araujo, Oriol Gomis-
    Bellmunt, Stephan Kusche, Florian Pöschke, Horst Schulte

Efficiency and Lifetime Analysis of Several Airborne Wind Energy Electrical Drive Concepts 3711
    Bakr Bagaber, Daniel Heide, Bernd Ponick, Axel Mertens

Design and Performance Analysis of Single-Phase Axial Flux Permanent Magnet Motor for
Coaxial Cascade 3722
    Chu Wang, Xiaowei Hu, Xiaoya Wang, Weiwei Geng, Qiang Li, Jingning Hou

Comparison of Pulse Current Capability of Different Switches for Modular Multilevel Converter-
Based Arbitrary Wave Shape Generator Used for Dielectric Testing of High Voltage Grid Assets 3729
    Dhanashree Ashok Ganeshpure, Ajeeth Phrassanna Soundararajan, Thiago Batista Soeiro,
    Mohamad Ghaffarian Niasar, Peter Vaessen, Pavol Bauer

Accurate Modeling of IGBT-Based Converters in PLECS 3740
    Anne Von Hoegen, Philipp Tillmann, Tetsuya Kojima, Rik W. De Doncker

Novel Analytical Method for Estimating the Junction-To-Top Thermal Resistance of Power
MOSFETs.. 3750
    José Miguel Sanz-Alcaine, Francisco Jose Perez-Cebolla, Carlos Bernal-Ruiz, Asier Arruti,
    Iosu Aizpuru

DC-Side Impedance for Handling Interoperability of Multi-Vendor Multi-Terminal HVDC
Systems... 3757
    Ashkan Nami, Adil Abdalrahman, Ying-Jiang Häfner, Malaya Kumar Sahu, Khirod Kumar
    Nayak

Utilizing the Electroluminescence of SiC MOSFETs as Degradation Sensitive Optical Parameter 3766
    Lukas A. Ruppert, Michael Laumen, Rik W. De Doncker

Characterization of GaN-On-AlN/SiC Transistors Towards Monolithic Integrability 3775
    Nick Wieczorek, Xiaomeng Geng, Carsten Kuring, Oliver Hilt, Frank Brunner, Mihaela Wolf,
    Joachim Würfl, Sibylle Dieckerhoff

Optimal Frequency for Dynamic Wireless Power Transfer 3786
    Mincui Liang, Khalil El Khamlichi Drissi, Christophe Pasquier

A Wide-Input-Voltage-Range 50W Series-Capacitor Buck Converter with Ancillary Voltage Bus for Fast Transient Response in 48V PoL Applications.. 3796
    Nameer Khan, James Xu, Gerard Villar Piqué, John Pigott, Henk Jan Bergveld, Alaa El Sherif, Olivier Trescases

Four-Level Boost Inverter Based on ANPC Topology with Switched-Capacitor Branch............................ 3804
    Robert Stala, Adam Penczek, Stanislaw Piróg, Aleksander Skala, Andrzej Mondzik, Zbigniew Waradzyn, Krishna Kumar Gupta, Pallavee Bhatnagar, Sanjay K. Jain, Kasinath Jena

Comparative Evaluation of Partially-Rated Energy Storage Integration Topologies for High Voltage Modular Multilevel Converters.. 3813
    Zoe Blatsi, Sebastian Neira, Stephen Finney, Michael M. C. Merlin

Influence of Current Collapse Due to V_{ds} Bias Effect on GaN-HEMTs I_d-V_{ds} Characteristics in Saturation Region... 3822
    Xuyang Lu, Arnaud Videt, Ke Li, Soroush Faramehr, Petar Igic, Nadir Idir

Deep-Learning Fault Detection and Classification on a UAV Propulsion System 3831
    Pierre-Yves Brulin, Fouad Khenfri, Nassim Rizoug

A Compact Solid State Transformer for Replacing Conventional Medium Power Transformer in Weight-Critical Applications.. 3838
    Leon Fauth, Felix Willer, Jens Friebe

Comparative Study of Single-Phase and Three-Phase DAB for EV Charging Application............................ 3846
    Nicola Blasuttigh, Hamzeh Beiranvand, Thiago Pereira, Marco Liserre

Dynamic Load Emulation for Automotive Power IC Robustness Validation 3855
    Alexander Ulbing, Daniel Kostynski, Markus Sievers

DAB Frequency Decoupling Control with Current Minimization 3862
    Simon Uicich, Jean-Yves Gauthier, Xuefang Lin-Shi, Bruno Allard, Arnaud Plat

Design and Performance Analysis of a Modified Proportional Multi-Resonant (PMR) Controller for Three-Phase Voltage-Source Inverters 3871
    Ahmad Ali Nazeri, Mahmoud Saeidi, Peter Zacharias

Proposition and Comparison of Several Solutions for High Induced Voltage Across Inactive Transmitting Coils in a Series-Series Compensation DIPT System.. 3883
    Wassim Kabbara, Tanguy Phulpin, Mohamed Bensetti, Antoine Caillierez, Serge Loudot, Daniel Sadarnac

Modeling and Measuring the Bearing Capacitance of Radially Loaded Bearings 3893
    Stefan Quabeck, Daniel C. Rodriguez, Rik W. De Doncker

Comprehensive Control of Matrix Converters in On-Board Electric Drive Applications............................ 3903
    Galina Mirzaeva

Power System Simulation Tool for Quick Benchmarking of Innovative MVDC Grids in E-Mobility Applications.. 3910
    Daniel Siemaszko, Philippe Noisette

An Artificial Intelligence Pipeline for Critical Equipment Thermal Conditioning System Design 3920
    Raik Orbay, Athanasios Tzanakis, Inko Marcaide, Jonas Löfgren, Torbjörn Thiringer, Thomas Bernichon

Aspects of Stability Issues of HVAC/HVDC Coupled Grids 3928
  Gianni Bakhos, Kosei Shinoda, Juan-Carlos Gonzalez-Torres, Abdelkrim Benchaib, Luigi Vanfretti, Seddik Bacha

Measurement of Coss-V Characteristic of the 1.7kV/900A SiC Power Module and Estimation of the Channel Current 3938
  Jacek Rabkowski, Fernando Gonzalez-Hernando, Mariusz Zdanowski, Irma Villar, Uxue Larrañaga

In-Slot Cooling of Electrical Machines using Traditional Techniques and Additive Manufacturing 3947
  Ahmed Hembel, Gokhan Cakal, Bulent Sarlioglu

Comparison of High-Power 2-Level and 3-Level Converters in Terms of Power Density, Costs and Performance 3957
  Ludwig Schlegel, Wilfried Hofmann

Autonomous Characterization of Lithium-Ion Battery Model Parameters Utilizing a Mathematical Optimization Methodology 3966
  Hamzeh Beiranvand, Helge Krüger, Sandra Hansen, Marco Liserre, Christian Werlig, Andreas Würsig

SOC Governed Algorithm for an EV Cascaded H-Bridge Connected to a DC Charger 3975
  Giulia Tresca, Andrea Formentini, Filippo Gemma, Federico Lusardi, Riccardo Leuzzi, Pericle Zanchetta

Shaping the Transition from Si-Based Power Devices to SiC MOSFETs and GaN HEMTs 3984
  Gerald Deboy

Reinventing Batteries Through Nanotechnology 3986
  Yi Cui

Advancing GaN Power ICs: Efficiency, Reliability & Autonomy 3987
  Dan Kinzer

Electrification Strategy of Volkswagen Group 3989
  Alexander Krick

Make it Fly — the Future of Sustainable Aviation 3991
  Tanja Neuland

The Instrumental but Extremely Challenging Role of Hydrogen Towards a Decarbonized Society 3992
  Stefan Linder

Short Circuit Behavior of Dual Three-Phase Permanent Magnet Synchronous Motors with Different Mutual Inductance in Electric Propulsion Application 3993
  Yinghui Yang, Georg Möhlenkamp

Hybrid Silicon-SiC Inverter – Combining the Best of Both Worlds 4003
  Hans-Günter Eckel, Felix Kayser, Pham Ha Trieu To

Robustness of SiC Trench MOSFETs 4004
  Christian Felgemacher

3D Predictive Fatigue Modeling of Power Modules 4005
  Ben Samples, Brandon Passmore

Heterogeneous Integration of Power Conversion using Power Supply on Chip and Power Supply in Package......... 4006
   Cian Ó Mathúna, Seamus O'Driscoll

Driving Innovations for Power Electronics with Integratable and Sustainable Magnetics........................... 4008
   Matt Wilkowski

Impact of Package Technology on the Switching Behavior of High-Voltage GaN FETs........................... 4011
   Sebastian Klötzer

Impact of Power Electronics on Battery Operation 4012
   Dirk Uwe Sauer

Trends in Power Electronics and Batteries for Electrified Vehicle Infrastructure....................................... 4013
   Torsten Leifert

Impact of High Frequency Current Pulses on Battery Ageing 4014
   Julia Kowal

Aircraft Electrification – System-Level Potentials for Aviation Decarbonization 4015
   Kathrin Ebner, Antoine Habersetzer, Arne Seitz

About Power Electronics Challenges in Aviation 4016
   Marco Bohllaender

Development of Electric Motors for Aircraft Applications... 4017
   Simon Wolfstädter

Powertrain Trends in Electric Trucks... 4018
   Luciana C. Afonso

Modulation Strategy Impact of BEV Inverters on the Voltage Ripple and the High-Voltage Traction System Stability 4019
   Cornelius Rettner

Zero Emission Trucks & Bodies 4020
   Martin Glaser

Integrating Offshore Wind & Hydrogen - An Operator's View 4021
   Florian Gremme

Status Quo and Future Prospects of Power Electronic Solutions for Electrolysis Plants 4022
   Sven Schumann

Modular Power Supply System for Large Scale Water Electrolyzers 4023
   Ralf Juchem, Klaus Rigbers

Properties of a Lithium-Ion Battery as a Partner of Power Electronics... 4025
   Alexander Blömeke, Katharina Lilith Quade, Dominik Jöst, Weihan Li, Florian Ringbeck,
   Dirk Uwe Sauer

Author Index

Design of Planar Coupled Inductor Applied to Zero-Current Switching Clamped Current Converter

Vinicius Freire Bezerra[*], Tobias Manthey[†], Montiê Alves Vitorino[‡], Jens Friebe[†]

[*]Institute of Dynamics and Vibration Research (IDS)
[†]Institute for Drive Systems and Power Electronics (IAL)
Leibniz University Hannover (LUH)
[‡]Industrial Electronics and Machine Drive Laboratory (LEIAM)
Federal University of Campina Grande (UFCG)
Email: bezerra@ids.uni-hannover.de; tobias.manthey@ial.uni-hannover.de;
vitorino@dee.ufcg.edu.br; friebe@ial.uni-hannover.de

Acknowledgement

Parts of this work were funded by the Minna-James-Heineman-Stiftung under reference number S0029/ 10110/2021 and also by the Ministry of Science and Culture of Lower Saxony and the Volkswagen Foundation. The authors are responsible for the content of this publication.

Keywords

≪Planar magnetics≫, ≪Magnetic coupling≫, ≪Soft switching≫, ≪ZCS converters≫, ≪Current Source≫.

Abstract

This work focuses on the design and measuremental verification of a current source converter (CSC) with planar coupled inductor in the DC-link. Soft switching is achieved with a pseudo resonance interval between the stray inductor of the MOSFET and the output capacitor. Investigations about the improvement of the inductor are also discussed.

Introduction

Voltage source converters (VSCs) and CSCs play important role in the industry and are continuously studied in academia for improvement, either by describing new topologies or investigating alongside techniques to the conventional arrangement. Each one has its distinct mode of operation, but usually, they have patterns that can be described in a dual way [1].

Among the common characteristics, some issues come with these structures. Regarding the CSCs, it is remarkable that a large DC-link inductor needs to be used in order to compensate the twice grid frequency oscillation in single-phase converters [2]. Also, switching stress and energy dissipation through conduction must be taken into account. In [3], a converter is proposed which achieves zero-current switching (ZCS) at high frequencies by adding an auxiliary circuit between the DC-link side. Moreover, the DC-link reactive element has been part of many studies and plenty ways to reduce it has been proposed. One approach to do this is by using planar inductors in which several papers have introduced ways to design it and interesting experimental results have been obtained. In [4], it is shown a design and modeling FEM-based method due to lack of approaches in simulation tools. In [5], analysis of different types of geometry are done and comparisons between number of turns, width of the track, diameter among other parameters are evaluated.

This work aims to study, investigate and design a single-phase CSC using a technique that involves pseudo-resonating the parasitic inductance of the switches with the output capacitor. With this, no additional switch or passive component need to be added and the soft-switching capability is achieved. These parasitic inductances, called stray inductances, are intrinsic to every semiconductor and its packaging and their presence usually causes undesired overshoot and oscillations, in addition to increasing switching losses [7]. In [6], investigations about the effect of the stray inductance on switching events for a SiC MOSFET is done via simulation and experimental prototype.

This principle of pseudo-resonance was addressed on the literature as in [8], in which is proposed a DC-DC converter topology with ease of control and zero-voltage switching capability, operating in high switching frequency. It is taken from results that quasi-resonant converters working high resonant concept are an interesting alternative, despite it aren't not suitable for higher power levels. In general, the concept of a resonant pole was proposed in a full bridge pseudo-resonant DC-DC converter and some desirable attributes were taken such as simple control, wide control range and constant frequency operation [8]. Also, in [9] is proposed a zero-voltage resonant-transition (ZVRT) with similar approach which uses high switching frequency and the intrinsic C_{oss} of the MOSFET to achieve ZVS.

With increasing switching frequency the power density of the passive components can be increased but also the design process becomes more complex [10]. Basic physical properties are discussed in [11] and serve as a fundamental basis for this design. With natural convection, better heat transfer can be achieved by choosing planar core shapes and allows higher power densities [10]. In this paper, the inductive DC-link is designed as a planar design and the advantages and disadvantages are compared with a conventional inductor design. FEM-based and artificial intelligence-based simulation softwares were used for the theoretical studies and a comparison was made between the measurements of the prototype and the simulation results. In particular, the comparison of the efficiency and temperature rise is to be analyzed.

Topology Description

In this section, the operating principle of the single-phase zero-current switching clamped current (ZCS-CC) converter is presented. The proposed converter and its waveforms are shown in Fig. 1 and 2, respectively. It can be seen that the basic structure is very similar to the conventional single-phase current source inverter (CSI). The main difference is taking account the stray inductance in each MOSFET-diode switch of the H-Bridge to allow the operation of the converter with soft-switching. Also, the topology works with two different types of aproach for each horizontal pair of MOSFETs - switched and clamped. As summarized in Table I, the converter operation mode can be divided into two modes. In mode 1, the switch s_3 is clamped off and s_4 is clamped on while the upper pair is switching in a complementary way. In the second mode, the upper pair is clamped while the bottom one switches. t_{on} indicates the time for when s_1 (or s_4, in case of mode 2) is turned on and therefore the output voltage v_o is increasing, t_{off} is the time when the same switch is turned off, t_{dt_1} and t_{dt_2} are the dead time for the switches s_1 and s_2 (mode 1) and T_s the switching period. Due to the architecture of the H-bridge and the switching characteristic, the converter can operate either as DC/DC or as DC/AC converter [7], [12].

Table I: Switching pattern of the two DC-DC operation modes.

	Mode 1	Mode 2
s_1	$t_{on} + t_{dt_1} + t_{dt_2}$	T_s
s_2	$t_{off} + t_{dt_1} + t_{dt_2}$	0
s_3	0	$t_{off} + t_{dt_1} + t_{dt_2}$
s_4	T_s	$t_{on} + t_{dt_1} + t_{dt_2}$

In order to simplify the analysis of the converter, it is assumed that the circuit operation is in steady-state, the output inductor is large enough to be considered as a current source, and the output filter capacitance is very small, providing sufficiently fast voltage changes in one switching period to enable the proposed

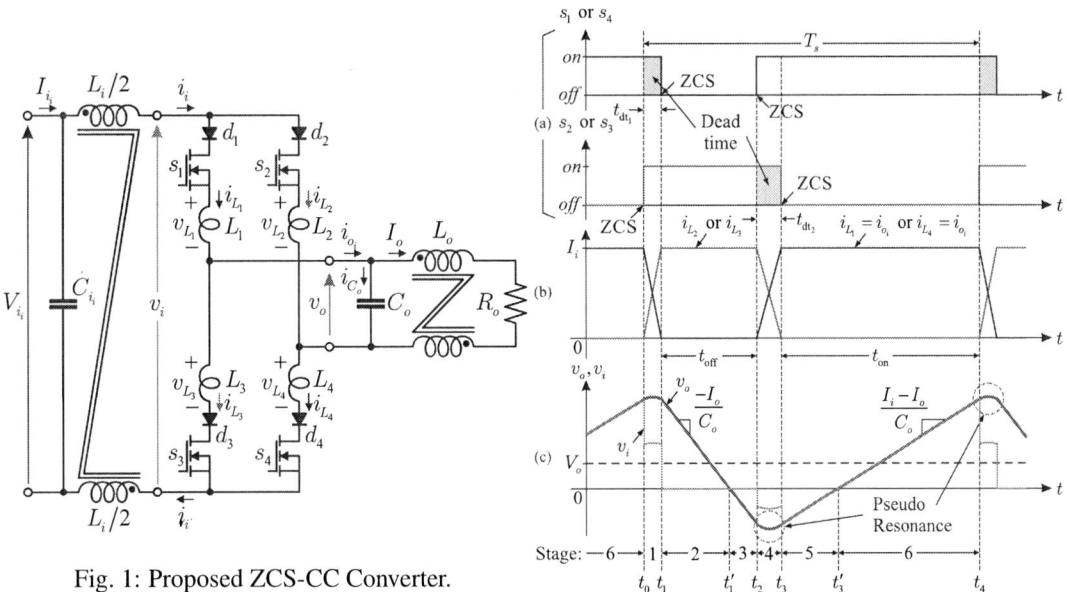

Fig. 1: Proposed ZCS-CC Converter.

Fig. 2: Main waveforms of the proposed converter.

operation, as shown in Fig. 2 (c).

The proposed converter has six stages of operation during each switching cycle. As mentioned before, the circuit uses one horizontal pair of MOSFETs to be clamped and another one for the switching operation. For the description of this section, the mode 1 will be considered.

Stage 1 $[t_0, t_1]$ - For $t \leq t_0$, s_1 is conducting the input current (I_i) and $i_{C_o} = I_i - I_o$, so the output voltage is increasing linearly with a slope of $(I_i - I_o)/C_o$. In $t = t_0$, s_2 is turned on with zero current due to the presence of the stray inductors, which do not allow sudden current variation. The current through L_2, i_{L_2}, grows more slowly compared to its current rise time (due to resonance). Even at this stage, the inductances resonate with the C_o capacitor. Since $L_1 = L_2$ is very small (as well as the C_o value), the resonance frequency $f_n = 1/(2\pi\sqrt{L_r \cdot C_o})$ is much higher than the switching frequency of the converter. This small resonance, present in the dead time, is what leads to the naming pseudo-resonant. During this interval, the voltage v_o flows through the two inductors equally with $v_o/2$, causing L_1 to discharge and L_2 to charge. Applying Kirchoff's Law:

$$v_{L_2} - v_C - v_{L_1} = 0 \qquad (1) \qquad -v_{L_1} = v_{L_2} = \frac{v_o}{2} \qquad (2)$$

Stage 2 $[t_1, t_1']$ - At $t = t_1$, the current i_{L_1} reaches zero and s_1 can be turned off at zero current. At that point, a freewheeling is created and all current from the source runs through s_2. In this interval, the output voltage, v_o, decreases linearly, since the capacitor starts to supply the load. At this moment, its voltage falls to a slope of $-I_o/C_o$.

Stage 3 $[t_1', t_2]$ - With s_2 still on, and with the capacitor-load combined, the current source load type begins to inject energy and the capacitor voltage reverses and becomes negative. The input voltage remains zero due to the presence of the freewheeling.

Stage 4 $[t_2, t_3]$ - At $t = t_2$, s_1 turn on with zero current and the same charge-discharge process of the inductors, as well as the pseudo-resonance, seen in stage 1, will happen again, with the inductor L_1 charging with $v_o/2$ and the inductor L_2 discharging with $-v_o/2$. So:

$$v_{L_2} + v_C - v_{L_1} = 0 \qquad (3) \qquad v_{L_1} = -v_{L_2} = \frac{v_o}{2} \qquad (4)$$

Stage 5 $[t_3, t_3']$ - In this stage, the current in the inductor L_2 reaches zero and s_2 can be opened at zero current. During this stage, a path from the source to the capacitor is created, through s_1, through which the capacitor will be charged and its voltage will increase linearly, by an angular coefficient of $(I_i - I_o)/C_o$; the input voltage is the opposite of the output voltage.

Stage 6 $[t_3', t_4]$ - From t_3', the period in which the output voltage returns to be positive and increases linearly, the capacitor remains being charged along the path created in the previous stage until the switching period is completed, returning to the point $t_4 = t_0$;

It can be observed, that zero-current switching occurs at the following times: $t_0 = t_4$, t_1, t_2 and t_3. Due to the stray inductances, the switches must be triggered only with zero current, otherwise the stored energy in them will be dissipated in the switches. To achieve soft switching, the voltage across the capacitor needs to be positive and negative in each switching period, so that the charging and discharging process of both capacitors is carried out completely, as shown in stages 1 and 4.

In practice, the value of $L_1 - L_4$ are not quite easy to control, which lead to unreliable results, especially regarding the output voltage value. As said before, this voltage need to have a ripple that covers positive and negative values so that the technique will properly work. In order to overcome this stability problem on the inductors values, a control for the length of the dead time can be done [13].

For this converter, the control of the output current can be done using a variable switching frequency PWM. Assuming that the dead times t_{dt_1} and t_{dt_2} are very small in proportion to T_s, the voltage waveform of the capacitor can be approximated to a triangular shape. The relationship between output current and input could be seen as for a buck converter, $I_o = D \cdot I_i$, where D is the duty cycle of s_1. The average output voltage is V_o.

Coupled Inductor Design

Planar inductors with PCB winding offer several advantages such as a high integration density, a defined structure and the associated reproducibility. However, the high parasitic coupling capacitance and low copper fill factor must be taken into account in the design process. The simulative design of the inductor is performed using Frenetic and GeckoMAGNETICs simulation software. Frenetic determines the results based on the input parameters and the artificial intelligence based algorithm and also offers several design proposals. Within the GeckoMAGNETICS simulation environment, there is more freedom regarding the geometric modeling of the winding and is thus more suitable for this specific application. Furthermore a 2D FEM simulation in ANSYS Maxwell is performed to avoid saturation effects.

For the evaluation of the planar design of the coupled inductor, an additional design is generated as a reference inductor using the simulation software Frenetic, which is initially considered as the optimal design with the lowest losses. Accordingly, the manual design of the planar inductor is performed using Gecko MAGNETICs and is based on the design methodology in [10]. This comparison is intended to illustrate the advantages and disadvantages of the planar design in terms of geometric dimensions as well as physical properties. The starting points for the design are, on the one hand, the current waveform shown in Fig. 3 as well as the voltage drop across the inductors and, on the other hand, the required inductance value for the current DC link of approx. $100\,\mu\text{H}$ resulting in $50\,\mu\text{H}$ for each inductor.

$$L = \frac{V_i \cdot (1-D)}{\Delta i_L \cdot f_{sw}}, \quad V_i = 24\,\text{V}, \quad D = 0.5, \quad \Delta i_L = 1\,\text{A}, \quad f_{sw} = 125\,\text{kHz} \tag{5}$$

Since the current converter can be used in both DC-DC and DC-AC operation, the requirements for the inductors in the respective operation must be considered separately. In DC-DC operation, sufficient energy should be stored for one switching period, whereas in DC-AC operation, the current ripple and output frequency determine the inductance value. The design procedure shown here refers specifically to DC-DC operation. In order to be able to investigate the influence of different switching frequencies within the topology, the N97 ferrite core material is chosen, which has a low power loss density for the selected operating range between $125\,\text{kHz}$ and $500\,\text{kHz}$. For the worst-case consideration of core losses and the amount of stored energy, the design process is performed for the switching frequency of $125\,\text{kHz}$.

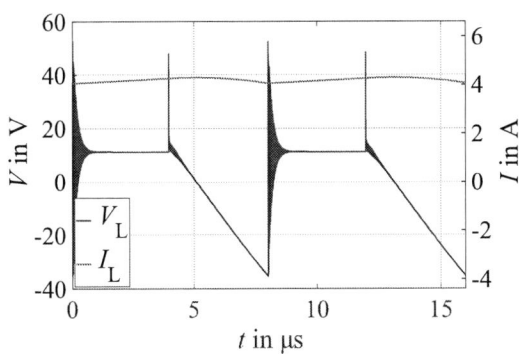

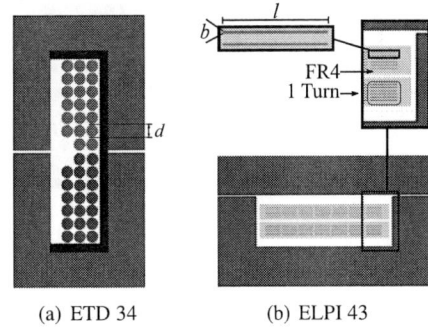

(a) ETD 34   (b) ELPI 43

Fig. 3: Voltage and current waveforms of one coupled inductor for simulative dimensioning and loss calculation of the inductor design.

Fig. 4: Schematic arrangement of the two coupled inductors inside one winding window. The windings are separated by color.

Due to the high number of turns and the associated low magnetic flux density excitation, the winding losses are dominant compared to the core losses. The low harmonic spectrum of the input current additionally leads to very low AC losses within the winding, so that the ohmic losses resulting from the DC resistance of the winding and the RMS-value of the current can be compared in a simplified way.

First simulative design results are shown in Fig. 4 - Fig. 7 and are summarized in Table II. The schematic illustration of the inductors shows a winding window of a core half, in which the windings of the two inductors contrast with each other in terms of color. One advantage of the conventional chokes is the significant high copper filling factor and the associated low DC resistance of the winding as well the smaller construction volume. To reduce the DC resistance of the PCB winding, a 6-layer PCB is chosen to parallel the winding.

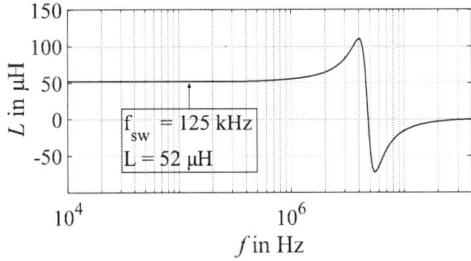

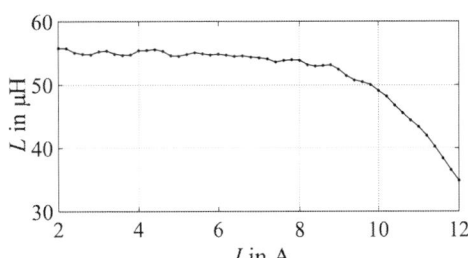

Fig. 5: Inductance measurement of one inductor with an impedance analyzer.

Fig. 6: Saturation current measurement.

A brief cost analysis was worked out on the basis of 1000 units. The result shows that the ferrite costs in the reference design are significantly lower, but due to the additional coil former and the higher copper costs compared to the PCB, the total costs are approximately the same.

Fig. 5 shows the inductance measurement with an impedance analyzer. The resulting value fits to the simulation results and also the frequency range from 125 kHz up to 500 kHz is far away from the resonant frequency causing by the inductance itself and coupling capacitance between both windings which is expected to be much higher than in the reference design due to the large area between both PCB winding. To take care of saturation effects it must be taken into account that both windings are sharing one core caused a doubling of the magnetic flux density. In Fig. 6 it can be seen that saturation effects starts at about 9 A, which is sufficient for this topology.

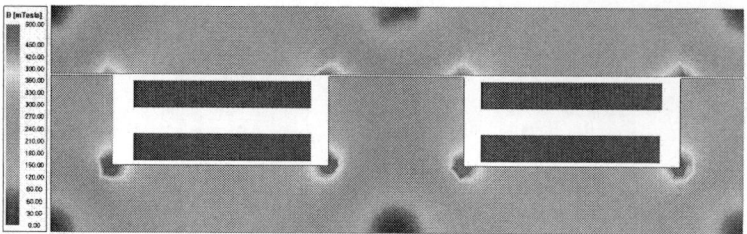

Fig. 7: 2D FEM simulation of the planar ELP core with an air gap of 0.12 mm to determine the magnetic flux density distribution in ANSYS Maxwell.

Table II: Simulation results of an reference inductor in comparison of a planar inductor. Parameters marked with * are refered to one inductor in the coupled inductor design. The cost calculation is an estimation for an order quantity of 1000 pieces.

Parameters	Reference inductor	Planar inductor
Core type	ETD 34/17/11	ELP/I 43/10/28
Core material	N97	N97
Number of turns*	20	7
Winding type	Round	PCB winding
Winding geometry	$d = 1.4\,\text{mm}$	$l = 1.6\,\text{mm}, b = 35\,\mu\text{m}$
Air gap	0.64 mm	0.12 mm
Core boxed volume	$23.11\,\text{cm}^3$	$32.54\,\text{cm}^3$
Weight	73 g	76 g
Copper filling factor	70.3 %	6.4 %
Inductance*	54.06 µH	57.84 µH
Core losses	-	-
Winding losses*	0.53 W	1.79 W
DC resistance* @ 20 °C	10.12 mΩ	46.9 mΩ
Temperature*	36.2 °C	41.7 °C
Core material cost	2.22 $	4.42 $
Coil former cost	1.07 $	-
Winding cost	2.14 $ (2.4 m winding length)	1.41 $ (PCB 35 cm x 50 cm)

Hardware Setup and Measurements

The hardware setup that was built for validation is shown in Fig. 8. A 200 V MOSFETs IPB107N20NA from Infineon is used in series with the 650 V diodes C6D10065E from Wolfspeed in order to achieve the reverse blocking capability. On the left side of the prototype the planar inductor is shown which consists of two 6-layer PCB, each for one winding. In addition a auxiliary wire make the connections between the two PCBs. Each PCB has a thickness of 1.6 mm, in which the insulation between the different layers is approximately 0.3 mm. At the main PCB, the power and control components are placed (top side) as well as filtering circuits for the gate driver power supply and heat sinks for the MOSFETs and diodes (bottom side). The control signals from the drivers are generated from a microcontroller STM32F411RE from STMicroelectronics and RECOM converters are used to provide an isolated 12 V voltage as a driver supply.

To achieve 100 W rated input power, an input current of 4.2 A and input voltage of 24 V was applied. All of the variables and the respective values of the experimental setup can be seen in the Table III as well as the experimental results for 100 W input power can be seen in Fig. 9. As can be seen in magenta, the output voltage (scaled by a 0.5 factor) reaches positive and negative values, totalizing 104 V of ripple, in

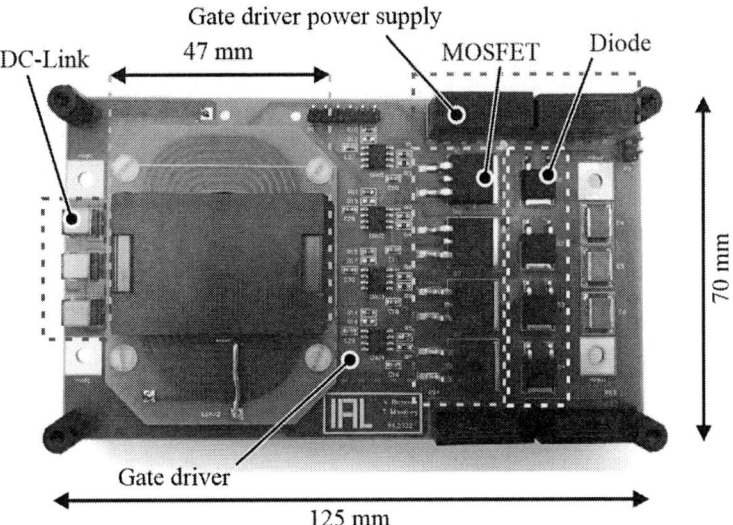

Fig. 8: Hardware prototype of the ZCS-CC with coupled planar inductors at the input side.

one switching period, enabling the proposed operation. The current waveforms in both active switches are shown in Fig. 9 (a) and Fig. 9 (b).

Table III: Variable values for the experimental setup.

Parameter	Variable	Value
Input voltage	V_{in}	24 V
Input current	I_i	4.2 A
Input inductor	$L_i/2$	57 μH
Output capacitor	C_o	220 nF
Switching frequency	f_s	125 kHz
Duty cycle	D	50 %
Deadtime	t_{dt}	200 ns
Output voltage ripple	v_o ripple	104 V
Efficiency	η	83 %

In the zoomed Fig. 9 (c), the zero-current turn-off transition can be seen. When the gate-source signal v_{gs2} turn on, the current on the switch s_1 decreases and the soft-switching on s_1 can be achieved as expected. However, in Fig. 9 (d), it could be seen a slightly different behaviour on the turn off moment of s_2. There is still a current through the switch, which leads to a pseudo-zcs on this component. Furthermore, an efficiency investigation on the prototype was done and the result can be seen in Fig. 10. Another investigation made is that since the output capacitance value is much smaller when compared to a capacitance value of a conventional DC-AC topology, it becomes extremely sensitive to enviromental changes (temperature changes) therefore the use of capacitors class 1 is more desirable.

A thermal photograph taken in steady state with an IR camera in Fig. 11 shows the heat distribution of the whole prototype. The estimates maximum temperature in the simulation has an deviation of only 3.9 K. Since s_4 is switched on continuously, comparatively high forward losses occur in the associated diode compared to the soft-switching diodes, which leads to a distinct temperature increase. For further optimisation, diodes with a lower forward voltage at nominal current can be selected.

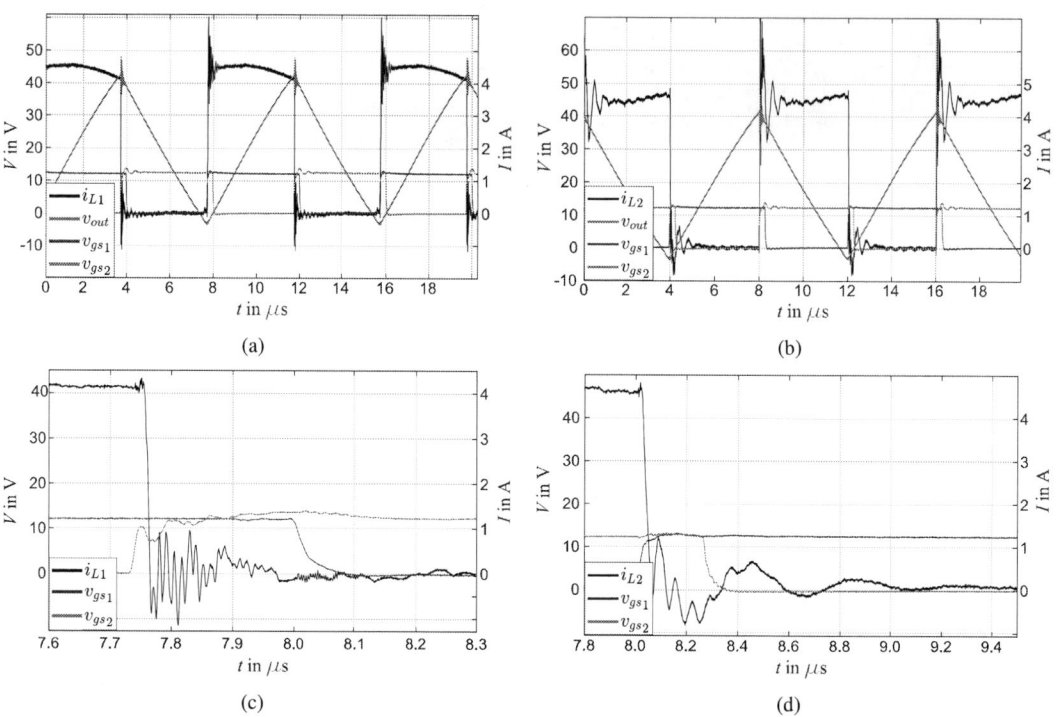

(a)

(b)

(c)

(d)

Fig. 9: Experimental results for 100 W and 125 kHz. Waveforms for: (a) S_1 and (b) S_2. Zoomed ZCS in (c) S_1 and (d) S_2. In black, both i_{L1} and i_{L2}, in magenta the output voltage (divided by two) and blue and green the gate signals v_{gs_1} and v_{gs_2}.

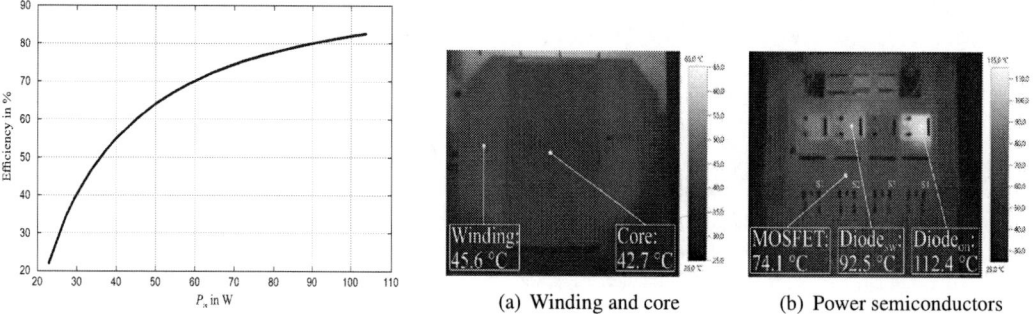

(a) Winding and core

(b) Power semiconductors

Fig. 10: Efficiency curve of the converter prototype for a maximum input power of 100 W.

Fig. 11: Thermal photograph taken with an IR camera.

Conclusion

An extensive topology description of the current source converter including switching state analysis is followed by a design methodology of a planar coupled inductor as well a measuremental verification. As part of the inductor design, a reference design was first generated for comparison with the planar inductor. A first design of the planar coupled inductor is presented and the advantages and disadvantages are discussed. A hardware prototype was built and analyzed in DC-DC operation. As in the theoretical approach expected, zero current switching is achieved in the MOSFETs and the switching transients are shown in greater detail. Furthermore an efficiency of about 83 % is achieved at an input power of 100 W.

References

[1] H. Ishikawa and Y. Murai, "A novel soft-switched PWM current source inverter with voltage clamped circuit," in IEEE Transactions on Power Electronics, vol. 15, no. 6, pp. 1081-1087, Nov. 2000, doi: 10.1109/63.892822

[2] M. A. Vitorino and M. B. d. R. Corrêa, "Compensation of DC link oscillation in single-phase VSI and CSI converters for photovoltaic grid connection," 2011 IEEE Energy Conversion Congress and Exposition, 2011, pp. 2007-2014, doi: 10.1109/ECCE.2011.6064033

[3] Zhihong Bai, Chushan Li and David Xu, "A zero-current switching (ZCS) current source converter for high-frequency PWM applications," IECON 2016 - 42nd Annual Conference of the IEEE Industrial Electronics Society, 2016, pp. 3213-3216, doi: 10.1109/IECON.2016.7793576

[4] A. Ammouri, T. Ben Salah and F. Kourda, "Design and modeling of planar magnetic inductors for power converters applications," 2015 7th International Conference on Modelling, Identification and Control (ICMIC), 2015, pp. 1-5, doi: 10.1109/ICMIC.2015.7409474

[5] A. Ammouri, H. Belloumi, T. B. Salah and F. Kourda, "Experimental analysis of planar spiral inductors," 2014 International Conference on Electrical Sciences and Technologies in Maghreb (CISTEM), 2014, pp. 1-5, doi: 10.1109/CISTEM.2014.7076937

[6] Liu, Yang, and Huai-Yu Ye. "Investigation on stray inductance of SiC MOSFET module." *2017 14th China International Forum on Solid State Lighting: International Forum on Wide Bandgap Semiconductors China (SSLChina: IFWS)*. IEEE, 2017.

[7] Lutz, Josef, et al. Semiconductor power devices: physics, characteristics, reliability. Springer Science & Business Media, 2011.

[8] Patterson, Oliver Desmond, and Deepakraj M. Divan. "Pseudo-resonant full bridge DC/DC converter." *IEEE Transactions on Power Electronics* 6.4 (1991): 671-678.

[9] Henze, C. P., H. C. Martin, and D. W. Parsley. "Zero-voltage switching in high frequency power converters using pulse width modulation." *APEC'88 Third Annual IEEE Applied Power Electronics Conference and Exposition*. IEEE, 1988.

[10] T. M. Undeland, J. Lode, R. Nilssen, W. P. Robbins and N. Mohan, "A single-pass design method for high-frequency inductors," in IEEE Industry Applications Magazine, vol. 2, no. 5, pp. 44-51, Sept.-Oct. 1996

[11] V. C. Valchev and V. A. D. Bossche, "Inductors and Transformers for Power Electronics (1st ed.)," CRC Press, 2005

[12] Erickson, Robert W., and Dragan Maksimovic. *Fundamentals of power electronics*. Springer Science & Business Media, 2007.

[13] V. F. Bezerra, M. A. Vitorino and J. Friebe, "Zero-Current Switching Clamped Current Converter," 2021 Brazilian Power Electronics Conference (COBEP), 2021, pp. 1-7, doi: 10.1109/COBEP53665.2021.9684011

Characterization of Online Junction Temperature of the SiC power MOSFET by Combination of Four TSEPs using Neural Network

Kanuj Sharma[a]*, Simon Kamm[b], Kevin Muñoz Barón[a] and Ingmar Kallfass[a]

[a]Institute of Robust Power Semiconductor Systems, University of Stuttgart
[b]Institute of Automation and Software Systems, University of Stuttgart
Stuttgart, Germany
Tel.: +49 / (0) 711/ 685 60833
Fax: +49 / (0) – 711/ 685 58700
*E-Mail: kanuj.sharma@ilh.uni-stuttgart.de
*URL: https://www.ilh.uni-stuttgart.de/

Keywords

«Condition monitoring», «Device characterization», «TSEP», «Silicon carbide», «Machine learning», «Neural network», «Wide bandgap»

Abstract

This paper presents an approach to combine multiple temperature-sensitive electrical parameters to improve the accuracy and precision of the junction temperature estimation of power transistors using the example of a silicon-carbide power MOSFET. Switching delays and the threshold voltage of the power transistor during turn-on and -off of a silicon-carbide power transistor are used as temperature-sensitive electrical parameters for the online junction temperature measurements. In order to improve the accuracy, a shallow fully-connected neural network is used as the means to combine the four measurements in one switching cycle of the transistor. The maximum measurement error of the junction temperature of the power transistor is reduced approximately 10-fold from 8.98 K to 0.92 K.

Introduction

The robustness and reliability of the power semiconductor devices in an application are critically affected by the junction temperature of wide-bandgap power transistors. The temperature swings and maximum temperature is the main cause of the failure in power electronics [1] which makes the knowledge of the operating temperature of the power transistor paramount. There are different methods depending on the usage and application that can be used to measure or estimate the junction temperature of the transistor. Some of them are physical contact where thermistors or thermocouples are positioned directly on or near the active chip area in order to get the temperature information on the junction of the power transistor. However, the issues like EMI and voltage isolation while implementing this method should be considered. There are also optical methods to measure the junction temperature of the transistor such as infrared sensor (IR). The exposure of a non-reflecting chip surface is required to have the two-dimensional temperature map of the transistor in order to use an optical method for the online junction temperature estimation [2]. Moreover, the optical measurements can lead to a bulky measurement setup and are difficult to implement in application. One of the most important methods to monitor the junction temperature of the power transistors is by measuring the change in the electrical characteristics of the transistor with temperature. These electrical parameters are called temperature-sensitive electrical parameters (TSEP) and have been widely used especially in the applications like the end of life estimation, condition monitoring, power cycling [3], and temperature control [2]. Some of the TSEPs and their dependencies are shown in Table 1.

The choice of the TSEP is highly dependent on the intended application usage. Some of the criteria to select a particular TSEP are temperature sensitivity, linearity, accuracy, invasiveness, complexity, and calibration need [4]. Most of the selection criteria are dependent on each other. For example, the non-linear behavior of the TSEP leads to high calibration needs in terms of measurement points. The accuracy of a TSEP measurement decreases with the decrease in temperature sensitivity of the electrical parameter and makes the circuit more complex. Similarly, the invasive TSEPs which affect the normal operation of the transistor are complex and require more attention during calibration.

Table I: TSEPs and their dependencies [5, 6]

TSEPs	Threshold voltage	Transient delay	Drain-source voltage	Internal gate resistance	On-resistance
Dependencies	T_J	T_J, V_{DS}, I_D, R_G	T_j, I_D	T_J	T_J, V_{DS}, I_D

In this paper, the transient delays and quasi threshold voltage, both of them during turn-on and -off, so altogether four TSEPs, are used as the means of monitoring the online junction temperature of the power device, and an improved circuit is proposed to measure the same which is independent of the DC-link voltage. Then the results are used to train a shallow fully connected neural network to improve the accuracy of the temperature measurements in an online application.

Measurement concept

The quasi-threshold voltage $V_{th,q}$ and transient delays during turn-on and off of the SiC power transistors as TSEPs are measured to extract the information on the junction temperature of the transistor. The measurement circuits for quasi-threshold voltage and transient delays are adapted from [5] and [7] respectively. The block diagrams of the measurement circuit are shown in Fig. 1 and 2. Whenever there is a change in current flowing through the transistor during turn-on and -off switching transient, there is a voltage drop ($V_{SS'}$) across the parasitic inductance between power and Kelvin source of the transistor. For the measurement of quasi-threshold voltage, the $V_{SS'}$ is used as a trigger during turn-on and turn-off transient of the transistor. To generate these triggers, a comparator is connected across the power and Kelvin source terminal of the transistor which generates an output as soon as the $V_{SS'}$ goes higher than a certain reference voltage. This comparator is designed in a way that the trigger is generated before the transistor reaches the Miller plateau region. The dependencies from other electrical parameters are introduced in the Miller plateau region, therefore, the trigger generation in or after the Miller plateau is avoided. The output of the comparator is connected to a D flip-flop. The D flip-flop is used so that the measurement circuit is triggered only once in one switching cycle to avoid the false triggering due to the ringing in $V_{SS'}$ [5]. The two different circuits are used to measure the quasi-threshold voltage during turn-on and off transient. The only difference between the two circuits is the polarity of the comparator connected between the power and Kelvin source.

A commercially available time-to-digital converter (TDC) is used to measure the turn-on delay ($t_{d,on}$) and turn-off delay ($t_{d,off}$). A TDC requires two signals: first to start and second to stop its internal clock. The pulse-width modulated (PWM) signal from a function generator, which is temperature-independent,

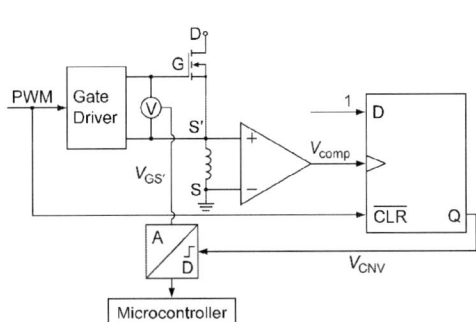

Fig. 1: Block diagram for quasi-threshold voltage measurements [5]

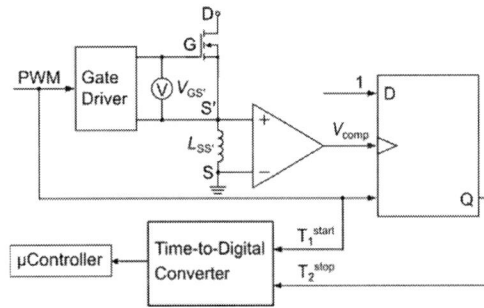

Fig. 2: Block diagram for quasi-threshold voltage measurements [7]

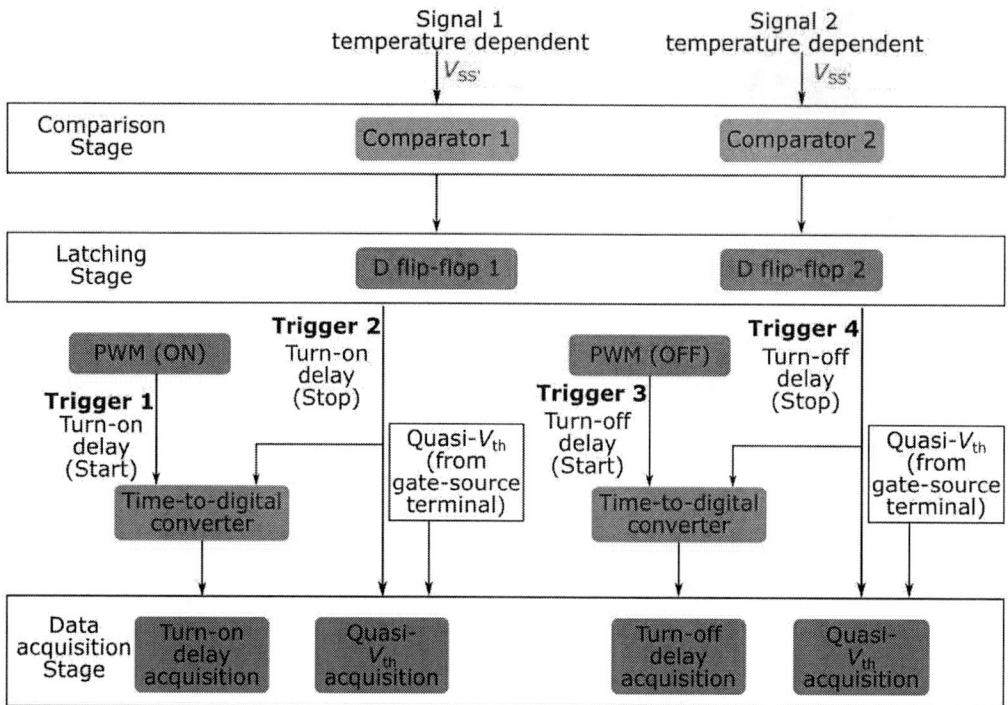

Fig. 3: Flow chart of all four TSEPs acquisition

is used as the first trigger to start the internal clock of the TDC. For the second trigger, which stops the TDC internal clock, the voltage drop across the parasitic inductance between Kelvin and the power source when the current flow changes in the transistor are used which is temperature-dependent [7]. The second trigger is the same which is used for $V_{th,q}$ acquisition. A flow chart is shown in Fig. 3 to explain the trigger generation and the acquisition of all four TSEPs.

Although measured TSEPs have very little dependencies on other electrical parameters, the novelty of this paper lies in the combination of TSEPs using machine learning (ML) algorithms to further improve the accuracy of the measurements. This method eliminates the influence of the load conditions on junction temperature measurements of the power transistor because the data acquisition takes place before the Miller plateau, moreover, the shallow fully connected neural network takes these conditions into account.

Online Measurements

The proposed measurement circuit is implemented on a commercially available SiC power module. The device under test used in this research is BSM120D12P2C005 from ROHM [8]. The gate driver circuits in a half-bridge configuration are designed and implemented in buck converter topology. The proposed junction temperature measurement method is applied on the low-side of the half-bridge while keeping the high-side off. An inductive load of 150 μH is used as an inductive load. The temperature is varied from 25 °C to 125 °C. The measurements are performed on SiC power MOSFET for quasi-threshold voltage and transient delay with varying DC-link voltage from 100 V to 400 V in 100 V step and load current of 5 A and 10 A during turn-on and -off transition of the silicon carbide power module. The 10 Ω external gate resistor in a double pulse test setup and the data is generated to train the ML model. During turn-on transition, the results show that the $t_{d,on}$ decreases linearly with the increase in temperature with the temperature sensitivity of around -275 ps/K. This negative temperature coefficient is due to its relation with the threshold voltage. As the DC-link voltage is increased from 100 V to 400 V, the change in temperature sensitivity has a negligible change of around -30 ps/K. The linearization of the raw data shows a root mean square error (RMSE) of 2.2 ns. Whereas for the threshold voltage during turn-on transition ($V_{\text{th,q}}^{\text{on}}$), the RMSE is 4 mV. The $V_{\text{th,q}}^{\text{on}}$ has a linear relationship with the temperature with a temperature sensitivity of around -7 mV/K. The negative temperature coefficient of $V_{\text{th,q}}^{\text{on}}$ is due to the increase in the carrier concentration and decrease in the bandgap. The measured value of the $V_{\text{th,q}}^{\text{on}}$ is decreased by 176 mV with the change in DC-link voltage from 100 V to 400 V whereas the change in temperature sensitivity is negligible. However, with the change in load current from 5 A to 10 A, the temperature sensitivity is increased to -11 mV/K and the threshold voltage is decreased by 667 mV. During the turn-off transition, the $t_{\text{d,off}}$ also shows a linear relationship with the temperature sensitivity of 350 ps/K. When the load current is increased from 5 A to 10 A, the $t_{\text{d,off}}$ is decreased by 25 ns but the temperature sensitivity remains similar. However, with the increase in DC-link voltage from 100 V to 400 V, the $t_{\text{d,off}}$ is increased by 13.6 ns. The quasi-threshold voltage during turn-off ($V_{\text{th,q}}^{\text{off}}$) also has a linear relationship with the temperature. It has a temperature sensitivity of -1.1 mV/K. The RMSE between the raw and the linearized data is around 20 mV. With the increase in load current from 5 A to 10 A, the acquired $V_{\text{th,q}}^{\text{off}}$ increases by around 50 mV with a little increase in temperature sensitivity to -1.5 mV/K whereas with the increase in DC-link voltage from 100 V to 400 V, the $V_{\text{th,q}}^{\text{off}}$ decreases by 41.7 mV. The results are summarized in Table II.

Table II: TSEPs and their temperature sensitivities

TSEPs	Temperature sensitivity	Root mean square error (RMSE)	Temperature error (in K)
Turn-on delay	-275 ps/K	2.2 ns	5.71
Threshold voltage during turn-on	-7 mV/K	4 mV	1.98
Turn-off delay	370 ps/K	1.85 ns	3.08
Threshold voltage during turn-off	-1.1 mV/K	20 mV	8.98

The error and variation in the measured value of all TSEPs are due to the parasitics introduced by the measurement circuit. A machine learning model can be useful to improve the overall error by combining multiple TSEPs.

Combination of TSEPs

ML algorithms are in general capable of learning complex input (TSEPs) to output (temperature) mappings. For the training procedure, training samples are needed, where the input-output relationship is included. For supervised learning, labels are needed which determine the target output. Since the relations between the TSEPs and temperature exist but are hard to combine which can lead to better accuracy, ML algorithms can be a promising approach to estimate the temperature based on the obtained TSEPs. For the first investigation, the correlation between the parameters and temperature is determined by the calculation of Pearson's linear correlation coefficient between all pairs of variables as shown in Fig. 4.

The figure shows a high correlation between the individual TSEPs as well as between the TSEPs and the temperature. Just $V_{\text{th,q}}^{\text{on}}$ shows a slightly lower correlation (-0.96) with the temperature compared to the other three TSEPs (-1, 1, -0.99). Two assumptions can be made based on this evaluation.

- The junction temperature of the power transistor can be estimated based on the given TSEPs data

- Individual TSEPs should also be used to estimate the junction temperature of the power transistor because of the high correlation between e.g. $V_{\text{th,q}}^{\text{on}}$, $t_{\text{d,on}}$, $t_{\text{d,off}}$ and the temperature

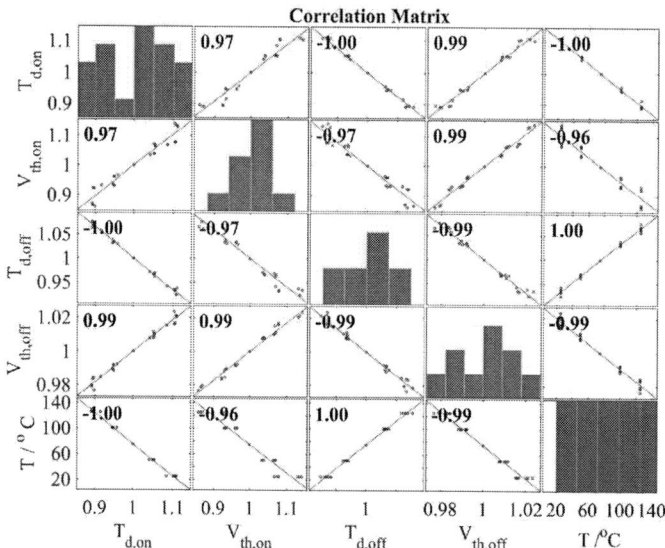

Fig. 4: Correlation coefficients between all TSEPs and temperature

A simple shallow fully connected neural network is used for estimating the junction temperature of the power transistor. In a fully connected layer, all neurons are connected to the outputs of the previous layer. Each connection has a weight parameter w_i, which is adjusted during training. Per neuron, one bias parameter b exists. The final output y_i of the neuron i in a fully connected neural network is given by equation 1 with the (non-linear) activation function σ.

$$y_i = \sigma(w_1 x_1 + \cdots + w_m x_m + b) \tag{1}$$

The network contains four layers, an input layer where the number of neurons is equal to the number of input parameters, two hidden layers with 10 and 5 neurons, and an output layer with one neuron for the junction temperature. The hidden layers use rectified linear unit (ReLU) activation function and the output layer uses a linear activation function. The basic architecture is shown in Fig. 5.

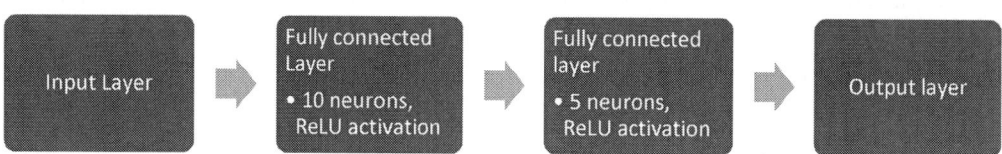

Fig. 5: Architecture of the shallow fully connected neural network

The network is trained for 100 epochs and the dataset is split randomly into 80% training data (32 measurements) and 20% test data (8 measurements). First, the neural network is trained with all four investigated TSEPs. Fig. 6 shows the prediction of the trained network on unseen test data. For each obtained parameter vector, estimation is performed. Therefore, the network can predict the temperature each time the TSEPs are sampled. The model shows promising results with a mean absolute temperature deviation of 0.92 K on the test data. The neural network is also trained with each TSEP as an input

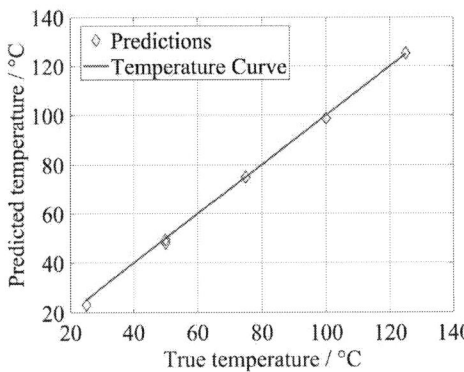
Fig. 6: Temperature predictions of the neural network (all TSEPs) on unseen test data

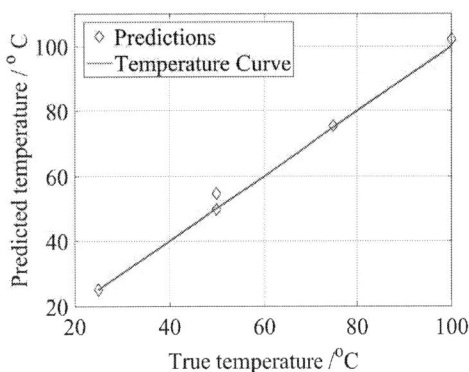
Fig. 7: Temperature predictions of the neural network ($t_{d,on}$) on unseen test data

before combining them. The network which is trained with $t_{d,on}$ as an input parameter shows best temperature prediction among models with other individual TSEP as input. Therefore, the prediction on unseen test data of $t_{d,on}$ is shown in Fig. 7. The model is also able to estimate the temperature well with a mean absolute temperature deviation of 1.05 K on the test data. Table III shows the improvement in the accuracy of measured TSEPs after processing with the ML algorithm.

Table III: TSEPs and their mean temperature deviation

Transition	TSEP	Mean absolute temperature deviation (in K)
Turn-on	Delay time	5.71
	Quasi-threshold voltage	1.98
Turn-off	Delay time	3.08
	Quasi-threshold voltage	8.98
Combined TSEPs		**0.92**

For further evaluation of the model learning behavior, the model for the combined TSEPs is trained with 20 to 100 epochs with a step size of 20. The obtained mean temperature deviation (in K) is shown in Table IV, with the mean results over five independent runs.

Table IV: Performance for different number of epochs with combined TSEPs

Number of Epochs	Mean absolute temperature deviation (in K)
20	3.05
40	2.39
60	1.43
80	**1.37**
100	1.57

The results of the table show, that already from 60 epochs on, good results can be obtained, while with 80 epochs the best mean performance over five independent runs are achieved. Therefore, 80 epochs are further used for model training. Until now, the data was split randomly into 80% training data (32 measurements) and 20% test data (8 measurements). As the next step, the model's capability of learning with a limited amount of training data shall be proved. Therefore, the ratio of training and test data will be changed from 20% training data to 80 % training in steps of 15%. The mean results of five independent runs for each split are shown in the following Table V.

The best results are obtained with the highest number of training data. However, already with 20 measurements as training data, better results are obtained than with the best single TSEP model (see Table III). From 20 measurements on, the performance of the model just slightly improves, so it can also be used with less training data than the previously selected 32. However, a decrease in performance has to be considered for this.

Table V: Performance for different amounts of training data with combined TSEPs

Number of measurements in training data	Mean absolute temperature deviation (in K)
8	5.72
14	2.84
20	1.88
26	1.59
32	**1.37**

For TSEPs noise and measurement, inaccuracy is always a big factor and one of the main reasons adoption of that measurement principle is difficult in the application. A small change in the measured parameter usually causes a huge estimation error. Therefore, the effect of adding noise to the measurement data is investigated in this paper. White Gaussian noise with mean zero and different values for the variance is added to the normalized signal (normalized to [0, 1]) and the final estimation performance of the model on test data over five independent runs is shown in Table VI. The variance was set to the values of 0.001, 0.005, 0.01, and 0.05.

Table VI: Performance for different variance values

Variance for noise generation	Mean absolute temperature deviation (in K)
0	**1.37**
0.001	2.66
0.005	3.51
0.01	4.96
0.05	18.82

The performance is drastically decreasing with added noise. With small variances (e.g. 0.005) the performance is in a similar range, but above 0.01 variance, the deviation of the estimation to the original value is becoming worse. Therefore, the model needs to be further adjusted for added noise in the measurements to become robust against measurement inaccuracies and noise.

It is assumed, that the accuracy of the temperature estimation can further be improved by using the additional temperature-sensitive parameter as an additional input. However, the results with the combined TSEPs are already promising and serve as a strong baseline for further work.

The temperature sensitivity after combining the TSEPs using neural networks cannot be determined because the neural network works as a black box with many non-linear behaviors. Each neuron in the model adds a non-linearity. However, the capability of the neural network as a universal function approximator offers the possibility to model arbitrary complex relationships and establish correlations between individual and combined TSEPs.

Conclusion

The junction temperature of the power transistor is measured via temperature-sensitive electrical parameters. Four TSEPs are considered here, namely the quasi-threshold voltage and the delay times during turn-on and turn-off. The data generated from the measurements are used to improve the accuracy of the measurement by training a shallow fully connected neural network. The maximum error in the measurement is reduced from 8.98 K to 0.92 K. The temperature sensitivities after combining the TSEPs are not possible to calculate since the used machine learning algorithm works as a black-box. However, there is a possibility to calculate second or third-order polynomial approximation using a symbolic regression algorithm in the machine learning model.

References

[1] E. Laloya, O. Lucia, H. Sarnago and J. M. Burdio, "Heat Management in Power Converters from State of the Art to Future Ultrahigh Efficiency Systems," *IEEE transactions on Power Electronics,* vol. 31, no. 11, pp. 7896-7908, 2016.

[2] J. Ruthardt, L. Schnabel, P. Ziegler, P. Marx, K. Sharma, M. Fischer, M. Nitzsche and J. Roth-Stielow, "Closed Loop Junction Temperature Control of Power Transistors for Lifetime Extension," in *IEEE Applied Power Electronics Conference and Exposition (APEC)*, New Orleans, LA, USA, 2020.

[3] K. Muñoz Barón, K. Sharma, M. Nitzsche, P. Ziegler, D. Koch and I. Kallfass, "Characterization of Threshold Voltage for Application-Oriented Power Cycling Conditions for Wide-Bandgap Power Devices," in *Internation Exhibition and Conference for Power Electronics, Intelligent Motion, Renewable Energy and Energy Management (PCIM)*, Nuremberg, Germany, 2020.

[4] Y. Avenas, L. Dupont and Z. Khatir, *Temperature Measurement of Power Semiconductor Devices by Thermo-Sensitive Electrical Parameters - A review,* vol. 27, 2016, pp. 3081-3092.

[5] K. Sharma, K. Muñoz Barón, J. Ruthardt, J. Hueckelheim, D. Koch, F. Muenzenmayer and I. Kallfass, "Characterisation of the Junction Temperature of SiC Power Devices via Quasi-Threshold Voltage as Temperature Sensitive Electrical Parameter," in *11th International Conference on Integrated Power Electronics Systems*, Berlin, 2020.

[6] L. Qiao, F. Wang, J. Dyer and Z. Zhang, "Online Junction Temperature Monitoring for SiC MOSFETs Using Turn-On Delay Time," in *IEEE Applied Power Electronics Conference and Exposition (APEC)*, New Orleans, USA, 2020.

[7] K. Sharma, K. Muñoz Barón, J. Ruthardt and I. Kallfass, "Online Junction Temperature Monitoring of Wide Bandgap Power Transistors using Quasi Turn-on Delay as TSEP," in *2021 IEEE 8th Workshop on Wide Bandgap Power Devices and Applications (WiPDA)*, Redondo Beach, CA, USA, 2021.

[8] ROHM Semiconductors, "SiC Power Module BSM120D12P2C005 Datasheet," ROHM semiconductors. [Online].

Novel Extended Robust Disturbance Observer for Improved Cogging Force Compensation in Permanent Magnet Linear Motors

Franz Luckert*, Axel Mertens**
WITTENSTEIN cyber motor GmbH
97999 Igersheim, Germany
Phone: +49 (0) 7931 493-18050
Fax: +49 (0) 7931 493-200
Email: franz.luckert@wittenstein.de
URL: https://www.cyber-motor.wittenstein.de/

**Institute for Drive Systems and Power Electronics
Leibniz University Hannover
30167 Hannover, Germany
Phone: +49 (0) 511 762-2471
Fax: +49 (0) 511 762-3040
Email: mertens@ial.uni-hannover.de
URL: https://www.ial.uni-hannover.de

Keywords

≪Linear drive≫, ≪Ripple minimization≫, ≪Neural network≫, ≪Industrial application≫, ≪Robust control≫

Abstract

This paper presents an extended robust disturbance observer for improved force ripple compensation of a permanent magnet synchronous linear motor. In extension, working from the output of the robust disturbance observer, the amplitudes of dominant harmonics are estimated by means of a harmonic-activated neural network. With the knowledge of the amplitude, the phase shift of an individual harmonic resulting from the robust disturbance observer can be corrected. This method improves the cogging force compensation by feedforward, especially when the frequency of the force ripple increases due to higher travelling speeds.

Introduction

Numerous industrial applications require linear motion with high demands on dynamic and positioning accuracy [1]. These include wafer scanners and wafer steppers used for photolithography in the manufacture of semiconductors. While wafer steppers have strict accuracy requirements regarding the final position of their movement, wafer scanners require a high positioning accuracy during their constant movement [2]. A drive that is ideally suited for linear positioning tasks is the permanent magnet synchronous linear motor (PMSLM). The PMSLM is a direct drive with high dynamics and virtually unlimited positioning range. However, its design leads to a force ripple, which has a detrimental effect on the positioning accuracy in the micrometer range.

Many efforts have already been made to reduce the cogging force of linear motors. In addition to design approaches, different control strategies have been investigated. A feedforward compensation based on offline determination of the cogging force was presented in [3]. However, due to manufacturing tolerances, the experimental determination of the first-order ripple forces must be performed for each

PMSLM individually. For this reason, the concept would be very costly for broad application. In [4], a neural-network-based feedforward-assisted feedback controller was investigated. [5] presents a disturbance observer based on an inverse model of the plant. Because of a lack of inhibiting ability for middle- and high-frequency force ripple components, the proposed disturbance observer cannot compensate force ripple completely. To deal with the higher harmonics of the cogging force, a compensation method based on inverse model iterative learning in combination with a robust disturbance observer (RDOB) is investigated in [6]. Neural networks for an inversion based feedforward controller design were investigated in [9]. While neural networks with an autoregressive exogeneous structure were not able to improve tracking performance due to being unable to track the general system dynamics, physicsguided neural networks increase performance compared to several other benchmark feedforward strategies like, for example, massacceleration feedforward. A force ripple compensation strategy using a fuzzy-adaptive Kalman filter was investigated in [10]. The fuzzy-adaptive Kalman filter algorithm is able to reduce the effort needed to estimate the noise covariance matrix by tuning the parameters. Due to the improved noise reduction, the force ripple and, thus, also the position tracking performance is significantly improved.

This paper is based on the RDOB detailed in [6]. The focus is placed on extending the RDOB to minimize the effect of phase shift at higher harmonics of the cogging force. Finally, an extended robust disturbance observer (ERDOB) is presented that is able to suppress the dominant higher harmonic components of the cogging force. For this purpose, the amplitudes of the dominant harmonic components are estimated by means of a harmonic-activated neural network (HANN). By knowing amplitudes of the dominant harmonics and position of the moving part, these components of the cogging force can be canceled by feedforward control.

This paper is divided into three sections. First, the problem is described and the underlying model is explained. Second, the structure of the ERDOB is presented and this is finally verified by experiments in the last section.

Modeling and problem description

The model of the PMSLM is known from the literature [4]. Since the current control loop is assumed to be fast enough, only the mechanical dynamics of the PMSLM are considered:

$$\ddot{x}(t) = \frac{f_i(i_q,t) - f_f(\dot{x},t) - f_c(x,t)}{m}. \tag{1}$$

$$\text{with } f_i(i_q,t) = k_F \cdot i_q(t). \tag{2}$$

The parameter m represents the mass of the glider. k_F is the force constant of the PMSLM and i_q is the current in the q-axis. x denotes the position of the glider and $\ddot{x}$ is its acceleration. Furthermore, the model contains the speed-dependent frictional force f_f and the position-dependent cogging force f_c, which are understood to be disturbance forces:

$$f_d(x,\dot{x},t) = f_f(\dot{x},t) + f_c(x,t). \tag{3}$$

As already mentioned, the cogging force is caused by the mechanical design of the PMSLM. Parameters such as the opening of the stator slots or the finite length of the glider generate spatially distributed periodic disturbance forces. According to [8], the cogging force can finally be modeled as a Fourier series expansion truncated after the N-th harmonic:

$$f_c(x,t) = F_0 + \sum_{n=1}^{N} F_{cA,n} \cdot \cos\left(\frac{\pi}{\tau_P} \cdot k_{c,n} \cdot x(t)\right) + \sum_{n=1}^{N} F_{cB,n} \cdot \sin\left(\frac{\pi}{\tau_P} \cdot k_{c,n} \cdot x(t)\right). \tag{4}$$

Here, $k_{c,n}$ denotes the period of the n-th harmonic of the cogging force as a multiple of the pole pitch τ_P. The parameters $F_{cA,n}$ and $F_{cB,n}$ describe the amplitudes of the even and odd parts of the harmonic.

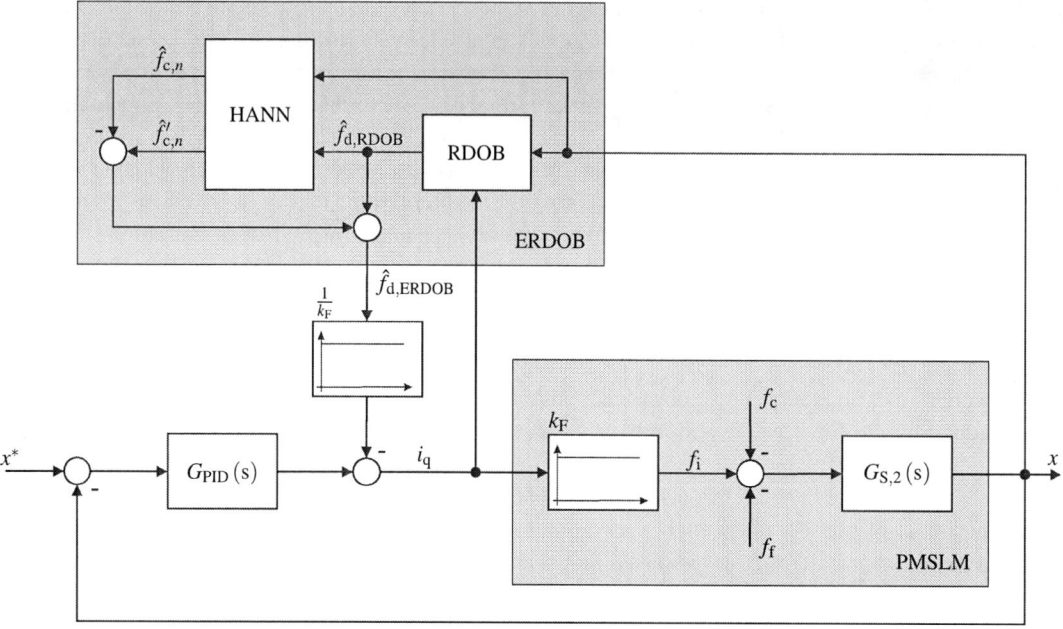

Fig. 1: The structure of the controlled system

Design of the extended robust disturbance observer

The ERDOB combines an RDOB and a HANN to minimize the effect of cogging force on the positionalaccuracy by using feedforward control. The structure of the control system is shown in Figure 1. The RDOB estimates the disturbance forces from the measured position x and known control value i_q. For this purpose, the positional value is multiplied with the inverse transfer function of the disturbance. Since the inverse plant model is double differentiating, it is multiplied by a second-order low-pass filter to make it proper and to obtain a practically realizable transfer function for the RDOB:

$$G_{\text{RDOB}}(s) = G_{\text{S},2}^{-1}(s) \cdot G_{\text{LP}}(s) = \frac{m \cdot s^2}{\xi_2 \cdot s^2 + \xi_1 \cdot s + \xi_0}. \tag{5}$$

The parameters ξ_0, ξ_1, ξ_2 are the coefficients of the low-pass filter. In Equation (1), it can be seen that not only the disturbance forces f_c and f_f, but also the control value f_i, affect the output x of the plant. To obtain solely the disturbance forces, the low-pass filtered control value f_i must be subtracted from the output of $G_{\text{RDOB}}(s)$:

$$\hat{F}_{\text{d,RDOB}}(s) = \hat{F}_{\text{c,RDOB}}(s) + \hat{F}_{\text{f,RDOB}}(s) = G_{\text{RDOB}}(s) \cdot X(s) - k_{\text{F}} \cdot G_{\text{LP}}(s) \cdot I_q(s). \tag{6}$$

$\hat{F}_{\text{d,RDOB}}(s)$, $\hat{F}_{\text{c,RDOB}}(s)$, $\hat{F}_{\text{f,RDOB}}(s)$, $X(s)$ and $I_q(s)$ denote the Laplace transformation of these variables. The final results are the estimated frictional force $\hat{f}_{\text{f,RDOB}}$ and the estimated cogging force $\hat{f}_{\text{c,RDOB}}$. In terms of estimating the cogging force, however, this robust disturbance force observer also has a distinct disadvantage. As already described at the beginning, the cogging force consists of several harmonics. Depending on their order and the speed of the glider, the estimated harmonics are phase shifted. As a result, these cannot be completely canceled out by feedforward control.

$\hat{F}_{\text{d,RDOB}}(s)$, $\hat{F}_{\text{c,RDOB}}(s)$, $\hat{F}_{\text{f,RDOB}}(s)$, $X(s)$ and $I_q(s)$ denote the Laplace transformation of these variables. The result is finally the estimated friction force $\hat{f}_{\text{f,RDOB}}$ and the estimated cogging force $\hat{f}_{\text{c,RDOB}}$. In terms of estimating the cogging force, however, this robust disturbance force observer also has a decisive disadvantage. As already described at the beginning, the cogging force consists of several harmonics. Depending on their order and the speed of the glider, the estimated harmonics are phase shifted. As a

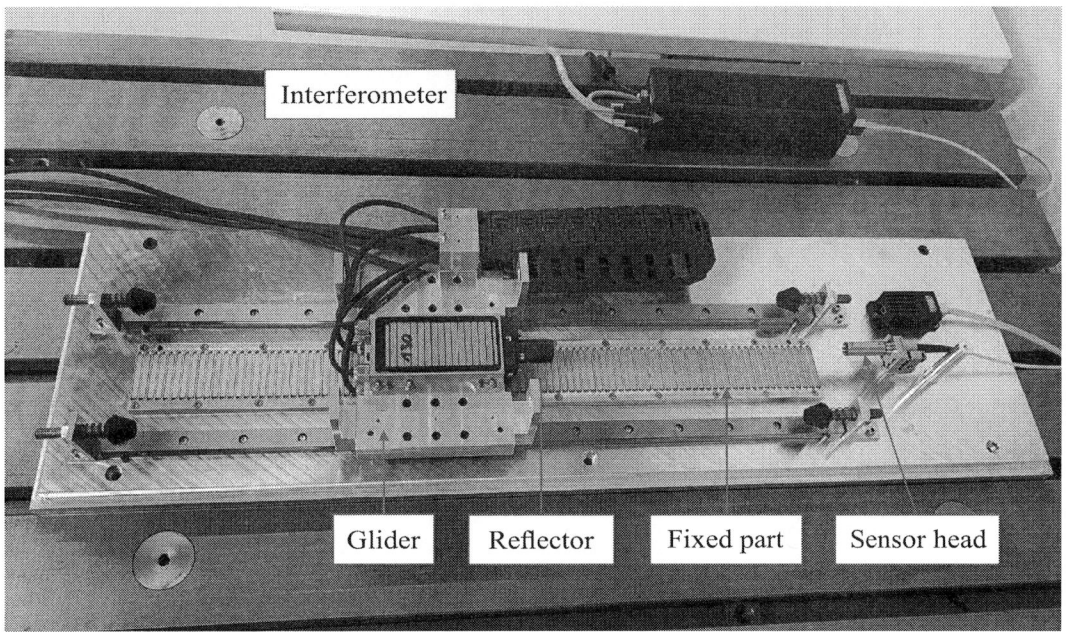

Fig. 2: The test setup for experimental investigation

result, these cannot be completely canceled out by feedforward control.

By performing an FFT on the positional error, the remaining influence of each harmonic can be determined. To improve the cogging force estimation, dominant harmonics can be canceled by feedforwarding them in the correct phase position. As the order of the desired harmonic is already known, only the amplitude and phase are needed. For estimating these two parameters, a HANN presented in [7] is used. The neural network for the desired harmonic is modeled according to Equation (4):

$$\hat{f}_{c,n}(x,t) = \hat{\Theta}_{A,n}(t) \cdot \sin\left(\frac{\pi}{\tau_P} \cdot k_{c,n} \cdot x(t)\right) + \hat{\Theta}_{B,n}(t) \cdot \cos\left(\frac{\pi}{\tau_P} \cdot k_{c,n} \cdot x(t)\right) \tag{7}$$

$$\text{with } \hat{\Theta}_{A,n}(t=0) = \hat{\Theta}_{A0,n}, \ \hat{\Theta}_{B,n}(t=0) = \hat{\Theta}_{B0,n}. \tag{8}$$

The index n denotes the order of the harmonic to be estimated. $\hat{\Theta}_{A,n}$ and $\hat{\Theta}_{B,n}$ are the amplitudes of its even and odd spectral components. For calculating the error, the previously estimated cogging force of the RDOB is used:

$$e(t) = \hat{f}_{c,n}(x,t) - \hat{f}_{d,\text{RDOB}}(x,t). \tag{9}$$

In principle, the error could also be calculated from the measured position, as with the RDOB. But in this case, the cogging force would pass through the strictly positive real transfer function $G_{S,2}(s)$ of the plant, which would require a more complex error network. In order to adapt the amplitudes $\hat{\Theta}_A$ and $\hat{\Theta}_B$, a gradient descent method is used as the learning law:

$$\frac{\mathrm{d}}{\mathrm{dt}}\hat{\Theta}_{A,n}(t) = -\eta \cdot e(t) \cdot \sin\left(\frac{\pi}{\tau_P} \cdot k_{c,n} \cdot x(t)\right) \tag{10}$$

$$\frac{\mathrm{d}}{\mathrm{dt}}\hat{\Theta}_{B,n}(t) = -\eta \cdot e(t) \cdot \cos\left(\frac{\pi}{\tau_P} \cdot k_{c,n} \cdot x(t)\right). \tag{11}$$

Here, η denotes the increment of the learning law. Finally, the HANN provides the amplitudes of the even and odd components of the n-th harmonic of the previously estimated and phase-shifted cogging

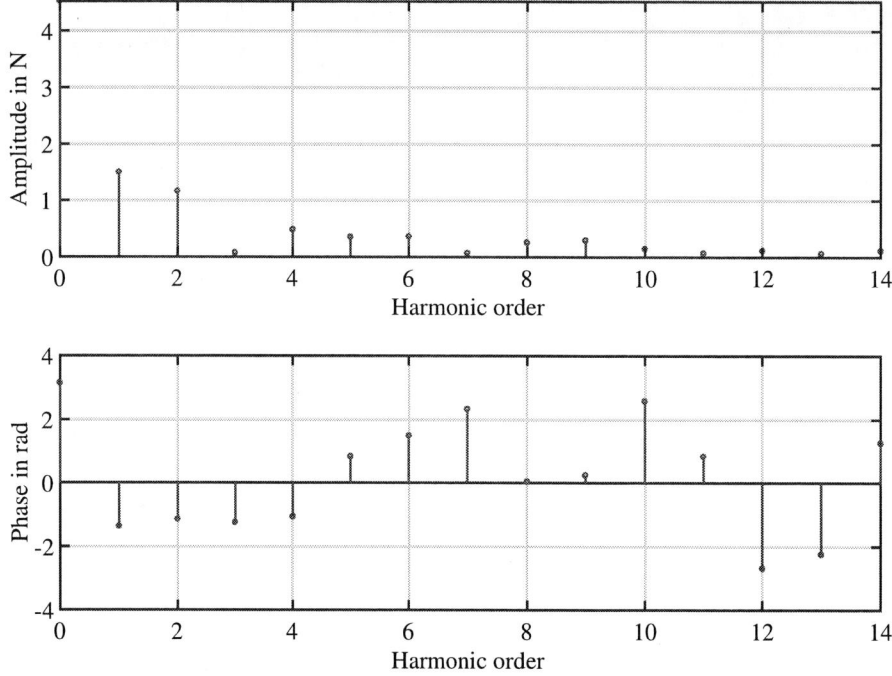

Fig. 3: Estimated disturbance force frequency characteristic

force $\hat{f}_{c,RDOB}$. Due to the fact that the frequency of the harmonic is known from its period $k_{c,n}$ and the glider speed $\dot{x}$, the phase shift $\varphi_{LP,n}$ caused by the low-pass filter can be calculated and corrected:

$$\hat{f}'_{c,n}(x,t) = \hat{\Theta}_{A,n}(t) \cdot \cos\left(\frac{\pi}{\tau_P} \cdot k_{c,n} \cdot x(t) - \varphi_{LP,n} \right) + \hat{\Theta}_{B,n}(t) \cdot \sin\left(\frac{\pi}{\tau_P} \cdot k_{c,n} \cdot x(t) - \varphi_{LP,n} \right). \quad (12)$$

Finally, the estimated disturbance force of the ERDOB is the calculated:

$$\hat{f}_{d,ERDOB}(x,t) = \hat{f}_{d,RDOB}(x,t) - \hat{f}_{c,n}(x,t) + \hat{f}'_{c,n}(x,t). \quad (13)$$

The estimated disturbance force can be used to compensate the force ripple by feedforward control. The validation of this concept will be presented in the next section.

Experimental results

To demonstrate the effectiveness of the proposed ERDOB, it is now validated by experiment. The experimental setup shown in Figure 2 uses a standard WITTENSTEIN PMSLM, which is not optimized for low cogging force. It is equipped with an Attocube interferometer providing a positional resolution of 1 nm. The newly designed control algorithm is implemented on a dSPACE rapid prototyping system with the DS1006 processor board, which has a sampling rate of 16 kHz.

In the first step, the plant of the experimental setup is analyzed and the dominant harmonics of the disturbance force are determined. In the second step, it is considered how well these can be suppressed by using the ERDOB.

Identification of the disturbance forces and system analysis

To determine the disturbance force, the glider is moved at a very low speed of 0.01 m/s by PID control. If we now apply an FFT to the output signal $\hat{f}_{d,RDOB}$ of the RDOB, we obtain the spectrum of the estimated

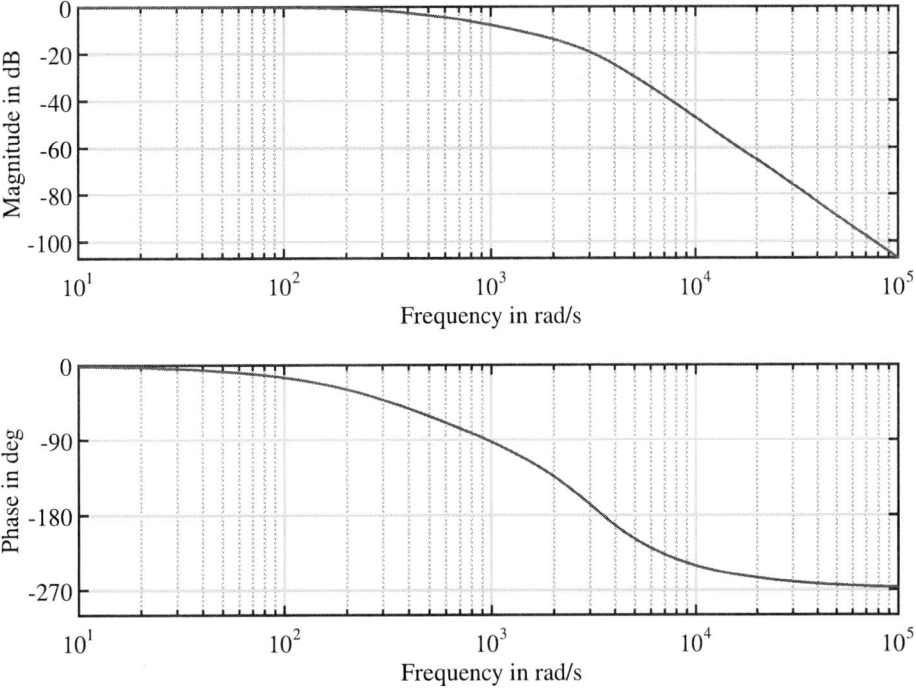

Fig. 4: Bode plot of the RDOB

disturbance force. The parameters of the RDOB are chosen to be $\xi_0 = 1$, $\xi_1 = 0.45\,\text{ms}$, $\xi_2 = 0.1\,\text{ms}^2$. The time constant for the additional low-pass filter is $T_\text{LP} = 1.59\,\text{ms}$. Since the transfer functions of $G_{\text{LP},1}(s)$ and $G_{\text{LP},2}(s)$ are known, the amplitudes of the single harmonics together with their phase shifts can finally be corrected. The results are shown in Figure 3. Beside the frictional force, the two harmonics with order $k_{\text{c},1} = 1$ and $k_{\text{c},2} = 2$ are dominant. The Bode plot of the RDOB in Figure 4 shows that a phase shift is already occurring for the first-order harmonic at a speed $v = 0.1\,\text{m/s}$, which results in a frequency of $\omega_{\text{c},1} = 57\,\text{rad/s}$. For the second-order harmonic, attenuation is already taking effect. Since the frequency increases proportional to the velocity, phase correction should already be applied to the first-order and second-order harmonics. The effect of these will be investigated in the next section.

Measurement

To evaluate the benefits of amplitude and phase corrections, the ERDOB is compared to a system in which the cogging force for feedforward control is estimated by using only the RDOB. The results are also compared to a system without cogging force feedforward, which is only PID controlled. For the investigation, a trapezoidal velocity profile is used and the glider is moved over a distance of 100 mm at a constant speed. For the system with cogging force estimation and feedforward control, a settling time of 0.1 s is allowed, after which the remaining positional error $\Delta x = x - x^*$ is then determined.

The HANN of the ERDOB is set up to estimate the two dominant harmonics $k_{\text{c},1} = 1$ and $k_{\text{c},2} = 2$ of the cogging force determined when analyzing the plant. For this purpose, the increment of the learning law is chosen to be $\eta = 50$.

Figure 5 a) shows the remaining positional error for a constant glider speed of $v = 0.1\,\text{m/s}$. The maximum positional error for the PID-controlled system is 4.503 μm. By feedforwarding the disturbance force estimated with the RDOB, the remaining positional error can be limited to 1.543 μm after the settling period. If the cogging force estimated with the proposed ERDOB is used for cogging force cancelation, the positional error can be reduced to 1.117 μm. Finally, the spectrum of the resulting positional error is shown in Figure 5 b). As can be seen here, the ERDOB does not have any impact on the

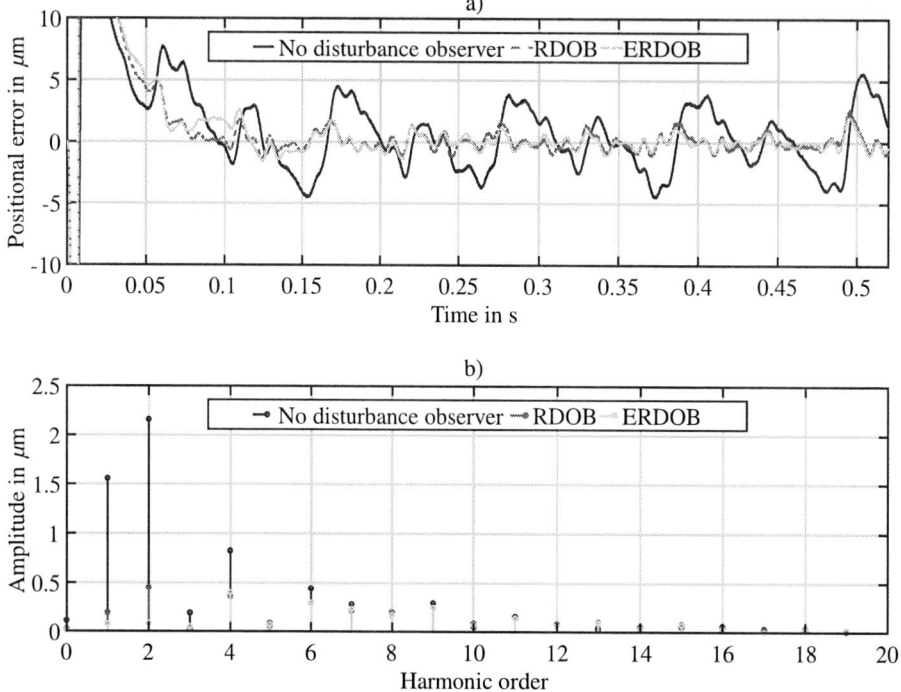

Fig. 5: a) Positional error for $v = 0.1\,\text{m/s}$; b) Positional error frequency characteristic for $v = 0.1\,\text{m/s}$

first-order harmonic of the cogging force compared to the RDOB. But it is able to reduce the impact of the second-order harmonic of the cogging force on the positional error.

Figure 6 a) shows the remaining positional error for the glider moving at a constant speed of $v = 0.2\,\text{m/s}$. The maximum positional error for the PID-controlled system is $5.152\,\mu\text{m}$. By feedforwarding the disturbance force estimated with the RDOB, the remaining positional error can be limited to $3.403\,\mu\text{m}$ after the settling period. By using the proposed ERDOB for cogging force estimation, the positional error can be reduced to $2.562\,\mu\text{m}$. Even at the speed $v = 0.2\,\text{m/s}$, the ERDOB was able to reduce the second-order harmonic in the positional error by suppressing the cogging force, as Figure 6 b) shows. The impact on the first-order harmonic is still very low as this is compensated well enough by the RDOB itself, due to low phase shift and low attenuation.

Lastly, a travelling speed of $v = 0.5\,\text{m/s}$ is considered. With this movement, the PID controller can only limit the positional deviation to $4.590\,\mu\text{m}$. If the disturbance forces, which are only estimated with the RDOB, are feedforward, the positional error after the settling period can be limited to $3.797\,\mu\text{m}$. By using the proposed ERDOB, the remaining positional error is only $2.3\,\mu\text{m}$. The results are shown in Figure 7 a). With the increased speed, the phase correction of the ERDOB becomes more important. Figure 7 b) shows that the RDOB is no longer able to suppress the second-order harmonic of the cogging force. But with the ERDOB, it can be reduced as well as the first-order harmonic by using the ERDOB appropriately.

The experiments have shown that the positional error can be reduced by using a disturbance observer. If no observer is used to suppress the harmonics of the cogging force, the positional error remains in the same range for all three glider speeds considered. With higher glider velocities, the positional error for the feedforward control with the RDOB increases together with the phase shift and attenuation of the two low-pass filters used in the disturbance force estimation. This was already predicted when the system was described and analyzed. The spectra of the positional errors in Figure 5 b), 6 b) and Figure 7 b) also show that the influence of the ERDOB rises with increasing speed. Thus, the first-order harmonic of the disturbance force was suppressed almost equally well by the RDOB and ERDOB at the speed $v = 0.2\,\text{m/s}$. But at the speed $v = 0.5\,\text{m/s}$, the ERDOB suppresses the first harmonic much better than

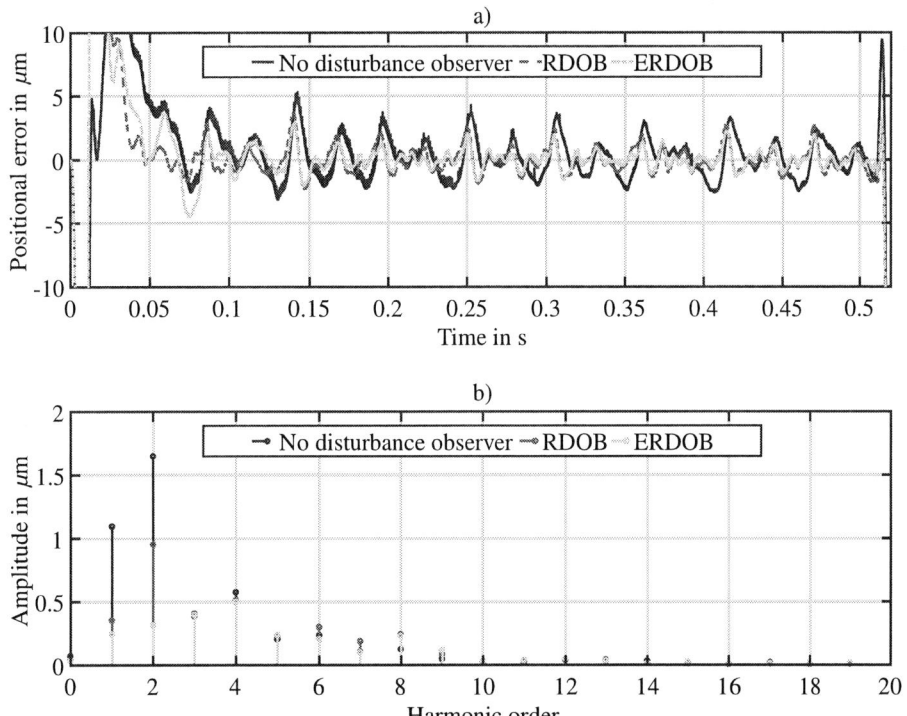

Fig. 6: a) Positional error for $v = 0.2\,\text{m/s}$; b) Positional error frequency characteristic for $v = 0.2\,\text{m/s}$

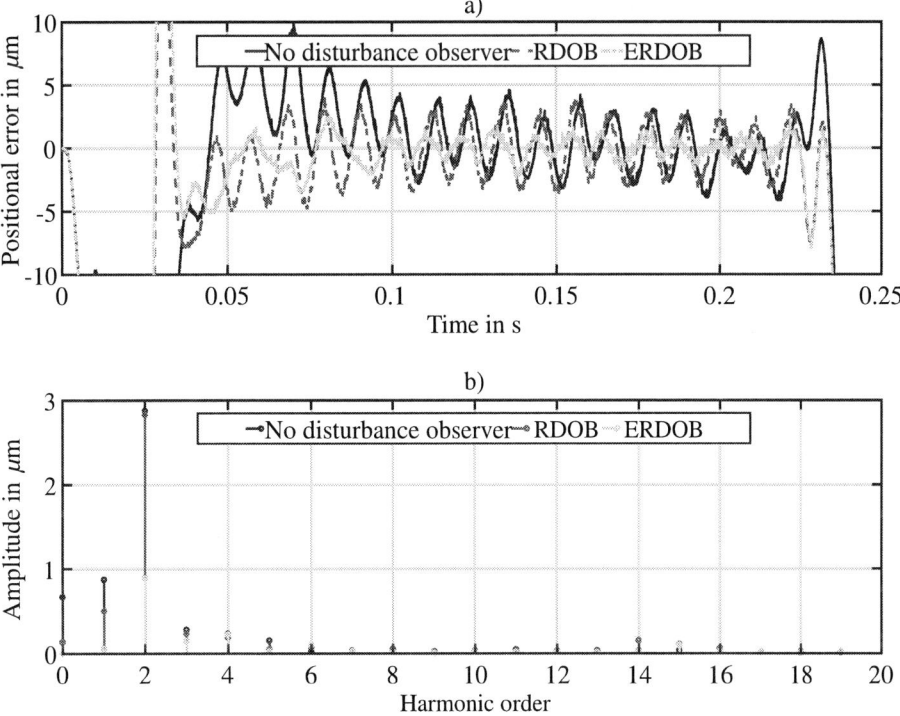

Fig. 7: a) Positional error for $v = 0.5\,\text{m/s}$; b) Positional error frequency characteristic for $v = 0.5\,\text{m/s}$

the RDOB. For the second-rder harmonic, the RDOB has already reached its limit and is no longer able to reduce the impact of this. However, the ERDOB is still able to reduce the impact of the second-order harmonic on the positional error. In summary, this section has shown that an improvement in positional accuracy can be achieved using the concept of ERDOB.

Conclusion

In this paper, the use of an ERDOB for motion control of a PMSLM has been demonstrated. The ability to feed forward dominant harmonics of the cogging force with correct phase positons has reduced the positional error significantly. Also, taking account of all other spectral components of the cogging force with the RDOB has further improved the system in terms of positional accuracy. Therefore, it is to achieve a sufficiently satisfying result possible without any design changes and with little modeling effort. The use of a feedforward structure has the consequence that the feedback controller can be designed with more emphasis on robustness. A further improvement of the system can still be achieved if, in addition to the phase position, the damping of the harmonic caused by the RDOB is also compensated.

References

[1] H.-H. Mu, Y.-F. Zhou, X. Wen, Y.-H. Zhou: "Calibration and compensation of cogging effect in a permanent magnet linear motor", Mechatronics Vol 19, 2009, pp. 577-585

[2] Hans Butler: "Position Control in Lithographic Equipment [Applications of Control]" PIEEE Control Systems Magazine, 2011, pp. 28-47

[3] P. Van Den Braembussche, J.Swevers, H.Van Brussel, P.Vanherck: "Accurate tracking control of linear synchronous motor machine tool axes", Mechatronics Vol 6, 1996, pp. 507-521

[4] G. Otten, T. J. A. de Vries, J. van Amerongen, A. M. Rankers, and E. W. Gaal: "Linear motor motion control using a learning feedforward controller", IEEE/ASME Trans. Mechatronics Vol 2, 1997, pp. 161-170

[5] K. K. Tan, T. H. Lee, H. F. Dou, S. J. Chin, and S. Zhao, "Precision motion control with disturbance observer for pulsewidth-modulated-driven permanent-magnet linear motors", IEEE Transactions on Magnetics, 2003, Vol 39, pp. 1813-1818

[6] F. Xuewei, Y. Xiaofeng, C. Zhenyu: "A New Linear Motor Force Ripple Compensation Method Based on Inverse Model Iterative Learning and Robust Disturbance Observer", Complexity Vol 2018, pp. 1-19

[7] D. Schroeder, Martin Buss: "Intelligente Verfahren", Berlin; Heidelberg: Springer Vieweg, 2017

[8] J. Malaiz, J. Lvine: "An Observer-based design for cogging forces cancellation in permanent magnet linear motors", Proceedings of the 48h IEEE Conference on Decision and Control (CDC) held jointly with 2009 28th Chinese Control Conference, 2009, pp. 6811-6816

[9] M. Bolderman, M. Lazar and H. Butler, "PhysicsGuided Neural Networks for Inversionbased Feedforward Control applied to Linear Motors", 2021 IEEE Conference on Control Technology and Applications (CCTA), 2021, pp. 1115-1120

[10] X. Zhang, T. Zhao and X. Gui, "Force Ripple Suppression of Permanent Magnet Linear Synchronous Motor Based on Fuzzy Adaptive Kalman Filter", 2021 13th International Symposium on Linear Drives for Industry Applications (LDIA), 2021, pp. 1-5

Improvement of a self-powered gate driver power supply

Mariana Raya[1], Oriol Aviñó[1], Sergio Busquets-Monge[2], Xavier Perpiñà[1], Miquel Vellvehi[1], Xavier Jordà[1]

1: MICROELECTRONICS NATIONAL CENTER (IMB-CNM, CSIC)
C/ dels Til·lers s/n, Campus UAB, Bellaterra, Spain.
2: UNIVERSITAT POLITÈCNICA DE CATALUNYA (UPC)
Campus Diagonal Sud, Edifici H, Av. Diagonal, 647, Barcelona, Spain
Tel.: +34 93 594 7700
Fax: +34 93580 0267
mariana.raya@imb-cnm.csic.es, oriol.avino@ imb-cnm.csic.es, sergio.busquets@upc.edu,
xavier.perpinya@imb-cnm.csic.es, miquel.vellvehi@imb-cnm.csic.es, xavier.jorda@imb-cnm.csic.es
https://www.imb-cnm.csic.es/en

Acknowledgments

This work was supported by the Spanish Ministry of Science, Innovation and Universities under Project HIPERCELLS (RTI2018-098392-B-I00), the Regional Government of the Generalitat de Catalunya (Grant 2017 SGR 1384), the Consejo Superior de Investigaciones Científicas (JAE-INTRO Grant JAEINT21_EX_0937 and PTI+ TransEner Platform) and the Agencia Estatal de Investigación (Grant FJC2019-040660-I).

Keywords

«Intelligent gate driver», «Smart Gate Drivers», «Driver concepts», «Power supply», «Multi-level converters»

Abstract

In this work, an improved self-powered gate driver power supply is proposed, analyzed by simulation and experimentally validated. This solution is based on the addition of a voltage regulator and it achieves a floating, constant and robust voltage to supply the gate driver and auxiliary circuits (protections, sensors, local control and communications) in switching cells for multilevel converters implementation. The obtained gate driver supply voltage is stable for a wide range of frequencies and auxiliary circuit current consumptions. Moreover, the main characteristics of the main transistor turn-on and turn-off are preserved while decreasing the power dissipation of the gate driver power supply circuit.

Introduction

Among all power topologies involved in AC conversion, multilevel converters have been attracting an increasing attention during the last years. Such a solution presents additional advantages compared with the traditional two-level VSI topologies widely spread in the industry. In 2003, the introduction of the Modular Multilevel Converter (MMC) topology [1, 2] drastically boosted the practical implementation of the multilevel approach into the industry for a wide power range. The MMCs are implemented by the interconnection of basic Switching Cells (SC) formed by power transistors with their antiparallel diodes connected in half-bridge configuration and a capacitor across this structure. This passive component on each SC slows down the initialization process and requires specific control strategies at converter level for assuring the correct voltage balancing among capacitors. This fact and the need for bulky and heavy branch inductors are two of the drawbacks of the MMC topology. Among all multilevel solutions, the multilevel active clamped (MAC) topology introduced in 2011 [3], though showing a slightly higher harmonic distortion and switching losses, presents several clear advantages in terms of modularity [3], integrability [3], redundancy [4], and fault-tolerant capability [4], justifying the complexity and over cost coming from the higher number of used active devices and drivers. The MAC converters are formed by the association of basic SC without any capacitor; only N-1 capacitors are required at the converter input (N-1 = converter levels) and they do not require any inductance, providing practically a "full Silicon" (or "full semiconductor") solution. In return, the number of devices required for an N-level converter increases as N^2 (and not proportionally with N as in MMCs), increasing the complexity of the MAC implementation. As an evolution of the MAC approach, a new paradigm has emerged in 2018 [5], the so-called Switching-Cell Array (SCA) converter topology, which consists of a matrix arrangement of $2N^2$ highly-optimized SCs that can be easily reconfigured to produce converter legs with different voltage and current ratings.

In such a scenario, SC modularity and compactness is a crucial feature in order to tackle with the implementation of a large number of controlled devices in the same converter. Nevertheless, there is a lack of power module solutions specifically designed as elementary building blocks for multilevel topologies to demonstrate their major performances under regular and faulty operating conditions, especially for the SCA case. The present work aims at solving one of the limiting factors to achieve smart and compact SC modular solutions for SCA implementation: the floating gate driver power supply (GDPS).

A usual solution is an external gate driver power supply (EGDPS) using compact dc-dc converters [6], however it becomes costly and inefficient when the number of switches increases. An alternative could be a bootstrap power supply [7][8], but high-power and low-power states are not galvanically isolated and there is a dependence between top-side and low-side switching.

This work will focus on the solution proposed in [9], shown in Fig. 1, which is a self-powered internal GDPS (IGDPS) for the gate driver of the main power switch S_m (a power MOSFET in this case). In [10], an optimization of the design in terms of losses and switching times was performed at high switching frequency (100 kHz), but the driver supply voltage presents significant variations (ripple) that make it unsuitable to power auxiliary circuitry such as short circuit, over-temperature and shoot-through protections, as well as for supplying digital local control and communication devices. In this sense, the present work proposes a significant improvement of this IGDPS scheme based on the implementation of a voltage regulator to stabilize the driver supply voltage. The improved solution proposed in this work (internal regulated GDPS, IRGDPS) has been exhaustively analyzed by simulation (LTspice), and experimentally validated using a buck converter (70V VDMOSFET, V_{bus}= 30V, L=0.4mH).

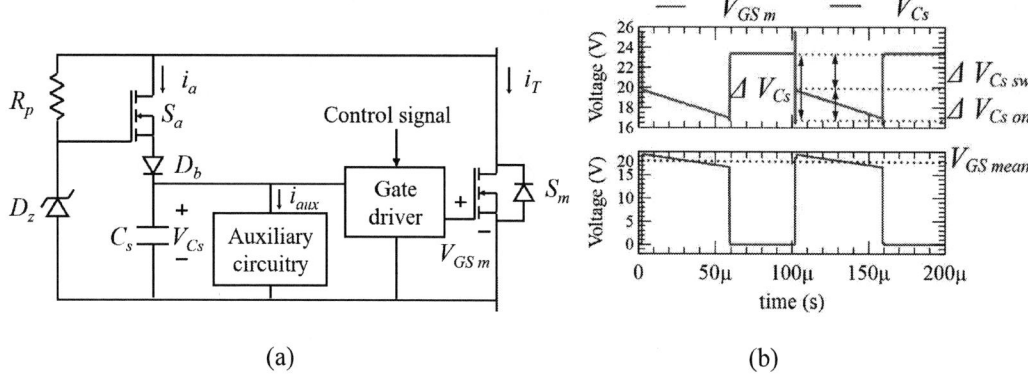

(a)　　　　　　　　　　　　　　　　　(b)

Fig. 1: (a) Internal gate driver power supply (IGDPS) circuit schematic. (b) Example of the capacitor voltage V_{Cs} and gate source voltage (V_{GSm}) of S_m of an IGDPS in a buck converter.

Internal gate driver power supply

In the IGDPS of Fig. 1 (a), the gate driver is powered by the capacitor voltage V_{Cs}. The capacitor C_s is charged during the turn-off of the main transistor S_m. Its maximum voltage ($V_{Cs\,max}$) will be the reverse voltage of the zener diode (V_z), minus the forward voltage drop of the diode D_b (V_{Db}) and the gate-source voltage drop of the auxiliary transistor S_a (V_{GSa}).

$$V_{Cs\,max} = V_z - V_{Db} - V_{GS\,a} \tag{1}$$

At the beginning of the S_m switching-off, the voltage at drain and source terminals of S_a start rising, then, S_a switches on and the current i_a charges the capacitor C_s. The capacitor stops charging when the cut off condition of S_a is achieved.

During the on-state of S_m, its gate voltage $V_{GS\,m}$ is V_{Cs} (gate driver supply voltage) minus the gate driver output stage voltage drop (ΔV_{driver}). Therefore, any variation in the C_s capacitor voltage during the on-state of S_m is directly transmitted to $V_{GS\,m}$ affecting the on-state resistance value of S_m. As mentioned before, $V_{Cs\,max}$ depends on V_z, V_{Db} and $V_{GS\,a}$, but its mean value depends also on the switching frequency (f) and the duty cycle (d) (which determine the C_s charging rate), as well as, its discharge process.

The diode D_b prevents the C_s from discharging through the auxiliary transistor when S_m is in on-state. On the other hand, C_s suffers two discharge processes: an abrupt discharge during the turn-on of S_m and a slow discharge during the S_m on-state. The abrupt discharge ($\Delta V_{Cs\,sw}$) associated to the abrupt V_{Cs} falling slopes shown in Fig. 1 (b) corresponds to the charge transfer between C_s and the S_m input gate capacitance during its turn-on. The subsequent slow discharge process is associated to the slow V_{Cs} falling slope shown in ($\Delta V_{Cs\,on}$) which corresponds to the current consumption of the gate driver and the auxiliary circuitry during S_m on-state ($I_{on} = I_{aux} + I_{driver}$) (shown in Fig. 1 (b)). During the off-state, the auxiliary circuitry current I_{aux} will slow down the C_s charge.

The voltage ripple (ΔV_{Cs}) and the S_m mean gate-source voltage ($V_{GS\,mean}$) can be estimated with the equations (2) and (3) respectively, where Q_g is the gate charge. These equations are only valid when the C_s capacitor is fully charged during the turn-off of S_m. This condition may not be true when the capacitance value is too large or the off time of S_m is not sufficiently long. The proper design of the circuit and its control must avoid these conditions during normal operation. As it can be seen in the equations, there is a dependence between the voltage ripple and the mean gate-source voltage of S_m. Besides, only some combinations of V_z-C_s can achieve a certain $V_{GS\,mean}$ which makes this circuit complex to dessign.

$$\Delta V_{Cs} = \Delta V_{Cs\,sw} + \Delta V_{Cs\,on} = \frac{Q_g}{C_s} + I_{on}\frac{d}{f \cdot C_s} \tag{2}$$

$$V_{GS\,mean} = V_{Cs\,max} - \left[\Delta V_{Cs\,sw} + \frac{\Delta V_{Cs\,on}}{2} + \Delta V_{driver}\right] \tag{3}$$

Fig. 2 shows ΔV_{Cs}, $\Delta V_{Cs\,sw}$, $\Delta V_{Cs\,on}$, the S_m dissipated power (P_m), and the total dissipated power (P_{total}) for a $V_{GS\,mean}$=15V (simulation parameters in **Table I**). For the estimation of the total dissipated power, only the elements with a significant power consumption have been considered (S_m, driver and S_a). Almost constant V_{Cs} values can be achieved with large capacitances ($\Delta V_{Cs} = 35$mV for C_s=10µF-V_Z=16V), but the circuit performs inefficiently showing the highest total power losses. For large C_s values, the capacitor continues charging after the turn-off of S_m while the auxiliary transistor S_a is conducting causing significant losses [9]. Moreover, there is less or no reduction of S_m power consumption in comparison to EGDPS. In summary, Fig. 3 shows that there is a trade-off between minimum voltage ripple and power losses in S_a when increasing C_s.

Table I: Simulation parameters for the IGDPS

Buck converter	$L = 0.4$mH,- V_{BUS}=30V	D_b	RSX051VYM30 (30V)
S_m	IPB015N04N (40V)	V_z	TFZ--B
R_G	3Ω	S_a	PMPB14XN (40V)
Driver	1ED44175N01B	R_p	215kΩ

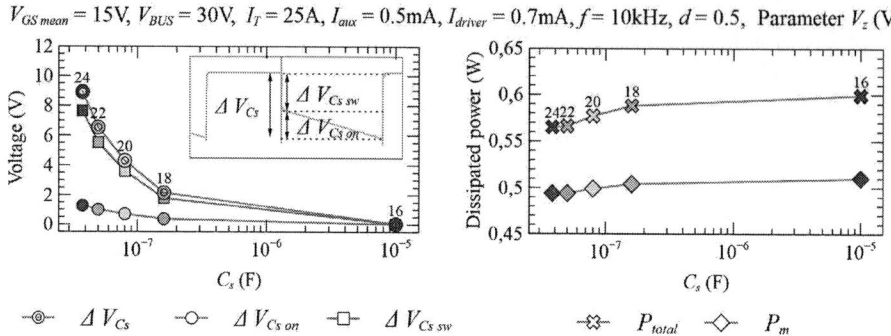

Fig. 2: Simulation of the IGDPS ΔV_{Cs} voltage drop and power dissipation (P_m: S_m power dissipation; P_{total}: driver, S_a and S_m power dissipation) for different combinations of V_z and C_s (24V – 38nF; 22V – 50nF; 20V – 80nF; 18V-160nF; 16V-10µF). The numbers over the symbols indicate zener voltages.

Once the capacitor and the zener diode have been selected (they can be considered as gate-driving design parameters), the $V_{GS\,m}$ and P_{total} can vary with the operating parameters since ΔV_{Cs} changes with I_{aux}, f and d for a given transistor S_m and its associated turn-on gate charge Q_g (equations (2) and (3)). As Fig. 3 (a) and Fig. 3 (b) show, $\Delta V_{Cs\,on}$ and $V_{GS\,mean}$ are more sensitive to I_{aux} and f variations when C_s is small. Increasing the S_m on-state semi-period (by decreasing f or increasing d) or increasing I_{aux} will lead to higher $\Delta V_{Cs\,on}$ values and a reduction in $V_{GS\,mean}$. If the capacitor is too small, at certain operation conditions such as low f or high I_{aux}, the minimum C_s voltage can be lower than the required minimum gate driver supply voltage and the circuit will not be operating correctly, therefore, in Fig. 3 (b) there are no simulations for V_z =24V – C_s = 38nF when I_{aux} is higher than 2mA.

P_{total} increases with I_{aux} (Fig. 3 (b)) due to a slower C_s charge, which causes the transistor S_a to be in on-state for a longer time. Similarly, the P_{total} increases with f or d (Fig. 3 (a)) because of a reduction of the C_s charging time that causes a lower $V_{GS\,mean}$ value. Those effects are more significant when the C_s value is higher. Fig. 4 shows the power dissipation as a function of the main current I_T. As it can be derived, the power consumption of S_a remains almost constant along the I_T sweep.

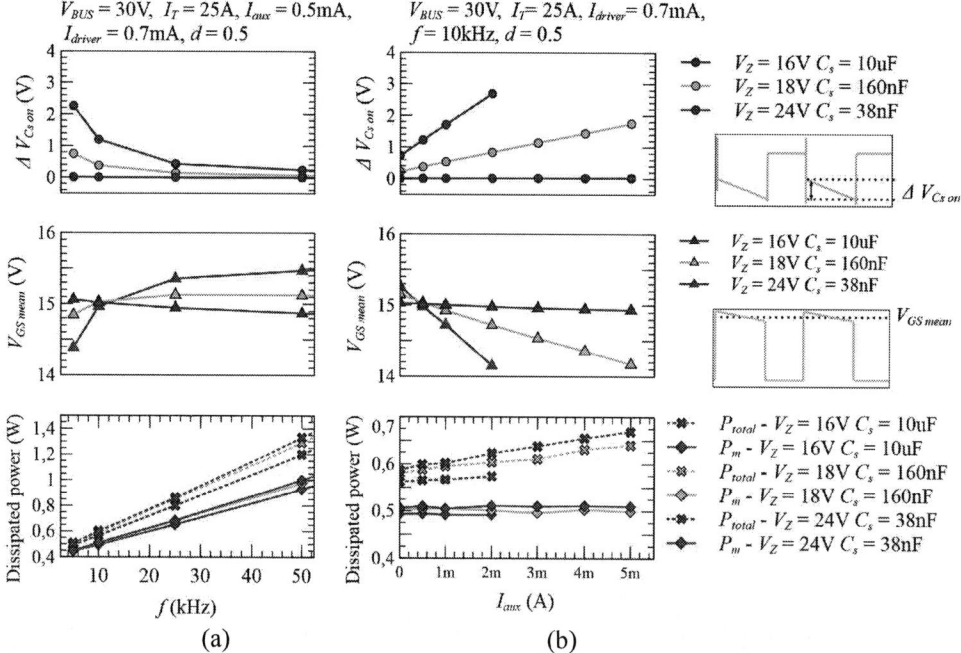

Fig. 3: Simulation of the IGDPS $\Delta V_{Cs\,on}$, $V_{GS\,mean}$ and power dissipation as a function of (a) I_{aux} and (b) f, for three combinations of V_z and C_s (24V – 38nF; 18V-160nF; 16V-10µF).

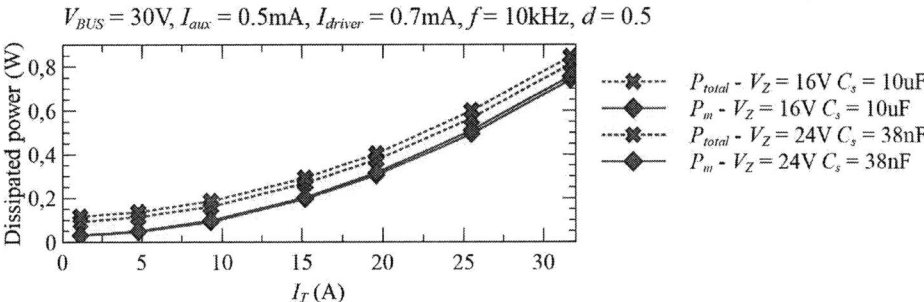

Fig. 4: Simulation of the (IGDPS) power dissipation (P_m: S_m losses; P_{total}: driver, S_a and S_m losses) as a function of main transistor current I_T for three combination of V_z and C_s (24V − 38nF; 16V-10μF).

Internal regulated gate driver power supply

Fig. 5 shows an improved version of the IGDPS circuit presented above, the Internal Regulated Gate Driver Power Supply (IRGDPS), which includes a voltage regulator between C_s and the gate driver. As mentioned before, the main reason for this modification is to achieve a constant and robust gate driver and auxiliary circuitry voltage supply. This approach facilitates the design of the gate driving and protection circuitry in elementary switching cells. The operation principle of the circuit concerning C_s charge remains unchanged, but the discharge process is controlled by the regulator. In the IRGDPS version, the voltage ripple at the gate driver supply voltage is drastically reduced for a wide range of C_s and switching frequency values and, in addition, the dissipation losses can also be significantly reduced in comparison with the lower ripple case of IGDPS (for the V_z-C_s combination 16V-10μF).

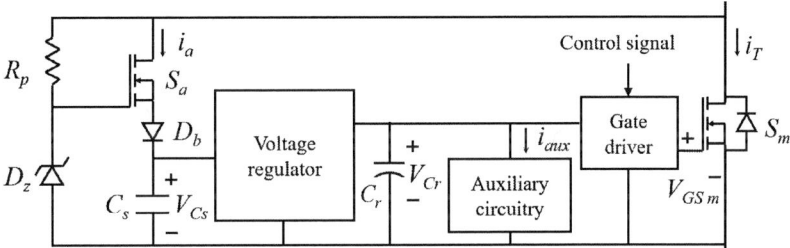

Fig. 5: Internal regulated gate driver power supply (IRGDPS) circuit schematic.

The selection of the voltage regulator is the main design issue of this circuit. A linear voltage regulator is easier to implement and requires less passive components but its losses and the increment of losses in the transistor S_a (due to the regulator high quiescent current) makes it unsuitable for the foreseen application. On the other hand, switching-mode voltage regulators can be considered for the IRGDPS due to their higher efficiency. **Table II** contains the most relevant components used in the simulation of the IRGDPS schematic. For the estimation of the total dissipated power, only the elements with a significant power consumption have been considered (S_m, driver, S_a and voltage regulator).

Unlike the IGDPS circuit, many combinations of capacitors and zener diodes can be implemented in order to achieve a suitable V_{Cs} supply voltage for the voltage regulator. In Fig. 6 (a), a C_s sweep has been performed for two different zener diodes, and it can be seen that the P_{total} (voltage regulator, driver, S_a and S_m) is reduced for zener blocking voltages close to V_{BUS}. Simulations reveal that the power consumption of S_a is drastically reduced while the driver and voltage regulator power consumption remains almost constant when increasing the zener blocking voltage.

A design constrain appears when V_{Cs} is lower than the minimum supply voltage of the voltage regulator, and this happens when C_s is too small, for example in Fig. 6 (a) for V_z=20V, C_s must be at least 500nF to operate correctly. Fig. 6 (b), (c) and (e) depict a comparison between the power dissipation of the IRGDPS circuit with $V_z = 27V$ and $C_s=100nF$ and the IGDPS circuit with $V_z = 16V$ and $C_s=10\mu F$ (which is the case with minimum voltage ripple) when varying f, I_T and I_{aux}. The obtained results show that the P_m of the IRGDPS is similar to the 16V-10µF combination of the IGDPS circuit and the P_{total} is significantly reduced for wide ranges of f, I_T and I_{aux}. In this topology, the f, d and I_{aux} variations do not affect the $V_{GS\,m}$ value. In addition, the introduction of the voltage regulator reduces the C_s discharge current. Hence, the P_{total} is less dependent against the operating parameters and the C_s value.

Table II: Simulation parameters for the IRGDPS

Buck converter	$L = 0.4mH,- V_{BUS}=30V$		D_b	RSX051VYM30 (30V)
S_m	IPB015N04N (40V)		V_z	TFZ--B
R_G	3Ω		S_a	PMPB14XN (40V)
Driver	1ED44175N01B		R_p	215kΩ
Voltage regulator	LT3991		C_r	22µF
			C_s	100nF

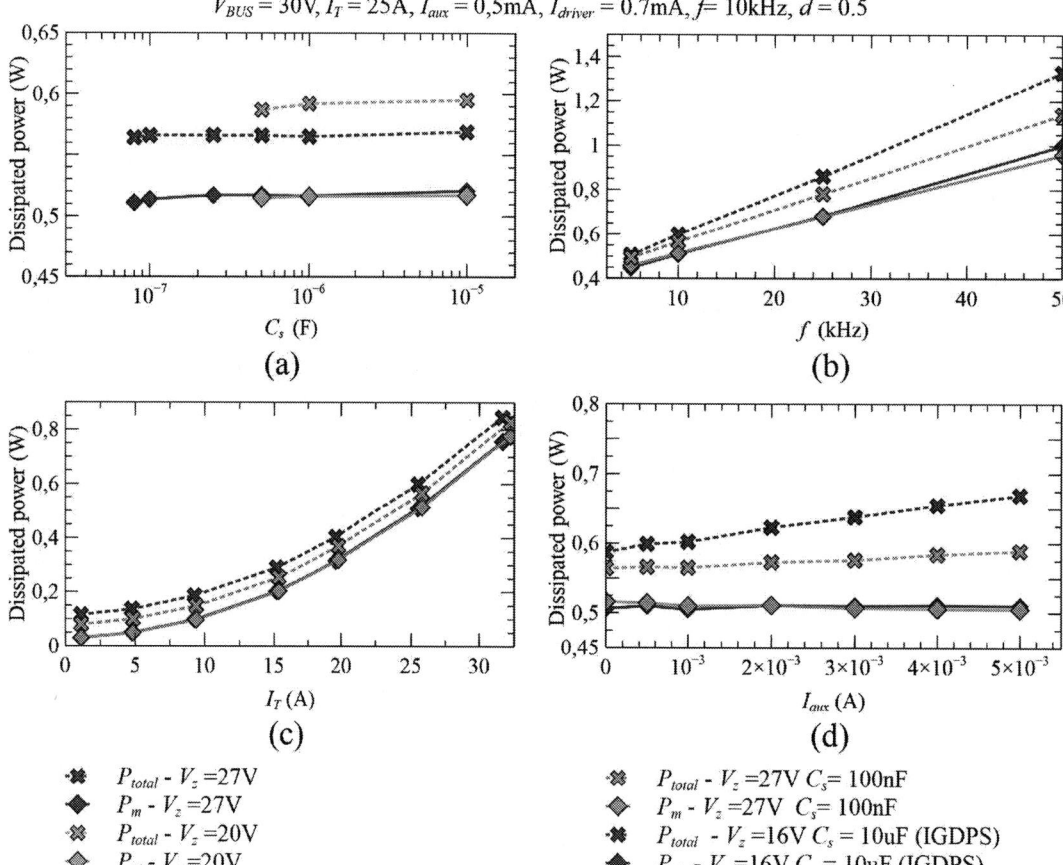

Fig. 6: Simulation of the IRGDPS power dissipation and comparison with IGDPS power losses (P_m: S_m power dissipation; P_{total}: voltage regulator, driver, S_a and S_m power dissipation) for: (a) C_s sweep, (b) f sweep, (c) I_T sweep, (d) I_{aux} sweep.

Experimental results

A conceptual validation and performance comparison among GDPS circuits (EGDPS, IGDPS and IRGDPS) has been experimentally studied in a test buck converter using a 70V MOSFET as the main transistor S_m. The circuit parameters are shown in **Table III** and the experimental setup is shown in the pictures of Fig. 7. As the main objective of these tests was to demonstrate the main characteristics and trends predicted by simulation of the proposed IRGDPS solution, a similar power MOSFET reference was used as S_m. In order to compare the performances of the different circuits under the same conditions, the IGDPS and IRGDPS were designed in order to obtain $V_{GS\,mean}$ =14V and the lowest voltage ripple.

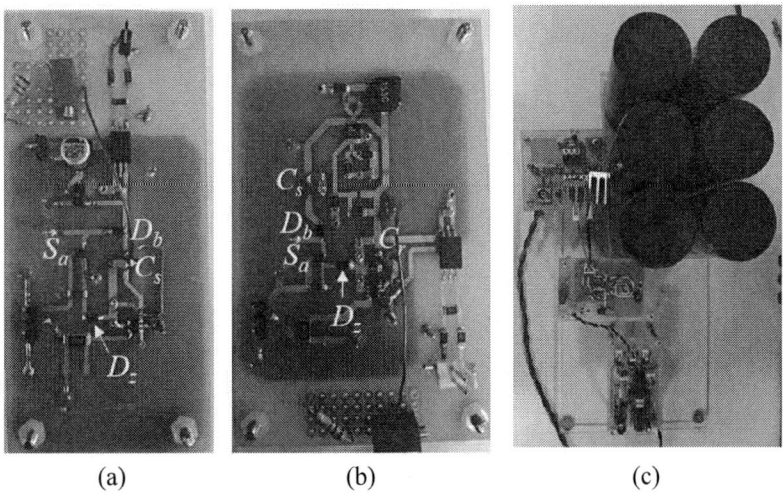

(a)   (b)   (c)

Fig. 7: Experimental setup pictures: (a) EGDPS and IGDPS, (b) IRGDPS, (c) Buck converter, IRGDPS and pre-driver stage.

Table III: Main components for the test buck converter and gate driver power supply circuits

	EGDPS	IGDPS	IRGDPS
Buck converter	L = 0.4mH, V_{BUS}=30V, Diode: MRB601000PT (100V, 60A)		
S_m	IXFH76N07-11 (70V, 76A)		
R_G	3Ω		
Driver	1ED44175N01B		
S_a	-	PMPB14XN (40V)	
D_b	-	SD101BW-E3-18 (50V)	
R_p	-	215kΩ	
V_z	-	16V	27V
C_s	-	10μF	430nF
C_r	-		22μF
Voltage regulator	-		LT3991

Driver and auxiliary circuitry supply voltage

Fig. 8 depicts the mean gate-source voltage $V_{GS\,m}$ and the driver supply voltage (V_{Cs} in IGDPS and V_{Cr} in IRGDPS) waveforms. At low frequency (Fig. 8 (a)), the IGDPS supply voltage shows a higher voltage drop during on-state of S_m ($\Delta V_{Cs\,on}$) which directly reduces the gate-source voltage. On the other hand, increasing the frequency (Fig. 8 (b)) has two effects on the capacitor voltage. First, it reduces the $\Delta V_{Cs\,on}$, and second, it reduces the maximum value of V_{Cs} (C_s is not fully charged). The IRGDPS is more robust working at low and high frequency (Fig. 8 (c) and (d)), in both cases, a minimum voltage ripple of V_{Cr} is achieved and the $V_{GS\,m}$ remains constant. Fig. 9 plots the mean gate source voltage ($V_{GS\,mean}$) and voltage drop of driver supply voltage (ΔV_{driver}) versus frequency for different auxiliary currents.

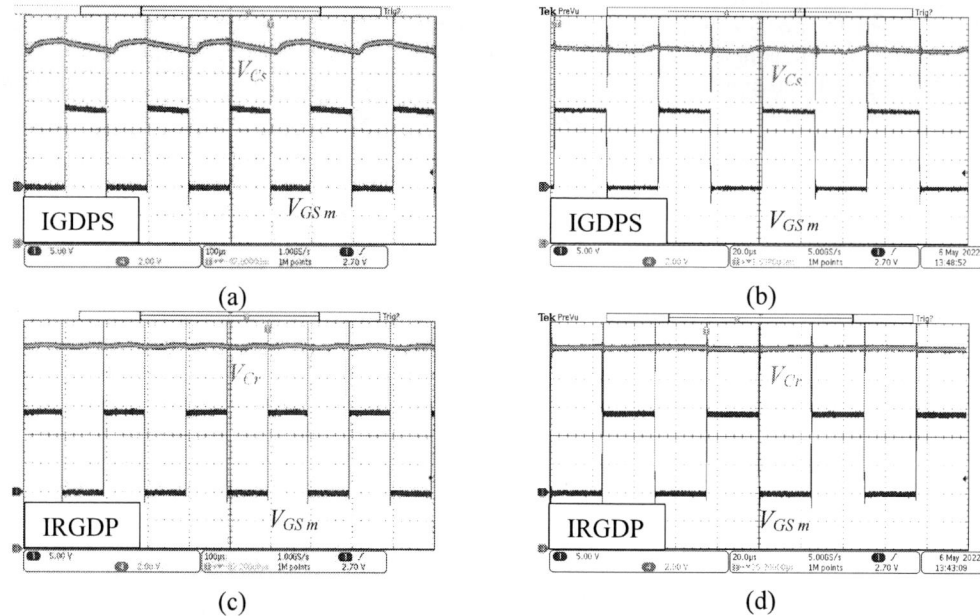

(a) (b)

(c) (d)

Fig. 8: Gate source voltage waveforms ($V_{GS\,m}$ in blue) and driver supply voltage (V_{driver} in green, V_{Cs} for IGDPS and V_{Cr} for IRGDPS) for: (a) IGDPS f = 5kHz, (b) IGDPS f = 20kHz, (c) IRGDPS f =5kHz, (d) IRGDPS f = 20kHz. All waveforms have been taken for V_{BUS} = 30V, I_T = 10A, I_{aux} = 14mA, d= 0.5.

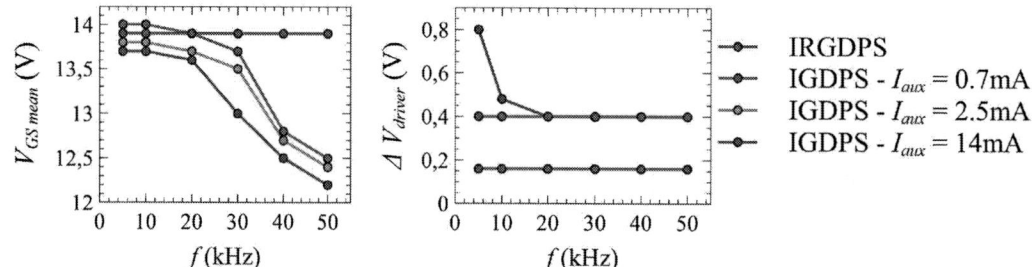

Fig. 9: $V_{GS\,mean}$ and ΔV_{driver} as a function of f, evidencing the superior behavior of the IRGDPS solution in a wide range of switching frequencies. All values have been taken for V_{BUS} = 30V, I_T = 10A, d= 0.5.

Power dissipation of the main transistor and gate driver power supply circuits

A comparative analysis of the power dissipation of S_m can be obtained from Fig. 10 (EGDPS), Fig. 11 (IGDPS) and Fig. 12 (IRGDPS) where the turn-off and turn-on processes are shown at different conditions and GDPS circuits. The duration of the whole switching process and power dissipation during the turn-off of the main transistor S_m of the self-powered solution are similar to the EGDPS solution (Fig. 10 (a), Fig. 11 (a), Fig. 12 (a) and (b)), although the IRGDPS presents less oscillations and a reduction of the power peak. Besides, in the IRGDPS option, when the auxiliary current increases the power dissipation and oscillations are slightly reduced. During the turn-on process with the EGDPS solution (Fig. 10 (b)), the gate-source voltage shows significant fluctuations caused mainly by the reverse recovery of the free-wheel diode of the buck converter. The self-powered circuits show also this behavior during the turn-on, possibly, enhanced by the turn-off of the auxiliary transistor S_a and the diode D_b (Fig. 11 (b), Fig. 12 (c) and (d)).

Finally, in order to study the power consumption distribution of the self-powered GDPS solutions, a thermography of the circuits working at different conditions is depicted in Fig. 13. An indirect method was selected to study the losses and their distribution among the different components since direct current measurement may alter their operation due to parasitic inductances. The thermographs show that as the auxiliary current increases, the dissipation of the GDPS increases, mainly at the auxiliary transistor S_a (hottest point) in both circuits. Nevertheless the lower S_a temperature obtained for the IRGDPS confirms the lower power dissipation already predicted by simulation for this solution.

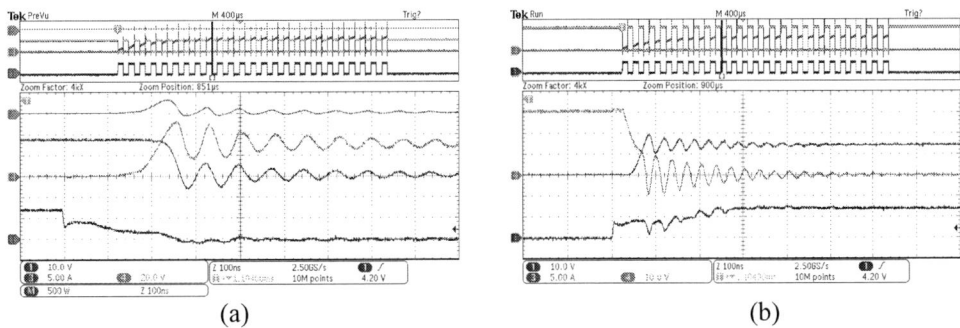

(a)   (b)

Fig. 10: Switching waveforms ($V_{GS\,m}$ in blue, I_T in violet, $V_{DS\,m}$ in green, P_m in red) of EGDPS: (a) Turn-off process, (b) Turn-on process. All waveforms have been taken for $V_{BUS} = 30$V, $f = 10$kHz, $d = 0.5$.

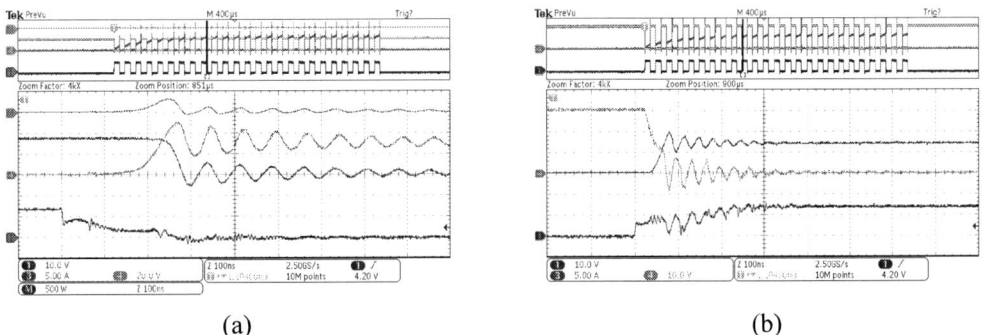

(a)   (b)

Fig. 11: Switching waveforms ($V_{GS\,m}$ in blue, I_T in violet, $V_{DS\,m}$ in green, P_m in red) of IGDPS: (a) Turn-off process $I_{aux} = 14$mA, (d) Turn-on process $I_{aux} = 14$mA. All waveforms have been taken for $V_{BUS} = 30$V, $f = 10$kHz, $d = 0.5$.

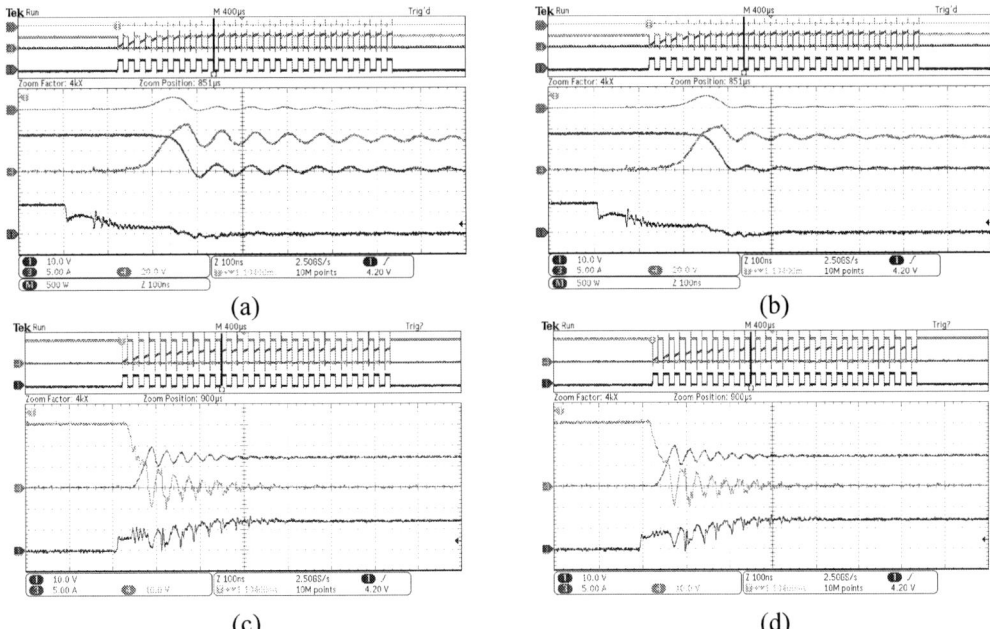

Fig. 12: Switching waveforms ($V_{GS\,m}$ in blue, I_T in violet, $V_{DS\,m}$ in green, P_m in red) of IRGDPS: (a) Turn-off process $I_{aux} = 0.7$mA, (b) Turn-off process $I_{aux} = 14$mA, (c) Turn-on process $I_{aux} = 0.7$mA, (d) Turn-on process $I_{aux} = 14$mA. All waveforms have been taken for $V_{BUS} = 30$V, $f = 10$kHz, $d = 0.5$.

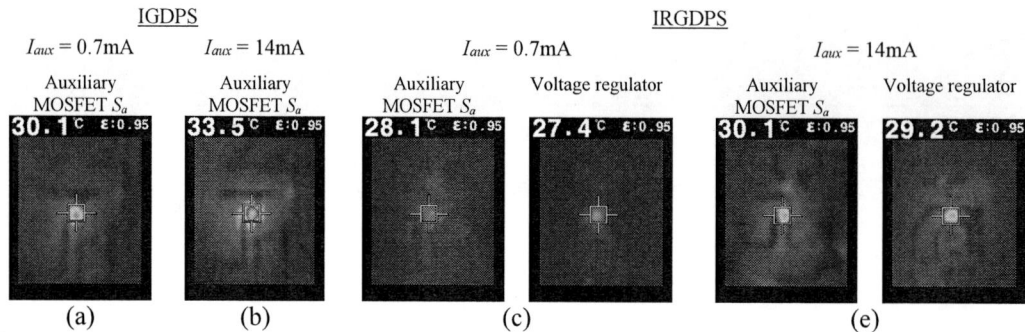

Fig. 13: Thermography of (a) IGDPS with I_{aux} = 0.7mA, (b) IGDPS with I_{aux} = 14mA, (c) IRGDPS with I_{aux} = 0.7mA, (d) IRGDPS with I_{aux} = 14mA. Lower dissipation of S_a is obtained for the IRGDPS.

Conclusion

The proposed IRGDPS circuit improves the main drawbacks of previous IGDPS schemes. It shows a big potential as a compact gate driver power supply solution for multilevel converters.

The simulation and experimental results have shown that the IRGDPS solution behaves as expected and its main operation conditions and constrains have been analyzed and compared with other options (IGDPS and EGDPS). Additional verifications have been carried out at different main current I_T and C_s capacitor values to verify that the experimental behavior corresponded to the simulated one.

The obtained supply voltage for the gate driver and auxiliary circuitry is stable for a wide range of frequencies and auxiliary current consumption. Power losses and switching times of the main transistor are unchanged while the consumption of the auxiliary switch required to ensure the C_s charge is reduced using the IRGDPS solution.

Present investigations are addressed towards the miniaturization of the proposed IRGPDS solution, its integration in smart power modules and the use of higher breakdown voltage devices and switching cells.

References

[1] A. Lesnicar and R. Marquardt, "An innovative modular multilevel converter topology suitable for a wide power range," 2003 IEEE Bologna Power Tech Conference Proceedings, 2003, pp. 6, Vol.3-, doi: 10.1109/PTC.2003.1304403.

[2] R. Marquardt, "Modular Multilevel Converter: An universal concept for HVDC-Networks and extended DC-Bus-applications," The 2010 International Power Electronics Conference - ECCE ASIA, 2010, pp. 502-507, doi: 10.1109/IPEC.2010.5544594.

[3] S. Busquets-Monge and J. Nicolas-Apruzzese, "A Multilevel Active-Clamped Converter Topology—Operating Principle," in IEEE Transactions on Industrial Electronics, vol. 58, no. 9, pp. 3868-3878, Sept. 2011, doi: 10.1109/TIE.2010.2098376.

[4] J. Nicolas-Apruzzese, S. Busquets-Monge, J. Bordonau, S. Alepuz and A. Calle-Prado, "Analysis of the Fault-Tolerance Capacity of the Multilevel Active-Clamped Converter," in IEEE Transactions on Industrial Electronics, vol. 60, no. 11, pp. 4773-4783, Nov. 2013, doi: 10.1109/TIE.2012.2222856.

[5] S. Busquets-Monge and L. Caballero, "Switching-Cell Arrays—An Alternative Design Approach in Power Conversion," in IEEE Transactions on Industrial Electronics, vol. 66, no. 1, pp. 25-36, Jan. 2019, doi: 10.1109/TIE.2018.2816002.

[6] J. L. Gálvez, X. Jordà, M. Vellvehi, J. Millán, M. A. José-Prieto, J. Martín. "Intelligent bidirectional power switch module for matrix converter applications". European Power Electronics Conference EPE, Aalborg (Denmark), 5-7 September 2007.

[7] B. A. Welchko, M. B. de Rossiter Correa, and T. A. Lipo, A threelevel MOSFET inverter for low-power drives, IEEE Trans. Ind. Electron., vol. 51, pp. 669-674, June 2004.

[8] F. Padilha, W. I. Suemitsu, M. D. Bellar, and P. M. Lourenco, Low cost gate drive circuit for three-level neutral-point-clamped voltagesource inverter, IEEE Trans. Ind. Electron., vol. 56, pp. 1196-1204, April 2009.

[9] R. Mitova, J. -. Crebier, L. Aubard and C. Schaeffer, "Fully integrated gate drive supply Around Power switches," in IEEE Transactions on Power Electronics, vol. 20, no. 3, pp. 650-659, May 2005

[10] S. Busquets-Monge, D. Boroyevich, R. Burgos y Z. Chen, Performance analysis and design optimization of a self-powered gate-driver supply circuit, 2010 IEEE International Symposium on Industrial Electronics, pp. 979-985, 2010.

Optimization and Scaling of a Compact High-Power IGCT Capacitor Charger Based on Simulation and Measurements with a 300 kW/3.3 kV Demonstrator

Felix Haag [1,2], Fabian Albrecht [1,2], Volker Brommer [1], Oliver Liebfried [1], Klaus F. Hoffmann [2]

[1]French-German Research Institute of Saint-Louis (ISL)
5 Rue du Général Cassagnou,
68300 Saint-Louis, France
Phone: +33 3 8969-5342
Email: felix.haag@isl.eu
URL: www.isl.eu

[2]Helmut Schmidt University
Holstenhofweg 85,
22043 Hamburg, Germany
Phone: +49 40 6541-4765
Email: felix.haag@hsu-hh.de
URL: www.hsu-hh.de/lek

Keywords

≪IGCT≫, ≪Pulsed power converter≫, ≪Capacitors≫, ≪DC-DC power converter≫, ≪Optimization≫

Abstract

This paper presents the efficiency and charging power optimization of a high pulsed-power inverting buck-boost converter used as a capacitor charger. The influence of converter parameters is investigated with a measurement-based semi-analytical model of the proposed topology. Additionally, the power scaling of such a pulsed power charging system is analyzed.

I Introduction

Pulsed power sources are used for the generation of short (ns-ms) very high power (MW-TW) pulses which are conventionally generated by high voltage capacitors [1]. Industrial applications like electromagnetic forming or heat treating use pulsed power. It is also needed in research for particle physics, electromagnetic acceleration or high power microwave generation. To charge the necessary energy buffers, e.g. high voltage capacitors, compact pulsed power chargers are required. A high efficiency is of special interest in order to reduce losses, as they increase the size of the cooling system.

Brommer et al. [2] analyzed a compact capacitor charger which is used as a basis for these investigations. It utilizes an inverting buck-boost topology in contrast to the H-bridge inverter which is commonly used in such charging systems [3], [4]. Albrecht et al. [5] discussed the different operation principles and showed that operation in Boundary Conduction Mode (BCM) is clearly superior for this specific setup.

The schematic of the charger which uses an IGCT switch and a Brooks-like storage inductor can be seen in Fig. 1 (a). Furthermore, a series diode and an input filter are connected to the input of the converter to protect the battery from high current pulses and negative current. In Fig. 1 (b) the three fundamental currents of BCM operation can be seen. The inductor current i_{LS} is the sum of the magnetizing loop (i_T) and demagnetizing loop (i_F) current.

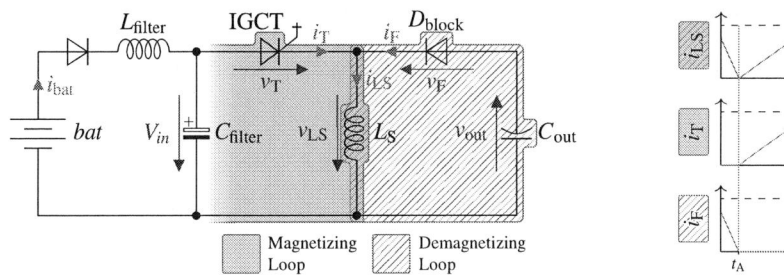

(a) Schematic with marked magnetizing and demagnetizing loop   **(b)** Fundamental converter currents

Fig. 1: Schematic of the capacitor charger and corresponding currents in the inverting buck-boost topology

The parameters of the converter's main components are given in Tab. I. A picture of the experimental setup is shown in Fig. 2. For an input voltage of 200 V its average charging power is > 300 kW with a power density of 4.07 kW/dm³ [5].

Tab. I: Component list of the converter with references to schematic in Fig. 1a and Fig. 2

	Symbol	Component	Information
①	L_S	Brooks-Like Coil	$L_S \approx 107\,\mu H$, $R_{LS,DC} = 3.15\,m\Omega$
②	C_{out}	Film Capacitor	$C_{out} \approx 872\,\mu F$
③	IGCT	ABB 5SHY 55L4500 [6]	$V_{DRM} = 4500\,V$, $I_{TGQM} = 5000\,A$
④	D_{block}	Fast Recovery Diode	Two series connected diodes
		ABB 5SDF 05D2505 [7]	$V_{RRM} = 2500\,V$
⑤	L_{filter}	Core-Less Inductor	$L_{filter} = 30\,\mu H$
⑥	C_{filter}	Electrolytic Capacitor	$C_{filter} = 75\,mF$
⑦		IGCT Gate-Unit Power Supply	$V_{supply} = 38\,V$
	V_{in}	Lithium Battery Bank [8]	$V_{in} = 200\,V$

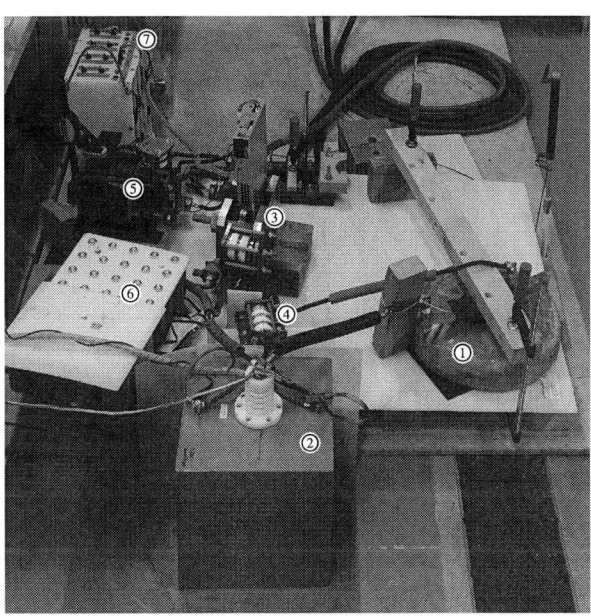

Fig. 2: Picture of the laboratory setup of the power converter

II Converter Simulation Model

In order to identify possible improvements to the proposed converter system without changes to the experimental setup, a simulation model is implemented using Python. This model is based on measurements and an analytical description of currents. With a deviation of the simulated losses of < 10 %, the simulated efficiency deviates less than two percentage points from measurement derived values. More information about the accuracy of the simulation can be found in the following section *Verification*.

Conduction Losses

The conduction losses are calculated based on an analytical description of the currents i_T (magnetizing loop) and i_F (demagnetizing loop). Equation (1) shows the definition of i_{LS} which is the sum of i_T and i_F (see Fig. 1 (b)). The total parasitic resistance in the magnetizing loop is defined as $R_{tot,m}$ and V_{in} defines the input voltage. Due to the input filter, V_{in} is not constant during one pulse. However, to simplify the mathematical model it is considered to be constant. The equation for the demagnetizing loop is obtained by solving the differential equation of a series resonant circuit with L_S as well as C_{out} and the corresponding initial conditions $i_{LS}(t_B) = \hat{I}$ and $v_{out}(t_A) = v_{out}(t_B) = V_{out,t_B}$. The voltage V_{out,t_B} describes the output voltage at the beginning of each demagnetizing cycle. The angular frequency ω is defined as $\omega = \left(\sqrt{L_S C_{out}}\right)^{-1}$. In contrary to magnetizing, the voltage drop across the parasitic loop resistance during demagnetizing is only a small fraction of the voltage v_{LS}. Therefore, it does not alter the waveform significantly and can be neglected.

$$
i_{LS}(t) = \begin{cases} \frac{V_{in}}{R_{tot,m}} \left(1 - \exp\left[-(t-t_A) \cdot \frac{R_{tot,m}}{L_S} \right] \right) & t_A \leq t < t_B \quad \text{(magnetizing loop)} \\ \hat{I} \cdot \cos(\omega[t-t_B]) - V_{out,t_B} \sqrt{\frac{C_{out}}{L_S}} \cdot \sin(\omega[t-t_B]) & t_B \leq t < t_{A'} \quad \text{(demagnetizing loop)} \end{cases} \tag{1}
$$

The total conduction losses are the sum of conduction losses in the magnetizing loop and the demagnetizing loop ($E_{loss,m}$ and $E_{loss,d}$). They are obtained by integrating the coil current i_{LS} combined with the loop resistances and threshold voltages of the respective loop (see Eq. (2)). The resistances of each loop are summarized in $R_{tot,m}$ and $R_{tot,d}$ - all threshold voltages in $V_{tot,m}$ and $V_{tot,d}$.

$$
E_{cond} = E_{loss,m} + E_{loss,d} = \underbrace{\int_{t_A}^{t_B} \left(V_{tot,m} \cdot i_{LS} + R_{tot,m} \cdot i_{LS}^2 \right) dt}_{E_{loss,m}} + \underbrace{\int_{t_B}^{t_{A'}} \left(V_{tot,d} \cdot i_{LS} + R_{tot,d} \cdot i_{LS}^2 \right) dt}_{E_{loss,d}} \tag{2}
$$

Each pulse is defined by a desired maximum inductor current $\hat{I}$ and a starting time t_A. The corresponding integration limits t_B and $t_{A'}$ can be obtained by solving (1) for the corresponding parameter t.

The storage inductor L_S is affected by skin and proximity effects. Therefore, its actual resistance is significantly higher than the DC value. In this simulation model, the actual resistance is calculated based on the measurement of $R_{LS}(\omega)$ and an FFT of the occurring current waveform, as proposed in [9]). Other solutions, like the ladder-model presented in [10], [11] are not accurate for all operation points.

Switching Losses

The total switching losses of the converter consist of the turn-on and turn-off losses of the switching semiconductor as well as the forward and reverse recovery losses of the blocking diodes (a series connection of two diodes is needed, as seen in Fig. 2). A previous paper has shown that operation in BCM results in a significant efficiency increase over Continuous Conduction Mode (CCM) [5].

The turn-on losses of the IGCT as well as forward and reverse recovery losses of the blocking diodes were measured and can be considered as neglectable. A series of measurements provided the turn-off losses of the IGCT for selected operating points. Linear interpolation is then used to approximate the turn-off energies between the individual measured points. Since the turn-off losses change when an RC-snubber is connected in parallel to the IGCT, the measurement series was conducted with and without a snubber circuit.

The turn-on losses increase with a snubber because the energy of the snubber capacitor is dissipated as heat in the snubber resistor and the IGCT. At turn-on, the stored energy of the snubber capacitor is minuscule ($< 50\,\mathrm{mJ}$) because the voltage at it is equal to V_{in}. Despite of the additional energy lost due to the snubber, the total turn-on losses can still be neglected.

Efficiency

Since all the major loss components are known, the energy related efficiency (per pulse) can be calculated with Eq. (3). The parameter $E_{\mathrm{LS}} = \frac{1}{2} L_{\mathrm{S}} \hat{I}^2$ is the energy in the storage inductor and $E_{\mathrm{loss,sw}}$ is the switching loss.

$$\eta_{\mathrm{pulse}} = \frac{E_{\mathrm{LS}} - E_{\mathrm{loss,d}}}{E_{\mathrm{LS}} + E_{\mathrm{loss,sw}} + E_{\mathrm{loss,m}}} \tag{3}$$

III Analysis of the Converter Setup

Unless otherwise noted, following values for power or efficiency refer to average values for a complete capacitor charging cycle up to $3.3\,\mathrm{kV}$. The setup used is the one shown in Fig. 2, with a battery input voltage of $200\,\mathrm{V}$. Measurements have been conducted with a combination of Rogowski coils and coaxial shunts as well as active and passive high-voltage probes (see Tab. IV for more information). Please take into account that especially losses derived from measurements in the mJ range (IGCT turn-on, reverse recovery, forward recovery) have associated inaccuracies.

Commutation Loop Inductance

The commutation loop inductance is the sum of parasitic inductances which hinder the commutation of the current from magnetizing loop to demagnetizing loop and vice versa. Typically, it is a result of the connections between components. A more extensive explanation can be found in [12]. The simplified commutation cell of the converter is shown in Fig. 3.

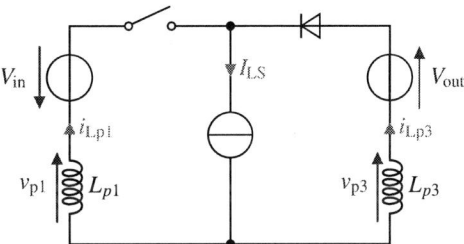

Fig. 3: Commutation cell of an inverting buck-boost converter

Since the commutation process is sufficiently fast, the current I_{LS} does not change throughout it. Therefore, the inductor L_{S} is represented by a constant current source. The Kirchhoff loop with V_{in} and V_{out} shows that the total commutation loop inductance is $L_{\mathrm{com}} = L_{\mathrm{p1}} + L_{\mathrm{p3}}$. Since I_{LS} is constant, any parasitic series-connected inductance does not influence the commutation process. Therefore, only the Kirchhoff loop with V_{in} and V_{out} benefits from connections with lower self-inductance and reduced length. Methods to achieve this have been proposed in [13]. By rearranging the components as well as shortening and rerouting the connections, the commutation loop inductance was reduced from more than $1900\,\mathrm{nH}$ to approximately $730\,\mathrm{nH}$. These values are determined based on IGCT turn-off measurements without a snubber circuit or a clamping network.

Switching Losses

The most dominant switching losses are the turn-off losses of the IGCT. Therefore, they are characterized for different turn-off voltages and currents i_{TGQ} of the IGCT. The following measurements were conducted with neither a snubber circuit nor a clamping network. Fig. 4 illustrates a typical turn-off measurement for $2500\,\mathrm{V}$ and $2500\,\mathrm{A}$. The turn-off energies of multiple operation points are visualized in Fig. 5. Contrary to [14] the turn-off energy does not scale linearly with current and blocking voltage.

For the used input voltage of $200\,\mathrm{V}$, the turn-on losses of the IGCT are less than 1 mJ. Forward recovery losses of the blocking diodes are below $500\,\mathrm{mJ}$ for this application. The blocking diodes' reverse recovery charge causes a slight discharge of the output capacitor, resulting in an energy loss of approximately

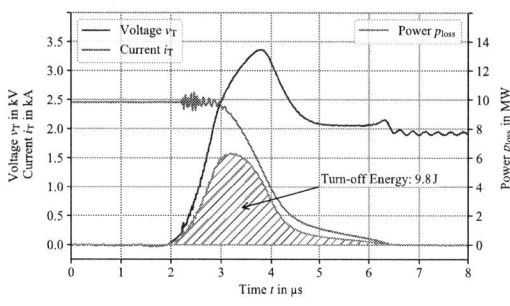

Fig. 4: IGCT turn-off measurement

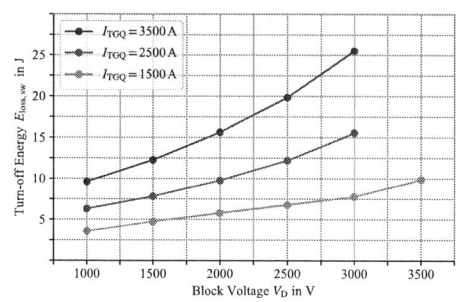

Fig. 5: Measured turn-off losses of the IGCT

5 J throughout one charging cycle up to 3.3 kV. The reverse recovery charge Q_{RR} increases with the parameter $-\frac{di_F}{dt}$. The measured maximum of 710 μC therefore occurs with the highest output voltage at the last pulse of a charging cycle. The relationship between the reverse recovery charge and the lost energy in the output capacitor is further explained in [5].

All small energy losses (≤ 5 mJ) listed above result in a less than 0.2 % of the total transferred energy over one capacitor charging cycle. Additionally, they are complicated to implement into the simulation model. Therefore, the only switching losses included in the model are the turn-off losses of the IGCT.

Conduction Losses

The values of threshold voltages and resistances of the components are given in Tab. I. Parasitic resistances of the connection cables are 1556 μΩ and 808 μΩ for the magnetizing and demagnetizing loop, respectively. The resulting conduction losses are dominated by the energy dissipation in the storage inductor. Its DC and AC resistance was measured via a micro-ohmmeter and an *LCR* meter. Moreover, a FEM simulation is used to verify these results. Parameters for the IGCT and diodes were obtained from the devices' datasheets.

Efficiency and Distribution of Losses

The simulation model combined with the obtained information of the experimental setup can be used to determine the efficiency for each possible pulse in the BCM operation. Additionally, the distribution of losses can be investigated. The former is of special interest because the converter does not operate in steady state. Therefore, every pulse during one charging cycle is unique and has to be calculated individually (based on the output voltage and desired $\hat{I}$). Information on the distribution of losses is of high interest since it reveals the components with the highest potential of optimization.

The following Fig. 6 shows the efficiency of pulses for different maximum conduction currents $\hat{I}$ and output voltages v_{out}. The dashed line crossing all peaks indicates the pulses of maximum efficiency for every output voltage. The influence of different input voltages is not considered, since it is fixed (see Tab. I).

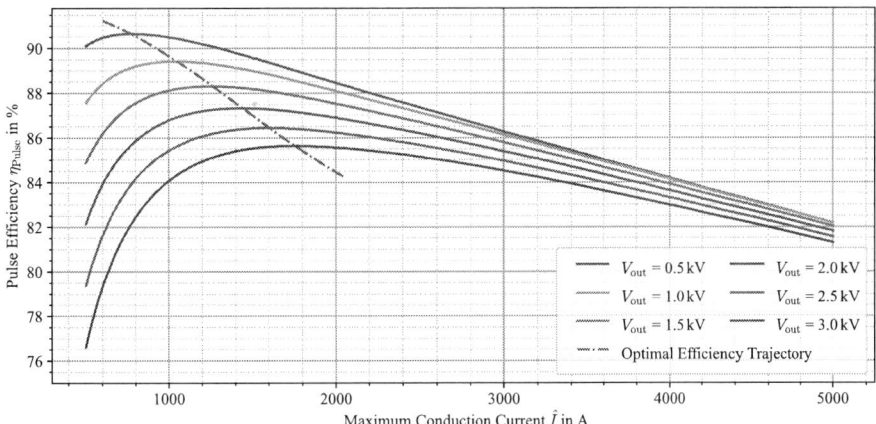

Fig. 6: Efficiency of the converter for different operation points

Fig. 6 illustrates that the pulse efficiency decreases with an increasing output voltage. This is due to the higher switching and reverse recovery losses of the corresponding semiconductors. Hence, the first pulse during a capacitor charging cycle has the highest efficiency. Towards higher maximum conduction currents the influence of V_{out} on the efficiency decreases. The maximum pulse efficiencies exceed 90 %. Pulses with higher maximum conduction current still have pulse efficiencies over 80 %. As shown in [5], power transfer rises approximately quadratically with $\hat{I}$. Therefore, operation along the optimal efficient trajectory drastically limits the power transfer of the converter.

Fig. 7 presents the simulated distribution of losses for the fastest possible charging cycle. The figure is a summary of multiple unique pulses needed to charge the output capacitor up to 3.3 kV. The power maximized control scheme used to achieve this is explained in the following section *Power Transfer*. A nested pie chart visualizes the distribution of losses. Its inner circle differentiates between the **type of loss** (switching and conduction). The conduction losses are furthermore split into losses occurring during the magnetizing ($t_A \leq t < t_B$) and demagnetizing phase ($t_B \leq t < t_{A'}$). The outer circle indicates in which component the losses are generated. Key parameters of the used setup and charging cycle are presented in the lower right.

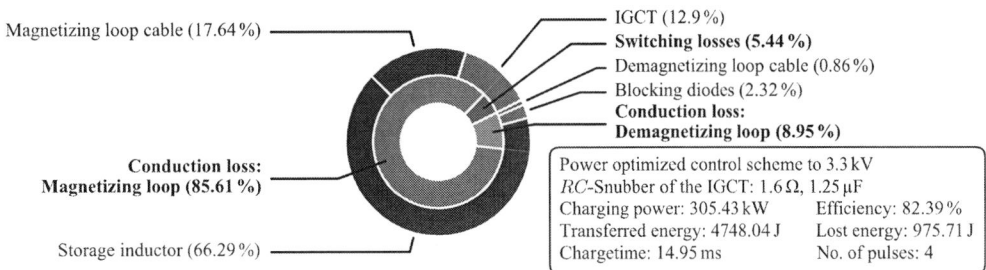

Fig. 7: Simulated distribution of losses during one charging cycle up to 3.3 kV; categorized by type of loss (**inner circle**) and by component (outer circle)

The following conclusion can be drawn from the simulation results summarized in Fig. 7: The total efficiency of the charging cycle is 82.39 %. Its average charging power is 305.43 kW. In total, four pulses are needed to charge the capacitor to 3.3 kV.

Note, that the switching losses of 5.4 % contribute the least to the overall losses. Even the losses generated in the IGCT turn out to be less than 50 % switching losses. The most optimization potential clearly lies in the reduction of the conduction losses. Since the output voltage is typically higher than the input voltage, the magnetizing phase is significantly longer than the demagnetizing phase. Therefore, the conduction losses during the magnetizing phase are higher than the losses during the demagnetizing phase.

IV   Verification

In order to verify the simulation model, a series of measurements were conducted.

According to Eq. (4), the output energy is determined by the difference in output voltage before and after a pulse in combination with the output capacitance.

$$E_{\text{Pulse,out}} = \frac{1}{2} C_{\text{out}} \left(V^2_{\text{out}, t_{A'}} - V^2_{\text{out}, t_B} \right) \tag{4}$$

A Fluke 233 multimeter in combination with a passive high-voltage probe is used to measure the output voltage.

The input energy is determined by integrating the input power over a given time interval. The input power is the mathematical product of the voltage v_{in} and the current i_{bat} marked in Fig. 1. An alternative approach would be to integrate the product of v_{in} and i_{T}. However, the first approach has the advantage that the amplitude of the measured current is significantly lower and also not as steep. This makes measurements easier and thus, improves accuracy.

A passive oscilloscope voltage probe is used to measure v_{in}, while the current is measured with a $1\,\text{m}\Omega$ coaxial shunt. Refer to Tab. IV for more information about the used metrology.

This means that the losses of the input filter are not resembled in the obtained measurements. The filter decreases the efficiency by approximately one percentage point. The efficiency is defined as the ratio between the output energy and the input energy for each pulse. It is presented in the following Fig. 8 for different output voltages and $\hat{I} = 2500\,\text{A}$. Measurements at other turn-off currents have a similar deviation.

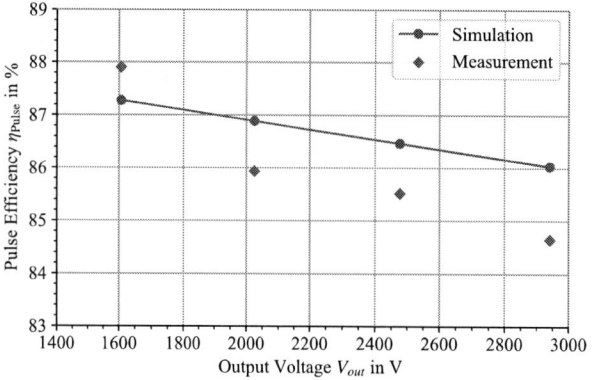

Fig. 8: Comparison between the simulation and measurements for 2500 A

It can be seen, that the simulated values match the measured values. The highest deviation in the simulated efficiency is ≈ 1.4 percentage points for a current of 2.5 kA and an output voltage of 2.9 kV. Simulated values for the efficiency are typically higher than the values derived from measurements. The value at 1.6 kV and 2.5 kA is the only exception for all 20 compared operation points. Note, that the measured values also have inaccuracies due to integration and measurement errors.

V Optimization

In order to increase the efficiency or power transfer of the converter, one can either change the experimental setup or alter the control strategy of the switching semiconductor. For the hardware changes the following factors are analyzed: Input voltage, storage coil inductance, commutation loop inductance (L_S), parasitic resistances. Additionally, optimal control strategies (Dynamic Pulse-Current Control in short DPCC) for both maximized power transfer and maximized efficiency are investigated.

Efficiency

As mentioned earlier, the majority of the converter's losses are dissipated in the magnetizing loop. The magnetizing losses $E_{loss,m}$ are approximately inversely proportional to the input voltage. Therefore, the best option to decrease them is an increase of V_{in}. This however is only practical up to a certain point, since a high ratio between the output and input voltage is the main reason a boost converter is used to charge the capacitor. A decrease of the total parasitic resistance through shorter interconnections or by redesigning the storage inductor using high-frequency litz-wire or foil conductors would further reduce the losses [15].

The commutation loop inductance also affects the efficiency of the converter. For each turn-off process, its total stored energy is dissipated. Since $L_S \gg L_{com}$, the transferred energy is significantly higher than the energy lost due to L_{com}. As mentioned in the subsection *Commutation Loop Inductance*, a noteworthy reduction of the commutation loop inductance was achieved by rearranging the components of the converter. The reduction from 1900 nH to 730 nH decreased the switching losses at $I_{TGQ} = 1.5\,kA$ and $v_T = 1\,kV$ by 20 % (from 4.4 J to 3.5 J). Since the switching losses are only a small fraction of the total losses, the increase in efficiency is only marginal.

Additionally, one can change the control strategy of the switching semiconductor to maximize efficiency. The maximal efficiency is achieved by adjusting the length of each pulse in accordance to the dashed line in Fig. 6. One full capacitor charging cycle with this efficiency optimized DPCC has an efficiency of 88.7 % and takes 41.6 ms (simulated), while the fastest possible charging cycle requires 14.95 ms and has an average efficiency of 82.39 %. However, for the efficiency optimized DPCC, $\hat{I}$ of each pulse is below 2000 A. The power transfer is therefore drastically reduced and the charging time inevitably increased.

Power Transfer

The power transfer of the converter increases approximately linear with the input voltage V_{in} and maximum conduction current $\hat{I}$ [5]. A higher input voltage results in a higher $\frac{di_{LS}}{dt}$ during the magnetizing phase and thus, a shorter pulse duration. Since the energy in the storage inductor rises quadratically with its current and the current increases linearly with time, a higher maximum conduction current also increases the power transfer. However, a higher $\hat{I}$ also results in a higher overvoltage generated by the commutation loop inductance during turn-off. Since this overvoltage has to be blocked by the IGCT, the current has to be limited if a certain output voltage is reached. Reducing this overvoltage is therefore a key element in order to increase the power transfer. Possible solutions for this are now presented in more detail.

The mentioned reduction of the commutation loop inductance (1900 nH to 730 nH) has reduced the overvoltage by up to 53 % (from 3650 V to 1725 V at $\hat{I} = 3000\,A$). This enables higher current pulses, especially at the end of each charging cycle. Increasing the power transfer from 83.1 kW to 166.3 kW.

A further power increase is possible by dynamically changing $\hat{I}$ throughout the charging cycle. There are two phases during one charging cycle. For the first pulses, the peak current $\hat{I}$ is equal to the maximum rating of the IGCT ($I_{TGQM} = 5\,kA$). Here, the specified maximum blocking voltage of the IGCT is higher than the sum of the input, output and overvoltage ($v_{DRM} > V_{in} + v_{out} + v_{Lcom}(\hat{I})$). If the charging state of the capacitor is increasing, the current $\hat{I}$ must be decreased in order to stay in the SOA of the IGCT. In this second phase, the voltage limit of the IGCT limits the maximum possible conduction current $\hat{I}$. Without the proposed regulation of $\hat{I}$ the maximum possible current of the last pulse has to be used throughout the whole charging cycle, thus reducing the power transfer drastically.

With the implementation of the presented power-optimized DPCC the average power transfer increases from 166.3 kW to 251.88 kW.

The implementation of an *RC*-snubber circuit (1.6 Ω, 1.25 μF) increases the possible power transfer from 251.88 kW to 305.43 kW by allowing higher currents without exceeding the IGCT's maximum voltage. Furthermore, the turn-off losses of the IGCT itself decrease by $\approx$ 20%. However, the total switching losses are typically not reduced if a snubber is implemented. Instead, the turn-on losses increase and the snubber's components also generate losses.

Summary

Table II presents the summary of the optimization possibilities. They are rated and sorted after relevance.

Tab. II: Rated overview of possible improvements

Relevance	Improvement	Power Transfer	Efficiency	Effort
1st	Power-Optimized DPCC	+ + +	----	----
2nd	Commutation Loop Inductance	+ + + +	+	---- ----
3rd	Input Voltage	+ + +	+ + +	---- ----
4th	Parasitic Cable Resistance	+	+	---- ----
5th	Storage Coil Inductance and Resistance	0	+ + +	---- ---- ----
6th	Efficiency-Optimized DPCC	– – –	+	–

VI Scaling of the Charging System

The benefit of a converter utilizing a ferrite-less storage inductor becomes apparent when the charger system is scaled towards a higher power. This can be realized by parallel converters or by a single, higher current converter that utilizes multiple semiconductors connected in parallel. For the latter, the coil's inductance is adapted so that the IGCT currents and switching frequency remain the same. Additionally, $\frac{L_S}{R_{DC}}$ is held constant so that the sum over all resistive inductor losses remains constant compared to the parallel converter system. This is possible by adjusting both the number of coils and the winding cross-section. The following Table III presents how the two approaches scale. The advantage of the high current converter lies in the inductive storage. If scaled appropriately, its volume stays constant while the overall losses increase by factor N in both cases [16]). In the approach of parallel converters, the size increases with N.

Tab. III: Scaling laws for parallel converters and a higher current converter. N resembles the power scaling factor

	Power Transfer	Semiconductors	Inductor Volume	Inductor Losses
Parallel Converters	N	N	N	N
High Current Converter	N	N	1	N

VII Conclusion

In this paper, the power transfer and the efficiency optimization of a high pulsed-power capacitor charger are investigated. A simple yet accurate simulation was set up and validated by experimental measurements. Consequently, the influence of parameters like the input voltage, commutation loop inductance and storage coil inductance on the performance were investigated. Simulation results show that conduction losses are significantly higher than switching losses and that most of the conduction losses are generated in the main inductor. Therefore, an improvement of the coil's L/R ratio and a higher input voltage are the most promising optimization approaches in terms of efficiency. The rated maximum blocking voltage of the IGCT switch was identified as "the" parameter which limits the power transfer rate. Methods to increase the power and simultaneously stay below the devices' voltage limitations are the decrease of the commutation loop inductance, the use of a snubber circuit or an improved control scheme with dynamically adjusted maximal pulse currents adapted to the charging state of the output capacitor. The power transfer could be increased from 80 kW to 300 kW without significant changes to the experimental setup.

In the case of scaling the charger system, the approach of operating a single coil at a higher current is more promising than multiple converters in parallel.

VIII Appendix

Tab. IV: Measurement Metrology

Name	Range	Bandwidth
Rogowski Coil [17]	$\pm 3\,\text{kA}$	16 MHz
Differential Voltage Probe [18]	$\pm 7\,\text{kV}$	70 MHz
1 mΩ Coaxial Shunt (Zirrgiebel)	–	170 MHz

References

[1] H. Bluhm and D. Rusch, *Pulsed Power Systems: Principles and Applications.* Berlin: Springer, 2006.

[2] V. Brommer, O. Liebfried, and S. Scharnholz, "A High-Power Capacitor Charger Using IGCTs in a Boost Converter Topology," *IEEE Transactions on Plasma Science*, vol. 41, no. 10, 2013.

[3] S. L. Holt, J. C. Dickens, J. L. McKinney, and M. Kristiansen, "A Compact 5 kV Battery-Capacitor Seed Source with Rapid Capacitor Charger," in *2009 IEEE Pulsed Power Conference*, IEEE, 62009, pp. 897–901.

[4] S. R. Jang, S. H. Ahn, H. J. Ryoo, and G. H. Rim, "Novel High Voltage Capacitor Charger for Pulsed Power Modulator," in *2010 IEEE International Power Modulator and High Voltage Conference*, IEEE, 52010, pp. 317–321.

[5] Fabian Albrecht, Felix Haag, Volker Brommer, Klaus F. Hoffmann, Oliver Liebfried, "The Relevance of Boundary Conduction Mode for High Pulse Power DC-DC Converters Using GCTs and IGCTs," in.

[6] ABB - Switzerland Ltd Semiconductors, "ABB IGCT 5SHY 55L4500," 2013.

[7] ABB - Switzerland Ltd Semiconductors, "ABB 5SDF 05D2505_5SYA1114," 2001.

[8] V. Brommer, O. Liebfried, and S. Scharnholz, "Développement d'une batterie Li-Ion pour des applications d'impulsion de forte puissance," 2016.

[9] C. R. Sullivan, "Optimal Choice for Number of Strands in a Litz-Wire Transformer Winding," no. Found in IEEE Power Electronics Specialists Conference, 1997.

[10] S. Kim and D. P. Neikirk, "Compact Equivalent Circuit Model for the Skin Effect," in *1996 IEEE MTT-S International Microwave Symposium Digest*, IEEE, 1996, pp. 1815–1818.

[11] S. Mei and Y. I. Ismail, "Modeling Skin and Proximity Effects with Reduced Realizable RL Circuits," 2004.

[12] J. Lutz, *Halbleiter-Leistungsbauelemente.* Berlin and Heidelberg: Springer, 2006.

[13] K. Moorthy, R. Sree, B. Aberg, M. Olimmah, L. Yang, *et al.*, "Estimation, Minimization, and Validation of Commutation Loop Inductance for a 135-kW SiC EV Traction Inverter," *IEEE Journal of Emerging and Selected Topics in Power Electronics*, vol. 8, no. 1, pp. 286–297, 2020.

[14] ABB - Switzerland Ltd Semiconductors, "Applying IGCTs," 2007.

[15] C. R. Sullivan and R. Y. Zhang, "Simplified Design Method for Litz Wire," in *2014 IEEE Applied Power Electronics Conference and Exposition - APEC 2014*, IEEE, 16.03.2014 - 20.03.2014, pp. 2667–2674.

[16] M. Brooks and H. M. Turner, "Inductance of Coils," 1912.

[17] PEM - Power Electronic Measurements Ltd., "Rogowski Coil - Specification Overview," 2010.

[18] Testek, "Testek TT-SI 9010 Instruction Manual for a 7 kV Differential Voltage Probe,"

Multilayer busbars for medium voltage ANPC converter dedicated to battery energy storage systems

Mamadou Lamine Beye, Luc Bimmel, Anthony Bier, Jérémy Martin
French Alternative Energies and Atomic Energy Commission – CEA
Department of Solar Technologies
73370 Le Bourget du Lac, France
Tel.: +33 (0)4 79 79 22 03
E-Mail: jeremy.martin@cea.fr

Acknowledgements

This project has received funding from the European Union's Horizon 2020 research and innovation programme under grant agreement No 864459.

Keywords

«Low inductive busbar», « 3.3 kV SiC power modules », «High power», «Medium voltage», «Energy storage system».

Abstract

The increase of energy storage system power leads to open a technological pass which is to increase the voltage level of battery racks. Available 3.3 kV Silicon Carbide (SiC) semi-conductors implemented in an ANPC topology allows tuning a 3.6 kV DC bus.Thus, researches are shifting to medium voltage systems in which battery racks are connected in series with a middle point grounded. SiC modules implementation requires low inductive busbars to achieve high efficiency when rising in switching frequency necessary to shrink the output filter. In this paper, a methodology for reducing the parasitic inductor of the busbars (< 20 nH) is presented.

Introduction

The energy transition leads to speed up the mass use of renewable energies, in particular solar and wind energy instead of fossil fuels [1]. However, to ensure the balance between electricity production and consumption, energy storage systems are combined with renewable energies generators [2]. These storage systems must also meet requirements in terms of efficiency and grid support. European Talent project was proposed to increase the voltage of BESS from conventional low voltage racks [3] 1 kV-1.4 kV to medium voltage racks (2×1 500 V with middle point grounded) achieving a high efficiency (> 99 %) and reducing the quantity of required power components raw materials. The ANPC converter operating in outer switching modulation mode (OSMM) presents the main advantage to use only small switching loops in inverter or rectifier mode allowing a switching speed increase [4]. This paper focuses on the ANPC converter design. The DC/DC converter analysis will be down by the authors in furthers articles.

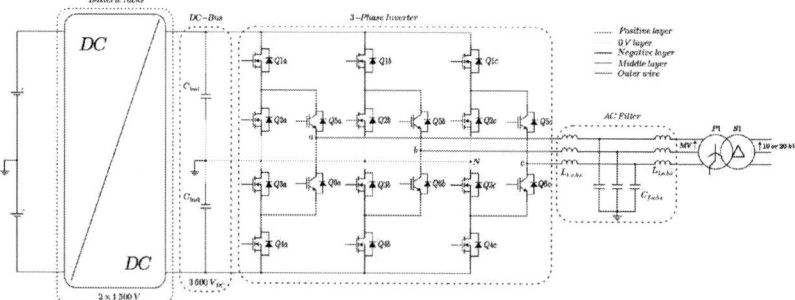

Figure. 1: Medium Voltage three-phase ANPC converter for battery energy storage systems

Several studies carried out to set up laminated busbars which have made it possible to reduce the parasitic inductance of busbars from a few hundred nH to around ten nH [5-6]. All of these papers use multilayer busbars that often requires algorithm to determine busbars sizing. In [7], authors bring an approach to build multi layers layers busbars from two layers busbars once assembled show interesting stray inductor reduction. The methodology presented, is applicable to an ANPC conversion PEBB using 3.3 kV power modules.

A mixed Si/SiC ANPC leg and its control principle are presented Fig. 2. A ANPC leg controlled in outer switch modulation mode consists of using a 3.3 kV IGBT power module operating at grid frequency (50 Hz) connected to the leg output while two SiC modules operating at the switching frequency (…kHz) are connected to poles. The implementation of the ANPC leg is done by an external assembly of 3×2L modules. This allows the manufacturer to take benefits of a wide choice of two-level power modules easy to replace.

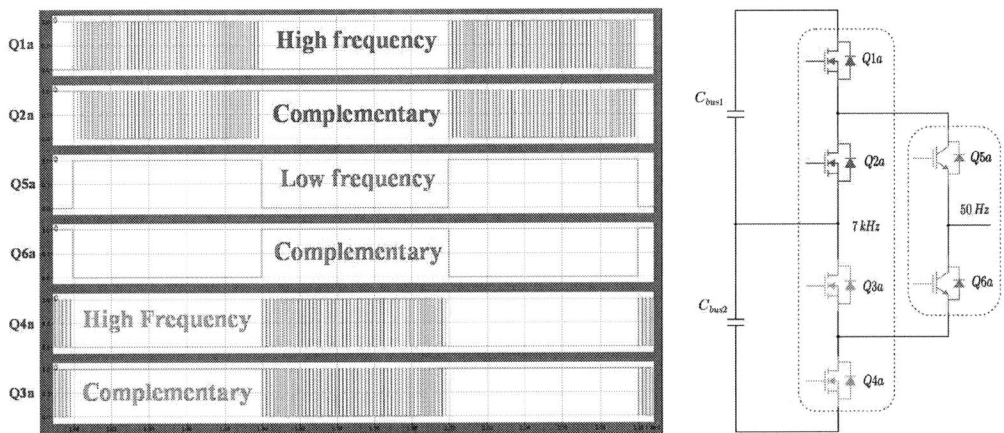

Fig. 2: Principle of operation of an ANPC leg (OSMM)

In order to build a 3MVA converter two 1.5 MVA ANPC PEBB will be connected in parallel. A switching frequency of 7 kHz seems an acceptable compromise to target an efficiency > 99% including output filter losses and DC/DC converter losses. An electro-thermal model was built under PLECS [8] to estimate the power modules losses from power modules datasheet. From switching and conduction characteristics of IGBT and MOSFET power modules, the simulation results show an efficiency range between 99% - 99.6%.

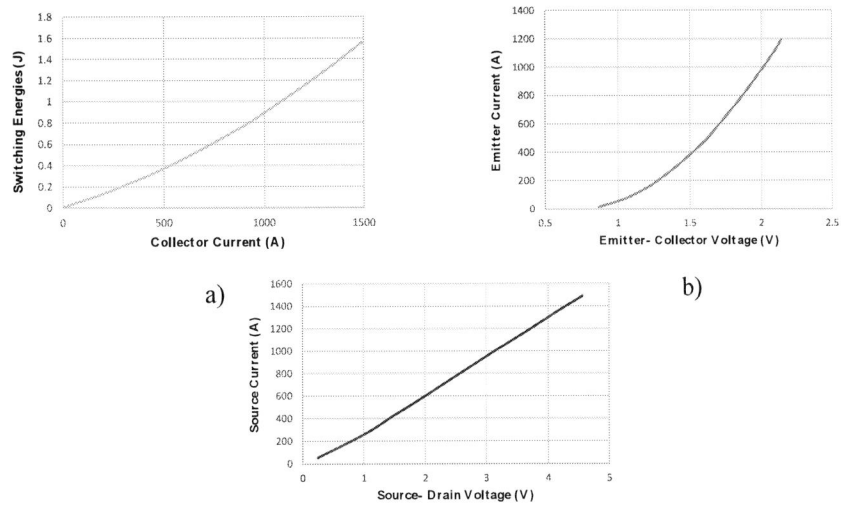

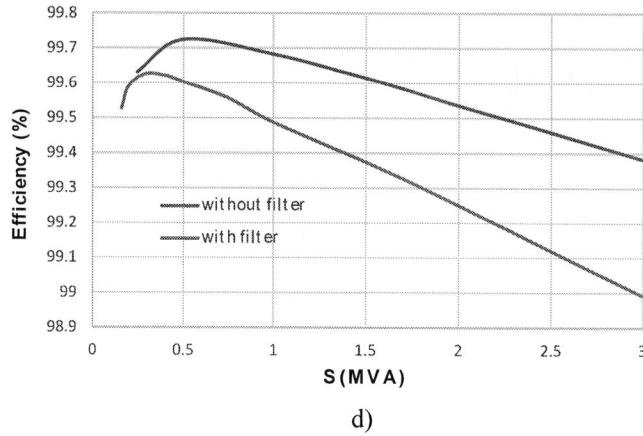

d)

Fig. 3: a) Half-Bridge switching energy characteristics, b) Free-Wheeling diode forward characteristics (MOSFET), c) Free-Wheeling diode forward characteristics (IGBT), d) Efficiency curves for 3 MVA three-phase ANPC converter (7 kHz)

In the section 1, the design methodology of the busbars will be described as well as a Q3D [9] based comparative studies between different layouts possibilities. Finally, the conclusion of this paper and the future work are presented in section 2.

1. Bus bar design and comparison for one phase Leg

As explained previously, the use of busbars is to ensure the connection between the power modules and the DC capacitor of the converter maximizing capacitive effect. Equations 1 to 4 give the characteristics of classic busbars formed by two conductive layers separated by an insulating layer [7]. L_C: is the parasitic inductance of each active layers, which depends on the length (l), the width (w), the thickness of the layers (e), the insulation layer thickness (d), the vacuum permeability (μ_0) and relative permeability (μ_r), Minimum busbars area (A_{min}). M_c is the mutual inductance between the two conductive layers; it depends on the same parameters as L_C. When the direction of current flow is the same in the two conductive layers the mutual inductance is positive, conversely it is negative. In order to reduce the total inductance of the bus bar (L_T) a design with a reverse flow direction is recommended. Figure 4, presents the flowchart of the design methodology.

$$L_c = 2.10^{-7}.l.\left[2.303 \log\left(\frac{2l}{w+d}\right) + 0.5 + 0.2235\left(\frac{w+d}{l}\right)\right] \qquad (1)$$

$$M_C = \frac{\mu_r}{2\pi}\left[l.\ln\left(\frac{\sqrt{d^2+l^2}+l}{d}\right) - \sqrt{d^2+l^2} + d\right] \qquad (2)$$

$$L_t = 2L_c - 2M_c \qquad (3)$$

$$A_{min} = w.e = \frac{400.I.(0.785).[1+0.05(N-1)].(1.10^{-6})}{0.123} \qquad (4)$$

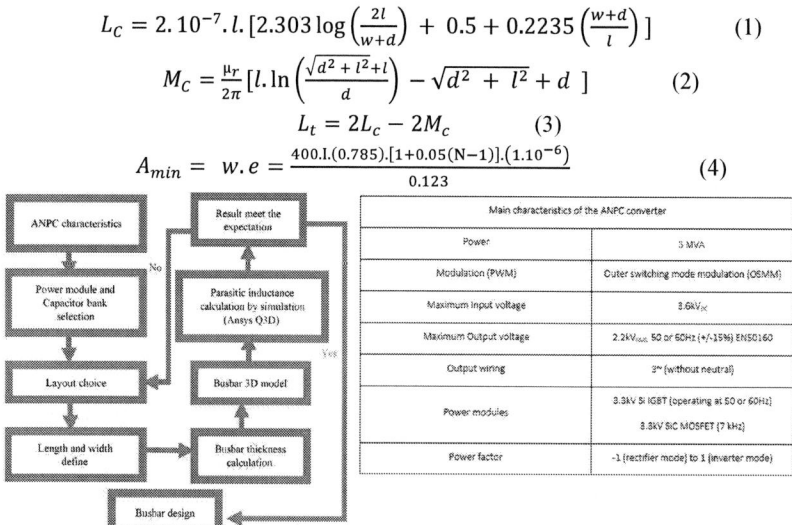

Figure. 4: Busbars algorithm design and ANPC converter characteristic

DC bus Capacitor Bank selection:

The capacitor bank ($C_{bat} = C_{bus1} + C_{bus2}$) is calculated to provide the peak I_{rms} current to the AC output of the converter. Its value can be determined by using equation 5 [10] where (f_s) is the switching frequency, (ΔV_{dc}) is the ripple voltage of the DC bus, fixed to 5%.

$$C_{batmin} \geq \frac{\sqrt{2} \times I_{rms\_max}}{2 \times f_s \times \Delta V_{dc}} \quad (5)$$

Based on this criteria and the converter parameter (Figure 4), the minimum value of C_{bat} is 1.2 mF.

ANPC Leg busbars layout comparisons

Figure 5 shows 3 layouts possibilities for inverter legs prototyping using two layer busbars and 3.3kV 2L power modules. Layout 1 consists of placing all power modules on a same line, so the capacitors are placed in front of MOS 1 and MOS 2 power modules (Figure 5b). This placement shows the advantage of using only one heatsink for the 3 power modules. However, placing the IGBT module (IGBT 1) between the negative layer (N) and positive layer (P) reduces the mutual inductance between these two layers. Layout 2 consists of placing the capacitors (C_{bus1} and C_{bus2}) in the center and power modules on the top and the bottom (figure 5c). One of the consequences of this placement is the use of one heatsink per power module; this can increase the costs/complexity of the converter. This placement would also increase busbar's surface area of layers, which may result in increased costs and volume.

Layout 3 consists in placing the MOS 1 and MOS 2 on the same plane line while IGBT 1 is placed back and between the two modules so as to form a triangle (Figure 5d). C_{bus1} and C_{bus2} are placed in front of the SiC power modules and MOS 1 and MOS 2. This type of placement makes it possible to reduce surface area of bus bar layers and therefore the material costs. In addition to this, to bring the MOSFET modules closer together has a positive impact because this increases the mutual inductor. Regarding the cooling of the power modules, depending on the technology used, two heatsinks will be necessary because of the non-alignment of the power modules.

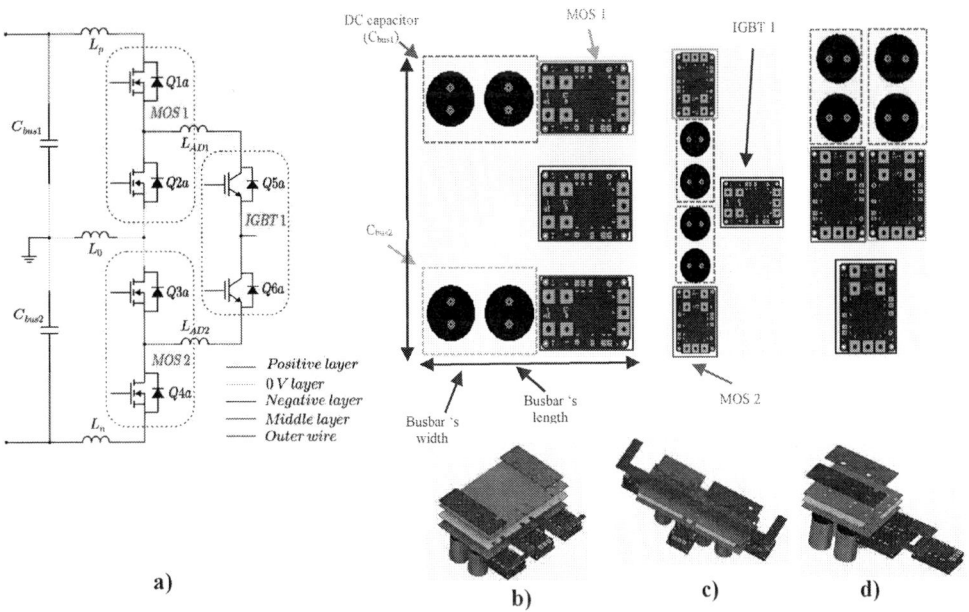

Figure. 5: Layout presentations, a) ANPC leg, b) **Layout 1**: line placement of power modules, c) **Layout 2**: Opposition placement of MOS 1 and MOS 2, d) **Layout 3**: triangular placement of power modules

Based on these comparative configurations, it can be concluded that layout 1 and 3 seems to be the most interesting configurations from a power density and cost point of view. However, taking into account the ANPC short switching loop inductance ($L_b = L_0 + L_p$) presented in Table 1, layout 3 is more interesting. These parasitic inductance values were obtained by extracting the 3D model of these different bus bars under Ansys Q3D.

Table 1: Parasitic inductances comparison, inverter leg busbars

	Layout 1	Layout 2	Layout 3
Length [mm]	427	459	259
Width [mm]	259	174	230
Thickness [mm]	3	3	3
L_p [nH]	7	7	4.8
L_0 [nH]	15.88	15.88	15.88

2. Experimental test

After these comparative studies, the paralleling 3 leg of inverter (500 kVA) is chosen in order to obtain 1.5 MVA ANPC power converter. The triangular layout placement of the power modules was chosen (figure 5d) since this solution shows the lowest parasitic inductance loop and busbar's volume (table 1a). Figure 6 presents the 3D model of the PEBB, the prototype ANPC converter phase leg as well as the test bench which made it possible to carry out preliminary tests presented in this part.

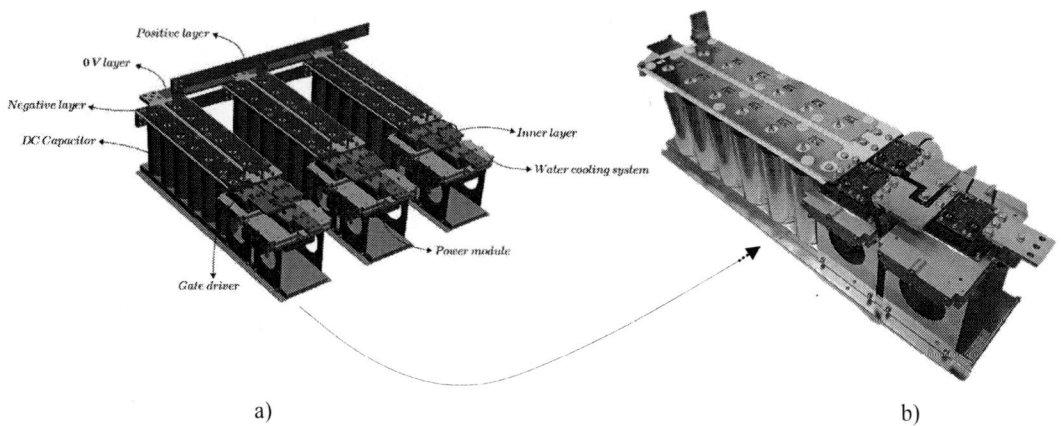

a)   b)

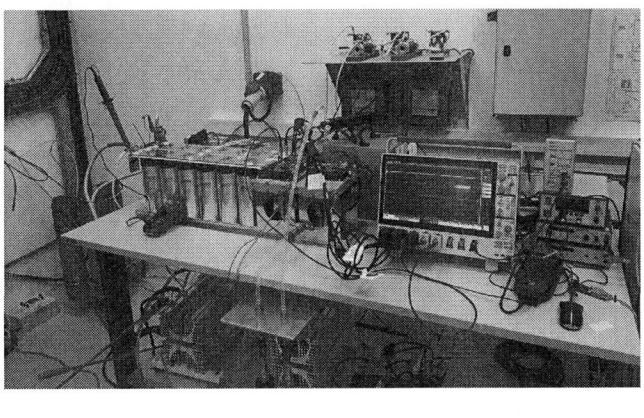

c)

Figure. 6: Inverter leg of the PEBB, a) 3D solid edge model, b) PEBB inverter leg prototype and c) the test bench of the DPT

To evaluate the parasitic inductance of the bar bus, double pulse tests (DPT) are carried out. The electrical schematic of the DPT is shown in Figure 8a. The total inductance of the ANPC short switching loop inductance ($L_{parT} = L_m + L_0 + L_p + L_{shunt} + L_{screw} + L_{cbus1}$) is determined using equation 6 as well as the electrical signals of Q2a (I_d and V_{ds}) measured during the turn off (Figure 8b). L_m is the parasitic of the SiC power module, L_C is the equivalent parasitic inductance of C_{bus1}, L_{screw}, is the parasitic inductance of the screw connection and L_{shunt} the shunt parasitic inductance. These parasitic inductances can be obtained from datasheet or can be calculated by estimation from the mechanical geometry.

$$L_{parT} = L_m + L_0 + L_{shunt} + L_{screw} + L_{cbus1} = \frac{\Delta V_{off\ overshoot}}{di/dt} \qquad (6)$$

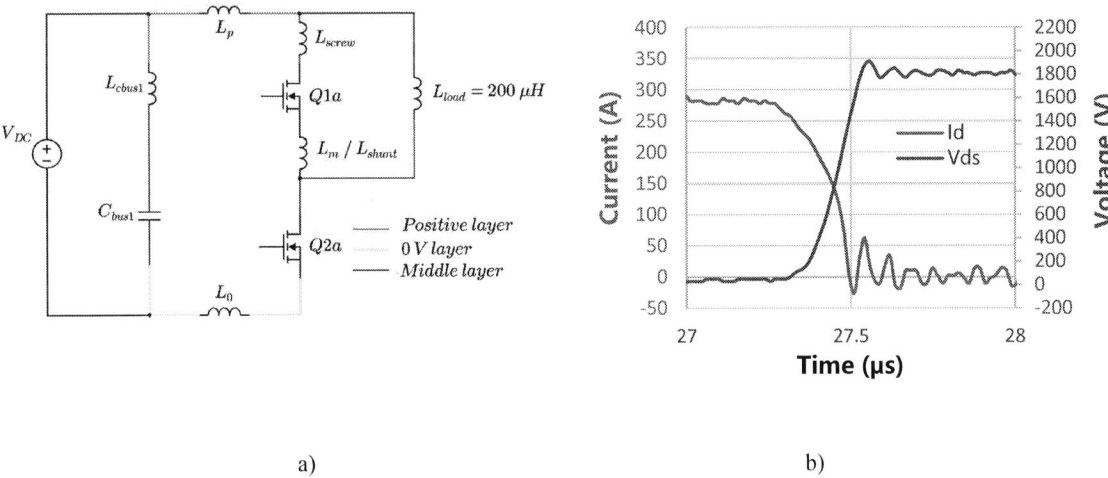

a)

b)

Figure. 7: Evaluation parasitic inductance (L_p and L_0) using Double Pulse Test of the PEBB, a) electrical scheme of the DPT, b) V_{ds} and I_d of Q2a using obtained by DPT

Table 2: Comparison with other works

References	This paper	[7]	[11]
Power [kVA]	500	500	750
V_{dc} (DC bus voltage) [kV]	3.6	1	2
L_{par} [nH]= $L_0 + L_p$	14.5	6.5	78

The value of the short switching parasitic inductance loop of the busbar ($L_0 + L_p$) obtained is listed in Table 2. It is compared with other work in the same field for ANPC or NPC converter prototyping. The parasitic inductance obtained is lower than the case [11] with a NPC converter, the dimensions of the converter are larger (1 000 × 700 × 500 mm) compared to the Talent converter. However, the study in [7] shows a smaller parasitic inductance. This example uses different kinds of power modules (HT − 3 000 series), with a lower DC busbar voltage (1 kV) than the one in Talent project (3.6 kV).

3. Conclusion

In this paper, a busbar design comparison for a high power and medium voltage ANPC converter has been presented. To validate the methodology, a comparative study is carried out on several types of placement for the power modules around the busbars and the capacitors. Simulation results show that the triangular configuration of power modules reduces the most parasitic inductance of the switching loops. The analysis of the switching waveform obtained by the experimental results shows a total parasitic inductor of 14.5 nH lower as the simulated value (20.68 nH).

Next investigations will focus on the SiC power modules switching losses measurements. Then gate resistors will be adjusted in order to increase the MOSFET switching speed in order to benefits the selected busbars and reduce the switching losses.

References

[1] Jaeger-Waldau, A., PV Status Report 2019, EUR 29938 EN, Publications Office of the European Union, Luxembourg, 2019, ISBN 978-92-76-12608-9, doi:10.2760/326629, JRC118058.

[2] H. Komatsu, T. Katayama, et N. Kawakami, « Development of Large-Capacity Converter for Battery Energy Storage Systems », in *2018 International Power Electronics Conference (IPEC-Niigata 2018 - ECCE Asia)*, Niigata, mai 2018, p. 1346-1350. doi: 10.23919/IPEC.2018.8507856.

[3] B. Burger, « For Solar Energy systems - ISE », Fraunhofer, 2019.

[4] Y. Jiao, « High Power High Frequency 3-level Neutral Point Clamped Power Conversion System », p. 302.

[5] D. Zhang, J. He, et S. Madhusoodhanan, « Three-Level Two-Stage Decoupled Active NPC Converter With Si IGBT and SiC MOSFET », *IEEE Trans. on Ind. Applicat.*, vol. 54, n° 6, p. 6169-6178, nov. 2018, doi: 10.1109/TIA.2018.2851561.

[6] L. Popova *et al.*, « Stray inductance estimation with detailed model of the IGBT module », in *2013 15th European Conference on Power Electronics and Applications (EPE)*, Lille, France, sept. 2013, p. 1-8. doi: 10.1109/EPE.2013.6631852.

[7] H. Gui *et al.*, « Methodology of Low Inductance Busbar Design for Three-Level Converters », *IEEE J. Emerg. Sel. Topics Power Electron.*, vol. 9, n° 3, p. 3468-3478, juin 2021, doi: 10.1109/JESTPE.2020.2999403.

[8] « Electrical Engineering Software | Plexim ». https://www.plexim.com/ (consulté le 11 juillet 2022).

[9] « Ansys Q3D Extractor | Q3D Simulation & Parasitic Extraction Tool ». https://www.ansys.com/fr-fr/products/electronics/ansys-q3d-extractor (consulté le 10 juillet 2022).

[10] Huiqing Wen, Weidong Xiao, Xuhui Wen, et P. Armstrong, « Analysis and Evaluation of DC-Link Capacitors for High-Power-Density Electric Vehicle Drive Systems », *IEEE Trans. Veh. Technol.*, vol. 61, nº 7, p. 2950-2964, sept. 2012, doi: 10.1109/TVT.2012.2206082.

[11] J. Wang, B. Yang, J. Zhao, Y. Deng, X. He, et X. Zhixin, « Development of a compact 750KVA three-phase NPC three-level universal inverter module with specifically designed busbar », in *2010 Twenty-Fifth Annual IEEE Applied Power Electronics Conference and Exposition (APEC)*, Palm Springs, CA, USA, févr. 2010, p. 1266-1271. doi: 10.1109/APEC.2010.5433338.

A Simulation Model for SiC MOSFET Switching Transients Controlled by an Adaptive Gate Driver with the Capability of Reducing Switching Losses and EMI across the Full Operating Range

[1]Zheming Li, [2]Robert W. Maier, [1]Mark-M. Bakran, [3]Franz-J. Niedernostheide, [3]Daniel Domes

[1]University of Bayreuth, Center of Energy Technology, Department of Mechatronics
[1]Universitätsstraße 30
[1]Bayreuth, Germany

[2]ZF Friedrichshafen AG
[2]Am Briefzentrum 2
[2]Bayreuth, Germany

[3]Infineon Technologies AG
[3]Am Campeon 1-15
[3]Neubiberg, Germany

Tel.: +49 / (0) – 921 55 7811
E-Mail: zheming.li@uni-bayreuth.de
URL: https://www.mechatronik.uni-bayreuth.de

Keywords

«Intelligent gate driver», «Device simulation», «EMC/EMI», «Switching losses», «Silicon Carbide (SiC)», «MOSFET».

Abstract

In this paper, the performance of an intelligent-gate-driver-based self-regulating gate control approach, which can reduce switching losses and EMI at SiC MOSFET turn-off and turn-on, is investigated by simulation and verified by measurements. Firstly, a MOSFET behavior model is presented and confirmed with double pulse measurement results of this approach. Based on this model, the performance of this approach in continuous operation is evaluated and compared with measurement in continuous operation. It is verified that there is a good match between measurement and simulation. The trade-off between switching losses and EMI is improved significantly by the proposed gate control approach compared to simple gate control with a single gate resistance.

1 Introduction

In recent years, wide bandgap semiconductors like SiC MOSFETs are in the process of replacing Si IGBTs in many power electronic applications. SiC MOSFETs show many advantages, such as low conduction and switching losses. But there are some challenges. Generally, fast switching transients are desired for low switching losses. However, the switching speed of SiC MOSFETs is limited by the drain-source overvoltage and the voltage oscillation, which are both induced by the commutation circuit stray inductance and the latter causes electromagnetic interference (EMI) emission [1-2]. Besides, the usually unwanted parasitic turn-on (PTO) effect [3-4], which occurs during the turn-on switching transients of a SiC MOSFET in a half-bridge configuration and results in a temporary shoot-through current of the phase, tends to increase at higher switching speed.

Active gate drivers (AGDs) are utilized to ensure a better compromise between switching losses and EMI emission of SiC MOSFET switching transients. A comprehensive overview of the state-of-the-art SiC devices AGD research is provided in [5]. The control strategies of AGD are classified into 3 categories, as shown in column 2-4 of Table I. The open-loop control shown in column 2 of Table I applies the same driver voltage profile to all scenarios. Yet optimum switching performance cannot be guaranteed in most operating conditions, as power electronic applications normally have a wide operating range. The direct feedback control shown in column 3 of Table I uses sensor circuits to detect the feedback signals. Their references are tracked by adapting the output of AGD. However, adjusting the gate control parameters of AGD during switching transients needs high speed, high-bandwidth electrical components like sensors, A/D converters, comparators, which are not economy friendly. Besides, even with these expensive components, their propagation delay and the parasitic effects of the AGD circuit may lead to insufficient switching performance. The model-based indirect control shown in column 4 of Table I estimates the feedback indices with an accurate switching trajectory model of power devices instead of direct measurement or describes the relation between operating parameters and gate control parameters with a look-up-table. Operating parameters like DC-link voltage, load current and junction temperature are sensed prior to each switching process to derive the optimum gate control parameters with the model or the look-up-table. No ultrafast components are required to detect the high frequent feedback signals. However, new problems such as high computation load of AGD, enormous calibration and measuring efforts are introduced. Moreover, measuring inaccuracies may lead to insufficient results.

Table I: Comparison between AGD control strategies in [5] and the proposed approach

AGD control strategies	Open-loop control [5]	Direct feedback control [5]	Model-based indirect control [5]	**Proposed approach**
Description	Applying constant gate control parameters for all operating conditions	Compensating errors between measured feedback signals and references via adjusting control parameters during actual switching transient	Building look-up-table/switching trajectory model to derive switching performance and gate control parameters	**Adjusting control parameters for the next switching event according to EMI evaluation of the actual one**
Feedback signals	No	du/dt, di/dt, $u_{DS}(t)$, $i_D(t)$, T_j etc.	U_{DC}, I_D, T_j etc.	**di/dt for EMI evaluation**
Operating condition dependence	Not considered	Considered	Considered	**Considered**
Requirements on driver-hardware	Low	High speed, high-bandwidth sensors, comparators, processors etc. with low propagation delay	Measurement of relevant operating parameters, high computation load of AGD, etc.	**Measurement and processing of di/dt-signal with low timing requirements**
Calibration effort	Low	Low	High	**Low**
Cost	Low	High	High	**Low**
Performance in operation	--	(--, ++) depending on switching speed	+	**Evaluated in this work**

An AGD-based self-regulating gate control approach is proposed in [6-7] and shown in column 5 of Table I. It consists of two methods, a boost-method for the optimization of a single operating point and a tracking method for the closed-loop control of varying operating conditions. The boost-method provides two levels of switching speed. Both levels are used successively in one switching process. It is verified in [6-7] by double pulse measurements that the boost-method can realize low switching losses and low switching oscillation at a certain operating point at both of SiC MOSFET turn-off and turn-on, when the gate control parameters are set as optimum. The control loop of the tracking method is closed not in the actual switching event, but over the next one. The drain-current slope di/dt is measured by the voltage drop from the power source to the kelvin source of SiC devices for oscillation evaluation by a simple comparator. The gate control parameters are adjusted for the next switching process according to the oscillation evaluation of the actual switching process. The adjustment is conducted after the actual and before the next switching process. Thanks to sufficient time between two successive switching events, the timing requirements for the di/dt-signal-processing are low. Furthermore, measurement of operating parameters or establishment of an accurate model or look-up-table is unnecessary. Therefore, hardware, calibration and measuring efforts are reduced significantly.

In this paper, a SiC MOSFET switching trajectory model is presented. The accuracy of this model is confirmed by double pulse measurements of the proposed gate control approach. Simulations of PWM controlled continuous operation are performed with this approach and compared with measurements performed under the same test condition. The performance of this approach is evaluated and compared to the performance of simple gate control with a single gate resistance.

2 Measurement setup and SiC MOSFET behavior simulation model

The devices under test (DUTs) are 1.7 kV SiC MOSFETs in TO-247 package. The rated current of a single chip is 37.5 A. The double pulse experiment is used to investigate the switching behavior of the SiC MOSFETs. All the experiments are performed as single chip measurements, where the commutation circuit stray inductance (L_σ) is scaled to emulate the switching behavior of a high power module, as shown in [8].

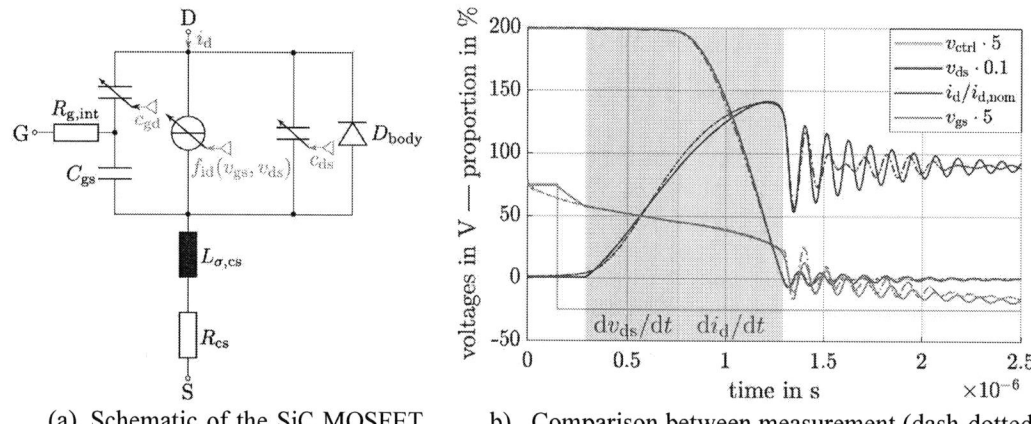

(a) Schematic of the SiC MOSFET
     simulation model

b) Comparison between measurement (dash-dotted)
    and simulation (solid) in a high inductive setup

Fig. 1: Schematic and results of the SiC MOSFET simulation model [9-10]

The SiC MOSFET behavior simulation model is shown in Fig. 1 (a). This model is proposed and explained in detail in [9-10]. With this model, the switching processes can be described with a small number of input parameters. The voltage-dependent MOSFET capacitances c_{ds}, c_{gd} and the constant capacitance C_{gs} can be received from capacity measurements or data sheet. The measurement of the common source elements $L_{\sigma,\,cs}$ and R_{cs} is explained in [9]. The central function of the model $i_d\,(v_{gs},\,v_{ds})$ can be acquired by measurements of transfer characteristics. Finally, the switching environment including the scaled commutation circuit stray inductance and gate resistances is needed. A comparison between a double pulse measurement shown in dash-dotted lines and the corresponding simulation shown in solid lines is shown in Fig. 1 (b). The accuracy of the model is verified.

3 Optimization of switching performance for a single operating point

In this section, the variant of the boost-method shown in [6] for the optimization of SiC MOSFET turn-off switching performance and the variant shown in [7] for the optimization of turn-on switching performance are briefly reviewed, respectively. The simulated and the measured switching trajectories are compared to verify the accuracy of the proposed simulation model.

3.1 Optimization of turn-off switching performance

The working principle of the boost-method implemented at SiC MOSFET turn-off is shown in Fig. 2 (a). The low boost gate resistance $R_{G,\,boost\text{-}off}$ accelerates the switching speed, whereas the high gate resistance $R_{G,\,normal}$ suppresses the drain-source voltage oscillation. The control signal (green), the output voltages of the normal-driver $U_{DR,\,normal}$ (blue) and the boost-driver $U_{DR,\,boost}$ (red) at SiC MOSFET turn-off are shown in Fig. 2 (a) right. When a turn-off switching process is initiated, $U_{DR,\,normal}$ switches from the positive driver voltage U_P to the negative U_N and maintains U_N until the next turn-on switching process. $U_{DR,\,boost}$ is normally U_P. It changes to U_N at the start of the turn-off switching process for a predefined time duration called boost-time t_B. The diode D_{off} ensures that $R_{G,\,boost\text{-}off}$ is only active and accelerates the switching speed during boost-time. A comparison between three switching processes performed without (dashed with high switching speed, dotted with low switching speed) and with the boost-method (solid) is shown in Fig. 2 (b). With the boost-method low switching losses and low oscillation are achieved simultaneously.

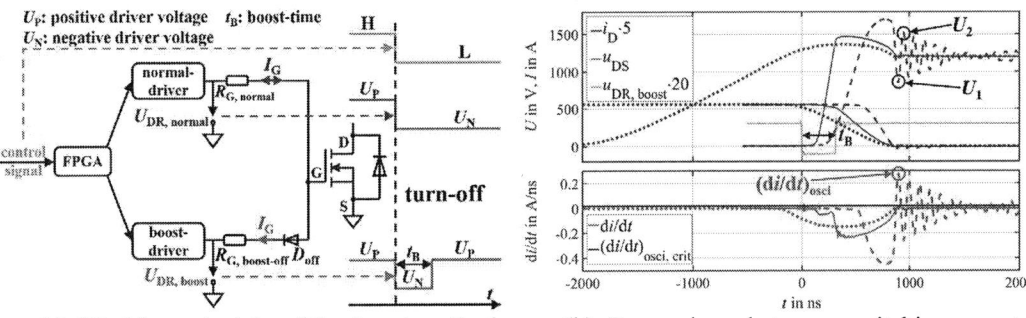

(a) Working principle of the boost-method implemented at SiC MOSFET turn-off with the control signal and output voltages of gate drivers

(b) Comparison between switching events with (solid) and without (dashed, dotted) the boost-method at $U_{DC} = 1.2$ kV, $I_D = 112$ A, $T_j = 25°C$

Fig. 2: A brief review of the boost-method implemented at SiC MOSFET turn-off [6]

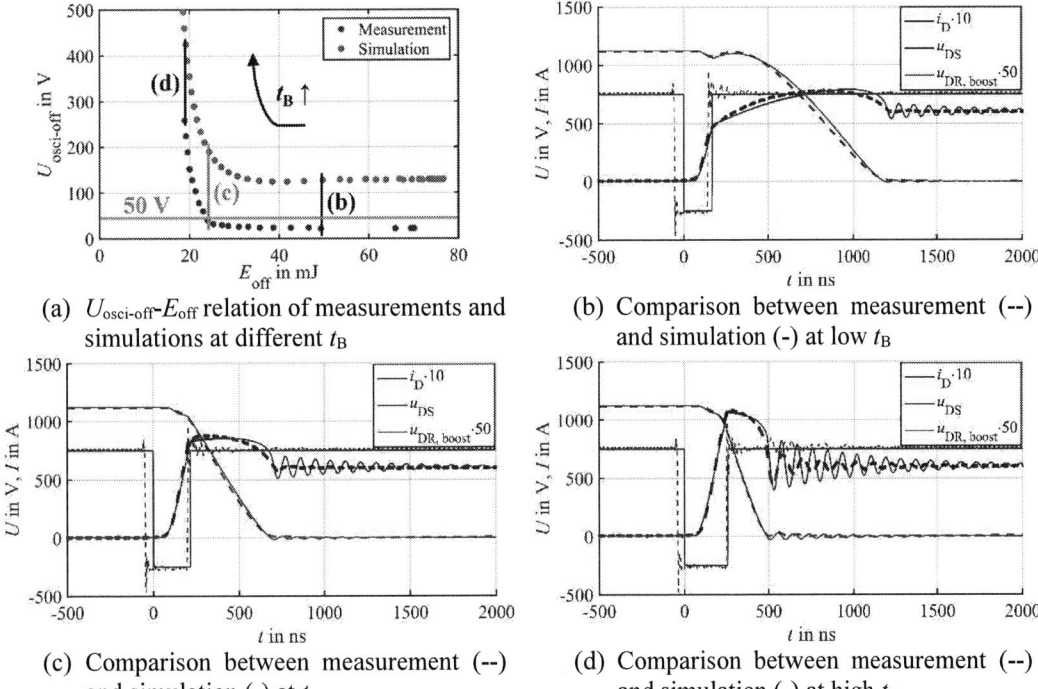

(a) $U_{osci\text{-}off}$-E_{off} relation of measurements and simulations at different t_B

(b) Comparison between measurement (--) and simulation (-) at low t_B

(c) Comparison between measurement (--) and simulation (-) at $t_{B, opt}$

(d) Comparison between measurement (--) and simulation (-) at high t_B

Fig. 3: Comparison between measurements and simulations of the boost-method implemented at SiC MOSFET turn-off at the test condition: $U_{DC} = 600$ V, $I_D = 112$ A, $T_j = 25°C$

Table II: Comparison between measurements and simulations shown in Fig. 3 (b) – (d)

Figure	Measurements shown in dashed lines			Simulations shown in solid lines		
	t_B/ns	$U_{osci\text{-}off}$/V	E_{off}/mJ	t_B/ns	$U_{osci\text{-}off}$/V	E_{off}/mJ
Fig. 3 (b)	195	22	49.4	165	128	49.8
Fig. 3 (c)	245	40	24.0	215	189	24.4
Fig. 3 (d)	285	259	18.9	255	424	19.2

To investigate the influence of the boost-time t_B on the switching performance, a test series of boost-switching processes are performed with different t_B under the test condition: $U_{DC} = 600$ V, $I_D = 112$ A, $T_j = 25°C$. The voltage oscillation indicator at SiC MOSFET turn-off $U_{osci\text{-}off}$ is quantitively defined as $|U_1 - U_2|$, as shown in Fig. 2 (b). The relation between $U_{osci\text{-}off}$ and the turn-off switching losses E_{off} in respect to t_B is shown in Fig. 3 (a) as discrete blue points. Each blue point represents a boost-switching measurement with a certain t_B. It can be seen that with increasing t_B, E_{off} decreases and $U_{osci\text{-}off}$ increases monotonously. A trade-off exists between $U_{osci\text{-}off}$ and E_{off}. The maximum t_B with an $U_{osci\text{-}off}$ of no more than a predefined threshold of the oscillation amplitude is defined as the optimum boost-time $t_{B,\,opt}$. When the threshold value is set as $U_{osci\text{-}off} = 50$ V, the boost-switching process with $t_{B,\,opt}$ is represented by the crossing point of the two green lines shown in Fig. 3 (a). The corresponding switching process is shown in dashed lines in Fig. 3 (c). The switching characteristics of the three measurements shown in dashed lines in Fig. 3 (b), Fig. 3 (c) and Fig. 3 (d) are listed in Table II left. $t_B < t_{B,\,opt}$ leads to high losses, as the switching process shown in dashed lines in Fig. 3 (b), while $t_B > t_{B,\,opt}$ leads to high oscillation amplitudes, as the switching process shown in dashed lines in Fig. 3 (d). Thus, t_B must be set as optimum. As the di/dt-signal is proportional to the voltage oscillation during switching transients, $U_{osci\text{-}off}$ can be determined by the voltage drop from the power source to the kelvin source for reduced measuring effort.

A simulation series of boost-switching processes is performed with different t_B at the same condition as the test series. The relation between the voltage oscillation indicator at SiC MOSFET turn-off $U_{osci\text{-}off}$ and the turn-off switching losses E_{off} is shown in Fig. 3 (a) as red points. Each red point represents a boost-switching simulation with a certain t_B. It can be seen that the $U_{osci\text{-}off}$-E_{off} relation of the test series and the $U_{osci\text{-}off}$-E_{off} relation of the simulation series are qualitatively identical with an oscillation offset. Three boost-switching simulations with different t_B are shown in solid lines in Fig. 3 (b), Fig. 3 (c) and Fig. 3 (d), respectively. The simulated boost-switching trajectories match well with the measured boost-switching trajectories. The switching characteristics of the three simulations are listed in Table II right. E_{off} of the compared measurements and simulations in Fig. 3 (b) – Fig. 3 (d) are almost equal, whereas t_B and $U_{osci\text{-}off}$ of the compared measurements and simulations are shifted by an offset. It is proven that the proposed simulation model can be used to qualitatively investigate the boost-switching performance.

3.2 Optimization of turn-on switching performance

The variant of the boost-method for the optimization of SiC MOSFET turn-on switching performance takes the advantages of the parasitic turn-on (PTO) effect to combine low switching losses and low EMI. The working principle of this variant is shown in Fig. 4 [7]. In case $I_D > 0$, the high side switch (HSS) performs a hard switching process and is called active switch (blue). The low side switch (LSS) acts as freewheeling device and is referred to as passive switch (red). The low boost gate resistance $R_{G,\,boost\text{-}off}$ accelerates the gate discharging, while the high gate resistance $R_{G,\,normal}$ decelerates it. The active turn-on of HSS is controlled by the turn-on gate resistance $R_{G,\,on}$, which is set to a low value for low switching losses. The control signals, the output voltages of the normal-drivers $U_{DR,\,normal}$ and the boost-drivers $U_{DR,\,boost}$ are shown in Fig. 4 right. Different from the variant of the boost-method implemented at SiC MOSFET turn-off (active turn-off), this variant is implemented at diode turn-off (passive turn-off). In order to avoid short circuit of the phase, a dead time t_D is set between the passive turn-off of LSS and the active turn-on of HSS. In other aspects, the two variants of the boost-method are similar.

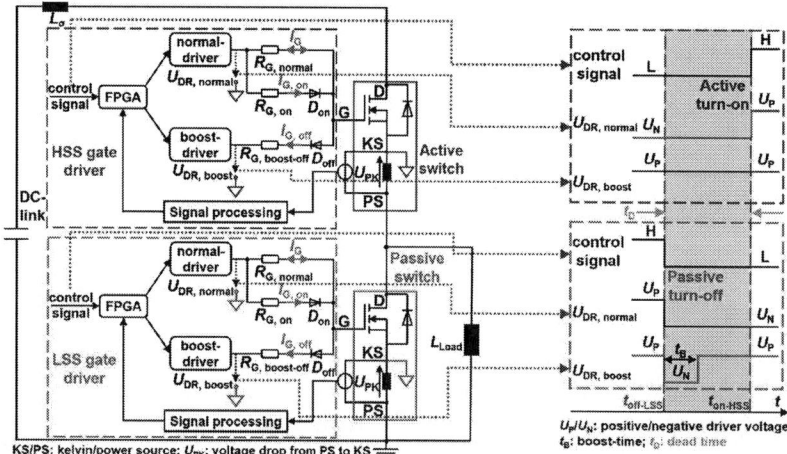

Fig. 4: Working principle of the boost-method for the optimization of SiC MOSFET turn-on switching performance with the control signals and output voltages of HSS/LSS gate drivers [7]

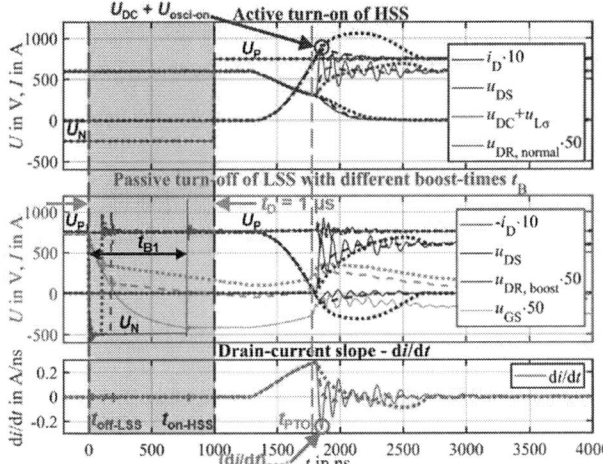

Test Nr.	1	2	3
Mark	Solid (-)	Dashed (--)	Dotted ($\cdots$)
t_B/ns	780	170	100
$U_{osci\text{-}on}$/V	330	42	90
E_{MOS}/mJ	8.4	9.1	11.7
E_{Dio}/mJ	0.5	1.1	9.0
E_{on}/mJ	8.9	10.2	20.7
$E_{on} = E_{MOS} + E_{Dio}$			

Fig. 5: Comparison between boost-method controlled switching processes with different t_B at the test condition: t_D = 1 µs, I_D = 75 A, U_{DC} = 600 V, T_j = 25°C [7]

The switching trajectories of three boost-method controlled measurements with different boost-time t_B is illustrated in Fig. 5 left. The HSS signals are shown in the top diagram, the LSS signals in the middle diagram and the drain-current slope di/dt in the bottom diagram. With reducing t_B the three switching processes are shown in solid, dashed and dotted lines, respectively. In the middle diagram of Fig. 5 left it can be seen that the adaption of the gate source voltage u_{GS} affected via t_B results in the same control effect as the adjustment of the negative driver voltage U_N. Lower t_B leads to higher u_{GS} at the start time of the PTO effect t_{PTO}, which results in stronger PTO. A high overvoltage at the LSS will induce an oscillation between the commutation circuit stray inductance L_σ and the LSS junction capacitance. The occurrence of the PTO dampens the overvoltage and reduces the switching oscillation. The switching characteristics of the three switching processes are demonstrated in Fig. 5 right. The voltage oscillation indicator at SiC MOSFET turn-on $U_{osci\text{-}on}$ is quantitively defined as the maximum voltage oscillation peak of L_σ, as shown in the top diagram of Fig. 5 left. The total switching losses at SiC MOSFET turn-on E_{on} is defined as the cumulated switching losses of the active MOSFET E_{MOS} and the passive body-diode E_{Dio}. The switching process shown in solid lines has a high $U_{osci\text{-}on}$ of 330 V due to insufficient PTO. The switching process shown in dotted lines has a low $U_{osci\text{-}on}$ of 90 V, but the shoot-through current caused by PTO is high, which leads to high E_{Dio} of 9 mJ. The switching process shown in dashed lines achieves low $U_{osci\text{-}on}$ of 42 V and low E_{Dio} of 1.1 mJ simultaneously.

To investigate the influence of the boost-time t_B on the switching performance, a test series of boost-switching processes are performed with different t_B at the test condition: I_D = 75 A, U_{DC} = 600 V, T_j = 25°C, t_D = 1 µs. The relation between the voltage oscillation indicator at SiC MOSFET turn-on $U_{osci\text{-}on}$ and the total switching losses at SiC MOSFET turn-on E_{on} in respect to t_B is illustrated in Fig. 6 (a) as discrete blue points. Each blue point represents a boost-switching measurement with a certain t_B. It can be seen that with decreasing t_B, $U_{osci\text{-}on}$ first decreases significantly with negligible increase of losses. Almost no losses are generated in the passive switch until $U_{osci\text{-}on}$ reaches its minimum. However, with further decreasing t_B, E_{on} rises dramatically due to excessive PTO. $U_{osci\text{-}on}$ increases not because of increasing oscillation, but caused by the increasing PTO current. The switching event shown in dotted lines in Fig. 5 left demonstrates this phenomenon. High t_B leads to high oscillation, while low t_B to high losses. Hence, t_B must be set appropriate. As E_{on} decreases with increasing t_B monotonously, the maximum t_B with an $U_{osci\text{-}on}$ of no more than a predefined oscillation threshold is defined as optimum boost-time $t_{B, opt}$. If the oscillation threshold is set as $U_{osci\text{-}on}$ = 50 V, the boost-switching process with $t_{B, opt}$ is marked with the green circle shown in Fig. 6 (a), which is corresponded to the switching process shown in dashed lines in Fig. 5 left and Fig. 3 (c).

A simulation series of boost-switching processes is performed with different t_B at the same test condition as the test series. The relation between $U_{osci\text{-}on}$ and E_{on} is shown in Fig. 6 (a) as red points. Each red point represents a boost-switching simulation with a certain t_B. It can be seen that the $U_{osci\text{-}on}$-E_{on} relation of the measurement series matches well with the $U_{osci\text{-}on}$-E_{on} relation of the simulation series with a slight difference of E_{on}. Three comparisons of switching trajectory between measurement and simulation with different t_B are shown in Fig. 6 (b), Fig. 6 (c) and Fig. 6 (d), respectively. Their switching characteristics are listed in Table III. The switching trajectories, the PTO effect and the influence of t_B on the boost-switching performance are well simulated by the proposed model.

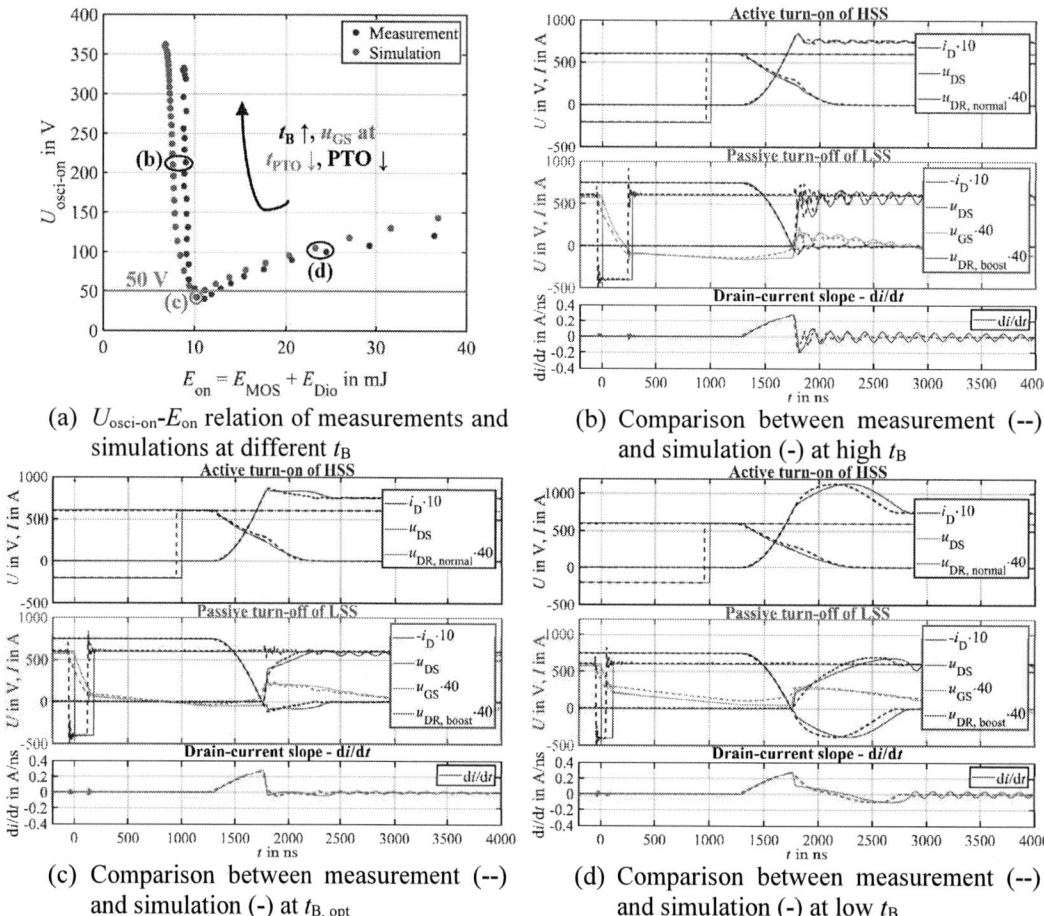

(a) $U_{osci\text{-}on}$-E_{on} relation of measurements and simulations at different t_B

(b) Comparison between measurement (--) and simulation (-) at high t_B

(c) Comparison between measurement (--) and simulation (-) at $t_{B,\,opt}$

(d) Comparison between measurement (--) and simulation (-) at low t_B

Fig. 6: Comparison between measurements and simulations of the boost-method implemented at diode turn-off at the test condition: $t_D = 1\ \mu s$, $I_D = 75$ A, $U_{DC} = 600$ V, $T_j = 25°C$

Table III: Comparison between measurements and simulations shown in Fig. 6 (b) – (d)

Figure	Measurements shown in dashed lines			Simulations shown in solid lines		
	t_B/ns	$U_{osci\text{-}on}$/V	E_{on}/mJ	t_B/ns	$U_{osci\text{-}on}$/V	E_{on}/mJ
Fig. 6 (b)	280	213	9.1	280	210	7.6
Fig. 6 (c)	170	42	10.2	180	42	10.2
Fig. 6 (d)	90	100	24.6	110	105	23.4

4 Optimization of switching performance in continuous operation

In this section, only the tracking method implemented at SiC MOSFET turn-off is discussed. Simulation and measurement of PWM controlled continuous operation are performed with the proposed gate control approach at the same condition. Their switching characteristics are compared. The performance of the proposed approach is compared with the performance of a simple gate control with a single gate resistor.

4.1 Review of the tracking method at SiC MOSFET turn-off

It is verified by measurements that the optimum boost-time $t_{B, opt}$ is strongly dependent on the operating condition, compare [6]. To realize an operating condition dependent gate control, the tracking method is developed. Its working principle is shown in Fig. 7. The basic idea of the tracking method is to trace varying $t_{B, opt}$ through the iteration of t_B according to the result of oscillation evaluation. During every boost-switching process, the di/dt-signal is measured and processed for oscillation evaluation. A switching process with an oscillation of no more than the predefined oscillation threshold is considered as oscillation-free. Therefore, $t_{B, opt}$ is the maximum t_B for an oscillation-free switching process. The tracking method control loop is closed over the next switching process. The boost-time of the actual switching process $t_{B, act}$ is not adjusted for the actual switching process, but the next one. As shown in Fig. 3 (a), the voltage oscillation indicator $U_{osci-off}$ increases monotonously with increasing t_B. If an oscillation is detected in the actual switching process, it means that $t_{B, act} > t_{B, opt}$. $t_{B, act}$ will be decremented with Δt_B for the next switching process. Otherwise, $t_{B, act} <= t_{B, opt}$. $t_{B, act}$ will be incremented with Δt_B for the next switching process. $t_{B, act}$ is assumed to be maintained within a narrow band around the varying $t_{B, opt}$ by the tracking method. A more detailed explanation of the tracking method is presented in [6].

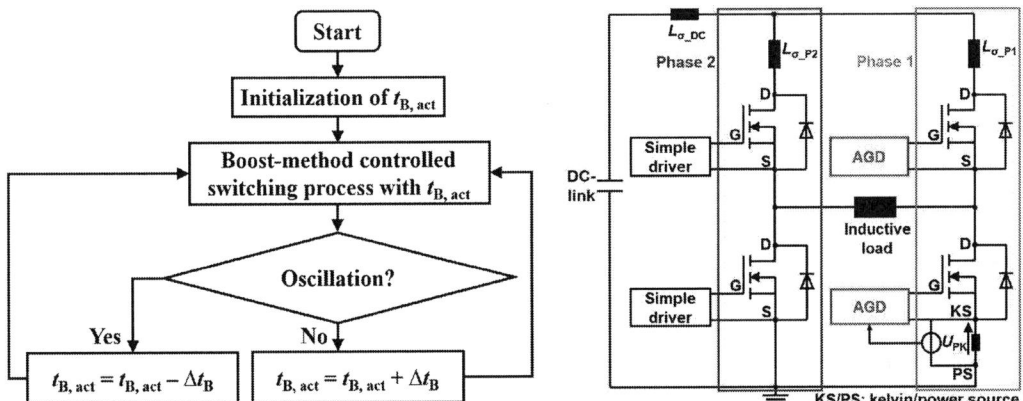

Fig. 7: Working principle of the tracking method   Fig. 8: Circuit diagram of the test setup

4.2 Implementation of the tracking method in continuous operation

The equivalent circuit diagram of the H-Bridge measurement setup is shown in Fig. 8. The AGDs (red) are connected to phase 1 (green). A test of PWM controlled continuous operation with a fundamental frequency $f_f = 60$ Hz, a switching frequency $f_{sw} = 10$ kHz, a dead time $t_D = 3$ µs and a finite number of periods $N_P = 5$ is performed under the test condition: $U_{DC} = 550$ V, $T_j = 25°C$. The drain-current course $I_D(t)$ is sinusoidal. The increment of the actual boost-time $t_{B, act}$ is set as $\Delta t_B = 5$ ns. In the performance evaluation, every turn-off switching process is extracted from the continuous operation and separately evaluated. Different switching characteristics of each turn-off switching process with $I_D > 0$ are evaluated and plotted as discrete points in different diagrams of Fig. 9 (a). For turn-off switching processes with $I_D < 0$, the actual boost-time $t_{B, act}$ is evaluated, the voltage oscillation indicator $U_{osci-off}$ and the turn-off switching losses E_{off} are always set as 0. U_{DC} and I_D are shown in the top diagram, $t_{B, act}$ in the second diagram, $U_{osci-off}$ in the third diagram and E_{off} in the bottom diagram. $t_{B, act}$, $U_{osci-off}$ and E_{off} of the measurement are shown in black. As the energy stored in the DC-link capacitors is consumed during continuous operation because of a power limitation of the DC voltage source, U_{DC} and the amplitude of sinusoidal I_D decrease with time as shown in the top diagram. It can be seen that $t_{B, act}$ increases monotonously at the start-up as $U_{osci-off}$ is much lower than the oscillation threshold. After the threshold is reached, $U_{osci-off}$ is maintained in close proximity to the threshold, as varying $t_{B, opt}$ caused by varying I_D, U_{DC} is traced by $t_{B, act}$ with an apparent concave trajectory. The behavior of the tracking method in the driving cycle meets the expectations.

A simulation of continuous operation is performed. The I_D- and U_{DC}-course of the measurement shown in the top diagram of Fig. 9 (a) are utilized as simulation condition to ensure a fair comparison. The switching characteristics of every simulated turn-off switching process are evaluated and plotted as discrete green points in Fig. 9 (a) in the same manner as the measurement. It can be seen that the E_{off}-courses of the simulation and the measurement are almost equal, whereas the $t_{\text{B, act}}$- and $U_{\text{osci-off}}$-courses of the simulation and the measurement show an offset, which is consistent with the comparisons of single operating point between simulations and measurements shown at the end of section 3.1. Fig. 9 (b) illustrates a comparison of the switching trajectories at $t = 24.62$ ms between the simulation shown in solid lines and the measurement shown in dashed lines. The simulated boost-switching trajectory matches well with the measured boost-switching trajectory. It is proven that the proposed SiC MOSFET behavior simulation model is capable of qualitatively investigating the proposed gate control approach consisting of the boost-method and the tracking method, even in continuous operation.

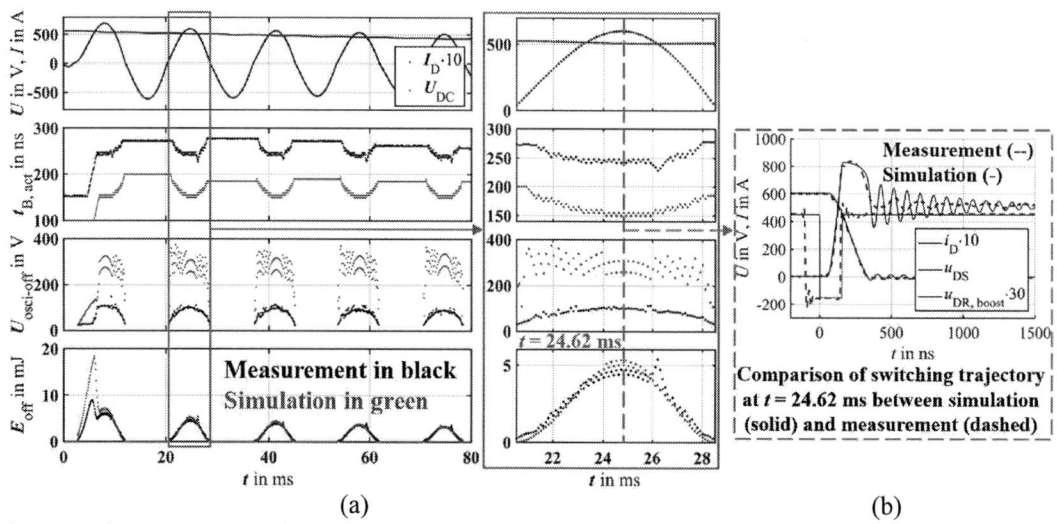

        (a)   (b)

Fig. 9: Performance comparison of PWM controlled continuous operation between measurement (black points in Fig. 9 (a), dashed lines in Fig. 9 (b)) and simulation (green points in Fig. 9 (a), solid lines in Fig. 9 (b)) at the condition: $f_{\text{f}} = 60$ Hz, $f_{\text{sw}} = 10$ kHz, $U_{\text{DC}} = 550$ V, $T_{\text{j}} = 25°$C, sinusoidal I_{D}

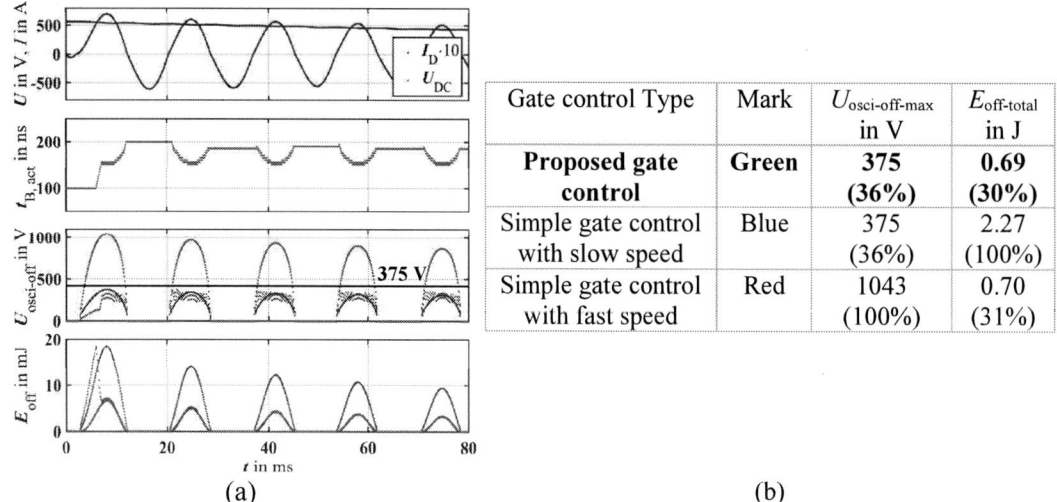

Gate control Type	Mark	$U_{\text{osci-off-max}}$ in V	$E_{\text{off-total}}$ in J
Proposed gate control	**Green**	**375 (36%)**	**0.69 (30%)**
Simple gate control with slow speed	Blue	375 (36%)	2.27 (100%)
Simple gate control with fast speed	Red	1043 (100%)	0.70 (31%)

        (a)   (b)

Fig. 10: Performance comparison of continuous operation between simulations of the proposed gate control (green) and simple gate control (fast in red, slow in blue) at the condition shown in Fig. 9

$$E_{off-total} = \sum_{n=1}^{N_S} E_{off-n}, N_S = \frac{f_{sw}}{f_f} \cdot N_P \tag{1}$$

To investigate the performance of the proposed gate control approach, two simulations of continuous operation are performed with different switching speeds at the same condition as the simulation shown in Fig. 9. Their gate control type is simple gate control with a single gate resistance. Fig. 10 demonstrates a performance comparison between these two simulations with simple gate control and the simulation with the proposed approach shown in Fig. 9. The switching characteristics of every simulated turn-off switching process are evaluated and plotted as discrete points in Fig. 10 (a) in the same way as Fig. 9 (a). The proposed approach is shown in green, the slow simple gate control in blue and the fast in red. The maximum voltage oscillation indicator of continuous operation $U_{osci\text{-}off\text{-}max}$ is defined to evaluate the oscillation magnitude. The total turn-off switching losses of continuous operation $E_{off\text{-}total}$ is calculated by means of Eq. (1). N_S is the total number of turn-off switching events of continuous operation, N_P the number of periods and $E_{off\text{-}n}$ the losses of the "n"-th turn-off switching process. It should be pointed out that the calculation of $E_{off\text{-}total}$ starts from the second period, as the transition of $t_{B,\,act}$ from start-up to $t_{B,\,opt}$ leads to an E_{off}-peak, as shown in the bottom diagram of Fig. 9 (a), Fig. 10 (a). It can be concluded from the table shown in Fig. 10 (b) that $E_{off\text{-}total}$ of the proposed gate control is 70% lower than $E_{off\text{-}total}$ of the slow simple gate control while $U_{osci\text{-}off\text{-}max}$ is equal for the two control techniques. $U_{osci\text{-}off\text{-}max}$ of the proposed gate control is 64% lower than that of the fast simple gate control without increased $E_{off\text{-}total}$. The performance of the proposed gate control in continuous operation is proven to be excellent.

5 Conclusion

A SiC MOSFET behavior model is proposed to simulate a boost-method for the switching performance optimization of a single operating point and a tracking method for the dynamic optimization of varying operating conditions. The simulated boost-switching trajectories are consistent with the measured ones of a SiC MOSFET at both turn-off and turn-on. The tracking method is investigated by simulations and measurements of PWM controlled continuous operation. The simulated behavior of the tracking method matches well with the measured behavior. The tracking method is capable of dynamically optimizing switching performance for varying operating conditions. It is proven by simulation that the performance of the proposed gate control approach consisting of the boost-method and the tracking method is significantly better than the performance of simple gate control in continuous operation. 70% of the switching losses or 64% of the oscillation amplitude can be reduced without increased oscillation amplitude or switching losses, respectively.

6 References

[1] E. U. Krafft, B. Laska, A. Nagel and J. Weigel: A new standard IGBT housing for high-power converters, EPE'15 ECCE Europe, pp. 1-11, 2015.

[2] A. Paredes, H. Ghorbani, V. Sala, E. Fernandez and L. Romeral: A new active gate driver for improving the switching performance of SiC MOSFET, 2017 IEEE APEC, pp. 3557-3563, 2017.

[3] M. R. Ahmed, R. Todd and A. J. Forsyth: Predicting SiC MOSFET behavior under hard-switching, soft-switching, and false turn-on conditions, IEEE Transactions on Industrial Electronics, vol. 64, no. 11, pp. 9001-9011, 2017.

[4] Z. Miao, Y. Mao, C.-M. Wang and K. D. T. Ngo: Detection of cross-turn-on and selection of off drive voltage for an SiC power module, IEEE Transactions on Industrial Electronics, vol. 64, no. 11, pp. 9064-9071, 2017.

[5] S. Zhao, X. Zhao, Y. Wei, Y. Zhao and H. A. Mantooth: A review of switching slew rate control for silicon carbide devices using active gate drivers, IEEE Journal of Emerging and Selected Topics in Power Electronics, vol. 9, no. 4, pp. 4096-4114, 2021.

[6] Z. Li, R. W. Maier, Mark-M. Bakran, D. Domes and F.-J. Niedernostheide: How to turn off SiC MOSFET with low losses and low EMI across the full operating range, PCIM Europe 2021, pp. 1263-1270, 2021.

[7] Z. Li, R. W. Maier, M. Walter and Mark-M. Bakran: A self-regulating gate control based on the parasitic turn-on effect for low losses and low EMI of SiC MOSFET, PCIM Europe 2022, 2022.

[8] R. W. Maier and Mark-M. Bakran: Switching SiC MOSFETs under conditions of a high power module, EPE'18 ECCE Europe, pp. 1-10, 2018.

[9] P. Hofstetter, R. W. Maier and Mark-M. Bakran: Influence of the threshold voltage hysteresis and the drain induced barrier lowering on the dynamic transfer characteristic of SiC power MOSFETs, 2019 IEEE APEC, pp. 944-950, 2019.

[10] P. Hofstetter and Mark-M. Bakran: Mitigating drain source voltage oscillation for SiC power MOSFETs in order to reduce electromagnetic interference, EPE'19 ECCE Europe, pp. 1-10, 2019.

Phase-Shift Modulation for Flying-Capacitor DC-DC Converters

Philipp Rehlaender, Frank Schafmeister, Joachim Böcker
Power Electronics and Electrical Drives, Paderborn University
Warburger Str. 100
Paderborn, Germany
Phone: +49 (0) 5251 60 2159
Fax: +49 (0) 5251 60 3443
Email: rehlaender@lea.upb.de
URL: http://lea.upb.de

Acknowledgments

The authors would like to thank Delta Energy Systems (Germany) GmbH for funding this research.

Keywords

≪High frequency power converter≫, ≪Switched-mode power supply≫, ≪DC-DC converter≫, ≪3-Level Inverter≫.

Abstract

Multi-level flying capacitor converters are an attractive solution for PFC and inverter stages due to their increased output frequency and the application of semiconductors of lower voltage. Due to the beneficial figure of merit of lower-voltage semiconductors switching losses and conduction losses can be reduced. Consequently, this paper analyzes the three-level flying capacitor (FC) module as a replacement for the conventional half bridge. It discusses flying-capacitor voltage control options when being employed for the conventional half-bridge LLC converter with a flying-capacitor inverter and proposes a simple-to-implement modulation to fully utilize the benefits of the three-level structure of the FC module. It explores potential applications to propose a flying capacitor phase shift converter and a three-phase three-level LLC resonant converter with a five-level line-to-line voltage.

1 Introduction

Multi-level flying capacitor (FC) converters are an attractive solution for higher-voltage applications because they can utilize semiconductors with a reduced blocking voltage. They are usually employed in power factor correction (PFC) stages or inverters to show outstanding performance [1–3]. By employing a multi-level voltage-source inverter to such systems, the individual switching frequency of the semiconductors can be kept low while the output frequency increases with the number of levels [4–6]. As an effect, the output filter can be designed for a much smaller inductance due to the increased effective frequency and number of voltage levels.

Additionally, the designer profits from the lower semiconductor blocking voltage since low-voltage power MOSFETs can be utilized that feature a better figure of merit (FOM) compared to high-voltage semiconductors [3, 7–9]. Due to the significantly better FOM of low-voltage devices, FC converters may also be used to DC-DC converters. While traditional DC-DC converters do not profit from the increased number of voltage levels, the use of low-voltage devices of an improved FOM can result in a better system performance.

Consequently, a FC cell (cf. Figure 1b) can be used to replace the traditional half-bridge leg (cf. Figure 1a). An exemplary application may be the LLC half bridge converter, which is visualized with an

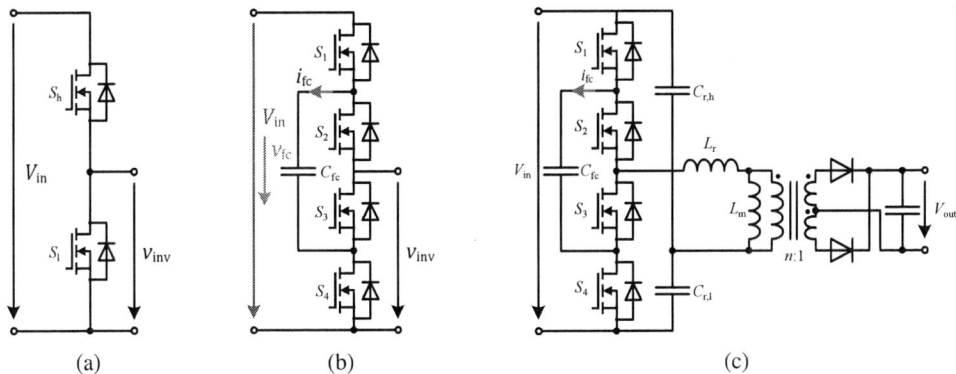

Fig. 1: (a)Traditional half-bridge module, (b) the flying capacitor module and (c) the flying capacitor inverter with a half-bridge LLC resonant converter with splitted resonant capacitor .

FC inverter in Figure 1c. However, in operation, the flying capacitor voltage v_{FC} must be controlled to half of the input voltage ($v_{FC} = V_{in}/2$) to ensure that every MOSFET is only stressed with half of the input voltage and avoid a dynamic increase of this voltage. Therefore, this paper analyzes the phase-shift modulation as a method to control the flying capacitor voltage and proposes an easy-to-implement three-level modulation to fully exploit the three-level structure of the FC converter enabling a quasi phase-shift modulation voltage pattern.

2 Flying-Capacitor DC-DC converters

In traditional applications of the FC converter (PFC and inverter applications), all levels of the modulations were utilized since the fundamental frequency of the filter was much lower than the switching frequency. However, in DC-DC converters (phase-shifted full bridge, LLC resonant converter, etc.), this is not the case since the inductor current changes polarity after half a switching period. To replace the conventional half-bridge with a FC module, a control mechanism must be found to control the FC voltage and avoid a run-off. The flying capacitor stage can be switched in four different states:

- $v_{inv} = V_{in}$ (ET$^+$, see Figure 2a),

- $v_{inv} = 0$ (ET$^-$, see Figure 2b),

- $v_{inv} = V_{in}/2$ (FW$^+$, see Figure 2c),

- $v_{inv} = V_{in}/2$ (FW$^-$, see Figure 2d).

In the following, the sequence of these states is analyzed to derive a balancing operation of the flying capacitor voltage and anaylze the modes of operation.

2.1 Phase-Shift Operation in Flying-Capacitor DC-DC Converters

When utilizing a phase-shift modulation to the FC module, the switching signals of the semiconductors S_1-S_4 (cf. Figure 1b) are phase shifted to the input signals of S_2-S_3 as depicted in Figure 3a. Compared to the traditional phase-shift modulation of the full-bridge inverter, this modulation cannot be as easily adapted since for either phase-shift modulation (Figure 3a, 3b), the flying capacitor current is either negative (cf. Figure 3a) or positive (cf. Figure 3b) resulting in a run-off of the flying capacitor voltage v_{FC}. Additionally, in the phase-shift modulation, the MOSFETs are stressed with different turn-off currents [10–12] as visualized in Figure 3a, 3b (current values named after [13]) and conduction losses [12].

The same voltage pulses of the phase-shift modulation can also be achieved by the asymmetrical phase-shift modulation or duty-cycle modulation depicted in Figures 3c and 3d [11, 14, 15]. This modulation

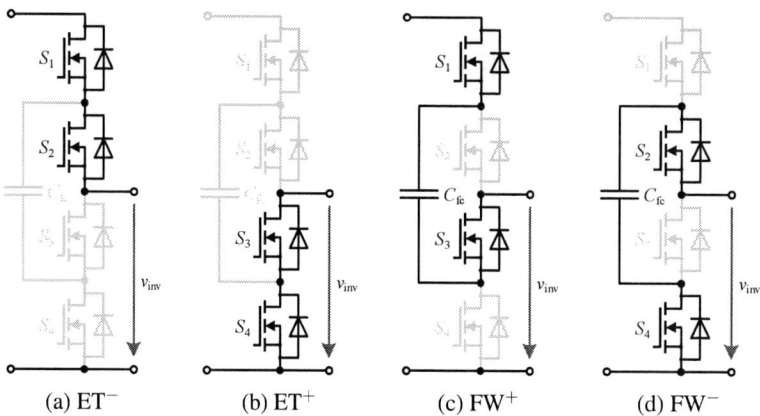

(a) ET⁻ (b) ET⁺ (c) FW⁺ (d) FW⁻

Fig. 2: Switching states of the flying-capacitor module.

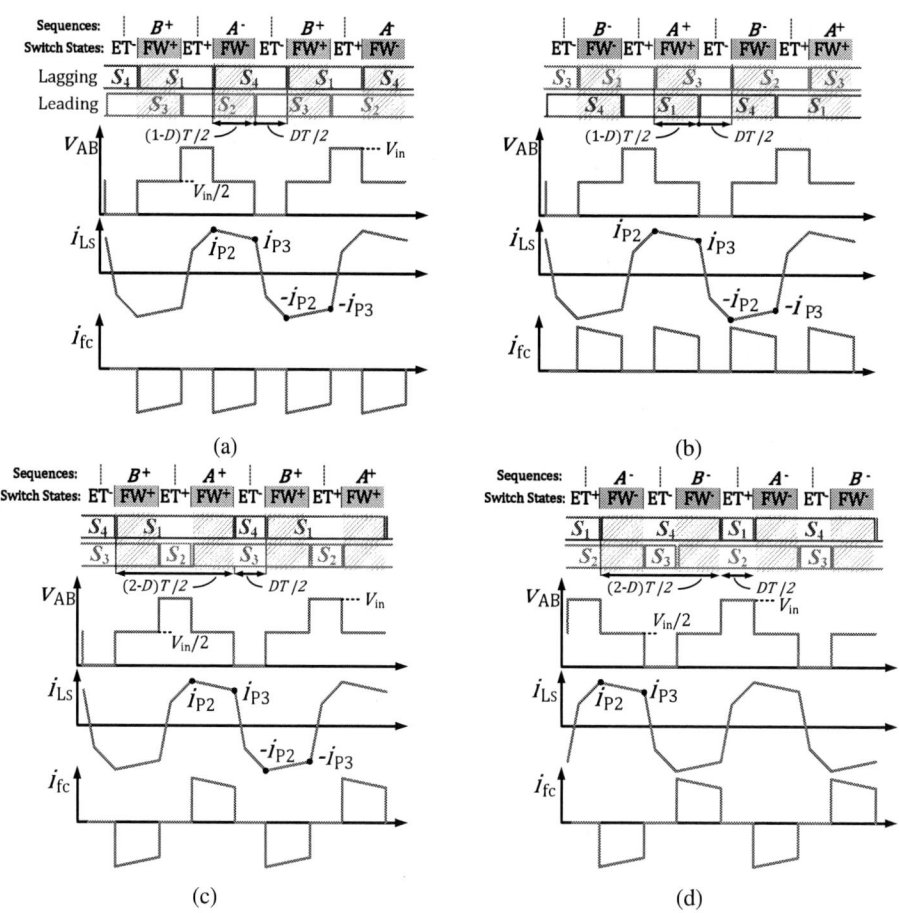

Fig. 3: Phase-shift modulation (a,b) and asymmetrical phase-shift modulation (c,d).

results in positive and negative flying-capacitor currents enabling a full utilization of the three levels of the flying capacitor module in a quasi phase-shift-modulation voltage pattern. However, the modulation results in unbalanced losses since two switches are turned on for $(2-D)T/2$ where $D \in [0,1]$ whereas the other two switches are only turned on for $DT/2$ resulting in unbalanced conduction losses. Additionally, the switches are also turned off at different current levels resulting in unbalanced switching losses.

In the conventional phase-shift modulation the switching intervals are repeated every period of the transformer current. If the switches S_2 and S_4 are operated as the leading leg, the intervals are repeated in the order

$$\mathbf{ET}^+(S_1, S_2) \to \mathbf{FW}^-(S_2, S_4) \to$$
$$\mathbf{ET}^-(S_3, S_4) \to \mathbf{FW}^+(S_1, S_3). \tag{1}$$

If the switches S_1 and S_3 are operated as the leading leg, the intervals are repeated in the reverse order:

$$\mathbf{ET}^+(S_1, S_2) \to \mathbf{FW}^+(S_1, S_3) \to$$
$$\mathbf{ET}^-(S_3, S_4) \to \mathbf{FW}^-(S_2, S_4). \tag{2}$$

Considering the preceding states four different sequences can be derived: the transition through a freewheeling interval from ET^+ to ET^- labeled type A sequence and the transition through a freewheeling interval from ET^- to ET^+, which are labeled type B sequence [12]. Both transitions can utilize either the positive or negative freewheeling state, which is emphasized through the superscript.

For the A type or B-type transition, the flying capacitor can be either charged or discharged (assuming steady state). The A^+ transition hereby results in a charging current whereas the A^- transition results in a discharging current:

$$A^+(i_{\mathrm{fc}} > 0)): \quad \mathbf{ET}^+(S_1, S_2) \to \mathbf{FW}^+(S_1, S_3) \to \mathbf{ET}^-(S_3, S_4), \tag{3}$$

and

$$A^-(i_{\mathrm{fc}} < 0): \quad \mathbf{ET}^+(S_1, S_2) \to \mathbf{FW}^-(S_2, S_4) \to \mathbf{ET}^-(S_3, S_4). \tag{4}$$

The two possible transitions from ET^- to ET^+ (labeled type-B sequence) can also result in a discharging current (B^+) or a charging current (B^-):

$$B^+(i_{\mathrm{fc}} < 0): \quad \mathbf{ET}^-(S_3, S_4) \to \mathbf{FW}^+(S_1, S_3) \to \mathbf{ET}^+(S_1, S_2) \tag{5}$$

and

$$B^-(i_{\mathrm{fc}} > 0): \quad \mathbf{ET}^-(S_3, S_4) \to \mathbf{FW}^-(S_2, S_4) \to \mathbf{ET}^+(S_1, S_2). \tag{6}$$

This analysis emphasizes that the description of above. The conventional phase-shift modulation either results in a continuous discharging current (Figure 3a) or continuous charging current (Figure 3b).

2.2 Flying-Capacitor Voltage Control for LLC Resonant Converters

While a full utilization of the traditional phase-shift (Figure 3a, 3b) modulation is not possible for the FC DC-DC converter, the introduction of a phase shift to the gate signals between S_1-S_4 and S_2-S_3 can be utilized to control the FC voltage in an LLC resonant converter that is traditionally operated with 50 % duty cycle and no phase shift. While ideally the FC current is zero for when there is no phase-shift, even a small delay in the operation may lead to a run-off of the FC voltage and can, therefore, result in a destruction of the switches. By operating S_2-S_3 as the lagging leg (cf. Figure 3a, 3b) if $v_{\mathrm{FC}} < V_{\mathrm{in}}/2$ and S_1-S_4 as the leading leg if $v_{\mathrm{FC}} > V_{\mathrm{in}}/2$ (in each with only a minor phase shift) the flying capacitor voltage can be controlled.

The control concept to stabilize v_{FC} to $V_{\mathrm{in}}/2$ is depicted in Figure 4a. For $\Delta D < 0.5$, the switches S_1-S_4 are leading with the switches S_2-S_3 lagging; for $\Delta D > 0.5$, the switches S_2-S_3 are leading while S_1-S_4 are lagging. Simulation results are depicted for an 2.4 kW LLC resonant converter in Figure 4b for $C_{\mathrm{fc}} = 2\,\mu\mathrm{F}$ with $V_{\mathrm{in}} = 800\,\mathrm{V}$ and $V_{\mathrm{out}} = 48\,\mathrm{V}$, $I_{\mathrm{out}} = 50\,\mathrm{A}$. The resonant parameters are $L_{\mathrm{r}} = 30\,\mu\mathrm{H}$, $L_{\mathrm{m}} = 300\,\mu\mathrm{H}$, $C_{\mathrm{r}} = 90\,\mathrm{nF}$ and $n = 7$. For $0 < t < 3\,\mathrm{ms}$, the control was turned off. After $t = 3\,\mathrm{ms}$, the control was turned back on resulting in a stabilization of the flying capacitor voltage. For $\Delta D < 0$, the switches S_1-S_4 are leading; for $\Delta D > 0$ they are lagging. The results show that the flying capacitor voltage can be well controlled to half of the input voltage V_{in}.

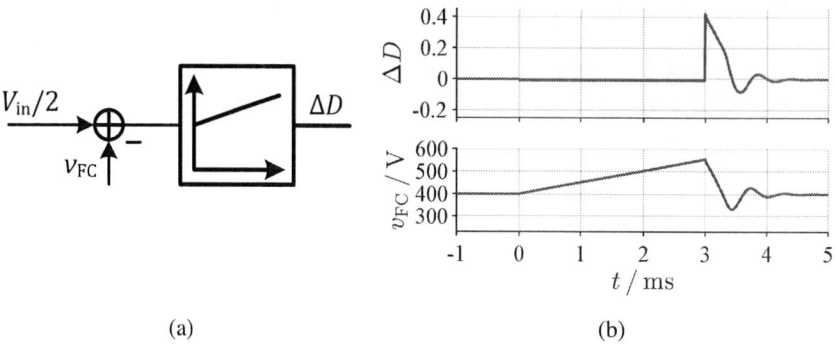

(a)   (b)

Fig. 4: (a) Control concept to adjust the flying capacitor voltage v_{FC}. For $\Delta D < 0$, the switches S_1-S_4 are leading, for $\Delta D > 0$, they are lagging; (b) simulation results proving the concept.

3 A Loss-Balancing Phase-Shift Modulation for Flying-Capacitor DC-DC Converters

The preceding section showed that the conventional phase-shift modulation is unsuitable for the application on a flying-capacitor DC-DC converter as it results in a continuous charge or discharge of v_{fc}. While the asymmetrical phase-shift modulation of Figures 3c and 3d may be an option to utilize quasi phase-shift modulation voltage shapes, it is a non-ideal solution as it results in unbalanced semiconductor losses. To utilize the full potential of the phase-shift modulation in LLC resonant converters, this chapter presents a multi-level modulation to achieve quasi-phase-shift modulation voltage shapes. Benefits of phase-shift modulation in LLC resonant converters are the potential of employing it for start-up purposes [16, 17], a low-gain loss reduction [18–21], an extension of the operation in wide voltage-transfer ratio applications [22–25] or in balancing when utilizing interleaving of multi-rail LLC converters [26–28]. To utilize the benefits of the phase-shift operation, several concepts have been presented in literature [29, 30] where both concepts employed additional switches or additional capacitors and diodes. However, this section shows that a modulation exists to achieve these voltage shapes without the need of additional components.

The alternating-asymmetrical modulation (a2PSM) is depicted in Figure 5. It uses the same pulse pattern that was used for a conventional full-bridge inverter in [10–12]. To the author's knowledge, no such modulation has yet been utilized to a multi-level inverter. The modulation consists of turn-on intervals of the length $(2-D)T/2$, $T/2$, $DT/2$ and $T/2$ where the pulses are phase-shifted from S_1-S_4 to S_2-S_3 by T. The modulation can be easily implemented on a DSP/microcontroller (e.g. in TI C2000 controllers using the action-qualifier control registers, AQCTL) by using an up-down counter. By employing two PWM units, the same synchronized up-down counter can be utilized where the compare values for one PWM unit (yellow lines) are $(2-D)N_{reg}/4$ and $N_{reg} - DN_{reg}/4$ while for the other PWM unit (dark red lines), they are $DN_{reg}/4$ and $(2+D)N_{reg}/4$. During up-count, the flying-capacitor current is negative while for down count, it is positive considering steady state. Exemplary simulation results are depicted in Figure 6. Figure 6a shows ideal simulation results with no delays in the PWM signals. A simulation with an exemplary 80 ns phase shift between the signals of S_1/S_4 and S_2/S_3 is depicted in Figure 6b. This modulation results in a non-ideal flying capacitor voltage v_{fc}. In average the voltage is $\bar{v}_{fc} = 452$ V, 52 V higher than the ideal voltage of $\bar{v}_{fc} = \frac{V_{in}}{2} = 400$ V. However, contrary to the conventional phase-shift modulation depicted in Figure 4b, this does not lead to a continuously increasing/decreasing flying capacitor voltage but results in a steady state offset in the voltage.

A control of the flying-capacitor voltage can be achieved by modifying the compare values for the up- and down count respectively by increasing or decreasing the freewheeling intervals during the respective counter slope. The upper two compare values are, thus, modified to $(2+D-\Delta D)N_{reg}/4$ and $N_{reg} - (D-\Delta D)N_{reg}/4$ whereas the lower two compare values are modified to $(2-D+\Delta D)N_{reg}/4$ and $(D-$

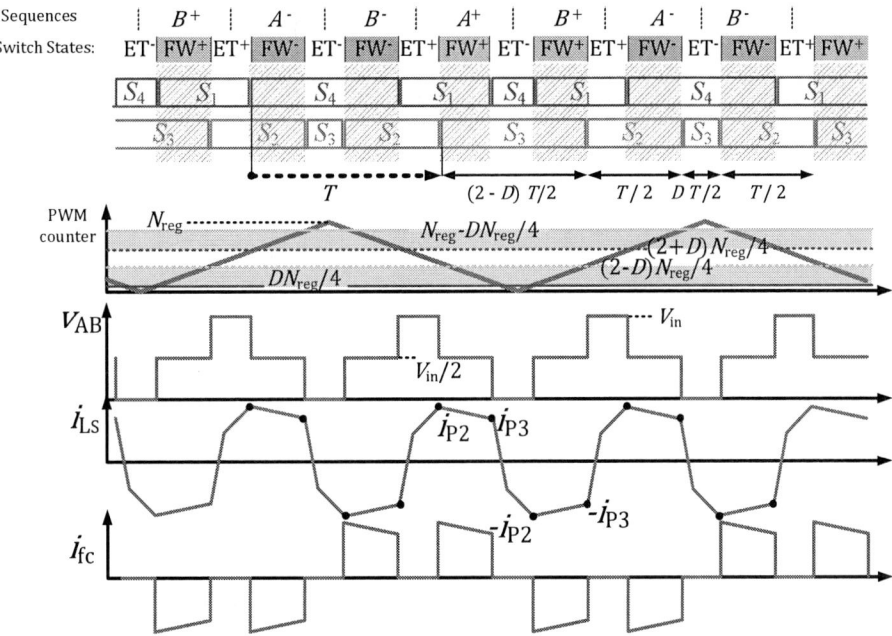

Fig. 5: Three-level alternating asymmetrical phase-shift modulation with balanced flying-capacitor currents.

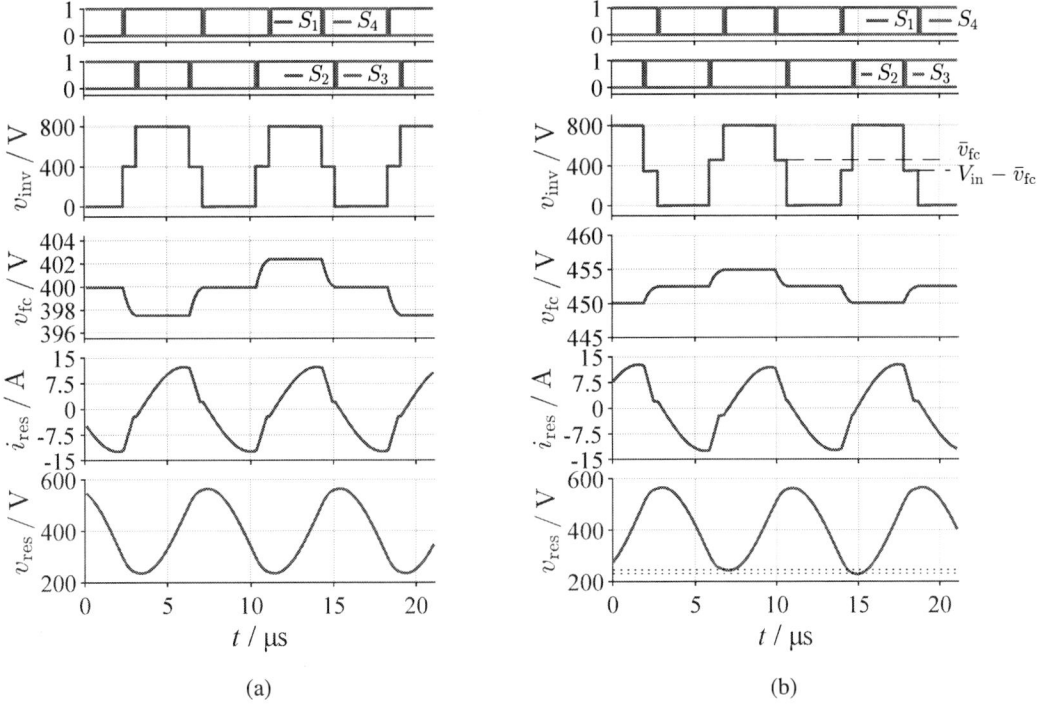

Fig. 6: 3-Level LLC converter simulation results. (a) balanced operation, (b) unbalanced operation with undesired phase shift of 80 ns for S_1 and S_4 leading to an undesired flying capacitor voltage.

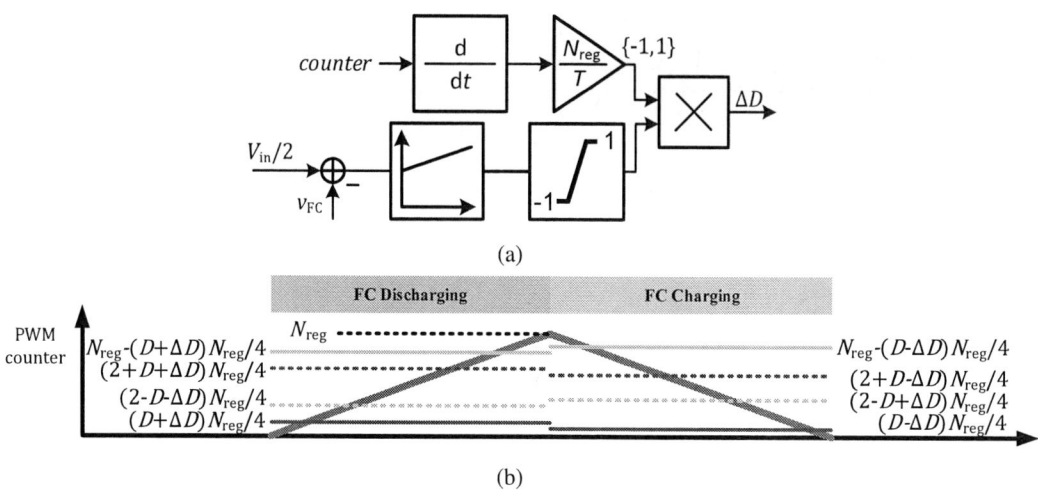

(a)

(b)

Fig. 7: (a) Control concept to adjust the flying capacitor voltage v_{FC} in the a2PSM. Depending on the counter slope, the compare values are altered; (b) Visualization of the adapted compare values to balance the flying-capacitor voltage by increasing the charging intervals.

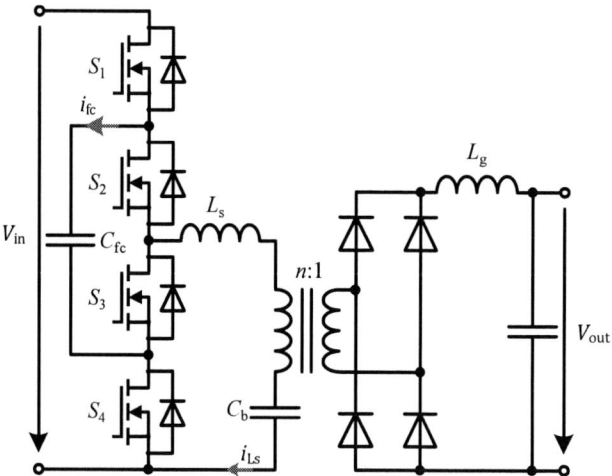

Fig. 8: Three-level phase-shift converter with a blocking capacitor C_b.

$\Delta D)N_{reg}/4$ where ΔD is determined as visualized in Figure 7a. The modified compare values are depicted in Figure 7b.

With the proposed modulation, it is possible to create a flying-capacitor phase-shift converter (*phase-shifted full bridge* as depicted in Figure 8. The blocking capacitor C_b is used to block half of the input voltage and is setup with a large capacitance. Simulation results are depicted in Figure 9 where Figure 9a shows a simulation with a introduced delay of 20 ns for S_1 and Figure 9b shows simulation results with the balance control resulting in the desired flying capacitor voltage.

4 Conclusion

Multi-level flying-capacitor converters are an attractive solution to exploit the beneficial figure of merit of semiconductors with a lower blocking voltage. Conventionally, they are employed in power-factor-correction stages or inverters. This paper proved that the flying-capacitor module can be successfully implemented to exchange the conventional half-bridge leg. It proved that the phase-shift modulation can be utilized to adjust the flying capacitor voltage of half-bridge LLC converter and proposed a modulation

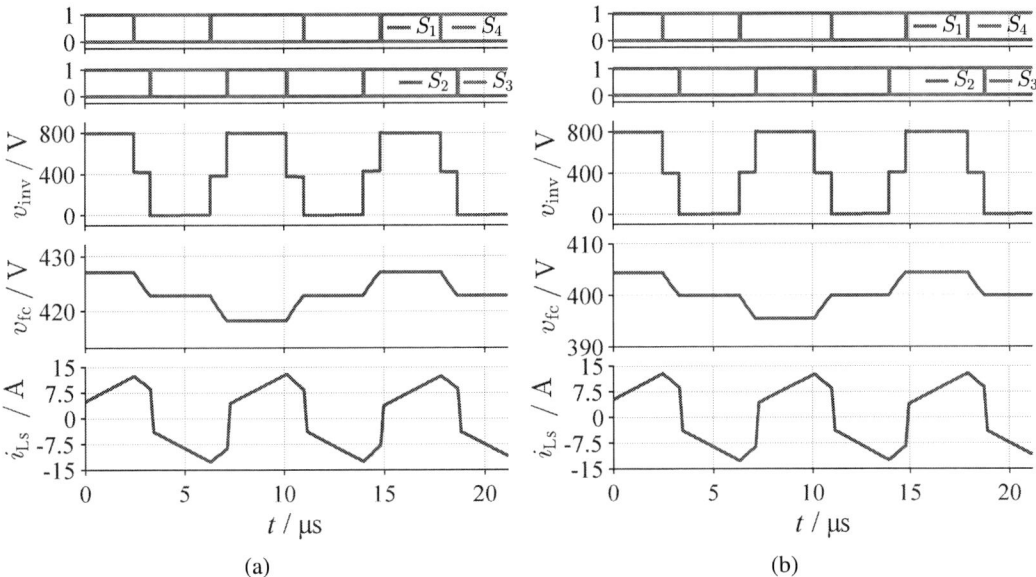

Fig. 9: 3-Level PSFB converter simulation results. (a) unbalanced operation with undesired delay of 20 ns for S_1 leading to an undesired flying capacitor voltage (b) balanced operation (controlled, $\Delta D = 0.01$) with undesired delay of 20 ns for S_1 leading to the desired operation.

to fully exploit the three-level structure of the FC module. Consequently, a FC phase-shift converter and a three-level LLC resonant converter were proposed along with a concept to control the voltage to the desired value.

References

[1]   Y. Lei, C. Barth, S. Qin, W.-C. Liu, I. Moon, A. Stillwell, D. Chou, T. Foulkes, Z. Ye, Z. Liao, and R. C. N. Pilawa-Podgurski. "A 2-kW Single-Phase Seven-Level Flying Capacitor Multilevel Inverter With an Active Energy Buffer". In: *IEEE Transactions on Power Electronics* 32.11 (2017), pp. 8570–8581.

[2]   J. Azurza Anderson, G. Zulauf, P. Papamanolis, S. Hobi, S. Miric, and J. W. Kolar. "Three Levels Are Not Enough: Scaling Laws for Multilevel Converters in AC/DC Applications". In: *IEEE Transactions on Power Electronics* 36.4 (2021), pp. 3967–3986.

[3]   J. Azurza Anderson, E. J. Hanak, L. Schrittwieser, M. Guacci, J. W. Kolar, and G. Deboy. "All-Silicon 99.35% Efficient Three-Phase Seven-Level Hybrid Neutral Point Clamped/Flying Capacitor Inverter". In: *CPSS Transactions on Power Electronics and Applications* 4.1 (2019), pp. 50–61.

[4]   G. P. Adam, O. Anaya-Lara, G. Burt, S. J. Finney, and B. W. Williams. "Comparison between flying capacitor and modular multilevel inverters". In: *2009 35th Annual Conference of IEEE Industrial Electronics*. IEEE, 3.11.2009 - 05.11.2009, pp. 271–276.

[5]   A. A. Sneineh, M.-y. Wang, and K. Tian. "A Hybrid Capacitor-Clamp Cascade Multilevel Converter". In: *IECON 2006 - 32nd Annual Conference on IEEE Industrial Electronics*. IEEE, 6.11.2006 - 10.11.2006, pp. 2031–2036.

[6]   H. Wang, M. Lu, Y. Deng, R. Zhao, and X. He. "Relationship between flying capacitor multilevel inverter PWM methods and switching loss minimized PWM method for flying capacitor multilevel inverter". In: *2004 IEEE 35th Annual Power Electronics Specialists Conference (IEEE Cat. No.04CH37551)*. IEEE, 20-25 June 2004, pp. 4418–4422.

[7]   G. Deboy. "Perspective of Loss Mechanisms for Silicon and Wide Band-Gap Power Devices". In: *CPSS Transactions on Power Electronics and Applications* 2.2 (2017), pp. 89–100.

[8]   J. W. Kolar. *X-Technologies - Power Electronics 4.0*. 26.04.2021.

[9]   J. W. Kolar. *X-Concepts The DNA of Future High-Performance Power Electronic Systems*. 13.11.2021.

[10] L. Mihalache. "A modified pwm control technique for full bridge ZVS DC-DC converter with equal losses for all devices". In: *Conference Record of the 2004 IEEE Industry Applications Conference, 2004. 39th IAS Annual Meeting*. IEEE, 3-7 Oct. 2004, pp. 1776–1781.

[11] J. Kucka and D. Dujic. "Equal Loss Distribution in Duty-Cycle Controlled H-Bridge LLC Resonant Converters". In: *IEEE Transactions on Power Electronics* 36.5 (2021), pp. 4937–4941.

[12] P. Rehlaender, R. Unruh, F. Schafmeister, and J. Böcker. "Alternating Asymmetrical Phase-Shift Modulation for Full-Bridge Converters with Balanced Switching Losses to Reduce Thermal Imbalances". In: *APEC '21*. Piscataway, NJ: IEEE Service Center, 2021.

[13] U. Badstuebner, J. Biela, and J. W. Kolar. "Design of an 99%-efficient, 5kW, phase-shift PWM DC-DC converter for telecom applications". In: *APEC '10*. IEEE, 2010, pp. 773–780.

[14] Y. Shen, H. Wang, F. Blaabjerg, X. Sun, and X. Li. "Analytical model for LLC resonant converter with variable duty-cycle control". In: *2016 IEEE Energy Conversion Congress and Exposition (ECCE)*. IEEE, 18.09.2016 - 22.09.2016, pp. 1–7.

[15] C. Wu and N. Guangfu. "Frequency conversion phase shift asymmetric duty cycle modulation method of series resonant full bridge converter". CN201711339609. 2018.

[16] Q. Chen, J. Wang, Y. Ji, and S. Liang. "Soft starting strategy of bidirectional LLC resonant DC-DC transformer based on phase-shift control". In: *2014 9th IEEE Conference on Industrial Electronics and Applications*. IEEE, 9.06.2014 - 11.06.2014, pp. 318–322.

[17] D. Yang, C. Chen, S. Duan, J. Cai, and L. Xiao. "A Variable Duty Cycle Soft Startup Strategy for ¡italic¿LLC¡/italic¿ Series Resonant Converter Based on Optimal Current-Limiting Curve". In: *IEEE Transactions on Power Electronics* 31.11 (2016), pp. 7996–8006.

[18] B. McDonald and F. Wang. "LLC performance enhancements with frequency and phase shift modulation control". In: *2014 IEEE Applied Power Electronics Conference and Exposition - APEC 2014*. IEEE, 16.03.2014 - 20.03.2014, pp. 2036–2040.

[19] Y.-K. Lo, C.-Y. Lin, M.-T. Hsieh, and C.-Y. Lin. "Phase-Shifted Full-Bridge Series-Resonant DC-DC Converters for Wide Load Variations". In: *IEEE Transactions on Industrial Electronics* 58.6 (2011), pp. 2572–2575.

[20] J.-H. Kim, C.-E. Kim, J.-B. Lee, Y.-D. Kim, H.-S. Youn, and G.-W. Moon. "A simple control scheme for improving light-load efficiency in a full-bridge LLC resonant converter". In: *IPEC '14*. IEEE, 2014, pp. 1743–1747.

[21] J.-H. Kim, C.-E. Kim, J.-K. Kim, J.-B. Lee, and G.-W. Moon. "Analysis on Load Adaptive Phase-Shift Control for High Efficiency Full-Bridge LLC Resonant Converter in Light Load Conditions". In: *IEEE Transactions on Power Electronics* (2015), p. 1.

[22] S. Liu, R. Ren, W. Meng, X. Zheng, F. Zhang, and L. Xiao. "Short-circuit current control strategy for full-bridge LLC converter". In: *2014 IEEE Energy Conversion Congress and Exposition (ECCE)*. IEEE, 14.09.2014 - 18.09.2014, pp. 3496–3503.

[23] N. Shafiei, M. Ordonez, M. Craciun, C. Botting, and M. Edington. "Burst Mode Elimination in High-Power LLC Resonant Battery Charger for Electric Vehicles". In: *IEEE Transactions on Power Electronics* 31.2 (2016), pp. 1173–1188.

[24] P. Rehlaender, F. Schafmeister, and J. Böcker. "Interleaved Single-Stage LLC Converter Design Utilizing Half- and Full-Bridge Configurations for Wide Voltage Transfer Ratio Applications". In: *IEEE Transactions on Power Electronics* 36.9 (2021), pp. 10065–10080.

[25] P. Rehlaender, T. Grote, S. Tikhonov, M. Schröder, F. Schafmeister, and J. Böcker. "A 3,6 kW Single-Stage LLC Converter Operating in Half-Bridge, Full-Bridge and Phase-Shift Mode for Automotive Onboard DC-DC Conversion". In: *PCIM Europe 2020*. Ed. by Mesago Messe Frankfurt GmbH. Stuttgart, 2020.

[26] H. Figge, F. Schafmeister, and T. Grote. "LLC balancing". US9263951B2. 2014.

[27] P. Rehlaender, S. Tikhonov, F. Schafmeister, and J. Böcker. "Dual Interleaved 3.6 kW LLC Converter Operating in Half-Bridge, Full-Bridge and Phase-Shift Mode as a Single-Stage Architecture of an Automotive On-Board DC-DC Converter". In: *2020 22nd European Conference on Power Electronics and Applications (EPE'20 ECCE Europe)*. IEEE, pp. 1–10.

[28] K. Murata and F. Kurokawa. "An Interleaved PFM ¡italic¿LLC¡/italic¿ Resonant Converter With Phase-Shift Compensation". In: *IEEE Transactions on Power Electronics* 31.3 (2016), pp. 2264–2272.

[29] M. S. Agamy and P. K. Jain. "A Single Stage Three Level Resonant LLC AC/DC Converter Using a Variable-Frequency-Phase-Shift Controller and a Voltage Balancing Auxiliary Circuit". In: *Twenty-First Annual IEEE Applied Power Electronics Conference and Exposition, 2006. APEC '06*. IEEE, March 19, 2006, pp. 411–416.

[30] Y. Guo, Z. Yang, Y. Yin, and H. Cao. "Digital Control of Hybrid Full Bridge Three-level LLC Resonant Converter Based on SiC MOSFET". In: *2018 IEEE International Power Electronics and Application Conference and Exposition (PEAC)*. IEEE, 4.11.2018 - 07.11.2018, pp. 1–6.

An EV Integrated Isolated DC Charger using a Six-Phase Synchronous Machine

Sukhjit S Ghumman, Mehanathan Pathmanathan, Peter W Lehn
Department of Electrical and Computer Engineering, University of Toronto
Toronto, Ontario, Canada
E-Mail: sukhjit.singh@mail.utoronto.ca
meha.pathmanathan@utoronto.ca
lehn@ece.utoronto.ca

Acknowledgements

This work was supported by the Natural Sciences and Engineering Research Council of Canada (NSERC) under Grant CRDPJ 513206-17.

Keywords

«Electric Vehicles», «DC Chargers», «Onboard Chargers», «Isolated Charging», «Integrated Charger» «Six-Phase Machine» «Multiphase Machine» «WFSM» «WRSM» «PMSM».

Abstract

This paper presents an integrated DC charger using a six-phase synchronous machine with two isolated neutrals. The synchronous motor provides isolation between the DC grid and the battery and does not require a separate transformer. Control techniques are developed which allow charging without generating electromechanical torque inside the machine. The concept is validated experimentally using a six-phase, wound field synchronous machine (WFSM), showing isolated DC charging with zero torque production. Simulation results using an equivalent permanent magnet synchronous machine (PMSM) model are provided, indicating the potential for this system as a means to achieve high-power, high-efficiency isolated DC charging.

Introduction

Multiphase machines have been explored in literature for high-power electromechanical energy conversion applications like wind turbines, electric ships, aerospace, etc. Specifically, six and nine-phase induction and synchronous machines have been studied, as off-the-shelf three-phase converters can be used. Multi-phase machines allow the splitting of current into more paths hence, multiple smaller power electronic converters (modular design), lower torque ripple compared to their three-phase counterparts, increased degrees of freedom, and can facilitate fault-tolerant operations [1]. Due to the aforementioned benefits, multiphase synchronous, induction, and switched reluctance machines have been considered in electric vehicles (EVs) for driving purposes. Recently, multiphase machines have found an interest in EV charging also, as the machine leakage inductances can be used as a filter during the power transfer [2]. In addition, these machines can integrate dissimilar energy sources like fuel cells, supercapacitors etc. [3].

Electric vehicle chargers can be broadly categorized into integrated and non-integrated variants. Non-integrated chargers (which can be off-board fast-charging stations or onboard chargers) usually include a galvanic isolation stage either at line frequency or through an isolated DC/DC converter. In contrast, integrated chargers usually do not provide the means for galvanic isolation [4]. The lack of a galvanic isolation stage (and the inherent large common-mode impedance it provides) can make it challenging for integrated chargers to meet the common-mode leakage current requirements from standards such as UL2231 and IEC61851 (though non-isolated standard-compliant integrated solutions are possible [5]).

Three different charging types exist, namely: 1-φ ac, 3-φ ac, and DC charging. Various integrated topologies using multi-phase machines have been published [2], but these topologies lack isolation between the input source and the battery unless additional transformers are used. A few solutions have been found in the literature, which allow isolation using multiphase machines. A 3-φ wound-field induction machine has been used in [6] for 3-φ charging, where the rotor acts as an input port and provides isolation. However, the torque is generated during charging and this method requires mechanical brakes to block the rotor during the charging process. A 3-φ ac charger based on a split winding 3-φ phase permanent magnet machine is presented in [7] but requires switches to split motor windings during driving and charging operation. Again, this method results in torque production. Therefore, a clutch is needed to decouple the traction motor from the EV powertrain during charging. [8] shows another 3-φ charger using a six-phase induction machine as a transformer, but uses switches in the rotor circuit to disconnect it during charging operation. Another topology has been published in [9] for a three-phase grid charging solution that provides isolation, where 1-φ ac and 3-φ ac grid charging is achieved using a six-phase PMSM. This topology requires additional grid-connected inductors and a three-phase diode bridge on the battery side during the charging process. Therefore, requires additional components and reconfigurations to charge the battery and provide isolation.

To date, DC charging solutions using multi-phase machines, especially integrated DC charging solutions with isolation have not yet been explored. Most EVs use synchronous machines (SM) as they provide high torque density compared to induction machines (IM). In synchronous machines, permanent magnet synchronous machines (PMSM) are more prevalent than wound field synchronous machines (WFSM) due to higher torque density and no additional hardware requirement for the field circuit [10]. This paper investigates machine-integrated DC charging solutions while using the synchronous machine as a medium of isolation. The proposed charger topology and control algorithm are validated experimentally using a six-phase WFSM, showing that isolated DC charging with zero torque production is achieved. Finally, simulation results with an equivalent PMSM model are shown indicating the improvements in efficiency and charging power that can be obtained with an equivalent six-phase PMSM.

Charger Topology

The proposed charger, shown in Fig. 1, uses two three-phase voltage source converters (converter1 and converter2) and a six-phase synchronous machine with two isolated three-phase windings (Y-configuration). Isolated neutrals of Y-connected three-phase windings have been leveraged to provide galvanic isolation between the DC grid and the battery. A two-pole contactor with (S_{1+} & S_{1-}) is the additional component required for reconfiguring the system between charging and driving operation. S_{1+} & S_{1-} remain closed during driving and open during charging operation and the DC-link of converter1 is used as an inlet from the DC grid.

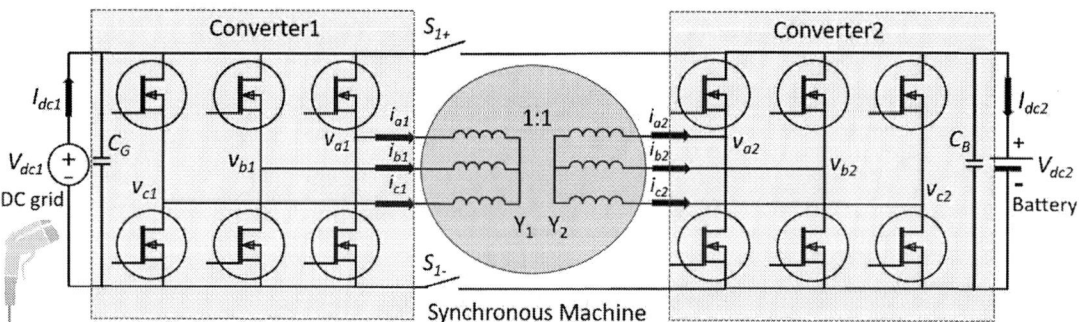

Fig 1: Integrated DC charger using a six-phase synchronous machine as a medium of isolation

The power injected from the DC grid goes through the converter1, which converts it to ac power and transfers it to the winding set2 magnetically, where converter2 converts it back to DC to charge the battery. If suitably oriented, the magnetic field created by stator windings does not introduce torque in the machine, as the torque-producing component does not get excited. The rotor field winding should be kept short-circuited during the charging process, as the proposed control algorithm will result in large voltages being induced across the field winding if it is left open-circuited. The charger has bidirectional power transfer capability, where it can be used for vehicle to vehicle charging (V2V), vehicle to grid (V2G) & V2X (for DC power) capability as well, although, this paper will only focus on DC grid to battery charging.

Machine modeling

The machine used for the proposed charger topology is a six-phase wound field synchronous machine (WFSM) with two isolated neutrals. The two-stator windings have a 0° phase shift between them, Fig 2(a). A Six-phase synchronous machine model using the double dq model ($d_1q_1 0_1$, $d_2q_2 0_2$) in the rotor reference frame is shown in Fig 2(b) and (c) for the charging mode [11], [12]. Zero sequence components will not be discussed here, as there will not be any path for them to flow due to the wye-connected windings. Machine stator voltage equations in the dq axes can be given by (1) – (4), where λ is flux linkage, ω_r is the rotor speed in electrical radians per second (zero, during charging), and p is the time derivative. The stator flux linkage equations are given in (4) - (8) where l_{ls1} and l_{ls2} are the stator leakage inductances, l_{md} and l_{mq} is the d and q-axis magnetizing inductances. As discussed in the previous section, the field winding circuit is short-circuited to avoid high voltages being induced.

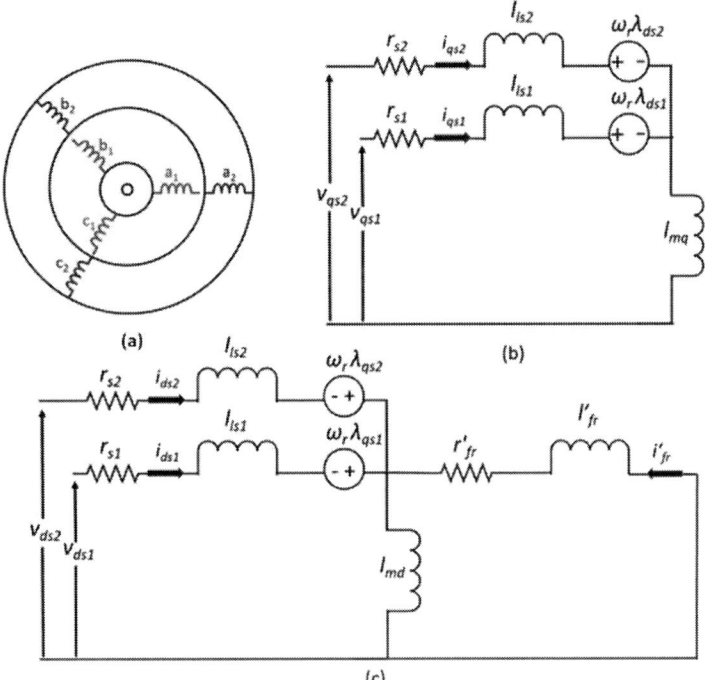

Fig. 2: (a) Six-phase synchronous machine with 0° phase shift between two sets of windings (rotor standstill) (b) q-axis (c) d-axis equivalent circuit referred to the winding set1

$$v_{ds1} = r_{s1}i_{ds1} + p\lambda_{ds1} - \omega_r\lambda_{qs1} \tag{1}$$

$$v_{ds2} = r_{s1}i_{ds2} + p\lambda_{ds2} - \omega_r\lambda_{qs2} \tag{2}$$

$$v_{qs1} = r_{s1}i_{qs1} + p\lambda_{qs1} + \omega_r\lambda_{ds1} \tag{3}$$

$$v_{qs2} = r_{s1}i_{qs2} + p\lambda_{qs2} + \omega_r\lambda_{ds2} \tag{4}$$

$$\lambda_{ds1} = l_{ls1}i_{ds1} + l_{md}(i_{ds1} + i_{ds2} + i'_{fr}) \tag{5}$$

$$\lambda_{ds2} = l_{ls2}i_{ds2} + l_{md}(i_{ds1} + i_{ds2} + i'_{fr}) \tag{6}$$

$$\lambda_{qs1} = l_{ls1}i_{qs1} + l_{mq}(i_{qs1} + i_{qs2}) \tag{7}$$

$$\lambda_{qs2} = l_{ls2}i_{qs2} + l_{mq}(i_{qs1} + i_{qs2}) \tag{8}$$

Charging Operation

The charging algorithm should ensure zero torque production and allow power to be transferred magnetically between two sets of three-phase windings. The following subsections explain the zero torque production, power flow, and control mechanisms during the charging operation.

Zero torque production

Equation (9) gives the electromagnetic torque developed by the machine in a rotor reference frame using the double dq model, where P represents the number of pole pairs. The rotor is assumed to be at standstill. Zero torque can be achieved if the stator currents of the q-axis are both set to zero.

$$T = \frac{3}{2}P\left[(i_{qs1} + i_{qs2})l_{md}(i_{ds1} + i_{ds2} + i'_{fr}) - (i_{ds1} + i_{ds2} + i'_{fr})l_{mq}(i_{qs1} + i_{qs2})\right] \tag{9}$$

Sinusoidal time-varying d-axes currents

The desired ac power can be transferred between the two sets of windings magnetically if time-varying d-axes currents are used. The time-varying d-axes currents in rotor reference will produce a pulsating magnetic field that does not rotate in space and accordingly does not produce electromagnetic torque. This pulsating magnetic field generated by the injection of a time-varying d-axis current in one set of windings will however induce voltages on the second set of windings (and also the field winding, which is why the field must be shorted during this process). The induced voltage on the second set of windings can be leveraged as a means to achieve isolated power transfer from one stator winding set to another. Fig. 3 shows the sample currents in winding set1 and set2 that generate no torque and only d-axis current can be allowed, therefore, the distribution of the current between the phases a, b, and c will depend on the rotor field angle.

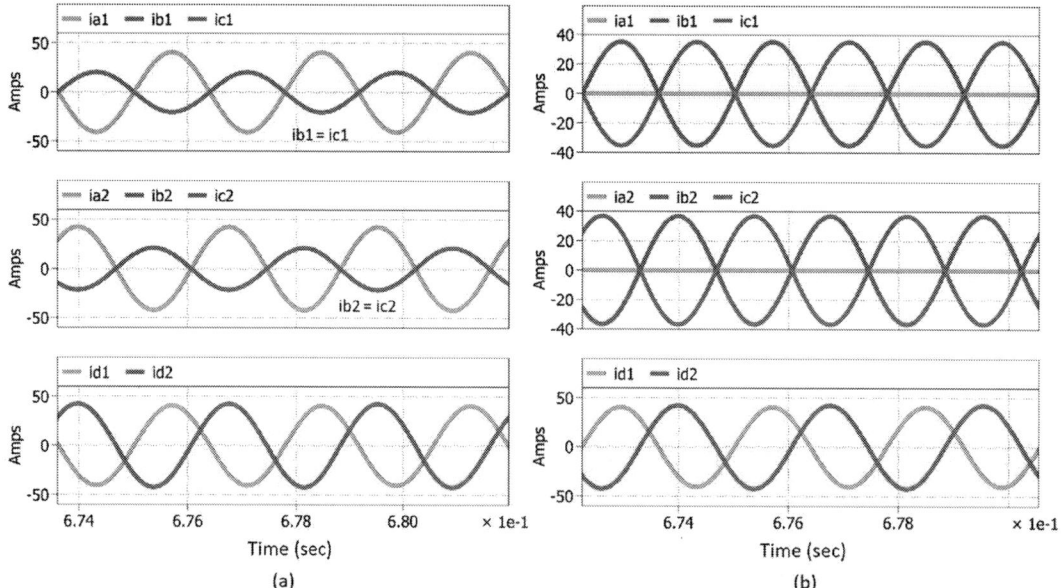

Fig. 3: Stator winding set1 and set2: abc & d-axes currents at rotor electrical angle: (a) 0° (b) 90°

Power flow model

Fig. 4. shows the equivalent T-model of the two d-axes of the synchronous machine during the charging process. In this image, Z_{ds1} represents the impedance of the series connection of l_{ls1} and r_{s1}, (Fig. 2c), Z_{ds2} is the impedance of the series connection of l_{ls2} and r_{s2}, while Z_{eq} is the parallel combination of d-axis magnetizing inductance l_{md} and field winding impedance (series connection of l'_{fr} , r'_{fr}). V_{ds1} and V_{ds2} represent the d-axes rms voltages of winding set1 and set2, whereas I_{ds1} and I_{ds2} represent the d-axes rms currents of the winding set1 and set2.

Please note, that unlike traditional dq frame models the quantities I_{ds1} and I_{ds2} are AC phasors and are given by (10) and (11). As such, they have a real part $I_{ds1,x}$ (defined as being in-phase with the winding set2 voltage phasor V_{ds2}), along with an imaginary part $I_{ds1,y}$ (defined as being orthogonal to V_{ds2}). The real part $I_{ds1,x}$ can be used to transfer real power from winding set1 to set2. Similarly, $I_{ds1,y}$ can be used to supply reactive power. I_{eq} and V_{eq} are the current through and the voltage across the impedance Z_{eq}.

P_1 and P_2 depict the power of the three-phase windings on the DC-grid and battery sides respectively, given by (13) and (14). Ideally, V_{ds1} and V_{ds2} voltages should be near to the rated winding voltage, for higher power flow to be achieved, provided the current I_{eq} does not become excessive. This suggests a high value of Z_{eq} is beneficial for power transfer, limiting reactive currents and system efficiency.

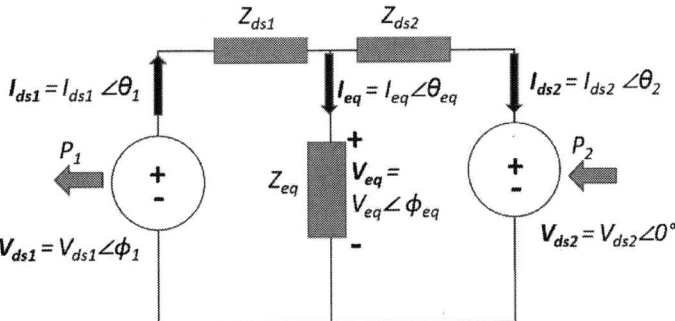

Fig. 4: Equivalent T-model of synchronous machine d-axes during the charging operation

$$I_{ds1} = I_{ds1,x} + jI_{ds1,y} \tag{10}$$

$$I_{ds2} = I_{ds2,x} + jI_{ds2,y} \tag{11}$$

$$I_{eq} = I_{ds1} - I_{ds2} \tag{12}$$

$$P_1 = Re\{V_{ds1}I^*_{ds1}\} \tag{13}$$

$$P_2 = Re\{V_{ds2}I^*_{ds2}\} \tag{14}$$

Controllers

Fig. 5 shows the control schematic for DC charging. Two PR controllers, one for i_{ds1} and another i_{qs1} are deployed. The i_{ds1} PR controller regulates real and reactive power exchange between the two sets of windings. The i_{qs1} PR controller ensures zero AC current flowing in the q-axis. Winding set2 voltages are directly assigned with $v_{ds2*} = \sqrt{2}V_{ds2}cos(\omega t)$, which is consistent with Fig. 4, and $v_{qs2*} = 0$.

A PI controller generates the required real part $I_{ds1,x}$ (of phasor current I_{ds1}) needed for dc power transfer, P_{ref}. The imaginary part $I_{ds1,y}$ can be chosen to provide approximately half of the magnetizing branch reactive current. Based on the two components of the required phasor current I_{ds1}, the time domain current reference $i_{ds1,ref}$ is fabricated. The modulating signals are generated by normalizing v_{abc1*} and v_{abc2*} to v_{dc1} and v_{dc2} respectively. The rotor angular position θ is taken from an encoder and is used for all rotating reference frame conversions.

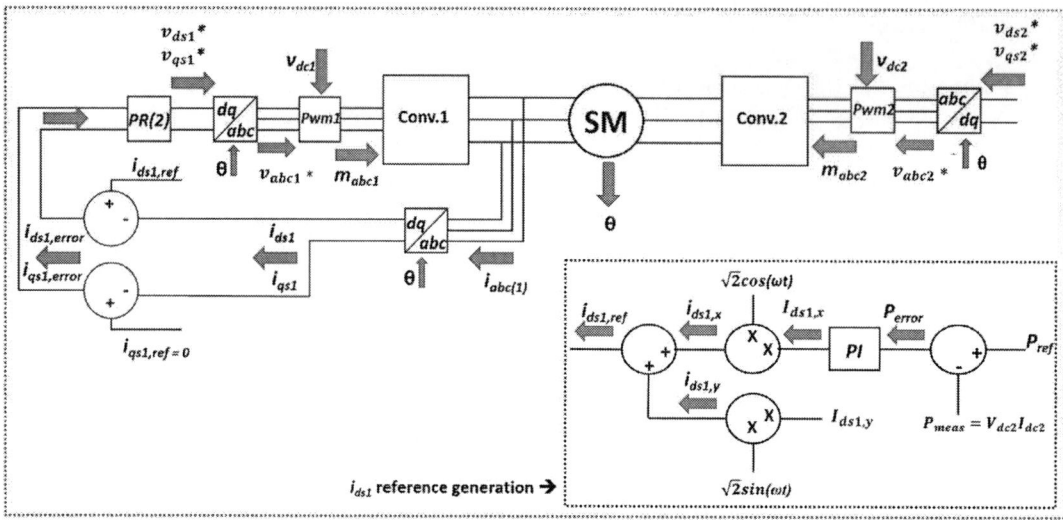

Fig. 5: DC charger control scheme for power transfer while ensuring zero torque production+

Results

The following subsections show the simulated and experimental results using the WFSM and then simulated results using an equivalent PMSM model are shown, emphasizing the higher power transfer using PMSM.

Simulation results

The proposed charger topology has been simulated using the six-phase WFSM machine. The machine used has ratings of 110V rms (line-line), and 40A rms per phase. The DC-link voltage on both inverters (battery and DC grid voltages) is set to 200V, with an inverter switching frequency of 10 kHz. The frequency of the d-axis reference current was set to 360 Hz.

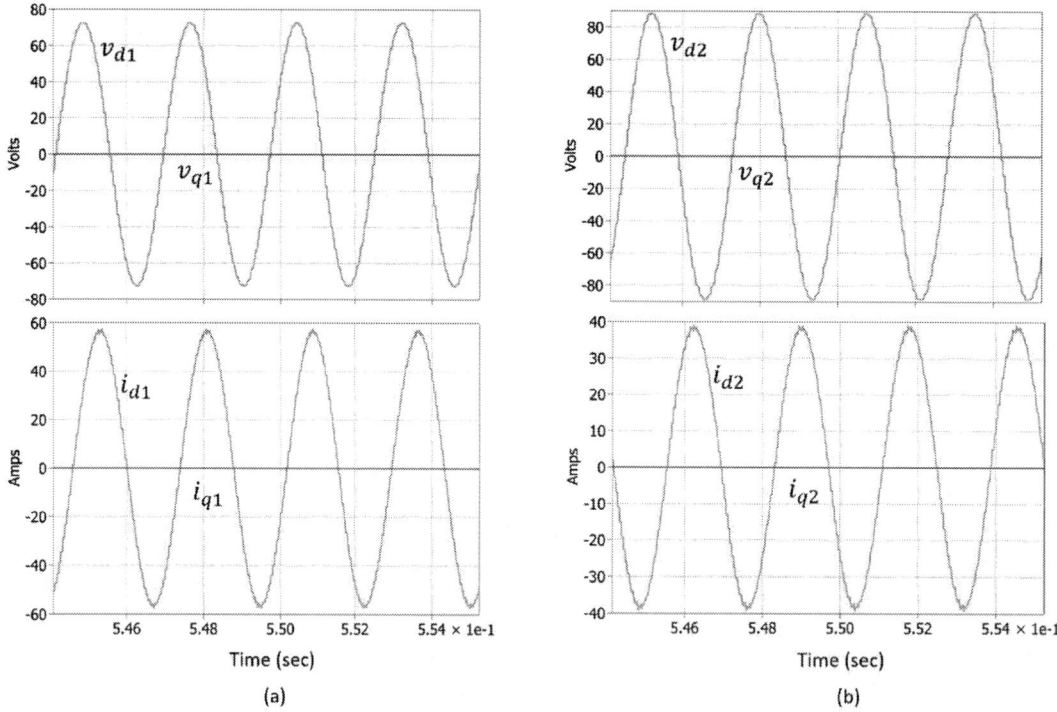

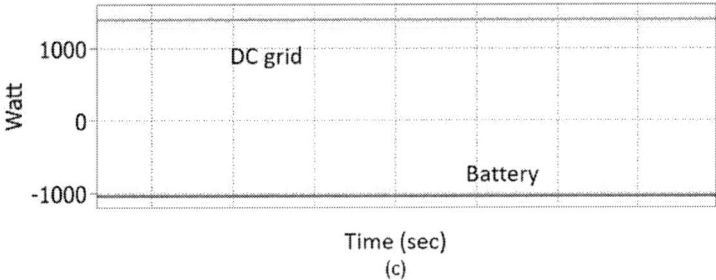

(c)

Fig. 6: dq-axes voltages and currents: (a)winding set1 (b)winding set2 (c) DC grid and battery power

Fig 6(a) and Fig 6(b) show the dq-axes voltages and currents of winding set1 and set2 respectively, at a rotor electrical angle of 0°. The q-axis voltages are not excited, therefore, the q-axis currents i_{q1} & i_{q2} remains zero. Hence, the electromagnetic torque is zero. Fig. 6(c) shows the DC power transferred from the DC grid to the battery. Power sent and received is 1394W and 1039W respectively.

Experimental results

The proposed charger topology has been experimentally validated using a six-phase WFSM with parameters shown in Table I. Fig. 7(a) shows the phase voltage and three-phase currents in both the winding sets in the abc frame at 0° rotor electrical angle. Fig. 7(b) shows the corresponding d and q axes currents. Currents i_{q1} & i_{q2} remains zero, indicating that the torque production is zero, based on the torque equation (9).

Table I: Six-phase WFSM parameters

Stator winding resistance ($r_{s1} = r_{s2}$, Fig. 2)	50.5 mΩ
Stator winding leakage inductance ($l_{ls1} = l_{ls2}$, Fig. 2)	0.2788 mH
Stator d-axis magnetizing inductance (l_{md}, Fig. 2)	1.1381 mH
Rotor resistance referred to winding set1 (r'_{fr}, Fig. 2)	65.3 mΩ
Rotor leakage inductance referred to winding set1 (l'_{fr}, Fig. 2)	0.3314 mH
Equivalent branch Impedance (Z_{eq} at 360 Hz, Fig. 4)	$(0.04245 + j0.605253)\Omega$

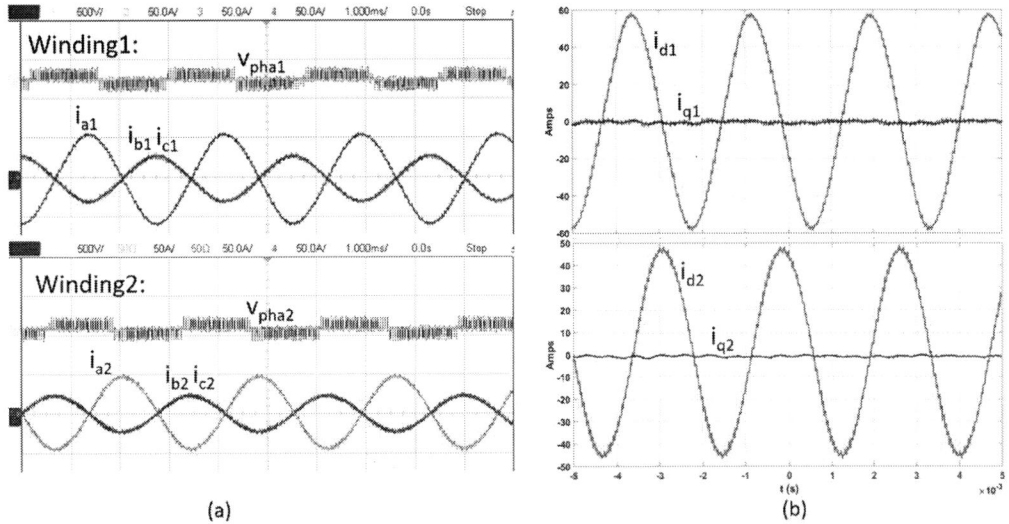

Figure 7. Winding set1and set2: (a) Phase voltage & abc currents (b) dq axes currents.

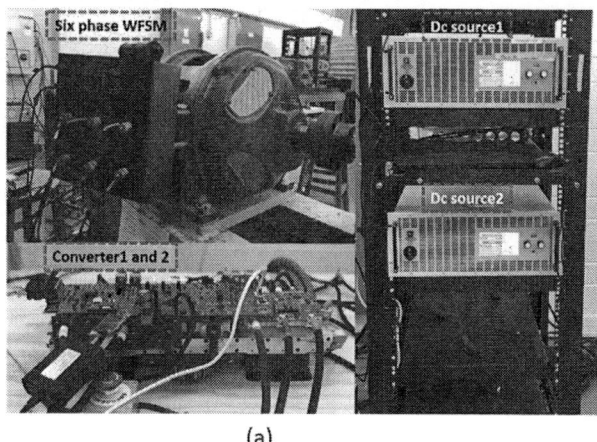

Fig 8: (a) Experimental setup (six-phase WFSM, dual inverter system, and two DC sources) (b) power analyzer results: machine line voltages, currents, power, and efficiency.

Fig. 8(a) shows the experimental setup used to validate the topology while Fig. 8(b) shows the power analyzer results for the power transfer during the charging operation, where line-line voltages and line currents on each winding set are shown. Power sent and received is 1330.9W and 808.4W respectively from each winding set. The lower efficiency is due to lower equivalent impedance Z_{eq} (in Fig. 4.) resulting in a higher magnitude of equivalent magnetizing branch current **Ieq**. The additional losses can be attributed to the machine winding ac resistance and core losses.

Simulation Results: PMSM Equivalent Model

The proposed charger topology can be implemented using a PMSM to increase the power transfer capability and particularly efficiency. The short-circuited field winding in the case of WFSM was needed to avoid a large induced voltage on the field but it results in a lower equivalent impedance Z_{eq} (in Fig. 4.). It is also associated with the rotor current conduction and I²R losses in the rotor circuit. Compared to WFSM, PMSM has no rotor field winding, which needs to be short-circuited, and accordingly will have a higher Z_{eq}. Tables I and II include the value of Z_{eq} for WFSM and PMSM respectively. As a result of higher Z_{eq}, higher power transfer is possible, and a higher efficiency since a smaller magnitude of equivalent magnetizing branch current **Ieq** will be produced for a given **Veq** (in fig. 4.). Equivalent six-phase PMSM parameters are shown in Table II

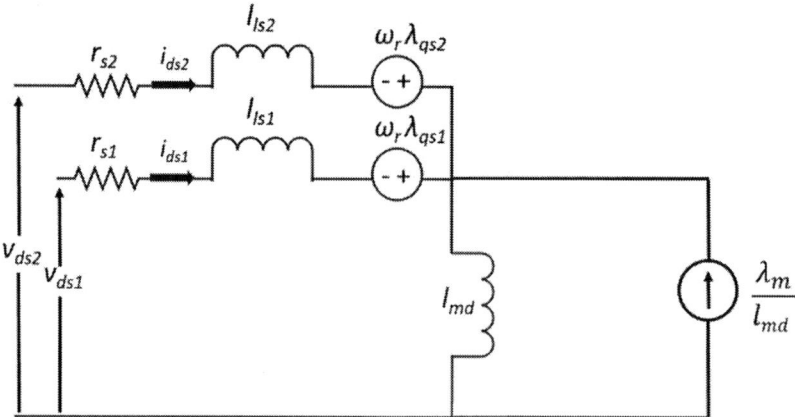

Fig 9: Six-phase PMSM d-axes equivalent circuit, permanent magnet flux linkage is modelled as a current source.

Fig. 9 shows the d-axis equivalent model for a six-phase PMSM, where the permanent magnet flux linkage is modelled as a DC source, and λ_m is the flux linkage due to permanent magnets. The machine stator voltage equations in the d-axis can be given by (15), (16), where λ is flux linkage, ω_r is the rotor speed in electrical radians per second (which is zero during charging), and p is the time derivative. The stator flux linkage equations in d-axis are given in (17), (18) where l_{ls1} and l_{ls2} are the stator leakage inductances, l_{md} is the d-axis magnetizing inductance. The q-axis model will not be discussed here and at zero rotor speed, rotor magnetic flux will not affect the q-axis operation. Fig 10(a) and Fig 10(b) show the dq-axes voltages and currents of winding set1 and set2 respectively, at a rotor electrical angle of 0°. The q-axis currents i_{q1} & i_{q2} are equal to zero. Hence, the electromagnetic torque is zero. Fig. 10(c) shows the DC power transferred from the DC grid to the battery. Power sent and received is 2106W and 2000W respectively. This corresponds to maximum theoretical efficiency of about 95%, compared to the WFSM simulation of Fig 6 where theoretical efficiency is approximately 75%.

$$v_{ds1} = r_{s1}i_{ds1} + p\lambda_{ds1} - \omega_r\lambda_{qs1} \tag{15}$$

$$v_{ds2} = r_{s1}i_{ds2} + p\lambda_{ds2} - \omega_r\lambda_{qs2} \tag{16}$$

$$\lambda_{ds1} = l_{ls1}i_{ds1} + l_{md}(i_{ds1} + i_{ds2}) + \lambda_m \tag{17}$$

$$\lambda_{ds2} = l_{ls2}i_{ds2} + l_{md}(i_{ds1} + i_{ds2}) + \lambda_m \tag{18}$$

Table II: Equivalent Six-phase PMSM parameters

Stator winding resistance ($r_{s1} = r_{s2}$, Fig. 2)	50.5 mΩ
Stator winding leakage inductance ($l_{ls1} = l_{ls2}$, Fig. 2)	0.2788 mH
Stator d-axis magnetizing inductance (l_{md}, Fig. 2)	1.1381 mH
Equivalent branch Impedance (Z_{eq} at 360 Hz, Fig. 4)	j3.1237 Ω

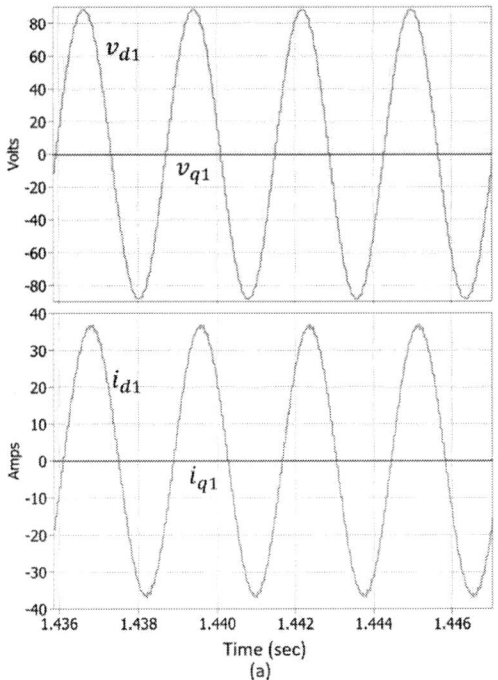

(a)

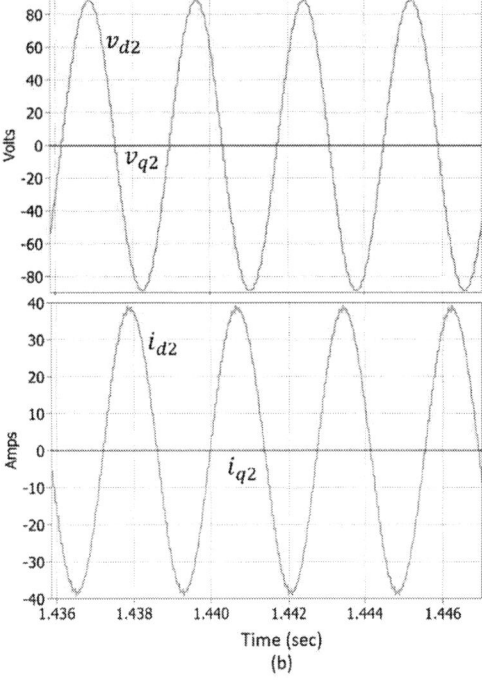

(b)

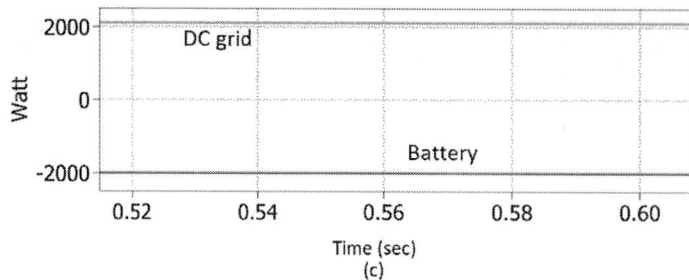

Fig. 10: dq-axes voltages and currents: (a)winding set1 (b)winding set2 (c) DC grid and battery power

Conclusions

This paper presented an isolated integrated DC charger topology based on a six-phase synchronous machine, which requires a minimum reconfiguration from driving and charging modes. The proposed control scheme produces zero average torque and allows power transfer magnetically between two isolated winding sets by injecting a time-varying d-axis current into the stator windings of the machine. The algorithm was experimentally validated with a six-phase wound field synchronous machine and was shown to allow isolated power transfer with zero torque production. The power transfer capability and efficiency of the proposed charger can be greatly increased by using a permanent magnet synchronous machine.

References

[1] A. Salem and M. Narimani, "A Review on Multiphase Drives for Automotive Traction Applications," in *IEEE Transactions on Transportation Electrification*, vol. 5, no. 4, pp. 1329-1348, Dec. 2019

[2] I. Subotic, N. Bodo, E. Levi, B. Dumnic, D. Milicevic and V. Katic, "Overview of fast on-board integrated battery chargers for electric vehicles based on multiphase machines and power electronics", *IET Electric Power Applications*, vol. 10, no. 3, pp. 217-229, March 2016

[3] Bojoi, A. Tenconi, F. Profumo and F. Farina, "Dual-Source Fed Multi-phase Induction Motor Drive for Fuel Cell Vehicles: Topology and Control," *2005 IEEE 36th Power Electronics Specialists Conference*, 2005

[4] M. Y. Metwly, M. S. Abdel-Majeed, A. S. Abdel-Khalik, R. A. Hamdy, M. S. Hamad and S. Ahmed, "A Review of Integrated On-Board EV Battery Chargers: Advanced Topologies, Recent Developments and Optimal Selection of FSCW Slot/Pole Combination," in *IEEE Access*, vol. 8, pp. 85216-85242, 2020

[5] C. Viana, S. Semsar, M. Pathmanathan and P. W. Lehn, "Integrated Transformerless EV Charger With Symmetrical Modulation," in *IEEE Transactions on Industrial Electronics*, vol. 69, no. 12, pp. 12506-12516, Dec. 2022

[6] F. Lacressonniere and B. Cassoret, "Converter used as a battery charger and a motor speed controller in an industrial truck," *2005 European Conference on Power Electronics and Applications*, 2005, pp. 7 pp.-P.7

[7] S. Haghbin, K. Khan, S. Zhao, M. Alakula, S. Lundmark and O. Carlson, "An Integrated 20-kW Motor Drive and Isolated Battery Charger for Plug-In Vehicles," in *IEEE Transactions on Power Electronics*, vol. 28, no. 8, pp. 4013-4029, Aug. 2013

[8] E. Hoevenaars, T. Illg and M. Hiller, "Novel Integrated Charger Concept Using an Induction Machine as Transformer at Standstill," *2020 IEEE Vehicle Power and Propulsion Conference (VPPC)*, 2020, pp. 1-5

[9] P. Pescetto and G. Pellegrino, "Isolated Semi Integrated On-board Charger for EVs Equipped with 6-phase Traction Drives," *IECON 2021 – 47th Annual Conference of the IEEE Industrial Electronics Society*, 2021, pp. 1-6

[10] K. T. Chau, C. C. Chan and C. Liu, "Overview of Permanent-Magnet Brushless Drives for Electric and Hybrid Electric Vehicles," in *IEEE Transactions on Industrial Electronics*, vol. 55, no. 6, pp. 2246-2257, June 2008

[11] R. F. Schiferl and C. M. Ong, "Six Phase Synchronous Machine with AC and DC Stator Connections, Part I: Equivalent Circuit Representation and Steady-State Analysis," in *IEEE Transactions on Power Apparatus and Systems*, vol. PAS-102, no. 8, pp. 2685-2693, Aug. 1983

[12] P. C. Krause, O. Wasynczuk, and S. D. Sudhoff, Analysis of Electric Machinery and Drive Systems, 2nd ed. Piscataway, NJ, USA: IEEE Press,2002

Configurable ISOP-IPOP DC-DC Converter for Universal Solid-State Transformer

Pramod Apte, Jens Friebe
Institute for Drive Systems and Power Electronics
Leibniz University Hannover
pramod.apte@ial.uni-hannover.de
URL: https://www.ial.uni-hannover.de

Lukas Fräger
BLOCK Transformatoren
Elektronik GmbH
lukas.fraeger@block.eu
URL: https://www.block.eu

Acknowledgments

This paper is funded by the German Federal Ministry of Economic Affairs and Climate Action (BMWK) pursuant to a decision of the German Parliament in the project STIM (Smart Transformers as Power Supply for the Future Mechanical Engineering Industry). Funding number: 03EN2010E. The authors are responsible for the content of this publication.

Keywords

≪Solid-State transformer≫, ≪Wide input voltage range≫, ≪Dual Active Bridge (DAB)≫

Abstract

This paper presents the investigation of configurable DC-DC converter for a Universal Solid-State Transformer with a wide input voltage range. The system design for the combination of the Active Front End (AFE) and the DC-DC converter is analyzed. A DC-DC converter solution with configurable series or parallel input connection is proposed.

Introduction

Universal Solid-State Transformers (USST) are designed to incorporate all typical low voltage (LV) grids. The USST accommodates all the industrial LV grid voltages like 120 V/230 V/277 V/290 V. The system schematic of the USST is shown in Fig. 1. As seen in Fig. 1, the DC-link 1 voltage is variable while the DC-link 2 voltage is set based on the DC grid. For this paper, a DC-link 2 voltage of 800 V is considered. The Active-Front End and the DC-DC converter of the USST are responsible for adjusting the voltage levels between the input AC grid and the output DC grid. Incorporating such a wide input voltage range leads to higher operation losses for certain operating conditions in both the AFE and the DC-DC converter. Different strategies to achieve the wide-gain operation and their pros and cons are explained in Section 1. To tackle wide range operation of the DC-DC converters, the switch is the topological configurations has been previously investigated [1] [2] [3]. In Section 2, this paper introduces a topological solution that uses configurable Input Series Output Parallel (ISOP) and Input Parallel Output Parallel (IPOP) topology for the DC-DC converter. Offline change of the input connection as series or parallel connection for a particular grid connection significantly increases the overall system efficiency without compromising the simplicity of design. A qualitative comparison of this implementation with the conventional implementation is put forward. Finally, the measurement results for a 5 kW DC-DC converter are presented.

Conventional System Implementation

The main objective of the USST is to be able to accommodate all the typical LV grid voltages as input. The conventional two-stage solution to adjust the voltage levels is to simultaneously use the voltage gain

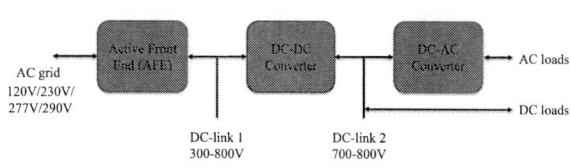

AC grid
120V/230V/
277V/290V

DC-link 1
300-800V

DC-link 2
700-800V

Fig. 1: Universal Solid-State Transformer schematic

Table I: AFE grid input voltage and pre-ferred output voltage levels for low losses

Grid Phase-Neutral RMS Voltage	Preferred DC-link voltage range
120 V	300-350 V
230 V	600-650 V
277 V	700-750 V
290 V	750-800 V

of the AFE (M_{AFE}) and the DC-DC converter ($M_{\text{DC-DC}}$) [4] [5]. The voltage gain of the AFE (M_{AFE}) is the term used in this paper to denote the DC-link voltages due to modulation of the AFE, $M_{\text{AFE}} = V_{\text{DC1}}/(\sqrt{6}V_{\text{AC,rms}})$. This section explains different scenarios of achieving wide-input voltage range by manipulating M_{AFE} and $M_{\text{DC-DC}}$. The variable that separates these scenarios is the DC-link 1 voltage (V_{DC1}). For this project, the DC-link 2 voltage (V_{DC2}) of 800 V is required. Hence, the maximum value of V_{DC1} voltage is also set to 800 V.

The highest and the lowest input grid voltages will be considered to investigate the system design. The standard North America grid voltage of 120 V at 60 Hz is the lower extreme voltage. The special industrial grid phase voltage of 290 V at 60 Hz is considered the highest input voltage. Other typical grid voltages like 230 V, 50 Hz will not significantly deviate the design of the USST. The adjustment of the voltage levels from input grid to DC link 2 is done by three important parameters: The AFE gain (M_{AFE}), the DC-DC converter gain ($M_{\text{DC-DC}}$) and the transformer ratio (n_r) of the DC-DC converter. The transformer ratio is a fixed hardware parameter while the other two can be varied during operation. By taking into account both the AFE and the DC-DC Converter, to achieve the wide-gain operation for the USST, the following strategies can be used:

- 1. Operate the AFE near unity gain. This leads to a highly efficient operation of the AFE [6]. V_{DC1} is set close to $\sqrt{6}V_{\text{AC,rms}}$ where $V_{\text{AC,rms}}$ is the rms phase-neutral voltage of the input grid. This leads to V_{DC1} between 294-710 V for the grid voltage range. Set the transformer ratio such that the DC-DC converter has symmetric buck and boost range. A wide-range operation of the DC-DC converter leads to high losses especially at the ends of the gain range. Additionally, most DC-DC converter topologies lose soft switching which often leads to worse EMI performance. Higher gain range also leads to higher current stress on active and passive components as well as larger magnetic components.

- 2. Operate the AFE with a constant output voltage of 800 V. This leads to an efficient performance for the higher grid voltage of 290 V as suggested in Table I. On the other hand, it leads to very high losses for low grid voltage of 120 V. Also, it leads to a larger volume of the inductors and bulkier filter design. Use $n_r = 1$ to operate the DC-DC converter with unity gain. A compact and efficient design of the DC-DC converter is possible in this region of operation.

- 3. With a variable DC link 1 voltage, optimize the transformer ratio n_r and distribute the losses between the AFE and the DC-DC converter. The range of DC link 1 voltage is set within 450-800 V. The AFE works near unity gain operation for 290 V, while the DC-DC converter needs to operate in step-down mode because of $n_r > 1$. For 120 V input grid voltage, V_{DC1} is set between 450-600 V. A detailed optimization is required to determine this voltage. The DC-DC converter components have lower peak ratings compared to strategy 1 but losses are high nevertheless. Similar to Strategy 1, deterioration in EMI performance occurs as the converter loses soft switching for buck and boost operation. Although this strategy leads to lower peak losses for the combined system of the AFE and the DC-DC converter compared to strategy 1, it still suffers from high losses for the lowest and the highest input grid voltages.

Table II: Loss trends for the AFE and DC-DC converter for different operational strategies

Input Grid Voltage	120 V		290 V	
	AFE	DC-DC	AFE	DC-DC
Strategy 1	Low	High	Low	High
Strategy 2	Very High	Low	High	Low
Strategy 3	Moderate	Moderate	Low	Moderate
Proposed Strategy	Moderate	Low	Low	Low

Configurable DC-DC Converter

All the above discussed strategies have their pros and cons. Overall, all the strategies suffer from significant losses in certain operating conditions than desired. For the USST application, once the input grid is connected, the voltage range is smaller compared to the design requirements as shown in Table I. The configurable ISOP-IPOP converter takes advantage of this detail. For the lower voltage grid (120 V), the AFE sets V_{DC1} to a fixed voltage of 400 V, while for the other higher voltage grids, V_{DC1} is set to 800 V. V_{DC1} is thus limited to two values, 400 V and 800 V. Fig. 2 shows the topology of the configurable DC-DC converter. The DC-DC converter for the designed USST consists of two individual dual-bridge converter modules which can either be connected in ISOP or IPOP configuration. When the switches S1 and S2 are ON and S3 is OFF, the configuration is IPOP whereas when S1 and S2 are OFF and S3 is ON the configuration is ISOP.

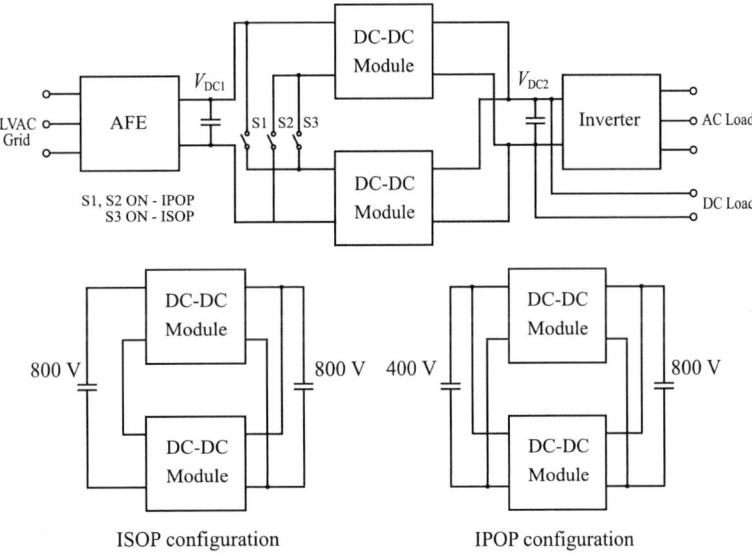

Fig. 2: Configurable ISOP-IPOP DC-DC converter and its two configurations

For low and high input grid voltages, V_{DC1} is set to 400 V and 800 V respectively by the AFE. V_{DC2} is fixed at 800 V and the n_r for the dual-bridge converter is set to 2. The combination of these DC-DC converters is then operated in IPOP and ISOP configurations for V_{DC1} of 400 V and 800 V respectively. Fig. 2 shows that, for both the configurations, each DC-DC converter is operated with fixed voltage ratio operation with input voltage of 400 V and output voltage of 800 V. For most of the DC-DC converter topologies, a fixed voltage ratio operation ensures high efficiency and low stresses on the components. Additionally, the AFE has significantly less losses compared to the conventional solutions discussed before. As an offline switch in configurations is possible, the switches S1,S2 and S3 can be realized using contactors, mechanical relays or even jumpers. Use of contactors or relays is simpler and more efficient compared to the use of bidirectional switches which add to the complexity of the converter. Use of jumpers is the most efficient and cheap solution, but needs human intervention which is often undesirable.

DC-DC Converter Module Topology

The DC-DC converter modules for the USST must exhibit galvanic isolation and bidirectional power transfer capability. For the ISOP/IPOP configuration, the DC-DC converter operates at unity gain. There are many DC-DC converter topologies relevant for this application. The Dual Active bridge (DAB) converter first introduced in [7] is one of the most popular topologies for SSTs. It features Zero Voltage Switching (ZVS) for both the full-bridges at fixed voltage ratio operation. It is also capable of buck and boost operation in both directions of power flow. Another popular topology is the Series Resonant Converter (SRC). SRC is one of most efficient converters for fixed voltage ratio operation especially when operated in sub-resonant or HC-DCM mode [8]. LLC converter is another often considered topology for soft switching and buck-boost capability. Also, it only works in buck mode in reverse direction when the resonant tank is placed on one side of the transformer. This loses a degree of freedom during non-ideal voltage conditions. The Series Resonant Dual-Active Bridge (SRC-DAB) is a resonant version of the DAB converter. It ensures ZVS on the primary bridge and ZCS with partial ZVS on the secondary bridge for unity gain operation [9]. The resonant topologies have added costs and volume because of the resonant capacitors. Additionally, the resonant topologies have a slower transition speed for bidirectional power. [10]. Overall, it is a practical choice dependent on the application, especially the voltage and power levels required for the USST. Although the resonant converters provide excellent efficiency, for high power applications, the costs far outweigh the benefits compared to the non-resonant DAB. Hence, for this project the DAB converter is considered.

Qualitative Comparison of Configurable DC-DC Converter with Single Wide-Input Voltage DC-DC Converter

There are multiple advantages to using the configurable DC-DC converter compared to a single wide-input voltage DC-DC converter. Due to the fixed voltage ratio operation of each module, the semiconductor losses are significantly lower due to lower peak currents and ubiquitous soft switching. Total losses in the semiconductors are usually limited by the cooling effort. Higher $R_{DS(on)}$ often co-relates to lower switching energy of the MOSFET. Soft-switching provides additional freedom to choose SiC MOSFETs with higher $R_{DS(on)}$ which further reduces the costs. As the DAB undergoes hard turn-off, lower switching energies would lead to lower losses. Due to the low switching losses, an increase in the switching frequency is possible. A larger switching frequency leads to smaller transformer volume as well as low current ripple on the DC-link. As shown in [11] use of DC-link capacitor for each module is better than for the complete DC-link irrespective of the configuration. Lower current ripple, lower voltage rating reduces the capacitor volume and cost. Additionally, the primary side of the DAB uses low voltage devices for half the power rating of the full converter. Lower voltage class switches are often cheaper. Lower peak currents also lead to lower conduction losses in the transformer. Soft switching for the whole power range leads to a much superior EMI performance too.

On the other hand, the number of semiconductor devices is doubled for this converter. Even though the transformer is small, the overall volume of two transformers is still higher than that of a single wide-input DC-DC converter. Although, the modules operate with fixed voltage ratio, additional control effort

Table III: Hardware prototype components

Component	Type
MOSFETs	Infineon IMZA65R027M1H (Prim)
	Infineon IMZ120R030M1H (Sec)
Transformer	E55/28/21 (2 sets), turns ratio 7:14
	$L_\sigma = 12.4\,\mu H$, $L_m = 400\,\mu H$
Litz Wire	1400 x 0.05 mm (Prim)
	200 x 0.1 mm (Sec)
Controller	Texas Instruments F28379D
Power Analyzer	Yokogawa WT1800

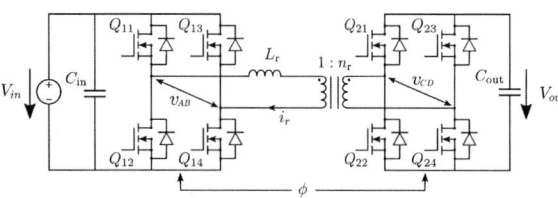

Fig. 3: Dual Active Bridge topology

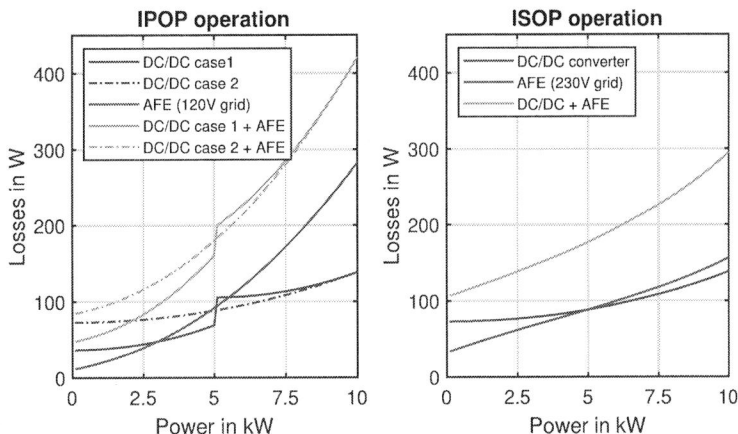

Fig. 4: Calculated losses for DC-DC converter and AFE for IPOP and ISOP configurations

is required to achieve power and voltage balance for the modules. However, a fixed input DC-link voltage reduces the complexity of control.

Design of Dual-Active Bridge Module

To analyze the performance of the configurable ISOP-IPOP converter, a fixed voltage ratio DAB module is designed. For a 10 kW SST, each DAB module is rated for 5 kW with input voltage of 400 V and output voltage 800 V. The primary bridge consists of 650 V SiC switches while the secondary consists of 1200 V SiC switches. For a fixed voltage ratio operation, all the switches undergo ZVS for the entire power range. The conduction losses and the turn-off losses are mainly influenced by the peak transformer current. A small leakage inductance is thus preferred as it results in lower peak currents and eventually lower semiconductor losses. Additionally, to reduce the volume of the converter only the leakage inductance of the transformer is used.

The modules are assumed to be identical and all parasitic components are neglected. The DAB modules are operated with a fixed frequency of 100 kHz using single phase shift (SPS) modulation. The power equation of the DAB converter for the SPS modulation is given by Eq. 1.

$$P_{12} = \frac{nV_1V_2\phi(\pi - |\phi|)}{2\pi^2 f_s L} \quad \text{where } \phi \text{ is the phase difference between } V_{AB} \text{ and } V_{CD} \tag{1}$$

In the IPOP configuration the individual DAB modules have the same operating conditions as the ISOP configuration if they equally share the power. Fig. 4 shows the calculated losses for the DAB modules for ISOP and IPOP configurations. In IPOP operation, case 1 implies that the second module is switched ON after 5 kW and case 2 implies equally power distribution for complete power range. In case of ISOP configuration, there is no added advantage to operate the two modules with different output power. On the other hand, it can be seen from Fig. 4 that for the IPOP configuration, use of a single module for partial load conditions significantly reduces the losses. The least loss operation involves use of a single module up to 5 kW and then equal power sharing after that.

Performance of the Combined DC-DC Converter and AFE System

To analyze the performance of the AFE, two different grid voltages of 120 V and 230 V are considered. Fig. 4 shows the loss plots for the AFE. The AFE has higher losses with the 120 V grid operation. This is primarily due to high currents which lead to higher conduction losses for the semiconductors. On the other hand, the benefits of lower grid voltage are seen for lower power values. As discussed in the previous sections, the losses from the DC-DC converter are nearly the same for both the configurations. The combined losses plots for the DC-DC converter and the AFE are also shown in Fig. 4. The IPOP configuration has the higher peak losses compared to ISOP. The difference in peak losses is not extreme

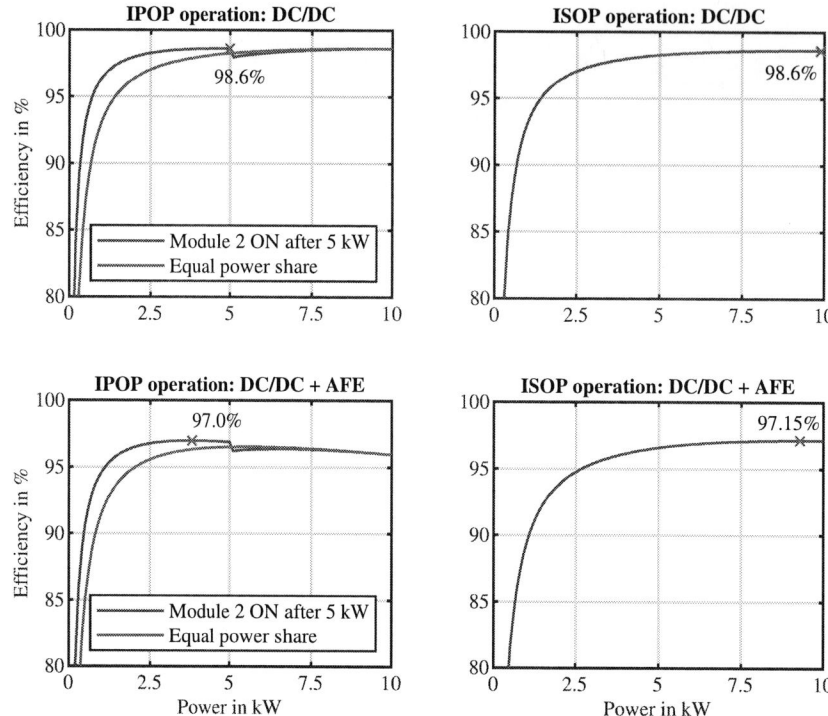

Fig. 5: Calculated efficiency plots and maximum efficiency for 10 kW DC-DC converter and combination of DC-DC converter and AFE for ISOP and IPOP configurations

as would have been the case for other SST strategies. Fig. 5 shows the efficiency of the DC-DC converter and the combined efficiency of the AFE and DC-DC Converter system. For both the grid voltages of 120 V and 230 V, the system has a peak efficiency of greater than 97 %. The higher grid voltages like 277 V and 290 V which were not considered for calculations actually lead to better efficiencies as the AFE would have a smaller boost gain to reach V_{DC1} of 800 V. In summary, for all the possible grid voltages, the AFE + DC-DC converter system has an excellent efficiency performance.

Measurement Results

Fig. 6 shows the 5 kW DAB module prototype with rated input voltage of 400 V and output voltage of 800 V. The hardware specifications of the prototype are mentioned in Table III. It consists of a single power PCB with both the full bridges of the DAB. The PWM signals and the SPS modulation are facilitated by the Texas Instruments F28379D Launchpad. The measurement waveforms at rated voltage and power are shown in Fig. 7. This prototype reached a peak efficiency of 97.95 % near full load which is very close to the calculated efficiency. More tests need to be conducted to validate the efficiency for the ISOP/IPOP configurations.

Conclusion

In this paper, different strategies to realize the USST were analyzed. A configurable DC-DC Converter is proposed as a solution for efficient operation of the USST. In contrast to a wide-input voltage DC-DC converter, a converter with two unique input voltages enables achieving high efficiency. A qualitative comparison of the configurable converter and a conventional wide-input voltage converter is presented. The performance of the AFE and DC-DC converter system is analyzed. A 5 kW DAB prototype with voltage ratio of 1:2 is built and tested for rated voltage and power to prove the high efficiency of the DC-DC converter.

Fig. 6: 5 kW DAB module prototype

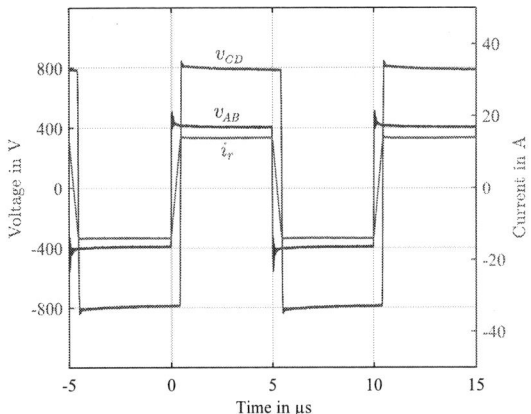

Fig. 7: Measurement waveforms for the DAB prototype with SPS modulation at $V_{in} = 400\,\text{V}$, $V_{out} = 798\,\text{V}$ and $P_{out} = 4985\,\text{W}$

References

[1] P. Sun, L. Zhou and K. M. Smedley, "A Reconfigurable Structure DC–DC Converter With Wide Output Range and Constant Peak Power," in IEEE Transactions on Power Electronics, vol. 26, no. 10, pp. 2925-2935, Oct. 2011, doi: 10.1109/TPEL.2011.2129576.

[2] Y. Shen, H. Wang, A. Al-Durra, Z. Qin and F. Blaabjerg, "A Structure-Reconfigurable Series Resonant DC–DC Converter With Wide-Input and Configurable-Output Voltages," in IEEE Transactions on Industry Applications, vol. 55, no. 2, pp. 1752-1764, March-April 2019, doi: 10.1109/TIA.2018.2883263.

[3] D. Lyu, T. B. Soeiro and P. Bauer, "Design and Implementation of a Re-configurable Phase-Shift Full-Bridge Converter for Wide Voltage Range EV Charging Application," in IEEE Transactions on Transportation Electrification, doi: 10.1109/TTE.2022.3176826.

[4] X. She, A. Q. Huang and R. Burgos, "Review of Solid-State Transformer Technologies and Their Application in Power Distribution Systems," in IEEE Journal of Emerging and Selected Topics in Power Electronics, vol. 1, no. 3, pp. 186-198, Sept. 2013, doi: 10.1109/JESTPE.2013.2277917.

[5] S. Falcones, X. Mao and R. Ayyanar, "Topology comparison for Solid State Transformer implementation," IEEE PES General Meeting, 2010, pp. 1-8, doi: 10.1109/PES.2010.5590086.

[6] L. Fräger, S. Langfermann, M. Owzareck and J. Friebe, "An Analytic Inverter Loss Model for Design and Operation Space Optimization," 2021 23rd European Conference on Power Electronics and Applications (EPE'21 ECCE Europe), 2021, pp. 1-10.

[7] M. N. Kheraluwala, R. W. Gascoigne, D. M. Divan and E. D. Baumann, "Performance characterization of a high-power dual active bridge DC-to-DC converter," in IEEE Transactions on Industry Applications, vol. 28, no. 6, pp. 1294-1301, Nov.-Dec. 1992, doi: 10.1109/28.175280.

[8] J. E. Huber and J. W. Kolar, "Analysis and design of fixed voltage transfer ratio DC/DC converter cells for phase-modular solid-state transformers," 2015 IEEE Energy Conversion Congress and Exposition (ECCE), 2015, pp. 5021-5029, doi: 10.1109/ECCE.2015.7310368.

[9] X. Li and A. K. S. Bhat, "Analysis and Design of High-Frequency Isolated Dual-Bridge Series Resonant DC/DC Converter," in IEEE Transactions on Power Electronics, vol. 25, no. 4, pp. 850-862, April 2010, doi: 10.1109/TPEL.2009.2034662.

[10] B. Zhao, Q. Song, W. Liu and Y. Sun, "Overview of Dual-Active-Bridge Isolated Bidirectional DC–DC Converter for High-Frequency-Link Power-Conversion System," in IEEE Transactions on Power Electronics, vol. 29, no. 8, pp. 4091-4106, Aug. 2014, doi: 10.1109/TPEL.2013.2289913.

[11] M. Sato et al., "High efficiency design for ISOP converter system with dual active bridge DC-DC converter," 2016 IEEE Applied Power Electronics Conference and Exposition (APEC), 2016, pp. 2465-2472, doi: 10.1109/APEC.2016.7468211.

Using System-on-Chip Boards for the Deployment of Controller for Verification and Prototyping

Adeel Jamal, Gerd Griepentrog
Technical University of Darmstadt
Fraunhoferstrasse 4, Darmstadt, Germany
Email: adeel.jamal@tu-darmstadt.de
URL: http://www.lea.tu-darmstadt.de

12.08.2022

Acknowledgments

This work was funded by the German Federal Ministry for Economic Affairs and Energy under the support code 0324101.

Keywords

≪System-on-Chip Boards≫, ≪PQ Controller≫, ≪Model-based Design Workflow≫, ≪Two-level Converter≫, ≪Controller Verification & Prototyping≫

Abstract

New control strategies for power converters are mainly tested using Hardware-in-the-Loop (HIL) based rapid control prototyping (RCP) platforms in academia and industry research. In this paper, different strategies to use the system-on-chip boards for power electronics application are presented with pros and cons of each technique. The hardware software co-design work flow utilizing Intel and Matlab's toolchain is elaborately described. The proposed approach to implement a controller for power electronics application, that is cost-effective, offers more freedom for customization and will help decrease the product's time-to-market is presented. The introduced approach will be used to experimentally verify the performance of the exemplary PQ Controller designed for Grid-tied converter with LC filter.

Introduction

Rapid Control Prototyping (RCP) platforms are used widely for testing, validation and prototyping of different controllers used in the drives of electric traction motors and Brushless DC (BLDC) motors for electric and hybrid vehicles [1]. Similarly, they are also employed in designing controllers for new converter topologies in medium and high voltage applications [2], [3] and DC-DC converter applications [4]. Depending on the manufacturer and hardware variant, RCP platforms are usually capable of running both Hardware-in-the-Loop(HIL) [5] and Processor-in-the-Loop(PIL) simulations [6]. dSpace™, Speedgoat™ and OPAT-RT Technologies™ are leading vendors for providing the RCP platforms but nearly all of the RCP control boards/platforms offered by them are quite expensive with less or close to zero freedom available for any hardware customization, for instance in terms of testing new communication protocols between power electronics' board hosting power stages or from ADC adapter card to control/DSP board. Vendors require customers to buy separate licenses for additional features such as new toolsets or analog/digital interfaces. In addition to buying the hardware for upgrading the performance of the processor of RCP platforms by appending more cores for high complexity controller design implementation, e.g. multiple step prediction based model predictive controller (MPC), customers are generally also required to pay for license for usage of extra cores which makes it rather more expensive if an upgrade is needed. Also, the need of perpetual support from the manufacturer throughout the product's life cycle should not be underestimated. The time required to transfer the controller from RCP to the ASIC/FPGA chip is relatively high as the design first needs to be converted to HDL code which could be synthesized and compiled. The availability of high performance, high resource SoC boards offer an alternative for control design deployment which is cheap and reliable. This approach will also drastically reduce time-to-market as the SoC Chips for which the entire Model-based Design workflow is employed can be directly used in the embedded SoC boards for end products with the same workflow still valid for programming it. dSpace RCP platform like MicroLabBox also utilizes the Model-based design approach to streamline the process of C and HDL (Hardware Description Language) code generation and integration but certain processes within the workflow cannot be modified or enhanced. MicroLabBox is used in the industry for automotive and control applications [5], [7]- [8]. Nowadays, SoC chips are rapidly penetrating the commercial products arena where hardware accelerators are required for fulfilling real-time processing needs. Most of the researchers working on power electronics based solutions are not catching up with the advancements in embedded technologies such as using high speed SoC/FPGA systems rather relying on RCPs for time saving in prototyping of controller designs. This paper attempts to elaborate the deployment of controller on an Intel SoC-FPGA De1-SoC board which hosts Cyclone V SoC that is powerful enough to support complex controllers. There will be some effort to master the toolchain required to deploy the design but once it is mastered, the complete step of transferring the control design in RCPs to HDL can be circumvented. If needed, the performance can be upgraded by just replacing the SoC board while the toolchain and workflow would remain the same if the chip belongs to the same family or minor changes of setting up the interface if the chip belongs to a different family.

In the first section, the model of plant and the structure of power controller [9] will be elaborated. Second section details different strategies to program the SoC Boards that can be used in power electronic applications. The Intel's toolchain and the complete workflow used to program the Cyclone V SoC chip hosted on De1-SoC board will be discussed in the third section. Paper ends with the discussion of results using the adopted toolchain and workflow with a brief conclusion.

Model of the Plant and Controller Structure

The plant consist of an LC filter that is directly coupled to the stiff grid voltage source via grid impedance R_g and L_g as shown in Figure 1. In general, the impedance of the grid at the Point of Common Coupling (PCC) varies depending on the configuration and connection scheme of the distribution grid at any particular instant of time. In the design and tuning of controller, grid impedance is assumed to be constant. Three phase DC/AC converter or Inverter applies a voltage at the inductor side of the LC-filter. Filter parameters are shown in Table I.

Table I: Chosen Parameters for LC-filter and Grid

Component	Value
Filter Inductance (L_f)	$850\,\mu H$
Filter Resistance (R_f)	$0.01\,\Omega$
Filter Capacitance (C_f)	$25\,\mu F$
Grid Resistance (R_g)	$10\,m\Omega$
Grid Inductance (L_g)	$80\,\mu H$

The voltage across the capacitor can be considered as disturbance while deriving for the gains of the controller. Therefore, the transfer function of the plant is;

$$G_{\text{plant}}(s) = \frac{i_{L_g}(s)}{u_{\text{in}}(s)} = \frac{1}{R_f + sL_f} \tag{1}$$

$$G_{\text{plant}}(s) = \frac{\tau_f}{L_f(1 + s\tau_f)}; \tau_f = \frac{L_f}{R_f} \tag{2}$$

Using System-on-Chip Boards for the Deployment of Controller for Verification and
Prototyping   JAMAL Adeel

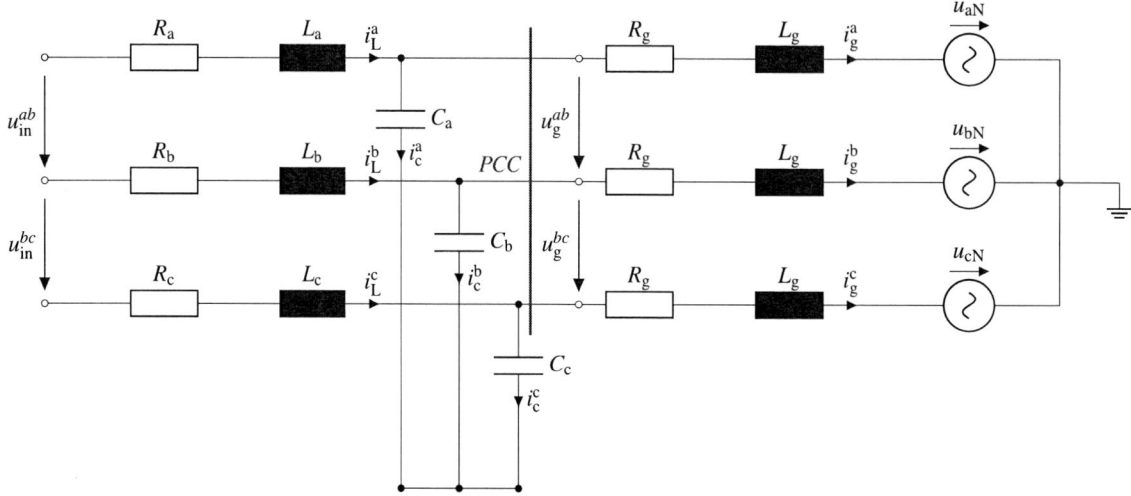

Fig. 1: Model of the plant consisting of LC filter and the Grid

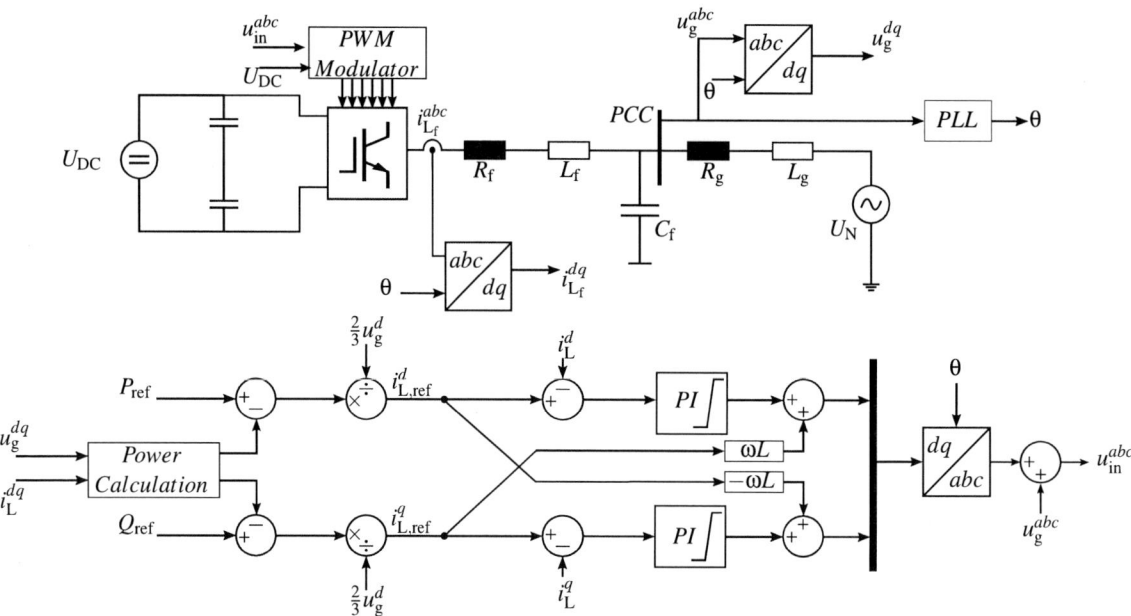

Fig. 2: PQ Controller Structure

The inverter can be modeled as a delay element;

$$G_{\text{delay}}(s) = \frac{1}{1 + 1.5sT_s} \tag{3}$$

Open-loop transfer function of the system including the PI controller with gain k_c and time constant τ_0 can be derived to get;

$$G_{\text{OL}} = \frac{k_c(s\tau_0 + 1)}{s\tau_0} \cdot \frac{1}{1 + 1.5sT_s} \cdot \frac{\tau_f}{L_f(1 + s\tau_f)} \tag{4}$$

Lastly, dominant time constant of the plant will be compensated by the controller according to the magnitude optimum criterion [10], hence time constant τ_0 and gain k_c becomes;

$$\tau_0 = \tau_f \tag{5}$$

$$k_c = \frac{L}{3T_s} \tag{6}$$

therefore the integral gain is;

$$k_{c,i} = \frac{L}{3T_s\tau_f} \tag{7}$$

The entire control scheme is being depicted in Figure 2. Inside the power calculation block in Figure 2, instantaneous power theory is applied to obtain the instantaneous power flows from the inverter to grid [11].

Strategies for Programming System-on-Chip

System-on-chips available from different manufacturers offer different levels of customization options, FPGA resources, toolchains and programming workflows. Modern SoC architecture comprises of processor system and FPGA therefore, there are multiple strategies to program the SoC corresponding to the requirements of application. Following are the different strategies that have been explored, tested and implemented but the list is not exhaustive in nature:

1. Linux/RTOS or any other compatible OS runs on ARM processor and FPGA is used for controller implementation. C/C++/Python program can be written for Linux to communicate with the AXI / AVALON bus connecting processor system to FPGA. FPGA can be programmed using the VHDL/Verilog program written using open-source hardware accelerators and built-in IP cores available from the manufacturer. Lower-level hardware understanding of the address mapping of the FPGA in the Hard Processor System [12](HPS[1]) address space and knowledge of the provided APIs of the manufacturer is required to implement this approach.

2. Sometimes, the FPGA design is provided by the manufacturer for lower level control of converter such as monitoring of current, voltage and temperature but if the FPGA is completely self-programmed or the FPGA design for lower level control of converter provided by manufacturer is not chip-locked [2]; then matlab's co-design workflow can be adopted to generate the HDL code automatically from the Simulink design file through the Matlab's HDL coder library while at the same time, the ARM controller, run by Matlab Linux OS firmware, is programmed through the C code generated by embedded coder library of Matlab.

3. If the FPGA design is chip-locked then there are two possibilities;
   - Linux/RTOS running on both cores using Asymmetric Multi-processing (AMP). One core is used for monitoring and transmission of control and measurement data to the host main controller which also provides the set points. Linux is limited for the execution of different threads, and assigning their priorities to only one core and is managed by Linux kernel. The

---

[1] As referred in the technical reference manual of Cyclone V chip.

[2] which means lower-level control is provided as a design-partition which can be appended in the end of FPGA's final design

other core just sits idle unless it is directed by the Linux to start executing instructions at a specific address in SDRAM memory space. It must be ensured that the second core doesn't intrude into the other's memory space on SDRAM by mistake after the end of execution of its own program. Closed loop controller executes on the second core which means that, the communication between them is within the framework of Linux. Communication between the ARM processor and FPGA can be via SDRAM memory. The AMP is not straight-forward and can be quite challenging to implement on Cyclone V chip because of the following reasons:

(a) Complexity associated with seperation of RAM, peripherals and caches between the two cores.

(b) Difficulty in ensuring cache coherence during transfer of data between cores.

(c) Lack of documentation or frameworks regarding AMP implementations on the Intel SoC FPGA (compared to OpenAMP for the Xilinx Zynq platform).

- One of the ARM cores is run by Linux, the other one is programmed in bare-metal[3]. Controller would be completely written in bare-metal and the delays are deterministic in nature which means the real-time execution can be guaranteed. Communication between the two cores can be on the principle of shared memory region based on interrupts. The shared memory has to be configured in Linux u-boot bootloader so that it doesn't deny the access to its memory region by default. The handling of communication and data exchange between the two cores is also not straightforward and it requires advanced considerations such as ensuring both programs running on distinct cores tend not to access the memory region at the same time hence it is recommended to be used if none of the other options is viable.

Hardware Setup, Intel's Toolchain and Workflow

The commercial version of the Siemens' S120 converter (Two level, silicon semiconductor switch) is adapted by replacing the built-in controller hosted on factory-programmed AISC chip by De1-SoC control platform [13]. To implement the customized controller algorithm, De1-SoC control platform is connected to the power electronics of the S120 converter via an interface card used for voltage level translation and isolation between digital and power electronics circuit. De1-SoC comprises of Cyclone V SoC chip which hosts dual core ARM Cortex-A9 processor [14] alongside FPGA on a single chip. FPGA and processor system are interconnected through high-speed AXI-Avalon bus as shown in Figure 3.

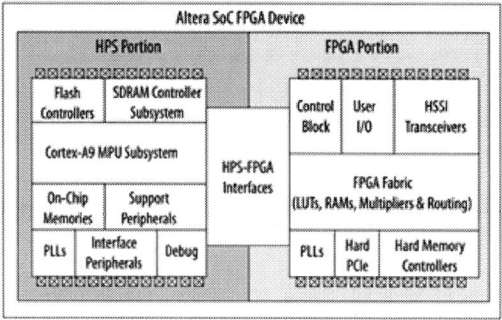

Fig. 3: Intel's SoC-FPGA Device Block Diagram [12]

The grid-connected converters demand real-time execution capability to avoid delays and other synchronization issues that could potentially lead to faults or temporary shutdown of the converter. Dual Core ARM Cortex-A9 is an application processor designed to be run using operating system (OS) by default [15]. Linux OS is configured to run on the processor in symmetric multi-processing mode i.e. the

---

[3]Bare-metal means that one has to program it by lower level c functions using APIs and library functions provided by the manufacturer.

Linux kernel and scheduler prioritizes the execution of threads on each core. Linux is not a Real-Time OS (RTOS) i.e. it doesn't guarantee a certain worst case maximum response time/latency to a triggering event. Similarly, Real-time patch of the Linux that is generally used to render some control on priority assignment of threads in Linux OS is also not so effective therefore the entire controller would be programmed on FPGA and all non-time critical blocks will be run on ARM processor using the MATLAB's hardware software (HW SW) co-design workflow concept.

The workflow that needs to be followed to program the FPGA and ARM processor separately is shown in Figure 4. For programming the FPGA, the HDL design must be finalized in Intel's Quartus software with the help of built-in IP Cores using Qsys/Platform designer. The generated .sof file can then be used to program the FPGA. Similarly, for programming the ARM A9 processor in bare-metal, ARM Debugger Studio 5 (DS-5) can be used. Intel's built-in hardware libraries and MATLAB's libraries can be utilized to generate the executable files for ARM processor. In Figure 4, a bird's eye view is presented, each step shown can entail further configurations and multiple sub-steps. The aforementioned entire toochain and workflow is handled by MATLAB if its HW, SW co-design workflow methodology is adapted. Following are the major steps that entails Matlab's co-design workflow;

1. Design the whole system that needs to be run on System-on-Chip in Simulink. All the blocks that are meant to be run on FPGA must be enclosed by a main subsystem block. All the blocks used must be compatible with HDL coder library for the HDL code generation. Where possible, HDL-optimized math operation blocks such as divide and square-root operation must be used as it ends up using less FPGA resources.

2. All the tasks and subsystems that are non-time critical e.g. monitoring systems must be placed outside the main subsystem block and it must be interfaced appropriately.

3. Using the HDL workflow advisor, the IP Core for the main subsystem block must be generated. The AXI bus interface code generation will be done by MATLAB if the specific pins of the SoC is defined for the main subsystem block interface.

4. Quartus project can be generated inside the HDL workflow advisor. The IP core can then be imported into the project followed by running analysis and synthesis to generate bitstream or .sof file.

5. Software interface model is to be generated with the help of HDL workflow advisor. This interface model can be used to interact with the System-on-Chip in MATLAB's real-time mode.

This is an iterative process and errors can appear at any stage of the compilation. It must be fixed step-by-step. It's possible that the design doesn't fit on the FPGA becasue of the over-usage of resources. There are different resource optimization techniques that can be employed to condense the design e.g. serialization of data. It should be kept in mind that improving the performance of the generated design is incremental and it takes time and many iterations to achieve a design that is final for deployment.

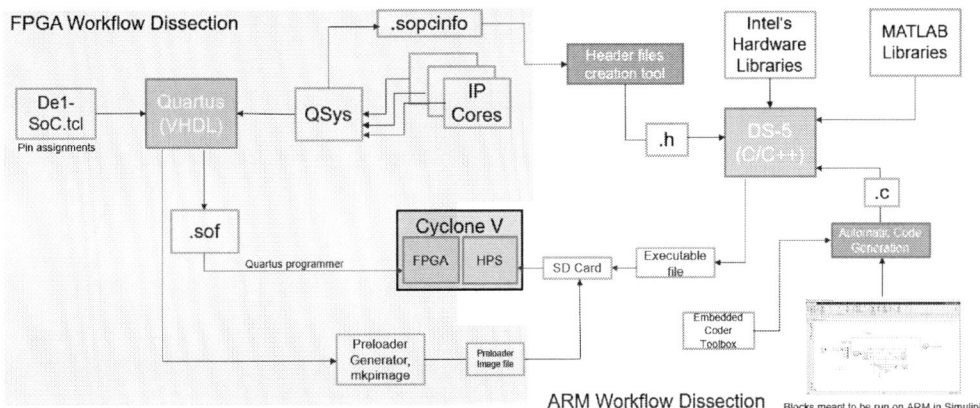

Fig. 4: Intel's SoC-FPGA Entire Toolchain and Workflow

Results and Discussion

In this section, the result of adopting the described workflow and applying the Intel's toolchain to implement the PQ Controller structure will be presented. The PQ-controller is fully implemented inside FPGA. All the debugging and data probes transmit the data from the FPGA to the ARM processor via 120-bit wide data bus via AXI/AVALON protocol. In Figure 5, the experimental result of applying different setpoints to the controller is presented. Initially, the active (P) and reactive power (Q) setpoints are zero. Setpoints for P and Q of 10 kVA and 7 kVA are applied at t = 10 ms and t = 35 ms respectively. The measured active and reactive power follows the setpoint with some time delay dependent on the time constant of the controller. The output current though the filter's inductor and the voltage at PCC is also shown in the figure. The controller tries to regulate the active and reactive power with very little deviation and the small ripple present in the measured active and reactive power waveform is due to the switching nature of the converter. More complex controller's with extensive calculations like MPC can also be employed by using the described methodology. This approach provides a lot of freedom and flexibility to implement the controller in different ways using FPGA resources.

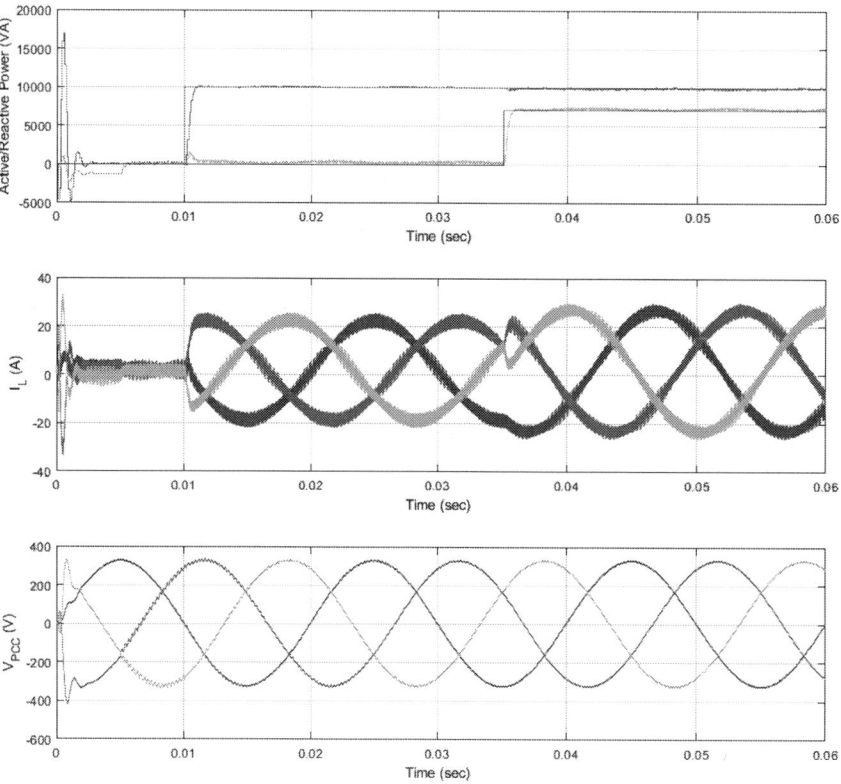

Fig. 5: Measured Active(P) and Reactive power(Q) following the reference

Conclusion

The proposed method and workflow has several advantages. It saves a lot of time in design prototyping. When designing controllers for converters in microgrid application, it provides flexibility to change and modify the controller with very little marginal effort. The whole process of design doesn't require a deep knowledge of VHDL/Verilog or C language, only elementary knowledge is sufficient which is a huge advantage for novice. The disadvantages are following: Matlab HDL coder and embedded coder licenses are required. Built-in resource optimization techniques offered from Matlab and HDL Coder are not robust and stable in complex design e.g. feedback controller design. The complete cycle of finalization of design in Simulink, synthesizing the design files and compilation can be tedious; as in the case, the design gets large, the compilation time increases proportionally and therefore the feedback on errors made in the design will be slow. The entire design process is iterative; as it can happen that after finalizing the whole design and carrying out functional testing, the design doesn't fit on FPGA in the last stage of the compilation and fitting process.

References

[1] H. Ross, Elektromobile Ideen verwirklichen, Published at: Elektronik automotive, 04/2012

[2] J. Dannehl, C. Wessels and F. W. Fuchs, "Limitations of Voltage-Oriented PI Current Control of Grid-Connected PWM Rectifiers With *LCL* Filters," in IEEE Transactions on Industrial Electronics, vol. 56, no. 2, pp. 380-388, Feb. 2009, doi: 10.1109/TIE.2008.2008774.

[3] Gholami-Khesht H, Davari P, Blaabjerg F. An Adaptive Model Predictive Voltage Control for LC-Filtered Voltage Source Inverters. Applied Sciences. 2021; 11(2):704. https://doi.org/10.3390/app11020704

[4] P. Karamanakos, T. Geyer and S. Manias, "Direct Voltage Control of DC–DC Boost Converters Using Enumeration-Based Model Predictive Control," in IEEE Transactions on Power Electronics, vol. 29, no. 2, pp. 968-978, Feb. 2014, doi: 10.1109/TPEL.2013.2256370.

[5] S. Khaldoune, P. -D. Maria, K. Abdelaziz and F. Maurice, "Hardware in loop methodologies for the control of dual-PMSM connected in parallel: FPGA implementation and experimentation," 2015 17th European Conference on Power Electronics and Applications (EPE'15 ECCE-Europe), 2015, pp. 1-10, doi: 10.1109/EPE.2015.7309461.

[6] Mangesh Kale, Narayani Ghatwai, Suresh Repudi, Processor-In-Loop Simulation: Embedded Software Verification & Validation In Model Based Development, Published at Design & Reuse

[7] T. Jaster, A. Rowe and Z. Dong, "Modeling and simulation of a hybrid electric propulsion system of a green ship," 2014 IEEE/ASME 10th International Conference on Mechatronic and Embedded Systems and Applications (MESA), 2014, pp. 1-6, doi: 10.1109/MESA.2014.6935601.

[8] M. Ruba, V. Chindris and D. Fodorean, "Design and experimental validation of a low voltage high current SRM for light electric vehicles," 2014 International Symposium on Power Electronics, Electrical Drives, Automation and Motion, 2014, pp. 118-123, doi: 10.1109/SPEEDAM.2014.6871910.

[9] B. Hammer, E. Lenz and U. Konigorski, "On the Design of Grid-Side Inverter Voltage Controllers," 2020 2nd IEEE International Conference on Industrial Electronics for Sustainable Energy Systems (IESES), 2020, pp. 227-232, doi: 10.1109/IESES45645.2020.9210662.

[10] D. Schröder, Elektrische Antriebe - Regelung von Antriebssystemen, 4th ed. 2015, isbn: 978-3-642-30096-7.

[11] Hirofumi Akagi, Edson Hirokazu Watanabe and Maur´ıcio Arede, "Instantaneous Power Theory and Applications to Power Conditioning", Second Edition, 2017, The Institute of Electrical and Electronics Engineers, Inc. Published 2017 by John Wiley & Sons, Inc.

[12] Cyclone V Hard Processor System, Technical Reference Manual, cv_5v4 2014.12.15, Altera 101 Innovation Drive San Jose, CA 95134

[13] https://www.terasic.com.tw/cgi-bin/page/archive.pl?Language=English&No=836

[14] https://developer.arm.com/ip-products/processors/cortex-a/cortex-a9

[15] ARM Cortex-A series Programmer's Guide, Version 4.0

[16] "Benchmark Systems for Network Integration of Renewable and Distributed Energy Resources", Final Report of Task Force C6.04.02, 2014

Utilizing the Reactive Current Control Capability of an MMC-Fed AC/DC Converter for Volt-Second Balancing in Medium Frequency Transformers

Kaveh Pouresmaeil, Maurice Roes, Jorge Duarte, Korneel Wijnands, Nico Baars, and George Papafotiou

Electromechanics and Power Electronics (EPE) group

Department of Electrical Engineering

Eindhoven University of Technology, the Netherlands

Email: k.pouresmaeil@tue.nl

Acknowledgments

This publication is part of the project NEON (with project number 17628 of the research programme Crossover which is (partly) financed by the Dutch Research Council (NWO)).

Keywords

≪Charging infrastructure for EVs≫, ≪Grid-connected converter≫, ≪Modular Multilevel Converters (MMC)≫, ≪AC-DC ≫, ≪Transformer ≫.

Abstract

The non-ideal behavior of power switches and/or circuit asymmetries in transformer-isolated converters can result in nonzero average voltage across the transformer terminals, which, in turn, can saturate the transformer. In this paper, a volt-second balancing scheme is developed for a Modular-Multilevel-Converter (MMC)-fed AC/DC converter to avoid transformer saturation.

Introduction

Rising adoption of Electric Vehicles (EV) and introduction of EVs with large battery capacities, such as electric trucks, demand Medium-Voltage (MV) connected ultra-fast charging stations [1, 2]. Here, a transformer is needed to provide the galvanic isolation and step down the MV grid voltage to a level that is suitable for EV battery charging. Typically, an MV-connected 50 Hz transformer at the required power level has a large volume, which can be reduced considerably by using a transformer that operates in the Medium-Frequency (MF) range. In this regard, MMC-based chargers are getting more popular [3]. Fig. 1 shows a typical structure of an MMC-based ultra-fast charger, where the MMC, followed by an MF transformer and a zero-voltage switched AC/DC converter, converts the grid voltage to an MF voltage wave [4].

The MMC-based charger comprises a transformer-isolated converter. In such a converter, preventing DC bias in the transformer magnetizing flux is critical, as it can take the transformer core outside its linear operation region. This bias can result from any mismatch in the applied volt-seconds to the transformer primary and secondary windings [5, 6]. Since the primary winding average current is often closed-loop controlled by the MMC, the applied voltage to the primary winding is free of a DC component [3]. On the other hand, the secondary winding is likely to be excited with a DC biased voltage through the Low-Voltage AC/DC converter (LVC). As a remedy, the average current of the secondary side can also be controlled to avoid DC bias, but this demands additional sensing circuity and control complexity. Passive methods are also used to avoid transformer saturation, such as the introduction of a DC-blocking capacitor, inclusion of an air gap in the transformer core, and overdimensioning the flux density, which

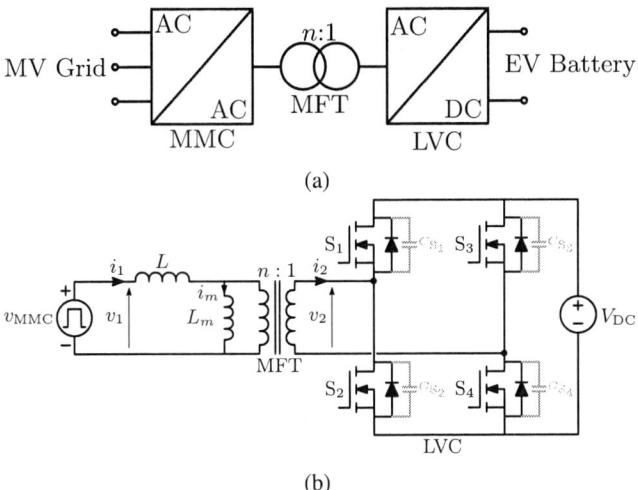

(a)

(b)

Fig. 1: (a) Block diagram of an MMC-based ultra-fast charger, (b) AC/DC converter with equivalent circuit model of the MMC and the MFT.

result in additional volume and/or power losses [5]. In [7], it is shown that a zero-voltage switched active bridge can compensate a small part of the net volt-seconds it undesirably applies to the transformer. The resulting DC component in the bridge current positively modifies the voltage waveform of the bridge within the switching transitions. This compensating effect becomes more pronounced when decreasing the switching current. This effect was exploited in [8] in a modified modulation strategy for a dual-active bridge (DAB) to restrict the switching current and enhance the inherent volt-second balancing effect of the bridge. This modulation strategy uses inner phase shift angles within each H-bridge in addition to the outer phase shift angle between primary and secondary bridges, thereby bringing about additional control effort.

In the MMC-fed LVC, the switching current of power devices can be easily modified by adjusting the reactive current drawn from the MMC [4, 9]. Here, the MMC acts like a controlled voltage source, which can be used to control the active and reactive terms of its terminal current separately. The objective of this paper is to employ the reactive term to control the switching current of the LVC, thereby minimizing the steady state volt-second imbalance. An analytical relationship between the transformer reactive current and resulting offset current through the transformer is derived, and the analysis is verified with simulations.

Volt-second balancing

The inherent volt-second balancing of the LVC can be analytically quantified. This in turn allows it to be enhanced by changing the available free parameters.

Inherent volt-second balancing

Fig. 2 shows the voltage and current waveforms of the LVC and magnetizing current of the transformer under both normal operation and in the presence of an offset current due to a timing error t_{err} in the switching cycle. During normal operation, the LVC switches with a constant duty cycle of 50% in a ZVS manner. In a zero-voltage switched bridge, the transitions of the AC-terminal voltage from $+V_{DC}$ to $-V_{DC}$ (or vice versa) are not instantaneous, and the transition time is determined by the commutation process of the power switches [8], which is governed by the charging and discharging of the output capacitances of the power switches in the bridge legs. If the switching current I_{sw} remains relatively constant during the switching transition, the commutation time is given to a good approximation by

$$t_{cmt} = \frac{CV_{DC}}{I_{sw}},$$ (1)

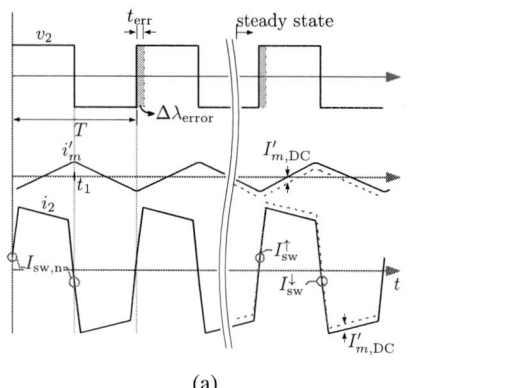

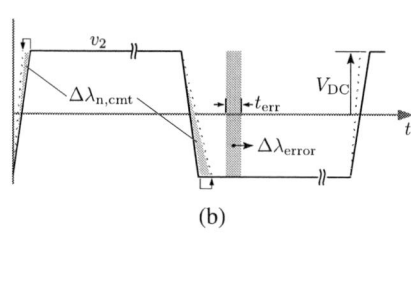

(b)

(a)

Fig. 2: (a) Nominal and DC-biased waveforms of the converter, where the solid lines and red-dotted lines indicate normal operation and operation under a DC bias, respectively, (b) the enlarged secondary voltage during switching transitions.

where C denotes the switch-node capacitance, the equivalent capacitance of the parallel connection of high- and low-side switch output capacitances (see Fig. 1b). As shown in Fig. 2b, during normal operation, the switching currents and therefore commutation times are the same for both rising and falling edges of the secondary voltage $v_2(t)$. Hence, given the half-cycle symmetry, the net volt-second area is zero in each switching cycle (see Fig. 2b). In case of a timing error t_{err} in the LVC due to the deviation of duty cycle from 50%, $v_2(t)$ will have an undesired net volt-second component $\Delta\lambda_{\mathrm{error}}$ equal to $2t_{\mathrm{err}}V_{\mathrm{DC}}$. Subsequently, a finite DC current component will flow through the LVC and magnetizing path of the transformer, providing that there is a nonzero resistance in the DC current path. As can be seen from Fig. 2b, in this case, the switching currents are no longer the same for the rising and falling edges of $v_2(t)$ and–providing the fact that the LVC still switches in a ZVS manner–they are, respectively,

$$
\begin{cases}
I_{\mathrm{sw}}^{\uparrow} = I_{\mathrm{sw,n}} - I'_{m,\mathrm{DC}} \\
I_{\mathrm{sw}}^{\downarrow} = I_{\mathrm{sw,n}} + I'_{m,\mathrm{DC}}
\end{cases},
\tag{2}
$$

where $I_{\mathrm{sw,n}}$ and $I'_{m,\mathrm{DC}}$ represent the switching current during normal operation and the secondary referred DC component of magnetizing current under DC-biased operation, respectively. Hereafter in this paper, the prime notation indicates secondary-referred primary quantities. According to (2), the commutation times will differ, or equivalently the switching transitions will be faster in one edge compared to the other edge of $v_2(t)$. As can be seen from Fig. 2b, this will create a net volt-second contribution $\Delta\lambda_{\mathrm{n,cmt}}$ (blue-shaded area) partially compensating the original undesired net volt-second component $\Delta\lambda_{\mathrm{error}}$ (red-shaded area). Using (1), the compensating net volt-second component is found to be

$$
\Delta\lambda_{\mathrm{n,cmt}} = \left(\frac{C V_{\mathrm{DC}}^2}{I_{\mathrm{sw,n}}} - \frac{C V_{\mathrm{DC}}^2}{I_{\mathrm{sw}}^{\uparrow}} \right) - \left(\frac{C V_{\mathrm{DC}}^2}{I_{\mathrm{sw,n}}} - \frac{C V_{\mathrm{DC}}^2}{I_{\mathrm{sw}}^{\downarrow}} \right).
\tag{3}
$$

Substitution of (2) into (3) yields

$$
\Delta\lambda_{\mathrm{n,cmt}} = -\frac{2 I'_{m,\mathrm{DC}} C V_{\mathrm{DC}}^2}{I_{\mathrm{sw,n}}^2 - I'_{m,\mathrm{DC}}}.
\tag{4}
$$

The other counteracting effect against the imbalanced net volt-second is the DC voltage drop over the resistance of the DC current path which is

$$
R_{\mathrm{DC}} = 2 R_{\mathrm{DS-on}} + R_{\mathrm{s,w}} + R_b,
\tag{5}
$$

where R_{DS-on}, $R_{s,w}$, and R_b represent the drain-source on resistance of the MOSFETs in the LVC, the secondary winding resistance, and the internal resistance of the EV battery pack, respectively. This resistance compensates for the volt-second imbalance by

$$\Delta\lambda_{n,R} = -V_R T$$
$$= -R_{DC} I'_{m,DC} T, \tag{6}$$

with T the switching period. Considering the (typically) low value of R_{DC}, the resulting offset current is so large that the transformer core can be driven into saturation.

At steady state, the volt-second balance of the circuit can be expressed as

$$\Delta\lambda_{error} + \Delta\lambda_{n,R} + \Delta\lambda_{n,cmt} = 0. \tag{7}$$

Substitution of (4) and (6) into (7) gives

$$2 t_{err} V_{DC} = -R_{DC} I'_{m,DC} T - \frac{2 I'_{m,DC} C V_{DC}^2}{I_{sw,n}^2 - I_{m,DC}'^2}. \tag{8}$$

Then, the offset current can be determined as a function of the normal switching current $I_{sw,n}$ and circuit parameters. Given (8), the compensating effect becomes more pronounced when decreasing the switching current during normal operation.

Switching current control

To ensure ZVS transitions in the LVC, the minimum commutation current of switching devices should be provided. The commutation current is equal to the sum of the inductive reactive currents given to the converter by the MMC and the magnetizing inductance of the transformer. Fig. 3a shows the simplified representation of the MMC-fed AC/DC converter, where the MMC is modeled as a controlled voltage source and linked to the LVC by the inductor L. Here, L represents the sum of the MFT's leakage inductance and arm inductance of the MMC. The terminal current of the MMC comprises an active term $i_d(t)$ and a reactive term $i_q(t)$ (see Fig. 3b). These terms can be controlled through v_{MMC}^{ref}, denoting the reference of the MMC terminal voltage. As a result, the reactive current of the MMC side $i_q(t)$ is adjustable [4, 9]. Apart from the MMC, the magnetizing inductance of the MFT contributes to the required commutation current in the LVC. Considering a negligible MFT leakage inductance compared to the arm inductance of the MMC, as can be seen from Fig. 3b, the secondary referred magnetizing current will have a triangular waveform with peak amplitude equal to

$$\hat{I}'_m = i'_m(t_1) = \frac{V_{DC} T}{4 L'_m} \tag{9}$$

where L'_m represents the secondary-referred magnetizing inductance of the MFT. As shown in Fig. 3b, the switching current of the LVC during normal operation is equal to the total reactive current delivered

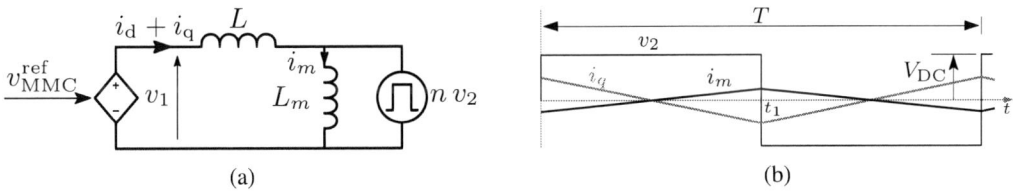

(a)   (b)

Fig. 3: (a) The simplified representation of the MMC-fed AC/DC converter, (b) the reactive currents through the LVC.

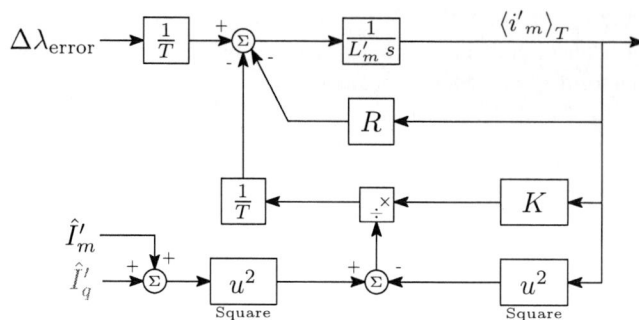

Fig. 4: Average dynamic representation of the LVC with control input $\hat{I}'_q$ for compensating the imbalance in the applied volt-seconds.

to the converter by the MMC and the MFT and given by

$$I_{\text{sw,n}} = |i'_q(t_1) - i'_m(t_1)| = \hat{I}'_q + \frac{V_{\text{DC}} T}{4 L'_m}. \tag{10}$$

The first term of (10) is adjustable by the MMC to change the normal switching current $I_{\text{sw,n}}$ of the LVC. Substitution of (10) into (8) yields

$$2 t_{\text{err}} V_{DC} = -R_{DC} I'_{m,\text{DC}} T - \frac{I'_{m,\text{DC}} \overbrace{2 C V_{\text{DC}}^2}^{K}}{(\hat{I}'_q + \frac{V_{\text{DC}} T}{4 L'_m})^2 - I'^{2}_{m,\text{DC}}}. \tag{11}$$

As a result, the normal switching current can be modified through controlling i_q to enhance inherent volt-second balancing of the LVC. This also can be represented by the average dynamic model of the system drawn in Fig. 4, where $\langle i'_m \rangle_T$ represents the average of the secondary referred magnetizing current over one switching period. As can be seen, both R_{DC} and $\Delta\lambda_{\text{n,cmt}}$ act as a negative feedback to limit the offset current.

Simulation results

To verify the theoretical analysis, simulations were conducted using PLECS with the circuit parameters given in Table I. Quite a large timing error $t_{\text{err}} = 50\,\text{ns}$ is considered in the LVC. In order to find the attenuation of the offset current with the proposed method, first, the offset current without inherent volt-second balancing is obtained by setting $\Delta\lambda_{\text{n,cmt}}$ to zero in (7,8), which yields an offset current equal to $\Delta\lambda_{\text{error}}/R_{\text{DC}} T$. Afterwards, using (8) the resulting offset current with the inherent volt-second balancing is derived versus the normal switching current. Subsequently, the attenuation of the offset current, with regard to $\Delta\lambda_{\text{error}}/R_{\text{DC}} T$, is plotted in Fig. 5. As can be seen, the higher the normal switching current, the

Table I: Circuit parameters

Description	Parameter	Value
Power rating	P	1 MW
Output DC voltage	V_{DC}	800 V
Operating frequency	f	1 kHz
Switch-node capacitance	C	13.2 nF
Dead time	t_d	500 ns
Magnetizing inductance	L_m	200 mH
Path resistance	R_{DC}	5.5 mΩ
Transformer turn ratio	$n:1$	10 : 1

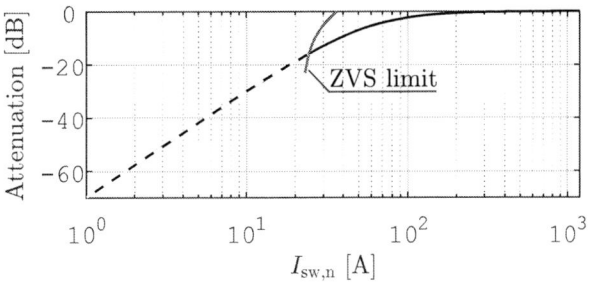

Fig. 5: The attenuation of the offset current versus the normal switching current.

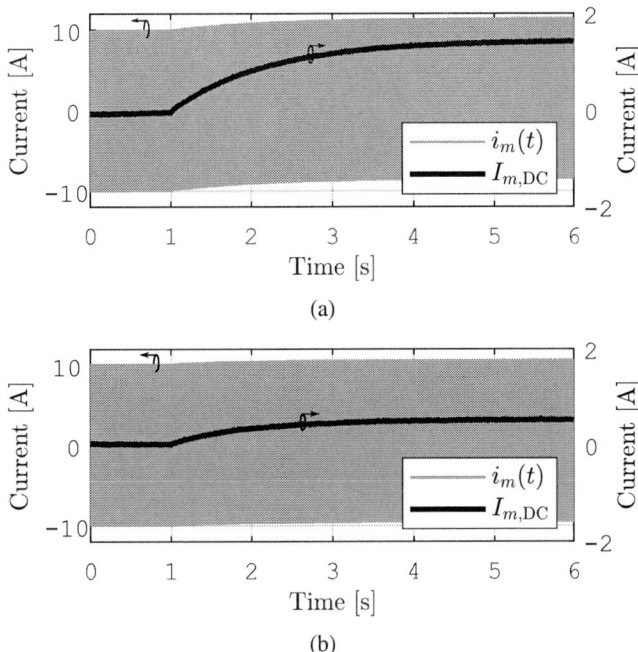

Fig. 6: Simulated magnetizing current i_m and its DC component $I_{m,\text{DC}}$ for (a) phase shift control and (b) the proposed method.

smaller the balancing effect of the bridge. Especially at elevated switching currents (which is a typical situation for phase shift control [4]), the inherent volt-second balancing effect of the bridge is negligible. In addition, the minimum desired offset current is determined by ZVS limit of power switches. Although this limit can be relaxed by increasing the dead time in the legs of the LVC, a very large dead time can bring about performance deterioration of the converter such as the duty cycle loss.

Fig. 6 depicts simulation results of the instantaneous transformer magnetizing current as well as its DC component, both with and without reactive current control. By means of the reactive current control, the switching current is set to be $I_{\text{sw,n}} = 40$ A, which is around two times the minimum required commutation current of power devices in the LVC. Initially, the LVC works ideally without a timing mismatch, so that there is no DC current through the transformer. At $t = 1$ s the timing error $t_{\text{err}} = 50$ ns is applied to the LVC. As can be seen from Fig. 6a, for large switching currents (phase shift control), the imbalance causes a large DC component in the transformer magnetizing current, which can take the transformer core into saturation. On the other hand, when the switching current during normal operation is limited by means of reactive current control, the resulting DC component of the magnetizing current is decreased by a factor of about 3, as shown in Fig. 6b. The further decrease in the offset current is achievable by reducing the normal switching current as shown in Fig. 5.

Conclusion

A volt-second balancing method is presented for an AC-DC converter interconnected to an (AC-AC) MMC by means of an MFT. This method utilizes the reactive current control capability of the MMC-connected AC/DC converter to boost the inherent volt-second balancing of the zero-voltage switched bridge and therefore mitigate the offset current through the transformer. The inherent volt-second balancing of the LVC is analytically quantified, which yields analytical relationship between the transformer reactive current, the parameters of the converter circuit, and the resulting offset current. In simulations, a large timing error is imposed on the LVC, and the resulting offset current is significantly reduced by the proposed method.

References

[1] M. Safayatullah, M. T. Elrais, S. Ghosh, R. Rezaii and I. Batarseh, "A Comprehensive Review of Power Converter Topologies and Control Methods for Electric Vehicle Fast Charging Applications," in IEEE Access, vol. 10, pp. 40753-40793, 2022

[2] N. S. Patil and A. Shukla, "Review and Comparison of MV grid-connected Extreme Fast Charging Converters for Electric Vehicles," 2021 National Power Electronics Conference (NPEC), 2021, pp. 1-6

[3] Y. P. Marca, M. G. L. Roes, J. L. Duarte and K. G. E. Wijnands: Isolated MMC-based ac/ac stage for ultrafast chargers, 2021 IEEE 30th International Symposium on Industrial Electronics (ISIE), 2021, pp. 01-08

[4] K. Pouresmaeil, J. L. Duarte, K. G. E. Wijnands, M. G. L. Roes, and N. H. Baars: Single-Phase Bidirectional ZVZCS AC-DC Converter for MV-Connected Ultra-Fast Chargers. In PCIM Europe 2022: International Exhibition and Conference for Power Electronics, Intelligent Motion, Renewable Energy and Energy Management Proceedings, 10 – 12 May 2022, Nuremberg, pp. 124-130

[5] G. Ortiz, L. Fässler, J. W. Kolar and O. Apeldoorn: Flux Balancing of Isolation Transformers and Application of "The Magnetic Ear" for Closed-Loop Volt–Second Compensation, in IEEE Transactions on Power Electronics, vol. 29, no. 8, pp. 4078-4090, Aug. 2014

[6] J. Voss, R. Mencher, P. Joebges, J. Mathé and R. W. D. Doncker, "Controller-in-the-Loop of a Transformer Saturation Control for High-Power Three-Phase Dual-Active Bridge DC-DC Converters," 2021 IEEE Energy Conversion Congress and Exposition (ECCE), 2021, pp. 1798-1804

[7] D. Costinett, D. Seltzer, D. Maksimovic and R. Zane: Inherent volt-second balancing of magnetic devices in zero-voltage switched power converters, 2013 Twenty-Eighth Annual IEEE Applied Power Electronics Conference and Exposition (APEC), 2013, pp. 9-15

[8] C. Winter, J. Riedel, Z. Mohzani, R. Mencher and S. Butzmann, "Enhancing Inherent Flux Balancing in a Dual-Active Bridge Using Adaptive Modulation," 2019 IEEE Applied Power Electronics Conference and Exposition (APEC), 2019, pp. 2202-2209

[9] S. Cui, J. Hu and R. W. De Doncker: Control and Experiment of a TLC-MMC Hybrid DC–DC Converter for the Interconnection of MVDC and HVDC Grids, in IEEE Transactions on Power Electronics, vol. 35, no. 3, pp. 2353-2362, March 2020

Cost Comparison for Different PV-Battery System Architectures Including Power Converter Reliability

Martijn Deckers[1],[*],[***], Leander Van Cappellen[*],[***], Glenn Emmers[*],[**], Fereshteh Poormohammadi[*],[**], and Johan Driesen[*],[**]

[*]EnergyVille - Thor Park 8301, 3600 Genk, Belgium
[**]ESAT/ELECTA KU Leuven - Kasteelpark Arenberg 10, 3001 Heverlee, Belgium
[***]IMO-IMOMEC Hasselt University - Wetenschapspark 1, 3590 Diepenbeek, Belgium
[1]martijn.deckers@kuleuven.be

Acknowledgments

This work has been supported by Flanders Innovation & Entrepreneurship and Flux50 under project DAPPER, HBC.2020.2144. Martijn Deckers is funded by a PhD grant of the Research Foundation Flanders (FWO), 1S87522N.

Keywords

≪Levelized cost of energy≫, ≪Renewable energy systems≫, ≪Reliability≫, ≪Aging≫, ≪Batteries≫.

Abstract

This paper compares the levelized energy cost of a commercial DC-coupled photovoltaic battery systems with a multiple input multiple output converter. The comparison is based on an electrothermal simulation allowing to include the actual converter efficiency and degradation in different use cases. The multiple input multiple output converter proofs to be less expensive and more reliable, however, the lower efficiency causes the final levelized energy cost to be higher.

Introduction

Currently, two categories of residential photovoltaic (PV) battery systems are commercially available, AC- and DC-coupled systems. The AC-coupled systems are mostly applied in a retrofit installation when the addition of a battery is wanted to an already operational PV system. This corresponds to the system in Fig. 1 a), the PV panels are connected to the AC grid using a DC/DC converter followed by a DC/AC converter and the battery has its own DC/DC and DC/AC converter in parallel, both of which need to be bidirectional to allow for either charging and discharging of the battery. The DC-coupled systems allows to eliminate one of the DC/AC converters by connecting the DC/DC converters to a common DC bus as shown in Fig. 1 b). Next to the use of less components, generally the efficiency improves as two conversion steps are eliminated when excess PV power is transferred to the battery [1]. Next to these two commercial system architectures, also a third topology is emerging. Here the number of converter modules is reduced further by combining two DC/DC converters into one Multiple Input Multiple Output (MIMO) converter as shown in Fig. 1 c). This decreases the number of components and conversion steps even further [2]. A good measure to compare these different systems architectures is the Levelized Cost Of Energy (LCOE) which reflects the cost of a kWh of energy consumed taking into account all the expenses and gains during the lifetime of the system [3].

The paper is organised as follows. First, the gaps in current literature are identified followed by the objectives of this work. Secondly, the method used to calculate the LCOE is elaborated starting from

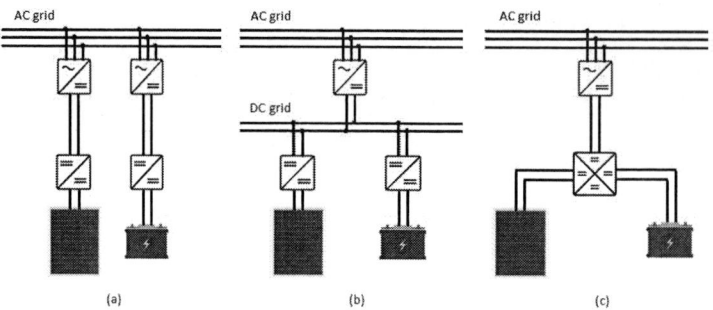

Fig. 1: System structure of (a) an AC coupled system, (b) a DC-coupled system and (c) a MIMO converter.

the main LCOE formula after which the individual terms such as cost, efficiency and reliability are discussed. Thirdly, the used converter architectures and their parameters are given. Fourthly, the results are presented after which the paper is summarised in the final section.

Scope of the paper

Reliability comparisons between commercial DC- and AC-coupled topologies can be found in literature [4]. To the authors knowledge these analyses do not yet exist for the newer MIMO topologies. These new systems could have great potential but more work is needed to quantify this. Also the LCOE analysis of a standard PV battery systems can be found in literature. To determine the power flow based on given mission profiles, mostly ideal battery dispatching is used which will also be done in this paper. However, frequently some additional simplifications are made which might have a big impact on the final results [3, 5, 6]. Summarised the following gaps are identified in the existing research:

- Currently no LCOE evaluation is available for MIMO converters in PV-battery systems;
- The efficiency is treated as a fixed value instead of being a function of the power flow;
- The component degradation is not taken into account or a fixed lifetime is assumed.

This paper aims to compare the LCOE of a commercial DC-coupled system with a MIMO architecture as presented in Fig. 1 b) and c). This LCOE will take into account the system construction costs, the operational efficiency and component degradation, potentially causing the need for system maintenance or replacement. The differences in LCOE between the two configurations will be highlighted together with the main causes for these discrepancies. A sensitivity analysis towards the cost, efficiency, lifetime and converter specifications will take into account modelling uncertainties and show how future changes might influence the result.

Levelized cost of energy (LCOE) calculation method

This section describes the method used for the LCOE calculation. After elaborating the base formula, also the individual terms such as manufacturing cost, reliability and efficiency will be discussed.

Base formula

The LCOE evaluates the costs and revenues generated with the setup during its operational lifetime, taking into account a certain discount rate r. Equation 1 gives the main expression for the LCOE, it is calculated in a similar way as shown in [7] with a few changes to allow the inclusion of variable efficiencies and degradation.

$$LCOE = \frac{C_{PV} + C_{inverter} + C_{E\ purchased} - C_{E\ sold}}{E_{load}} \tag{1}$$

The cost of the inverter system $C_{inverter}$ is the main scope of the paper. The different architectures use different components, leading to a different manufacturing cost $C_{inverter\ manufacturing}$ and the degradation during its lifetime may differ, leading to discrepancies in operation and maintenance cost $C_{inverter\ o\&m}$ as given in equation 2. The cost of the purchased energy $C_{E\ purchased}$ and the sold energy $C_{E\ sold}$ can be set depending on different scenarios. The battery is always dispatched in the most economic way satisfying the required load energy E_{load}. The cost of the PV panels C_{PV} will not be the focus of this paper as it is the same for all the different system architectures.

$$C_{inverter} = C_{inverter\ manufacturing} + \sum_{n=0}^{N} \frac{C_{inverter\ o\&m}}{(1+r)^n} \tag{2}$$

$$C_E = \sum_{n=0}^{N} \frac{C_{E\ n}}{(1+r)^n} \tag{3}$$

An important aspect in this paper is that the conversion efficiencies $\eta_{storage->out}$, $\eta_{PV->storage}$ and $\eta_{PV->out}$ will not be taken constant but instead are a function of the actual energy-flows based on the electrothermal simulation.

$$E_{load} = E_{out} + E_{purchased} - E_{sold} \tag{4}$$

$$E_{out} = \sum_{n=0}^{N} (E_{out\ load} + E_{out\ sold}) = \sum_{n=0}^{N} (\eta_{PV->out}\ E_{PV\ out} + \eta_{storage->out}\ E_{storage\ out}) \tag{5}$$

$$\sum_{n=0}^{N} E_{PV} = \sum_{n=0}^{N} (E_{PV\ out} + E_{PV\ storage}) \tag{6}$$

$$\sum_{n=0}^{N} E_{storage\ in} = \sum_{n=0}^{N} (\eta_{PV->storage}\ E_{PV\ storage}) \tag{7}$$

Summarised, LCOE differences between system architectures can be caused by: 1) Differences in initial investment due to the dependence on the number of components and needed component ratings. 2) The power flow dependent conversion efficiencies between PV panels, battery and output energy which are dependent on the number of conversion steps. 3) Differences in system reliability which are dependent on the component ratings and the component stress.

Converter manufacturing cost

The manufacturing costs $C_{DCcoupled}$ and C_{MIMO} of the two converter architectures need to be estimated. For this a base cost C_{base} is calculated, taking the switching devices, diodes, inductors, heat sinks and cooling fans costs $C_{Component\ n}$ into account. The used cost values are based on the single unit price of a component n with the correct rating. In Fig. 2 an example is shown from the search for heat sinks and their price in function of the thermal resistance. Other components such as capacitors, PCB, casing, etc. are excepted to be similar for both architectures and can be modelled as an additional cost $C_{additional}$ which is the same for both architectures. At this point, additional parameters such as labour, transportation, engineering, etc. are not taken into account. To acquire a realistic price estimate for both systems, the current market price C_{market} for a DC-coupled PV systems with similar ratings is chosen as the cost $C_{DCcoupled}$ in this analysis. The cost of the MIMO converter C_{MIMO} is then calculated based on C_{market} and the relative difference between the estimated base costs C_{base} of the two converters ΔC keeping the ratio constant.

$$C_{base} = \sum_{n=0}^{N} C_{Component\ n} + C_{additional} \tag{8}$$

$$\Delta C[-] = \frac{C_{base\ MIMO}}{C_{base\ DCcoupled}} - 1 \tag{9}$$

$$C_{DCcoupled} = C_{market} \tag{10}$$

$$C_{MIMO} = C_{market}(1 + \Delta C) \tag{11}$$

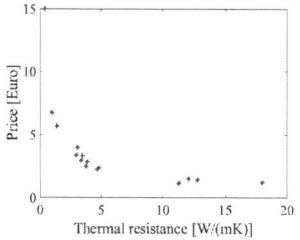

Fig. 2: The cost of a heat sink in function of the required thermal resistance.

Converter efficiency and degradation modelling

Electrothermal simulation

The electrothermal simulation is done based on an electrical and a thermal model. For the electrical model, a set of equations is written based on the schematics in Fig. 4, relating the voltages and currents. For the MIMO converter multiple sets of equations are used for each operational condition. Different sets of equations are used for both the continuous and discontinuous operation modes of the converters. The thermal model uses a lumped network of thermal resistances which can be found in the component data-sheets. Thermal capacitances will not be considered, meaning that the system is always in steady state. The coupling between thermal and electrical simulation is done using a temperature dependent drain to source resistance $R_{DS\ on}$. An iteration between the electrical and thermal model results in a lookup table relating the converter power to the component die temperatures and losses. This allows to convert the mission profile into temperature profiles and to calculate the efficiency.

Degradation modelling

For the degradation modelling, the main focus is placed on the switching devices. The failure rates of the capacitors improved drastically with the use of film capacitors. When the switch is operated within specifications, package failures caused by thermal stress contribute most to the degradation. The main failure modes are bondwire degradation and die solder layer delamination. To calculate the amount of damage a certain temperature cycle invokes, an empirical model is used. The number of cycles is counted using the rainflow counting algorithm. When the failure criteria are met, components need to be replaced leading to increased operation and management costs in the LCOE [4, 8, 9]. Here the assumption is made that when a component fails, the entire power electronic system is replaced. A frequently used thermal degradation model is the Bayerer model given in Equation 12. Here the number of cycles in a lifetime N_f is linked to the junction temperature swing ΔT_j, the minimum junction temperature $T_{j\ min}$ and the heating time t_{on}. Additionally, the bondwire configuration is taken into account by including the bond wire current I and the thickness of the bondwire D. The device voltage class V estimates the device thickness [10, 11]. Often a set of default fitting parameters is used called the CIP08 model [13] which is used in this paper as well. This model was developed for standard Si power modules while the components used in this paper are SiC components. Even when the failure modes are very similar, care should be taken when this lifetime model is applied to different technologies as proven in [12]. Because of this, the lifetime results will be normalised for comparison and a base lifetime of 15 year will be used in the LCOE calculations. This means the DC-coupled system gets a default lifetime of 15 years and the lifetime model will determine the difference that needs to be added or subtracted to get the MIMO converter lifetime.

$$N_f = A\ (\Delta T_j)^{\beta_1}\ e^{\frac{\beta_2}{T_{jmin}+273}}\ t_{on}^{\beta_3}\ I^{\beta_4}\ V^{\beta_5}\ D^{\beta_6} \tag{12}$$

The total workflow is shown in Fig. 3. In step one the temperature profile of each component is obtained based on the lookup tables and the power flows determined by the converter control. Applying rainflow counting to this temperature profile and using the reliability model, the expected time of failure is obtained in step two. A Monte Carlo analysis can be used to take into account uncertainties and translate this one point in time into a failure probability distribution. In step three the integral is taken to get the unreliability function. This function can easily be converted to the reliability function. Integrated, this gives the Mean Time To Failure (MTTF) which is used to estimate the maintenance costs.

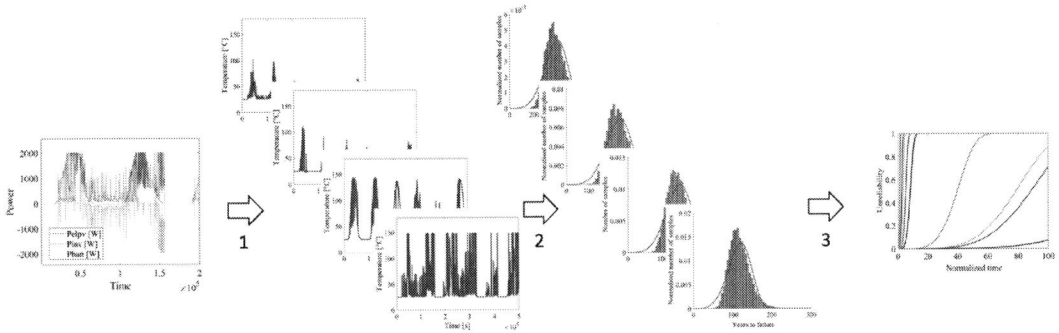

Fig. 3: Workflow to calculate the unreliability of the different switching devices based on a mission profile.

Converter design, specifications and control

In this section the converter system architectures are described in more detail. The used specifications and control methodology will be given as well.

Converter topologies

In Fig. 4 the converter schematics used in the simulation are given. The circuits are selected to be the most basic topologies achieving the required functionality such as the ability to have a bidirectional power flow to the battery. The PV converter is a basic boost converter and the battery converter is a synchronous boost converter [4]. The MIMO converter is synthesised from the combination of boost converters to achieve the same capabilities making this the simplest MIMO converter that can achieve the required functionality [14].

Converter specifications and control

The converter specifications used are summarised in the Table I. The power limits and voltages are the same for both the DC-coupled and MIMO converter. The inductances are converter specific as the DC-coupled requires two inductors while the MIMO converter only has one. The inductors values are chosen to ensure a similar current ripple for both the DC-coupled and MIMO converter.

In this work a policy case with different injection and consumption tariffs is implemented. As a result, the used control aims to maximise self consumption. This means that the produced PV power is used to satisfy the load and excess energy is transferred to the battery. When PV production is smaller than the load, energy from the battery is used to supply the difference. Within this control scheme, the limits of the different power converters in the system are also taken into account, possibly leading to curtailment

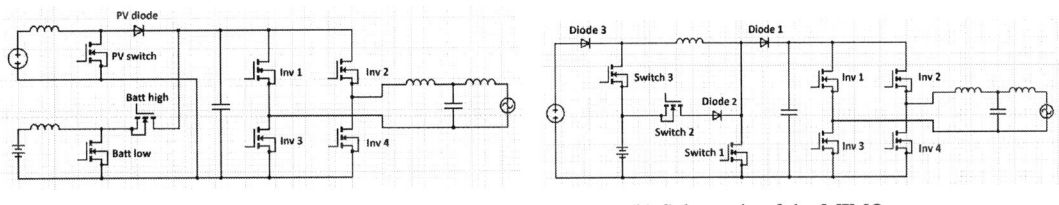

(a) Schematic of the DC-coupled converter   (b) Schematic of the MIMO converter

Fig. 4: Converter schematics used in the simulation.

of the PV by deviating from MPP and limiting the power transferred from or to the battery. The battery also has limited energy storage capacity. When the battery is full, excess PV energy will be sold to the grid and when it is empty, energy will be bought from the grid if needed. Fig. 5 shows a section form the load profile, electrical PV energy and the battery state of charge. The curtailment of the PV production can be seen as well as the maximum and minimum battery state of charge.

Table I: General converter specifications and converter components.

Parameter	Symbol	Value	Unit
General converter parameters			
Maximum PV power	Ppv_{max}	2000	W
Maximum PV voltage	Vpv_{max}	350	V
minimum PV voltage	Vpv_{min}	40	V
Battery voltage	$Vbatt$	350	V
Battery charging power	$Pbatt_{max}$	2000	W
Battery discharging power	$Pbatt_{min}$	-2000	W
Battery energy content	$Qbatt$	6000	Wh
Inverter maximum power	$Pinv_{max}$	2000	W
DC bus voltage	$Vbus$	450	V
Grid RMS voltage	$Vgrid$	230	Vrms
Grid frequency	$fgrid$	50	Hz
Switching frequency	fs	100000	HZ
DC-coupled converter parameters			
PV converter inductance	Lpv	0.0024	H
Battery converter inductance	$Lbatt$	0.0027	H
MIMO converter parameters			
MIMO inductance	L	0.0015	H
Used components			
SiC mosfet	IMW65R107M1H		
SiC diode	AIDW40S65C5		
PV panel	UL-305M-60		

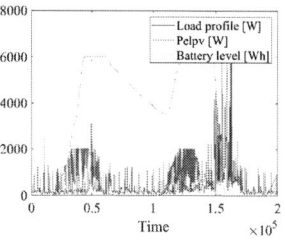

Fig. 5: Battery state of charge based on the load profile and the PV energy production.

In order to get a realistic power flow, real load and irradiance data are needed. For this work second resolution load data from HTW Berlin University of Applied Sciences [15] and second resolution irradiance data from the University of New South Wales [16] was used.

Results and discussion

In this section the simulation results are discussed. First the lifetime, converter manufacturing cost and efficiency are treated after which the final LCOE is calculated using these parameters. At the end a small sensitivity analysis on the LCOE will be conducted to examine the effects of potential modelling errors in the simulation.

The comparison of a DC-coupled and MIMO converter has been conducted for two main design cases. In the first case, the size of the heat sink is determined by the component that dissipated the most heat. Based on the maximum allowable die temperature the required heat sink was calculated and the same sink was then given to all the components. Because of this, the thermal cycling amplitudes are different for all components and the lifetime is mostly determined by the amount of dissipation rather than the number of cycles. In the second case, the required heat sink was calculated for each component independently, leading to smaller heat sink sizes for the non-restrictive components. This means that all components have similar thermal swings and the lifetime will be determined by the amount of cycles it undergoes.

Converter manufacturing cost

The manufacturing converter cost is evaluated for both the case with identical heat sinks and the case with optimised heat sinks. The used simulation values as well as the thermal resistances of the heat sinks in the optimised case can be found in Table II. The cost of the heat sinks are determined based on Fig 2. The diode and switch costs are equal for all devices because of the very similar required maximum ratings.

The MIMO converter appears to be the least expensive option as can be seen in Fig. 6. The cost of the DC-coupled converter with optimised sinks is set to 1. The main contributor to the cost is the inductor of which the MIMO only needs one. The MIMO converter has more components and a bit more losses leading to larger heat sinks expenses. Notwithstanding, the overall analyses remains in the advantage of the MIMO converter.

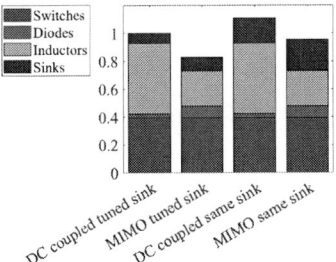

Fig. 6: Normalised cost of the DC-coupled and MIMO converter.

Table II: Optimised heat sink values and cost estimation parameters.

DC-coupled		MIMO		Simulation parameters	
Component	Thermal resistance	Component	Thermal resistance	Parameter	Value
Batt low	36.8000 W/(mK)	Switch 1	3.2600 W/(mK)	C_{base}	800 Euro
Batt high	20.5800 W/(mK)	Switch 2	7.9900 W/(mK)	$C_{E\ offtake}$	20 c/kWh
PV switch	3.2500 W/(mK)	Switch 3	20.9000 W/(mK)	$C_{E\ injection}$	-16 c/kWh
PV diode	23.8400 W/(mK)	Diode 1	23.8600 W/(mK)	r	0.05 [-]
Inv 1-4	23.1900 W/(mK)	Diode 2	18.4000 W/(mK)	C_{switch}	10 Euro
		Diode 3	6.9200 W/(mK)	C_{diode}	5 Euro
		Inv 1-4	23.1900 W/(mK)		

Lifetime

The resulting unreliability curves from the identical sink case can be seen in Fig. 7a and 7b. The MTTF of all the components are also summarised in Table III. The MTTF of the DC-coupled topology is normalised to 1. From this it can be seen that the MIMO topology is most reliable. The main restricting components are PV switch and switch 1. PV switch handles the PV power flow while switch 1 handles both the PV power flow and battery discharging. The reason for this higher reliability of switch 1 is the smaller boost ratios between the PV and battery voltage while in the DC-coupled converter everything has to be boosted to the DC bus voltage. The same analysis is also done with an infinite battery size. The reliability of the inverter components decreases as well as those involved in charging and discharging the battery.

Fig. 7c and 7d give the reliability curves for the case with optimised sinks. Now the lifetime will no longer be determined by the temperature swing amplitudes but rather by the number of cycles. The inverter switches are the most limiting for the lifetime which is to be expected as they are operational when either PV energy goes to the load or the battery discharges. The second most lifetime restricting components are batt low and switch 3. This shows that the battery discharging causes a lot af thermal cycles. Switch 3 is more reliable because it does not have to switch when the battery keeps discharging between switching periods lowering the amount of thermal cycles. PV switch and switch 1 now have very high MTTF because their low reliability was purely caused by the large thermal cycle amplitude compared to the other components.

The same lifetime model is applied to the diodes giving an idea of their contribution to the system failure. Because of the high losses caused by the forward voltage they are susceptible to large temperature swings and can be limiting for the lifetime as well. The model was not made for these types of components but the similar packaging allows to make a first estimate.

Table III: MTTF of DC-coupled and MIMO converter components while using identical and optimised heat sinks.

Same sink				Optimised sink			
DC-coupled		MIMO		DC-coupled		MIMO	
Component	MTTF	Component	MTTF	Component	MTTF	Component	MTTF
Tot	1	Tot	2.08	Tot	1	Tot	1.01
Batt low	455.20	Switch 1	2.08	Batt low	2.52	Switch 1	159.74
Batt high	94.62	Switch 2	32.26	Batt high	5.10	Switch 2	87.56
PV switch	1.00	Switch 3	119.71	PV switch	77.66	Switch 3	5.23
PV diode	436.03	Diode 1	142.39	PV diode	8.24	Diode 1	3.20
Inv 1-4	58.51	Diode 2	121.81	Inv 1-4	1.37	Diode 2	8.14
		Diode 3	10.84			Diode 3	39.18
		Inv 1-4	58.51			Inv 1-4	1.37

Efficiency

The efficiency difference is compared in Fig. 8. The MIMO converter experiences less inductor loss as well as less losses in the switches as a result of the reduced voltage ratios. The forward voltages of the diodes cause a large loss increase resulting in a slightly lower average efficiency for the MIMO converter.

LCOE

Finally the resulting LCOE for the optimised sink case is shown in Fig. 9. It is clear that the converter efficiency is very important because of the large amount of energy that is processed by the installation during its lifetime. The additional cost of installing the system on location is not taken into account so the real LCOE will be higher, however, the relative difference remains the same. Table IV shows the final results achieved. Although the MIMO converter has a lower manufacturing cost and higher reliability, the slightly lower efficiency leads to a higher LCOE. The influence of the inverter is excluded in the MTTF of the optimised sink case to focus the results on the differences in the system.

In Fig. 10 a sensitivity analysis is included on the impact of the relative difference in cost, efficiency and MTTF. The current estimate is indicated with a dashed line. The parameters of the DC-coupled system are kept constant while the parameters of the MIMO converter are varied. The influence of the MTTF shows discrete steps as a prolonged lifetime might make it possible to buy one unit less during the system lifetime. At the end the inverter cost saturates as only one unit is purchased in the beginning.

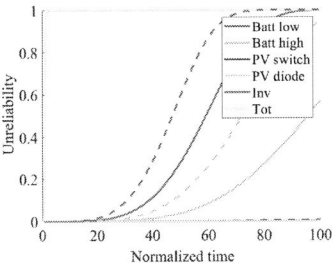

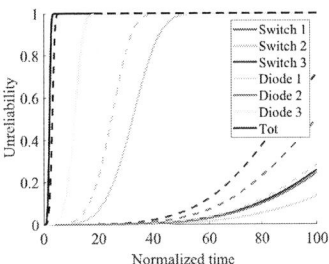

(a) Unreliability of the DC-coupled system with finite (full line) and infinite battery (dashed line). All the heat sinks have the same size.

(b) Unreliability of the MIMO system with finite (full line) and infinite battery (dashed line). All the heat sinks have the same size.

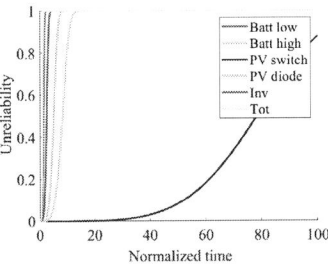

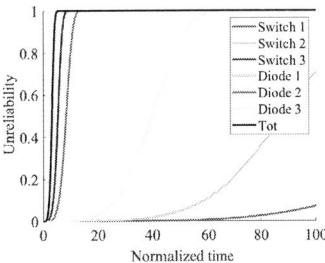

(c) Unreliability of the DC-coupled system where all the heat sinks are optimised to the same maximum temperature.

(d) Unreliability of the MIMO system where all the heat sinks are optimised to the same maximum temperature.

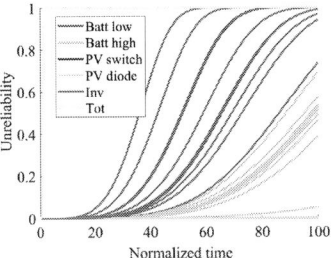

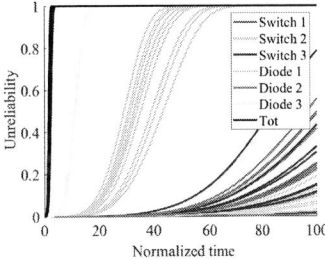

(e) Unreliability of the DC-coupled system for different load profiles. All the heat sinks have the same size.

(f) Unreliability of the MIMO system for different load profiles. All the heat sinks have the same size.

Fig. 7: Results of the lifetime study.

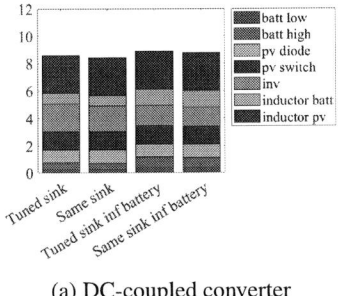

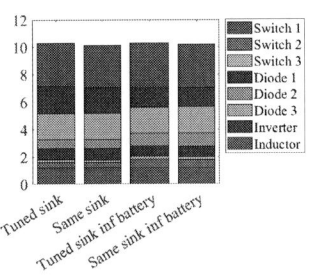

(a) DC-coupled converter

(b) MIMO converter

Fig. 8: Power loss [W] for optimised and identical heat sinks with either the rated or an infinite battery.

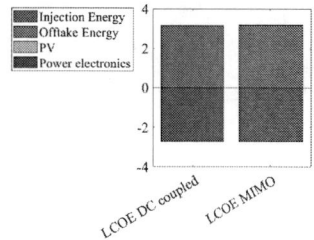

(a) Composition of the LCOE with both the contribution of energy injection and offtake.

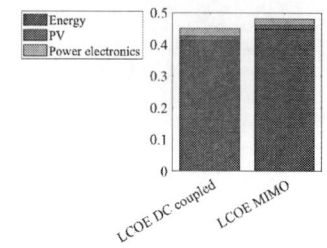

(b) Composition of the LCOE with net energy.

Fig. 9: LCOE [c/kWh] of the DC-coupled and MIMO converter with optimised heat sinks.

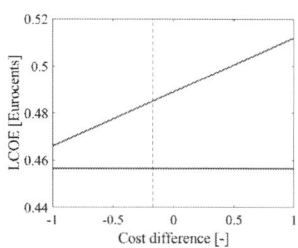

(a) Relative component cost difference.

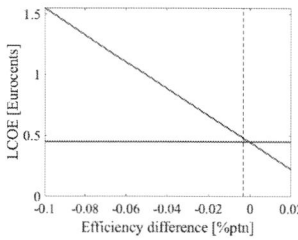

(b) Relative efficiency difference.

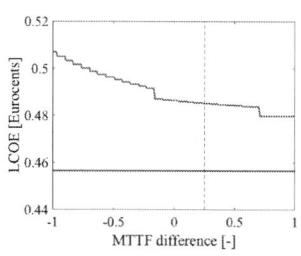

(c) Relative MTTF difference.

Fig. 10: Sensitivity analysis of the MIMO converter LCOE (red line), the DC-coupled converter values (blue line) are kept constant.

Table IV: Summary of the results from the LCOE analysis.

	Same sink		Optimised sink	
	DC-coupled	MIMO	DC-coupled	MIMO
Cinverter [-]	1	0.9542	1	0.8298
MTTF [-]	1.0030	2.0841	0.9981	1.2453
Efficiency [-]	0.9771	0.9738	0.9765	0.9734
LCOE [c/kWh]	0.4504	0.4778	0.4565	0.4851

Conclusion

This paper compared the LCOE of a DC-coupled and MIMO PV battery system in two different design cases. Next to the classical analysis taking into account the investment cost and fixed operational efficiency, this paper conducted a full electrothermal simulation of both systems. As result, a power flow dependent efficiency could be introduced together with mission profile dependent degradation. This influences the operation and maintenances expenses introducing new dependencies into the LCOE. The MIMO converter was shown to be more reliable and less expensive. However, the reduced efficiency of the MIMO converter due to the use of multiple diodes caused the resulting LCOE to be higher than the achieved results with the classical DC-coupled system. Future work will be devoted to improve the efficiency while keeping the reliability high and the manufacturing costs low. The sensitivity analysis showed that with only a slight improvement in efficiency, a big impact on the final LCOE can be made.

References

[1] S. Ravyts, M. D. Vecchia, G. V. den Broeck, and J. Driesen, "Review on building-integrated photovoltaics electrical system requirements and module-integrated converter recommendations," *Ener-*

gies, vol. 12, p. 1532, apr 2019.

[2] S. Danyali, S. H. Hosseini, and G. B. Gharehpetian, "New extendable single-stage multi-input dc-dc/ac boost converter," *IEEE Transactions on Power Electronics*, vol. 29, pp. 775–788, feb 2014.

[3] A. Pena-Bello, E. Barbour, M. Gonzalez, M. Patel, and D. Parra, "Optimized PV-coupled battery systems for combining applications: Impact of battery technology and geography," *Renewable and Sustainable Energy Reviews*, vol. 112, pp. 978–990, sep 2019.

[4] M. Sandelic, A. Sangwongwanich, and F. Blaabjerg, "Reliability evaluation of PV systems with integrated battery energy storage systems: DC-coupled and AC-coupled configurations," *Electronics*, vol. 8, p. 1059, sep 2019.

[5] L. Ayompe, A. Duffy, S. McCormack, and M. Conlon, "Projected costs of a grid-connected domestic pv system under different scenarios in ireland, using measured data from a trial installation," *Energy Policy*, vol. 38, pp. 3731–3743, jul 2010.

[6] F. L. Franco, A. Morandi, P. Raboni, and G. Grandi, "Efficiency comparison of dc and ac coupling solutions for large-scale pv bess power plants," *Energies*, vol. 14, p. 4823, aug 2021.

[7] Y. Khawaja, D. Giaouris, H. Patsios, and M. Dahidah, "Optimal cost-based model for sizing grid-connected PV and battery energy system," in *2017 IEEE Jordan Conference on Applied Electrical Engineering and Computing Technologies (AEECT)*, IEEE, oct 2017.

[8] W. Van De Sande, S. Ravyts, O. Alavi, P. Nivelle, J. Driesen, and M. Daenen, "The sensitivity of an electro-thermal photovoltaic dc–dc converter model to the temperature dependence of the electrical variables for reliability analyses," *Energies*, vol. 13, no. 11, 2020.

[9] W. V. D. Sande, S. Ravyts, A. Sangwongwanich, P. Manganiello, Y. Yang, F. Blaabjerg, J. Driesen, and M. Daenen, "A mission profile-based reliability analysis framework for photovoltaic DC-DC converters," *Microelectronics Reliability*, vol. 100-101, p. 113383, sep 2019.

[10] P. D. Reigosa, H. Wang, Y. Yang, and F. Blaabjerg, "Prediction of bond wire fatigue of igbts in a pv inverter under a long-term operation," *IEEE Transactions on Power Electronics*, vol. 31, no. 10, pp. 7171–7182, 2016.

[11] R. Bayerer, T. Herrmann, T. Licht, J. Lutz, and M. Feller, "Model for power cycling lifetime of igbt modules - various factors influencing lifetime," in *5th International Conference on Integrated Power Electronics Systems*, pp. 1–6, 2008.

[12] F. Hoffmann and N. Kaminski, "Power cycling capability and lifetime estimation of discrete silicon carbide power devices," *Materials Science Forum*, vol. 1004, p. 977, 07 2020.

[13] L. Ceccarelli, R. M. Kotecha, A. S. Bahman, F. Iannuzzo, and H. A. Mantooth, "Mission-profile-based lifetime prediction for a sic mosfet power module using a multi-step condition-mapping simulation strategy," *IEEE Transactions on Power Electronics*, vol. 34, no. 10, pp. 9698–9708, 2019.

[14] H. Wu, K. Sun, S. Ding, and Y. Xing, "Topology derivation of nonisolated three-port dc–dc converters from dic and doc," *IEEE Transactions on Power Electronics*, vol. 28, no. 7, pp. 3297–3307, 2013.

[15] T. Tjaden, J. Bergner, J. Weniger, and V. Quaschning, "Representative electrical load profiles of residential buildings in germany with a temporal resolution of one second," 12 2015.

[16] B. Zheng, J. Fletcher, A. Lennon, Y. Jiang, and P. Burr, "Solar irradiance measurements, with one second resolution, and estimated power ramp rate of photovoltaic modules with integrated capacitor energy storage. unsw, sydney.dataset.," 2019.

Insulation Design and Analysis of a Medium Voltage Planar PCB-based Power Bus Considering Interconnects and Ancillary Circuit Integration

Joshua Stewart, *Student Member, IEEE*, Rolando Burgos, *Senior Member, IEEE*, Dushan Boroyevich, *Life Fellow, IEEE*
Center for Power Electronics Systems (CPES) – Virginia Tech
Blacksburg, VA, USA
City, Country
E-Mail: joshuastewart@vt.edu

Acknowledgements

This research was funded in part by the U.S. Office of Naval Research (ONR) under Award Number N00014-16-1-2956. The information, data, or work presented herein was funded in part by the Advanced Research Projects Agency-Energy (ARPA-E), U.S. Department of Energy, under Award Number DE-AR0001727-1519. The views and opinions of authors expressed herein do not necessarily state or reflect those of the United States Government or any agency thereof.

Keywords

«Bus bar», «High power density systems», «Medium voltage converter», «Modular Multilevel Converters (MMC) », «Partial discharge»

Abstract

This paper presents a design method for a medium voltage (MV) PCB-based bus, focusing on interconnects and considerations for the integration of converter level ancillary circuits. A 6 kV power electronics building block (PEBB) is used as a case study to analyze the design of its PCB bus. Surface mounted balancing resistors and interconnects for power terminals are integrated to further increase the PEBB's power density. PCB-embedded structures, referred herein as shield pads, are introduced as a method to control the peak electric field (E-field) intensity in air near critical terminals and other devices. Additional features to relax the requirements for insulation design within the converter were also incorporated to fully leverage the design flexibility offered from a PCB bus. The final bus demonstrated a partial discharge inception voltage (PDIV) of 11.04 kV.

Introduction

The introduction of wide bandgap (WBG) power semiconductors devices, such as silicon carbide (SiC) and gallium nitride (GaN), have presented new opportunities in power electronic converter design. These WBG devices offer faster turn-on and turn-off times, higher blocking voltage, and operate at higher junction temperature allowing reductions in passive components and ease cooling requirements [1]. It is essential to consider power loop inductance to ensure components are not subject to an overvoltage as the device turns off. To achieve low inductance, it is common to place parallel conductors serving as a power bus in close proximity to achieve mutual inductance cancellation. Not considering apertures, inductance of a planar bus is directly proportional to conductor length and width and is inversely proportional to the conductor spacing [2]. However, as the conductor spacing is reduced, electric field intensity (E-Field) is increased. Low inductance planar busses have been designed to reduce impedance [2], [3] but did not address high voltage concerns. A medium voltage planar bus was designed for operation in a 6 kV converter but failed to meet partial discharge (PD) requirements [4]. Dielectric thickness is commonly selected considering average E-field intensity and with an added safety factor. However, finite element analysis (FEA) simulations performed in COMSOL Multiphysics indicate a peak E-field intensity more than an order of magnitude larger where the conductor edges interface the dielectric, also known as the triple point. A design method was introduced in [5] which uses geometric techniques to reduce the peak E-field intensity and optimize insulation thickness to avoid overdesign. In this work, a multilayer PCB is designed using geometric techniques for E-field control to optimize conductor clearances on a layer-by-layer basis to ensure insulation reliability while achieving a low inductance design for operation in a 6 kV 500 kW modular converter utilizing Wolfspeed 10 kV XHV-6 or XHV-9 MOSFET modules with the overall target of >10 kW/l power density. In addition to the bus insulation design, full system integration of the power cell will be considered as the larger purpose is to improve total power density by integrating all necessary power stage components into a highly modular power electronics building block (PEBB). Power levels can be customized by assembling individual PEBBs various series or parallel configurations to achieve higher voltage or current levels respectively, depending on end use requirements. This work is a continuation of [5] which eliminates bulky decoupling capacitors and considers integration of PEBB level ancillary circuitry such as the voltage sensor, temperature sensor, controller, and power supplies for these circuits [6]-[8].

In this paper, Section 1 describes the PEBB-level system architecture. Section 2 describes the method for electric field control used for the overall bus and terminals. The insulation design strategy for terminals, pads, and ancillary components is presented in Section 3. Finally, testing results are provided in Section 4.

1. Power Cell Architecture

Each PEBB includes all components necessary for the power cell to operate independently, or as part of a larger converter assembly. In a modular multilevel converter- (MMC) type assembly, PEBBs can be stacked in series to get a higher voltage. To achieve true modularity, each PEBB must be designed to meet the maximum voltage requirement while operating in the MMC. Each PEBB is cooled by a set of push-pull fans powered from an earth-grounded power supply. Since the heatsink is connected to the midpoint (MID) of each power cell, the maximum common mode voltage for the MMC assembly is the MMC bus voltage plus half of the cell voltage (since the heatsink is connected to MID). In this application, the target MMC bus voltage is 24 kV leading to an insulation system designed for 27 kV. The insulation system is divided into two primary zones: PEBB-level and converter-level (referring to the MMC). Components included in the converter-level insulation zone are the heatsink, fans and wireless auxiliary power converter (WPT). Other components have a maximum differential voltage of 6 kV and are included in the PEBB-level insulation design. All ancillary circuits are referenced to MID making this a local ground. The 6 kV 500 kW half-bridge PEBB is shown in Fig. 1 with key components

Fig. 1: 6 kV, 500 kW SiC MOSFET-based PEBB with key components indicated; 12 kW/l

indicated. A simplified circuit schematic is shown in Fig. 2. All ancillary circuits are powered from the WPT directly or indirectly. First, the dc-link capacitors are charged from a precharge circuit. Both of these circuits are located under the capacitors for improved power density. The precharge circuit is able to connect across the positive (+DC) and negative (-DC) dc-link capacitor terminals from below the bus. All other powered ancillary circuits are located above the bus which requires a power wire routed from below the bus to above it. Power from the WPT is then split and fed to a current-transform (CT)-based auxiliary power supply (GDPS) and a UPS. The gate drivers, voltage sensor, and temperature sensor are powered from the GDPS. Power to the controller and discharge circuit is passed through the UPS. Three parallel XHV-9 power modules are located below the bus and mounted to a single heatsink. Output from the modules connect to a differential mode (DM) inductor. PEBB power terminals connect to the -DC and the output of the DM inductor.

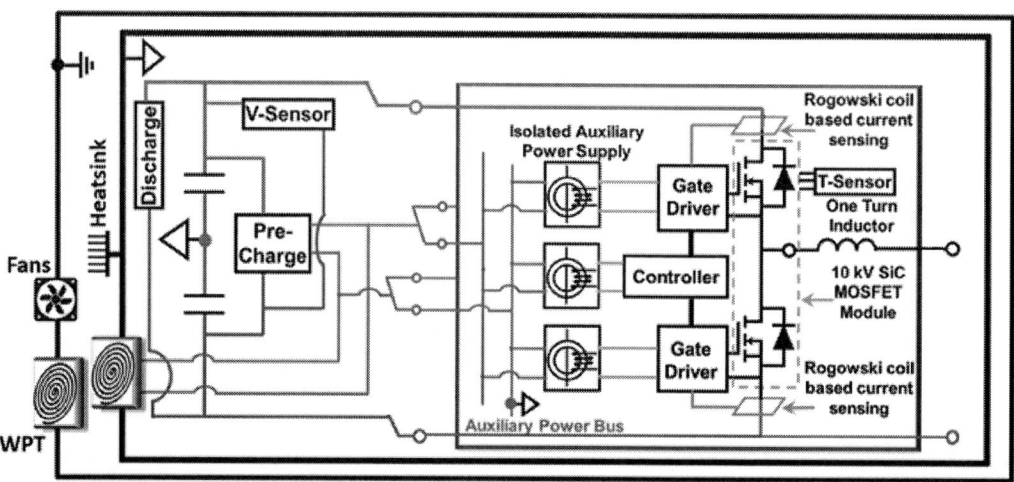

Fig. 2: Simplified schematic of full-bridge PEBB with voltages indicated w.r.t local ground. Dark blue) -3 kV V; Brown) 0 V; Red) +3 kV

2. High Voltage PCB Insulation Design

The design flow shown in Fig. 3 was implemented to optimize conductor clearance requirements near component terminals to ensure E-field intensity was within the design target throughout the PCB as well as in air. It should be noted that the terminal modeled in this section can be used for components that have enough space from the edge of the pad, so they do not influence the E-field around this section of the board. A method to account for surface mount components, larger boards with higher voltage, or to simply add an additional safety margin is presented in Section 2. As a continuation from [5], the layer stackup is such that a MID layer is placed as the outermost conductor layer on both sides of the bus followed by +DC and -DC. The +DC and -DC are separated by another MID layer centered between them. In the final stackup, +DC and -DC consisted of multiple parallel layers to increase the current carrying

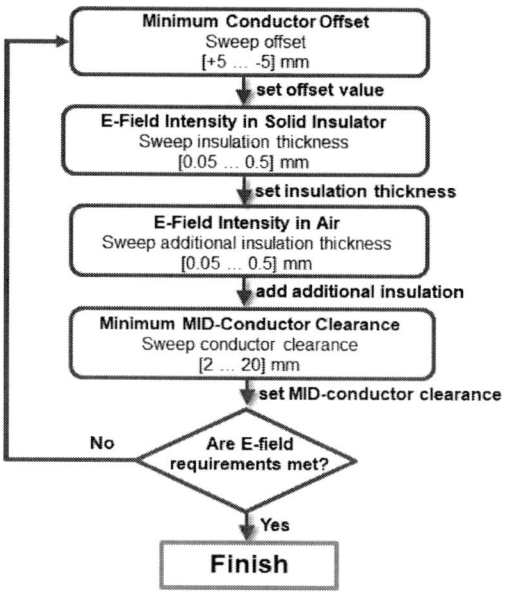

Fig. 3: Design flow for E-field constrained PCB design

capability. The 2D model used for E-field analysis at a terminal connected to the -DC layer is shown in Fig. 4. The material selected was Isola 370HR for its high electrical strength of 54 kV/mm (1350 V/mil) and glass transition temperature of 180 °C. To account for aging [9], a derating factor for FR4 is applied so that the maximum allowable field strength inside of the PCB will be derated. A common value for air breakdown is 3 kV/mm. The actual value at which air begins to ionize may be lower as air conditions vary. The limit for this design is 2 kV/mm in air to ensure its usefulness in applications without corona or surface

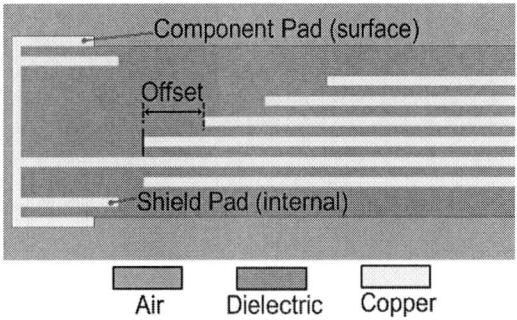

Fig. 4: Example of PCB cross section at component terminal.

discharge. Before the design flow was initiated, the +DC and all MID conductor edges were isolated from the -DC terminal to ensure the only factors influencing the results were the actual conductor offset and insulation thickness. The minimum conductor offset determined from Step 1 indicates the importance of incorporating an offset between layer edges as the highest field strength occurred with the case where no offset was applied. Introducing a 1 mm offset reduced the peak field intensity by 13%. After setting the offset, insulation thickness was increased, and the clearance was set to meet the field intensity requirement within the PCB.

To meet field intensity requirements in air, three options were investigated. These included adding dielectric 1) to all layers 2) between the two outer MID and +/-DC layers and 3) adding an additional core between the outer MID layers and air. Adding the additional core between the outer MID and air proved to be the most effective option as this forced the high fields into the dielectric instead of air.

3. Ancillary Circuits and Power Terminals

To achieve a highly modular design, ancillary circuits necessary for the power converter operation should be included in a single PEBB enclosure. The target power density is >10 kV/l power density, which requires components to be placed in close proximity to each other. With the bus structure implemented, the outer most bus layers are at the MID and serve as local ground. This allows the local controller, power module temperature sensors, voltage sensor, and their power supplies to be referenced to the same local ground and placed in close proximity to the bus with low E-field between. Shared terminals for the voltage sensor measurement and bus discharge were created using the standard approach presented in Section 2. E-field simulations were performed in ANSYS Maxwell 3D to determine spacing for components with high differential voltage such as the power terminals. Interconnects were designed to keep the peak E-field below 2 kV/mm.

For cold start operation, the PEBB is designed with a 10 kV 12 W custom pre-charge circuit. To charge the 32.5 µF capacitors, power dissipated through balancing resistors should be limited. Twelve surface mount device (SMD) resistors rated at 1 MΩ/1 W were placed in series between the dc-link capacitor terminals to provide voltage balancing while dissipating 3 W at full voltage. A small radius on SMD pads causes a high E-field intensity in that region. To alleviate this issue, shielding pads were placed on layers below the SMD pads to concentrate fields inside of the PCB dielectric rather than air; see Fig. 5.

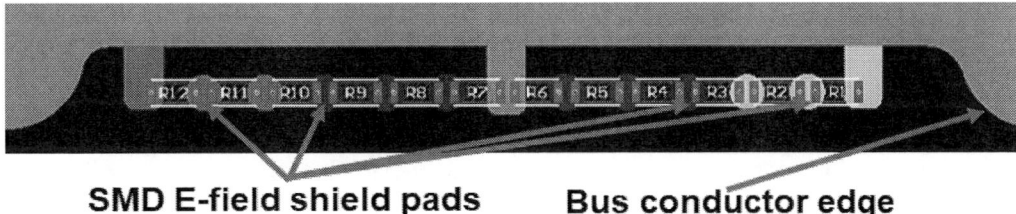

Fig. 5: PCB embedded pads for SMD component electric field control

For thicker boards, optimizing the height and width of this shield pad can reduce the peak field intensity in air by an order of magnitude while preserving overall insulation reliability [10]. Although they can be placed at any height that meets the maximum field strength requirements, placing them on existing layers used for one of the other bus conductors can reduce the final board cost due to a lower layer count.

Power from the WPT must be routed from below the bus to the top. Although the wire is approximately 48 V w.r.t. local ground, placing it near the board edges can still cause field enhancements due to the embedded +DC and -DC layers. Additionally, thermocouples referenced to MID must connect from the temperature sensor above the bus to the heatsink below. These wires have very small radii and present challenges like the WPT power wire. Leveraging the flexibility offered from a PCB, plated-through-holes (PTH) connected to MID can be used as shielded feedthroughs for wires and cables that need to pass from one side of the bus to the other. The hole diameter is dependent on the size and number of wires passing through any one PTH. Although this method can be used to pass wires at any voltage common to one of the bus plan voltages, there is no requirement to size the PTH outer annular ring since the outermost conductor layers are also MID. A 3D rendering of the PCB bus is shown in Fig. 6.

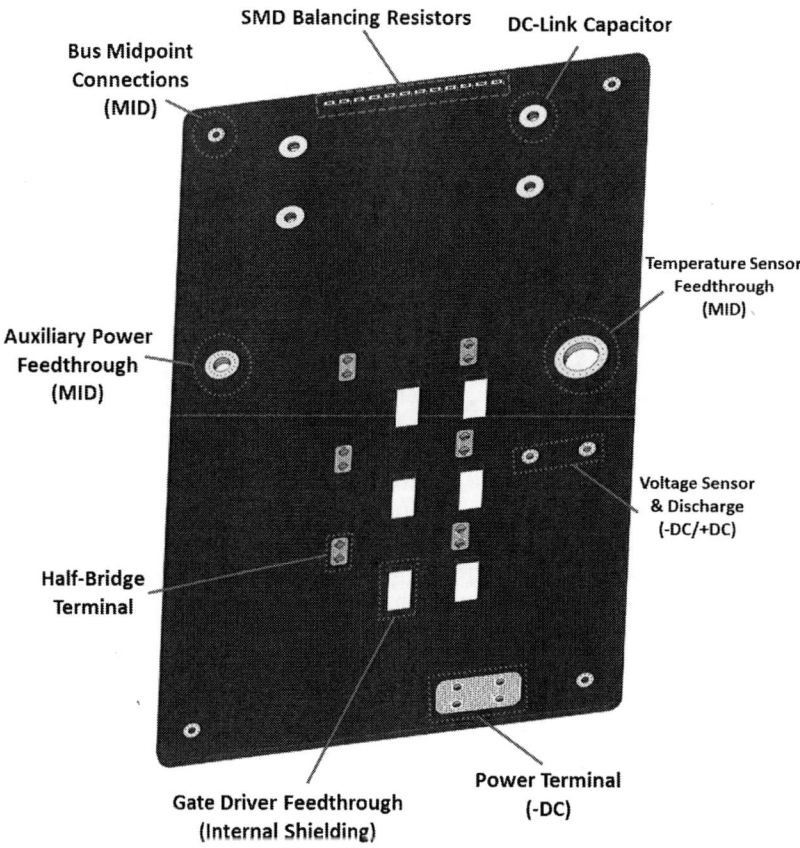

Fig. 6: 3D rendering of 6 kV PCB bus with terminals and feedthroughs labeled

4. Experimental Results

The bus insulation quality was tested using an MPD 600, 1 nF 100 kV coupling capacitor, and a measurement impedance for PD measurement and analysis. A 60 Hz sinusoidal excitation was applied using a Phenix 6CP100/50-7.5. High voltage (HV) and reference (REF) were connected across +DC, -DC, and MID in various combinations to test the insulation across each adjacent layer and well as +DC and -DC. Individual discharge events are plotted as a function of magnitude and time, then superimposed on top of the voltage excitation signal. The pattern created is the phase resolved partial discharge (PRPD) diagram. The PRPD pattern can help determine the type of discharge events that are likely occurring, e.g., PD or corona discharge. Corona discharge occurs in air near the insulator due to a metal part ionizing the air. PD occurs of the insulation. This can near the triple point or anywhere within the insulation due to defects such as voids, delamination, or contaminants.

The PDIV for each layer combination is shown in Table I. All combinations were required to ensure that each insulation layer was tested adequately. Due to the varying nature of PD measurements, six measurements were taken for each connection and averaged. Although the PDIV for Connection 6 much higher than the other measurements, this is expected since the total insulation between +DC and -DC is much thicker than any of the other layer combinations (MID is left floating). In the final application, the maximum voltage between any two adjacent layers is 3 kV. The minimum measured PDIV is about twice the operating voltage. Fig. 7 shows the PRPD for Connection 6. Here, the tested insulation is inside of the PCB between +DC and MID layers as well as the air insulation primarily at +DC and MID terminals. The voltage was kept at the PDIV for 3 minutes to record the PRPD pattern. This pattern indicates discharge occurs first in air which is consistent with what is expected based on simulation results.

Table I: Bus PDIV for various layer combinations

Connection	High Voltage	Reference (GND)	Floating	PDIV (kV)
1	-DC	MID	+DC	7.96
2	+DC	MID	-DC	6.31
3	+DC	-DC	MID	11.04
4	+DC & -DC	MID	-	6.03
5	+DC & MID	-DC	-	6.42
6	-DC & MID	+DC	-	5.77

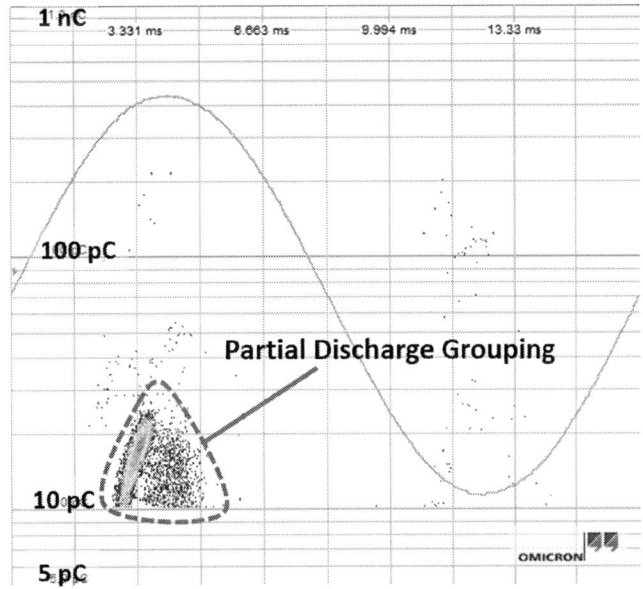

Fig. 7: PRPD pattern for insulation test between layers +DC and MID.

Conclusion

In this work, an MV PCB-based dc bus is presented, demonstrating PD free operation up to 11.04 kV. As the PCB fabrication process is mature, insulation quality is superior due to lower defects when compared to a conventional planar laminated bus. Due to the inherently low inductance, overall system weight and cost can be reduced by eliminating decoupling capacitors and placing the dc-link capacitors directly on the bus. The converters power density can be increased by placing ancillary circuitry, such as discharge and measurement circuits, in the area previously occupied by decoupling capacitors. To fully utilize the customizability offered by a PCB bus and eliminate long wire runs with a high potential, terminals should be added near circuitry that would otherwise need to connect to the dc capacitor terminals via long traces for power or measurement purposes. Additionally, shield feedthroughs connected to any potential can be placed where necessary for cables to pass from above to below the bus without creating field enhancements near the board edges.

References

[1] A. Anurag, S. Acharya and S. Bhattacharya, "Evaluation of Extra High Voltage (XHV) power module for Gen3 10 kV SiC MOSFETs in a mobile utility support equipment based solid state transformer (MUSE-SST)," *2019 10th International Conference on Power Electronics and ECCE Asia (ICPE 2019 - ECCE Asia)*, Busan, Korea (South), 2019, pp. 1-7

[2] Cree, "Design considerations for designing with Cree SiC modules part 2. Techniques for minimizing parasitic inductance," App Note CPRW-AN13, 2013

[3] J. Stewart, J. Neely, J. Delhotal and J. Flicker, "DC link bus design for high frequency, high temperature converters," *2017 IEEE Applied Power Electronics Conference and Exposition (APEC)*, Tampa, FL, 2017, pp. 809-815 [4] Vanderkeyn Ralf W.: Example of fast switching component, EPE Journal Vol. 20 no 5, pp. 48- 56

[4] J. Wang *et al.*, "Design and Testing of 6 kV H-bridge Power Electronics Building Block Based on 10 kV SiC MOSFET Module," *2018 International Power Electronics Conference (IPEC-Niigata 2018 -ECCE Asia)*, Niigata, Japan, 2018, pp. 3985-3992

[5] J. Stewart, Y. Xu, R. Burgos and M. Ghassemi, "Design of a Multilayer PCB Bus for Medium Voltage DC Converters," *2019 IEEE Electric Ship Technologies Symposium (ESTS)*, Washington, DC, USA, 2019, pp. 329-336

[6] S. Mocevic *et al.*, "Power Cell Design and Assessment Methodology Based on a High-Current 10-kV SiC MOSFET Half-Bridge Module," in *IEEE Journal of Emerging and Selected Topics in Power Electronics*, vol. 9, no. 4, pp. 3916-3935, Aug. 2021, doi: 10.1109/JESTPE.2020.2995386.

[7] K. Sun, J. Wang, R. Burgos, D. Boroyevich, J. Stewart and N. Yan, "Design and Multiobjective Optimization of an Auxiliary Wireless Power Transfer Converter in Medium-Voltage Modular Conversion Systems," in IEEE Transactions on Power Electronics, vol. 37,

[8] Y. Rong, J. Wang, Z. Shen, R. Burgos, D. Boroyevich and S. Zhou, "Distributed Control and Communication System for PEBB-based Modular Power Converters," 2019 IEEE Electric Ship Technologies Symposium (ESTS), 2019, pp. 627-633, doi: 10.1109/ESTS.2019.8847807.

[9] R. Tarzwell, K. Bahl, "*High voltage printed ciruict design & manufacturing notebook*," Sierra Proto Express, design guide, Nov. 4 2004.

[10] J. Stewart, I. Cvetkovic and R. Burgos, " Design and Analysis of a 24 kV PCB Bus for the Low Impedance Interconnect of a Multiphase PEBB-based Converter," *2022 IEEE International Power Modulator and High Voltage Conference (IPMHVC)*, 2022.

Modular multilevel converter control with using a general space vector PWM method in medium voltage hydro power application

Chengjun Tang, Torbjörn Thiringer
Chalmers University of Technology
Hörsalsvägen 11
Gothenburg, Sweden
Email: chengjun.tang@chalmers.se, torbjorn.thiringer@chalmers.se

Acknowledgments

This work has received funding from the European Union's Horizon 2020 research and innovation programme under grant agreement No. 764011.

Keywords

≪Space Vector PWM≫, ≪Modular Multilevel Converters (MMC)≫, ≪Converter control≫, ≪Efficiency≫.

Abstract

This paper studies a generalized space vector PWM (SVPWM) method for modulating the modular multilevel converter (MMC) for a medium voltage hydropower application. In addition to the modulation of the MMC, the circulating current control and the submodule capacitor voltage balancing are included in the study. The simulation and experimental results show the feasibility of using the generalized SVPWM method for controlling the MMC. Furthermore, the loss study shows that the switching loss in the MMC can be reduced with 28% when a modification of the generalized SVPWM method is utilized, thus, the total efficiency of the converter can be increased.

Introduction

As the requirement of reducing greenhouse emissions is more and more urgent, the total installed capacity of renewable energy, such as wind power and solar power, are growing. However, renewable energy also brings fluctuations to the energy generation due to intrinsic characteristics. Therefore, to facilitate the integration of renewable energy into the grid, flexible energy storage methods are needed; and the century-old pumped storage hydropower technology can be one of those methods.

Traditionally, the machine and turbine/pump of pumped storage hydropower stations are working under fixed speed, which is associated with the grid frequency. Later, it is found that the overall efficiency of the system can be increased when variable speed operation is employed [1]. Then, the double-fed induction machine (DFIM) technology is widely used in pumped storage hydropower stations. However, DFIM is more suitable for high capacity power stations. For those with the installed capacity below 100 MW, the converter-fed synchronous machine (CFSM) technology is preferred [2]. The key component in the CFSM technology is the converter, which is the interface between the machine and the grid.

Due to the high voltage level in the system, it is natural to think that multilevel converter topology can be employed. The MMC [3], which belongs to the multilevel converter family, has been widely used in the HVDC applications [4], and it is a good candidate for the CFSM hydropower application. For HVDC applications, the voltage levels of an MMC can be a few hundred, thus the nearest level modulation

(NLM) is most suitable [4]. For pumped storage hydro power, since the voltage level is not that high, to optimize the performance of the converter, SVPWM can be utilized [5].

In [6], it is shown that multilevel converter can be controlled with a generalized SVPWM method, however, the application is for cascaded H-bridge (CHB) converter. The SVPWM scheme in [7] is for MMC, while the control of circulating current and submodule voltage balancing are not included. In [5], the circulating current control and submodule voltage balancing are given along with the SVPWM method, however, the controllers structure are complicated, and the controller parameters needs to be tuned.

The purpose of this paper is to study a generalized SVPWM method for the MMC modulation in a hydropower application. In the meanwhile, a simple circulating current control and a submodule voltage balancing control are being studied to show the possibility of integrating different control aspects for the MMC with the generalized SVPWM method. At the end of this study, a modification of the generalized SVPWM is investigated to show the potential of decreasing the switching loss of the converter.

The general SVPWM method

SVPWM has been widely used in some multilevel converter applications, for instance, in the applications where three-level NPC converters are used, and it can be employed for a converter with even higher voltage levels. However, with the increase of the voltage level, the voltage vectors increases drastically. For a $n+1$ level converter, the SVPWM has $(n+1)^3 - n^3$ voltage vectors [6]. Fig. 1 shows the space vector plane for a five-level converter.

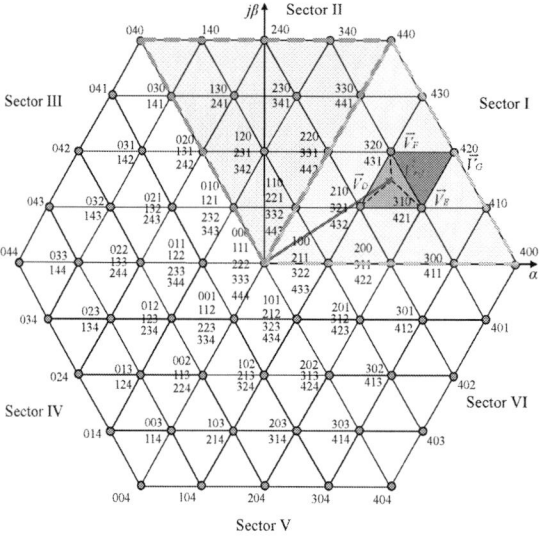

Fig. 1: Space vector plane for a five-level converter

In the same manner as in the two-level SVPWM, the primary task of the SVPWM for a multilevel converter is to find the three closest voltage vectors and its corresponding dwell time which can be used to synthesize the reference voltage vector. For example, the reference voltage $\overrightarrow{V_{ref}}$ in Fig. 1 can be synthesized by small voltage vectors $\overrightarrow{V_D}$, $\overrightarrow{V_E}$, and $\overrightarrow{V_F}$. Some of the small voltage vectors in Fig. 1 contain more than one switching state, for example, the small voltage vector $\overrightarrow{V_D}$ contains three switching states: [210], [321], and [432]; and these switching states in one small voltage vector are called redundant states. The challenge of using SVPWM for a multilevel converter is to select the appropriate switching states to synthesize the reference voltage vector.

One way to tackle this challenge is to use predefined look-up tables for the switching states in the controller, however, the complex lookup table for storing the switching states results in a large burden on the controller memory. In [6], a generalized SVPWM method for the multilevel converter is proposed, and in [8], this method is being successfully utilized for modulating a five-level NPC converter. The process of this generalized SVM method is being briefly described in the following paragraphs, and the details of the overall process can be found in [8].

Firstly, based on the angle of $\overrightarrow{V_{ref}}$ in the space vector plane, the sector number where $\overrightarrow{V_{ref}}$ is located can be found. If $\overrightarrow{V_{ref}}$ is located in a sector other than Sector I, then the angle conversion which listed in Table I is utilised to converter $\overrightarrow{V_{ref}}$ to Sector I, and this step is the same as in the two-level SVPWM.

Table I: Reference voltage angle transformation

Sector	Angle θ	Angle θ'
1	$(0, \pi/3]$	$\theta' = \theta$
2	$(\pi/3, 2\pi/3]$	$\theta' = -\theta + 2\pi/3$
3	$(2\pi/3, \pi]$	$\theta' = \theta - 2\pi/3$
4	$(\pi, 4\pi/3]$	$\theta' = -\theta - 2\pi/3$
5	$(4\pi/3, 5\pi/3]$	$\theta' = \theta + 2\pi/3$
6	$(5\pi/3, 2\pi]$	$\theta' = -\theta$

Secondly, $\overrightarrow{V_{ref}}$ is being normalized with respect to $2V_{dc}/3$ to get v^*, then v^* is being transferred from $\alpha\beta$ coordinate system to a new 60° coordinate system ($\alpha'\beta'$) according to

$$
\begin{cases}
v_{\alpha'} = v^* cos\theta - \hat{V} sin\theta / \sqrt{3} \\
v_{\beta'} = v^* sin\theta \cdot 2/\sqrt{3}
\end{cases}
\tag{1}
$$
$$
\tag{2}
$$

where v^* is the normalized voltage vector and θ is the vector angle. Once $v_{\alpha'}$ and $v_{\beta'}$ are calculated, the following vectors of the parallelogram $DEFG$ which contains the reference vector in the space vector plane can be calculated,

$$
\begin{bmatrix} V_D \\ V_E \\ V_F \\ V_G \end{bmatrix} = \begin{bmatrix} floor(v_{\alpha'}) & floor(v_{\beta'}) \\ ceil(v_{\alpha'}) & floor(v_{\beta'}) \\ floor(v_{\alpha'}) & ceil(v_{\beta'}) \\ ceil(v_{\alpha'}) & ceil(v_{\beta'}) \end{bmatrix}
\tag{3}
$$

Fig. 2 shows one-sixth of the five-level SVPWM plane when mapping from $\alpha\beta$ coordinates to $\alpha'\beta'$ coordinates.

Thirdly, based on the new coordinates in the 60° $\alpha'\beta'$ coordinate system, the triangle which contains $\overrightarrow{V_{ref}}$ is decided. There are two types of triangles in general, one is type DEF, and another one is type EFG. If $\overrightarrow{V_{ref}}$ falls in the triangle DEF, then the voltage vectors that used to synthesize $\overrightarrow{V_{ref}}$ are being switched in the following order, $\overrightarrow{V_D} \rightarrow \overrightarrow{V_E} \rightarrow \overrightarrow{V_F} \rightarrow \overrightarrow{V_D}'$, where $\overrightarrow{V_D}'$ is the redundant switching states of $\overrightarrow{V_D}$. If $\overrightarrow{V_{ref}}$ is insider triangle EFG, then the sequence of the switching states are $\overrightarrow{V_E} \rightarrow \overrightarrow{V_F} \rightarrow \overrightarrow{V_G} \rightarrow \overrightarrow{V_E}'$. The dwell time of the corresponding voltage vectors can be calculated in the 60° $\alpha'\beta'$ coordinate system.

At the end, once the switching states and corresponding dwell time are acquired, based on a transformation matrix, the switching states can be acquired in the normal $\alpha\beta$ coordinate system, while the dwell time of each voltage vector is the same.

Modification of the generalized SVPWM for the MMC

The schematic diagram of the MMC is shown in Fig. 3. For a $n+1$ level MMC, each arm of the MMC consists of a n number of submodules and one arm inductor L_0. The presence of the arm inductor in the circuit can prevent inrush current [9], and it can limit the circulating current as well.

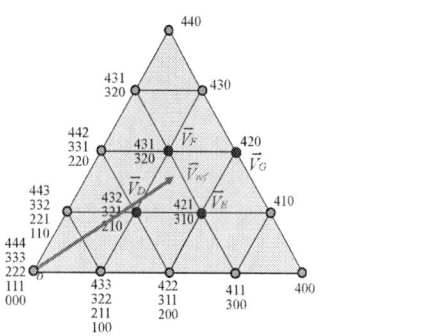

(a) Space vector plane, αβ coordinates

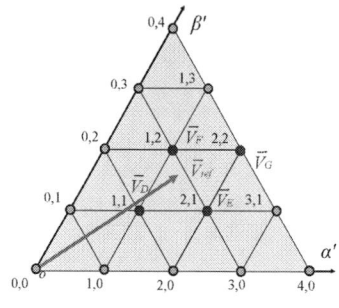

(b) Space vector plane, 60° α'β' coordinates

Fig. 2: SVM vectors converting from αβ coordinates to α'β'

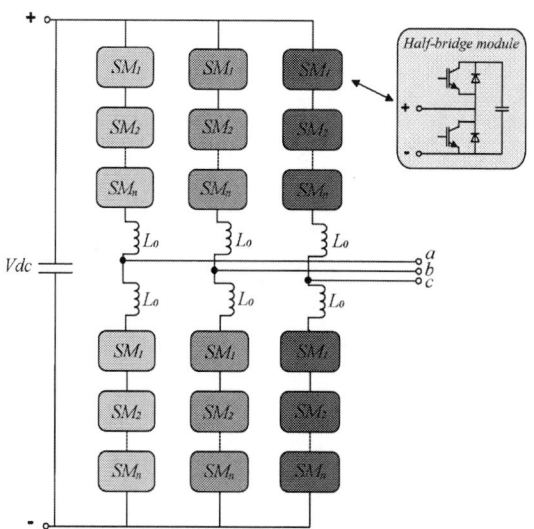

Fig. 3: Schematic diagram of the MMC topology

The studied submodule is a half-bridge IGBT submodule, which has shown the lowest losses and the lowest cost in HVDC applications [10]. The half-bridge submodule consists of two IGBT-diode switch pairs and one capacitor, and the two IGBTs in the half-bridge SM are being switched in a complementary mode. The submodule can be inserted into or bypassed from the circuit, depending on the switching states of the two IGBTs.

For the multilevel NPC converter, the aforementioned SVPWM method can be used directly, whereas, for the MMC, a modification of the SVPWM is needed due to the need of a "dual-modulator" for both the upper arm and the lower arm of the MMC. Fig. 4 shows the two voltage references for the upper arm ($\overrightarrow{V_{ref}}$) and lower arm ($\overrightarrow{V_{ref}}'$) of the MMC when using the SVM modulation. The two voltage references rotate in the same direction with a phase-shift of 180°.

By using the aforementioned generalized SVPWM method, the small voltage vectors for synthesizing the upper arm reference voltage can be determined. For the lower arm, there are two ways to determine the small voltage vectors. The first way is to use the voltage vector [nnn] minus the corresponding upper arm small voltage vectors, where n is the number of submodules in each arm. For example, for the five-level MMC with 4 submodules in each arm in Fig. 4, if the small voltage vectors for $\overrightarrow{V_{ref}}$ are [310] $\rightarrow$

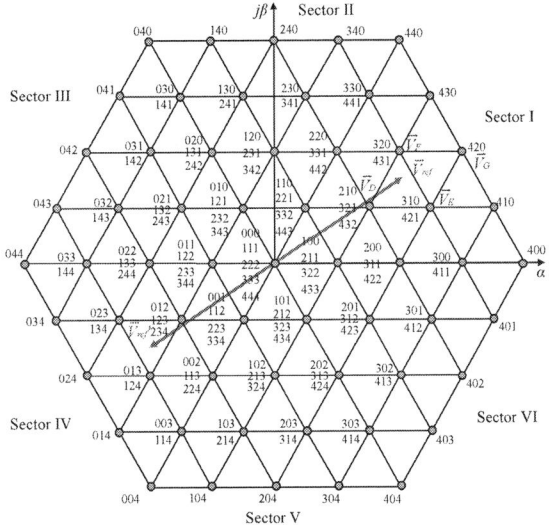

Fig. 4: Dual voltage reference of SVM for MMC

$[320] \rightarrow [321] \rightarrow [421]$, then the corresponding small voltage vectors for $\overrightarrow{V'}_{ref}$ would be $[134] \rightarrow [124]$ $\rightarrow [123] \rightarrow [023]$ in Sector IV. However, the drawback of this method is the output phase voltage level would be limited to $n+1$ level, thus, the THD in the output current and voltage would be higher. The second way is to use the redundant switching states of the lower arm small voltage vectors. For example, for the same upper arm $\overrightarrow{V_{ref}}$, instead of using vectors $[134] \rightarrow [124] \rightarrow [123] \rightarrow [023]$, the lower arm small voltage vectors can be $[023] \rightarrow [123] \rightarrow [124] \rightarrow [134]$. It can be seen that the average number of the total inserted submodules in one phase during one switching period T_{sw} would be equal to n, which is 4 for the five-level MMC.

Circulating current control

The circulating currents flow internally among the converter legs, thus, the converter output voltage and current are not being affected [9]. However, the presence of the circulating currents increases the losses of the power switches in each arm, and that reduces the efficiency of the converter. Thus, it is important to suppress the amplitude of the circulating currents in the MMC.

The circulating currents in the MMC are presented in the form of a negative sequence with the frequency of twice the fundamental frequency, and a double-line frequency circulating current controller [11] can be used to control the circulating currents, and the structure of the controller is shown in Fig. 5.

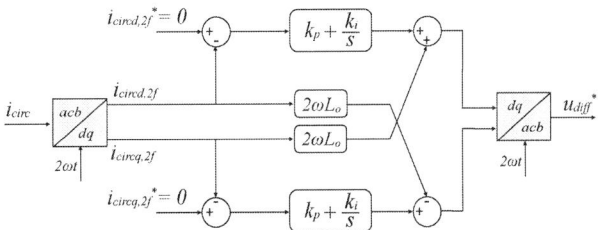

Fig. 5: The double line frequency circulating current controller

The circulating currents are firstly transformed from acb to dq coordinates. The PI controller is designed in dq coordinates, and the parameters of k_p and k_i are acquired by using the loop shaping method to have a first-order response of the closed-loop system.

Submodule capacitor voltage balancing

For each submodule capacitor, the voltage of it is alternating around the nominal value V_{dc}/n, which is caused by the charging and discharging processes due to the energy exchange. However, the submodule capacitor voltages can be unbalanced and deviate from the nominal value during the operation if no active balancing strategy is applied.

Various submodule voltage balancing methods have been proposed by different researchers, such as in [12], [13] and [14]. In this study, the classical sorting and balancing algorithm, which is based on the submodule capacitor voltages comparison and the polarity of the arm currents, is employed at the modulation stage [15],[16]. The balancing principle is, if the arm current is positive, then the arm submodule capacitor voltages are firstly sorted in ascending order. After that, from the submodule with the lowest capacitor voltage, a certain amount of submodules are inserted into the circuit; the inserted number of the submodule is based on the modulation signal. If the arm current is negative, a similar process is applied, while the order of the sorting for the submodule capacitor voltages is descending. Consequently, the inserted submodule capacitor will be charged or discharged based on the polarity of the arm currents, and the voltage of the corresponding submodule will be increased or decreased, thus, all the submodule capacitor voltages are balanced.

Simulation results

To demonstrate the feasibility of the studied generalized SVPWM method, a simulation model which contains a five-level MMC at the grid side was simulated in MATLAB Simulink and PLECS, and the parameters of the simulation model are listed in Table II. The simulation results are shown in Fig. 6.

Table II: Submodule and arm parameters of the simulated five-level MMC

Parameters	Value	Unit
Power S	6	MVA
Power factor $cos(\varphi)$	1	/
DC-link voltage V_{dc}	12	kV
Grid phase peak voltage $V_{g,peak}$	5.5	kV
Grid frequency f	50	Hz
PWM frequency modulation ratio m_f	23	/
RL filter Resistance R_f	1.5	mΩ
RL filter inductance L_f	10	mF
Number of submodules in each arm N_{SM}	4	/
Submodule voltage E_{SM}	3	kV
Submodule capacitance C_{SM}	1.41	mF
Arm inductance L_{arm}	5	mH

The voltage and current waveforms in Fig. 7a and Fig. 7b show that the generalized SVPWM method works well in the MMC, the line-to-line voltages contain high number of voltage levels and the phase currents are close to sinusoidal shape. The circulating current in Fig. 7c is controlled with the objective to have a minimum oscillation. The submodule capacitor voltages in Fig. 7d are balanced around 3 kV with using the sorting and balancing algorithm. However, it can also be observed that the submodule capacitor voltages are not perfectly symmetrical around 3 kV compared with using PWM method. This is due to the u^*_{diff} signal from the circulating current controller is being subtracted from the modulation signal $(V_{dc}/2 - u_{ref})$, and sometimes this gives a modulation index m_a smaller than 0 or larger than 1, while the actual m_a is being fixed to 0 if $m_a <= 0$ or to 1 if $m_a >= 1$. Due to the fact of two symmetrical arms in the MMC, this change of m_a can cause the modulation signals for the two arms unbalanced, thus, distorts the submodule capacitor voltages.

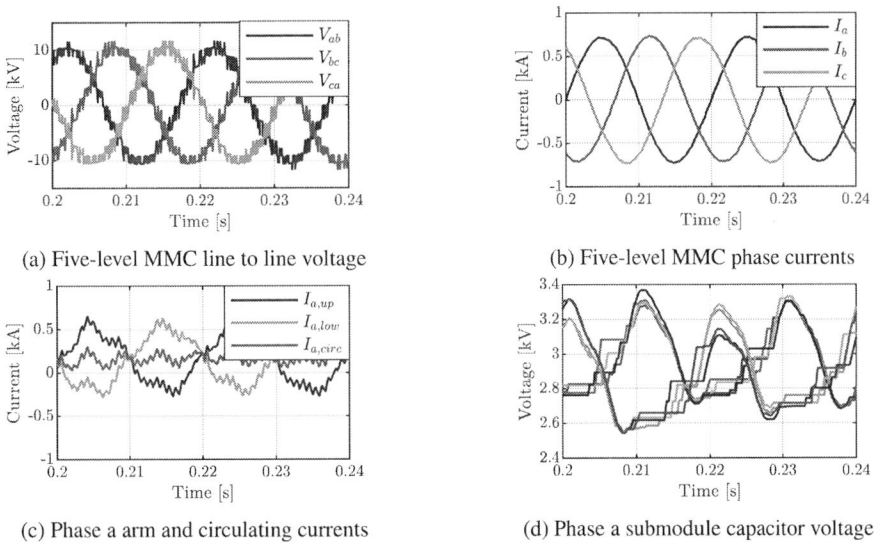

Fig. 6: Five-level MMC converter with SVPWM modulation, 100% power

For HVDC applications where the submodules are being switched under the frequency close to fundamental frequency, the switching loss of the converter is not that high, however, for the studied application, the submodules are being switched with 287.5 Hz, which means the switching loss will account for the majority part in the total losses. In addition to this, the submodule capacitor voltages sorting and balancing algorithm will impose more switching events on the submodules, thus, increase the switching loss as well. To decrease the switching loss in the MMC, one way is using the discontinuous SVPWM, which means only five segment of switching states are needed in one switching period. In this study, it is found when using the discontinuous SVPWM for the MMC, that the phase voltage should be maintained with $n + 1$ level; if $2n + 1$ level operation is utilized, the overall system will become unstable, and the simulation will diverge eventually. The simulation results of using discontinuous SVPWM are shown in Fig.7.

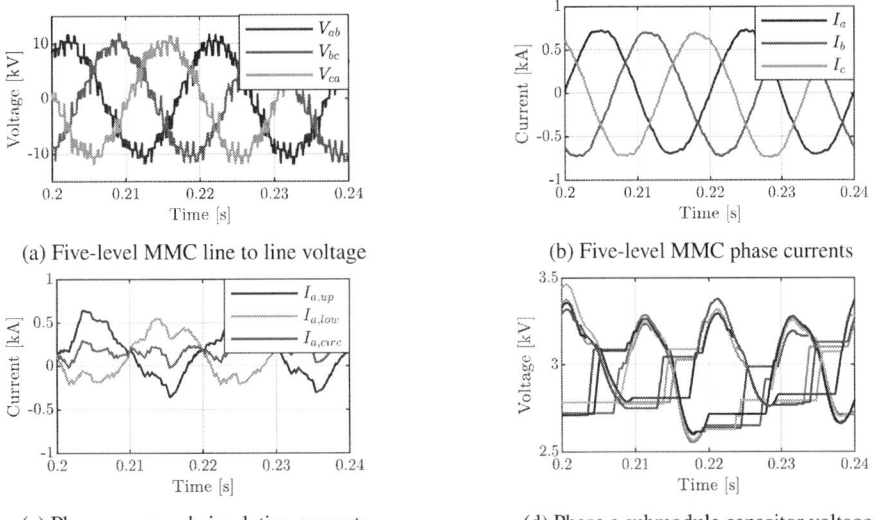

Fig. 7: Five-level MMC converter with discontinuous SVPWM modulation, 100% power

It can be clearly seen that due to the $n+1$ level operation, the current THD in the discontinuous SVPWM case is higher than that in the normal SVPWM case. The circulating current controller and the submodule capacitor voltages balancing algorithm work fine in the discontinuous SVPWM case, while the performance of them are slightly worse when comparing with the normal SVPWM case.

The losses results of the normal SVPWM and discontinuous SVPWM operation are shown in Fig. 8. The value of the losses are being normalized with respect to the nominal power (6 MW) of the system.

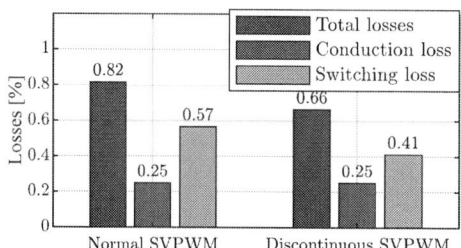

Fig. 8: Loss comparison between normal SVPWM and discontinuous SVPWM

It can be seen that when using the discontinuous SVPWM for modulating the five-level MMC under the defined conditions, the switching loss can be decreased from 0.57% to 0.41%, while the conduction loss are almost the same. The 28% reduction of switching loss is as expected, since the discontinuous SVPWM method requires less switching events for the operation of the MMC, however, the slightly worse waveforms that can be seen in Fig.7 means that the investments on the filter for the discontinuous SVPWM method would be higher in order to have a good voltage and current waveform quality. If the annual power deliver mission profile of the studied application is known, then, a justification of whether the discontinous SVPWM should be employed can be conducted based on the total investment and the total saved energy, i.e., the life-cycle cost.

Experimental results

The experiment focuses on the waveform verification of the SVPWM method for modulating the MMC converter, and the MMC experimental setup is shown in Fig 9.

Fig. 9: Lab set-up of the MMC: 1) DC power supply 2) MMC 3) Three-phase RL load

The lab set-up consists of three main parts: a 400 V DC power supply; a five-level MMC; a three-phase RL load. The five-level MMC contains three B-Box RCP controllers, 24 IGBT submodules, and 6 arm inductors. The used IGBT submodule is shown in Fig. 10.

The submodule includes 4× IGBT power switches on the board, forming a "H-bridge", while only two of the switches are used in the five-level MMC to form the half-bridge submodule. Apart from the power switches, the submodule also includes $9 \times PanasonicEET$ capacitors to form the submodule capacitor

Fig. 10: The IGBT submodule inside the converter cabinet

with a capacitance of 5 mF. The driver circuit, the optical PWM input, and the onboard measurements of the submodule are shown in Fig. 10 as well.

The experimental results are shown in Fig. 11, and a closed-loop current controller is utilized to control the current to follow the reference current.

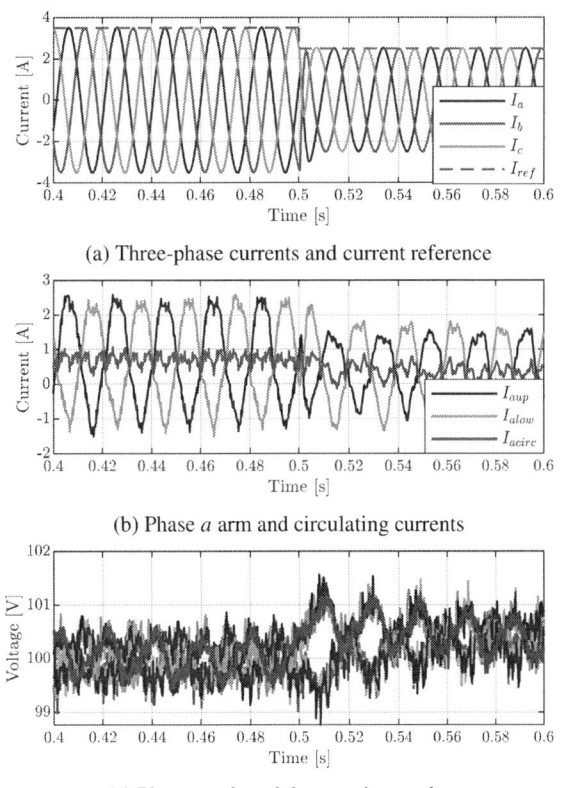

(a) Three-phase currents and current reference

(b) Phase a arm and circulating currents

(c) Phase a submodule capacitors voltage

Fig. 11: Experimental results of SVM operation with current step: 3.5 A to −2.5 A

It can be seen that for the current step at $t = 0.5$ s, the phase current amplitude can follow its reference value, the circulating currents are minimized, and the submodule capacitor voltages are balanced. This means that the studied generalized SVPWM can be utilized for modulating the MMC.

Conclusion

This paper studies a generalized SVPWM method for modulating the MMC, and in addition to that, the circulating current control and the submodule capacitor voltage balancing of the MMC are included in the study. The simulation and experimental results show the feasibility of using the generalized SVPWM method for modulating the MMC. However, the waveform qualities of using SVPWM are not perfect due to the distortion of circulating current control signal on the reference modulation signal. On the other hand, this generalized SVPWM shows the flexibility of using discontinuous SVPWM to decrease 28% of the switching loss inside the converter, which can benefit the thermal design of the MMC for the studied application.

References

[1] M. Valavi and A. Nysveen, "Variable-speed operation of hydropower plants: A look at the past, present, and future," *IEEE Industry Applications Magazine*, vol. 24, no. 5, pp. 18–27, 2018.

[2] T. Holzer and A. Muetze, "Full-size converter operation of hydro power generators: A state-of-the-art review of motivations, solutions, and design implications," *e & i Elektrotechnik und Informationstechnik*, vol. 136, no. 2, pp. 209–215, 2019.

[3] A. Lesnicar and R. Marquardt, "An innovative modular multilevel converter topology suitable for a wide power range," in *2003 IEEE Bologna Power Tech Conference Proceedings,*, IEEE, vol. 3, 2003, 6–pp.

[4] S. Debnath, J. Qin, B. Bahrani, M. Saeedifard, and P. Barbosa, "Operation, control, and applications of the modular multilevel converter: A review," *IEEE transactions on power electronics*, vol. 30, no. 1, pp. 37–53, 2014.

[5] Y. Deng, Y. Wang, K. H. Teo, M. Saeedifard, and R. G. Harley, "Optimized control of the modular multilevel converter based on space vector modulation," *IEEE Transactions on Power Electronics*, vol. 33, no. 7, pp. 5697–5711, 2017.

[6] S. Wei, B. Wu, F. Li, and C. Liu, "A general space vector pwm control algorithm for multilevel inverters," in *Eighteenth Annual IEEE Applied Power Electronics Conference and Exposition, 2003. APEC'03.*, IEEE, vol. 1, 2003, pp. 562–568.

[7] Y. Deng, K. H. Teo, C. Duan, T. G. Habetler, and R. G. Harley, "A fast and generalized space vector modulation scheme for multilevel inverters," *IEEE Transactions on Power Electronics*, vol. 29, no. 10, pp. 5204–5217, 2013.

[8] G. Mademlis, "Medium Voltage Generation System with Five-level NPC Converters for Kite Tidal Power," Licentiante Thesis, Chalmers University of Technology, 2019.

[9] B. Wu, M. Narimani, and Knovel, *High-power converters and AC drives*, Second edition., ser. IEEE Press series on power engineering. Hoboken, New Jersey: Wiley : IEEE Press, 2017.

[10] K. Sharifabadi, L. Harnefors, H.-P. Nee, S. Norrga, and R. Teodorescu, *Design, control, and application of modular multilevel converters for HVDC transmission systems*. John Wiley & Sons, 2016.

[11] Q. Tu, Z. Xu, and L. Xu, "Reduced switching-frequency modulation and circulating current suppression for modular multilevel converters," *IEEE Transactions on Power Delivery*, vol. 26, no. 3, pp. 2009–2017, 2011.

[12] M. Hagiwara and H. Akagi, "Control and experiment of pulsewidth-modulated modular multilevel converters," *IEEE transactions on power electronics*, vol. 24, no. 7, pp. 1737–1746, 2009.

[13] F. Deng and Z. Chen, "A control method for voltage balancing in modular multilevel converters," *IEEE Transactions on Power Electronics*, vol. 29, no. 1, pp. 66–76, 2013.

[14] K. Ilves, L. Harnefors, S. Norrga, and H.-P. Nee, "Analysis and operation of modular multilevel converters with phase-shifted carrier pwm," *IEEE Transactions on Power Electronics*, vol. 30, no. 1, pp. 268–283, 2014.

[15] S. Rohner, S. Bernet, M. Hiller, and R. Sommer, "Modulation, losses, and semiconductor requirements of modular multilevel converters," *IEEE transactions on Industrial Electronics*, vol. 57, no. 8, pp. 2633–2642, 2009.

[16] E. Solas, G. Abad, J. A. Barrena, S. Aurtenetxea, A. Carcar, and L. Zajac, "Modular multilevel converter with different submodule concepts-part i: Capacitor voltage balancing method," *IEEE Transactions on Industrial Electronics*, vol. 60, no. 10, pp. 4525–4535, 2013.

A technical overview of single-stage three-port dc-dc-ac converters

Sebastian Neira[1,2], Zoe Blatsi[1], Michael M.C. Merlin[1] and Javier Pereda[2]
[1]THE UNIVERSITY OF EDINBURGH
[2]PONTIFICIA UNIVERSIDAD CATÓLICA DE CHILE
[1]Colin Maclaurin Road
[1]Edinburgh, United Kingdom
Phone: +44 (0)131 650 5646
Fax: +44 (0)131 650 6554
Email: s.neira@ed.ac.uk
URL: https://www.eng.ed.ac.uk/research/institutes/ies

Keywords

≪DC-AC converters≫, ≪Double-input converter≫, ≪Hybrid power integration≫, ≪Emerging technology≫, ≪Split-Source inverter≫.

Abstract

Multiport dc-ac converters are becoming a pivotal technology to integrate renewable energy sources, energy storage systems and ac grids/loads. This article presents a classification and comparative analysis of the recently proposed single-stage three-port topologies. The study leads to the definition of three major groups, based on the operation principle to perform the dc-dc-ac power conversion. Each group is analysed in terms of the rated power, voltage boost characteristic and components count. Furthermore, a comparative study is performed to provide the best suited applications for each category.

Introduction

Hybrid power systems integrating renewable energy sources (RES), energy storage systems (ESS) and ac grids/loads have risen rapidly in the later years. In this context, systems with two dc elements connected to an ac port have become relevant because of their multiple applications, such as photovoltaic systems with integrated ES and hybrid EV powertrains [1, 2]. Therefore, there is a growing interest in developing multi-functional three-port power conversion systems with high power density and efficiency [3]. Thus, single-stage three-port topologies have been proposed due to their reduction of power processing stages and required components. These topologies perform the tasks of two or three separated converters, allowing the power flow control between three sources/loads with an optimised power processing stage. Reference [4] presents a throughout analysis of three-port dc-dc converters focused on PV-ESS applications and concludes that the extension to three-port dc-dc-ac converters is an essential topic to further investigations.

Single-stage three-port dc-dc-ac power converters have emerged during the last decade, looking to interface the elements of different hybrid power systems. The motivation behind these topologies is to reduce the required components, thus improving the efficiency and reliability of the power conversion system [5]. Additionally, they can provide extra features as embedded boost capabilities, continuous current at the dc elements or galvanic isolation with the ac port [6, 7]. However, they also present some limitations, as restricted power regulation capabilities and the need for advanced control and modulation techniques. Nevertheless, better-suited topologies for different applications can be determined, considering the mentioned limitations at the design stage. Thus, a comparative analysis of these single-stage three-port topologies is fundamental to classify them according to their features and possible applications.

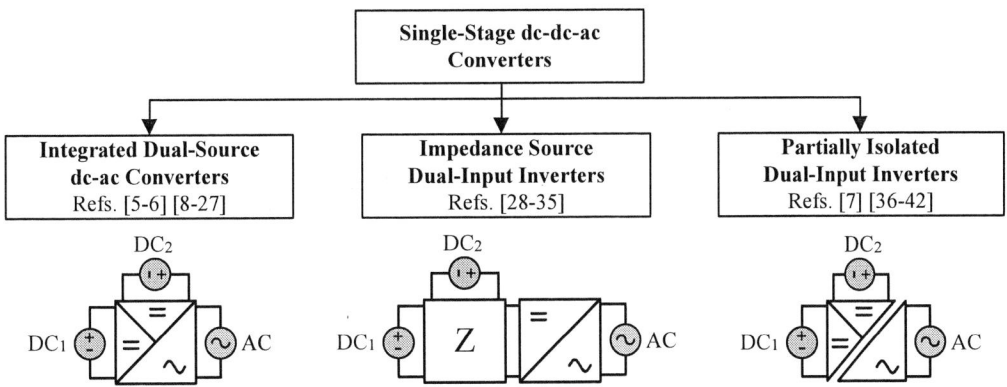

Fig. 1: Classification of single-stage three-port dc-dc-ac converters

This paper will present a technical overview of the single-stage three-port dc-dc-ac converters proposed for emerging hybrid power systems. The objective is to introduce a comparative analysis between the different converters based on the topology characteristics, features, limitations and applications. The study also classifies the different topologies based on the operating principle used to minimise the power processing stages. The latter looks for standardising the existing solutions to facilitate the state-of-art analysis for future proposals. The following sections present the proposed classification for the single-stage three-port dc-dc-ac converters, the relevant application cases and a comparative study considering the features of the analysed the topologies.

Topologies classification

Fig. 1 presents the proposed classification for the single-stage three-port dc-dc-ac topologies published in the last decade. Three major groups are defined based on the topology feature allowing the connection of the two dc ports with the ac side. This section will present a description of the proposed categories, with focus on the topologies and operation principles.

Integrated dual-source dc-ac converters

The first category is called integrated dual-source dc-ac converters, it contains three-port topologies that use as a basis a conventional dc-ac converter. Fig. 2 shows the overall topology of these solutions for a single-phase case, although they also work in three-phase systems. The topologies in this category integrates a dc-dc conversion within the legs of the conventional dc-ac converter, thus enabling the power flow with the second dc port. The interface connecting the second dc port allows to define three major groups within this category.

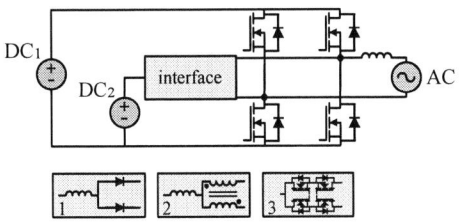

Fig. 2: Integrated dual-source dc-ac converter examples

The first subcategory (Fig. 2, interface 1) correspond to the split-source inverters [8–17], which embed a boost stage on one or more legs of the dc-ac converter. These topologies allows to transfer power from a low voltage dc source to both dc_1 and ac ports, using the legs of the converter to control two currents at the same time (L_{dc2}, L_{ac}). This is possible due to a degree of freedom (DoF) given by the switching states of the converter, as the zero states for the ac side (both upper or lower switches activated) create a different path for the current of L_{dc2}. Thus, a multi-variable control to enable the power regulation between the three ports is performed using modified modulation schemes. The minimum additional components to generate this topology are one dc inductor and one diode per leg, thus the components count is reduced compared to a classical two-stage implementation. The use of diodes limits the power flow to be unidirectional from port dc_2, and also exposes these diodes to the fast switching of the converter legs.

The second subcategory (Fig. 2, interface 2) include the buck-boost three-port converters (TPC) [6, 18–21] that integrate a buck-boost conversion to each leg of the dc-ac converter. These topologies have similar features to the split-source subcategory, but they add bidirectional power flow between all three ports. Topology in [6] uses one inductor per leg, enabling the simultaneous operation as a dc-ac converter between dc_1 and ac ports and as interleaved buck-boost converter between dc_1 and dc_2. The operation is also based in a multi-variable controller exploiting the DoF available to regulate the power flow between the sources and loads of the system. The minimum additional components for the operation of this topology is one inductor per leg, eliminating the need for extra semiconductor devices for the connection of the second dc port. Although current at port dc_2 is purely dc, the inductors' currents include an ac component at the frequency of the ac port. Therefore, the size of these passive components increase for low frequency applications. However, including magnetic coupling between the windings reduces the resultant ac flux and consequently the required size of these components.

The last subcategory (Fig. 2, interface 3) consists on the dual-port asymmetrical multilevel converters (DP-AMI) [5, 22–27], which use modified three-level converters connecting the second dc port to the neutral point. Thus, these converters (NPC or T-type) replace the connection to the mid-point of dc_1 port by a connection to the new dc_2 element. This group also uses the DoF given by the switching states of the topologies to perform bidirectional power flow between the three ports. Furthermore, the operation is more flexible due to the increased number of states given by the additional semiconductor devices. These topologies use at least twice the number of switching devices compared to the first two subcategories, but there is no need for extra passive components. The main limitation of this group is the limited power regulation capability for port dc_2, which leads to the use of an auxiliary dc-dc converter to extend this operation range. This limitation is generated by the unbalanced operation of the three-level converters and it varies with the voltage ratio between both dc ports.

Table I shows the components comparison for three examples of the subcategories described in the present subsection. The split-source inverter shows the higher voltage boost capability for port dc_2, reaching maximum gain values of 4 to 5 [10]. However, the modulation scheme should be carefully selected to minimise low-frequency ac ripple at the current of this second dc port that can be detrimental for batteries or fuel-cells. The buck-boost three-port converter presents the most flexible power range, as the power flow for port dc_2 is symmetrical for both source and sink scenarios. The current at dc_2 is free of low-frequency oscillations due to the interleaved operation of the interface inductors. Nevertheless, these inductors will have ac components at the frequency of the ac port, leading to higher volume and losses. The latter can be addressed at the design stage by including magnetic coupling to reduce the ac component currents on the windings. Finally, the dual-port AMI shows the best power quality at the ac side, due to the increase number of available switching states. Additionally, the volume is reduced as there is no need for extra passive components. However, the power regulation capability is reduced at port dc_2, complicating its use with elements that require wide bidirectional power flow capabilities as batteries or supercapacitors.

Table I: Summary for three-phase implementations of integrated dual-source dc-ac converters

Reference	Topology	Transistors	Diodes	Inductors	Power flow
[8]	Split-source inverter	6	3	1	Unidirectional
[6]	Buck-boost TPC	6	0	3	Bidirectional
[25]	Dual-port AMI	12	0	0	Bidirectional

Impedance source dual-input inverters

The second category includes the impedance source converters modified to incorporate a second dc input on the impedance network. These topologies use one of the available capacitors in the Z- or qZ-source network to connect a dc element exchanging power with both dc input and ac output. Fig. 3 shows the overall structure of these converters for a single-phase configuration, which can be extended to a three-phase. These converters work actuating on two control inputs: the modulation index of the dc-ac converter and the shoot-through ratio that allows boosting the voltage of the dc ports. Therefore, these two DoF are used to regulate the currents at the ac and one of the dc ports, using the remaining one to address the power balance between the three elements.

The Z-source inverter [28] includes a second dc source in one of the capacitors of the impedance network, allowing bidirectional power flow between all three ports. Additionally, it can provide boost capabilities using the shoot-through state of the dc-ac converter. However, the boost ratio should be kept between 1 to 2 as higher values generate excessive losses. References [29, 30] use two impedance networks and coupled inductors to enhance the boost capabilities, but at expenses of doubling the amount of required passive elements. The main limitation of the Z-source inverters is the discontinuous current at the dc elements, which has to be addressed by large size input capacitors.

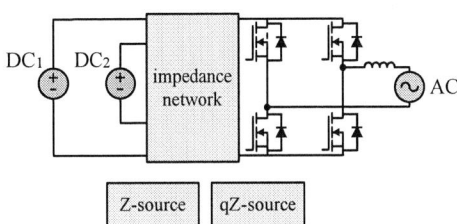

Fig. 3: Impedance source dual-input inverter examples

The second type of converters in this category correspond to the quasi-Z-source inverters, which uses a modified impedance network creating a common dc rail between dc_1 and ac ports [31–35]. This topology follows the same operational principle of the Z-source, but features continuous dc currents at both ports and consequently reduce passive components sizing. The reported qZ-source converters provide limited voltage boost capabilities by using both dc ports with the impedance network to generate a higher voltage at the ac port.

Table II presents the components comparison for three examples of converters in this category. The switching devices count is kept low, as the needed DoF is given by the impedance network itself (with the shoot-through state). Voltage boost capability can be achieve with a dual impedance network, but the passive components count is significantly incremented. The qZ-source inverter shows the best operative conditions for the dc elements, and it is the most adequate choice when there is no requirement for a high-ratio voltage boost.

Table II: Summary for three-phase implementations of impedance source dual-input dc-ac converters

Reference	Topology	Transistors	Diodes	Inductors	Capacitors	Voltage Boost
[28]	Z-source	7	0	2	2	Limited
[29]	Z-source	7	4	2*	8	High
[31]	qZ-source	6	1	2	2	Limited

Inductors are replaced with three-winding transformers in this case.

Partially isolated dual-input inverters

The last category consists of the partially isolated dual-input inverters, which feature galvanic isolation between the dc ports with the ac side (Fig. 4). These converters are based on microinverter topologies, as the Flyback [36–39] and the dual active bridge (DAB) [7, 40–42], including an extra dc port on the primary side of the transformer. Additionally, they use a current source converter on the secondary to interface directly the high-frequency signals with the low-frequency grid. Therefore, these converters manage to connect the three ports without requiring extra unfolding or power conditioning stages. Modulation techniques as the dual-phase shift or triple-phase shift give the DOFs required to control the power flow between the elements. Furthermore, partially isolated dual-input converters have embedded voltage boost capabilities with the transformer. Also, some of them can work with soft-switching increasing the overall efficiency of the solution.

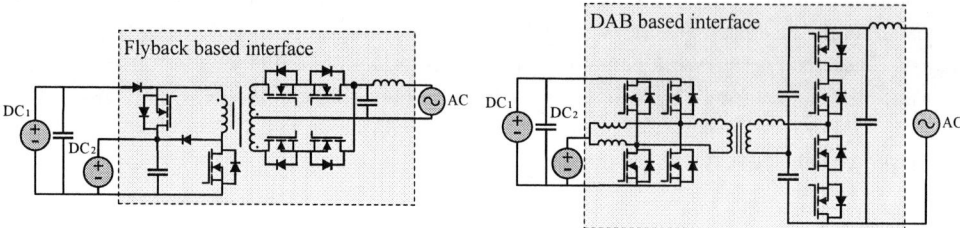

Fig. 4: Partially isolated dual-input inverter examples

These converters have a reduced volume due to the use of a high-frequency link that decreases the sizing of the required passive components. Therefore, they are suitable for its use as PV microinverters with energy storage support, enhancing the overall operation of the system. The main limitaion for this category is the elevated number of components and the complexity of the control and modulation techniques, compared to the previous categories.

Applications

Three-port dc-dc-ac converters present features that enable their use in different applications relevant for the present and future electrical system. Among these applications, the ones that can be highly benefited with single-stage conversion systems are:

- On- and off-grid hybrid PV-BES systems
- Hybrid EV powertrains
- On-grid low voltage PV or BES with embedded boost stage
- Hybrid fuel-cell BES systems
- PV green hydrogen generation

Hybrid PV-BES systems are already commercially available, and most of the solutions use power conversion systems with two or more power processing stages. Thus, the development of single-stage three-port converters can help to improve the efficiency and reliability of the system by reducing the number of components. The same principle applies for other commercially available systems, such as dual-PV systems. Additionally, single-stage three-port converters can also connect individual low-voltage elements (PVs or BES) to the grid, using the embedded boost capabilities present in some of the described topologies.

Hybrid EV powertrains are another application field that can be benefited by using single-stage three-port converters. These systems require to manage power interactions between the vehicle battery, an auxiliary dc source, as supercapacitors or fuel-cells, and the electrical machine performing the traction. Therefore, the reduction of power conversion stages can decrease the overall losses and volume of the solution, which are critical parameters in the design of such systems. Furthermore, single-stage three-port topologies can also be beneficial for newer applications related to the use of hydrogen. For example, fuel-cell systems could enhance its operation by using a second dc energy storage element as complement. Moreover, the generation of green hydrogen using PV energy is also an interesting field to explore the use of three-port topologies in an efficient and reliable manner.

Figure 5 showcase the listed applications that can enhance its operation using single-stage three-port topologies. The integrated dual-source converters present features, such as bidirectional power flow and boost capabilities, that allow their use in several applications. Moreover, the reduced component count of these topologies can decrease the total volume and failure rate of the power conversion stage, generating a positive impact on applications such as hybrid PV-BES and new hydrogen applications. However, these topologies use the redundant switching states of the converter to regulate the power flow, generating oscillations in the common-mode voltage. Therefore, their use in hybrid EV powertrains should be further analysed, as these oscillations could be detrimental to electric motors. Impedance source dual-input converters also show bidirectional power flow and boost capabilities for the dc ports, with a reduced count of active devices. Additionally, common-mode voltage oscillations can be reduced

Application	Needed Features	Useful Features	Suitable Topologies
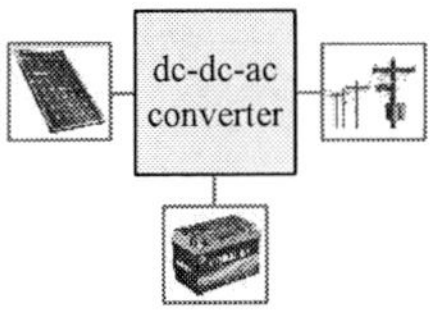 dc-dc-ac converter	- Bidirectional power flow at battery port. - Low-frequency ripple rejection at battery port.	- Extended power regulation capability to address variability of resource. - Boost capability to reduce serial connections PV or battery arrays.	- Buck-boost TPC & dual-port AMI. - qZ-source dual-input converter. - Partially isolated dual-input converters.
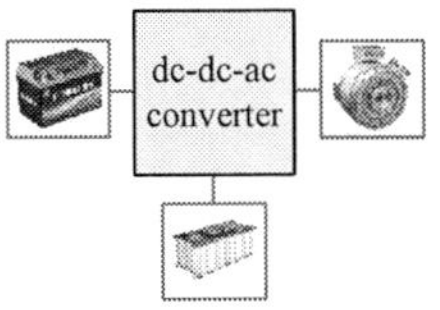 dc-dc-ac converter	- Bidirectional power flow for the three ports. - Low-frequency ripple rejection at battery port.	- Extended power regulation capability to address dynamic operation. - Reduced common-mode voltage variations to protect the motor.	- Z-source dual-input converter. - qZ-source dual-input converter.
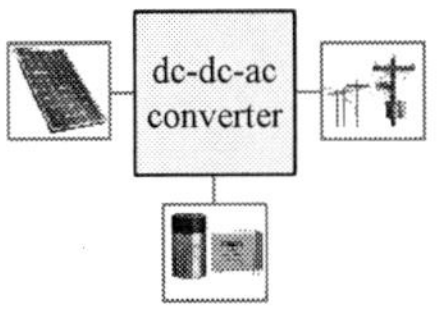 dc-dc-ac converter	- High voltage gain.	- Extended power regulation capability to address variability of resource. - Reduced leakage currents.	- Split-source inverters. - Partially isolated dual-input converters.
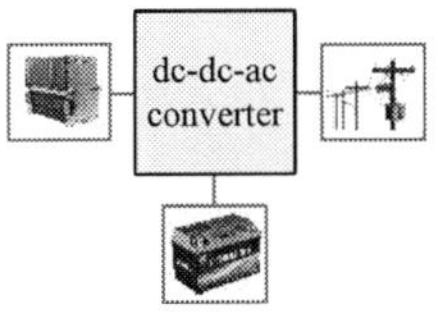 dc-dc-ac converter	- Bidirectional power flow at battery port. - Low-frequency ripple rejection at dc ports.	- Continuous steady current at fuel-cell port. - Boost capability to connect to the ac grid.	- Buck-boost TPC & dual-port AMI. - qZ-source dual-input converter.
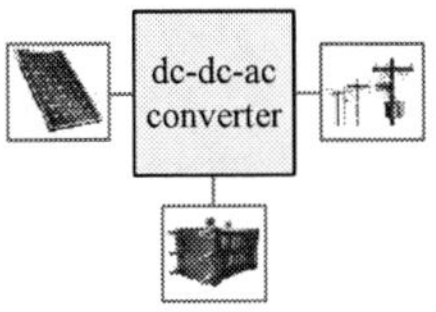 dc-dc-ac converter	- Continuous steady current at electrolyzer port. - Low-frequency ripple rejection at electrolyzer port.	- Extended power regulation capability to address variability of resource. - Boost capability to connect to the ac grid.	- Buck-boost TPC & dual-port AMI. - qZ-source dual-input converter.

Fig. 5: Three-port converter applications and key features for their operation.

through modulations techniques, enabling its use for hybrid EV powertrains. Finally, partially isolated dual-input inverters are best suited for low-power PV or hybrid PV-BES systems connected to the ac grid due to the use of high-frequency transformers. The transformer perform voltage boost and also provide galvanic isolation, maximising the impedance of the leakage current path. Furthermore, these topologies present high efficiencies as soft-switching techniques can be implemented for the active devices.

Discussion

This section present a comparative analysis based on the features of the three-port topologies previously described. The main parameters to perform the comparison are: Operation principle and control system requirements, power ratings and power flow capabilities, voltage boost capabilities and quantity of active and passive components.

The operation principle of the single-stage three-port topologies is to use the available DoFs to perform the task that normally would be carried on by a second conversion stage. Thus, the three categories condense dc-dc and dc-ac conversion functionalities into an individual power converter. The first category (Integrated dual-source) uses the redundant states of a standard dc-ac converter to control a second current at a dc port in addition to the current at the ac port. Two main control schemes have been proposed in previous literature: The first scheme uses a nested loop with single-variable controllers for each current, together with a modified modulation scheme allowing to make use of the redundant switching states; And the second scheme uses a multi-variable controller acting directly on the switching states based on

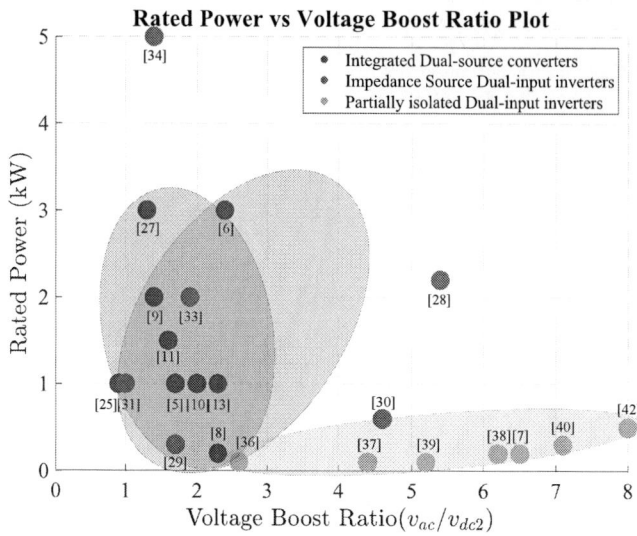

Fig. 6: Rated power against voltage boost capability for dc-dc-ac converters.

the required currents. The second category (Impedance source dual-input) uses the shoot-through state enabled by the impedance network to perform buck-boost functionality at the same time of the dc-ac conversion. Reported controllers for the topologies in this category include single-variable controllers acting on the shoot-through ratio to regulate the dc-dc conversion and on the modulation index to regulate the dc-ac power transfer. The last category (Partially isolated dual-input) uses a high-frequency link, working with the phase-shift angles to perform the dc-dc-ac power conversion. Publications for these topologies show that double- or triple-phase shift modulations can be used to regulate the power flow, using multi-variable controllers to obtain the duty cycles and phase-shifts.

The power rating of the different categories is a main parameter to take into account for a comparative study, as it defines the possible applications for the different categories. Additionally, the voltage boost capability is also a key feature to analyse, as single-stage dc-dc-ac converters have been proposed to replace two-stage solutions to connect low-voltage dc sources to the ac grid. Figure 6 shows a plot that relates these two variables for topologies in the three defined categories. The integrated dual-source converters present a maximum boost ratio of 2.4, as for greater values the power range gets too reduced and losses increase. The impedance source dual-input converters could reach higher voltage gains, but at expenses of increasing the passive components count. The plot shows that the first two categories cover a similar area of higher rated power but limited voltage boost ratio. On the contrary, the third category presents very high voltage ratios but low power rating due to the high-frequency isolation stage. Voltage gains up to 8.0 are reported in literature for this category, with rated powers less than 0.5 kW.

Finally, Table III summarises the features and limitations of the defined categories using one topology as example for each of them.

Conclusion

This article presents a categorisation for the emerging single-stage dc-dc-ac converters, defining three main groups based on the operation principle of the studied topologies. Each group uses different degrees of freedom to perform dc-dc and dc-ac power transfer simultaneously, without requiring additional power converters. The components count differs between the defined categories, as they include either passive or active components to enable the use of the required degrees of freedom. The reported converters have been proposed for low- and medium-power applications and for different voltage boost ratios between the dc ports and the ac one. Moreover, it is shown that each category covers a different range considering the rated power and voltage gain ratio. The first two categories (Integrated dual-source

Table III: Single-stage three-port dc-dc-ac topologies comparison

Reference	Category	Topology	Features	Limitations
[8]	Integrated dual-source	Split-source	Single-stage boost	Unidirectional power flow Low dc-link utilisation
[6]	Integrated dual-source	Buck-boost TPC	Bidirectional power flow Minimum device count	Inductors losses Common-mode voltage
[5]	Integrated dual-source	DP-AMI	Boost capabilities Reduced losses	Reduced power range
[28]	Impedance source dual-input	Z-source	Bidirectional power flow	Reduced dc voltage range
[31]	Impedance source dual-input	Quasi-Z-source	Reduced active devices count Continuous dc currents	Unidirectional power flow Reduced power range
[7]	Part. Isolated dual-source	DAB based	Soft-switching capabilities Galvanic isolation	Low power capability High active devices count

converters and Impedance source dual-input converters) are suitable for applications with higher power requirements and lower voltage boost needs. The remaining category (Partially isolated dual-input converters) is adequate for low-power applications with high-voltage boost requirement.

Hybrid dc-ac power systems have become an enabling technology for the integration of renewable energies and EVs to the electrical grid. Therefore, the optimisation of power conversion systems to interface multiple dc and ac elements has gained relevance in the last years. In this context, this article aims to help in the standardisation of three-port dc-dc-ac converters to facilitate the state-of-art study for new proposals. Three categories are defined and analysed considering the operation principle, rated power and components count. Firstly, Integrated dual-source converters perform the three-port power regulation by using dc-ac converter legs to perform a dc-dc conversion at the same time. These converters offer limited voltage boost capability (ratios up to 2.4) and present rated powers in the order of 1-5 kW. Thus, they are suitable for applications such as hybrid PV-BES or hydrogen fuel-cells with BES systems. Secondly, Impedance source dual-input converters use an passive components network to enable the connection of the two dc ports with an ac grid or load. This category presents voltage gain ratios up to 5.4 between the dc and ac ports and it is also reported for a range of 1-5 kW in rated power. The main applications for these topologies are hybrid energy storage systems, either connected to the grid or driving electric motors in EVs. Lastly, Partially isolated dual-input converters use a high-frequency isolated link with two dc sources at the primary and an ac port at secondary. The converters in this category have the highest voltage gain ratios (up to 8.0), but the power is limited to less than 0.5 kW. Consequently, these converters are most suitable for PV micro-inverter operation with complementary BES.

A possible direction for the future research in single-stage dc-dc-ac converters include increasing the rated power of the topologies. Commercially available multi-stage solutions for applications such as hybrid EV powertrains, hybrid PV-BES or FC-BES systems are rated for powers from 10 to more than

100 kW. Therefore, studies should be performed to analyse the losses increment and the size of passive components required to extend the power capabilities of the single-stage three-port topologies to match the available multi-stage solutions.

References

[1] I. Batarseh and K. Alluhaybi, "Emerging Opportunities in Distributed Power Electronics and Battery Integration: Setting the Stage for an Energy Storage Revolution," in IEEE Power Electronics Magazine, vol. 7, no. 2, pp. 22-32, June 2020.

[2] M. Ehsani, K. V. Singh, H. O. Bansal and R. T. Mehrjardi, "State of the Art and Trends in Electric and Hybrid Electric Vehicles," in Proceedings of the IEEE, vol. 109, no. 6, pp. 967-984, June 2021.

[3] A. K. Bhattacharjee, N. Kutkut and I. Batarseh, "Review of Multiport Converters for Solar and Energy Storage Integration," in IEEE Transactions on Power Electronics, vol. 34, no. 2, pp. 1431-1445, Feb. 2019.

[4] N. Zhang, D. Sutanto, K. M. Muttaqi, "A review of topologies of three-port DC–DC converters for the integration of renewable energy and energy storage system," in Renewable and Sustainable Energy Reviews, vol. 56, pp. 388-401, April 2016.

[5] H. Wu, L. Zhu, F. Yang, T. Mu, and H. Ge, "Dual-DC-Port Asymmetrical Multilevel Inverters With Reduced Conversion Stages and Enhanced Conversion Efficiency," IEEE Trans. Ind. Electron., vol. 64, no. 3, pp. 2081–2091, Mar. 2017.

[6] S. Neira, J. Pereda, and F. Rojas, "Three-Port Full-Bridge Bidirectional Converter for Hybrid DC/DC/AC Systems," IEEE Trans. Power Electron., vol. 35, no. 12, pp. 13077–13084, Dec. 2020.

[7] A. K. Bhattacharjee and I. Batarseh, "An Interleaved Boost and Dual Active Bridge-Based Single-Stage Three-Port DC–DC–AC Converter With Sine PWM Modulation," IEEE Trans. Ind. Electron., vol. 68, no. 6, pp. 4790–4800, Jun. 2021.

[8] H. Ribeiro, A. Pinto, and B. Borges, "Single-stage DC-AC converter for photovoltaic systems," in 2010 IEEE Energy Conversion Congress and Exposition, Atlanta, GA, pp. 604–610, Sep. 2010.

[9] J. Kan, S. Xie, Y. Wu, Y. Tang, Z. Yao, and R. Chen, "Single-Stage and Boost-Voltage Grid Connected Inverter for Fuel-Cell Generation System," IEEE Trans. Ind. Electron., vol. 62, no. 9, pp. 5480–5490, Sep. 2015.

[10] A. Abdelhakim, P. Mattavelli, P. Davari, and F. Blaabjerg, "Performance Evaluation of the Single Phase Split-Source Inverter Using an Alternative DC–AC Configuration," IEEE Trans. Ind. Electron., vol. 65, no. 1, pp. 363–373, Jan. 2018.

[11] A. Abdelhakim, P. Mattavelli and G. Spiazzi, "Three-Phase Three-Level Flying Capacitors Split-Source Inverters: Analysis and Modulation," IEEE Trans. Ind. Electron., vol. 64, no. 6, pp. 4571-4580, June 2017.

[12] L. An, T. Cheng and D. D. Lu, "Single-Stage Boost-Integrated Full-Bridge Converter With Simultaneous MPPT, Wide DC Motor Speed Range, and Current Ripple Reduction," IEEE Trans. Ind. Electron., vol. 66, no. 9, pp. 6968-6978, Sept. 2019.

[13] S. S. Lee, A. S. T. Tan, D. Ishak, and R. Mohd-Mokhtar, "Single-Phase Simplified Split-Source Inverter (S3I) for Boost DC–AC Power Conversion," IEEE Trans. Ind. Electron., vol. 66, no. 10, pp. 7643–7652, Oct. 2019.

[14] M. S. Hassan, A. Abdelhakim, M. Shoyama, J. Imaoka and G. M. Dousoky, "Three-Phase Split-Source Inverter-Fed PV Systems: Analysis and Mitigation of Common-Mode Voltage," IEEE Trans. Power Electron., vol. 35, no. 9, pp. 9824-9838, Sept. 2020.

[15] C. Yin, W. Ding, L. Ming, and P. C. Loh, "Single-Stage Active Split-Source Inverter With High DC Link Voltage Utilization," IEEE Trans. Power Electron., vol. 36, no. 6, pp. 6699–6711, Jun. 2021.

[16] M. S. Hassan, A. Abdelhakim, M. Shoyama, J. Imaoka and G. M. Dousoky, "Parallel Operation of Split-Source Inverters for PV Systems: Analysis and Modulation for Circulating Current and EMI Noise Reduction," IEEE Trans. Power Electron., vol. 36, no. 8, pp. 9547-9564, Aug. 2021.

[17] Y. Elthokaby, I. Mohamed and N. Abdel-Rahim, "Model Predictive Control for Three-Phase Split-Source Inverter," 2020 22nd European Conference on Power Electronics and Applications (EPE'20 ECCE Europe), Lyon, France, pp. 1-10, Sep. 2020.

[18] J. Pereda, S. Neira, "Multi-port DC-DC-AC power converter," INAPI PCT/CL2021/050033, April 2021.

[19] S. Neira, A. Lizana and J. Pereda, "A Novel Three-Port NPC Converter for Grid-Tied Photovoltaic Systems with Integrated Battery Energy Storage," 2020 IEEE 11th International Symposium on Power Electronics for Distributed Generation Systems (PEDG), Dubrovnik, Croatia, pp. 104-109, Nov. 2020.

[20] C. Perera, J. Salmon and G. J. Kish, "Multiport Converter With Enhanced Port Utilization Using Multitasking Dual Inverters," IEEE Open Journal of Power Electronics, vol. 2, pp. 511-522, 2021.

[21] M. Aly, E. A. D. Ibrahim, S. Kouro, E. M. Ahmed, T. A. Meynard and J. Rodriguez, "Model Predictive Control-Based Three-Port Common Ground Photovoltaic-Battery Grid-Connected Inverter," IECON 2021 – 47th Annual Conference of the IEEE Industrial Electronics Society, Toronto, Canada, pp. 1-6, Nov. 2021.

[22] L. Dorn-Gomba, P. Magne, B. Danen, and A. Emadi, "On the Concept of the Multi-Source Inverter for Hybrid Electric Vehicle Powertrains," IEEE Trans. Power Electron., vol. 33, no. 9, pp. 7376–7386, Sep. 2018.

[23] H. Wu, L. Zhu, and F. Yang, "Three-Port-Converter-Based Single-Phase Bidirectional AC–DC Converter With Reduced Power Processing Stages and Improved Overall Efficiency," IEEE Trans. Power Electron., vol. 33, no. 12, pp. 10021–10026, Dec. 2018.

[24] J. Wang, H. Wu, T. Yang, L. Zhang and Y. Xing, "Bidirectional Three-Phase DC–AC Converter With Embedded DC–DC Converter and Carrier-Based PWM Strategy for Wide Voltage Range Applications," IEEE Trans. Ind. Electron., vol. 66, no. 6, pp. 4144-4155, June 2019.

[25] J. Wang, K. Sun, D. Zhou, and Y. Li, "Virtual SVPWM-Based Flexible Power Control for Dual-DC Port DC–AC Converters in PV–Battery Hybrid Systems," IEEE Trans. Power Electron., vol. 36, no. 10, pp. 11431–11443, Oct. 2021.

[26] C. Liu, C. Xu, J. Ruan, L. Jin, L. Bao and Y. Deng, "A Vector Control Strategy for a Multi-Port Bidirectional DC/AC Converter With Emphasis on Power Distribution Between DC Sources," IECON 2019 - 45th Annual Conference of the IEEE Industrial Electronics Society, Lisbon, Portugal, pp. 1579-1584, Dec. 2019.

[27] J. Wang, K. Sun, H. Wu, L. Zhang, J. Zhu and Y. Xing, "Quasi-Two-Stage Multifunctional Photovoltaic Inverter With Power Quality Control and Enhanced Conversion Efficiency," IEEE Trans. Power Electron., vol. 35, no. 7, pp. 7073-7085, July 2020.

[28] S. Hu, Z. Liang, and X. He, "Ultracapacitor-Battery Hybrid Energy Storage System Based on the Asymmetric Bidirectional Z -Source Topology for EV," IEEE Trans. Power Electron., vol. 31, no. 11, pp. 7489–7498, Nov. 2016.

[29] M. G. Varzaneh, A. Rajaei, A. Jolfaei and M. R. Khosravi, "A High Step-Up Dual-Source Three-Phase Inverter Topology With Decoupled and Reliable Control Algorithm," IEEE Trans. Ind. Applications, vol. 56, no. 4, pp. 4501-4509, July-Aug. 2020.

[30] M. G. Varzaneh, A. Rajaei, M. Forouzesh, Y. P. Siwakoti and F. Blaabjerg, "A Single-Stage Multi-Port Buck-Boost Inverter," IEEE Trans. Power Electron., vol. 36, no. 7, pp. 7769-7782, July 2021, doi: 10.1109/TPEL.2020.3042338.

[31] Y. Liu, B. Ge, H. Abu-Rub, and F. Z. Peng, "Control System Design of Battery-Assisted Quasi-Z-Source Inverter for Grid-Tie Photovoltaic Power Generation," IEEE Trans. Sustain. Energy, vol. 4, no. 4, pp. 994–1001, Oct. 2013.

[32] W. Liang, Y. Liu, B. Ge, H. Abu-Rub, R. S. Balog, and Y. Xue, "Double-Line-Frequency Ripple Model, Analysis, and Impedance Design for Energy-Stored Single-Phase Quasi-Z-Source Photovoltaic System," IEEE Trans. Ind. Electron., vol. 65, no. 4, pp. 3198–3209, Apr. 2018.

[33] A. Lashab, D. Sera, J. Martins, and J. M. Guerrero, "Dual-Input Quasi- Z -Source PV Inverter: Dynamic Modeling, Design, and Control," IEEE Trans. Ind. Electron., vol. 67, no. 8, pp. 6483–6493, Aug. 2020.

[34] V. Castiglia, R. Miceli, F. Blaabjerg and Y. Yang, "A Comparison of Two-Stage Inverter and Quasi-Z-Source Inverter for Hybrid Energy Storage Applications," 2020 22nd European Conference on Power Electronics and Applications (EPE'20 ECCE Europe), Lyon, France, pp. 1-10, Sep. 2020.

[35] W. Liang, Y. Liu and J. Peng, "A Day and Night Operational Quasi-Z Source Multilevel Grid-Tied PV Power System to Achieve Active and Reactive Power Control," IEEE Trans. Power Electron., vol. 36, no. 1, pp. 474-492, Jan. 2021.

[36] H. Hu, Q. Zhang, X. Fang, Z. J. Shen and I. Batarseh, "A single stage micro-inverter based on a three-port flyback with power decoupling capability," 2011 IEEE Energy Conversion Congress and Exposition, Phoenix, USA, pp. 1411-1416, 2011.

[37] S. Harb, H. Hu, N. Kutkut, I. Batarseh, and A. Harb, "Three-port micro-inverter with power decoupling capability for Photovoltaic (PV) system applications," in 2014 IEEE 23rd International Symposium on Industrial Electronics (ISIE), Istanbul, Turkey, pp. 2065–2070, Jun. 2014.

[38] K. Alluhaybi, H. Hu, and I. Batarseh, "Design and Implementation of Dual-input Microinverter for PV-Battery Applications," in 2020 IEEE Applied Power Electronics Conference and Exposition (APEC), New Orleans, LA, USA, pp. 1467–1474, Mar. 2020.

[39] M. Hadi Zare, M. Mohamadian and R. Beiranvand, "A Single-Phase Grid-Connected Photovoltaic Inverter Based on a Three-Switch Three-Port Flyback With Series Power Decoupling Circuit," IEEE Trans. Ind. Electron., vol. 64, no. 3, pp. 2062-2071, March 2017.

[40] M. Safayatullah, S. Ghosh, S. Gullu and I. Batarseh, "Model Predictive Control for Single-Stage Grid-Tied Three-Port DC-DC-AC Converter Based on Dual Active Bridge and Interleaved Boost Topology," IECON 2021 – 47th Annual Conference of the IEEE Industrial Electronics Society, Toronto, Canada, pp. 1-6, Nov. 2021.

[41] A. Bhattacharjee and I. Batarseh, "An Interleaved Boost and Dual Active Bridge Based Three Port Microinverter," 2020 IEEE Applied Power Electronics Conference and Exposition (APEC), New Orleans, LA, USA, pp. 1320-1326, Mar. 2020.

[42] X. Chen and I. Batarseh, "A Modified Three-Port Bidirectional LLC Resonant Converter for Renewable Power Systems," 2021 IEEE Energy Conversion Congress and Exposition (ECCE), Vancouver, Canada, pp. 2067-2073, Oct. 2021.

Common-Mode EMI Noise Modeling of Three-Level T-Type Inverter for Adjustable Speed Drive Systems

Vefa Karakasli, Abdelmoumin Allioua and Gerd Griepentrog
Technical University of Darmstadt
Institute for Power Electronics and Control of Drives
Fraunhoferstr. 4, 64283 Darmstadt, Germany
Email: vefa.karakasli@lea.tu-darmstadt.de
URL: https://www.lea.tu-darmstadt.de/

Acknowledgments

This work was supported by the German Research Foundation (DFG) under the grant GR 4487/7-1.

Keywords

≪Adjustable speed drive≫, ≪Electromagnetic Interference (EMI)≫, ≪EMC/EMI ≫, ≪MOSFET≫, ≪T-type inverter≫

Abstract

This article studies the EMI noise modeling of a three-level T-type inverter ($3LT^2I$) for adjustable speed drive (ASD) systems. A methodology to generate the common-mode (CM) voltages in the time-domain is proposed based on SiC MOSFETs switching behavior, such as the variable rise and fall times. The frequency spectrums of the time-domain voltages are used with the CM noise paths to predict the CM current of the $3LT^2I$. The CM noise model of the $3LT^2I$ consists mainly of two CM noise sources. The effect of these noise sources on the CM current is investigated. The proposed EMI noise prediction method is validated by the comparison between the simulation and experimental results of the CM voltage and CM current spectrums.

Introduction

The $3LT^2I$ is becoming a popular topology because of its advantages, such as low total harmonic distortion (THD) and switching losses compared to a two-level voltage source inverter (2L-VSI) [1]. It also provides a better EMI behavior in the medium and low power ranges [2]. Wide-bandgap semiconductors, such as silicon carbide (SiC) MOSFETs, are increasingly used within $3LT^2I$ since they have significant operational advantages, in particular switching at higher current and voltage, with low switching losses and higher operating temperatures. However, the fast switching speed of SiC MOSFETs results in high-voltage slew rates (dv/dt), which leads to increasing EMI challenges, higher bearing currents in motors and deterioration in the insulation of motor windings [3–5]. Standards such as EN 50121 and IEC 61800 set the limits of EMI emissions to ensure that the system is electromagnetically compatible with its environment. An accurate EMI prediction model must be developed in order to evaluate the EMI performance of the $3LT^2I$ with parasitic capacitances. The prediction model will help designers to take early precautions in the design phase against the EMI noises.

The EMI noise modeling of the 2L-VSI has been widely conducted in the literature [6–9]. While each phase of a 2L-VSI has only one voltage varying node, the three-level neutral point clamped (3L-NPC) has rather three and the $3LT^2I$ has two voltage varying nodes. The increase in voltage-varying nodes complicates the EMI modeling of three-level inverter topologies. Most EMI models of the 3-level inverter

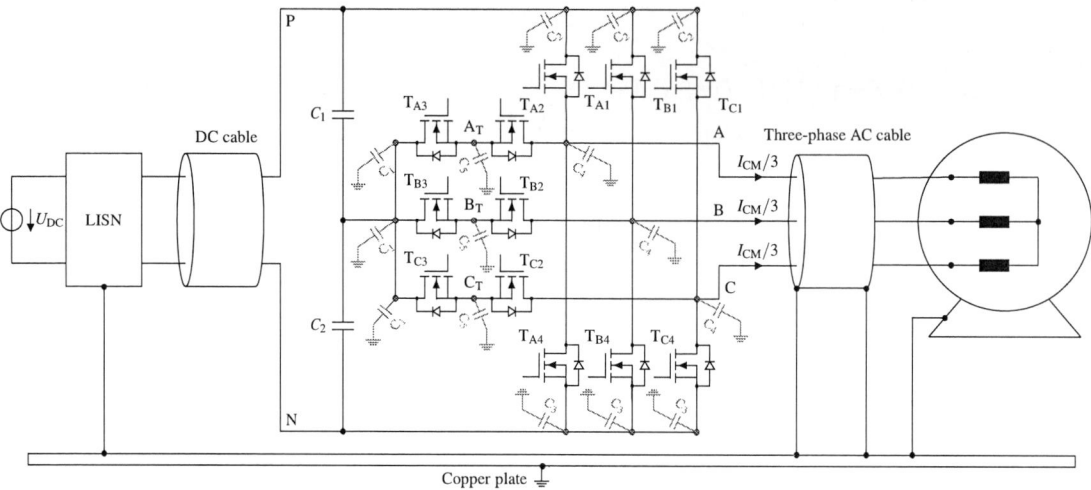

Fig. 1: Topology of a 3LT^{2}I for an ASD system with line impedance stabilization network (LISN) and parasitic capacitances.

– mainly the 3L-NPC – are focused on photovoltaic or grid applications [10–14]. However, no study has been conducted on the EMI modeling of the 3LT^{2}I for an ASD system, which is shown in Fig. 1.

The voltage spectrum used in the frequency-domain model can be obtained from the simulated or measured time-domain voltage waveforms using Fourier transform. In some papers, the switching waveforms are modeled as trapezoidal waves [9, 15]. In [13, 16], the switching waveform accuracy is improved by adding the ringing effect of the switching devices. Although the real switching waveform has variable commutation times, these models proposed to rather use fixed commutation times, which reduces the accuracy of the frequency-domain simulations. Furthermore, the model in [17] proposes an approach based on the switching behavior of SiC MOSFET, such as the variable rise and fall times. Nevertheless, the ringing effect is not considered in [17].

In this paper, a frequency-domain CM noise model is proposed for a 3LT^{2}I used in ASD systems. First, the CM noise sources for the 3LT^{2}I are derived, and a methodology to generate time-domain CM voltages is proposed based on SiC MOSFETs switching behavior. Then, the CM noise propagation paths of the ASD system are presented. Finally, the simulation and experimental results are provided to validate the proposed EMI modeling for the 3LT^{2}I.

Common-Mode Model for Three-Level T-Type Converter

The EMI noises can be simulated either in time-domain or in frequency-domain [9]. Frequency domain models are often used to simulate the EMI noise due to their simplicity and reduced computational effort [6]. Therefore, the frequency-domain EMI approach will be used in this article. Moreover, it should be noted that the asymmetry due to an unbalanced impedance in the three-phase system and the nonlinear operation of the inverter leads to a coupling between CM and differential-mode (DM) noises [18–20]. This type of noise is called the mixed-mode effect. For the sake of simplicity, the decoupled CM noise model in the frequency-domain is used in the proposed method, as it has also been considered in [6, 7, 10, 11].

Common-Mode Voltage Source Modeling

The CM model of three-phase 3LT^{2}I begins from analyzing a single-phase leg with parasitic capacitances, as shown in Fig. 2(a). The parasitic capacitance between the single-phase leg circuit and the grounded heatsink are presented by $C_1 - C_5$. Discrete SiC MOSFETs are used in the 3LT^{2}I circuit. The parasitic capacitances of the drain of SiC MOSFETs to the heatsink (C_{MH}) and the parasitic capacitances of the printed circuit board (PCB) traces to the heatsink ($C_{PCB,H}$) have an important effect on the EMI

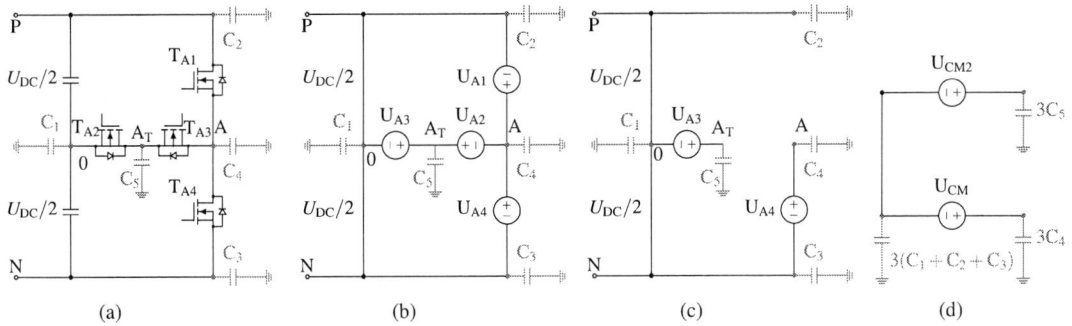

Fig. 2: Extraction of $3\text{LT}^2\text{I}$ CM model. (a) A phase of the $3\text{LT}^2\text{I}$ with parasitic capacitances. (b) Replacement of the switching devices by voltage sources. (c) Elimination of U_{A1} and U_{A2}. (d) The 3-phase CM model.

modeling of $3\text{LT}^2\text{I}$. C_1 represents the parasitic capacitance between the midpoint of dc-link capacitors and the ground. C_2 is the combined parasitic capacitances of the drain of T_{A1} to the heatsink and the node P to the heatsink. C_3 is the node N to the heatsink parasitic capacitance. Furthermore, the parasitic capacitance of phase A and C_{MH} are summed up and presented as C_4. Since the bidirectional switches located at the midpoint leg has the common drain configuration, the parasitic capacitance of the common drain is determined by

$$C_5 = 2C_{MH} + C_{PCB,A_T}. \tag{1}$$

In general, semiconductors are replaced either by voltage or current sources within EMI prediction models, with the so-called substitution theorem [10, 11]. The switching devices are substituted with voltage sources to reflect the CM voltage behavior appropriately, as shown in Fig. 2(b). It is important to understand that these noise sources have the same time-domain waveform of the switching devices. The dc capacitors are short-circuited in Fig. 2(b) due to their small ac impedances. It must be ensured that the voltage sources are not placed in parallel to avoid a closed loop of the voltage sources [11,14]. Thus, the voltage source of the switch T_{A1} and T_{A2} are eliminated, as shown in Fig. 2(c). The switching voltages U_{A3} and U_{A4} equal U_{A_T0} and U_{A0}, respectively. Finally, the CM EMI model of the $3\text{LT}^2\text{I}$ is obtained by extending the single-phase model in Fig. 2(c) to Fig. 2(d), where the CM voltages can be expressed as follow:

$$U_{CM} = \frac{U_{A0} + U_{B0} + U_{C0}}{3}, \tag{2}$$

$$U_{CM2} = \frac{U_{A_T0} + U_{B_T0} + U_{C_T0}}{3}. \tag{3}$$

The voltage source model in [17], which is for a 2L-VSI, will be modified for the $3\text{LT}^2\text{I}$, and the ringing effect will be added. Fig. 3 shows the proposed algorithm for the generation of time-domain voltage sources. The gate signals of the $3\text{LT}^2\text{I}$ are generated based on the pulse-width modulation (PWM) technique. The PWM technique in [21], which proposes a modulation method for the $3\text{LT}^2\text{I}$ based on two-level space vector PWM (SVPWM), is implemented. The fundamental frequency of reference voltage (f_{fund}), the switching frequency (f_{sw}), the sampling frequency (f_{samp}), the dead time (t_{dt}) and the modulation index (m) are the input parameters of the SVPWM technique. It is worth mentioning that other PWM techniques can also be used. The generated PWM signals are implemented to the $3\text{LT}^2\text{I}$ circuit in which the ideal switching devices are used to obtain the output pulse voltages. On the output of the $3\text{LT}^2\text{I}$, a three-phase star or delta RL load is connected, through which the values depend on the reactive and active power of the motor. The ideal source voltages U_{A0}, U_{B0} and U_{C0} in (2) together with U_{A_T0}, U_{B_T0} and U_{C_T0} in (3) are extracted as the output voltages ($U_{out,ideal}$) in the diagram.

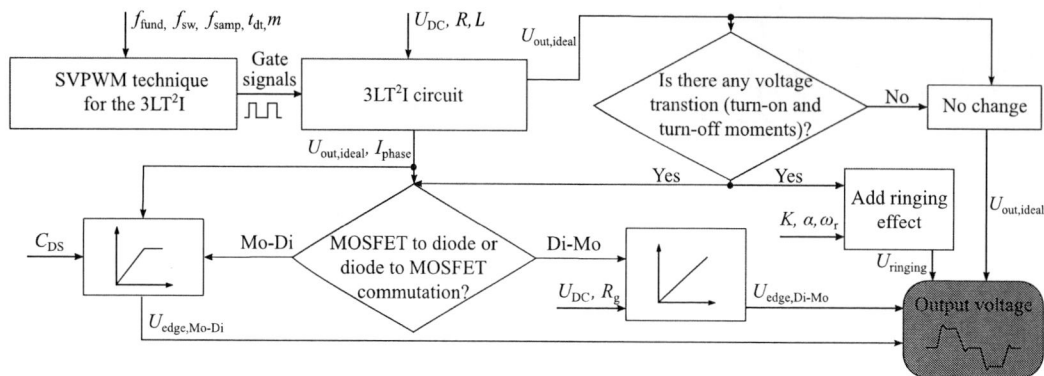

Fig. 3: The diagram for the generation of time-domain voltages of the 3LT²I.

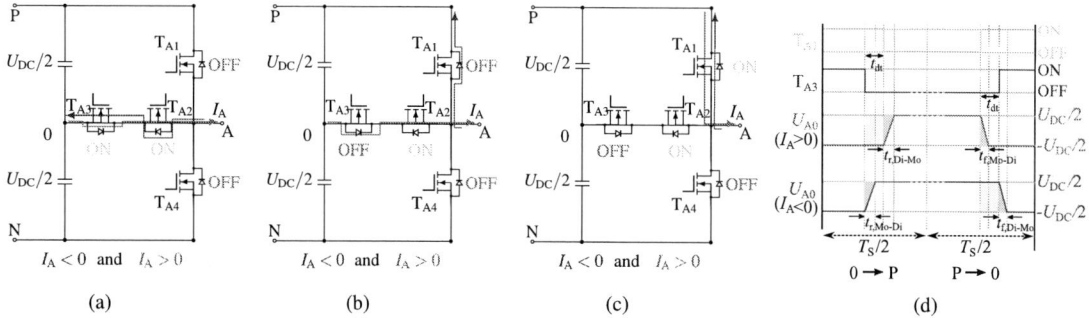

Fig. 4: 0→P transition of a single-phase leg of the 3LT²I. (a) Steady-state before 0→P transition. (b) During the dead-time period. (c) Steady-state after 0→P transition. (d) Gate signals (T_{A1} and T_{A3}) and output voltage waveform (U_{A0}).

According to applied PWM signals and output phase currents, two commutation types occur: 1) MOSFET to diode commutation and 2) diode to MOSFET commutation. There are four possible output voltage state transitions of the 3LT²I: P→0, 0→P, 0→N, and N→0. The semiconductors' rise and fall times during these transitions must be considered to model the voltage sources accurately. In Fig. 4, the switching behavior of a single-phase leg of the 3LT²I during 0→P transition is analyzed under different current directions. A dead-time is set between the switch T_{A1} and T_{A3} gate signals in order to avoid shoot-through fault, as illustrated in Fig. 4(d). When the phase current is negative ($I_A < 0$), the MOSFET to diode commutation takes place during the dead-time period, as can be seen in Fig. 4(a) and 4(b). First, I_A charges the drain-source capacitance of T_{A3}, and the drain-source voltage of T_{A3} starts to increase while the current flows through the midpoint leg remains almost constant [17]. The internal diode of T_{A1} becomes forward biased after U_{A0} reaching $\frac{U_{DC}}{2}$. A similar MOSFET to diode commutation is observed in P→0 transition in the case of the positive phase current ($I_A > 0$), as shown in Fig. 4(d). The voltage rise time (t_r) and fall time (t_f) of the MOSFET to diode commutation are defined by the MOSFET drain-source capacitance (C_{DS}) as:

$$t_{\text{edge,Mo-Di}} = \frac{C_{DS}\frac{U_{DC}}{2}}{I_A} = \frac{Q_{DS}}{I_A} \tag{4}$$

where Q_{DS} is the MOSFET charge [17]. A linear model for the capacitance is assumed in (4). Thus, the edges with variable duration occur because of the MOSFET to diode commutation in the 3LT²I. On the other hand, the diode to MOSFET commutation occurs after the dead-time period in 0→P transition when the phase current is positive, as given in Fig. 4(b) and 4(c). The t_r and t_f of the diode to MOSFET commutation mainly depend on the gate resistance (R_g) and U_{DC}, also the plateau voltage (U_P) and the reverse transfer capacitance (C_{rss}) [22]. In the measurement results, it is observed that t_r and t_f are

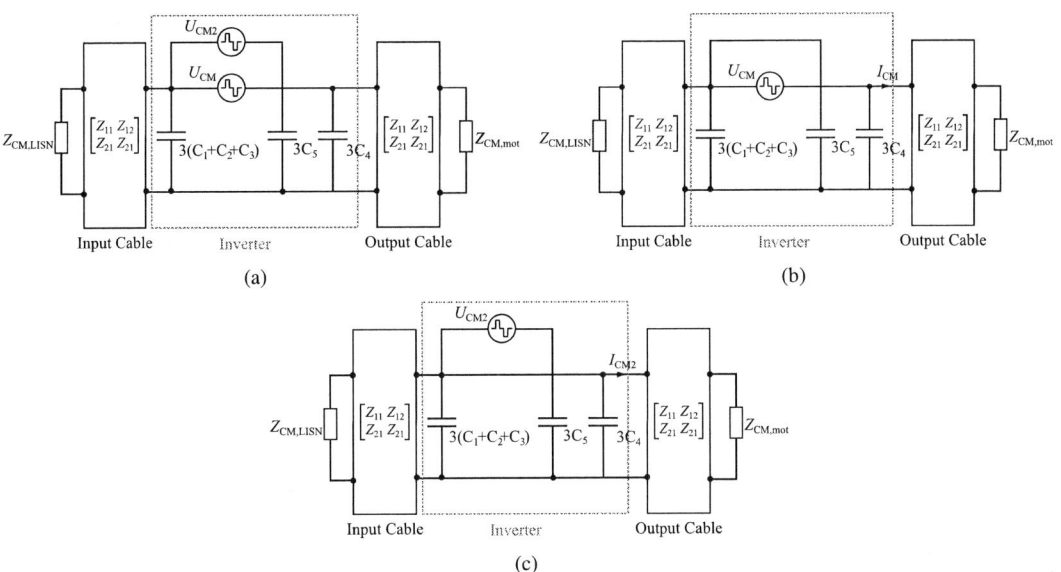

Fig. 5: The CM prediction model of an ASD system. (a) Extracted CM noise model. (b) Separate analysis of U_{CM}. (c) Separate analysis of U_{CM2}.

almost constant during the diode to MOSFET commutation. In the simulation of time-domain voltage, a constant value, which is determined from measurement results, is used for t_{r} and t_{f} of the diode to MOSFET commutation.

Finally, a damped exponential sinusoidal wave is added to the trapezoidal output voltage based on the following mathematical function:

$$U_{\mathrm{ringing}} = K\exp(-\alpha t)\sin(\omega_{\mathrm{r}} t) \tag{5}$$

where K is the magnitude of the ringing, α is the damping coefficient, and ω_{r} is the ringing frequency [13].

Final Common-Mode Model

The extracted CM model of the $3\mathrm{LT}^2\mathrm{I}$ in Fig. 2(d) is connected to the other noise paths of the ASD system, as shown in Fig. 5(a). The two-port impedance matrices of input and output cables are used to ensure that the model accurately characterizes the cable properties. CM impedance of the ADS system components and the parasitic capacitances of the $3\mathrm{LT}^2\mathrm{I}$ can be determined by direct measurements or simulation models. In the presented model, the measurement of CM impedance of the ADS system components and the parasitic capacitances of the drain of SiC MOSFETs to the heatsink are used. Furthermore, a finite element method (FEM) tool is used to find the parasitic capacitances between PCB and the heatsink.

There are two voltage sources in Fig. 5(a) and their spectrum forms are acquired from the time-domain voltages, which are generated based on the diagram in Fig. 3. It should be noted that the response of a linear circuit having two independent sources can be obtained by applying superposition principle. Thus, the contributions of CM current generated due to U_{CM} and U_{CM2} can be dissected, as shown in Fig. 5(b) and Fig. 5(c). The CM current at the input of the inverter (I_{CM} and/or I_{CM2}) is then analyzed in the ASD systems.

Simulation and Measurement Results

A 20-kW $3\mathrm{LT}^2\mathrm{I}$ is used to validate the proposed CM voltage and current models for the ASD system in Fig. 1. The input voltage is set at 300 V, the switching frequency is 24 kHz and the modulation index is

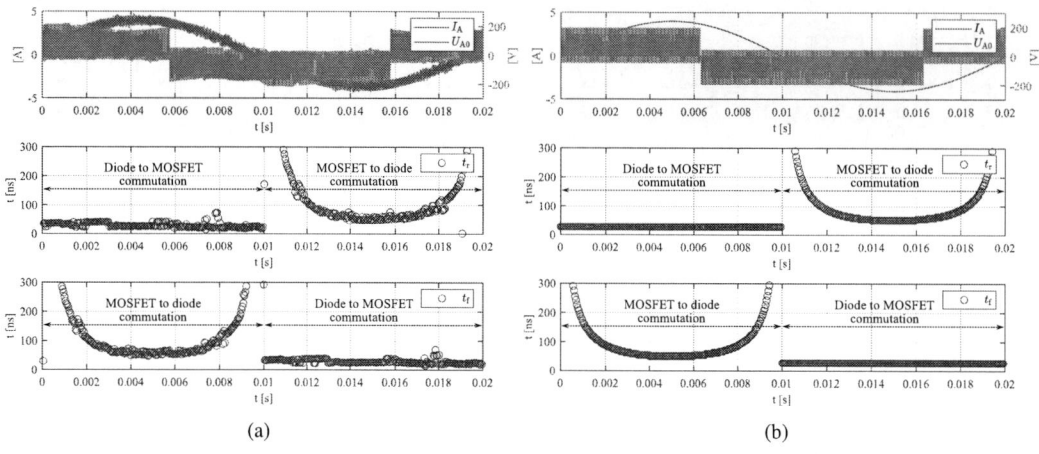

(a)   (b)

Fig. 6: (a) Measured and (b) simulated phase current (I_A), phase voltage (U_{A0}), the rise time (t_r) and fall time (t_f).

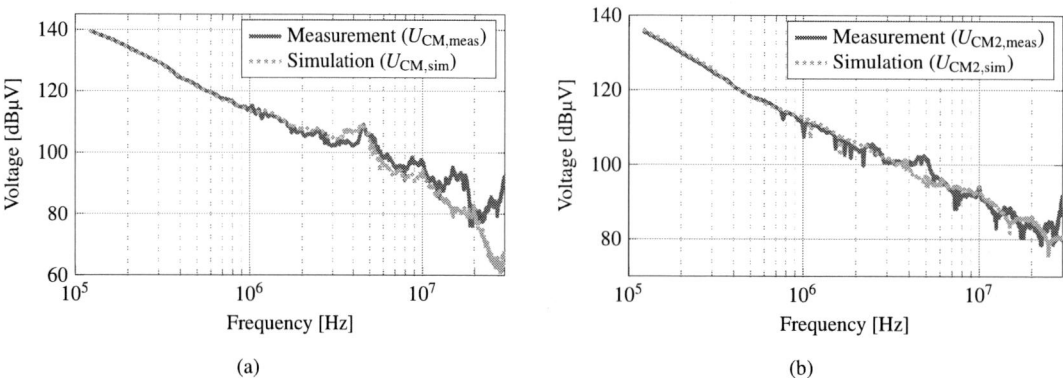

(a)   (b)

Fig. 7: Comparison between measured and simulated CM voltage sources. (a) U_{CM}. (b) U_{CM2}.

0.8. A Keysight E5061B vector network analyzer (VNA) with a test fixture 16047E is used to measure the SiC MOSFETs drain to the heatsink parasitic capacitances, the motor CM impedance and 2-port cable impedances in Fig. 5(a).

Fig. 6 shows the measured and simulated phase A current and voltage with respective commutation times (t_r and t_f). For each switching even, the value of the corresponding commutation time is displayed as a single circle in the graph. Fig. 6(a) shows that the duration of the measured diode to MOSFET commutation are concentrated around 30 ns. However, the MOSFET to diode commutations in the corresponding figure have variable t_r and t_f, which is from 50 ns to the dead time. The variable phase current is the main reason for the variable commutation times, as it was defined in (4). Furthermore, the measured and simulated results in Fig. 6 show that the proposed time-domain voltage model replicates the SiC MOSFETs switching behaviors, such as t_r and t_f of both commutation types.

The short window discrete Fourier transform (SWDFT) is applied to get the frequency domain peak value of the voltage and current spectrums, as described in [17]. Fig. 7 shows that the prediction of CM voltage has good agreement with the measured CM voltage. In the high-frequency range (10-30 MHz), there are 5-20 dBµV mismatches, and it can be explained by the non-linear switching behavior of SiC MOSFET, which was ignored in the proposed model.

Next, the CM current contributions generated due to U_{CM} and U_{CM2} will be investigated. The measured voltage spectrums in Fig. 7 are used to maximize the accuracy of the CM current in the simulation. The proposed CM current prediction model in Fig. 5(b) and Fig. 5(c) are simulated in the frequency-domain.

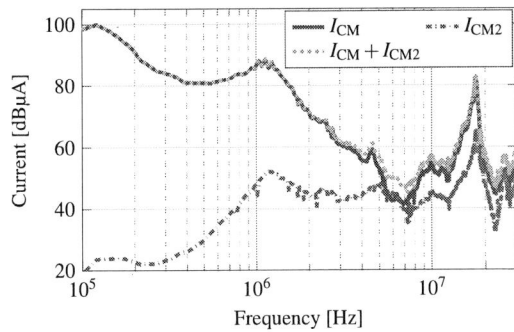

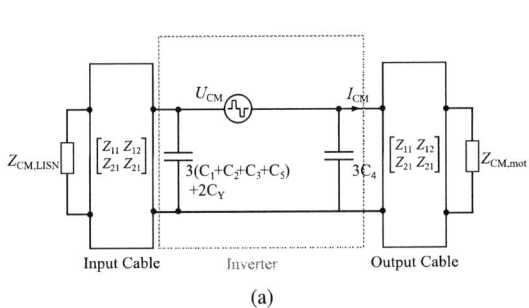

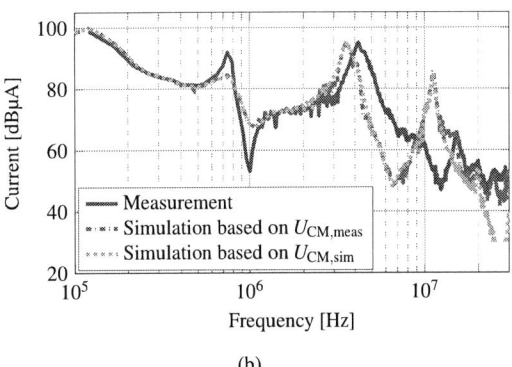

Fig. 8: Comparison between simulated I_{CM} and I_{CM2}.

Fig. 9: Comparison between measured and simulated CM currents.

Fig. 10: (a) CM noise model after adding Y-capacitors between dc-link capacitor points and the heatsink. (b) Comparison between measured and simulated CM currents after adding Y-capacitors.

Fig. 8 shows separately calculated CM currents (I_{CM} and I_{CM2}) along with the overall CM current (I_{CM} + I_{CM2}) which is calculated based on the superposition principle. It can be observed that there is almost no effect of I_{CM2} on the overall CM current until 5 MHz. Although I_{CM2} relatively increases the overall CM current at high frequencies, I_{CM} is still the dominant CM current. Therefore, the CM current generated as a consequence of U_{CM2} can be ignored, and the model in Fig. 5(b) can be used as the final CM EMI noise model.

Fig. 9 shows the comparison between the measured and the simulated CM current results. First, the measured voltage spectrum of Fig. 7(a) ($U_{CM,meas}$) is used in the CM noise model. Even though some differences are observed in the high-frequency range, simulation and measurement results are in good agreement. The high-frequency mismatches between the measurement and simulation based on $U_{CM,meas}$ are due to the neglected stray inductance of the dc capacitors, the measured impedance errors and the mixed mode noise effect. Second, the simulated voltage spectrum in Fig. 7(a) ($U_{CM,sim}$) is used to find the CM current. The results in Fig. 9 show that the proposed CM current model is capable of sufficient accuracy when the simulated CM voltage based on SiC MOSFETs switching behavior is used. Furthermore, two Y-capacitors (C_Y) are connected between dc-link capacitors points (P and N) and the heatsink in order to analyze the proposed method under different situations. The CM noise model with the added Y-capacitors is given in Fig. 10(a), and the measured CM current is compared with the simulation results in Fig. 10(b). The change of the CM current is adequately covered in the new situation.

Conclusion

This article studied the CM voltage and current modeling of a $3LT^2I$. First, the noise sources and propagation paths of the $3LT^2I$ were extracted. Then, a circuit model was developed by connecting the inverter model to the other noise propagation paths of the ASD system. The simulation result of the CM voltage and current of the model showed a good agreement with measurements, although some differences were observed in the high-frequency range due to the simplifications.

References

[1] U.-M. Choi, J.-S. Lee, and K.-B. Lee, "New modulation strategy to balance the neutral-point voltage for three-level neutral-clamped inverter systems," IEEE Transactions on Energy Conversion, vol. 29, no. 1, pp. 91–100, 2014.

[2] M. Schweizer and J. W. Kolar, "Design and implementation of a highly efficient three-level T-type converter for low-voltage applications," IEEE Transactions on Power Electronics, vol. 28, no. 2, pp. 899–907, 2012.

[3] H. Chen, Y. Yan, and H. Zhao, "Extraction of common-mode impedance of an inverter-fed induction motor," IEEE Transactions on Electromagnetic Compatibility, vol. 58, no. 2, pp. 599–606, 2016.

[4] J. Luszcz and K. Iwan, "AC motor transients and EMI emission analysis in the ASD by parasitic resonance effects identification," in 2007 European Conference on Power Electronics and Applications. IEEE, 2007, pp. 1–9.

[5] G. Vidmar and D. Miljavec, "A universal high-frequency three-phase electric-motor model suitable for the delta-and star-winding connections," IEEE Transactions on Power Electronics, vol. 30, no. 8, pp. 4365–4376, 2015.

[6] H. Bishnoi, P. Mattavelli, R. Burgos, and D. Boroyevich, "EMI behavioral models of DC-fed three-phase motor drive systems," IEEE Transactions on Power Electronics, vol. 29, no. 9, pp. 4633–4645, 2014.

[7] F. Costa, C. Vollaire, and R. Meuret, "Modeling of conducted common mode perturbations in variable-speed drive systems," IEEE Transactions on Electromagnetic Compatibility, vol. 47, no. 4, pp. 1012–1021, 2005.

[8] D. Han, S. Li, Y. Wu, W. Choi, and B. Sarlioglu, "Comparative analysis on conducted CM EMI emission of motor drives: WBG versus Si devices," IEEE Transactions on Industrial Electronics, vol. 64, no. 10, pp. 8353–8363, 2017.

[9] B. Revol, J. Roudet, J.-L. Schanen, and P. Loizelet, "EMI study of three-phase inverter-fed motor drives," IEEE Transactions on Industry Applications, vol. 47, no. 1, pp. 223–231, 2010.

[10] H. Zhang, L. Yang, S. Wang, and J. Puukko, "Common-mode EMI noise modeling and reduction with balance technique for three-level neutral point clamped topology," IEEE Transactions on Industrial Electronics, vol. 64, no. 9, pp. 7563–7573, 2017.

[11] J. Wang, X. Liu, Y. Xun, and S. Yu, "Common mode noise reduction of three-level active neutral point clamped inverters with uncertain parasitic capacitance of photovoltaic panels," IEEE Transactions on Power Electronics, vol. 35, no. 7, pp. 6974–6988, 2020.

[12] S. Gulur, V. M. Iyer, and S. Bhattacharya, "Improved common mode noise models for three level t-type neutral point clamped converters," in 2018 IEEE Energy Conversion Congress and Exposition (ECCE). IEEE, 2018, pp. 6398–6403.

[13] H. Hizarci, U. Pekperlak, and U. Arifoglu, "Conducted emission suppression using an EMI filter for grid-tied three-phase/level T-type solar inverter," IEEE Access, vol. 9, pp. 67 417–67 431, 2021.

[14] F. A. Kharanaq, A. Emadi, and B. Bilgin, "Analytical EMI modeling of an active neutral point clamped inverter," in IECON 2021–47th Annual Conference of the IEEE Industrial Electronics Society. IEEE, 2021, pp. 1–5.

[15] E. Gubia, P. Sanchis, A. Ursúa, J. López, and L. Marroyo, "Frequency domain model of conducted EMI in electrical drives," IEEE Power Electronics Letters, vol. 3, no. 2, pp. 45–49, 2005.

[16] Y. Xiang, X. Pei, W. Zhou, Y. Kang, and H. Wang, "A fast and precise method for modeling EMI source in two-level three-phase converter," IEEE Transactions on Power Electronics, vol. 34, no. 11, pp. 10 650– 10 664, 2019.

[17] D. Drozhzhin, V. Karakasli, and G. Griepentrog, "Comprehensive analysis of converter output voltage for conducted noise simulation," in 2019 International Symposium on Electromagnetic Compatibility-EMC EUROPE. IEEE, 2019, pp. 42–47.

[18] J. Xue and F. Wang, "Mixed-mode EMI noise in three-phase DC-fed PWM motor drive system," in 2013 IEEE Energy Conversion Congress and Exposition. IEEE, 2013, pp. 4312–4317.

[19] D. Drozhzhin and G. Griepentrog, "Simulation of conducted noise of an AC drive by means of mixed mode 6-port networks," in 2018 International Symposium on Electromagnetic Compatibility (EMC EUROPE). IEEE, 2018, pp. 1–6.

[20] V. Karakasli, G. Griepentrog, J. Wei, and D. Drozhzhin, "Mixed-mode effect on motor common mode current," The Applied Computational Electromagnetics Society Journal (ACES), vol. 35, no. 11, p. 1374–1375, Nov. 2020.

[21] J. H. Seo, C. H. Choi, and D. S. Hyun, "A new simplified space-vector PWM method for three-level inverters," IEEE Transactions on power electronics, vol. 16, no. 4, pp. 545–550, 2001.

[22] B. Agrawal, M. Preindl, B. Bilgin, and A. Emadi, "Estimating switching losses for SiC MOSFETs with non-flat miller plateau region," in 2017 IEEE Applied Power Electronics Conference and Exposition (APEC). IEEE, 2017, pp. 2664–2670.

A Condition Monitoring Scheme for Semiconductor Devices in Modular Multilevel Converters With Cascaded H-Bridge Submodules

Mohsen Asoodar, Mehrdad Nahalparvari, Christer Danielsson*, Hans-Peter Nee
KTH Royal Institute of Technology, School of Electrical Engineering & Computer Science
SE-100 44 Stockholm, Sweden
* Hitachi Energy
Mäster Ahls gata 16, 72212, Västerås, Sweden
E-mails: asoodar@kth.se, mnah@kth.se,
christer.danielsson@hitachienergy.com, hansi@kth.se

Keywords

≪Cascaded H-Bridge≫, ≪Condition monitoring≫, ≪Degradation≫, ≪Device characterisation≫, ≪Diagnostics≫, ≪Estimation technique≫, ≪FACTS≫, ≪Modular Multilevel Converters (MMC)≫, ≪Reliability≫. ≪Semiconductor device≫, ≪Thermal model≫

Abstract

In this paper, a novel online semiconductor device monitoring scheme is presented. The condition monitoring (CM) scheme is based on measuring the ON-state voltage drop of semiconductor devices, and tracking the changes in their ON-state resistance. The proposed solution measures the ON-state voltage of semiconductor devices at a controlled and readily measurable temperature. This allows for accurate CM of semiconductors as it decouples temperature related and degradation related changes in the ON-state voltage. The temperature decoupling is achieved using natural switching redundancies available to modular multilevel converter systems. Hence, the proposed CM scheme does not interfere with the output voltages and currents generated by the converter.

Introduction

Modular multilevel converters (MMCs) have long been the preferred topology for high-voltage and high-power grid-connected applications. This is mainly due to their high efficiency, and high quality voltage and current outputs [1–6]. Specifically, static synchronous compensators (STATCOMs) that are based on the cascaded H-bridge modular multilevel converter (CHB-MMC) are proven to be effective in providing stability to the power grid, increasing the overall power quality of the power network, regulating the grid voltage, and in some cases, regulating the grid frequency as well [1, 3, 5, 7, 8].

CHB-MMC STATCOMs utilize a string of series connected submodules to improve the harmonic performance of the output voltage and current, and reduce potential issues caused by electromagnetic interferences. In STATCOM applications, the submodules are typically based on the full-bridge topology to allow operation in all four operating quadrants. Each CHB-MMC submodule is usually equipped with a dc capacitor unit. These dc capacitors are typically in the millifarad range, resulting in stored energy levels in the tens to hundreds of kilojoules [9, 10]. Hence, in case of a semiconductor device failure, the entire energy of the dc link capacitor will discharge in the semiconductors. If left unprotected, the semiconductor modules can explode leading to a catastrophic failure in the submodule [11, 12]. Therefore, having knowledge of the health state of the semiconductors is very valuable, and can help avoid such events. Both the dc capacitor and the semiconductors are subject to high currents, and are prone to thermal degradation. Accurate online condition monitoring (CM) methods for dc capacitors in CHB-MMCs have been addressed in the literature [9, 13]. Additionally, in [14] a general monitoring of semiconductors in MMCs has been presented, where the ON-state collector-emitter voltage ($v_{CE,ON}$) is used as the monitoring parameter. The study presents a solution for parameter estimation in noisy environments, but does not address the effect of temperature on $v_{CE,ON}$. In IGBTs, $v_{CE,ON}$ can vary both as a result of degradation, and also when subjected to temperature variations. Hence, in order to accurately monitor the condition of semiconductors, these two effects must be decoupled [15]. Temperature-sensitive electrical parameters (TSEP) have shown to accurately estimate the junction temperature of semiconductor devices [16–19]. These methods typically require an initial measurement characterization, and may not be suitable when the device parameters change due to degradation [20]. Although tracking the junction temperature itself is also a suitable method for CM [21], it is complicated to directly measure the junction temperature of semiconductor devices online. Consequently, a more accurate method is needed for decoupling CM parameters from temperature variations.

The ON-state resistance of semiconductors varies as a result of gate-oxide and package-related degradation. On the other hand, changes in the threshold voltage is shown to be mainly a result of gate oxide related issues [21, 22]. Package-related degradation can be in the form of bond wire lift-off, and solder fatigue [23, 24], both of which result in an increased effective resistance of the semiconductors [23–26]. The ON-state voltage drop can be monitored online using added circuitry that isolates the measurement circuit from high voltages when the device is in OFF-state [27]. The measured ON-state resistance can then be extracted from the slope of the $v_{CE} - i_C$ curve using recursive estimation algorithms [23].

In conventional 2-level and 3-level converters, it is difficult to keep the temperature of the semiconductors constant during normal operation of the converter. This is because there are not enough switching redundancies that allow for isolation of specific semiconductors while the converter is in operation. On the contrary, modular multilevel converters exhibit many redundant switching patterns that result in the same modulated voltage. This paper shows how these redundancies enable accurate CM of semiconductors at a known temperature. Although this study focuses on accurate estimation of the ON-state resistance of semiconductors, the proposed solution can be used to measure any other CM parameter of semiconductors at a constant temperature.

System Description

The CM scheme proposed in this paper is mainly intended for CHB-MMC STATCOMs. This family of STATCOMs are offered in various topologies [7, 28, 29]. The common feature in all these topologies is the existence of multiple arms consisting of series-connected full-bridge submodules. The proposed CM algorithm can be implemented on each arm separately. Hence, the proposed solution can be used for all CHB-MMC topologies. In this paper, a single-phase grid-connected STATCOM consisting of only one arm is considered. In STATCOMs there are typically a few redundant submodules per arm in order to continue the operation in case of failure of one or more submodules [30]. Moreover, the full voltage rating of STATCOMs is only needed when they are operating in full capacitive mode of operation. Therefore, in the majority of operational scenarios, a few of the submodules are redundant, which provide redundant switching. Of the many switching redundancies, one simple solution is to temporarily bypass one submodule and operate the converter using the remaining submodules in the system. In this study, this specific redundancy is used to isolate the semiconductor device of interest, and estimate its ON-state resistance online.

This study is conducted on a single-phase STATCOM with full-bridge submodules, as depicted in Fig. 1(a), and with the system parameters summarized in Table I. A simplified schematic diagram of the full-bridge submodule is shown in Fig .1(b), where the parameter s_i, $i \in \{1, 2, 3, 4\}$ represents the switching command of the corresponding switch. Hence, $s_i = 1$ and $s_i = 0$ correspond to the ON-state and OFF-state of the switch s_i, respectively.

STATCOMs can control the voltage at the point of common coupling (PCC) through injection of reactive power [31]. In order to achieve a desired voltage at the PCC, a reference current is injected into the network. The injected current together with the arm inductance define the required voltage that is generated by the converter. This behaviour is summarized in the following Kirchhoff's voltage law (KVL)

$$v_{\text{PCC}} = v_{\text{conv}} - L_{\text{arm}}\frac{\mathrm{d}i_{\text{arm}}}{\mathrm{d}t} = v_{\text{grid}} + L_{\text{grid}}\frac{\mathrm{d}i_{\text{arm}}}{\mathrm{d}t}, \tag{1}$$

where L_{arm} and i_{arm} represent the arm inductance and arm current, respectively.

Table I: System parameters

Parameter	Symbol	Value	Unit
Nominal power	S_{base}	33	MVA
Nominal arm current (RMS)	i_{base}	0.6	kA
Submodule capacitance	C_{cell}	10	mF
Carrier frequency	f_c	87.5	Hz
Number of submodules	N	22	–

The STATCOM shown in Fig. 1(a) is operated in current control mode using a proportional resonant controller as depicted in Fig. 1(c). Moreover, the converter is equipped with an overall energy control using the sum voltage of all submodule dc capacitors. All capacitor voltages are also controlled to their mean value using a voltage balancing control scheme. The modulation scheme used for this study is the phase-shifted pulse width modulation (PS-PWM); however, the proposed method can be used together with any other modulation scheme as well. The parameters opted for this study are based on the 5SNA0800N330100 HiPak module from Hitachi Energy.

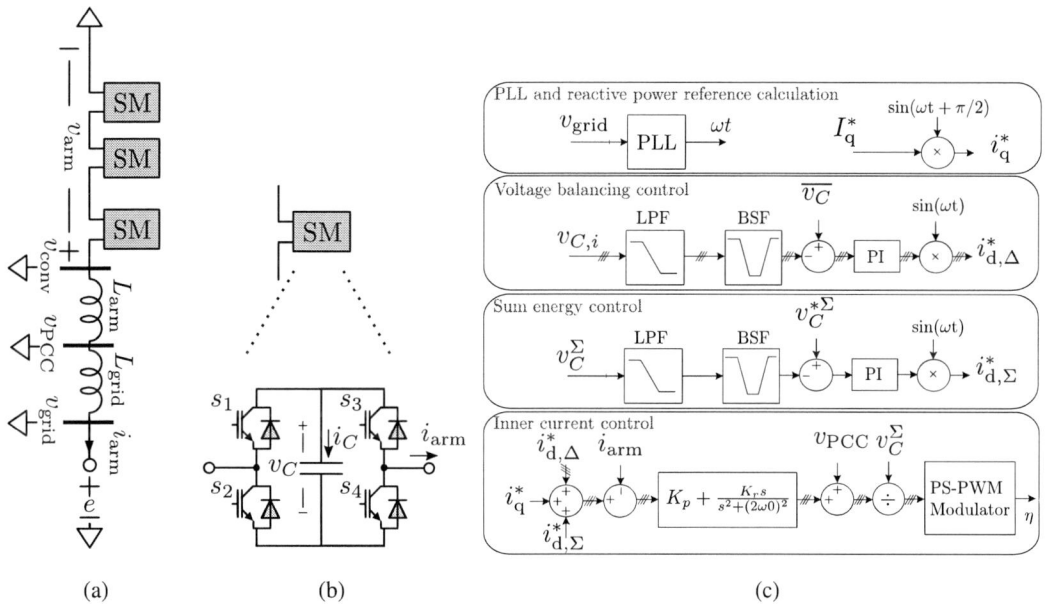

Fig. 1: (a) Single phase CHB-MMC STATCOM. (b) Full-bridge submodule. (c) Closed-loop control system.

Online Condition Monitoring Scheme

The proposed online monitoring scheme is based on online curve tracing of semiconductor devices in CHB-MMCs. The procedure for this curve tracing is as follows:

1. Bypass the selected submodule though (s_1, s_3) or (s_2, s_4) semiconductors.
2. Re-arrange the phase-shifted carriers in the remaining submodules in order to maintain a low total harmonic distortion (THD).
3. Actively discharge the selected submodule to a desired voltage.
4. Bypass the selected submodule through (s_1, s_3) or (s_2, s_4) semiconductors and wait for the non-conducting semiconductor to cool down to a known temperature.
5. Periodically insert the non-conducting semiconductor for short durations.

6. Measure the ON-state voltage during each short ON-state duration.
7. Recharge the submodule to its nominal value.
8. Change the modulation to its initial state where all submodules are actively operated.

These steps are explained in the following subsections.

Submodule Isolation

The PS-PWM scheme in MMCs provides a modulated signal with low harmonic content. In full-bridge MMCs, this low THD is achieved when carrier signals of the submodules are shifted by $180°/N$, where N is the number of submodules in each arm. In order to maintain a low THD, the carriers must be rearranged after a submodule is bypassed. Hence, after bypassing a submodule, the carrier signals of the active submodules are phase shifted by $180°/(N-1)$ instead. In this study, the transition of the carrier phases is done linearly and over a predefined time period of one millisecond.

After bypassing the submodule and rearranging the carrier signals, the junction temperature of its non-conducting switches will cool down to the case temperature of the device in a matter of seconds. This is due to the small thermal capacity of industrial-type semiconductor device modules. Unlike the junction temperature, measuring the case temperature or heat sink temperature is trivial. This is mainly because the heat sink is accessible, and it is not expected to have fast temperature changes. After allowing the junction temperature of the device to settle to the temperature of the heat sink, it is possible to extract its ON-state voltage at a known junction temperature. The proposed procedure for conducting this measurement is to temporarily toggle from one bypass switch group to the other, sample the ON-state voltage, and toggle back to the initial state. For example, if s_4 in Fig. 1(b) is to be monitored, initially the switch group (s_1, s_3) are ON and the switch group (s_2, s_4) are OFF. Then, for the purpose of sampling the ON-state voltage of s_4, (s_1, s_3) are turned OFF and (s_2, s_4) are turned ON for a short period. This procedure is shown in Fig. 2. Repeating this process at different arm current levels allows for online curve tracing of the semiconductor at a known temperature.

After a cell is bypassed, the sum energy and voltage balancing controllers of Fig. 1(c) must be manipulated accordingly. For the sum energy control, the direct voltage of the bypassed cell must be replaced with its reference value in v_C^Σ. Moreover, the direct voltage of the bypassed submodule must be replaced with the average voltage of the remaining active submodules in $\overline{v_C}$. These manipulations allow for the active submodules to operate independently from the bypassed submodule.

Although it is possible to conduct the sampling procedure of Fig. 2 when the bypassed submodule is at rated voltage, the switching losses that occur as a result of $v_{CE,ON}$ sampling can increase the junction temperature and reduce the accuracy of the measurements. Hence, it is beneficial to reduce the submodule voltage prior to sampling $v_{CE,ON}$. After bypassing, the remaining $N-1$ active submodules are continuously operated in current control mode, while the bypassed submodule can be independently operated. In order to discharge the submodule of interest, it can be temporarily forced to provide a modulated sinusoidal voltage that is $180°$ out of phase with respect to the arm current. As a result of this modulation, the selected submodule gradually discharges. In order to allow selective conducting and non-conducting switch groups of (s_1, s_3) and (s_2, s_4), complete discharge of the submodule should be avoided.

The temporary discharge operation is shown in Fig. 3, where the capacitor voltage of one selected submodule is lowered actively. The procedure is divided into six phases. In P_1, the submodule of interest is bypassed. During P_2, the submodule is actively discharged. In P_3, the ON-state voltage sampling according to Fig. 2 is conducted. Once the data acquisition is finalized, the submodule is actively recharged in P_4, temporarily bypassed in P_5, and actively operated in normal condition in P_6. It is best to conduct the sampling procedure of Fig. 2 at different current levels. Doing so replicates a similar procedure to how commercial curve tracers extract the $v_{CE} - i_C$ data curves. In order to not heat up the semiconductor during the data acquisition process, it is essential to keep the duration of the sampling pulse short. The length of this pulse can vary depending on the size of the semiconductor module. In this study, a sampling pulse of 50 μs is considered. Once the online data acquisition for the two non-conducting semiconductors is finalized, the function of the upper and lower semiconductors can be switched, and the second set of previously conducting semiconductors can now be curve traced. Applying this method one submodule at a time allows for accurate online monitoring of semiconductors at readily measurable temperatures.

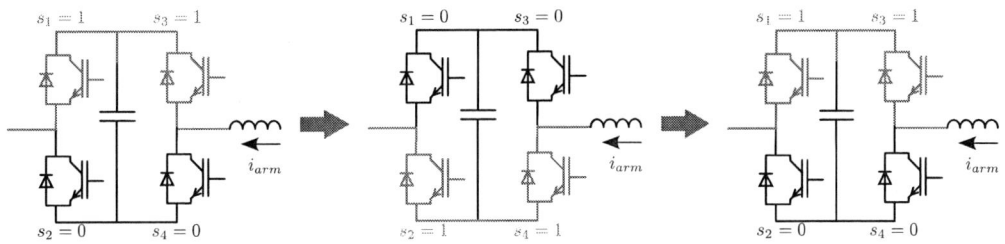

Fig. 2: Procedure for sampling $v_{CE,\text{ON}}$ in a bypassed submodule

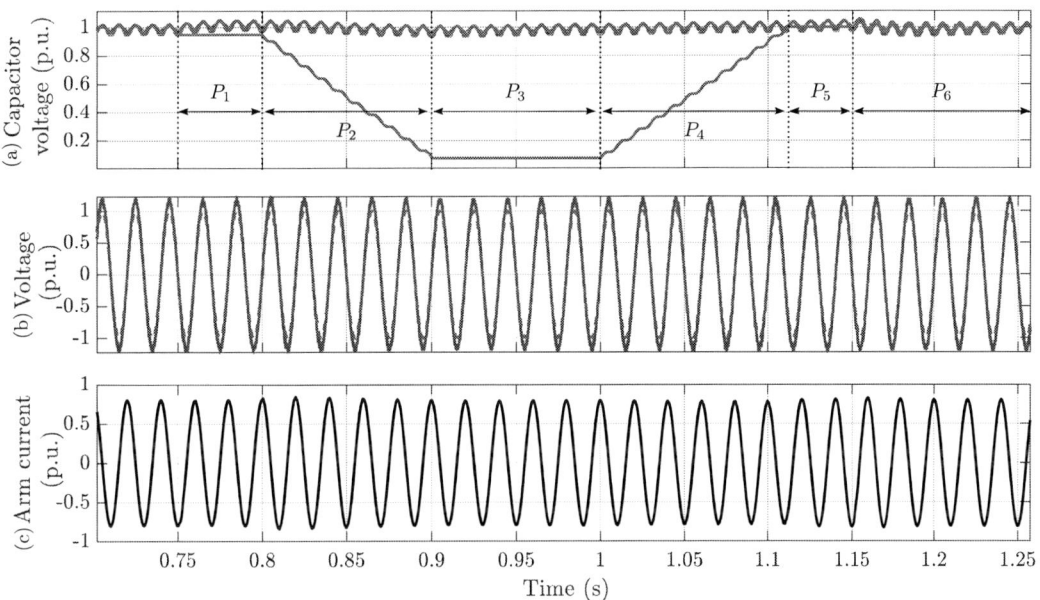

Fig. 3: Selective submodule bypass (P_1), discharge (P_2), sampling (P_3), and recharge (P_4), during normal operation of the converter. a) Capacitor voltages with selected module discharged to 10% nominal value. b) Grid voltage (dashed line) and converter voltage (solid line). c) Arm current of the converter at 0.8 p.u., capacitive.

Loss and Thermal Analysis

In this study, the $v_{CE} - i_C$ curve of the semiconductor is extracted from the data sheet and placed as a lookup table inside the simulation model in MATLAB/Simulink as shown in Fig. 4(a). Hence, the ON-state voltage of each semiconductor module changes in the simulation according to the data sheet parameters. In the simulation model, the semiconductor data are sampled at a predefined frequency. The sampling frequency should not be an integer multiple of the fundamental frequency of the network. With this choice, the ON-state voltages are naturally sampled at different points in the converter's current waveform. As a result, a $v_{CE} - i_C$ curve is traced online. Due to the small duration of the sampling pulse, small power losses are expected in the semiconductor. The losses of the semiconductor under test (SUT) are divided into switching losses and conduction losses. The conduction and switching losses are defined as

$$P_{\text{cond}}(t, T_j) = v_{CE,\text{ON}}(t, T_j).i_{CE}(t) \tag{2}$$

$$E_{\text{sw}}(t, T_j) = E_{\text{sw}}(T_{j,\text{ref}}, v_{CE,\text{ref}}, i_C).(1 + k_{T_1}.(T_j(t) - T_{\text{ref}})).(\frac{v_{CE}}{v_{\text{ref}}})^{k_v}, \tag{3}$$

where k_{T_1} and k_v are extracted from the data sheet [32].

The ON-state voltage $v_{CE,\text{ON}}$ in (2) can be simplified as,

$$v_{CE,\text{ON}}(t, T_j) = v_{CE,0}(1 + k_{T_2}.(T_j(t) - T_{\text{ref}})) + r_{CE}(1 + k_{T_3}.(T_j(t) - T_{\text{ref}})).i_{CE}(t), \tag{4}$$

where k_{T_2} and k_{T_3} represent the temperature coefficient of the intercept and slope of the $v_{CE} - i_C$ curve, respectively.

According to (3), a lower voltage over the semiconductor results in lower switching losses. This is the main reason for reducing the voltage of the submodule prior to sampling the SUT. Moreover, due to the short period of the sampling instant, the conduction losses are negligible. The conduction and switching losses are implemented in the MATLAB/Simulink simulation environment. In the simulation model, the TSEPs of (2) and (3) are dynamically updated based on their calculated temperature from the previous simulation time step.

The thermal model of the 5SNA0800N330100 IGBT is shown in Fig. 5(a) with parameters summarized in Table II. The thermal model from junction to case is typically provided in the data sheet in the Foster representation. The thermal model and parameters shown in this paper are based on the equivalent Cauer model using the Foster model parameters. A heat sink with a coolant temperature of 25°C is assumed in this study. The coolant temperature is typically higher in practice. Nevertheless, the purpose of this study is to show that the junction temperature does not vary significantly from the initially stabilized coolant temperature as a result of sampling with the proposed method. Consequently, the hypothesis holds regardless of the initial coolant temperature.

Table II: Thermal model parameters of the 5SNA0800N330100

Symbol	Value	Unit
$R_{th,1}$	2.5	K/kW
$R_{th,2}$	4.7	K/kW
$R_{th,3}$	1.9	K/kW
$R_{th,4}$	0.7	K/kW
$C_{th,1}$	1.8	J/K
$C_{th,2}$	9.5	J/K
$C_{th,3}$	192	J/K
$C_{th,4}$	4913	J/K

Simulation results in Fig. 5(b) reveal that for a sampling pulse duration of 50 μs, and a sampling frequency of 10.37 Hz, the increase in semiconductor chip temperature is negligible. Hence, it can be assumed that all sampled points are achieved at a fixed temperature of the semiconductor's heat sink. The sampled voltage and current data can then be used to identify the $v_{CE} - i_C$ curve of each semiconductor online. This is shown in Fig. 4(b). In this simulation, the effect of noise is omitted; however, in the presence of large noise levels, it is possible to fit a linear curve to the resistive part of the $v_{CE} - i_C$ curve and extract the ON-state resistance of the device [14].

Conclusion

In this paper, a novel method of online health monitoring of semiconductors in CHB-MMCs is presented. The proposed method has the advantage of curve tracing semiconductors at a known and easily

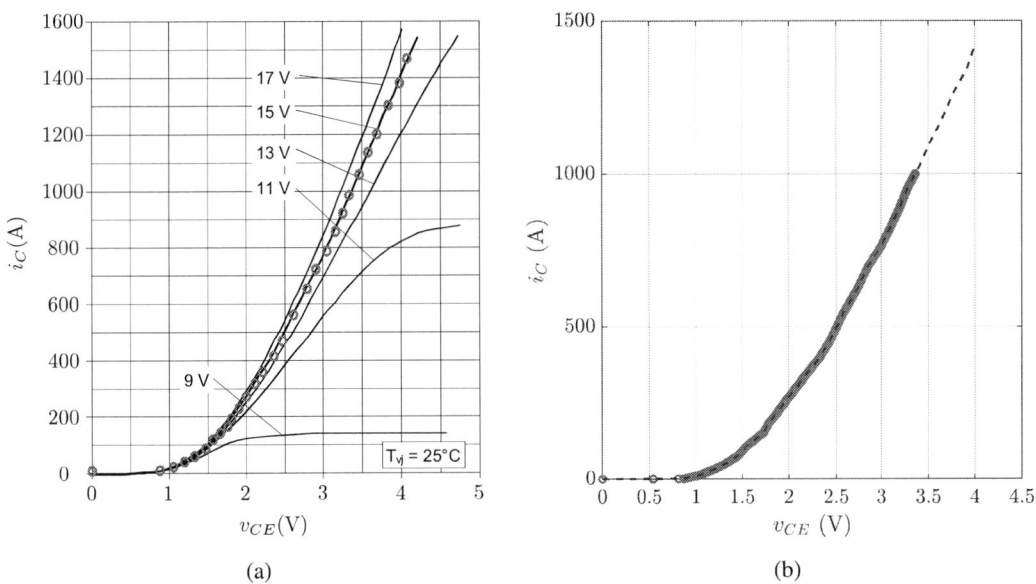

(a)

(b)

Fig. 4: (a) $v_{CE} - i_C$ characteristics of 5SNA0800N330100 extracted from the datasheet (b) $v_{CE} - i_C$ characteristics of 5SNA0800N330100 at VGE = 15 V, and T_j = 25 °C from datasheet (blue) and through online estimation (red).

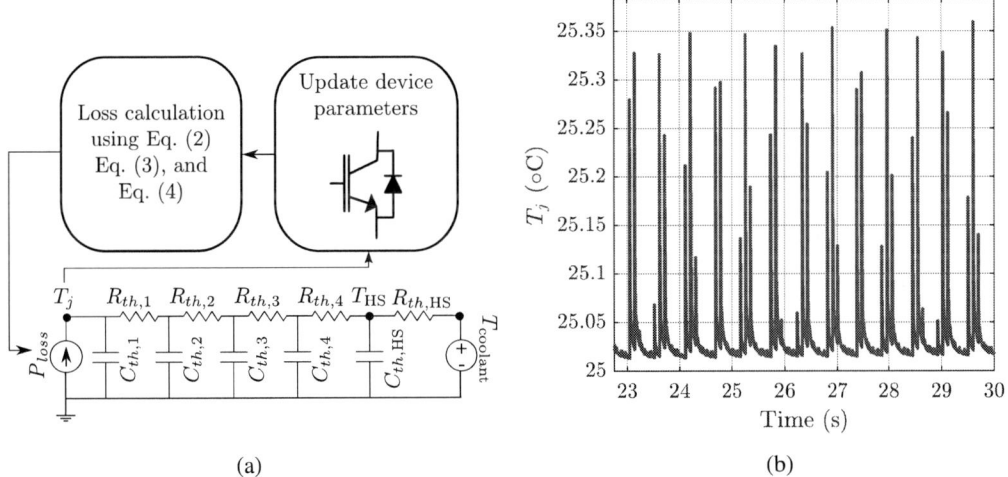

(a)

(b)

Fig. 5: (a) Thermal model of 5SNA0800N330100. (b) Semiconductor junction temperature variations when a peak arm current of 1 kA, a reduced submodule voltage of 200 V, a sampling frequency of 10.37 Hz, and a sampling period of 50 μs is considered.

measurable temperature. Therefore, variations in the ON-state resistance of semiconductors caused by degradation can be separated from that caused by temperature variations. Although the proposed method has been investigated for CHB-MMCs in STATCOM applications, it can be applied to all MMC-based topologies that comprise full-bridge submodules.

References

[1] S. Sirisukprasert, A. Huang, and J.-S. Lai, "Modeling, analysis and control of cascaded-multilevel converter-based STAT-COM," in *IEEE Power Engineering Society General Meeting*, vol. 4, 2003, pp. 2561–2568.

[2] F. Peng and J. Wang, "A universal STATCOM with delta-connected cascade multilevel inverter," in *IEEE 35th Annual Power Electronics Specialists Conference*, vol. 5, 2004, pp. 3529–3533.

[3] F. Z. Peng and Jin Wang, "A universal STATCOM with delta-connected cascade multilevel inverter," in *IEEE 35th Annual Power Electronics Specialists Conference*, vol. 5, 2004, pp. 3529–3533.

[4] S. Chou, B. Wang, S. Chen, C. Lee, P. Cheng, H. Akagi, and P. Barbosa, "Average power balancing control of a STATCOM based on the cascaded H-bridge PWM converter with star configuration," in *IEEE Energy Conversion Congress and Exposition*, 2013, pp. 970–977.

[5] N. Thitichaiworakorn, M. Hagiwara, and H. Akagi, "Experimental verification of a modular multilevel cascade inverter based on double-star bridge cells," *IEEE Trans. Ind. App.*, vol. 50, no. 1, pp. 509–519, Jan.-Feb. 2014.

[6] S. Sinsel, I. Ramsay, W. Hörger, A. Janke, and M. A. Alegria, "Multilevel STATCOMs — a new converter topology that opens up the market," in *IEEE PES T&D Conference and Exposition*, 2014, pp. 1–5.

[7] T. Soong and P. W. Lehn, "Evaluation of emerging modular multilevel converters for BESS applications," *IEEE Trans. Power Deliv.*, vol. 29, no. 5, pp. 2086–2094, Oct. 2014.

[8] ——, "Control of energy storage enabled modular multilevel converters with reduced storage requirements," in *15th Workshop on Control and Modeling for Power Electronics (COMPEL)*, 2014, pp. 1–7.

[9] M. Asoodar, M. Nahalparvari, C. Danielsson, R. Söderström, and H.-P. Nee, "Online health monitoring of dc-link capacitors in modular multilevel converters for FACTS and HVDC applications," *IEEE Trans. Power Electron.*, vol. 36, no. 12, pp. 13 489–13 503, Dec. 2021.

[10] K. Ilves, S. Norrga, L. Harnefors, and H.-P. Nee, "On energy storage requirements in modular multilevel converters," *IEEE Trans. Power Electron.*, vol. 29, no. 1, pp. 77–88, Jan. 2014.

[11] D. Li, F. Qi, M. Packwood, A. Islam, L. Coulbeck, X. Li, Y. Wang, H. Luo, X. Dai, and G. Liu, "Explosion mechanism investigation of high power IGBT module," in *19th International Conference on Thermal, Mechanical and Multi-Physics Simulation and Experiments in Microelectronics and Microsystems (EuroSimE)*, 2018, pp. 1–5.

[12] M. Billmann, D. Malipaard, and H. Gambach, "Explosion proof housings for IGBT module based high power inverters in HVDC transmission application," in *Proc. PCIM Eur. Conf., Nuremberg*, 2009, pp. 352–257.

[13] C. Liu, F. Deng, Q. Yu, Y. Wang, F. Blaabjerg, and X. Cai, "Submodule capacitance monitoring strategy for phase-shifted carrier pulsewidth-modulation-based modular multilevel converters," *IEEE Trans. Ind. Electron.*, vol. 68, no. 9, pp. 8753–8767, Sep. 2021.

[14] S. Chen, S. Ji, L. Pan, C. Liu, and L. Zhu, "An on-state voltage calculation scheme of MMC submodule IGBT," *IEEE Trans. Power Electron.*, vol. 34, no. 8, pp. 7996–8007, Aug. 2019.

[15] V. Smet, F. Forest, J.-J. Huselstein, A. Rashed, and F. Richardeau, "Evaluation of V_{ce} monitoring as a real-time method to estimate aging of bond wire-IGBT modules stressed by power cycling," *IEEE Trans. Ind. Electron.*, vol. 60, no. 7, pp. 2760–2770, Jul. 2013.

[16] H. Luo, W. Li, F. Iannuzzo, X. He, and F. Blaabjerg, "Enabling junction temperature estimation via collector-side thermo-sensitive electrical parameters through emitter stray inductance in high-power IGBT modules," *IEEE Trans. Ind. Electron.*, vol. 65, no. 6, pp. 4724–4738, Jun. 2018.

[17] N. Baker, S. Munk-Nielsen, F. Iannuzzo, and M. Liserre, "IGBT junction temperature measurement via peak gate current," *IEEE Trans. Power Electron.*, vol. 31, no. 5, pp. 3784–3793, May 2016.

[18] Y. Avenas, L. Dupont, and Z. Khatir, "Temperature measurement of power semiconductor devices by thermo-sensitive electrical parameters—a review," *IEEE Trans. Power Electron.*, vol. 27, no. 6, pp. 3081–3092, Jun. 2012.

[19] V. Sundaramoorthy, E. Bianda, R. Bloch, D. Angelosante, I. Nistor, G. Riedel, F. Zurfluh, G. Knapp, and A. Heinemann, "A study on IGBT junction temperature (Tj) online estimation using gate-emitter voltage (Vge) at turn-off," *Microelectron. Rel.*, vol. 54, no. 11, pp. 2423–2431, Nov. 2014.

[20] F. Yang, E. Ugur, and B. Akin, "Evaluation of aging's effect on temperature-sensitive electrical parameters in SiC MOS-FETs," *IEEE Trans. Power Electron.*, vol. 35, no. 6, pp. 6315–6331, Jun. 2020.

[21] H. Oh, B. Han, P. McCluskey, C. Han, and B. D. Youn, "Physics-of-failure, condition monitoring, and prognostics of insulated gate bipolar transistor modules: A review," *IEEE Trans. Power Electron.*, vol. 30, no. 5, pp. 2413–2426, May 2015.

[22] E. Ugur, C. Xu, F. Yang, S. Pu, and B. Akin, "A new complete condition monitoring method for SiC power MOSFETs," *IEEE Trans. Ind. Electron.*, vol. 68, no. 2, pp. 1654–1664, Feb. 2021.

[23] M. A. Eleffendi and C. M. Johnson, "In-service diagnostics for wire-bond lift-off and solder fatigue of power semiconductor packages," *IEEE Trans. Power Electron.*, vol. 32, no. 9, pp. 7187–7198, Sep. 2017.

[24] G. J. Riedel and M. Valov, "Simultaneous testing of wirebond and solder fatigue in IGBT modules," in *8th International Conference on Integrated Power Electronics Systems*, 2014, pp. 1–5.

[25] U.-M. Choi, F. Blaabjerg, S. Jørgensen, S. Munk-Nielsen, and B. Rannestad, "Reliability improvement of power converters by means of condition monitoring of IGBT modules," *IEEE Trans. Power Electron.*, vol. 32, no. 10, pp. 7990–7997, Oct 2017.

[26] J. M. Anderson and R. W. Cox, "On-line condition monitoring for MOSFET and IGBT switches in digitally controlled drives," in *IEEE Energy Conversion Congress and Exposition*, 2011, pp. 3920–3927.

[27] Y. Peng and H. Wang, "A comparative study on converter-level on-state voltage measurement circuits for power semiconductor devices," in *IEEE Energy Conversion Congress and Exposition (ECCE)*, 2021, pp. 3638–3643.

[28] P.-H. Wu, H.-C. Chen, Y.-T. Chang, and P.-T. Cheng, "Delta-connected cascaded H-bridge converter application in unbalanced load compensation," *IEEE Trans. Ind. App.*, vol. 53, no. 2, pp. 1254–1262, Nov. 2017.

[29] C. D. Townsend, R. A. Baraciarte, Y. Yu, D. Tormo, H. Z. de La Parra, G. D. Demetriades, and V. G. Agelidis, "Heuristic model predictive modulation for high-power cascaded multilevel converters," *IEEE Trans. Ind. Electron.*, vol. 63, no. 8, pp. 5263–5275, Aug. 2016.

[30] W. Song and A. Q. Huang, "Fault-tolerant design and control strategy for cascaded H-Bridge multilevel converter-based STATCOM," *IEEE Trans. Ind. Electron.*, vol. 57, no. 8, pp. 2700–2708, Aug. 2010.

[31] M. Nahalparvari, M. Asoodar, L. Bessegato, S. Norrga, and H.-P. Nee, "Modeling and shaping of the dc-side admittance of a modular multilevel converter under closed-loop voltage control," *IEEE Trans. Power Electron.*, vol. 36, no. 6, pp. 7294–7306, Jun. 2021.

[32] Y. Zhang, H. Wang, Z. Wang, Y. Yang, and F. Blaabjerg, "Simplified thermal modeling for IGBT modules with periodic power loss profiles in modular multilevel converters," *IEEE Trans. Ind. Electron.*, vol. 66, no. 3, pp. 2323–2332, Mar. 2019.

Particular Requirements on Drive Inverters for Safe and Robust Operation on an Open Industrial DC Grid

Simon Puls[1], Jan-Niklas Koch[2], Martin Ehlich[3], Holger Borcherding[4]

[1,3]Lenze SE	[2,4]OWL University of Applied Sciences and Arts
Breslauer Str. 3	Campusallee 12
32699 Extertal, Germany	32657 Lemgo, Germany
+49 5154 82-1528 /-2066	+49 5261 702-5597 /-5217
simon.puls@lenze.com	jan-niklas.koch@th-owl.de
martin.ehlich@lenze com	holger.borcherding@th-owl.de
http://www.lenze.com	http://www.th-owl.de

Acknowledgments

The Authors gratefully acknowledge financial support by the German Federal Ministry of Economic Affairs and Climate Action (BMWK) via grant Numbers 03ET7558A to N (Project acronym: "DC-INDUSTRIE") and 03EI6002A to Q (Project acronym: "DC-INDUSTRIE2").

Supported by:

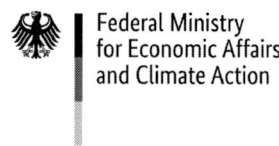

on the basis of a decision by the German Bundestag

Keywords

≪Lifetime of DC-link capacitor≫, ≪Fault handling strategy≫, ≪Short circuit≫, ≪DC voltage control≫, ≪Grid-connected inverter≫.

Abstract

Today DC offers far-reaching advantages over AC. Therefore, many devices have been equipped with an internal DC link for years. In the field of energy supply, the use of DC technology is also growing and is state of the art e.g. in offshore, high-voltage, automotive and data center applications. The spread of industrial open DC grids is currently starting and is completely different due to the requirements: The DC grid itself and energy flows in an industrial environment are highly dynamic and bidirectional. Due to the low impedance electrical connection of the DC links of many devices, stored energies in fault cases as well as ripple currents during operation place particular requirements on the devices.

Introduction

DC grids offer a series of advantages in the industrial sector, especially for voltage source inverter-based electric drives. For this reason, proprietary DC grids in the cabinet have been state of the art for many years. Hence there is a need for extended, open DC grids in the industrial environment. The elimination of the rectifier and the direct electrical coupling makes it possible to exchange energy, e.g. generative

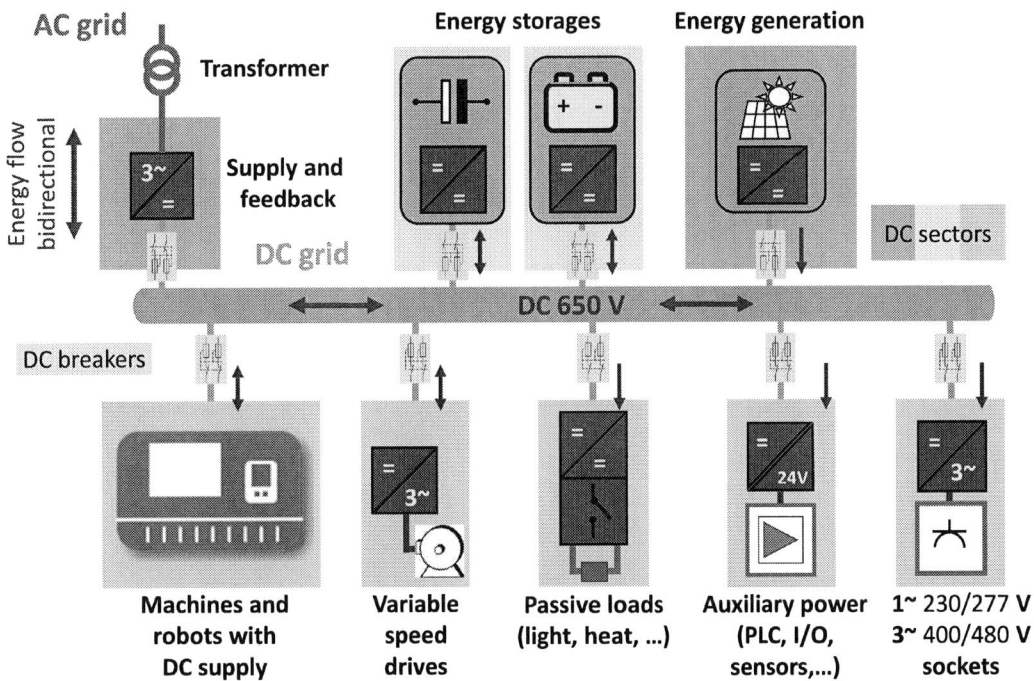

Fig. 1: Concept for an open industrial DC grid [2]

energy of drives in braking operation. Consequently braking resistors can be completely eliminated and 20 - 30 % energy can be saved. The potential savings increase significantly during intermittent operation, for example in intralogistics for storage and retrieval machines or industrial robots with short and fast accelerations and decelerations [1].

Due to the fact that devices - especially electrical drives - in industrial environments often operate very dynamically and the flow of electrical power is correspondingly highly dynamic, the DC grid itself must also offer a high degree of dynamics. In relation to the nominal power of the DC grid, this results in a high capacity and low inductance, which also causes the short-circuit power to increase dramatically in the event of a fault [3] and transient overvoltages to reach high maximum levels [4]. Furthermore, due to the low impedance connection of the DC links of the devices, ripple currents can flow between individual devices and additionally load them. Devices must be designed for these grid-side influences to ensure safe and robust operation for the intended lifetime.

Concept for an Open Industrial DC Grid

The concept of DC-INDUSTRIE [2] is to connect industrial devices, such as electric drives, to each other via an open DC grid. The AC/DC conversion, which is always necessary with controlled industrial drives, can be carried out at a central point to the AC grid. By eliminating the need for uncontrolled rectifiers, the overall efficiency of plants can be increased and the emitted disturbance of harmonics significantly reduced. A further advantage of DC grids is the easy coupling of storage units and decentralized energy generators, such as photovoltaic systems. If the DC grid is controlled with intelligent grid management, it can react flexibly to grid voltage changes and failures. This increases the availability of the plants. The idea of DC-INDUSTRIE is to integrate devices for dedicated machine functions in load sectors. Each load sector is connected to the DC grid with a DC connection box. The task of the DC connection box is to protect all devices within the load sector from disturbances in the DC grid (overcurrent, over- and undervoltage) and to disconnect the load sector from the DC grid in case of short circuit faults etc. in the inner circle. For this task, very fast DC switches in semiconductor- or hybrid-technology are needed [5].

Infeed sections are managed as single load sectors. The DC grid can be operated at a star point grounded AC transformer or designed as an isolated DC grid (typically with an Active Infeed Controller).

Fig. 1 illustrates the concept as an overview. In addition to drive inverters (centralized or decentralized) and power supply units for coupling to the AC grid, innovative switching and protective devices as well as electromagnetic compatibility (EMC) input filters are required.

Grid-side Influences on Devices

In the industrial environment, devices are exposed to many influences that can affect their lifetime or function. While environmental influences are limited by appropriate housings or mounting locations, for example, electrical influences from the mains supply must also be limited. This is necessary so that the expected lifespan is maintained and the function is ensured - i.e. a device functions safely and robustly on the DC grid. The devices in the industrial DC grid are usually very similar in design, see Fig. 2. The most important components, regardless of the type of converter, are
- an input filter (inductance),
- a DC link (capacitance),
- and a power output stage (semiconductor-based switches).

The capacitances in the DC link, which are usually electrolytic capacitors in drive inverters, age over time [6]. Semiconductor-based switches, on the other hand, whether IGBT, SiC, etc., fail abruptly in the event of overload [7].

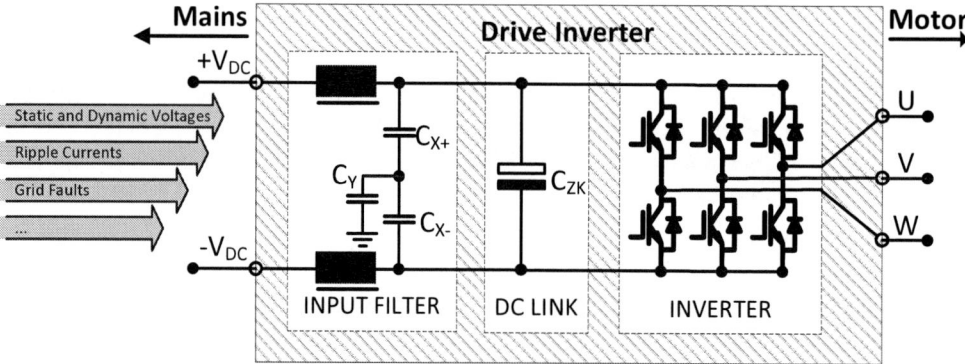

Fig. 2: Grid-side Influences on Devices and a symbolized Drive Inverter with its main components: Input Filter, DC Link and Inverter (left to right)

The most important electrical influences from the grid are:
- Voltage ranges, especially transient overvoltages [8]
- Ripple currents between individual devices [9]
- Device and grid faults [10]

Requirements on Drive Inverters for Safe and Robust Operation

To ensure that devices function safely and robustly when operated in the DC grid, the possible influences mentioned before must not lead to any significant impairments. If the influences and the devices are matched to each other, this can be ensured.

Voltage: Ranges and Transient Overvoltages

The voltage with which the devices are supplied is the decisive factor in the selection of components. A distinction must be made between the supply voltage, which is comparatively static and lies within

Table I: Specified static (right column S4) and dynamic voltage ranges (columns S1 to S3b) [2]

Upper voltage limit Ux for nominal voltage 540 V / 650 V	Voltage band	S1: t < 100 µs	S2: 100 µs ≤ t ≤ 1 ms	S3a: 1 ms ≤ t ≤ 5 s	S3b: 5 s ≤ t ≤ 60 s	S4: t > 60 s
U6: 1500 V	B7	A7				
U5: 1200 V	B6	A6	A7			
U4: 880 V	B5	A6	A7	A7	A7	
U3b: 800 V	B4	A4	A5	A5	A7	A7
U3a: 750 V	B3h	A3	A3	A3	A4	A5
U2: 485 / 600 V	B3l	A3	A3	A3	A3	A3
U1: 400 V	B2	A4	A4	A2	A2	A1
	B1	A4	A2	A1	A1	

Time →

fixed limits, and possible transient overvoltages as well as undervoltages. The reason for this is that different components have different sensitivities. Capacitors, especially electrolytic capacitors, are comparatively robust to transient overvoltages. In other words, voltages that exceed the supply voltage for a few microseconds by a factor of up to two do not necessarily lead to damage. This is mainly due to their parasitic properties, such as the R_{ESR}, R_{Leak} and L_{ESL}, and their ability to absorb energy. Semiconductors, on the other hand, cannot handle overvoltages beyond their rated voltage. In the case of 1200 V rated types used in industrial applications, even slight voltages in addition to this lead to a defect [7]. Static voltages, i.e. voltages that are permanently present, have comparatively little effect on semiconductors. On the other hand, too small voltages lead to increased currents in order to provide the desired power for e.g. a drive. The increased currents lead to further heating in all components involved. In particular, the semiconductors with their small thermal capacity can be overloaded and fail.

Table I shows in detail possible voltages in relation to the time for which they can be present. In the right column S4 : t > 60 s, the voltages and voltage limits that are permanent are indicated:
- > 800 V: Overvoltage (with active protection)
- 750 V to 800 V: Overvoltage (without active protection)
- 485 V to 750 V: Normal working voltage
- < 485 V: Undervoltage

Since transient voltages can also occur in a grid, caused for example by load changes, these are considered normal operating conditions. Accordingly, devices that are operated on the grid must also withstand voltages exceeding their nominal values without negative influence. Table I lists ranges for smaller time intervals in the remaining columns from right to left. Column S1 : t < 100 µs, for example, indicates that voltages up to 1500 V for less than 100 µs are allowed in principle and must be assumed to exist. In order to avoid damage to devices influenced by voltages, they must be designed and components dimensioned accordingly. An inductive component in the input filter in combination with a capacitive DC link capacitor, for example, forms a low-pass filter which can prevent a high, but only short-acting transient overvoltage from affecting semiconductors.

Fig. 3 illustrates an example of transient voltage events triggered by the switching off of a simple load in the DC grid. Even under real conditions, the maximum value of the voltage can easily reach twice the initial value. On the other hand, this example also shows the time frame in which the event and thus a potential danger is over again: In about 10 µs after the switch-off event, the voltage has reached the initial value again.

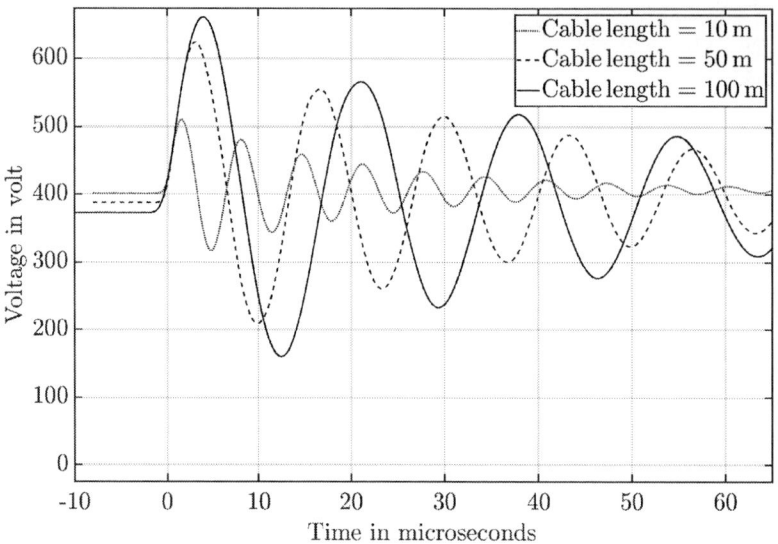

Fig. 3: Measurement of overvoltage events due to a Load Dump with variations in cable length

Ripple Currents between Individual Devices

Due to the fact that the devices are operated on the open industrial DC grid without an input rectifier on the grid (see Fig. 2), equalizing currents can occur between the devices. Only the input filters of the devices and the cable connections separate the individual DC links from each other. A device with a high AC current ratio in the DC link can serve itself on the DC link of another device. In this case, the other device is additionally stressed and can be overloaded if the ripple current is too high. The result is premature aging of a device forced by the mains side.

In order not to overload the components of individual devices, they must be able to handle a mutual ripple current load. This is because these currents cannot be completely avoided. In the case of the DC system concept, there is an agreement that each device has a minimum DC link capacitance: $C_{ZK} \geq 25\,\mu F/kW$. This ensures, on the one hand, that devices emit only limited amounts and, on the other hand, what devices must be able to withstand. A possible solution may also be to connect external capacitors to the DC grid at certain points, e.g. on a rail system for mobile equipment [11].

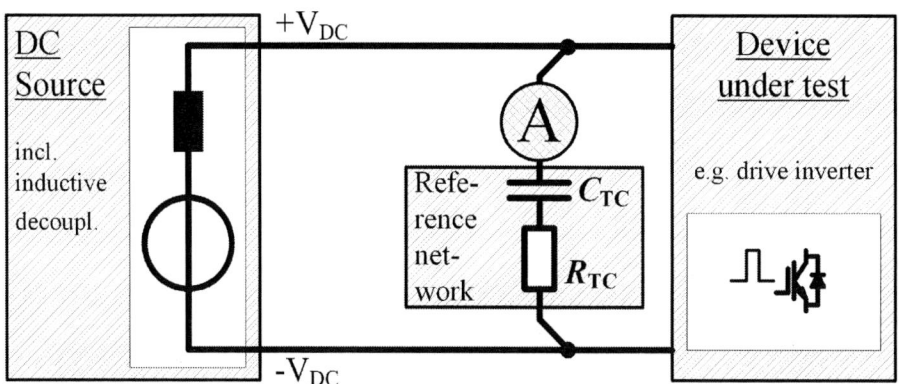

Fig. 4: Ripple current measurement method

Increased ripple currents emanate in particular from devices which use film capacitors as DC link storage.

Due to their increased resistance to ripple current loads, these can be designed smaller in the individual device. For operation in a combined system, on the other hand, the use of such a device means a load on other devices in the system.

To detect a potential overload of electrolytic capacitors, a simple measuring method can be used. Fig. 4 shows the basic setup. Using a simple multimeter, the RMS current is measured that flows into a series circuit of film capacitor and resistor. This circuit is inserted into the DC grid at $+V_{DC}$ and $-V_{DC}$ and can test both individual devices and devices in a combination. With an appropriate design, the circuit emulates the impedance of a real DC link electrolytic capacitor and does not influence the devices due to comparatively small values. With a capacitance of $C_{TC} = 1\,\mu F$ and a resistance of $R_{TC} = 20\,\Omega$, the maximum current is $I_{RMS} = 40\,mA$ for individual devices and $I_{RMS} = 28\,mA$ for devices in combination.

Capacitors play a special role when considering possible loads due to ripple currents, which, unlike other influences, can be permanent and unnoticeable during operation. And in the lifetime, these are a significant part.

The resulting lifetime L_X of an electrolytic capacitor can be calculated by:

$$L_X = L_0 \cdot 2^{\frac{T_0-T_a}{10k}} \cdot K_i^{\left(1-\left(\frac{I_a}{I_0}\right)^2\right)\cdot\frac{\Delta T_0}{10K}} \cdot \left(\frac{U_0}{U_a}\right)^n \quad (1)$$

Each individual factor is based on heat generation.

The single factors of equation (1) are:

L_0 Nominal lifetime

$2^{\frac{T_0-T_a}{10k}}$ Temperature factor

$K_i^{\left(1-\left(\frac{I_a}{I_0}\right)^2\right)\cdot\frac{\Delta T_0}{10K}}$ Ripple current factor

$\left(\frac{U_0}{U_a}\right)^n$ Voltage factor

Table II: Parameters for the lifetime calculation equation (1)

L_X	resulting lifetime	L_0	nominal lifetime	K_i	safety factor (2 - 4)
I_a	actual ripple current	I_0	nominal ripple current	n	exponent (3 - 5)
T_a	actual ambient temp.	T_0	nominal temperature	ΔT_0	overtemp. core (5 - 10 K)
U_a	actual voltage	U_0	rated voltage		

The ripple currents include a wide range of frequencies. Mainly these are multiples of the semiconductor switching frequency, which is 4 kHz, 8 kHz or sometimes 16 kHz with IGBTs today. These multiples must be taken into account with regard to the frequency F_x and the amplitude I_x in the equation (2). In this case, I_a is the frequency-weighted equivalent ripple current, which is calculated as follows:

$$I_a = \sqrt{\left(\frac{I_1}{F_1}\right)^2 + \left(\frac{I_2}{F_2}\right)^2 + ... + \left(\frac{I_n}{F_n}\right)^2} \quad (2)$$

In general, it can be stated that the lifetime of capacitors is greater when they are operated in the DC grid than in the AC grid. This is due in particular to the fact that the rectifier's recharging currents are not present.

Device and Grid Faults

In contrast to AC devices, faults in the DC grid or in other devices play a special role and can place a particular load on connected devices [10]. If a short circuit occurs in the grid or in a nearby device, the DC links of the connected devices are immediately discharged. This results in very high currents, which can be up to a thousand times higher than the operating current in real devices [12].

During such a discharge of the DC link capacitance (see Fig. 5), not only the capacitor is stressed. All conductive tracks and possibly existing connectors are also overloaded, whereby these develop contact

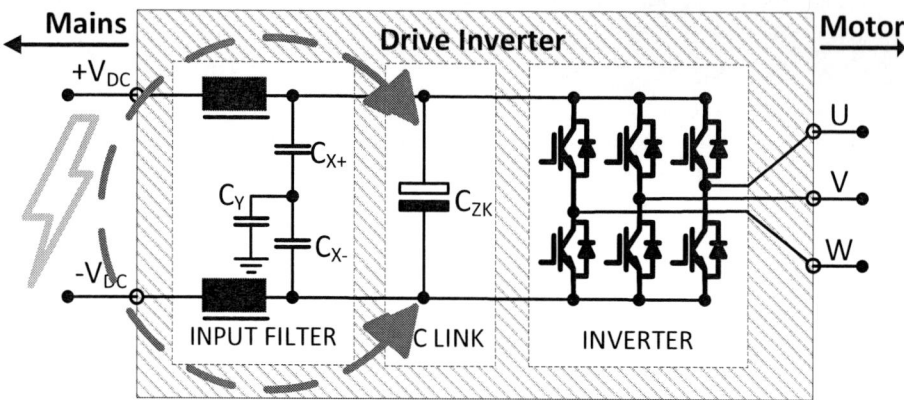

Fig. 5: Drive Inverter with the main components and an external short-circuit fault and resulting current flow

resistances or even break down completely. Measurements already carried out show however: Especially the inductance in the input filter (see Fig. 2), which usually has only a few turns, can get the entire voltage in the worst case: 0 V at the DC terminal and the full voltage in the DC link. The result may be voltage flashovers and thus local thermal overloads, as well as component failure.

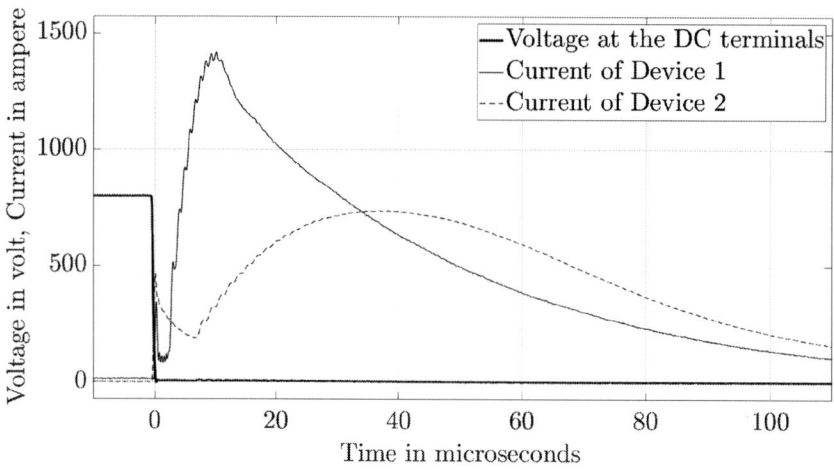

Fig. 6: Measured voltage and current curves of a discharge process of two drive inverters with different DC link capacitors: Film capacitor in Device 1 and electrolytic capacitor in Device 2

Fig. 6 shows the measurement of an external discharge of the DC links of two different devices of the same power class ($P_{nom} \approx 1\,kW$). Device 1 uses a film capacitor as the DC link capacitor and accordingly shows a rapidly increasing current curve. This is due to the low attenuation and the comparatively low impedance of the film capacitor. Device 2 has an electrolytic capacitor in the DC link. Due to parasitic effects, the current rise is limited and the amplitude is also smaller. The maximum current, on the other hand, is maintained longer. In both cases, the discharge current reaches a maximum of more than 1000 times the nominal operating current.

Conclusion

Devices in the industrial environment are subject to high demands on their lifetime: They must be achieved. For this to work, the electrical influences on the device must be known when selecting the components. The most important influencing factors here are: Static and dynamic voltages, ripple currents flowing between individual devices, and fault events in the DC grid that lead to sudden discharges. All these influencing factors can have a negative impact on the lifetime. Damage and premature aging can be avoided by the correct design and selection of components such as semiconductors, capacitors and also the inductances of the input filter.

The paper shows the permissible voltage ranges of DC INDUSTRIE in relation to time Devices can be easily designed for this with suitable input filters and DC link capacitors. Furthermore, the determination of ripple current loads is possible by a simple measurement and a test circuit, so that overloads can be detected already during initial operation. Devices must be designed for fault events such as short-circuits that can occur during operation and must not be impaired. Since these are short-term events, devices can also be designed for them without obstacles. Therefore, these listed influences do not stand in the way of a wide use of devices, such as drive inverters, in an open industrial DC grid.

References

[1] H. Borcherding, J. Austermann, T. Kuhlmann, B. Weis and A. Leonide, "Concepts for a dc network in industrial production" IEEE Second International Conference on DC Microgrids," Nuremberg, 2017, pp. 227 234

[2] ZVEI & Consortium DC-INDUSTRIE2, "System concept DC-INDUSTRIE2," Published on April 4th, 2022

[3] S. Puls, M. Ehlich, J. Austermann, H. Borcherding, F. Blank, "The Influence on Drive Inverters under the Effects of Short Circuits in an Open Industrial DC grid," IEEE 21nd European Conference on Power Electronics and Applications, Genua, 2019

[4] S. Puls, J. Hegerfeld, J. Austermann, H. Borcherding, "Transient Overvoltage Protection Solutions for Drive Inverters operating on an Open Industrial DC Grid," PCIM Europe digital days 2020; International Exhibition and Conference for Power Electronics, Intelligent Motion, Renewable Energy and Energy Management, Germany, 2020, pp. 1-8.

[5] K. Askan, M. Bartonek, K. Weichselbaum, "Power Module for Low Voltage DC Hybrid Circuit Breaker," IEEE Third Internatinal Conference on DC Microgrids, Matsue, 2019

[6] S. Puls, J. Austermann and H. Borcherding, "Lifetime Calculation for Capacitors in Industrial Micro DC grids," 2019 IEEE Third International Conference on DC Microgrids (ICDCM), Matsue, Japan, 2019, pp. 1-6

[7] C. Oeder, N. Foerster and T. Duerbaum, "Implementation of an Adaptive Dead Time in Resonant Converters," PCIM Europe 2018; International Exhibition and Conference for Power Electronics, Intelligent Motion, Renewable Energy and Energy Management, Nuremberg, Germany, 2018, pp. 1-8.

[8] S. Puls, J. Austermann and H. Borcherding, "Potential Hazards of Transient Overvoltages in an Industrial DC Grid and Basic Protective Measures," 2021 22nd IEEE International Conference on Industrial Technology (ICIT), 2021, pp. 625-630, doi: 10.1109/ICIT46573.2021.9453597.

[9] S. Puls, S. Warkentin, J. Austermann and H. Borcherding, "Characteristics and Possible Resonant Oscillations in an Open Industrial DC grid," PCIM Europe digital days 2021; International Exhibition and Conference for Power Electronics, Intelligent Motion, Renewable Energy and Energy Management, 2021, pp. 1-8.

[10] S. Puls, F. Blank, J. Höflsauer, O. Grünberg and H. Borcherding, "Effects of Component Failures in Drive Inverters during Parallel Operating on an Open Industrial DC Grid," 2021 23rd European Conference on Power Electronics and Applications (EPE'21 ECCE Europe), 2021, pp. 1-9.

[11] J. -N. Koch, R. Otte and H. Borcherding, "Highly Efficient SiC-based Active Infeed Converter for Industrial DC Conductor Systems," PCIM Europe digital days 2021; International Exhibition and Conference for Power Electronics, Intelligent Motion, Renewable Energy and Energy Management, 2021, pp. 1-8.

[12] S. Puls et al.: The Influence on Drive Inverters under the Effects of Short Circuits in an Open Industrial DC grid, 2019, IEEE 21st European Conference on Power Electronics and Applications (EPE), Genua

Investigation about Operation and Performance of Gate Drivers for Power Electronics Converters for Cryogenic Temperatures

Mustafeez-ul-Hassan, Yuxuan Wu, Vyacheslav Solovyov and Fang Luo
Stony Brook University
100 Nicolls Rd,
Stony Brook, NY 11794, USA
Tel.: +1– 479.404.9966.
E-Mail: Mustafeez.hassan@stonybrook.edu; yuxuan.wu@stonybrook.edu;
vyacheslav.solovyov@stonybrook.edu and fang.luo@stonybrook.edu
URL: https://www.stonybrook.edu

Acknowledgements

The authors would like to acknowledge the financial support lended by NASA to carry out this research under ULI: Development of the Cryogenic Hydrogen-Energy Electric Transport Aircraft (CHEETA) Design Concept under award number 80NSSC19M0125. Furthermore, NSF CAREER award numbered 1846917 titled Semiconductor-Based EMI Mitigation Architecture for Future Power Electronics Systems is also acknowledged.

Keywords

«Cryogenic», «Smart Gate Drivers», «More Electric-Aircraft»

Abstract

Utilization of power electronics converters under cryogenic temperatures (CT) offers higher power efficiency and volumetric density. This becomes possible due to reduced channel resistance and switching energies for Si and GaN-based devices. Therefore, both quantities, constituting a figure of merit for power converters; offer lower conduction losses and increased switching speeds with the reduction in temperatures. This leads to increased power efficiency, and lower volume of the passive components involved. Since reliable and efficient operation of a power electronics converter depends upon the operational characteristics of switching devices; gate drivers play a critical role as they not only decide, but also help optimize the operating conditions of such devices. To take full advantage of CT, not only proper working but also the quality of switching performance of gate driving circuits is extremely important. This paper presents operation and performance of numerous commercially off-the-shelf (COTS) gate drivers under cryogenic operating conditions. Gate drivers (GD) selected for the analysis are capable enough to drive high-speed wide band gap (WBG) devices. As part of the experimentation process, whole GD board was designed while selecting all the auxiliary components to be compatible with CT. The paper compares five different GDs and presents successful operation of two of them at CT of 77 K.

Introduction

Power electronics is the key enabling technology for electromechanical drives, transportation, renewable energy systems, and power grids. Cryogenic power electronics technology (CPET) is the next step to obtain higher power efficiency, higher power density, and superior performance for various applications. CPET finds its applications in deep space and terrestrial applications, medical diagnostics, all-electric vehicles, propulsion motors, and superconducting magnetic energy storage systems [1]. Utilization of cryogenically cooled environment offers opportunity to design and manufacture lightweight, and highly efficient power electronics converters. Such cryogenic power conversion systems offer significant improvements in performance, including the reduction in semiconductor losses, higher switching speed, and performances [2].

Power converters are majorly composed of three components as shown in Fig. 1 (a): semiconductor devices, cooling materials, and passive components [3]. These components dictate the efficiency and power density of the overall conversion system, and therefore their appropriate selection is important to ensure high density and efficiency. A framework to select and optimize a converter together with its components is provided in [4]. Considering the operation of converters at low temperatures, numerous literatures have characterized the performance of different passive components for low-temperature applications [5]. Similarly, performance of different semiconductor devices, such as Si MOSFETs and GaN high electron mobility transistors (HEMT), has been reported to improve with reduction in temperature [6].

Although appropriate devices for low temperatures have been identified and their static characterization performed, yet dynamic characterization and full converter development at lower temperatures has not been reported. [6] reports the dynamic characterization of a 650 V GaN until 133 K whereas [7] reports dynamic characterization until 93 K. Similarly, some of the converters developed recently include a three-level active neutral point clamped converter (3L-ANPC) with series cascading of Si devices [8]. Development of a single-phase three-level flying capacitor multi-level inverter using 200 V GaN devices was reported until 133 K in [9]. From the literature, it can be observed that development of power electronics converters with high voltage GaN (> 600 V) devices is not reported at cryogenic temperatures. The reason for this may be attributed to non-availability of gate driving and auxiliary components at such low temperatures. Gate driver (GD) circuitry is used to drive the semiconductor devices as depicted in Fig. 1 (b), and comprise of an isolated auxiliary power supply, integrated circuits (ICs) and passive components. GDs are extremely crucial for the optimized performance of semiconductor switches as both the conduction and switching losses are related to the magnitude and nature of gate driving voltages. Although efforts have been made to successfully operate such components at low temperatures as reported in [10], however not even a single successful operation at a temperature as low as 77 K is reported.

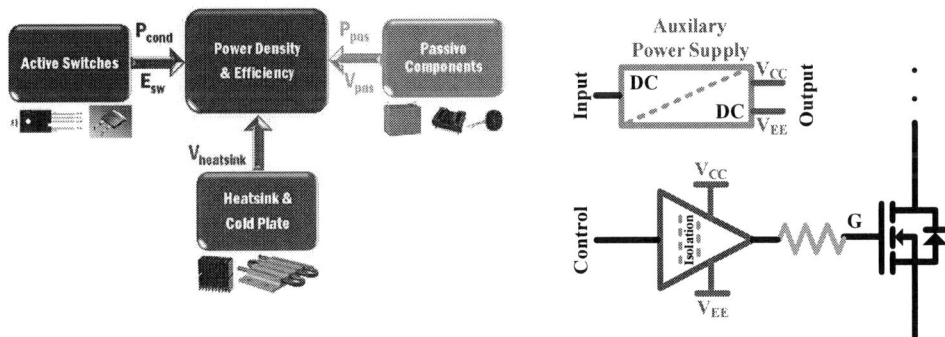

Fig. 1: (a) Building block of a power electronics converter   Fig. 1: (b) Basic GD schematics

Keeping in view the significance of gate driving components for CPET, this paper focuses on analyzing the performance of numerous commercially available gate drivers at cryogenic temperatures (CT). Section two of the paper discusses selection of gate driver and its associated components suitable for CT, whereas third section presents development of hardware and setting up of a cryogenic chamber. Discussion on experimentation and results of gate drivers at liquid nitrogen temperature (LNT), i.e., 77 K is carried out in the second last section. The paper concludes afterward.

Selection of Gate Driver and Associated Components

Considering the advantages associated with low-temperature operation of semiconductor devices; both gate drivers and switching devices should preferably be placed together and inside a cold environment [12]. This will not only make the connections short and compact but will also offer noise immune and faster dynamic performances, as [13] placed both the gate driver and power circuitry inside the chamber while using Si devices for a DC-DC converter. To have the whole gate driving circuitry function properly at low temperatures, all the constituent components, for example, ICs and passive components must be carefully selected and designed.

Gate Driver Integrated Circuit (IC)

Selection of gate driver is critical as it drives the gate-source of any transistor and makes or breaks its channel to establish the current. Keeping in view the faster turn-on and turn-off together with lower operating temperatures, the gate driver ICs should not only ensure higher sourcing/sinking of gate current, but also need to have complementary metal-oxide semiconductor (CMOS) technology. CMOS-based integrated circuits are preferred over other types because of their improved performance in switching speed with reduction in temperature. On the contrary, other technologies, for example, bipolar junction transistor (BJT) suffers from a significant decrease in current gain caused by the reduction in carrier lifetime at reduced temperatures [12], [14]-[15].

To evaluate the performance of GDs, numerous commercially off-the-shelf (COTS) drivers capable enough to drive GaN devices were selected; since these devices offer reduced conduction losses, improved switching performance, and constant breakdown voltage [2]. Combined with CMOS technology and availability in the market, GD must also offer extremely low isolation barrier capacitance, higher common-mode transient immunity (CMTI), and higher current sourcing/sinking capability. Therefore, a thorough survey was conducted, and a brief comparison of key parameters for candidate GD is provided in Table I.

Table I: Key Parameters for Candidate Gate Drivers

PROPERTY	SI8271	SI8235AD	UCC21540A	MAX22702	ADUM4221
CMTI	200 V/ns	45 V/ns	100 V/ns	300 V/ns	150 V/ns
C_{IO}	0.5 pF	1.4 pF	1.2 pF	1 pF	2.2 pF
PEAK CURRENT	4 A	4 A	4 A/6 A	4 A/5.7 A	4 A
FABRICATION PROCESS	--	--	BiCMOS	BiCMOS	Monolithic transformer
TECHNOLOGY	SiC	Si	--	SiC	--

Selection of Auxiliary Power Supply (APS)

In addition to gate driving ICs, numerous auxiliary power supplies (APS) were also selected and tested. For the APS, 5 to 9 V conversion ratio was considered as to offer bipolar +6 V, -3 V to drive the gate of GaN devices, alongside lowest values of isolation capacitance (C_{ISO}). Lower values of C_{ISO} matter as it leads to leakage currents between primary and secondary sides for higher switching speeds. R05P205S and RV-0509S were the APS utilized in most of the test combinations. Since most of the APS use ferrite-based cores, they are highly likely to fail at CT because of reduction in their permeability and increase in core losses [5].

Selection of Auxiliary Components

Operating power electronics circuits at low temperatures need all associated components to be temperature resilient as well. Therefore, for the design of whole circuitry, thin film-based resistors and negative-positive zero (NP0) capacitors were employed since these components have been reported to show negligible variation in performance with reduction in temperature [1], [2].

Development of Hardware and Cryogenic Test Setup

Development of Gate Driving boards

As part of hardware development, separate printed circuit boards (PCBs) were designed for all the gate driver ICs, whereas a similar buffer sage IC, ISO7820x from Texas Instruments was utilized as the first stage. During the PCB development and prototyping, board shape and layout, buffer IC, all the passive components, and soldering materials were kept constant as to only see the influence of GD IC's performance and operation. Furthermore, initially, two-stage isolation in APS was implemented with the R05P205S COTS converter. Resulting four layered PCBs and assembled gate driving boards are shown in Fig. 2 and Fig. 3 respectively. The resulting PCBs were 61mm wide and 45 high, whereas normal FR-4 material together with Sn42/Bi57.6/Ag0.4 solder paste was used to attach different components.

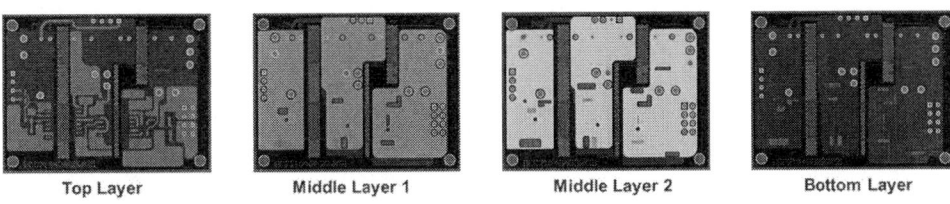

| Top Layer | Middle Layer 1 | Middle Layer 2 | Bottom Layer |

Fig. 2: Symmetric PCB layout for all five gate driver boards

Fig. 3: Development of different gate driver boards

Development of Double Pulse Test (DPT) Platform

In addition to five symmetrical gate driver boards, a DPT platform was also built so that successfully tested gate drivers and auxiliary components can be employed in driving a GaN-based converter at lower temperatures. For the DPT, a two-layered PCB structure having GaN systems GS66516T-MR device was utilized together with a 2 GHz, 101 mΩ current shunt labelled SSDN-10 to monitor the device current. Based on the initial testing observations, gate driver boards were modified while mal-operating components were removed and replaced. Hence, finalized gate driver boards (version 2) together with DPT board containing a current shunt and switching device were developed and are shown in Fig. 4.

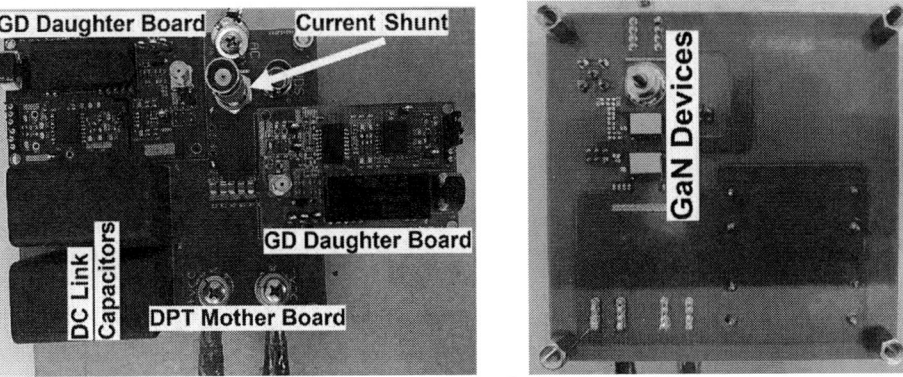

Fig. 4: Development of DPT platform together with (finalized) gate drivers

Development of Cryogenic Test Setup

For the cryogenic testing, a setup having an 18-inch stainless steel chamber integrated with A230 CryoMech single-stage cryopump was used. The chamber was evacuated to 10^{-6} Torr and the assembled gate driver boards were thermally connected to a thick copper plate by a thermally conductive epoxy, StyCast 2850FT. The copper plate was tightened with the baseplate of the chamber, where the cold/base-plate temperature was measured by a Si-diode cryogenic temperature sensor (LakeShore 420 model). The GD assembly was encompassed by a copper radiation shield and ten layers of Mylar multi-layer insulation. All the connections with source voltage, control signals, and measurement were established using feedthroughs. For the measurements, long jumper wires were involved to bring the signal outside the chamber and displayed on Tektronix TBS 2000 B oscilloscope. Complete test setup containing the cryo-chamber alongside measurement equipment is shown in Fig. 5. The control pulses were generated using a DSP microcontroller F28379d which was also placed outside the chamber.

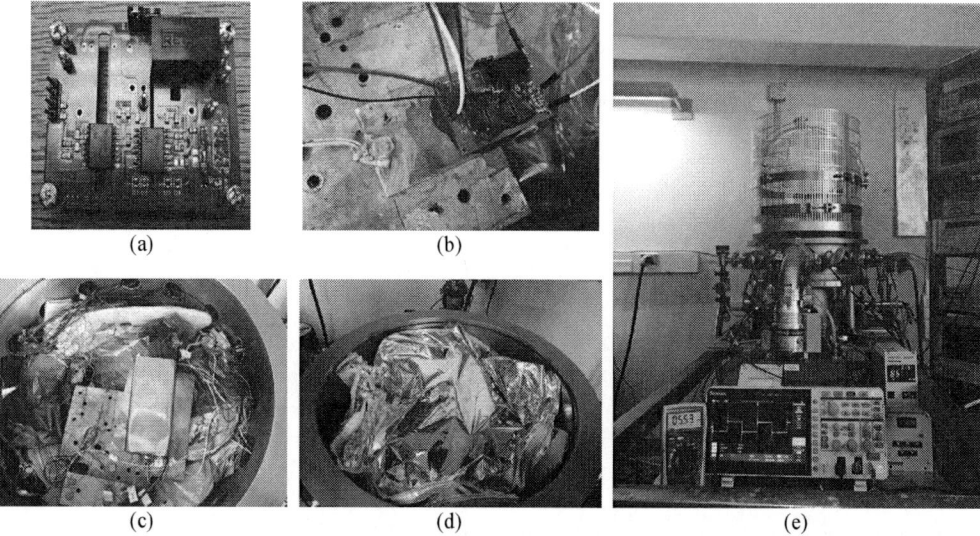

Fig. 5: (a) GD at RT (b) GD board connected to base plate of chamber and temperature sensor installed (c) GD covered with copper plate inside chamber (d) GD board covered with multi-layer insulation (e) Overall cryogenic chamber setup from outside containing oscilloscope, cryogenic temperature sensor.

Experimentation and Results

Auxiliary Power Supply

Numerous APS were configured on the gate driver boards assembled, but almost all of them failed except for RV-0509S. RV-0509S was tested numerous times, and was characterized during lowering the temperature to LNT, as well as bringing it back to room temperature. Although the power supply operated fine, yet there were changes in output voltages generated and input currents drawn. This trend has been depicted in Fig. 6 (a) where output voltage drops while the input current rises. The curve shown is for constant output load, which means the efficiency of isolated DC-DC converter in the APS drops as the temperature goes down. This goes well in accordance with the projected increase in core loss and decrease in permeability for ferrite cores, therefore overall efficiency of APS drops [5]. Additionally, the change in output voltage is also related to open-loop control of most of the COTS APS, and therefore an investigation about closed-loop control with variable temperature can be carried out as future work.

Gate Driving Boards

Using the experimental setup shown earlier, all five gate drivers were characterized, and unique temperature performance was observed for different ICs. Even though the gate drivers have been rated from -40 C-125 C, yet they were tested for lower temperatures as not even a single gate driver is available COTS for such applications. From the experiments, two of the gate drivers MAX22702 and Si8235AD stopped working around ~ 130 K, mainly because of their fabrication technology being BiCMOS. However, UCC21540A showed better immunity to temperature and performed well until ~ 93 K. On the contrary, ADUM4221 has a monolithic transformer technology and therefore it performed well until ~ 77 K. Similarly, Si8271 was also found to be operating fine till LNT. It is worth mentioning that numerous samples of a particular IC were characterized, and the results were found to be consistent with each iteration of experimentation. Furthermore, the buffer IC ISO7820x was found to be consistent in all the boards tested. Additionally, both the successfully operating gate drivers were configured to switch under loaded and unloaded conditions at different switching frequencies. A silicon carbide power MOSFET C3M0016120K was used as a dummy load to verify the operability of the gate drivers. Fig. 6 (b) shows the operation of gate driver at 77 K with an operating frequency around 150 kHz.

During the process of characterization, the gate drivers were found to have variation in rise and fall times with the change in temperature. As can be seen from Fig. 6 (c), it can be observed that overshoot for Si8271 increases at first and then seems to further slowdown. This process can be related to a decrease in resistance of high output transistor of gate driver with temperature, hitting a minimum at 193 K, and starts increasing afterwards. However, looking at the turn-off time in Fig. 6 (e), the trend for undershoot looks similar which means the resistance of low output transistor keeps decreasing. Similar analysis can be carried out for ADUM4221, which shows an increase in resistance of high output transistor until 193 K and starts reducing afterwards. This trend can be inferred from Fig. 6 (d). Whereas the low output transistor impedance seems to decrease until 153 K and increases onwards as inferred from Fig. 6. (f). This variation in resistance of output transistors directly affect the rise and fall times, together with direct influence on amount of current being sourced/sinked.

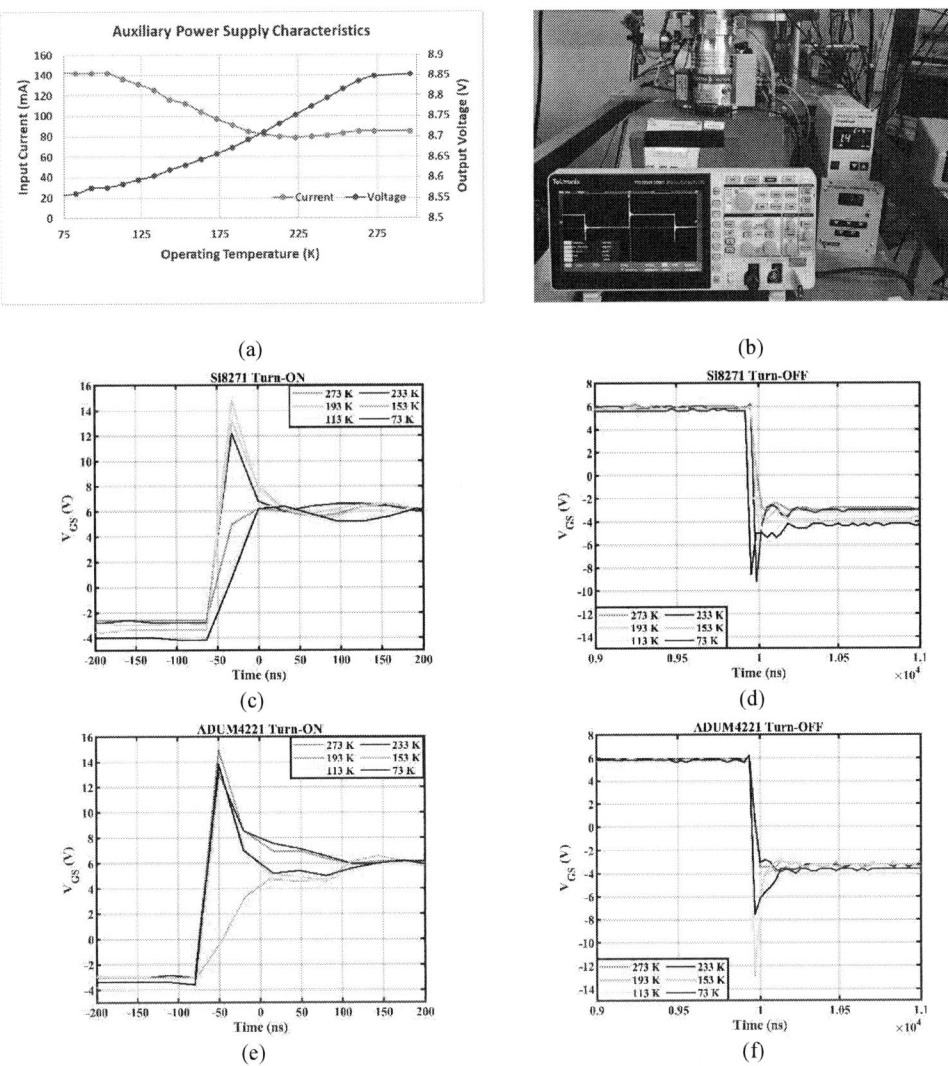

Fig. 6: (a) Performance of APS RV-0509S with temperature (b) GD boards operating fine at LNT (c) turn-on of Si8271 (d) turn-off of Si8271(e) turn-on of ADUM4221 (f) turn-off of ADUM4221

Double Pulse Tests at LNT and Device Dynamic Characterization

Having characterized the operating performance of two gate drivers, a double pulse test (DPT) setup was established and DPT experiments were only conducted at RT and LNT. The characterization gets limited to only two temperatures as DPT board cannot be placed inside the cryo-chamber because of connections involving long leads and feedthroughs of the chamber. The challenge behind doing DPT inside the chamber is related to measurement issues. Therefore, a Dewar having liquid nitrogen was used to insert DPT and gate driving boards together as shown in Fig. 7 (a).

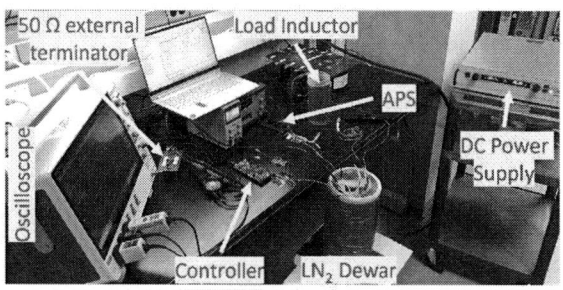 

Fig. 7: (a) LN$_2$ Setup for DPT (b) LN$_2$ dewar

After characterizing the setup at RT, both DPT and GD board were inserted inside LN$_2$. Both the boards were subjected not only to LNT, but they were also given sudden thermal shock while inserting in LN$_2$ Dewar. For the measurements, active probe THDP0200 was used, and it was directly inserted inside the Dewar. Furthermore, V-I alignment with deskew adjustment was conducted before doing the LNT measurements. The results from both RT and LNT measurements are presented in Fig. 8 for device turn off behaviors. As can be seen from Fig. 8 (a), turn-off voltage at 400 V is not only advanced in time, has higher overshoot, but also has more sinusoidal high frequency oscillations at LNT temperature (labelled as A). Moreover, the response during initial time of turning off is also different at LNT (labelled as B). Similarly, turn-off current waveform has also been shown to advance, amplify (labelled as C) and have different profile during the starting times of turn-off process. Turn-off current comparison can be seen from Fig. 8 (b).

From the measurements, it can be observed that switching speeds at lower operating temperatures increase, therefore resulting in reduction in overlap times for current and voltage. This reduction in overlap has a direct influence on calculation of switching energies which can be much lower for LNT.

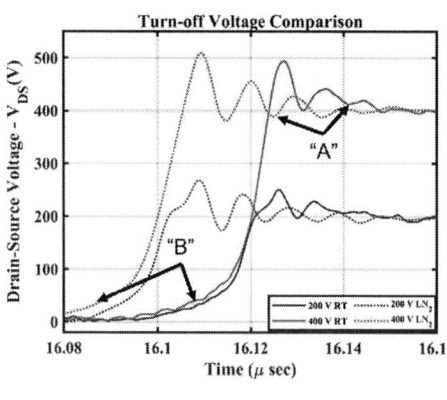 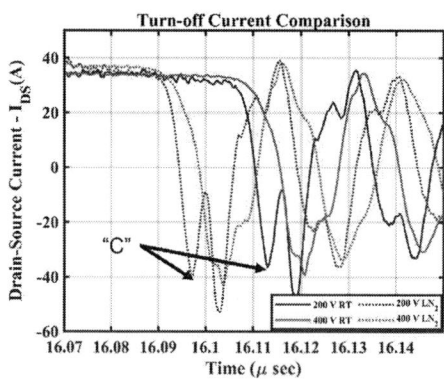

(a)   (b)

Fig. 8: Switching Loss variation (a) Turn-on (b) Turn-off

Conclusion

In this paper, numerous commercially available gate drivers (GD) for wide band gap devices (WBG), particularly for GaN HEMTs, were characterized for low temperature applications. GDs from different manufacturers with different fabrication processes and semiconductor technology were evaluated to find the one which not only works appropriately, but its performance does not get degraded a lot. From the experimentation, two GD and a buffer integrated circuit (IC) were found to operate well until 77 K. As part of investigation, most of the commercially available auxiliary power supplies were found to fail operating except for RV-0509S. The paper characterizes the operating performance of which with temperature.

The paper also demonstrated the change in resistance of high and low output transistors of gate drivers. This change clearly gets reflected onto rise and fall times, and therefore it directly influences the over/under voltages applied at the gate source of power transistors. These over/under voltages can directly impact the switching characteristics together with failure modes of devices being switched. Furthermore, this can in resistance can also have a direct impact onto turn-on/turn-off dV/dt of power transistors. These challenges can be extremely influential for low temperature applications and need to be covered as part of future research.

References

[1] K. Rajashekara and B. Akin, "A review of cryogenic power electronics - status and applications," 2013 International Electric Machines & Drives Conference, 2013, pp. 899-904.

[2] H. Gui et al., "Review of Power Electronics Components at Cryogenic Temperatures," in IEEE Transactions on Power Electronics, vol. 35, no. 5, pp. 5144-5156, May 2020.

[3] Y. Chen, Z. Yuan and F. Luo, "A Model-Based Multi-Objective Optimization for High Efficiency and High-Power Density Motor Drive Inverters for Aircraft Applications," NAECON 2018 - IEEE National Aerospace and Electronics Conference, 2018, pp. 36-42.

[4] Mustafeez-ul-Hassan, Z. Yuan, A. I. Emon and F. Luo, "A Framework for High Density Converter Electrical-Thermal-Mechanical Co-design and Co-optimization for MEA Application," 2021 IEEE Energy Conversion Congress and Exposition (ECCE), 2021, pp. 3120-3125.

[5] R. Chen et al., "Core Characterization and Inductor Design Investigation at Low Temperature," 2018 IEEE Energy Conversion Congress and Exposition (ECCE), Portland, OR, 2018, pp. 4218-4225.

[6] R. Ren et al., "Characterization and Failure Analysis of 650-V Enhancement-Mode GaN HEMT for Cryogenically Cooled Power Electronics," in IEEE Journal of Emerging and Selected Topics in Power Electronics, vol. 8, no. 1, pp. 66-76, March 2020.

[7] Z. Zhang et al., "Characterization of Wide Bandgap Semiconductor Devices for Cryogenically-Cooled Power Electronics in Aircraft Applications," 2018 AIAA/IEEE Electric Aircraft Technologies Symposium (EATS), 2018, pp. 1-8.

[8] H. Gui, Z. Zhang, R. Chen, R. Ren, J. Niu, H. Li, Z. Dong, C. Timms, F. Wang, L. M. Tolbert, B. J. Blalock, D. Costinett, and B. B. Choi, "Development of high-power high switching frequency cryogenically cooled inverter for aircraft applications," IEEE Transactions on Power Electronics, vol. 35, no. 6, pp. 5670–5682, 2020.

[9] C. B. Barth, T. Foulkes, O. Azofeifa, J. Colmenares, K. Coulson, N. Miljkovic, and R. C. N. Pilawa-Podgurski, "Design, operation, and loss characterization of a 1-kw gan-based three-level converter at cryogenic temperatures," IEEE Transactions on Power Electronics, vol. 35, no. 11, pp. 12 040–12 052, 2020.

[10] Y. Wei, M. M. Hossain, R. Sweeting and A. Mantooth, "Functionality and Performance Evaluation of Gate Drivers under Cryogenic Temperature," 2021 IEEE Aerospace Conference (50100), 2021, pp. 1-9.

[11] Hui Li, D. Liu and C. A. Luongo, "Investigation of potential benefits of MOSFETs hard-switching and soft-switching converters at cryogenic temperature," in IEEE Transactions on Applied Superconductivity, vol. 15, no. 2, pp. 2376-2380, June 2005.

[12] B. Patra et al., "Cryo-CMOS Circuits and Systems for Quantum Computing Applications," in IEEE Journal of Solid-State Circuits, vol. 53, no. 1, pp. 309-321, Jan. 2018.

[13] F. F. Perez-Guerrero, K. Venkatesan, B. Ray and R. L. Patterson, "Low temperature performance of a closed loop three level buck converter," Proceedings of the IEEE 1999 International Conference on Power Electronics and Drive Systems. PEDS'99 (Cat. No.99TH8475), 1999, pp. 58-62 vol.1.

[14] W. F. Clark, B. El-Kareh, R. G. Pires, S. L. Titcomb and R. L. Anderson, "Low temperature CMOS-a brief review," in IEEE Transactions on Components, Hybrids, and Manufacturing Technology, vol. 15, no. 3, pp. 397-404, June 1992.

[15] F. Torres, J. Pastrana, R. Colon, E. Delgado and B. Ray, "Low temperature characterization of CMOS operational amplifier ICs," IECEC 96. Proceedings of the 31st Intersociety Energy Conversion Engineering Conference, 1996, pp. 546-552 vol.1, doi: 10.1109/IECEC.1996.552942.

Synchronization Angle determination in DVCSFO of DFIM naval propulsion

Youssef Drimizi, Maria Pietrzak-David, Pascal Maussion
LAPLACE, UNIVERSITE DE TOULOUSE, INP, UPS, CNRS, Toulouse, France
Phones : +33 5 34 32 24 07, +33 5 34 32 23 59, +33 5 34 32 23 64,
Fax : +33 5 61 63 75 88
E-Mails : Drimizi@laplace.univ-tlse.fr, Maria.David@laplace.univ-tlse.fr,
Pascal.Maussion@laplace.univ-tlse.fr,

Keywords

«Variable speed drive», «Induction Motor», «Flux Model», «Reliability», «Power Sharing»

Abstract

This paper presents the **D**ual **V**ector **C**ontrol with **S**tator **F**lux **O**rientation (**DVCSFO**) of Doubly Fed Induction Machine (DFIM). This DFIM operates in motor mode and is intended for naval propulsion. This machine, without permanent magnets, is fed by two identical PWM VSIs. It has been specially designed to share its power equally between the stator and the rotor. The previous studies have discussed the cooling of all system carried out by the water, for the stator and rotating rotor of DFIM and also for two IGBT VSIs. The stator and rotor fluxes are controlled in rotating «d,q» reference frame and must be maintained at their nominal value. Therefore, the DVCSFO strategy requires precise knowledge of all transformation angles and especially the Synchronization Angle when an incremental encoder is used as a speed sensor. Unlike the squirrel cage asynchronous machine, this DFIM actuator can be blocked if the synchronization angle is neglected in Park transformations.

Introduction

In modern naval propulsion concepts, the electrical machines have a crucial role, moving towards "the more electric" or even "all electric" ship and must respect the environmental demand "Clean-Sea" and "Clean-Ship". The authors propose the DFIM for naval propulsion application, where this machine works in motor mode and offers some advantages as listed below:

- high degree of freedom,
- excellent operation at very low speed and even when stopped,
- natural structural redundancy in case of VSI failure,
- active power dispatching between stator P_s and rotor P_r sides (Fig.1),
- total losses of two PWM VSIs are lower than the same power squirrel cage induction machine supplied by only one PWM VSI,
- wide rotation speed range with nominal torque.

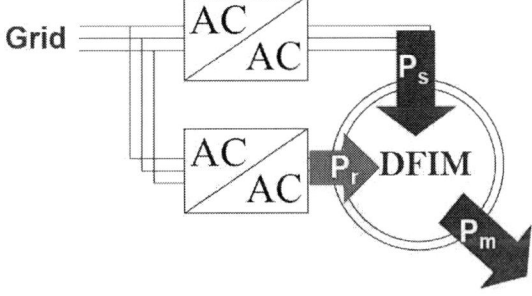

Fig. 1: Active power dispatching between the stator and rotor armatures of DIFIM

In this paper the authors discuss an important problem of Real Time control realization of a Multi-Pole DFIM propulsion. The multipolar solution is applied to reduce the volume and the weight of the motor, operates in an embedded system.

The DVCSFO strategy was chosen to guarantee an optimal DFIM drive system operation according to a real naval propulsion specification.

As the system operating and performances depend on position of the stator and rotor fluxes, therefore to reach the control performances and its expectations, the synchronization angle must be cancelled before any start up. This angle is often ignored in FOC of classical squirrel cage induction machine. The synchronization angle precise information is required in the first step of DVCSFO strategy and the main key of DFIM starting and achieving performances.

DIFIM propulsion presentation

The studied DFIM naval propulsion structure, shown in the Fig. 2, is composed of:
- AC grid,
- two AC/AC converters (two rectifiers and two PWM VSIs),
- DFIM propulsion,
- symmetrical three blade propeller load,
- transformer between the AC grid and the rectifier to respect the voltage level on the rotor and stator sides.

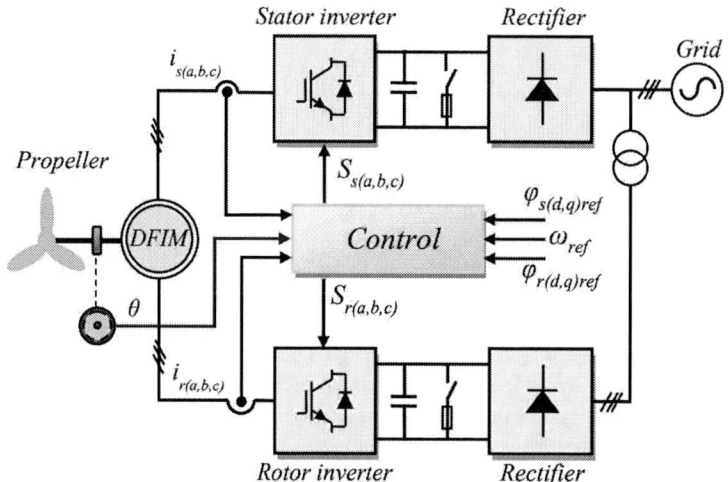

Fig. 2: General diagram of DFIM naval propulsion

DFIM actuator modelling

We have chosen for this study the flux model of the DFIM [3] defined in rotating d,q reference frame given in the Fig. 3 and described by equations of electrical part of the machine (1) and (2), with the stator flux orientation such as $\Phi_{sd}=\Phi_s$ and $\Phi_{sq}=0$.

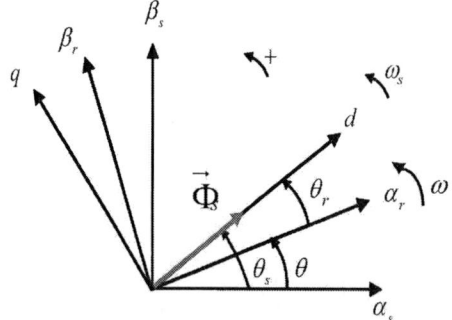

Fig. 2: Vector diagram chosen for DFIM modelling

The electrical equations for the stator (1) and rotor (2) of DIFIM can be written as follows:

$$V_{sd} = \frac{R_s}{\sigma L_s}\Phi_{sd} + \frac{d\Phi_{sd}}{dt} - \omega_s\Phi_{sq} - \frac{R_s M_{sr}}{\sigma L_r L_s}\Phi_{rd}$$

$$V_{sq} = \frac{R_s}{\sigma L_s}\Phi_{sq} + \frac{d\Phi_{sq}}{dt} + \omega_s\Phi_{sd} - \frac{R_s M_{sr}}{\sigma L_r L_s}\Phi_{rq}$$

(1)

$$V_{rd} = \frac{R_r}{\sigma L_r}\Phi_{rd} + \frac{d\Phi_{rd}}{dt} - \omega_r\Phi_{rq} - \frac{R_r M_{sr}}{\sigma L_r L_s}\Phi_{sd}$$

$$V_{rq} = \frac{R_r}{\sigma L_r}\Phi_{rq} + \frac{d\Phi_{rq}}{dt} + \omega_r\Phi_{rd} - \frac{R_r M_{sr}}{\sigma L_r L_s}\Phi_{sq}$$

(2)

When the motor is non saturated, the linear relation can be expressed between the fluxes and currents as below (3):

$$\begin{cases} \Phi_{sd} = L_s I_{sd} + M_{sr} I_{rd} \\ \Phi_{sq} = L_s I_{sq} + M_{sr} I_{rq} \\ \Phi_{rd} = L_r I_{rd} + M_{sr} I_{sd} \\ \Phi_{rq} = L_r I_{rq} + M_{sr} I_{sq} \end{cases}$$

(3)

Finally, the mechanical equation is expressed by (4)

$$J_t \frac{d\Omega}{dt} = T_{em} - f_t\Omega - T_L$$

$$T_{em} = \frac{N_p M_{sr}}{L_s L_r \sigma}(\Phi_{sq}\Phi_{rd} - \Phi_{sd}\Phi_{rq})$$

(4)

where $V_{sd}, V_{sq}, V_{rd}, V_{rq}$ - direct and quadrature stator and rotor voltages

$I_{sd}, I_{sq}, I_{rd}, I_{rq}$ - direct and q quadrature stator and rotor currents

$\Phi_{sd}, \Phi_{sq}, \Phi_{rd}, \Phi_{rq}$ - direct and quadrature stator and rotor fluxes

R_s, R_r - stator and rotor phase resistances,

L_s, L_r - stator and rotor cyclic inductances

M_{sr} - mutual cyclic stator/rotor inductance

$\sigma = 1 - M_{sr}^2/L_s/L_r$ - dispersion ratio,

ω_r, ω_s - rotor and stator velocities.

θ_s - stator angle between α_s axis of the stationary reference frame and d axis,

θ_r - angle between the α_r axis of the rotor reference frame and d axis

θ - electrical angle between the stator and the rotor

J_t - DFIM and load total inertia

f_t - total viscous friction coefficient

N_p - number of pairs of poles

T_{em} - electromagnetic torque

T_L	- load torque
Ω	- mechanical rotation speed (rad/s)
$\omega = N_p\,\Omega$	- electrical rotation speed (rad/s)

We can dress the general diagram of DIFIM actuator (Fig. 4) using the transfer functions of its electrical modes expressed as follows:

Axis d

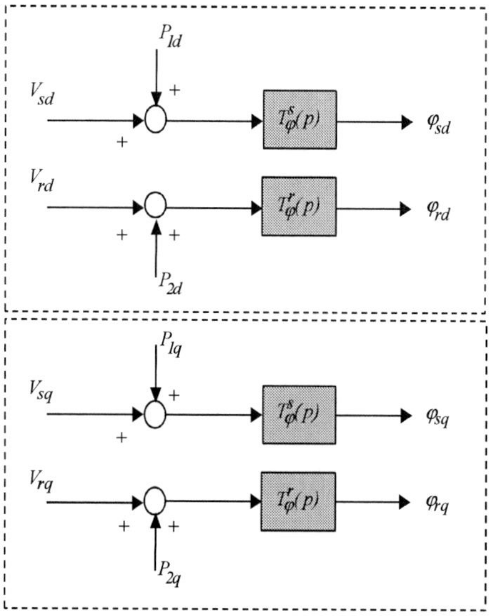

For index i=d, q

$$T_\varphi^s(p) = \frac{V_{siref} + P_{1i}}{\Phi_{si}} \qquad T_\varphi^s(p) = \frac{\dfrac{L_s\sigma}{R_s}}{p\dfrac{L_s\sigma}{R_s}+1}$$

$$T_\varphi^r(p) = \frac{V_{riref} + P_{2i}}{\Phi_{ri}} \qquad T_\varphi^r(p) = \frac{\dfrac{L_r\sigma}{R_r}}{p\dfrac{L_s\sigma}{R_r}+1}$$

(5)

The coupling terms are expressed as below:

$$\underline{P} = \begin{pmatrix} P_{1d} \\ P_{1q} \\ P_{2d} \\ P_{2q} \end{pmatrix} = \begin{pmatrix} \dfrac{R_s M_{sr}}{L_r L_s \sigma}\Phi_{rd} + \omega_s\Phi_{sq} \\[2mm] \dfrac{R_s M_{sr}}{L_r L_s \sigma}\Phi_{rq} - \omega_s\Phi_{sd} \\[2mm] \dfrac{R_s M_{sr}}{L_r L_s \sigma}\Phi_{sd} + \omega_s\Phi_{rq} \\[2mm] \dfrac{R_s M_{sr}}{L_r L_s \sigma}\Phi_{sq} - \omega_s\Phi_{rd} \end{pmatrix}$$

(6)

Fig. 3: DFIM diagram of its electrical modes

We add also the diagram of mechanical mode shown in the fig. 5, where T_L corresponds to load torque created by propeller, T_{em} is electromagnetic torque issue from DFIM and defined by equation (4) given previously. The direct and quadrature stator and rotor fluxes are expressed in (3). For non-saturated DFIM we obtain the linear relations between the stator and rotor fluxes and currents.
The mechanical equation can by written by (4), with:

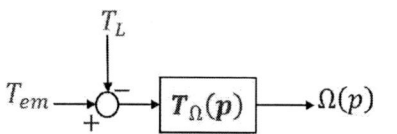

Fig. 4: DFIM diagram of its mechanical mode

The transfer function of mechanical mode of this DFIM actuator, can be formulated by:

$$T_\Omega(p) = \frac{\dfrac{1}{ft}}{1+\dfrac{Jt}{ft}p}$$

(7)

DVCSFO strategy of DFIM propulsion drive

Now we present the control diagram for the DFIM propulsion which is defined from the inverse model of the system. So, the general scheme of this control is given in the Fig. 6. We observe that first, the fluxes Φs and Φr have to be installed. This magnetization step must be preceded by the alignment of the two fluxes, stator and rotor. In this phase the synchronization angle will be determined precisely. This method is made automatically and will be presented in the next part of this paper.

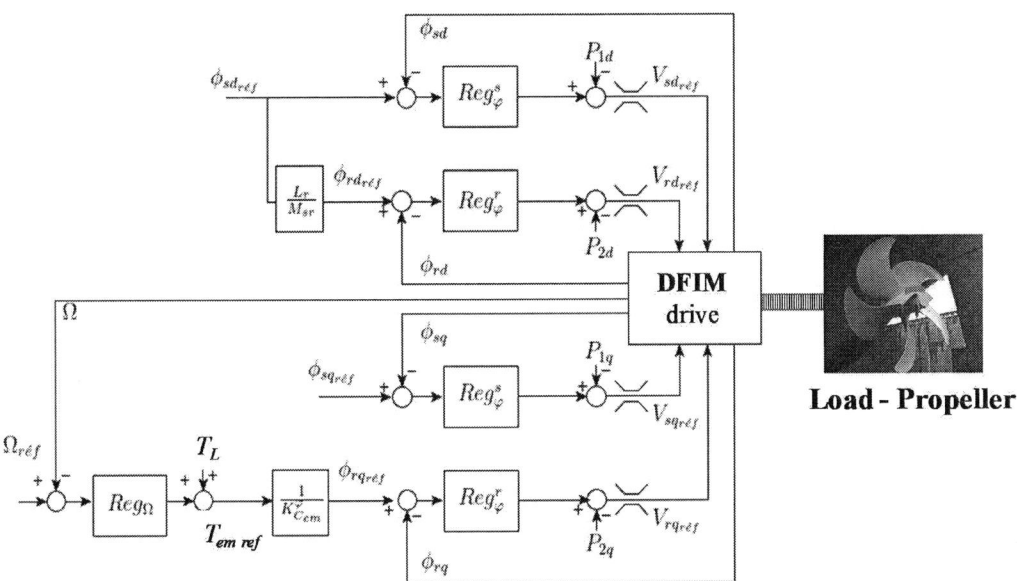

Fig. 5: General diagram for DVCSFO strategy of DIFIM propulsion

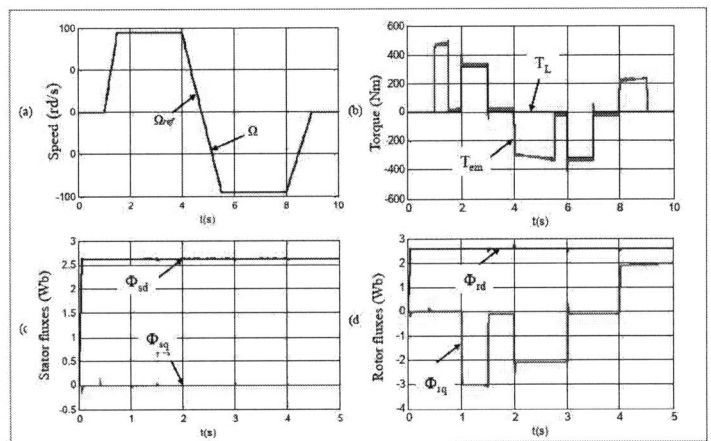

Fig. 6: 30kW DIFIM propulsion behavior during the operation test in four quadrants torque-speed plane

So, after the magnetization phase with installation of the two stator and rotor fluxes, the DFIM is started and stabilizes its rotation speed according to the imposed reference value. A reversal of the rotation direction is then ensured. The constant speed is also guaranteed in the opposite direction. Then, the DFIM returns to zero speed according to the imposed reference value.

It must be noted that this satisfactory DIFIM working will be problematic in real ship condition. In all laboratory tests, the Park transformations can be made with perfect angles. As it was write previously, the starting condition of this machine requires the perfect alignment of two stator and rotor fluxes. When this condition is not respected the DFIM starting is impossible.

The authors propose the generic and reliable method to guarantee the good DFIM starting in real operating conditions and especially delicate for the multi-pole, high voltage and high power DFIM propulsion.

Synchronization angle determination for reliable DFIM propulsion

The procedure for determining the synchronization angle of a DFIM having multi-poles must be carried out before its starting. As this machine can have several rotor shaft positions and therefore the different initial rotor angle defined as Θ_r.

We recall that in the vector control carried out in the d,q reference frame the Park transformation are very important and depends on the stator and rotor flux positions Θ_s and Θ_r as shown in Fig. 3. According to the control strategy conditions, these positions can be defined from the frequency or angular self-piloting laws. Due to the realization adapted to our digital control architecture we prefer to apply the angular self-piloting illustrated in Fig. 8

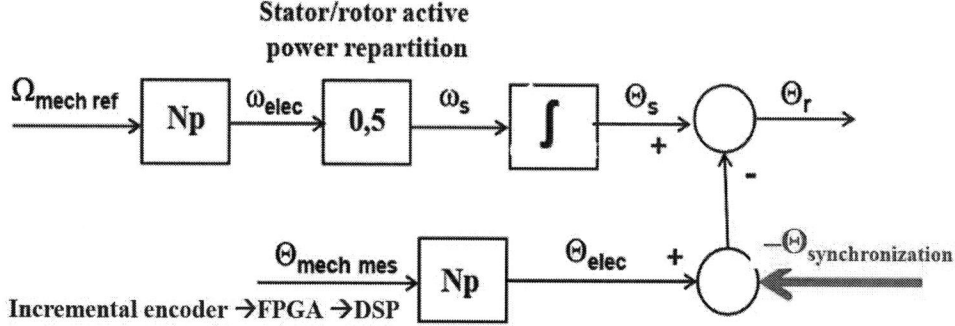

Fig. 7: Principle of angular self-piloting with correction term of synchronization angle

It can be noted that according to the DVCSFO strategy, the stator flux orientation with "d" axis requires the precise definition of its position Θ_s. So, from the reference value of mechanical rotation speed $\Omega_{mecaref}$ expressed in rad/s, we calculate the reference of electrical rotation speed ω_{elec} Next, using the active stator and rotor power repartition ratio equal to 0,5 we obtain the stator velocity ω_s. Its integration gives the stator flux position Θ_s. In squirrel cage induction machine the rotor flux position Θ_r is obtained by simple subtracting the electric position Θ_{elec} from Θ_s.

When the initial conditions for all angles are equal to zero, the synchronization angle is also equal to zero and the control algorithm DVCSFO is realized correctly. In the DFIM real operating conditions and for multi-poles DFIM, the synchronization angle has to be known precisely to introduce it as correction term for Θ_r calculation. Finally, we use the modified angular self-piloting shown in the Fig. 8. One observe that rotor flux position Θ_r is not obtained by direct subtracting the measured electrical position $\Theta_{elec\ mes}$ from stator flux position Θ_s, but this initially known synchronization angle, has to be introduced as correcting term $\Theta_{synchronization}$ (red indication in the Fig.8).

In fine, for 3-phase, 6 poles 30kW DFIM to determine this initial electric rotor position we have to consider its rotor structure. The synchronization angle of fluxes will be find by pre-supplying the first stator phase with DC voltage $V_{s1}=+V$, two other stator phases are fed also with DC voltages $V_{s2}=V_{s3}=-V/2$. The rotor will be supplied in the same way with DC voltages but theirs signs will be opposite, i.e. $V_{r1}=-V$ and $V_{r2}=V_s=+V/2$. For 3 pairs of DFIM poles the initial rotor position can be different and situated in one of three sectors (red indication). Each sector of $2\pi/3$ rad (120°) corresponds to one pair of poles (Fig. 9)

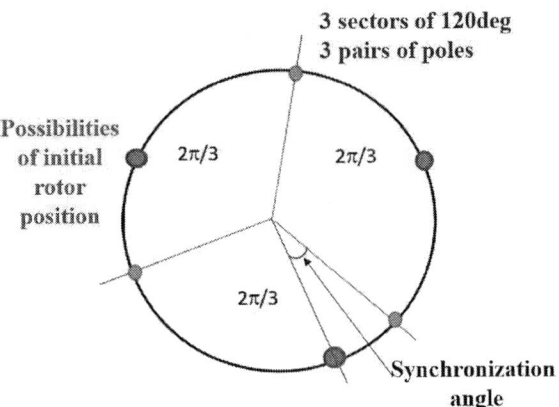

Fig. 8: Synchronization angle can be find inside of 120° sector corresponding to each pair of poles

First, we have to verify if the machine is fed to operate in motor mode. So, if the sequences of the 3 phases of the stator and the rotor are correct, then the DFIM has to be placed in a squirrel cage induction machine topology successively on the stator side or on the rotor side. For these topologies the DFIM must rotate in the opposite direction. It can be noted that this verification is made applying real time system and software reconfigurations, i.e. programmable short circuits in the stator or in the rotor to obtain the squirrel cage induction motor operation.

The synchronization angle measurement is made by fixing the rotor in any position, next we send 3 DC modulating signals Vs1, Vs2 and Vs3 to stator PWM modulator as it was described previously. We apply also to rotor PWM modulator 3 modulating signals. These DC supply provoke that the rotor and stator fluxes are aligned. The modulators are realized with FPGA solution operating with fixed point format, whereas the modulating signals are sent from DSP working with floating point format. So, the special registers allow to adapt these formats of the data issued: from DSP sent to FPGA and from FPGA sent to DSP.

So, to define the synchronization angle, first we must determine the number of sector where the rotor was stopped, next the angle value inside this sector has to be defined. If an angle of $2\pi rad$ (or of 360 degrees) is coded by 2048, each sector is characterized by g=2048/3 (or 120degrees per sector).

Consequently, after a reading of FPGA data, "N" which can be a value found between 0 and 2048, is transferred to DSP. This value must be divided by "g" to obtain the final result containing the sector number X,00 and 0,xx corresponds to angle inside this sector. To find this angle expressed in degrees we calculate 0,xx*120°. To determine Θ_r this angle must be expressed in doubly precision format compatible to DSP algorithm and is the synchronization angle to introduce it as correction term in modified angular self-piloting.

In any case, there is always a non-zero angle which should be compensated in the control strategy, it due to the fact of the mechanical shaft elasticity, torsion impact and the accuracy of the speed sensor.

DVCSFO strategy – task management

This DVCSFO strategy is implemented on a digital system containing a DSP card, TMS320C6713 from TEXAS INSTRUMENT, associated with the EP1K 100QC 208-2 FPGA card from ALTERA and an ADC/DAC interfaces card (THS10064 / TLC7628CDW) used to receive the current and voltage measurements. The Fig. 10 shows the principal task management of this control strategy where all algorithmic part is coded in DSP and interruptions, data acquisition, dead time and also modulation are programmed in FPGA.

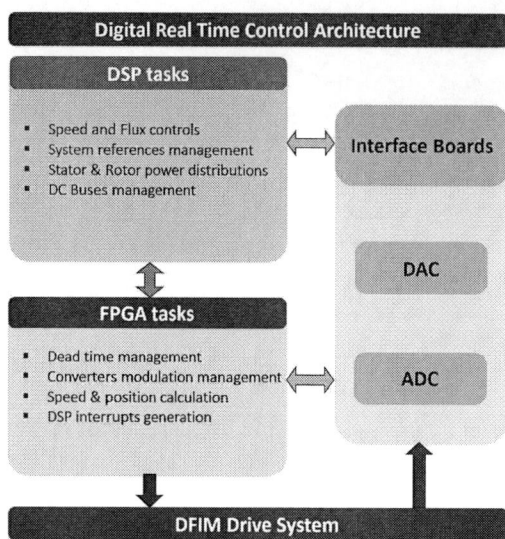

Fig. 9: Task management of DVCSFO strategy for DFIM propulsion

Conclusion

In this paper the authors present very important aspect concerning the synchronization angle which influences the quality of DVCSFO strategy applied to DFIM naval propulsion. This solution is interesting because the applied machine is Special design DFIM drive was designed to accept the active power distribution equivalently between stator and rotor. Consequently, the stator and rotor PWM VSIs have the same structure and the same sizing. Multi-poles, high voltage and high power DFIM drive allowing the best solution for this embedded system with weight and volume reduction. This introduces some difficulties in the control of the machine.

To guarantee the high reliability in all operating point the quality of proposed strategy has to be required. So, all phases of this control must be sure, i.e. starting, magnetization and four quadrants of torque/speed plane operation according to the specific naval conditions the synchronization angle is known precisely. We focused this presentation on the "synchronization angle" determination conditioning all transformations of control variables, necessary to good working of proposed DVCSFO. We affirm that the starting of this DFIM propulsion is possible when the synchronization angle is known precisely. This angle determination is not simple when the DFIM drive has many poles and when we work using the incremental speed sensor placed on the load side of this drive system.

To solve the synchronization angle problem, we recommend to use an absolute encoder instead of an incremental encoder or a sensorless control strategy.

References

[1] Y. Kawabata , E. Ejiogu et T. Kawabata . Vector-controlled double-inverter-fed wound-rotor induction motor suitable for high-power drives, IEEE Transactions on Industry Applications, 35.5 , Sept. 1999, pp. 1058–1066.

[2] G. Poddar et V. T. Ranganathan . Direct torque and frequency control of double-inverter-fed slip-ring induction motor drive. IEEE Transactions on Industrial Electronics 51.6 , Dec. 2004. P. 1329–1337. ISSN : 0278-0046. DOI : 10.1109/TIE.2004.837897.

[3] F. Bonnet, Contribution à l'Optimisation de la Commande d'une Machine Asynchrone à Double Alimentation utilisée en mode Moteur, PhD, Institut National Polytechnique de Toulouse, 2009

[4] M. Debbou, A. Damdoum, M. Pietrzak-David, Optimal Sliding Mode Control for DFIM Electric Marine Thruster, International Conference on Electrical Systems for Aircraft, Railway, Ship Propulsion and Road Vehicles **and** International Transportation, ESARS'2016, Toulouse, France

[5] J. Gillet, M. Pietrzak-David, F. Messine,, Optimization of the control of a doubly fed induction machine, 13th International Workshop on Optimization and Inverse Problems in Electromagnetism, September 10 – 12, 2014, Delft, Netherlands

[6]. P. Han; M. Cheng; R. Luo, Design and Analysis of a Brushless Doubly-Fed Induction Machine With Dual-Stator Structure, IEEE Transactions on Energy Conversion, P. 1132-1141, Volume: 31, Issue: 3, Sept. 2016.

Power control of LCR-DAB converter with phase shift in fixed switching frequency

Seung-Hyuk Baek[1], Jaehong Lee[2], Seung-Hwan Lee[2], and Sungmin Kim[3]

[1]Power ICT Research Center, Korea Electrotechnology Research Institute
Ansan, South Korea

[2]School of Electrical and Computer Engineering, University of Seoul
Seoul, South Korea

[3]Department of Electrical and Electronic Engineering, Hanyang University
Ansan, South Korea

Tel.: +82 / (031) 8040–4270

E-Mail: seunghb@keri.re.kr

Acknowledgements

This work was supported by the Korea Institute of Energy Technology Evaluation and Planning (KETEP) and the Ministry of Trade, Industry & Energy(MOTIE) of the Republic of Korea (No. 20202020800060)

Keywords

«Dual Active Bridge Converter», «DC-DC power converter», «Bi-directional converters», «Solid-State Transformer», «Wireless Power Transmission»

Abstract

This paper proposes the power control method of the Loosely Coupled Resonant Dual-Active-Bridge (LCR-DAB) converter using Phase-Shift Modulation (PSM) of a conventional DAB converter. In this paper, to apply PSM to the LCR-DAB converter, the impedance characteristics were analyzed to design an efficient operating frequency for the LCR-DAB converter. In addition, the power equation of the LCR-DAB converter was derived to control the power using the phase, which is the control variable of the PSM. Experiments were performed on the 840W prototype LCR-DAB converter to verify the validity of the proposed power control method.

Introduction

According to the increase of grid-connected distributed energy resources, the need for large-capacity power conversion systems is being emphasized. To increase the power density of a high-power converter, research on a Solid-State Transformer (SST) that converts power through converters with high-frequency transformers instead of line frequency transformers is being actively conducted [1-7]. Since the insulation performance of the SST is determined by the high-frequency transformer of the Dual-Active-Bridge (DAB) converter in the SST, the design of the high-frequency transformer has been becoming very important. To achieve the high-power density of SST, the size of the high-frequency transformer needs to be as small as possible. However, the high voltage insulation of the transformer requires enough volume of the transformer. To solve this problem, the Loosely Coupled Resonant DAB (LCR-DAB) converter using Wireless Power Transfer (WPT) coils with a low coupling coefficient instead of the high-frequency transformer was studied [8-9]. The LCR-DAB converter can secure high insulation performance because there is a physical gap between the two WPT coils. And, to enhance the efficiency of the power transfer, the series capacitors are employed with the WPT primary and secondary coils, respectively. In this paper, the impedance characteristics of the WPT coil and the series capacitor has been analyzed and the power transferred between two active bridge converters have been derived according to the output voltage of the two converters. Based on the impedance characteristics and the transferred power, the simple power control method of the LCR-DAB converter is newly presented in this paper. The proposed power control method uses the fixed frequency Phase-Shift Modulation (PSM), which is the power control method of the conventional Dual-Active-Bridge (DAB) converter having only a high-frequency transformer. The PSM based fixed frequency power control method can determine the transferred power between primary and secondary power by the phase angle difference of the primary and secondary output voltage. Since the proposed method uses only a single parameter, phase angle difference, the power control can be simply implemented. In this paper, the frequency characteristics for selecting the operating frequency of the LCR-DAB converter are explained, and the validity of the proposal is verified through the power control experiment of the 840W proto-type LCR-DAB converter.

Feature of the LCR-DAB converter

Figure 1 shows the SST applied to the MV (Medium Voltage)/LV (Low Voltage) system that converts the distribution system voltage of 22.9kV to 220V. If this system is designed to have an electrical breakdown voltage of 1.5 times the grid voltage, an electrical breakdown voltage of 35 kV or higher is required. Since the high-frequency transformer of the conventional DAB converter has a relatively small volume, however, it is difficult to secure the high insulation performance required for a higher breakdown voltage than 35kV. Figure 2 depicts the proto-type LCR-Coil of the LCR-DAB converter used in this paper. As shown in the figure, the distance between the two WPT coils in the LCR-Coil designed in this paper is 3cm. If the material inside the distance is air, the electrical breakdown voltage between the primary and secondary coils of the LCR-coil is approximately 60 [kV]. Therefore, in case an LCR-DAB converter is used instead of a high-frequency transformer of the conventional DAB converter, high insulation performance can be easily secured. Figure 3 shows the configuration of the LCR-DAB converter. As shown in the figure, The LCR-DAB converter is configured with the additional compensation circuit for efficient power transfer, such as the general Wireless Power Transmission (WPT) system. The compensation circuit is generally composed of a resonant circuit using capacitors, and the LCR-DAB converter in figure 3 is the Series-Series (SS) compensation circuit with the capacitor in series in the coil. The SS compensation circuit is one of the compensation circuits mainly used in the WPT system because it has the characteristics of the voltage source or the current source depending on the operating frequency of the converter [10]. Therefore, in this paper, the SS compensation circuit was applied as a method to compensate for the coupling coefficient.

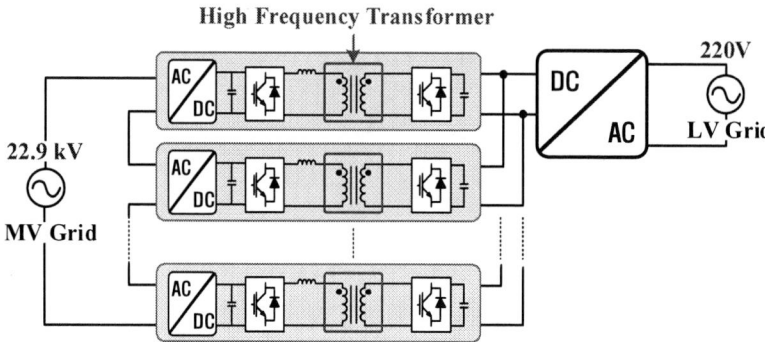

Fig. 1: SST applied to MV/LV system

Fig. 2: LCR-coil applied to 840[W] class LCR-DAB converter

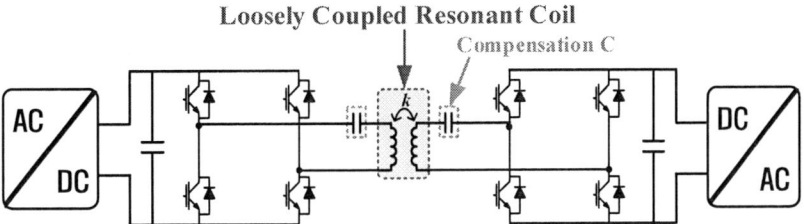

Fig. 3: Configuration of the LCR-DAB converter

Table I: Parameters of the proto-type LCR-DAB converter

Parameters	Value
V_P, V_S (Priamry & secondary DC voltage)	100 [V]
C_P, C_P (Series capacitance)	72 [nF]
R_P, R_S (Coil resistance)	70 [mΩ]
L_P, L_S (Series inductance)	100 [uH]
k (Coupling coefficient)	0.26

Impedance characteristics of the LCR-DAB Converter with SS compensation circuit

The conventional DAB converter having a high-frequency transformer can transfer power between primary and secondary converters by phase angle difference of the primary and secondary output voltages. Because the power transferred can be simply determined by the phase angle difference namely phase-shift, the power can be controlled easily. In a similar manner, the power control of the LCR-DAB converter can be performed through phase-shift of the secondary converter output voltage at the fixed operating frequency. However, the electric circuit characteristics of the WPT coils and the series capacitor in the LCR-DAB converter change according to the phase-shift. Therefore, to determine the operating frequency of the LCR-DAB converter, it is necessary to analyze the characteristics of the input impedance according to the phase shift. In the case of LCR-DAB converters which are consisted of the WPT coils and series capacitors, the input impedances have three of resonant frequencies. Each resonance frequency can be calculated through Short/Open circuit analysis from the equivalent circuit of the LCR-DAB converter [5]. Figure 4 shows the equivalent circuit of the LCR-DAB converter. Both wireless coils of the LCR-DAB converter can be equivalent to T-model for electrical analysis as shown in the figure 4. In figure 4, V_P, V_S are converter output voltages, θ is phase-shift angle of V_S relative to V_P, I_P, I_S are converter currents, R_P, R_S are Equivalent Series-Resistances (ESR) of the wire, C_P, C_S are capacitances of compensation capacitors, L_P, L_S are self-inductances, and k is coupling coefficient between the primary coil and the secondary coil, respectively.

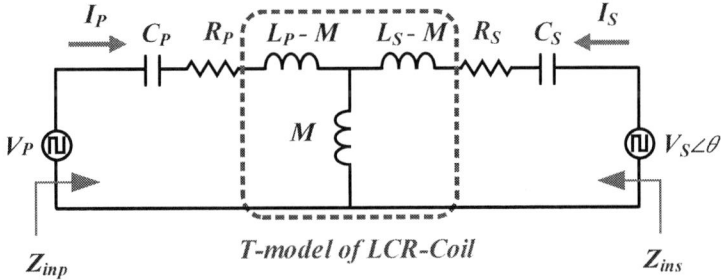

Fig. 4: The equivalent circuit of the LCR-DAB converter

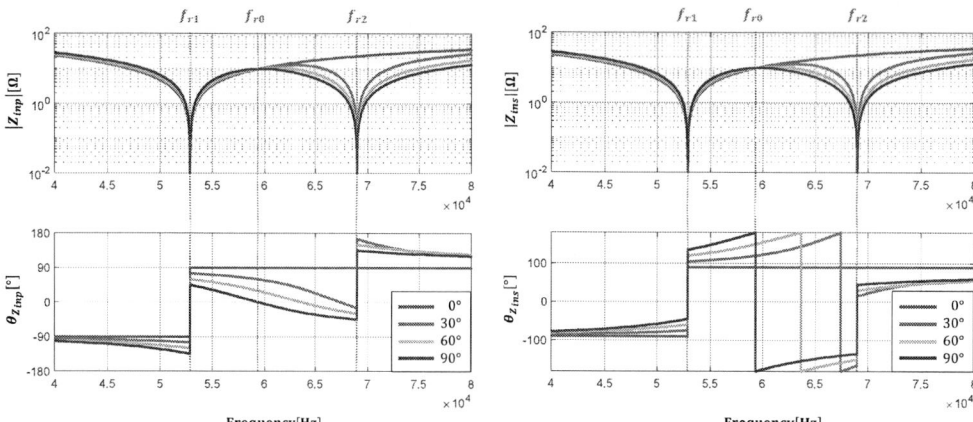

Fig. 5: Frequency responses of input impedances by phase-shift

In addition, Z_{inp} and Z_{ins} represent the input impedance seen from primary converter and secondary converter, respectively. From figure 4, if $L_P=L_S$, $C_P=C_S$, and $R_P=R_S=0[\Omega]$, each resonant frequency of input impedances can be expressed as the following equations.

$$f_{r0} = \frac{1}{2\pi\sqrt{L_P C_P}} = \frac{1}{2\pi\sqrt{L_S C_S}} \tag{1}$$

$$f_{r1}, f_{r2} \cong \frac{f_{r0}}{\sqrt{1 \pm k}} \tag{2}$$

In equations (1) and (2), f_{r0} is the series resonance frequency of the short circuit in the LCR-DAB converter equivalent circuit, and f_{r1}, f_{r2} are the two resonance frequencies of the open circuit. Figure 5 shows the frequency response characteristics of the input impedance according to the phase-shift. The specifications in table 1 were applied to the input impedance calculations. Figure. 5 indicates that the Q at f_{r1} decreases and the Q at f_{r2} increases as the phase-shift increases. Therefore, in the frequency band between f_{r1} and f_{r0}, the input impedance increases as the phase-shift increases, and in the frequency band between f_{r0} and f_{r2}, the input impedance decreases as the phase-shift increases. On the other hand, at the series resonance frequency f_{r0}, the input impedance magnitude is always constant regardless of the phase-shift.

Selection of operating frequency of LCR-DAB converter to apply fixed frequency PSM

The proposed power control method of the LCR-DAB converter would control the transferred power through only phase-shift. Therefore, because the switching frequency needs to be fixed for the simplicity of the control structure, the switching frequency should be determined from the viewpoint of efficiency. In figure 5, in the frequency band between f_{r1} and f_{r0}, it was confirmed that the magnitude of the input impedances increased as the phase-shift increases. However, since the input impedance in this frequency band is relatively small, the magnitude of the average current in the phase-shift range ($-90[°]\sim90[°]$) is large. Figure 6 shows the average input current according to the phase-shift range in case the LCR-DAB converter is operated with a frequency between f_{r1} and f_{r0}. In the figure, it can be confirmed that the case of f_{r0} has the lowest average input current in the entire phase-shift range. On the other hand, in the frequency band between f_{r0} and f_{r2}, as the phase shift increases, the magnitude of the input impedance decreases. Therefore, in this frequency band, the magnitude of the input current increases as the phase shift increases. In addition, in this frequency band, as the phase shift increases, the decrease in input impedance is relatively large. Therefore, even when the phase shift is less than 90°, the impedance phase becomes 0°. This means that when the

LCR-DAB converter operates at a frequency higher than f_{r0}, the rated power is output even when the phase shift is less than 90°. Consequently, when the LCR-DAB converter operates at a frequency between f_{r0} to f_{r2}, the usable range of the phase shift, which is the only control variable, is reduced, thereby increasing the control sensitivity. Figure 7 shows the output power for the phase shift in the case of the LCR-DAB converter operated at a frequency between f_{r0} and f_{r0}-f_{r2}(65kHz). In the case of f_{r0} in the figure, the output power increases according to the phase-shift, and the rated power is output under the condition that the the the phase-shift is 90[°]. On the other hand, at a frequency between f_{r0} and f_{r2}(65kHz), the rated power can be transferred at the phase-shift of about 40 [°], and as the phase-shift increases, the transferred power would be larger tha the rated power. When the LCR-DAB converter operates at f_{r0} with the PSM method, the magnitude of input current is minimized as compared with other frequencies, and the range of phase-shift angle capable of transmitting rated power is extended to 90°, thereby minimizing control sensitivity. In addition, the magnitude of input impedance is constant, and the input impedance have inductive characteristics in the entire phase-shift range. Therefore, the current magnitude would not only be constant in the whole power range but the ZVS operation is also achieved in the whole power range.

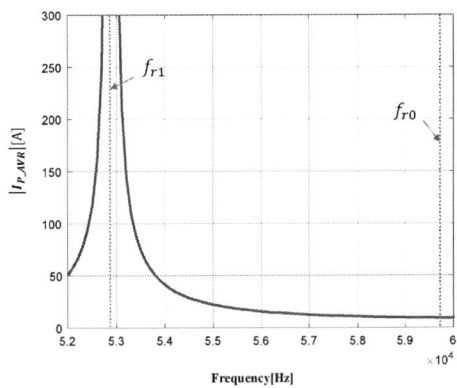

Fig. 6: Average input current magnitude for frequencies between f_{r1} and f_{r0}

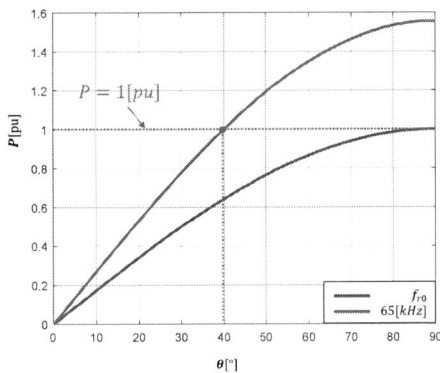

Fig. 7: Output power for phase-shift in case of f_{r0} and 65kHz

Proposed power control of LCR-DAB converter with fixed frequency PSM

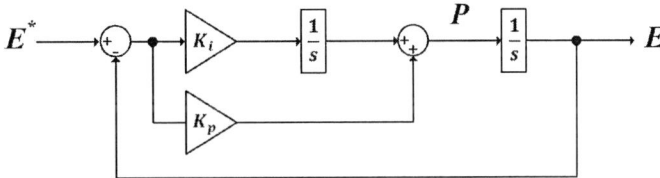

Fig. 8: Power controller structure

Typically, the conventional DAB converter controls the amount of power flowing through the converter to control the secondary DC-link voltage [11]. Figure 8 shows the structure of the voltage controller applied in this paper, and the energy, which is the input variable of the power controller, was calculated from the DC-link voltage using the following equation:

$$E = \frac{1}{2}CV^2 \tag{3}$$

The reference value is the stored energy when the secondary DC link capacitor voltage is the command voltage (V_{DCS}^*) and the feedback value is the stored energy in the secondary DC link capacitor of which value is V_{DCS}. In equation (3), C is the capacitances of DC-link. To apply PSM to the control structure shown in Fig. 8, an operator is required to convert power, which is the output of the power controller, into the phase-shift. Therefore, it is necessary to derive a power equation having the phase-shift as a variable. The power equation of the LCR-DAB converter can be derived by analyzing the LCR-DAB converter equivalent circuit in Figure 4. From the equivalent circuit of the LCR-DAB converter, assuming that $L_P=L_S$, $C_P=C_S$, and ESR = 0Ω, the power equation of the LCR-DAB converter operating at f_{r0} can be simply expressed as follows:

$$P_P = -\frac{1}{2}Re\{V_P I_P\} \cong -\frac{V_P^2}{Z_{inp0}} = -\frac{V_P V_S}{2\omega_{r0}M}\sin\theta \tag{4}$$

$$P_s = -\frac{1}{2}Re\{V_s I_s\} \cong -\frac{V_S^2}{Z_{ins0}} = \frac{V_P V_S}{2\omega_{r0}M}\sin\theta \tag{5}$$

In equations (4) and (5), Z_{inp0} and Z_{ins0} are the input impedance when the LCR-DAB converter operates at f_{r0}, ω_{r0} (= $2\pi f_{r0}$)is the angular frequency, and M (= $k\sqrt{L_P L_S}$) is the mutual inductance, respectively. From the derived power equation, it can be seen that the sign of the power and the sign of the phase-shift are opposite in the LCR-DAB converter, unlike the conventional DAB converter. Therefore, in the case of an LCR-DAB converter, when power is transferred from the primary side to the secondary side, the output voltage of the primary side converter is led the output voltage of the secondary side converter, and in the opposite case, the primary side output voltage is lagged than the secondary side output voltage. From Equations (4) and (5), assuming that the magnitude of the phase-shift is small, the operator for deriving the phase-shift from the output of the power controller can be expressed as follows.

$$\theta = -\frac{2\omega_{r0}M}{V_P V_S}P_P \quad (Power\ flow\ P \rightarrow S) \tag{6}$$

$$\theta = \frac{2\omega_{r0}M}{V_P V_S}P_S \quad (Power\ flow\ S \rightarrow P) \tag{7}$$

Figure 9 shows the power control structure of the LCR-DAB converter proposed in this paper. Figure 9 shows the control structure when power is transmitted from the primary side to the secondary side, and Equation (6) is applied to calculate the phase-shift. On the other hand, when power is transmitted from the secondary side to the primary side, power control can be performed by calculating the phase-shift using Equation (7)

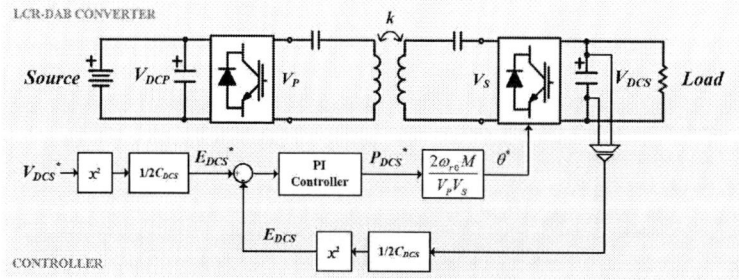

Fig. 9: Proposed power controller structure of the LCR-DAB converter

Results of experiment

To verify the proposed power controller of the LCR-DAB converter, the 840[W] proto-type LCR-DAB converter power control experiments were performed. Figure 10 shows the experiment set-up, and the specifications of the system used for the experiment are described in Table 1. Figure 11 shows the voltage and current waveforms when there is no load and when the rated load is applied in each direction. In Figure 11, as confirmed by Equations (4) and (5), since the sign of power and the sign of phase are opposite, it can be confirmed that the phase-shift increases in the opposite direction as the load increases. In addition, since the magnitude of the input impedance is constant regardless of the phase-shift in case of the operating at the series resonance frequency (f_{r0},), it can be confirmed that the magnitude of the input current is almost constant under all load conditions. Figure 12 shows the load current and the secondary DC link voltage when the rated load is applied in the step manner. In figure 12, it can confirm that the DC link voltage is well controlled even with a momentary load change. Figure 13 shows the efficiency and input-output loss for the output power of the LCR-DAB converter. The maximum efficiency is about 96.8 [%] was measured when power is transmitted from the secondary side to the primary side, and the loss was measured in the form of a decrease as the load increased. Since the LCR-DAB converter performs the ZVS operation in all load conditions, the loss of the converter is dominated by the resistance loss due to the converter current. Therefore, ideally, since the magnitude of the converter output current is constant regardless of the load when operating at the series resonance frequency, a constant loss should occur even when the load changes. However, due to the error of the system parameter, the current increased as the load increased, and the loss was measured in the form of a decrease.

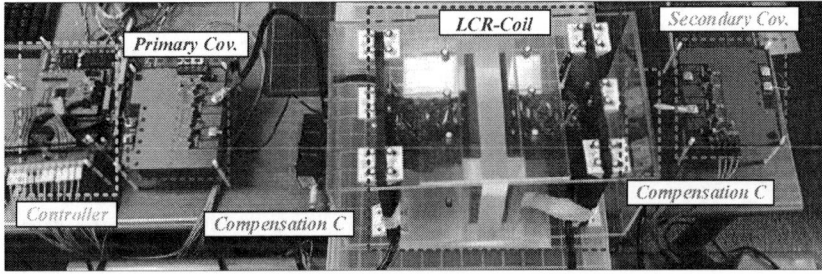

Fig. 10: Experiment set-up of 840[W] LCR-DAB converter

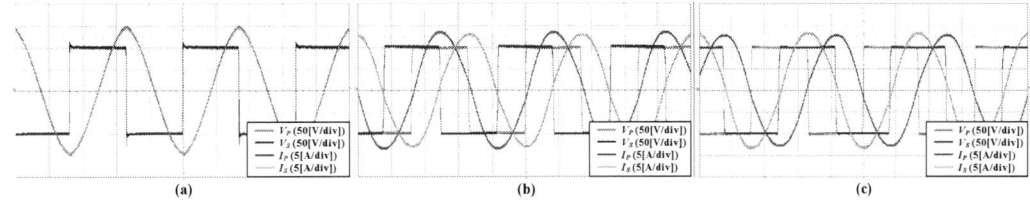

Fig. 11: (a) No-load, (b) Rated load (positive direction), (c) Rated load (negative direction)

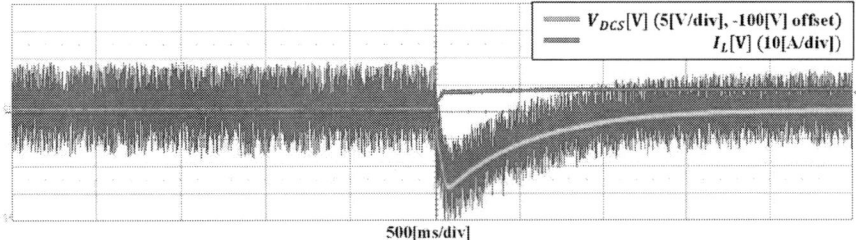

Fig. 12: The load current and the secondary side DC link voltage for the step rated load

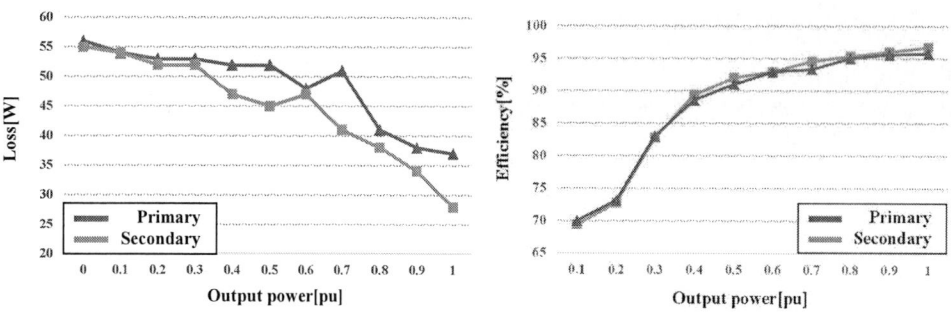

Fig. 13: Efficiency and loss for output power

Conclusion

In this paper, the power control method of applying the fixed frequency PSM that can perform with a relatively simple structure to the LCR-DAB converter was proposed. To this end, the selection of the most suitable compensation topology and efficient switching frequency for the LCR-DAB converter was described. The validity of the proposal was verified through the power control experiment of the 840[W] proto-type LCR-DAB converter to which the selected efficient frequency was applied.

References

[1] J. E. Huber and J. W. Kolar: Applicability of Solid-State Transformers in Today's and Future Distribution Grids, IEEE Transactions on Smart Grid, vol. 10, no. 1, pp. 317- 326

[2] R. P. Londero, A. P. C. d. Mello and G. S. da Silva: Comparison between conventional and solid state transformers in smart distribution grids, IEEE PES Innovative Smart Grid Technologies Conference, pp. 1- 6

[3] A. Joshi and S. Nath: Efficiency Comparison of Solid-State Transformer and Low-Frequency Power Transformer, 3rd International Conference on Energy, Power and Environment: Towards Clean Energy Technologies, pp. 1- 6

[4] B. Zhao, Q. Song, W. Liu, and Y. Sun: Overview of dual-active-bridge isolated bidirectional DC–DC converter for high-frequency-link power-conversion system, IEEE Transactions on Power Electronics, vol. 29, no. 8, pp. 4091- 4106

[5] X. She, A. Q. Huang and R. Burgos: Review of Solid-State Transformer Technologies and Their Application in Power Distribution Systems, IEEE Journal of Emerging and Selected Topics in Power Electronics, vol. 1, no. 3, pp. 186- 198

[6] J. E. Huber and J. W. Kolar: Solid-State Transformers: On the Origins and Evolution of Key Concepts, IEEE Industrial Electronics Magazine, vol. 10, no. 3, pp. 19- 28

[7] J. E. Huber and J. W. Kolar: Volume/weight/cost comparison of a 1MVA 10 kV/400 V solid-state against a conventional low-frequency distribution transformer, 2014 IEEE ECCE, pp. 4545- 4552

[8] J. Lee, J. Roh, S. -H. Lee, S. Kim and M. -Y. Kim: A Novel Solid-State Transformer with Loosely Coupled Resonant Dual-Active-Bridge Converters, in 2020 IEEE ECCE, pp. 3972- 3978

[9] J. Lee, J. Roh, M. Y. Kim, S. -H. Baek, S. Kim and S. -H. Lee: A Novel Solid-State Transformer With Loosely Coupled Resonant Dual-Active-Bridge Converters, IEEE Transactions on Industry Applications, vol. 58, no. 1, pp. 709- 719

[10] S. Cho, I. Lee, S. Moon, G. Moon, B. Kim and K. Y. Kim: Series-series compensated wireless power transfer at two different resonant frequencies, IEEE ECCE Asia Downnunder, pp. 1052- 1058

[11] C. Mi, H. Bai, C. Wang, S. Gargies: Operation, Design and Control of Dual H-bridge-based Isolated Bidirectional DC-DC Converter, IET Power Electronics, vol. 1, no. 4, pp. 507- 517

A Simplified Braking Method for Direct Matrix Converter-Fed PMSM Drives with Consideration of Avoiding Regenerative Energy

Jun Xie, Dustin Henneberg, Martin Suberski, Thomas Ellinger,
Uwe Rädel and Jürgen Petzoldt
Technische Universität Ilmenau
Power Electronics and Control Group
Ilmenau, Germany
Phone: +49 367769-1553
Fax: +49 367769-1469
Email: jun.xie@tu-ilmenau.de
URL: http://www.tu-ilmenau.de

Keywords

≪Direct matrix converter≫, ≪AC-AC converter≫, ≪Permanent magnet motor≫, ≪Variable speed drive≫, ≪Regenerative power≫

Abstract

This paper presents a simplified braking method for direct matrix converter (DMC) - fed permanent magnet synchronous motor (PMSM) drives with special consideration of avoiding the regenerative energy from feeding back to the grid. As a four-quadrant converter, the DMC has the bidirectional energy-flow capability, which is seen as a great benefit. However, the regenerative energy feeding back to the grid is not always needed, in some cases might be not allowed or not possible. These could happen due to the grid load status, strict quality specifications of the grid current or grid failure, where the DMC needs to be decoupled from the grid. In order to avoid the regenerative energy temporarily or completely, a three-phase diode bridge (B6) with chopper resistance on the motor side is employed together with the traditional B12-clamp circuit, which is used for overvoltage protection. Furthermore, the motor torque during braking could be controlled without large fluctuation and within the required limitation. The principle of the proposed strategy is described in detail. Simulation results are used to verify the feasibility of the proposed strategy.

Introduction

The three-phase DMC shown in Fig. 1 currently attracts many research interests in power electronics and power systems [1, 2]. As an alternative topology of direct AC-AC conversion without any bulky energy storage elements in an intermediate link, the DMC shows great advantages with its compact size, bidirectional energy-flow capability and longtime durability, especially in adjustable speed drives applications.

Compared with the conventional AC-DC-AC converter, which also has the bidirectional energy-flow capability as shown in Fig. 2 (a), the DMC has no energy storage components in the DC-link, which is a great benefit in volume, since no additional circuit is required. However, the harmonics involved in the grid current could be an issue for DMC in applications, where strict quality specifications are required.

On the other hand, the regenerative energy could be temporarily undesired according to the grid load status, especially for micro grid with many loads in parallel, where the power flow in the grid must be controlled properly to keep the grid stable. It could also happen, when there is error in grid and the

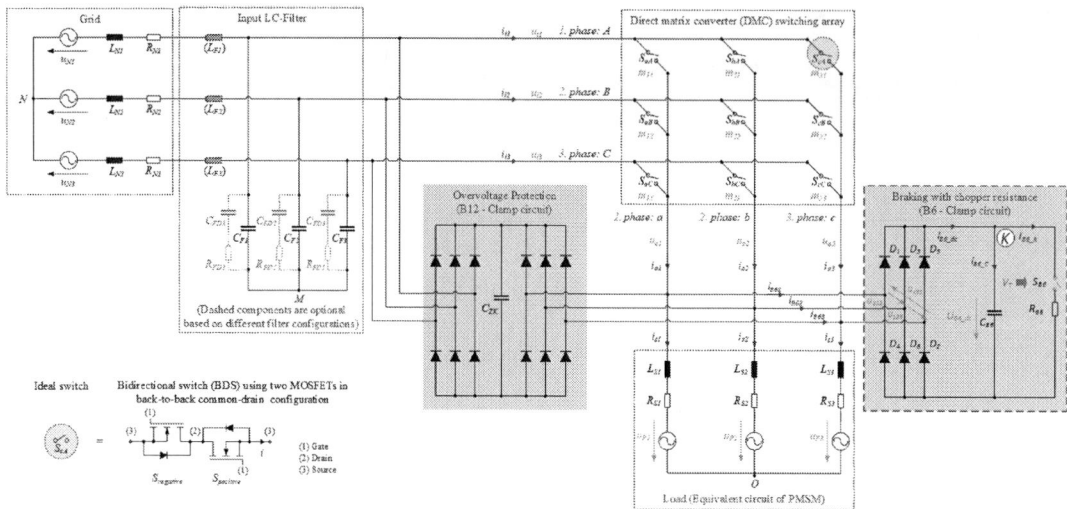

Fig. 1: System topology of three-phase to three-phase DMC-fed PMSM drives with braking circuit for avoiding regenerative energy

DMC needs to be decoupled from the grid, in this case, the regenerative energy is not able to feed back to the grid. For these reasons, it is important to have a control method to avoid the regenerative energy temporarily or completely. So far, not much researches are made in this potential application and more research interests are expected in this field.

In [6–8], continue studies for avoiding regenerative energy are presented with three different methods in the special application of electric aircraft, which is a great foundation for further study. In [6], B6-diode bridge with one chopper resistance on the grid side is as input power clamp (IPC) method employed as shown in Fig. 2 (b). In [7], three-phase chopper resistance on the grid side is implemented as bi-directional switch (BDS) method as shown in Fig. 2 (c). Both methods rely on the detection of regenerative energy. There are two methods presented for detecting regenerative energy. The first method is the power comparison (PC) method, in which the input and output power are estimated separately and the result of comparison is then used as reference for duty cycle calculation of the related chopper switches. The second method is the input voltage reference (IVR) method, in which the input capacitor voltage is used as reference. In [8], the standard B12-clamp circuit is used with a chopper resistance in the DC-link as shown in Fig. 2 (d). In [9], further studies are made with special consideration of gird faults in regeneration in order to properly decide the halt sequence of the DMC.

In conclusion, all these methods above show good results and have specific advantages and some short-comings to some extent. This paper proposed a simplified method for braking of DMC-fed PMSM drives with consideration of avoiding regenerative energy. As shown in Fig. 1, in normal drive operation, the chopper resistance R_{B6} in the B6-diode bridge is shut down by chopper switch S_{B6}. When the PMSM drive brakes, the switching matrix of DMC is shut down by disabling the gate signals and the chopper switch S_{B6} starts to switching according to the duty cycle V_t, which is determined by the proposed control strategy. The energy stored in PMSM drives will be consumed by the chopper resistance R_{B6} to avoid the regenerative energy back to the grid. Meanwhile, the braking moment mi in this period could be controlled within the required limitation as long as the duty cycle V_t of the chopper switch S_{B6} still inner the upper limit of 1.

Compared with previous research, the main features of the proposed strategy are as follows:
- It provides a simplified braking method by employing B6-diode bridge with one chopper resistance on motor side and shutting down the switching matrix of DMC during regeneration. The DC-link current in B6-diode bridge $i_{B6,dc}$ is used as reference without calculating power on both sides of DMC, which needs less computation effort.

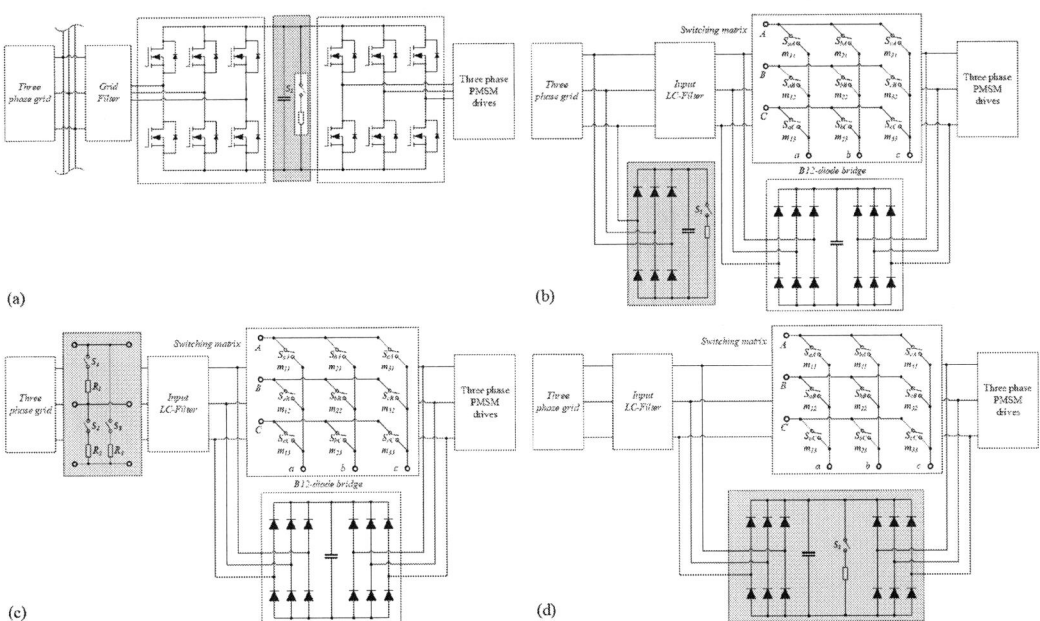

Fig. 2: Summary of methods for avoiding regenerative energy: (a) Conventional AC-DC-AC converter. (b) B6-diode bridge with one chopper resistance on grid side. (c) Three-phase chopper resistance on grid side. (d) Standard B12-clamp circuit with chopper resistance.

- It prevents grid failure to further reflect to the motor drive side though the coupled switching matrix of DMC and keeps the harmonics, which are generated through the B6-diode bridge, away from the grid.
- It is able to control the motor torque within the required limitation and without large fluctuation during braking period, which helps to ensure the motor braking safe and smooth.

The rest of this paper is organized as follows. In section II, a brief introduction of the system topology is presented. The proposed strategy is subsequently addressed in section III. Simulation results verifying the performance of the proposed strategy are demonstrated in section IV. Finally, this study is summarized in section V.

System topology

The system topology of three-phase to three-phase DMC-fed PMSM drives with avoiding regenerative energy is shown in Fig. 1, which consists of power supply, input LC filter, DMC switching array, PMSM drive system, the B12-clamp circuit for overvoltage protection and the B6-diode bridge for braking.

A. Topology of DMC

The DMC is composed of nine bidirectional switches (BDS) as a 3×3 matrix, which connects the grid to the PMSM drives. Each BDS should have the capability to block voltage and conduct current in both directions. There are many configurations for the realization [12]. One possibility could be two power semiconductor switches in back-to-back arrangement, which is shown in Fig. 1. The power semiconductor switch used could be anti-parallel insulated gate bipolar transistor (IGBT) or metal–oxide–semiconductor field-effect transistor (MOSFET).

The input LC-Filter is necessary to reduce high-frequency current harmonics and voltage fluctuation. Many topologies of input filter for DMC have been proposed [10]. In this study, the damped LC filter is employed, in which the filter capacitor C_f is parallel with a RC damping circuit as shown in Fig. 1. It is worth noticing that, if the grid impedance $(R_N + jX_{L_N})$ can be precisely determined, it is possible to use grid inductance L_N instead of the normal filter inductance L_f within the permissible resonance frequency range. The output filter is commonly neglected due to the inductive nature of PMSM drive system.

The B12-clamping circuit is frequently implemented as a standard circuit for overvoltage protection, which makes up the deficiencies of DMC for absence of passive free-wheeling paths. It connects the input and output sides of DMC using 12 fast recovery diodes. The double B6 diode bridge are connected by a DC-link capacitor C_{ZK}, which is determined proportional to the energy stored in PMSM drives [11].

The B6-diode bridge is employed to avoid the regenerative energy by motor braking. The chopper resistance R_{B6} is switched at duty cycle V_t, which is determined by the control strategy. The DC-link capacitor of the B6-diode bridge C_{B6} is used for voltage measurement and to absorb the high frequency components of the chopper resistance current $i_{B6,R}$.

B. Direct modulation strategy of DMC

The direct modulation strategy, namely the optimized Venturini's method, is employed for normal operation of the PMSM drives in the study, because of its convenience and straightforward of understanding the basics of DMC.

The switching status S_{ij} of each BDS in the 3×3 switching array can be assigned as 1 to represent ON state and 0 to represent OFF state. In this manner, to avoid short circuit of input phases (A,B,C) or open circuit of output phases (a,b,c), the possible switching states to satisfy the constraint (1) is well known to be $3^3 = 27$ combinations.

$$\sum_{j=A,B,C} S_{ij} = 1, \; \forall i = \{a,b,c\} \tag{1}$$

The mathematical relationship between input and output of DMC can be described in (2) with the 3×3 instantaneous transfer matrix of switching state S_{ij}.

$$\begin{bmatrix} u_{o1} \\ u_{o2} \\ u_{o3} \end{bmatrix} = \begin{bmatrix} S_{aA} & S_{aB} & S_{aC} \\ S_{bA} & S_{bB} & S_{bC} \\ S_{cA} & S_{cB} & S_{cC} \end{bmatrix} \begin{bmatrix} u_{i1} \\ u_{i2} \\ u_{i3} \end{bmatrix} \quad , \quad \begin{bmatrix} i_{i1} \\ i_{i2} \\ i_{i3} \end{bmatrix} = \begin{bmatrix} S_{aA} & S_{aB} & S_{aC} \\ S_{bA} & S_{bB} & S_{bC} \\ S_{cA} & S_{cB} & S_{cC} \end{bmatrix}^T \begin{bmatrix} i_{o1} \\ i_{o2} \\ i_{o3} \end{bmatrix} \tag{2}$$

where $\vec{u}_i = \begin{bmatrix} u_{i1} & u_{i2} & u_{i3} \end{bmatrix}^T$ and $\vec{i}_i = \begin{bmatrix} i_{i1} & i_{i2} & i_{i3} \end{bmatrix}^T$, $\vec{u}_o = \begin{bmatrix} u_{o1} & u_{o2} & u_{o3} \end{bmatrix}^T$ and $\vec{i}_o = \begin{bmatrix} i_{o1} & i_{o2} & i_{o3} \end{bmatrix}^T$ are the input and output voltage and current vectors respectively.

Suppose the sinusoidal input voltage vector $\vec{u}_i$ and output current vector $\vec{i}_o$ of DMC are expressed as (3):

$$\vec{u}_i(t) = \begin{bmatrix} \hat{U}_i cos(\omega_i t) \\ \hat{U}_i cos(\omega_i t - \frac{2\pi}{3}) \\ \hat{U}_i cos(\omega_i t + \frac{2\pi}{3}) \end{bmatrix} \quad , \quad \vec{i}_o(t) = \begin{bmatrix} \hat{I}_o cos(\omega_o t + \phi_o) \\ \hat{I}_o cos(\omega_o t - \frac{2\pi}{3} + \phi_o) \\ \hat{I}_o cos(\omega_o t + \frac{2\pi}{3} + \phi_o) \end{bmatrix} \tag{3}$$

By adding third harmonics of the input and output angular frequencies into the desired output voltage, a maximum voltage transfer ratio q of $\frac{\sqrt{3}}{2}$ (or 86.7%) is achieved, which is known as the intrinsic limitation of DMC [12]. The desired output voltage vector $\vec{u}_o$ and input current vector $\vec{i}_i$ in average value are described as (4).

$$\vec{u}_o(t) = q\hat{U}_i \begin{bmatrix} cos(\omega_o t) - \frac{1}{6}cos(3\omega_o t) + \frac{1}{2\sqrt{3}}cos(3\omega_i t) \\ cos(\omega_o t - \frac{2\pi}{3}) - \frac{1}{6}cos(3\omega_o t) + \frac{1}{2\sqrt{3}}cos(3\omega_i t) \\ cos(\omega_o t + \frac{2\pi}{3}) - \frac{1}{6}cos(3\omega_o t) + \frac{1}{2\sqrt{3}}cos(3\omega_i t) \end{bmatrix} , \quad \vec{i}_i(t) = \begin{bmatrix} \hat{I}_i cos(\omega_i t + \phi_i) \\ \hat{I}_i cos(\omega_i t - \frac{2\pi}{3} + \phi_i) \\ \hat{I}_i cos(\omega_i t + \frac{2\pi}{3} + \phi_i) \end{bmatrix} \tag{4}$$

where f_i and f_o are the input and output frequencies with $\omega_i = 2\pi f_i$ and $\omega_o = 2\pi f_o$; q is the voltage transfer ratio with $q = \frac{\hat{U}_o}{\hat{U}_i}$; $\hat{U}_i$ and $\hat{U}_o$, $\hat{I}_i$ and $\hat{I}_o$ are the input and output voltage and current amplitude, respectively; ϕ_i and ϕ_o are the input and output phase displacement angle.

By employing unity input power factor with $cos(\phi_i) = 1$ into the direct modulation method, the duty cycle $m_{ij}(t)$ of each switch is expressed in a simplified form in (5).

$$
\begin{aligned}
m_{ij}(t) = & \frac{1}{3}\left\{ 1 + 2q \cdot cos\left(\omega_i t - (j-1)\frac{2\pi}{3}\right)\left[cos\left(\omega_o t - (i-1)\frac{2\pi}{3}\right) - \frac{1}{6}cos(3\omega_o t)\right] \right. \\
& \left. + \frac{7}{12}cos\left(2\omega_i t + (j-1)\frac{2\pi}{3}\right) - \frac{1}{12}cos\left(4\omega_i t - (j-1)\frac{2\pi}{3}\right) \right\}
\end{aligned}
\tag{5}
$$

where $\forall i = \{1,2,3\}$ corresponds to output phase $\{a,b,c\}$ and $\forall j = \{1,2,3\}$ corresponds to input phase $\{A,B,C\}$. With $\vartheta_i = \omega_i t$ and $\vartheta_o = \omega_o t$, the entire transfer matrix can be expressed using duty cycles $m_{ij}(t)$ with the vectors defined in (6) and (7) as shown in (8).

$$
\begin{bmatrix} x_1 \\ x_2 \\ x_3 \end{bmatrix} = \begin{bmatrix} cos(\vartheta_o) - \frac{1}{6}cos(3\vartheta_o) \\ cos(\vartheta_o - \frac{2\pi}{3}) - \frac{1}{6}cos(3\vartheta_o) \\ cos(\vartheta_o + \frac{2\pi}{3}) - \frac{1}{6}cos(3\vartheta_o) \end{bmatrix} \quad , \quad \begin{bmatrix} y_1 \\ y_2 \\ y_3 \end{bmatrix} = \begin{bmatrix} cos(\vartheta_i) \\ cos(\vartheta_i - \frac{2\pi}{3}) \\ cos(\vartheta_i + \frac{2\pi}{3}) \end{bmatrix}
\tag{6}
$$

$$
\begin{bmatrix} z_1 \\ z_2 \\ z_3 \end{bmatrix} = \frac{7}{36} \begin{bmatrix} cos(2\vartheta_i) \\ cos\left(2(\vartheta_i - \frac{2\pi}{3})\right) \\ cos\left(2(\vartheta_i + \frac{2\pi}{3})\right) \end{bmatrix} - \frac{1}{36} \begin{bmatrix} cos(4\vartheta_i) \\ cos\left(4(\vartheta_i - \frac{2\pi}{3})\right) \\ cos\left(4(\vartheta_i + \frac{2\pi}{3})\right) \end{bmatrix}
\tag{7}
$$

$$
\begin{bmatrix} m_{11} & m_{12} & m_{13} \\ m_{21} & m_{22} & m_{23} \\ m_{31} & m_{32} & m_{33} \end{bmatrix} = \begin{bmatrix} \frac{1}{3} + \frac{m}{\sqrt{3}}(x_1 y_1 + z_1) & \frac{1}{3} + \frac{m}{\sqrt{3}}(x_1 y_2 + z_2) & \frac{1}{3} + \frac{m}{\sqrt{3}}(x_1 y_3 + z_3) \\ \frac{1}{3} + \frac{m}{\sqrt{3}}(x_2 y_1 + z_1) & \frac{1}{3} + \frac{m}{\sqrt{3}}(x_2 y_2 + z_2) & \frac{1}{3} + \frac{m}{\sqrt{3}}(x_2 y_3 + z_3) \\ \frac{1}{3} + \frac{m}{\sqrt{3}}(x_3 y_1 + z_1) & \frac{1}{3} + \frac{m}{\sqrt{3}}(x_3 y_2 + z_2) & \frac{1}{3} + \frac{m}{\sqrt{3}}(x_3 y_3 + z_3) \end{bmatrix}
\tag{8}
$$

where $m = \frac{q}{\sqrt{3}/2}$ is the modulation index with a range from 0 to 1.

C. Model of PMSM drive system

Since no output filter is required, the motor voltages and currents are actually the output voltages $\vec{u}_o$ and currents $\vec{i}_o$ of DMC. The continuous-time model of the PMSM is given in dq-coordinate in (9):

$$
\begin{cases} u_d = R_s i_d - \omega_e \cdot \psi_q + \frac{d\psi_d}{dt} \\ u_q = R_s i_q + \omega_e \cdot \psi_d + \frac{d\psi_q}{dt} \end{cases} \quad , \quad \begin{cases} \psi_d = L_d i_d + \psi_m \\ \psi_q = L_q i_q \end{cases}
\tag{9}
$$

The electromagnetic torque m_i and mechanical dynamics of PMSM can be estimated in (10):

$$
m_i = \frac{3}{2}P_p(\psi_d i_q - \psi_q i_d) \quad , \quad m_i - m_l - m_r = \frac{1}{J}\frac{d\omega_m}{dt} \quad , \quad m_r = k_r \omega_m \quad , \quad \omega_e = P_p \cdot \omega_m
\tag{10}
$$

where u_d and u_q, i_d and i_q, ψ_d and ψ_q, L_d and L_q represent stator voltages $\vec{u}_o$, stator currents $\vec{i}_o$, magnet flux and stator inductance in dq-coordinate, respectively; ψ_m represents the magnet flux linkage of PMSM; R_s is the resistance of stator; P_p is the number of pole pairs; J is the moment of inertia; ω_e and ω_m are the electrical and mechanical angular frequency of PMSM, respectively; m_l is the load moment and m_r is the frictional moment, which is proportional to the mechanic angular frequency with the friction coefficient k_r.

Proposed braking method with avoiding regenerative energy

In normal conditions, where the regenerative energy is allowed back to the grid. The chopper switch S_{B6} in the B6-clamp circuit is switched off. The system is protected by the B12-clamp circuit to avoid over-voltage. The DMC-fed PMSM drive operates in four-quadrant and the energy flows in both directions of DMC, which depends on the motor operation to be acceleration or braking.

In special cases, the regenerative energy back to the grid is not allowed, temporarily or completely. In practice, especially when the PMSM drive is fully loaded, the following features of the drive system are frequently desired:
- The regenerative energy could be fully avoided from feeding back to the grid.

- The braking torque of the motor drive is controllable and is able to keep relative constant without large fluctuation, so that the braking process is smooth and safe. Meanwhile, the braking torque should not exceed the required limitation.

To achieve this, in this proposed principle, the switching matrix of DMC will be shut down during braking, so that the regenerative energy produced by braking operation of the PMSM drives will be dissipated by the braking resistor R_{B6} in the B6-diode bridge, instead of flowing back to grid.

During braking process, the motor moment reference, which is the DC-link current $i_{B6,dc}$, will be controlled by setting the duty cycle V_t of the chopper switch S_{B6} based on the required moment limitation. The simplified circuit for analysing the braking process is shown in Fig. 3.

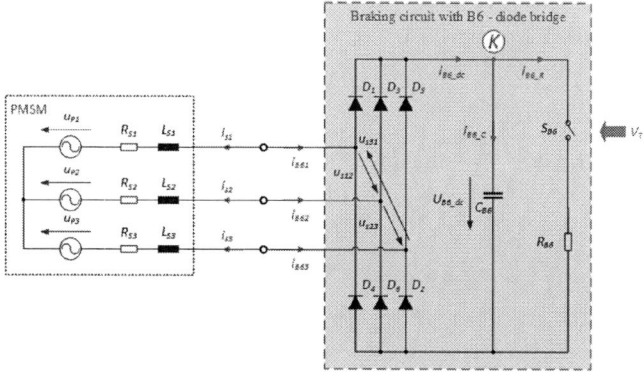

Fig. 3: Analyse of the braking circuit using B6-diode bridge for DMC-fed PMSM drives

It can be seen from Fig. 3 that the current of the chopper resistance $i_{B6,R}$ can be described in (11).

$$i_{B6,R} = \frac{U_{B6,dc}}{R_{B6}} \cdot V_t \quad , \quad (0 \leq V_t \leq 1) \tag{11}$$

where V_t is the duty cycle of chopper switch S_{B6}, which is limited between 0 and 1; R_{B6} is the chopper resistance.

If Kirchhoff's current law is employed in node K, then (12) can be obtained.

$$i_{B6,dc} = i_{B6,C} + i_{B6,R} = C_{B6} \cdot \frac{d(U_{B6,dc})}{dt} + \frac{U_{B6,dc}}{R_{B6}} \cdot V_t \quad , \quad (0 \leq V_t \leq 1) \tag{12}$$

Since the switching matrix is switched off during braking in this case and the three-phase motor currents are no longer represented by the three-phase output currents of DMC, in this strategy, the DC-link current in B6-diode bridge $i^*_{B6,dc}$ is employed in the outer current control loop as reference value to control the motor moment mi within the required limitation and without large fluctuation.

The output of the current control loop is used as reference voltage $U^*_{B6,dc}$ for the inner voltage control loop. In the voltage control loop, the capacitor voltage $U_{B6,dc}$ is controlled to drop smoothly, which is corresponding to the current motor speed. The output of the voltage control loop will be current of the chopper resistance $i_{B6,R}$, which is used to determine the duty cycle V_t of the chopper switch S_{B6} as can be derived from (11).

The block diagram of the closed cascade control loops is shown in Fig. 4.

The discrete-time integrator is employed in the control using Forward-Euler-Method. The expression for the approximation is shown in (13).

$$\frac{1}{s} = T_p \cdot \frac{z^{-1}}{1 - z^{-1}} \tag{13}$$

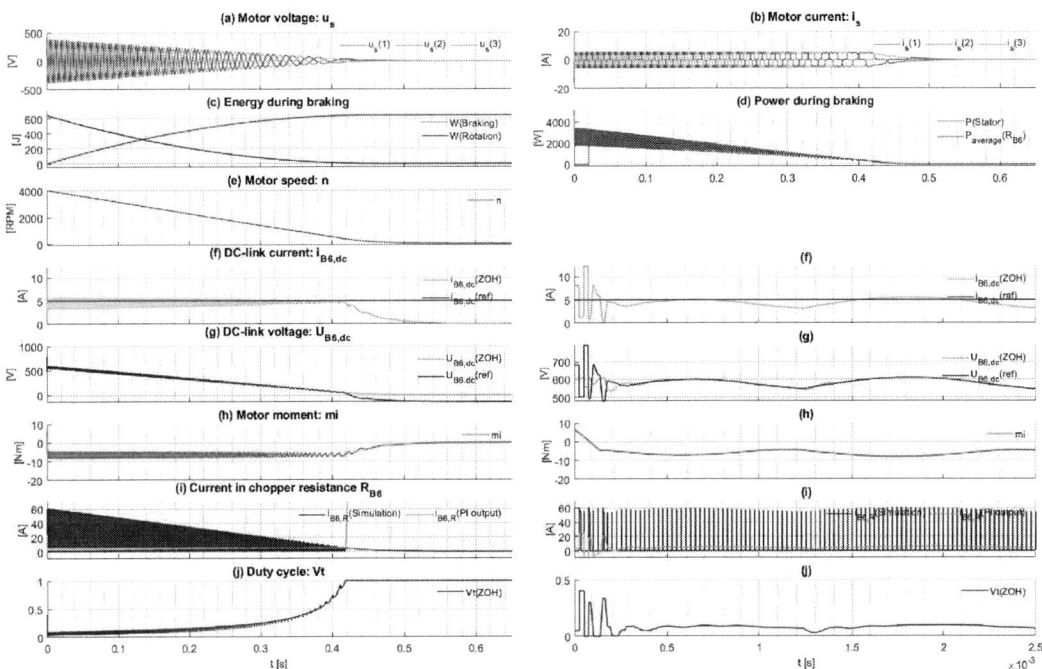

Fig. 4: Closed cascade control loops of the proposed braking method with avoiding regenerative energy

Simulation results

Simulation results in Fig. 5 is used to verify the feasibility of the proposed method using MATLAB. The parameters of the system are shown in Table I.

Fig. 5: Simulation results: (a) Motor voltage: $\vec{u}_s$. (b) Motor current: $\vec{i}_s$. (c) Energy during braking. (d) Power during braking. (d) Motor speed: n. (f) DC-link current: $i_{B6,dc}$. (g) DC-link voltage: $U_{B6,dc}$. (h) Motor moment: mi. (i) Current in chopper resistance R_{B6}. (j) Duty cycle V_t of chopper switch S_{B6}.

Table I: System parameter in simulation

Symbol	Value	Symbol	Value
C_{B6}	$5\mu F$	R_{B6}	10Ω
P_p	2	L_d, L_q	$4mH$
R_s	0.4Ω	J	$0.007kg \cdot m^2$
ψ_m	$0.4V/(rad \cdot s^{-1})$	I_{max}	$70A$
f_p	$40kHz$		

In the simulation, the PMSM drive starts to brake by $t = 0$ from $4000RPM$ as shown in (e). The DC-link current $i^*_{B6,dc}$ in B6-diode bridge is set to be $5A$ as reference value for braking moment limitation and it can be well controlled through adjusting duty cycle V_t of the chopper switch S_{B6} as shown in (f) and (j). As the motor brakes, the motor speed decreases (g), as well as the motor voltage (a). This can also be indicated through the DC-link capacitor voltage $U_{B6,dc}$ as shown in (g).

The motor moment mi during braking is relative constant without large fluctuation as shown in (h), which shows a good result for braking moment limitation and also verifies the feasibility of the proposed control method.

Conclusion

In this paper, a simplified braking method with special consideration of avoiding regenerative energy is presented. By employing B6-diode bridge with one chopper resistance on motor side and shutting down the switching matrix of DMC during motor braking, the regenerative energy can be fully avoided from feeding back to the grid. Meanwhile, the braking torque is controlled within the required limitation and without large fluctuation, so that the braking process is smooth and safe. Simulation results confirm the feasibility of the proposed strategy.

References

[1] Alesina, A. and Venturini, M.: Solid-state power conversion: A Fourier analysis approach to generalized transformer synthesis, IEEE Transactions on Circuits and Systems, Vol 28 no 4, pp. 319-330, 1981.

[2] Alesina, A. and Venturini, M.G.B.: Analysis and design of optimum-amplitude nine-switch direct AC-AC converters, IEEE Transactions on Power Electronics, Vol 4 no 1, pp. 101-112, 1989.

[3] J. Rodriguez, M. Rivera, J.W. Kolar, P.W. Wheeler: A Review of Control and Modulation Methods for Matrix Converters, IEEE Transactions on Industrial Electronics, vol. 59, pp. 58-70, 2012.

[4] L. Empringham, J. W. Kolar, J. Rodriguez, P. W. Wheeler, J. C. Clare: Technological Issues and Industrial Application of Matrix Converters: A Review, IEEE Transactions on Industry Electronics, vol. 60, no. 10, pp. 4260-4271, Oct. 2013.

[5] J. Mahlein, J. Igney, J. Weigold, M. Braun, O. Simon: Matrix Converter Commutation Strategies With and Without Explicit Input Voltage Sign Measurement, IEEE Transactions on Industry Electronics, vol. 49, no. 2, pp. 407-414, April 2002.

[6] M. Imayavaramban, Patrick W. Wheeler, L. Empringham, Jon C. Clare: Input power clamp for controlling regeneration in matrix converter drive, IET, pp. 515-519, 2008.

[7] M. Imayavaramban, Patrick. W. Wheeler: Avoiding Regeneration with a Matrix Converter Drive, IEEE pp. 2529-2534, 2007.

[8] M. Imayavaramban, Patrick. W. Wheeler: Third method for regenerative braking in matrix converter drive: More electric aircraft, IEEE, pp. 1-6, 2017

[9] Tsuyoshi Nagano, Jun-ichi Itoh: Halt Sequence for Matrix Converter to Suppress Increase of Snubber Capacitor Voltage during Motor Regeneration, IEEE, pp. 6261-6266, 2016.

[10] She, Hongwu and Lin, Hua and Wang, Xingwei and Yue, Limin: Damped input filter design of matrix converter, PEDS, pp. 672-677, 2009.

[11] Costa, L. A. and Fan, B. and Burgos, R. and Boroyevich, D. and Chen, W. and Blasko, V.: The Fast Overvoltage Protection Consideration and Design for SiC-based Matrix Converters, APEC, pp. 1567-1574, 2020.

[12] Wheeler, P.W. and Rodriguez, J. and Clare, J.C. and Empringham, L. and Weinstein, A.: Matrix converters: a technology review, IEEE Transactions on Industrial Electronics, Vol 49 no 2, pp. 276-288, 2002.

Inverter-machine parametric co-design for energy efficient electric drives

Jaedon Kwak, Alberto Castellazzi
Kyoto University of Advanced Science
Ukyo Ward, Yamanouchi Gotandacho, 18
Kyoto, Japan
Tel.: +81 / 7033013693.
E-Mail: 2020md02@kuas.ac.jp

Keywords

«Optimal efficiency drive», «Wide Bandgap devices», «Multi-level inverters», «Silicon Carbide (SiC)», «Hybrid Electric Vehicle (HEV)»

Abstract

This paper presents the development of a unified parametric machine-inverter design framework targeted at energy efficiency optimization that is taking into account the most frequent operational conditions and the integral of the power losses over time. A case study is conducted with 400V-30kW ISG in hybrid electric vehicle application. For 3L-ANPC with wide band gap device, various switching frequencies, and different type of topologies and PWM strategies are compared with respect to energy loss in both inverter and machine. The machine design methodology is presented considering interactive effect between inverter and machine. As a result, proposed design has an advantage of energy loss compared to conventional design. The activity will contribute to drive progress beyond state-of-the-art in key application domains.

Introduction

Electrical machine drive systems(ED) are a key point for electric vehicle(EV) and hybrid electric vehicle(HEV) applications. To achieve satisfactory operational range, energy consumption must be low: that is, the ED must be efficient and lightweight. With this goal, many studies have been conducted to increase energy efficiency in the fields of inverters and machines. In general, conventional design of 3-phase inverter for ED uses 2-level half bridge topology with Si-IGBT for their power device. With high demand collector-emitter current, it can be challenging to adopt high switching frequency due to the high heat-generation of the Si-IGBT. Wide-band-gap(WBG; silicon carbide, SiC; gallium nitride, GaN) based power conversion is well known to enable increased power density and efficiency as a result of higher current density of the semiconductors, smaller module footprints, higher switching frequencies and operational temperature [1-3]. Compared to 2-level inverters, multilevel inverter topologies such as advanced neutral point clamped(ANPC) inverter offers several advantages in terms of efficiency and current harmonics [4-6]. In machine design for ED, permanent magnet synchronous machine(PMSM) is wide spread and vastly used due to its high power density and efficiency [7], [8]. These design parameters not only affect the energy efficiency of each inverter and machine, but also affects each other mutually. In [9] and [10], authors investigate that machine's iron loss and magnet eddy current loss are influenced by the PWM waveform depending on the inverter switching frequency. By increasing switching frequency with SiC inverter, the losses and temperature of motor can be reduced in [11]. Though design variables of each inverter and machine can affect mutually, research has hitherto largely targeted the machine and the inverter independently. Besides, few studies have considered integrated design considering interactive effects at initial design stage. For this reason, unified parametric machine-inverter co-design must be conducted for minimizing energy consumption of system. System performance is optimized taking into account the prevalent load, that is the statistically most relevant operational condition from application driving cycle.

In this paper, unified framework enabling for energy efficient optimization is proposed with a case study of 400Vdc-30kW integrated starter generator(ISG) and their drive system in plug-in hybrid vehicle. A parametric design method using some of major design variables of machine and inverter is presented. A co-simulation considering inverter-machine mutual effect is conducted in order to calculate energy consumption of system during driving cycle, represented by US06.

Approach and methodology

Calculation energy loss consumption during driving cycle

Initial design specification of 400Vdc-30kW belt driven ISG model is shown in Table I. To calculate energy loss consumption during driving cycle, US06 driving cycle is used. Vehicle profile and machine profile of US06 is presented in Fig. 1.

Table I: Specification of 400Vdc-30kW ISG model

	Item	Specification
	Power / Torque	30kW (motoring & generating) / ±50Nm
	DC-link voltage / max. phase current	400Vdc / 180Arms
	Max. speed	18,000 rpm
	Pulley ratio	3:1
Inverter	Inverter topology	2L- half bridge 3-phase inverter with Si-IGBT
Inverter	Switching frequency	10 kHz
Machine	Motor type	IPMSM
Machine	Pole / Slot	8 / 48
Machine	Outer Diameter / Stack Length [mm]	130mm / 55mm
Machine	Winding type	Rectangular wire / hair-pin winding
Machine	Magnet	NdFeB

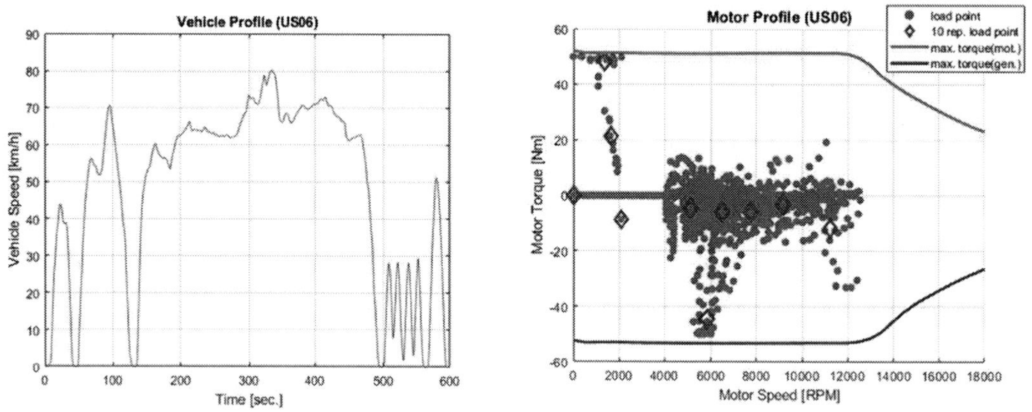

(a)   (b)

Fig. 1: US06 driving cycle : (a) vehicle profile (b) ISG profile with S-T curve and 10 representative load point

In [12], The subregions of each load point can be devided and representative equivalent load points are calculated by equation (1)-(3), where N_i represents the number of points in the ith subregion, Emi is the energy of ith region, n_{mci} and T_{mci} represents speed and torque in the energy gravity center point.

$$E_{mi} = \sum_{j=1,2,\cdots}^{N_i} E_{mij} \tag{1}$$

$$n_{mci} = \frac{1}{E_{mi}} \sum_{j=1,2,\cdots}^{N_i} E_{mij} n_{mij} \tag{2}$$

$$T_{mci} = \frac{1}{E_{mi}} \sum_{j=1,2,\cdots}^{N_i} E_{mij} T_{mij} \tag{3}$$

The number of subregions influences the accuracy of the representative points. In this paper, 10 representative points are sufficient to represent the entire cycle, which is shown in Table II.

Table II: 10 representative points of US06 driving cycle

No.	Time [sec.]	Speed [rpm]	Torque [Nm]
1	1.2	1640.9	21.2
2	2.7	1342.4	48.3
3	8.6	5815.1	-44.5
4	12.9	11215.9	-11.9
5	22.4	2094.8	-8.6
6	36.6	9186.7	-3.3
7	42.0	7789.6	-5.9
8	77.9	6490.6	-6.0
9	195.0	5105.2	-4.5
10	200.7	0.0	0.0

Design parameters

Given machine type and inverter, design optimization can exploit the following parameters:

1) Number of levels(N_L): higher N_L yields voltage waveforms closer to pure sinus, improving machine efficiency due to better harmonics, and decreasing voltage endurance level for each power device; however, it can imply lower inverter efficiency due to larger switch number, on top of a more complex design.
2) Switching frequency(f_s): higher f_s reduces output current harmonics, and losses of machine can be decreased, especially at high-speed region. But it causes high inverter switching loss.
3) PWM strategy: depending on inverter topology, current ripple and inverter losses are influenced by PWM strategies. Besides, current ripple impacts on machine losses.
4) Machine pole number(n_p) and stator number of series turns(n_T): higher n_p increase power density with low magnetic reluctance; it requires higher f_s to keep stable controllability. n_T impact on power density as well as stator inductance, that is related to current ripple.

The governing design equations for the inverter loss that comprised conduction and switching loss can be expressed as

$$P_{cond,IGBT} = I(\theta)V_{ce} + I(\theta)^2 R_{ce} \tag{4}$$

$$P_{cond,\text{Diode}} = I(\theta)V_f + I(\theta)^2 R_f \tag{5}$$

$$P_{cond,\text{MOS}} = I(\theta)^2 R_{ds} \tag{6}$$

$$P_{sw,\text{IGBT}} = \left[E_{on,\text{IGBT}}(I, T, V_{DC}) + E_{off,\text{IGBT}}(I, T, V_{DC}) \right] \cdot f_s \tag{7}$$

$$P_{sw,\text{MOS}} = \left[E_{on,\text{MOS}}(I,T,V_{DC}) + E_{off,\text{MOS}}(I,T,V_{DC}) \right] \cdot f_s \qquad (8)$$

$$P_{rec,\text{Diode}} = E_{rec,diode}(I,T,V_{DC}) \cdot f_s \qquad (9)$$

$I(\theta)$ is the current flowing through switches, V_{ce} and V_f are the initial saturation voltage drop of IGBT and diode, R_{ce}, R_f, and R_{ds} are the on-state resistance of IGBT, diode, and MOSFET. $E_{on,\text{IGBT}}(I,T,V_{DC})$ $E_{off,\text{IGBT}}(I,T,V_{DC})$, $E_{on,\text{MOS}}(I,T,V_{DC})$, and $E_{off,\text{MOS}}(I,T,V_{DC})$ represent the energy per unit switching on and off state of IGBT and MOSFET, and $E_{rec,diode}(I,T,V_{DC})$ is the reverse recovery energy per unit switching period of diode.

For the machine, the copper loss and iron loss can be expressed as

$$P_{copper} = 3i_S^2 \cdot R_{copper} \qquad (10)$$

$$P_{iron} = P_{hyst} + P_{eddy} = k_{minor}k_h(f,B_m) \cdot f \cdot B_m^2 + \sum k_e\,(f,B_m) \cdot f_n^2 \cdot B_{m\_n}^2 \qquad (11)$$

$$k_{minor} = 1 + k\frac{1}{B_m}\sum \Delta B_n \qquad (12)$$

with R_{copper} the stator resistance of one phase, that is dependent on frequency due to AC skin and proximity effect [13], and i_S the stator current; $k_{h,e,minor}$ are parameters for hysteresis, eddy-current losses and B_m is flux density, f_n and B_n is harmonic component of frequency and flux density [9].

Design with co-simulation

Inverter design

To investigate multi-level topology, 3-level ANPC inverter are taken into account. By rationally selecting the zero-state loop, the loss balance of each devices in 3L-ANPC can be achieved. In addition, the flexible commutation mode of the ANPC topology also provides the possibility for hybrid configuration of power devices [4-6]. In this paper, two different topologies are presented shown in Fig. 2.

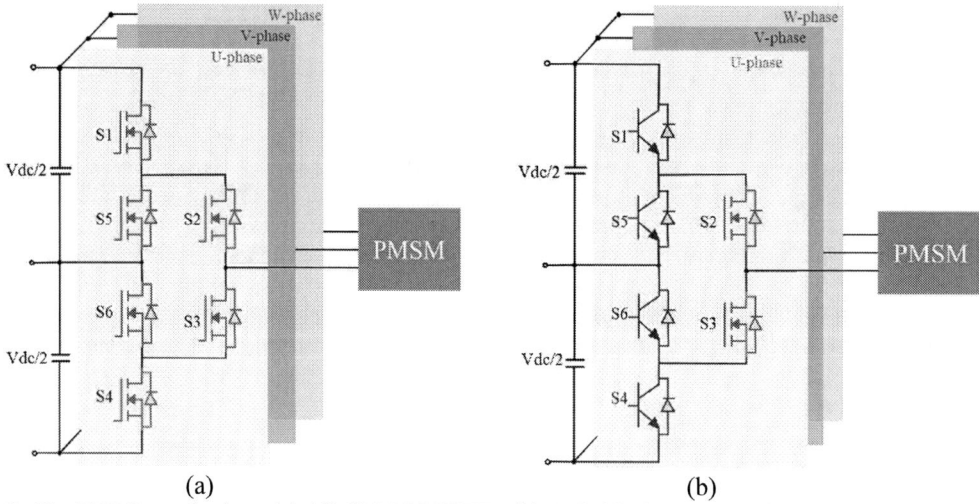

(a)   (b)

Fig. 2: 3L-ANPC topologies : (a) All SiC-MOSFET; (b) Hybrid Si-IGBT & SiC-MOSFET

The proposed PWM strategies are presented in Fig. 3 and Fig. 4. In PWM1, the current flows in two parallel paths at zero-states: S5-S2 and S6-S3, that leads the low conduction losses at zero-states [1]. Only two devices(S2, S3) switches at switching frequency and the others switch at line frequency, that leads the low switching losses with 4 devices except S2, S3 in PWM2 [4], [6]. This strategy is suitable for hybrid topology due to little switching loss on the 4 IGBT devices.

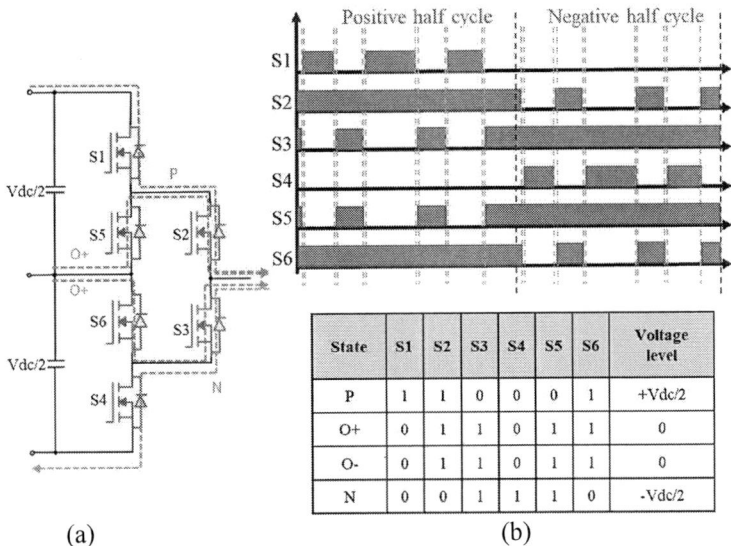

(a)   (b)

Fig. 3: PWM1 strategy: (a) current loop for each state; (b) switching signal and states

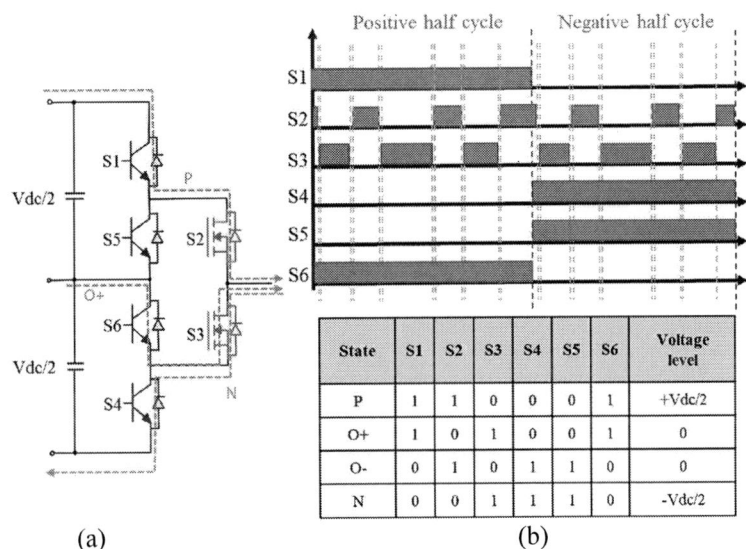

(a)   (b)

Fig. 4: PWM2 strategy: (a) current loop for each state; (b) switching signal and states

The simulation model is composed of a closed loop with PI current controller, PWM strategies, 3level-ANPC model, and PMSM model. For PMSM model, the tabulated parameters interpreted as FEM based at the corresponding load point are used. Inverter model have power devices refer to Table III. The losses are compared with a combination of two topologies and two PWM strategies at two different load points in Table II; 1) load 2: a relatively large current excitation at low speed, and 2) load 4: a relatively small current excitation at a high speed.

Table III: Power device specification

Topology	Device	Part No.	V_{ce} / V_{ds}	I_c / I_d	Remarks
2L-HB	Si-IGBT/Diode	Infineon - IKQ75N120CT2	1200	75	3-parallel
3L-ANPC	Si-IGBT/Diode	Infineon – AIKQ120N60CT	600	120	2-parallel
	SiC-MOSFET	Wolfspeed - C3M0015065D	650	120	2-parallel

Speed[rpm]	Torque[Nm]	Current [A]	Voltage[V]
1342.4	48.3	235.0	55.4

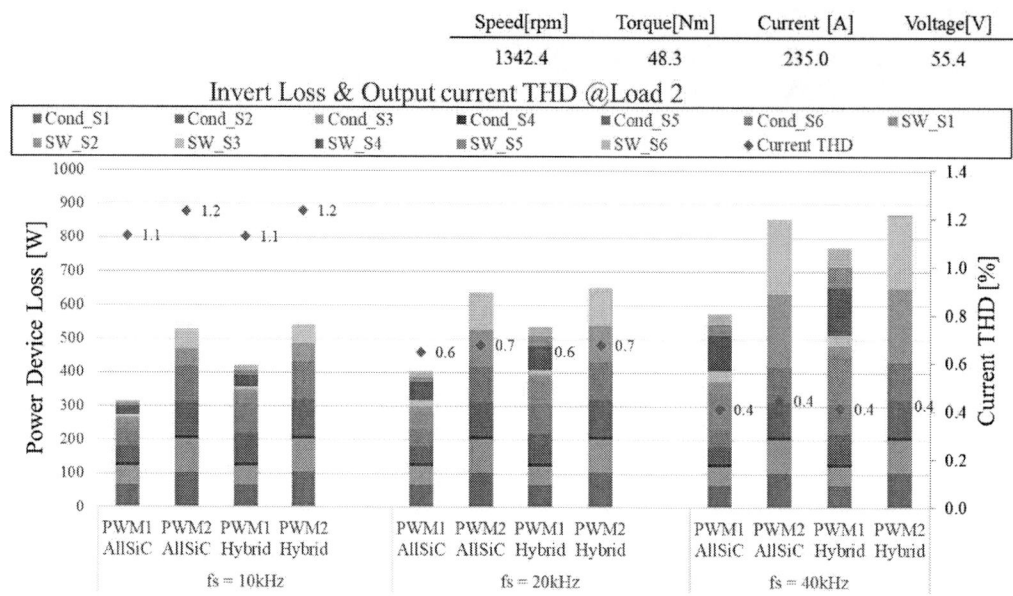

Fig. 5: Comparison of 3L-ANPC inverter losses at load2 in Table II

Speed[rpm]	Torque[Nm]	Current [A]	Voltage[V]
11,215.9	-11.9	57.9	291.2

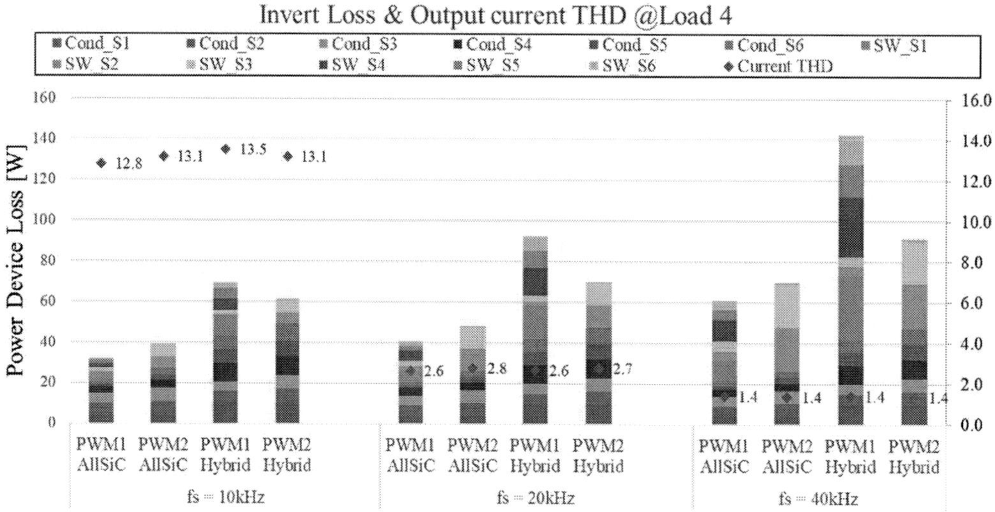

Fig. 6: Comparison of 3L-ANPC inverter losses at load4 in Table II

The results of comparison of the inverter loss are illustrated in Fig. 5, and Fig. 6. The quality of the output current waveform that is represented by total harmonic distortion(THD) is dependent on f_s, and not greatly effected by 2 PWM strategies. THD of output current is calculated by equation (13) where I_n is amplitude of current harmonics.

$$THD = \frac{\sqrt{I_2^2 + I_3^2 + I_4^2 + \cdots}}{I_1} \tag{13}$$

In low speed and high current region that is such as load 2 in Table II, inverter losses of PWM1 are smaller than one of PWM2 with both topologies as Fig. 5. In contrast, PWM2 is more efficient in hybrid topology in high speed and low current region such as load 4 in Table II, as illustrated in Fig. 6. As results, PWM2 has an advantage in high-speed region where switching loss is dominant, while PWM1 is more efficient in low speed where conduction loss is dominant.

For the machine, especially in high region such as load 4, the losses are affected by the fs, that is the harmonic component, THD of output current waveform as Fig. 7.

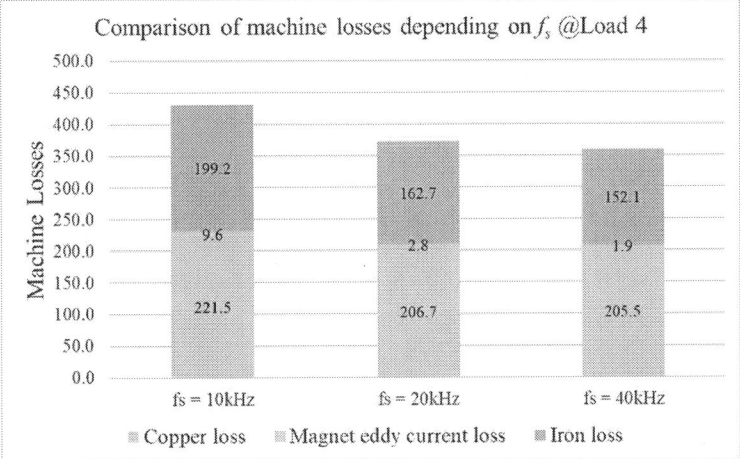

Fig. 7: Comparison of machine losses depending on f_s in PWM1 hybrid topology at load 4 in Table II

Machine design considering inverter parameter

As far as the machine is concerned, the discussion can actually be easily extended to include additional interrelation between f_s and other design parameters. For the sake of illustration, here the possibility to increase the machine number of poles when f_s is increased is considered. When designing the machine, the maximum number of poles, n_{p_max} can be expressed as follows.

$$n_{p\_max} = 120 \times \frac{f_s}{N_{max} \times m_f} \tag{14}$$

N_{max} is the highest rotational speed of the machine, in rpm, and m_f the frequency modulation index, that is, the ratio of switching to fundamental electrical frequency. Thus, by increasing f_s, n_{p_max} can also be increased. In [14] and [15], increasing pole number of PMSM offers advantages of power density and efficiency. As the number of poles increases, the fundamental flux per pole Φ_1 decreases at the same air-gap flux density B_g as equation (15) where D_r is rotor diameter, and l_{stk} is effective stack length. Thinner yoke thickness are allowed due to reduced magnetic flux per pole and shorter magnetic flux paths. That means high power density at given size by increasing air-gap flux density.

$$\Phi_1 = B_g \cdot \frac{\pi D_r l_{stk}}{n_p} \tag{15}$$

For the number of series turns per phase n_T, expressed by equation (16)-(18).

$$E_{ph} = \frac{P_{max}}{3 i_{s\_max}} \text{ (@base speed)} \tag{16}$$

$$E_{ph} = \frac{2\pi}{\sqrt{2}} \cdot f k_{w1} n_T \Phi_1 = \frac{\pi^2}{\sqrt{2}} \frac{\omega_m}{60} k_{w1} n_T B_g D_r l_{stk} \tag{17}$$

$$n_T = \frac{P_{max}}{3 i_{s\_max}} \cdot \frac{\sqrt{2}}{\pi^2} \cdot \frac{60}{\omega_m k_{w1} B_g D_r l_{stk}} \tag{18}$$

E_{ph} is the phase electro-motive force at base speed, P_{max} and i_{s_max} are the maximum value of power and phase rms current, k_{w1} is the fundamental harmonic winding factor, and ω_m is the mechanical base speed of machine for rpm. Furthermore, n_T affects output current harmonics. The stator inductance is proportional to square of n_T, and that impact on output current ripple [1].

$$L = \frac{V_{DC}/(N_L-1)}{8 \cdot f_s \cdot \Delta i_s} \tag{19}$$

Hair-pin winding technology is beneficial in terms of power density, thermal performance, and electromagnetic force with high slot fill factor and low slot openings [16]. In contrast, limited n_T is feasible for given number of poles and slots since even number of layer per slot are essential to hair-pin winding. For given same ratio of pole/slot combination, by increasing n_p, not only the power density increase, but higher n_T close to the optimal value following equation (18) can be designed shown in Table IV.

Table IV: Number of series turns per phase in hair-pin design depending on pole/slots

pole	slot	Layer per slot	Parallel circuit	Series turns per phase
8	48	4	1	32
	48	6	1	48
	48	8	2	32
10	60	4	1	40
	60	6	2	30
	60	8	2	40
12	72	4	1	48
	72	6	2	36
	72	8	2	48

Improved design and results

In this section, improved design is proposed in the way covered in the previous section. In Table V, design variables of machine design are described. Despite the increase in n_T, increasing n_p leads to decrease copper weight with lower phase current; that means proposed model has more higher power density.

Table V: Comparison of machine design

Item	Current Design	Improve Design
Image		
Pole / Slot	8 / 48	10 / 60
Outer diameter / Stack length [mm]	130 / 55	
Number of turns per phase	32	40
Magnet weight [p.u]	1.0	1.0
Copper weight [p.u]	1.0	0.90

For the inverter variable, the optimal switching frequency is 30kHz as the target of less than 5% of THD in the entire operation load point. Despite the loss reduction using 3L-ANPC topology, the use of higher switching frequency compared to 10kHz that is the value of original design has resulted in a slight increase in inverter loss. However, in terms of machine losses, the energy loss is improved. This is particularly evident in high-speed region where improvements in current waveforms are distinct. As a result, using All-SiC topology offers 20.7% reduction of total energy loss during driving cycle, even in hybrid topology, it has total 5.6% of energy savings shown in Table VII.

Table VI: Comparison of inverter and machine loss at each load point

Load point	Time [sec.]	Speed [rpm]	Torque [Nm]	Inverter Loss [W]			Machine Loss [W]		
				Original	Improve (All SiC)	Improve (Hybrid)	Original	Improve (All SiC)	Improve (Hybrid)
1	1.2	1640.9	21.2	211.5	219.6	376.2	522.1	510.1	511.7
2	2.7	1342.4	48.3	888.6	975.0	1504.9	2561.7	2380.1	2374.3
3	8.6	5815.1	-44.5	671.6	836.7	1229.8	2186.2	2090.7	2090.4
4	12.9	11215.9	-11.9	85.6	123.5	193.6	582.2	402.8	404.6
5	22.4	2094.8	-8.6	57.1	71.1	129.3	102.1	99.4	99.9
6	36.6	9186.7	-3.3	20.4	23.4	42.0	222.8	112.1	112.8
7	42	7789.6	-5.9	38.4	48.5	84.4	286.9	134.8	134.3
8	77.9	6490.6	-6.0	38.9	48.9	86.0	215.1	114.8	114.5
9	195	5105.2	-4.5	27.4	32.7	59.4	102.6	68.7	68.0
10	200.7	0.0	0.0	0.0	0.0	0.0	0.0	0.0	0.0

Table VII: Comparison of energy loss during driving cycle

Load point	Time [sec.]	Speed [rpm]	Torque [Nm]	Energy loss [kJ]		
				Original	Improve (All SiC)	Improve (Hybrid)
1	1.2	1640.9	21.2	0.9	0.9	1.1
2	2.7	1342.4	48.3	9.3	9.1	10.5
3	8.6	5815.1	-44.5	24.6	25.2	28.6
4	12.9	11215.9	-11.9	8.6	6.8	7.7
5	22.4	2094.8	-8.6	3.6	3.8	5.1
6	36.6	9186.7	-3.3	8.9	5.0	5.7
7	42	7789.6	-5.9	13.7	7.7	9.2
8	77.9	6490.6	-6.0	19.8	12.7	15.6
9	195	5105.2	-4.5	25.4	19.8	24.8
10	200.7	0.0	0.0	0.0	0.0	0.0
Total energy loss during driving cycle [kJ]				114.7	90.9 (20.7%↓)	108.3 (5.6%↓)

Conclusion

This paper has proposed the development of a comprehensive framework for joint inverter-machine co-design. The impact analysis is studied with important design parameters such as switching frequency, the number of level with 3L-ANPC topology, PWM strategies and machine pole number and turns. With simulation model based on analytic approach, the optimum design variable can be selected. A key figure-of-merit is energy efficiency considering total energy loss consumption of driving cycle. The effect of these design variables on energy loss vary depending on the application and driving condition. However, this methodology can be applied to the design of other EDs for EVs and HEVs.

References

[1] E. Gurpinar and A. Castellazzi: Tradeoff Study of Heat Sink and Output Filter Volume in a GaN HEMT Based Single-Phase Inverter, IEEE Transactions on Power Electronics, Vol. 33 no 6, pp. 5226-5239, June 2018, doi: 10.1109/TPEL.2017.2730038.

[2] S. Ozdemir, F. Acar and U. S. Selamogullari: Comparison of silicon carbide MOSFET and IGBT based electric vehicle traction inverters, 2015 International Conference on Electrical Engineering and Informatics (ICEEI), 2015, pp. 1-4, doi: 10.1109/ICEEI.2015.7387215.

[3] A. Allca-Pekarovic, P. J. Kollmeyer, P. Mahvelatishamsabadi, T. Mirfakhrai, P. Naghshtabrizi and A. Emadi: Comparison of IGBT and SiC Inverter Loss for 400V and 800V DC Bus Electric Vehicle Drivetrains, 2020 IEEE Energy Conversion Congress and Exposition (ECCE), 2020, pp. 6338-6344, doi: 10.1109/ECCE44975.2020.9236202.

[4] Z. Feng, X. Zhang, S. Yu and J. Zhuang: Comparative Study of 2SiC&4Si Hybrid Configuration Schemes in ANPC Inverter, IEEE Access, Vol. 8, pp. 33934-33943, 2020, doi: 10.1109/ACCESS.2020.2974554.

[5] A. Kersten, E. Grunditz and T. Thiringer: Efficiency of Active Three-Level and Five-Level NPC Inverters Compared to a Two-Level Inverter in a Vehicle, 2018 20th European Conference on Power Electronics and Applications (EPE'18 ECCE Europe), 2018, pp. P.1-P.9

[6] L. Zhang et al.: Evaluation of Different Si/SiC Hybrid Three-Level Active NPC Inverters for High Power Density, IEEE Transactions on Power Electronics, Vol. 35, no. 8, pp. 8224-8236, Aug. 2020, doi: 10.1109/TPEL.2019.2962907.

[7] Kim, DM., Jung, YH., Cha, KS. et al.: Design of Traction Motor for Mitigating Energy Consumption of Light Electric Vehicle Considering Material Properties and Drive Cycles, International Journal of Automotive Technology, Vol. 21, pp. 1391–1399, doi: 10.1007/s12239-020-0131-7

[8] K. T. Chau, C. C. Chan and C. Liu: Overview of Permanent-Magnet Brushless Drives for Electric and Hybrid Electric Vehicles, IEEE Transactions on Industrial Electronics, Vol. 55, no. 6, pp. 2246-2257, June 2008, doi: 10.1109/TIE.2008.918403.

[9] S. Xue et al.: Iron Loss Model for Electrical Machine Fed by Low Switching Frequency Inverter, IEEE Transactions on Magnetics, vol. 53, no. 11, pp. 1-4, Nov. 2017, Art no. 2801004, doi: 10.1109/TMAG.2017.2696360.

[10] K. Yamazaki and S. Watari: Loss analysis of permanent-magnet motor considering carrier harmonics of PWM inverter using combination of 2-D and 3-D finite-element method, IEEE Transactions on Magnetics, Vol. 41, no. 5, pp. 1980-1983, May 2005, doi: 10.1109/TMAG.2005.846278.

[11] K. Yamaguchi, K. Katsura and T. Jikumaru: Motor loss and temperature reduction with high switching frequency SiC-based inverters, 2017 IEEE 5th Workshop on Wide Bandgap Power Devices and Applications (WiPDA), 2017, pp. 127-131, doi: 10.1109/WiPDA.2017.8170534.

[12] Dong Wei, Hongwen He, Jianfei Cao: Hybrid electric vehicle electric motors for optimum energy efficiency: A computationally efficient design, Energy, Vol. 203, 2020, doi: 10.1016/j.energy.2020.117779

[13] J. W. Chin, K. S. Cha, J. C. Park, D. M. Kim, J. P. Hong and M. S. Lim: Investigation of AC Resistance on Winding Conductors in Slot According to Strands Configuration, IEEE Transactions on Industry Applications, Vol. 57, no. 1, pp. 316-326, Jan.-Feb. 2021, doi: 10.1109/TIA.2020.3033815.

[14] G. Artetxe, J. Paredes, B. Prieto, M. Martinez-Iturralde, and I. Elosegui: Optimal Pole Number and Winding Designs for Low Speed–High Torque Synchronous Reluctance Machines, Energies, Vol. 11, no. 1, p. 128, Jan. 2018, doi: 10.3390/en11010128.

[15] D. Misu, M. Matsushita, K. Takeuchi, K. Oishi and M. Kawamura: Consideration of optimal number of poles and frequency for high-efficiency permanent magnet motor, 2014 International Power Electronics Conference (IPEC-Hiroshima 2014 - ECCE ASIA), 2014, pp. 3012-3017, doi: 10.1109/IPEC.2014.6870113.

[16] G. Berardi, S. Nategh, N. Bianchi and Y. Thioliere: A Comparison Between Random and Hairpin Winding in E-mobility Applications, IECON 2020 The 46th Annual Conference of the IEEE Industrial Electronics Society, 2020, pp. 815-820, doi: 10.1109/IECON43393.2020.9255269.

Bidirectional Cuk Converter in Partial-Power Architecture with Current Mode Control for Battery Energy Storage System in Electric Vehicles

J.S. Artal-Sevil[1], J. Anzola[2], V. Ballestín-Bernad[1], I. Aizpuru[2]
[1]Department of Electrical Engineering.
Polytechnic School of Engineering and Architecture, EINA.
University of Zaragoza. Spain.
[2]Electronics and Computer Science Department.
Goi Eskola Politeknikoa-Orona Ideo-Fundazioa eraikina
Mondragon Unibertsitatea. Spain.
E-mail: {jsartal, ballestin}@unizar.es, {janzola, iaizpuru}@mondragon.edu

Acknowledgements

The authors would like to thank the support of Government of Aragon and the European Union for the project T28_20R, "building Aragon from Europe". This work was supported in part by the Spanish MINECO under Grant RTC-2015-3358-5. The authors also want to thank the support of the IDI project, "study and analysis of partial-power processing architectures for the development of an on-board charger in electric and hybrid vehicles" with reference OTRI-2020/0416.

Keywords

«Partial-Power Processing Converter (PPC)», «non-isolated bidirectional Partial-Power architectures», «DC-DC switched-mode power supplies», «high power», «DC-fast charges», «battery energy storage system (BESS)», «on-board charger».

Abstract

This paper presents a partial-power processing architecture intended for an on-board charger. This module is integrated into a Battery Energy Storage System (BESS). This model allows us to easily control the charge-discharge current of the LiFePO4 battery, as well as the current injection on the DC-bus (V2G). The architecture used in the partial-power processing is based on the non-isolated bidirectional Cuk converter, with average current mode control. The purpose has been to compare both topologies, Full-Power and Partial-Power, to observe the advantages and disadvantages that each on-board charger design offers. Partial architecture has some advantages such as high power density, small size, decrease stress on devices, as the DC-DC converter only processes a fraction of the total power. Thus, the purpose of this paper has been to analyze and explore the usefulness of the non-isolated bidirectional Cuk converter with partial-power processing architecture in battery charging systems in electric vehicles. The effectiveness of the strategy has been validated by the *Matlab/Simulink* software simulation.

1. Introduction

The incorporation of the partial-power philosophy allows us to use smaller and cheaper converters since power losses are reduced. These design concepts are an attractive solution and have attracted the attention of the research community in recent years [1]. The full-power converter processes all the energy supplied to the load, while the partial-power converter only processes a fraction of the power.

This architecture has already been implemented in numerous applications, such as the integration of photovoltaic systems [2] [3], battery charging systems in electric vehicles [4], DC-power supply [5], active balancing of PV-arrays [6] or MPPT search algorithms in TEG systems [7], spacecraft [8], etc., in order to improve system performance. In short, the partial converter can improve the efficiency of the entire system while reducing its cost. It is an advantage of this architecture. Although, there are also opinions that doubt the performance improvement in non-isolated topologies [9].

This partial-power topology is being implemented in higher power applications. As an example, Iyer *et al.* [4], [10] propose a fast-charging station for different battery electric vehicles (BEV) and plug-in hybrids (PHEV) based on partial-power processing architecture. Likewise, Xue *et al.* [11] and Artal-Sevil *et al.* [12] develop a low-cost bidirectional fractional DC-DC converter, intended for a high-power battery energy storage system (BESS). Its main objective was to reduce the power processed by the converter, in order to increase the overall efficiency of the system. While Anzola *et al.* [13] describe a charging unit based on partial-power processing for extremely fast charging stations for electric vehicles. Similarly, Mira *et al.* [14] present the analysis of a DC-DC mode switching power supply in a partial-power processing configuration. The presented model is based on the Dual Active Bridge (DAB) architecture and constitutes a unidirectional charge converter. This topology is also being used in systems that introduce energy storage systems. In [15] a partial-power processing architecture for a hybrid electric vehicle (HEV) based on Fuel-Cells is presented. The supply system includes an active buffer, in partial-power topology, to supply the power peaks demanded by the vehicle's traction.

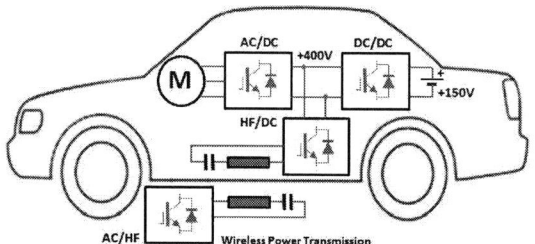

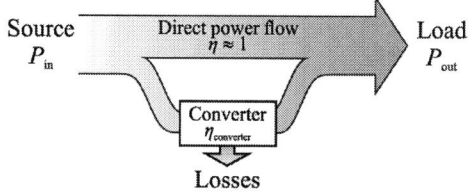

Fig. 1.   On-board charger diagram in a wireless power transfer charging system. Application of the partial-power processing architecture on a non-isolated bidirectional Cuk DCDC converter.

Fig. 2.   Concept of power flow in partial-power converter.

On the other hand, vehicle to grid (V2G) is an emerging technology that is being analyzed for electric vehicles (EV), see Fig. 1. The energy stored in battery packs can be an alternative to power demand peaks on the grid [16]. This technology requires the development of smart on-board chargers, integrated into the vehicle, that have the ability to manage the power flow [17] [18]. This paper proposes partial-power processing architecture for the design of a smart on-board charger. The main objective is to validate the advantages of the partial-power architecture in these applications. For this purpose, multiple simulations have been carried out in *Matlab-Simulink*. Likewise, in [19] [20] the design of a high-density battery charger with different converter architectures based on partial-power processing has been described. These designs stand out for their high efficiency.

This paper is organized as follows. Section 1 shows a brief introduction associated with the problem addressed. Section 2 presents the mathematical analysis of the partial-power converter. Section 3 provides a description of the system configuration. Section 4 shows the different simulation results obtained with the *Matlab-Simulink* software. Finally, the conclusions and some brief considerations are described in Section 5.

2. Proposed Cuk Converter in Partial-Power Processing Architecture

In Fig. 2 the concept of power flow in partial-power architecture is shown. The proposed converter has been derived from the classic Cuk converter. To simplify the steady-state analysis of the proposed architecture, some assumptions are made, such as ideal switching devices and inductors, with no delay in the switching process. For a sampling time T_{SW}, the switch is ON for $D \times T_{SW}$ and OFF for the period $(1 - D) \times T_{SW}$. Therefore, depending on the state of the switch, two modes of operation can be identified.

Mode 1: ON-state $[0 \leq t \leq D(t)T_{SW}]$.

At the initial time t = 0, when switch SW_1 turns on, diode D_1 turns off. The inductor currents (i_{L1}, i_{L2}) increase from their respective initial value. The voltage and current equations are as follows:

$$v_{L1}(t) = L_1 \frac{\partial i_{L1}}{dt} = v_G \; ; \quad v_{L2}(t) = L_2 \frac{\partial i_{L2}}{dt} = v_{C1} - v_C \tag{1}$$

$$i_{C1}(t) = C_1 \frac{\partial v_{C1}}{dt} = i_{L2} \; ; \; i_C(t) = C \frac{\partial v_C}{dt} = i_{L2} - i_O \tag{2}$$

$$v_O = v_G - (-v_C) \tag{3}$$

Mode 2: OFF-state $[D(t)T_{SW} \le t \le T_{SW}]$.

When switch SW_1 turns off at t = D·T$_{SW}$, diode D_1 turns on. A similar analysis gives the equations for voltage and current as follows:

$$v_{L1}(t) = L_1 \frac{\partial i_{L1}}{dt} = v_G - v_{C1} \; ; \; v_{L2}(t) = L_2 \frac{\partial i_{L2}}{dt} = -v_C \tag{4}$$

$$i_{C1}(t) = C_1 \frac{\partial v_{C1}}{dt} = -i_{L1} \; ; \; i_C(t) = C \frac{\partial v_C}{dt} = i_{L2} - i_O \tag{5}$$

The average inductor voltage (v_{L1}, v_{L2}) must be zero at steady-state. Hence, analyzing the volt-sec balance for inductors (L_1, L_2) during a switching period T_{SW}, we obtain:

$$v_{L1}(t) = 0 \rightarrow v_G \cdot D(t) + \{v_G - v_{C1}\} \cdot \{1 - D(t)\} = 0 \tag{6}$$

$$v_{L2}(t) = 0 \rightarrow \{v_{C1} - v_C\} \cdot D(t) - v_C \cdot \{1 - D(t)\} = 0 \tag{7}$$

As input voltage (v_G) is assumed to be constant over a switching period. Solving for steady-state output voltage v_O then,

$$v_O = v_G \frac{1}{1-D} \tag{8}$$

Likewise, the instantaneous value in the duty cycle ratio $D(t)$ depends on the input voltage v_G, and can be expressed as

$$D = 1 - \frac{v_G}{v_O} \tag{9}$$

Similarly, the average capacitor current (v_C, v_{C1}) must be zero in steady-state. Hence, analyzing the current-sec balance for the capacitors (C, C_1) during a T_{SW} switching period, we obtain:

$$i_{C1}(t) = 0 \rightarrow i_{L2} \cdot D(t) + (-i_{L1}) \cdot \{1 - D(t)\} = 0 \tag{10}$$

$$i_C(t) = 0 \rightarrow \{i_{L2} - i_O\} \cdot D(t) + \{i_{L2} - i_O\} \cdot \{1 - D(t)\} = 0 \tag{11}$$

Solving for the current in the inductors (i_{L1}, i_{L2}) in steady state then,

$$i_{L1} = i_{L2} \frac{D}{1-D} \; ; \quad i_{L2} = i_O \tag{12}$$

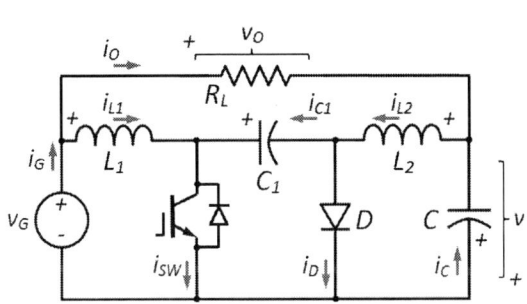

Fig. 3.   Cuk converter in partial-power processing architecture.

Fig. 4.   Experimental testing of the Cuk converter in partial-power processing architecture.

And calculating the current i_G supplied by the voltage source v_G, we obtain (13).

$$i_G = i_{L1} + i_O = i_O \frac{1}{1-D} \tag{13}$$

Also, it is possible to determine the input resistance R_G of the converter. In this case, it is given by,

$$R_G = \frac{v_G}{i_G} = \frac{v_O}{i_O}(1-D)^2 = R_L(1-D)^2 \qquad (14)$$

3. On-Board Charger based on Partial-Power Processing Architecture

Traditionally, the load (LiFePO$_4$ battery in this case) is connected to the output (v_C) on the DC-DC converter. In other words, in the full-power structure, this converter processes all the power supplied to the battery from the DC-bus, see Fig. 3. Meanwhile, in the partial-power architecture, the load is connected in series between the DC-bus and the Cuk converter output (v_C). Figure 5 shows the partial-power processing architecture diagram in the Cuk converter. This converter topology is bidirectional, non-isolated, and has two inductors. The charge-discharge current (i_{Li}) in the LiFePO$_4$ battery, or the current injection on the grid (V2G), can be controlled by the duty cycle (D) in the converter.

For example, during the current charging process in the LiFePO$_4$ battery, the G_{vFPC} voltage gain ($G_{vFPC} = v_{OUT}/v_{IN}$) in continuous conduction mode (CCM) for the Cuk converter operating as full-power architecture (FPC), is given by (15). Whereas the G_{vPPC} voltage gain for the Cuk converter operating in partial-power architecture (PPC) is given by (16).

$$G_{vFPC} = \frac{v_O}{v_{IN}} = \frac{v_{Li}}{v_{DC}} = \frac{D}{1-D} \qquad (15)$$

$$G_{vPPC} = \frac{v_O}{v_{IN}} = \frac{v_{DC}}{v_{C2}} = \frac{v_{DC}}{v_{DC} - v_{Li}} = \frac{1}{1-D} \qquad (16)$$

where v_{IN} is the input voltage and v_O is the output voltage of the Cuk converter in each case; v_{Li} and v_{DC} represent the voltage at the LiFePO$_4$ battery terminals (BESS) and the voltage on the DC-bus respectively, and D is the duty cycle in Cuk converter module.

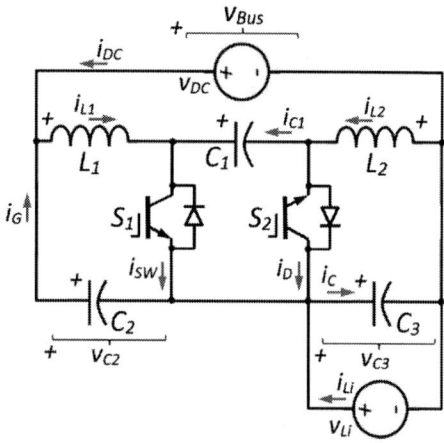

Fig. 5. Schematic diagram of the bidirectional Cuk converter in partial-power processing architecture applied to an on-board charger.

TABLE I. PARAMETERS ASSOCIATED WITH THE BIDIRECTIONAL CUK PARTIAL-POWER CONVERTER MODEL.

Parameter	Symbol	Value
DC-bus voltage	v_{DC}	+150V
Module internal resistance	r_{DC}	0,1Ω
LiFePO$_4$ battery voltage	v_{Li}	+45V to +55V
Internal battery resistance	r_{Li}	0,025Ω
Cuk inductors	L_1, L_2	5mH
Cuk capacitor	C_1	500µF
Switching frequency	f_{SW}	10kHz
Switch resistance S$_1$, S$_2$	R_{on}	0,01Ω
Schottky forward voltage	v_F	0,3V
Snubber network	R_S, C_S	10kΩ; 10nF
Input/Output Capacitor	C_2, C_3	500µF

Note that in the full-power topology, the output voltage can be higher or lower than the input voltage as the duty cycle increases (D). Whereas in the partial-power architecture, the output voltage is always higher than the input voltage depending on the value adopted by the duty cycle (D). In this assumption, operating with eq. 16, the relationship between the DC-bus voltage (v_{DC}) and the battery voltage (v_{Li}) is obtained as,

$$v_{DC}(1-D) = v_{DC} - v_{Li}; \quad v_{Li} = D \cdot v_{DC} \qquad (17)$$

4. Simulation Results

In order to study the characteristics of a partial-power architecture applied to a battery energy storage system, a non-isolated bidirectional Cuk converter has been modelled in partial-power processing mode.

The analyzed topology is shown in Fig. 5. The simulation software used has been *Matlab-Simulink*. At the same time, an average current mode control has been developed. Likewise, Table I contains some of the parameters used in the development of the model.

Cuk converter in partial-power topology switches at a frequency $f_{SW} = 10\text{kHz}$. This structure uses two coils L_1, L_2 with the value of 5mH, and a capacitor C_1 with the value of 500µF. A small internal resistance in series with each voltage source ($r_{DC} = 0,1\Omega$ and $r_{Li} = 0,025\Omega$) represents the power losses during charge/discharge processes in the bidirectional Cuk converter. Figure 6 shows the currents associated with each coil (i_{L1}, i_{L2}) as well as the current i_{Li} during the LiFePO4 battery charging process. The sign of the current i_{Li} is considered negative ($i_{Li} < 0$) during the charging process, while i_{Li} is considered positive ($i_{Li} > 0$) during the LiFePO4 battery discharge process. The signs in the current flows can be seen in the diagram of the Cuk partial-power converter, see Fig. 5.

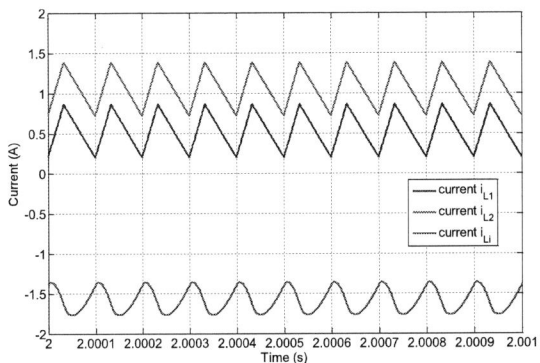

Fig. 6. Steady-state response of the bidirectional Cuk converter in partial-power architecture: current mode control. i_{L1}, i_{L2}, and i_{Li} currents during the LiFePO4 battery charging process ($i_{Li} = -1,50\text{A}$).

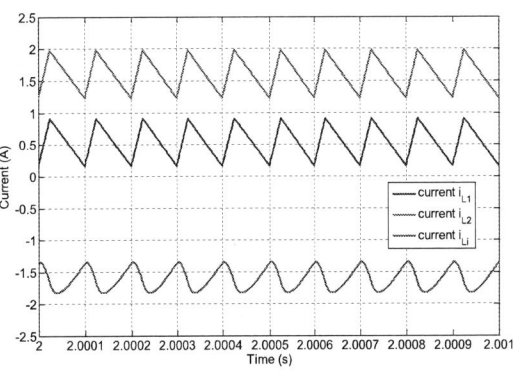

Fig. 7. Steady-state response of the bidirectional Cuk converter in full-power architecture: current mode control. i_{L1}, i_{L2}, and i_{Li} currents during the LiFePO4 battery charging process ($i_{Li} = -1,50\text{A}$).

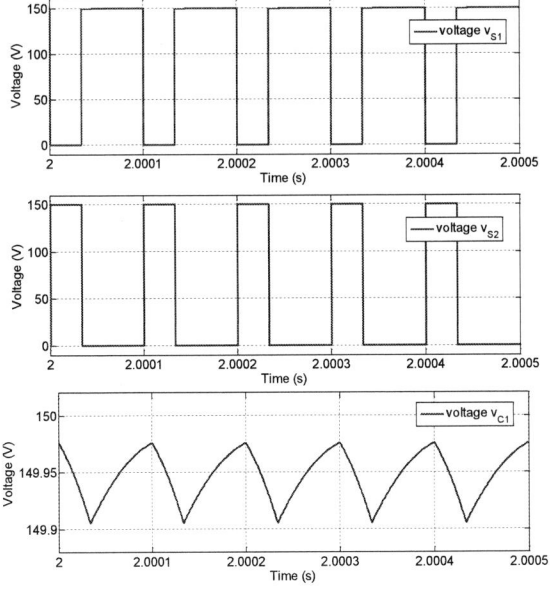

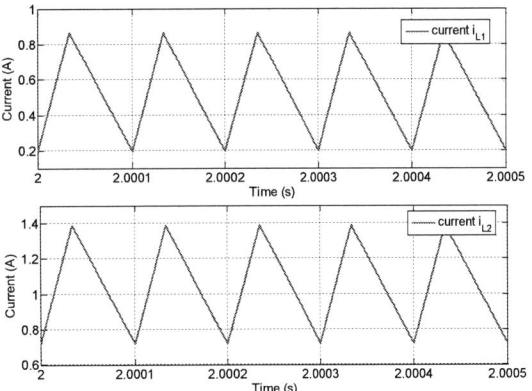

Fig. 8. Steady-state response of the bidirectional Cuk converter in partial-power architecture: v_{S1}, i_{L1}, v_{S2}, i_{L2}, and v_{C1} voltages and currents during the LiFePO4 battery charging process ($i_{Li} = -1,50\text{A}$).

In the Cuk converter in partial-power architecture, with average current mode control, during the battery charging process ($i_{Li} = -1,50\text{A}$) the following average currents are obtained in the inductors: $i_{L1} = +0,524\text{A}$, $i_{L2} = +1,042\text{A}$, see details in Fig. 6. While the current ripple in the inductors corresponds to $\Delta i_{L1} = \pm0,334\text{A}$, $\Delta i_{L2} = \pm0,334\text{A}$. Under these conditions, the capacitor voltage v_{C1} is equal to the DC-bus voltage (v_{DC}), $v_{C1} = +150\text{V}$. In Fig. 7 the currents i_{L1}, i_{L2} corresponding to the Cuk topology coils in full-converter mode are observed. The purpose is to compare the results between both topologies (Cuk partial-converter and Cuk full-converter), considering the same charging current in the LiFePO4 battery.

In this way, it is possible to analyze the advantages provided by the partial-power processing architecture applied to battery energy storage systems (BESS).

In the Cuk converter in full-power architecture, with average current mode control, during the battery charging process ($i_{Li} = -1,50A$) the following average currents are obtained in the inductors: $i_{L1} = +0,5354A$, $i_{L2} = +1,596A$, see details in Fig. 7. While the current ripple in the inductors corresponds to $\Delta i_{L1} = \pm 0,375A$, $\Delta i_{L2} = \pm 0,375A$. Under these conditions, the capacitor voltage v_{C1} is not similar to the DC-bus voltage (v_{DC}), $v_{C1} = +200V$. In this case, the duty cycle for the Cuk full-power topology is $D_{S1}|_{FP} = 0,2502$. In view of the simulation results, the stress on the different devices is higher in the full-power converter than in the partial-power converter for the same requested power, see Figs. 6 and 7.

Figure 8 represents the different voltages (v_{S1}, v_{S2}, and v_{C1}) in the semiconductor devices and capacitor respectively, as well as the currents in the inductors (i_{L1}, i_{L2}) of the Cuk converter in partial-power architecture during the current charging process in LiFePO$_4$ battery. Under these conditions, the following values were obtained: DC-bus voltage $v_{DC} = +150V$, capacitor voltage C_2 ($v_{C2} = +99,91V$), average DC-bus current $i_{DC} = 0,5239A$, while the average current and voltage in the LiFePO$_4$ battery (i_{Li}, v_{Li}) are $i_{Li} = 1,565A$, $v_{Li} = +50,04V$, respectively. In this case, for the Cuk partial-power architecture, the duty cycle results $D_{S1}|_{PP} = 0,3338$.

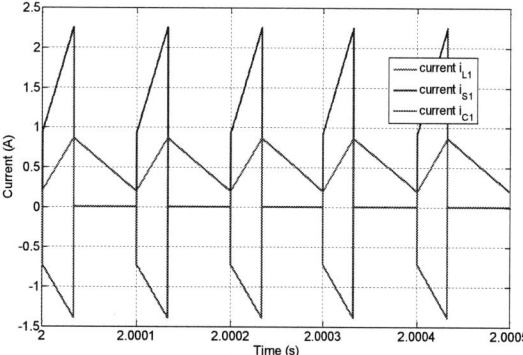

Fig. 9. Steady-state response of the bidirectional Cuk converter in partial-power architecture: i_{L1}, i_{S1}, and i_{C1} currents during the LiFePO$_4$ battery charging process ($i_{Li} = -1,50A$).

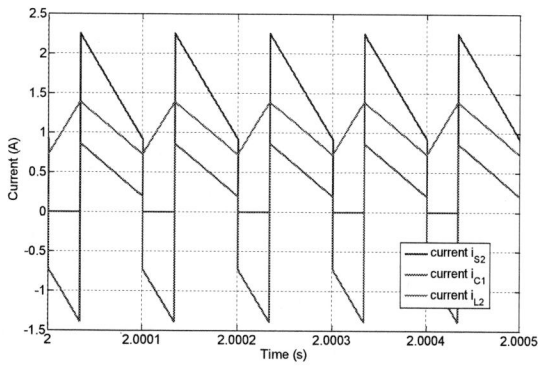

Fig. 10. Steady-state response of the bidirectional Cuk converter in partial-power architecture: i_{L2}, i_{S2}, and i_{C1} currents during the LiFePO$_4$ battery charging process ($i_{Li} = -1,50A$).

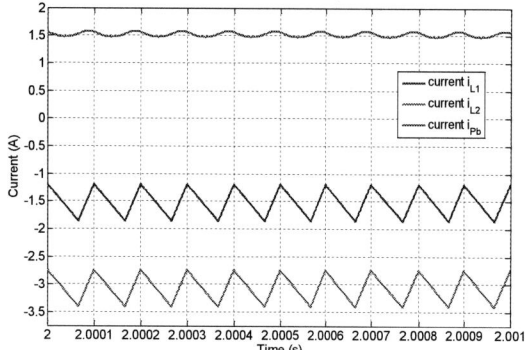

Fig. 11. Steady-state response of the bidirectional Cuk converter in partial-power architecture: current mode control. i_{L1}, i_{L2}, and i_{DC} currents during the LiFePO$_4$ battery discharging process. Power flow goes from LiFePO$_4$ battery to DC-bus ($i_{DC} = +1,50A$); vehicle-to-grid (V2G) operation.

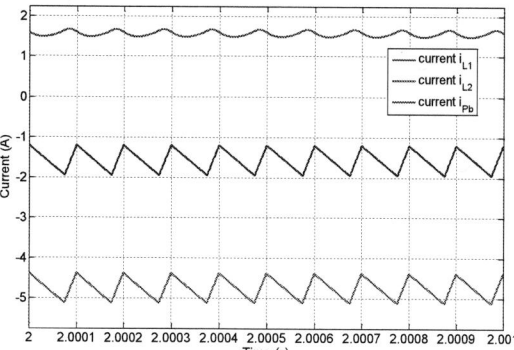

Fig. 12. Steady-state response of the bidirectional Cuk converter in full-power architecture: current mode control. i_{L1}, i_{L2}, and i_{DC} currents during the LiFePO$_4$ battery discharging process. Power flow goes from LiFePO$_4$ battery to DC-bus ($i_{DC} = +1,50A$); vehicle-to-grid (V2G) operation.

Figure 9 shows the current waveforms in the inductor L_1 (i_{L1}), switch S_1 (i_{S1}), and the capacitor C_1 (i_{C1}), during the LiFePO$_4$ battery charging process. Similarly, applying Kirchhoff's law to the initial node (see Fig. 5), we obtain:

$$i_{L1}(t) = i_{S1}(t) + i_{C1}(t) \tag{18}$$

While in Fig. 10 the current waveforms in the inductor L_2 (i_{L2}), switch S_2 (i_{S2}), and the capacitor C_1, (i_{C1}), during the LiFePO$_4$ battery charging process, are observed.

$$i_{S2}(t) = i_{L2}(t) + i_{C1}(t) \tag{19}$$

Under these conditions (LiFePO$_4$ battery charging current i_{Li} = -1,50A), the average value of the current in the different switches (i_{S1}, i_{S2}) is: i_{S1} = 0,524A; i_{S2} = 1,042A. Likewise, these values support those obtained in the average currents of the L_1 and L_2 converter coils.

Similarly, the simulation results corresponding to the bidirectional Cuk converter in partial-power architecture during the LiFePO$_4$ battery discharge process are presented in Fig. 11, that is, the power flow goes from the battery to the DC-bus. This case corresponds to the power flow injection in the DC-bus, vehicle-to-grid (V2G) operation.

Figure 11 shows the currents i_{L1}, i_{L2} corresponding to the inductors of the Cuk converter in partial-power topology together with the current injected into the DC-bus (i_{DC} = +1,50A). As mentioned above, in the case of the DC-bus, the sign of the i_{DC} current is considered positive ($i_{DC} > 0$) during the power injection process (V2G operation), that is, in the LiFePO$_4$ battery discharge process. Meanwhile, the i_{DC} current is considered negative ($i_{DC} < 0$) during the LiFePO$_4$ battery charging process. See this sign convention on the Cuk converter diagram (Fig. 5).

In the Cuk converter in partial-power architecture, with current mode control, during current injection on the DC-bus (i_{DC} = +1,50A), the following average currents are obtained in the inductors: i_{L1} = -1,524A, i_{L2} = -3,062A, see Fig. 11. Furthermore, the current ripple in the inductors corresponds to Δi_{L1} = ±0,334A, Δi_{L2} = ±0,334A. Under these conditions, the capacitor voltage v_{C1} is equal to the DC-bus voltage (v_{DC}), v_{C1} = +150,2V.

In this case the following values were obtained: DC-bus voltage v_{DC} = +150,3V; capacitor voltage C_2 v_{C2} = +100,3V and LiFePO$_4$ battery voltage v_{Li} = +49,89V. Likewise, the average currents obtained in the DC-bus (i_{DC}) and in the LiFePO$_4$ battery (i_{Li}) were: i_{DC} = +1,524A, i_{Li} = +4,582A.

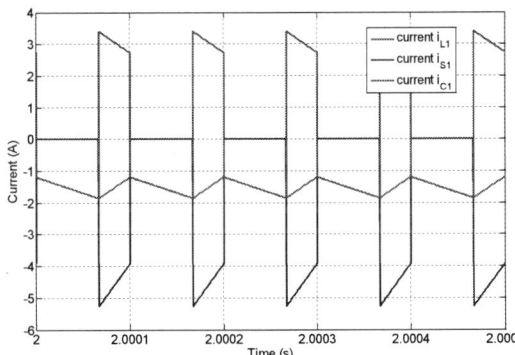

Fig. 13. Steady-state response of the bidirectional Cuk converter in partial-power architecture: i_{L1}, i_{S1}, and i_{C1} currents during the current injection process on the DC-bus. Power flow goes from LiFePO$_4$ battery to DC-bus (i_{DC} = +1,50A); vehicle-to-grid (V2G) operation.

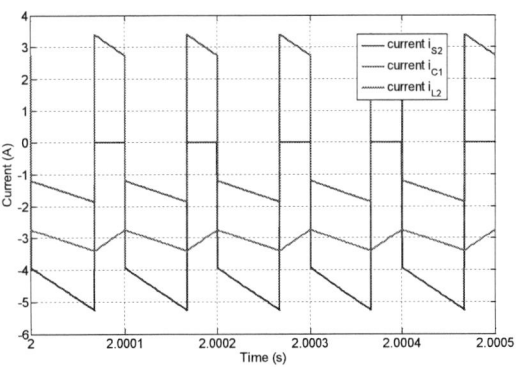

Fig. 14. Steady-state response of the bidirectional Cuk converter in partial-power architecture: i_{L2}, i_{S2}, and i_{C1} currents during the LiFePO4 battery discharging process. Power flow goes from LiFePO$_4$ battery to DC-bus (i_{DC} = +1,50A); vehicle-to-grid (V2G) operation.

Figure 12 shows the behavior of the bidirectional Cuk converter in the full-power architecture, with control in average current mode, during current injection on the DC-bus (i_{DC} = +1,50A). In this case, the following average currents were obtained in the inductors: i_{L1} = -1,556A, i_{L2} = -4,698A; see details in Fig. 12. While the current ripple in the inductors corresponds to Δi_{L1} = ±0,375A, Δi_{L2} = ±0,375A. Under these conditions, the capacitor voltage C_1 becomes v_{C1} = +200V. Now, in this case, the duty cycle of the Cuk converter in full-power architecture is $D_{S2|FP}$ = 0,7502. As in the previous assumption, in view of the simulation results obtained in the current injection to the DC-bus, the stress of the devices is higher in the full-power architecture than in the partial-power architecture, considering the same power requirements (see details in Figs. 11 and 12).

Figure 13 shows the current waveforms in inductor L_1 (i_{L1}), switch S_1 (i_{S1}), and capacitor C_1 (i_{C1}) during the current injection process on the DC-bus in partial-power topology. That is, the power flow goes from the LiFePO$_4$ battery to DC-bus (i_{DC} = +1,50A); vehicle-to-grid (V2G) operation. Likewise, Fig. 14 shows the current waveforms in inductor L_2 (i_{L2}), switch S_2 (i_{S2}), and capacitor C_1 (i_{C1}), during this LiFePO$_4$ battery discharge process. Under these conditions, the average current value in the different switches (i_{S1}, i_{S2}) was: i_{S1} = -1,556A; i_{S2} = -4,698A. Now, in this case, the duty cycle of the bidirectional Cuk converter in partial-power architecture is $D_{S2|PP}$ = 0,6679.

5. Conclusions

This paper presents the analysis of a non-isolated bidirectional Cuk converter in partial-power processing architecture. The topology has been applied to an on-board charger in an electric vehicle, supported by the Battery Energy Storage System (BESS). Furthermore, an average current mode control has been developed and implemented in this application. The purpose has been to compare both architectures (full-power and partial-power converter) to validate their advantages in the on-board charger design.

In view of the simulation results obtained, the stress on the different devices is greater in the full-power architecture than in the partial-power architecture, for the same power requirements. In addition, the partial-power topology only processes a fraction of the total power (the voltage difference between the DC-bus and the battery energy storage system), reducing power losses in the overall system. This allows us to reduce the nominal power in the DC-DC converter, with respect to the full-power topology, and at the same time causes a reduction in its cost, volume, and weight.

Finally, the simulation model has been developed with *Matlab-Simulink* software. The control system based on the average current has responded satisfactorily to the charge-discharge current requirements of the LiFePO$_4$ battery, as well as the current injection on the DC-bus.

References

[1] J. Anzola, I. Aizpuru, A. Arruti Romero, A. Alacano Loiti, R. Lopez-Erauskin, J.S. Artal-Sevil and C. Bernal, "Review of Architectures Based on Partial Power Processing for DC-DC Applications" IEEE Access. IEEExplore Digital Library. Vol. 8, June 2020. pp.: 1-14.

[2] J.R.R. Zientarski, M.L. da Silva, J. Renes Pinheiro and H. Leães Hey, "Evaluation of Power Processing in Series-Connected Partial-Power Converters". IEEE Journal of Emerging and Selected Topics in Power Electronics. IEEExplore Digital Library. Vol.: 7, issue: 1, March 2019; pp. 343–352.

[3] J.W. Zapata, S. Kouro, G. Carrasco, H. Renaudineau and T.A. Meynard, "Analysis of Partial Power DC–DC Converters for Two-Stage Photovoltaic Systems". IEEE Journal of Emerging and Selected Topics in Power Electronics. IEEExplore Digital Library. Vol.: 7, issue: 1, March 2019; pp. 591–603.

[4] V.M. Iyer, S. Gulur, G. Gohil, and S. Bhattacharya, "An Approach Towards Extremev_S_ Fast Charging Station Power Delivery for Electric Vehicles with Partial Power Processing". IEEE Transactions on Industrial Electronics. IEEExplore Digital Library. In press.

[5] T. Kanstad, M.B. Lillholm and Z. Zhang, "Highly Efficient EV Battery Charger Using Fractional Charging Concept with SiC Devices". IEEE Applied Power Electronics Conference and Exposition (APEC'19). IEEExplore Digital Library. March 2019, Anaheim, CA, (USA); pp. 1601–1608.

[6] S. Mallangada, M.O. Badawy and Y. Sozer, "A Novel Differential Power Processing Architecture for a Partially Shaded PV String Using Distributed Control". IEEE Energy Conversion Congress and Exposition (ECCE). IEEExplore Digital Library. September 2018, Portland (USA); pp. 6220–6227.

[7] J.S. Artal-Sevil, C. Bernal-Ruiz, J. Beyza and V.M. Bravo, "Evaluation of a Thermoelectric Generation system based on Differential-Power Processing architecture under non-uniform temperature conditions" IEEE International Autumn Meeting on Power, Electronics and Computing (ROPEC'20). IEEExplore Digital Library. November 2020, Ixtapa (Mexico); pp.: 1-6.

[8] L. Chen, H. Wu, P. Xu, H. Hu and C. Wan, "A high step-down non-isolated bus converter with Partial Power conversion based on synchronous LLC resonant converter". IEEE Applied Power Electronics Conference and Exposition (APEC'15). IEEExplore Digital Library. Charlotte, NC, USA. March 2015; pp.: 1950-1955.

[9] T. Suntio and A. Kuperman, "Comments on "An efficient Partial Power Processing DC/DC converter for distributed PV architectures"" IEEE Transactions on Power Electronics. IEEExplore Digital Library. vol. 30, no. 4. April 2015; pp.: 2372.

[10] V.M. Iyer, S. Gulur, G. Gohil, and S. Bhattacharya, "Extreme Fast Charging Station Architecture for Electric Vehicles with Partial Power Processing". IEEE Applied Power Electronics Conference and Exposition (APEC'18). IEEExplore Digital Library. March 2018, San Antonio Texas (USA); pp. 659–665.

[11] F. Xue, R. Yu and A. Huang, "Fractional converter for high efficiency high power battery energy storage system". IEEE Energy Conversion Congress and Exposition (ECCE'17). IEEExplore Digital Library. Cincinnati, Ohio (USA). October 2017; pp. 5144–5150.

[12] J.S. Artal-Sevil, C. Bernal-Ruiz, J. Anzola, I. Aizpuru, A. Bono-Nuez and J.M. Sanz-Alcaine, "Partial Power Processing architecture applied to a Battery Energy Storage System" IEEE Vehicle Power and Propulsion Conference (VPPC'20). IEEExplore Digital Library. November-December 2020; Gijón (Spain) pp.: 1-6.

[13] J. Anzola, I. Aizpuru, A. Arruti, A. Alacano, R. Lopez, J.S. Artal-Sevil and C. Bernal-Ruiz, "Partial Power Processing Based Charging Unit for Electric Vehicle Extreme Fast Charging Stations". IEEE Vehicle Power

and Propulsion Conference (VPPC'20). IEEExplore Digital Library. November 2020, Gijon (Spain); pp.: 1-6.

[14] M.C. Mira, Z. Zhang, K.L. Jørgensen and A.E.M. Andersen, "Fractional Charging Converter With High Efficiency and Low Cost for Electrochemical Energy Storage Devices". IEEE Transactions on Industry Applications. IEEExplore Digital Library. Vol.: 55, issue: 6 Nov 2019; pp. 7461–7470.

[15] J.S. Artal-Sevil, V. Ballestín-Bernad, A. Coronado-Mendoza and J.L. Bernal-Agustín, "Design and Analysis of a Partial-Power Converter with an Active Power-Buffer for a Fuel Cell-based Hybrid Electric Vehicle". IEEE Vehicle Power and Propulsion Conference (VPPC'21). IEEExplore Digital Library. Gijón (Spain), October 2021; pp.: 1-6.

[16] Y. Benomar, M. El Baghdadi, O. Hegazy, Y. Yang, M. Messagie and J. Van Mierlo, "Design and modeling of V2G inductive charging system for light-duty Electric Vehicles". International Conference on Ecological Vehicles and Renewable Energies (EVER17). IEEExplore Digital Library. April 2017. MonteCarlo (Mónaco); pp. 1–7.

[17] A. Farjah, E. Bagheri, A.R. Seifi and T. Ghanbari, "Main and auxiliary parts of battery storage, aimed to fast charging of electrical vehicles" Annual Power Electronics, Drives Systems and Technologies Conference (PEDSTC'18). IEEExplore Digital Library. February 2018; Tehran (Iran) pp.: 277-282.

[18] J.S. Artal-Sevil, V. Ballestín-Bernad, J. Anzola and J.A. Domínguez-Navarro, "High-Gain Non-isolated DC-DC Partial-Power Converter for Automotive Applications" IEEE Vehicle Power and Propulsion Conference (VPPC'21). IEEExplore Digital Library. October 2021; Gijón (Spain) pp.: 1-6.

[19] Y. Cao, M. Ngo, N. Yan, D. Dong, R. Burgos and A. Ismail, "Design and Implementation of an 18 kW 500 kHz 98.8% Efficiency High-density Battery Charger with Partial Power Processing". IEEE Journal of Emerging and Selected Topics in Power Electronics. IEEExplore Digital Library. August 2021; pp.: 1-14.

[20] K. Zheng, W. Zhang, X. Wu and L. Jing, "Optimal Control Method and Design for Modular Battery Energy Storage System Based on Partial Power Conversion". IEEE Access. IEEExplore Digital Library. Vol. 9, September 2021; pp.: 133376-133386.

Design Space Exploration for a Capacitive 36V, 4A, 4:1 DCDC Converter with GaN Switches Using a Performance-Cost-Matrix Including Uncommon Topologies

Adrian Gehl, Malte Kempchen, Simon Disselkamp, Markus Olbrich, Bernhard Wicht
LEIBNIZ UNIVERSITÄT HANNOVER, INSTITUTE OF MICROELECTRONIC SYSTEMS
Appelstr. 4
30167 Hannover, Germany
Phone: +49 / (511) – 762 19670
Fax: +49 / (511) – 762 19694
Email: adrian.gehl@ims.uni-hannover.de
URL: https://www.ims.uni-hannover.de

Keywords

≪DC-DC converter≫, ≪Switched capacitor≫, ≪Wide bandgap≫, ≪Modelling≫, ≪Design optimization≫

Abstract

A systematic approach to compare capacitive DCDC converters regarding performance and cost is presented. All possible topologies for a 4:1 voltage conversion ratio are generated using an algorithm developed in this research. Selected topologies are implemented benefiting from GaN transistors. Experimental results over various operating conditions up to 36 V input, 4 A load and 2 MHz switching frequency confirm theory and simulation. Two uncommon topologies demonstrate superior cost and load performance. The parameter space supports various industrial and automotive applications.

Introduction

Capacitive or switched-capacitor (SC) DCDC converters have become a common solution for small and medium power applications due to their high power density at large voltage conversion ratios [1]. Compared to traditional inductive DCDC converters they can omit a bulky inductor. SC DCDC implementations found in the literature usually rely on a few well-known topologies, such as the Series-Parallel, Dickson, Fibonacci or Ladder topology [2, 3]. However, especially for larger voltage conversion ratios, which require multiple flying capacitors [4], there exist uncommon topologies that have not been investigated in prior art. Therefore, a systematic approach to compare the performance of all possible SC DCDC topologies for a given voltage conversion ratio (VCR) is required. In a second step, a cost metric for the actual implementation of the SC DCDC converters can be derived, yielding a performance-cost-matrix from which a certain topology for a specific use case can be selected. Good understanding of performance and cost/complexity trade-offs in fixed-ratio topologies is crucial when designing multi-VCR SC DCDC converters [5].

This paper is organized as follows: First, constraints for the SC DCDC topologies are chosen, so that a practical comparison is possible with reasonable effort. To achieve this comparison on a PCB level a universal hardware platform is introduced, using the same components for each SC DCDC topology. This hardware platform also allows for a comparable cost metric. Simulations allow to derive a performance figure-of-merit (FOM) for each topology. Based on a cost-performance matrix, six different topologies are chosen to be implemented on PCB level. Measurements finally validate the simulation results.

EPE'22 ECCE Europe

Topology Synthesis

To achieve a voltage conversion, the flying capacitors in the power stage of a SC DCDC converter are re-arranged in two or more switching phases. As an example, the implemented 4:1 topologies are shown in Fig. 1.

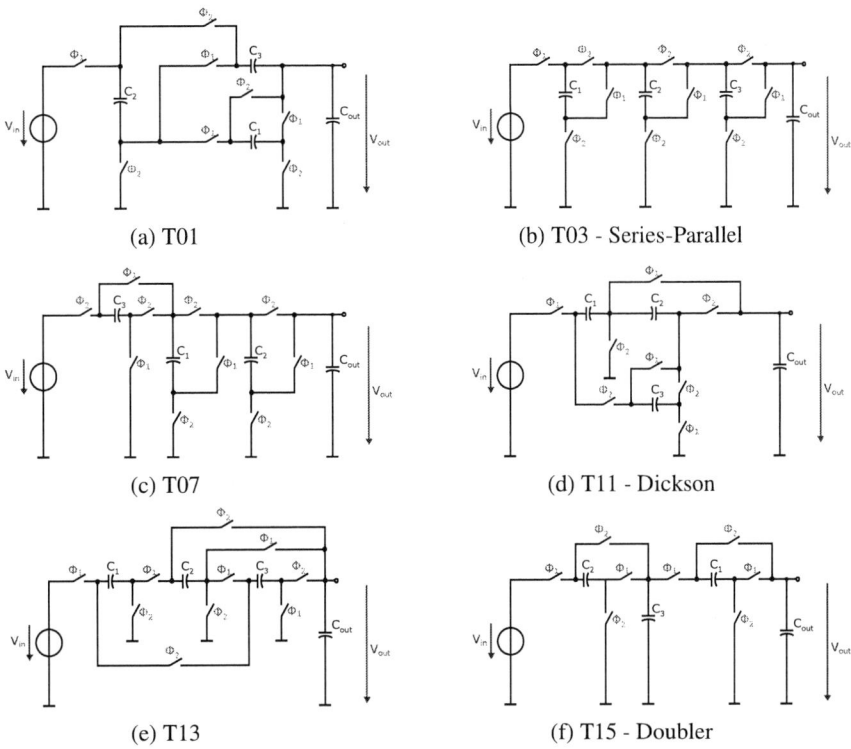

(a) T01

(b) T03 - Series-Parallel

(c) T07

(d) T11 - Dickson

(e) T13

(f) T15 - Doubler

Fig. 1: Implemented SC DCDC converter topologies

To realize a desired VCR, a certain minimum number of flying capacitors and switching phases is required [6]. However, with increasing number of flying capacitors and switching phases, the possibilities to achieve the desired VCR also grow significantly. Hence, it must be ensured to generate all possible topologies for given VCR, flying capacitor and switching phase constraints to enable a comprehensive comparison of the different implementations. A software algorithm based on the charge flow analysis introduced in [1] is developed as part of this research. The algorithm computes all possible SC DCDC architectures under given constraints and checks for realizability by applying Kirchhoff's laws. For the topology analysis, in this work a VCR of 4:1 is chosen. The number of flying capacitors and switching phases is set to three and two, respectively. These constraints give a total of 15 realizable SC DCDC architectures which include the well-known series-parallel, Dickson and cascaded / doubler implementations (Fig. 1).

PCB implementation

To perform a fair comparison the synthesized topologies are implemented using the same hardware components as shown in Table I. For the power switches EPC2014C GaN-FETs with a rated V_{DS} of 40 V are chosen while the flying capacitors are standard SMD ceramic types. Due to their low C_{OSS} GaN-FETs enable high switching frequencies resulting in low output resistance. This way, high efficiency can be achieved at load currents up to 4 A. The power switches are driven by isolated gate drivers (Infineon 2EDF7275F) since their source potentials can float. Because of the partially floating source potential, the gate drive supplies in a SC DCDC are of particular interest. For the drivers which cannot be supplied by bootstrapping or directly from the low-side supply, isolated DCDC converters (Analog Devices ADuM5028) are used.

Table I: Selected Components for PCB implementation

Component	Name
Power switches	EPC 2014C
Flying capacitors	1uF 50V MLCC
Isolated gate drivers	Infineon 2EDF7275F
Isolated DCDC converters	Analog Devices ADuM5028

An example PCB implementation is shown in Fig. 2. The gate drivers and power switches are placed on the opposite sides of the PCB to keep the gate loops short which is crucial for GaN-FETs. In addition, this enables a compact powerstage layout to decrease PCB parasitics.

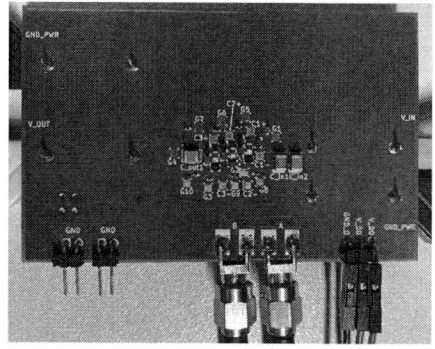 

(a) Powerstage with GaN-FETs and flying capacitors

(b) Gate drivers and gate drive supply

Fig. 2: SC DCDC PCB implementation

Performance and Cost Considerations, Methodology, FOM Development

The common SC DCDC converter model in Fig. 3 consists of an ideal transformer with the VCR of the topology and the output resistance R_{out} which models the intrinsic losses in the powerstage.

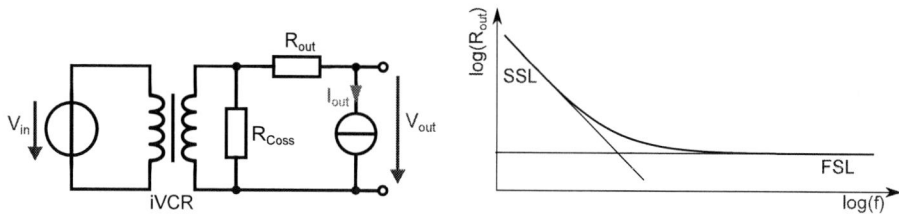

Fig. 3: SC DCDC converter model

The output resistance depends on the switching frequency and can be calculated according to eq. (1).

$$R_{out} = (V_{in} \cdot iVCR - V_{out})/I_{out} \tag{1}$$

At lower frequencies (Slow Switching Limit, SSL) the charge redistribution losses of the flying capacitors dominate. At higher frequencies the voltage swing across the flying capacitors becomes neglectable and the output resistance approaches the fast-switching limit (FSL) which is determined by the equivalent DC resistance of the powerstage. This FSL resistance also sets the maximum load current for given input and output voltages. The resistance R_{Coss} models the charging losses in the output capacitances of the power switches. In order to accurately model the efficiency of each topology, other loss mechanisms must be taken into account. Fig. 4 shows a breakdown of the different losses for topology T01 in (see Fig. 1a).

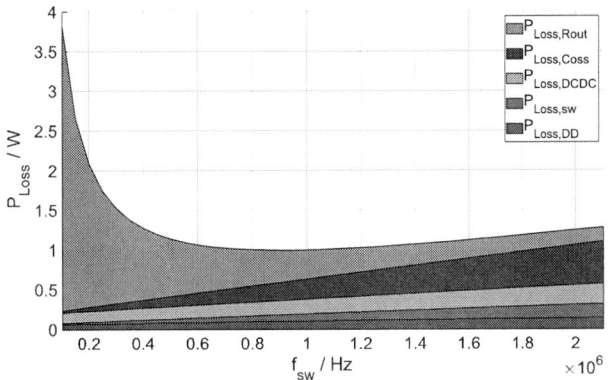

Fig. 4: Losses in SC DCDC Topology T01

At low switching frequencies, the R_{out} losses are dominant while at higher switching frequencies all frequencies dependent losses increase. $P_{Loss,Coss}$ are the losses during charging / discharging of the power switch output capacitances which are not negligible especially at higher input voltages. For accurate modeling, the voltage dependence of C_{OSS} must be taken into account: For the 4:1 SC DCDC topologies, the drain-source voltage of the power switches varies between $1/4 \cdot V_{in}$ and $3/4 \cdot V_{in}$, which corresponds to a C_{OSS} of 130 pF to 180 pF for the EPC2014 GaN-FETs. $P_{Loss,DCDC}$ accounts for the losses in the isolated DCDC converters, these are mainly topology dependent since the topologies require different numbers of isolated gate drive supplies. $P_{Loss,sw}$ and $P_{Loss,DD}$ indicate the gate charge losses of the power transistors and the low-side logic supply, respectively. From these losses the efficiency plots as shown in Fig. 5 can be derived.

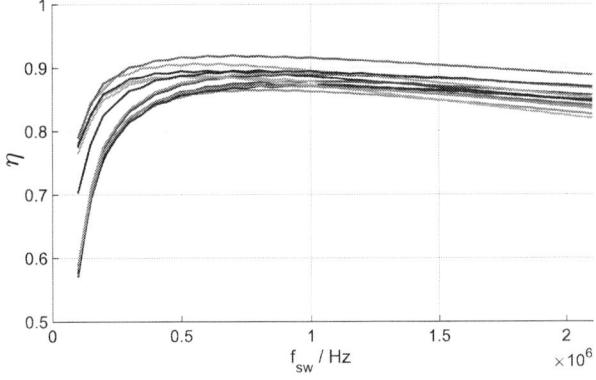

Fig. 5: Simulated efficiencies for all 15 topologies, $V_{in} = 36\,\text{V}$, $I_{load} = 1\,\text{A}$

Moreover, a performance figure-of-merit (FOM) can be introduced according to eq. (2)

$$\text{FOM} = \max(\eta)/\min(R_{out}) \tag{2}$$

with its maximum indicating the best performance. The FOM includes load current capability represented by the minimum Rout and the plain maximum efficiency. To obtain the SC DCDC topologies which are beneficial for actual implementation the gate drive complexity / cost must also be considered. Some topologies need more power switches or more isolated DCDC converters for gate drive supply and have therefore a larger bill of material. A power switch which is directly supplied by the low-side supply has significantly lower cost than a floating power switch that is supplied by an isolated DCDC converter. The cost considerations also apply directly for an integrated SC DCDC powerstage where cost is determined by silicon area. Here, the main area contributors are usually capacitors, so a topology which requires fewer floating voltage domains is beneficial since a floating voltage domain requires either bootstrapping or a dedicated charge pump.

Using the accumulated component costs, each SC DCDC topology can be entered into the cost-performance matrix as shown in Fig. 6 where the best topology would be located top left.

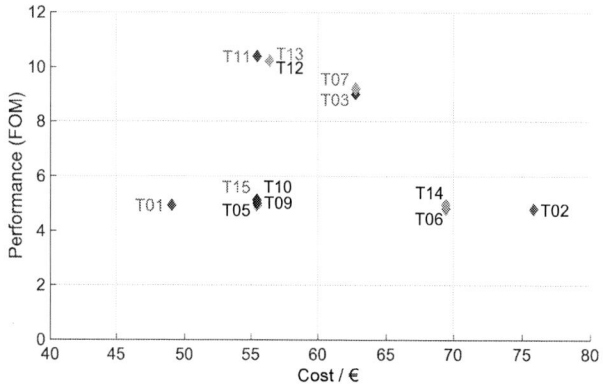

Fig. 6: Cost-Performance-Matrix of the 4:1 SC DCDC topologies (implemented topologies marked in red)

Experimental Results

From the complete set of 15 topologies, six were chosen to be implemented on a PCB (see schematics in Fig. 1, also marked in red in Fig. 6). These include the Series-Parallel (T03), Dickson (T11) and Cascaded / Doubler (T15) architecture as well as the three uncommon topologies T01, T07 and T13.

Fig. 7 shows the measured vs. simulated output resistances and efficiencies exemplarily for all implemented topologies. The deviation between simulation and measurement at low switching frequencies, especially for T11-Dickson, is caused by the non-linear flying capacitors where the actual capacitance changes significantly with the capacitor voltage. This is a concern at low switching frequency since the voltage swing over the flying caps at SSL is large (Fig. 8). In addition, T11-Dickson has larger steady-state voltages across the flying capacitors as compared to T01. The occurrence of resonances can be observed caused by parasitic inductances mainly from PCB traces, which is not included in the simulation model. Figure 9 shows the efficiency of each topology vs. load.

The measurements confirm the cost-performance matrix of Fig. 6. T01, T11-Dickson and T13 build the pareto front of all topologies. T01 has mediocre performance while by far the lowest cost. T11-Dickson has a slight performance advantage over T13, mainly due to its higher peak efficiency, while T13 has lower minimum R_{out}. This behavior is also confirmed by an efficiency measurement with fixed input and output voltages (Fig. 9) where T13 and T11-Dickson can provide significantly more load current of more than 4.1 A than the other topologies and also show the highest peak efficiencies.

Table II: Comparison between selected SC DCDC topologies

	T01	T11 - Dickson	T13
Peak efficiency	○	++	+
Max. load current	○	+	++
Cost	++	○	−
Max. C_{fly} voltage	$0.5 \cdot V_{in}$	$0.75 \cdot V_{in}$	$0.5 \cdot V_{in}$
Max. switch voltage	$0.5 \cdot V_{in}$	$0.5 \cdot V_{in}$	$0.75 \cdot V_{in}$

Design Space Exploration for a Capacitive 36V, 4A, 4:1 DCDC Converter with GaN
Switches Using a Performance-Cost-Matrix Including Uncommon Topologies

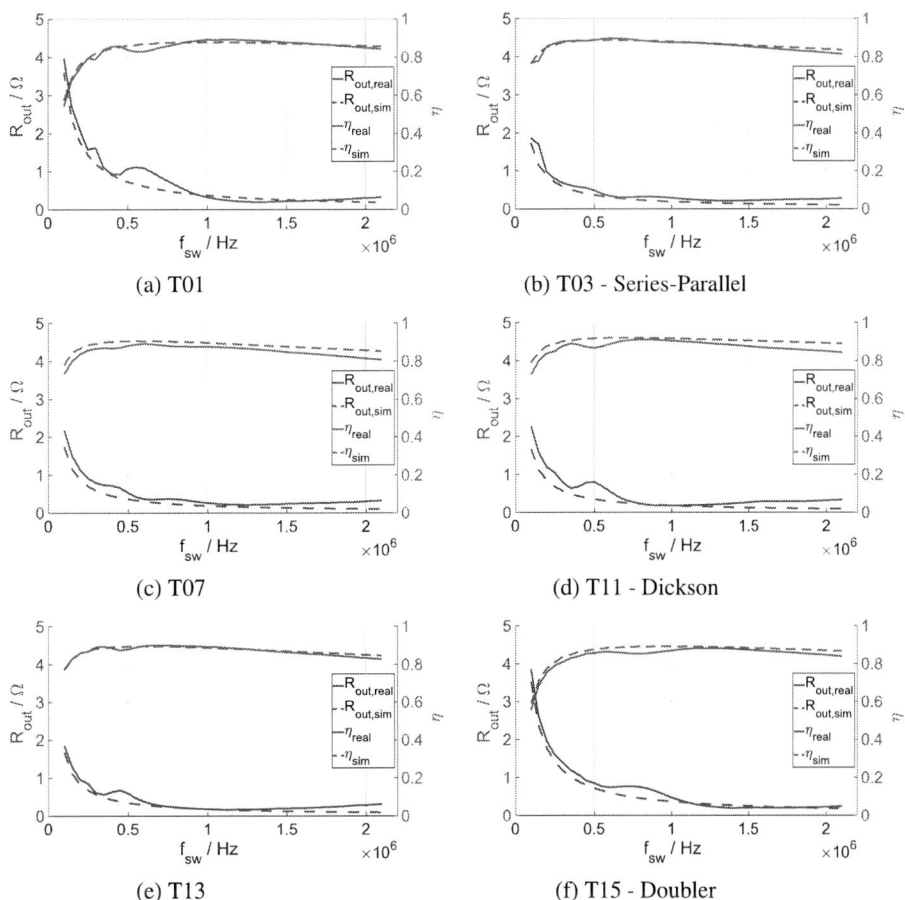

(a) T01

(b) T03 - Series-Parallel

(c) T07

(d) T11 - Dickson

(e) T13

(f) T15 - Doubler

Fig. 7: Simulated vs. measured output resistances and efficiencies

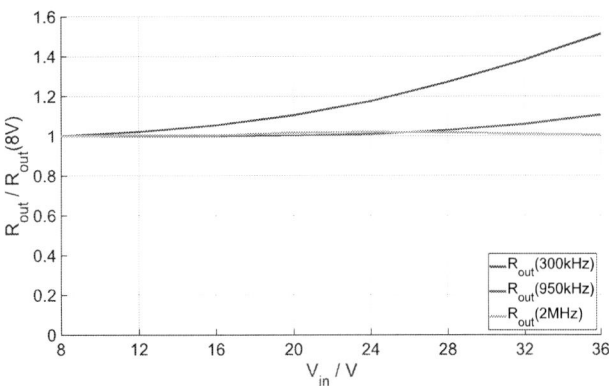

Fig. 8: Output resistance of T11-Dickson vs. input voltage for different switching frequencies

Table II summarizes the results for T01, T11-Dickson and T13 and includes the maximum voltage ratings for flying capacitors and power switches which is an important information for a final DCDC converter design since it determines the maximum input voltage for a given set of components. For an integrated solution, the limits are set by the semiconductor technology, so the topology selection becomes even more important.

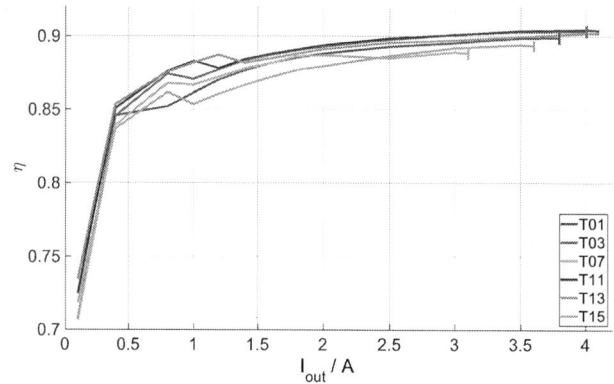

Fig. 9: Efficiencies vs. load current for $V_{in} = 36\,V$, $V_{out} = 8.3\,V$

Conclusion

This work presents a comprehensive analysis of 4:1 SC DCDC converter topologies with three flying capacitors and GaN power switches. These topologies include not only the well-known Dickson, Series-Parallel and Doubler / Cascaded architectures, but also all other possible topologies for the given constraints are generated using a self-developed software algorithm. The cost-performance matrix shows that the Dickson topology has benefits in terms of efficiency at reasonable implementation costs. However, also uncommon topologies turn out to be attractive. The topology T013 has an advantage due to its minimal output resistance but also requires a slightly more expensive gate drive supply. For applications which do not have the highest performance demands the topology T01 could also be of interest, since it has significantly lower implementation cost. The measurement results are in line with the simulations, also showing practical limitations of the different topologies, such as different maximum voltage ratings of the power switches and flying capacitors.

The presented methodology can be easily adapted for other voltage conversion ratios and more flying capacitors which might lead to even more uncommon SC DCDC architectures since the parameter space grows rapidly with increasing number of flying capacitors and larger voltage conversion ratios.

References

[1] M. D. Seeman and S. R. Sanders, "Analysis and Optimization of Switched-Capacitor DC–DC Converters," IEEE Trans. Power Electron., vol. 23, no. 2, pp. 841–851, 2008, doi: 10.1109/TPEL.2007.915182.

[2] J. Liu and S. Gregori, "Switched-Capacitor Boost-Buck Ladder Converters With Extended Voltage Range in Standard CMOS," IEEE Trans. Circuits Syst. I, vol. 67, no. 12, pp. 4593–4606, 2020, doi: 10.1109/TCSI.2020.3028268.

[3] D. Gunasekaran and F. Z. Peng, "Design of GaN based ultra-high efficiency, high power density resonant Dickson converter for high voltage step-down ratio," in ECCE 2019: IEEE Energy Conversion Congress & Expo : Baltimore, MD, Sept. 29-Oct. 3, Baltimore, MD, USA, 2019, pp. 845–852.

[4] M. S. Makowski and D. Maksimovic, "Performance limits of switched-capacitor DC-DC converters," in Proceedings of PESC '95 - Power Electronics Specialist Conference, Atlanta, GA, USA, Jun. 1995, pp. 1215–1221.

[5] D. Lutz, P. Renz, and B. Wicht, "12.4 A 10mW fully integrated 2-to-13V-input buck-boost SC converter with 81.5% peak efficiency," in 2016 IEEE International Solid-State Circuits Conference (ISSCC), San Francisco, CA, USA, Jan. 2016 - Feb. 2016, pp. 224–225.

[6] M. S. Makowski, "Realizability Conditions And Bounds On Synthesis Of Switched-capacitor DC-DC Voltage Multiplier Circuits," IEEE Trans. Circuits Syst. I, vol. 44, no. 8, 1997.

A fast control for a three-switch multi-input DC-DC converter

Simone Cosso, Andrea Formentini, Mario Marchesoni, Massimiliano Passalacqua, Luis Vaccaro

UNIVERSITY OF GENOVA, DEPARTMENT OF ELECTRICAL, ELECTRONIC,
TELECOMMUNICATIONS ENGINEERING AND NAVAL ARCHITECTURE
Via all' Opera Pia 11A,16145
Genova, Italy
simone.cosso@edu.unige.it, andrea.formentini@unige.it, marchesoni@unige.it,
massimiliano.passalacqua@unige.it, luis.vaccaro@unige.it

Keywords: «DC-DC Converter» «Feed-Forward» «DCM» «CCM» «Power Converters»

Abstract

In this paper, a double-input bidirectional DC-DC converter is taken into account. The converter allows to use less switches than a traditional solution and, moreover, it guarantees a higher efficiency. The modulation strategy proposed in the technical literature allows the converter to work in Discontinuous Conduction Mode at low-load and, therefore, to increase the efficiency in comparison to the CCM. However, such a control reduces the dynamics of the converter. To improve the transient response, a Feed-Forward approach is proposed in this paper. Since the converter model in DCM is complex and highly nonlinear, a simplified model is considered. The effectiveness of the proposed approach is proved with experimental results on a converter prototype.

Introduction

The increasing attention on air pollution and greenhouse emissions led to increasing attention on renewable energy sources and in electric mobility. Regarding the power electronics sector, this aspect increases the studies on DC-DC converters. As a matter of fact, DC-DC converters are used to interface renewables and storages with a common DC-bus [1-4]; moreover, the storage systems are connected with DC-DC converters with the motor inverter DC-link in hybrid and electric vehicles [5-9]. When several DC sources are used, it could be useful to use multi-input or multi-output converters. Various solutions are proposed in the technical literature both with galvanic insulation [10-14] and without galvanic insulation [15-19].

Regarding bidirectional multi-input single-output converters, a new double-input converter is shown in [20]. The proposed converter uses only three switches, whereas in a conventional solution (i.e., two half-bridge converters in parallel connection) four switches are needed. Moreover, the converter proposed in [20] guarantees lower losses in comparison to two half-bridges converters, as shown in [20] and analyzed in detail in [21]. However, in low-load working conditions, the converter works in Continuous Conduction Mode (CCM) with the control strategy proposed in [20].

To improve the efficiency in this working region by exploiting Discontinuous Conduction Mode (DCM), a new modulation strategy is presented in [22]. Moreover, in [20] the current path at high loads is also optimized and a loss reduction is achieved also in this working area.

However, the efficiency increase is obtained at the cost of dynamics reduction; as a matter of fact, in DCM the converter becomes strongly nonlinear. Indeed, as highlighted in the paper, in CCM a small variation in the control output (i.e., switch duty cycles) causes a high variation in the converter currents, whereas, in DCM a high variation in the control output causes a low variation in the converter currents. For this reason, an optimal control tuning in the CCM region causes a very slow response in the DCM region, whereas an optimal control tuning in the DCM region causes instability in the CCM region. Since the converter is strongly nonlinear in the DCM region, also the CCM-DCM bound depends on several variables and, therefore, variable gains lead analogously to instability.

In this paper, a Feed-Forward (FF) control is proposed to improve the converter speed response. A simplified converter model is studied and it is used to calculate the duty cycle values for the desired current references. The proposed control is integrated into the control proposed in [20]. The effectiveness of this solution is verified with experimental tests on a converter prototype.

Converter structure and control

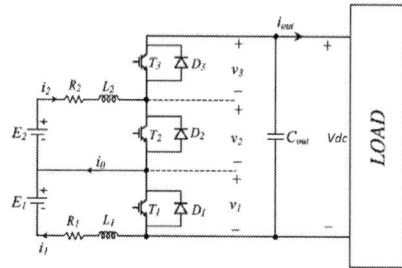

Fig. 1. Three-switch double-input bidirectional DC-DC converter.

The three-switch converter is shown in Fig. 1. The converter is a double-input bidirectional boost, i.e., the sum of the two-input voltages E_1 and E_2 has to be lower than the output voltage Vdc.
In Fig. 2 the control developed in [22] is shown. In order to allow DCM at low loads, five different working conditions are identified. Control outputs, gains K_1 and K_2 varies according to the working condition as shown in Fig. 2.
Please note that i_1 and i_2 are the current instantaneous values whereas I_1 and I_2 are the current average values during the duty cycle.
The five working conditions depends on current reference and are listed in the followings:

1. $I_1 > 0, I_2 > 0$
2. $I_1 > 0, I_2 < 0$
3. $I_1 < 0, I_2 > 0$
4. $I_1 < 0, I_2 < 0, I_1 > I_2$
5. $I_1 < 0, I_2 < 0, I_1 < I_2$

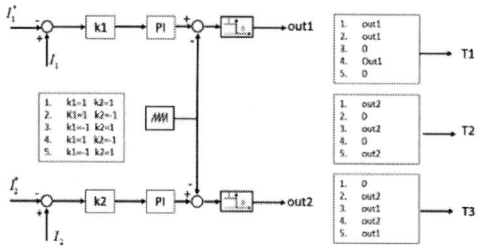

Fig. 2. The control scheme for the three-switch converter.

If both currents are positive and the converter works in CCM, the relation between I_1 and m_1 (i.e., the duty cycle of switch 1), and I_2 and m_2 can be modelled by the transfer functions in equations (1):

$$\frac{I_1}{m_1} = \frac{V_{dc}}{(R_1 + sL_1)(sT_p + 1)}; \quad \frac{I_2}{m_2} = \frac{V_{dc}}{(R_2 + sL_2)(sT_p + 1)} \qquad (1)$$

Where V_{dc}, R_1, R_2, L_1 and L_2 are shown in Fig. 1, whereas T_p is the converter delay, which can be considered 1.5 the switching frequency.
The control PI gains can be tuned from equations (1) (e.g., using the phase margin criterion). Even if a fast response is obtained in CCM, the converter dynamics worsens significantly in DCM. Indeed, the tuning in CCM (i.e., starting from equations (1)) leads to PI gains which are too low for the DCM region.
In order to fasten the response, a FF approach is carried out in this paper.

The feed-forward inputs guarantee a fast response, with a small error, which is then annulled by the PI with a time constant that can be relatively high. For the sake of brevity, the FF equations are shown just for condition 1. However, the same method can be applied also to the other four conditions. Please note that in the test bench control platform the FF control is implemented for all the five conditions in order to perform the experimental tests.

Feed-Forward approach

Fig. 3 shows the four possible current paths in condition 1 (i.e., when both I_1 and I_2 are positive). For each path, one can calculate the current slopes (i.e., the current derivatives), which depends on the voltage and inductance values. The current slopes for each path are reported in Table I, together with the related switch configuration.

In the low-load region, the transfer function between the switch duty cycles and the currents vary according to several variables (e.g., the current signs, the current amplitude and based on which current is higher). In addition to that, in path D the transfer function becomes very complex to calculate and difficult to implement in the control.

For this reason, a simplified approach is considered in this study and the relation between duty cycles and current references are approximated with an heuristic approach.

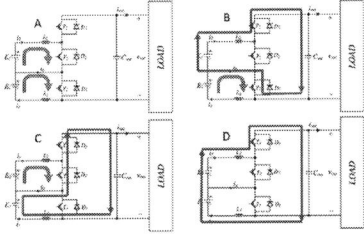

Fig. 3. Current paths in condition 1 (I_1 and I_2 >0)

TABLE I. CURRENT SLOPES
AND IGBT CONFIGURATION IN CONDITION 1

Path	IGBT config.	Current slope I1	Current slope I2
A	T_1 on T_2 on T_3 off	$\dfrac{V_1}{L_1}$	$\dfrac{V_2}{L_1}$
B	T_1 on (or T_1 off and $i_1 < i_2$) T_2 off T_3 off	$\dfrac{V_1}{L_1}$	$\dfrac{V_{dc} - V_2}{L_2}$
C	T_1 off T_2 on (or T_2 off and $i_2 < i_1$) T_3 off	$\dfrac{V_{dc} - V_2}{L_1}$	$\dfrac{V_2}{L_1}$
D	T_1 off T_2 off T_3 off	$\dfrac{V_{dc} - V_1 - V_2}{L_1 + L_2}$	$\dfrac{V_{dc} - V_1 - V_2}{L_1 + L_2}$

The instantaneous current waveforms in condition 1 are shown in Fig. 4. It is possible to note that when all switches are off, path C occurs at first; then, when I_1 and I_2 reach the same instantaneous value, the two currents decrease overlapping (i.e., path D occurs).

During path D, current slope is $\dfrac{V_{dc}-V_1-V_2}{L_1+L_2}$, whereas it is $\dfrac{V_{dc}-V_2}{L_2}$ and $\dfrac{V_{dc}-V_1}{L_1}$ in path B and C, respectively. Differently, when both switches T1 and T2 are closed, current path A occurs and both currents increase independently.

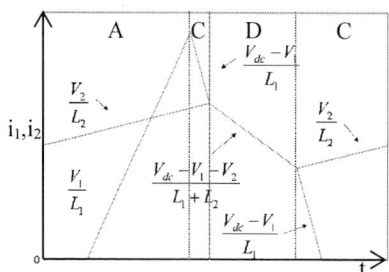

Fig. 4. Condition 1, $I_1>0$ and $I_2>0$ waveforms. i_1 (red), i_2 (blue).

From Fig. 4 one can note that the mathematical relation between the duty cycles and the current references is hard to obtain and unsuitable to implement in the control. For this reason a simplified method is proposed in the followings.

The relation between duty cycles and currents is evaluated in the two limit conditions: when there is no overlap (NOV) and when there is a total overlap (TOV), i.e., when $I_1=I_2$. Once this two conditions are evaluated, one can obtain the FF duty cycle values in case of partial overlap, interpolating the NOV and TOP values. As it is shown later, a parameter K to define the overlap ratio is defined.

In the simplified model, six different conditions can be considered:

1) $I_1 < I_2, I_1 > 0, I_2 > 0$ (NOV and TOV)

2) $I_1 < I_2, I_1 < 0, I_2 > \hat{I}_2$ (NOV)

3) $I_1 < I_2, I_1 < 0, I_2 < \hat{I}_2$ (NOV and TOV)

4) $I_1 > I_2, I_1 > 0, I_2 > 0$ (NOV and TOV)

5) $I_1 > I_2, I_1 > \hat{I}_1, I_2 > 0$ (NOV)

6) $I_1 > I_2, I_1 < \hat{I}_1, I_2 < 0$ (NOV and TOV)

$\hat{I}_1$ and $\hat{I}_2$ are defined later in the paper and they are always positive quantities. For this reason, in the 2nd and the 5th conditions the currents have opposite signs, therefore the overlap is not possible. Please note that these are the six conditions to define the FF and should not be confused with the five conditions used to implement the control shown in the previous paragraph.

Feed-forward equations for condition 1

NOV, equation for m2

In Fig. 5 (left) the ideal DCM waveform for i_2 in condition 1 ($I_1 > 0$ and $I_2 > 0$) is shown. Please note that if both currents are in DCM, path D always occur and the real current waveforms are shown in Fig. 5 (right). However, considering the ideal case is useful to obtain the equation to use in the interpolation process.

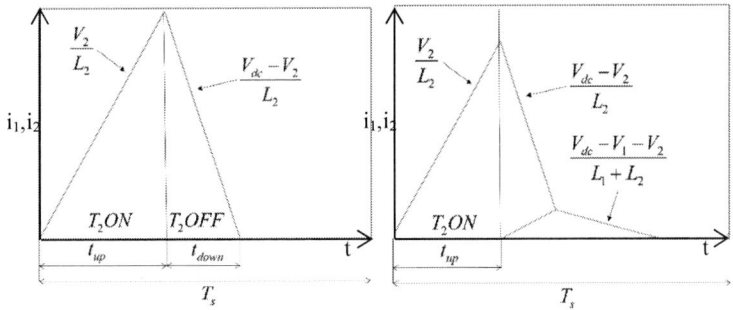

Fig. 5. Ideal waveform of i_2 (left), real waveform of i_2 (right)

The quantities t_{up}, I_{peak} and t_{down} are defined in equations (2), (3) and (4) and the calculation of the average current I_2 is evaluated in equation (5). From equations (2), (3), (4), (5) one can obtain equation (6), which is valid if i_2 is in DCM.

$$t_{up} = m_2 \cdot Ts \tag{2}$$

$$I_{peak} = \frac{V_2}{L_2} \cdot t_{up} \tag{3}$$

$$t_{down} = \frac{I_{peak}}{\frac{V_{dc} - V_2}{L_2}} \tag{4}$$

$$I_2 = \frac{I_{peak} \cdot (t_{up} + t_{down})}{2 \cdot T_s} \tag{5}$$

$$m_2 = \sqrt{\frac{2 L_2 (V_{dc} - V_2)}{V_2 T_s V_{dc}}} \sqrt{I_2} \tag{6}$$

When i_2 is in CCM, equation (7) is verified. Replacing (6) in (7) one obtains the condition in equation (8), which is valid for the CCM zone, whatever I_2 amplitude. Thus, in case of NOV when both current are positive and $I_1 < I_2$ (ideal case), the FF contribution for m_2 is given by equation (6), with the condition that $m_2 \in \left[0; \dfrac{V_{dc} - V_2}{V_{dc}} \right]$.

$$\frac{V_2}{L_2} m_2 = \frac{V_{dc} - V_2}{L_2} (1 - m_2) \tag{7}$$

$$m_2 = \frac{V_{dc} - V_2}{V_{dc}} \tag{8}$$

NOV, $I_1 > \hat{I}_1$, equation for m_1

In Fig. 6, i_1 waveform (red) is plotted together with i_2 waveform (blue). The quantities t_{up}, I_{peak} and t_{down} are defined in equations (9), (10), (11) and the calculation of the average current I_2 is evaluated in equation (12). Please note that in order to simplify the notation, avoiding long subscripts, the definition of t_{up}, I_{peak} and t_{down} is valid until the end of the subparagraph and the same variable are redefined in each subparagraph.

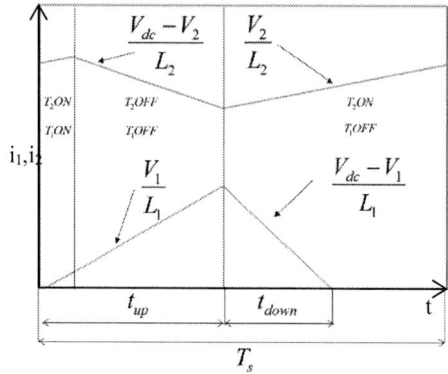

Fig. 6. i_1(red) and i_2(blue) waveforms during a switching period.

$$t_{up} = (m_1 + 1 - m_2)T_s \qquad (9)$$

$$I_{1peak} = \frac{V_1}{L_1} t_{up} \qquad (10)$$

$$t_{down} = \frac{I_{1peak}}{\dfrac{(V_{dc} - V_1)}{L_1}} \qquad (11)$$

$$I_1 = \frac{I_{1peak}(t_{up} + t_{down})}{2T_s} \qquad (12)$$

$$m_1 = \sqrt{\frac{2L_1(V_{dc} - V_1)}{T_s V_1 V_{dc}}} \sqrt{I_1} + m_2 - 1 \qquad (13)$$

When i_2 is in CCM, equation (14) is verified. Replacing (13) in (14), (15) is obtained, which is valid for the CCM zone, whatever I_2 amplitude. Thus, in case of NOV, when both currents are positive and $I_1 < I_2$, the FF contribution for m_1 is given by equation (15), with the condition that $m_1 = \dfrac{V_{dc} - V_1}{V_{dc}}$.

$$\frac{V_1}{L_1}(m_1 - m_2 + 1) = \frac{(V_{dc} - V_1)}{L_1}(m_2 - m_1) \qquad (14)$$

$$m_1 = \frac{V_{dc} - V_1}{V_{dc}} \qquad (15)$$

Please note that when m_1 is zero, I_1 is higher than zero, as it can be noticed in Fig. 7. $\hat{I}_1$ is I_1 value when $m_1 = 0$.

$\hat{I}_{1NOV}$ is I_1 value while $m_1 = 0$ and there is no overlap, i.e., i_2 is in CCM. Substituting $m_1 = 0$ and $m_2 = \dfrac{V_{dc} - V_2}{V_{dc}}$ in equation (13) one obtains equation (16). Since I_1 tends to zero while I_2 tends to zero, $\hat{I}_{1TOV} = 0$. Therefore, the FF value is given by the interpolation between $\hat{I}_{1NOV}$ and $\hat{I}_{1TOV}$.

$$\hat{I}_{1NOV} = \frac{T_s V_1 V_2^2}{2L_1 V_{dc}(V_{dc} - V_1)} \qquad (16)$$

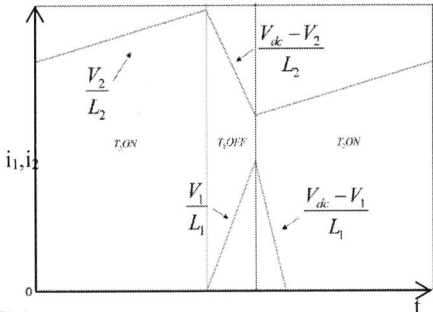

Fig. 7. i_1(red) in DCM and i_2(blue) in CCM.

NOV, $I_1 < \hat{I}_1$, equation for m_1

In Fig. 8, the current waveforms when $I_1 < \hat{I}_1$ and there is NOV are shown. Please note that in this condition, I_1 is controlled by m_3. t_1, $i_{1peakNeg}$, t_2, t_3, $I_{1peakPos}$ and t_4 are defined in equation (17), (18), (19), (20), (21) and (22), whereas the calculation of the I_1 is shown in equation (23).

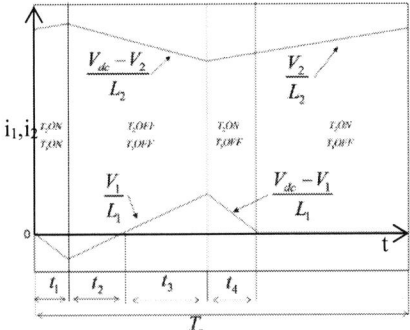

Fig. 8. i_1(red) and i_2(blue) in $I_1 < \hat{I}_1$ condition.

$$t_1 = m_3 T_s \tag{17}$$

$$I_{1peakNeg} = t_1 \frac{V_{dc} - V_1}{L_1} \tag{18}$$

$$t_2 = \frac{I_{1peakNeg}}{\frac{V_1}{L_1}} \tag{19}$$

$$t_3 = 1 - m_2 - t_2 \tag{20}$$

$$I_{1peakPos} = \frac{V_1}{L_1} t_3 \tag{21}$$

$$t_4 = \frac{I_{1peakPos}}{\frac{V_{dc} - V_1}{L_1}} \tag{22}$$

$$I_1 = \frac{I_{1peakPos}(t_3 + t_4) - I_{1peakNeg}(t_1 + t_2)}{2T_s} \tag{23}$$

Substituting (17), (18), (19), (20), (21) and (22) in (23), one obtains (24).

$$m_3 = \frac{V_1(1-m_2)}{2(V_{dc}-V_1)} - \frac{L_1}{T_s V_{dc}(1-m_2)} I_1 \qquad (24)$$

It is worth to note that equation (24) is valid only when there is NOV, therefore $m_2 = \dfrac{V_{dc}-V_2}{V_{dc}}$. Substituting m_2 expression in (24) one obtains equation (25).

$$m_3 = \frac{V_1 V_2}{2V_{dc}(V_{dc}-V_1)} - \frac{L_1}{T_s V_2} I_1 \qquad (25)$$

TOV, equations for m_1 and m_2

In the previous subparagraphs the condition in which there is not current overlap (NOV) has been shown. Now, the other condition, i.e., when $i_1=i_2$ and there is a total overlap (TOV), is taken into account. The current waveforms in this condition are shown in Fig. 9. Neglecting the small time in which current slopes have opposite signs, one can define t_{2up}, I_{2peak}, t_{2down} and I_2 as in equations (26), (27), (28), and (29).

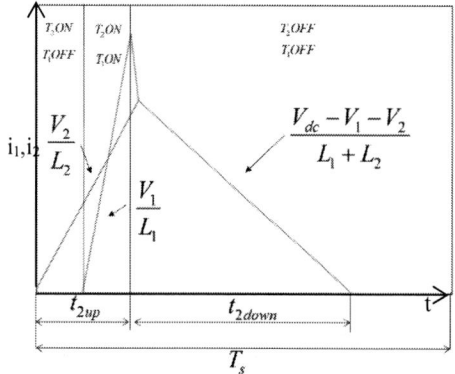

Fig. 9. i_1(red) and i_2(blue) waveform when $I_1=I_2$.

$$t_{2up} = m_2 T_s \qquad (26)$$

$$I_{2peak} = \frac{V_2}{L_2} t_{2up} \qquad (27)$$

$$t_{2down} = \frac{I_{2peak}}{\dfrac{V_{dc}-V_1-V_2}{L_1+L_2}} \qquad (28)$$

$$I_2 = \frac{I_{2peak}(t_{2up}+t_{2down})}{2T_s} \qquad (29)$$

Combining the previous equations, one obtains equation (30).

$$m_2 = \sqrt{\frac{2L_2^2(V_{dc}-V_1-V_2)}{V_2 T_s(L_1 V_2 - L_2 V_1 + L_1 V_{dc})}}\sqrt{I_2} \qquad (30)$$

Equation (30) is valid when i_2 is in DCM. On the contrary, equation (31) is verified when i_2 is in CCM. Substituting (26), (27) and (28) in (31) equation (32) is obtained and solving for m_2 results in equation (33). Thus, the FF contribution for m_2 in this condition is given by equation (30) with the limitation $m_2 \in \left[0; \dfrac{L_2(V_{dc}-V_1-V_2)}{(L_1 V_2 - L_2 V_1 + L_2 V_{dc})}\right]$.

$$t_{2up} = t_{2down} \tag{31}$$

$$\frac{V_2 m_2}{L_2} = \frac{V_{dc} - V_1 - V_2}{L_1 + L2}(1 - m_2) \tag{32}$$

$$m_2 = \frac{L_2(V_{dc} - V_1 - V_2)}{(L_1 V_2 - L_2 V_1 + L_2 V_{dc})} \tag{33}$$

t_{1up}, I_{1peak}, t_{1down} and I_1 are defined in equations (34), (35), (36) and (37). Substituting (34), (35) and (36) in (37) one obtains equation (38).

$$t_{1up} = m_1 T_s \tag{34}$$

$$I_{1peak} = \frac{V_1}{L_1} t_{1up} \tag{35}$$

$$t_{1down} = \frac{I_{1peak}}{\dfrac{V_{dc} - V_1 - V_2}{L_1 + L_2}} \tag{36}$$

$$I_1 = \frac{I_{1peak}(t_{1up} + t_{1down})}{2T_s} \tag{37}$$

$$m_1 = \sqrt{\frac{2L_1^2}{T_s V_1}\left(\frac{Vdc - V_1 - V_2}{L_2 V_1 - L_1 V_2 + L_1 V_{dc}}\right)}\sqrt{I_1} \tag{38}$$

Since with a TOV $i_1 = i_2$, if $\dfrac{V_2}{L_2} < \dfrac{V_1}{L_1}$, then $m_1 \in [0; m_2]$. On the contrary if $\dfrac{V_2}{L_2} > \dfrac{V_1}{L_1}$, then $m_1 \in \left[0; \dfrac{L_1(V_{dc} - V_1 - V_2)}{(L_2 V_1 - L_1 V_2 + L_1 V_{dc})}\right]$ and $m_2 \in [0; m_1]$.

Interpolation Process

The interpolation process is used to find feed-forward outputs while there is a partial overlap. The current ripples when both currents are in CCM and without overlap are defined in equations (39) and (40). If the currents are in DCM and there is no overlap, it is easy to verify that the ripple is given by equations (41) and (42). Therefore, a parameter K, which measures the overlap ratio, is defined in equation (43). Since in equations (6), (13), (30) and (38) the current is under square root, the duty cycles can be calculated as in equation (44).

$$di_1 = \frac{V_1(V_{dc} - V_1)T_s}{L_1 V_{dc}} \tag{39}$$

$$di_2 = \frac{V_2(V_{dc} - V_2)T_s}{L_2 V_{dc}} \tag{40}$$

$$di_{1real} = di_1^{(sign)}\sqrt{\frac{I_1}{di_1/2}} \qquad di_{1real} \in [-d_{i1}; d_{i1}] \tag{41}$$

$$di_{2real} = di_2^{(sign)}\sqrt{\frac{I_2}{di_2/2}} \qquad di_{2real} \in [-d_{i2}; d_{i2}] \tag{42}$$

$$K = \left|\frac{2(I_2 - I_1)}{|di_{1real} + di_{2real}|}\right| \in [0;1] \tag{43}$$

$$m_x = \sqrt{K}\, m_{x(NOV)} + (1 - \sqrt{K})\, m_{x(TOV)} \tag{44}$$

Experimental results

The proposed FF approach is tested on a converter prototype, which is shown in Fig. 10. A 12V battery and a 22V supercapacitor are employed as sources, whereas there is no load on the DC-link, therefore one source charges the other one. A voltage-loop is implemented to keep the DC-link at 150 V. Therefore I_1 reference is the voltage PI output, whereas I_2 reference can be set arbitrarily. The control is implemented of Dspace Microlabbox.

Currents waveform while I_2 reference is changed from +5V to -5V are shown in Fig. 11, both for the control with and without FF. Moreover, in Fig. 12 a test with a sinusoidal reference for I_2 at 3Hz is shown both with and without FF. Please note that in order to plot the current reference, Fig. 12 is obtained exporting the data from Dspace and post-processing them with Matlab.

In both cases one can note a significant increase in dynamics with the FF control.

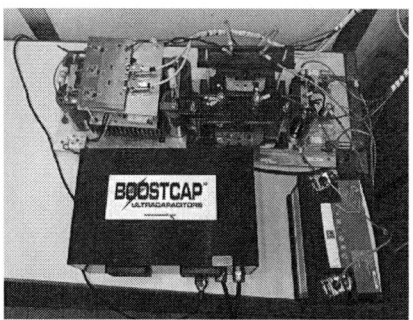

Fig. 10 Experimental test bench.

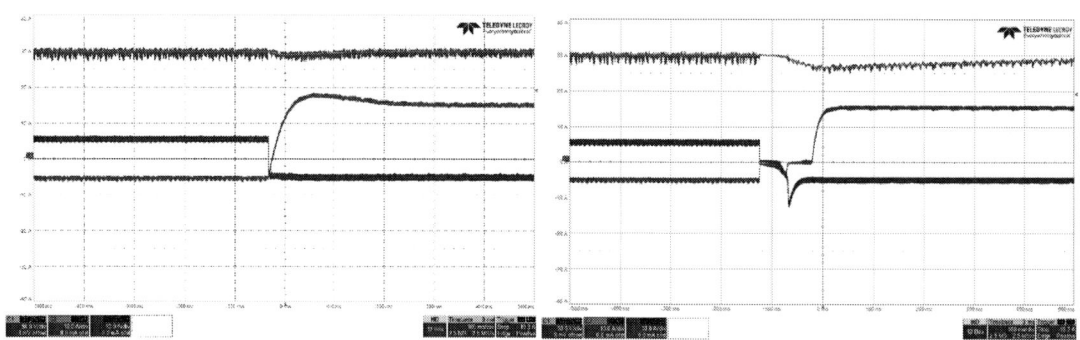

Fig. 11. i_1(magenta), i_2(blue) and v_{dc}(yellow) during a I_2 variation from +5 A to -5A. FF (left), NO-FF (right).

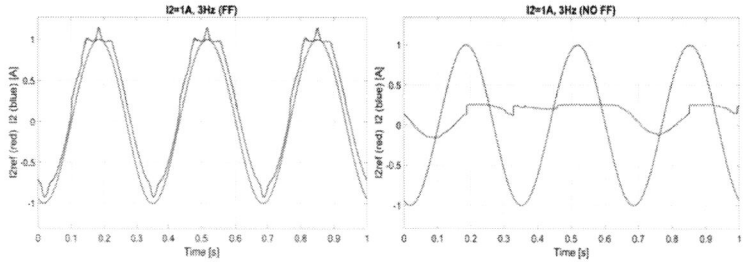

Fig. 12 Experimental current I_2(blue) and reference (red) at 1 A, 3 Hz. FF (left), NO-FF (right).

Conclusions

In this paper a double-input bidirectional DC-DC converter is taken into account. The modulation proposed in the technical literature to work in DCM allows loss reduction, however it worsen the converter dynamics. To improve the transient response, a Feed-Forward control is proposed in this paper and added to the converter control. Since the converter model in DCM is highly nonlinear and complex, simplified relations are considered for the Feed-Forward. The effectiveness of the proposed strategy is validated with experimental results on a converter prototype comparing the transient response with and without Feed-Forward.

REFERENCES

[1] H. Fakham, D. Lu, and B. Francois, "Power Control Design of a Battery Charger in a Hybrid Active PV Generator for Load-Following Applications," *IEEE Transactions on Industrial Electronics,* vol. 58, no. 1, pp. 85-94, 2011.

[2] T. Hirose and H. Matsuo, "Standalone Hybrid Wind-Solar Power Generation System Applying Dump Power Control Without Dump Load," *IEEE Transactions on Industrial Electronics,* vol. 59, no. 2, pp. 988-997, 2012.

[3] H. Cha, J. Choi, and P. N. Enjeti, "A Three-Phase Current-Fed DC/DC Converter With Active Clamp for Low-DC Renewable Energy Sources," *IEEE Transactions on Power Electronics,* vol. 23, no. 6, pp. 2784-2793, 2008.

[4] Z. Liang, R. Guo, J. Li, and A. Q. Huang, "A High-Efficiency PV Module-Integrated DC/DC Converter for PV Energy Harvest in FREEDM Systems," *IEEE Transactions on Power Electronics,* vol. 26, no. 3, pp. 897-909, 2011.

[5] M. Passalacqua, D. Lanzarotto, M. Repetto, and M. Marchesoni, "Conceptual design upgrade on hybrid pow ertrains resulting from electric improv ements," *International Journal of Transport Development and Integration,* Article vol. 2, no. 2, pp. 146-154, 2018.

[6] G. Su and L. Tang, "A Multiphase, Modular, Bidirectional, Triple-Voltage DC–DC Converter for Hybrid and Fuel Cell Vehicle Power Systems," *IEEE Transactions on Power Electronics,* vol. 23, no. 6, pp. 3035-3046, 2008.

[7] M. B. Camara, H. Gualous, F. Gustin, and A. Berthon, "Design and New Control of DC/DC Converters to Share Energy Between Supercapacitors and Batteries in Hybrid Vehicles," *IEEE Transactions on Vehicular Technology,* vol. 57, no. 5, pp. 2721-2735, 2008.

[8] C. Lai, Y. Cheng, M. Hsieh, and Y. Lin, "Development of a Bidirectional DC/DC Converter With Dual-Battery Energy Storage for Hybrid Electric Vehicle System," *IEEE Transactions on Vehicular Technology,* vol. 67, no. 2, pp. 1036-1052, 2018.

[9] D. Lanzarotto, M. Passalacqua, and M. Repetto, "Energy comparison between different parallel hybrid vehicles architectures," *International Journal of Energy Production and Management,* Article vol. 2, no. 4, pp. 370-380, 2017.

[10] R. Wai, C. Lin, and Y. Chang, "High Step-Up Bidirectional Isolated Converter With Two Input Power Sources," *IEEE Transactions on Industrial Electronics,* vol. 56, no. 7, pp. 2629-2643, 2009.

[11] P. C. Heris, Z. Saadatizadeh, and E. Babaei, "A New Two Input-Single Output High Voltage Gain Converter With Ripple-Free Input Currents and Reduced Voltage on Semiconductors," *IEEE Transactions on Power Electronics,* vol. 34, no. 8, pp. 7693-7702, 2019.

[12] H. Tao, J. L. Duarte, and M. A. M. Hendrix, "Three-Port Triple-Half-Bridge Bidirectional Converter With Zero-Voltage Switching," *IEEE Transactions on Power Electronics,* vol. 23, no. 2, pp. 782-792, 2008.

[13] H. Matsuo, T. Shigemizu, F. Kurokawa, and N. Watanabe, "Characteristics of the multiple-input DC-DC converter," in *Proceedings of IEEE Power Electronics Specialist Conference - PESC '93,* 1993, pp. 115-120.

[14] Z. Qian, O. Abdel-Rahman, and I. Batarseh, "An Integrated Four-Port DC/DC Converter for Renewable Energy Applications," *IEEE Transactions on Power Electronics,* vol. 25, no. 7, pp. 1877-1887, 2010.

[15] G. Chen, Z. Jin, Y. Deng, X. He, and X. Qing, "Principle and Topology Synthesis of Integrated Single-Input Dual-Output and Dual-Input Single-Output DC–DC Converters," *IEEE Transactions on Industrial Electronics,* vol. 65, no. 5, pp. 3815-3825, 2018.

[16] R. Wai, C. Lin, and B. Chen, "High-Efficiency DC–DC Converter With Two Input Power Sources," *IEEE Transactions on Power Electronics,* vol. 27, no. 4, pp. 1862-1875, 2012.

[17] Y. Liu and Y. Chen, "A Systematic Approach to Synthesizing Multi-Input DC–DC Converters," *IEEE Transactions on Power Electronics,* vol. 24, no. 1, pp. 116-127, 2009.

[18] A. Ganjavi, H. Ghoreishy, and A. A. Ahmad, "A Novel Single-Input Dual-Output Three-Level DC–DC Converter," *IEEE Transactions on Industrial Electronics,* vol. 65, no. 10, pp. 8101-8111, 2018.

[19] O. Ray, A. P. Josyula, S. Mishra, and A. Joshi, "Integrated Dual-Output Converter," *IEEE Transactions on Industrial Electronics,* vol. 62, no. 1, pp. 371-382, 2015.

[20] M. Marchesoni and C. Vacca, "New DC–DC Converter for Energy Storage System Interfacing in Fuel Cell Hybrid Electric Vehicles," *IEEE Transactions on Power Electronics,* vol. 22, no. 1, pp. 301-308, 2007.

[21] M. Marchesoni, M. Passalacqua, and L. Vaccaro, "A refined loss evaluation of a three-switch double input DC-DC converter for hybrid vehicle applications," *Energies,* Article vol. 13, no. 1, pp. 1-13, 2020, Art. no. 204.

[22] M. Passalacqua, M. Marchesoni, and L. Vaccaro, "A New Modulation Strategy for Exploiting Discontinuous Conduction Mode in a Double-Input Three-Switch Bidirectional DC-DC Converter," *IEEE Transactions on Industrial Electronics,* vol. 68, no. 11, pp. 10815-10825, 2021.

Impact on the torque and on the copper losses under fault-tolerant Control of 5-phase PMSG

A. Dieng
LER- Université Cheikh Anta Diop de Dakar, IREENA – UNIVERSITY OF NANTES
Tel : 00 221 776665138
abdoulayendaw.dieng@ucad.edu.sn

Keywords

Permanent magnet synchronous generator, Torque Control, Control strategy, Open-circuit faults

Abstract

Here an analysis of the impact on the torque and on the copper losses under fault operation is done. The goal of each adopted torque control strategy is to generate current references to maximize the average torque and to minimize the copper losses. An itemized analysis is done.

Introduction

Five-phase Permanent magnet synchronous generator PMSG seems to be an interesting solution in the context of renewable energy source energy [1]-[2]. It offer many advantages. For exemple when it associated to AC-DC converter the converted power and the operating of the energy conversion chain are increased. In this paper an analysis of the impact on the torque and on the copper losses under fault-tolerant Control of 5-phase PMSG is done. The goal of each adopted torque control strategy is to generate current references to maximize the average torque and to minimize the copper losses. Torque control strategy of five-phase permanent magnet synchronous machine, under fault operation, have already been proposed in [3]-[10].

This work focuses on a detailed theoretical analysis of the impact on the torque and on the copper losses under fault operation but the current control performances of the five-phase PMSG (Fig. 1) achieved thanks to a robust and accurate AC current controller under one-phase is permanently open is presented. Figure 1 shows the control scheme of the 5-phase PMSG under the e-phase is permanently open. The choice of the faulty phase is the e-phase and it is given as example.

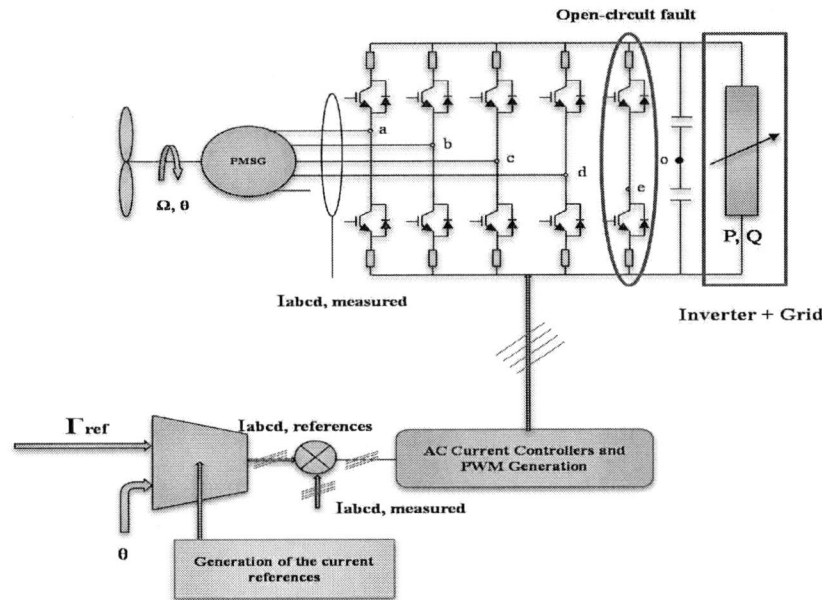

Fig. 1: Control scheme of the 5-phase PMSG under fault operation

Electrical model of the 5-phase PMSG under fault operation mode in the abcde frame

For the 5-phase PMSG, under fault operation, the used model is given by [4] for one open phase and two open phases.

One open phase

Under normal operation, in the abcde frame, the electrical equations of the generator can be written in the following matrix form:

$$[E] = [R][I] + [L]\frac{d}{dt}[I] + [V] \tag{1}$$

Where $[R] = \begin{bmatrix} r & 0 & 0 & 0 & 0 \\ 0 & r & 0 & 0 & 0 \\ 0 & 0 & r & 0 & 0 \\ 0 & 0 & 0 & r & 0 \\ 0 & 0 & 0 & 0 & r \end{bmatrix}$, $L = \begin{bmatrix} L_1 & L_2 & L_3 & L_3 & L_2 \\ L_2 & L_1 & L_2 & L_3 & L_3 \\ L_3 & L_2 & L_1 & L_2 & L_3 \\ L_3 & L_3 & L_2 & L_1 & L_2 \\ L_2 & L_3 & L_3 & L_2 & L_1 \end{bmatrix}$, $[V] = \begin{bmatrix} V_a \\ V_b \\ V_c \\ V_d \\ V_e \end{bmatrix}$, $[E] = \begin{bmatrix} E_a \\ E_b \\ E_c \\ E_d \\ E_e \end{bmatrix}$

and $[I] = [I_a \quad I_b \quad I_c \quad I_d \quad I_e]^t$

[R], [L] are respectively the resistance matrix and the inductance matrix. The inductance matrix is always real, circular and symmetrical whatever the winding [16]. [E], [V], [I] are respectively the EMF vector, the voltage vector and the current vector.

Let us assume that the e-phase is continually open, after a fault, $I_e = 0$. The new current vector becomes $[I] = [I_a \quad I_b \quad I_c \quad I_d]^t$

The voltage equation of the e-phase, in the abcde frame, is given by:

$$V_e = E_e - \left(L_2\frac{dI_a}{dt} + L_3\frac{dI_b}{dt} + L_3\frac{dI_c}{dt} + L_2\frac{dI_d}{dt} \right) \tag{2}$$

As the neutral point of the machine is not connected, $I_a + I_b + I_c + I_d = 0$, then:

$$V_a + V_b + V_c + V_d + V_e = 0 \tag{3}$$

The voltage across each healthy phase can be expressed as a function of the output voltages of the inverter as follows:

$$\begin{bmatrix} V_a \\ V_b \\ V_c \\ V_d \end{bmatrix} = \begin{bmatrix} V_{ao} - V_{No} \\ V_{bo} - V_{No} \\ V_{co} - V_{No} \\ V_{do} - V_{No} \end{bmatrix} \tag{4}$$

N is the neutral point of the machine.

Using (3) and (4) the voltage V_{No} are obtained:

$$V_{No} = \frac{1}{4}(V_{ao} + V_{bo} + V_{co} + V_{do} + V_e) \tag{5}$$

Using (4) and (5), the voltage equation across each healthy phase becomes:

$$\begin{bmatrix} V_a \\ V_b \\ V_c \\ V_d \end{bmatrix} = \begin{bmatrix} V_{ao}\text{-}V_{No} \\ V_{bo}\text{-}V_{No} \\ V_{co}\text{-}V_{No} \\ V_{do}\text{-}V_{No} \end{bmatrix} = \frac{1}{4}\begin{bmatrix} 3 & -1 & -1 & -1 \\ -1 & 3 & -1 & -1 \\ -1 & -1 & 3 & -1 \\ -1 & -1 & -1 & 3 \end{bmatrix}\begin{bmatrix} V_{ao} \\ V_{bo} \\ V_{co} \\ V_{do} \end{bmatrix} - \frac{1}{4}V_e \tag{6}$$

Let us consider $\begin{bmatrix} V_a' \\ V_b' \\ V_c' \\ V_d' \end{bmatrix} = \frac{1}{4}\begin{bmatrix} 3 & -1 & -1 & -1 \\ -1 & 3 & -1 & -1 \\ -1 & -1 & 3 & -1 \\ -1 & -1 & -1 & 3 \end{bmatrix}\begin{bmatrix} V_{ao} \\ V_{bo} \\ V_{co} \\ V_{do} \end{bmatrix}$

By substituting (2) in (6), (6) becomes:

$$\begin{bmatrix} V_a \\ V_b \\ V_c \\ V_d \end{bmatrix} = \begin{bmatrix} V_a' \\ V_b' \\ V_c' \\ V_d' \end{bmatrix} - \frac{1}{4}\left(E_e - \left(L_2\frac{dI_a}{dt} + L_3\frac{dI_b}{dt} + L_3\frac{dI_c}{dt} + L_2\frac{dI_d}{dt}\right)\right) \tag{7}$$

Where $\begin{bmatrix} V_a' \\ V_b' \\ V_c' \\ V_d' \end{bmatrix}$ is the new voltage vector

Based on (1) and (7) the new electrical equation, for the remaining healthy phases, under the e-phase open-circuit fault can be written as follows:

$$[E'] = [R'][I] + [L']\frac{d}{dt}[I] + [V'] \tag{8}$$

Where : $[R'] = \begin{bmatrix} r & 0 & 0 & 0 \\ 0 & r & 0 & 0 \\ 0 & 0 & r & 0 \\ 0 & 0 & 0 & r \end{bmatrix}$, $[L'] = \begin{bmatrix} L_1 + \frac{1}{4}L_2 & L_2 + \frac{1}{4}L_3 & L_3 + \frac{1}{4}L_3 & L_3 + \frac{1}{4}L_2 \\ L_2 + \frac{1}{4}L_2 & L_1 + \frac{1}{4}L_3 & L_2 + \frac{1}{4}L_3 & L_3 + \frac{1}{4}L_2 \\ L_3 + \frac{1}{4}L_2 & L_2 + \frac{1}{4}L_3 & L_1 + \frac{1}{4}L_3 & L_2 + \frac{1}{4}L_2 \\ L_3 + \frac{1}{4}L_2 & L_3 + \frac{1}{4}L_3 & L_2 + \frac{1}{4}L_3 & L_1 + \frac{1}{4}L_2 \end{bmatrix}$

[R'], [L'] are respectively the new resistance matrix and the new inductance matrix.
Hence the new fictitious equivalent EMF vector is given by:

$$[E'] = \begin{bmatrix} E_a' \\ E_b' \\ E_c' \\ E_d' \end{bmatrix} = \begin{bmatrix} E_a + \frac{1}{4}E_e \\ E_b + \frac{1}{4}E_e \\ E_c + \frac{1}{4}E_e \\ E_d + \frac{1}{4}E_e \end{bmatrix}$$

Now the expression of the total generator's electromagnetic torque can be deduced:

$$\Gamma = \frac{1}{\Omega}\left(E_a'I_a + E_b'I_b + E_c'I_c + E_d'I_d\right) \tag{9}$$

Where Ω is the mechanical angular speed.
Finally the same approach can be applied to any open phase. It can be noticed that following the open phase the inductance matrix [L'] changes

Two open phases

The faulty phases is selected arbitrary. Let us assuming that the d-phase ($I_d = 0$) and the e-phase ($I_e = 0$) are permanently open. The new current vector becomes $[I] = [I_a \quad I_b \quad I_c]^t$
The voltage equations of the d-phase and the e-phase, in the abcde frame, are given by:

$$V_d = E_d - \left(L_3 \frac{dI_a}{dt} + L_3 \frac{dI_b}{dt} + L_2 \frac{dI_c}{dt}\right) \tag{10}$$

$$V_e = E_e - \left(L_2 \frac{dI_a}{dt} + L_3 \frac{dI_b}{dt} + L_3 \frac{dI_c}{dt}\right) \tag{11}$$

As the neutral point of the machine is not connected, $I_a + I_b + I_c = 0$, then:

$$V_a + V_b + V_c + V_d + V_e = 0 \tag{12}$$

The voltage across each healthy phase can be expressed as a function of the output voltages of the inverter as follows:

$$\begin{bmatrix} V_a \\ V_b \\ V_c \end{bmatrix} = \begin{bmatrix} V_{ao} - V_{No} \\ V_{bo} - V_{No} \\ V_{co} - V_{No} \end{bmatrix} \tag{13}$$

Using (12) and (13) the voltage V_{No} are obtained:

$$V_{No} = \frac{1}{3}(V_{ao} + V_{bo} + V_{co} + V_e + V_d) \tag{14}$$

Using (13) and (14), the voltage equation across each healthy phase becomes:

$$\begin{bmatrix} V_a \\ V_b \\ V_c \end{bmatrix} = \begin{bmatrix} V_{ao} - V_{No} \\ V_{bo} - V_{No} \\ V_{co} - V_{No} \end{bmatrix} = \frac{1}{3}\begin{bmatrix} 2 & -1 & -1 \\ -1 & 2 & -1 \\ -1 & -1 & 2 \end{bmatrix}\begin{bmatrix} V_{ao} \\ V_{bo} \\ V_{co} \end{bmatrix} - \frac{1}{3}V_d - \frac{1}{3}V_e \tag{15}$$

Let us consider $\begin{bmatrix} V_a^{'} \\ V_b^{'} \\ V_c^{'} \end{bmatrix} = \frac{1}{3}\begin{bmatrix} 2 & -1 & -1 \\ -1 & 2 & -1 \\ -1 & -1 & 2 \end{bmatrix}\begin{bmatrix} V_{ao} \\ V_{bo} \\ V_{co} \end{bmatrix}$

By substituting (10) and (11) in (15), (15) becomes:

$$\begin{bmatrix} V_a \\ V_b \\ V_c \end{bmatrix} = \begin{bmatrix} V_a^{'} \\ V_b^{'} \\ V_c^{'} \end{bmatrix} - \frac{1}{3}\left(E_d - \left(L_3 \frac{dI_a}{dt} + L_3 \frac{dI_b}{dt} + L_2 \frac{dI_c}{dt}\right)\right) - \frac{1}{3}\left(E_e - \left(L_2 \frac{dI_a}{dt} + L_3 \frac{dI_b}{dt} + L_3 \frac{dI_c}{dt}\right)\right) \tag{16}$$

Where $\begin{bmatrix} V_a^{'} \\ V_b^{'} \\ V_c^{'} \end{bmatrix}$ is the new voltage vector

Based on (1) and (16) the new electrical equation, for the remaining healthy phases, under two-phase open-circuit fault can be written as follows:

$$[E^{'}] = [R^{'}][I] + [L^{'}]\frac{d}{dt}[I] + [V^{'}] \tag{17}$$

Where : $[R^{'}] = \begin{bmatrix} r & 0 & 0 \\ 0 & r & 0 \\ 0 & 0 & r \end{bmatrix}$, $[L^{'}] = \begin{bmatrix} L_1 + \frac{1}{3}L_3 + \frac{1}{3}L_2 & L_2 + \frac{2}{3}L_3 & L_3 + \frac{1}{3}L_2 + \frac{1}{3}L_3 \\ L_2 + \frac{1}{3}L_3 + \frac{1}{3}L_2 & L_1 + \frac{2}{3}L_3 & L_2 + \frac{1}{3}L_2 + \frac{1}{3}L_3 \\ L_3 + \frac{1}{3}L_3 + \frac{1}{3}L_2 & L_2 + \frac{2}{3}L_3 & L_1 + \frac{1}{3}L_2 + \frac{1}{3}L_3 \end{bmatrix}$

$[R^{'}]$, $[L^{'}]$ are respectively the new resistance matrix and the new inductance matrix.

Hence the new fictitious equivalent EMF vector is given by:

$$[E^{'}] = \begin{bmatrix} E_a^{'} \\ E_b^{'} \\ E_c^{'} \end{bmatrix} = \begin{bmatrix} E_a + \frac{1}{3}E_d + \frac{1}{3}E_e \\ E_b + \frac{1}{3}E_d + \frac{1}{3}E_e \\ E_c + \frac{1}{3}E_d + \frac{1}{3}E_e \end{bmatrix}$$

Now the expression of the total generator's electromagnetic torque is now given by:

$$\Gamma = \frac{1}{\Omega}\left(E_a^{'}I_a + E_b^{'}I_b + E_c^{'}I_c\right) \tag{18}$$

Finally, the same approach can be applied to any combination of two open phases. In the remainder the case of one-phase open-circuit fault is studied. The choice of the faulty phase is the e-phase and it is given as example. The proposed control strategy can be applied to any open phase.

Torque control strategy of the phase the 5-phase PMSG

Assuming the *e*-phase is open due to the failure of power devices, torque ripples appear.
In order to reduce the torque ripples and to minimize the copper losses, the chosen torque control strategy consists to impose optimal current references in the remaining four phases. The reactive power is equal to zero when the EMF and current vectors of the machine are collinear:

$$\frac{E'_a}{I_{aref}} = \frac{E'_b}{I_{bref}} = \cdots = \frac{E'_d}{I_{dref}} \tag{19}$$

Generation of the current references under fault operation mode in the abcde frame

In fault operation mode, the copper losses are minimal when the current and EMF vectors, corresponding to the healthy phases, are collinear. When one or i (i≤3) phases of the machine is open due to the failure of power devices, the current references of the healthy phases are expressed in the following form:

$$I_{zref} = \frac{E_z}{\sum_{z=a}^{e}(s_z E_z^2)} \Gamma_{dref}\Omega \tag{20}$$

Where

z = a, b, c, d, e, $s_z = 1$ for the healthy phases and $s_z = 0$ for the open phases.
E_z : EMF
Γ_{dref} : the reference torque under fault operation
When the neutral of the machine is not connected of the midpoint to the DC bus or if the sum of the imposed currents must be zero, the expression of the current references of the healthy phases becomes:

$$I_{zref} = \frac{E'_z}{\sum_{z=a}^{e} E'^2_z} \Gamma_{dref}\Omega \tag{21}$$

Where

$E'_z = s_z E_z - \frac{1}{n-i}\sum_{z=a}^{e} s_z E_z$ and i the number of open phases
The Fourier analysis of the EMF of the machine under consideration is summarized in Table 1. Table 1 summarizes the normalized magnitude of each harmonic of the considered machine.

Table 1 Fourier analysis of the EMF profile

EMF Harmonic	1	3	7	9
Magnitude/Fundamental %	100%	30%	0.2%	0.7%

The Fourier analysis of the EMF shows that the ninth harmonic and the seven harmonic are very low and can be neglected.

Robust current controller

The next figure shows a basic scheme where the chosen current controller has been used in DC/AC converters. It runs in sliding mode and offers an accurate current control. At high and low frequencies, it operates in different ways, i.e. at high frequency it operates for frequency switching control and at low frequency it operates for current control [11]-[12].

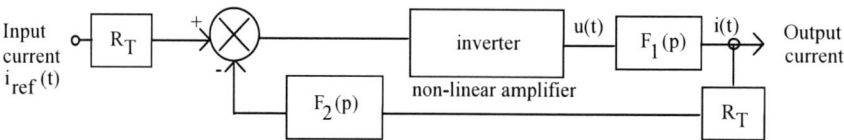

Figure 3: Scheme of the inverter current control loop

Simulations and Experimental results

Now, simulation and experiments are done to show the current control performances under open-circuit fault to validate the torque control strategy. Therefore the *e*-phase is permanently open. The choice of the faulty phase is the *e*-phase and it is given as example.

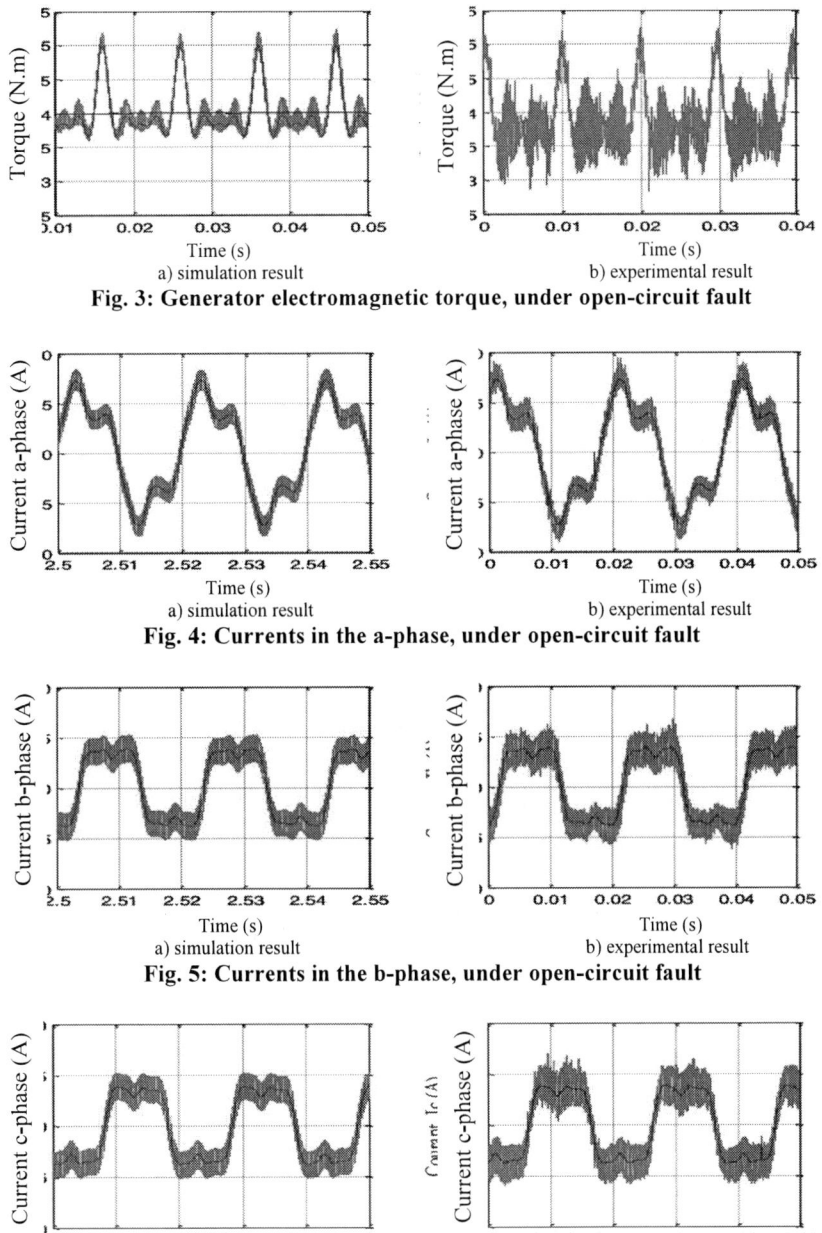

Time (s)
a) simulation result

Time (s)
b) experimental result

Fig. 3: Generator electromagnetic torque, under open-circuit fault

Time (s)
a) simulation result

Time (s)
b) experimental result

Fig. 4: Currents in the a-phase, under open-circuit fault

Time (s)
a) simulation result

Time (s)
b) experimental result

Fig. 5: Currents in the b-phase, under open-circuit fault

Time (s)
a) simulation result

Time (s)
b) experimental result

Fig. 6: Currents in the c-phase, under open-circuit fault

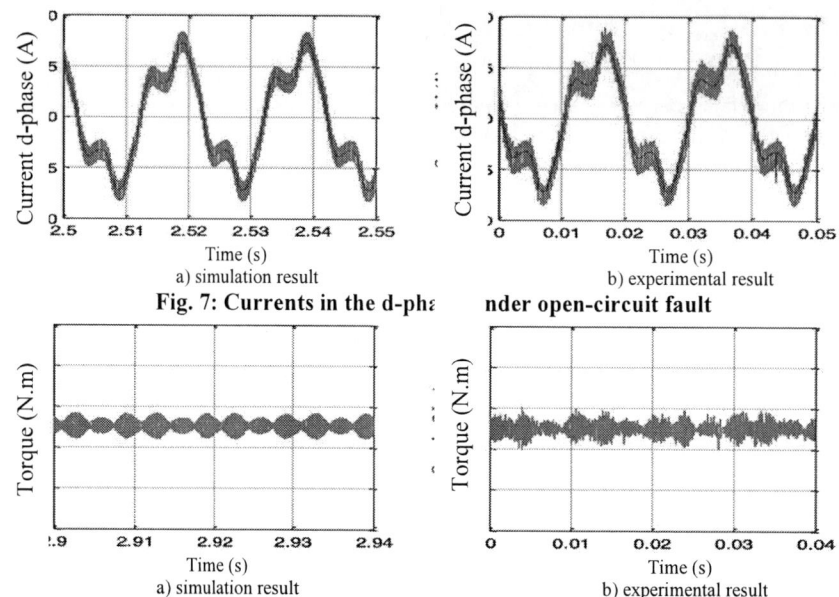

a) simulation result

b) experimental result

Fig. 7: Currents in the d-pha nder open-circuit fault

a) simulation result

b) experimental result

Fig. 8: Generator electromagnetic torque, under open-circuit fault, new current references applied

Fig. 3 shows the generator electromagnetic torque under fault operation when the same currents under normal operation and their magnitudes are maintained. The generator torque is pulsating and the average value decreases. Under fault operation (the e-phase is open), the new current references are imposed in the four healthy phase. The generator torque is kept constant (Fig. 8) thanks to the good performance of the robust current controllers which accurately track its references (Fig. 4, Fig. 5, Fig. 6 and Fig. 8).

Impact on the torque and on the copper losses

In this part, a theoretical analysis of the impact on the torque and on the copper losses, under fault operation, is studied. This impact depend to the configurations of the adopted torque control strategy. Three cases are considered, one open phase (for exemple phase e), two open adjacent phases (phases d and e) and two open non-adjacent phases (phases c and e).

The expression of the generator electromagnetic torque, under fault operation, is given by :

$$\Gamma_d = \frac{1}{\Omega} \sum_{z=a}^{e} (s_z E_z I_z) \tag{4}$$

Where

z = a, b, c, d, e, $s_z = 1$ for the healthy phases and $s_z = 0$ for the open phases.

The expression of the copper losses is given by :

$$P_{jd} = r < (\sum_{z=a}^{e} (s_z I_z)^2) > \tag{5}$$

Where

z = a, b, c, d, e, $s_z = 1$ for the healthy phases and $s_z = 0$ for the open phases.

Using (2) and (5) or (3) and (5), the expression of the copper losses can be rewritten as a function of the torque reference in degraded mode [7]. Thus, at given copper losses, the torque reference in degraded mode can be calculated.

Table 2 shows, for each configurations studied, the impact on the copper losses when it is not limited and the torque reference in degraded mode is equal to 5 N.m. In normal operation mode, for this same value torque reference, the copper losses obtained in simulation are:

- neutral of the machine is not connected of the midpoint to the DC bus, $P_{jN} = 24,47\ W$

- neutral of the machine is connected of the midpoint to the DC bus and the homopolar EMF is exploited, $P_{jN} = 24,05\ W$

Table 2 Impact on the copper losses

Configurations	$\dfrac{P_{jd}}{P_{jN}}$ %
One open phase Neutral not connected to the midpoint of the DC bus	+36
Two open adjacent phases Neutral not connected to the midpoint of the DC bus	+1663 (**)
Two open non-adjacent phases Neutral not connected to the midpoint of the DC bus	+79
One open phase Neutral connected to the midpoint of the DC bus Sum of the current references is not equal to zero	+25
Two open adjacent phases Neutral connected to the midpoint of the DC bus Sum of the current references is not equal to zero	+70
Two open non-adjacent phases Neutral connected to the midpoint of the DC bus Sum of the current references is not equal to zero	+69

In the case of two open adjacent phases, at constant torque, the copper losses increase by +1663 (Table 2) which is every high and can have very serious consequences for the machine. This increase is due to the amplitude of the current references which is very high. The adopted control strategy is not good. By connecting the machine neutral to the midpoint of the DC bus and by imposing the sum of the current references is not egal to zero, the copper losses increase by + 70% (Table 2).

Table 3 Impact on the copper losses

Configurations	$\dfrac{\Gamma_{dref}}{\Gamma_{ref}}$ %
One open phase Neutral not connected to the midpoint of the DC bus	-14
Two open adjacent phases Neutral not connected to the midpoint of the DC bus	-76 (**)
Two open non-adjacent phases Neutral not connected to the midpoint of the DC bus	-25
One open phase Neutral connected to the midpoint of the DC bus Sum of the current references is not equal to zero	-10
Two open adjacent phases Neutral connected to the midpoint of the DC bus Sum of the current references is not equal to zero	-23
Two open non-adjacent phases Neutral connected to the midpoint of the DC bus Sum of the current references is not equal to zero	-23

Table 3 shows, for each considered configurations, the impact on the electromagnetic torque when the copper losses in degraded mode are kept equal to the copper losses in normal mode. In the case of one open phase, the torque reference decreases by -14% (Table 3) if the machine neutral is not connected. When it is connected to the midpoint of the DC bus and the sum of the imposed current references is not equal to zero, the torque reference decreases by -10% (Table 3). When two adjacent phases are opened with the machine neutral not connected, the torque reference decreases by -76% (Table 3). This is no good for the machine.

Conclusion

In this paper a theoretical analysis of the impact on the torque and on the copper losses, under fault operation, has been done. Firstly the current control performances under open-circuit fault to validate the torque control strategy is presented. Simulation and experimental results prove the effectiveness of the torque control strategy. Finally, the itemized analysis shows the impact on the copper losses and on the torque in fault operation.

References

[1] E Levi, Multiphase electric machines for variable-speed applications, IEEE Trans. Ind. Electron., vol. 55, no. 5, pp. 1893–1909, May 2008.

[2] N. Bianchi, S. Bolognani, and M. D. Pre, "Strategies for the fault-tolerant current control of a five-phase permanent-magnet motor," IEEE Trans. Ind. Appl., vol. 43, no. 4, pp. 960–970, Jul./Aug. 2007.

[3] Cong X, Tao G, Peng Z and Haiping X, A Fault-Tolerant FOC Strategy for Five-Phase SPMSM With Minimum Torque Ripples in the Full Torque Operation Range Under Double-Phase Open-Circuit Fault, IEEE Transactions on Industrial Electronics, Vol.67, Issue 11, pp 9059 - 9072, November 2020.

[4] A Dieng, M F Benkhoris, M Ait-Ahmed and J C Le Claire, Modeling and Optimal Current Control of Five-Phase PMSG - PWM Rectifier SET Non-Sinusoidal EMF under Open-Circuit Faults, 21st European Conference on Power Electronics and Applications, EPE 2019 ECCE Europe, 8915578, pp. 1-9, 2019

[5] A Dieng, J C Le Claire, M F Benkhoris and M Ait-Ahmed, Torque control strategy for fault –tolerant "five-phase PMSG-PWM rectifier" set for marine current turbine applications, Journal of electrical engineering, Vol 19, N°4, pp. 1-9, 2019,

[6] A Hosseyni, R Trabelsi, M F Mimouni and A Iqbal, Fault tolerant control strategy of a five-phase permanent magnet synchronous motor drive, 2015 16th International Conference on Sciences and Techniques of Automatic Control and Computer Engineering (STA), Tunisie, Dec. 2015, ISBN 978-1-4673-9234-1.

[7] X Kestelyn and E Semail, A vectorial approach for generation of optimal current references for multiphase permanent-magnet synchronous machines in real time, IEEE Trans. Ind. Electron., vol. 58, No. 11, pp. 5057–5065, November 2011.

[8] F Baudart, B. Dehez, F. Labrique and E Matagne, Optimal current waveforms for permanent magnet synchronous machines with any number of phases in open circuit, Mathematics and Computers in Simulation, Elsevier, Vol. 90, pp.1-14, April 2013.

[9] S. Dwari and L. Parsa, Fault-tolerant control of five-phase permanent-magnet motors with trapezoidal back EMF, IEEE Trans. Ind. Electron., vol. 58, no. 2, pp. 476-485, Feb. 2011.

[10] F. Baudart, B. Dehez, E. Matagne, D. Telteu-Nedelcu, P. Alexandre and F. Labrique, Torque control strategy of polyphase permanent-magnet synchronous machines with minimal controller reconfiguration under open-circuit fault of one phase, IEEE Trans. Ind. Electron., vol. 59, no. 6, pp. 2632-2644, Jun. 2012.

[11] J.C. Le Claire, «Power Electronic Converters – PWM Strategies and Current Control Techniques», Chapter 14: «Current Control using Self-Oscillating Currents Controllers», ISTE, London, United Kingdom, 2011, WILEY, Hoboken, USA, 2011, pp 417-447.

[12] J.C. Le Claire, S. Siala, J. Saillard, R. Le Doeuff, «A new Pulse Modulation for Voltage Supply Inverter's Current Control», in 8th European Conference on Power Electronics and Applications, Lausanne, Switzerland, September 1999, CD-ROM ref. ISBN 90-75815-04-2.

Weighting Factor Design for FS-MPC in VSCs: A Brain Emotional Learning-Based Approach

Mohammad Sadegh Orfi Yeganeh
Technical University of Denmark
2800 Kgs. Lyngby, Denmark
morfi@dtu.dk

Arman Oshnoei
Aalborg University
Aalborg, Denmark
aros@et.aau.dk

Saeed Peyghami
Aalborg University
Aalborg, Denmark
sap@energy.aau.dk

Nenad Mijatovic
Technical University of Denmark
2800 Kgs. Lyngby, Denmark
nm@dtu.dk

Tomislav Dragicevic
Technical University of Denmark
2800 Kgs. Lyngby, Denmark
tomdr@dtu.dk

Frede Blaabjerg
Aalborg University
Aalborg, Denmark
fbl@energy.aau.dk

Keywords

Brain emotional learning (BEL) - Finite set model predictive control (FS-MPC) - Total harmonic distortion (THD) - Uninterruptible power supply (UPS).

Abstract

Finite set model predictive control (FS-MPC) has been identified as one of the most favorable controllers for power electronic applications due to its capability over real-time solutions to multiple objectives and constraints. However, the main challenge in the FS-MPC is the choice of appropriate weighting factors in the cost function to reach the best switching state of the inverter. This study proposes an approach based on brain emotional learning (BEL) to provide online tuning of weighting factors in FS-MPC of a power converter, which prevents the dependency of the converter control system on the various uncertainties coming from operating conditions and loading conditions. The proposed BEL approach is fully model-free, indicating that the weighting factors are adjusted without previous knowledge of the system model and parameters. Simulation and experimental results validate the proposed control scheme's effectiveness under different load conditions.

1. Introduction

Among different converter topologies, voltage source converters (VSCs) are the most widely spread in practice. Many advanced control techniques and topologies for VSCs have been proposed over the past years, aiming to mitigate some of the well-known limitations of classical linear control approaches [1] and [2]. Power converters and motor drives play an important role in power electronics technology and various industrial applications such as renewable energy sources, electric vehicles, HVDC transmission systems, and uninterruptible power supply systems (UPS).

Finite control set model predictive control (FS-MPC) has been extensively utilized in both academia and industry in the last decades. This control method bears many advantages such as there is no need for a modulation stage, constraints can be included in the main cost function, the easy inclusion of non-linearities in the model, and low complexity that makes the implementation much easier. In contrary to the conventional control techniques, the FS-MPC is capable of obtaining faster dynamic response and higher frequency bandwidth [3-5].

The performance of FS-MPC is deeply influenced by the weighting factors, the tuning of which is still a challenge to be undertaken [6]. Recently, model-free intelligent controllers such as fuzzy logic and neural network have been developed to decrease the sensitivity to modeling inaccuracy. The main characteristic of intelligent controllers is the model-free design that enables them to manage model non-linearity, complexity, and uncertainty in power electronic applications [7]. In this regard, refs. [8] and [9] have employed an artificial neural network method in the off-line mode for weighting factor design of FS-MPC in UPS and motor drives applications. This method, however, demands many computations to cover all the assortments of the possible coefficients. Also, the conducted analysis for identifying the optimal values of weighting factors is dependent on operating conditions, which may give rise to a flawed performance of the control system. A model based on the emotional learning in the human brain's limbic system was developed in [10]. The brain emotional learning (BEL) model is an effective controller for fast decision-making, particularly in uncertain states. The online learning capacity, minimum computational intricacy, and, most notably, no demand for prior knowledge of system dynamics make the BEL a distinctive controller over other intelligent controllers. Also, it is simple, with fewer tuning parameters in emotional controllers, and it does not need a further iterative procedure for updating parameters or learning [11]. The BEL is growingly being utilized in electrical motors [12], control engineering [13], and intelligent devices [14]. It is illustrated in [10] that the BEL can present more effective solutions than neural networks and fuzzy logic in controlling the synchronous machines in power systems.

Motivated by the previous discussion, this paper proposes a model-free and adaptive approach based on BEL to adjust the weighting factors appearing in the FS-MPC objective function. The proposed BEL-based FS-MPC is applied to a benchmark UPS VSC system. The proposed strategy is totally model-free, suggesting that the weighting factors are tuned without previous knowledge of the parameters and system model. The method offers superior performance in varying operating point conditions. To achieve optimum results with the BEL method working based on the knowledge of the experts, considering desirable scaling factors are the necessary part, which is also addressed in this paper. Both simulation and experimental validations are provided to demonstrate the effectiveness of the proposed control scheme.

2. Finite set model predictive control principle

In this paper, a BEL-based FS-MPC approach is proposed to adjust the voltage and current of the VSC, and reduce the switching losses. Fig. 1 illustrates the BEL-based FS-MPC that is operated on a VSC for UPS application. In this way, a proper control command is obtained based on the prediction from the converter model and a cost objective function. The converter model is a two-level three-phase VSC, in which there are eight switching states in total.

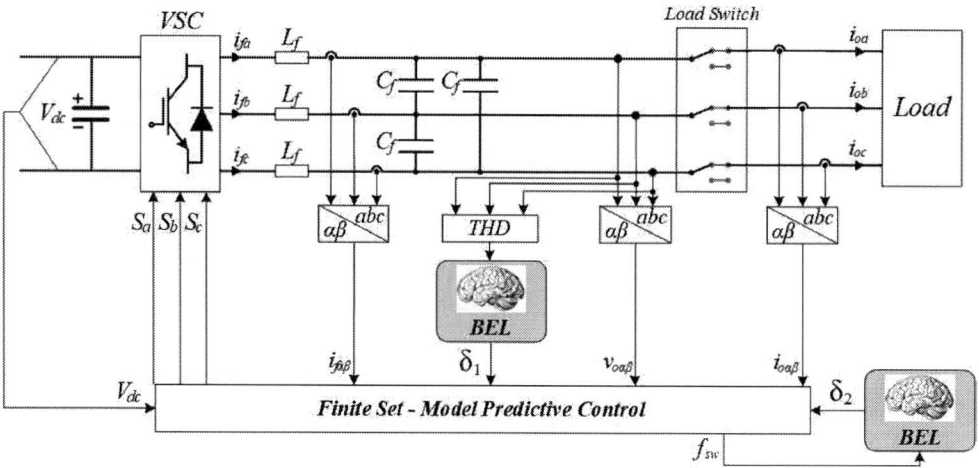

Fig. 1: A schematic of the proposed control structure and power section

To eliminate the harmonics of the output voltage and current, a three-phase LC filter is connected to the load (see Fig. 1). $R_f, C_f,$ and L_f are the resistance, capacitance, and inductance of the LC filter, respectively. The output current (i_o), the filter current (i_f), and output voltage (v_o) are presented in vectors as follows:

$$i_o = [i_{ou} \quad i_{ov} \quad i_{ow}]^T \tag{1}$$

$$i_f = \begin{bmatrix} i_{fu} & i_{fv} & i_{fw} \end{bmatrix}^T \tag{2}$$

$$v_o = \begin{bmatrix} v_{ou} & v_{ov} & v_{ow} \end{bmatrix}^T \tag{3}$$

Three-phase variable vectors are transferred to a two-dimensional vector ($\alpha\beta$ stationary reference frame) by employing the Clarke transformation (T) as follows:

$$T = \frac{1}{3} \begin{bmatrix} 1 & e^{j\frac{2}{3}\pi} & e^{j\frac{4}{3}\pi} \end{bmatrix} \tag{4}$$

Finally, the output voltage and current of the converter can be expressed in the state-space form as follows:

$$\frac{d}{dt} \begin{bmatrix} i_f \\ v_f \end{bmatrix} = A \begin{bmatrix} i_f \\ v_f \end{bmatrix} + B \begin{bmatrix} v_i \\ i_o \end{bmatrix} \tag{5}$$

$$A = \begin{bmatrix} -\dfrac{R_f}{L_f} & -\dfrac{1}{L_f} \\ \dfrac{1}{C_f} & 0 \end{bmatrix} \tag{6}$$

$$B = \begin{bmatrix} \dfrac{1}{L_f} & 0 \\ 0 & -\dfrac{1}{C_f} \end{bmatrix} \tag{7}$$

where A is the system matrix, B is the control matrix, and v_i is the input voltage.

The MPC control technique works based on predicting v_f and i_f, and then applying a proper magnitude for v_i in the objective function. Therefore, the main objective function consists of the prediction error (v_e) with a weighting factor (δ_1), the current limitation ($\mathcal{E}_{lim}$), the number of switching efforts (SW) with a weighting factor (δ_2), and a minimizer for the voltage derivative (v_{reg}), which is given as below

$$v_e(k) = v_f(k+1) - v_{ref}(k) \tag{8}$$

$$\mathcal{E}_{lim}(k) = \begin{cases} 0, & if\ |i_f(k)| \leq i_{max} \\ \infty, & if\ |i_f(k)| > i_{max} \end{cases} \tag{9}$$

$$SW(k) = \sum |u(k) - u(k-1)| \tag{10}$$

$$v_{reg}(k) = (C_f.w_{ref}.v_{f\beta}(k+1) - i_{f\alpha} + i_{o\alpha})^2 + (C_f.w_{ref}.v_{f\alpha}(k+1) - i_{f\alpha} + i_{o\beta})^2 \tag{11}$$

$$CF: \delta_1.v_e(k) + \mathcal{E}_{lim}(k) + \delta_2.SW(k) + v_{reg}(k) \tag{12}$$

$$f_{sw} = \sum_{k=1}^{1/T_s} \frac{|\Delta S_a(k)| + |\Delta S_b(k)| + |\Delta S_c(k)|}{6} \tag{13}$$

where v_{ref} and w_{ref} are the voltage and frequency of the reference signal, respectively. As it can be seen, the system performance is highly influenced by the weighting factors (δ_1 and δ_2), which should be adjusted optimally. In this paper, the BEL method is proposed to adjust the weighting factors and thereby improve the performance of the converter control system. Performance evaluation criteria include the converter's switching frequency and the total harmonic distortion (THD) of the output voltage. The design process of the BEL-based regulation scheme is discussed in the next section.

3. BEL-based regulation scheme

The BEL is used as a model-free method in a range of control engineering applications. The BEL can learn quick-auto, and thus it is proper for robust and fast decision-making in nonlinear systems, especially in systems with uncertainty. In [15], the BEL controller has superior performance compared with PI and fuzzy logic controllers in both online and offline simulations for PMSM drive systems in different test conditions. This method is constituted by the Amygdala, which is in charge of emotional learning; the orbitofrontal cortex, sensory cortex, and Thalamus [10]. The model is provided with two inputs including sensory input (*SI*) and emotional signal (*ES*). Preprocessing of SI signals such as filtering or noise reduction is performed by the Thalamus. The sensory cortex receives the Thalamus output and then submits it to the Amygdala and Orbitofrontal cortex. A simplified structure of the BEL used in this work is depicted in Fig. 2. *A* and *O* networks, respectively, express the functional blocks associated with the amygdala and orbitofrontal cortex. The BEL output is obtained based on the subtraction of network *A* and network *O* outputs as follows.

$$M = \sum_z A_z - \sum_z O_z \tag{14}$$

The output of A network is computed as

$$A_{th} = \max(S_z) \tag{15}$$

$$A_z = S_z G_z \tag{16}$$

where A_{th} is a neuron that receives maximum sensory signals from the thalamus directly. Similarly, the output of the OC network is given by:

$$O_z = S_z H_z \tag{17}$$

The weights in AD and OC networks are updated by using (18) and (19)

$$\Delta G_z = \alpha[S_z \max(0, ES - \sum_z A_z)] \tag{18}$$

$$\Delta H_z = \gamma[S_z (\sum_z O_z - ES)] \tag{19}$$

where α and γ are BEL's learning rates. The schematic diagram of the proposed BEL-based regulation scheme is depicted in Fig. 3. The critical functionality of this structure is to minimize the THD and switching frequency. The BEL output is supplementary regulation coefficients to update the weighting factors of the FC-MPC. This adaptive feature improves the converter's robustness against different uncertainties and a vast range of working states. Since the FS-MPC includes two weighting factors, thus two BEL structures are designed. To attain favorable performance of the BEL-based approach, constituting an empirical relation between SI, ES, and output (δ_1 and δ_2) is essential. The *SI* and *ES* inputs for the first BEL to find the optimal value of δ_1 are (20) and (21), respectively.

$$SI = \mu_1 THD + \mu_2 \int THD \, dt \tag{20}$$

$$ES = \mu_3 THD + \mu_4 \int THD dt + \mu_4 \delta_1 \tag{21}$$

Similarly, for the second BEL to find the optimal value of δ_2, the *SI* and *ES* inputs are written as follows:

$$SI = \lambda_1 f_{sw} + \lambda_2 \int f_{sw} \, dt \tag{22}$$

$$ES = \lambda_3 f_{sw} + \lambda_4 \int f_{sw} dt + \lambda_4 \delta_1 \tag{23}$$

The weighting coefficients appearing in (20), (21), (22), and (23) are determined through a trial and error process by simulations. The functions *SI* and *ES* are chosen as the outputs of a PI block in response to the THD and f_{sw} signals. In other words, those weights are the same as proportional and integral coefficients appearing in the PI controller.

Two scaling coefficients (SCs) are added to the coordinator body to normalize the outputs. A particle swarm optimization algorithm tunes the SCs by minimizing the following performance index.

Performance index:

$$\min F = \int_{t=0}^{T_s} t(\Delta v_o{}^2)dt \tag{24}$$

Decision variables:

$$SF_{Ai,min} \leq SF_{Ai} \leq SF_{Ai,max} \tag{25}$$

$$SF_{Pi,min} \leq SF_{Pi} \leq SF_{Pi,max} \tag{26}$$

where Δv_o represents the output voltage deviation; and T_s is the time length of Δv_o. As (24) implies, the integral of time multiplied by squared error (ITSE) is used to obtain the optimal solution.

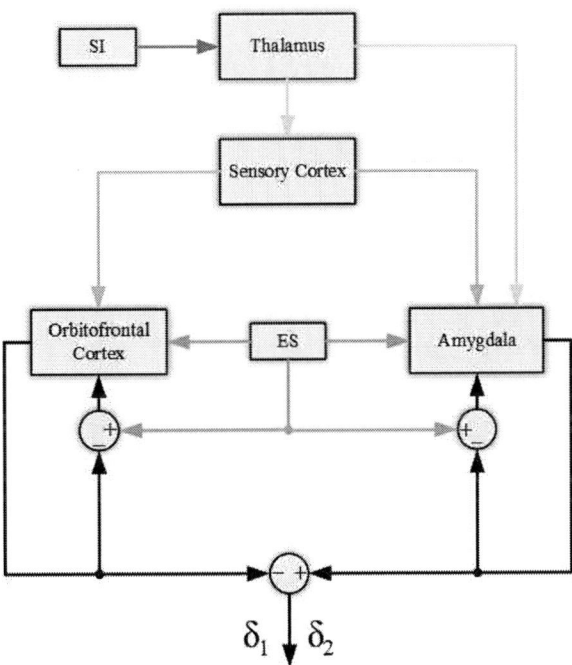

Fig. 2: Structure of BEL approach

4. Simulation and experimental results

The proposed weighting factor design strategy has been verified experimentally with a test system. As mentioned in Fig. 1, the experimental setup consists of a full-bridge three-phase VSC (SEMITEACH IGBT, 20kW), and a DC power supply (Delta Elektronika SM1500-CP-30), an LC filter, impedances, measurements, and resistive loads are employed and illustrated in Fig. 3. In addition, a dSpace MicroLab Box DS-1202 is used in the control section. Details of the utilized parameters are provided in Table I.

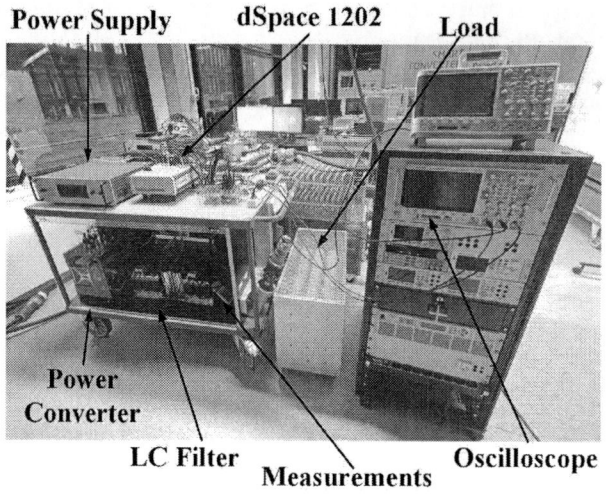

Fig. 3: A photograph of the experimental motor drive test setup

Table I. Parameters of the test system

Component Parameters		
Parameters	Symbol	Value
DC power supply	V_{dc}	260 V
Nominal voltage magnitude	V_{ref}	100 V
Nominal frequency	f_{ref}	50 Hz
Sampling time	T_s	25 μs
Capacitance of LC filter	C_f	10 μF
Inductance of LC filter	L_f	2.2 mH
Load 1	R_{Load1}	58 Ω
Load 2	R_{Load2}	116 Ω
Load 3	R_{Load3}	12 Ω
Control Parameters		
Switching Frequency Section (BEL Parameters)	α	2
	β	-1
THD Section (BEL Parameters)	α	1
	β	1

Based on standard (IEC 62040-3), the UPS should follow different tests on performance, components, and main parts [16]. To verify the dynamic performance of the proposed control technique, a two-step load change has been carried out on the UPS setup. Therefore, the controller has been tested for three different resistive loads. In this way, for the first test, at $t = 2.0$ s, load 1 (58 Ω) is paralleled with load 2 (116 Ω), thus the equal resistance is around 39 Ω. According to Fig. 4 (a), the FS-MPC controller can perform properly and control the voltage magnitude by well tracking the reference value. The current waveform is presented in Fig. 4 (b), and it has been increased due to impedance reduction on the load side. Fig. 4 (a) and (b) present a fast and proper performance of the proposed control strategy. The magnitudes of THD and switching frequency of the proposed control technique are improved by utilizing the BEL controller, which can be seen in a comparison with Table III in [9]. Another achievement of the proposed solution of this study is designing the weighting factors in an online approach, which can overcome the defects of the state-of-the-art ([8] and [9]). Based on Fig. 4 (c) and (d), the optimal value of the switching frequency and THD magnitudes are floating and they are around 6.3 kHz and 1.2% respectively. By utilizing the mentioned values of the control parameters in Table I, the weighting factors of the cost functions for the switching frequency and THD are obtained and presented in Fig. 4 (e) and (f). The value of the switching frequency weighting factor is around 4.8 most of the time, and the value of the THD weighting factor has a variation around 1.1.

For the second test, at $t = 2.0$ s, load 3 (12 Ω) is paralleled with load 2 (116 Ω), thus the equal resistance is around 11 Ω. According to Fig. 5 (a) and (b), the FS-MPC controller can perform properly and control the voltage and current by employing the BEL controller for the weighting factor design. In this test, the load variations are higher than in the first test, so there are more changes in both weighting factors at $t = 2.0$ s in comparison with the previous test. The figures also reveal that the BEL-based method regulates the weighting coefficients such that a stable operation of the VSC is achieved.

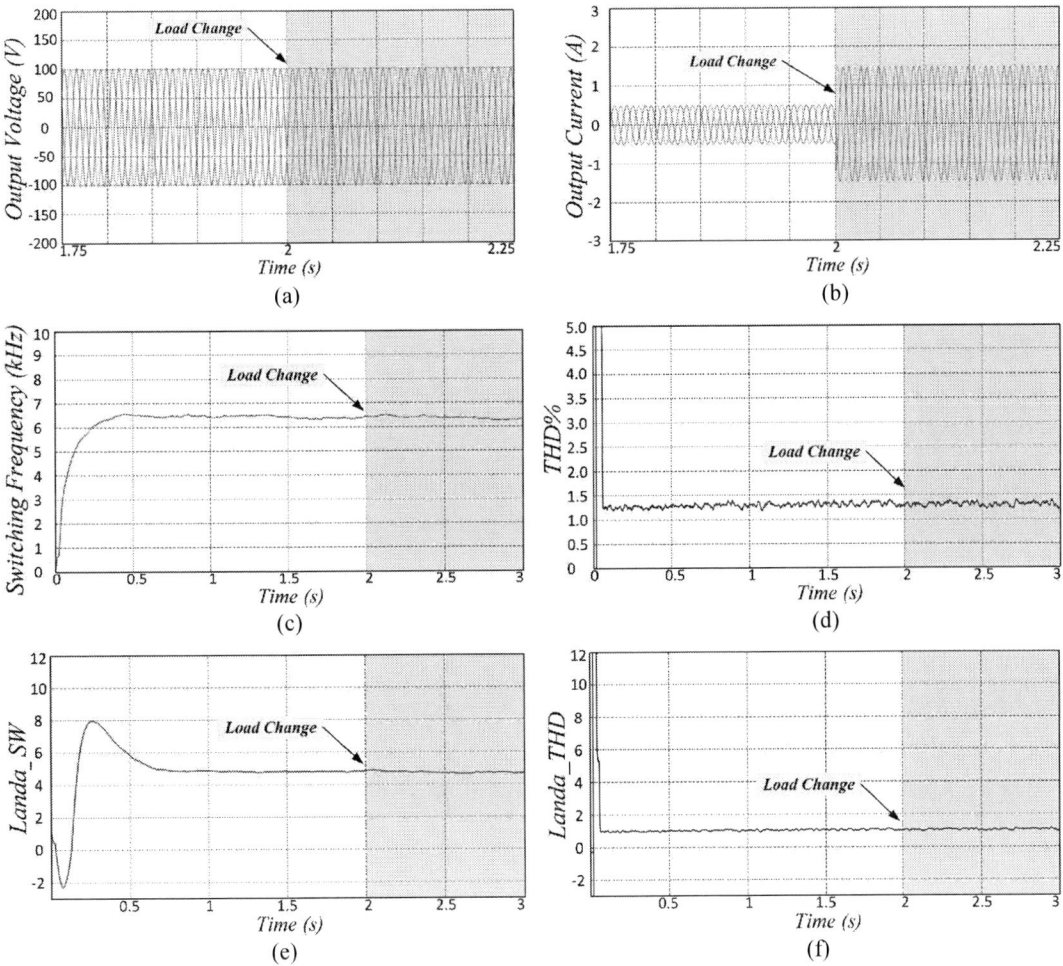

Fig. 4: Simulation results of the specified first test, (a) Output voltage, (b) Output current, (c) Switching frequency, (d) Total harmonic distortion of the output voltage, (e) Switching frequency weighting factor (δ_1), and (f) THD weighting factor (δ_2)

The proposed control strategy's robust and optimal design of the weighting factors has been demonstrated experimentally for the first test in Fig. 6 (a) and (b). The experimental results of the power converter's output voltage and current validate the obtained simulation results.

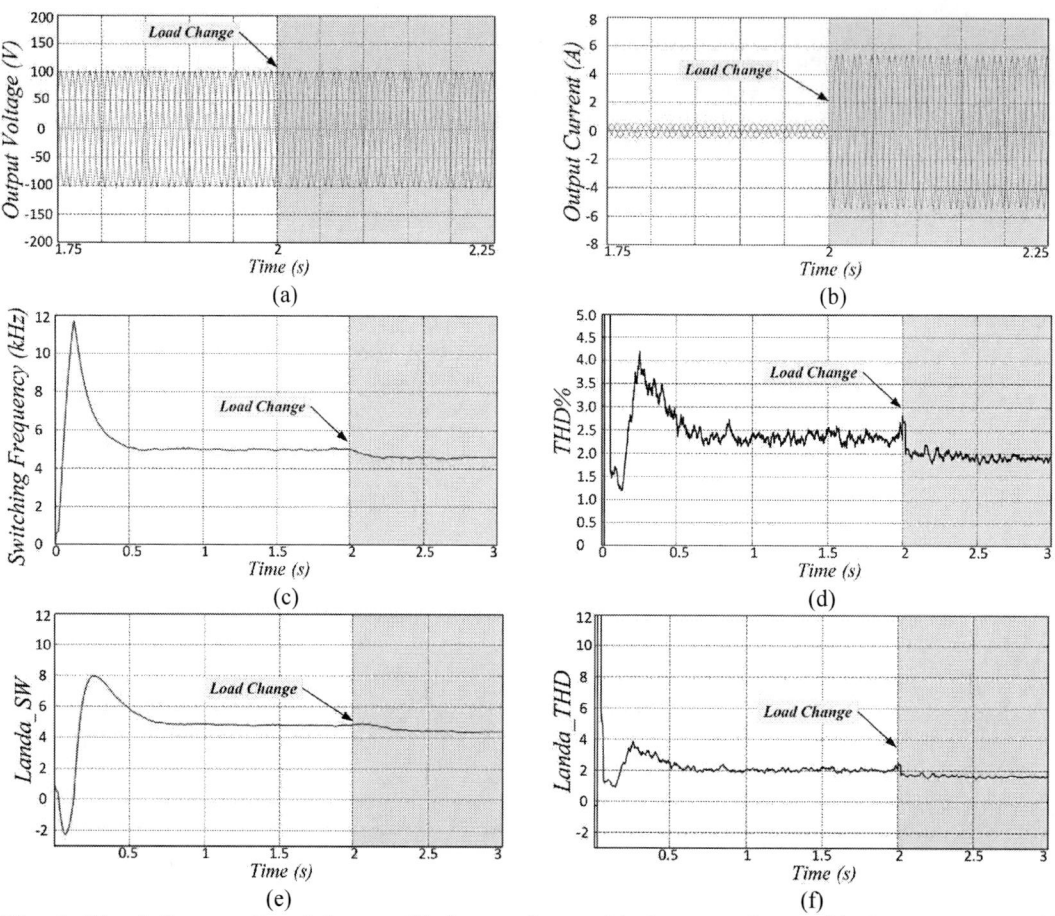

Fig. 5: Simulation results of the specified second test, (a) Output voltage, (b) Output current, (c) Switching frequency, (d) Total harmonic distortion of the output voltage, (e) Switching frequency weighting factor (δ_1), and (f) THD weighting factor (δ_2).

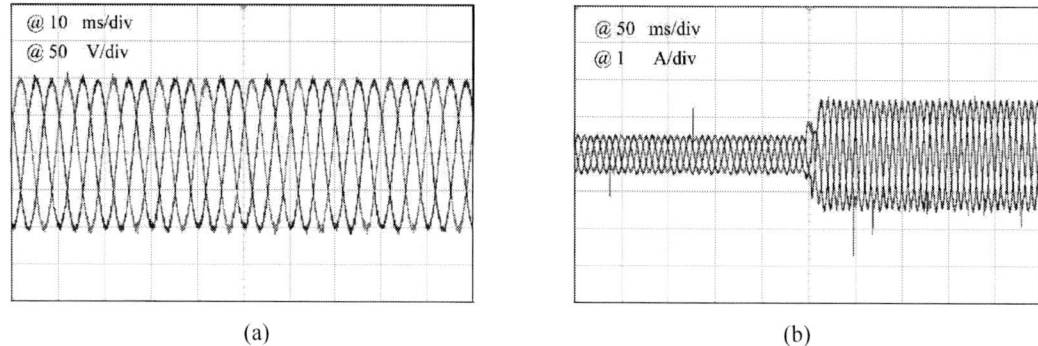

Fig. 6: Experimental results of the specified first test, (a) Output voltage, and (b) Output current of the power converter

Conclusion

In this study, a real-time solution by employing brain emotional learning method was proposed to design the weighting factors of the FS-MPC. The approach's effectiveness was illustrated on a UPS VSC setup. Minimizing the switching frequency and THD were the main objectives behind the BEL-based FS-MPC method. The proposed method's key features are the online learning capacity, minimum computational complexity, and there is no need for prior knowledge of the VSC dynamics. Finally, the simulation and experimental results demonstrated the effectiveness of the proposed strategy under different operational conditions.

References

[1] Yeganeh, M.S.O., Davari, P., Chub, A., Mijatovic, N., Dragicevic, T. and Blaabjerg, F., "A Single-Phase Reduced Component Count Asymmetrical Multilevel Inverter Topology," IEEE Journal of Emerging and Selected Topics in Power Electronics, 9(6), pp.6780-6790, 2021.

[2] Heydari, R., Khayat, Y., Amiri, A., Dragicevic, T., Shafiee, Q., Popovski, P. and Blaabjerg, F., "Robust high-rate secondary control of microgrids with mitigation of communication impairments," IEEE Transactions on Power Electronics, 35(11), pp.12486-12496, 2020.

[3] Khayat, Y., Heydari, R., Naderi, M., Dragicevic, T., Shafiee, Q., Fathi, M., Bevrani, H. and Blaabjerg, F., "Decentralized frequency control of AC microgrids: an estimation-based consensus approach," IEEE Journal of Emerging and Selected Topics in Power Electronics, 9(5), pp.5183-5191, 2020.

[4] Yeganeh, M.S.O., Mijatovic, N. and Dragicevic, T., "Dynamic Performance Optimization of Single-Phase Inverter based on Model Predictive Control," In 2021 IEEE International Conference on Predictive Control of Electrical Drives and Power Electronics (PRECEDE) (pp. 235-240). IEEE, 2021.

[5] Mardani, M.M., Mijatovic, N. and Dragicevic, T., "Optimal Model Predictive Controller for Grid-Connected Voltage Source Converters," In 31st IEEE International Symposium on Industrial Electronics. IEEE., 2021.

[6] Ding, D., Yeganeh, M.S.O., Mijatovic, N., Wang, G. and Dragicevic, T., "Model Predictive Control on Three-Phase Converter for PMSM Drives with a Small DC-link Capacitor," In 2021 IEEE International Conference on Predictive Control of Electrical Drives and Power Electronics (PRECEDE) (pp. 224-228). IEEE, 2021.

[7] Oshnoei, A., Sadeghian, O., Mohammadi-Ivatloo, B., Blaabjerg, F. and Anvari-Moghaddam, A., "Data-Driven Coordinated Control of AVR and PSS in Power Systems: A Deep Reinforcement Learning Method," In 2021 IEEE International Conference on Environment and Electrical Engineering and 2021 IEEE Industrial and Commercial Power Systems Europe (EEEIC/I&CPS Europe) (pp. 1-6). IEEE, 2021.

[8] Novak, M., Xie, H., Dragicevic, T., Wang, F., Rodriguez, J. and Blaabjerg, F., "Optimal cost function parameter design in predictive torque control (PTC) using artificial neural networks (ANN)," IEEE Transactions on Industrial Electronics, 68(8), pp.7309-7319, 2020.

[9] Dragičević, T. and Novak, M., "Weighting factor design in model predictive control of power electronic converters: An artificial neural network approach," IEEE Transactions on Industrial Electronics, 66(11), pp.8870-8880, 2018.

[10] Khezri, R., Oshnoei, A., Yazdani, A. and Mahmoudi, A., "Intelligent coordinators for automatic voltage regulator and power system stabiliser in a multi-machine power system," IET Generation, Transmission & Distribution, 14(23), pp.5480-5490, 2020.

[11] Saeed, M.U., Sun, Z. and Elias, S., "Semi-active vibration control of building structure by Self Tuned Brain Emotional Learning Based Intelligent Controller," Journal of Building Engineering, 46, p.103664, 2022.

[12] Yazdani, A.M., Mahmoudi, A., Movahed, M.A., Ghanooni, P., Mahmoudzadeh, S. and Buyamin, S., "Intelligent speed control of hybrid stepper motor considering model uncertainty using brain emotional learning," Canadian Journal of Electrical and Computer Engineering, 41(2), pp.95-104, 2018.

[13] Hsu, C.F. and Lee, T.T., "Emotional fuzzy sliding-mode control for unknown nonlinear systems. International Journal of Fuzzy Systems," 19(3), pp.942-953, 2017.

[14] Iranpour, E. and Sharifian, S., "An FPGA implemented brain emotional learning intelligent admission controller for SaaS cloud servers," Transactions of the Institute of Measurement and Control, 39(10), pp.1522-1536, 2017.

[15] Qutubuddin, M. D., and Narri Yadaiah. "Modeling and implementation of brain emotional controller for permanent magnet synchronous motor drive," Engineering Applications of Artificial Intelligence 60, pp.193-203, 2017.

[16] "IEC-International Electrotechnical Commission," IEC 62040-2, 2016.

A Strategy for Smooth Microgrid Transitions without Phase Misalignment and Voltage Mismatch

Gabriel Silva Rocha[1], Amiron Wolff dos Santos Serra[1], Cesar Augusto Santana Castelo Branco[1], Hercules Araujo Oliveira[1], Jose Gomes de Matos[1], and Luiz Antonio de Souza Ribeiro[1]

[1]Institute of Electrical Energy – IEE
Federal University of Maranhao – UFMA
Sao Luiz, Brazil
iee.ufma@gmail.com

Acknowledgements

The authors would like to thank the support and motivation provided by Federal University of Maranhao (UFMA) and Equatorial Energia.

Keywords

«Microgrid», «Seamless transfer», «Synchronization», «Grid-forming converters», «Voltage Source Converters (VSCs)»

Abstract

In this work, it is proposed a modified control structure to provide smooth transitions of control modes of Grid-forming Converter without phase misalignment and voltage mismatch. A background of the traditional structure is provided and a carefully description of the method is introduced. At the results, a comparison with methods in literature is carried out and it was possible to prove that with the correct phase and voltage compensations, one can almost eliminate the transients in the grid voltage due to the switching of control modes of the Grid-forming Converter and transition of the MG operating mode.

Introduction

The integration of distributed energy resources (DER) with power electronics, energy storage system (ESS), and local loads is starting be widespread due to the diffusion of Microgrids (MG) in today's power systems. The control and management strategy of a MG depends on its objectives and modes of operations. A MG can operate either connected or disconnected from the utility grid in order to enable the grid-connected or islanded modes. The decision about which operating mode is driven by innumerous reasons, such as economic and environment issues, reliability, power quality, among others. In grid-connected mode, frequency and voltage magnitude are regulated by the main grid and the MG is responsible for the generation and consumption management. In islanded mode, the MG should be able to generate enough active and reactive power to meet local demands, frequency and voltage control, and operate within specified limits regulated by international standards.

Since a MG can operate either connected or disconnected from the utility grid, it needs to transit between the operating modes. The integration of MG at the power system started raising the questions about how secure and reliable the transitions can be to guarantee its continuity of operation. The transition between the operating modes of the MG can happen intentionally, where the islanding can be planned and controlled, producing small transients in the continuity of the MG islanded operation, or unintentionally, when islanding occurs without any predictability, making it impossible to perform previous adjustments to the MG, causing severe transients and hindering the success of its continued operation in islanded mode [1].

According to IEEE 2030-7 standard [2], the implementation approach for the transition from grid-connected to islanded mode is left to the MG designer or operator. For synchronization and grid-connected operation, IEEE 1538-1 standard stablish the limits of voltage, frequency, and phase mismatch for the reconnection at the PCC, but they lack the methodology to cover the synchronization process. During the transitions of MG, two main problems may occur: a phase misalignment resulted by the sudden change of the reference frame in the transformations of the control structure and the transients caused by the switching of the operating modes of the VSI. Although the change from grid-forming to grid-feeding operation of the inverter could be a problem, the main contribution to the transients in the MG during the transition process is due to the phase misalignment and voltage mismatch [3].

In literature, there are several methods used provide smooth transitions of MG. In Arafat [4], a dispatch unit has been utilized, which is responsible for compensation of power variations at the islanding and synchronization by adjusting its output power. In Pilehvar [5], a Smart Inverter is proposed, where the term 'smart' is due to a very carefully switching of phase during transitions modes of a droop-controlled inverter to avoid phase misalignment, large voltage, and power oscillations. In Talapur [6] , a modified PLL loop for a grid-forming converter has been introduced to avoid the aforementioned issues at the islanding, but it increased the complexity of the control structure due to several switching of modes in the structure. Alves [7] proposed an improved synchronization loop with maximum mismatch variation per unit, which could provide a smooth transition from islanded to grid-connected operation. However, the islanding was not studied, and the loop tends to fail at the worst-case scenario of synchronization. Wang [3] proposed a method to estimate the difference phase angle between the grid and VSI reference, but it was only used to synchronization and intentional islanding. In Tran [8] it was proposed a modified PLL structure that was possible to work as a phase detector (PD), providing references of phase during islanding and grid-connected operation without mismatch. However, the unintentional islanding was not investigated, and a smooth synchronization loop was not proposed.

Most of VSI in MG utilize PLLs for grid synchronization, but PLL are known to possess a poor dynamic performance, and a difficult and time-consuming tunning process, which commonly leads to large frequency transients and phase jumps [9] [10]. Besides, a constant voltage excitation for correct phase estimation is another drawback of PLLs. To overcome the problems of PLL, a windowing factor scheme has been proposed in [9] for grid synchronization. However, the estimation strategy is very complex and the tunning process was not cited. Recent trends in MG control with respect to seamless transition have been towards developing faster and advanced PLL or PLL-less solutions to attain grid synchronization [11].

Therefore, a good strategy and procedure for MG transition is essential for traditional controls to avoid large transients and provide smooth transition between the modes of operations of a MG. In this paper it is proposed a modified control structure to provide smooth transitions of modes of MG without high frequency variations, phase misalignment and voltage amplitude mismatch.

VSI Control Structure

The notion of control is central in MGs, since this is what distinguishes a MG from a distribution system with DER, so that they appear to the main grid as a controlled and coordinated unit [12]. The correct operation of the controls of the VSI is necessary for a proper operation the entire MG. In a MG, DERs with a dispatchable energy source, such a battery energy storage system (BESS), use different inverter operation modes depending upon the MG state. The operating mode of the MG determine whether the converter is working as a grid-forming or grid-feeding converter. In grid connected mode, the inverter operates in grid-feeding mode, controlling the active and reactive power that the primary source delivers to the grid, called PQ control. In islanded mode, the inverter operates in grid-forming mode, sharing power with other inverters and controlling the voltage and frequency of the MG, called Vf control. Fig. 1 illustrates the block diagram of the traditional control structure of a VSI that can operate in both modes: grid-forming and grid-feeding.

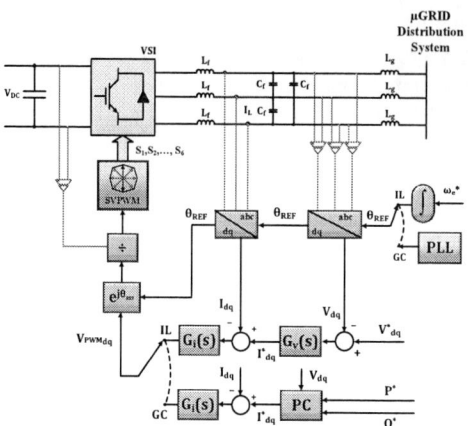

Fig. 1: Traditional control structure of a grid-forming converter with dual mode of operation (adapted from [8]).

The references of power and voltage can be provided by local controllers. The voltage controller (VC), represented by the block G_v and current controller (CC), represented by the block G_i, in Fig. 1 are implemented in the synchronous reference frame (SRF) and are Transfer Functions (TF) of Proportional-Integrative (PI) controllers. A power controller (PC) is added for current reference generation based on the references of power at grid-feeding operation.

There are three main problems with the structure provided in Fig. 1. First, under grid-feeding operation, the reference of phase used by the dq transform to the SRF is provided by the main grid, usually estimated by a Phase Locked Loop (PLL), while under grid-forming operation, the reference is generated by an internal oscillator in the VSI. The use of an internal generator is due to security reasons, since PLL needs voltage excitation for correct generation of phase, which is not possible during a grid shutdown. In the transition from grid-feeding to grid-forming operation, a sudden change of phase reference (θ_{REF}) in the VSI switching operation is a big problem faced during the islanding of microgrids due to the high transients of voltage, frequency and current even with no power flow between grid and MG. Second, at the time of reconnection, the lack of smooth method for synchronization of the grid and MG phases tends to provide high frequency fluctuations in the system. Third, the VSIs are programmed to control the voltage of the PCC, whereas for a smooth MG reconnection, a method to compensate the voltage drop along the MG distribution system, as illustrated in the Fig. 2, needs to be devised in order to avoid voltage distortion along the process and inrush of currents during the reconnection.

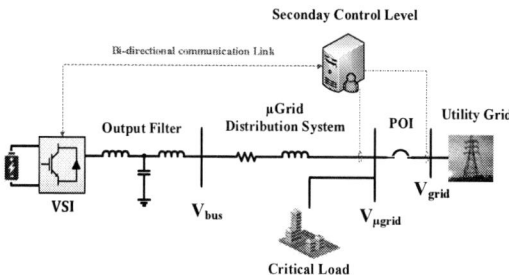

Fig. 2: Typical infrastructure of a Microgrid.

More details of the traditional structure of VC and CC presented in the Fig. 1 can be found in [13]. At the Fig. 2, the interface between the MG and the utility grid is called Point of Interconnection (POI).

Phase estimation and Synchronization Algorithm

The smooth transition between the operation modes of the converter can be summarized on the aim to avoid sudden change of operating point of the VSI. To accomplish this task, the phase reference for the transformations should not change during this process. Therefore, the difference phase angle between the grid voltage and the VSI output needs to be calculated and compensate in the internal reference of the VSI. Other techniques, described in Tran [8] and Alves [7], tried to smoothly change the references of the transformations by integrating the angular frequency over time, slightly increasing or decreasing the angular frequency to avoid sudden change of reference and make the phases match each other. In this work, it is proposed the adjusting of the difference phase angle ($\Delta\theta$) to make the phases match and θ_{REF} can be calculate by (1).

$$\theta_{REF} = \theta_{VSI} + \Delta\theta \tag{1}$$

Where θ_{REF} is the phase reference of the MG and θ_{VSI} is the internal reference of the grid-forming power converter.

If the grid voltage (V_{grid}) and VSI voltage reference (V_{VSI}) vectors rotate at the same angular frequency, they can be treated as stationary vectors. The vector cross product is determined by (2).

$$|V_{grid} \times V_{VSI}| = |V_{grid}||V_{VSI}| \sin \varepsilon_\theta = V_{grid_\alpha} V_{VSI_\beta} - V_{VSI_\alpha} V_{grid_\beta} \tag{2}$$

Where the phase error (ε_θ) between the grid phase (θ_{grid}) and the reference of phase (θ_{REF}) of the MG is given in (3).

$$\varepsilon_\theta = \theta_{grid} - \theta_{REF} \tag{3}$$

Where the α, β subscripts mean the alpha and beta components of the voltage vectors. Therefore, from [14], it is possible to derive the expression of the sine of the phase angle difference between the grid and the inverter voltage vectors. Rearranging the terms, it is possible to obtain (4).

$$|V_{grid}| \sin \varepsilon_\theta = V_{grid_\alpha} \frac{V_{VSI_\beta}}{|V_{VSI}|} - \frac{V_{VSI_\alpha}}{|V_{VSI}|} V_{grid_\beta} \tag{4}$$

From the small signal analysis, the $\sin \varepsilon_\theta \cong \varepsilon_\theta$ if $\varepsilon_\theta \cong 0$. Therefore, (4) can be linearized for small difference angles and be solved continuously, compensating the error with PI controllers. At the steady state, the difference between phase reference of the MG and main grid reference should be zero. The normalized components of the VSI voltage in (4) is obtained by the sine and cosine components of the phase reference. A drawback of this equation is that it only holds for balanced and undistorted grid conditions, since under these circumstances the resulting voltage vector does no vary with amplitude and frequency. To eliminate influence of negative sequences components, the positive sequence component extractor (PSCE) [14] can be used to filter it out of the $\alpha\beta$ components of the grid voltage. The proposed Phase Angle Estimation Block (PAEB) is illustrated in Fig. 3.

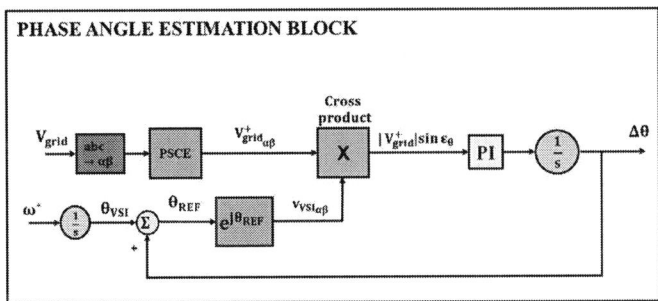

Fig. 3: Control structure of the phase angle detection algorithm.
The closed loop TF is expressed in (5).

$$\frac{\Delta\theta}{\theta_{grid}} = \frac{2\zeta\omega_n s + \omega_n^2}{s^2 + 2\zeta\omega_n s + \omega_n^2} \tag{5}$$

Where, $\omega_n = \sqrt{K_I \times |V_{grid}^+|}$ and $\zeta = (K_p/2)\sqrt{|V_{grid}^+|/K_I}$.

The block resembles the heterodyning process utilized to estimate rotor position in induction machines [15]. For synchronization, an additional loop is required in order to provide smooth variation of the difference phase angle.

Instead of estimating the phase angle at each time, the block estimates the difference phase angle between the grid voltage and the internal reference of the power converter. The advantage of the proposed block is that one can have more control over the phase, which is a mandatory requirement for low-inertia power converters, and avoids discontinuous difference phase angle ($\theta_{grid} - \theta_{REF}$) at the transitions. The integrator in the structure has the ability of memory. Therefore, at the time of islanding, the last value of the phase angle can be used by the VSI to avoid a discontinuity in phase and, hence, a sudden change of reference in the transformations of the control structure. At the time of synchronization, the phase angle of the block can be smoothly varied to avoid high frequency variation in the MG. In Fig. 4, it is illustrated the proposed smooth synchronization unit (SSU) to provide seamless reconnection of the MG.

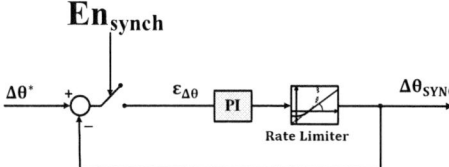

Fig. 4: Control structure of the smooth synchronization algorithm.

The main objective of the SSU is limit the first derivative of the difference phase angle from the PAEB, according to (6), to provide a safe reconnection, avoiding sudden change of reference and high frequency fluctuations in the system.

$$\left|\frac{d(\Delta\theta)}{dt}\right| \leq \lambda \tag{6}$$

Where λ is the maximum variation of the frequency in the MG during the synchronizing process, in rad/s. Limiting the module of the first derivative means to define a maximum variation of frequency (λ) around an operating point, typically 50 or 60 Hz. To guarantee the allowable range of frequency, the IEEE 1547-3 stablishes the variation of ± 1.2 Hz as a continuous operation of the MG. Therefore, the synchronization is carried out without great frequency changes to the system. For the correct operation of the smooth synchronization loop, proper tunning needs to be done. The proportional gain, K_p, is defined as (7).

$$K_p = \frac{\Delta\theta}{e_{ss}} \tag{7}$$

In order to remove steady-state error, e_{ss}, integral gain should be employed and it is evaluated by (8). The value of e_{ss} should be small, typically, in order of 0.1.

$$K_i = \frac{2\lambda}{\Delta\theta} \tag{8}$$

The proportional and integral gains are adaptive. Before the synchronization starts, the gains are calculated based on the estimated $\Delta\theta$ from PAEB, after that the gains are kept constant and the process initiates.

By using these gains in (7) and (8), the block implements a trapezoidal variation of the frequency for synchronization, which minimize variation and abrupt frequency changes over critical loads. Also, this method possesses a simpler and practical implementation than the one presented in [16].

Control Structure

For correct synchronization and islanding process of the MG, hierarchical control needs to be employed, which, in turns, refers to secondary control level. The secondary control level seeks to manage different operating modes of the MG and prescribe the compensations signals (ΔV and $\Delta\theta$) for primary control level in order to improve the dynamic response of the MG [17]. In Fig. 5 (a), it is shown the controllers used by the secondary control to correctly manage the operation of the MG.

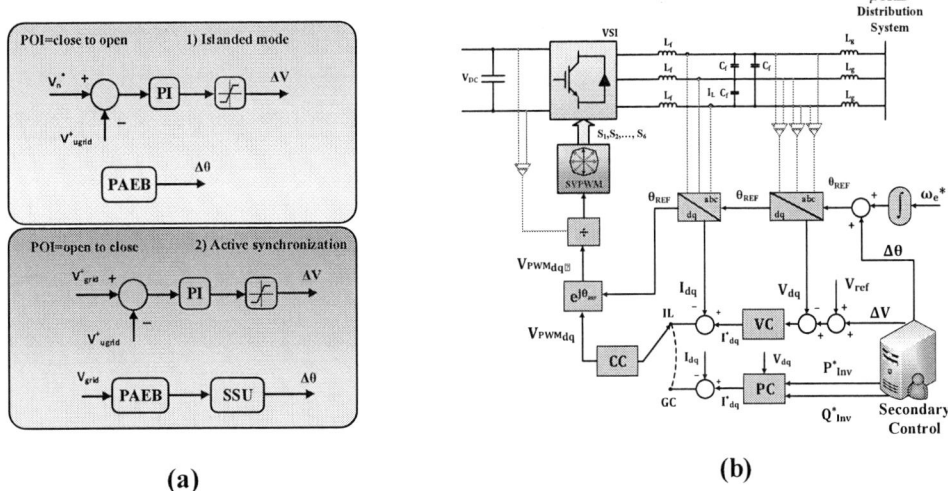

(a)   **(b)**

Fig. 5: (a) Secondary control structure of synchronization and islanding algorithm; (b) Modified control structure in the grid-forming power converter.

At islanded mode, the secondary control is responsible for voltage amplitude correction to the nominal value and the PAEB compensates the phase in order to avoid misalignment during the transitions and switching of mode of the MG and VSI. During the transition from islanded to grid-connected, the SSU, the PAEB and a smooth voltage controller is employed to smoothly vary phase and voltage amplitude of the MG to eliminates high frequency variation and voltage distortion.

The structure of the controller for the grid-forming converter is illustrated in Fig. 5 (b) and possesses the following modification: generation of phase is internal of VSI and is only compensated by the secondary control for operation synchronized to the grid; the voltage reference is compensated by the secondary control based on measurements at the PCC in order to compensate voltage drops along the system; to avoid misalignment of the VSI voltage reference (V_{VSI}) and voltage sags and swells during the switching of operation of the power converter, only the current reference is changed, which is provided by the VC, at islanded operation, while, at the grid-connected mode, by the PC.

The method employed for the PC was the Balanced Positive Sequence Control (BPSC) [18], which provides balanced and positive sequence components to the output current, thus, contributing to the MG power quality. However, this method comes at the cost of a greater oscillatory output power during grid-fault conditions compared with others power control methods [19]. The VC structure was the same implemented in Fig. 5 (b) with feedforward of the current capacitor [20] and state-feedback decoupling [21] to improve MG stability during islanding, specially, unintentional.

Performance Evaluation and Analysis

In order to evaluate the performance of the proposed controllers, comparison with others method in literature were carried out and also the worst-case scenarios, either for synchronism (180 degrees of displacement) or islanding (unintentional), were considered. The system considered for analysis is the same illustrated in Fig. 2, where the parameters are summarized in Table I.

Table I: Parameters for the MG system

Description	Simbology	Values
Current Controller Gains	K_{pc}, K_{ic}	$11.3, 2.63 \times 10^3$
Voltage Controller Gains	K_{pv}, K_{iv}	$0.0157, 0$
Hierarchical voltage control gains	K_{phc}, K_{ihc}	$0, 25$
Filter parameters	L_f, C_f, L_g	$1.8\ mH, 25\ \mu F, 1.8\ mH$
Frequency	f_e	$60\ Hz$

Maximum frequency variation	λ	7.54 rad/s
Rated RMS line-to-line Voltage	V_{line}	380 V

Microgrid reconnection

In this section, the proposed synchronization algorithm is compared to one used in literature [16], illustrated in Fig. 6 (a). A frequently method of MG synchronization utilized in literature is based on sine of phase error between the grid and MG reference, $\sin(\theta_{grid} - \theta_{REF})$. The sine is calculated, since the difference of phases provided is a discontinuous and oscillating function, therefore smooth synchronization cannot be done with it. This method attempts to compensate a difference of phases by minimizing the sine. However, for a 180 degrees of phase difference, no proper synchronization is carried out, as shown in Fig. 6 (b).

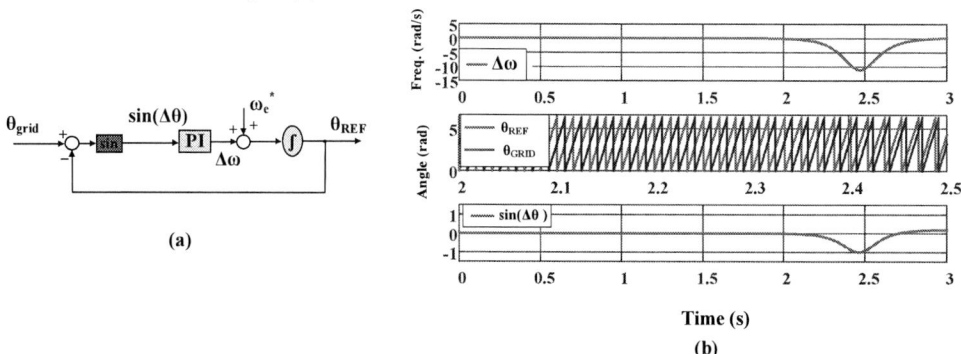

Fig. 6: (a) Classical control structure of synchronization algorithm; (b) Variation of frequency and angle in the MG with synchronization starting at the time of 0 seconds.

The main problems of this loop are the unconstrained frequency variations and low bandwidth response. Better performance could be attained increasing the bandwidth. However, this would mean in a higher oscillation and peak in frequency response of the MG. This is a crucial drawback of this loop, since no proper performance is obtained in the worst-case scenario of displacement. The proposed smooth synchronization loop in Fig. 4 mitigates this problem, insofar as the angle of displacement, $\Delta\theta$, is utilized for synchronization and the proposed block for angle estimation can correctly estimate the angle of displacement advanced or delayed. Also, it permits lesser variations of frequency in the MG. The Fig. 7 shows the performance of the proposed synchronization loop.

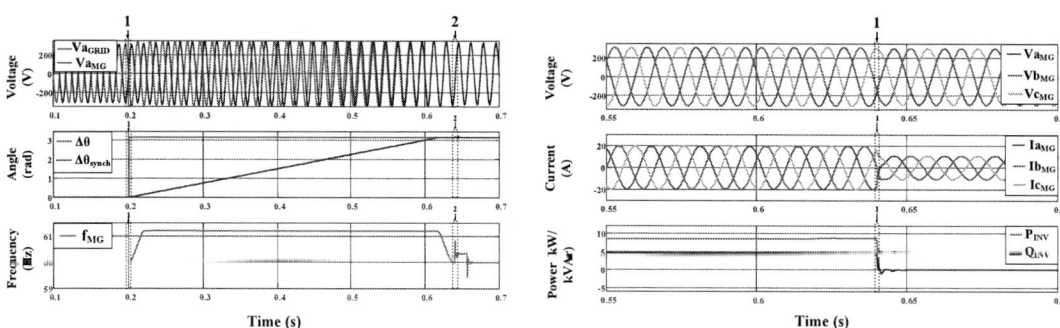

1: At the time of 0.2s, the synchronization starts and the grid voltage is measured for angle and amplitude estimation.

2: At the time of 0.64s, the microgrid is synchronized in both phase and amplitude, then the PCC switch closes and the VSI changes to current source operation.

(a)

1: At the time of 0.64s, the microgrid is synchronized in both phase and amplitude, then the PCC switch closes and the VSI changes to current source operation.

(b)

Fig. 7: (a) Smooth synchronization process; (b) Voltage, VSI power and current outputs during transition of modes of VSI.

At time of 0.2 s, the synchronization starts. The standard allows a reconnection to the grid if the voltage is within a maximum of 110 % of the nominal voltage. Therefore, the grid voltage was programmed to experience an increase of 0.1 pu in the voltage amplitude to evaluate the performance of synchronization. As can be seen, the control permits a smoothly adjustment the references of the VSI without distortion at the MG. Also, the increase in frequency in the MG present a maximum of 1.2 Hz, which is the limit stablished for continuous operation of a MG. At the time of 0.64 s, the MG is synchronized both in amplitude and phase, then the command of closing the POI is sent to the transfer switch and VSI to changes its operation mode. A small oscillation in frequency is experienced in the MG at the time of 0.64 s due to power injection changes from islanded to grid-connected operation and switching of modes of operation. In islanded operation, the VSI is injecting about 8.5 kW and 5 kVAr of active and reactive power, respectively, in the MG distribution system to supply the loads. When the MG migrates for grid-connected operation, the VSI is set to inject only active power, which was pre-defined as 5 kW.

Microgrid disconnection

The unintentional islanding is the worst-case scenario of disconnection of a MG. The decision of the disconnection is usually driven due to grid power quality degradation. The most common degradation of the utility grid power quality happens due to grid faults. When a grid fault occurs at the utility grid, the MG should be disconnected and operate in islanded mode. However, if the transients are highly severe, the system may not be capable of recovering from such conditions and can become unstable, causing a MG blackout and a drop of reliability of the system. The performance of proposed controller was compared to a similar in literature during a phase to ground short circuit at a point near the POI, but in the utility grid, as illustrated in Fig. 8. For a fair comparison, the same bandwidth was utilized, given the similarity in VSI low level control structure. The differences in the proposed method and the one in [6] sum up to the grid phase estimation, the power controller and the feedforward of the output current.

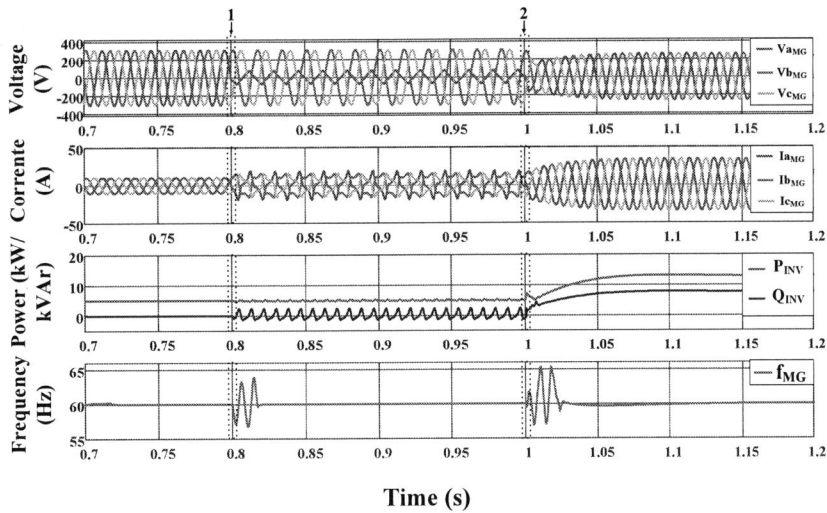

Time (s)

1: At the time of 0.8s , the a phase to ground fault happens at the grid side.
2: At the time of 1s, the islanding command is sent to the inverter, then the PCC switch opens and the VSI changes to voltage source operation.

Fig. 8: Voltage, current of VSI, power and frequency in the MG amidst a grid fault with the method proposed by **[6]**.

At the time of 0.7 s, the power drawn by the load is changed to 16 kW and 12 kVAr nominal of active and reactive power, respectively, to simulate a stochasticity with that the controllers need to handle. At 0.8 s, a grid fault occurs at the grid side subjecting the system to a high unbalanced condition. As can be noted, this causes a high distortion of the current injected by the inverter. In [6], the power controller is a simpler one, which did not take into consideration unbalanced conditions and distortion at the MG voltage. As an advantage, the authors proposed a modified PLL structure to avoid phase misalignment during MG islanding. However, the lack of the feedforward of the output current degrades the performance of the controller, so as to during the islanding, the VSI controls takes more time to recover the system and restore the frequency. Also, the load acts as a perturbation and its influence is very high in the output voltage of the VSI, so that a huge gap of the MG voltage at the POI is encountered from the nominal voltage value. The authors did not propose a secondary control to correct the difference of voltage amplitude to the nominal at the POI caused by the influence of the load and the drops along the system. In the case of a real MG, the stochasticity of the loads would bring about voltage fluctuations in the system. The performance of the proposed controllers is illustrated in Fig. 9 during the unintentional islanding with the same scenario and controller gains.

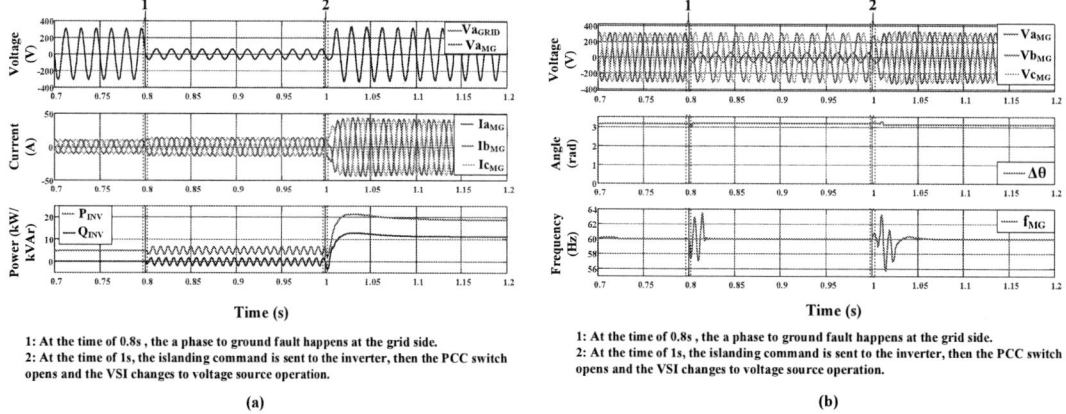

1: At the time of 0.8s , the a phase to ground fault happens at the grid side.
2: At the time of 1s, the islanding command is sent to the inverter, then the PCC switch opens and the VSI changes to voltage source operation.

(a)

1: At the time of 0.8s , the a phase to ground fault happens at the grid side.
2: At the time of 1s, the islanding command is sent to the inverter, then the PCC switch opens and the VSI changes to voltage source operation.

(b)

Fig. 9: (a) Voltage, current and VSI output power in the MG amidst a grid fault with the method proposed in this article; (b) Voltage, phase angle and frequency in the MG during the unintentional islanding.

As can be seen in the Fig. 9 (a), the BPSC maintains sinusoidal current components at the output current of the inverter, which it is very important in order to avert damaging the hardware of the VSI, contribute to MG power quality and comply with IEEE standards. Even though, it comes at the cost of a more oscillatory power output, but they are very small, so that the influence on power quality and frequency oscillation is negligible. The MG restoration of voltage and frequency is also very fast, owing to modified controllers employed in the VSI control structure, which improve the dynamic response and ability to resist against perturbations. Besides, the hierarchical control structure, described in Fig. 5, keeps the voltage amplitude to the nominal value.

In the Fig. 9 (b), one sees that the angle estimated by the PAEB hardly changes under the abnormal conditions faced by the MG. This guarantees that the MG does not faced misalignment of phase during the islanding, which would avoid sudden change of reference at the VSI control structure. In turns, providing a seamless intentional islanding and a smooth unintentional one, since the transitory are great minimized.

Conclusion

In this work, it was presented a methodology to avoid phase misalignment and voltage mismatch at the MG transitions. It was shown that the estimation of phase angle has the advantage of avoiding the discontinuous phase, which facilitates the synchronization, even in the worst-case scenarios, where the most techniques fails. By using the SSU, it was able to provide a smooth variation of phase by limiting the variation in frequency in order to comply the IEEE standards and averting large frequency variations due to an instantaneous change of reference. The control structure proposed also was able to increase the resilience of MG amidst abrupt grid conditions, such as faults, load changes, and provide smooth variation of amplitude voltage for synchronization and islanded operation in order to eliminate voltage distortion in the MG.

References

[1] D. Ioris, A. B. Almeida and P. T. Godoy, "A Microgrid Islanding Performance Study Considering Time Delay in Island Detection," *IEEE PES,* 2020.

[2] IEEE Standard Association, *IEEE 2030.7 - IEEE Stand. for the Specification of Microgrid Controllers,* 2017.

[3] J. Wang, A. Pratt and M. Baggu, "Integrated Synchronization Control of Grid-Forming Inverters for Smooth Microgrid Transition," *IEEE PES,* August 2019.

[4] M. N. Arafat, A. Elrayyah and Y. Sozer, "An Effective Smooth Transition Control Strategy using Droop Based Synchronization for Parallel Inverters," *IEEE Trans. on Industry Applications,* vol. 51, no. 3, pp. 2443-2454, 2015.

[5] M. S. Pilehvar and B. Mirafzal, "Smart Inverter for Seamless Recconection of Isolated Residential Microgrid to Utility Grid," in *IEEE EPEC,* 2020.

[6] G. G. Talapur, H. M. Suryawanshi, L. Xu and A. B. Shitole, "A Reliable Microgrid With Seamless Transition Between Grid Connected and Islanded Mode for Residential Community With Enhanced Power Quality," *IEEE Trans. on Industry Applications,* vol. 54, no. 5, pp. 5246-5453, 2018.

[7] A. G. P. Alves, L. G. B. Rolim, R. F. S. Dias and P. T. P. Santos, "VSC plug-and-play operation using online grid parameter estimation for PI self-tuning," *IET Power Electronics,* vol. 13, no. 18, pp. 4359-4367, 2020.

[8] T.-V. Tran, T.-W. Chun, H.-H. Lee, H.-G. Kim and E.-C. Nho, "PLL-Based Seamless Transfer Control Between Grid-Connected and Islanding Modes in Grid-Connected Inverters," *IEEE Trans. on Power Electronics,* vol. 29, no. 10, pp. 5218 - 5228, 2014.

[9] S. Kumar and B. Singh, "Seamless Operation and Control of Single-Phase Hybrid PV-BES-Utility Synchronized System," *IEEE Trans. on Industry Applications,* vol. 55, no. 2, pp. 1072-1082, 2019.

[10] M. K. Ghartemani, S. A. Khajehoddin, P. K. Jain and A. Bakhshai, "Problems of Startup and Phase Jumps in PLL Systems," *IEEE Trans. on Power Electronics,* vol. 27, no. 4, pp. 1830 - 1838, 2011.

[11] S. D'Silva, M. Shadmand, S. Bayhan and H. Abu-Rub, "Towards Grid of Microgrids: Seamless Transition between Grid-Connected and Islanded Modes of Operation," *IEEE Open Journal of the Industrial Electronics Society,* pp. 66-81, 2020.

[12] N. Hatziargyriou, Microgrids: Architectures and Control, 1nd ed., Wiley-IEEE Press, 2014.

[13] M. Ganjian-Aboukheili, M. Shahabi, Q. Shafiee and J. M. Guerrero, "Seamless Transition of Microgrids Operation From Grid-Connected to Islanded Mode," *IEEE Trans. on Smart Grid,* vol. 11, no. 3, pp. 2106-2114, 2020.

[14] P. Rodriguez, A. Luna, R. S. M. Aguilar, I. E. Otadui, R. Teodorescu and F. Blaabjerg, "A Stationary Reference Frame Grid Synchronization System for Three-Phase Grid-Connected Power Converters Under Adverse Grid Conditions," *IEEE Trans. on Power Electronics,* vol. 27, no. 1, pp. 99-112, 2012.

[15] L. A. S. Ribeiro, M. W. Degner, F. Briz and R. D. Lorenz, "Comparison of Carrier Signal Voltage and Current Injection for the Estimation of Flux Angle or Rotor Position," in *IEEE Industry Applications Conference,* 1998.

[16] M. N. Arafat, S. Palle, Y. Sozer and I. Husain, "Transition Control Strategy Between Standalone and Grid-Connected Operations of Voltage-Source Inverters," *IEEE Trans. on Industry Applications,* vol. 48, no. 5, pp. 1516-1524, 2012.

[17] X. Hou, Y. Sun, J. Lu, X. Zhang, L. H. Koh, M. Su and J. M. Guerrero, "Distributed Hierarchical Control of AC Microgrid Operating in Grid-Connected, Islanded and Their Transition Modes," *IEEE Access ,* vol. 6, pp. 77388 - 77401, 2018.

[18] P. Rodriguez, A. V. Timbus, R. Teodorescu, M. Liserre and F. Blaabjerg, "Flexible Active Power Control of Distributed Power Generation Systems During Grid Faults," *IEEE Trans. on Industrial Electronics,* vol. 54, no. 5, pp. 2583-2592, 2007.

[19] A. F. Cupertino, L. S. Xavier, E. M.S.Brito, V. F. Mendes and H. A. Pereira, "Benchmarking of power control strategies for photovoltaic systems under unbalanced conditions," *International Journal of Electrical Power & Energy Systems,* vol. 106, pp. 335-345, 2019.

[20] J. Lu, M. Savaghebi and J. M. Guerrero, "Feedforward Control Strategy for the state-decoupling Stand-alone UPS," in *EPE ECCE,* 2017.

[21] F. Bosio, L. A. S. Ribeiro, F. D. Freijedo, M. Pastorelli and J. M. Guerrero, "Effect of State Feedback Coupling and System Delays on the Transient Performance of Stand-Alone VSI With LC Output Filter," *IEEE Trans. on Industrial Electronics,* vol. 63, no. 8, pp. 4909-4917, 2016.

Subtle Design and Performance Comparison of WF-FSM and DC-VRM for Large-Scale Direct-Drive Wind Power Generation

Udochukwu B. Akuru
TSHWANE UNIVERSITY OF TECHNOLOGY
Department of Electrical Engineering
Pretoria, South Africa
E-Mail: AkuruUB@tut.ac.za

Maarten J. Kamper
STELLENBOSCH UNIVERSITY
Department of Electrical and Electronic Engineering
Cape Town, South Africa
E-Mail: kamper@sun.ac.za

Zi-Qiang Zhu
THE UNIVERSITY OF SHEFFIELD
Department of Electronic and Electrical Engineering
Sheffield, U.K.
E-Mail: z.q.zhu@sheffield.ac.uk

Acknowledgements

The work is supported in part by the Centre for Renewable and Sustainable Energy Studies (CRSES), Stellenbosch University, South Africa and Tshwane University of Technology, South Africa.

Keywords

«AC machine», «Brushless drive», «Design», «Finite-element analysis», «Wind-generator systems».

Abstract

We attempted the design and performance comparison of field-excited flux modulation machines viz., wound-field flux switching machine versus DC Vernier reluctance machine at large-scale power. Based on simple conceptual topologies and performance comparison using finite element analysis, preliminary findings at ~5.5 MW show potentials, while in-depth optimization is required for stiffer conclusions.

Introduction

Renewable energy is associated with notable credentials which include technological revolution, reduction in fossil fuel usage, climate change mitigation and economic empowerment, among others. Over the years, wind power has become a global reckoning force among new renewable energy electricity sources. Currently, the global cumulative total installed wind power generating capacity stands at 845 GW, with a 13.5 % growth experienced in the global wind power market in 2021 [1].

Considering the proliferation of wind power generation, wind generators are critical because they are integral drivetrain components in wind turbines [2]. In modern times, variable-speed wind turbines are preferred to fixed-speed because of extended operating power capture range [3]. The leading industrial wind generator technology is the variable-speed doubly fed induction generator (DFIG) using a three-stage high-speed gearbox (3G) drivetrain, meaning a smaller wind generator. However, DFIGs are besieged with low reliability due to its large-sized gearbox and brushed generators [3, 4, 5]. Due to large gearbox reliability constraints in high-speed wind generators, recent innovations have led to single-stage (1G) or two-stage (2G) geared medium-speed wind generators. In effect, medium-speed wind generators yield a middle-point in terms of the size for both generator and gearbox. In some studies, the medium-speed drivetrain has been investigated for different wind generator concepts such as for DFIG, permanent magnet synchronous generator (PMSG) and wound-field flux switching machine (WF-FSM), providing the best tradeoff in terms of low cost of energy compared to other drivetrains [5-7].

In recent times, there is also the so-called direct-drive or gearless wind generator technologies which are gaining prominence because of total elimination of the gearbox. Direct-drive wind generators operate at very low speed and without gearboxes, meaning that the generators tend to increase dramatically in size and thus producing very high electromagnetic torque [3, 8]. To this end, direct-drive wind generators are prone to reduced generator torque density. Another issue with direct-drive generators is an enlarged airgap which is meant to reduce manufacturing challenges but at the same time, increases the cost and efficiency constraints of the generator because of the need for more excitation. To tackle some of these challenges, the mainstay wind generator technology for the direct-drive systems are PM machines which uses high-energy permanent magnets (PMs) for increased power density [9, 10]. But as identified in the recently published 5–year INNWIND.EU project which focused on design concepts for futuristic 10–20 MW offshore wind turbine drivetrains, PM drives based on a magnetic pseudo-direct drive (PDD) could be competitive in terms of lowering levelised cost of electricity (LCOE) but are notwithstanding vulnerable to the unpredictable market prices of PMs, among others [11].

Generally, it is believed that the future of wind turbines will be large-scale and offshore based [12]. This is because onshore systems have matured fully over the years with power rating and size limitations, exacerbated by huge logistics; whereas offshore wind turbines are not that limited [2]. This makes the variable-speed direct-driven PMSGs as the mainstay technology for this to happen. Notwithstanding, it is important to note that large offshore wind turbines operate remotely in harsher conditions compared to onshore systems; hence the cost of manufacturing, logistics and maintenance make them extremely difficult to manage. To this end, the important factors to consider in determining the appropriate wind generator technology for futuristic wind turbines should be electromagnetic output, mechanical design, thermal management, cost/size and reliability [3, 11].

Meanwhile, wound-rotor or electrically excited synchronous generator (WRSG) are being experimented for direct-drive wind generating systems. Such wind generator technologies have the advantage of having no PMs (i.e., no demagnetization threats and low-cost) as well as supplying adjustable reactive power and generator output voltage based on a variable field current excitation source but are known to be oversized and with reduced efficiency [2]. In addition, WRSGs often than not, require brushes and slip rings for their DC electromagnet exciters.

To this end, there are novel and emerging wound-field wind generator technologies which eliminate the need for brushes and slip rings to power their DC electromagnets, while offering a very robust rotor structure. They operate based on the principle of minimum reluctance or flux modulation effect to produce torque [13]. These machines typically comprise an armature, a field exciter and a flux modulator. Some examples of these brushless machines are the so-called stator-mounted flux modulation machines such as the DC field-excited double salient machines (DC-DSMs), DC field-excited flux reversal machines (DC-FRMs), wound-field flux switching machines (WF-FSMs) and DC-excited Vernier reluctance machines (DC-VRMs) [14-17], among others.

Currently, the so-called wound-field flux-modulation synchronous machines (e.g., WF-FSM and DC-VRM), has not received much consideration, unlike the conventional PMSG or WRSG [18, 19], for applications in MW wind power generation. WF-FSM for applications in wind generator drives have been demonstrated at industrial-scale power levels for geared medium-speed wind generators [17]. To the best of the authors, the DC-VRM have only been exhibited for small-scale direct-drive wind generators [20]. So far, it remains to be seen what the WF-FSM and DC-VRM would yield in terms of performance for MW-scale direct-drive wind generator systems. The study in this paper is a delicate contemplation of such research adventure of which the WF-FSM and DC-VRM will be evaluated and compared for large-scale power direct-drive wind power generator systems. The study will be based on finite element analyses leading to small-scale experimental validation. For the proposed study, the overarching prioritization of these machine variants is due to their less emphasis on rare-earth materials for wind generator designs, as well as because of their robust design and high-power density compared to traditional wound-rotor machines [21, 22].

Topology Design Selection and Presentation

In this study, the three-phase 12/10 stator coils/rotor poles radial-flux concentrated winding machine topology is selected for both the WF-FSM and DC-VRM due to simplicity and optimum stator/rotor pole combinations [23, 24]. The generic machine models are presented in Fig. 1 as developed in 2D FEA software. As earlier indicated, both machines are benchmarked for direct (gearless) wind generator drivetrain at rotor speed of 12 r/min based on specifications laid out for a 5 MW direct-drive WRSG in [19].

The steady-state dq-axes generator modelling of both wound-field machine variants considered in this study, which fast-tracks the machine performance in FEA and where $I_d = 0$ control can be applied, has been undertaken previously in [25]; hence, it is not rehearsed here. Moreover, the in-house 2D FEA software utilised for this study applies position-stepping magnetostatic solutions which makes it highly efficient to undertake the proposed electromagnetic design analysis in MW-scale [26]. The research methodology for the study is summarized as shown in Fig. 2.

The technical specifications for the proposed machines are presented as shown in Table I. The specific dimensional parameters are sized approximate to sizing routines detailed in [27] and [20] for the WF-FSM and DC-VRM, respectively. Due to large-scale designs, approximations on end-windings are considered for the end-winding field and armature resistance. To facilitate a fair comparison of both generators, similar values have been imposed on the rated generator technical and dimensional characteristic data such as output power, stator outer diameter, airgap length and split ratio (ratio between stator airgap diameter to stator outer diameter), among others. To this end, the subtleness of the design process of the proposed MW-scale direct-drive wind generators is expressed in terms of the non-optimised designs, as well as inconsideration of the generator structural and thermal design aspects. The electromagnetic and design performance analysis now follows in the next section.

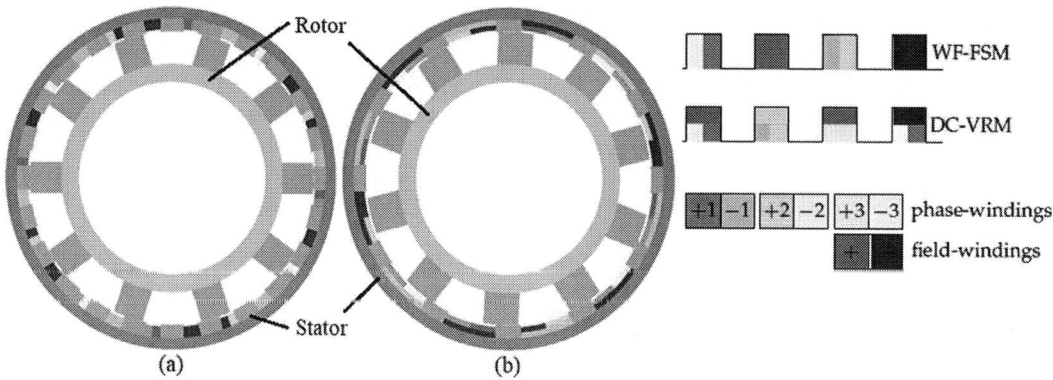

Fig. 1: Selected three-phase 12/10 stator coil/rotor poles machine topologies drawn in in-house FEM software: (a) WF-FSM, and (b) DC-VRM.

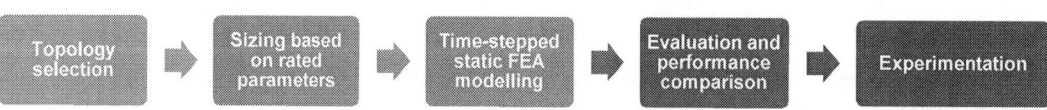

Fig. 2: Workflow of the research process.

Comparative Analysis of Electromagnetic and Design Performance

The no-load and on-load characteristics of the proposed machines are given first, thereafter the on-load characteristics. Figs. 3 and 4 presents the field lines and flux density vector diagrams of the designed machines under no-load conditions. As can be seen, high saturation, although much lower in the WF-FSM, is exhibited on the stator of both machines because of their stator-mounted configuration, besides the fact that optimization is yet to be undertaken. The airgap flux density waveforms of the designed machines under no-load conditions are shown in Fig. 5. The flux density in the WF-FSM design is much lower because of a higher split ratio which tends to minimize the DC loading in spite of a fixed current density compared to the DC-VRM.

In terms of the on-load performance, a summary is provided in Table II. The rated on-load average torque, as well as the generated power under rated conditions, are indicated; the latter approximates to 5.7 MW for both machines. The torque ripple values are seen to be slightly above 10%, which is very rough but considering that the machines are not optimized yet. The power factor of the DC-VRM is much lower compared to the WF-FSM, granted that the former naturally exhibits poor power fact due to its higher armature reaction effects in the Vernier machines' family [28]. For such direct drive designs, power factor is very critical because it Very interestingly, the evaluated efficiencies of these machines are quite high, especially for such wound-field machines. Although mechanical and other scant rotational losses have been ignored, the nearly lossless designs are quite remarkable considering that field winding losses are included. Based on the masses of the machines, which dwarfs 220 tons 7.5 MW, 10 r/min, direct-drive conventional wound-field generator as indicated in [19], it could have been expected that the core losses will dominate. But alas, the proposed machines end up with rated frequency of 2 Hz for evaluating the stator core losses, and even lower for the rotor [29]! Due to much lower pole number for the considered topology. Besides, the high efficiency is also indicative that the machines are oversized as indicated by their masses which may not be practical for the proposed wind turbine application, but with optimization, and selection of higher stator/rotor pole combination, the torque density of the machines can be potentially improved.

Table I: Design specifications of proposed direct-drive brushless wound-field wind generators

	WF-FSM	DC-VRM
Power	≥ 5 MW	
Line voltage	≈ 6 kV	
Speed	12 r/min	
Phase current density	1. 5A/mm^2	
Field current density	2.5 A/mm^2	
Stator outer diameter	8 m	
Airgap length	5 mm	
Shaft diameter	3 m	
Stack length	1.5 m	
Split ratio	0.8	
Slot fill factor	0.5	
Split ratio	0.85	0.8

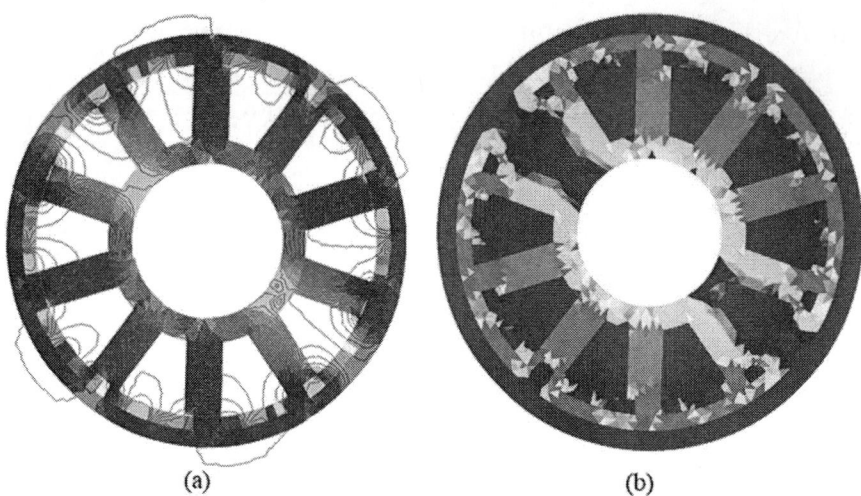

Fig. 3: WF-FSM: (a) flux lines, and (b) flux density map.

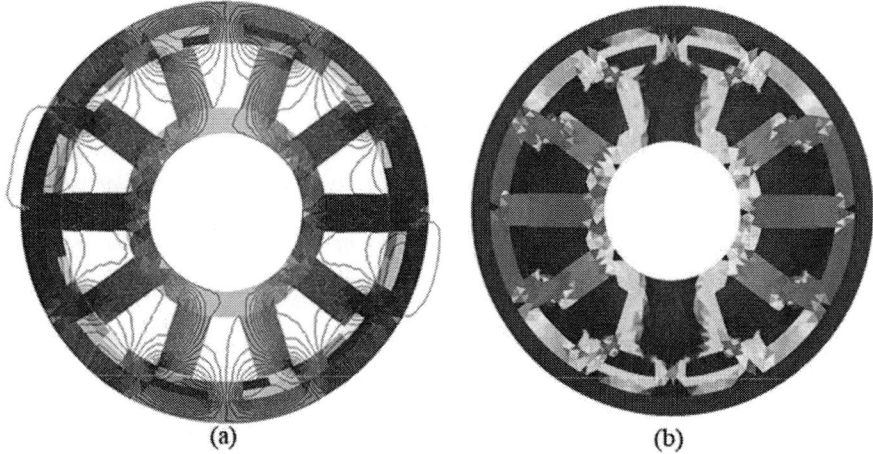

Fig. 4: DC-VRM: (a) flux lines, and (b) flux density map.

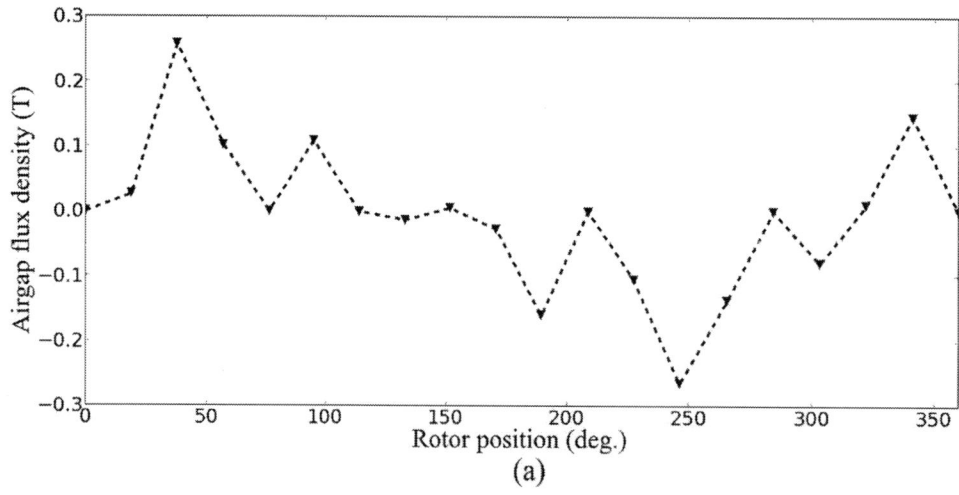

(a)

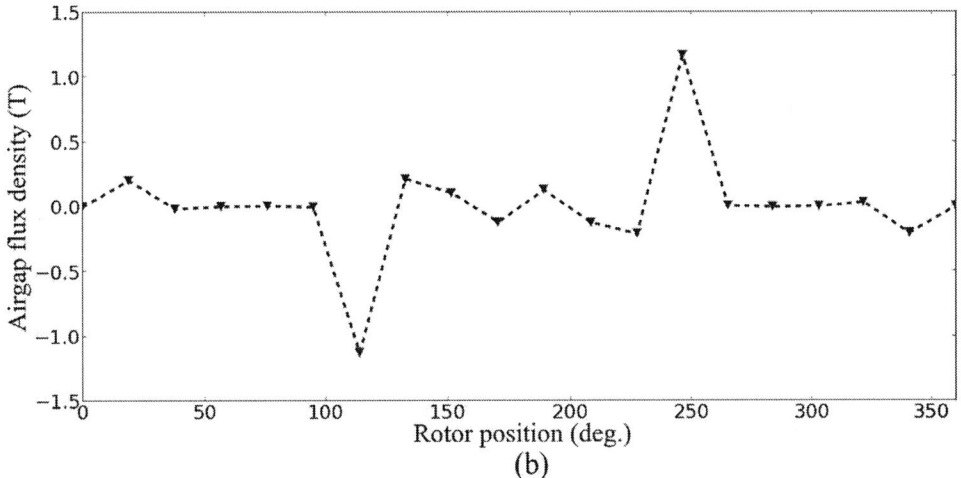

(b)

Fig. 5: No-load airgap flux density waveforms at rated field currents and 12 r/min:
(a) WF-FSM (I_f = 0.97 kA), and (b) DC-VRM (I_f = 1.28 kA).

Table II: Comparative performance evaluation

	WF-FSM	DC-VRM
Power (MW)	5.8	5.7
Line voltage (kV)	6	6
Phase current (kA)	0.58	0.73
Field current (kA)	0.97	1.28
Torque (MNm)	4.6	4.5
Cogging torque (%)	11.32	12.6
Torque ripple (%)	10.7	12.5
Power factor	0.96	0.74
Core loss (kW)	11.0	9.2
Field Cu loss (kW)	5.60	4.04
Phase Cu loss (kW)	2.01	1.39
Efficiency (%)	99.7	99.7
Active mass (tons)	324	344
Torque/litre (kNm/m^3)	61.1	59.9
Loss density (kW/m^3)	0.24	0.22

Experimental Results

In this section, small experimental prototypes of the WF-FSM and DC-VRM are investigated at no-load to serve as modest proof of concepts. The machine design and rated parameters of both prototypes are presented in Table III. Both prototype machines are built for wind generator drives [17, 20]. Fig. 6 shows the experimental test benches for the prototype machines. The comparison of the no-load line voltages between FEA and measurement for both prototype machines is shown in Fig. 7. The disagreement between simulated and measured results is due to complications encountered during fabrication of the prototypes, especially for the WF-FSM, as well as fringing effects due to underestimation of so-called on-load induced DC winding induced voltages in predicted values [30].

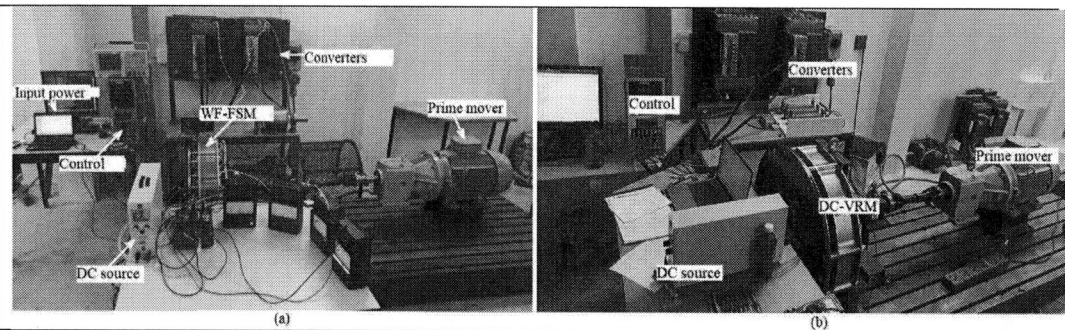

Fig. 6: Experimental set-up for prototype generators: (a) WF-FSM, and (b) DC-VRM.

Table III: Design and performance data for prototype generators

	WF-FSM	DC-VRM
Stator outer diameter (mm)	600	700
Rotor outer diameter (mm)	414.7	454.1
Stack length (mm)	104	109.6
Air gap thickness (mm)	0.7	0.8
Number of rotor poles	10	10
Number of stator slots	24	12
Rated speed (r/min)	360	200
Rated frequency (Hz)	60	33.3
Rated phase current (A)	9.8	27.5
Rated field current (A)	7.8	28.7
Rated line voltage (V)	700	384
Rated terminal power (kW)	10	14.6

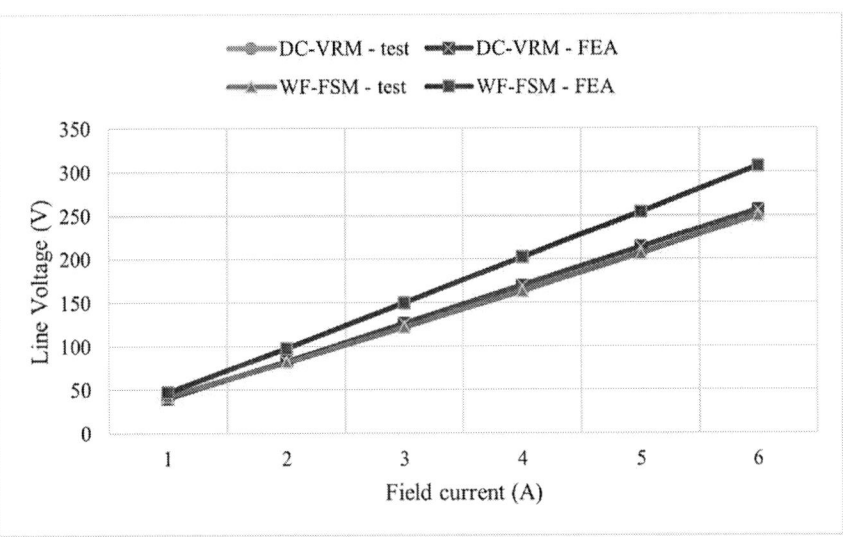

Fig. 7: Comparison of no-load voltages, WF-FSM@360 r/min and DC-VRM@300 r/min.

Conclusion and Future Work

In this study, conceptual WF-FSM and DC-VRM are designed, and their performance compared based on finite element analyses. Preliminary findings show the potentials of the design adventure, but the need for in-depth optimization is required for deeper conclusions to be made. At 5.5 MW, the resolved machines appear oversized for a 12/10 pole, 12 r/min designs, but appearing almost lossless at 99.7% efficiency. It must be reiterated that both machines have not been optimized, hence, the evaluated cogging torque and torque ripple ranged from 10% to 13%. Also, the torque per litre at approximately 60 kNm/m³ and the active mass in excess of 300 tons are beyond standard ranges, due to the oversimplification of the airgap length, current density and selected stator/rotor pole design.

However, the design feasibility of the proposed machines for wind generator drives are validated based on existing small-scale prototypes. In future, additional studies would be undertaken to consolidate the design candidature of the highlighted MW-range stator-mounted brushless machines for direct-drive wind power systems through the following:

- Parametric variation on pole numbers, airgap length and current density on the overall generator performance.
- Multi-objective optimisation studies prioritising optimum torque density and torque performance.

References

[1] REN21, Renewables 2020 global status report Paris, 2020.
[2] Keysan O.: Future electrical generator technologies for offshore wind turbines, Wind Power, E&T IET Services, Aug 16, 2017.
[3] Chen H, Zuo Y, Chau K. T., Zhao W. and Lee C. H. T.: Modern electric machines and drives for wind power generation: A review of opportunities and challenges. IET RPG Journal, 15: pp. 1864-1887, 2021.
[4] Polinder H. Ferreira J. A., Jensen B. B., Abrahamsen A. B., Atallah K. and McMahon R. A. Trends in wind turbine generator systems, IEEE JESTPE Journal, Vol. 1 no 3, pp. 174–185, Sept. 2013.
[5] Zhu Z. Q. and Hu J.: Electrical machines and power-electronic systems for high-power wind energy generation applications: Part I—market penetration, current technology and advanced machine systems, COMPEL Journal, Vol. 32 no 1, pp.7–33, 2012.
[6] de Vries E.: The evolution of wind turbine drive systems, WINDPOWER Monthly, 26 April 2012.
[7] Akuru U. B. and Kamper M. J.: Design and investigation of low-cost PM flux switching machine for geared medium-speed wind energy applications, EPC&S Journal, Vol. 46 no 9, pp. 1082-1090, 2018.
[8] Li H., Chen Z. and Polinder H.: Research report on numerical evaluation of various variable speed wind generator systems, Project UpWind, Technical Report, 2006.

[9] Blaabjerg F. and Ionel D. M.: Renewable Energy Devices and Systems – State-of-the-Art Technology, Research and Development, Challenges and Future Trends, EPC&S journal, Vol. 43 no 12, pp. 1319-1328, 2015.

[10] Khan F., Sulaiman E. and Ahmad Md Z.: Review of switched flux wound-field machines technology, IETE Technical Review, Vol. 34 no 4, pp. 343-352, 2017.

[11] Jensen P. H., Chaviaropoulos T. and Natarajan A.: LCOE reduction for the next generation offshore wind turbines, INNWIND.EU PROJECT, October 2017.

[12] Bensalah A., Benhamida M. A., Barakat G. and Amara Y.: Large wind turbine generators: State-of-the-art review, ICEM 2018, pp. 2205-2211.

[13] Li D., Qu R. and Li J.: Topologies and analysis of flux-modulation machines, ECCE 2015, pp. 2153-2160.

[14] Udosen, D., Kalengo, K., Akuru U. B., Popoola O. and Munda J. L.: Non-Conventional, Non-Permanent Magnet Wind Generator Candidates, Wind Journal, Vol. 2, 429-450, 2022.

[15] Liu W., Wang H., Wang Y., Ren L. and Xiao L.: New approach to suppress torque ripple and improve torque output for wound-excited doubly salient machine, IECON 2016, pp. 2857-2861, 2016.

[16] Liu X. and Zhu Z. Q.: Electromagnetic performance of novel variable flux reluctance machines with DC-Field coil in stator, IEEE TransMag Journal, Vol. 49 no 6, pp. 3020-3028, Jun. 2013.

[17] Akuru U. B. and Kamper M. J.: Intriguing behavioral characteristics of rare-earth-free flux switching wind generators at small- and large-scale power levels, IEEE TIA journal, Vol. 54 no 6, pp. 5772-5782, Nov.-Dec. 2018.

[18] Stuebig C., Seibel A., Schleicher K., Haberjan L., Kloepzig M. and Ponick B.: Electromagnetic design of a 10 MW permanent magnet synchronous generator for wind turbine application, IEMDC 2015, pp. 1202-1208.

[19] Boldea I., Tutelea L. and Blaabjerg F.: High power wind generator designs with less or no PMs: An overview, ICEMS 2014, pp. 1-14.

[20] Akuru U. B., Kamper M. J. and Mabhula M.: Optimisation and design performance of a small-scale dc vernier reluctance machine for direct-drive wind generator drives, ECCE 2020, pp. 2965–2970.

[21] Pavel C. C., Lacal-Arántegui R., Marmier A., Schüler D., Tzimas E., Buchert M., Jenseit W. and Blagoeva D.: Substitution strategies for reducing the use of rare earths in wind turbines, Resources Policy, Vol. 52, pp. 349-357, 2017.

[22] Lee C. H. T., Chau K. T., Liu C. and Chan C. C.: Overview of magnetless brushless machines, IET EPA Journal, Vol. 12 no 8, pp. 1117-1125, 2018.

[23] Cheng M., Hua W., Zhang J. and Zhao W.: Overview of stator–permanent magnet brushless machines, IEEE TIE Journal, Vol. 58 no 11, pp. 5087–5101, Nov. 2011.

[24] Liu, X. and Zhu Z. Q.: Stator/rotor pole combinations and winding configurations of variable flux reluctance machines, IEEE TIA Journal, Vol. 50 no 6, pp. 3675-3684, Nov./Dec. 2014.

[25] Mabhula M., Akuru U. B. and Kamper M. J.: Cross-coupling inductance parameter estimation for more accurate performance evaluation of wound-field flux modulation machines, Electronics Journal, Vol. 9 no 11, August 2020.

[26] SEMFEM. Available: www0.sun.ac.za/semfem/index.html

[27] Akuru U. B. and Kamper M. J.: Formulation and multi-objective design optimisation of wound-field flux switching machines for wind energy drives, IEEE TIE Journal, vol. 65 no 2, pp. 1828-1836, 2017.

[28] Spooner E. and Haydock L.: Vernier hybrid machines, IEE Proc. EPA Journal, Vol. 150 no 6, 655-662, Nov. 2003.

[29] Fukami T., Avoki H., Shima K., Momiyama M., and Kawamura M.: Assessment of core losses in a flux-modulating synchronous machine, IEEE TIA Journal, Vol. 48 no 2, pp. 603-611, Mar./Apr. 2012.

[30] Wu Z., Zhu Z-Q., Wang C., Mipo J-C., Personnaz S. and Farah P.: Analysis and Reduction of On-Load DC Winding Induced Voltage in Wound Field Switched Flux Machines, IEEE TIE Journal, Vol. 67 no 4, pp. 2655-2666, April 2020.

Analysis and Implementation of different non-isolated Partial-Power Processing Architectures based on the Cuk Converter

J.S. Artal-Sevil[1], J. Anzola[2], V. Ballestín-Bernad[1], J.L. Bernal-Agustín[1]
[1]Department of Electrical Engineering.
Polytechnic School of Engineering and Architecture, EINA.
University of Zaragoza. Spain.
[2]Electronics and Computer Science Department.
Goi Eskola Politeknikoa-Orona Ideo-Fundazioa eraikina
Mondragon Unibertsitatea. Spain.
E-mail: {jsartal, ballestin, jlbernal}@unizar.es, janzola@mondragon.edu

Acknowledgements

The authors would like to thank the support of Government of Aragon and the European Union for the project T28_20R, "building Aragon from Europe". This work was supported in part by the Spanish MINECO under Grant RTC-2015-3358-5. The authors also want to thank the support of the IDI project, "study and analysis of partial-power processing architectures for the development of an on-board charger in electric and hybrid vehicles" with reference OTRI-2020/0416.

Keywords

«Partial-Power Processing Converter (PPC)», «non-isolated bidirectional partial-power architectures», «DC-DC switched-mode power supplies», «high power», «Converter Modelling», «DC-fast charges», «battery energy storage system (BESS)», «Converter Design and Optimization», «on-board charger», «Hybrid Cuk converter».

Abstract

This paper presents the analysis and study of different partial-power processing architectures based on the Cuk converter. Thus, the non-isolated topologies analyzed have been considered operative as voltage source converter (VSC). Likewise, the study of the converters is based on the continuous conduction mode (CCM). The partial-power topology has some advantages, such as high power density, small size, the decreased voltage in semiconductor devices, etc., this is because the DC-DC converter only processes a fraction of the total power. The purpose has been to compare the different topologies, Full-Power, Partial-Power, and hybrid-converters, in order to observe the advantages and drawbacks in each case. Among the parameters to be studied is the voltage stress in semiconductors as well as the voltage gain in each architecture. The effectiveness of these new converter architectures has been validated by the *Matlab/Simulink* software simulation. In this way, different simulation results are presented and analyzed throughout the paper. The purpose of this paper has been to study and explore the usefulness of the non-isolated Cuk converter with partial-power processing architecture for industrial power applications.

1. Introduction

The introduction of partial-power processing techniques can allow us to design and optimize converters. In this way, it is possible to implement smaller and cheaper power converters because now the power losses decrease. In recent years, these new design concepts have attracted the attention of engineers, converter designers, and researchers as an attractive solution for the development of industrial applications [1], [2]. While full-power converters process all the power flow that is supplied to the load, partial-power converters only process a fraction of this power flow, see Fig. 1. In this way, its power losses are lower than the full-power topology, meanwhile, the voltage and current stress in its semiconductor components also decrease.

The partial-power architecture has already been implemented in numerous industrial applications, such as battery charging systems in electric vehicles [3], power flow control in hybrid systems based on Fuel-Cell [4], energy distribution subsystems in wind turbines [5], integration of photovoltaic systems [6] [7], DC-power supply [8], active balancing of PV-arrays [9] or MPPT search algorithms in TEG systems [10], spacecraft [11], etc., in order to improve system performance. In summary, the partial-power converter can improve the efficiency of the whole system, while the equipment design costs are reduced. Although, there are also opposing opinions that doubt the improvement of performance in non-isolated partial-power topologies [12]. It is clear that isolated partial-power processing architectures have more advantages, but this does not justify that non-isolated partial-power topologies cannot be used in industrial applications where isolation is not a design condition, or simply in those cases in which an improvement in system efficiency is sought.

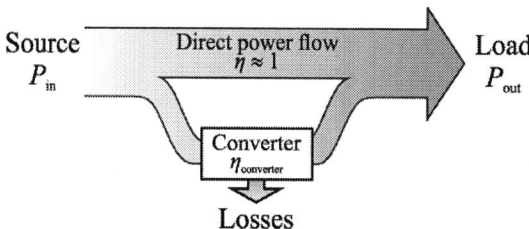

Fig. 1. Concept of power flow in partial-power converter. The converter only processes a fraction of the power supplied to the load. In this case, as an advantage the partial-power processing, the power losses are lower than in the full converter.

Partial-power architecture has some advantages over full converters, such as high power density, small size, and decreased voltage stress in semiconductor devices. These new topologies are being applied in high-power automotive applications. Some examples of applications developed are the fast-charging station for battery electric vehicles (BEV) and plug-in hybrids (PHEV) [13-15], DC-power distribution for LVDC/MVDC systems [16], a DC-DC mode switching power supply [17], a high-power battery energy storage system (BESS) [18] [19], etc.

In this paper, the authors propose different partial-power processing architectures based on the Cuk converter. These DC-DC converter topologies are intended for industrial application design or auxiliary systems in electric vehicles such as smart on-board chargers, auxiliary power supplies, etc. The main objective is to validate the advantages and drawbacks of the partial-power processing strategy in converter design. To increase the conversion efficiency and the voltage gain, different topologies of step-up converters based on the Cuk model have been proposed. Thus, to contrast these ideas, multiple simulations using *Matlab-Simulink* software have been carried out.

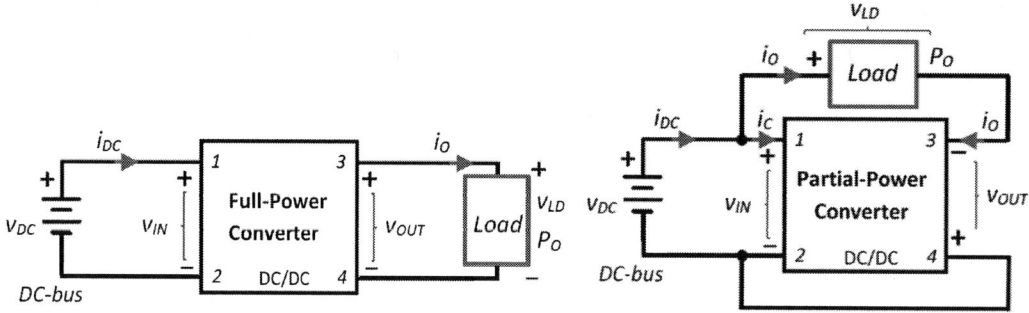

Fig. 2. Block diagram of the power conversion system. Comparison between full-power converter (traditional DC-DC converter system) and partial-power processing architecture (series-connected DC-DC converter system). In this case, the partial-power converter processes a fraction of the load's power.

This document is organized as follows. Section 1 shows a brief introduction associated with the problem addressed. Section 2 presents the mathematical analysis of the Cuk converter in partial-power processing architecture. Different hybrid topologies based on the Cuk converter are analyzed in Section 3. Meanwhile, section 4 shows the simulation results obtained in each case analyzed. The software used has been *Matlab/Simulink*. Finally, the conclusions and some brief considerations are described in Section 5.

2. Cuk Converter in Partial-Power Processing Architecture

Figure 2 shows the basic concept of power flow in partial-power architecture. The diagram shows a comparison between the full-power converter (traditional DC-DC converter) and the partial-power converter. Its objective is to reduce the power processed by the converter, in order to increase the overall efficiency of the system. The proposed converter has been derived from the typical Cuk converter, see Fig. 3.

To simplify the steady-state analysis of the proposed architecture, some assumptions are made, such as ideal switching devices and inductors, with no delay in the switching process. For a sampling time T_{SW}, the switch is ON-state for $D \times T_{SW}$ and OFF-state for $(1 - D) \times T_{SW}$. Therefore, depending on the state of the switch, two modes of operation can be identified.

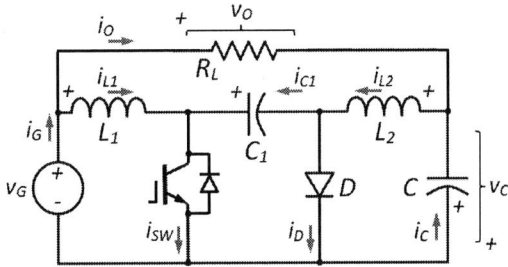

Fig. 3. Cuk converter in partial-power processing architecture.

Mode 1: ON-state $[0 \le t \le D(t)T_{SW}]$.

At the initial time t = 0, when switch SW_1 turns on, diode D_1 turns off. The inductor currents (i_{L1}, i_{L2}) increase from their respective initial value. The voltage and current equations are as follows:

$$v_{L1}(t) = L_1 \frac{\partial i_{L1}}{dt} = v_G \; ; \; v_{L2}(t) = L_2 \frac{\partial i_{L2}}{dt} = v_{C1} - v_C \tag{1}$$

$$i_{C1}(t) = C_1 \frac{\partial v_{C1}}{dt} = i_{L2} \; ; \; i_C(t) = C \frac{\partial v_C}{dt} = i_{L2} - i_O \tag{2}$$

$$v_O = v_G - \left(-v_C\right) \tag{3}$$

Mode 2: OFF-state $[D(t)T_{SW} \le t \le T_{SW}]$.

When switch SW_1 turns off at t = D×T$_{SW}$, diode D_1 turns on. A similar analysis gives the equations for voltage and current as follows:

$$v_{L1}(t) = L_1 \frac{\partial i_{L1}}{dt} = v_G - v_{C1} \; ; \; v_{L2}(t) = L_2 \frac{\partial i_{L2}}{dt} = -v_C \tag{4}$$

$$i_{C1}(t) = C_1 \frac{\partial v_{C1}}{dt} = -i_{L1} \; ; \; i_C(t) = C \frac{\partial v_C}{dt} = i_{L2} - i_O \tag{5}$$

The average inductor voltage (v_{L1}, v_{L2}) must be zero at steady-state. Hence, analyzing the volt-sec balance for inductors (L_1, L_2) during a switching period T_{SW}, we obtain:

$$v_{L1}(t) = 0 \rightarrow v_G \cdot D(t) + \left\{v_G - v_{C1}\right\} \cdot \left\{1 - D(t)\right\} = 0 \tag{6}$$

$$v_{L2}(t) = 0 \rightarrow \left\{v_{C1} - v_C\right\} \cdot D(t) - v_C \cdot \left\{1 - D(t)\right\} = 0 \tag{7}$$

As input voltage (v_G) is assumed to be constant over a switching period. Solving for steady-state output voltage v_O then,

$$v_O = v_G \frac{1}{1 - D} \tag{8}$$

In this way, the voltage gain "M" of the Cuk converter in partial-power processing architecture will be

$$\frac{v_O}{v_G} = \frac{1}{1 - D} = M \tag{9}$$

Likewise, the instantaneous value in the duty cycle ratio $D(t)$ depends on the input voltage v_G, and can be expressed as

$$D = 1 - \frac{v_G}{v_O} = 1 - \frac{1}{M} \tag{10}$$

Similarly, the average capacitor current (v_C, v_{C1}) must be zero in steady-state. Hence, analyzing the current-sec balance for the capacitors (C, C_1) during a T_{SW} switching period, we obtain:

$$i_{C1}(t) = 0 \rightarrow i_{L2} \cdot D(t) + (-i_{L1}) \cdot \{1 - D(t)\} = 0 \tag{11}$$

$$i_C(t) = 0 \rightarrow \{i_{L2} - i_O\} \cdot D(t) + \{i_{L2} - i_O\} \cdot \{1 - D(t)\} = 0 \tag{12}$$

Solving for the current in the inductors (i_{L1}, i_{L2}) in steady-state then,

$$i_{L1} = i_{L2} \frac{D}{1-D} ; \quad i_{L2} = i_O \tag{13}$$

Fig. 4. Experimental testing of the Cuk converter in partial-power processing architecture.

And calculating the current i_G supplied by the voltage source v_G, we obtain (14).

$$i_G = i_{L1} + i_O = i_O \frac{1}{1-D} \tag{14}$$

3. Improved Cuk Converter in Partial-Power Architecture

In this section, some modifications to the typical Cuk converter have been introduced. The objective is to obtain the voltage gain in the partial-power processing architecture in each of the cases analyzed. As in the previous discussion, in order to simplify the steady-state analysis, some assumptions are made, such as ideal switching devices and inductors.

Figure 5 shows the double inductor Cuk converter in partial-power processing architecture. In this case, the inductor L_1 has been replaced by a network of diodes and coils. Again, for a T_{SW} sampling time, the switch is ON-state for $D \times T_{SW}$ and OFF-state for $(1 - D) \times T_{SW}$. Therefore, depending on the state of the switch, two modes of operation can be identified.

Mode 1: ON-state $[0 \le t \le D(t)T_{SW}]$.

At the initial time t = 0, when switch SW_1 turns on, diode D_1 turns off. Likewise, diodes D_{N1} and D_{N2} are in conduction while diode D_{N3} remains in cut-off. The inductor currents (i_{L1}, i_{L2} and i_{L3}) increase from their respective initial value. In this case, the voltage and current equations are as follows:

$$v_{L1}(t) = L_1 \frac{\partial i_{L1}}{dt} = v_G ; \quad v_{L2}(t) = L_2 \frac{\partial i_{L2}}{dt} = v_G \tag{15}$$

$$v_{L3}(t) = L_3 \frac{\partial i_{L3}}{dt} = v_{C1} - v_C \tag{16}$$

$$i_{C1}(t) = C_1 \frac{\partial v_{C1}}{dt} = i_{L3} ; \quad i_C(t) = C \frac{\partial v_C}{dt} = i_{L3} - i_O \tag{17}$$

$$v_O = v_G - (-v_C) \tag{18}$$

Analysis and Implementation of different non-isolated Partial-Power Processing
Architectures based on the Cuk Converter

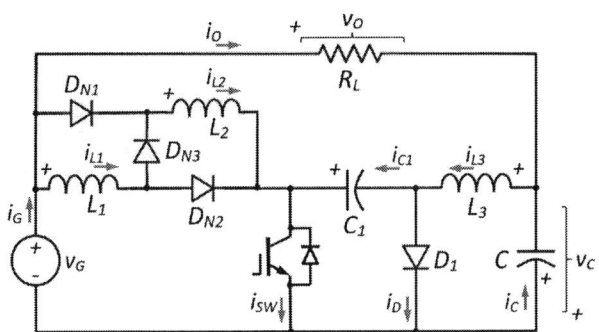

Fig. 5. Improved Cuk converter in partial-power processing architecture. Now the model has a double inductor at the input of the converter.

Mode 2: OFF-state $[D(t)T_{SW} \leq t \leq T_{SW}]$.

When switch SW_1 turns off at $t = D \times T_{SW}$, diode D_1 turns on. Now the D_{N1} and D_{N2} diodes turn off, meanwhile, the D_{N3} diode turns on. A similar analysis gives the equations for voltage and current as follows:

$$v_{L1}(t) = L_1 \frac{\partial i_{L1}}{dt} ; \quad v_{L2}(t) = L_2 \frac{\partial i_{L2}}{dt} \tag{19}$$

$$v_{L1}(t) = v_{L2}(t) ; \quad v_{L1}(t) = \frac{1}{2}\left(v_G - v_{C1}\right) \tag{20}$$

$$v_{L3}(t) = L_3 \frac{\partial i_{L3}}{dt} = -v_C \tag{21}$$

$$i_{C1}(t) = C_1 \frac{\partial v_{C1}}{dt} = -i_{L1} ; \quad i_C(t) = C \frac{\partial v_C}{dt} = i_{L3} - i_O \tag{22}$$

The average inductor voltage (v_{L1}, v_{L2} and v_{L3}) must be zero at steady-state. Hence, analyzing the volt-sec balance for inductors (L_1, L_3) during a switching period T_{SW}, we obtain:

$$v_{L1}(t) = 0 \rightarrow v_G \cdot D(t) + \frac{1}{2}\left\{v_G - v_{C1}\right\} \cdot \left\{1 - D(t)\right\} = 0 \tag{23}$$

$$v_{L3}(t) = 0 \rightarrow \left\{v_{C1} - v_C\right\} \cdot D(t) - v_C \cdot \left\{1 - D(t)\right\} = 0 \tag{24}$$

Fig. 6. Experimental testing of the improved Cuk converter in partial-power processing architecture.

As input voltage (v_G) is assumed to be constant over a switching period. Solving for steady-state output voltage v_O then,

$$v_O = v_G \frac{1 + D^2}{1 - D} \tag{25}$$

In this case, a quadratic converter has been obtained, where the voltage gain "M" is,

$$\frac{v_O}{v_G} = \frac{1 + D^2}{1 - D} = M \tag{26}$$

Therefore, for the same value of the duty cycle $D(t)$, the improved Cuk converter with a double inductor provides a higher voltage gain.

Another additional alternative to this Cuk architecture is the incorporation of the double capacitor C_1. Figure 7 shows the double capacitor Cuk converter in partial-power processing architecture. In this case, the capacitor C_1 has been replaced by a network of diodes and capacitors. Again, for a T_{SW} sampling time, the switch is ON-state for $D \times T_{SW}$ and OFF-state for $(1 - D) \times T_{SW}$. Therefore, depending on the state of the switch, two modes of operation can be identified.

Mode 1: ON-state $[0 \leq t \leq D(t)T_{SW}]$.

At the initial time t = 0, when switch SW_1 turns on, diodes D_1 and D_2 turn off. The inductor currents (i_{L1}, i_{L2}) increase from their respective initial value. In this case, the voltage and current equations are as follows:

$$v_{L1}(t) = L_1 \frac{\partial i_{L1}}{dt} = v_G ; \quad v_{L2}(t) = L_2 \frac{\partial i_{L2}}{dt} = 2v_{C1} - v_C \tag{27}$$

$$i_{C1}(t) = C_1 \frac{\partial v_{C1}}{dt} = i_{L2} ; \quad i_C(t) = C \frac{\partial v_C}{dt} = i_{L2} - i_O \tag{28}$$

$$v_O = v_G - \left(-v_C\right) \tag{29}$$

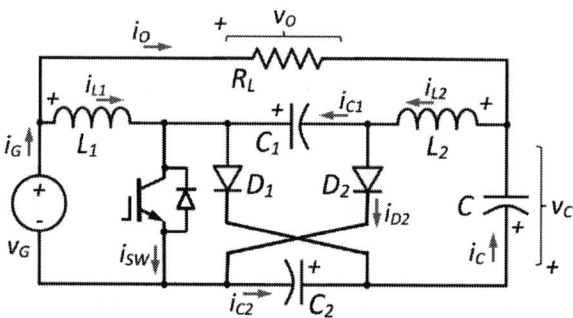

Fig. 7. Improved Cuk converter in partial-power processing architecture. Now the model has a double capacitor in the middle of the Cuk converter.

Mode 2: OFF-state $[D(t)T_{SW} \leq t \leq T_{SW}]$.

When switch SW_1 turns off at t = D×T$_{SW}$, diode D_1 and D_2 turn on. A similar analysis gives the equations for voltage and current as follows:

$$v_{L1}(t) = L_1 \frac{\partial i_{L1}}{dt} = v_G - v_{C1} ; \quad v_{C1} = v_{C2} \tag{30}$$

$$v_{L2}(t) = L_2 \frac{\partial i_{L2}}{dt} = v_{C1} - v_C \tag{31}$$

$$i_{C1}(t) = C_1 \frac{\partial v_{C1}}{dt} = \frac{1}{2}\left(i_{L2} - i_{L1}\right) \tag{32}$$

$$i_C(t) = C \frac{\partial v_C}{dt} = i_{L2} - i_O \tag{33}$$

$$i_{D1} = i_{D2} = \frac{1}{2}\left(i_{L1} + i_{L2}\right) \tag{34}$$

The average inductor voltage (v_{L1}, v_{L2}) must be zero at steady-state. Hence, analyzing the volt-sec balance for inductors (L_1, L_2) during a switching period T_{SW}, we obtain:

$$v_{L1}(t) = 0 \rightarrow v_G \cdot D(t) + \left\{v_G - v_{C1}\right\} \cdot \left\{1 - D(t)\right\} = 0 \tag{35}$$

$$v_{L2}(t) = 0 \rightarrow \left\{2v_{C1} - v_C\right\} \cdot D(t) + \left\{v_{C1} - v_C\right\} \cdot \left(1 - D(t)\right) = 0 \tag{36}$$

As input voltage (v_G) is assumed to be constant over a switching period. Solving for steady-state output voltage v_O then,

$$v_O = v_G \frac{2}{1-D} \tag{37}$$

In this new partial-power topology, the voltage gain "M" obtained in the improved Cuk converter is,

$$\frac{v_O}{v_G} = \frac{2}{1-D} = M \tag{38}$$

Another additional topology is described in Fig. 8. In this case, the dual capacitor Cuk converter is shown in a partial-power processing architecture. Now, the initial capacitor C_1 has been replaced by a network of diodes (D_{N1}, D_{N2}, D_{N3}) and capacitors (C_1, C_2). Again, for a T_{SW} sampling time, the switch is ON-state for $D \times T_{SW}$ and OFF-state for $(1 - D) \times T_{SW}$. Therefore, depending on the state of the switch, two modes of operation can be identified.

Mode 1: ON-state $[0 \le t \le D(t)T_{SW}]$.

At the initial time t = 0, when switch SW_1 turns on, diodes D_{N1} and D_{N2} turn on, while D_{N3} turns off. The inductor currents (i_{L1}, i_{L2}) increase from their respective initial value. In this case, the voltage and current equations are as follows:

$$v_{L1}(t) = L_1 \frac{\partial i_{L1}}{dt} = v_G ; \quad v_{L2}(t) = L_2 \frac{\partial i_{L2}}{dt} = v_{C1} - v_C \tag{39}$$

$$i_{C1}(t) = C_1 \frac{\partial v_{C1}}{dt} = \frac{1}{2}i_{L2} ; \quad i_C(t) = C \frac{\partial v_C}{dt} = i_{L2} - i_O \tag{40}$$

$$v_O = v_G - \left(-v_C\right) \tag{41}$$

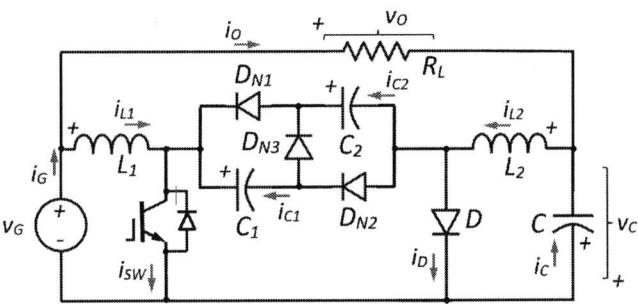

Fig. 8. Dual capacitor Cuk converter in partial-power processing architecture.

Mode 2: OFF-state $[D(t)T_{SW} \le t \le T_{SW}]$.

When switch SW_1 turns off at t = $D \times T_{SW}$, diode D_{N1} and D_{N2} turn off, while D_{N3} turns on. A similar analysis gives the equations for voltage and current as follows:

$$v_{L1}(t) = L_1 \frac{\partial i_{L1}}{dt} = v_G - 2v_{C1} ; \quad v_{C1} = v_{C2} \tag{42}$$

$$v_{L2}(t) = L_2 \frac{\partial i_{L2}}{dt} = -v_C \tag{43}$$

$$i_{C1}(t) = C_1 \frac{\partial v_{C1}}{dt} = -i_{L1} ; \quad i_{C1} = i_{C2} \tag{44}$$

$$i_C(t) = C \frac{\partial v_C}{dt} = i_{L2} - i_O \tag{45}$$

$$i_{D3} = i_{L1} \tag{46}$$

The average inductor voltage (v_{L1}, v_{L2}) must be zero at steady-state. Hence, analyzing the volt-sec balance for inductors (L_1, L_2) during a switching period T_{SW}, we obtain:

$$v_{L1}(t) = 0 \rightarrow v_G \cdot D(t) + \left\{v_G - 2v_{C1}\right\} \cdot \left\{1 - D(t)\right\} = 0 \tag{47}$$

$$v_{L2}(t) = 0 \rightarrow \left\{v_{C1} - v_C\right\} \cdot D(t) - v_C \cdot \left\{1 - D(t)\right\} = 0 \tag{48}$$

As input voltage (v_G) is assumed to be constant over a switching period. Solving for steady-state output voltage v_O then,

$$v_O = v_G \frac{2-D}{2(1-D)} \tag{49}$$

In this additional partial-power architecture, the voltage gain "M" obtained in the dual capacitor Cuk converter is,

$$\frac{v_O}{v_G} = \frac{2-D}{2(1-D)} = M \tag{50}$$

Table I shows a comparison between the voltage gains in different full-power and partial-power architectures. It should be noted that in the Cuk full-power topology, the output voltage v_O can be higher or lower than the input voltage v_G, as the duty cycle (D) increases. Whereas in Cuk partial-power architecture, the output voltage v_O is always higher than the input voltage v_G depending on the value adopted by the duty cycle (D). The non-isolated partial-power converter used has shown good results in terms of efficiency and power density. Traditionally, the load is connected to the output (v_C) of the DC-DC converter. In other words, in the full-power topology, this converter processes all the power supplied to the load. Meanwhile, in partial-power architecture, the load is connected in series between the input voltage (v_G) and the Cuk converter output (v_C).

TABLE I. COMPARISON DIFFERENT PROPOSED CUK CONVERTERS IN FULL-POWER AND PARTIAL-POWER ARCHITECTURE.

Converter model	Voltage-Gain in Full-Power	Voltage-Gain in Partial-
Buck-Boost Converter	$\dfrac{D}{(1-D)}$	$\dfrac{1}{(1-D)}$
Hybrid-Boost Converter [20]	$\dfrac{3-D}{(1-D)}$	---
Typical Cuk Converter	$\dfrac{D}{(1-D)}$	$\dfrac{1}{(1-D)}$
Improved Cuk Converter (2L)	$\dfrac{D(1+D)}{(1-D)}$	$\dfrac{1+D^2}{(1-D)}$
Improved Cuk Converter (2C)	$\dfrac{1+D}{(1-D)}$	$\dfrac{2}{(1-D)}$
Dual Capacitor Cuk Converter	$\dfrac{D}{2(1-D)}$	$\dfrac{2-D}{2(1-D)}$

TABLE II. PARAMETERS ASSOCIATED WITH THE NON-ISOLATED CUK PARTIAL-POWER CONVERTER MODEL.

Parameter	Symbol	Value
DC-bus voltage	v_G	+25V
Module internal resistance	r_{DC}	0.1Ω
Cuk inductors	L_1, L_2	2.5mH
Cuk capacitor	C_1, C_2	500μF
Switching frequency	f_{SW}	20kHz
Switch resistance S_1, S_2	R_{on}	0.01Ω
Schottky forward voltage	v_F	0.3V
Snubber network	Rs, Cs	100kΩ; 1nF
Output Capacitor	C	250μF
Load resistance	R_L	50Ω

4. Simulation Results

In order to study the properties and characteristics of the improved non-isolated Cuk converter in applied partial-power architecture, the topology shown in the previous figures has been modeled. The simulation software used has been *Matlab-Simulink*. At the same time, an average voltage mode control has been developed. Likewise, Table II contains some of the parameters used in the model.

Figure 8 represents the different voltages (v_{S1}, v_{D1}) in the semiconductor devices and the voltage and current in the capacitor (v_{C1}, i_{C1}) respectively, as well as the inductor currents (i_{L1}, i_{L2}) of the non-isolated Cuk converter in the partial-power processing architecture. These waveforms correspond to the topology shown in Fig. 3. In this case, the switching frequency f_{SW} corresponds to f_{SW} = 20kHz.

Under these conditions, the following values were obtained: DC-bus voltage v_G = +25V, output voltage v_O = +100V, capacitor voltage C_1 (v_{C1} = +100V), average DC-bus current i_G = 8.212A, while the average currents of the inductors i_{L1} = 6.212A and i_{L2} = 2A and the value of the current in the load R_L is i_O = 2A. Meanwhile, the average switch current i_{S1} and the average diode current i_D are i_{S1} = 6.212A and i_D = 2A, respectively. In this case, for the Cuk partial-power topology, the duty cycle results $D_{S1}|_{PP}$ = 0,756.

Figs. 9 and 10 show a comparison of the peak currents i_{L1}, i_{S1} and i_{C1} for the Cuk converter in full-power and partial-power topology. In both cases, an output voltage v_O = +100V has been considered. Now in the full-power topology, the duty cycle corresponds to $D_{S1}|_{FP}$ = 0,807. As can be seen in the figures, for the same power supplied to the R_L load, the peak currents are higher in the full-power topology than in the partial-power architecture. In the extended paper, more oscillograms will be shown comparing the different architectures discussed in the previous sections. In this way, the advantages and drawbacks of each of the architectures presented will be observed. The limitation of pages in the abstract has limited the simulation results shown.

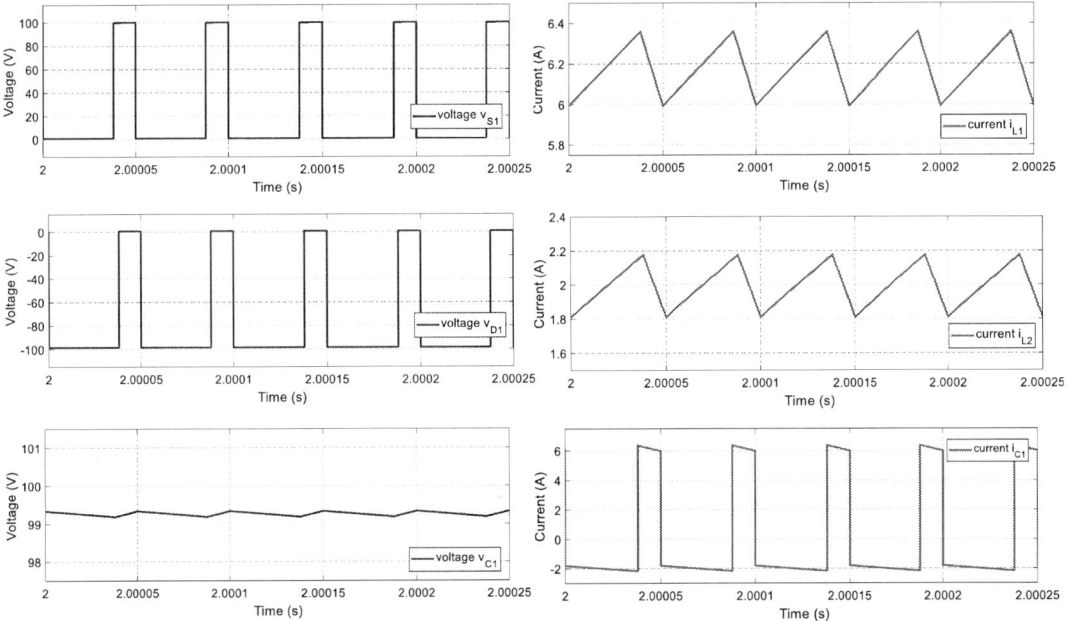

Fig. 9. Steady-state response of the non-isolated Cuk converter in partial-power architecture: v_{S1}, i_{L1}, v_{S2}, i_{L2}, v_{C1} and i_{C1}. Voltages and currents to provide a converter output voltage $v_O = +100V$ (duty cycle D = 0.756).

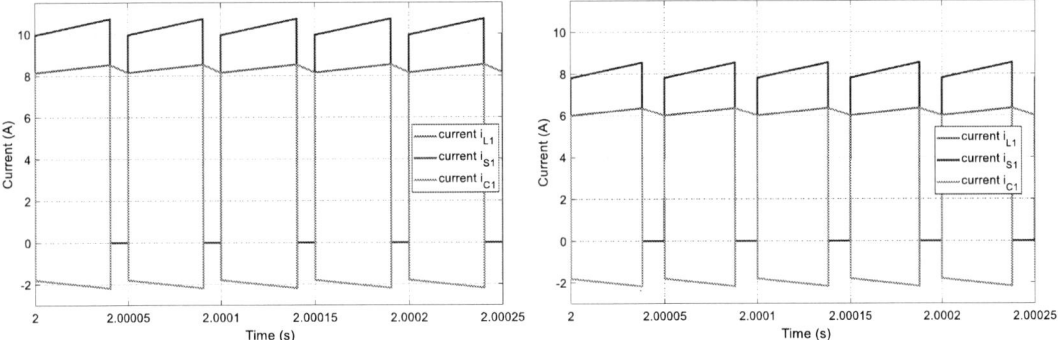

Fig. 10. Steady-state response of the non-isolated Cuk converter in full-power topology. Currents i_{L1}, i_{S1} and i_{C1} to provide a converter output voltage $v_O = +100V$ (duty cycle $D_{FP} = 0.807$).

Fig. 11. Steady-state response of the non-isolated Cuk converter in partial-power topology. Currents i_{L1}, i_{S1} and i_{C1} to provide a converter output voltage $v_O = +100V$ (duty cycle $D_{PP} = 0.756$).

5. Conclusions

This paper presents the analysis of different non-isolated Cuk converters in partial-power processing architecture. The purpose has been to compare both architectures (full-power and partial-power converter) to validate their advantages. In the same way, the improved Cuk converter has been studied, analyzing its use in partial-power architecture. The different simulation models have been developed with the *Matlab-Simulink* software. The results obtained support the different equations of the mathematical model studied.

In view of the simulation results obtained, the stress on the different devices is higher in the full-power architecture than in the partial-power architecture, for the same power supplied to the load. The partial-power topology only processes a fraction of the supplied power, reducing power losses and increasing system efficiency. In this way, it is possible to reduce the cost, volume, and weight of the converter.

References

[1] J. Anzola, I. Aizpuru, A. Arruti Romero, A. Alacano Loiti, R. Lopez-Erauskin, J.S. Artal-Sevil and C. Bernal, "Review of Architectures Based on Partial-Power Processing for DC-DC Applications". IEEE Access. IEEExplore Digital Library. Vol. 8, June 2020; pp.: 1-14.

[2] H. Jeong, H. Lee, Y.C. Liu and K.A. Kim, "Review of Differential Power Processing Converter Techniques for Photovoltaic Applications". IEEE Transaction on energy Conversion. IEEExplore Digital Library. Vol. 34, issue:1, October 2018; pp.: 351-360.

[3] V.M. Iyer, S. Gulur, G. Gohil, and S. Bhattacharya, "An Approach Towards Extreme Fast Charging Station Power Delivery for Electric Vehicles with Partial-Power Processing". IEEE Transactions on Industrial Electronics. IEEExplore Digital Library. Vol. 67, issue: 10, October 2020; pp.: 8076-8087.

[4] J.S. Artal-Sevil, J.A. Domínguez, C. Bernal-Ruiz and A. Coronado-Mendoza, "Power Flow Control through Differential Power Processing to improve reliability in hybrid systems based on PEM-Fuel Cell". International Conference on Ecological Vehicles and Renewable Energies (EVER'20). IEEExplore Digital Library. September 2020, MonteCarlo (Mónaco); pp.: 1-12.

[5] M. Pape and M. Kazerani, "Turbine Startup and Shutdown in Wind Farms Featuring Partial-Power Processing Converters". IEEE Open Access Journal of Power and Energy. IEEExplore Digital Library. Vol. 7, July 2020; pp.: 254-264.

[6] J.R.R. Zientarski, M.L. da Silva, J. Renes Pinheiro and H. Leães Hey, "Evaluation of Power Processing in Series-Connected Partial-Power Converters". IEEE Journal of Emerging and Selected Topics in Power Electronics. IEEExplore Digital Library. Vol.: 7, issue: 1, March 2019; pp. 343–352.

[7] J.W. Zapata, S. Kouro, G. Carrasco, H. Renaudineau and T.A. Meynard, "Analysis of Partial-Power DC–DC Converters for Two-Stage Photovoltaic Systems". IEEE Journal of Emerging and Selected Topics in Power Electronics. IEEExplore Digital Library. Vol.: 7, issue: 1, March 2019; pp. 591–603.

[8] T. Kanstad, M.B. Lillholm and Z. Zhang, "Highly Efficient EV Battery Charger Using Fractional Charging Concept with SiC Devices". IEEE Applied Power Electronics Conference and Exposition (APEC'19). IEEExplore Digital Library. March 2019, Anaheim, CA, (USA); pp. 1601–1608.

[9] S. Mallangada, M.O. Badawy and Y. Sozer, "A Novel Differential Power Processing Architecture for a Partially Shaded PV String Using Distributed Control". IEEE Energy Conversion Congress and Exposition (ECCE). IEEExplore Digital Library. September 2018, Portland (USA); pp. 6220–6227.

[10] J.S. Artal-Sevil, C. Bernal-Ruiz, J. Beyza and V.M. Bravo, "Evaluation of a Thermoelectric Generation system based on Differential-Power Processing architecture under non-uniform temperature conditions" IEEE International Autumn Meeting on Power, Electronics and Computing (ROPEC'20). IEEExplore Digital Library. November 2020, Ixtapa (Mexico); pp.: 1-6.

[11] L. Chen, H. Wu, P. Xu, H. Hu and C. Wan, "A high step-down non-isolated bus converter with Partial-Power conversion based on synchronous LLC resonant converter". IEEE Applied Power Electronics Conference and Exposition (APEC'15). IEEExplore Digital Library. Charlotte, NC, USA. March 2015; pp.: 1950-1955.

[12] T. Suntio and A. Kuperman, "Comments on "An efficient Partial-Power Processing DC/DC converter for distributed PV architectures"" IEEE Transactions on Power Electronics. IEEExplore Digital Library. vol. 30, no. 4. April 2015; pp.: 2372.

[13] F. Hoffmann, J. Person, M. Andersen, M. Liserre, F.D. Freijedo and T. Wijekoon, "A multiport Partial-Power processing converter with energy storage integration for EV stationary charging". IEEE Journal of Emerging and Selected Topics in Power Electronics. IEEExplore Digital Library. August 2021; pp.: 1-13.

[14] S. Rivera, J. Rojas, S. Kouro, P.W. Lehn, R. Lizana, H. Renaudineau and T. Dragicevic, "Partial-Power Converter Topology of Type II for Efficient Electric Vehicle Fast Charging". IEEE Journal of Emerging and Selected Topics in Power Electronics. IEEExplore Digital Library. October 2021; pp.: 1-10.

[15] J. Anzola, I. Aizpuru, A. Arruti, A. Alacano, R. López, J.S. Artal-Sevil and C. Bernal, "Partial-Power Processing Based Charging Unit for Electric Vehicle Extreme Fast Charging Stations". IEEE Vehicle Power and Propulsion Conference (VPPC'20). IEEExplore Digital Library. November 2020; pp.: 1-6.

[16] Y. Cao, M. Ngo, N. Yan, Y. Bai, R. Burgos and D. Dong, "DC Distribution Converter with Partial-Power Processing for LVDC/MVDC Systems". International Conference on DC Microgrids (ICDCM'21). IEEExplore Digital Library. July 2021; Arlington (USA); pp.: 1-8.

[17] M.C. Mira, Z. Zhang, K.L. Jørgensen and A.E.M. Andersen, "Fractional Charging Converter With High Efficiency and Low Cost for Electrochemical Energy Storage Devices". IEEE Transactions on Industry Applications. IEEExplore Digital Library. Vol.: 55, issue: 6 Nov 2019; pp. 7461–7470.

[18] J.S. Artal-Sevil, C. Bernal-Ruiz, J. Anzola, I. Aizpuru, A. Bono-Nuez and J.M. Sanz-Alcaine, "Partial Power Processing architecture applied to a Battery Energy Storage System" IEEE Vehicle Power and Propulsion Conference (VPPC'20). IEEExplore Digital Library. November-December 2020; Gijón (Spain) pp.: 1-6.

[19] J.S. Artal-Sevil, V. Ballestín-Bernad, J. Anzola and J.A. Domínguez-Navarro, "High-Gain Non-isolated DC-DC Partial-Power Converter for Automotive Applications" IEEE Vehicle Power and Propulsion Conference (VPPC'21). IEEExplore Digital Library. October 2021; Gijón (Spain) pp.: 1-6.

[20] M. Chen, K. Li and A. Ioinovici, "A family of high DC gain step-up non-isolated converters based on a new hybrid passive switching cell" 18th European Conference on Power Electronics and Applications (EPE'16 ECCE Europe). IEEExplore Digital Library. September 2016, Karlsruhe (Germany); pp.: 1-9.

GaN HEMT and SiC Diode Commutation Cell based Dual-Buck Single-Phase Inverter with Premagnetized Inductors and Negative Gate Driver Turn-off Voltage

Tobias Brinker, Hendrik Gräber, and Jens Friebe
Leibniz University Hannover
Institute for Drive Systems and Power Electronics
Hanover, Germany
Email: Tobias.Brinker@ial.uni-hannover.de
URL: http://www.ial.uni-hannover.de

Acknowledgments

Parts of this work were funded by the German Federal Ministry for Economic Affairs and Climate Action under Grant No. 03EE1057A (Voyager-PV) on the basis of a decision by the German Bundestag and also by the Ministry of Science and Culture of Lower Saxony and the Volkswagen Foundation. The authors are responsible for the content of this publication.

Keywords

≪Single phase system≫, ≪Switching losses≫, ≪Magentic device≫, ≪Gallium Nitride (GaN)≫, ≪Silicon Carbide (SiC)≫.

Abstract

This paper presents a highly efficient single-phase dual-buck inverter topology for the use in photovoltaic (PV) micro inverters with specific advantages for the application of switching cells composed of Gallium Nitride enhancement-mode high electron mobility transistors (GaN E-HEMTs) and Silicon Carbide (SiC) diodes. For this inverter topology, the opportunity to use negative gate driver turn-off voltage without affecting the reverse conduction loss is identified as a significant advantage in terms of efficiency and cross talk immunity. However, the low magnetic utilization of the filter inductors is a well-known drawback. Therefore, permanent magnet premagnetization is proposed as a way to alleviate this issue.

Introduction

Single-phase photovoltaic (PV) micro inverters require high efficiency and compact size as well as high reliability, making them an interesting field of application for 650 V GaN E-HEMTs. Their inherent advantages include high switching speed, high breakdown voltage and low on-resistance [1]. Despite all these, GaN E-HEMTs suffer from false turn-on phenomena due to their low threshold voltage and reverse conduction loss when switched-off during dead-time [1]. A highly efficient single-phase dual-buck inverter topology, with specific advantages for the use of GaN E-HEMTs by avoiding conventional half-bridge circuits, is presented and discussed in this paper. The topology allows high efficiency and reliability but suffers from low magnetic utilization and bulky inductor dimensions [2]. For further optimization, specific benefits regarding the use of GaN E-HEMTs and Silicon Carbide (SiC) Schottky diodes will be introduced and discussed.

For typical full-bridge inverters, the influence of the gate driver turn-off voltage $V_{\text{GS,off}}$ on transistor loss has been investigated in [3]. With the reduction of $V_{\text{GS,off}}$, a negative effect on the reverse conduction loss and a positive effect on the turn-off loss has been observed, prevailing slightly for higher currents.

In contrast to the full bridge topology, dead-times can be omitted in the dual buck topology due to the specific switching cells, consisting of a SiC Schottky diode and GaN E-HEMT. Therefore, the opportunity to use negative $V_{GS,off}$ without affecting the reverse conduction loss is identified as a significant advantage in terms of turn-off loss with the side effect to prevent false turn-on phenomena.

The dual buck inverter topology and its modulation scheme leads to unipolar excitation of the filter inductors during a single half-period of the output voltage [4]. For optimization of this low magnetic utilization, permanent magnetic premagnetization has been proposed as a suitable method in [5]. On this basis, an alternative configuration of permanent magnet and core is proposed and compared in this work.

This paper is structured as follows: The first section presents the dual buck inverter topology and its modulation scheme. Then the impact of negative $V_{GS,off}$ on switching losses is described analytically. For validation, double pulse tests with a hardware prototype were performed. Furthermore, the improvement of the low magnetic utilization by permanent magnetic premagnetization is discussed and the alternative concept is presented. Conclusions are given in the final section.

Dual-Buck Inverter Topology

Fig. 1 shows the proposed dual buck single-phase inverter topology and the active power driving scheme. The circuit is composed of two parallel buck converters (T1-L1-T2-D2 and T3-L2-T4-D1), whose outputs are connected to the load with opposite polarity [6, 7]. Each buck converter is modulating the output current i_{ac} for one polarity of the grid voltage v_{ac}. For the positive (negative) polarity, v_{ac} is modulated by T2 (T4) with switching frequency f_{sw}, while T1 (T3) controls the polarity of v_{ac} with low ac frequency f_{ac}. This enables an arrangement of transistors optimized for either low conduction losses (T1, T3) or low switching losses (T2, T4).

The inverter is capable of providing active and reactive power, as can be derived from the switching states depicted in Fig. 2 [6, 8]. In the figure, modulating inductor currents are indicated by the blue lines and freewheeling currents are indicated by the orange lines. In quadrants I and III, active power is fed into the grid and the switching scheme shown in Fig. 1 applies. For the reactive power quadrants (II and IV) the transistors (T1, T3) operated with f_{ac} are both turned-off, while the switching scheme of T2 or T4 does not need to be adjusted. To reduce diode conduction losses, T2 and T4 can be switched complementary in order to bypass the low ac frequency current from the diodes (dashed lines).

For the well-known issue of leakage currents in single-phase inverters, the common mode voltages over the module capacitances are relevant [4, 9]. These voltages can be measured between DC+ or DC- and the grounded neutral terminal (N) of the transformerless inverter. For active power operation (quadrants I and III), N is always connected to either the high- or low-side of the DC link (DC+, DC-, respectively). During the positive half-wave of v_{ac} (quadrant I) T1 clamps the DC+ potential to v_{ac}, while T3 clamps it to the grounded neutral terminal during the negative half-wave (quadrant III). Therefore, no discontinuities of the voltage between DC+ or DC- and ground are possible and leakage currents are considered low. [9]

For reactive power operation (quadrants II and IV), the potential of N depends on the current flow direction since T1 and T3 are turned-off. For quadrant II, D2 clamps it to v_{ac}, while D4 clamps it to DC- in quadrant IV. Regarding the voltage between DC+ or DC- and ground, a discontinuity appears when changing from any active power state to any reactive power state. As a result, peak leakage currents

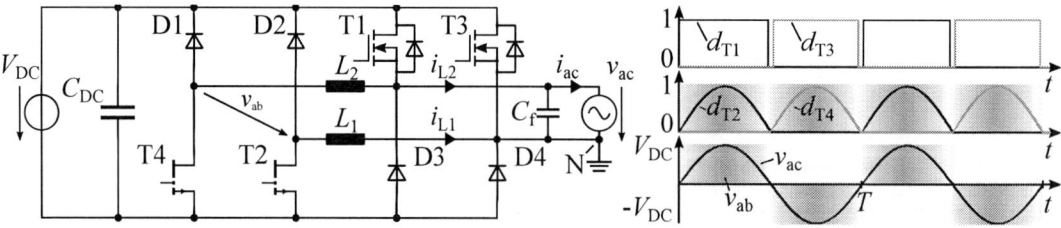

Fig. 1: Circuit diagram of dual buck inverter with control signals for active power operation.

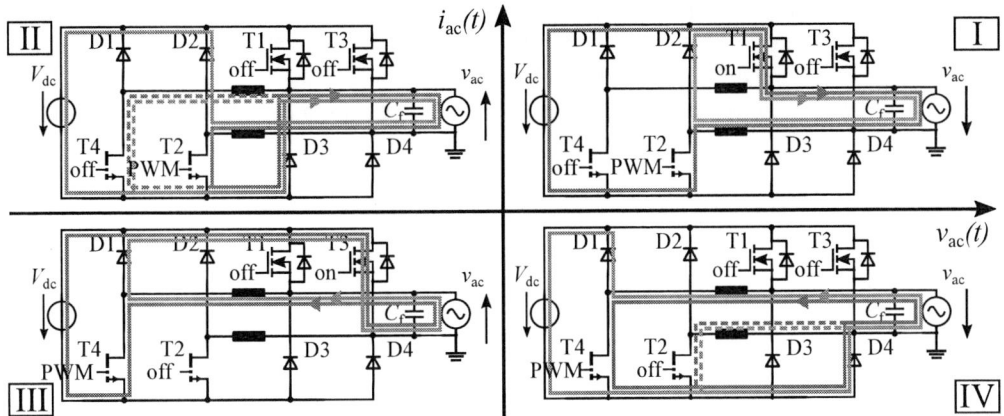

Fig. 2: Overview of the active and reactive power switching states. Modulating inductor currents are shown by the blue lines and freewheeling currents are shown by the orange lines. Dashed lines indicate additional current paths to reduce diode conduction losses.

can occur at the transition between active and reactive power states. In contrast to the conventional full-bridge inverters, the dual-buck inverter is not recommended for the use in transformerless PV inverters when reactive power capability is required. For the investigations in this paper, only active power states are considered.

Hardware Prototype

For the measurements, a prototype with adjustable $V_{GS,off}$ and exchangeable inductors has been assembled (see Fig. 3). Components were selected according to their specific characteristics (see Table I). Conduction loss optimized Silicon superjunction MOSFETs are used for the ac frequency operated transistors T1 and T3 as their switching losses are too small to be considered [7]. Due to their dynamic performance, GaN E-HEMTs and and SiC Schottky diodes are used for the high-frequency (HF) stage (T1,T4,D1-D4). Both of these are wide-bandgap devices that do not exhibit any bipolar behavior, such as reverse recovery charge Q_{rr}. Since there are no conventional half-bridge circuits, it is possible to omit dead-times for T2 and T4. Nevertheless, T1 and T3 still require a dead-time to prevent short circuiting the grid, but are only operated slowly (f_{ac}). As described in the section above, there are no reverse conduction losses for T2 and T4, which allows to use negative $V_{GS,off}$ without losing efficiency. This is expected to reduce switching loss [10] and to prevent false turn-on phenomena [1].

For the prototype a driver supply with adjustable turn-on and turn-off voltage is used. It is based on a

Table I: Hardware prototype components and operational parameters.

Parameter	Symbol	Value
GaN HEMTs	T2, T4	GS66504B
Si MOSFETs	T1, T3	IPT60R022S7XTMA1
SiC Schottky diodes	D1-D4	STPSC4H065DLF
Gate driver IC	–	Si8271 series
Adj. driver supply	–	RP-1209D, TPS7A39
Filter capacitance	C_f	250 nF (C0G)
DC-link capacitance	C_{DC}	3x 180 µF, 2x 2 µF
DC-link voltage	V_{DC}	400 V
Ext. turn-on res. (GaN)	$R_{G,on}$	3 Ω
Ext. turn-off res. (GaN)	$R_{G,off}$	0 Ω
Gate turn-on voltage (GaN)	$V_{GS,on}$	6 V

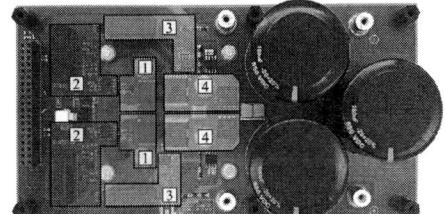

Fig. 3: Photograph of the hardware prototype.
1: GaN HEMT and SiC diode switching cells.
2: Adjustable gate driver supplies for GaN HEMTs.
3: High-side driver supplies for Si MOSFETs.
4: Si MOSFET and SiC diode commutation cells.

bidirectional low dropout regulator with the ability of regulating the negative rail down to zero volts [11]. Conventional high-side drivers with isolated driver supplies are used for all transistors of this prototype. No isolated gate driver supply is required for the low-side gate drivers [5]. The absence of di/dt in reverse conduction also allows normally-off depletion mode GaN cascodes or super junction MOSFETs to be considered, since the reverse recovery of their slow body diodes is irrelevant here [12].

Impact of Negative Gate Driver Turn-off Voltage

The effects of a negative $V_{GS,off}$ on the switching characteristics were investigated with the inverter prototype configured for an inductively clamped double pulse test according to Fig. 4. The double pulse test is usually used to characterize the turn-on and turn-off energies of a transistor in dependence of the drain current. For this purpose, a high bandwidth measurement of the commutation current as well as the drain source voltage of the device under test (DUT) is required. Installing a suitable current sensor would insert additional inductance to the commutation cell and affect the measurement result. For this reason, the evaluation of the switching energies is omitted here and instead qualitative conclusions about the switching energies are drawn on the basis of slew rate measurements of v_{DS}. Thereby, an increased slew rate is associated with lower switching energy.

The results obtained from the slew rate measurements are presented in Fig. 5 and indicate faster switching transients and therefore lower switching losses for negative $V_{GS,off}$ during turn-off, while for turn-on transients no significant effect can be observed. In the figure, the $R_{DS(on)}$ for a junction temperature of 150° is used as an indicator for the the usable operating current range preventing saturation of the GaN E-HEMT (GS66504B [13]). The current range of the hardware prototype is highlighted to show that negative turn-off voltages do not affect the efficiency in typical inverter operation current range. In the following sections, the turn-on and turn-off sequences are examined to explain the effects more in detail.

Turn-on

The turn-on slope depends indirectly on the drain current i_{T4} via the Miller plateau voltage v_{pl} and is rather determined by the gate driver, the transconductance g_{fs} and the output capacitance C_{OSS} of the device under test (DUT) and the output capacitance of the diode $C_{OSS,D}$ [14, 15, 16, 17]:

$$v_{pl} = V_{th} + \frac{i_{T4}}{g_{fs}} \tag{1}$$

$$\frac{dv_{DS}}{dt} = \frac{-g_{fs}(v_{GS} - v_{pl})}{C_{OSS} + C_{OSS,D}}. \tag{2}$$

As indicated in (1) v_{pl} increases for higher drain currents. For the turn-on process, this leads to a slight reduction in slew rate (see (2)) and consequently higher switching energies for higher drain currents, regardless of the gate driver turn-off voltage [15]. This explains the downward trend in turn-on speed for increasing drain currents.

Another cause for the slight enhancement in turn-on speed can be identified in the interaction of the parasitic gate inductance and the negative gate turn-off voltage [10]. While the GaN E-HEMT is switched-off, the gate-source capacitance C_{GS} is charged to $V_{GS,off}$. During the turn-on process, C_{GS} is then first

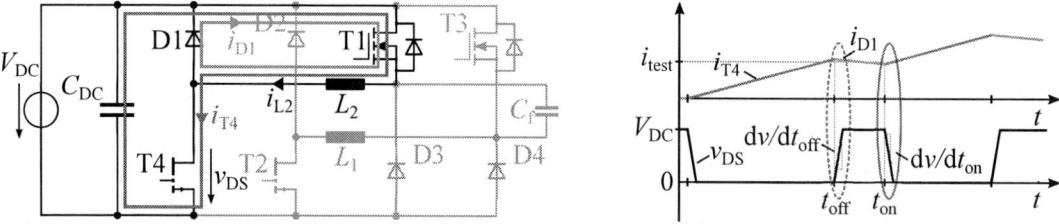

Fig. 4: Double pulse test setup of the prototype board and basic operation waveform. Slew rates dv/dt are used as indicators for switching energy. T4 is the DUT, while T1 is constantly switched-on.

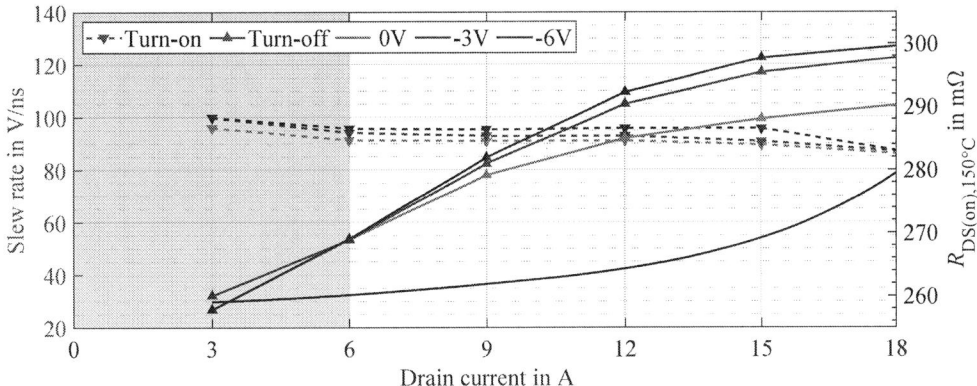

Fig. 5: Turn-on and turn-off slew rates $(10\% - 90\%)$ of v_{DS} for different $V_{GS,off}$ with respect to the drain current I_{T4}. The inverter current range is highlighted. $R_{(DS(on))}$ indicates saturation region of GaN HEMT.

charged up to V_{th}. As the threshold voltage is reached, the drain-source voltage of the semiconductor drops rapidly, causing a capacitive displacement current (Miller current) that effectively counteracts the gate current and stops the charge process of C_{GS} until C_{GD} is discharged (Miller plateau). At the same time, the nonlinear voltage dependent output capacitance C_{OSS} rises rapidly for lower V_{DS}. When a negative turn-off voltage is applied, the parasitic gate inductance is exposed to a larger voltage-time area, which results in a higher gate current being present when v_{GS} reaches the miller plateau [10]. This higher current can compensate the Miller current more effectively, keeping the gate of the semiconductor more stable above V_{th} and thus accelerating the turn-on process and in turn reduces the turn-on losses.

Turn-off

In contrast to the turn-on slope, the gate voltage can actually drop below v_{th} during the turn-off transient, at which time no 2DEG current flows at all. Therefore, (2) has to be separated for turn-off as a piecewise equation [14]:

$$\frac{dv_{DS}}{dt} = \begin{cases} \frac{i_{T4}}{C_{OSS}+C_{OSS,D}} & v_{GS} \leq V_{th} \\ \frac{g_m(v_{pl}-v_{GS})}{C_{OSS}+C_{OSS,D}} & v_{GS} > V_{th} \end{cases} \tag{3}$$

For $v_{GS} > v_{th}$ the turn-off process can exemplary be described as follows [1, 14, 15]: First, the gate current i_G starts to discharge the input capacitance C_{GS}. The load current i_{T4} continues to flow through 2DEG, with no current flowing into the output capacitance C_{OSS}. When v_{GS} drops down to v_{pl}, 2DEG starts to pinch-off and i_{T4} redirects from 2DEG into C_{OSS}. In this case, the Miller plateau voltage shift with increasing currents accelerates the turn-off transition (see (3)). During this period, the gate driver partially controls the turn-off transition by keeping v_{gs} below V_{pl}. When v_{GS} drops below V_{th}, 2DEG is completely pinched-off and entire i_{T4} charges C_{OSS} until it reaches V_{DC}. During this interval, the gate driver does not affect the turn-off slew rate at all. [16, 17]

After v_{GS} falls below V_{th}, 2DEG is completely pinched-off and the entire i_{T4} charges C_{OSS} until it reaches V_{DC}. For GaN E-HEMTs, applying a negative gate driver turn-off voltage reduces v_{pl} [16, 17]. In principle, v_{GS} can be immediately dropped below v_{th} by selecting a proper negative gate driver turn-off voltage. Then no Miller plateau is visible and the turn-off transition is completely controlled by the drain current, potentially accelerating the switching transient and lowering the switching energy.

From the experimental data provided in Fig. 5 an increase in turn-off slew rate can be observed for negative $V_{GS,off}$ and higher drain currents. For lower drain currents in the range below approx. 9 A, no significant deviation of the slew rate from zero volt can be observed for negative $V_{GS,off}$. For drain currents above 9 A, the 2DEG is apparently pinched off by v_{GS} for each investigated negative turn-off voltage of the gate driver, since no significant change in the slew rate can be observed between negative $V_{GS,off}$.

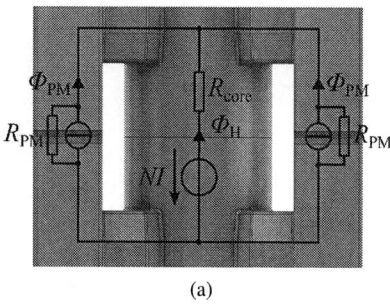 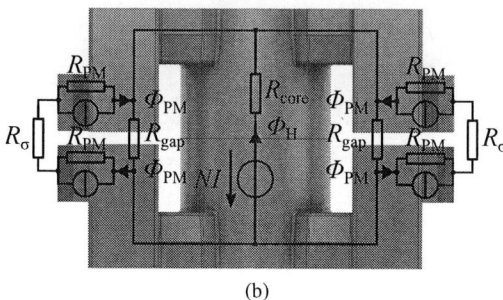

(a)   (b)

Fig. 6: (a) Serial concept of premagnetization with PM inserted into the air gaps and (b) parallel concept with PM located in the vicinity of the air gaps. Magnetic north and south poles of the PM are marked in red and green, respectively.

Permanent Magnet Premagnetized Filter Inductors

From Fig. 2 can be derived, that in each switching state, current only can flow in one direction through one of the two inductors, leading to a poor magnetic utilization, which depicts the main drawback of the topology. Although, this significantly affects the weight and volume of the system, it does not affect the efficiency [6, 8]. Due to their low utilization, the inductors can be designed with at least twice the specific core losses during switching operation of their corresponding half-bridge. As a direct consequence of this, the inductor design of both inductors will lead to a higher switching frequency of which the saturation flux will limit the design instead of the thermal limit [5]. A optimization possibility is offered by PM premagnetization [5]. Single inductor variants of the inverter have been introduced, but suffer from higher conduction loss due to additional switches or diodes in the power loop [8].

The concept of PM premagnetization is as follows: One or more PM is used to insert a negative flux into the core to extend the saturation flux density for unipolar excitation. As a result, a smaller core volume can be used and thus a higher magnetic utilization can be achieved.

The most conventional and intuitive approach of PM premagnetization is illustrated in Fig. 6a and relies on inserting PM into the air gap of the power inductor [5]. This imposes several intrinsic constraints on the available magnetic cross section A_{PM}, demagnetization by the flux ϕ generated by the coil, and eddy current losses in the PM material at high switching frequencies [18]. However, this approach provides the most compact volume and can be used with standard cores. Other variants of this concepts include inserting PM into all air gaps or in the central leg only.

To prevail the limitations mentioned above, PM can be placed in the vicinity of the air gaps according to Fig. 6b. In this configuration, the PM may not be crossed by the magnetic flux generated by the coil, resulting in lower eddy current losses. Eddy currents are rather generated by deviations in the PM reluctance which are caused by the change of the PM operating point, but are smaller than those caused

Fig. 7: Parameters of conventional and premagnetized inductors with 40 % copper filling factor.

Parameter	Air gap (no PM)	Parallel concept	Serial concept
Core type	PQ32/20	PQ26/20	PQ26/20
Effective magnetic cross section	154.2 mm^2	122.6 mm^2	122.6 mm^2
Core box volume	15 cm^2	12.57 cm^2	10.57 cm^2
Permanent magnet material	-	Neofer 41/100p	Neofer 41/100p
Number of turns	31	26	26
Air gap width (central/outer leg)	0.8/0.8 mm	0/1 mm	0/1 mm
Dimensions	(32x22x21.3) mm	(31.5x19x21) mm	(26.5x19x21) mm
Core box volume	15 cm^3	12.57 cm^3	10.57 cm^3

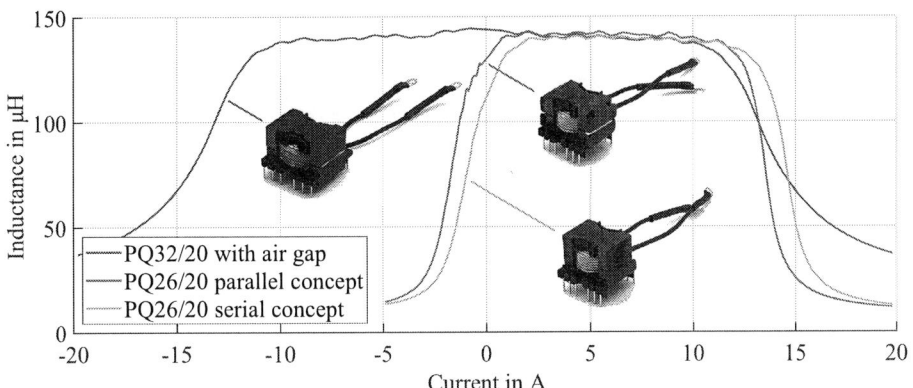

Fig. 8: L-I curves of conventional PQ32/20 inductors and premagnetized PQ26/20 inductors.

by direct induction. There are also no limitations in space in terms of magnetic cross-section and PM length. However, when using standard cores, this configuration results in an increased box volume due to the external PMs. [18]

The dimensioning of the magnetic circuit is not different from conventional air gapped inductors and is dominated by the relucatance of the air gaps (see Fig. 6). For a complete shift of the L-I characteristic the relation given in (4) can be used. The dimensioning of the combined magnetic circuit can be performed following the methodology presented in [19], considering the lumped magnetic circuits given in Fig. 6.

$$B_{PM} = \frac{L \cdot I_{sat}}{N A_{PM}} \tag{4}$$

Table 7 compares a conventional PQ32/20 inductor with two premagnetized configurations based on smaller PQ26/20 cores. Polymer-bonded (PA11) rare earth magnets (NdFeB) of Neofer 41/100p type were used for both configuration. These magnets are characterized by a residual flux density B_r, that fits to the saturation flux density of the TDK N95 ferrite core material. They also have good mechanical properties and are easy to shape. All inductors were set up with the same copper filling factor of 40 % and same litz wire (100x0.1 mm). The air gap is adjusted in order to reach the inductance of 140 μH. Fig. 8 shows the measured L-I curves, which show almost equal saturation currents of the premagnetized and conventional inductors.

For unipolar excitation, as it appears for the investigated inverter topology, a premagnetized inductor with up to 30 % smaller core volume can be used. Another advantage of using smaller cores is the reduced number of turns, which improves winding losses.

Conclusion

In this paper, two design improvement options for a specific single-phase dual-buck inverter topology using GaN E-HEMTs and SiC Schottky diodes were identified and investigated. For this purpose, an analysis of the switching states for both, active and reactive power operation has been performed. This revealed a low utilization of the magnetic components and the possibility to omit dead-times for the GaN E-HEMTs and SiC Schottky diode based switching cells. The absence of dead-times eliminates reverse conduction losses when the GaN E-HEMT is switched-off. This gives the opportunity to use a negative gate driver turn-off voltage $V_{GS,off}$ without affecting the reverse conduction losses.

A prototype with adjustable $V_{GS,off}$ and exchangeable inductors has been assembled and double pulse tests were performed. Instead of evaluating of the switching energies, qualitative conclusions about the switching energies are drawn on the basis of slew rate measurements of v_{DS}. From the experimental data an increase in turn-off slew rate can be observed for negative $V_{GS,off}$, but only for higher drain currents, while no significant effect on slew rate could be observed. Thus, it could be shown that a negative $V_{GS,off}$

cannot affect the efficiency, since a significant reduction of switching loss only becomes evident beyond the current range of a typical inverter configuration.

Permanent magnet (PM) premagnetization has been proposed to overcome poor magnetic utilization. Therefore, two PM configurations were proposed and compared based on their specific characteristics and design limitations. A comparison of premagnetized and conventional PQ core-based inductors on the basis of L-I curve measurements showed that a premagnetized inductor with a 30 % smaller PQ26/20 core can be used instead of the conventional PQ32/20 inductor for the investigated inverter topology.

References

[1] Jianchun Xu, Yajie Qiu, Di Chen, Juncheng Lu, Ruoyu Hou, Peter Di Maso, "An Experimental Comparison of GaN E-HEMTs versus SiC MOSFETs over Different Operating Temperatures," *GaN Systems Inc.*, Online: http://www.gansystems.com, accessed: 10/28/2021.

[2] P. Zacharias, "Inverter capable of providing reactive power," U.S. Patent 8 638 581 B2, Oct. 9, 2009.

[3] H. Qin, W. Wang, Z. Peng, A. Liu, and S. Bai, "Characterization and optimization of gate driver turn-off voltage for eGaN HEMTs in a phase-leg configuration," Energy Reports, vol. 8. Elsevier BV, pp. 908–919, Apr. 2022. doi: 10.1016/j.egyr.2021.11.093.

[4] W. Li, Y. Gu, H. Luo, W. Cui, X. He and C. Xia, "Topology Review and Derivation Methodology of Single-Phase Transformerless Photovoltaic Inverters for Leakage Current Suppression," in IEEE Transactions on Industrial Electronics, vol. 62, no. 7, pp. 4537-4551, July 2015.

[5] J. Friebe, S. Lin, L. Fauth and T. Brinker, "Premagnetized Inductors in Single Phase dc-ac and ac-dc Converters," *2019 IEEE 15th Brazilian Power Electronics Conference and 5th IEEE Southern Power Electronics Conference (COBEP/SPEC)*, 2019, pp. 1-6.

[6] S. V. Araujo, P. Zacharias and R. Mallwitz, "Highly Efficient Single-Phase Transformerless Inverters for Grid-Connected Photovoltaic Systems," in *IEEE Transactions on Industrial Electronics*, vol. 57, no. 9, pp. 3118-3128.

[7] Y. Wang, L. Yang, Z. Meng, G. Li and P. Chen, "Power loss distribution analysis for a high frequency dual-buck full-bridge inverter," *2017 IEEE Transportation Electrification Conference and Expo*, (ITEC Asia-Pacific), 2017, pp.1-6.

[8] L. Zhou and F. Gao, "Dual buck inverter with series connected diodes and single inductor," *2016 IEEE Applied Power Electronics Conference and Exposition (APEC)*, 2016, pp. 2259-2263.

[9] D. Barater, E. Lorenzani, C. Concari, G. Franceschini, and G. Buticchi, "Recent advances in single-phase transformerless photovoltaic inverters," IET Renewable Power Generation, vol. 10, no. 2. Institution of Engineering and Technology (IET), pp. 260–273, Feb. 2016. doi: 10.1049/iet-rpg.2015.0101.

[10] L. Will, S. Sprunck and P. Zacharias, "Impact of Negative Turn-Off Voltage On Turn-On Losses in GaN E-HEMTs," *PCIM Europe digital days 2020; International Exhibition and Conference for Power Electronics, Intelligent Motion, Renewable Energy and Energy Management*, 2020, pp. 1-5.

[11] Texas Instruments Incorporated, "TPS7A39 Dual, 150-mA, Wide VIN Positive and Negative LDO Voltage Regulator," Datasheet, SBVS263A, 2017.

[12] T. Kerekes, R. Teodorescu, P. Rodríguez, G. Vázquez and E. Aldabas, "A New High-Efficiency Single-Phase Transformerless PV Inverter Topology," in *IEEE Transactions on Industrial Electronics*,vol.58,no.1,pp.184-191,Jan.2011.

[13] GaN Systems Inc, "GS66504B Bottom-side cooled 650 V E-mode GaN transistor," Datasheet, Rev 200402, 2020.

[14] E. A. Jones, Z. Zhang and F. Wang, "Analysis of the dv/dt transient of enhancement-mode GaN FETs," *2017 IEEE Applied Power Electronics Conference and Exposition (APEC)*, 2017, pp. 2692-2699.

[15] B. Sun, Z. Zhang and M. A. E. Andersen, "Switching transient analysis and characterization of GaN HEMT," *2018 3rd International Conference on Intelligent Green Building and Smart Grid (IGBSG)*, 2018, pp. 1-4.

[16] Z. Wang, Y. Wu, J. Honea and L. Zhou, "Paralleling GaN HEMTs for diode-free bridge power converters," *2015 IEEE Applied Power Electronics Conference and Exposition (APEC)*, 2015, pp. 752-758.

[17] L. Lu, G. Liu and K. Bai, "Critical transient processes of enhancement-mode GaN HEMTs in high-efficiency and high-reliability applications," in *CES Transactions on Electrical Machines and Systems*, vol.1, no.3, pp.283-291, Sept.2017.

[18] A. R. Aguilar and S. Munk-Nielsen, "Design, analysis and simulation of magnetic biased inductors with saturation-gap," *2014 16th European Conference on Power Electronics and Applications*, 2014, pp. 1-11.

[19] S. Lin, J. Friebe, S. Langfermann and M. Owzareck, "Premagnetization of High-Power Low-Frequency DC-Inductors in Power Electronic Applications," PCIM Europe 2019; *International Exhibition and Conference for Power Electronics, Intelligent Motion, Renewable Energy and Energy Management*, 2019, pp. 1-7.

Determination of Optimal Associated Discrete Circuit Switch Model Parameters for Real-Time Simulation of Dual-Active Bridge Converters

Marija Stevic and Ravinder Venugopal
OPAL-RT Germany GmbH
Bucher Str. 100
90408 Nuremberg, Germany
Phone: +49 (0)911 38-44-52-02
Email: marija.stevic@opal-rt.com
URL: https://www.opal-rt.com/de/

Acknowledgments

The presented research and its results have been obtained in the project "DC-Sek: Control and automation of hybrid AC/DC distribution networks, in consideration of cyber-physical security aspects", funded by the German Federal Ministry of Education and Research (BMBF, funding code 03SF0594) under the roof of the Flexible Electrical Networks Research Campus.

Keywords

≪Real-time simulation≫, ≪Dual Active Bridge Converter≫, ≪Discrete-time≫, ≪Field Programmable Gate Array (FPGA)≫, ≪DC-DC power converter≫.

Abstract

Real-time simulation of a Dual-Active Bridge (DAB) converter based on the Associated Discrete Circuit (ADC) switch model is considered in this work. The ADC switch model allows for efficient simulation as its topology is fixed regardless of the switch state. The main disadvantages of the ADC-based simulation of power electronics converters are virtual power losses and artificial oscillations following switching events, which might significantly deteriorate simulation fidelity. A suitable parameter for constant conductance of the ADC switch model should be selected to minimize virtual power losses and artificial oscillations. However, a constant conductance parameter that minimizes virtual power losses might result in a low degree of simulation fidelity in terms of transformer current for the case of the DAB converter. This paper demonstrates with a numerical example the inconsistency of the objectives when selecting an optimal constant conductance parameter for simulation of the DAB converter. To address the challenge, a Loss Compensation Algorithm (LCA) is utilized to minimize virtual power losses, while a method to determine a suitable ADC switch model parameter that ensures specified accuracy of the transformer current is derived analytically. Time-domain simulation results of DAB converter are provided for validation of the proposed method.

Introduction

Hardware-in-the-Loop (HiL) concept is widely used for testing and validation of devices in the context of research and innovation activities. Ongoing developments of novel control concepts for power electronics converters increasingly rely on the HiL testing to bridge the gap between simulations and the real-world conditions. The HiL testing of controllers allows for significantly wider range of testing conditions compared to hardware prototypes and simplifies testing procedures as safety measures are relaxed. Typically, in addition to the simulation of a power electronics converter that is controlled by an external controller, a RTS simulates the system surrounding the power electronics converter, that often represents a multi-converter system. High degree of fidelity of real-time simulation is of critical importance to ensure valid HiL testing. Applications of real-time simulation of Dual-Active Bridge (DAB) converters are in the context of multi-converter systems, such as Multi-Terminal Direct-Current (MTDC) grids [1] or Solid-State Transformers (SST) [2]. Real-time simulations are based on fixed-step solvers and the simulation time-step is determined considering the fastest system dynamics, including switching dynamics. To meet these timing requirements Field Programmable Gate Array (FPGA) platform is mostly required. One possible approach is to derive a mathematical model of a power electronics converter of interest and determine the custom implementation for FPGA platform while focusing on efficiency and system specifics of interest [3, 4]. The approach considered in this paper is a generic solver with the objective to provide the tradeoff between the simulation efficiency and model flexibility.

Suitable switch model is one of the key factors to enable model flexibility in terms of converter topologies, as well as model scalability in terms of number of converters to be simulated within the defined simulation time-step and with feasible memory requirements. Associated Discrete-Circuit (ADC) switch model [5] is proposed as a solution enabling fixed system matrix regardless of states of the switches and thus reducing memory and online computation demands. The equivalent circuit of the ADC switch model consists of a conductance and a current source connected in parallel, that represent an inductance for a switch in closed state and a capacitance for a switch in open state. If parameters are selected to obtain equal conductance parameter values for both switch states, the system matrix is constant for all combinations of switch states regardless of converter topologies. However, the ADC switch model introduces simulation inaccuracies in terms of virtual power losses and artificial oscillations following switching events. The degree of these inaccuracies depends on the conductance parameter of the ADC switch model that should be selected with respect to a specific application and simulation scenarios to ensure required simulation fidelity.

Several methods for selecting a suitable constant conductance parameter of ADC switch model to meet simulation fidelity requirements have been proposed in literature. This paper provides analysis of the ADC-based simulation of DAB converter topologies and discusses its simulation fidelity specific the the DAB topology. In this context, a suitable procedure to determine optimal conductance parameter is derived specifically for ADC-based real-time simulations of DAB converters. Validation is carried out and simulation fidelity is quantified based on a reference model with ideal switch model representation.

The paper is organized as follows. First, ADC switch model and the role of conductance parameter are introduced. Next, a single-phase DAB converter is introduced as a case study. Challenges specific to simulation of DAB converter topologies based on ADC switch model are described. Following this, a procedure for determination of optimal constant conductance parameter for simulation of DAB converters is derived. Validation and assessment are provided based on time-domain simulations of the described case study.

Associated discrete circuit switch model

The ADC switch model is proposed in the context of Fixed Admittance Matrix Nodal Method (FAMNM) for simulation of power electronics converters. If an ideal switch model is utilized, the admittance matrix varies depending on the states of the switches. The objective is to enable efficient simulation for a relatively large number of switches and independently of power electronics converter topologies. The ADC switch model allows for fixed admittance matrix, regardless of the switch states.

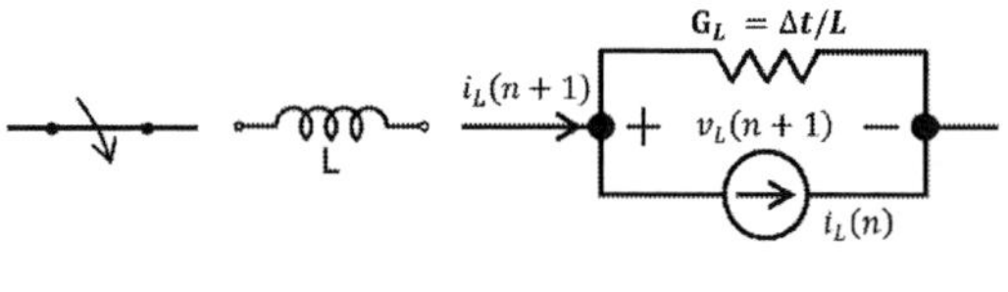

Fig. 1: ADC model of a switch in on-state 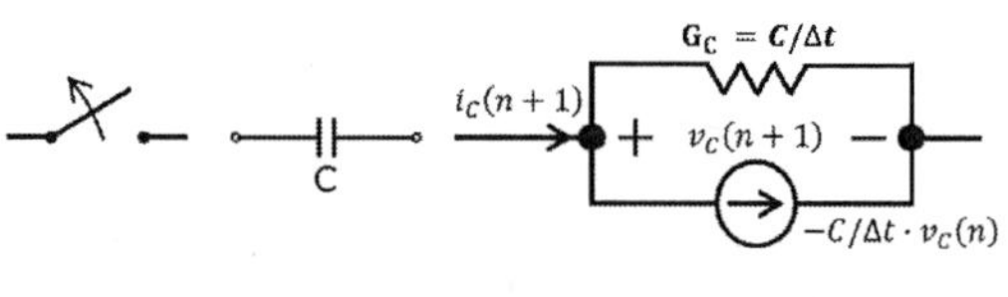   Fig. 2: ADC model of a switch in off-state

A switch is represented by an inductance in on-state and a capacitance in off-state [5]. The associated discrete circuits for an inductance and capacitance are illustrated in Fig. 1 and Fig. 2. For both states of the switch, an associated discrete circuit consists of a current source, connected in parallel to a conductance. Following the Euler Backward numerical integration method, the values of the conductance can be calculated based on the simulation time-step T_s and values of inductance or capacitance, while the value for current source is calculated based on the current or voltage quantity from the previous simulation time-step and thus represents the history of the element.

Fixed admittance matrix regardless of the states of the switches can be ensured if the value of the conductance in the ADC model of the switch is the same for both states of the switch. Namely, if conductance L_{sw} of the switch model in on-state and capacitance C_{sw} of the switch model in off-state are selected in a way that holds $G_s = T_s/L = C/T_s$, then the conductance has a constant value despite switching events and is known as a constant conductance parameter. The constant conductance parameter plays an important role in the overall power electronics converter model and might therefore significantly impact the simulation fidelity.

If a power electronics converter model with an ideal switch representation is considered as a reference benchmark, there are two effects of an ADC switch model that impact the simulation fidelity. First, the ADC switch model introduces virtual power losses. Namely, following a switching event, a capacitance (or an inductance) is removed from the circuit and replaced by an inductance (or a capacitance). In a standard FAMNM solvers the energy stored in a capacitance (or an inductance) disappears upon removing the element. As a result, power losses can be observed in the circuit and these losses are known as virtual because they are not related to actual power electronics converter behavior. The virtual power losses depend on the constant conductance parameter and the simulation time-step. Second, the ADC switch model introduces a parasitic inductance element of the value $L_{sw} = T_s/G_s$ when the switch is in the on-state and a parasitic capacitance element of the value $C_{sw} = T_s \cdot G_s$ when the switch is in off-state. As a result, artificial oscillations occur following the switching events and inaccurate dynamics of voltages and currents might be observed. In addition, if these parasitic elements are not sufficiently small compared to elements of the rest of the circuit, the dynamics of the overall circuit might be affected. The described behavior of the ADC switch model should be considered when selecting the constant conductance parameter value to ensure the simulation fidelity. In the following section a brief literature overview of methods and considerations for selecting suitable constant conductance parameter of ADC switch model is provided.

Constant conductance parameter of the ADC switch model

The ADC switch model provides substantial advantage for real-time simulation of power electronics converters; however, it potentially introduces simulation inaccuracies. The degree of the simulation inaccuracy depends on constant conductance parameter and the simulation time-step. The initial measure to ensure the simulation fidelity is to reduce the simulation time-step to the lowest value feasible for real-time execution of the system to be simulated. The size of the simulation time-step required to meet system and switching frequency dynamics is typically in the sub-microsecond range. Determination of the suitable size of the simulation time-step for real-time simulation of a single-phase DAB converters is described in [6] and it is out of the scope of this work. Here, the focus is on determining the suitable constant conductance parameter for a defined simulation time-step.

One approach to determining the constant conductance parameter value is to select parasitic inductance L_{sw} and parasitic capacitance C_{sw} of the ADC switch model based on the parameters of the physical switch [5]. Nevertheless, the actual parameters would typically require an extremely small simulation time-step, in the range of a few nanoseconds to several tens of nanoseconds, which is typically not suitable for a real-time execution of a moderate size of the system, even when a Field-Programmable Gate Array (FPGA) hardware is utilized.

Inaccuracies introduced by the ADC switch model can be determined by comparing the voltage and current quantities of the circuit simulated by employing the ADC switch model and the corresponding voltage and current quantities obtained from the simulation based on the ideal switch model. This allows the selection of constant conductance parameter values based on a trial-and-error approach, where multiple simulation runs are carried out for various values of the constant conductance parameter. In case of a multi-converter systems, different values of the constant conductance parameter should be considered for each of the converters. This approach provides reliable results as the reference model is utilized, however, it is time consuming.

One systematic approach for selecting the constant conductance parameter value is to minimize the virtual power losses. To do so, the power electronics converter topology and switching operation must be analyzed to derive the total energy loss depending on the switch parameters and current and voltage values of inductance and capacitance prior to switching events. Considering an arm of a two-level inverter, the optimal constant conductance parameter is equal to $G_s = I_{rms}/V_{dc}$, where I_{rms} is the effective value of the load current and V_{dc} represents the dc-link voltage level [7]. However, selection of a constant conductance parameter that minimizes virtual power losses does not guarantee high degree of simulation fidelity of all converter quantities. Typical parameters of a DAB converter often result in the described inconsistency, which is shown in this work.

A method to select optimal constant conductance parameter based on the eigenvalues of the admittance matrix of the simulated system is proposed in [8]. The constant conductance parameters are determined based on an optimization algorithm with an objective to minimize the Euclidean distance between the eigenvalues of an admittance matrix obtained with ideal switch models and the eigenvalues of an admittance matrix obtained with ADC switch models. More precisely, a sum of Euclidean distances between the corresponding eigenvalues for all combination of switch states is considered as the objective function.

While the described method provides a holistic approach to selecting constant conductance parameter of an arbitrary converter topologies, this work focuses specifically on the DAB converter topology. In the context of DAB converters, a simple rule for determining optimal constant conductance parameter is derived and a high degree of simulation fidelity is achieved.

Simulation case study

The challenges of ADC-based simulations of DAB converter topologies are introduced based on a single-phase DAB converter topology illustrated in Fig. 3. The converter consists of two single-phase H-bridges, a bridge 1 with a dc-link voltage V_1 and a bridge 2 with a dc-link voltage V_2, interconnected by a single-phase two-winding power transformer with a total inductance L_{tot} and transformation ratio of 1 : n. Utilization of ADC-based switch model for real-time simulations of DAB power converters introduces several challenges that must be addressed to achieve high degree of simulation fidelity. To demonstrate the degree of impact of constant conductance parameter on the simulation fidelity, time-domain simulations of the DAB converter are conducted.

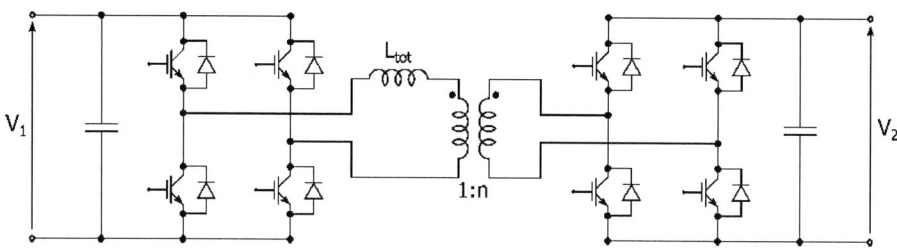

Fig. 3: Schematic of the single-phase dual-active bridge converter

Three simulation environments are utilized for simulation fidelity assessment and validation of the method for selecting a suitable constant conductance parameter. All simulation environments utilize fixed-step solver with Euler Backward numerical integration method. First, time-domain simulations of a DAB converter are conducted in an offline simulation environment to generate reference simulation results. Offline simulation environment utilized in this work is based on SimPowerSystems library in MATLAB/Simulink. Switching model of a DAB converter is based on the Resistive Switch Model (RSM), where a switch is represented with a two-valued resistance, $R_{on} \approx 0$ when the switch is in on-state and $R_{off} \gg 0$ when the switch is in off-state [9]. The parameters of the RSM are selected to ensure simulation results of negligible difference compared to the simulation with ideal switch model. Second, time-domain simulations in a simulation environment based on ADC switch model are conducted with various values of constant conductance parameter G_s. The ADC-based simulation environment that is utilized is electrical Hardware Solver (eHS) from OPAL-RT, that is a generic FPGA-based electromagnetic transient solver [10]. The eHS solver is accompanied with the LCA algorithm [11] that minimizes virtual power losses and artificial oscillations in ADC-based simulations. Two simulation environments are used, that are the eHS solver without the LCA algorithm and the eHS solver with LCA algorithm.

The parameters of a single-phase DAB converter used as a case study are listed in the Table I. The DAB converter is operated in open loop based on single-phase shift modulation [12], where the input and output bridges operate at a duty cycle of 50% and the converter is controlled through the phase-shift angle. The reference for the phase-shift angle used in the simulation is $\varphi = 18°$. The simulation time-step of $T_s = 500\,\text{ns}$ is selected as suitable for high-fidelity simulation the DAB converter [6].

Table I: Parameters of DAB converter

Parameter	Value
dc-link capacitance of primary side bridge	$C_1 = 2\,\text{mF}$
dc-link voltage of primary side bridge	$V_1 = 5\,\text{kV}$
dc-link capacitance of secondary side bridge	$C_2 = 2.5\,\text{mF}$
dc-link voltage of secondary side bridge	$V_2 = 5\,\text{kV}$
Transformer leakage inductance	$L_{tot} = 750\,\mu\text{H}$
Transformer winding resistance	$50\,\text{m}\Omega$
Switching frequency	$2\,\text{kHz}$

ADC-based simulation of a DAB converter

Time-domain simulation results for the parameters given in the previous section in offline simulation environment and in eHS simulation without LCA are illustrated in Fig. 4 and Fig. 5.

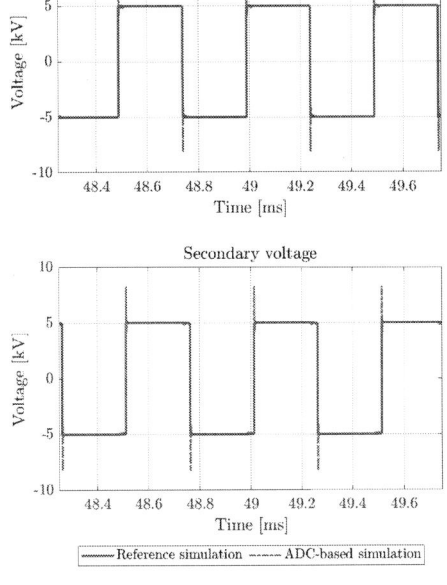

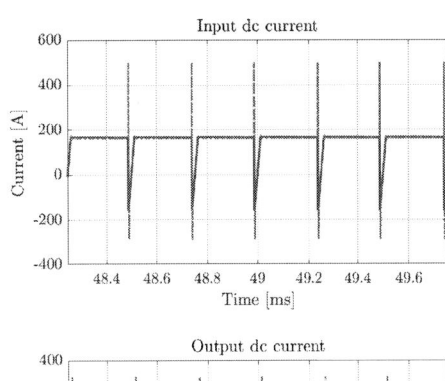

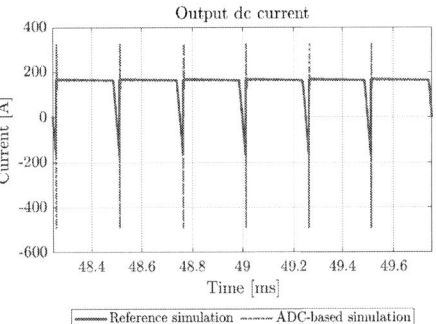

Fig. 4: Primary and secondary voltage in offline simulation and eHS simulation without LCA.

Fig. 5: Input and output dc current in offline simulation and eHS simulation without LCA.

The constant conductance parameter value of $G_s = 0.1\,\mathrm{S}$ is selected. Fig. 4 shows primary and secondary voltages of the DAB converter and indicates that voltage overshoots and artificial oscillations due to the ADC switch model are relatively large. Input and output dc currents shown in Fig. 5 demonstrate the same effect.

Another consequence of the ADC switch model are virtual power losses due to removal of the parasitic inductance or capacitance following the switching event. Table II gives values of power at the input and output of the DAB converters, as well as relative power losses. The difference between the power losses in the reference simulation and in the ADC-based simulation represents virtual power losses caused by the ADC switch model and it does not represent the actual converter behaviour. Therefore, these results might be misleading.

Table II: Power losses in eHS simulation without LCA

Parameter	P_{in}	P_{out}	P_{loss}
Reference simulation	752.8 kW	747.6 kW	0.69%
eHS without LCA	744.2 kW	716.8 kW	3.68%

In addition to the simulation fidelity issues related to virtual power losses and artificial oscillations, fidelity of the primary and secondary currents is affected by the parasitic inductances L_{sw} of the switches in on-state. Namely, the apparent inductance in the converter model in the ADC-based simulation is equal to the transformer leakage inductance L_{tot} increased by the parasitic inductances L_{sw} of all switches in on-state that are on the same current path. For the case of a single-phase DAB converter, the number of switches in on-state that should be considered when calculating the apparent inductance in the circuit is four.

Since the parasitic inductance is defined with $L_{\text{sw}} = T_s/G_s$, the degree of the simulation fidelity depends on the constant conductance parameter G_s for the given simulation time-step. Fig. 6 shows primary currents in offline simulation and eHS simulation without LCA for the constant conductance parameter values of $G_s = 0.01\,\mathrm{S}$, $G_s = 0.03\,\mathrm{S}$ and $G_s = 0.1\,\mathrm{S}$. The simulation results show that the impact of the parasitic inductances on simulation fidelity in terms of the transformer current significantly depends on the constant conductance parameter.

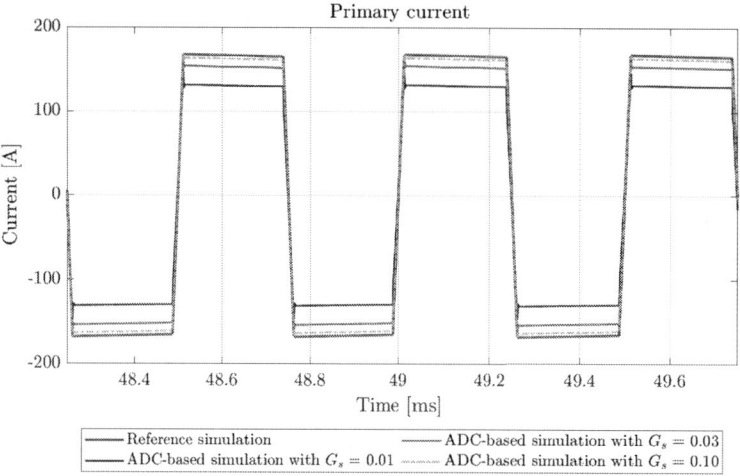

Fig. 6: Primary current in offline simulation and eHS simulation without LCA for Gs values of $G_s = 0.01\,\mathrm{S}$, $G_s = 0.03\,\mathrm{S}$ and $G_s = 0.1\,\mathrm{S}$.

To summarize, there are multiple objectives to be met when selecting a constant conductance parameter for ADC-based simulation of DAB converters. Virtual power losses, artificial oscillations and degree of simulation fidelity of transformer current depend on the constant conductance parameter. To characterize the dependency of virtual power losses and degree of simulation fidelity in terms of transformer current, time-domain simulations in eHS environment without LCA are conducted for the conductance parameter G_s in the range of values from $0.1\,\text{S}$ to $0.4\,\text{S}$. Simulation results are compared to the reference simulation in offline simulation environment and relative power losses and relative error of the primary Root Mean Square (RMS) current are calculated and illustrated in Fig. 7.

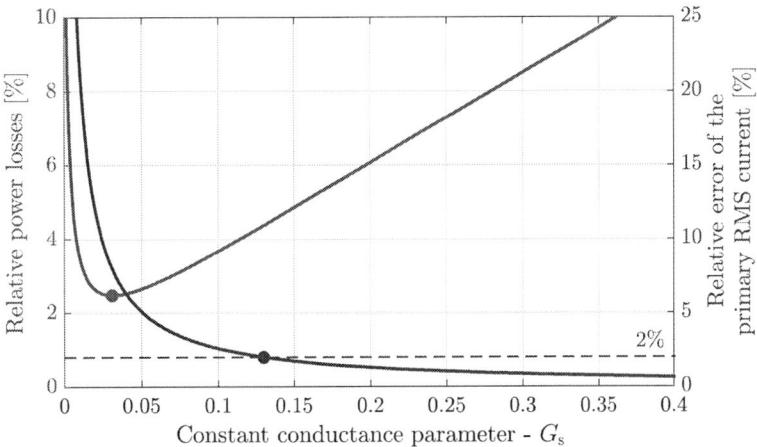

Fig. 7: Relative power losses and error of the primary RMS current.

The characteristic of relative power losses depending on the constant conductance parameter of the ADC switch model, in Fig. 7 shown in blue, is of typical shape for a converter topology consisting of 2-level arms. To determine the conductance parameter that minimizes virtual power losses, an approach derived in [7] for a two-level inverter can be followed. Following this method, ADC switch models are configured with constant conductance parameter values of $G_s = I_{rms}/V_1 = 0.03\,\text{S}$. This result is consistent with the optimum obtained based on time-domain simulations and indicated with blue dot in Fig. 7.

However, the conductance parameter that minimizes the virtual power losses results in the relative error of primary RMS current above 5% as it is shown in Fig. 7 in red. Primary currents shown in Fig. 6 indicate significant difference of the current obtained in reference simulation and in eHS simulation with the conductance parameter of $G_s = 0.03\,\text{S}$, that is calculated to minimize virtual power losses. This is the consequence of a relatively high ratio between the parasitic inductance of the ADC switch and the transformer leakage inductance. As it can be observed in Fig. 6 and Fig. 7, to achieve higher degree of simulation fidelity in terms of the transformer currents, a higher value of constant conductance parameter should be selected. For example, to achieve a relative error of the primary RMS current of 2%, the constant conductance parameter should fulfill the condition $G_s \geq 0.13\,\text{S}$. However, this in turn results in relative power losses above 4%. The following section addresses the challenge of selecting optimal conductance parameter for ADC-based simulation of DAB converters in the context of conflicting objectives as described above.

ADC-based simulation with Loss Compensation Algorithm

Loss Compensation Algorithm (LCA) [11] is developed to minimize virtual power losses and artificial oscillations in ADC-based simulation of power electronics converters. The idea behind LCA is to properly initialize current and voltage values of the inductance and capacitance of the ADC switch model following switching events. Therefore, energy stored in the capacitance or inductance of the ADC switch model does not disappear and voltage and current states are not initialized with zero values each time the switch state is changed. The LCA is fully embedded in the eHS [10] and implemented for FPGA-based real-time simulation.

Assessment of the LCA in the context of simulation of DAB converters is illustrated in Fig. 8. First, relative power losses are significantly reduced and the values range between 0.75% and 0.95%. Note that relative power losses in the reference simulation are $P_{\text{loss}} = 0.69\%$, as given in Table II. Therefore, LCA ensures high degree of simulation fidelity in terms of relative power losses almost independently from conductance parameter values.

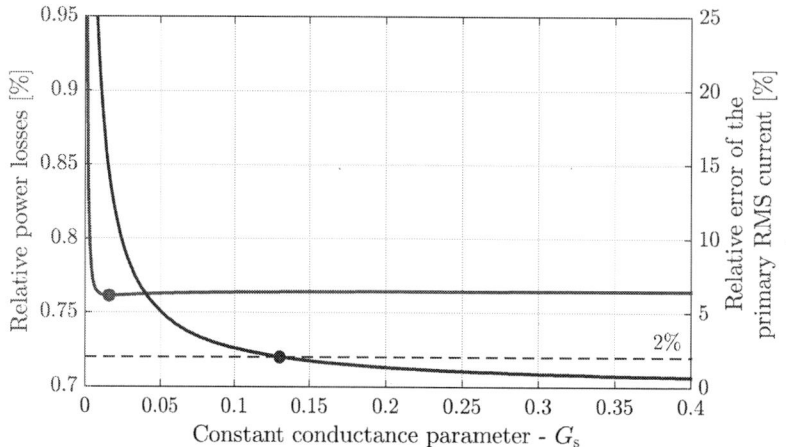

Fig. 8: Relative power losses and error of the primary RMS current with LCA.

Relative error of the primary RMS current is shown in red in Fig. 8. As expected, LCA does not improve this error, since this is the consequence of the parasitic inductance of the ADC switch model, which is the same regardless of the LCA. However, since relative power losses significantly reduced across the entire range of constant conductance parameter, the parameter can be selected to met the requirement for the relative error of the primary RMS current. The following subsection provides a simple rule to determine a suitable ADC switch model conductance parameter that meets a maximum error constraint of the transformer current.

Determination of the ADC switch model parameter

The current in a single-phase DAB converter with the same input V_{in} and output voltage output V_{out} levels is defined by Eq. 1.

$$I = \frac{V_{\text{in}} \left(1 - \frac{\varphi}{\pi}\right) \frac{\varphi}{\pi}}{2 f_{\text{sw}} L} \tag{1}$$

where φ denotes the phase-shift angle expressed in radians, L represents the total leakage inductance and f_{sw} refers to the switching frequency. As already described, the main source of fidelity degradation in terms of relative error of primary RMS current originates from the parasitic inductance of the ADC switch model. To characterize the degree of fidelity degradation in this context, the derivative of the converter current given by Eq. 1 with respect to the inductance value is derived as given in Eq. 2.

$$\frac{\mathrm{d}I}{\mathrm{d}L} = -\frac{V_{\text{in}} \left(1 - \frac{\varphi}{\pi}\right) \frac{\varphi}{\pi}}{2 f_{\text{sw}}} \cdot \frac{1}{L^2} \tag{2}$$

Next, the relative error of the DAB converter current $I_{\text{err}}^{\text{rel}} = |\Delta I / I|$ can be expressed by Eq. 3.

$$\left| \frac{\Delta I}{I} \right| = \left| -\frac{\Delta L}{L} \right| \tag{3}$$

The absolute error of the apparent inductance is equal to the total inductance of four parasitic inductances of ADC switches connected in series and is given by Eq. 4.

$$\Delta L = 4 \cdot L_{\mathrm{sw}} = 4 \cdot \frac{T_{\mathrm{s}}}{G_{\mathrm{s}}} \tag{4}$$

If a constraint for maximum of the relative error of RMS current is defined $I_{\mathrm{err,max}}^{\mathrm{rel}} = 2\%$, the condition for parameters to satisfy this constrainet is given by Eq. 5.

$$\frac{1}{L} \cdot 4 \cdot \frac{T_{\mathrm{s}}}{G_{\mathrm{s}}} \cdot 100\% \leq I_{\mathrm{err,max}}^{\mathrm{rel}} \tag{5}$$

Finally, for a defined simulation time-step, the condition for constant conductance parameter is expressed with Eq. 6.

$$G_{\mathrm{s}} \geq \frac{4T_{\mathrm{s}}}{L} \cdot \frac{100\%}{I_{\mathrm{err,max}}^{\mathrm{rel}}} \tag{6}$$

In this context, the smallest value of G_{s} that satisfies the constraint for the relative error of the current should be selected. The reason is to minimize the size of the capacitor in off-state and therefore minimize its impact on the rest of the system.

Simulation results

Conductance parameter that satisfies the constraint given by Eq. 6 is equal to $G_{\mathrm{s}} \geq 0.13\,\mathrm{S}$, which is consistent with the fidelity assessment provided in Fig. 8. Simulation results in eHS environment and LCA with the specified value of the conductance parameter are provided in Fig. 9 and Fig. 10. Voltages and currents indicate high degree of simulation fidelity.

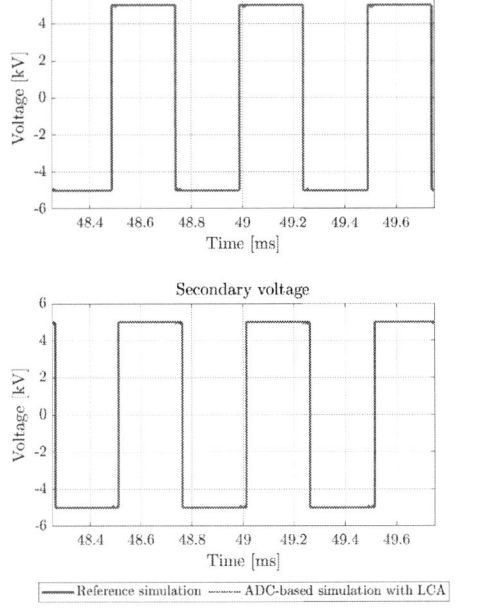

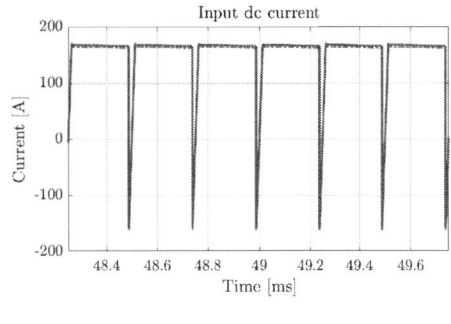

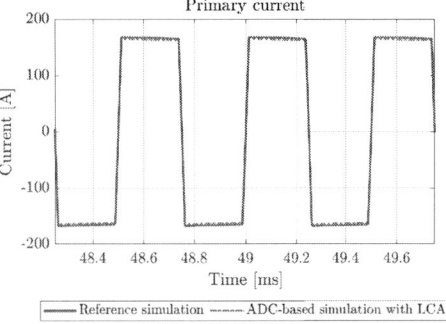

Fig. 9: Primary and secondary voltage in offline simulation and eHS simulation with LCA.

Fig. 10: Input and output dc current in offline simulation and eHS simulation with LCA.

Conclusion and outlook

The ADC switch model has been introduced and its advantages for FPGA-based real-time simulation due to its simplicity and computation efficiency have been outlined. The importance of constant conductance parameter value for simulation fidelity has been described. Moreover, the challenges specific to simulation of DAB converter topologies based on ADC switch model have been discussed. In this context, the specific characteristics of the DAB converter must be accounted for when determining the suitable constant conductance parameters.

As demonstrated based on the numerical example, the method to select constant conductance parameter values based on the minimization of the virtual power losses of each of the H-bridges might result in parasitic inductance values that are relatively large compared to the transformer leakage inductance. To address this challenge, a Loss Compensation Algorithm (LCA) is utilized to minimize virtual power losses, while a method to determine a suitable ADC switch model conductance parameter that meets a maximum error constraint of the transformer current is derived analytically. Finally, the procedure for determination of optimal conductance parameter values of ADC switch models for the case of DAB converters is validated based on the concrete example and time-domain simulations.

References

[1] Stefan P. Engel et al. "Dynamic and Balanced Control of Three-Phase High-Power Dual-Active Bridge DC–DC Converters in DC-Grid Applications". In: *IEEE Transactions on Power Electronics* 28.4 (2013), pp. 1880–1889.

[2] Hrishikesan V. M., Chandan Kumar, and Marco Liserre. "An MVDC-Based Meshed Hybrid Microgrid Enabled Using Smart Transformers". In: *IEEE Transactions on Industrial Electronics* 69.4 (2022), pp. 3722–3731.

[3] Gabriele Arena et al. "A Cost-Effective Hardware in the Loop Implementation of Dual Active Bridge for Fast Prototyping of Electric Vehicles Charging Controls". In: *2021 23rd European Conference on Power Electronics and Applications (EPE'21 ECCE Europe)*. 2021, P.1–P.10.

[4] Ming Jia, Philipp Joebges, and Rik W. De Doncker. "A Fast and Robust Model of Dual-Active Bridge Converters in Real-Time Simulation". In: *2020 22nd European Conference on Power Electronics and Applications (EPE'20 ECCE Europe)*. 2020, pp. 1–11.

[5] P. Pejovic and D. Maksimovic. "A method for fast time-domain simulation of networks with switches". In: *IEEE Transactions on Power Electronics* 9.4 (1994), pp. 449–456.

[6] Marija Stevic et al. "Challenges in Real-Time Simulation of Smart Transformers". In: *2022 IEEE 13th International Symposium on Power Electronics for Distributed Generation Systems (PEDG)*. 2022, pp. 1–6.

[7] Xizheng Guo et al. "FPGA-based hardware-in-the-loop real-time simulation implementation for high-speed train electrical traction system". In: *IET Electric Power Applications* 14.5 (2020), pp. 850–858.

[8] R. Razzaghi et al. "A method for the assessment of the optimal parameter of discrete-time switch model". In: *Electric Power Systems Research* 115 (2014), pp. 80–86. ISSN: 0378-7796.

[9] Ramin Mirzahosseini and Reza Iravani. "Small time-step FPGA-based real-time simulation of power systems including multiple converters". In: *IEEE Transactions on Power Delivery* 34.6 (2019), pp. 2089–2099.

[10] Christian Dufour et al. "General-purpose reconfigurable low-latency electric circuit and motor drive solver on FPGA". In: *IECON 2012 - 38th Annual Conference on IEEE Industrial Electronics Society*. 2012, pp. 3073–3081.

[11] Christian Dufour. *Method and system for reducing power losses and state-overshoots in simulators for switched power electronic circuit*. US Patent 9,665,672. 2017.

[12] Nie Hou and Yun Wei Li. "Overview and Comparison of Modulation and Control Strategies for a Nonresonant Single-Phase Dual-Active-Bridge DC–DC Converter". In: *IEEE Transactions on Power Electronics* 35.3 (2020), pp. 3148–3172.

Integrated motor drive: A multidisciplinary approach

Betty Lemaire-Semail[1], Nadir Idir[1], Eric Semail[1], Souad Harmand[2]
Univ. Lille, Arts et Metiers Institute of Technology, Centrale Lille, Junia,
EA 2697 – L2EP - Laboratoire d'Electrotechnique et d'Electronique de Puissance, F-59000
Lille, France
Email: betty.semail@univ-lille.fr
[2] Université de Valenciennes et du Hainaut Cambraisis, CNRS, UMR 8201 LAMIH

Acknowledgements: this work has been achieved within the framework of CE2I project (Convertisseur d'Energie Intégré Intelligent). CE2I is co-financed by European Union with the financial support of European Regional Development Fund (ERDF), French state and the French Region of Hauts-de-France.

Keywords: "Multiphase drive", "Thermal design", "Wide bandgap devices", "Fault tolerance"

Abstract: integrated drives are becoming a target for manufacturers aiming at addressing the increasing market of electrical embedded systems, such as e-transportation systems. Anyway, the integration of the power converter inside the electrical machine induces challenges if compactness is required. Additionally to the research of the best choices for the power electronics, electromagnetic, and thermal components, interconnection issues also arise, leading to cope simultaneously with several physical challenges. This paper gives an example of a multidisciplinary approach to propose smart integrated drive solution dedicated to middle power embedded systems.

I. Introduction

Integrated drives with only two DC cables are quite attractive and have been developed for transportation systems since 20 years by academics and manufacturers with Si components [1-4]. The integration of the inverter in the electrical machine avoids long AC cables and therefore EMI (Electromagnetic Interference). A common cooling for the inverter and motor can thus be considered. Nevertheless, maximum temperature (around 150°C) of Si semi-conductors prevents from high compactness and efficiency in harsh thermal environment of transportation with transient operations. Up to now, many current industrial solutions consist in putting an existing inverter packaging close to an existing machine [5-7].

In order to propose optimized solutions in terms of both carbon cost and economical cost for the electrified transportation mass-market up to 50-150kW, new topologies, with a more intimate integration of semi-conductor components in the machine, have been developed [8-10]: for a given compactness criterion, such as robustness and functional reliability, recyclability, efficiency should be considered simultaneously for the choice of a topology. But it is with GaN and SiC semi-conductors that a strong evolution can be expected since the maximum temperatures of the motor and the components are now close to each other [11], which is favouring the compactness and a common cooling. All these specifications may not be addressed independently from each other, as they are closely imbricated. To address these questions, several proposals exist in the state of the art, some of them as prototypes [12-15], others as proofs of concept [16-17], others at a design stage, and some also at the commercial stage [5],[10],[18].

These proposals differ from each other by the system structure [5] – the inverter is located on one side of the machine [6], [16], or split into several parts and put around the circumference of the machine [4], [15], or again located on one axis side of the machine [3], [8-10], [12-14]. The number of legs for the VSI and phases for the machine is often a design parameter according to the power requirements, and the value of voltage DC bus. From that, the choice of the wide band gap components is deduced, among GaN and SiC technologies. High voltage applications (DC bus over 300V) [12] [14], [16-17] can already benefit from SiC component with a significant level of maturity. Nevertheless, high voltage solutions require expensive Battery Management System, dielectric constraints for the motor windings [19], and human security levels. For DC voltages under 48V (normalized maximum voltage for human), GaN components are potentially well adapted with small size packaging but the maturity for high currents is still ongoing due to constraint associated with thermal dissipation.

This paper aims at giving an example of the multidisciplinary approach leading to an original topology of integrated drive [27], dealing together with the thermal, power electronics and electromagnetism issues. The targeted market concerns rather the low-voltage automotive one with power up to 40-50 kW but the topology could be extended to higher voltage level with SiC components for high torque applications (truck, marine, plane). It relies on the experience of a project on "smart integrated drives" (CE2I project) performed together by three laboratories with regional funding.

The outline of the paper is as following: in the second part, the general approach and the available technological choices are described regarding the requirements. The third paragraph presents in more details the different parts of the integrated drive and the first performances.

II. General approach for smart integrated drive design

The "CE2I" project (French acronym for smart integrated drive) is achieved by four academic laboratories which target to design and realize integrated drive prototypes in an affordable power range of about 40kW, that is clearly relevant for mass-market low-voltage terrestrial mobility applications, such as light automotive and also motorbikes. But before reaching this power, intermediary steps have been performed to test the different technological possibilities.

The choice of the drive topology presented in the paper is an innovative multiphase tooth-concentrated open-winding drive with two multiphase inverters as shown in figure 1. In order to guarantee a high torque/volume ratio, PMSM is chosen. With a 48V DC bus, it is possible, thanks to open-winding choice, to impose 48V for each phase which is an advantage in comparison with a classical star configuration (24v in that case). Two inverters allow to keep easily a symmetrical spatial repartition of the coil connections, identical on both sides of the machine, in comparison with asymmetry due to internal connection between elementary coils. Moreover, the symmetry will be complete if the two inverters are located respectively on front side of the machine. The inverters can be supplied either by only one DC source or by two independent ones to warrant a higher reliability.

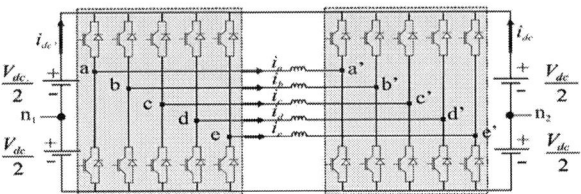

Figure 1: Multi-phase open winding PMSM and its inverters

The symmetrical topology is an advantage to design a relevant cooling system: the thermal dissipation should be the same on both axial sides of the machine and hot spots may be avoided by that way.

By increasing the number of phases, the aim is to avoid the use of transistors working in parallel. Thereby, necessary oversizing with three-phase solutions for reaching required reliability can be alleviated. With multiphase solution, the loss of one transistor can be tolerated without oversizing if a given loss of power is accepted. With transistors working in parallel, the transistors must be sized in order to support the death of one of them if snowball effect must be avoided.

Regarding the choice of the transistors, it highly depends on the power range. With wide band gap transistors, the chosen integrated solution can benefit of higher switching frequencies to simultaneously reduce the size of passive components such capacitors but also the heat associated to the component losses. Last but not least, for GaN transistors, as the maximum temperature is of the same order than the windings' ones, a common thermal cooling system could be shared.

The cooling system can be liquid if external pump already exists, as in numerous automotive, or with air in lighter solutions such as scooter or other light vehicles. In the last case, non-linear cooling of the machine by heat pipes has a particular interest in order to allow transient operations (such as acceleration or breaking): when the heat is increasing too much during a transient, the heat pipes start to operate. Their thermal résistance is decreasing slowing down the rise of the temperature winding. Moreover, heat pipes fulfil compactness requirements, and do not need extra external liquid cooling system. In the rotor, rotating heat pipes can also be chosen to protect the magnets in case of an increase of the rotor losses. In that case, the machine structure has to be designed accounting for the heat pipes' location.

It can be noticed that for high power solutions, eg for trucks (P> about150kW), such a kind of topology can be used with profit, but with SiC components and high DC bus voltage.

III. Details on technological choices and first results

1) Annular Voltage Source Inverter

To take into account the open phase winding of the machine (figure 1), the voltage source inverter (VSI) is split into two inverters located on both sides of the axis as shown in figure 2. The main constraints which have to be considered for the power electronics design is the PCB surface which is limited and the thermal pad of the GaN transistors which must be on the top side of the packaging regarding the PCB location near the machine [1], [20].
For a 5-phase machine, two 5-leg inverters are installed on both sides as shown in figure 2.

Figure 2: location of the two inverters in the machine

Top-side cooled 100V enhancement mode GaN transistors (GS61008T- GaN Systems) have been chosen, according to their high switching frequency, their low $R_{ds(on)}$ (7mΩ), and their high current (90A@25°C). Their PCB footprint is very low (7.0x4.0mm², with 0.54mm thickness) which meets the surface constraint available for the converter.
The geometry of one inverter side is shown in figure 3: an annular shape has been chosen in order to adapt itself to the machine geometry. The new PCB structure has been studied in simulation in order to determine all parasitic elements as well as the thermal behaviour of the active parts; it is important to determine parasitic inductances of power and gate loops which have the more influence on the transistors switch losses [21-22]. The calculation of the total power losses of the transistors are used to design the cooling systems of the converter.

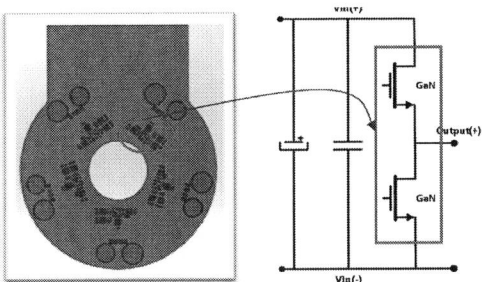

Figure 3: PCB geometry including devices placement of the 5 legs and the basic structure of one GaN inverter leg

As a first step towards the final full integrated drive with a single cooling system, the double 5-leg inverter has been designed with its own thermal environment, and use "classical" dissipators thermally connected to the top size of the transistors, thanks to spreaders and thermal interface material (TIM) system [23]. One of the realized inverter is shown in figure 4. To minimize the parasitic inductances of the power switch loops, the connections of the 5-legs of the inverter are carried out by busbar. The wire connections shown in figure 4 are only used to determine the conduction losses of GaN transistors.

Figure 4: Control side of 5-leg GaN inverter

The inverter has been tested first separately from the machine, on a passive resistive and inductive load, and gave the expected results on the voltage control.

2) 5-phase Internal Permanent Magnet Synchronous Machine (IPMSM)

The first machine prototype chosen for the project was intended for mild-hybrid 48V application. It is a 20slots/14poles tooth concentrated windings IPMSM with five phases in open winding configuration. This choice leads to several advantages: in general, a multiphase machine is tolerant to fault from the semi-conductor components of the inverter. For high compacity in integrated drive the safe margins of the components must be small and faults must be taking into consideration at the drive design stage. For a given torque the current per phase can be reduced, avoiding to work with transistors in parallel and providing a better dispatching in space of the VSI legs with also a better repartition of the thermal effects. Regarding the choice of the phase number -here 5-, as it is not a multiple of 3, additional advantages exist such as higher torque density with non-sinusoidal emf while ensuring low torque ripples in normal and fault operations. For fault operations, specific fast sensitive fault detections and more efficient reconfiguration strategies can be implemented [24-26].
The open-winding structure is suited to the low voltage DC bus (48V) since it allows roughly to double the voltage applied to each phase in comparison with star connexion. It is also well adapted to the solution with two VSI, one at each side of the machine.
In a first step, a "classical" 5-phase PMSM has been tested and equipped with the two annular VSI presented before (figure 6). The connection between the two parts of the VSI and the end of the phases

has been realized thanks to busbars. Mechanical adaptation rings have been designed in order to fix the PCBs to the end side of the machine (figure 5).

Figure 5: exploded view of the annular inverter	Figure 6: Views of one 5-phase PMSM prototype with the two VSI parts on each rotating axis side

To go further towards integration, and especially towards integration of the cooling system, another prototype has been designed and realized. It is also a multi-phase PMSM with twelve coils allowing 5-phase or 10-phase configurations depending on the disc connection between the coils and the VSIs. Thanks to the symmetrical topology and tooth concentrated winding, statoric heat pipes can be inserted symmetrically in the bottom of each slot (Figure 7) and condensers placed just over copper coil (Figure 8) with very short end-winding

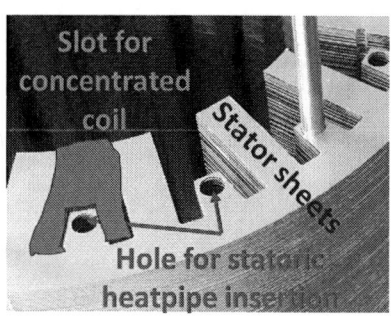 

Figure 7: open slots of stator sheet with hole for heat pipe insertion

Figure 8: Condenser of the heat pipe over short end-winding coils (stator)

In figures 9 and 10 more global views of the topology are given.

  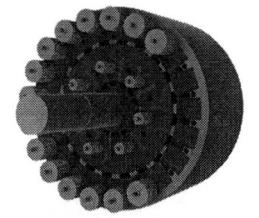
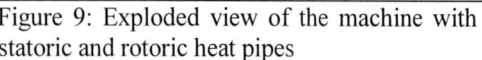

Figure 9: Exploded view of the machine with statoric and rotoric heat pipes	Figure 10: Prototype with twenty coils and internal radial magnets without heat pipe left with heat pipe right

3) Thermal cooling system

Thermal simulations have been performed on the VSI equipped with GaN transistors as well as on the machine core in order to determine the cooling requirements regarding the maximum power range of the integrated drive [27]. Special attention has been paid to take into account the location of the transistors. First, nodal thermal model of a single GaN transistor and its associated TIM, spreader and heat sink has been established. The cooling device proposed for the two GaN transistors (1-leg inverter) is shown in Fig. 11.

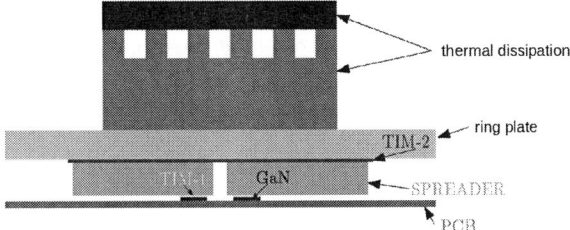

Figure 11: Cooling system of the 1-leg inverter

Then, the typical geometry of the annular VSI and the 5 legs have been considered in order to study their thermal interaction depending on the choice of the Pulse Width Modulation. Figure 12 gives an example of the thermal simulation of the 5-legs for two configurations: when 2-legs or 3-legs are ON-state.

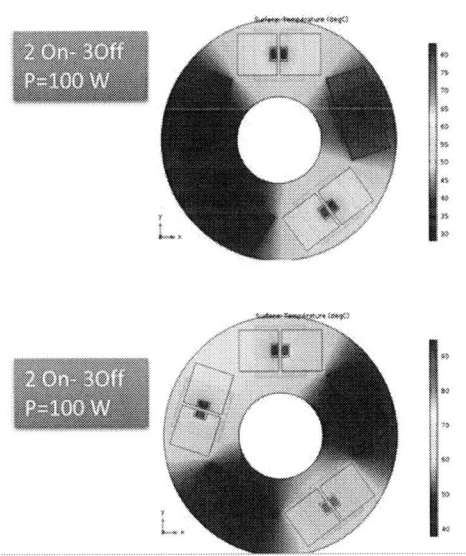

Figure 12: thermal simulations of different conducting situations for the 5-legs inverter

As for the machine cooling, simulations have also been performed using coupled FEM to determine the heating sources and thermal software to calculate the temperature of the sensitive areas [28]. From these simulations, the heat pipes solution has been proposed, which can be really integrated to the machine core, as it is shown in Figures 3 and 13. The tests on stator heat pipes gave the expected results, whereas the first experimental trials on the rotor heat pipes have shown the accurate operation of the heat pipes, despite the centrifugal effect due to the rotation speed. The revolving heat pipes used for the cooling of the rotor are hollow copper tubes, with an outer diameter of 4 mm and a wall thickness of 0.4 mm. The length of the evaporator (part inserted in the rotor) is 40mm and the condenser part (equipped with annular fins) is 15.5mm. This heat pipe uses water as a working fluid.

A specific test bench has been developed in order to determine the thermal conductance of these heat pipes as a function of the rotation speed and the applied heat power at evaporator (Figure 14).

 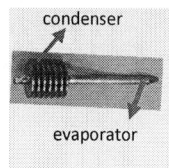

Figure 13: Rotor of the 5-phase PMSM equipped with heat pipes. Structure of the heat pipe

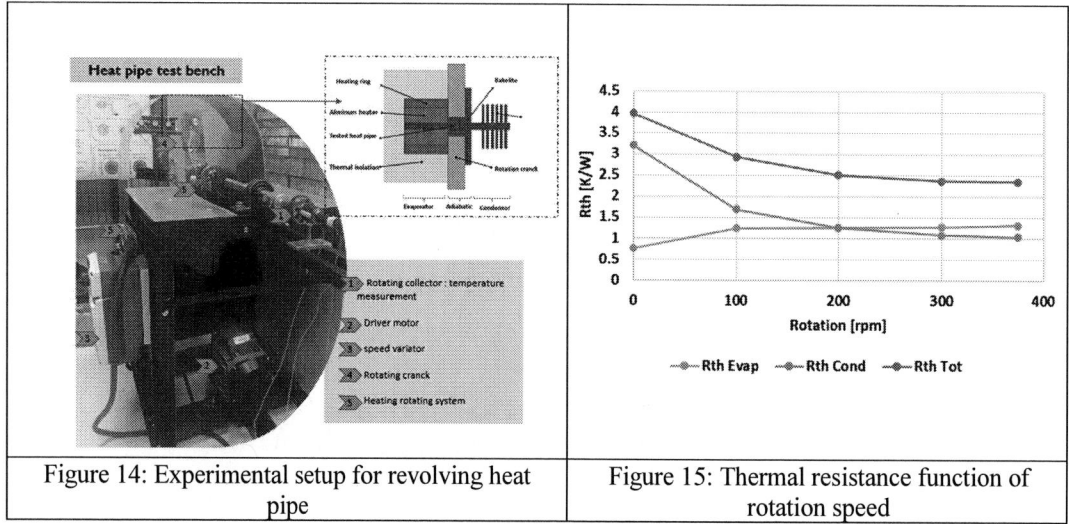

Figure 14: Experimental setup for revolving heat pipe	Figure 15: Thermal resistance function of rotation speed

The thermal conductance, which is the inverse of thermal resistance (figure 15), was then used in the global modeling with the nodal method and showed the interest of these revolving heat pipes to improve the cooling of the machine. Thereby, the contribution of heat pipes for a given operating point has been studied. In this case, Figure 16 shows the temperature distribution with and without the heat pipes. We can distinguish an extreme operating case with the losses of 35 W at the rotor and 1010 W at the stator, the maximum temperature is estimated at 131.5 °C according to the nodal modelling.

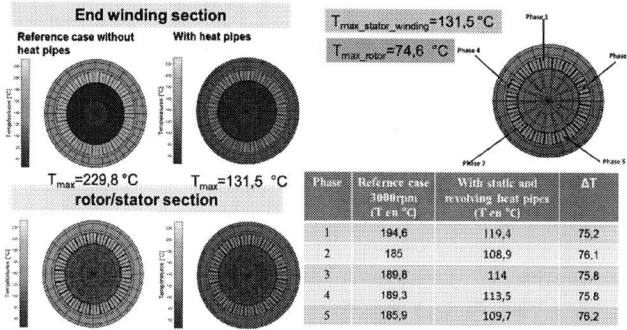

Figure 16: Thermal performance comparison between a machine without heat pipe and machine with heat pipes

As the statoric copper heat pipes have been inserted in stator iron sheets (figure 7), it has been verified at first by FE software and then by measurements of losses that no significant supplementary losses were induced in the heat pipes despite their proximity to electromagnetic fields at the bottom of slots (figure

7). The insulation with resin between iron sheets and each heat pipe prevents effectively from huge induced eddy currents predicted by FE software in case of no insulation of heat pipe.

IV. Conclusion

This paper describes the simultaneous studies which have to be performed in order to design an integrated drive: on the power electronics side, on the point of view of the machine design and control, and of course on the thermal side, regarding the crucial importance of the cooling system within the compacity constraints. Elements of the design method of the integrated inverter in a multiphase PMSM has been presented. With the choice of an original symmetrical topology, heat pipes for cooling can be inserted close to loss locations in a machine which is adapted thus to transient operations. Considering the number of phases as a parameter of design allows more flexibility for the VSI design while giving tolerance to a transistor fault.

The presented configuration of integrated drive is based on the location of two multi-leg inverters on each axial side of the machine. The number of chosen legs is depending on space and thermal constraints of the VSIs. To increase the power density of the converter and meet with the volume constraints, GaN transistors are used because of their small dimensions of their packaging and high switching frequency.

A 10-phase configuration with two separate supplies and two controls is ongoing.

This multidisciplinary approach is illustrated thank to the work done in an academic project named "CE2I" which gathers laboratories from 3 universities.

V. References

[1] T. M. Jahns, H. Dai, "The Past, Present, and Future of Power Electronics Integration Technology in Motor Drives", CPSS Trans. on Power Electronics and Applications, vol. 2, no. 3, Sept 2017.

[2] N. R. Brown, T. M. Jahns and R. D. Lorenz, "Power Converter Design for an Integrated Modular Motor Drive," 2007 IEEE Industry Applications Annual Meeting, 2007, pp. 1322-1328

[3] D. Richard and Y. Dubel, "Valeo StARS Technology: A Competitive Solution for Hybridization," 2007 Power Conversion Conference - Nagoya, Nagoya, 2007, pp. 1601-1605. https://www.youtube.com/watch?v=VJSEQaxAh4c

[4] J. J. Wolmarans, M. B. Gerber, H. Polinder, S. W. H. de Haan, J. A. Ferreira, and D. Clarenbach, "A 50 kW integrated fault tolerant permanent magnet machine and motor drive," in Proc. IEEE Power Electron. Specialists Conf., Jun. 2008, pp. 345–351.

[5] W. Lee, S. Li, D. Han, B. Sarlioglu, T. A. Minav and M. Pietola, "A Review of Integrated Motor Drive and Wide-Bandgap Power Electronics for High-Performance Electro-Hydrostatic Actuators," Trans. on Transp. Elect., vol. 4, no. 3, pp. 684-693, Sept. 2018.

[6] Doerr, J., Attensperger, T., Wittmann, L. & al., » The New Electric Axle Drives from Audi ». ATZ Elektron Worldw 13, 16–23 (2018).

[7] J. Reimers, L. Dorn-Gomba, C. Mak and A. Emadi, "Automotive Traction Inverters: Current Status and Future Trends," in IEEE Transactions on Vehicular Technology, vol. 68, no. 4, pp. 3337-3350, April 2019, doi: 10.1109/TVT.2019.2897899.

[8] P. Brockerhoff, W. Schön, P. Blaha, P. Václavek and Y. Burkhardt, "Disc inverter in highly integrated 9-phase drivetrain for E-mobility," EPE'15 ECCE-Europe, pp. 1-9. https://motorbrain.eu/

[9] X. Deng, S. Lambert, B. Mecrow and M. A. S Mohamed, "Design Consideration of a High-Speed Integrated Permanent Magnet Machine and its Drive System," ICEM, 2018, pp. 1465-1470,

[10] S. Runde, A. Baumgardt, O. Moros, B. Rubey and D. Gerling, "ISCAD — Design, control and car integration of a 48 volt high performance drive," in CES Transactions on Electrical Machines and Systems, vol. 3, no. 2, pp. 117-123, June 2019, https://molabo.eu/#maincontent

[11] A. K. Morya et al., "Wide Bandgap Devices in AC Electric Drives: Opportunities and Challenges," in IEEE Transactions on Transportation Electrification, vol. 5, no. 1, pp. 3-20, March 2019.

[12] L. de Lillo, B. Ahmadi, L. Empringham, M. Johnson, J. Espina and R. Abebe, "Next Generation Integrated Drive, NGID: A Novel Approach to Thermal and Electrical Integration of High Power Density Drives in Automotive Applications," IEEE ECCE, Portland, OR, 2018, pp. 1228-1232.

[13] Z. Gao et al., "A GaN-Based Integrated Modular Motor Drive for Open-Winding Permanent Magnet Synchronous Motor Application," 2018 1st Workshop on Wide Bandgap Power Devices and Applications in Asia (WiPDA Asia), 2018, pp. 73-79, doi: 10.1109/WiPDAAsia.2018.8734587.

[14] N. Bianchi, S. Calligaro, G. Maldini, M. Marson, M. Iurich and R. Petrella, "Integrated High-Frequency SiC Based Modular Multi Three-Phase PMSM Drive for Automotive Range Extender," 2021 IEEE Energy Conversion Congress and Exposition (ECCE), Vancouver, BC, Canada, 2021, pp. 4838-4845.

[15] A. H. Mohamed, H. Vansompel and P. Sergeant, "Electrothermal Design of a Discrete GaN-Based Converter for Integrated Modular Motor Drives," in IEEE Journal of Emerging and Selected Topics in Power Electronics, vol. 9, no. 5, pp. 5390-5406, Oct. 2021

[16] Hemsen J & all, « Innovative and Highly Integrated Modular Electric Drivetrain », World Electric Vehicle Journal. 2019; 10(4):89; https://cordis.europa.eu/project/id/769953

[17] Y. Chen et al., "Mild hybridisation of turboprop engine with high-power-density integrated electric drives," in IEEE Transactions on Transportation Electrification, doi: 10.1109/TTE.2022.3160153.

[18] T. M. Jahns and B. Sarlioglu, "The Incredible Shrinking Motor Drive: Accelerating the Transition to Integrated Motor Drives," in IEEE Power Electronics Magazine, vol. 7, no. 3, pp. 18-27, Sept. 2020

[19] X. Ju et al., "Voltage Stress Calculation and Measurement for Hairpin Winding of EV Traction Machines Driven by SiC MOSFET," in IEEE Transactions on Industrial Electronics, vol. 69, no. 9, pp. 8803-8814, Sept. 2022.

[20] Wang, Y. Li, Y. Han, "Integrated modular motor drive design with GaN power FETs", IEEE Transactions on Industry Applications, Vol. 51, N°. 4, July/August 2015.

[21] F. Salomez, S. Vienot, B. Zaidi A. Videt, T. Duquesne,E. Semail, N. Idir, "Design of an integrated GaN inverter into a multiphase PMSM", VPPC 2020.

[22] L. Pace, N. Defrance, A. Videt, N. Idir, J-C Dejaeger, "Extraction of packaged GaN power transistors parasitics using S-parameters", IEEE Transactions on Electron Devices, Vol. 66, N°. 6, pages. 2583-2588, 04/2019.

[23] S. Pickering, P. Wheeler, F. Thovex, K. Bradley, "Thermal Design of an Integrated Motor Drive", 32nd Annual Conference on IEEE Industrial Electronics, IECON 2006.

[24] M. Trabelsi, N.K. Nguyen, E. Semail, "Real-time Switches Fault Diagnosis based on Typical Operating Characteristics of Five-Phase Permanent Magnet Synchronous Machines", IEEE Transactions on Industrial Electronics, Vol 63, N°8, Aug. 2016 pp. 4683-4694

[25] N.K. Nguyen, F. Meinguet, E. Semail, X. Kestelyn, "Fault-Tolerant Operation of an Open-End Winding Five-Phase PMSM Drive with Short-Circuit Inverter Fault", IEEE Transactions on Industrial Electronics, Vol 63, N°1, pp 595-   605, Jan. 2016,

[26] D. T. Vu, N. K. Nguyen and E. Semail, "Fault-tolerant Control for Non-sinusoidal Multiphase Drives with Minimum Torque Ripple," in IEEE Transactions on Power Electronics, Vol 37, n°6, june 2022,pp 6290-6304

[27] R. Boubaker, S. Ouenzerfi, S. Harmand "Experimental Comparison of Air Cooling Solutions Adapted to GaN Systems", International Conference on Mechanics and Energy (ICME) 2018, Hammamet (Tunisia), December 2018.

[28] S. Ouenzerfi, H. Zahr, M. Trabelsi, R. Boubaker, E. Semail, S. Harmand « 3-D Multi-Nodal Thermal Modelling For Fault-Tolerant Machine » , IEEE ICIT 2019, 20th Ieee International Conference on Industrial Technology, Melbourne (Australia), February 2019.

[29] E. Semail, S. Harmand, N. Idir, B. Lemaire-Semail, «Integrated Polyphase Electric Machine», Patent WO2022136804_A1, 30/06/2022, https://worldwide.espacenet.com/publicationDetails/biblio?locale=en_EP&II=0&date=20220630&CC=WO&NR=2022136804A1&ND =3&KC=A1&rnd=1657892810098&adjacent=true&FT=D

Hardware in the Loop Test of an Electric Aircraft Powertrain

Sebastian Mönninghoff, Moritz Scholjegerdes, Kay Hameyer
INSTITUTE OF ELECTRICAL MACHINES RWTH (IEM)
Schinkelstraße 4
52062 Aachen, Germany
Tel: +49 241 80 97648
E-Mail: sebastian.moenninghoff@iem.rwth-aachen.de
URL: https://www.iem.rwth-aachen.de

Keywords

«Airplane», «Electrical Machine», «Test Bench», «Modelling»

Abstract

Electric machines are attracting attention as potential alternatives to conventional aircraft powertrains. During the development process of an electric aircraft it is often necessary to test and analyze the entire powertrain or components before all interacting systems are physically available. The function and behavior of subsystems, which interact with the powertrain, have to be emulated at a test bench. The powertrain is subject to varying external conditions during flight, which effect the function of the subsystems. Environmental conditions and the characteristic behavior of the aircraft have to be considered. This paper studies the development of a hardware in the loop test (HIL test), which allows to emulate flight conditions on a test bed during powertrain testing. The HIL test is subsequently implemented and developed for the example of the touring motor glider FVA 30. A duty cycle is derived from an aircraft flight test case and used to simulate the aircraft behavior. The results of the HIL test indicate, that the aircraft will meet the specified flight performance goals, but also hint at high temperatures inside the v-shaped tail section, which requires further study and possibly additional cooling effort.

Introduction

Testing aircraft components prior to full system integration offers the possibility to study aircraft subsystems and their interactions at early design stages. Thereby cost and risks of ground or flight tests can be reduced by identifying design issues early. The HIL test discussed here is developed exemplarily for the touring motor glider FVA 30.

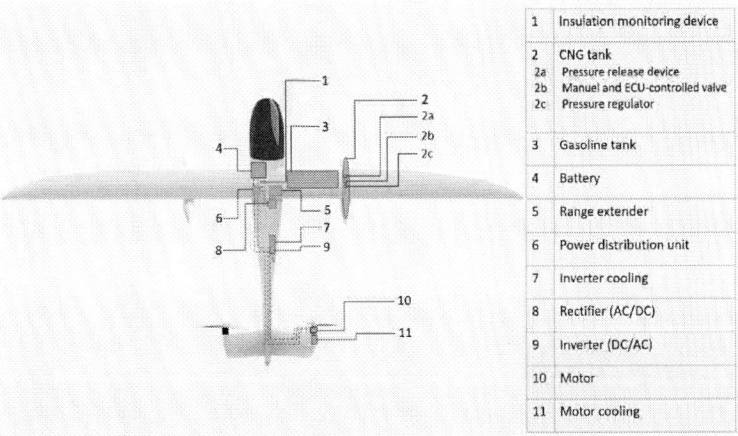

Fig. 1: Components and basic topology of the FVA 30 project [1].

The FVA 30 is a two-seated, hybrid-electric aircraft research platform. It is developed and built by the *Flugwissenschaftliche Vereinigung Aachen (1920) e.V.* (FVA) [1]. The FVA represents an aerospace society, operated and organized by students from RWTH Aachen University. The FVA 30 is based on the motor glider e-Genius of the University of Stuttgart [2]. An overview of the aircraft is shown in figure 1. A particular feature of the FVA 30 is the serial hybrid electric powertrain, which is constructed employing a range extender, a battery, two inverters and two electric machines. The electric machines are mounted to the tips of the v-shaped tail section and are fed by inverters, which are located in the middle section of the aircraft. The FVA has developed the necessary modifications to the structure.

As a constructive boundary condition, the range extender has to be able to run on both, natural gas as well as on gasoline. For this purpose, the natural gas is stored in high-pressure tanks inside the wing pods. The gasoline is stored in integral tanks inside the wings. The combined use of the range extender and the battery enables a range of 650 km. The take-off can be performed without using the range extender, since the aircraft is capable of climbing 1000 m while providing full power to the electric powertrain for 15 minutes by using the battery. [1]

Topology of the HIL Test

The components, which are physically available for the HIL test, include one electric machine, one power electronic inverter, the power cables and a mockup of one tail section side. The measurement campaign includes duty cycle tests, thermal tests, validation of the wiring concept and the prediction of flight performance. During flight operation, the powertrain is subject to varying external conditions, which have to be considered while conducting the tests. Since the inverters, the electric machines and their electric connections are redundant and symmetric, only one side of the powertrain is physically tested at the test bench. The second side of the electric powertrain is accounted for by assuming, that both sides of the powertrain exhibit identical behavior.

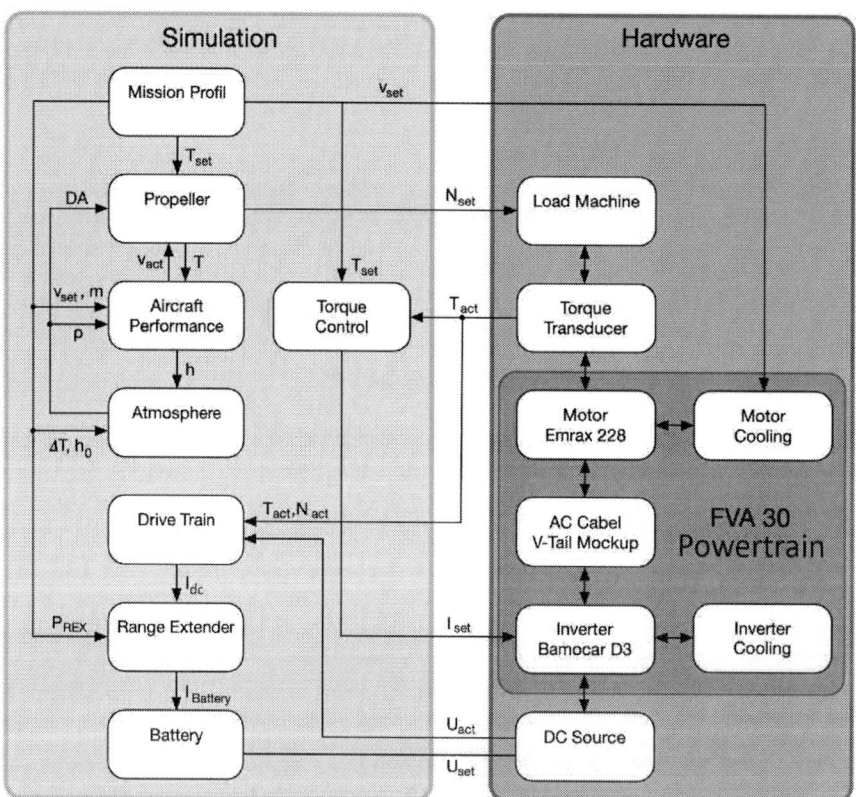

Fig. 2: Topology of the HIL test.

In [3] a framework is proposed, which allows the SIL/HIL testing of aircraft hard- and software components, which are either physically available or simulated. The framework in [3] uses an unmanned aerial vehicle as a test case and the simulator X-Plane to model the aircraft's aerodynamic state. The HIL test proposed in this paper implements a flight performance module directly, which makes use of available measurements of aircraft polars. The atmosphere model allows to simulate environmental conditions encountered during flight. Furthermore the HIL test proposed in this paper focuses on thermal validation of the powertrain concept of a manned aircraft. Cooling systems of the inverter and the electric machine are taken into account and modelled at the test bench. The battery temperature is simulated in the scope of this work, since it is not available physically.

The simulated data can be used to emulate the physical interfaces of the HIL test. A topology for the HIL test is proposed as shown in figure 2. The emulated physical interfaces are the shaft load, the electric connection between the inverter and the battery, the control bus of the inverter and the cooling system of all components.

To emulate the load of the shaft mechanically, an electric machine is employed. The battery behavior is emulated by a dc source, while the control bus of the inverter receives input data from the control hardware at the test bench. The cooling air flow at the test bench is emulated by fans. A mission profile as a duty cycle, is used to set the desired torque and the power provided by the range extender at multiple operating points. All of the aforementioned systems have to be controlled by the HIL test's control hardware to dynamically emulate the physical interfaces in real-time according to the current state of the virtual aircraft.

Physical Interface: Shaft

The proposed HIL test uses the torque T, defined by a mission profile, the state of the aircraft and the atmospheric conditions to calculate a resulting rotational speed of the propeller. Due to operational safety considerations the load machine controls the rotational speed of the HIL test, while the electric machine under test controls the torque.

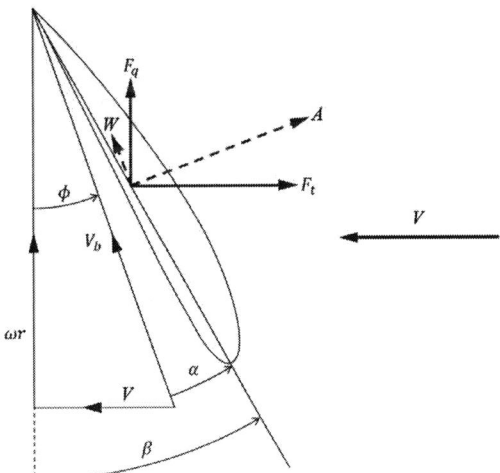

Fig. 3: Forces and velocities at propeller. [4]

The propeller converts the mechanical power of the electric machine into a propulsive force. A propeller has one or more blades rotating around the propeller axis. According to [4] the propeller blades act similar to an airfoil. The airfoils are provided with airflow with the velocity v_b and thus generate lift A and drag W as shown in figure 3. A and W can be transformed into the thrust force F_T and the force F_Q. F_Q results in a drag torque. The resulting blade airspeed v_b is calculated from the aircraft airspeed v_∞, the radius r and the angular frequency ω with equation 1 according to [4].

$$v_b = \sqrt{v_\infty^2 + (\omega * r)^2} \tag{1}$$

The simulation of the propeller is implemented with precomputed maps to determine rotational speed, thrust and efficiency in real-time. The maps are created for multiple operating points of the propeller, which is modelled with the blade element momentum theory [4], [5]. The propeller module receives a desired torque T_{Set}, the current airspeed of the aircraft v_{act} and the density altitude DA as inputs. The resulting rotational speed N is determined by the precomputed lookup tables. The desired torque T_{Set} is defined by the mission profile, while the current airspeed v_{act} is provided by the aircraft performance model. The air density has to be considered, when determining the current state of the propeller. It is provided as the density altitude DA by the atmosphere model.

The flight performance module is used to calculate the airspeed v_{act} and the altitude h of the aircraft in real-time. Figure 4 shows the forces acting on an aircraft during flight. According to [6] three coordinate systems are used to describe the state of the aircraft. In the aerodynamic coordinate system, the axis x_a is aligned with the direction of the aircraft airspeed vector v. In the geodetic coordinate system, the perpendicular axis z_g is parallel to the gravitational vector. The aircraft fixed coordinate system is bound to the aircraft geometry.

The lift force A acts in the opposite direction of z_a and the drag force W acts in the opposite direction of x_a in the aerodynamic coordinate system. The weight force G acts along the gravitational vector z_g in the geodetic coordinate system. As a simplification, it is assumed that the propeller thrust T is aligned parallel to x_a in the aerodynamic coordinate system. The flight performance module receives the thrust of the propeller F, the density of the surrounding air ρ, the desired airspeed v_{set} and the current aircraft mass m as input.

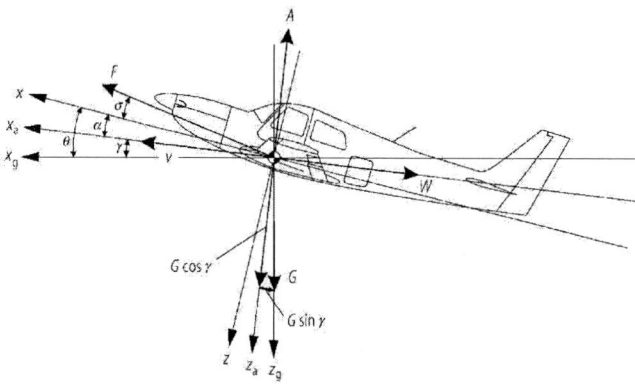

Fig. 4: Angles and Forces acting on an aircraft. [6]

$$0 = A - G * \cos(\gamma) \tag{2}$$
$$0 = F - W - G * \sin(\gamma) \tag{3}$$
$$\gamma = \arcsin\left(\frac{v}{w}\right) \tag{4}$$

In a quasi-stationary flight state without acceleration, a balance of forces is assumed. The relation between the lift A and the weight force G can be described with the equation 2, [6]. The weight force G results from the aircraft mass m and the gravity g. The relation between thrust F, drag W and weight force G is given by equation 3. Aircraft lift and weight are described by coefficients, which are known from measurements of the e-Genius prototype. The thrust is provided by the propeller model. The trajectory angle γ can be determined by the airspeed v and the vertical speed w with equation 4. [4].

The flight attitude control controls the virtual aircraft around the lateral axis. Two PID controllers are used. The first controller controls the climb angle γ over the load multiple n within the limits 0.9 to 1.18. During take-off, n is set to 1 until the rotation speed is reached. The target value for the angle of climb

γ_{set} is calculated with the equation 5 from the thrust F and the drag W_{set}, which would result in a stationary flight state ($n = 1$) at the desired airspeed v_{set} [4].

$$\gamma_{set} = \arcsin\left(\frac{F-W_{set}}{G}\right) \tag{5}$$

The second PID controller uses the airspeed v as reference variable and a difference to γ_{set} as controlled variable. This design allows the speed to be controlled and also limits both n and γ to realistic values. The aircraft performance model and the propeller model depend on atmospheric conditions. The international standard atmosphere (ISA) standardizes these conditions. The basis of the ISA atmosphere is defined at a height of 0 m above mean sea level. The air is assumed to have a standardized temperature of 15 °C, an air density of 1.225 kg/m3 and a pressure of 1013.25 hPa in this height [7].

The international standard atmosphere distinguishes between the air layers troposphere, tropopause and stratosphere [7]. For small aircraft such as the FVA-30 without pressurized cabin, generally only flight altitudes of about 3000 m are permitted without additional oxygen [8]. It is thus further assumed, that all flights take place in the troposphere, which reaches up to an altitude of 11000 m. The flight altitude above mean sea level is converted into the density altitude DA. The density altitude corresponds to the theoretical altitude above a pressure level of 1013.25 hPa. Based on the density altitude, the density is then determined by ISA using an interpolated lookup table.

Physical Interface: Battery

The hybrid power supply of the FVA 30 consists of an internal combustion engine, which is used as a range extender and a Lithium-ion battery. The combustion engine is only designed to provide sufficient power for cruising flight. With respect to the weight, this enables a lighter combustion engine with only 25 kW electric output power. The Lithium-ion battery provides the additional power for take-off and climb, but also enables safe take-off and flying without range extender. Here, the battery is designed for an overload output power of up to 95 kW during take-off and 87 kW during climb.

The drivetrain module calculates the total current I_{dc} necessary to provide the torque T. It takes the current voltage of the battery U, the measured rotational speed N and the measured torque T as input. The current I_{dc} is provided by the battery and the range extender. The range extender module calculates the current, which is drawn from the battery $I_{Battery}$ with the range extender power P_{REX} and the total current I_{dc} as input. The output of the battery module is used to set the voltage of the dc source U_{set}. In the context of this work the battery is emulated by a dc Source at the test bench.

The state of charge SOC results from the charge and the capacity of the cell with the equation 6. The charge of the cell Q_C depends on the initial charge $Q_C(t_0)$ and results from the integration of the cell current I_C over time according to equation 7.

$$SOC = \frac{Q_C}{C_C} \tag{6}$$

$$Q_C(t) = Q_C(t_0) + \int_{t_0}^{t} I_C(t)dt \tag{7}$$

The *Peukert* effect describes the influence of the discharge current on the storage capacity of the battery. In the data sheets of the batteries a nominal capacity C_N is usually given, which can be taken with the corresponding nominal current I_N. Equation 8 can be used to describe the extractable charge Q in dependence of the current I and the Peukert number k. [9]

$$Q = C_N * \left(\frac{I_N}{I}\right)^{k-1} \tag{8}$$

The temperature T_{bat} is obtained by integrating the power dissipation $P_L(t)$ over time, the mass m and the specific heat capacity c_p with equation 9. [10]

$$T_{bat}(t) = T_{bat}(t_0) + \int_{t_0}^{t} \frac{P_L(t)}{c_p * m} dt \qquad (9)$$

Physical Interface: Data Bus

The data bus of the inverter is used to control the electric machine. With the desired torque T_{set} and the measured rotational speed N the initial i_d-, i_q-current combination is determined from a lookup table. A PID controller is used within the torque control module depicted in figure 2 to account for the remaining difference between the desired torque and the measured torque. The inverter is controlled via a CAN bus.

Physical Interface: Cooling

The electric machine and the inverter use separate liquid cooling systems. In both cases a pump circulates the coolant through the component and a radiator.

The radiator of the electric machine is installed directly in the nacelle behind the propeller, so that an air flow is provided by the propeller. To emulate an equivalent air flow on the test bench an additional fan is used, which is directly mounted to the radiator as shown in figure 5. The speed of the air flow is measured by anemometer measurements in a circular duct on the suction side of the fan. Due to the known velocity profile of the air flow in the circular duct the mass flow can be calculated. The fan is set to generate the airflow expected during flight operation of the aircraft. The electric machine will be mounted on a steel mount in the aircraft. On the test bench however, the machine is mounted on an aluminum base plate with higher heat conductivity and a test bench mount with higher heat capacity than the real structure. To reduce the influence of the test bench mount on the heat dissipation mechanisms, the contact area between the electric machine and the aluminum plate is designed to be as small as possible. The plate is also insulated by sheets of foam to reduce convection at the plate surface.

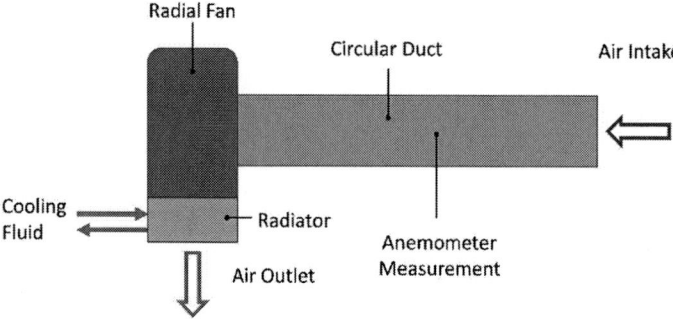

Fig. 5: Cooling setup of electric machine.

The inverter is installed further forward in the fuselage and needs a fan, which ensures sufficient air flow through the inverter radiator even when the aircraft is rolling at low speed. The inverter is equipped with a liquid cooling circuit. On the test bench the same setup is used as in the aircraft according to the current state of the aircraft design. The speed of the fan is fixed.

The cooling of the wiring was identified as a potentially critical aspect during the aircraft design process. Due to the restricted range of the center of gravity of the aircraft the inverters have to be placed beneath the wing section of the aircraft. The two electric machines are mounted on the v-shaped tail section of the aircraft. The electric wiring is routed from beneath the wing section through the whole tail section of the aircraft to the electric machines. Joule losses lead to increased temperatures in the cables and the surrounding structure. Since there is no airflow available in the aircraft body, natural convection is expected to be the dominant cooling mechanism in the vicinity of the wiring. The surrounding fiber reinforced polymer structure imposes an upper temperature limit of 80 °C on the wiring [11]. The HIL test is set up in such a way, that the wiring can be studied under realistic thermal conditions. A mockup of the v-shaped tail section is used at the test bench to encase the wiring. The upper opening of the mockup is sealed to create worst case conditions with impeded natural convection.

Measurements

Figure 6 shows the setup at the test bench. To emulate the physical interface of the shaft, the electric machine of the FVA 30 is coupled to a load machine with a torque transducer in between the machines. The measured torque and rotational speed are used as input for the HIL test software. The load machine is controlled by the HIL test software to emulate the behavior of the propeller.

Fig. 6: Mechanical connection of the shaft with torque and speed sensors.

The electric machine of the FVA 30 is supplied by the inverter, which is used in the aircraft, while the load machine is supplied by a test bench inverter. The inverter of the FVA 30 is provided with electric power by a dc source. The inverter current is measured and used as input data for a model, which characterizes the battery. The voltage of the dc source is controlled by this model to simulate the battery behavior.

A mockup of the v-shaped tail section was constructed and equipped with temperature sensors to monitor the temperature in the glass fiber structure of the plane. This setup enables realistic measurements of temperatures occurring during the duty cycle.

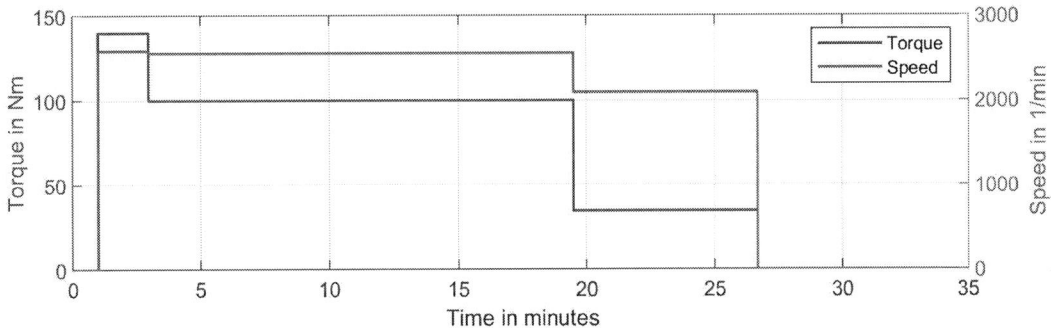

Fig. 7: Assumed duty cycle and operating conditions for the laboratory measurement.

To simulate the maximum expected load on the powertrain, the duty cycle shown in figure 7 is used. It simulates the take-off followed by an ascend until the battery is depleted. The time t_{vl} between the end of the take-off power phase and the arrival at cruise altitude depends on the aircraft simulation and is determined during the test run.

Figure 8 shows component temperatures, which are measured during the test run. The battery temperature is simulated, since it is emulated by a dc source and not physically available. The electric machine's coolant pump is operated at 8.1 l/min and the fan between 360 and 410 m³/h during the tests. During the measurements the ambient temperature is measured at 24 °C, which results in a correction of 14 K to account for the highest expected temperature under real operating conditions. Measurements shown in figure 8 show, that the maximum end-winding temperature of the electric machine is about 74 °C while operated at maximum continuous power. Corrected for warm days this results in 88 °C, which is well below the manufacturer's limit of 120 °C. The electric machine coolant temperature stabilizes at 44 °C.

The coolant pump of the inverter is operated at 7 l/min and the fan between 210 and 250 m³/h during the tests. The temperatures of the inverter IGBTs stabilize at 48 °C, which corresponds to a compensated

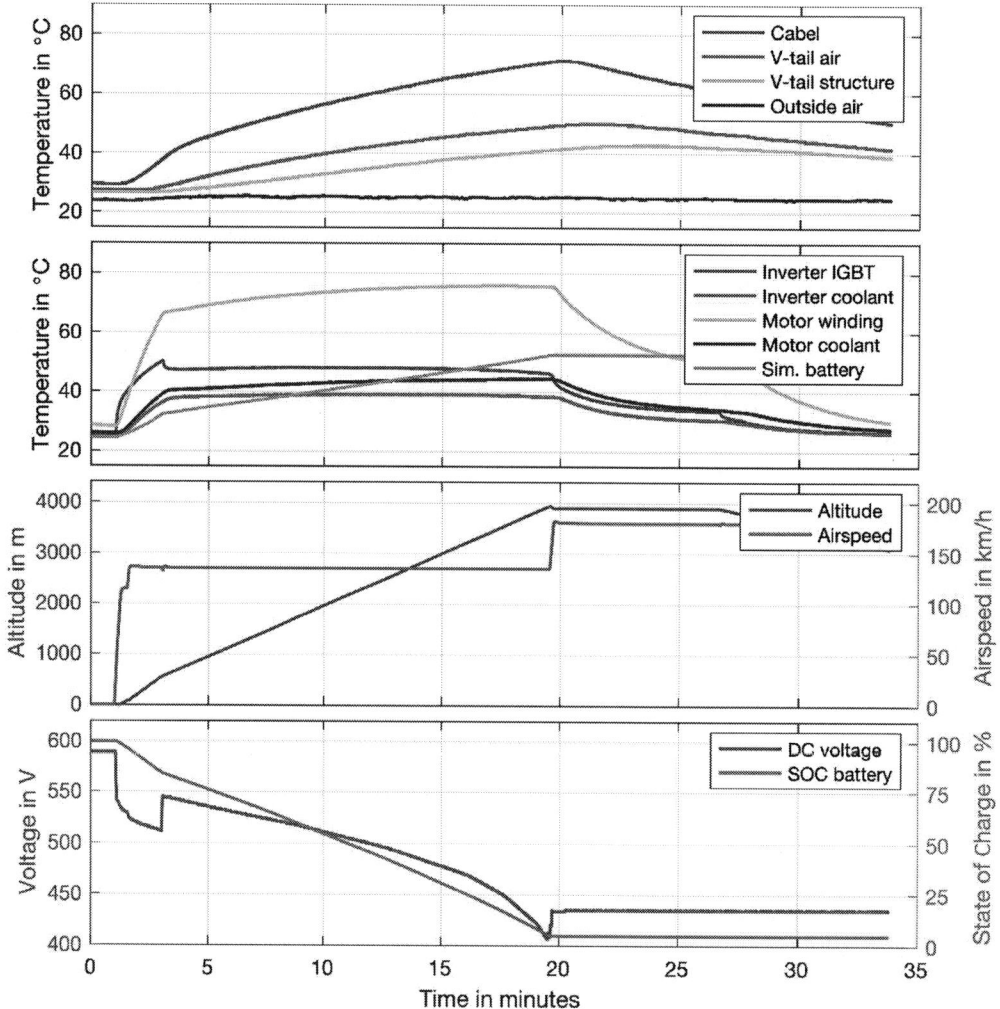

Fig. 8: Temperatures and simulated state of the aircraft.

temperature of 62 °C. This temperature is below the 75 °C of the inverter overtemperature warning. The coolant settles at 39 °C. The corrected temperature of 53 °C is below the 65 °C maximum inlet temperature of the inverter cooler. The air in the tail section reaches a temperature of 52 °C, which corresponds to corrected 66 °C. The actual heating of the tail section's inner side reached its maximum at 43 °C and corrected 57 °C. The material of the tail section has to be heat treated to withstand this temperature [11]. All components in the tail section must be able to withstand the treatment. This should be considered separately during the aircraft design process when selecting materials for the tail section.

During the duty cycle the aircraft is able to ascend to 4000 m above the initial altitude. The simulated state of charge of the battery and the temperatures of all components show, that the electric powertrain design fulfills the requirements set by the duty cycle.

Conclusions

The HIL test proposed and discussed in this paper is implemented at a test bench and used to validate the powertrain design of the FVA 30. The HIL test offers the possibility to test the electric powertrain with realistic duty cycles. The cooling of the components is set up to mimic the situation in the final aircraft design. This enables the determination of the thermal powertrain performance under realistic conditions and the prediction of overall flight performance of the aircraft.

The measurement results show, that the cooling for inverter and electric machine are sufficiently dimensioned. The HIL test is also used to validate the voltage range of the battery. The battery, supported by the range extender, provides sufficient energy to climb to almost 4000 m above the initial altitude. The voltage level is sufficient to achieve the required rotational speed throughout the entire duty cycle.

The HIL test measurement results demonstrate, that the electric losses in the connecting cables lead to increased temperatures in the tail section. The material of the tail section itself is warmed up considerably. In the further design process, these findings have to be considered in the fuselage design, so that an unacceptable temperature increase in the material can be avoided. In further tests the use of an active ventilation of the tail section and its influence on the temperatures could be interesting.

References

[1] Moxter T., Enders W., Kelm B., Scholjegerdes M., Koch C., Garbade M., Dahmann P.: Investigation of Alternative Propulsion Concepts for Small Aircraft with the Hybrid Electric Motor Glider FVA 30, Proc. Conf. DLRK 2020, 2021.

[2] Schumann L.: Reduktion des Energiebedarfs mittels eines batterieelektrischen Antriebs am Beispiel eines Kleinflugzeugs, Universität Stuttgart, 2018.

[3] Bole B., Teubert C., Chi Q.C., Hogge E.: SIL/HIL replication of electric aircraft powertrain dynamics and inner-loop control for V&V of system health management routines, Annual Conference of the PHM Society, 2013.

[4] Gudmundsson S.: General Aviation Aircraft Design, Applied Methods and Procedures. Butterworth-Heinemann, 2013.

[5] Glauert H.: The Elements of Aerofoil and Airscrew Theory, Cambridge University Press, 1983.

[6] Grote K. H., Bender B., Gohlich D.: Dubbel - Taschenbuch für den Maschinenbau. Springer-Verlag, 2018.

[7] Scheiderer J.: Angewandte Flugleistung: eine Einführung in die operationelle Flugleistung vom Start bis zur Landung. Springer Science and Business Media, 2008.

[8] Mies J.: AOPA Germany Safety Letter, Sauerstoffmangel, AOPA Germany, 2018.

[9] Omar N., Daowd M., Bossche P. v. D., Hegazy O., Smekens J., Coosemans T., Mierlo v. J.: Rechargeable energy storage systems for plug-in hybrid electric vehicles-Assessment of electrical characteristics. Energies, 5(8), 2012.

[10] VDI-Gesellschaft Verfahrenstechnik und Chemieingenieurwesen: VDI-Wärmeatlas. Springer Vieweg, 2013.

[11] Hexion, Specialty Chemicals, Epoxy and Phenolic Resins Division Epoxy Resins: EPIKOTE Resin MGS LR 285, 2010.

A Multi-port Smart Transformer for Green Airport Electrification

Giampaolo Buticchi[1], Giovanni De Carne[2], Thiago Pereira[3], Kangan Wang[3],
Xiang Gao[4], Jiajun Yang[1], Youngjong Ko[5], Zhixiang Zou[6], Marco Liserre[3]
University of Nottingham Ningbo China[1], Karlsruhe Institute of Technology[2],
University of Kiel[3], Siemens Energy[4],
Pukyong National University[5], Southeast University[6]
Ningbo, China; Karlsruke, Germany; Kiel, Germany; Germany;
Busan, South Korea; Nanjing, China
Email: buticchi@ieee.org

Keywords

≪smart transformer≫, ≪green airport≫, ≪multi-port power electronics≫.

Abstract

Green transportation and renewable energy production have attracted a global attention due to the needs of decreasing the environmental impact and still sustain increased energy needs. In this framework, the aircraft and airports are facing a profound renovations towards green technologies, among which the electrical ones are playing a central role. This paper explores how a Smart Transformer can upgrade the existing airport power system, enabling an efficient interface for renewable energy, electric vehicles and the future hybrid/electric aircraft, substituting the ground power units and enabling a smarter behavior of the electrical grid.

Introduction

The renewable energy exploitation has pushed the industry and the academia towards the development of power electronics solutions that have profoundly changed the electrical grids [1]. In particular, the increased adoption of power electronics interfaced sources and loads into the electrical grid has drastically changed the power flow conditions, the power quality and has been challenging the grid stability. In order to sustain an increased distributed generation and a growing demand at the same time (i.e., electric vehicle charging stations), several concepts have been proposed to make the electrical grid "smarter". One of the solutions which have become increasingly popular is the Smart Transformer (ST) [2], a solid-state transformer which can replace the existing power transformer as ac-ac converter, with the additional benefits of electronic control, dc connectivity, smart maintenance [3] and ancillary services. In this way, a gradual upgrade of the electrical grid becomes possible, with increased benefits as the penetration of the STs increases.

The transportation sector is used to be one of the primary pollution resources since fossil fuel is the dominant energy source in this sector. In order to reduce the emission of greenhouse gas, SO_x, and NO_x, transportation electrification has been proposed. The electric vehicle (EV) is considered as a promising solution. To make convenient charging access, Germany has installed over 20,000 charging facilities, and around 15% of them is fast charging facilities [4, 5]. As for maritime transportation, which contributes about $5 - 8\%$ of global emissions, vessel cold-ironing at berth has been promoted worldwide to increase the efficiency of energy consumption. Cold ironing is a process that uses shoreside power to provide the demand of vessels while they are docked. By plugging in the vessel at port allows shutting down all diesel generators on board of the vessel. The Siemens has built Germany's largest shore power system at the port of Kiel. With $16MVA$ power capability, the port is capable of supplying two ships simultaneously

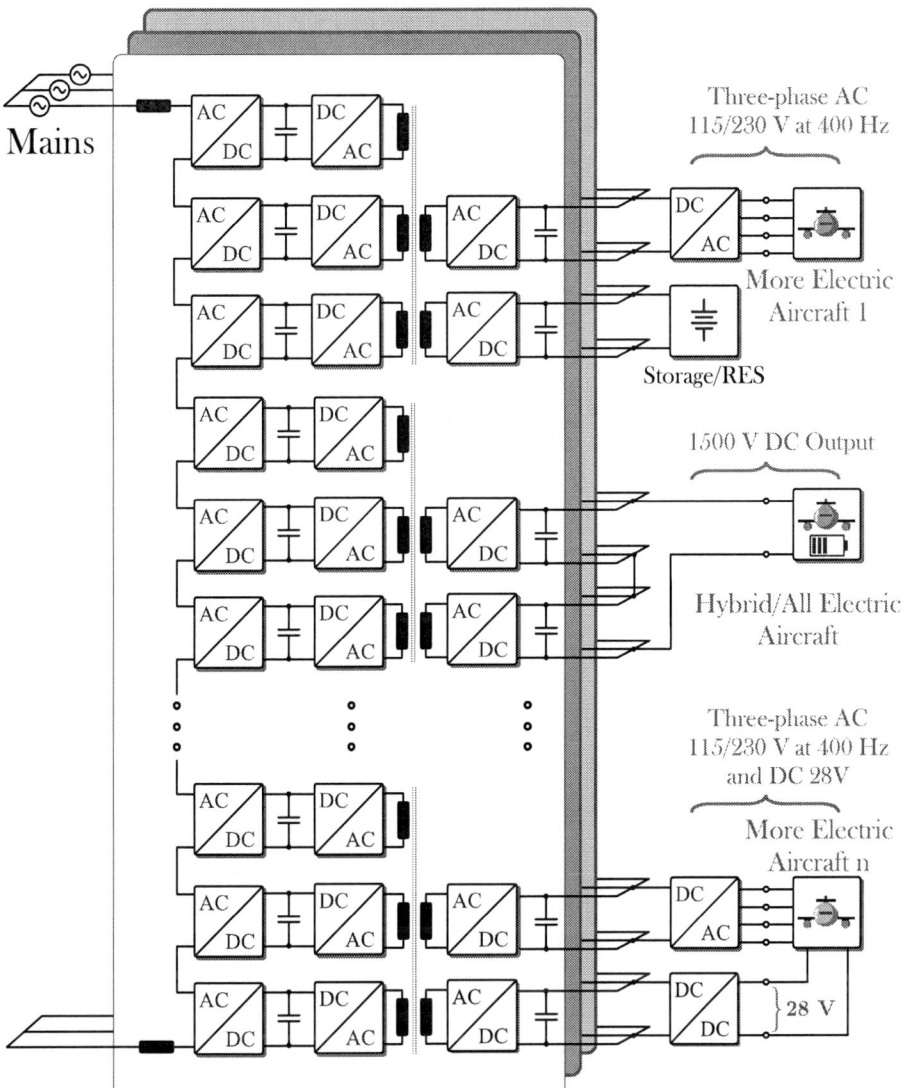

Fig. 1: Example of an architecture for the Smart Transformer based on Cascaded H-bridge (CHB) and a multi-port power converter structure which allows for the independent control of the output ports.

and will help to reduce more than 8000 tons of CO_2 annually [6]. Other researches have also confirmed the effectiveness of cold ironing in terms of reduction of CO_2 in different countries [7].

The ST can further benefit the green transportation ambition. For one side, it provides together with traditional ac connection also the dc connection. Consequently, the equipment that requires dc connection, such as charging facilities, can be directly connected to the ST, decreasing the complexity of the operation. In the meantime, some obstacles to host high power charging facilities, such as degrading the voltage quality in the ac grid, can be avoided. Furthermore, the onboard grid on vessels can use either 50 or 60 Hz ac power supply, requiring frequency conversion in the harbor if the frequency doesn't match. The ST can decouple the MV and LV grids connected to its both sides. This feature helps adjust the frequency easily between 50 and 60 Hz, solving frequency conversion between vessels and harbor.

An airport is a complex system with peculiar characteristics in terms of energy needs and construction. In addition to the energy for lighting, heating and airport logistics, there is an important contribution of the parked airplanes, whose electrical system gets connected to the airport one. In the short term, the electrification of the ground transport and the usage of low-carbon energy sources (i.e., renewable)

Simulation Parameters			
Power level	320 kW	Switching Frequency	5 kHz
DC/DC Topology	5-Active Bridge	Phase connection	Custom
DC Link Capacitance	3 mF	Transformer Leakage Inductance	15 uH
Voltage Control Bandwidth	300 Hz	Droop Coefficient	0.02 Ohm
Low-voltage current I_{LV}	300 A	PV Power level	0-10 kW

are seen as promising [8]. However, improving the energy management through smart metering, demand management and planning is an important long-term goal to achieve [8]. The improvement of the scheduling to reduce delays and the pollution has also been addressed [9]. As reported in [10] and [11], both the Copenhagen and Kansai airport rely on electrical ground power for the parked aircraft, to allow the main engine to be switched off. In the UK, Manchester and Stansted were already featuring renewable energy plants in 2014 [12]. In a plan to renovate the Tabriz International Airport, solar energy production is envisaged for the flat roofs of the airports [13]. Combined Heat Power (CHP) plants can also be adopted for the simultaneous production of electricity and heat, improving the efficiency. CHP and solar generations were two technologies used for the Rome International Airport [14]. In the next China Civil Aviation (CCA) 5-year plan, a great push towards greener airports and electrical technologies is also envisaged [15] to reduce the environmental impact.

Proposed Architecture

Several airports have undergone an upgrade of the electrical grids towards redundant medium voltage distribution.

Figure 1 shows a possible example of a smart transformer architecture which can fulfill the application requirements. A Cascaded H-bridge is adopted due to the wide industrial acceptance for medium voltage applications and due to the reduced need of having a MV-rated DC Link (in the range of tens of kV) which would make a Modular Multilevel Converter (MMC) solution more interesting. At the core of the structure is the DC/DC converter, which can be built in several configurations of active bridges. Although the initial proposals for ST were based on the Dual Active Bridge (DAB), which could offer bidirectional power transfer with soft-switching feature, the single-input single-output nature of the DAB limits the designer choices. For this reason, Multiple Active Bridges (MAB), which are the extension of the DAB with multi-winding transformers, have been investigated [16] due to their advantages in terms of arbitrary power flow among the different ports and for the potential weight saving compared to the DAB solutions due to better usage of the transformer core (up to 30% [16]).

Figure 1 shown a possibility of employing a MAB with five ports to build up several DC ports to power different aircraft kinds, since the conventional aircraft need 400 Hz, 115 V AC supply, whereas some More Electric Aircraft could be powered directly by a low voltage DC [17, 18] (270 V or 540 V). Considering the recent trends of electrification, hybrid/electric aircraft solutions are currently under investigation or prototyping, whereas for full electric aircraft there are already available products (Bye Aerospace eFlyer 800, eight-seater, Eviation Alice, nine-seater, and Pipistrel Alpha-Electro). Due to the high-power needs of electric propulsion (in the range of MW) even for regional aircraft, the power distribution will most likely shift from ac to dc and toward higher voltage in the kV range [19, 20].

Simulation Results

A proof-of-concept simulation targeting the DC/DC converter is set up in a Simulink/PLECS environment. A 300 kW power electronics operating at an equivalent DC voltage of 3 kV is considered as the target. Exemplified parameters are given in TABLE I. The target of the simulations is to prove that the control can stably regulate the DC links even in strongly asymmetrical power conditions.

Figure 2a shows a proposal of intertwined multiple active bridge (IMAB) configuration for the DC/DC converter, with the following characteristics:

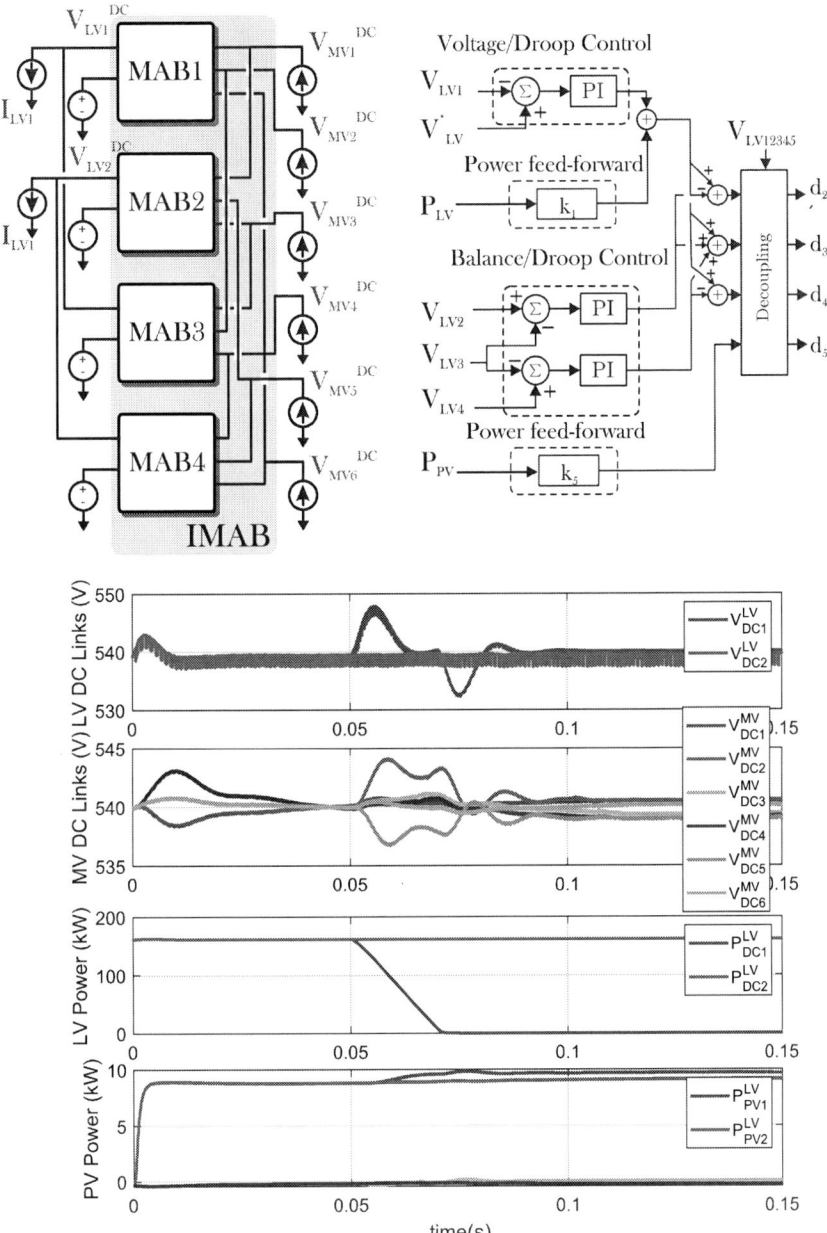

Fig. 2: Simulation Results. Proposed architecture based on the intertwined connection of Multiple Active Bridge (5 ports) converters (a), phase-shift control based on PI with droop characteristic and power feedforward, simulation results of the converter during a step change in the PV power at t = 0 s and a power ramp-down from a low-voltage port at t = 0.05 s.

1. Each MAB has its own control system separated from the others.

2. Each MV DC Link is connected to at least two MAB ports.

3. There must be a path connecting the MAB ports allowing for a redistribution of the power.

A simple control scheme based on the phase-shift modulation can be implemented and is reported in Figure 2b. The main feature are the possibility of regulating the LV port by drawing power from the MV ports and an equalizer control for the MV ports, which allows the CHB to draw unbalanced power from each AC cell. This control is based on an evolution of the distributed droop regulation with virtual resistors in [21] with the addition of the real-time decoupling of [22]. Additionally, a power feed-forward for the LV port is added to reduce the voltage transient. For the sake of simplicity, just a feed-forward control is employed for the Storage/PV port, although the implementation of a closed-loop control based on the state of charge or MPPT is straightforward. When it comes to the tuning of control parameters, the Symmetrical Optimum (SO) as one of popular tuning techniques, can be a proper choice. The SO was firstly proposed by Kessler [23], aiming at enhancing the stability, robustness and optimizing the transient performance of systems. With SO, the open loop phase margin of system can be associated with the PI parameters based on the linearized plant and feedback [24]. For the sake of simplicity, only the SO-based tuning process for the output voltage control is presented in the following content. It is noted that the same tuning procedure can be applied for the balancing controller. Fig. 3 shows a simplified transfer function block scheme of output voltage loop, where the output voltage controller is given as

$$G_v(s) = \frac{K_p s + K_i}{s} \tag{1}$$

The sampling period of controller and the PWM delay is modeled together as a first-order transfer function with cut-off frequency of ω_d, given as

$$G_d(s) = \frac{1}{s/\omega_d + 1} \tag{2}$$

And the control-to-output-current gain A is given as

$$A = \frac{V_i}{2N f_s L_{lk}} (1 - 2d) \tag{3}$$

Where V_i is the input voltage of converter, N is the turn ratio of transformer (secondary side voltage to primary side voltage), f_s is the switching frequency, L_{lk} is the leakage inductance and d is the phase shift. With the above equations, the open loop gain of the output voltage loop can be calculated as

$$G_{ol}(s) = A G_v(s) G_d(s) = \frac{A K_i \frac{s}{\omega_v} + 1}{s^2 C_o \frac{s}{\omega_d} + 1} \tag{4}$$

Where ω_v is the controller bandwidth. The gain crossover frequency ω_o can be approximated as

$$\omega_o = \sqrt{\omega_d \omega_v} \tag{5}$$

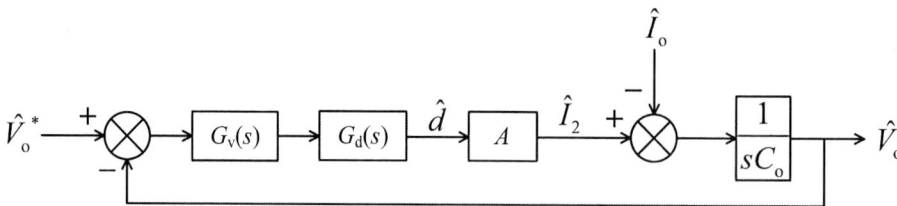

Fig. 3: Simplified transfer function block scheme of output voltage control

Then, it exists

$$|G_{\text{ol}}(j\omega_{\text{o}})| = \frac{AK_{\text{i}}}{\omega_{\text{o}}^2 C_{\text{o}}} \sqrt{\frac{\frac{\omega_{\text{o}}^2}{\omega_{\text{v}}^2}+1}{\frac{\omega_{\text{o}}^2}{\omega_{\text{d}}^2}+1}} = 1 \tag{6}$$

By rearranging the above equation, K_{i} can be expressed as

$$K_{\text{i}} = \frac{\omega_{\text{v}}\sqrt{\omega_{\text{v}}\omega_{\text{d}}}C_{\text{o}}}{A} \tag{7}$$

And therefore, K_{p} can be obtained as

$$K_{\text{p}} = \frac{\sqrt{\omega_{\text{v}}\omega_{\text{d}}}C_{\text{o}}}{A} \tag{8}$$

In the simulations, the CHB cells are emulated by current sources which draw an equal current and a PI control which regulates the total sum of the MV DC Links. It is worth noting that this represents a worst-case scenario, since the capability of power redistribution offered by the CHB are not exploited.

Two conditions are tested: the increase of the power drawn by the PV cells at t = 0 s (from 0 to 10 kW) and the power ramp-down of one of the LV ports at t = 0.05 s from 180 kW to zero (for example, an aircraft disconnecting). As can be seen, although the IMAB is working in asymmetrical condition (MAB 1 is working with a phase shift of opposite sign with respect to the other MABs), the distributed control allows for a good voltage tracking. A small steady-state error in the power of the PV happens due to the feed-forward nature of the PV port control. The performance of the control could be further improved by changing the phase-shift modulation [25] to ensure the minimum current circulation within the MAB or to implement a centralized controller [26] which would allow for an optimized transient response at the expense of an increase computational burden.

Conclusion

This paper has proposed a Smart Transformer as the core of a future green airport, allowing for the integration of electric loads, renewable energy systems and electrical grid, allowing for multiple isolated dc ports in a modular structure. Simulations proved that the proposed structure based on intertwined multiple active bridge allows for the power redistribution even during strongly asymmetrical power processing. A proof-of-concept distributed control has shown to achieve acceptable performance in terms of voltage transients.

References

[1] G. Buticchi, C.-S. Lam, R. Xinbo, M. Liserre, D. Barater, M. Benbouzid, O. Gomis-Bellmunt, C. A. Ramos-Paja, C. Kumar, and R. Zhu, "The role of renewable energy system in reshaping the electrical grid scenario," *IEEE Open Journal of the Industrial Electronics Society*, vol. 2, pp. 451–468, 2021.

[2] M. Liserre, G. Buticchi, M. Andresen, G. De Carne, L. F. Costa, and Z.-X. Zou, "The smart transformer: Impact on the electric grid and technology challenges," *IEEE Industrial Electronics Magazine*, vol. 10, no. 2, pp. 46–58, 2016.

[3] M. Liserre, G. Buticchi, J. I. Leon, A. Marquez Alcaide, V. Raveendran, Y. Ko, M. Andresen, V. G. Monopoli, and L. Franquelo, "Power routing: A new paradigm for maintenance scheduling," *IEEE Industrial Electronics Magazine*, vol. 14, no. 3, pp. 33–45, 2020.

[4] BMWi-Germany, "Masterplanladeinfrastruktur der bundesregierung," accessed on Jan.31.2021.

[5] Statistisches-Bundesamt-Germany, "Anzahl der ladestationen für elektrofahrzeuge in deutschland," accessed on Jan.31.2021. [Online]. Available: https://de.statista.com/statistik/daten/studie/460234/ umfrage/ladestationen-fuer-elektroautos-in-deutschland-monatlich/

[6] J. Kemper, "Siemens builds germany's largest 'power outlet' for ships for port of kiel," accessed on Aug. 11, 2020. [Online]. Available: https://www.metec-tradefair.com/en/News/Business_News/Siemens_builds_Germany's_largest_'power_outlet'_for_ships_for_Port_of_Kiel

[7] F. Ballini and R. Bozzo, "Air pollution from ships in ports: The socio-economic benefit of cold-ironing technology," *Research in Transportation Business and Management*, vol. 17, pp. 92–98, 2015, energy Efficiency in Maritime Logistics Chains.

[8] F. Greer, J. Rakas, and A. Horvath, "Airports and environmental sustainability: a comprehensive review," *Environmental Research Letters*, vol. 15, no. 10, p. 103007, oct 2020. [Online]. Available: https://doi.org/10.1088/1748-9326/abb42a

[9] L. Adacher, M. Flamini, and E. Romano, "Airport ground movement problem: Minimization of delay and pollution emission," *IEEE Transactions on Intelligent Transportation Systems*, vol. 19, no. 12, pp. 3830–3839, 2018.

[10] G. Baxter, P. Srisaeng, and G. Wild, "Sustainable airport waste management: The case of kansai international airport," *Recycling*, vol. 3, no. 1, pp. 6–0, 2018. [Online]. Available: https://www.mdpi.com/2313-4321/3/1/6

[11] ——, "An assessment of airport sustainability, part 2—energy management at copenhagen airport," *Resources*, vol. 7, no. 2, 2018. [Online]. Available: https://www.mdpi.com/2079-9276/7/2/32

[12] 2014. [Online]. Available: https://www.aoa.org.uk/wp-content/uploads/2014/09/AOA-Sustainable-Airports-Report.pdf

[13] A. Azami, H. Sevinc, and N. Akbarzadeh, "Bipv approach in modeling and re-designing of tabriz international airport, iran," in *2018 International Conference on Photovoltaic Science and Technologies (PVCon)*, 2018, pp. 1–9.

[14] M. Falvo, F. Santi, R. Acri, and E. Manzan, "Sustainable airports and nzeb: The real case of rome international airport," in *2015 IEEE 15th International Conference on Environment and Electrical Engineering (EEEIC)*, 2015, pp. 1492–1497.

[15] "Outline of action for the construction of china's civil aviation type 4 airport [in chinese]." [Online]. Available: http://www.gov.cn/zhengce/zhengceku/2020-03/25/content_5495472.htm

[16] T. Pereira, F. Hoffmann, R. Zhu, and M. Liserre, "A comprehensive assessment of multiwinding transformer-based dc–dc converters," *IEEE Transactions on Power Electronics*, vol. 36, no. 9, pp. 10 020–10 036, 2021.

[17] G. Buticchi, S. Bozhko, M. Liserre, P. Wheeler, and K. Al-Haddad, "On-board microgrids for the more electric aircraft—technology review," *IEEE Transactions on Industrial Electronics*, vol. 66, no. 7, pp. 5588–5599, July 2019.

[18] G. Buticchi, P. Wheeler, and D. Boroyevich, "The more-electric aircraft and beyond," *Proceedings of the IEEE*, pp. 1–15, 2022.

[19] D. Golovanov, D. Gerada, G. Sala, M. Degano, A. Trentin, P. H. Connor, Z. Xu, A. La Rocca, A. Galassini, L. Tarisciotti, C. N. Eastwick, S. J. Pickering, P. Wheeler, J. C. Clare, M. Filipenko, and C. Gerada, "4mw class high power density generator for future hybrid-electric aircraft," *IEEE Transactions on Transportation Electrification*, pp. 1–1, 2021.

[20] A. Trentin, G. Sala, L. Tarisciotti, A. Galassini, M. Degano, P. H. Connor, D. Golovanov, D. Gerada, Z. Xu, A. La Rocca, C. N. Eastwick, S. J. Pickering, P. Wheeler, J. C. Clare, and C. Gerada, "Research and realization of high-power medium-voltage active rectifier concepts for future hybrid-electric aircraft generation," *IEEE Transactions on Industrial Electronics*, vol. 68, no. 12, pp. 11 684–11 695, 2021.

[21] G. Buticchi, M. Andresen, M. Wutti, and M. Liserre, "Lifetime-based power routing of a quadruple active bridge dc/dc converter," *IEEE Transactions on Power Electronics*, vol. 32, no. 11, pp. 8892–8903, 2017.

[22] G. Buticchi, L. F. Costa, D. Barater, M. Liserre, and E. D. Amarillo, "A quadruple active bridge converter for the storage integration on the more electric aircraft," *IEEE Transactions on Power Electronics*, vol. 33, no. 9, pp. 8174–8186, 2018.

[23] C. Kessler, "Das Symmetrische Optimum, Teil I," *at - Automatisierungstechnik*, vol. 6, no. 1-12, pp. 395–400, Dec. 1958, publisher: Oldenbourg Wissenschaftsverlag. [Online]. Available: https://www.degruyter.com/document/doi/10.1524/auto.1958.6.112.395/html

[24] R. Teodorescu, M. Liserre, and P. Rodríguez, *Grid Converters for Photovoltaic and Wind Power Systems*. Chichester, UK: John Wiley & Sons, Ltd, Jan. 2011. [Online]. Available: http://doi.wiley.com/10.1002/9780470667057

[25] G. Buticchi, D. Barater, L. F. Costa, and M. Liserre, "A pv-inspired low-common-mode dual-active-bridge converter for aerospace applications," *IEEE Transactions on Power Electronics*, vol. 33, no. 12, pp. 10 467–10 477, 2018.

[26] F. Savi, J. Harikumaran, D. Barater, G. Buticchi, C. Gerada, and P. Wheeler, "Femtocore: An application specific processor for vertically integrated high performance real-time controls," *IEEE Open Journal of the Industrial Electronics Society*, vol. 2, pp. 479–488, 2021.

Improvement of EMI Filter Attenuation Using Shielding

Mohammad Ali, Rehnuma Bushra, Jens Friebe, Axel Mertens
LEIBNIZ UNIVERSITY HANNOVER
Institute for Drive Systems and Power Electronics
Welfengarten 1
30167 Hannover, Germany
Tel: +49 / (0) - 511 762 3778
Fax: +49 / (0) - 511 762 3040
E-mail: mohammad.ali@ial.uni-hannover.de
URL: www.ial.uni-hannover.de

Acknowledgments

This work was supported by Forschungsvereinigung Antriebstechnik e.V. (FVA) within Project FVA 637V.

Keywords

≪Electromagnetic Interference (EMI) Filters≫, ≪Shielding Modeling and Methods≫, ≪Mutual Couplings≫, ≪Filter Optimization≫, ≪EMI≫.

Abstract

The parasitic couplings between EMI filter components strongly impact the high-frequency attenuation of an EMI filter. One of the most effective ways to reduce the magnetic and capacitive couplings between components and improve the high-frequency attenuation is placing a conductive shield between filter components. Therefore, this paper analyzes and investigates the shielding enclosure's size and volume effect on EMI filter performance in common-mode (CM) and differential-mode (DM) configurations using copper shields positioned over a CM choke.

I. Introduction

Power electronic (PE) converters have become an indispensable technology in the field of power processing and conversion due to their high efficiency, high reliability, high power density, and low losses in the power semiconductor devices [1], [2]. On top of this, the introduction of silicon carbide (SiC) wide-bandgap (WBG) power semiconductors in the form of the metal–oxide–semiconductor field-effect transistor (MOSFET) has strengthened these advantages further. These WBG semiconductors offer faster switching than other conventional silicon (Si) semiconductors, such as insulated-gate bipolar transistors (IGBTs), resulting in a substantial reduction in switching losses and transmission losses. The high switching frequency and low losses help to reduce the size and weight of the PE converters [3]. However, the fast switching of WBG semiconductor devices comes along with very short transient times. These may induce voltage transients up to $50 \frac{V}{ns}$, which is approximately ten times higher than with the conventional Si semiconductors [4]. As a result, controlling the conducted emissions from power converters is becoming increasingly important [5], [6]. However, the physical size of the EMI filters needs to be considered as well.

As modern electronics become smaller, EMI filters must match this trend. The size of a passive EMI filter can be minimized by positioning the filter components very close to each other. However, this may raise parasitic inductive couplings between EMI filter components (inductors and capacitors) [7], [8], [9], [10]. As a result, the performance of EMI filters in the high-frequency region may suffer due to increased parasitic couplings. Several techniques can be used to lessen mutual couplings between EMI filter components that are placed close to each other. Placing shielding between EMI filter components is one of the most effective approaches for mitigating couplings between components [11], [12].

A common-mode (CM) choke is one of the major components of an EMI filter and provides high common-mode (CM) inductance to attenuate CM noise. Its leakage inductance acts like a differential-mode (DM) inductance and helps to attenuate DM noise. As the high-frequency performance of an EMI filter is heavily degraded by the magnetic coupling between the CM choke and capacitors, it is important to analyze the effect of shielding a CM choke.

This paper discusses the effect of placing shielding over a CM choke in an EMI filter to improve the high-frequency attenuation of the EMI filter. An extensive study is carried out to investigate the effect of copper shielding with different geometries, varying positions in an EMI filter, and various input and output filters. 3D simulations are conducted in ANSYS Q3D (FEM-MOM) with and without shielding to examine the magnetic coupling between two CM chokes and filter components and to determine the effects of the presence of the shielding structures on the performance of the CM choke in the EMI filter attenuation. Though it is mentioned in [11] that a metallic box over a CM choke is inadequate because horizontal eddy currents may exacerbate coupling with nearby capacitors, the dimensions of the metal box are not mentioned in the paper. This paper contends that the dimensions of the copper shielding box must be large enough to get effective shielding behavior and better high-frequency EMI filter attenuation. Moreover, the importance of the effective placement of a shielding enclosure in an EMI filter is analyzed in this paper.

II. Numerical Work, Experimental Analysis and Models of EMI Filters
A. Magnetic Field Distribution of Common-mode Choke with and without Shielding

The presence of a shielding box significantly changes the magnetic flux distribution of a CM choke. Figure 1a and Figure 1b show the magnetic flux distribution in ANSYS Q2D with and without shielding. For both simulations, the area of the ground surface is kept the same. Figure 1b shows that the magnetic field distribution of a shielded CM choke is comparatively less concentrated outside the shield. The field distribution of the normal common-mode choke can be seen in Figure 1a. The shielding also helps to reduce the coupling between the CM choke and the adjacent components, which are discussed in the following sections.

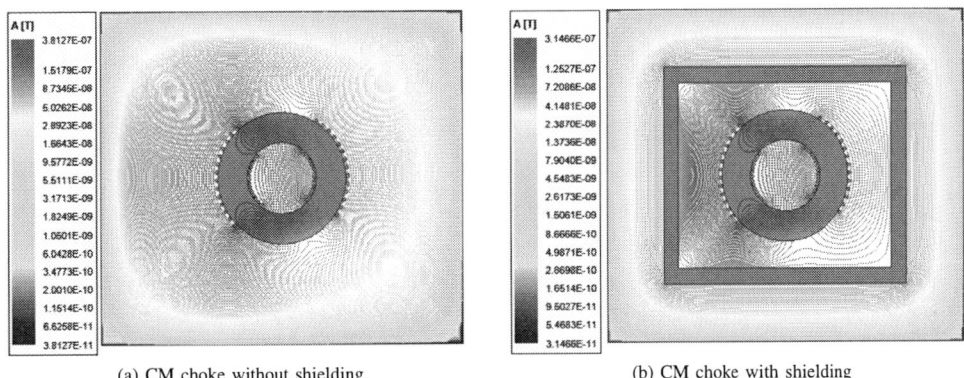

(a) CM choke without shielding
(b) CM choke with shielding

Fig. 1. Effect of shielding box on the magnetic flux of a CM choke, shown in ANSYS Q2D: (a) Magnetic flux distribution of the simple CM choke without copper shielding box; (b) Magnetic flux distribution of the CM choke with copper shielding box.

B. Effect of Shielding Box With Two Adjacent CM Chokes Placed Horizontally

This section studies the magnetic coupling between two CM chokes and the attenuation with and without a copper shielding box. Figs. 2a, 2b, 2c, and 2d show the simulation model and the magnetic coupling coefficients between windings of two horizontally placed CM chokes with and without a shielding box in both CM and DM configurations. The mutual coupling is extracted in ANSYS Q3D with the center points of the chokes positioned 40 mm apart. The blue arrows indicate the current directions in these figures, and the red arrows denote the magnetic flux directions. These figures distinguish between couplings before and after introducing a shielding box over a CM choke in both the CM and DM configurations. We have placed two CM chokes horizontally on a two-layer PCB to perform the measurement, as shown in Fig. 3a. A copper shielding box (Length $L=47$ mm, Width $W = 39$ mm, Height $H=37$ mm) is placed over one CM choke. The thickness of the box is 0.15 mm, and it is soldered to the ground conductor, which can be observed in Fig. 3b. The electrical circuit diagrams for two CM chokes in the CM and DM configurations are shown in Figs. 3c and 3d, respectively.

The CM attenuation, shown in Fig. 4a, also indicates a slight increase at high frequencies. The shielding significantly affects a CM choke's low-frequency performance, as shown in Fig. 4a. However, the large CM impedance of the CM choke will cause a considerable voltage drop across the windings for CM currents, which capacitively couples to any nearby conductive surface that is connected to the ground, for example, the copper shielding box. This phenomenon helps to provide a low-impedance path for displacement current through the conductive shielding box to the ground. Therefore, the improvement observed in the high-frequency CM attenuation in Fig. 4a is due to the capacitive coupling between the CM choke and the grounded copper shielding box. As the magnetic field lines induced by DM currents are mostly closed outside the core, the magnetic field inside the core is weaker. This alternating magnetic field of the CM choke caused by DM currents induces eddy currents in the conducting shielding box placed over the CM choke. These induced eddy currents reduce the net magnetic flux inside the windings, lowering the leakage inductance of the CM choke, which can be seen from the DM attenuation shown in Fig. 4b. As the leakage inductance decreases, the low-frequency DM attenuation decreases [11]. However, the decrease is not very significant.

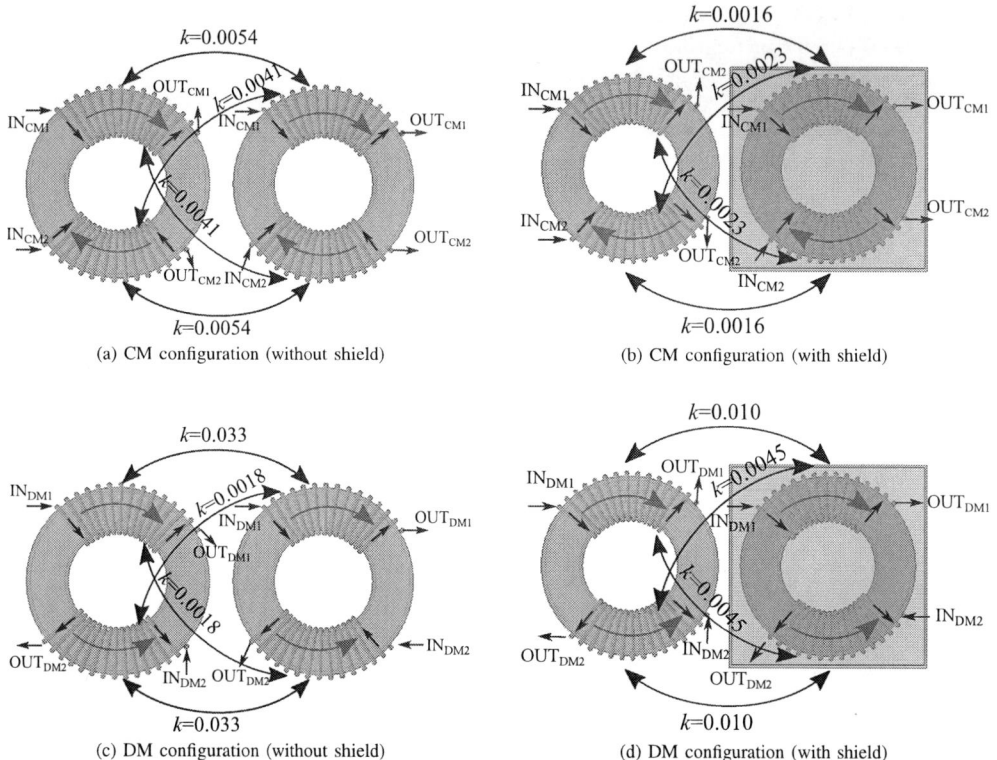

Fig. 2. Influence of shielding box on the magnetic coupling coefficients between two CM chokes in the CM and DM configurations, where two adjacent CM chokes are placed horizontally.

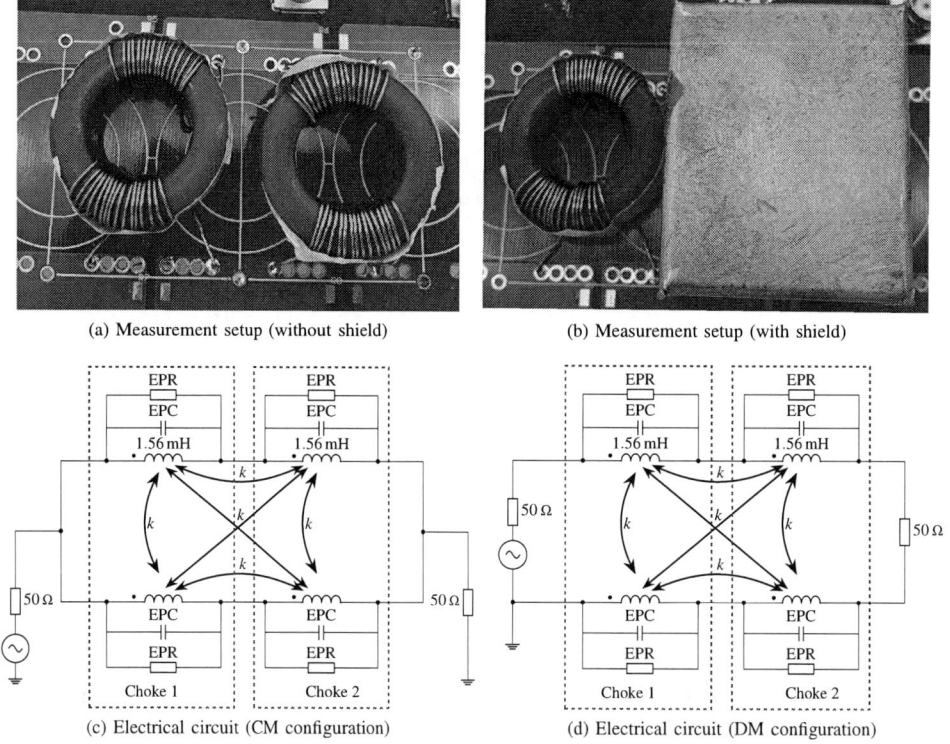

Fig. 3. (a, b) Measurement setup for two CM chokes without and with shielding box; (c, d) Electrical circuit diagrams with magnetic couplings.

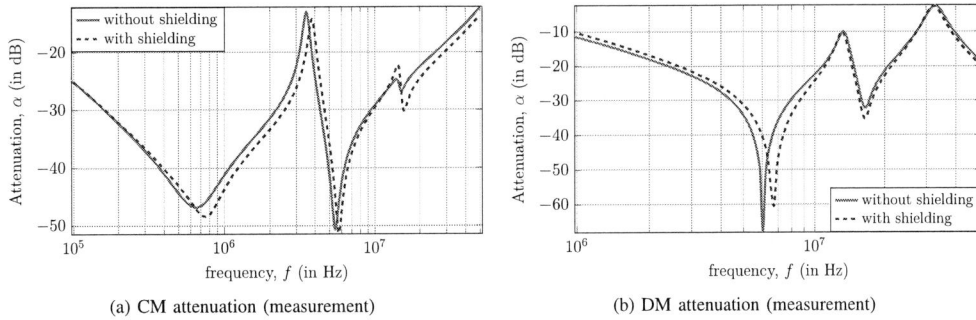

(a) CM attenuation (measurement) (b) DM attenuation (measurement)

Fig. 4. Influence of shielding on the CM and DM attenuation of the two CM chokes.

C. Effect of the Size of Shielding Box

When designing the shielding box, its size plays an important role in the effectiveness of the shielding. The size of a shielding box is rarely discussed. The shielding effectiveness of a box is strongly dependent on its volume. In addition, the resonant frequencies of the shielding box also depend on its volume. The shielding effectiveness is largely influenced by the electrical parameters and the resonant frequency [13]. At the resonant frequency, the electric field strength inside the shielding box increases, resulting in poorer shielding effectiveness. In addition, a small shielding box produces high capacitive coupling between the conductive box and the CM choke, which can be seen in Fig. 5. The field distribution between a CM choke and a copper shielding box is plotted in the XY plane using ANSYS Q2D Extractor. For simplicity, the copper wire wound around the cross-section of the ferrite core is represented by the orange circle segments in these figures. The blue rectangle depicts the copper shielding. Fig. 5b shows that the strength of the electric field between the copper windings and the box is comparatively higher with a smaller shield box.

The CM choke with a shielding box is simulated in the XYZ plane using ANSYS Q3D to extract the mutual capacitive couplings between the CM choke and the copper shielding box, as shown in Figs. 6a and 6b. In Figs. 6c and 6d, it can be seen that with a smaller height of the shielding box, the magnetic capacitive coupling between the copper windings and the box is comparatively higher. However, the height of the shielding box does not affect the magnetic coupling coefficient between two CM chokes, as shown in Fig. 7.

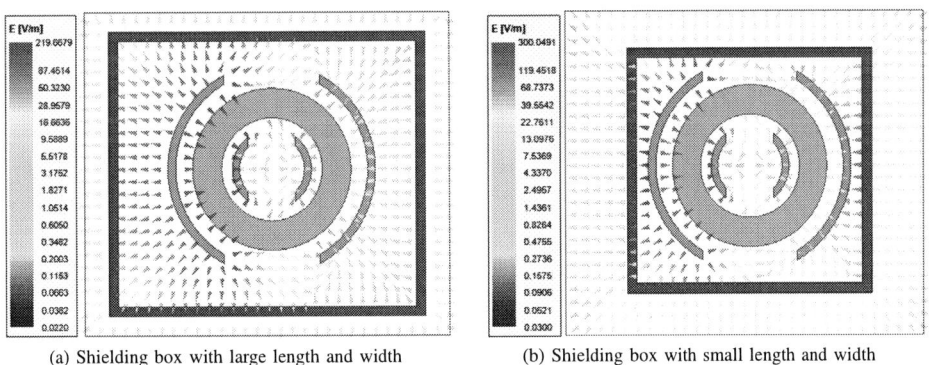

(a) Shielding box with large length and width (b) Shielding box with small length and width

Fig. 5. Electric field strength between the copper winding and the copper shielding box (2D simulation).

Since a shielding box for use on a PCB usually has both a fixed width and a fixed length in order to keep the PCB compact, its height is the only factor that be adjusted to change its volume. The high capacitive coupling due to a small shielding box also does not improve filter attenuation at higher frequencies. Fig. 8a and Fig. 8b show the experimental setup for two filters with higher and lower shielding boxes, where the CM chokes are built with a block core. The DM inductance of these chokes is improved by installing a block core since the magnetic coupling between the windings is reduced in this way [14], [15]. The length and width of the boxes are the same as above. However, the height of the small shielding box is 20 mm, and the height of the large shielding box is 37 mm. The CM and DM attenuations of the filter are shown in Fig. 8c and Fig. 8d, respectively. These figures show that the attenuation of the filter, regardless of the CM or DM configuration, remains the same when the height of the shielding box is small. However, the larger shielding box improves the attenuation by increasing the effectiveness of the shield and reducing the capacitive coupling between the shielding box and the CM choke. When using a higher shielding box, a 2 to 5 dB improvement in attenuation above 100 kHz can be achieved in the CM configuration. An approximately 10 to 15 dB improvement in attenuation can be achieved in the DM configuration in the frequency range from 150 kHz to 15 MHz.

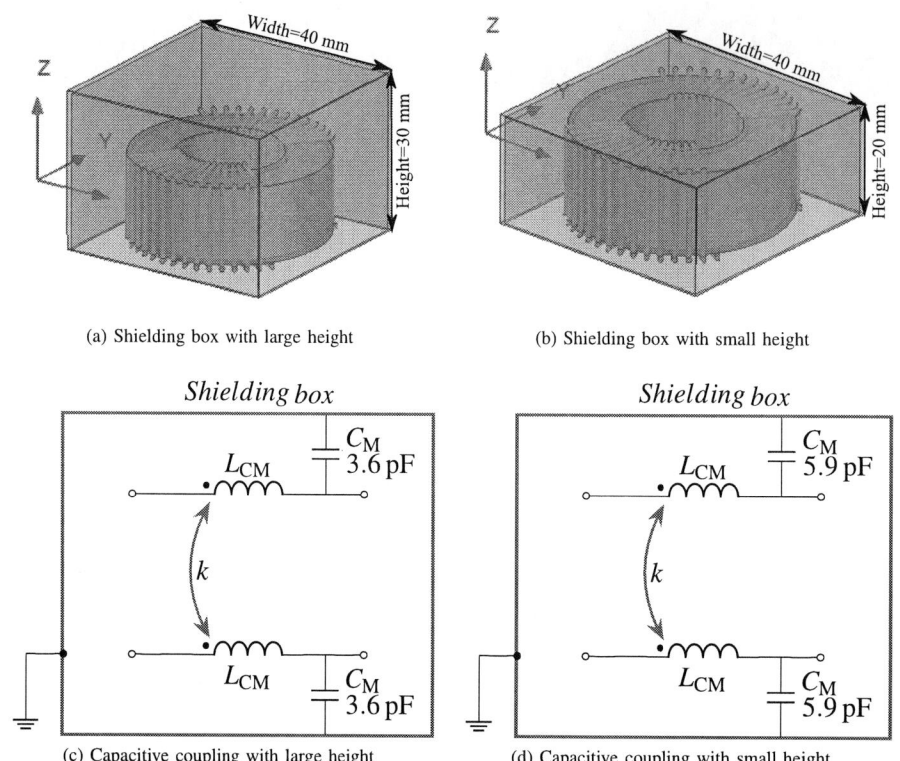

(a) Shielding box with large height   (b) Shielding box with small height

(c) Capacitive coupling with large height   (d) Capacitive coupling with small height

Fig. 6. Mutual capacitive coupling C_M between the copper windings and the copper shielding box (3D simulation).

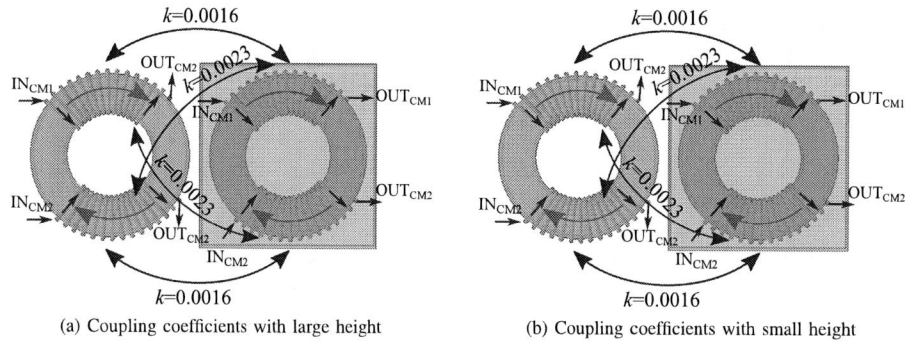

(a) Coupling coefficients with large height   (b) Coupling coefficients with small height

Fig. 7. Magnetic coupling coefficients between two CM chokes with large and small shielding box heights (3D simulation).

D. Positioning of the Shielding Box in the EMI Filter

When used in an EMI filter, the perfect positioning of the shielding box is one of the most important considerations. Since, in an EMI filter, the magnetic coupling between the CM choke and the Y capacitors has the most significant influence on the high-frequency attenuation of the filter, it is most beneficial to place a shield between these to reduce the coupling. As the CM choke is usually the largest component of an EMI filter, placing a shield over the CM choke also helps to reduce many couplings associated with the choke. In a multi-stage EMI filter, multiple chokes are typically used. In this case, it is essential to place a shield over the CM choke at a suitable position to obtain maximum attenuation.

Figs. 9a and 9b show a multi-stage EMI filter with two CM chokes. In this filter, a shielding box is placed over one choke at a time. The distance of the right choke from the X and Y capacitors is smaller than the distance of the left choke. Therefore, the right choke was more susceptible to inductive and capacitive coupling. The simulation model in Fig. 9c shows the coupling between the CM chokes and the Y-capacitors. Only the strongest coupling is shown in the figure. The magnetic coupling between the brown CM choke and the Y capacitors is dominant over the coupling between the blue CM choke and its adjacent Y capacitors.

Fig. 10a and 10b show that the left shielding (over the blue common-mode choke) has no significant effect on the overall performance of the filter because the other components are mounted at some distance from the

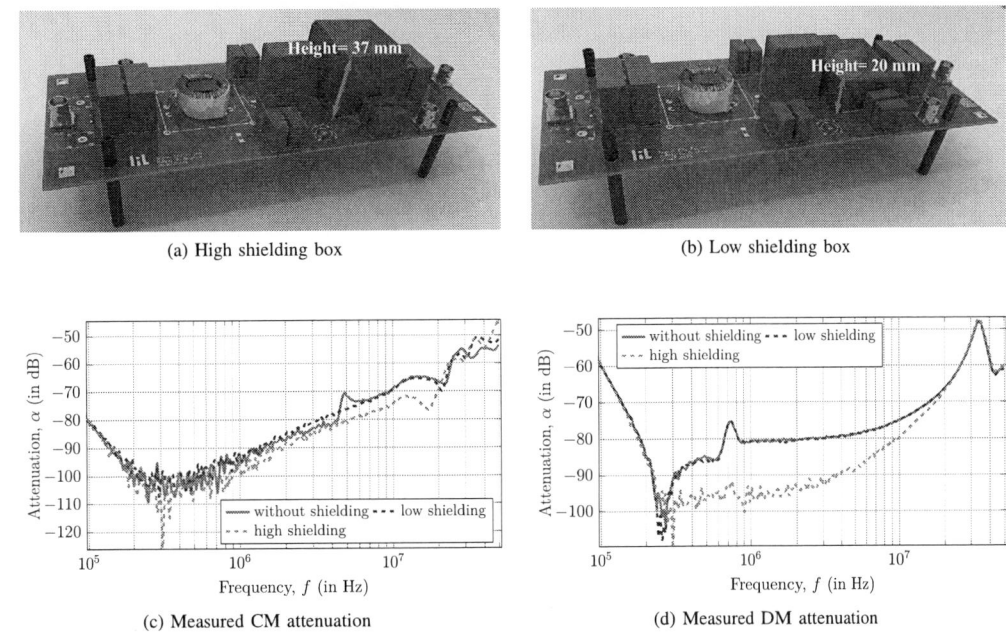

(a) High shielding box   (b) Low shielding box

(c) Measured CM attenuation   (d) Measured DM attenuation

Fig. 8. Influence of the shielding box height on the CM and DM EMI filter attenuations.

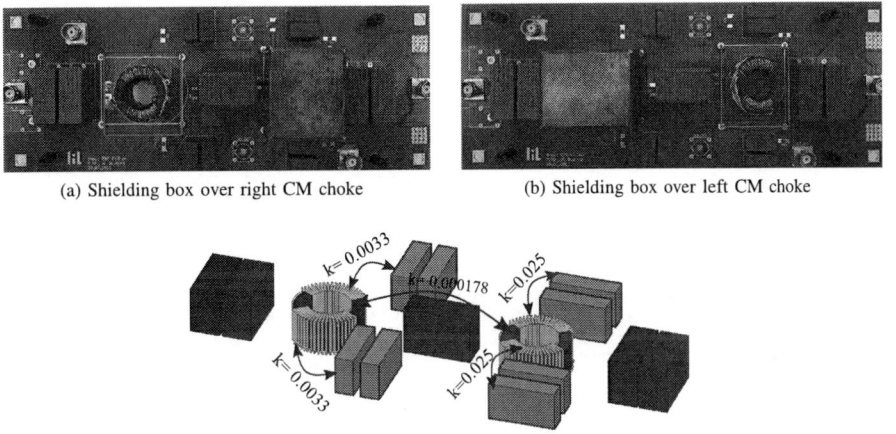

(a) Shielding box over right CM choke   (b) Shielding box over left CM choke

(c) Coupling coefficients between components

Fig. 9. Effect of different positions of the shielding box: (a) Shielding box over the left CM choke; (b) Shielding box over the right CM choke; (c) Coupling between CM chokes and Y capacitors obtained from Q3D simulation.

choke, which naturally reduces parasitic coupling. The right shielding (over the red CM choke) can achieve about 5 dB to 7 dB better attenuation above 10 MHz in the CM configuration, while the low-frequency CM attenuation and overall DM attenuation remain unaffected. Besides, the unwanted resonance in the CM attenuation in the frequency range from 2 MHz to 3 MHz, which is caused by the higher coupling between the choke and capacitors, also disappears. So, in summary, shielding reduces parasitic coupling and can help to attenuate more of the noise at higher frequencies.

E. Effect of Shielding in Different EMI Filters

To check the performance of different EMI filters, we have designed different input filters with a shielding box. Fig. 11a shows the EMI input filter A in which one choke is mounted on the core axis horizontal and another choke is mounted on the core axis vertical, in order to reduce the coupling between them, since there is less magnetic coupling flux in the perpendicular arrangement. This increases the size of the EMI filter, and its largest dimension is about 199 mm. To reduce the size of the filter, the best option would be to arrange two chokes vertically. However, the perpendicular arrangement of two chokes increases the mutual coupling between them; as mentioned in [9], the installation of an EMI shielding box helps in this case. As shown in Fig. 12a, the installation of the shielding box helps to the coupling and reduce the dimensions of the filter. The PCB

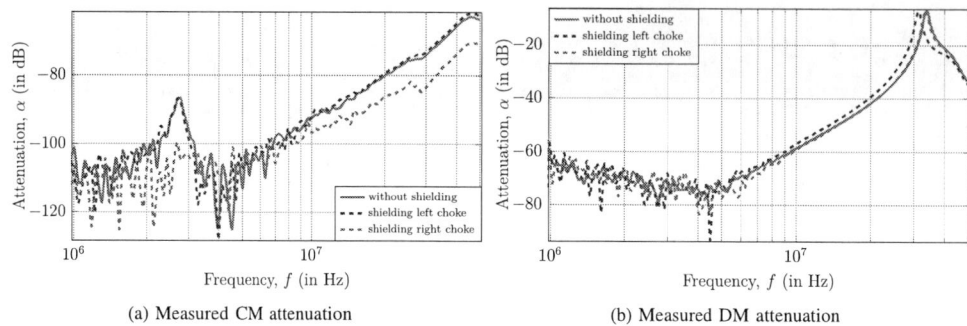

(a) Measured CM attenuation

(b) Measured DM attenuation

Fig. 10. Effect of different positions of the shielding box on the measured CM and DM attenuations.

length of the EMI input filter B is 14 mm smaller than that of the EMI input filter A, which contributes to a reduction in both the ground inductance [10] and the coupling between the two chokes, while the high-frequency performance of EMI filter B is better than that of EMI filter A.

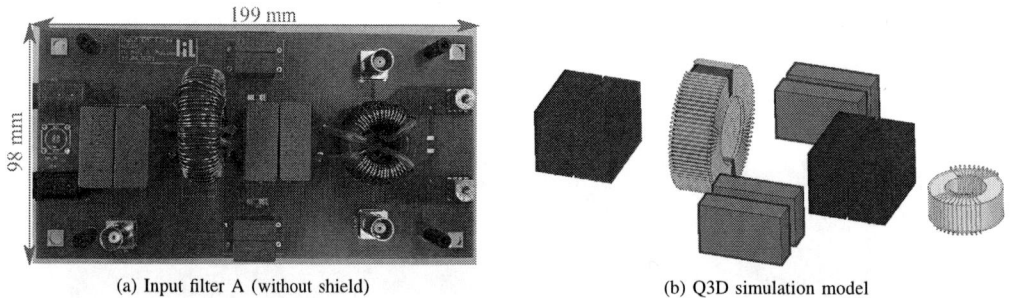

(a) Input filter A (without shield)

(b) Q3D simulation model

Fig. 11. (a) Measurement setup for two CM chokes without copper shielding box, where the chokes are arranged orthogonally; (b) ANSYS Q3D simulation model for input filter A.

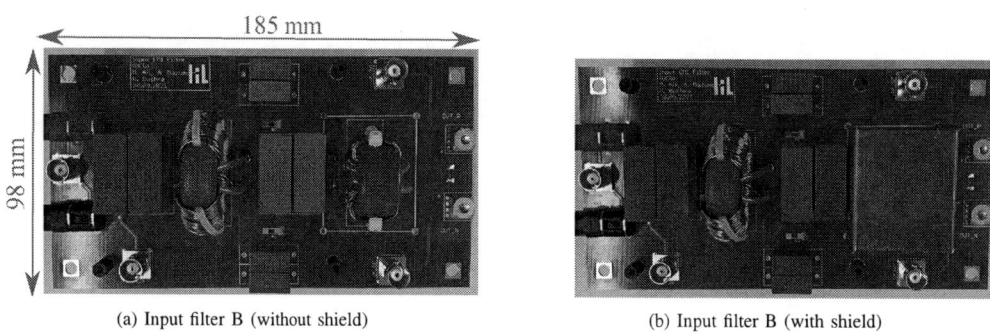

(a) Input filter B (without shield)

(b) Input filter B (with shield)

Fig. 12. (a, b) Measurement setup for two CM chokes without and with copper shielding box, where both chokes are arranged on the core axis vertical.

The input filter A is simulated using ANSYS Q3D (FEM-MOM) to extract the parasitics of components and the electromagnetic couplings between the filter components, as shown in Fig. 11b. The simulation model of the filter is built according to the geometric dimensions of the physical filter model. Fig. 13 shows the equivalent electrical network model developed for filter A with external parasitics and mutual couplings. Here, the ESL (equivalent series inductance) reduction technique applied to two parallel C_X and C_Y capacitors reduces the ESL [8], [9]. The magnetic couplings between components are shown by the red and magenta double-headed arrows. The second CM choke is surrounded by a magenta box because the shielding box is placed over the second choke. The shielding box reduces the dominant magnetic couplings between the C_Y capacitors and the second CM choke and between the first and second chokes, which is marked by the magenta double-headed arrows.

Fig. 14 compares the measurement and simulation results of filter A's CM and DM configurations. It can be seen that a good agreement exists between the measurement and simulation results up to 20 to 30 MHz.

A comparison of the measured CM and DM attenuations with an optimized arrangement of the CM chokes (input filter A) and with and without shielding of the choke (input filter B) is shown in Fig. 15. For the CM attenuation shown in Fig. 15a, input filter B with a shielding box shows 8 dB to 10 dB better attenuation than input filter A in the frequency range from 500 kHz to upwards. For the DM attenuation shown in Fig. 15b, input filter B exhibits 5 dB to 10 dB better attenuation in the frequency range from 400 kHz to 5 MHz. However, according to [9], input filter A without a shielding box exhibits better attenuation at higher frequencies due to the optimized placement of the chokes.

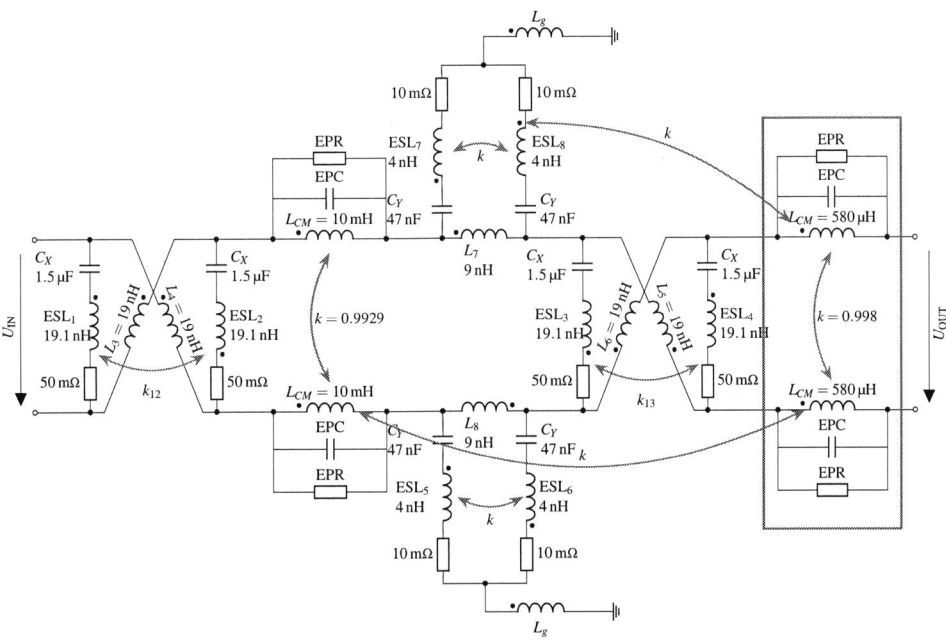

Fig. 13. An equivalent electrical circuit diagram for filter A and filter B.

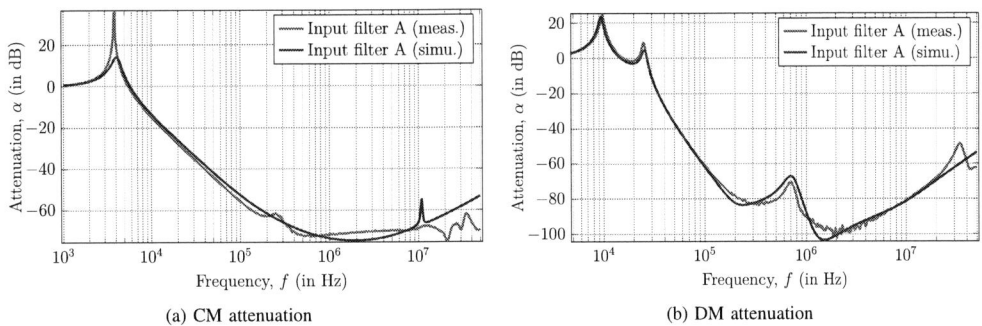

(a) CM attenuation

(b) DM attenuation

Fig. 14. Comparison of the measured and simulated CM and DM attenuations with optimized arrangement of the CM chokes (input filter A).

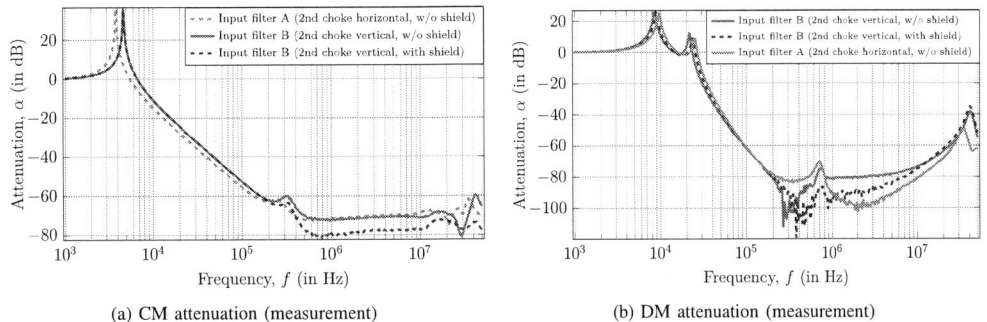

(a) CM attenuation (measurement)

(b) DM attenuation (measurement)

Fig. 15. Comparison of the measured CM and DM attenuations with optimized arrangement of the CM chokes (input filter A) and with and without shielding of the choke (input filter B).

III. Conclusion

The study presented in this paper highlights the effect of the shielding enclosure's size and volume on EMI filters in common-mode (CM) and differential-mode (DM) configurations. An EMI shield can effectively improve the attenuation of an EMI filter by reducing the parasitic coupling between filter components while reducing the physical size of the EMI filter. A larger volume shielding box is more effective than a smaller shielding box in improving the effectiveness. Besides, positioning the shielding box in an EMI filter is one of the most important considerations when seeking to reduce dominant couplings between components and improve the high-frequency attenuation. Finally, the design requirements of EMI filters with a shielding box have been discussed.

REFERENCES

[1] A. Merkert, T. Krone and A. Mertens, "Characterization and Scalable Modeling of Power Semiconductors for Optimized Design of Traction Inverters with Si- and SiC-Devices," *in IEEE Transactions on Power Electronics*, vol. 29, no. 5, pp. 2238–2245, 2014.

[2] J. Biela ,M. Schweizer, S. Waffler and J. W. Kolar, " SiC versus Si-Evaluation of Potentials for Performance Improvement of Inverter and DC–DC Converter Systems by SiC Power Semiconductors," *in IEEE Transactions on Industrial Electronics*, Vol. 58, No. 7, 2011.

[3] Di Han J. Noppakunkajorn and B. Sarlioglu, "Comprehensive efficiency, weight, and volume comparison of SiC and Si-based bidirectional DC-DC converters for hybrid electric vehicles," *in IEEE Trans. Veh. Technol*, vol. 63, no. 7, pp. 3001–3010, Sep. 2014.

[4] D. Han, C. T. Morris, W. Lee and B. Sarlioglu, "Three-phase common mode inductor design and size minimization," *IEEE Transportation Electrification Conference and Expo (ITEC)*, 2016.

[5] M. Ali, J. K. Müller, J. Friebe and A. Mertens, "Analysis of Switching Performance and EMI Emission of SiC Inverters under the Influence of Parasitic Elements and Mutual Couplings of the Power Modules," *European Conference on Power Electronics and Applications*, 2020.

[6] M. Ali, J. Friebe and A. Mertens, "Simplified Calculation of Parasitic Elements and Mutual Couplings of Wide-bandgap Power Semiconductor Modules," *in 22nd European Conference on Power Electronics and Applications (ECCE Europe)*, IEEE, Sep. 2020.

[7] E. Hoene, A. Lissner, S. Weber, S. Guttowski, W. John and H. Reichl, "Simulating Electromagnetic Interactions in High Power Density Inverters," *in IEEE Power Electronics Specialists Conf.*, 2005.

[8] M. Ali, J. Friebe and A. Mertens, "3-D Electromagnetic Modeling and Analysis of Electromagnetic Field Couplings of EMI Filter Capacitors," *in 12th International Conference on Power Electronics-ECCE Asia*, IEEE, 24-27 May 2021.

[9] M. Ali, J. Friebe and A. Mertens, "Design and Optimization of Input and Output EMI Filters under the Influence of Parasitic Couplings," *in 23rd European Conference on Power Electronics and Applications (ECCE Europe)*, IEEE, 6-10 Sep. 2021.

[10] M. Ali, T. Brinker, J. Friebe and A. Mertens, "Analysis of EMI Filter Attenuation under the Influence of Parasitic Elements of Components and their Mutual Coupling," *in 23rd European Conference on Power Electronics and Applications (ECCE Europe)*, IEEE, 6-10 Sep. 2021.

[11] C. Domínguez-Palacios, P. González-Vizuete, M. A. Martín-Prats and J. B. Mendez, "Smart Shielding Techniques for Common Mode Chokes in EMI Filters," *in IEEE Transactions on Electromagnetic Compatibility, vol. 61, no. 4, pp. 1329-1336*, Aug. 2019, doi: 10.1109/TEMC.2019.2918863.

[12] A. Asmanis, G. Asmanis, D. Stepins and L. Ribickis, "Modeling of EMI filters with shields placed between the filter components," *International Symposium on Electromagnetic Compatibility - EMC EUROPE*, 2016, doi: 10.1109/EMCEurope.2016.7739252.

[13] K.-H. Gonschorek and R. Vick, "Electromagnetic Compatibility for Device Design and System Integration," Springer Heidelberg Dordrecht London New York, pp. 153-157, 2009.

[14] J. Borsalani, A. Dastfan, J. Ghalibafan, "An Integrated EMI Choke With Improved DM Inductance," *in IEEE Transactions on Power Electronics*, 2021.

[15] M. Ali, R. Bushra, J. Friebe and A. Mertens, "Design and Potential of EMI CM Chokes with Integrated DM Inductance," *in 24th European Conference on Power Electronics and Applications (ECCE Europe)*, IEEE, 5-9 Sep. 2022.

Implementation of onsite Junction Temperature Estimation for a SiC MOSFET Module for Condition Monitoring

Farzad Hosseinabadi[1,2], Shahid Jaman[1,2], Sachin Kumar Bhoi[1,2], Md. Mahamudul Hasan[1,2], Sajib Chakraborty[1,2], (Member, IEEE), Mohamed El Baghdadi[1,2], (Member, IEEE), Omar Hegazy[1,2*], (Senior Member, IEEE)

[1]MOBI-EPOWERS Research Group, ETEC Department, Vrije Universiteit Brussel (VUB), Pleinlaan 2,1050 Brussels, Belgium;
[2]Flanders Make, 3001 Heverlee, Belgium

*Omar.hegazy@vub.be

Acknowledgements

This work was supported by HiEFFICIENT project. This project has received funding from the ECSEL Joint Undertaking (JU) under grant agreement no. 101007281. The JU receives support from the European Union's Horizon 2020 research and innovation programme and Austria, Germany, Slovenia, Netherlands, Belgium, Slovakia, France, Italy, and Turkey. We also acknowledge Flanders Make for the support to our research group.

Keywords

«Junction Temperature Estimation», «Reliability», «Condition Monitoring», «SiC MOSFET, « $R_{ds,\text{on}}$ ».

Abstract

This paper presents an advanced methodology for mapping junction temperature (T_j) based on the drain to source resistance ($R_{ds,\text{on}}$) of a SiC MOSFET module to monitor the power electronics converter health condition. Capturing real-time measurement of on-state drain-source voltage ($V_{ds,on}$), drain-source current ($I_{ds,on}$) and baseplate temperature, and taking advantage of a fast edge computing device, a significant correlation can be established between T_j and $R_{ds,\text{on}}$. Due to having a linear correlation and simple circuity in comparison to other junction temperature estimation methods, e.g., the internal gate resistance method (R_{Gint}), gate threshold voltage method ($V_{g,th}$) and short-circuit current method (I_{sc}), the output results of the proposed paper can be effortlessly implemented in a simple microcontroller that can monitor the health condition of a SiC MOSFET module in terms of bond wire fatigue and metallization reconstruction. To validate the proposed method, a synchronous boost converter is prototyped and tested. The experimental results depict the effectiveness of the proposed method.

Introduction

Nowadays, Power Electronic Converters (PECs) are utilized to efficiently convert electrical energy between power generators, storage systems, and power consumers. The power electronics industry comes across a growing appeal for cost reduction, increasing density, upgrading their products fast and continuously, and reducing time to market, which has brought severe reliability challenges that should be addressed. Field data shows that the PECs in some applications like renewable energy is considered one of the primary causes of malfunctions with the contribution of 13% to the overall failure rate, while in the PECs, semiconductor devices are the most failure-prone devices which are responsible for an overall 31% of failures in power electronics converters [1]-[2]. Thus, it is required to identify stress factors in power electronics devices. One of these factors is junction temperature (T_j) and junction temperature swing (ΔT_j), which are associated with wear-out and solder fatigue of the semiconductor module and can be measured onsite. Using real-time methods, junction temperature can be converted to the parameter that can predict the end of life (EOL) and also be the basis for derating estimation.

Regarding derating, it is worth mentioning that it will be fruitful to manage the processed power or speed up the cooling process if the module is experiencing abnormal temperature through a condition monitoring routine. Therefore, implementing condition monitoring systems via appropriate thermal management can enhance the lifetime of semiconductor modules [2].

In order to benefit from thermal management and condition monitoring, it is necessary to estimate onsite junction temperature. So far, several approaches have been proposed in the literature to estimate junction temperature [3]-[7]. The first one is implementing temperature sensors directly in module packages. In this method, a negative temperature coefficient sensor (NTC) or a p-n diode is implemented on the die or direct bonded copper (DBC) in the module and thus can estimate junction temperature [3]. The NTC temperature sensor can measure junction temperature several millimeters away from the die; thus, the thermal model is needed for junction temperature estimation. Also, junction temperature can be measured via a p-n diode built on the die which its instantaneous forward voltage changes according to temperature variations. On the other hand, temperature measurement by means of NTC and the p-n diode will not be highly accurate and fast. Also, most semiconductor modules lack integrated temperature sensors which worsens the situation.

Temperature-sensitive electrical parameters (TSEP) such as the internal gate resistance method (R_{Gint}), gate threshold voltage ($V_{g,th}$), short-circuit current (I_{sc}), turn-OFF transition time, and turn-ON delay are indirect temperature measurement methods with appropriate time resolution. Internal gate resistances in MOSFET and IGBT modules depend on junction temperature variations. A variation on this resistor can be calculated by applying test pulse currents with delay to switching pulses. However, a specially modified substrate is required to measure this resistance [4]-[5].

To benefit temperature estimation via measuring change in the gate threshold voltage, it is necessary to sense the $V_{g,th}$ in a short time interval, which requires high-speed data acquisition [6]. In the short circuit approach, the corresponding elements like auxiliary switches should be added to the converter for performing any short circuit tests for junction temperature estimation. It is necessary to ensure only considered device-under-test (i.e., MOSFET) are subjected to short circuit tests. Also, precise protection circuits are needed to guarantee that short circuit current will not pass safety limitations, limiting the applicability of this approach [7]. Turn-OFF transition time and turn-ON delay also are used for junction temperature estimation; however, it should be mentioned they need sensitive and accurate measurement circuits which cannot be easily implemented [8]. Also, system-level approaches like identification of change of low-order harmonics in the output of a voltage source inverter are used for junction temperature estimation, which is out of the scope of this paper [9]. The temperature within semiconductors can also be estimated using optical methods via measuring stimulated, naturally emitted, and reflected radiation, which has been explained in [10].

To fulfill the above-mentioned research gap, this paper focuses on an advanced method for junction temperature estimation and validates it experimentally. In this approach, as shown in Fig. 1, by heating up heatsink temperature and turning-ON semiconductor for a short duration and sensing drain-source voltage ($v_{ds,on}$) and drain-source current ($i_{ds,on}$), drain-source resistance ($R_{ds,on}$) can be mapped onto junction temperature. The rest of the paper discusses the proposed test bench preparation, calibration approach, and characterizing tests of a SiC MOSFET module. Then, experimental results are presented, followed by the conclusion in the last section.

Measurement circuit (Sensorics)

The schematic of the analog circuit used for measuring $v_{ds,on}$, $i_{ds,on}$ and $R_{ds,on}$ is shown in Fig. 2. The sensorics comprise a PTC sensor and its amplifier, Current Transducer (LA 55-P/SP23) and its amplifiers, and several op-amps in different combinations for sensing ON-state drain-source voltage and blocking OFF-state voltage. The outputs of the sensorics are connected to a Microcontroller with 12-bit ADC. The other topologies for sensing on-state voltage of semiconductor modules have been discussed in [11]-[14].

Test Bench Preparation and Calibration for Mapping Routine

A synchronous boost converter is prototyped to verify the proposed approach using a SiC half-bridge MOSFET module (CAS120M12BM2) mounted on a temperature-controlled heatsink. For heating up the heatsink, six parallel heating resistors (450Ω, 150 W) are connected to the heatsink, and one positive temperature coefficient (PTC) sensor is pasted on the heatsink close to the module to measure baseplate temperature. The junction temperature of the MOSFET module is considered equal to the baseplate temperature measured by the PTC sensor because of generating a short ON time signal (i.e., 100 ms) from 2 A to 10A with 1 A slope and utilizes natural air circulation for cooling the MOSFET module. As Fig. 1 shows, to start calibration, first, the temperature of the baseplate is increased up to a higher temperature (i.e., 80 °C), and then the control circuit pauses heating the heatsink and starts generating turn-ON pulses. There is a 200ms delay between each pulse to guarantee that any residual temperature perturbation has been eliminated [8].

Meanwhile, using current and voltage sensors and a 12-ADC unit, the microcontroller is able to measure $v_{ds,on}$ and $i_{ds,on}$ and thus can calculate $R_{ds,on}$. Also, using a PTC thermocouple, the microcontroller can measure baseplate temperature in each interval. It is worth mentioning that switching pulses are generated when baseplate temperature decreases by 5 °C, and the process continues until the baseplate temperature becomes equal to room temperature (25°C). Finally, a 3-D map to estimate $R_{ds,on}$ is generated based on extracted data.

Condition Monitoring

After completing the above-mentioned commissioning test, a 3-D $R_{ds,on}$ map is calculated, and thus it is programmed in edge, fog and cloud devices. In the actual operational condition, by sensing ON-state drain-source current and voltage and hence calculation of drain-source resistance of the MOSFET and also feeding edge/fog/cloud device on drain-source resistance and drain-source current and using implemented 3-D $R_{ds,on}$ map, the junction temperature is estimated. T_j data are valuable for thermal and lifetime control techniques like derating and increasing cooling effort.

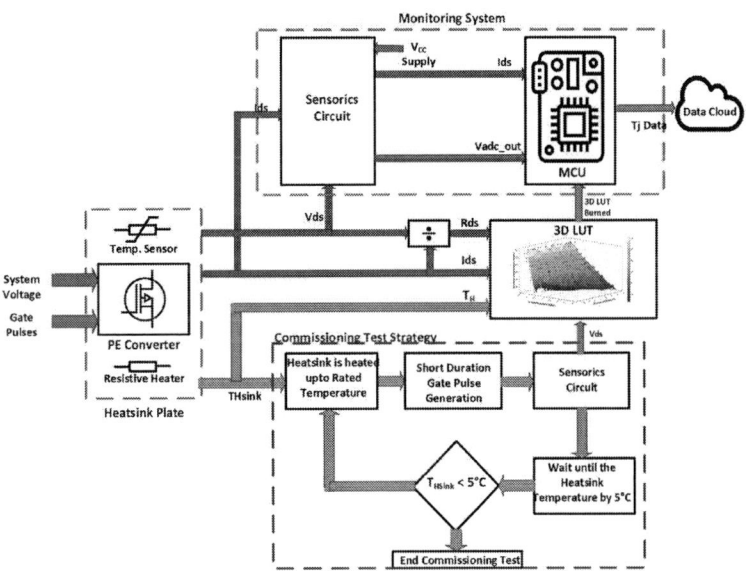

Fig. 1. Commissioning and junction temperature estimation algorithm.

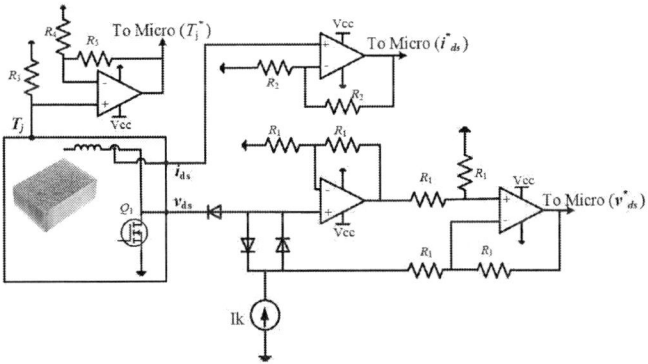

Fig. 2. Sensorics schematic of the T_j estimation.

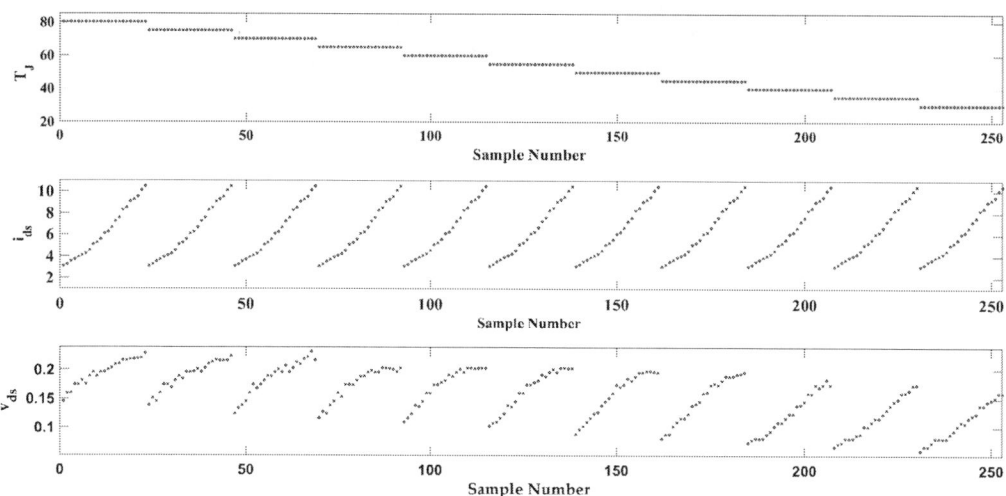

Fig. 3. Data collected during the commissioning test. From top to bottom: Junction temperature, drain-source current and drain-source voltage.

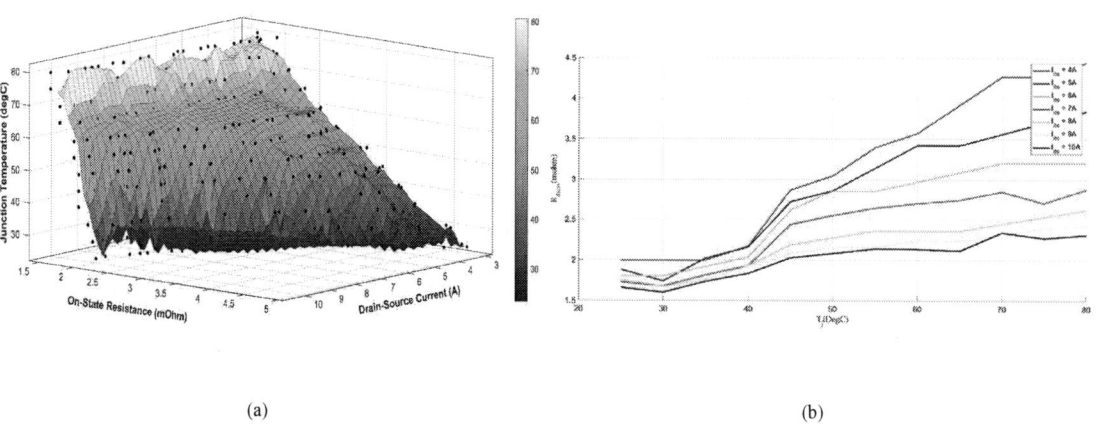

(a)   (b)

Fig. 4. Look-up tables gained via commissioning test (a) 3-D Look-up table, (b) 2-D Look-up table

Results and Discussion

After conducting commissioning tests, the results are reported in the form of T_j, $R_{ds,\text{on}}$ and $v_{ds,\text{on}}$. As Fig. 3 shows, the results are collected whenever junction temperature decreases by 5 °C until the temperature reaches 25 °C. Also, Fig. 4 represents 3-D and 2-D look-up tables of a SiC MOSFET module from calibration tests for current below 10A, respectively. These look-up tables have the capability of being mapped to the microcontroller for use as a junction temperature estimation tool for condition monitoring purposes. In Fig. 4(a), the x-axis is T_j, while y-axis is $i_{ds,\text{on}}$ and z-axis is $R_{ds,\text{on}}$. Also, Fig. 5 illustrates the gate pulses (V_{gs}) generated by the dSPACE unit and drain-source current ($i_{ds,\text{on}}$) measured by Agilent current probe (1146B). During the test, the time interval between pulses is kept at about 200ms. The experimental test setup is illustrated in Fig. 6. In addition, the results are interpolated using the polynomial function, which is shown as Equation (1)

$$T_j(i_{ds}, R_{dson}) = p_{00} + p_{10}i_{ds} + p_{01}R_{dson} + p_{20}i_{ds}^2 + p_{11}i_{ds}R_{dson} + p_{02}R_{dson}^2 + p_{30}i_{ds}^3 + p_{21}R_{dson}i_{ds}^2 + p_{12}i_{ds}R_{dson}^2 \qquad (1)$$

where, $p00 = 369$, $p10 = -98.56$, $p01 = -176.5$, $p20 = 8.603$, $p11 = 53.56$, $p02 = 20.36$, $p30 = -0.1636$, $p21 = -3.069$, $p12 = -5.436$.

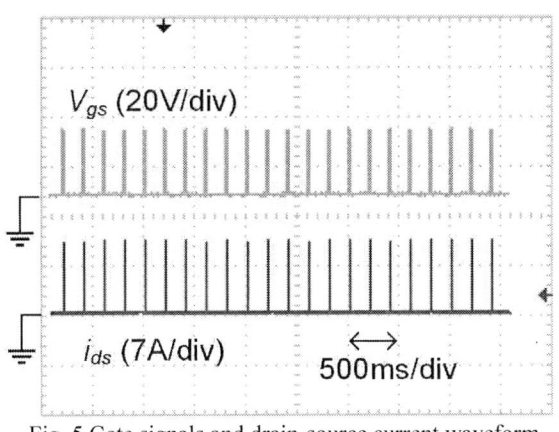

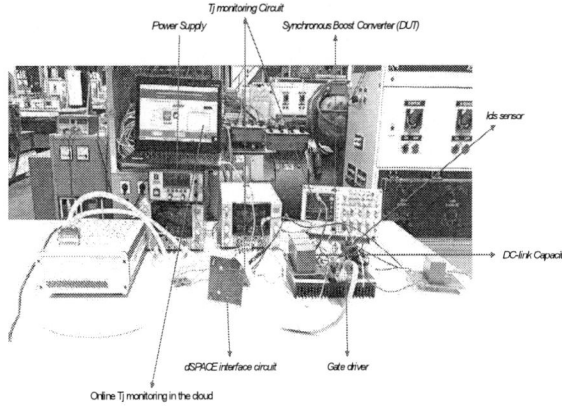

Fig. 5 Gate signals and drain-source current waveform   Fig. 6 Test set up

The proposed methods have some advantages and disadvantages as follows. Also, in Table 1, a comparison between some parameters used for junction temperature estimation is presented.

- This method does not require any especial equipment like a semiconductor parameter analyzer and can be performed on a power electronics converter, and implementation of the circuity is not costly in comparison with converter implementation. Also, discrepancies between modules manufactured by the same company will not affect the test process.
- There is no need for any thermocouple sensor inside the SiC MOSFET module.
- Temperature can be estimated very fast. Also, results can be easily implemented in a Microcontroller.
- The temperature estimation process does not influence the normal operation of the PECs.
- But due to semiconductor degradation, the test should be conducted for continuous accurate temperature estimation after receiving each level of degradation [11].
- Variation in gate threshold voltage might influence the test, which should be investigated [6],[11].
- Last but not least, this method gives information about average temperature and cannot identify hotspots on the chip [10].

Table 1. Comparison between different TSEPs [14]-[15]

TSEP	Online	Sensitivity	Effect On Converter	Linearity	Integrability	Complexity
$V_{ce,on}$	Yes	$\pm$ (2-3) mV/°C	Yes	Good	Yes	Less
R_{Gint}	Yes	(1-3) mΩ/°C	No	Good	Yes	High
I_{sat}	Yes	-	No	Exponential	No	High
I_{sc}	Yes	17% of $I_{sc,max}$/°C	Yes	Good	No	High
$V_{g,th}$	No	-(2-10) mV/°C	-	Good	No	High
R_{ds}	Yes	(3-7) mΩ/°C	No	Good	Yes	Less

CONCLUSION

This paper has focused on the feasibility of using drain-source resistance as a TSEP for tracking the junction temperature of the SiC MOSFET module for onsite condition monitoring. During the calibration test, the heatsink is heated to around 80 °C and left until its temperature reaches 25°C. Meantime, baseplate temperature, drain-source voltage, and drain-source current are measured, and also by employing fast edge computing, 3-D and 2-D look-up tables are programmed. Since the heatsink is cooled naturally and the interval between two turn-ON pulses is short (i.e., 200 ms), it is estimated that the instantaneous baseplate temperature is equal to junction temperature precisely.

On the other hand, due to the aging process and thus an increase in drain-source resistance, the SiC module requires to be calibrated after a specific time. In future work, it will be worthwhile to investigate using accelerated aging tests, counting algorithms, and accumulative damage models to update drain-source resistance after a specific level of degradation.

REFERENCES

[1] H. Wang, F. Blaabjerg, K. Ma, and R. Wu, "Design for reliability in power electronics in renewable energy systems - status and future," in *4th International Conference on Power Engineering, Energy and Electrical Drives*, May 2013, no. May, pp. 1846–1851, doi: 10.1109/PowerEng.2013.6889108.

[2] Chakraborty, S., Hasan, M. M., Paul, M., Tran, D. D., Geury, T., Davari, P., Blaabjerg, F., El Baghdadi, M., & Hegazy, O. (2021). Real-Life Mission Profile Oriented Lifetime Estimation of a SiC Interleaved Bidirectional HV DC/DC Converter for Electric Vehicle Drivetrains. *IEEE Journal of Emerging and Selected Topics in Power Electronics.* https://doi.org/10.1109/JESTPE.2021.3083198.

[3] On-Chip Sensor Built-In IGBT Modules for Driving xEV Motors. Available at: https://www.fujielectric.com/company/tech/pdf/64-04/FER64-04-186-2018.pdf

[4] N. Baker, S. Munk-Nielsen, M. Liserre, and F. Iannuzzo, "Online junction temperature measurement via internal gate resistance during turn-on," in *2014 16th European Conference on Power Electronics and Applications*, Aug. 2014, vol. 33, no. 0, pp. 1–10, doi: 10.1109/EPE.2014.6911024.

[5] W. Brekel, T. Duetemeyer, G. Puk, and O. Schilling, "Time resolved in situ Tvj measurements of 6.5 kV IGBTs during inverter operation," *Proc. PCIM Eur.*, pp. 806–813, 2009.

[6] F. Stella, O. Olanrewaju, Z. Yang, A. Castellazzi, and G. Pellegrino, "Experimentally validated methodology for real-time temperature cycle tracking in SiC power modules," *Microelectron. Reliab.*, vol. 88–90, no. July, pp. 615–619, 2018, doi: 10.1016/j.microrel.2018.07.072.

[7] Z. Xu, F. Xu, and F. Wang, "Junction temperature measurement of IGBTs using short-circuit current as a temperature-sensitive electrical parameter for converter prototype evaluation," *IEEE Trans. Ind. Electron.*, vol. 62, no. 6, pp. 3419–3429, 2015, doi: 10.1109/TIE.2014.2374575.

[8] N. Degrenne, J. Ewanchuk, E. David, R. Boldyrjew, and S. Mollov, "A review of prognostics and health management for power semiconductor modules," Proc. Annu. Conf. Progn. Heal. Manag. Soc. PHM, pp. 242–252, 2015.

[9] D. Xiang, L. Ran, P. Tavner, S. Yang, A. Bryant, and P. Mawby, "Condition monitoring power module solder fatigue using inverter harmonic identification," IEEE Trans. Power Electron., vol. 27, no. 1, pp. 235–247, 2012, doi: 10.1109/TPEL.2011.2160988.

[10] D. L. Blackburn, "Temperature measurements of semiconductor devices - A review," *Annu. IEEE Semicond. Therm. Meas. Manag. Symp.*, vol. 20, pp. 70–80, 2004, doi: 10.1109/stherm.2004.1291304.

[11] F. Stella, G. Pellegrino, and E. Armando, "Three-phase SiC inverter with active limitation of all MOSFETs junction temperature," *Microelectron. Reliab.*, vol. 110, no. 0, p. 113659, Jul. 2020, doi: 10.1016/j.microrel.2020.113659.

[12] R. Gelagaev, P. Jacqmaer, and J. Driesen, "A fast voltage clamp circuit for the accurate measurement of the dynamic ON-resistance of power transistors," IEEE Trans. Ind. Electron., vol. 62, no. 2, pp. 1241–1250, 2015, doi: 10.1109/TIE.2014.2349876.

[13] M. Guacci, D. Bortis, and J. W. Kolar, "On-State Voltage Measurement of Fast Switching Power Semiconductors," CPSS Trans. Power Electron. Appl., vol. 3, no. 2, pp. 163–176, 2018, doi: 10.24295/cpsstpea.2018.00016.

[14] Choi, U. M., Blaabjerg, F., & Jørgensen, S. (2018). Power Cycling Test Methods for Reliability Assessment of Power Device Modules in Respect to Temperature Stress. *IEEE Transactions on Power Electronics*, 33(3), 2531–2551. https://doi.org/10.1109/TPEL.2017.2690500

[15] Hu, K., Liu, Z., Yang, Y., Iannuzzo, F., & Blaabjerg, F. (2020). Ensuring a Reliable Operation of Two-Level IGBT-Based Power Converters: A Review of Monitoring and Fault-Tolerant Approaches. *IEEE Access*, 8, 89988–90022. https://doi.org/10.1109/ACCESS.2020.2994368.

Energy Storage Systems for Airborne Wind Generators

Bakr Bagaber and Axel Mertens
LEIBNIZ UNIVERSITY HANNOVER
Institute for Drive Systems and Power Electronics
Hannover, Germany
Phone: +49 (0) 511-762-3766
Email: bakr.bagaber@ial.uni-hannover.de
URL: www.ial.uni-hannover.de

Acknowledgments

This work was supported by the German Ministry of Economics and Technology (BMWi) – 0324217D.

Keywords

≪Wind-generator systems≫, ≪Energy Storage≫, ≪Flywheel≫, ≪Ultra capacitors≫, ≪Batteries≫,

Abstract

Airborne wind energy systems are known for their volatile output power profile, which can be improved by incorporating energy storage systems for filtering. Lithium iron phosphate batteries, ultracapacitors, and flywheels are compared in this paper under various wind conditions using different power filtering strategies. The comparison considers the energy storage capacity, cost, aging, and reserve storage for ancillary grid services. The results suggest that Lithium batteries have a competitive edge in terms of cost/kWh, whereas steel flywheels provide the cheapest overall solution. The impact of wind turbulence on the system cost was found to be limited. Moreover, reducing the power filtering quality did not contribute to any cost savings. The study also reveals the importance of the holistic cost analysis approach to recognize the trade-offs between savings in the energy storage cost and the rest of the system.

Introduction

Airborne wind energy systems (AWES) are a new class of wind generators promising to harness wind at high altitudes above 200 meters in a cost-effective way [1]. Several concepts of AWES are under research, among which the pumping cycle (PC) type has already reached a market commercialization stage [2]. The system consists of a flying soft kite connected to a ground-based electrical machine by means of a strong tether; a control pod attached to the kite is used for steering. The principle of operation can be understood with the help of Fig. 1a. The cycle is initialized by positioning the kite at a suitable altitude against the wind direction in phase one. In phase two, reel-out (generation) starts by maneuvering the kite in a crosswind direction at a certain wind window angle (ϑ). The crosswind principle exploits the lift-to-drag ratio (E) to induce high apparent wind speed -and lift force- which maximizes the extraction of power [3]. Once the tether is almost entirely winched out, the third phase starts (transition phase). The kite is steered towards the zenith position ($\vartheta = 90$). During this transfer phase, the machine decelerates rapidly towards zero while the torque remains high. Eventually, the fourth phase (reel-in) begins. The electric machine accelerates as a motor to pull the kite into lower elevations. Reel-in is usually accomplished under maximum speed at extensive angles ($\vartheta > 90$) to reduce the tether force. This unique load cycle produces varying output power, as depicted in Fig. 1b, which brings about serious challenges regarding the thermal cycling and lifetime of the machine-side converter [5]

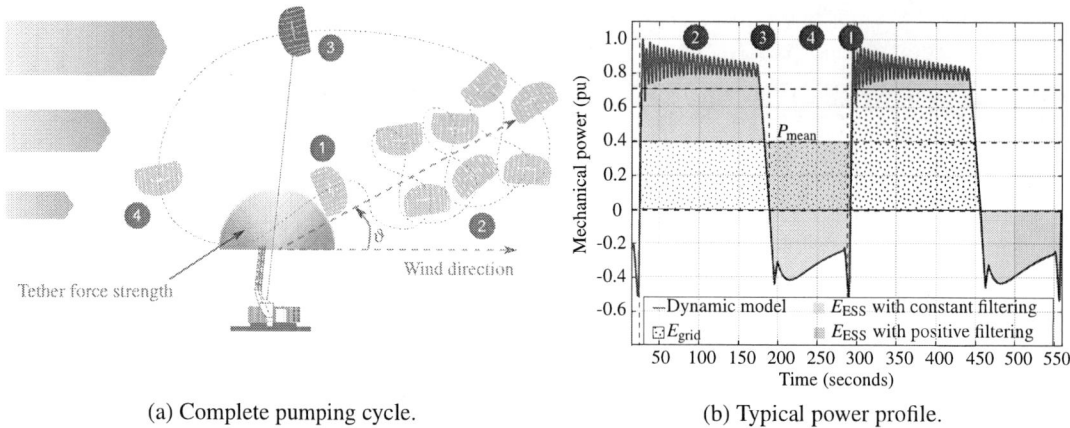

(a) Complete pumping cycle. (b) Typical power profile.

Fig. 1: principle of operation of a typical PC-AWES: (1) Start/restart, (2) Reel-out (generation), (3) Transfer (generation), (4) Reel-in (consumption), (modified from [4]). The power is normalized to the nominal reel-out power. The shaded area represents the ESS capacity requirements; the dotted area represents the energy transferred directly to the grid.

and compatibility with existing grid codes [6]. Additionally, due to the low inertia of the kite system, it is highly affected by wind gusts which could cause up to 50% overloads during the reel-out phase. The grid-side converter should therefore be over-dimensioned to handle up to 8 times the average cycle power P_{mean}, especially under high wind conditions.

The previous study [6] solved the grid conformity issue by incorporating an energy storage system (ESS). The sizing of two different ESSs, namely lithium iron phosphate (LFP) batteries and ultracapacitors (UC), was conducted for different scenarios. It was found that LFP batteries are more cost-effective than UC in fulfilling the fault ride-through (FRT) requirements. The ESS was also sized for filtering the power injected into the grid. It was concluded that LFP batteries are also better suited than UC for this purpose, although they had to be oversized due to their limited charging currents. However, the previous investigation was based on an average mathematical model for the kite, and it considered only ideal power curves under constant wind speed, as shown in Fig. 1a. Furthermore, the aging mechanisms of the ESS were not considered. In this work, the previous study is extended by using a detailed dynamic model of the AWES and the incorporation of aging models for the ESSs.

On the one hand, LFP batteries are characterized by their high energy density, moderate power density, and operational lifetime. On the other hand, UC are known for their outstanding high power density and cycle lifetime, but they suffer from much lower energy density than lithium batteries and they cost much more [7, 8]. Therefore, alternative ESS, namely flywheels, are also considered in this work. A Flywheel energy storage (FWES) system possesses trade-off characteristics between LFP batteries and UC. They have a slightly better energy density than UC and a comparable power density to LFP [7]. Their cost is also relatively lower than UC, and depending on the application, they could provide a lower cost/cycle count than LFP batteries, making them particularly suitable for AWES [7].

The first section of this paper presents a quick introduction to the system model. An explanation of the investigation methodology then follows. The system sizing results with different power filtering strategies and wind conditions are discussed afterward. Finally, the key findings are summarized.

System Modeling

This section describes the modeling approach of the AWES, the electrical drive, and the ESS in detail.

Pumping Cycle AWES Model

The dynamic model of the pumping cycle AWES can be explained with the help of Fig. 2. The kite position is described using three state variables in the polar coordinate system, the wind window angle

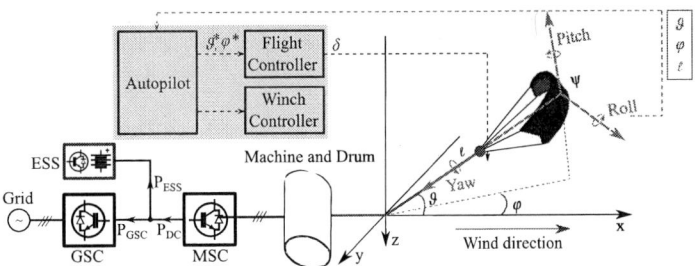

Fig. 2: Overview of the system model.

(ϑ), the azimuth angle (φ), and the tether length (l) [9, 3], whereas a fourth state variable (ψ) defines the orientation of the kite around the roll-axis.

Assuming a mass-less kite and the tether in a state of aerodynamic equilibrium with a homogeneous wind field along the x-axis, the dynamic and kinematic equations of motion can be simplified into a first-order system of equations,

$$\dot{\psi} = g_k \cdot v_a \cdot \delta + \dot{\varphi} \cdot \cos \vartheta \tag{1}$$

$$\dot{\vartheta} = \frac{v_w}{l} \cdot (E \cdot \cos \vartheta \cdot \cos \psi - \sin \vartheta) - \frac{\dot{l}}{l} \cdot E \cdot \cos \psi \tag{2}$$

$$\dot{\varphi} = -\frac{v_w \cdot E \cdot \cos \vartheta - \dot{l} \cdot E}{l \cdot \sin \vartheta} \sin \psi \tag{3}$$

$$\dot{l} = v_{winch}, \tag{4}$$

where (v_w) is the wind speed, (v_a) is the kite's apparent wind speed, and (E) is the glide ratio.

The control system of the AWES comprises a central autopilot, which with the help of a flight controller, defines the optimal flight trajectory of the kite. The steering dynamics are simplified in Equation (1) by the turn-rate law (TRL) $g_k v_a \delta$, where (g_k) is an empirical system parameter that quantifies the maneuvering response of the kite, and (δ) is a non-dimensional control input to the pod controller. A winch controller calculates the optimal winching speed for maximum energy yield according to the method described in [3]. An overview of the kite system parameters is available in Table I.

Modeling the Drivetrain

The electrical machine and the machine-side converter (MSC) are modeled by their rated efficiencies according to Table I.

Modeling the Energy Storage System

The LFP battery equivalent circuit model (ECM) is depicted in Fig. 3a; the model is based on the work of [11]. The model can be described as follows,

$$U_t = U_{oc}(SoC, T) - I \cdot \left[R_s(SoC, T, SoA) + ZARC(SoC, T, SoA) \right] \tag{5}$$

$$SoC = SoC_0 - \frac{1}{Q(I, T, SoA)} \cdot \int_0^t I \cdot dt, \tag{6}$$

where (U_{oc}) is the open-circuit voltage of the battery cell, (R_s) is the series resistance representing the ohmic losses, ($ZARC$) is the complex impedance of the cell represented by a resistor and a parallel constant phase element, (Q) is the cell's capacity, (I) is the cell current, (SoC) is the state of charge, and (U_t) is the cell's terminal voltage. All the equivalent-circuit parameters are either function of the SoC, the state of aging (SoA), and the temperature of the cell core (T). The core temperature is calculated

Table I: System Parameters. The reference cost for the LFP battery, the UC, and the composite FWES is available in [8]. The reference cost for the steel FWES is taken from [10].

Component	Parameter	Symbol	Value	Unit
Kite	Reel-out power	P_{mech}	1.8	MW
	Projected area	A	300	m^2
	Aerodynamic coefficient	C_R	1	-
	Glide ratio	E	5	-
Drivetrain	Winch efficiency	η_{Winch}	95	%
	PMSM efficiency	η_{PMSM}	94	%
	MSC efficiency	η_{MSC}	98	%
	GSC	-	198	$\frac{\$}{kW}$
LFP cell	Nominal capacity	Q_{nom}	2.5	Ah
	Nominal voltage	U_{nom}	3.3	V
	Discharge current	I_{dis}	70	A
	Max. charging current	I_{char}	10	A
	Rated series resistance (25°C, 80% SoC)	R_s	4.4	$m\Omega$
	ZARC resistances (25°C, 80% SoC)	$R_{1,2}$	3.7, 218	$m\Omega$
	ZARC general capacitances (25°C, 80% SoC)	$Q_{1,2}$	8.3, 961	$\frac{s^n}{m^2 \cdot \Omega}$
	Power conversion system efficiency	η_{PC}	94	%
	Energy storage cost (now)	-	189	$\frac{\$}{kWh}$
	Energy storage cost (in 10 years)	-	96	$\frac{\$}{kWh}$
UC cell	Rated capacitance	C_{nom}	3000	F
	Nominal voltage	U_{nom}	2.7	V
	Max. cont. current	I_{max}	210	A
	Internal resistance	R_{int}	0.29	$m\Omega$
	Voltage factor	k	917.985	$\frac{F}{V}$
	Power conversion system efficiency	η_{PC}	94	%
	Energy storage cost	[8]	32000	$\frac{\$}{kWh}$
FWES	Steel material density	ρ	7780	$\frac{kg}{m^3}$
	Maximum yield strength	σ_{yield}	1.1	GPa
	Shape factor	S	0.3	-
	Poisson's ratio	v	0.3	-
	Fixed $\frac{r_o}{r_{in}}$ ratio	-	0.15	-
	Housing thickness	-	0.05	m
	Speed range	-	3-6	kRPM
	FWES efficiency	η_{FWES}	90	%
	Energy storage cost (Composite)	-	37266	$\frac{\$}{kWh}$
	Power convertion system efficiency	η_{PC}	89	%
	Bearings cost	-	3.45	$\frac{\$}{kWh}$
	Miscellaneous items cost (of equipment cost)	-	20%	%
	Vaccume pump cost	-	6000	$
Common costs	Power conversion system (LFP, UC)	-	113	$\frac{\$}{kW}$
	Power conversion system (FWES)	-	168	$\frac{\$}{kW}$
	C&C (LFP)	-	96	$\frac{\$}{kWh}$
	C&C (UC, FWES) of (of equipment cost)	-	20%	$
	Fixed O&M (LFP, UC, Steel, Composite)	-	8, 1, 6.9, 5.6	$\frac{\$}{kWh\text{-}year}$
	Variable O&M (of annual energy throughput)	-	0.0003	$\frac{\$}{kWh}$

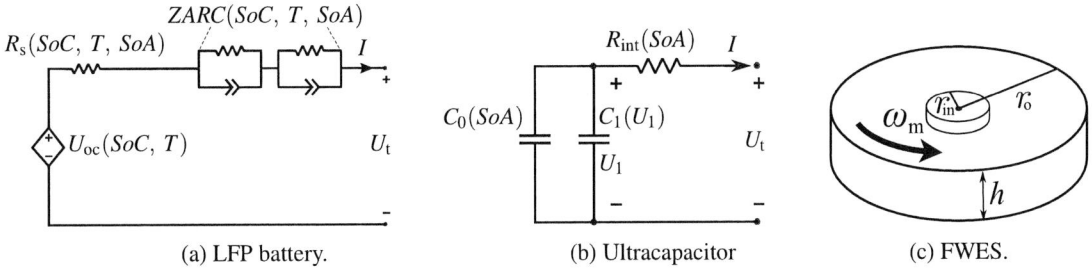

Fig. 3: Model structure of different ESS.

according to [12],

$$T = T_{amb} + q \cdot R_{th} \tag{7}$$

$$q = I^2 \cdot R_s(SoC, T, SoA) + I \cdot T \cdot \frac{\partial U_{oc}}{\partial T}, \tag{8}$$

where (T_{amb}) is the ambient temperature, (R_{th}) is the core-to-ambient thermal resistance of the cell, (q) is the heat dissipation inside the cell, and $(\frac{\partial U_{oc}}{\partial T})$ is the voltage-temperature coefficient. The first term of Equation (8) represents the joule heat, while the second term represents the reversible heat of the reaction [13]. The reference battery cell parameters are listed in Tabel I.

The degradation of the battery cell is represented by the so-called state-of-aging (SoA) parameter, which combines the calendar and cycle aging mechanisms of each component of the ECM of Fig. 3a independently. According to [11], the SoA can be represented by,

$$SoA_{calendar} = f(SoC, T, t) \tag{9}$$

$$SoA_{cycle} = f(SoC, T, DoD, NC) \tag{10}$$

$$SoA = SoA_{calendar} + SoA_{cycle}, \tag{11}$$

where (t) is the storage time, (DoD) is the depth of discharge, and (NC) is the number of cycles. The detailed aging equations have been dropped out for conciseness; the reader can refer to [11] for a more comprehensive discussion.

The UC ECM is depicted in Fig. 3b according to the modeling approach in [14, 15]. The governing equations are,

$$U_t = U_1 - I \cdot R_{int}(SoA) \tag{12}$$

$$U_1 = U_{1,0} - \frac{1}{C_0(SoA) + C_1} \int_0^t I \cdot dt \tag{13}$$

$$C_1 = k \cdot U_1 \tag{14}$$

where (C_0) and (C_1) are the fixed and voltage-dependent capacitances, respectively, (R_{int}) is the equivalent internal resistance of the cell, and (k) is the voltage dependency factor. The UC lifetime degradation follows the method described in [16]. This method extends the classical Eyring's law for calendar aging by incorporating the impact of the load current on the aging acceleration. The SoA is therefore expressed as,

$$SoA = e^{Log(2) \cdot \frac{T - T_{ref}}{T_0}} \cdot \left(e^{Log(2) \cdot \frac{U_t - U_{t,ref}}{U_{t,0}}} + B \right) \cdot e^{n_{RMS} \cdot \frac{\bar{I}_{RMS}}{C_0}} \cdot t, \tag{15}$$

where (B) is a correction factor for low SoC high storage temperature T condition, (n_{RMS}) is the acceleration factor, and $\bar{I}_{RMS}$ is the average RMS current over one complete load-cycle. The UC cell temperature can be calculated similarly using Equation (7). A comprehensive list of the UC reference cell parameters is available in Table I.

Finally, the Flywheel is modeled as a rotating cylinder with a shaft at the center, as depicted in Fig. 3c. The stored mechanical energy (E_m) can be described as,

$$E_m = \frac{1}{4} \cdot \pi \cdot h \cdot \rho \cdot (\omega_{m,\,max}^2 - \omega_{m,\,min}^2) \cdot (r_o^4 - r_{in}^4), \tag{16}$$

where (r_o, r_{in}) is the cylinder's outer and inner radius, respectively, (h) is the height, (ρ) is the material density, and ω_m is the rotational speed. Whereas the maximum specific energy of the FWES is calculated according to,

$$\frac{E_m}{m} = \frac{S \cdot \sigma}{\rho} \tag{17}$$

where (m) is the cylinder's mass, (S) is the Shape factor, and (σ_{yield}) is the maximum allowable material yield strength, which in this work is selected as the yield strength (σ_{yield}) of the material. The DoD, the disk radius, and the maximum allowed material stress (σ_{max}) influence The FWES's allowable number of cycles before breakdown (N_f) [17, 18] as described below,

$$\text{Log}(N_f) = 11.62 - 3.75 \cdot \text{Log}10\left(1.45 \cdot e^{-7} \cdot \sigma \cdot DoD^{0.44} - 80\right) \tag{18}$$

$$\sigma_{max} = \frac{1}{4}\left[(1-v) \cdot r_{in}^2 + (3+v) \cdot r_o^2\right] \cdot (\rho \cdot \omega_m^2) \tag{19}$$

where v is Poisson's ratio. Two types of FWES are investigated in this work. The first is a steel alloy FWES selected because of its competitive cost and higher safety. The second is a composite FWES characterized by higher energy density. The FWES parameters are summarized in Table I.

ESS Sizing Methods

Following the modeling approach of the last section, two sizing methods are developed and implemented. The first method is based on optimum sizing of the LFP battery and the UC using a particle swarm optimization (PSO) algorithm. The second methods is for sizing the FWES using an iterative process given specific design rules.

LFP and UC Packs Sizing Method

A battery or an UC pack usually comprises several cells connected in series to build up a specific voltage and in parallel to deliver a certain amount of current. To find the optimum number of series and parallel cells (N_s, N_p) of an ESS, a PSO [19] algorithm is used in this work. It is a population-based optimization algorithm for finding the global optima of a given problem under certain constraints.

The multi-objective optimization problem can be described by the following general form [15],

$$
\begin{aligned}
\text{Minimize} \quad & f_{obj}(Z, \zeta) \\
\text{Subject to} \quad & H_{e,i}(Z, \zeta) = 0, \quad i = 1, 2, ..., N_e \\
& H_{ine,j}(Z, \zeta) \leq 0, \quad j = 1, 2, ..., N_{ine},
\end{aligned} \tag{20}
$$

where (ζ) is the input vector, (Z) is the optimization variable vector, (f_{obj}) is the objective function to be minimized, ($H_{e,i}$) and ($H_{ine,j}$) are the i-th and j-th equality and inequality constraints, respectively. The following has been considered,

$$
\begin{aligned}
f_{obj} &= \min(N_s \cdot N_p) \\
\zeta &= [P_{DC}] \\
Z &= [N_s, N_p, SOC_0] \\
H_{e,i} &= [\text{Lifetime}] \\
H_{ine,j} &= [E_{ESS} \geq E_{min}, \ 10\% \leq SoC \leq 100\%, ... \\
& \quad ..., I_{cell} \leq |I_{max}|, \ U_t \leq U_{t,\,max}, \ 400\,\text{V} \leq U_{DC} \leq 750\,\text{V}],
\end{aligned} \tag{21}
$$

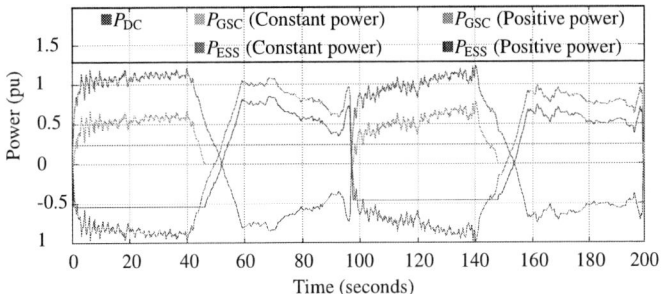

Fig. 4: Power profile of the GSC and the ESS under different power filtering strategies for an average wind speed of $14 \frac{m}{s}$ in open land conditions (c_1). The values are normalized to the average reel-out power.

where (E_{ESS}) is the ESS capacity, and (E_{min}) is the minimum capacity to meet the power filtering requirements. In order to accelerate the convergence of the solution, the system ordinary differential equations (ODE) describing the LFP and UC models are solved using the forward Euler numerical integration method with a 20 ms simulation time-step. Furthermore, the complex impedance term of the LFP model ($ZARC$) is approximated by six RC components following the method in [20]. Based on the estimated ESS capacity, the cost is calculated according to the price suggestions of [8]. The cost of the power electronic DC-DC converter is acquired from personal correspondence with multiple manufacturers.

Flywheel Sizing Method

For sizing the steel alloy FWES, the Equations 16 to 19 are solved iteratively until the design's energy storage and lifetime requirements are fulfilled, given the σ_{max} and $\frac{r_o}{r_{in}}$ dimensional constraints of Table I. Based on the selected design parameters (cylinder dimension), the FWES's cylinder, shaft, and housing masses are calculated. Finally, the cost is estimated according to the prices suggested in [10]. Regarding the composite material FWES, the cost is calculated based on the estimates found in [8], which are calculated from the rated power and energy capacity of the FWES according to the prices in [8].

Power Filtering Strategies

The sizing of the ESS is carried out for two different power filtering strategies. The goal is to identify the grid power quality's impact on the system's overall cost. The "Constant power filtering" strategy achieved full grid power filtering, where the GSC's power remains constant and equal to the average pumping cycle power. Fig. 1b and Fig. 4 explain the relationship between the GSC and ESS power and energy. This strategy trades off higher P_{grid} quality and smaller-sized GSC with more stress on the ESS power requirements.

The "Positive power filtering" allows the ESS to take only so much energy during the reel-out phase to supply the machine during the reel-in phase. The goal is not to consume any power from the grid and avoid the high energy rates dictated by the grid operators. This strategy trades off the quality of P_{grid} filtering by relatively lower ESS power during the reel-out phase.

ESS Sizing Results

ESS capacity and cost

The average results for different wind conditions and power filtering strategies are depicted in Fig. 5. The first column of grouped bars represents the sizing results for the LFP battery based on the power and energy requirements (P/E) as dictated by the kite's mechanical power and the filtering strategy. The second column represents the LFP sizing if the system is designed to last for ten years, which indicates the capacity results. On the other hand, the battery is expected to be replaced once during the project lifetime of 20 years; the replacement expenditures are included in the system cost taking into account the expected reduction in capital cost of LFP cells ten years from now. The remaining columns represent

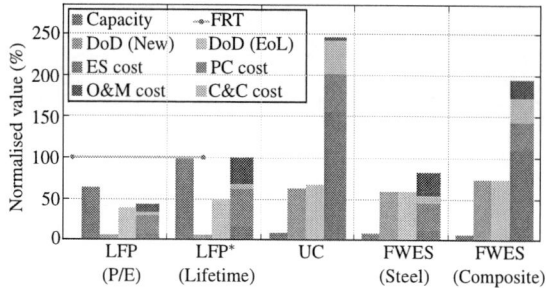

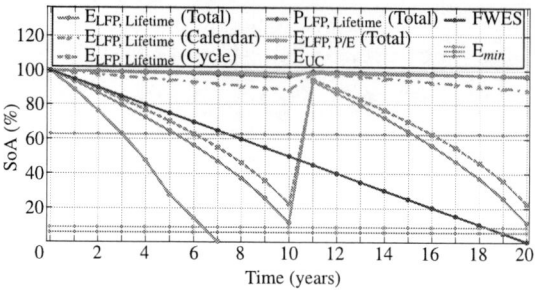

Fig. 5: Energy storage capacity and breakdown of costs for different ESS. The results indicate the average value of different wind conditions and power filtering strategies. Values with an asterisk (*) indicate the base value for normalization. The base value for the capacity is 777 kWh, and 1.23 M$ for the cost under constant filtering. The DoD is not normalized.

Fig. 6: SoA of different ESS. Capacity and power fade for the LFP and the UC indicate a gradual reduction in performance as the system ages. The FWES SoA indicator only reflects the expected system lifetime assuming the performance remains unchanged.

the sizing and cost estimates for the UC and steel and composite material (carbon-fiber) FWES under P/E and lifetime sizing criteria, respectively. Sizing the UC for lifetime was unnecessary since it can inherently withstand many discharge cycles, as will be discussed later in this chapter.

Looking at the results, we can deduce the following. In terms of energy storage capacity, sizing the LFP battery for ten years requires an increase in the capacity by 36% compared to the P/E sizing criteria. However, the sizes of all other ESS are only 8.4%-6.5% of the ten-year LFP battery size. So clearly, the LFP battery needs to be hugely oversized to meet the same P/E and lifetime requirements compared with UC and FWES, primarily due to the LFP's limited power density and feasible cycle count. The ESS capacity also impacts the maximum expected DoD of a newly installed system. The ten-year LFP would have a DoD of only 6%, which leaves plenty of room for additional utilization of the battery capacity for other purposes, such as providing frequency response support for the grid. On the contrary, the DoD for the UC and FWES is about 63% on average, which leaves far fewer reserves for supporting the grid.

The total cost of each system is broken down into energy storage (ES) costs, power conversion system (PC) costs, construction and commissioning (C&C) costs, and operation and maintenance (O&M) costs. The ES costs reflect the costs of the electrochemical cells, the battery management system and other ancillary electronics, and the manufacturing of the battery and UC packs. In the case of the FWES, it represents the costs of the rotating cylinder, the shaft, the steel housing, the vacuum pump, and the assembly costs. The PC costs represent the cost of the DC-DC power converter in the case of electrochemical energy storage systems and the cost of a permanent magnet synchronous machine and a three-phase inverter in the case of FWES. The C&C costs represent the installation costs on-site, whereas O&M costs represent the fixed and variable maintenance costs during the 20 years lifetime. The cost of equipment replacement is included in the cost with which it is associated.

Based on this cost breakdown, the expenses of each ESS are evaluated and depicted against each other in Fig.5, from which we can deduce the following. Although overly sized, the lifetime-sized LFP is still 245% cheaper than the UC option and 195% cheaper than the composite FWES solution. The only solution that provides lower costs than the LFP battery would be steel FWES; it is roughly 21 % cheaper. Additionally, the costs of the LFP battery and the steel FWES are governed by PC costs followed by O&M costs. In contrast, the ES costs do not account for more than 17% for the LFP battery and 12% for the steel FWES. These findings agree with the system specifications, which require very high-rated power but much less storage capacity.

ESS State of Aging

The comparison between the different ESS is further extended in this section to include aging and performance degradation during the expected project lifetime. Fig. 6 depicts the SoA results as a function of the operational years. The P/E sized LFP total energy capacity fade is depicted in light blue. The capacity drops annually until it reaches the minimum required capacity to fulfill the power filtering goal, roughly after 6.5 years. Therefore, the lifetime sizing was used instead; it is indicated by the solid orange curve, where the LFP capacity approaches E_{min} after ten years before being replaced by another battery pack. The capacity fade is a function of both calendar and cycle aging. In this investigation, the ambient temperature remains constant at 25°C. Therefore the calendar aging contribution is relatively limited, as shown in the figure. On the contrary, the cycle aging mechanism is the main contributor to LFP degradation because of the vast number of annual cycles. The capacity degradation of LFP causes the DoD to increase year after year. Eventually, the DoD could be as high as 49% after ten years, which renders the LFP less helpful in providing ancillary services for the grid towards the end of its lifetime.

Another form of LFP degradation is associated with the pulsed power capability (P_{LFP}) of the cell, as depicted in Fig. 6. This depreciation is, however, limited to no more than 4 %, which means the LFP would still be able to fulfill the fault ride-through requirements of the grid codes even towards the end of its life. The minimum LFP size required to fulfill the FRT (depicted by the green line in Fig. 5) is associated with the limited power density of the LFP and the peak power delivery during grid faults [6]; therefore, the capacity fade is irrelevant in this case as long as the pulsed power capability remains high.

The UC aging is prolonged and mainly contributed by calendar aging, and the capacity drop does not exceed 3% by the end of its lifetime. Here again, the calendar aging is limited by the seamlessly moderate ambient temperature of 25^{circ}C, if higher temperatures were considered, we could expect faster decline rates. The FWES SoA is assumed to be linear, where after 20 years, the accumulated stress on the disk could increase the failure rate. However, unlike the other ESS, the FWES performance is assumed to remain constant during its lifetime and no capacity or power fade is expected.

Impact of Wind Turbulence and Power Filtering Strategy

From the previous discussion, we concluded that LFP batteries and steel FWES are the most competitive candidates for AWES application in terms of capacity and cost. In this section, we try to find out the impact of different wind conditions and power filtering strategies on the overall system cost, including the cost of the ESS and the GSC.

Four different wind conditions [21] are considered in the investigation, and the results are depicted in Fig. 7. Compared with the ideal condition under constant wind speed, the wind turbulence in open land areas (c_1) produces the highest spikes in the kite's output power, it also generates the highest amount of energy, roughly around 18% more than the constant wind condition. The case for agricultural land areas (c_2) generates the next highest values, followed by offshore wind conditions (c_0). The overall system cost follows the power and energy trends; however, the percentage increase in cost is equal to 8% for the LFP and 9% for the FWES compared to the ideal wind condition. The price increase is mainly associated with the cost of the ESS since the GSC only constitutes a small percentage of the overall cost.

The impact of power filtering strategies on the system cost is shown in Fig. 8. Three scenarios are depicted; the first replicates the absence of an ESS, so the overall cost consists of that of the GSC, which has to be sized for the maximum kite power (P_{DC}). The second scenario represents constant output power filtering; the GSC conducts constant power as depicted by the straight yellow line in Fig. 4. Therefore, it is only sized at 13 % of the "No ESS" case. The remaining power has to go through the ESS, as depicted by the red line in Fig. 4. The cost of the ESS plus the GSC is 312% more expensive than that of the "No ESS" scenario for the LFP, and 253% for the FWES.

The positive power filtering strategy results are depicted in the last bar group of Fig. 8. The ESS power and energy demand are slightly less than that of the first strategy. Therefore, the GSC conducts more power and is roughly 300% more expensive than the constant power filtering strategy. It is, however, 37.5% cheaper than without an ESS. Regarding the LFP system cost, although the batteries have to carry

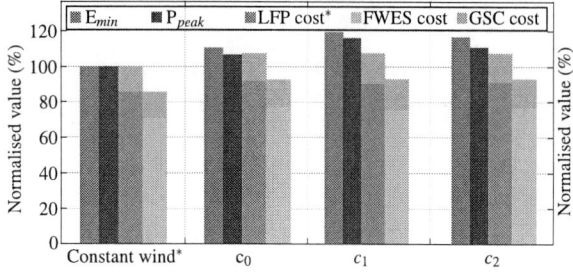
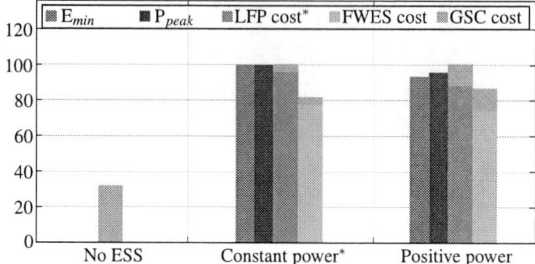

Fig. 7: Impact of wind turbulence on the ESS and GSC loading and cost. The results indicate the average values of different ESS, the GSC, and power filtering strategies.

Fig. 8: Impact of power filtering strategy on the ESS capacity and cost. The results indicate the average values of different ESS, the GSC, and wind conditions.

7 % less energy and 4 % less power than the constant filtering case, the overall cost remains unchanged. The cost saving in the LFP system are compromised by the more expensive GSC. The FW case is even more counter intuitive, as one would expect the reduction in the ESS power and energy demand should result in a cost reduction. However, because the energy storage cost constitutes only a fraction of the FW cost, the slight savings achieved are less than the extra costs asociated with the oversized GSC. These findings intensify the holistic approach's importance in understanding such a system's cost trade-offs.

Conclusion

This work investigated the sizing of lithium-iron phosphate batteries, ultracapacitors, steel, and composite flywheels for power filtering of pumping cycle airborne wind energy generators. Two power filtering strategies targeting either constant grid power (complete filtering) or positive grid power (partial filtering) were considered. Additionally, the implication of the installation site in terms of different wind classes was also taken into account. The energy storage systems were compared in capacity, usefulness for ancillary grid services, aging/performance impact, and cost.

The results reveal that, on the one hand, lithium-iron-phosphate battery systems provide the lowest cost/kWh at 1690$/kWh for the complete project, making it the best candidate for installations requiring the provision of ancillary services to the grid. Here, the extra energy reserve can be handy, especially for small island installations where such systems are meant to be used. On the other hand, steel flywheels provide the cheapest cost/kW at 410$/kW; this is due to the decoupling of the energy storage and power costs for such systems, where the power cost is mainly dictated by the cost of the electrical machine and power electronics converter. They also provide the lowest overall cost, 21% lower than lithium batteries, and they do not suffer from capacity or power fade during their expected lifetime. However, they are much smaller in capacity and do not allow for additional grid services at a competitive cost. The ultracapacitors and composite flywheels are generally 250%-300% more expensive than batteries and thus deemed too expensive for any commercial realization.

The impact of site locations and the associated wind turbulence class was also investigated in this work. The findings suggest that peak power can increase by as much as 19% compared to ideal wind conditions, but the associated cost increase is no more than 8%-9% for batteries and steel flywheels systems. This is because batteries are already oversized to meet the lifetime requirements of the project, so they can inherently withstand higher loads, and the extra cost is linked with oversizing the DC-DC converter system. The case is similar to flywheels, where the extra power only impacts the power conversion system, which has a cost-competitive edge, as discussed earlier.

The power filtering strategy results are even more counterintuitive. One would expect lower power filtering quality to result in cost savings; however, this is not the case. The results indicate that reducing the filtering quality reduces the energy storage system cost. However, it also increases the cost of the grid-side converter. In the case of lithium batteries, it ends up with a similar cost to the constant power filtering strategy, and in the case of the flywheel increases the overall system cost by 6%.

References

[1] M. Diehl, U. Ahren, and R. Schmehl, *Airborne Wind Energy.* Springer, Berlin, Heidelberg, 2013.

[2] S. P. GmbH, "Kite Power For Mauritius." https://skysails-power.com/kite-power-for-mauritius/.

[3] M. Erhard and H. Strauch, "Flight control of tethered kites in autonomous pumping cycles for airborne wind energy," *Control Engineering Practice*, vol. 40, pp. 13–26, July 2015.

[4] S. P. GmbH, "SkySails Power GmbH Image Brochure." https://skysails-group.com/downloads/.

[5] B. Bagaber, P. Junge, and A. Mertens, "Lifetime Estimation and Dimensioning of the Machine-Side Converter for Pumping-Cycle Airborne Wind Energy System," in *2020 22nd European Conference on Power Electronics and Applications (EPE'20 ECCE Europe)*, Sept. 2020.

[6] B. Bagaber and A. Mertens, "Fault Ride-Through Performance of Pumping Cycle Airborne Wind Energy Generators with the Support of Optimally Sized Energy Storage System," in *2021 IEEE 12th Energy Conversion Congress & Exposition - Asia (ECCE-Asia)*, pp. 1144–1150, May 2021.

[7] X. Luo, J. Wang, M. Dooner, and J. Clarke, "Overview of current development in electrical energy storage technologies and the application potential in power system operation," *Applied Energy*, vol. 137, pp. 511–536, Jan. 2015.

[8] K. Mongird, V. Viswanathan, P. Balducci, J. Alam, V. Fotedar, V. Koritarov, and B. Hadjerioua, "An Evaluation of Energy Storage Cost and Performance Characteristics," *Energies*, vol. 13, p. 3307, June 2020.

[9] M. Erhard and H. Strauch, "Control of Towing Kites for Seagoing Vessels," *IEEE Transactions on Control Systems Technology*, vol. 21, pp. 1629–1640, Sept. 2013.

[10] M. M. Rahman, E. Gemechu, A. O. Oni, and A. Kumar, "The development of a techno-economic model for the assessment of the cost of flywheel energy storage systems for utility-scale stationary applications," *Sustainable Energy Technologies and Assessments*, vol. 47, p. 101382, Oct. 2021.

[11] D.-I. Stroe, *Lifetime Models for Lithium Ion Batteries Used in Virtual Power Plants.* PhD thesis.

[12] C. Forgez, D. Vinh Do, G. Friedrich, M. Morcrette, and C. Delacourt, "Thermal modeling of a cylindrical LiFePO4/graphite lithium-ion battery," *Journal of Power Sources*, vol. 195, pp. 2961–2968, May 2010.

[13] Y. Cai, Y. Che, H. Li, M. Jiang, and P. Qin, "Electro-thermal model for lithium-ion battery simulations," *Journal of Power Electronics*, vol. 21, pp. 1530–1541, Oct. 2021.

[14] R. Faranda, M. Gallina, and D. Son, "A new simplified model of Double-Layer Capacitors," in *2007 International Conference on Clean Electrical Power*, (Capri, Italy), pp. 706–710, IEEE, May 2007.

[15] A. Ostadi and M. Kazerani, "A Comparative Analysis of Optimal Sizing of Battery-Only, Ultracapacitor-Only, and Battery–Ultracapacitor Hybrid Energy Storage Systems for a City Bus," *IEEE Transactions on Vehicular Technology*, vol. 64, pp. 4449–4460, Oct. 2015.

[16] T. Kovaltchouk, B. Multon, H. Ben Ahmed, J. Aubry, and P. Venet, "Enhanced Aging Model for Supercapacitors Taking Into Account Power Cycling: Application to the Sizing of an Energy Storage System in a Direct Wave Energy Converter," *IEEE Transactions on Industry Applications*, vol. 51, pp. 2405–2414, May 2015.

[17] X. Li, B. Anvari, A. Palazzolo, Z. Wang, and H. Toliyat, "A Utility-Scale Flywheel Energy Storage System with a Shaftless, Hubless, High-Strength Steel Rotor," *IEEE Transactions on Industrial Electronics*, vol. 65, pp. 6667–6675, Aug. 2018.

[18] "Military Handbook: Metallic Materials And Elements For Aerospace Vehicle Structures," 1998.

[19] J. Kennedy' and R. Eberhart, "Particle Swarm Optimization," p. 7.

[20] S. Buller and R. W. de Doncker, *Impedance Based Simulation Models for Energy Storage Devices in Advanced Automotive Power Systems.* No. 31 in Aachener Beiträge Des ISEA, Aachen: Shaker, 2003.

[21] T. Haas and J. Meyers, "AWESCO Wind Field Datasets [Data set]," *Zenodo*, 2019.

Design Interactions of AC- and DC-Side Filters for Traction Drives with SiC Inverters

Hedieh Movagharnejad, Benjamin Knebusch, Axel Mertens, Bernd Ponick
Institute for Drive Systems and Power Electronics
Leibniz University of Hannover
Hannover, Germany
Tel.: +49 / (511) – 762-14338
Fax: +49 / (511) – 762-3040
E-Mail: Hedieh.Movagharnejad@ial.uni-hannover.de, Benjamin.knebusch@ial.uni-hannover.de, Mertens@ial.uni-hannover.de, Ponick@ial.uni-hannover.de
URL: https://www.ial.uni-hannover.de

Acknowledgements

The authors gratefully acknowledge financial support from the German Federal Ministry of Education and Research (BMBF) under grant number 16EMO0252 (Project acronym: UmSiChT).

Keywords

« EMC/EMI», « Passive filters», « High frequency power converter», « High power density systems», « Electrical drive».

Abstract

This paper proposes a straightforward comprehensive design methodology for selecting and dimensioning parameters of common AC- Side filters and investigates the interactions between AC- and DC-side filters to avoid any over dimensioning in a traction drive system. A test setup has been implemented to validate the simulative investigations.

1. Introduction

Traction drives with silicon carbide (SiC) inverters offer several advantages such as lower losses and higher efficiency at partial load, increased vehicle range, higher power density and lower noise. However, the high switching frequencies related to SiC inverters - in addition to the desired switching loss reduction - also lead to steep voltage gradients (du/dt) of up to over $50 \, kV/\mu s$. Possible consequences are damage to the electrical motor (winding and bearings), accelerated ageing of insulation materials due to resulting partial discharges, and electromagnetic interference (EMI) problems [1]-[5]. For these reasons, SiC components can only be used to a limited extent with today's converter and machine structures and their voltage gradients must be reduced. This significantly decreases the efficiency and associated advantages. Besides, the cost increases due to larger chip sizes.

In order to counteract the problem mentioned, the use of output AC filters, such as a du/dt filter to limit the slope of output voltages or a sine-wave filter to generate sinusoidal output voltages, can be considered. Moreover, in order to reduce the generated high frequency (HF) common-mode (CM) current and stress on the motor bearings, HF-CM chokes can be applied between the inverter output and the motor [6]-[8]. Furthermore, to suppress the EMI noise generated due to the fast variation of the switching node voltage, EMI filters are connected between the fast switching converters and the power supply [9].

This paper proposes a comprehensive design methodology for typical output AC filters and investigates the interactions of input DC and output AC filters to obtain an optimized design procedure for filter sizing in traction drive systems. This design process simultaneously considers several important parameters and selects optimal solutions with respect to size and weight.

2. Analyzed System Structure

The system, which is considered in this paper is shown in Fig. 1.

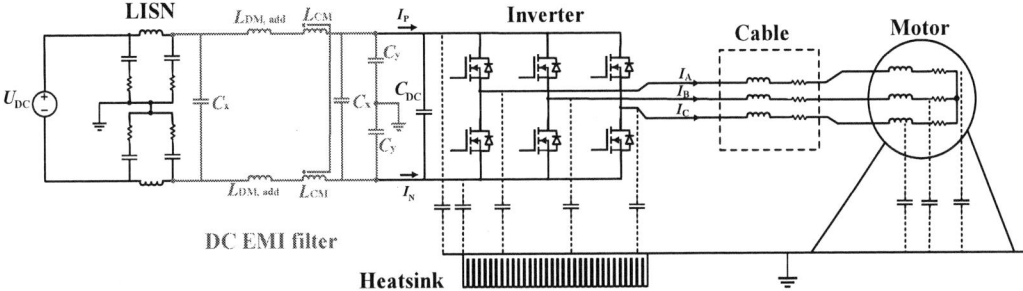

Fig. 1: Structure of the considered system

The considered system comprises a DC source, a DC line impedance stabilization network (LISN), and a three-phase inverter with SiC MOSFETs, which is connected via a cable to the induction motor. In Fig. 1, the red part belongs to the DC EMI filter, which is connected between LISN and DC side of the inverter. In [9], a comprehensive design methodology for dimensioning multi-stage DC EMI filters has been proposed. This paper focuses on the design approach for selecting an appropriate AC filter topology and its effect on the DC EMI filter design.

3. AC-Filters Design Approach

3.1. Sine-Wave Filter

Fig. 2 (a) shows an exemplary sine-wave filter, which is connected between the inverter output and the motor. The sine-wave filter, which forms a passive second-order low-pass filter, comprises a capacitor ($C_{F,s}$) to limit the steep voltage gradients, and an inductor ($L_{F,s}$) to limit the di/dt and correspondingly reduce the capacitor charging current. In this arrangement, a feedback from the capacitors' star point to the DC- Rail is considered to make a low impedance path for CM current and leading CM current into the DC- Rail. This can reduce the CM voltage at the motor terminals.

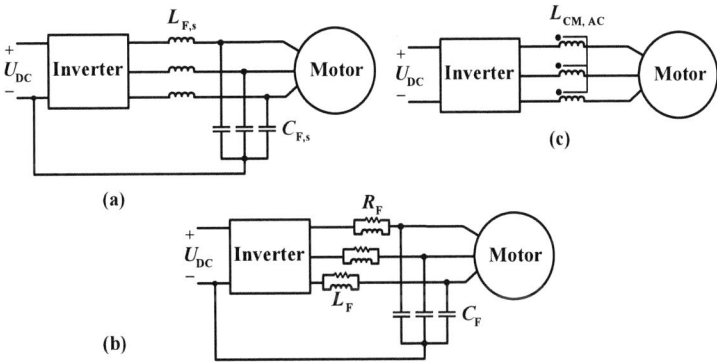

Fig. 2: AC filter topologies (a): sine-wave filter, (b): du/dt filter, (c): CM-Choke

Sine-wave filter operation depends mainly on the $L_{F,s}$ and $C_{F,s}$ values, as well as on the DC input voltage (U_{DC}) and the switching frequency (f_s). The choice of selecting filter parameters is not straightforward and depends on many important factors such as filter current ripple, desired voltage ripple on the filter output or motor terminals, a trade-off between the inductor value and thus the capacitor ripple current, losses and the size of the filter. Hence, in this paper, it is tried to propose and summarize a straightforward design procedure for selecting the filter component parameters.

Fig. 3 (left) presents a flowchart for the design of sine-wave filters. In this design methodology, with input data such as DC link voltage, switching frequency, fundamental frequency (f_0), rated output power (P_{out}), and load characteristics, the design process can be started. By designing sine-wave filters, some considerations should be made. The fundamental frequency voltage drop on the inductor should not be too large. Moreover, the cut-off frequency of the filter should be selected so that the HF noise generated will be attenuated. Furthermore, the consumed reactive power of the filter capacitors should be limited. The considered conditions are as follows:

- Fundamental frequency voltage drop across the filter < 10 % of the input voltage,
- 10* fundamental frequency < filter cut-off frequency < 0.5*switching frequency,
- Capacitors reactive power < 5 % of the inverter output power,
- Minimum inductance of the filter to be determined from maximum allowable current ripple.

Following the step-by-step design procedure, the values of $L_{F,s}$ and $C_{F,s}$ can be extracted. The size of a sine-wave filter is mainly defined by the size of its inductor. Therefore, from several valid designs, the one resulting in the smallest inductance is selected. It should be mentioned that by increasing the switching frequency, the size of the sine-wave filter can be reduced. The design input data and extracted filter parameters according to the design procedure can be found in the table in Fig. 3 (right).

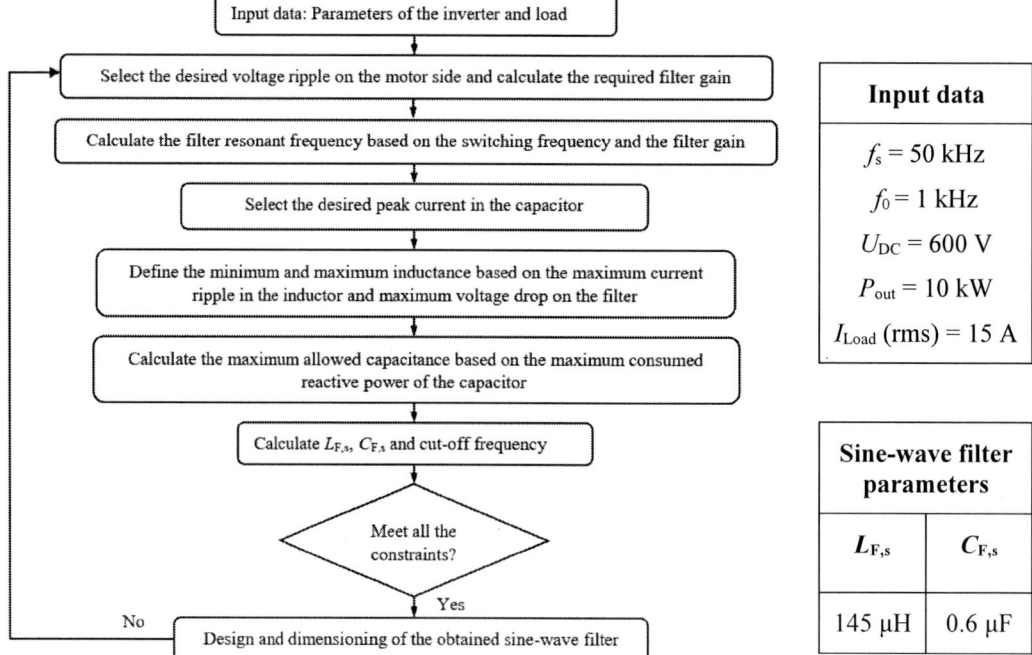

Fig. 3: Sine-wave filter design flowchart (left), Design input data and extracted sine-wave filter parameters (right)

For the conducted disturbances in the range of 150 kHz to 30 MHz, the FFT analysis of the CM and the differential mode (DM) disturbance spectrum in the frequency domain was performed. The IEC CISPR

25 standard was used to check compliance with EMI standards. Fig. 4 shows the U_{CM} interference spectrum at the LISN and the motor for the case without and with a sine-wave filter. It should be mentioned that the AC sine-wave filter does not influence the DC side DM voltage, since the DC and AC sides are decoupled due to the DC link capacitor. The simulations have been done at $U_{DC} = 600$ V and $f_s = 50$ kHz. Considering the U_{CM} disturbance spectrum at the LISN, it is clear that adding a sine-wave filter on the AC side with DC feedback can significantly reduce the disturbance at the LISN especially in the lower frequency range, which can notably reduce the demand for DC EMI filter. Based on the DC EMI filter design methodology proposed in [9], the required L_{CM} for the one-stage DC EMI filter is around 4.3 mH for the case without any AC filter, and this value reduces significantly to 154 µH if the aforementioned sine-wave filter is added to the AC side. Hence, it is recommendable to design a sine-wave filter in the system before the DC EMI filter design to avoid over-dimensioning of the DC side EMI filter. Moreover, a remarkable CM noise reduction can be detected at the motor terminals.

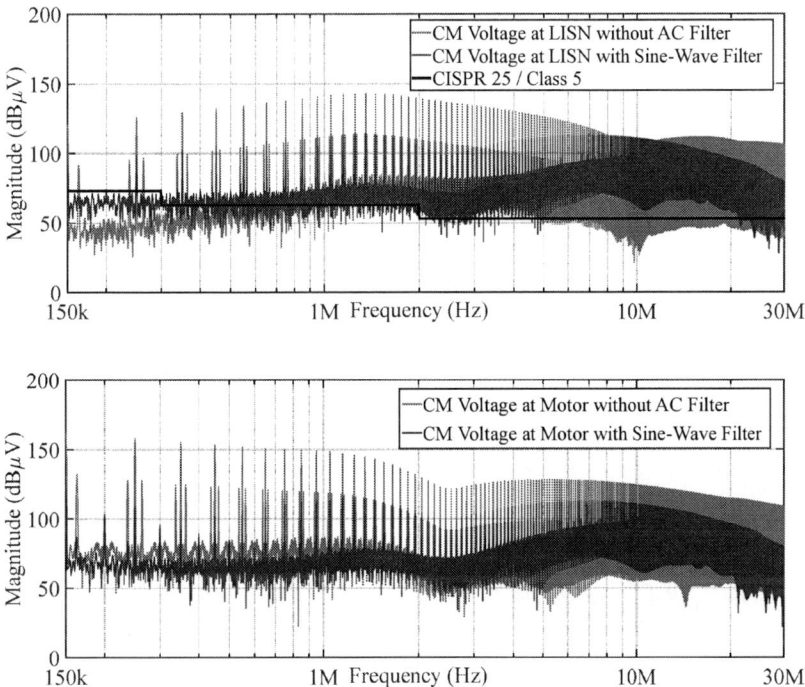

Fig. 4: Simulation results at $U_{DC} = 600$ V, $f_s = 50$ kHz, $n = 0$/min; U_{CM} at LISN and motor: without and with sine-wave filter

3.2. du/dt Filter

Another option for an AC filter is a du/dt filter, which comprises also an additional damping resistor, and can be seen in Fig. 2 (b). du/dt filters are passive second-order low-pass filters, and the frequency response can be adjusted by selecting the inductance L_F and the capacitance C_F. The filter resonant frequency is selected to be higher than the switching frequency. The energy of the resonance oscillation between L_F and C_F is dissipated in the damping resistors R_F. In order to design a du/dt filter, a design methodology has been considered mainly based on [7], which starts with listing of all available values for R_F, L_F and C_F and formation of all permutations of components. After this, a systematic limiting process of component parameters will be performed by some conditions, as follows:

- C_F > effective capacitance of the motor: In order to ensure the filtering effect during the operation of an electrical machine, the filter must provide a lower impedance path for the HF noise signals than the electrical machine. To achieve this, C_F must be larger than the effective capacitance of the motor.

- Desired du/dt at the filter output and filter time constant (τ) calculation:

  $\tau_{linear} = 0.8 \dfrac{U_{DC}}{du/dt}$ & $\tau = \sqrt{L_F C_F}$. After that, the selection can be limited by $\tau_{linear} \langle \tau \langle 1.5 * \tau_{linear}$.

- Damping degree: $D > 1$: Other permutations can be eliminated, which cannot meet the requirement $D = \dfrac{1}{2R_F} \sqrt{\dfrac{L_F}{C_F}}$.

Consequently, possible component values will be determined and the expected du/dt from their step responses can be achieved and further limitations can be done if the solutions do not meet the requirements. Moreover, power dissipation in the damping resistors can be calculated (losses per phase: $P_{R,avg} = C_F . U_{DC}^2 . f_s$) and the permutations with the least total volume and power losses will be chosen. The losses, which should be dissipated by the resistors, are linearly dependent on the switching frequency. Although the components required to implement a du/dt filter are very small compared to the sine-wave filter, it should be mentioned that resistors with higher power dissipation are required for each increase in the switching frequency. Here, a du/dt of 12 V/ns at the filter output was considered as a condition for the investigations. Table 1 shows the best possible combination results that enable the requirements to be met at $U_{DC} = 600$ V and $f_s = 50$ kHz.

Table 1: Extracted du/dt filter parameters

at $U_{DC} = 600$ V, $f_s = 50$ kHz	$L_F = 2$ µH	$C_F = 1.5$ nF	$R_F = 16\ \Omega$	$D = 1.1$

To investigate the interaction of du/dt filters on the DC side, the U_{CM} interference spectrum has been plotted at the LISN and the motor (Fig. 5).

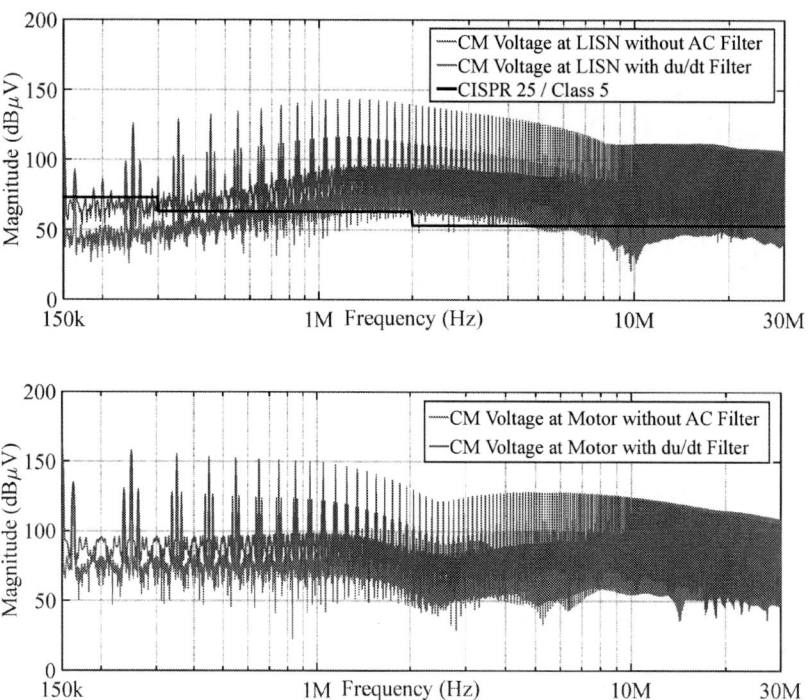

Fig. 5: Simulation results at $U_{DC} = 600$ V, $f_s = 50$ kHz, $n = 0$/min; U_{CM} at LISN and motor: without and with du/dt filter

It can be seen from the results that the presence of a du/dt filter does not affect the DC side EMI filter requirement, as it shows its effect only on the interference reduction in the HF range (MHz range).

This can be seen also in the results at the motor terminals. The insertion of a du/dt filter between the inverter and the motor can effectively decelerate the steep voltage edges but does not affect the EMI disturbances much.

3.3. CM-Choke

Another type of AC filter that can be considered is a CM core (Fig. 2 (c)). AC CM cores are placed over the line power cables between the inverter output and the motor input. In this configuration, the cores operate as a CM-choke. These types of cores are provided to reduce the HF CM current and correspondingly lower the stress on the motor bearings. Moreover, there is no fundamental frequency voltage drop across the chokes. It should be mentioned that these cores do not eliminate switching frequency noise.

In order to select an appropriate CM core and the corresponding required inductance value of the CM-choke, it is recommendable to investigate the whole system in simulations, to observe the peak CM currents at the motor terminals, and to define how much this CM current should be reduced. Then, the approximate value for the CM-choke inductance can be extracted, and based on this value and expected saturation characteristics of the materials, as well as core dimensions, an appropriate core can be chosen and inserted into the system.

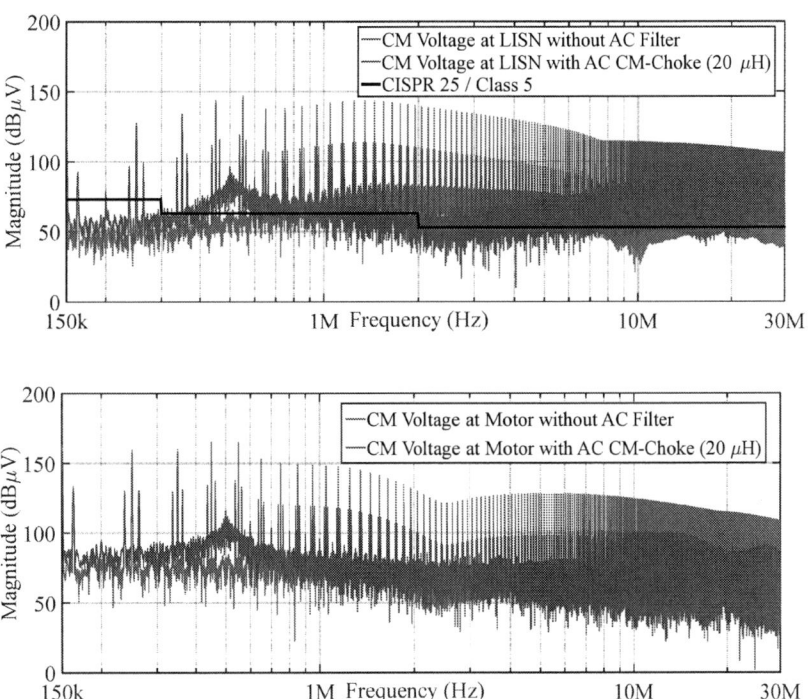

Fig. 6: Simulation results at U_{DC} = 600 V, f_s = 50 kHz, n = 0/min; U_{CM} at LISN and motor: without and with AC CM-choke

Fig. 6 shows the simulation results for the case without and with AC CM-choke. In this investigation, a CM inductance of $L_{CM,AC}$ = 20 μH has been chosen. By insertion of this CM-choke, a notable reduction in the CM current at the motor terminals can be seen (Fig. 7). The CM current reduces from its maximum value of 19.5 A to 6.5 A if $L_{CM,AC}$ = 20 μH is added. Besides, it can be seen from the CM voltage spectrum at the LISN, that there is no effective reduction in the CM noise in the low frequency range. But it should be mentioned that by inserting higher values of $L_{CM,AC}$, the LISN side CM voltage spectrum can be reduced remarkably and correspondingly the DC EMI filter design requirements can be affected. Hence, it is recommendable to design an AC CM-choke before the DC side EMI filter to avoid any over-dimensioning. In addition, a notable reduction of the CM noise spectrum in the MHz region can be detected at the motor terminals.

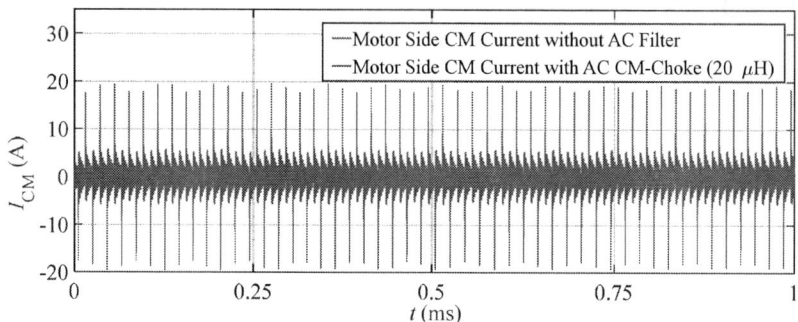

Fig. 7: Simulated I_{CM} at the motor terminals (U_{DC} = 600 V, f_s = 50 kHz, n = 0/min)

Table 2 shows a summary of the design interactions between DC EMI filter and various AC-side filters and their features.

Table 2: Design Interactions between DC- and AC-Side Filters

AC-side filter topology	Influence on DC-side EMI filter requirement	Feature
Sine-wave filter	+++	Safely inhibits du/dt, over voltages and bearing currents
du/dt filter	No influence	Reduces du/dt, but strongly affects efficiency
CM-choke	+	Reduces CM currents effectively

4. Test Setup

In order to validate the simulation results, a test setup was built as a 10 kW demonstrator. The setup (Fig. 8) consists of DC voltage source, LISN, 10 kW inverter with SiC-MOSFET half-bridge modules, and induction motor. The setup uses the Institute for Drive Systems and Power Electronics' (IAL) ControlCube, which incorporates a Xilinx Zynq 7000 (SoC) with an ARM Dual Core Cortex A9 and an Artix-7 FPGA as programmable logic (PL).

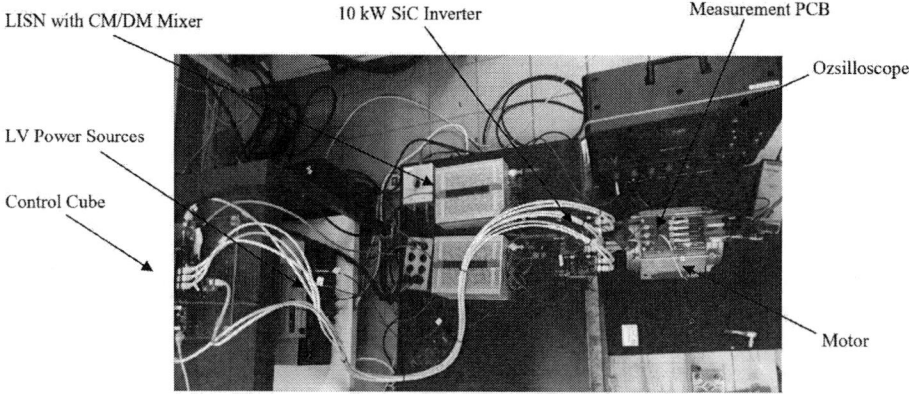

Fig. 8: Test bench setup

Fig. 9 represents the measurement results, driven at $U_{DC} = 600$ V, $f_s = 20$ kHz and $n = 0$/min. Fig. 9 (top) shows the U_{CM} interference spectrum at the LISN for the case without any AC filter and for the case with a du/dt filter. Here, the du/dt filter is implemented based on the design parameters of Table 1.

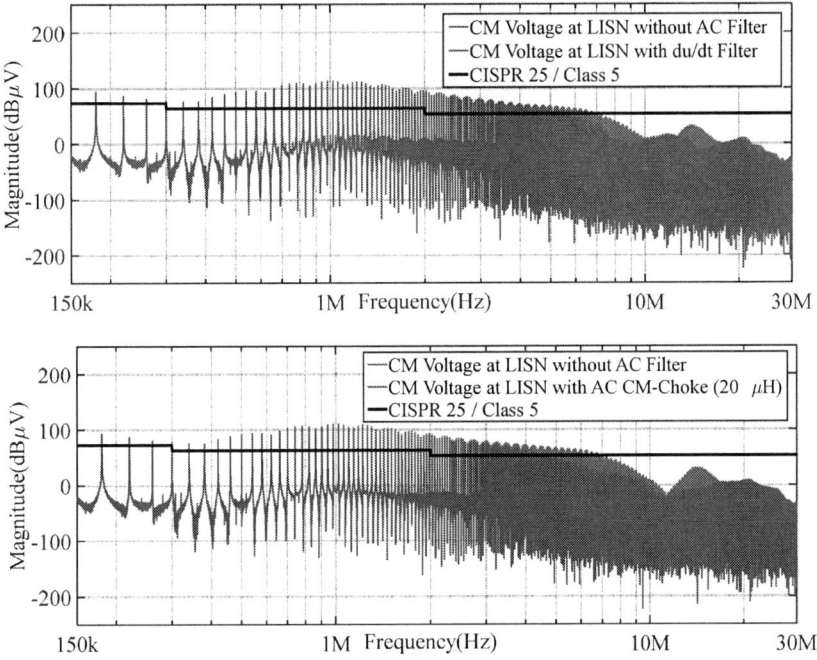

Fig. 9: Measurement results at $U_{DC} = 600$ V, $f_s = 20$ kHz, $n = 0$/min; U_{CM} at LISN: without and with du/dt filter (top), without and with AC CM-choke ($L_{CM,AC} = 20$ µH) (bottom)

As it was expected, it can be seen that an insertion of a du/dt filter does not affect the U_{CM} interference spectrum in the low frequency range, and thus does not influence the DC-side EMI filter design. A reduction can be seen in the spectrum from 4 MHz. Moreover, Fig. 9 (bottom) shows the same results for the case without and with an AC-side CM-choke. Here, two CM-cores (inductance at 10 kHz: 10 µH) in series are inserted between the inverter output and the motor input. A small reduction - also in the lower frequency range - can be detected in the interference spectrum after adding the AC CM-chokes. Insertion of such cores can affect the DC-side EMI filter design depending on how large the CM-choke inductance is.

Fig. 10 and Fig. 11 represent the transient waveforms for two cases, namely the case with $n = 0$/min and the case with $n = 1000$/min. Each group of graphs present the case without any filter, the case with a du/dt filter and the case with an AC CM-choke of 20 µH between the inverter output and the motor input. In these results, phase to ground voltages ($u_{U\_GND}(t)$, $u_{V\_GND}(t)$, $u_{W\_GND}(t)$), $u_{CM}(t)$, motor star point to ground voltage ($u_{SP\_GND}(t)$) and $i_{CM}(t)$ at the motor terminals are shown. Besides, the CM voltage at the LISN $u_{CM\_LISN}(t)$ can also be seen. For each group of the results, the absolute peak values of the related waveforms can be seen in the corresponding tables.

In the case $n = 0$/min, where the maximum CM current occurs, it can be seen from the results that inserting an AC CM-choke has reduced the CM current from its maximum value of 6.9 A to 2.5 A. Adding an AC CM-choke is more effective compared to a du/dt filter in both, I_{CM} and U_{CM} reduction. The maximum I_{CM} value at the motor is 3.7 A, in case a du/dt filter is added.

Looking at Fig. 11, it can be seen that the maximum I_{CM} value decreases to 2.8 A for the case without any filter at $n = 1000$/min. Both du/dt filter and CM-choke reduce this I_{CM} value to 1.6 and 1.5 A accordingly. No remarkable reduction in the CM voltage has been detected in this condition for various AC filter cases.

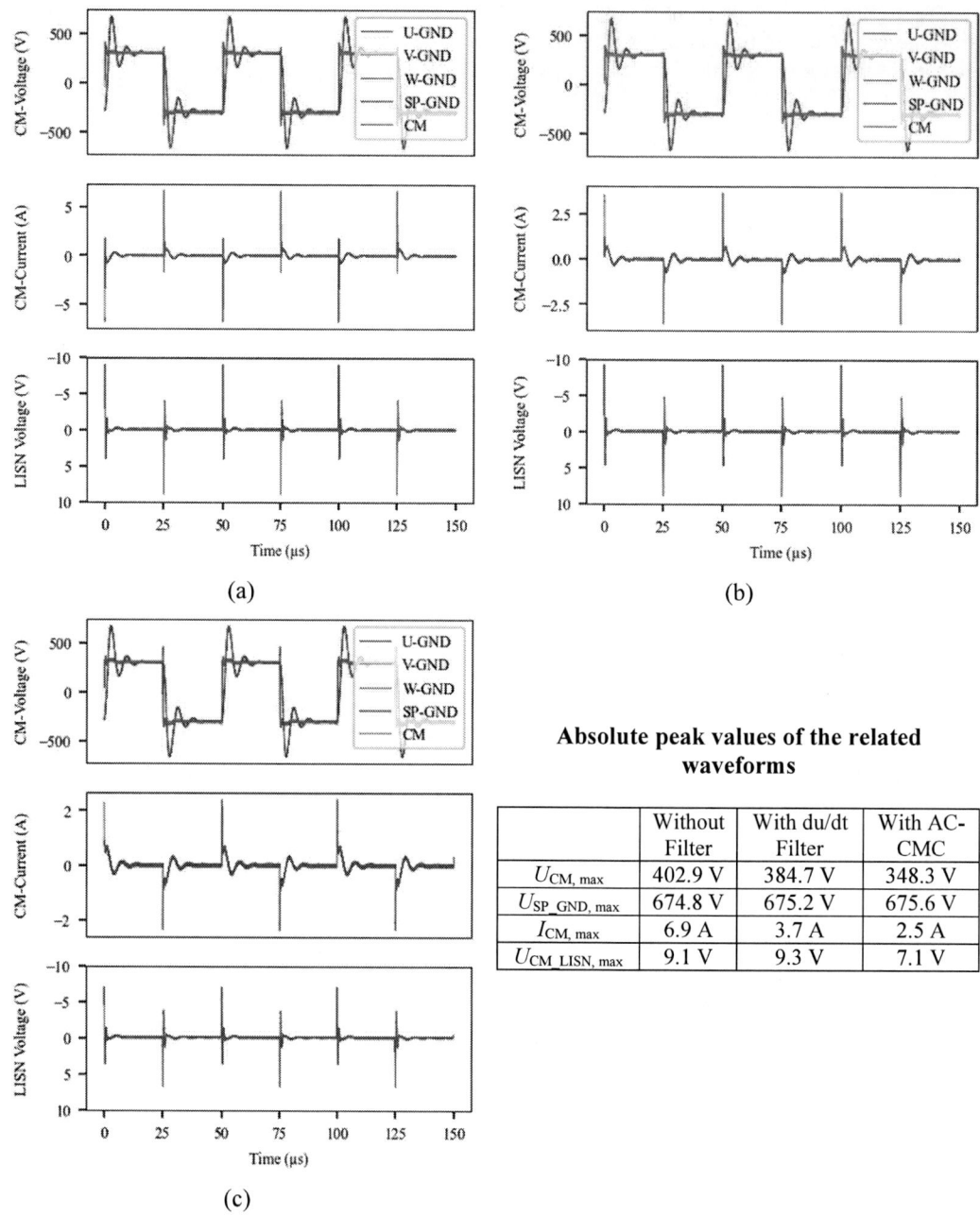

Absolute peak values of the related waveforms

	Without Filter	With du/dt Filter	With AC-CMC
$U_{CM, max}$	402.9 V	384.7 V	348.3 V
$U_{SP\_GND, max}$	674.8 V	675.2 V	675.6 V
$I_{CM, max}$	6.9 A	3.7 A	2.5 A
$U_{CM\_LISN, max}$	9.1 V	9.3 V	7.1 V

Fig. 10: Measurement results at $U_{DC} = 600$ V, $f_s = 20$ kHz, $n = 0$/min: without AC-filter (a), with du/dt filter (b), with AC CM-choke ($L_{CM,AC} = 20$ µH) (c)

5. Summary and Conclusion

In this paper, a systematic design methodology for selecting and dimensioning AC-side filters for traction drive systems is proposed. Moreover, the design interactions of AC- and DC-side filters are investigated and a design priority is advised. This straightforward design process can be useful for designers to avoid over-dimensioning and to achieve an optimum filter design with a higher power density. It can be concluded that adding an AC-side filter can affect the DC-side filter design and considerations should be taken before. From the investigations, it can be concluded that adding an AC-side sine-wave filter affects the DC EMI filter design in a positive way and can help to have smaller DC filter dimensions. Moreover, by generating sinusoidal output voltages at the motor terminals, the steep voltage gradients, over voltages and bearing currents will be inhibited safely. A du/dt filter - due to its effect in the high-frequency range - does not influence the DC EMI filter design, but reduces the steep du/dt effectively at the motor terminals. du/dt filters are not an interesting option at higher switching frequencies because of the high losses, which must be dissipated by the resistors. In addition, inserting an AC CM-choke can reduce the CM currents at the motor terminals effectively and, depending on the CM-choke inductance value, the DC-side EMI filter design can be affected in a positive way. A test setup has been implemented which validates the simulative investigations and the results comply with the simulation results.

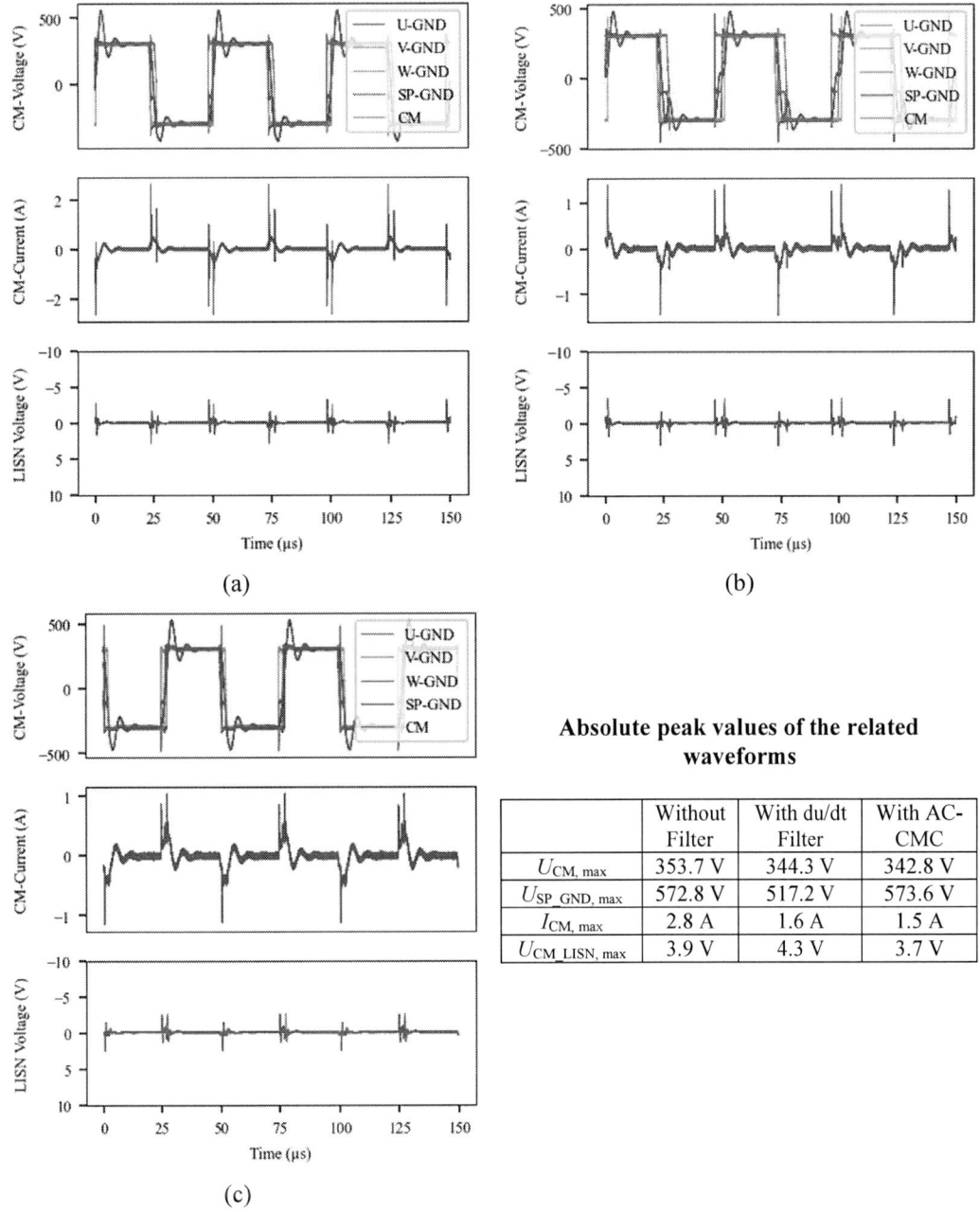

Absolute peak values of the related waveforms

	Without Filter	With du/dt Filter	With AC-CMC
$U_{CM, max}$	353.7 V	344.3 V	342.8 V
$U_{SP\_GND, max}$	572.8 V	517.2 V	573.6 V
$I_{CM, max}$	2.8 A	1.6 A	1.5 A
$U_{CM\_LISN, max}$	3.9 V	4.3 V	3.7 V

Fig. 11: Measurement results at $U_{DC} = 600$ V, $f_s = 20$ kHz, $n = 1000$/min: without AC-filter (a), with du/dt filter (b), with AC CM-choke ($L_{CM,AC} = 20$ µH) (c)

References

[1] A. Mütze, "Bearing Currents in Inverter-Fed AC Motors," Ph.D. dissertation, Technische Universität Darmstadt, 2004.

[2] H. Tischmacher, S. Gattermann, M. Kriese and E. Wittek, "Bearing wear caused by converter-induced bearing currents," *IECON 2010 - 36th Annual Conference on IEEE Industrial Electronics Society*, pp. 784-791, 2010.

[3] M. Kriese, E. Wittek, S. Gattermann, H. Tischmacher, G. Poll, and B. Ponick, "Influence of bearing currents on the bearing lifetime for converter driven machines," *International Conference on Electrical Machines*, pp. 1735-1739, 2012.

[4] A. Hoffmann and B. Ponick, "Method for the Prediction of the Potential Distribution in Electrical Machine Windings Under Pulse Voltage Stress," *IEEE Transactions on Energy Conversion*, vol. 36, no. 2, pp. 1180–1187, Jun. 2021.

[5] M. Ali, J.-K. Müller, J. Friebe, and A. Mertens, "Analysis of switching performance and EMI emission of SiC inverters under the influence of parasitic elements and mutual couplings of the power modules," *22nd European Conference on Power Electronics and Applications (EPE'20 ECCE Europe)*, 2020.

[6] N. Hanigovszki, J. Landkildehus and F. Blaabjerg, "Output filters for AC adjustable speed drives," *APEC 07 - Twenty-Second Annual IEEE Applied Power Electronics Conference and Exposition*, Anaheim, CA, USA, 2007.

[7] J.-K. Müller, "Untersuchungen zu Ausgangsfiltern in Siliziumcarbid-Antriebswechselrichtern," Ph.D. dissertation, Leibniz Universität Hannover, TEWISS Verlag ISBN 978-3-95900-525-8, 2021.

[8] J.-K. Müller, T. Brinker, J. Friebe and A. Mertens, "Output dv/dt filter design and characterization for a 10 kW SiC inverter," *IECON 2018 - 44th Annual Conference of the IEEE Industrial Electronics Society*, pp. 2122-2127, 2018.

[9] H. Movagharnejad and A. Mertens, "Design methodology for dimensioning EMI filters for traction drives with SiC inverters," *23rd European Conference on Power Electronics and Applications (EPE'21 ECCE Europe)*, 2021.

Investigation of an Interleaved Current-Fed Single Active Bridge DC-DC Converter for PV Applications

Lucas Vinícius de Araújo Gomes*, Tobias Manthey[†], Montiê Alves Vitorino*, Jens Friebe[†]
*Industrial Electronics and Machine Drive Laboratory (LEIAM)
Federal University of Campina Grande (UFCG)
[†]Institute for Drive Systems and Power Electronics
Leibniz University Hannover
Email: lucas.araujo@ee.ufcg.edu.br; tobias.manthey@ial.uni-hannover.de;
vitorino@dee.ufcg.edu.br; friebe@ial.uni-hannover.de

Acknowledgement

Parts of this work were funded by the Minna-James-Heineman-Stiftung under reference number S0029/10110/2021 and also by the Ministry of Science and Culture of Lower Saxony and the Volkswagen Foundation. The authors are responsible for the content of this publication.

Keywords

≪DC-DC converter≫, ≪Dual Active Bridge (DAB)≫, ≪Soft switching≫, ≪Design optimization≫, ≪Singel Active Bridge≫

Abstract

The increase in efficiency of photovoltaic inverters is the subject of many researches and in particular losses of semiconductors and magnetics have to be reduced. This topology follows the approach of a combined interleaved boost converter and a series resonant converter (SRC). In addition, it differs from the dual active bridge topology due to the non-synchronous rectification with less control effort.

Introduction

For the application of boosting a PV voltage level, a usual topology, which is composed of two stages, the interleaved boost converter, and the series resonant converter, is simplified to an interleaved current-fed single active bridge (CF-SAB). The amount of active components is significantly reduced by half , while also reducing the overall conduction and switching losses in these elements. Also, a DC blocking capacitor is no longer necessary in the configuration due to the control approach. There are also many studies about similar topologies, such as the CF-DAB (current-fed dual active bridge). In [1]-[4], this topology and various operating points are discussed, comparing the results for different phase shifts between the two active bridges, as well as using different duty cycles in each bridge and mapping the power and zero voltage switching (ZVS) boundaries for every operating point. Different approaches to control these converters are presented in many studies, like using the voltage across the clamping capacitor [3], complex control systems [4], [5], or discrete-time modelling [6]. Related topologies were used to achieve DC-DC conversion in a similar way, like using additional switches and capacitors to avoid voltage spikes and to achieve ZVS [7], using double half-bridges [8], in which one leg is active and one is composed by only two capacitors as well as a three-port converter [9], using two active bridges of three legs each. This paper discusses the approach for loss reduction by combining the two full bridges of the boost converter and the series resonant converter into one shared full bridge. The differences are worked out and the advantages and disadvantages are discussed. A comprehensive analysis of the overall system is given in the following chapter and a first design of the prototype is presented afterwards.

Topology Description

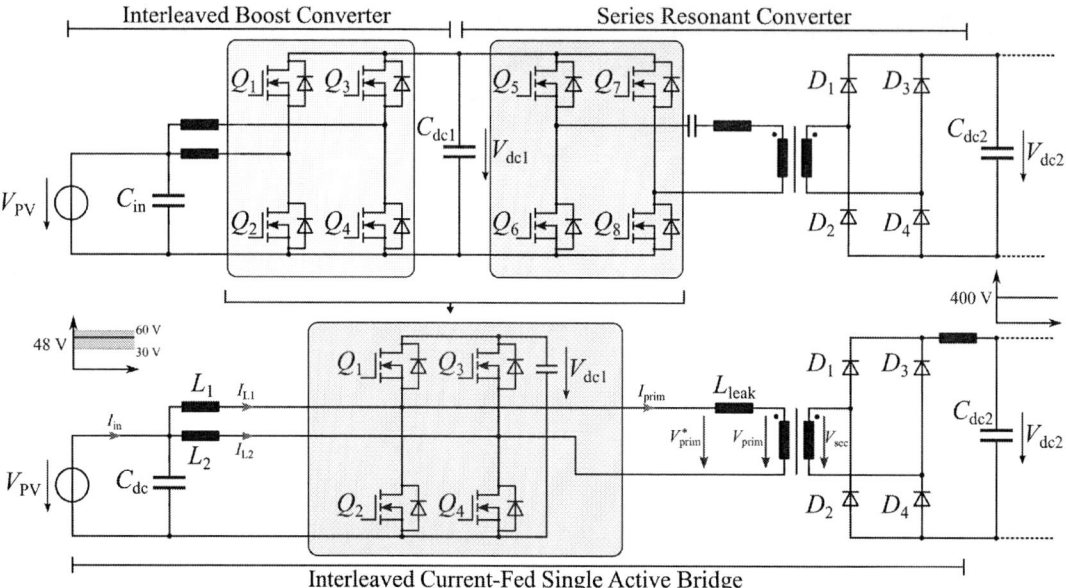

Fig. 1: Circuit description of the conventional two stage topology (top) in comparison to the combined interleaved current-fed single active bridge (bottom).

In Fig. 1 a comparison of the proposed topology to the conventional two stage topology is given. In this study, two operating points with the same output conditions are analyzed to show the degrees of freedom regarding the duty cycle and phase shift. First, a fixed phase shift of 180° is selected to reduce the input current ripple, followed by a phase shift of 120° to achieve a switching state where both low side switches are turned on at the same time which is initially not possible at a phase shift of 180°, since a duty cycle above 50 % would no longer meet the output requirements, even with the lowest input voltage of 30 V. For this research a constant input voltage of 48 V is selected, which is within the typical output voltage range of PV modules. The output voltage is set to 400 V at 1 kW output power and will be achieved by the boost characteristic of the proposed topology as well as the transformer winding ratio of $n = 8$. The resonant components can be eliminated since zero voltage switching (ZVS) can be achieved with the help of the boost inductors. In summary, the number of required MOSFETs is halved, switching and conduction losses are reduced and the more complex resonant component design is not required.

In Fig. 2 and Fig. 3 the main waveforms of the topology are shown. The switches Q_1 and Q_2 are complementary to each other as well as Q_3 and Q_4. It is possible to change the duty cycle D and the phase shift ϕ between the switching signal of the two legs, unlike previous studies [4], [5], in which the phase-shift is made between the active bridges of the primary and secondary side. The proposed converter has four possible switching states, however, which are not reached at every operating point within one period. The equivalent circuit diagram of each switching state is shown in Fig. 4 and only the active paths are highlighted. The four stages of operation are described as follows.

Stage I: Q_1 and Q_4 are active

Starting from the boost converter characteristic, the input voltage U_{PV} is applied to L_2 through the switched-on MOSFET Q_4, which increases the current I_{L2} through the inductor, and the energy will be stored in the magnetic field. Since the capacitor C_{dc1} is discharging, shown in Fig. 4 (a), the current I_{prim} of the primary side of the transformer is increasing. The voltage V_{prim}^* across the transformer is defined by the boost voltage V_{dc1}, which is in parallel during this state.

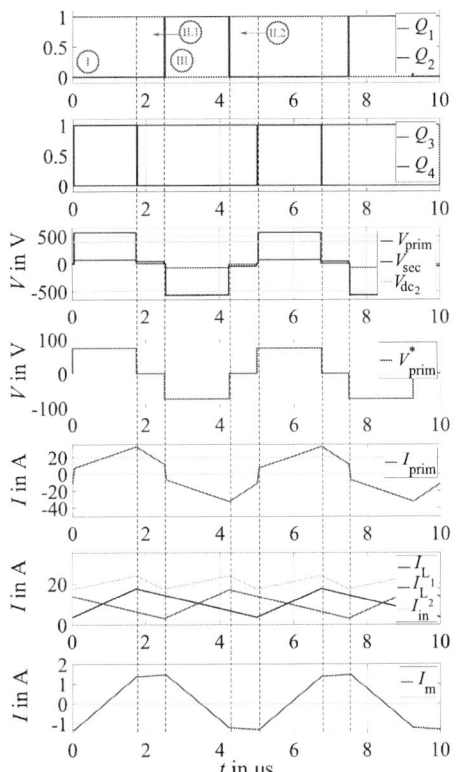 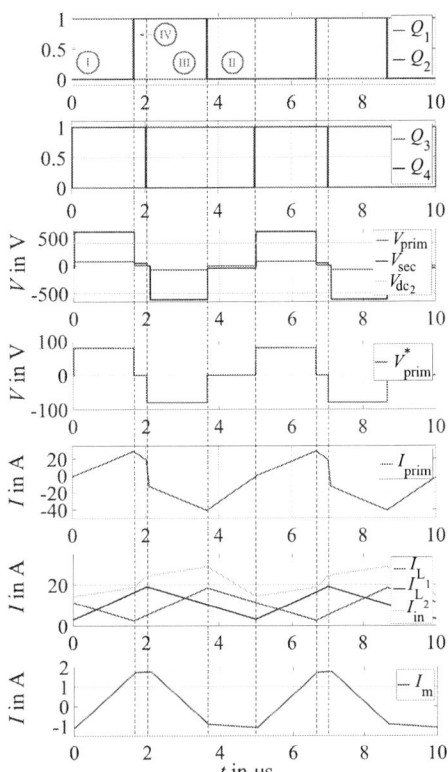

Fig. 2: Main waveforms of the proposed topology taken from the simulation @ $V_{PV} = 48\,V$, $V_{dc2} = 400\,V$, $P_{out} = 1\,kW$, $\phi = 180°$ and $D = 0.35$.

Fig. 3: Main waveforms of the proposed topology taken from the simulation @ $V_{PV} = 48\,V$, $V_{dc2} = 400\,V$, $P_{out} = 1\,kW$, $\phi = 120°$ and $D = 0.40$.

Stage II: Q_1 and Q_3 are active

State II can be divided into states II.1 and II.2 depending on the current direction of I_{prim} which is in the first stage positive and due to the previous state of II.2 negative. Both inductors L_1 and L_2 maintain the current flow due to the previously stored energy and are decreasing over time which has the same effect on the primary side current. Since both winding endings are connected to V_{dc1} the voltage drop V^*_{prim} is zero as a result of which the magnetizing current remains nearly constant.

Stage III: Q_2 and Q_3 are active

State III corresponds to the inverted state II. The voltage at the transformer V^*_{prim} has reversed, causing the magnetic field to be excited in the negative direction by the magnetizing current. C_{dc1} is discharged again after being charged in the previous state. It can be observed that there exists a small voltage drop across the leakage inductance L_{leak}, which leads to a small plateau in the secondary-side voltage V_{sec}. Vice versa, the current and voltage characteristics of L_1 are comparable to the properties of a traditional boost converter.

Stage IV: Q_2 and Q_4 are active

State IV only appears if phase shift and duty cycle allow the two low side gate signals to overlap. However, this is the case with a phase shift from about 120° or less. In comparison to switching state II, the voltage at the transformer is also zero in this case. Because the magnetic field in the transformer is no longer symmetrical, there is an offset in the magnetizing current.

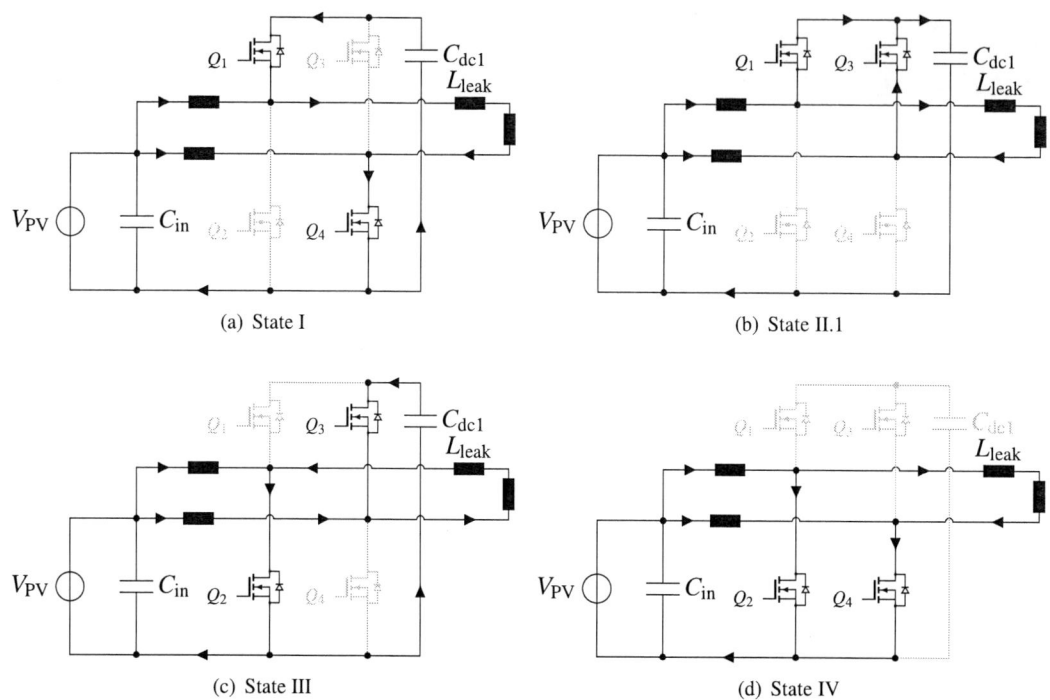

Fig. 4: The various switching states of the converter.

Since it is possible to achieve the same initial conditions with different operating points, symmetrical loading of the components is recommended. The asymmetry observed in Fig. 3 leads to a DC current offset in the amount of 8 A. Likewise, the higher input current ripple leads to an additional load on the input capacitors and the associated shorter service life.

Zero Voltage Switching Analysis

Before analyzing the zero voltage switching possibilities of the presented topology, the soft switching conditions of the conventional topology from Fig. 1 will be analyzed. Here, soft switching for the respective stage must be differentiated.

In the case of the boost converter, ZVS at turn-on without an additional network can only be realized by using a sufficiently low inductance value so that the current ripple reaches its negative value before switching on in order to discharge the output capacitor C_{OSS} of the low-side switches. The high-side switches are soft-switched by topology. Likewise, research on ZVS was carried out with the help of auxiliary circuits [10], [11], [12].

In resonant topologies like the SRC ZVS can be achieved by setting the switching frequency above or slightly below the resonant frequency [13]. In the first case, the transformer current is discharging C_{OSS} due to the lagging behavior of the current. In the second case, only the magnetizing current is present during the turn-on transition. The magnetizing inductance should be small enough to achieve a sufficiently high enough peak value of the magnetizing current to discharge C_{OSS} during the dead-time, resulting in a more complex transformer design even if the leakage inductance will be used as resonant inductance.

For visualizing the possibility of soft-switching, Fig. 5 is representing one leg of the converter as well as the voltage and current waveforms. In order to achieve ZVS, the antiparallel diode of the MOSFET must conduct at the moment the switch will be turned on so that only the forward voltage of the diode in the switching moment is present. That means, if the switch Q_1 is turned on, the current I_{Leg} must be lower than zero, respectively when Q_2 is turned on, the current needs to be higher than zero to realize ZVS. As

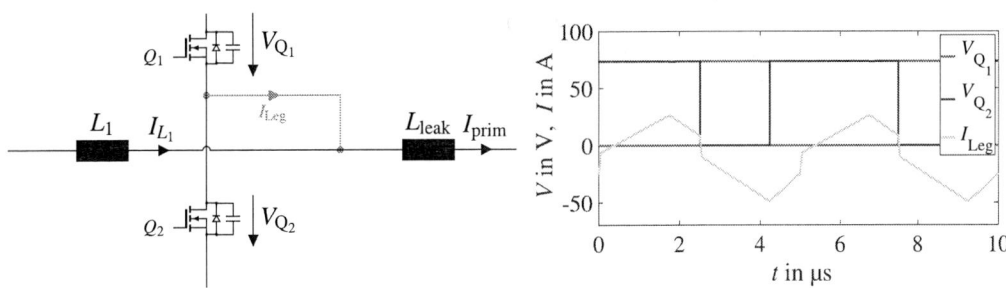

(a) Circuit diagram analysis of one half bridge.

(b) Waveforms analysis of the drain-source voltage drop and leg current with $D = 0.35$ and $\phi = 180°$. ZVS is achieved in all switches.

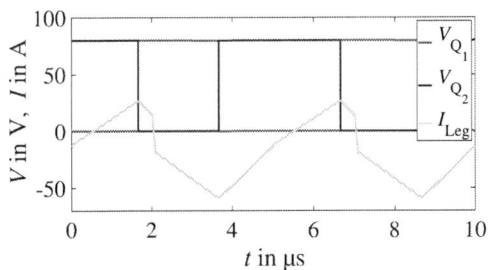

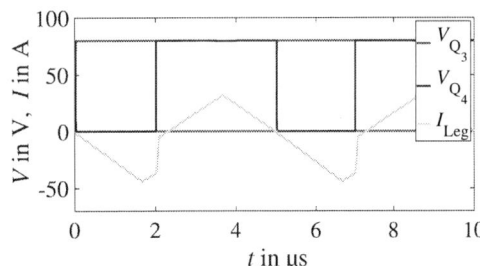

(c) Waveforms analysis of the drain-source voltage drop and leg current with $D = 0.40$ and $\phi = 120°$. ZVS is only achieved in Q_1, Q_2 and Q_3 (c).

(d) Waveforms analysis of the second leg, with its drain-source voltage drop and current, with $D = 0.40$ and $\phi = 120°$. ZVS is not achieved in Q_4.

Fig. 5: Depiction of soft switching analysis. ZVS in Q_2 is achieved when I_{Leg} is positive before the turn-on transition to discharge the output capacitor and same switching behavior for Q_1 vice versa.

long as the conditions in equation 1 and 2 are given for the respective switch, ZVS is achieved. Since switching state III preceeds the switch-on moment of the Q1 at a phase shift of 180°, the primary current I_{prim} in equation 1 has a negative polarity and the inductor current I_{L1} is at its maximum. Therefore, the switching operation of Q_1 is always soft-switching and Q_3 vice versa. Fig. 5 (d) shows an example in which the ZVS condition for Q_4 from equation 2 is not fulfilled, so that due to the phase shift the leg current I_{Leg} in the switching moment is zero.

$$Q_1 : \quad I_{\text{prim}} - I_{L1} < 0 \qquad\qquad Q_2 : \quad I_{\text{prim}} - I_{L1} > 0 \tag{1}$$

$$Q_3 : \quad -I_{\text{prim}} - I_{L2} < 0 \qquad\qquad Q_4 : \quad -I_{\text{prim}} - I_{L2} > 0 \tag{2}$$

Transformer Design

For the design of the transformer, the influence of the magnetizing inductance on the primary current is investigated in Fig. 6 for full load and half load conditions to achieve discontinuous conduction mode (DCM). The plateau which occurs in DCM operation is mainly defined by the magnetizing current and can lead to increased DC losses in the circuit. To achieve ZVS in the SRC, the inductance would be chosen to be $4.5\,\mu\text{H}$ to achieve a large enough magnetizing current to discharge C_{OSS} [13]. Since this criterion does not have to be fulfilled in the proposed topology, a significantly higher magnetising inductance can be chosen, which reduces the maximum amplitude of the primary current.

Based on the waveforms in Fig. 2 and Fig. 3 a transformer is designed by using the simulation software Frenetic, which artificial intelligence based algorithm offers several design recommendations. As a result, a PQ40 core with 3C97 ferrite material was chosen with overall losses of about $6.7\,\text{W}$. All parameters a given in Table I and the winding scheme is shown in Fig. 7. In order to reduce the winding

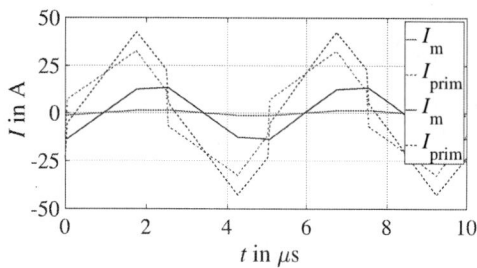
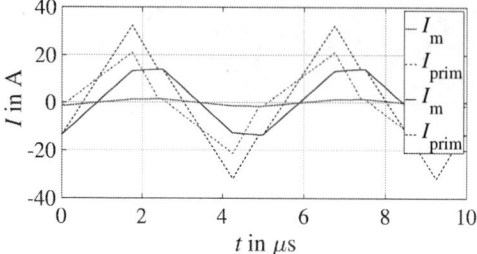

(a) Continuous conduction mode @ 1 kW output power

(b) Discontinuous conduction mode @ 500 W output power and $D_{Q2,4} = 0.35$

Fig. 6: Influence of the magnetizing inductance for a future design of the transformer. Red: $L_m = 45\,\mu H$, blue: $L_m = 4.5\,\mu H$.

Table I: Simulation results of the transformer design.

Core type	PQ 40	
Core material	Ferrite 3C97	
Air gap	0.64 mm	
Copper filling factor	78.1 %	
Number of turns	6 (prim.)	48 (sec.)
Winding type	Litz wire	
Winding geometry	512 x 0.05 mm	100 x 0.1 mm
Parallel windings	3 (prim.)	none (sec.)
Inductance	49 µH	
Core losses	0.55 W	
Winding losses	1.96 W (prim.)	4.17 W (sec.)
DC resistance @ 20 °C	3.3 mΩ (prim.)	129 mΩ (sec.)

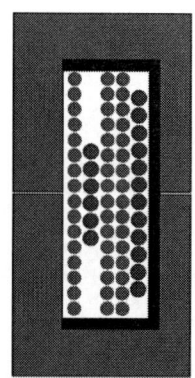

Fig. 7: Winding arrangement in one winding window of the transformer. The primary winding is marked in blue.

losses on the primary side, three primary windings are connected in parallel. Due to the lower current on the secondary side, this is not necessary. A small air gap is necessary to avoid saturation effects.

Component Specification and Hardware Setup

The hardware setup is shown in Fig. 8 and represents only the primary side of the overall prototype, which contain one half-bridge with boost inductor. For a compact design and at the same time easy exchange of the components or to measure different assembly options, each half bridge with its boost inductor is designed separately, so that two of the boards shown are necessary. The gate driver allows stand-alone operation but also enables control by a microcontroller. A sensing resistor prevents an unwanted high current rise within a half bridge as well as providing a symmetrical current distribution in each half bridge. A solder pad is provided for connecting one primary side winding end of the transformer. The secondary side of the transformer is connected to a common full-bridge diode rectifier, which is not shown here. Regarding the voltage ratings of the power semiconductors, 100 V Si MOSFETs from Infineon's OptiMOS series are selected and 650 V SiC Schottky diodes for the rectifier stage.

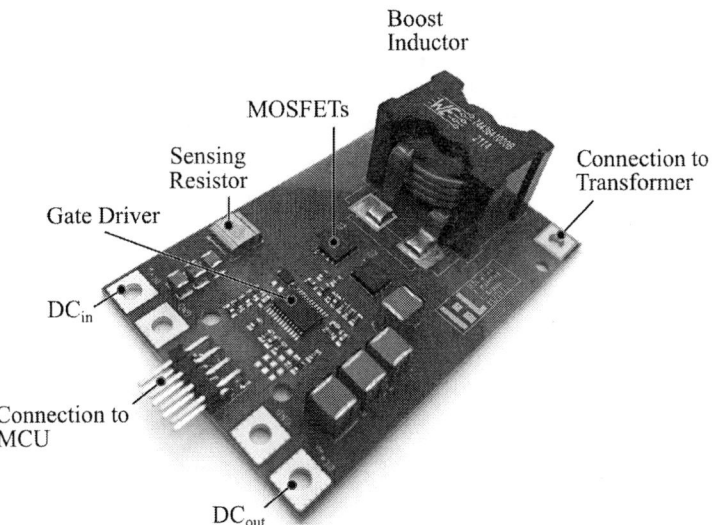

Fig. 8: Photograph of the hardware prototype. The two solder pads highlighted in blue are identical to the blue path of the current I_{Leg} marked in Figure 5 (a). Two of these PCBs form the overall system of the primary side.

Conclusion

This paper presents the investigation of a current-fed single active bridge as a combined topology of boost converter and series resonant converter at 1 kW output power and a boost operation from 48 V up to 400 V. The focus is on the analysis of different switching states and the possibilities for controlling the presented topology on the basis of two different operating points. In addition, a condition for soft-switching possibilities was developed and contrasted with the two stage topology.

Although no harmonics could be detected in the simulation model, oscillations occurred in the hardware setup, which spread to the transformer voltage, drain-source voltages as well as gate signals, it was not possible to show measurements of the operating points shown in this paper until the end of this work.

References

[1] D. Sha, X. Wang and D. Chen, "High-Efficiency Current-Fed Dual Active Bridge DC–DC Converter With ZVS Achievement Throughout Full Range of Load Using Optimized Switching Patterns," in IEEE Transactions on Power Electronics, vol. 33, no. 2, pp. 1347-1357, Feb. 2018.

[2] Y. Shi, R. Li, Y. Xue and H. Li, "Optimized Operation of Current-Fed Dual Active Bridge DC–DC Converter for PV Applications," in IEEE Transactions on Industrial Electronics, vol. 62, no. 11, pp. 6986-6995, Nov. 2015.

[3] D. Sha, X. Wang, K. Liu and C. Chen, "A Current-Fed Dual-Active-Bridge DC–DC Converter Using Extended Duty Cycle Control and Magnetic-Integrated Inductors With Optimized Voltage Mismatching Control," in IEEE Transactions on Power Electronics, vol. 34, no. 1, pp. 462-473, Jan. 2019.

[4] Y. Shi, R. Li, Y. Xue and H. Li, "High-Frequency-Link-Based Grid-Tied PV System With Small DC-Link Capacitor and Low-Frequency Ripple-Free Maximum Power Point Tracking," in IEEE Transactions on Power Electronics, vol. 31, no. 1, pp. 328-339, Jan. 2016.

[5] P. Vu, D. T. Anh, and H. D. Chinh, "A Novel Modeling and Control Design of the Current-Fed Dual Active Bridge Converter under DPDPS Modulation", Eng. Technol. Appl. Sci. Res., vol. 11, no. 2, pp. 7054–7059, Apr. 2021.

[6] A. Pal and S. Kapat, "Accurate Discrete-Time Modeling of an Interleaved Current-Fed Dual Active Bridge DC-DC Converter," 2019 IEEE Applied Power Electronics Conference and Exposition (APEC), pp. 1616-1621, 2019.

[7] H. Xiao and S. Xie, "A ZVS Bidirectional DC–DC Converter With Phase-Shift Plus PWM Control Scheme," in IEEE Transactions on Power Electronics, vol. 23, no. 2, pp. 813-823, March 2008.

[8] X. Liu, H. Li and Z. Wang, "A Fuel Cell Power Conditioning System With Low-Frequency Ripple-Free Input Current Using a Control-Oriented Power Pulsation Decoupling Strategy," in IEEE Transactions on Power Electronics, vol. 29, no. 1, pp. 159-169, Jan. 2014.

[9] Z. Wang and H. Li, "An Integrated Three-Port Bidirectional DC–DC Converter for PV Application on a DC Distribution System," in IEEE Transactions on Power Electronics, vol. 28, no. 10, pp. 4612-4624, Oct. 2013.

[10] G. Yao, A. Chen and X. He, "Soft Switching Circuit for Interleaved Boost Converters," in IEEE Transactions on Power Electronics, vol. 22, no. 1, pp. 80-86, Jan. 2007.

[11] Je-Hyun Yi, Paul Jang, Sang-Woo Kang and Bo-Hyung Cho, "Soft-switching synchronous interleaved boost converter with an auxiliary coupled inductor," 2016 IEEE 8th International Power Electronics and Motion Control Conference (IPEMC-ECCE Asia), pp. 3330-3335, 2016.

[12] M. Abbasi, N. Mortazavi and A. Rahmati, "A novel ZVS interleaved boost converter," The 5th Annual International Power Electronics, Drive Systems and Technologies Conference (PEDSTC 2014), pp. 535-538, 2014.

[13] T. Manthey, T. Brinker and J. Friebe, "Design of an Isolated DC-DC Converter for PV Micro-Inverters with Planar Transformer and PCB Integrated Winding," 2021 IEEE Energy Conversion Congress and Exposition (ECCE, pp. 554-560), 2021.

Real-Time Thermal Characterization of Power Semiconductors using a PSO-based Digital Twin Approach

Johannes Kuprat[1], Yoann Pascal[2], Marco Liserre[1]

[1] Chair of Power Electronics, Kiel University, Kaiserstrasse 2, 24143 Kiel, Germany

[2] Fraunhofer Institute for Silicon Technology ISIT, Fraunhoferstr. 1, 25524 Itzehoe, Germany

Phone: +49 (0) 431-880 6105

Fax: +49 (0) 431-880 6103

Email: jk@tf.uni-kiel.de, yoann.pascal@isit.fraunhofer.de, ml@tf.uni-kiel.de

URL: https://www.pe.tf.uni-kiel.de

Keywords

≪Modelling≫, ≪Thermal Model≫, ≪Optimization≫, ≪Diagnostics≫.

Abstract

Thermal impedance is essential for assessing the state-of-health of power semiconductors and to use thermal observers. This work proposes a Particle-Swarm-Optimization-based Digital Twin approach to extract the thermal impedance for online monitoring. A proof of concept of the approach is achieved in a real-time simulation with a digital reference model by showing the convergence to the given parameter set. Further, the convergence of the algorithm to a fixed parameter set is validated in the laboratory.

Introduction

The field of power electronics is of utmost importance due to its necessity for the integration of renewable energy sources and electric vehicle charging stations into the electrical grid [1]. A relevant topic in the field of power electronics is the reliability of the devices [2], especially the one of semiconductors and capacitors [3]. The thermal characteristics of the devices are a keystone for assessing their state-of-health via condition monitoring [4] and to use thermal observers [5] as well as active thermal control [6].

The concept of Digital Twins (DTs) is of growing relevance for academia and industry [7]. There are various definitions of DTs [7], in this work a DT refers to a real-time digital replicate of a physical system, which takes into account measurements and historical data to optimize the accuracy of the digital model. Recently, different approaches for the implementation of DT replicating the electrical behavior of power electronic converters have been proposed, which are based on Particle-Swarm-Optimization (PSO) [8], neural networks [9], polynomial chaos expansion [10], and bayesian optimization [11]. Prior, real-time thermal simulations have been proposed for reliability evaluation [12–16].

Furthermore, [17] proposed using DT approaches for the description of the converter reliability and [18] implemented a real-time thermal DT based on an extended Kalman filter. However, the demonstrated thermal DT was implemented with the assumption of temperature measurements at all nodes of the equivalent circuit, so the problem is decoupleable and the algorithm optimizes only one RC element of a Cauer network for each pair of temperature measurements. However, this does not enable to access also not measurable temperatures within power semiconductor modules. This work proposes a PSO-based DT approach for thermal real-time identification, which is capable of optimizing up to three RC elements of a Cauer network with only one pair of temperature measurements (Fig. 1).

This work is structured as follows: the proposed PSO-based thermal DT approach is introduced in the next section. Afterwards, the results of the Real-Time Simulation (RTS) with a digital reference model

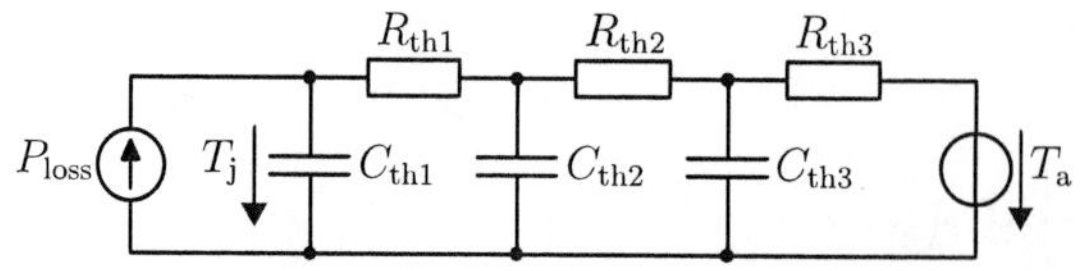

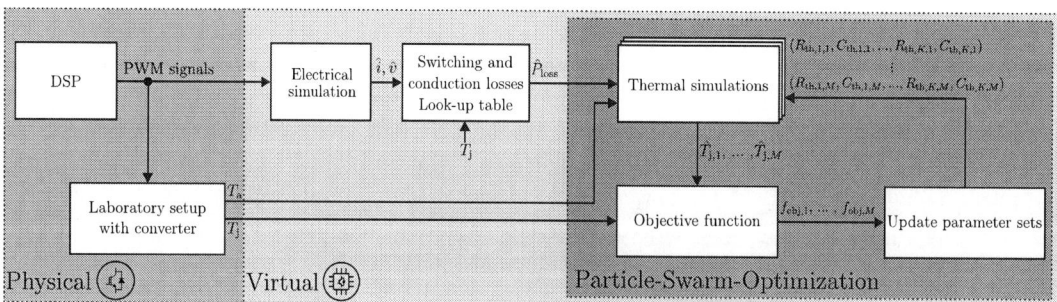

Fig. 1: Thermal modeling between junction and case temperature by a Cauer network with three RC elements.

are presented. Then the performance of the DT on the laboratory setup is shown. Thereafter, critical topics for the implementation of the PSO-based thermal DT in field applications are discussed. At the end the conclusion of the work is given.

PSO-based Thermal Digital Twin Approach

The PSO-based thermal DT is a RTS of a physical converter which is executed in parallel to its real operation. The PSO optimizes the thermal model based on measurements of junction and ambient temperatures (T_j and T_a) on the physical converter. As shown in Fig. 2, the PWM signals given by a Digital Signal Processor (DSP) are provided to the physical converter, which is in this case a buck converter, and to the electrical RTS of this buck converter at the same time. The voltage and current of this electrical RTS are then used to calculate the switching and conduction losses based on look-up tables, which also depend on the measured T_j. The estimated losses and the measured T_a are the inputs for the thermal RTSs.

There are M thermal simulations, which are executed in parallel. Each of these thermal simulations represents a particle of the PSO, which is characterized by a set of parameters

$$p_m = \{R_{th1,m}, C_{th1,m}, R_{th2,m}, C_{th2,m}, R_{th3,m}, C_{th3,m}\} \ \forall m \in [1,M] \tag{1}$$

describing a Cauer network with three RC elements between junction and ambient (Fig. 1). There are different execution rates in the virtual space, the one of the RTS $T_{exe,RTS}$ and that of the PSO $T_{exe,PSO}$. Depending on the considered converter it can be beneficial to have a higher execution rate for the electrical RTS $T_{exe,RTS,el}$ and a lower execution rate for the thermal RTS $T_{exe,RTS,th}$ to achieve higher accuracy with the limited computing capabilities. The performance of each particle is evaluated in each thermal RTS step based on the measured T_j and the junction temperature estimated by the particle $\hat{T}_{j,m}$. During one $T_{exe,PSO}$ period there are N executions at $T_{exe,RTS,th}$ and the PSO takes into account the average of the evaluations with the objective function:

$$f_{obj,m} = \frac{\sum_{n=1}^{N} \sqrt{(\hat{T}_{j,m,n} - T_{j,n})^2 + \frac{c_d}{s^{-2}} \left(\frac{\Delta \hat{T}_{j,m,n}}{T_{exe,RTS,th}} - \frac{\Delta T_{j,n}}{T_{exe,RTS,th}} \right)^2}}{N} \ \forall m \in [1,M] \ . \tag{2}$$

Herein c_d is the coefficient of the derivative part and for the implementation in the laboratory a filter needs to be applied to find the derivatives without distortion by the measurement noise.

Fig. 2: Scheme of a PSO-based thermal DT.

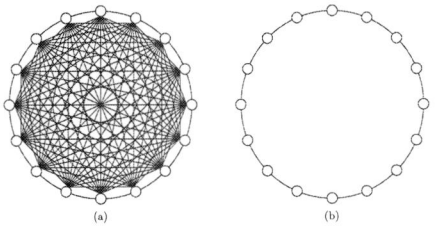

Fig. 3: Global topology (a) and local topology (b) for the PSO.

Table I: Coefficients and settings of the RTS and PSO in the real-time implementation.

Coefficient/Setting	Value
$T_{\text{exe,RTS,el}}, T_{\text{exe,RTS,th}}$	$10\,\mu\text{s}, 800\,\mu\text{s}$
$T_{\text{exe,PSO}}$	$2\,\text{s}$
ω, c_1, c_2, c_3	$0.7, 0.5, 0.5, 0.01$
c_{d}, M	$0.1, 16$

Based on $f_{\text{obj},m}$ in this PSO execution and the best $f_{\text{obj},m}$ of previous PSO executions for each particle it is decided whether the particles best parameter set $p_{\text{best},m}$ should be updated or not. Further, there are different information topologies for the PSO (Fig. 3). In the global topology (Fig. 3 (a)) all particles are connected and converge towards the globally best parameter set g_{best} whereas in the local topology (Fig. 3 (b)) each particle is only connected with its neighbors and converges towards the locally best parameter set $l_{\text{best},m}$. The global topology provides faster convergence, but is prone to premature convergence, therefore the local topology is used in this work. The velocities of all dimensions of each particle v_m in step k of the PSO are calculated according to:

$$
\begin{aligned}
v_m(k) =& \omega v_m(k-1) + c_1 r_1 \left(p_{\text{best},m}(k-1) - p_m(k-1)\right) \\
& + c_2 r_2 \left(l_{\text{best},m}(k-1) - p_m(k-1)\right) + c_3 r_3 p_m(k-1) \; \forall m \in [1,M] \; .
\end{aligned}
\tag{3}
$$

Here ω is the inertia weight, r_1 and r_2 are randomly generated numbers between 0 and 1, r_3 is a randomly generated number between -1 and 1, and c_1, c_2, c_3 are coefficients of the PSO. Where, the first term multiplied with the coefficient c_1 describes the orientation on the best parameter set $p_{\text{best},m}$ the specific particle ever had, judged on the achieved objective function value. The second term multiplied with the coefficient c_2 orients on the parameter set $l_{\text{best},m}$ of the locally best neighbor particle. The last part multiplied with the coefficient c_3 inserts an own vibration of the particles, which prevents the swarm from getting stuck in local minima of the objective function. The parameters of each particle are updated by adding the velocity to the parameters of the last period:

$$
p_m(k) = p_m(k-1) + v_m(k) \; \forall m \in [1,M] \; .
\tag{4}
$$

Real-Time Implementation

The capability of the thermal DT to find the proper Cauer network parameters is demonstrated based on a real-time simulation with a digital thermal reference model. This proof of concept is investigated with a simplified Cauer network, which describes the thermal behavior between a single junction, which

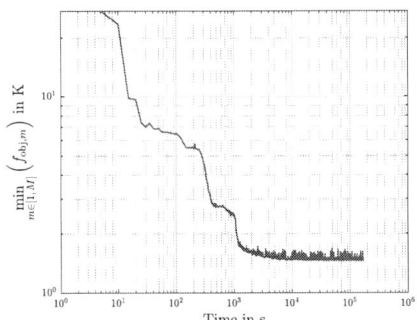

Fig. 4: Evolution of the objective function in the real-time implementation.

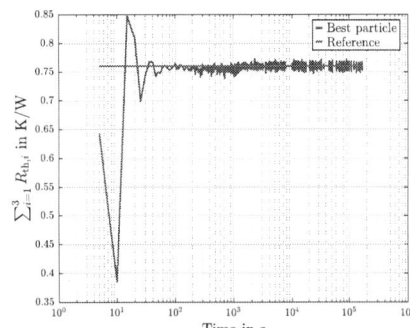

Fig. 5: Convergence of the sum of thermal resistances of the best particle in the real-time implementation.

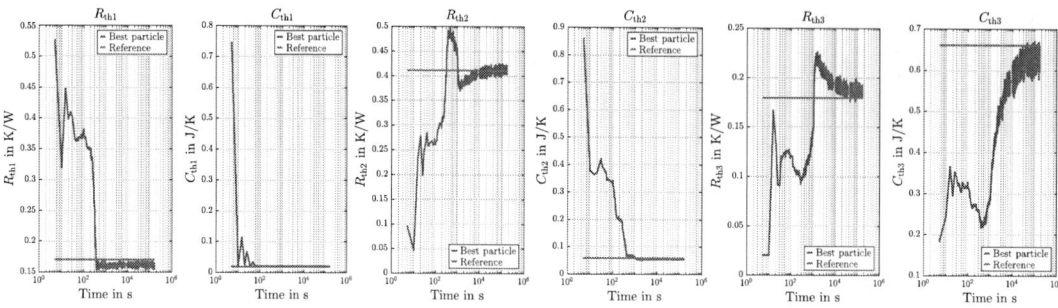

Fig. 6: Convergence of the Cauer network parameters of the best particle in the real-time implementation.

represents the IGBT and the diode of the buck converter, and the ambient temperature with three RC elements. A discussion on considering multiple heat sources is given in the second last section. The real-time simulation was executed on a Typhoon HIL402 system and the chosen coefficients and settings of the RTS and PSO are listed in table I. The converter is operated with a square-wave input for the duty cycle whose period is equal to $T_{\text{exe,PSO}}$, so that one full thermal cycle is considered in each PSO step.

The evolution of the objective function (2) can be seen in Fig. 4. Herein, always the value of the best particle in each PSO step is used, this also applies for the parameters shown in the following. Among the parameters the fastest convergence can be achieved for the sum of thermal resistances (Fig. 5), because it has great influence on the average junction temperature, which affects the first part of the objective function (2) strongly. As it can be seen on the single parameters (Fig. 6), the closer the RC element is to the junction, the faster its parameters are converging, because changing them has a greater impact on the junction temperature. The parameters of the last RC element have only slight impact on the junction temperature and can only converge to the values of the thermal reference model after the other RC elements have already reached the values of thermal reference parameters.

Laboratory Results

The laboratory setup is shown in Fig. 7 and the connection scheme of the equipment in Fig. 9. Herein, the digital signal processor (DSP) (TMS320F28379D from Texas Instruments) provides the gate signals to the physical converter (with an open FP25R12KE3 power module from Infineon) and to the HIL system

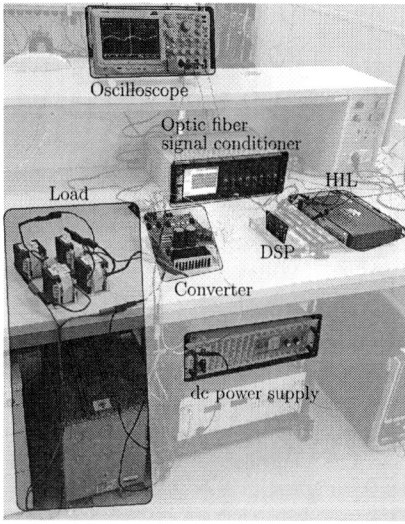

Fig. 7: Laboratory setup for thermal DT.

Fig. 8: Optic fiber temperature sensor placement in the open IGBT module.

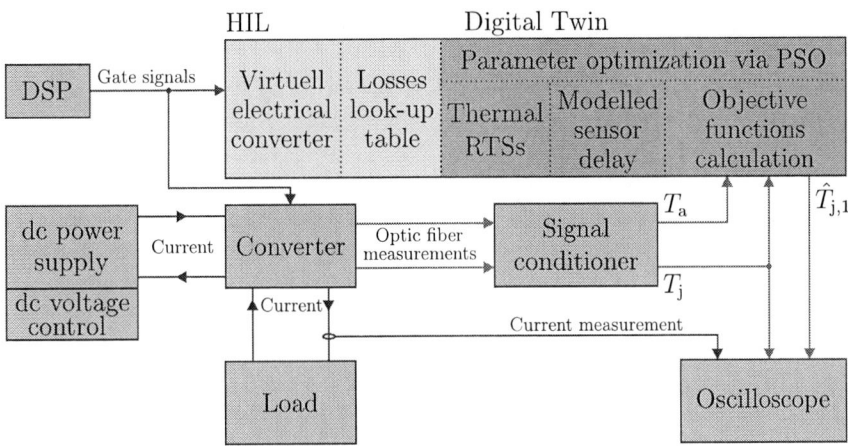

Fig. 9: Connection scheme of equipment in the laboratory setup.

(Typhoon HIL402), which executes the PSO-based thermal digital twin, at the same time. The converters dc link voltage of 200 V is controlled by the dc power supply (EA-PSI 9750-20 from Elektro-Automatik) and the converter feeds a load composed of a resistor (18 Ω) and an inductor (1.8 mH). The temperature of the IGBT in the open power module (Fig. 8) as well as the ambient temperature are measured with optic fiber temperature sensors (OTG-A from Opsens Solutions), which are connected to the optic fiber signal conditioner (PSC-D-N-N-N from Opsens Solutions). The signal conditioner scales the measured temperatures to voltages between $-5\,\text{V}$ and $5\,\text{V}$, which are given to the HIL system and scaled back to the temperatures there. The $-5\,\text{V}$ to $5\,\text{V}$ scaled measured junction temperature as well as the estimated

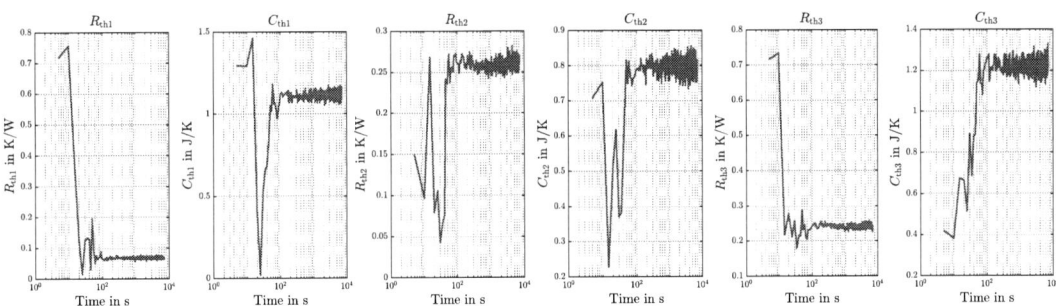

Fig. 10: Convergence of the Cauer network parameters of the best particle in the laboratory test.

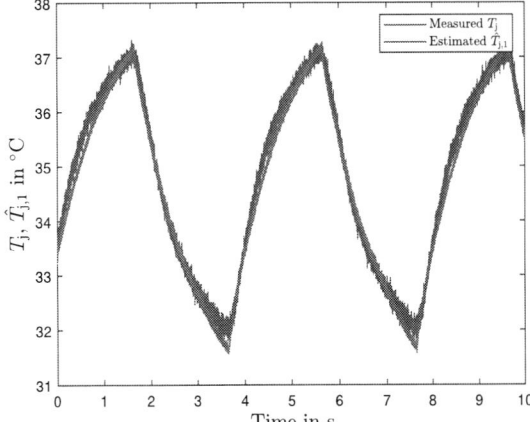

Fig. 11: Measured junction temperature (blue) and estimated junction temperature of particle one (red).

junction temperature of particle one are captured with an oscilloscope (DPO 3014 from Tektronix) and scaled back in Matlab afterwards. For the laboratory test the coefficients and settings of table I are used such as for the real-time implementation, except for a change in $T_{\text{exe,PSO}}$ and accordingly the period for the square-wave input for the duty cycle to $4\,\text{s}$. The convergence of the parameters to a fixed parameter set in the laboratory test can be seen in Fig. 10. Herein, giving reference parameters is not possible because the real thermal behavior can never be fully captured by the model, however the PSO based DT adapts the parameters of the Cauer networks of all the particles to find the optimal representation of the real behavior by the model according to the objective function. The measured junction temperature and the estimated junction temperature of particle one after convergence to the fixed parameter set can be seen in Fig. 11.

Discussion of Field Application

A critical point for the usage of thermal digital twins in field applications is the need for modeling temperatures of multiple dies. As long as the heat dissipation of the different dies is coupled negligibly, the approach can be applied to each die independently. However, in semiconductor modules the thermal cross-coupling effects between the different dies are significant [19]. Therefore, the whole thermal network of all the dies, which considers the thermal cross-coupling between them, needs to be estimated by each of the particles of the PSO. An often used thermal model to consider cross-coupling effects is the linear accumulation of self heating and all cross-coupled heating by other dies [20–23]

$$
\begin{bmatrix} \hat{T}_{\text{j,I1}}(t) \\ \hat{T}_{\text{j,D1}}(t) \\ \hat{T}_{\text{j,I2}}(t) \\ \hat{T}_{\text{j,D2}}(t) \end{bmatrix} = \begin{bmatrix} \dot{Z}^{\text{I1}}_{\text{th,self}}(t) & \dot{Z}^{\text{I1}\leftarrow\text{D1}}_{\text{th,cross}}(t) & \dot{Z}^{\text{I1}\leftarrow\text{I2}}_{\text{th,cross}}(t) & \dot{Z}^{\text{I1}\leftarrow\text{D2}}_{\text{th,cross}}(t) \\ \dot{Z}^{\text{D1}\leftarrow\text{I1}}_{\text{th,cross}}(t) & \dot{Z}^{\text{D1}}_{\text{th,self}}(t) & \dot{Z}^{\text{D1}\leftarrow\text{I2}}_{\text{th,cross}}(t) & \dot{Z}^{\text{D1}\leftarrow\text{D2}}_{\text{th,cross}}(t) \\ \dot{Z}^{\text{I2}\leftarrow\text{I1}}_{\text{th,cross}}(t) & \dot{Z}^{\text{I2}\leftarrow\text{D1}}_{\text{th,cross}}(t) & \dot{Z}^{\text{I2}}_{\text{th,self}}(t) & \dot{Z}^{\text{I2}\leftarrow\text{D2}}_{\text{th,cross}}(t) \\ \dot{Z}^{\text{D2}\leftarrow\text{I1}}_{\text{th,cross}}(t) & \dot{Z}^{\text{D2}\leftarrow\text{D1}}_{\text{th,cross}}(t) & \dot{Z}^{\text{D2}\leftarrow\text{I2}}_{\text{th,cross}}(t) & \dot{Z}^{\text{D2}}_{\text{th,self}}(t) \end{bmatrix} * \begin{bmatrix} P_{\text{loss,I1}}(t) \\ P_{\text{loss,D1}}(t) \\ P_{\text{loss,I2}}(t) \\ P_{\text{loss,D2}}(t) \end{bmatrix} + \begin{bmatrix} T_{\text{a}} \\ T_{\text{a}} \\ T_{\text{a}} \\ T_{\text{a}} \end{bmatrix} . \quad (5)
$$

Each of these thermal impedances is typically described by a third- or fourth-order thermal network, which leads to a high number of parameters [24]. Having a high dimensional search area can impede the convergence of the PSO-based thermal DT algorithm. Thus, the approach would benefit from thermal models, which include cross coupling effects and provide a low number of parameters.

Conclusion

The thermal behavior of power semiconductors is highly important for assessing their state-of-health as well as for the implementation of measures to improve their reliability, such as active thermal control. This work proposed a Particle-Swarm-Optimization-based Digital Twin approach to extract the thermal impedance of power semiconductors for real-time applications. By using a digital reference model in a real-time simulation, it has been shown that the PSO-based thermal Digital Twin can find the parameters of a Cauer network with three RC elements. This proves that the approach is able to converge to thermal networks with possible physical meaning. The convergence of the algorithm to a fixed parameter set is validated in the laboratory and the estimation of the junction temperature after convergence is shown.

References

[1] F. Blaabjerg, R. Teodorescu, M. Liserre, and A. Timbus, "Overview of control and grid synchronization for distributed power generation systems," *IEEE Transactions on Industrial Electronics*, vol. 53, no. 5, pp. 1398–1409, 2006.

[2] H. Wang, M. Liserre, F. Blaabjerg, P. de Place Rimmen, J. B. Jacobsen, T. Kvisgaard, and J. Landkildehus, "Transitioning to physics-of-failure as a reliability driver in power electronics," *IEEE Journal of Emerging and Selected Topics in Power Electronics*, vol. 2, no. 1, pp. 97–114, 2014.

[3] J. Falck, C. Felgemacher, A. Rojko, M. Liserre, and P. Zacharias, "Reliability of power electronic systems: An industry perspective," *IEEE Industrial Electronics Magazine*, vol. 12, no. 2, pp. 24–35, 2018.

[4] M. Andresen, J. Kuprat, V. Raveendran, J. Falck, and M. Liserre, "Active thermal control for delaying maintenance of power electronics converters," *Chinese Journal of Electrical Engineering*, vol. 4, no. 3, pp. 13–20, 2018.

[5] S. Kalker, L. A. Ruppert, C. H. Van der Broeck, J. Kuprat, M. Andresen, T. A. Polom, M. Liserre, and R. W. De Doncker, "Reviewing thermal monitoring techniques for smart power modules," *IEEE Journal of Emerging and Selected Topics in Power Electronics*, pp. 1–1, 2021.

[6] J. Kuprat, C. H. van der Broeck, M. Andresen, S. Kalker, M. Liserre, and R. W. De Doncker, "Research on active thermal control: Actual status and future trends," *IEEE Journal of Emerging and Selected Topics in Power Electronics*, vol. 9, no. 6, pp. 6494–6506, 2021.

[7] F. Tao, H. Zhang, A. Liu, and A. Y. C. Nee, "Digital twin in industry: State-of-the-art," *IEEE Transactions on Industrial Informatics*, vol. 15, no. 4, pp. 2405–2415, 2019.

[8] Y. Peng, S. Zhao, and H. Wang, "A digital twin based estimation method for health indicators of dc-dc converters," *IEEE Transactions on Power Electronics*, vol. 36, no. 2, pp. 2105–2118, 2021.

[9] A. Wunderlich and E. Santi, "Digital twin models of power electronic converters using dynamic neural networks," in *2021 IEEE Applied Power Electronics Conference and Exposition (APEC)*, 2021, pp. 2369–2376.

[10] M. Milton, C. De La O, H. L. Ginn, and A. Benigni, "Controller-embeddable probabilistic real-time digital twins for power electronic converter diagnostics," *IEEE Transactions on Power Electronics*, vol. 35, no. 9, pp. 9850–9864, 2020.

[11] S. Chen, S. Wang, P. Wen, and S. Zhao, "Digital twin for degradation parameters identification of dc-dc converters based on bayesian optimization," in *2021 IEEE International Conference on Prognostics and Health Management (ICPHM)*, 2021, pp. 1–9.

[12] M. Musallam, C. Buttay, M. Whitehead, and M. Johnson, "Real-time compact electronic thermal modelling for health monitoring," in *2007 European Conference on Power Electronics and Applications*, 2007, pp. 1–10.

[13] M. Musallam and C. M. Johnson, "Real-time compact thermal models for health management of power electronics," *IEEE Transactions on Power Electronics*, vol. 25, no. 6, pp. 1416–1425, 2010.

[14] H. Chen, B. Ji, V. Pickert, and W. Cao, "Real-time temperature estimation for power mosfets considering thermal aging effects," *IEEE Transactions on Device and Materials Reliability*, vol. 14, no. 1, pp. 220–228, 2014.

[15] T. K. Gachovska, B. Tian, J. L. Hudgins, W. Qiao, and J. F. Donlon, "A real-time thermal model for monitoring of power semiconductor devices," *IEEE Transactions on Industry Applications*, vol. 51, no. 4, pp. 3361–3367, 2015.

[16] W. Wang, H. R. Wickramasinghe, K. Ma, and G. Konstantinou, "Real-time co-simulation for electrical and thermal analysis of power electronics," in *2019 9th International Conference on Power and Energy Systems (ICPES)*, 2019, pp. 1–5.

[17] L. Felsberger, B. Todd, and D. Kranzlmueller, "Power converter maintenance optimization using a model-based digital reliability twin paradigm," in *2019 4th International Conference on System Reliability and Safety (ICSRS)*, 2019, pp. 213–217.

[18] B. Rodriguez, E. Sanjurjo, M. Tranchero, C. Romano, and F. Gonzalez, "Thermal parameter and state estimation for digital twins of e-powertrain components," *IEEE Access*, vol. 9, pp. 97 384–97 400, 2021.

[19] A. Stippich, M. Neubert, A. Sewergin, and R. W. De Doncker, "Significance of thermal cross-coupling effects in power semiconductor modules," in *2016 IEEE 2nd Annual Southern Power Electronics Conference (SPEC)*, 2016, pp. 1–6.

[20] B. Zahn, "Steady state thermal characterization of multiple output devices using linear superposition theory and a non-linear matrix multiplier," in *Fourteenth Annual IEEE Semiconductor Thermal Measurement and Management Symposium (Cat. No.98CH36195)*, 1998, pp. 39–46.

[21] U. Drofenik, D. Cottet, A. Müsing, J.-M. Meyer, and J. W. Kolar, "Modelling the thermal coupling between internal power semiconductor dies of a water-cooled 3300v/1200a hipak igbt module," in *Proceedings of Power Conversion and Intelligent Motion Conference*, 2007, pp. 1–8.

[22] A. S. Bahman, K. Ma, and F. Blaabjerg, "Thermal impedance model of high power igbt modules considering heat coupling effects," in *2014 International Power Electronics and Application Conference and Exposition*, 2014, pp. 1382–1387.

[23] M. Shahjalal, M. R. Ahmed, H. Lu, C. Bailey, and A. J. Forsyth, "An analysis of the thermal interaction between components in power converter applications," *IEEE Transactions on Power Electronics*, vol. 35, no. 9, pp. 9082–9094, 2020.

[24] Y. Zhang, Z. Wang, H. Wang, and F. Blaabjerg, "Artificial intelligence-aided thermal model considering cross-coupling effects," *IEEE Transactions on Power Electronics*, vol. 35, no. 10, pp. 9998–10 002, 2020.

Self-Sensing Design and Control for an Induction Machine with an Additional Short-Circuited Rotor Coil

Stefan Luecke* and Axel Mertens
Leibniz University Hannover
Institute for Drive Systems and Power Electronics
Welfengarten 1
Hanover, Germany
Phone: +49 (0) 511-762 2514
Fax: +49 (0) 511-762 3040
Email: stefan.luecke@stud.uni-hannover.de
URL: http://www.ial.uni-hannover.de

Acknowledgments

This work was funded by the Deutsche Forschungsgemeinschaft (DFG, German Research Foundation) – project identification number 424944120.

Keywords

≪Induction motor≫, ≪Design≫, ≪Impedance measurement≫, ≪Sensorless control≫

Abstract

In order to control an electrical machine at lower speeds, several control methods require a saliency. This paper presents an induction machine design to achieve a saliency for high frequency injection signals. The saliency is characterized for a prototype and a control method for the whole operation range is shown.

Introduction

Self-sensing control methods for induction machines (IMs) often use electromotive force (EMF)-based techniques in the medium and upper speed range and saliency-based methods applying a high frequency (HF) injection at low or zero speed. For the control, a certain level of saliency is required. However, IMs have a low saliency inherently. A machine design with good self-sensing ability is desired where other properties, like torque ripple, rated torque and efficiency remain unchanged. A demodulation technique applying a rotational HF injection is proposed in [1]. There, an IM design is used, which modulates the rotor slot leakage by changing the heights or widths of the slot openings. For this approach, a simulative design study concerning self-sensing and power conversion properties is done in [2]. Experimental results for prototypes with spatial modulation in the rotor slot opening height and width are shown in [3]. However, there is a trade off between self-sensing ability and power conversion, because the demands on the rotor slot opening design are opposed. An IM design using additional sensing windings on the stator is proposed in [4]. However, this approach requires additional terminals and an external excitation source. Another option for the enhancement of the saliency is the integration of an additional short-circuited coil in one axis of the rotor, which has been investigated for synchronous machine so far, e.g. in [5], [6], [7] and [8]. Thereby, the inductance is damped in this axis for high frequencies and thus,

---

*was an employee at the Institute for Drive Systems and Power Electronics from 2014 to 2019 and currently works at IAV GmbH.

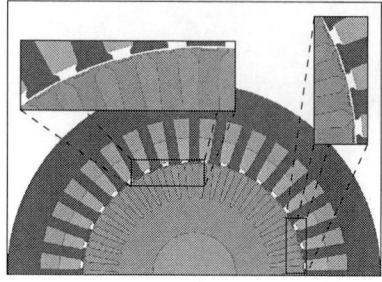

(a) Rotor with additional coil   (b) Schematic sketch

Fig. 1: Prototype with additional short-circuited rotor coil of a 4-pole, 1,5 kW IM

an inductive saliency occurs. In addition, the resistive saliency is amplified with this approach and the saliency enhancement has been experimentally validated, for instance in [8].

This paper evaluates the inductive saliency and the resistive saliency for an IM prototype with an additional short-circuited coil. Furthermore, a self-sensing control based on a flux observer structure and using a combined EMF-based and saliency-based evaluation is suggested.

Induction machine with an additional short-circuited rotor coil

The generation of a rotor-fixed saliency is based on the fact that a short-circuited winding damps a time-varying magnetic field, because the current, which is induced in the winding, counteracts the cause according to Lenz's law. If a winding is only placed in one axis of the rotor, the magnetic field is only damped in one axis. This leads to unequal effective inductances in the axes. The damping effect of the additional winding is frequency-dependent determined by its time constant, which is given by its inductance and resistance. Thus, the added rotor coil can be designed to be primarily effective in the high frequency range. The principle is described for this special rotor design in [9] in more detail. A prototype of a four-pole IM was built with four and five additional conductors per pole pitch. The additional slots are placed in the middle of the rotor teeth and the conductors are connected via two short-circuit rings at the rotor ends. A picture of the rotor and the cross section of the prototype's design are shown in Fig. 1a and Fig. 1b, respectively.

Impedance measurement

The saliency of the prototype is characterized using different methods. First, the impedance is measured with the impedance analyzer 6500B from Wayne Kerr and the network analyzer Bode100 from OMICRON LAB. During the measurement, the IM is at standstill and the measurement is repeated for several rotor positions. In Fig. 2 the results for a single frequency (here 2.5 kHz) and rotor positions covering one pole pair are shown. A β-excitation is applied using the setup shown in Fig. 2a. The inductance L_β and resistance R_β concerning this excitation over the mechanical rotor angle δ_R are presented in Fig. 2b and Fig. 2c. A variation of both quantities can be seen, which is caused by the additional coil. In order to evaluate the frequency behavior, Fig. 3 shows the inductance L_β and resistance R_β for selected rotor positions. With the impedance analyzer from Wayne Kerr and the network analyzer Bode100 the impedance can be measured down to 20 Hz and 1 Hz, respectively. It can be seen that the inductance depends on the rotor position. In addition, a reversal of the direction of the inductance difference can be noticed in the range from 300 to 700 Hz. The increase of the resistance with frequency results from the skin effect. In the range from around 1 to 3 kHz a clear spread of the resistance values can be observed, so that a high resistive saliency is present in this range. In order to characterize the inductive and resistive saliency, the salience ratios

$$S_L = \frac{L_\Delta}{L_\Sigma} \quad \text{and} \quad S_R = \frac{R_\Delta}{R_\Sigma} \tag{1}$$

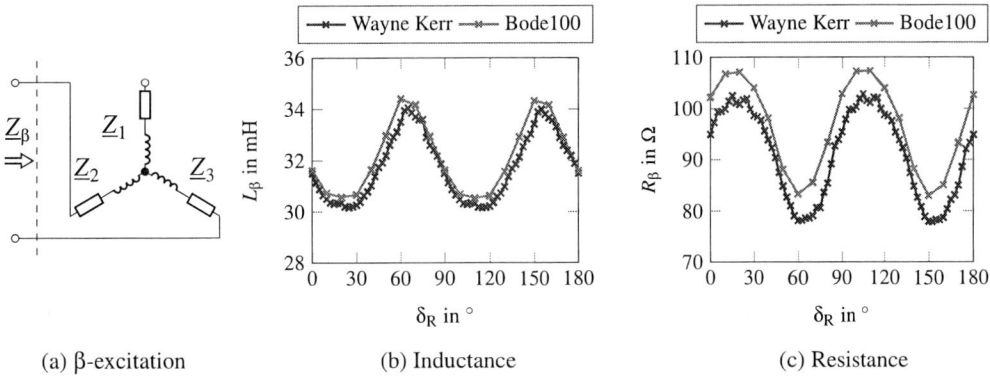

(a) β-excitation   (b) Inductance   (c) Resistance

Fig. 2: Impedance measurement: β-excitation setup (a), results obtained with the impedance analyzer (Wayne Kerr) and the network analyzer (Bode100) for the rotor position-dependent inductance (a) and resistance (b) at a frequency of 2.5 kHz

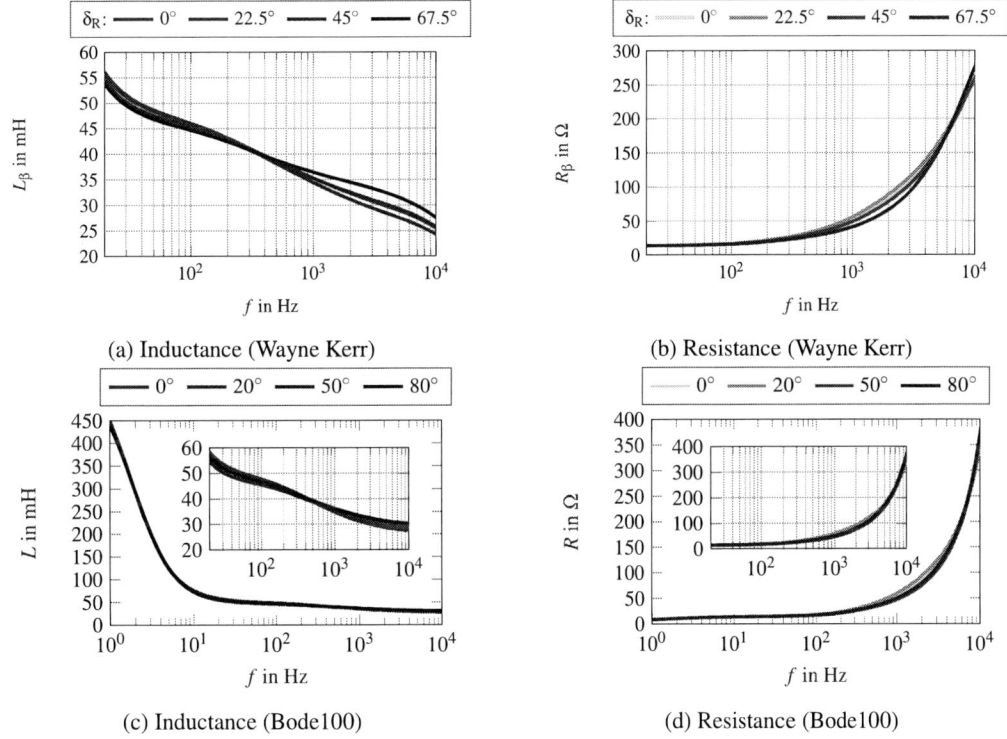

(a) Inductance (Wayne Kerr)   (b) Resistance (Wayne Kerr)

(c) Inductance (Bode100)   (d) Resistance (Bode100)

Fig. 3: Frequency dependency of the inductance (a)/(c) and the resistance (b)/(d) for certain rotor positions using the β-excitation setup, 1. row: measurement with the impedance analyzer form Wayne Kerr, 2. row: measurement with the network analyzer Bode100

are introduced, where L_Σ/R_Σ are the mean values of the inductance/resistance and L_Δ/R_Δ are the amplitudes of the 2th harmonic concerning one pole pair. Fig. 4 shows the behavior of S_L and S_R over frequency. A maximum inductive and resistive saliency of around 7% and 16% are reached, respectively. The following values are used to characterize the saliency behavior:

- The maximal saliency for low frequencies S_{LF}. This value is calculated by taking the maximum of S in a frequency range for 1 to 100 Hz.
- The maximal saliency for high frequencies S_{HF}. This value is calculated by taking the maximum of S in a frequency range above 100 Hz.

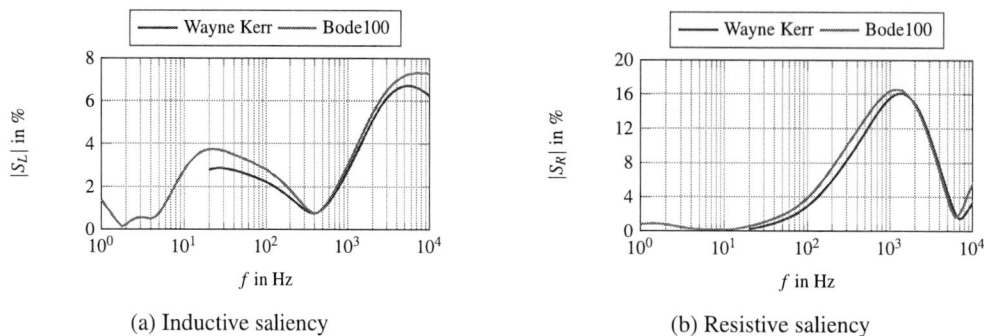

(a) Inductive saliency   (b) Resistive saliency

Fig. 4: Inductive saliency (a) and resistive saliency (b) over the frequency based on the impedance measurement

Table I: Comparison of characteristic saliency values

	S_{LF} in %	S_{HF} in %	$f_{S_{\text{HF}}}$ in kHz
inductive (Wayne Kerr)	3.33	7.16	1.75
inductive (Bode 100)	3.92	7.47	1.94
resistive (Wayne Kerr)	2.88	16.50	0.50
resistive (Bode 100)	3.54	16.55	0.41

- The corner frequency of the HF saliency $f_{S_{\text{HF}}}$. At this frequency the saliency value of $-3\,\text{dB}$ (70.7 %) of S_{HF} is reached.

Table I summarizes the characteristic values obtained by the impedance measurement for the inductive and resistive saliency.

Measurement based on evaluation of current gradients

As a second method, the current gradients during a pulse width modulation (PWM) period are evaluated to determine the inductance matrix. The inverter, which supplies the IM, is operated with a constant duty cycle for one half bridge. The duty cycles of the other two half bridges remain zero. Thereby, different DC stator current levels with a certain current ripple due to the switching can be applied in the IM. The current during a PWM period can be divided into an active switching state (ASS) and a passive switching state (PSS). The following equations

$$\bar{\vec{u}}_{\text{S,ASS},\alpha\beta} = R_{\text{S}} \cdot \bar{\vec{i}}_{\text{S,ASS},\alpha\beta} + \mathbf{L}^{\star}_{\sigma,\alpha\beta} \cdot \overline{\frac{\mathrm{d}\,\vec{i}}{\mathrm{d}t}}_{\text{S,ASS},\alpha\beta} \quad \text{and} \tag{2}$$

$$\bar{\vec{u}}_{\text{S,PSS},\alpha\beta} = R_{\text{S}} \cdot \bar{\vec{i}}_{\text{S,PSS},\alpha\beta} + \mathbf{L}^{\star}_{\sigma,\alpha\beta} \cdot \overline{\frac{\mathrm{d}\,\vec{i}}{\mathrm{d}t}}_{\text{S,PSS},\alpha\beta}. \tag{3}$$

can be derived. The voltage due to the back EMF can be neglected, since the IM is operated at standstill and a DC current is applied. The bar over a magnitude denotes its average value during the respective switching state. The difference of (2) and (3)

$$\underbrace{\bar{\vec{u}}_{\text{S,ASS},\alpha\beta} - \bar{\vec{u}}_{\text{S,PSS},\alpha\beta}}_{\Delta\bar{\vec{u}}_{\alpha\beta}} = \mathbf{L}^{\star}_{\sigma,\alpha\beta} \cdot \underbrace{\left(\overline{\frac{\mathrm{d}\,\vec{i}}{\mathrm{d}t}}_{\text{S,ASS},\alpha\beta} - \overline{\frac{\mathrm{d}\,\vec{i}}{\mathrm{d}t}}_{\text{S,PSS},\alpha\beta} \right)}_{\Delta\overline{\frac{\mathrm{d}}{\mathrm{d}t}\vec{i}}_{\alpha\beta}}. \tag{4}$$

can be used to eliminate the resistance dependency. Finally, the inductance matrix can be calculated with

$$\mathbf{L}^{\star}_{\sigma,\alpha\beta} = \begin{pmatrix} \Delta\bar{\vec{u}}_{\alpha\beta,\text{A1}} & \Delta\bar{\vec{u}}_{\alpha\beta,\text{A2}} \end{pmatrix} \cdot \begin{pmatrix} \Delta\overline{\frac{\mathrm{d}}{\mathrm{d}t}\vec{i}}_{\alpha\beta,\text{A1}} & \Delta\overline{\frac{\mathrm{d}}{\mathrm{d}t}\vec{i}}_{\alpha\beta,\text{A2}} \end{pmatrix}^{-1} \tag{5}$$

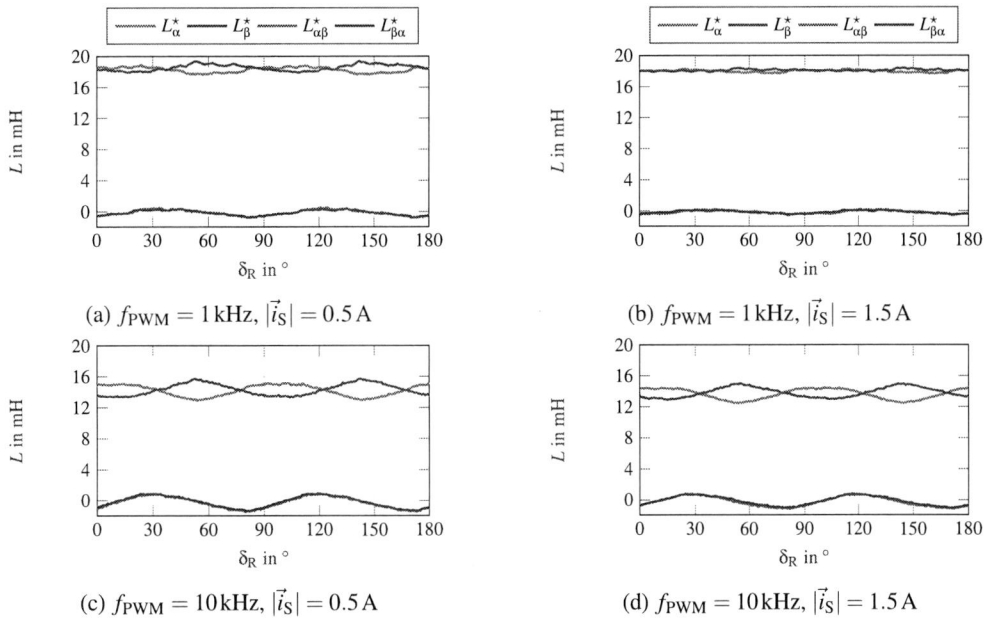

(a) $f_{\text{PWM}} = 1\,\text{kHz}$, $|\vec{i}_S| = 0.5\,\text{A}$

(b) $f_{\text{PWM}} = 1\,\text{kHz}$, $|\vec{i}_S| = 1.5\,\text{A}$

(c) $f_{\text{PWM}} = 10\,\text{kHz}$, $|\vec{i}_S| = 0.5\,\text{A}$

(d) $f_{\text{PWM}} = 10\,\text{kHz}$, $|\vec{i}_S| = 1.5\,\text{A}$

Fig. 5: Rotor position-dependent elements of $\mathbf{L}^{\star}_{\sigma,\alpha\beta}$ obtained at different switching frequencies f_{PWM} and stator current levels $|\vec{i}_S|$

Table II: Inductive saliency values S_L obtained at different switching frequencies f_{PWM} and stator current levels $|\vec{i}_S|$

$|\vec{i}_S|$ \ f_{PWM}	1 kHz	2.5 kHz	5 kHz	10 kHz
0.5 A	2.57 %	5.59 %	6.76 %	7.09 %
1.5 A	1.02 %	4.42 %	6.39 %	6.87 %

by using two excitations A1 and A2. During A1 and A2 the first and second half bridge is switching, respectively. Fig. 5 shows the results for two different stator current levels and two switching frequencies. $L^{\star}_{\alpha}$, $L^{\star}_{\beta}$, $L^{\star}_{\alpha\beta}$ and $L^{\star}_{\beta\alpha}$ are the elements of $\mathbf{L}^{\star}_{\sigma,\alpha\beta}$. The rotor position is changed during the identification process in $1°$-steps. The phase current and voltages are measured with the oscilloscope HD08108 from Lecroy. In order to get a better resolution for the current gradient evaluation, the DC link voltage is small. Table II summarizes the results of the inductive saliency obtained for four different frequencies. With a rising current level, the inductive saliency values S_L decreases. The saturation effect is more dominant at smaller frequencies, because at higher frequencies the effect of the additional rotor coil becomes stronger.

Comparison

The results obtained by the impedance measurement and the evaluation of the current gradients are compared in Table III for a frequency of 2.5 kHz. Such a value can be applied using a rotational or alternating HF injection. In order to compare the results for a similar IM operation condition, the value for $|\vec{i}_S| = 0.5\,\text{A}$ is used. Because of the β-excitation setup the measured inductances are twice as high as the inductances received by the second method. Therefore, the inductances obtained by the first method are divided by two for comparison reason. Regarding the maximal saliency for high frequencies S_{HF}, the values 7.16 %, 7.47 % and 7.09 % are obtained. The values match well.

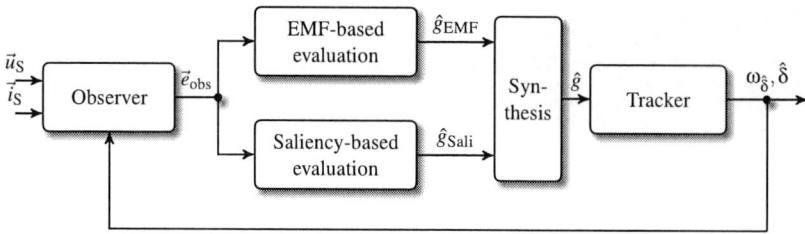

Fig. 6: Block diagram of rotor speed and position estimation method for the proposed self-sensing control

Self-sensing control method for the whole operation range

A self-sensing control method for the IM based on the gradient descent method is suggested, which is capable to operate the IM in the complete speed range. The basic block diagram of the electrical rotor angle and the rotor angle velocity estimation is shown in Fig. 6. The key features are the observer (Fig. 7a) and the two evaluations which calculate the gradients based on the EMF and the saliency (Fig. 7b and Fig. 7c). Further, the tracker consists of a PI controller and an integrator. The gradients are combined depending on three speed intervals. If the absolute speed is below the limit n_{Sali}, the saliency-based evaluation is active. If the absolute speed is above the limit n_{EMF}, the EMF-based evaluation is active and in between these limits the gradients are weighted linearly. An arbitrary HF injection is needed if the saliency-based evaluation is active at low speed. A similar control method is used for the synchronous machine in [10].

The self-sensing control is tested at two test stands. Test stand A has a 20 kW, six-pole IM with no additional short-circuited rotor coil and test stand B includes a 1.5 kW, four-pole IM with an additional short-circuited rotor coil. Fig. 8a shows a reference speed step from 1500 rpm to 1600 rpm obtained at test stand A at rated torque while the EMF-based evaluation is active. Low-frequency oscillation can be

Table III: Comparison of the results of the impedance measurement and evaluation of the current gradients for the frequency 2.5 kHz

Method	L_Σ in mH	L_Δ in mH	S_L in %
Impedance measurement (Wayne Kerr)	15.9	0.97	6.09
Impedance measurement (Bode 100)	16.1	0.95	5.93
Current gradient	16.0	0.90	5.59

Fig. 7: Block diagrams of key features of the proposed estimation

(a) Observer

(b) EMF-based evaluation

(c) Saliency-based evaluation

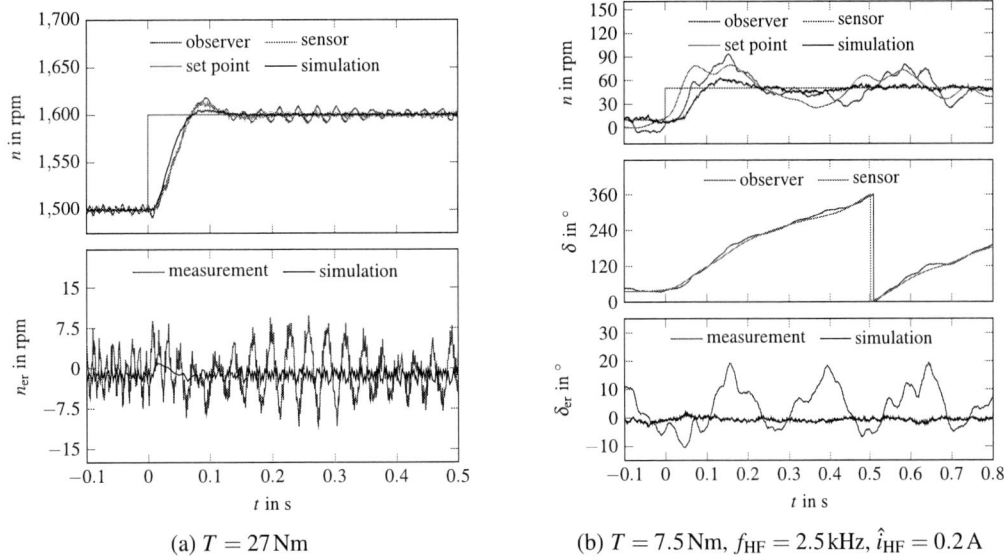

(a) $T = 27\,\text{Nm}$ (b) $T = 7.5\,\text{Nm}$, $f_{\text{HF}} = 2.5\,\text{kHz}$, $\hat{i}_{\text{HF}} = 0.2\,\text{A}$

Fig. 8: Measurement results and comparison with simulation for self-sensing speed control for different speed ranges using the EMF-based evaluation at test stand A (a) and using saliency-based evaluation at test stand B (b)

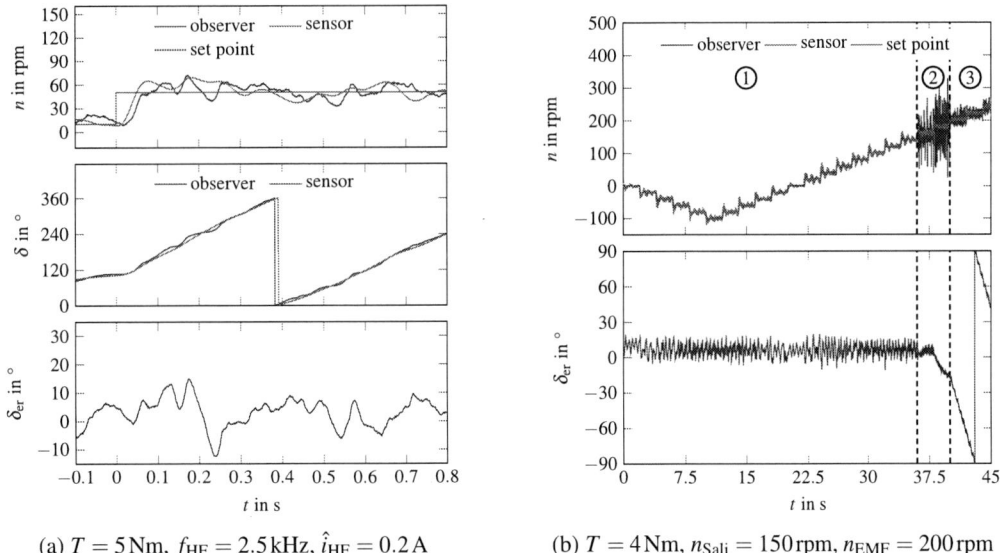

(a) $T = 5\,\text{Nm}$, $f_{\text{HF}} = 2.5\,\text{kHz}$, $\hat{i}_{\text{HF}} = 0.2\,\text{A}$ (b) $T = 4\,\text{Nm}$, $n_{\text{Sali}} = 150\,\text{rpm}$, $n_{\text{EMF}} = 200\,\text{rpm}$

Fig. 9: Measurement results for self-sensing speed control at test stand B using the saliency-based evaluation (a) and using the EMF- or/and saliency-based evaluation, respectively (b)

noticed on the speed error n_{er}. However, the mean value of the speed error is close to zero. The simulation forecasts the speed dynamic well. The overshoot is higher than predicted. A reference speed step from 10 rpm to 50 rpm is shown in Fig. 8b. The waveforms are obtained at test stand B at a torque of 7.5 Nm (75 % of the rated torque) during the saliency-based evaluation is active. The speed follows the reference value. However, significant oscillations can be seen on the speed. The root cause is the oscillating rotor angle estimation error. The oscillations are not visible in the simulation. In order to evaluate the results, it has to be considered that the saliency is in the range of 6 % at the used injection frequency of 2.5 kHz. This is not a high value for a saliency-based self-sensing control. Further, the saturation has a negative effect. Appling a smaller torque, the control performance becomes better shown in Fig. 9a.

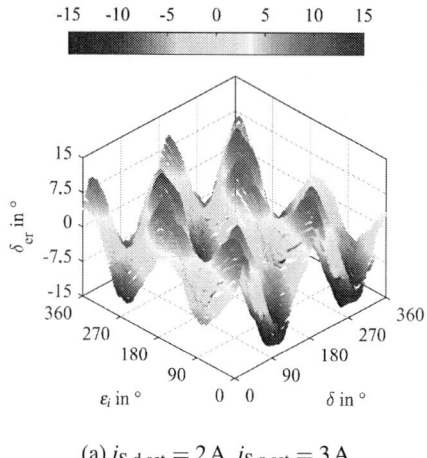

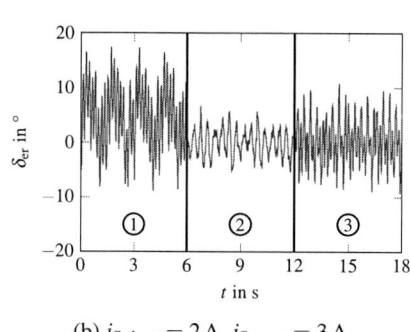

(a) $i_{S,d,set} = 2\,A$, $i_{S,q,set} = 3\,A$ (b) $i_{S,d,set} = 2\,A$, $i_{S,q,set} = 3\,A$

Fig. 10: Rotor angle estimation error identification depending on electrical rotor angle δ and the angle of the stator current space vector ε_i (a) and experimental results for rotor angle estimation error compensation obtained during current control operation (1: no compensation, 2: indirect compensation, 3: direct compensation)

Fig. 9b shows the experimental results, in which all speed ranges used for the synthesis of the gradients are passed through. The waveforms are obtained at a torque of 4 Nm. The performance of the speed control is degraded in the interval, in which both evaluations are active. In this range, the HF injection cannot be disabled, since the saliency-based evaluation is active. However, the HF injection disturbs the EMF-based evaluation. In order to identify the rotor angle estimation error, the IM is operated in current control mode using a speed sensor, the load machine is speed-controlled and the rotor angle estimation runs is active. Fig. 10a shows identified rotor angle estimation error. Two maxima exist over one electrical period of the rotor angle and the angle of stator current space vector. Two compensation methods are applied to reduce the rotor angle estimation error. The rotor angle can be corrected after the estimation (indirect compensation) or the rotor angle, which is fed back to the observer, is directly corrected within the estimation (direct compensation). Fig. 10b shows the experimental results for the compensation. The IM is operated with a speed sensor in order to avoid an interaction of self-sensing control on the estimation.

Conclusion

A modified rotor design with an additional short-circuited coil is proposed. The effect of the additional coil on the impedance and resistance is evaluated using two measurement techniques. For the investigated prototype, a maximal inductive saliency of around 7 % is achieved. The maximal resistive saliency is around 16 %. Further, a self-sensing control for the whole operation range using a EMF- and a saliency-based evaluation is suggested. Simulation and experimental results demonstrate the validity of the proposed method.

References

[1] Jansen P. L. and Lorenz R. D.: Transducerless position and velocity estimation in induction and salient AC machines, IEEE Trans. Ind. Appl. Vol. 31 no. 2, pp. 240-247, 1995

[2] Brown I. P. and Lorenz R. D.: Induction machine design methodology for self-sensing: Balancing saliencies and power conversion properties, IEEE Trans. Ind. Appl. Vol. 47 no. 1, pp. 79-87, 2011

[3] Mingardi D., Bianchi N., Alberti L., and Zeni R.: Analysis and test of the sensorless capability of induction motors with created saliency, IEEE Trans. Ind. Appl. Vol. 52, no. 3, pp. 2186-2193, 2016

[4] Joy M. T. and Bcker J.: Sensorless control of induction motor drives using additional windings on the stator, IEEE 9th International Symposium on Sensorless Control for Electrical Drives (SLED), pp. 162-167, 2018

[5] Faggion A. , Bianchi N. , and Bolognani S.: Ringed-pole permanent-magnet synchronous motor for position sensorless drives, IEEE Trans. Ind. Appl. Vol. 47 no. 4, pp. 1759-1766, 2011

[6] Graus J. and Hahn I.: Modelling and optimization of a short-circuited rotor winding of a PMSM for saliency tracking, 5th International Symposium on Sensorless Control for Electrical Drives (SLED), 2014

[7] Quattrone F. and Ponick B.: Active differential inductance control of permanent magnet synchronous machines using short-circuited rotor coils, IEEE Vehicle Power and Propulsion Conference (VPPC), 2015.

[8] Quattrone F. and Ponick B.: Evaluation of a permanent magnet synchronous machine with a rotor coil for improved self-sensing performance at low speed, XXII International Conference on Electrical Machines (ICEM), pp. 1680-1685, 2016

[9] Luecke S. and Mertens A.: Self-sensing control of induction machines using an additional short-circuited rotor coil, IEEE International Symposium on Sensorless Control for Electrical Drives (SLED), 2017

[10] Wiedmann K. and Mertens A.: Self-sensing control of PM synchronous machines including online system identification based on a novel mras approach, 3rd IEEE International Symposium on Sensorless Control for Electrical Drives (SLED), 2012

Calculating the tractive power and power conversion efficiency of battery electric vehicles using a global navigation satellite system and a road elevation database

Shinichi Domae[1], Alberto Castellazzi[1], Hamzeh J. Jaber[1], Tenghui Dong[2], Taketsune Nakamura[2]

[1]KYOTO UNIVERSITY OF ADVANCED SCIENCE
18 Yamanouchi Gotanda-cho, Ukyo-ku,
Kyoto, 615-8577, Japan
+81-75-496-6254
domae.shinichi@kuas.ac.jp
[2]KYOTO UNIVERSITY
1-30, Goryo-Ohara, Nishikyo,
Kyoto, 615-8245, Japan
+81-75-383-3117
dong.tenghui.7v@kyoto-u.ac.jp

Acknowledgements

This research is supported by the discretionary budget of the President of Kyoto University of Advanced Technologies.

Keywords

«Data analysis», «Electric vehicle», «Efficiency»

Abstract

The energy conversion efficiency of battery electric vehicles (BEVs), which is an essential index in the development of motor systems, remains unknown during real driving. This paper presents a calculation method of the tractive power and energy conversion efficiency of BEVs during driving. The progress of the global navigation satellite system (GNSS) and the widespread use of road elevation database enable a tractive power measurement during driving. We have developed the calculation method to evaluate an advanced motor-drive for BEVs under development without using chassis dynamometers. The calculation method showed almost the same energy conversion efficiencies on the three different routes. The results demonstrate the universality of the method and raise the possibility of benchmarking of the power conversion efficiency of any of various BEVs.

Introduction

The energy conversion efficiency of battery electric vehicles (BEVs)—an essential index to expand the driving ranges of BEVs and to reduce the global energy consumption—remains unknown during driving. Chassis dynamometers, which have been used to evaluate the tractive powers of vehicles, cannot measure actual tractive powers during driving and involve higher costs. A tractive power and energy conversion efficiency calculation method with the longitudinal dynamic model has been developed to evaluate an advanced motor-drive for BEVs under development without using chassis dynamometers. The main input data for the calculation includes the locations, speeds, and altitudes of vehicles. The progress of the global navigation satellite system (GNSS) enables accurate measurement of location and speed, whereas the widespread use of road elevation databases allows access to precise altitude information. The method can calculate the tractive powers and energy conversion efficiencies during driving and requires lower costs. Many works have been reported regarding energy consumption estimations for BEVs with the longitudinal dynamic model but their main purposes are precise predictions of driving ranges [1]. The paper presents a calculation method for BEV power train evaluations that can substitute the measurement with Chassis dynamometers. Almost same energy conversion efficiencies on the three different routes demonstrate sufficient universality achievable with the calculation method. The advanced motor system consisting of a 3-phase Y-inverter and a Halbach motor [2], which will be equipped to a small BEV, will be evaluated by the reported method in the near future.

Experiments

Tractive powers and energy conversion efficiencies are calculated with the longitudinal dynamic model which is well accepted and universal for any vehicles [3]. The major external forces acting on a vehicle include the following four forces: the rolling resistance of tires on road surfaces, F_{roll}, the aerodynamic drag during traveling at a particular speed in the air, F_{aero}, the grading resistance due to gravitation force, F_{grad}, and the acceleration resistance, F_{accel}. The traction force, F_{tract}, can be obtained as the sum of the four external force as follows. These forces are shown in Figure 1.

$$F_{tract} = F_{roll} + F_{aero} + F_{grad} + F_{accel} \tag{1}$$

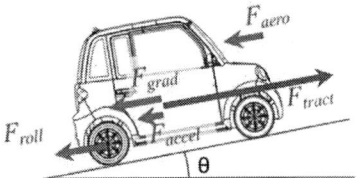

Fig. 1: Longitudinal dynamics of a vehicle

F_{roll}, F_{aero}, F_{grad}, and F_{accel} are expressed as follows,

$$F_{roll} = \mu mg \cos \theta \tag{2}$$

$$F_{aero} = \frac{1}{2} \rho C_d A (v + v_w)^2 \tag{3}$$

$$F_{grad} = mg \sin \theta \tag{4}$$

$$F_{accel} = (m + m_f) \frac{dv}{dt} \tag{5}$$

where μ is the rolling resistance coefficient, m is the vehicle mass, g is the gravitational acceleration, θ is the road inclination angle, ρ is the air density, C_d is the aerodynamic drag coefficient, A is the vehicle frontal area, v is the vehicle speed, v_w is the component of wind speed, m_f is the fictive mass of rolling inertia, and $\frac{dv}{dt}$ is the acceleration of the vehicle. The motor torque, T_m, and the motor speed, N_m (rpm), are calculated with the traction force and vehicle speed as shown below.

$$T_m = F_{tract} \frac{D_p}{2 A_r} \tag{6}$$

$$N_m = 60 A_r \frac{v}{\pi D_p} \tag{7}$$

where D_p is the diameter of the tire and A_r is the reduction ratio. The tractive power, P_{tract}, is calculated as a product of traction force and vehicle speed as below.

$$P_{tract} = F_{tract} v \tag{8}$$

The battery output power, P_{bat}, can be calculated with the battery output current and the battery output voltage obtained via a controller area network (CAN) bus. The power consumed by non-traction loads (auxiliary loads) are subtracted from P_{bat}. The brake booster, power steering, and wind wiper were the main auxiliary loads. The air conditioner or heater were not used during driving to avoid any unanticipated influences on the calculation. The energy conversion efficiency, η, can be calculated as

$$\eta = P_{tract} / P_{bat} \tag{9}$$

All the parameters for the model are shown in **Table I**. The locations and speeds of vehicles are measured by GNSS. The Quasi-Zenith Satellite System (QZSS) is available as GNSS in Japan since 2018. The sampling rate of the used GNSS system is 1 Hz. The road inclination angles, θ, are calculated by dividing the changes in elevation (m) per second by the horizontal vehicle velocities (m/sec) as shown in Fig. 2. The elevation information for each location measured by GNSS every second is obtained with using an open database of Geospatial Information Authority of Japan (GSI). The altitude of the driving route measured by GNSS and that obtained at the GSI database are compared in Figure 3. The altitude measured by GNSS (orange line) is instable because GNSS has larger error in vertical direction than in horizontal direction. The altitude obtained from the GSI database (blue line) is found to be much more stable. The use of the elevation database enables an accurate calculation of the road inclination angles. Air densities, ρ, are calculated by measuring the temperature, the relative humidity, and the atmospheric pressure during driving. Rolling resistance coefficient, μ, and aerodynamic drag coefficient, C_d, were obtained by coast down tests. Wind velocity, v_w, is assumed to be zero because it is relatively low and the measurement is practically difficult. Fictive mass of rolling inertia, m_f, is also assumed to be zero. The traction forces were calculated every second to synchronize with the sampling rate of the used GNSS system of 1 Hz.

Table I: Vehicle longitudinal dynamics equation parameters

Symbol	Meaning	Value (unit)
g	Gravitational Acceleration	9.81 (m/s²)
θ	Road Angle	Variable (rad)
m	Vehicle Mass	616 + Passengers' Weight (kg)
v	Vehicle Speed	Variable (m/s)
μ	Rolling Resistance Coefficient	0.0122
ρ	Air Density	Variable (kg/m³)
C_d	Aerodynamic Drag Coefficient	0.495
A	Vehicle Frontal Area	1.7 (m²)
v_w	Wind Velocity	Assumed To Be Zero (m/s)
m_f	Fictive Mass of Rolling Inertia	Assumed To Be Zero (kg)
D_P	Tire Diameter	0.507 (m)
A_R	Reduction Ratio	6.07

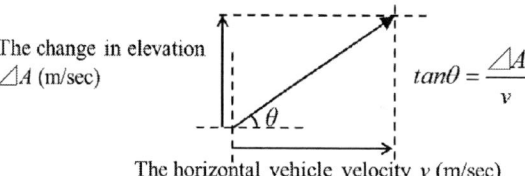

Fig. 2: The calculation of the road inclination angle θ

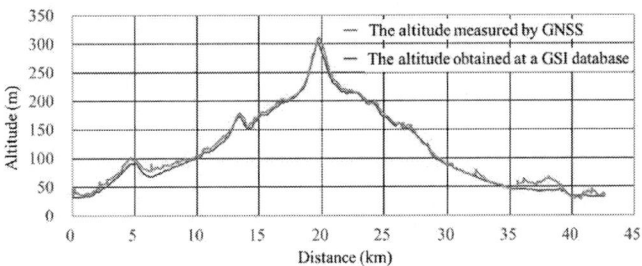

Fig. 3: The altitude of the driving route measured by GNSS and that obtained at a GSI database

The used test vehicle is a lightweight electric vehicle, TAJIMA-JIAYUAN. The vehicle, which has a simple structure and a simple control system, allows easy replacement of the Y-inverter and the Halbach motor under development with the present inverter and motor. The main characteristics of the test vehicle are shown in **Table II**. The maximum speed is 60 [km/h] and the regeneration energy is not available.

Table II: Test vehicle characteristics

Parameters	Value (unit)
Length/ Width/ Height	2,495 x 1,295 x 1,930 (mm)
Curb Weight	616 (kg)
Electric Motor Type	Induction Motor
Max. Power	16 (kW)
Max. Torque	101 (Nm)
Max. speed	60 (km/h)
Battery Capacity	10 (kWh)
Regeneration Energy	Not available

Results and Discussions

Torque-speed

An example of the traction force, F_{tract}, and the tractive power, P_{tract}, are shown in Fig. 4. In Fig. 4 (a), the vehicle speed increases from 10 km/h to nearly 50 km/h and the road angle increases to about 2 degrees with time. In Fig. 4 (b), F_{tract} plotted on the black line graph and F_{roll}, F_{aero}, F_{grad}, and F_{accel} are plotted on the stacked column graphs of blue, orange, gray, and yellow, respectively. F_{roll} is almost constant. F_{aero} is proportional to the square of the velocity. F_{grad} increases with the road angle. F_{accel} changes corresponding to the acceleration and deceleration of the vehicle. In Fig. 4 (c), P_{tract} plotted on the black line graph and P_{roll}, P_{aero}, P_{grad}, and P_{accel}, which are work done per second against F_{roll}, F_{aero}, F_{grad}, and F_{accel}, are plotted on the stacked column graphs of blue, orange, gray, and yellow, respectively. As shown in this example, the calculation enables to visualize the changes of P_{tract} with the changes of the vehicle speed and road angle. Furthermore, it enables in-situ monitoring of P_{tract} if road elevation information is available during driving.

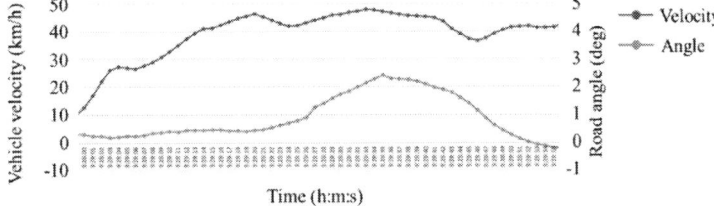

(a) Vehicle velocity ad Road angle

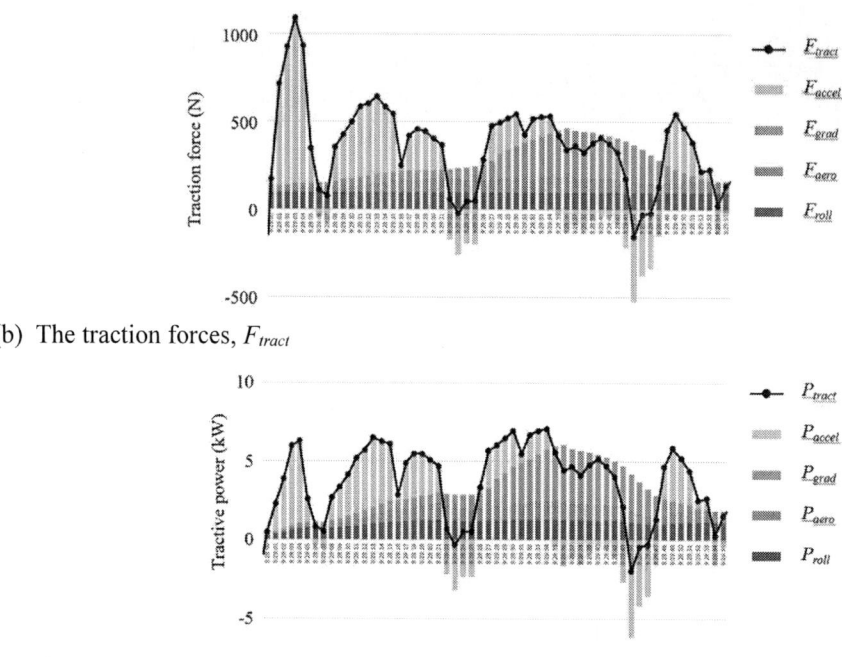

(b) The traction forces, F_{tract}

(c) The tractive powers, P_{tract}

Fig. 4: An example of the calculated traction forces F_{tract} and tractive powers P_{tract}

Fig. 5 (a) shows another example of the calculation: a scatter plot of the motor torque, T_m, and the motor speed, N_m, when the vehicle started at a green traffic signal. The numbers in the chart indicate the time by seconds after starting. The vehicle locations on the road are represented by yellow dots in Fig. 5 (b). N_m increases from 0 rpm to 3500 rpm which is equivalent to a car velocity from 0 km/h to 55 km/h. T_m increases to the maximum of 70 Nm to accelerate the vehicle and then decreases to 20 Nm which is required to keep speed constant against the rolling resistance and the aerodynamic drag. In general, the plot draws a trajectory shown as a blue arrow in Fig. 5 (a) and becomes almost constant.

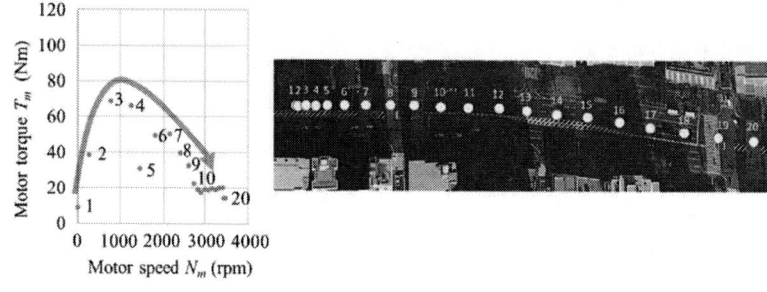

(a) Scatter plot of T_m and N_m (b) Vehicle location

Fig.5: Scatter plot of T_m and N_m when the vehicle started at a green traffic signal

Test runs were done on the three different routes in Kyoto (Route 1, Route 2, and Route 3) shown in Fig. 6 (a). Each route has different road gradients, and the Route 1 runs almost on a mountain road, the Route 2 runs in the city area and the Route 3 is in the middle as shown in the altitude profiles in Fig. 6 (b). The distances and numbers of test run of the routes are shown in **Table III**. Highways were not included in the routes because the test vehicle, which has the maximum speed of 60 km/h, is not allowed to run any highways.

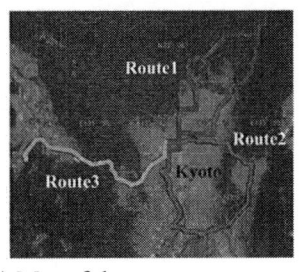

(a) Map of the routes

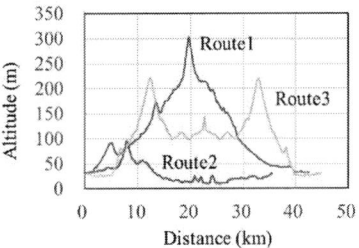

(b) Altitude profiles of the routes

Fig. 6: Maps and altitudes of the test courses

Table III: Information of routes

	ROUTE 1	ROUTE 2	ROUTE 3
Distance (km)	42.6	48.8	35.3
Number of Test Runs	12	12	12

Fig. 7 (a), (b), and (c) show scatter plots of T_m and N_m on Route 1, Route 2, and Route 3. Data for zero P_{bat} are removed from the plots because the vehicle is in coasting or friction braking when P_{bat} is zero. The regeneration braking is not available for the test vehicle. The scatter plots on the three routes show almost same tendency: T_m spread widely when N_m is smaller than around 2000 rpm and it becomes dense when N_m is larger than around 2000 rpm. Fig. 8 (a), (b), and (c) show the contour maps of the distribution of N_m and T_m on Route 1, Route 2, and Route 3, which are converted from the scatter plots shown in Fig. 7. The contour maps on the three routes have different shapes. Route 1 has the sharpest peak at around T_m of 18 Nm and N_m of 3000 rpm. This indicates that the vehicle was driven at almost constant torque and speed. Route 2 has a relatively broad peak and a tail spreading in the low speed and high torque region. Route 3 is in between. The difference of the shape among the routes should be due to traffic jams. In Route 1, the test vehicle runs at almost constant speed on mountain roads without traffic jams. In Route 2, the vehicle accelerates and decelerates repeatedly in the heavy traffic city area. As shown in Fig. 5 (a), higher torque is required in acceleration.

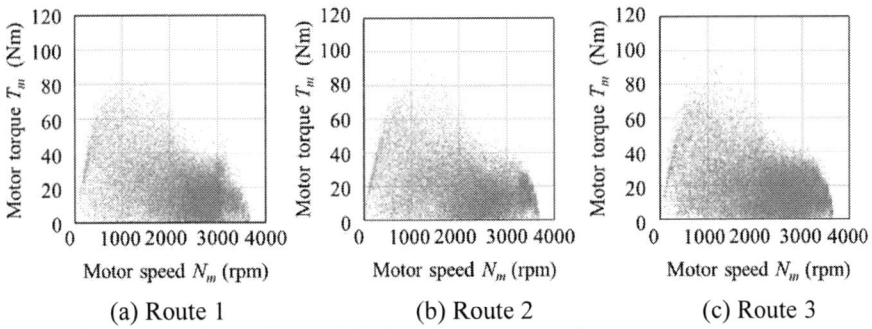

(a) Route 1   (b) Route 2   (c) Route 3

Fig. 7: The scatter plots of N_m and T_m for all runs on the three routes

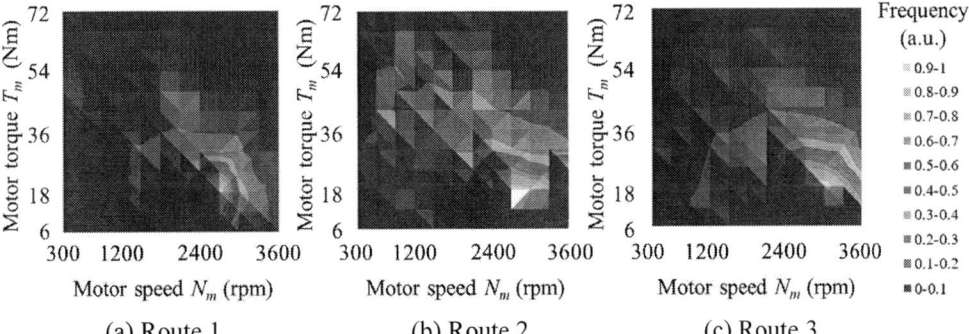

(a) Route 1   (b) Route 2   (c) Route 3

Fig. 8: The contour maps of the distribution of N_m and T_m

Energy conversion efficiency

Fig. 9 (a), (b), and (c) show the energy conversion efficiency maps on the T_m-N_m planes measured on Route 1, Route 2, and Route 3. Fig. 9 use the same data sets as Fig. 7 and Fig. 8. The battery output powers, P_{bat}, are combined with the tractive powers, P_{tract}, to calculate the energy conversion efficiencies, η, for positive P_{bat}. The T_m-N_m plane is divided into small squares and the average energy conversion efficiency is calculated in each square after removing the outliers statistically. In Fig. 9, the three routes show almost the same efficiency maps even though the road slopes and the T_m-N_m distributions are different. Fig. 10 (a) and (b) show the close energy conversion efficiencies among routes at the motor torque T_m of 18 Nm and 30 Nm. The results demonstrate the sufficient universality to calculate the efficiency of the motor system achievable regardless of the routes. Furthermore, the results raise the possibility of benchmarking of power conversion efficiency of any of various BEVs on any vehicle locations on the globe using the road elevation information. The reported method will be used to evaluate the advanced motor system under development.

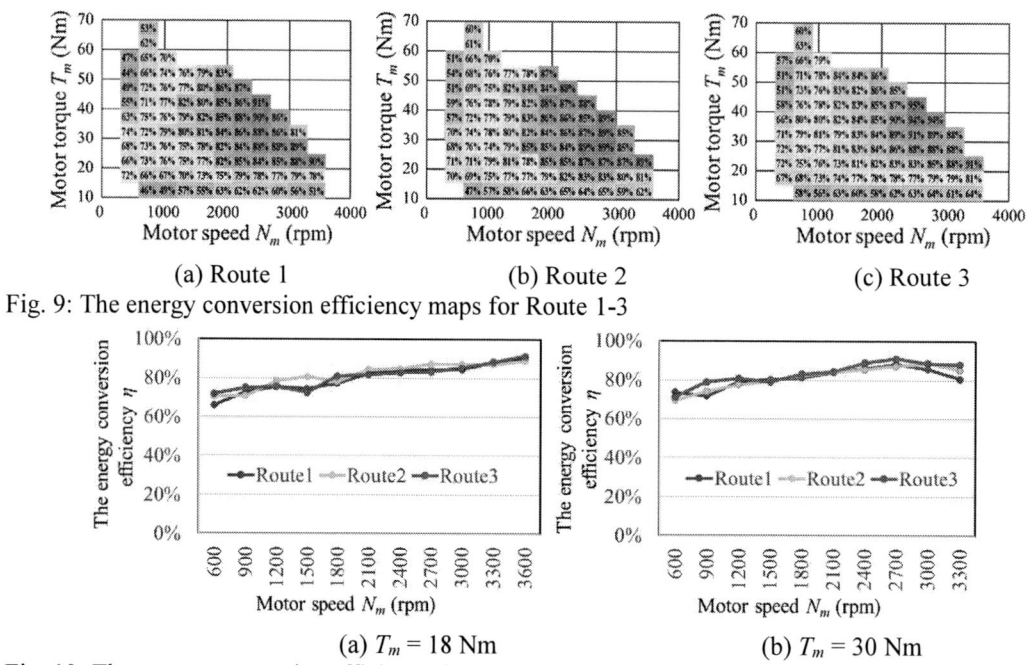

(a) Route 1   (b) Route 2   (c) Route 3

Fig. 9: The energy conversion efficiency maps for Route 1-3

(a) T_m = 18 Nm   (b) T_m = 30 Nm

Fig. 10: The energy conversion efficiency for Route 1-3

Motor system

The advanced motor system consisting of a 3-phase Y-inverter and a Halbach motor is under development. An interleaved three-phase Y-inverter is introduced to the motor-drive [4]. The Y-inverter has some innovative features such as sinusoidal output voltage synthesis as opposed to square wave modulated, no high voltage DC-link, and integrated voltage boost capability. In addition, the discontinuous pulse width modulation (D-PWM) is adopted with the state-of-the-art wide-bandgap semiconductor power devices. The electrical design has been described in [2]. Fig. 11 shows the proposed three-phase Y-inverter's circuit diagram and a photograph of one-phase Y-inverter.

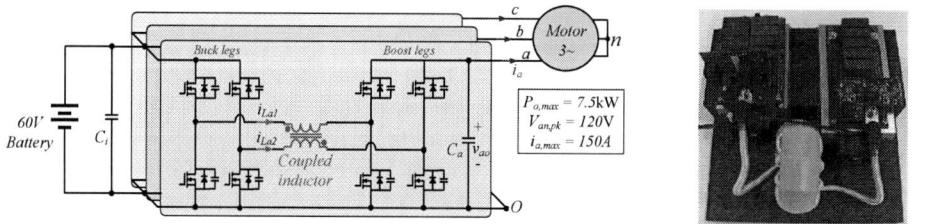

(a) The three-phase Y-inverter's circuit diagram   (b) A photo of one-phase Y-inverter

Fig. 11: The circuit diagram and a photograph of Y-inverter under development

The motor is a single-rotor Halbach motor, which is designed to operate at high voltage and high rotational speed with more than double power density as an equivalent traditional permanent magnet synchronous machine (PMSM) design. Fig. 12 shows the basic structure and a photograph of the single-rotor Halbach motor. It is designed to ensure very high efficiency at the most diverse and over as broad a range as possible of operational conditions, torque and speed. [2, 5, 6].

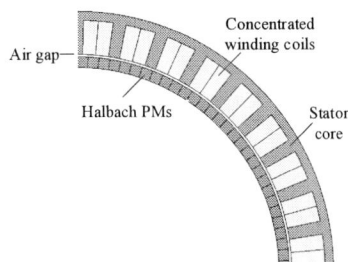

(a) The basic structure of the single-rotor Halbach motor   (b) A photo of Halbach motor

Fig. 12: The basic structure and a photograph of the single-rotor Halbach motor

The required torques for the motor system under development were defined based on the T_m-N_m distribution discussed above. The inverter and motor design have been proceeded iteratively to optimize the joint performance and the power density for up to 5000 rpm and the full torque capability. The Y-inverter and the Halbach motor will replace the present inverter and motor of the test vehicle, and the energy conversion efficiency will be investigated by the reported calculation method in the near future.

Conclusions

A tractive force and energy conversion efficiency calculation method with the longitudinal dynamic model has been developed. The location and speed of a vehicle is obtained through GNSS and the elevation information is derived from an elevation database of GSI, which enables an accurate calculation of the road inclination angles. The visual information of dynamic behaviors and energy conversion efficiency of the vehicle can be obtained by the calculation method. The universality of the method is demonstrated by the almost same efficiencies on three different routes. The advanced motor system under development, which will be equipped to a small BEV, will be evaluated by the reported method in the near future. The calculation method discussed in this paper raises the possibility of benchmarking of the power conversion efficiency of any of various BEVs on any locations on the globe.

References

[1] Chen, Y.: A Review and Outlook on Energy Consumption Estimation Models for Electric Vehicles, SAE J. STEEP 2(1):79-96, 2021

[2] Tran H.: Low-voltage-battery powered hybrid-GaN/SiC Y-Inverter dual-rotor Halbach motor drive for light electric vehicles, in Proc. ICEMS 2021

[3] Ehsani M.: Modern Electric, Hybrid Electric, and Fuel Cell Vehicles, CRC Press, 2005

[4] Antivachis M.: Three-Phase Buck-Boost Y-Inverter with Wide DC Input Voltage Range, in Proc. APEC 2018

[5] Kwak J.: Multi-parameter machine-inverter co-design for energy-efficient electric drives, in Proc. ICEMS2021

[6] Dong T.: High Power-Density High-Efficiency Electric Drive Design with Halbach-Rotor PMSM and WBG-Based High-Frequency Inverter, in Proc. ICEMS2021

PCB Layer Optimization of Planar Medium Frequency Transformer for On-Board EV Chargers

Fabian Groon, Hamzeh Beiranvand, Thiago Pereira, Görkem Can, Marco Liserre
Chair of Power Electronics, Kiel University
Kaiserstraße 2, 24143 Kiel
Kiel, Germany
Tel.: +49 431 880-6100
Fax: +49 431 880-6103
E-Mail: fagr@tf.uni-kiel.de, hab@tf.uni-kiel.de, tp@tf.uni-kiel.de, goerkem-can@hotmail.de,
ml@tf.uni-kiel.de
URL: https://www.pe.tf.uni-kiel.de/en

Acknowledgements

We acknowledge the Zentrale Innovationsprogramm Mittelstand (ZIM), which is a funding program of the Federal Ministry for Economic Affairs and Climate Action in Germany for providing financial support (Förderkennzeichen: KK5302202PR1).

Keywords

« Planar magnetics», «Transformer», «Power density optimization», «Efficiency», «Power converters for EV»

Abstract

Planar medium frequency transformer (MFT) is a promising solution for on-board electric vehicle (EV) chargers, to achieve high power density and high efficiency. The trend towards higher power densities and higher efficiencies exposes a number of limitations on conventional litz wire transformers, especially for increasing the current density. Litz wire current density is limited by the temperature, due to poor thermal management capabilities. PCBs, however, have better thermal management capabilities, which allows for higher current densities. This paper optimizes planar MFT windings focusing on maximizing current density and the simplicity of the implementation. The losses, power density and thermal constraints are investigated and a pareto-front is created based on optimal solutions. High efficiency and power densities are achieved from 2D/3D FEM simulations. A solution within thermal constraints is selected and a prototype is built based on similar ratings. Tests with different current densities are carried out on the prototype and the temperatures are compared. The results verify that planar MFT with high current densities are feasible solutions for high efficiency and high power density MFTs.

Introduction

Compact and efficient power electronic converters are a prerequisite for E-mobility and particularly for on-board electric vehicle (EV) chargers [1], [2]. IEC60950 recommends that the off-grid power electronics converters might provide galvanic isolation in grid-connected mode operation for providing the required level of safety. Even though there are multiple ways for achieving galvanic isolation [3], magnetic transformer based isolation is much more popular owing to its reliable and low-loss operation. Isolated DC-DC converters with planar transformer are interesting solutions to satisfy the safety requirements as well as high efficiency and power density for on-board EV chargers [4]. A unique feature of the PCB windings is their ability to carry high current densities [5]. While, conventional litz wires suffer from poor thermal management capabilities. Litz strands insulator breakdown could easily propagate inside the entire winding and consequently it leads to medium frequency transformer (MFT) failure [6].

Planar transformers are used in a wide range of applications such as data centers [7] and on-board EV chargers where the power ratings are normally higher than 1 kW [8]. The existing trade-off between core losses and planar winding losses at different frequencies is addressed in [9] where the influence of the number of turns per PCB layer on the transformer total losses is also evaluated. Figure-of-merit method has been employed in [5] to achieve an integrated design of planar transformer and inductor in a DAB converter. However, the design variables are limited to magnetic field density B_m and current density J for a fixed integrated geometry of cores and windings. Moreover, low-power MHz range frequency planar transformers, in which volume is reduced effectively [10], are not suitable for high-power applications. Generally, the motivation to increase the frequency of the transformers is to reduce the volume according to the Faraday's Law of induction where the core volume reduces proportional to the inverse of frequency, i.e. $A_C \propto 1/f$. Increasing the switching frequency results in the higher parasitic capacities which might lead to lose of soft-switching and therefore impacts on the optimum design of the transformer [11]. Therefore, the correlation between current density and copper area, i.e. $A_{Cu} \propto 1/J$ is a motivation to increase the current density to reduce the copper volume. Nevertheless, thermal constraints severely restrict the increase of J and a careful design is demanded. Therefore, optimizing the current density is considered as an objective in this paper.

This paper presents an $\eta\rho$-brute-force optimization for planar transformer focusing on maximizing the current density which results in the reduced required copper area and therefore weight and volume of the used copper can be saved which is desirable for EV on-board charging. Analytical analysis is carried out to correlate the transformer parameter and thermal limitation to the Pareto fronts in the $\eta\rho$-plane. Considering level 1 on-board chargers, a 10 kW 400:400 V planar MFT is designed at 50 kHz for all simulation as well as experiment scenarios. Comprehensive analytical, 2D/3D FEM simulations, and experimental results show that high efficiency and power densities can be achieved at current densities up to 6 times of a conventional wounded MFTs. This paper is organized as follows. Section II explains the theoretical design approach. Loss equations and thermal equations are given to design the transformer. Section III described the optimization control variables and the creation of the $\eta\rho$-plane. Analysis for different numbers of layers per track, PCBs and a final pareto front with all solutions are given. The results of the pareto front are experimentally validated in section IV, where different currents and voltages are applied for thermal measurement at different current densities. Finally, Section V concludes this paper.

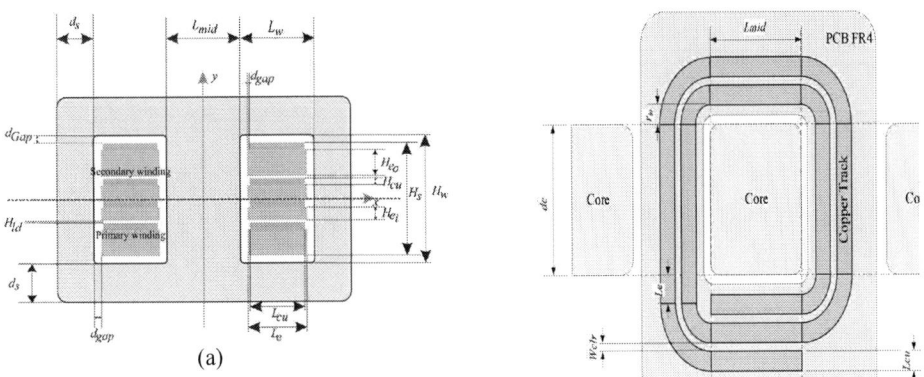

Fig. 1: Geometry of the MFT planar transformer: (a) core front view, and (b) one-layer PCB construction.

Theoretical Explanation of the Design

Fig. 1 shows the core and winding geometry of the planar MFT. Windings are constructed using PCB boards. Therefore, the number of turns can be identified by using the PCBs. The number of PCBs and the number of tracks per PCB determine the number of turns which must satisfy the Faradays law of the induction [12]. Theoretical basis to achieve a compact and thermally stable design of planar MFT are developed in the following. Power density (ρ) and efficiency (η) are defined as follows:

$$\rho = \frac{P_N}{V_T} \xrightarrow{(V_T = V_w + V_c)} \rho = \frac{P_N}{V_w + V_c} \quad \text{and} \quad \eta = \frac{P_N - P_{Cu} - P_{fr}}{P_N} \tag{1}$$

where P_N is the nominal output power, P_{Cu} is the winding losses and P_{fr} represents the core losses of the planar MFT. V_T, V_w and V_c are respectively total, winding and core volumes. Winding and core losses are analytically described by the following equation in [13]:

$$P_{Cu} = \rho_W V_W k_u J^2 \quad \text{and} \quad P_{fr} = V_c k_c f^\alpha B_{max}^\beta \tag{2}$$

Where ρ_W and k_u are the electrical resistivity of the conductor and the utilization factor of the window. P_{Cu} is explicitly given as a function of current density J. Steinmetz equation is used to analytically describe the core losses in equation (2), where k_c, α and β are constant given by the core manufacture. The power rating of the MFT planar transformer is related to the current density and geometrical dimensions of the winding as well as magnetic properties and magnetic core geometry by equation (3):

$$\frac{2P_N}{k_P} = k_V f B_{max} J k_f k_u A_c A_W \tag{3}$$

Where A_c and A_W are the magnetic core cross-section and core window area.

Thermal constraints

Temperature rise of the transformer, i.e. the difference between the hot spot and the ambient temperature can be estimated from the dissipated heat from the total surface area (A_T) of the transformer as in [14]. Therefore, thermal limits are also related to the $\eta\rho$ -plane as follows:

$$P_{Cu} + P_{fr} = h_t A_t \Delta T \tag{4}$$

Where ΔT and h_t are respectively the thermal transfer of the planar transformer. Then equation (4) can be inserted into equation (1):

$$\eta = \frac{\rho - h_t \frac{A_T}{V_T} \Delta T}{\rho} \tag{5}$$

To relate everything to the $\eta\rho$-plane the right side is divided by the transformer volume V_T. For further simplifications the thermal limit can be described with the thermal resistance R_θ.

$$h_t A_t \Delta T = \frac{\Delta T}{R_\theta} \xrightarrow{(1)} \eta = \frac{\rho - \frac{1}{R_\theta V_T} \Delta T}{\rho} = \frac{P_N - \frac{\Delta T}{R_\theta}}{P_N} \tag{6}$$

The thermal resistance for natural convection can be estimated with the core volume (V_C) [13]:

$$R_\theta = \frac{0.06}{\sqrt{V_c}} \xrightarrow{(1)} \eta = \frac{P_N - \frac{\Delta T \sqrt{V_c}}{0.06}}{P_N} \tag{7}$$

Winding losses

To optimize current density the winding losses needs to be considered. In planar MFT the winding losses can be estimated with the relation of AC resistance to DC resistance [15]:

$$\frac{R_{ac}}{R_{dc}} = \frac{\zeta}{2}\left[\frac{\sinh(\zeta) + \sin(\zeta)}{\cosh(\zeta) - \cos(\zeta)} + (2m - 1)\frac{\sinh(\zeta) - \sin(\zeta)}{\cosh(\zeta) + \cos(\zeta)}\right] \tag{8}$$

Where $\zeta = H_{Cu}/\delta$ is the relation of copper thickness H_{Cu} and skin depth δ and m is the magnetomotive force (MMF) ratio at the layer.

For planar transformer the DC resistance can be assumed with the mean length of the track (MLT):

$$R_{dc} = \rho_{Cu}\frac{MLT}{H_{Cu} * L_{Cu}} \tag{9}$$

The MLT can be estimated with the following equation:

$$MLT = 2(d_C + 3L_{Cu} + L_{Cu}(N_{Tracks} - 1)) + L_{mid} \tag{10}$$

Where N_{Tracks} is the number of tracks per layer. The equation for MLT estimation is only a rough estimation and the accuracy of the equation depends on the design of the track. The total winding losses then results of the sum of the AC resistance of all n layers multiplied by current through the winding:

$$P_{Cu,} = \sum_{i=1}^{n} R_{ac,i}I_{rms}^2 \tag{11}$$

Where I_{rms} is the rms current through the winding, which is for a 400:400 V transformer the same in both windings.

$\eta\rho$-Brute-Force Optimization and Results

Comprehensive $\eta\rho$-brute-force optimization is carried out for the specifications mentioned in the introduction section. The simplest possible winding non-interleaved topology is considered. For the simulation different design scenarios are selected. Optimization control variables are copper thickness, current density, magnetic flux, number of PCB layers and number of tracks (turns) per PCB. Moreover, magnetic core geometry is freely swept according to the minimum and maximum planar core dimensions available in the market. The current density varied from 3 to 16 A/mm² in steps of 1 A/mm², the copper thickness from 0.105 mm to 0.28 mm in steps of 0.035 mm and the magnetic flux density from 0.1 to 0.3 T in steps of $(2^k/4) \times 0.1$ T with k from 0 to 3. Fig. 1 shows geometry of core and winding and their positions respective to each other.

To simply show the impact of optimization control variables on the objectives, i.e. η and ρ, a series of 2D-FEM simulations using Ansys Electronics is carried out for different number of PCBs and also tracks at different current densities. Fig. 2 (a), (b) and (c) show the impact of the number of PCB layers and tracks for J=16 A/mm². In this design, only the top layer is considered for copper tracks for maximizing the simplicity of implementation. As it can be seen maximum efficiency can be achieved for a specific number of tracks per layer and also there is a trade-off between efficiency and power density versus the number of PCBs. The overall results with variation of J, B_m and core and winding geometries for a fixed number of PCBs (i.e. 4) and tracks (i.e. 2) is shown in Fig. 2 (d).

Fig. 3 (a) illustrates the overall Pareto front obtained from the $\eta\rho$-brute-force optimization. Thermal limitations which are explained theoretically in $\eta\rho$-plane by equation (4) – (7). The Pareto front with thermal limitation is illustrated in Fig. 3 (b). Only the remaining solutions within thermal limitation are shown. Considering a minimum acceptable 99.5% for efficiency, a power density beyond 30 kW/l can be achieved. Utilizing a cooling system capable for removing larger amount of heat, lower efficiencies with higher power densities and higher current densities can be employed. Two solutions in the Pareto front are selected and the power losses distribution in the core and their current density are shown in Fig. 3 (c) and (d). For the same selections, detailed design parameters are presented in table I. As can be seen from this table both the designs are achieved at high current densities. These current densities are 3 times higher than a typical value for litz wires as in [16].

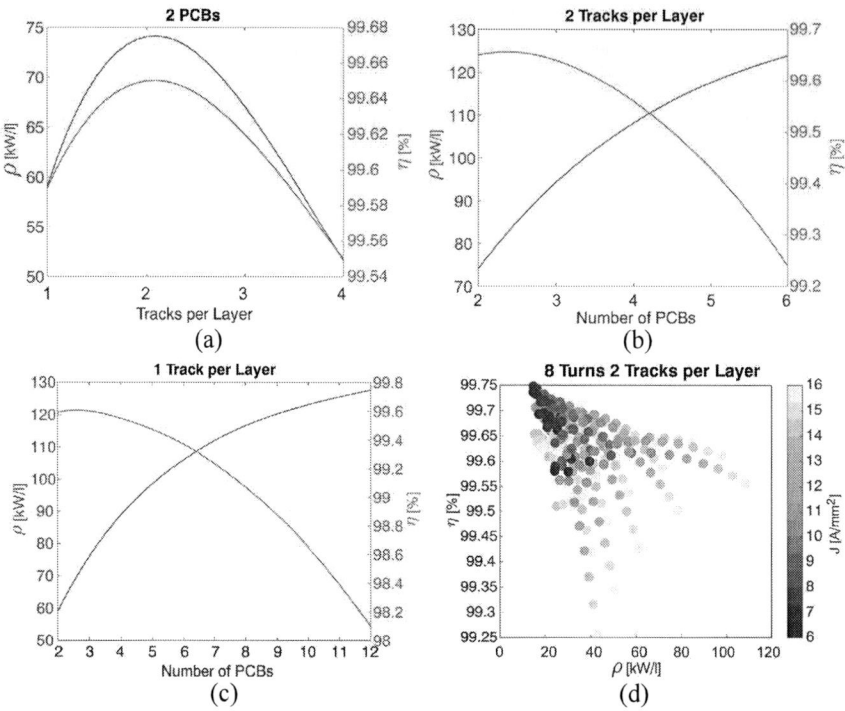

Fig. 2: Analysis for different number of tracks and PCBs at J=16 A/mm^2: (a) variation of efficiency and power density versus the number of tracks per layer, (b) variation of efficiency and power density with 2 tracks per layer versus the number of PCBs, and (c) variation of efficiency and power density with 1 track per layer versus the number of PCBs and (d) variation of copper thickness H_{Cu}, B_m and J.

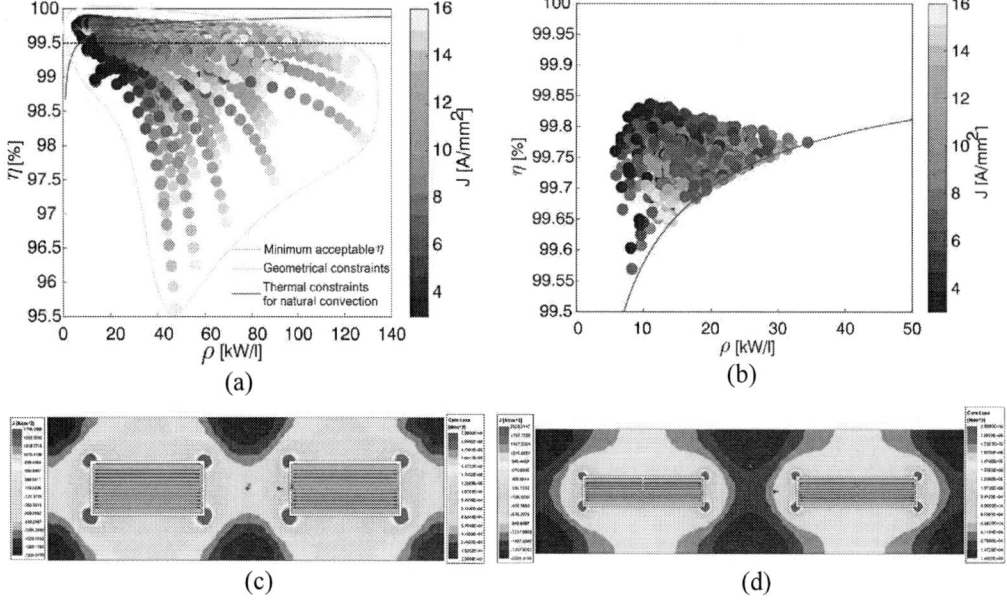

Fig. 3: Analysis for different number of tracks and PCBs: (a) comprehensive $\eta\rho$-brute-force optimization to form the Pareto front, (b) remaining Pareto front within thermal constraints, (c) selected 2D FEM design for 6 PCB layer per winding and 1 track per layer, (d) selected 2D FEM design for 4 PCB layer per winding and 2 tracks per layer.

Table I: Optimization results for 2 selected solution respect to $\eta\rho$ trade-off

Parameters	Variable	6 PCBs 1 Tracks	4 PCBs 2 Tracks
Volume [dm³]	V_T	0.2909	0.4484
Power density [kW/l]	ρ	34.38	22.30
Current density [A/mm²]	J	**7**	**8**
Efficiency [%]	η	99.77	99.72
Number of turns	T	6	8
Core width [mm]	d_C	135.9	123.62
Core length [mm]	L_C	78.43	112.92
Core height [mm]	H_C	27.29	32.12
Core window length [mm]	L_W	22.41	32.26
Core window height [mm]	H_W	10.49	7.93
Winding length [mm]	L_{Wi}	20.41	31.26
Winding height [mm]	H_{Wi}	4.425	3.47
Copper thickness [mm]	H_{Cu}	0.175	0.21
Copper width per track [mm]	L_{Cu}	20.41	14.88

Interleaving

For further current density optimization interleaving is considered. Interleaving the windings lowers the winding losses. The ratio of AC to DC resistance is shown in Fig. 4 (a). The ratio increases with increase of ζ, which with a constant δ is an increase of the copper thickness H_{Cu}. The higher the MMF the higher the ratio of resistance. The ratio of the AC to DC resistance can be interpreted as the losses in the windings. This is due to the fact, that the change of dc-resistance with the length L_{Cu} or the thickness H_{Cu} of the copper is lower than the change in the ratio due to the MMF. In Fig. 4 (b) the distribution of the MMF is shown for an example of 4 turns. When the windings are non-interleaved the MMF increases up to 4. Interleaving every layer will result in a constant MMF of 1.

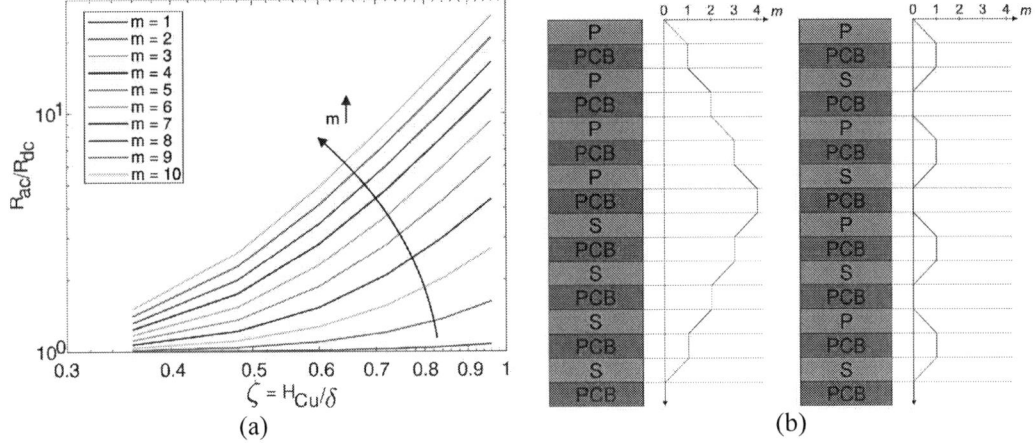

Fig. 4: (a) Ratio of AC to DC resistance as a function of ζ with the swept copper thickness H_{Cu} and (b) example of distribution of the MMF for 4 turns.

The effect of interleaving is shown for the example of 6 turns and 1 track per Layer in Fig 5. Fig. 5 (a) shows the pareto front of the non-interleaved structure and Fig. 5 (b) shows the pareto front of every layer interleaved structure (P-S-P-S-P-S-P-S). It is visible that the overall efficiency increases with interleaving the windings. The higher the copper thickness H_{Cu} the higher the increase of efficiency. This leads to an overall availability of higher current densities within thermal constraints.

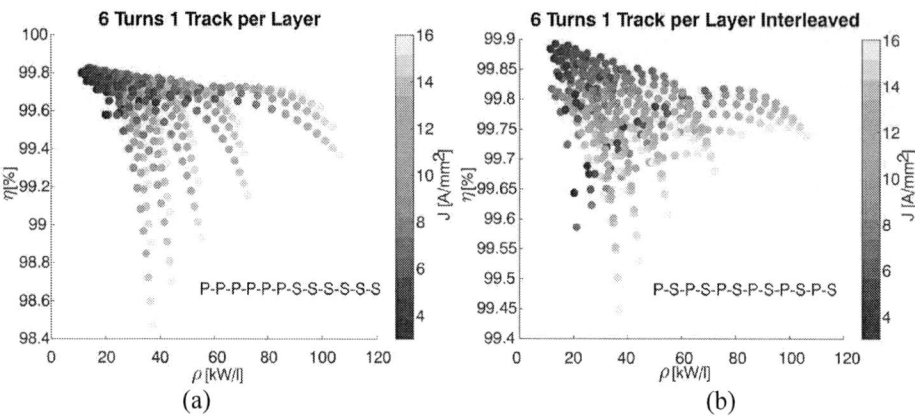

(a)   (b)

Fig. 5: Analysis of the effect of interleaving on the losses on the example of a 6 turn 1 track per layer transformer (a) non-interleaved and (b) interleaved P-S-P-S-P-S-P-S.

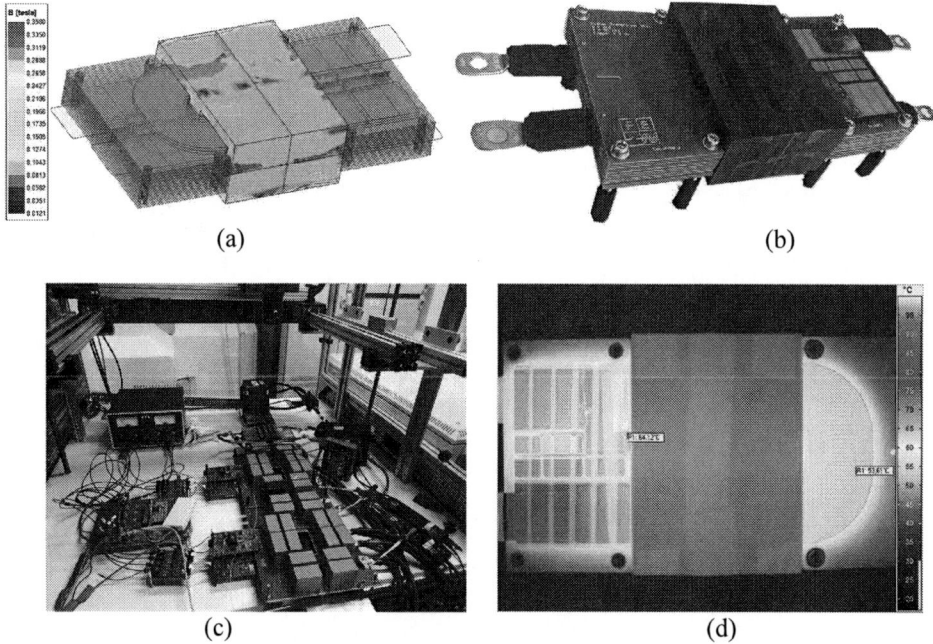

Fig. 6: Experimental results: (a) 3D FEM analysis of an example design to verify the performance, (b) developed hardware, (c) DAB converter prototype operated at 50 kHz and (d) Planar transformer at nominal current density in short circuit test.

Experimental validation

Current density optimization is a trade-off between core losses and copper losses. In the following work different current densities are analyzed and compared to show the impact of current density in planar MFT. Fig. 6 includes the planar MFT prototype and the DAB converter setup for operating at 50 kHz. Fig. 6 (a) shows the 3D Ansys simulation of a simple planar MFT design with 1 track per layer and total of 6 turns where the magnetic field density is at linear range when excited with nominal quantities and approximately distributed uniformly. Fig. 6 (b) shows the built prototype with similar ratings as in table I and Fig. 6 (c) shows the planar MFT connected to the DAB converter. The ratings of the build prototype are given in table II. The design given in table II was selected, because lower copper thickness results in higher current density, to prove that high current densities are feasible for planar transformers. The winding structure of the build prototype is P-P-P-S-S-S-S-S-S-P-P-P. The temperature of the copper tracks at nominal current density is shown in Fig. 6 (d).

Table II: Parameters of the designed planar MFT

Parameters	Variable	6 PCBs 1 Track
Volume [dm³]	V_T	0.796
Power density [kW/l]	ρ	12.56
Current density [A/mm²]	J	**12**
Number of turns	T	6
Core width [mm]	d_C	75
Core length [mm]	L_C	102
Core height [mm]	H_C	40.6
Core window length [mm]	L_W	36
Core window height [mm]	H_W	26.6
PCB width [mm]	d_{PCB}	192.1
Copper thickness [mm]	H_{Cu}	0.07
Copper width per track [mm]	L_{Cu}	29.8

Fig. 7 shows the recorded temperature at different current densities in short circuit test (a) and with applied voltage (b). In the short circuit test the core temperature is due to the heat of the windings. At the nominal current density, the hotspot temperature reaches 64 °C. The temperature is increasing with the square of the current. In Fig. 7 (b) the temperatures are given with applied voltage, where the phase shift of the converter is hold constant. The current is increased through applying higher voltages. The temperatures at load is higher than in short circuit test, due to higher temperature in the core. It is visible, that in short circuit test and test at load that the temperature curves are very similar and the temperature difference between the track and hotspot to the core are in both case close together. This means, that with lower temperature in the core, higher current density within acceptable temperatures can be achieved. The build prototype achieves a current density of 5 times of litz wires and can achieve a current density up to 6 times of litz wire with lower applied voltage.

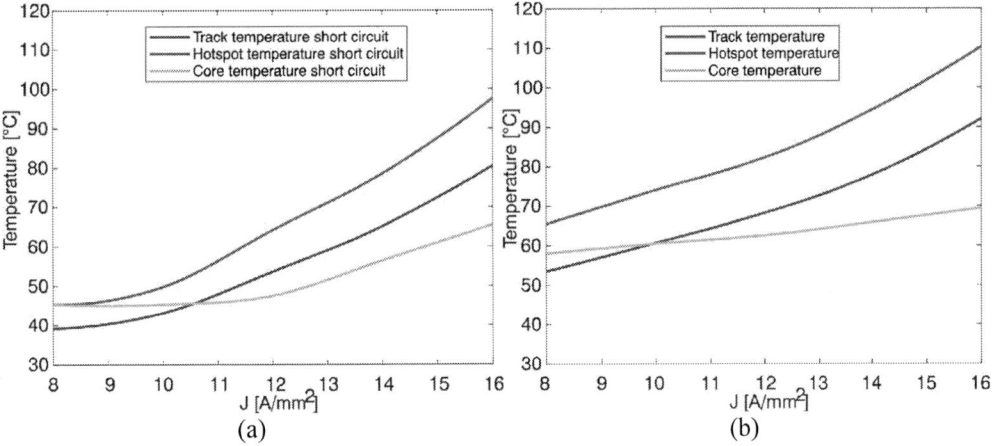

Fig. 7: Experimental results of temperature measurement on different current densities: (a) temperature during short circuit test and (b) temperature with increasing voltage at higher current densities.

Conclusion

This paper presents a comprehensive optimization of planar MFT windings to achieve high efficiency and power density for on-board EV chargers. The behavior of the planar MFT design in $\eta\rho$-plane is evaluated theoretically and analytical equations for explaining the variation of efficiency versus power density as well as thermal constraints. Extensive $\eta\rho$-brute-force optimization is carried out for ratings of 50 kHz 400 V 10 kW based on 2D-FEM. Simulation results show the feasibility of designs with efficiencies and power densities beyond 99.5% and 30 kW/l, respectively. Moreover, a high level of current density, up to 4 times of a conventional litz wire, can be applied to the planar MFT windings without significantly increasing the temperature rise. Interleaving the windings at higher copper thickness can increase the current density further. Simulation results show that at 50 kHz, 400 V and 10 kW current densities up to 10 A/mm² are feasible at natural convection. For solutions with forced convection higher current densities up to 16 A/mm² results in the best trade-off between power density and efficiency. The experiments show that high current densities within acceptable temperatures are feasible and that with a trade-off between current density and applied voltage, the current density can be increased further.

References

[1] He P., Mallik A., Sankar A. and Khaligh A.: Design of a 1-MHz high-efficiency high-power-density bidirectional GaN-based CLLC converter for electric vehicles, IEEE Trans. Veh. Technol. vol. 68 no. 1 pp. 213–223, 2018.

[2] Son W.-J., Lee J. H., Ann S., Byun J. and Lee B. K.: Design Methodology of a Planar Transformer in LLC Converters for High Power Density On-Board Chargers, in 2021 IEEE 12th Energy Conversion Congress & Exposition-Asia (ECCE-Asia) pp. 1613–1617, 2021.

[3] Yao C.: Semiconductor galvanic isolation based onboard vehicle battery chargers, The Ohio State University, 2018.

[4] Park Y., Chakraborty S. and Khaligh A.: DAB Converter for EV On-Board Chargers Using Bare-die SiC MOSFETs and Leakage-Integrated Planar Transformer, IEEE Trans. Transp. Electrification, 2021.

[5] Ramadhan R., Suk C., Kim S., Choi S., Yu B. and Park S.: Integrated Planar Transformer Design of 3 kW LDC for Electric Vehicles, in Proceedings of the KIPE Conference, pp. 157–159, 2020.

[6] Jaritz M., Hillers A. and Biela J.: General analytical model for the thermal resistance of windings made of solid or litz wire, IEEE Trans. Power Electron. vol. 34 no. 1 pp. 668–684, 2018.

[7] Fei C., Lee F. C. and Li Q.: High-efficiency high-power-density LLC converter with an integrated planar matrix transformer for high-output current applications, IEEE Trans. Ind. Electron. vol. 64 no. 11 pp. 9072–9082, 2017.

[8] Li B., Li Q. and Lee F. C.: A novel PCB winding transformer with controllable leakage integration for a 6.6 kW 500kHz high efficiency high density bi-directional on-board charger, in 2017 IEEE Applied Power Electronics Conference and Exposition (APEC) pp. 2917–2924, 2017.

[9] Ouyang Z., Thomsen O. C. and Andersen M. A.: Optimal design and tradeoff analysis of planar transformer in high-power DC–DC converters, IEEE Trans. Ind. Electron. vol. 59 no. 7 pp. 2800–2810, 2010.

[10] Fu M., Fei C., Yang Y., Li Q. and Lee F. C.: Optimal design of planar magnetic components for a two-stage GaN-based DC–DC converter, IEEE Trans. Power Electron. vol. 34 no. 4 pp. 3329–3338, 2018.

[11] Pereira T. A. et al.: Optimal design of planar transformer for GaN based phase-shifted full bridge converter, in 2020 IEEE Applied Power Electronics Conference and Exposition (APEC) pp. 2241–2248, 2020.

[12] Beiranvand H., Rokrok E. and Liserre M.: Vf-constrained ηρ-pareto optimisation of medium frequency transformers in ISOP-DAB converters, IET Power Electron. Vol. 13 no. 10 pp. 1984–1994, 2020.

[13] Hurley W. G. and Wölfle W. H.: Transformers and inductors for power electronics: theory, design and applications. John Wiley & Sons, 2013.

[14] Mogorovic M. and Dujic D.: Sensitivity analysis of medium-frequency transformer designs for solid-state transformers, IEEE Trans. Power Electron. vol. 34 no. 9 pp. 8356–8367, 2018.

[15] Ferreira J.: Improved analytical modeling of conductive losses in magnetic components, IEEE Trans. Power Electron., vol. 9, no. 1, pp. 127–131, 1994.

[16] Bahmani A.: Design and optimization considerations of medium-frequency power transformers in high-power DC-DC applications. Chalmers Tekniska Hogskola (Sweden), 2016.

Fault Current Capability Assessment of Low-Voltage side Inverters in Smart-Transformers

Thiago Pereira, Luis Camurca, Francisco Santos, Marco Liserre
Chair of Power Electronics, Kiel University
Kaiserstraße 2, 24143 - Kiel, Germany
Email: tp@tf.uni-kiel.de

Acknowledgments

This work was supported in part by the German Federal Ministry for Economic Affairs and Energy (BMWi) within Project KielFlex Kiel als Vorbild für die Errichtung von Ladeinfrastruktur in einem flexiblen Stromnetz zur Umsetzung einer Emissionsreduktion im Transportsektor under Grant 01MZ18002D and within Priority Programme "Energy Efficient Power Electronics 'GaNius' (DFG SPP 2312).

Keywords

≪Smart-Transformer≫, ≪Overload≫, ≪Overcurrent≫, ≪Thermal model≫ ≪DC-AC converter≫

Abstract

The Smart Transformer (ST) arises as one promising solution for the modern electric grid by providing ancillary services to support AC and DC distribution grid. In this context, as the classical low-frequency transformer (LFT), the ST is prone to experience overload conditions caused by faults and peak loads. However, unlike the LFT, the overcurrent capability of the ST is further limited by the thermal time constant of the power semiconductors. Thus, since the overload operation has been only partially investigated, this paper proposes and presents a comprehensive analysis of the LV-side power converter under normal and overload conditions to estimate its overcurrent capability. For this purpose, the LV-side inverter is assessed in terms of power losses considering continuous and discontinuous modulation strategies and different power semiconductor technologies along with multiple types of cooling systems.

Introduction

The ST is a complex system comprised usually of several power converter stages and, as such, the ST is subject to a multitude of different failure modes that has to handle, e.g., internal faults (e.g. semiconductor failures, thermomechanical failures, control errors, and insulation breakdowns), MV/LV short circuit, switching transients, and non-ideal load [1–3]. Unlike the classical LFTs, STs have limited overcurrent capabilities, where the employed power semiconductors represent the main limiting factor due to their low admissible overcurrent ratios (around $1.5 \times$ the nominal current for some minutes and $4 \times$ for some milliseconds, while the LFT can withstand $3 \times$ the rated current for several minutes [1,4]). In other words, the safety limit of power semiconductors devices is defined by the maximum operating junction temperature $(T_{j,max})$, which indicates that the operation of ST is possible only when the power devices operate with T_j lower than $T_{j,max}$. Above $T_{j,max}$, thermomechanical failures might happen [5].

For instance, in case of a fault at the LV side, the LV-side inverter should be the last stage of the ST to trip cf. Fig. 1 (a), according to the selectivity requirements. Consequently, the LV-side inverter has to handle higher current values for some time interval until the LV-side protection devices trips (i.e. breakers, and fuse as the last resources). The current values and time interval depend on the adopted protection device and they are usually defined by the time-current curves (e.g. tripping a 1.0 MVA and 0.5 MVA fuse demands 12 s for $5 \times$ the nominal current and 20 s for $2 \times$ the nominal current, respectively [1]). Therefore, regardless of the adopted protection devices, without prior knowledge of the overcurrent capability information and its maximum allowable limits, the suitable design of the ST's LV-side design, including the protection scheme, is hardly achieved [6,7].

In the literature, there is an increasing number of works focused on the ST operation [1–4, 8]. Nevertheless, the overcurrent capability of an LV-side inverter has been only partially investigated for ST applications [1, 4, 9]. Understanding that these critical operation modes are essential for inserting ST in future grids, this paper assesses the fault current capability of the LV-side inverters considering different modulation schemes and power semiconductor technologies (Si IGBT and SiC MOSFET). For this purpose, theoretical analysis of the power losses and thermal behavior are systematically addressed in a methodology considering the inherent characteristic of the power devices.

The paper is organized as follows. Section II describes the ST architecture and the investigated scenarios. Then, Section III summarizes the employed methodology and also the adopted assumptions for the overcurrent capability assessment. Next, Section IV and V present the power losses analysis of the power devices and the developed thermal network model, respectively. After that, the obtained results are discussed in Section VI. Finally, the outcomes are provided in Section VII.

LV Side Inverter of the Smart Transformer - System Description

In terms of temperature rise, the ST has the potential to handle more than the designed rated power, depending on the power semiconductor devices and overall design system [2, 4, 8]. However, it is also necessary that the system itself might provide such power. Therefore, it is considered that the MVAC stage provides the rated power (1.0 pu) while the LVDC will support the LVAC with the additional power by means of the storage system, cf. Fig.1 (a). Hence, in this scenario, the overload operation will affect only the LV-side inverter of the ST, such that the other stages are not considered in the next analysis. The DC-AC inverter, or 2L-inverter cf. Fig. 1 (b), considered in this paper is a four-leg inverter due to the requirement of a 3-phase 4-wire system in the distribution grid and for the fault identification (i.e. by using the neutral wire). The overall system's specifications are described in Table I.

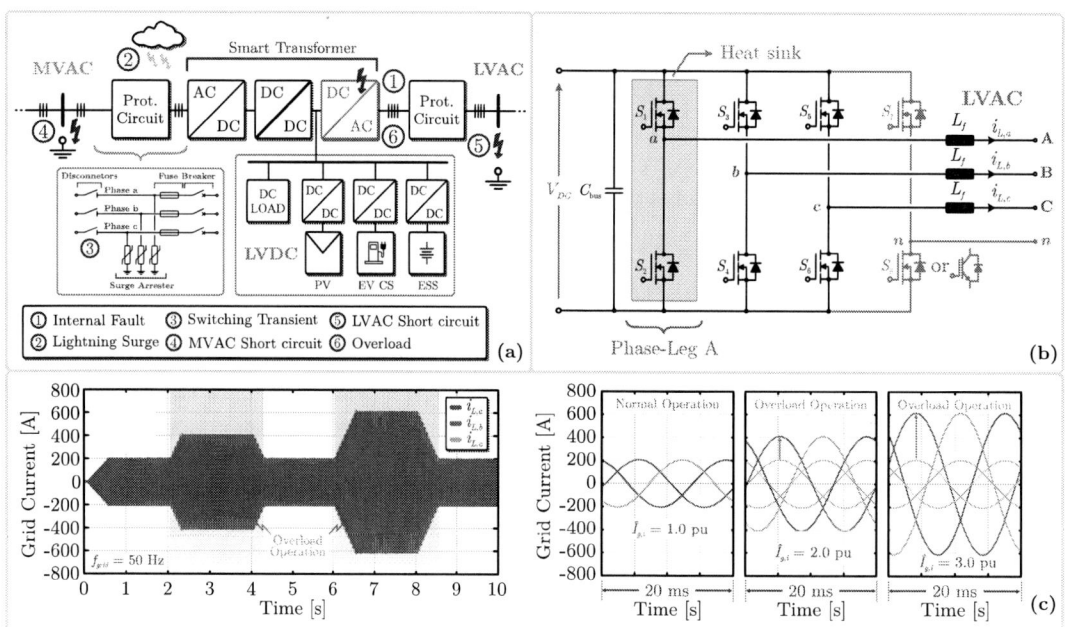

Fig. 1: (a) Electric distribution grid considering the possible abnormal operations on the smart transformer's stage [1]; (b) the four-leg 2L-inverter, and (c) current profile considering the normal operation (1.0 pu) and two different overload conditions (for 2.0 pu and 3.0 pu).

Table I: Electrical specification of the whole System considering the nominal operation.

Power	Voltage (LVAC)	Current (LVAC)	V_{DC} - LVDC	$f_{grid} = 1/T$	$f_{sw} = 1/T_{sw}$	L_f	$\cos(\varphi)$
100 kVA	400 V	145 A (1.0 pu)	800 V	50 Hz	5 kHz	1.2 mH	1.0

Overcurrent Capability Assessment of the LV side Inverter

The overcurrent capabilities of the 2L-inverter, cf. Fig. 1 (b)-(c), will be investigated by using analytical models taking into account the continuous and discontinuous pulse width modulation (CPWM and DPWM, respectively) strategies, cf. Fig. 1 [10–17]. The semiconductor loss models are based on analytical conduction/switching loss equations considering their equivalent polynomial approximations obtained from the datasheets. The thermal models are based on the equivalent electrical circuits to derive the temperature's evolution. The analysis is carried for two conditions: (i) in normal and (ii) in overload operation (grid current higher than 1.0 pu). Thereby, the focus of the comparison is on the junction temperatures T_j and on the power losses of the power semiconductors. In this context, Simulink/PLECS environment is used to validate the developed models. Further, since the analysis involves complex thermal effects with limited information, the following assumptions are made to proceed with the analysis:

- The phase current presents no current ripple and hence it is perfectly sinusoidal.
- The operation of the 2L-inverter is symmetrical and balanced.
- The phase current is defined by $i_L(\omega t) = \hat{I}_L \sin(\omega t - \varphi)$, where $\hat{I}_L$ is the peak current value and φ is the phase angle (which can particularly define the power factor, i.e. $\cos\varphi$).
- The 2L-inverter is composed of three half-bridge power modules and three individual heat sinks.
- The switching energy of the power devices is phase-current dependent.
- Although power modules for high power applications are usually made of multiple chips in parallel, the analysis is performed considering a single chip (according to the available data).
- The model is only valid for junction temperatures between $25\,°C$ (which is defined as the ambient temperature T_{amb}) and $T_{j,max}$ (which can be either $150\,°C$ or $175\,°C$).
- The heat flow has a one-dimensional propagation.

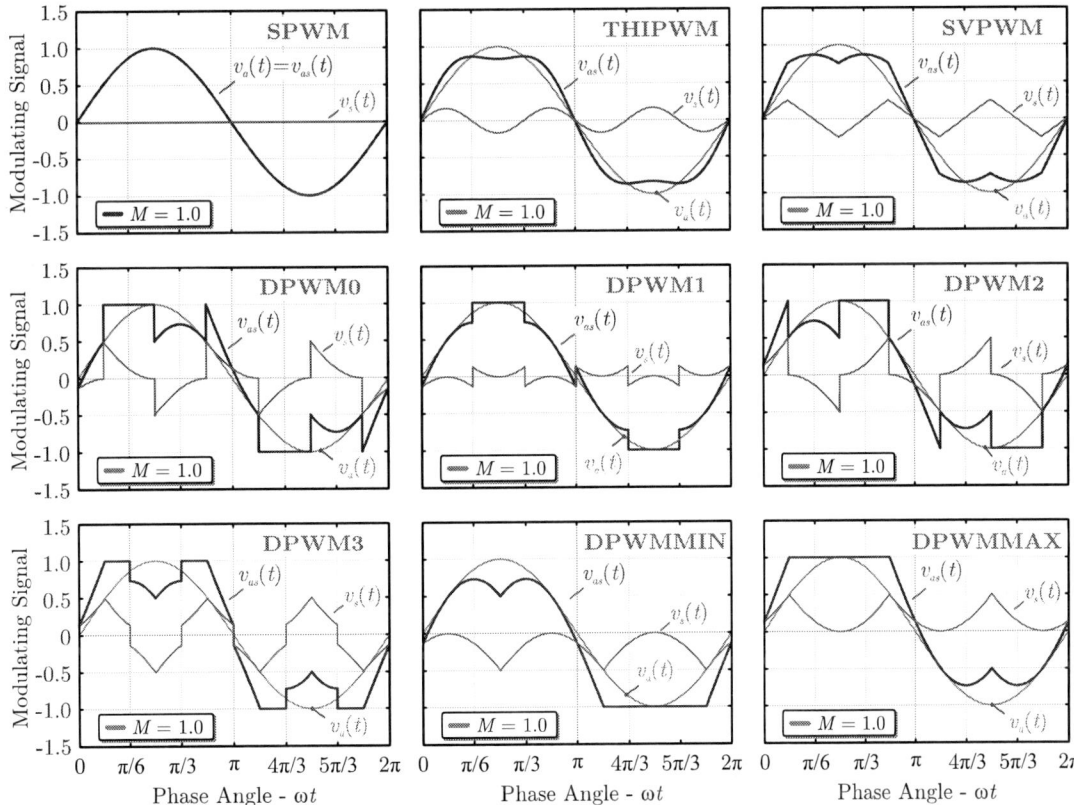

Fig. 2: Modulation strategies for phase a considering a modulation index $M = 1.0$, where v_a is the sinusoidal reference and $v_{as}(t)$ is the modulation functions, which is defined as the sum of $v_a(t)$ and the zero-sequence signal $v_s(t)$, i.e. $v_{as}(t) = v_a(t) + v_s(t)$ [10].

Power Losses Analysis of the Power Semiconductor Devices

The losses in the power devices of the 2L-inverter are comprised of both conduction losses and switching losses. These losses are highly dependent upon the power semiconductor technologies due to the different thermal dependence (i.e. Si IGBT or SiC MOSFET) and the modulation strategy due to the different duty-cycle patterns, as illustrated in Fig. 2. Therefore, the average value of the conduction losses P_{cond} for the most common power semiconductor devices can be mathematically estimated by (1).

$$P_{cond} = \frac{1}{T} \int_0^T v(i, T_j) i(\tau) d\tau \approx \frac{1}{2\pi} \begin{cases} \int_{\varphi}^{\varphi+\pi} d(\theta) \left[V_{CE}(i_L, T_j) + R_{on}(i_L, T_j) i_L(\theta) \right] i_L(\tau) d\theta, & \forall \text{ IGBT} \\ \int_{\varphi+\pi}^{\varphi+2\pi} d(\theta) \left[V_F(i_L, T_j) + R_D(i_L, T_j) i_L(\theta) \right] i_L(\theta) d\theta, & \forall \text{ APD} \\ \int_{\varphi}^{\varphi+2\pi} d(\theta) \left[R_{DS,on}(i_L, T_j) i_L(\theta) \right] i_L(\theta) d\theta, & \forall \text{ MOSFET} \end{cases} \quad (1)$$

Whereas the on-state voltage $v(i(\tau), T_j)$ is approximated around a phase-current value by means of the following thermal-dependent parameters: V_{CE} and V_F represent the forward voltage of the IGBT and anti-parallel diode (APD) respectively; R_{on} and R_D defines the dynamic resistance of the IGBT and APD respectively; and $R_{DS,on}$ represents on-resistance of the MOSFET. These parameters are defined by applying a first-order curve fitting of the on-state voltage. In Fig. 3 (a)-(d), the aforementioned electrical parameters are presented. As can be seen, they are highly dependent on the junction temperature T_j and phase current, which is assumed to be sinusoidal. Further, $d(\theta)$ indicates the duty-cycle function of a specific modulation strategy in one phase-current period. The resulting expressions obtained by using (1) are listed in Appendix (cf. Table V and VI) and depicted in Fig 4 (a)-(b).

It should be noticed that the DPWMMAX and DPWMMIN methods have unequal thermal stress on the power devices. For DPWMMAX, the high-side device of the half-bridge power module has higher conduction losses than the low-side one, while for DPWMMIN occurs the opposite. Thus, both modulation strategies are not considered in the analysis, since the overcurrent capability is inherently reduced. For the other modulation strategies, the conduction losses remain slightly similar, cf. Fig 4 (a)-(b).

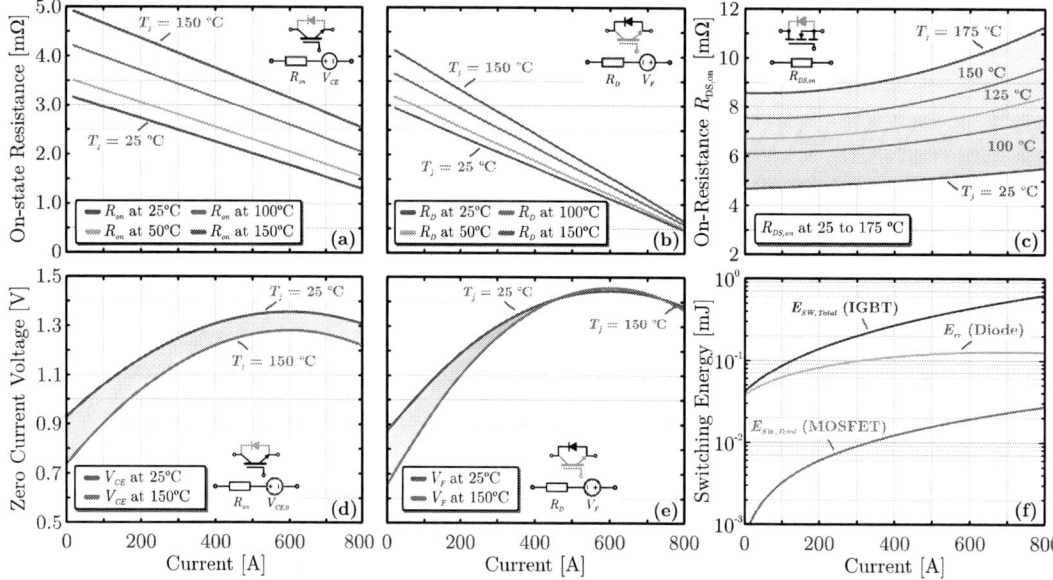

Fig. 3: Thermal-dependent parameters of the IGBT half-bridge module FF400R17KE4 and the SiC MOSFET half-bridge module CAB400M12XM3 considering the approximation of the on-stage voltage around the operating current value. Whereas (a) R_{on}, (b) R_D, and $R_{DS,on}$ represent respectively the on-resistance of the IGBT, APD and MOSFET; while (d) V_{CE} and (e) V_F represent the forward voltage of the IGBT and APD. (c) Switching energy losses extracted from the reference power module datasheets for 800 V, where $E_{sw,Total} = E_{on} + E_{off}$.

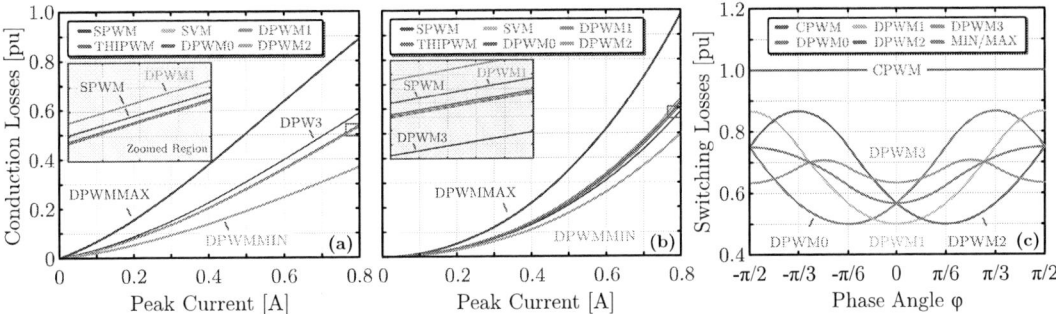

Fig. 4: Normalized conduction losses considering the following power modules for the continuous and discontinuous modulation strategies (where $P_{base} = 1.2\,\mathrm{kW}$): (a) IGBT half-bridge module FF400R17KE4, and (b) SiC MOSFET half-bridge module CAB400M12XM3. Normalized switching losses with respect to the switching losses of CPWM (i.e. $P_{sw,CPWM}$) for: (c) the continuous and discontinuous modulation strategies.

For the calculation of the switching losses P_{sw} considering different modulation strategies (cf. Fig.1), it is assumed a linear dependency of the switching energy losses w_S with respect to the phase current $i_L(t)$, i.e. $w_S \propto k_s i_L$ (where k_s is the constant related to the voltage across the device and to turn-on/off times) [13, 18–21]. Hence, the average value of the switching loss can be calculated according to (2).

$$P_{sw} = \left[\frac{V_{DC}(t_{on}+t_{off})f_{sw}}{4\pi}\right]\int_0^{2\pi}F_i(\theta)d\theta \approx \frac{E_{sw}}{2\pi}\int_0^{2\pi}F_i(\theta)d\theta, \quad \text{where} \quad F_i(\theta) = \begin{cases} 0, & |v_{as}| \geq 1.0 \\ |i_L|, & |v_{as}| < 1.0 \end{cases} \quad (2)$$

In (2), $F_i(\theta)$ is the switching current function, which defines generically the characteristic of the continuous and discontinuous modulation strategies. The switching current function is equal to zero during the intervals in which $v_{as}(t)$ is clamped. Otherwise, $F_i(\theta)$ is equal to the absolute value of the phase current $|i_L|$. For instance, by using the CPWM strategies, the phase currents are commutated within each switching period T_{sw} during an entire fundamental period T [10, 22]. Hence, the switching losses are the same and independent of the phase angle φ, which yields $P_{sw,CPWM} = (\hat{I}_L E_{sw})/\pi$. On the other hand, for the DPWM strategies, the switching losses are influenced by the modulation method and phase angle φ, as shown in Fig. 4 (c). As a result, by selecting DPWM strategies, it is noticed that the switching losses can be reduced up to 50 % as compared to the CPWM, cf. Fig. 4 (c).

Thermal Network Model of Half-Bridge Power Modules

A thermal network model is developed to estimate the junction temperature of the half-bridge power modules taking into account the power losses, as shown in Fig. 5 (a). For this reason, a thermal network model is required to represent the transient thermal behavior of the power semiconductor during the normal and overload conditions for different modulation strategies. as a result, the thermal network model allows the estimation of the overcurrent capability of the 2L-inverter by means of the maximum junction temperature $T_{j,max}$ and the respective time interval to reach this value (e.g. 150 °C).

Previous investigations have shown the performance advantage of the Cauer model for the thermal behavior analysis when more elements are considered along with the device's $R_{th}C_{th}$ network (e.g. TIM and heatsink) [23, 24]. In addition, Cauer models are meaningful to describe the internal structure of the device and therefore this model was selected to develop the thermal network model, cf. Fig. 5 (b) and Table II. The complete model based on the state-space representation of the Cauer model is implemented to estimate the T_j of each chip by means of (3) and (4) (cf. Appendix). Whereas, $T_{j,n+1}$ is the chip's junction temperature ($n = 0, 1, 2, 3$), T_{TIM} and T_{HS} are the temperatures on the TIM and heat sink surface respectively, and ΔT_s is the sampling period, which should be $\leq 1.0\,\mu s$.

The heat sink plays an important role to ensure a specific T_j of the power devices by supporting the heat exchange with the air (or water). For selecting a suitable cooling system, it is essential to know its thermal performance along with the thermal properties of the power semiconductor devices and the requirements given by the operation (e.g. power at normal and overload operation, T_{amb}, and $T_{j,max}$). Therefore, several types of heat sinks are included in the analysis, as shown in Fig. 5 (c)-(f).

Table II: Thermal model of the adopted power semiconductor devices considering the $R_{th}C_{th}$ network.

Device	$R_{th,1}$	$R_{th,2}$	$R_{th,3}$	$R_{th,4}$	$C_{th,1}$	$C_{th,2}$	$C_{th,3}$	$C_{th,4}$
IGBT	0.0050 K/W	0.0117 K/W	0.0429 K/W	0.0036 K/W	0.0371 J/K	0.3840 J/K	0.6328 J/K	155.30 J/K
APD	0.0152 K/W	0.0691 K/W	0.0166 K/W	0.0052 K/W	0.1346 J/K	0.3831 J/K	7.00 J/K	211.39 J/K
MOSFET	0.0588 K/W	0.0398 K/W	0.0524 K/W	-	0.1526 J/K	0.3364 J/K	2.4430 J/K	-

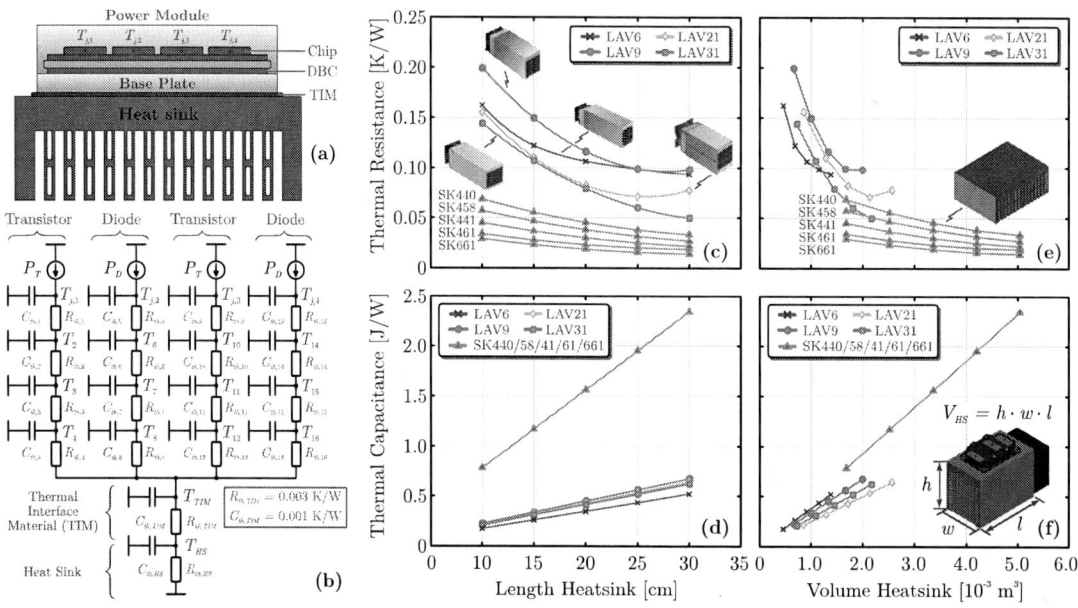

Fig. 5: (a) Cross-section view of a half-bridge power module placed on a heat sink. (b) Thermal network model of the half-bridge considering the thermal interface material (TIM) and the heat sink. Thermal resistance R_{th} and thermal capacitance C_{th} values for different heat sinks in terms of: (a)-(d) length in cm; and (e)-(f) volume in m^3.

$$T_{j,n+1}[k+1] = T_{j,n+1}[k] + \left(\frac{P_{loss}[k]\Delta T_s}{C_{th,4n+1}} \right) + \Delta T_s \left(\frac{T_{4n+2}[k] - T_{j,n+1}[k]}{R_{th,4n+1}C_{th,4n+1}} \right) \text{ for } n = 0,1,2,3$$

$$T_{TIM}[k+1] = T_{TIM}[k] + \left(\frac{\Delta T_s T_{HS}[k]}{R_{th,TIM}C_{th,TIM}} \right) + \Delta T_s \sum_{n=1}^{4} \left(\frac{T_{j,4n}[k] - T_{TIM}[k]}{R_{th,4n}C_{th,TIM}} \right)$$

$$T_{HS}[k+1] = \left(\frac{\Delta T_s T_{amb}[k]}{R_{th,HS}C_{th,HS}} \right) + \left[1 - \frac{\Delta T_s (R_{th,TIM} + R_{th,HS})}{R_{th,TIM}C_{th,HS}C_{th,HS}} \right] T_{HS}[k] + \left(\frac{\Delta T_s T_{TIM}[k]}{R_{th,TIM}C_{th,HS}} \right)$$

(3)

Overcurrent Capability Analysis - Results and Discussion

In order to perform the overcurrent capability analysis among the different modulation strategies, power semiconductor devices, and heat sinks, the 2L-inverter was evaluated under multiple overload conditions (i.e. current values from 1.0 pu to 4.0 pu), considering the specifications of Table I. As can be seen, Fig. 6 and Fig. 7 exhibit the junction temperatures in steady-state for the SiC MOSFET and IGBT power module, respectively. Regardless of the modulation strategies, the power module based on SiC MOSFET has an overcurrent capability limited to 3.0 pu, while the IGBT counterpart has the overcurrent capability extended to 4.0 pu when the DPWM strategies are applied (and the proper cooling system is adopted).

Due to the fact that total losses of the power devices depend on the junction temperature, the results illustrated in Fig. 5 and Fig. 6 provide only the junction temperature values after reaching the thermal steady-state for each overload condition. Nevertheless, as part essential of the overcurrent capability analysis, the related time interval to achieve these temperature values should be defined by monitoring the T_j evolution with respect to the time. as a result, it is possible to estimate the time interval in which the device can withstand the overload conditions.

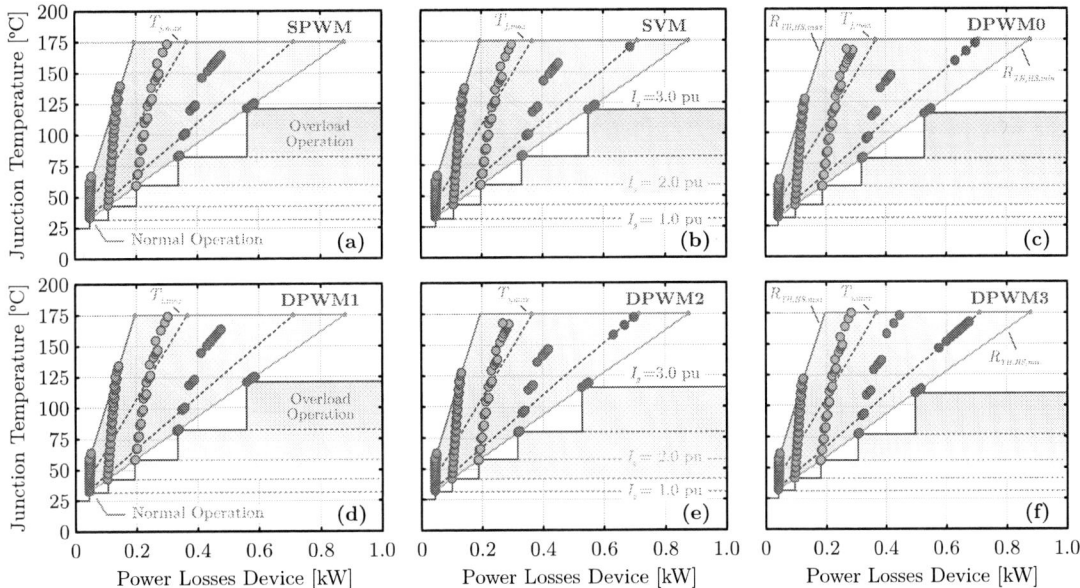

Fig. 6: Steady-state behavior of the junction temperature on the SiC MOSFET CAB400M12XM3 considering the influence of modulation strategies, semiconductor technologies, and cooling systems for different overload conditions. Where the thermal resistance varies within a range cf. Fig. 5, i.e. $R_{th,HS} \in [0.005\,\text{K/W}, 0.30\,\text{K/W}]$.

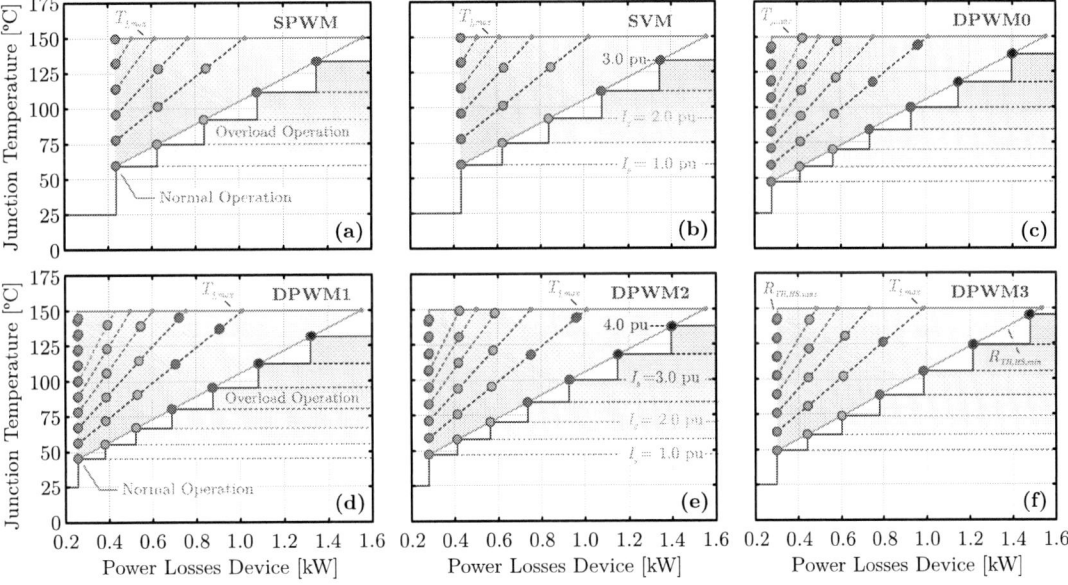

Fig. 7: Steady-state behavior of the junction temperature on the IGBT FF400R17KE4 considering the influence of modulation strategies, semiconductor technologies, and cooling systems for different overload conditions. Where the thermal resistance varies within a range cf. Fig. 5, i.e. $R_{th,HS} \in [0.005\,\text{K/W}, 0.30\,\text{K/W}]$.

For this purpose, in order to reduce the number of data, the number of cooling systems was reduced to only three different types, as listed in Table III. These parameters were extracted from Fig. 5 (c)-(f) to ensure $T_j \leq 75\,^\circ\text{C}$ at normal operation (without increasing drastically V_{HS}). As a result, different heat sinks are required for each power device due to their inherent thermal characteristic. Thus, based on these values of $R_{th,HS}$ and $C_{th,HS}$, all junction temperatures are monitored systematically under three overload conditions (2.0 pu, 2.5 pu, and 3.0 pu), considering SPWM and DPWM1 strategies due to the similarity with the other ones, cf. Fig. 8 and Fig. 9. The time is accounting between the overload instant (i.e. 100 s) and the instant which the temperature reach $T_{j,max}$ (i.e. 150 °C for IGBT and 175 °C for SiC MOSFET).

Table III: Thermal resistance $R_{th,HS}$ and capacitance $C_{th,HS}$ of the adopted heat sinks for the analysis, cf. Fig. 5.

Power Device	Si IGBT Power Module			SiC MOSFET Power Module		
Heat sink	Type I	Type II	Type III	Type I	Type II	Type III
Thermal Resistance $R_{th,HS}$	0.018 K/W	0.024 K/W	0.030 K/W	0.024 K/W	0.122 K/W	0.200 K/W
Thermal Capacitance $C_{th,HS}$	1562 J/K	1170 J/K	780 J/K	1170 J/K	260 J/K	206 J/K

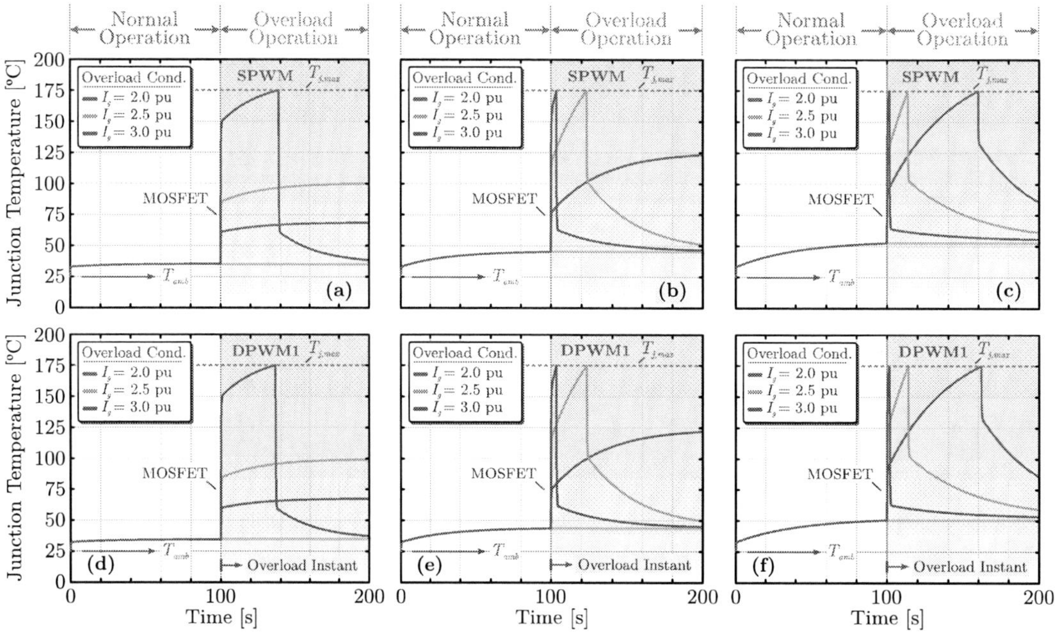

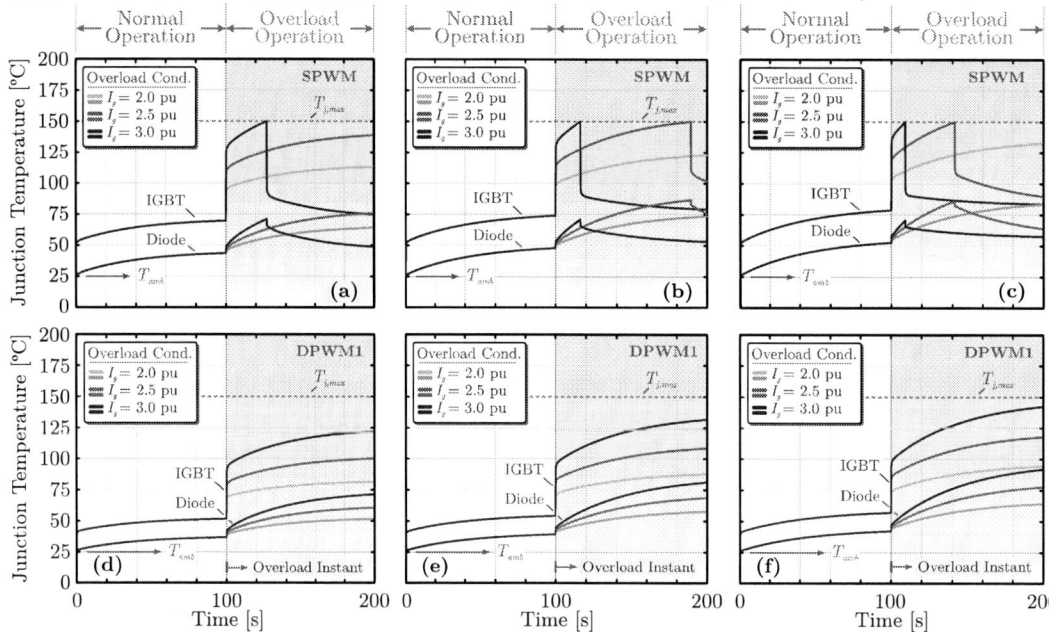

Fig. 8: Dynamic behavior of the junction temperature on the SiC MOSFET considering the influence of continuous and discontinuous modulation strategies along with different heat sinks: (a)-(d) $R_{th,HS} = 0.024$ K/W; (b)-(e) $R_{th,HS} = 0.122$ K/W; and (c)-(f) $R_{th,HS} = 0.200$ K/W. Where T_j is the average value, i.e. $\Delta T_j = 0$.

Fig. 9: Dynamic behavior of the junction temperature on the IGBT/APD FF400R17KE4 considering the influence of continuous and discontinuous modulation strategies along with different heat sinks: (a)-(d) $R_{th,HS} = 0.024$ K/W; (b)-(e) $R_{th,HS} = 0.122$ K/W; and (c)-(f) $R_{th,HS} = 0.200$ K/W. Where T_j is the average value, i.e. $\Delta T_j = 0$.

For the dynamic results, as can be seen in Fig. 8 and Fig. 9, the power module based on SiC MOSFET has an overcurrent capability almost independent of the CPWM and DPWM strategies (as already discussed). For the heat sink Type I (lowest value of $R_{th,HS}$), the power module can handle safely overloads conditions below 2.5 pu. However, for an overload condition equal to 3.0 pu, the SiC MOSFET power can operate for approximately 36.5 s before failing. With the modification of the cooling system for Type II and III, the trend is maintained. In this case, since $R_{th,HS}$ increases, the overcurrent capability is further reduced for both CPWM and DPWM strategies, cf. Fig. 8. On the contrary, the power module based on Si IGBT has an overcurrent capability highly dependent on the modulation strategy, since the switching losses are significantly higher for this power semiconductor technology (as demonstrated in Fig. 3 (e)). Therefore, the DPWM strategies can provide a notable reduction of the switching losses and the overcurrent capability can also be enlarged, as shown in Fig. 9 for the DPWM1. It should be noted that the select cooling system for the IGBT power modules has an influence on these results since the $R_{th,HS}$ values are lower as compared to those used with the SiC MOSFET power modules. Therefore, Table IV summarizes the overcurrent capability analysis by means of the time interval for reaching the $T_{j,max}$. Additionally, the analysis is extended to overload conditions of 4.0 pu, as presented in Table IV. Yet, it can be concluded that the obtained results follow the trend of the overcurrent values mentioned in the literature, i.e. some minutes for an overload of 1.5 pu and some milliseconds for 4.0 pu. The main difference is that the obtained values in this paper are numerically estimated.

Table IV: Fault Current Capability Results for the LV side Inverter.

Modulation Strategy	Power Device	Si IGBT Power Module			SiC MOSFET Power Module		
	Heat sink	Type I	Type II	Type III	Type I	Type II	Type III
SPWM	Overload Condition of 2.0 pu	> 100 s	> 100 s	> 100 s	> 100 s	> 100 s	59.82 s
	Overload Condition of 2.5 pu	> 100 s	89.22 s	41.42 s	≈ 100 s	23.12 s	13.60 s
	Overload Condition of 3.0 pu	27.31 s	16.22 s	9.21 s	38.72 s	3.41 s	1.71 s
	Overload Condition of 4.0 pu	95 ms	79 ms	66 ms	57 ms	45 ms	38 ms
DPWM1	Overload Condition of 2.0 pu	> 100 s	> 100 s	> 100 s	> 100 s	> 100 s	61.90 s
	Overload Condition of 2.5 pu	> 100 s	> 100 s	> 100 s	> 100 s	23.40 s	13.91 s
	Overload Condition of 3.0 pu	> 100 s	> 100 s	≈ 100 s	36.52 s	3.51 s	1.81 s
	Overload Condition of 4.0 pu	34.24 s	23.32 s	16.27 s	55 ms	45 ms	39 ms

Comparatively, the SiC MOSFET devices have lower power losses for normal operation (in particular switching losses are the lowest portion of the losses), while the power losses of the IGBT devices are slightly higher (with the switching as the critical portion). Furthermore, from the thermal network model, the overall losses of the IGBT power module are distributed over two transistors and two APDs, while for the MOSFET counterpart the losses are mainly concentrated in two transistors. This means that the heat flow is better distributed among the $R_{th}C_{th}$ network of the IGBT power module and hence the temperature variation tends to be slower. On the other hand, for the SiC MOSFET, the temperature will be more sensitive to the power variation, which results in a higher temperature rise and hence high T_j.

Conclusion

In this paper, the LV-side inverter of the ST is assessed in terms of its overcurrent capability taking into account the continuous and discontinuous modulations strategies, the most common power semiconductor technologies, and a cooling system with a wide range of characteristics. A systematic methodology was developed to allow the overcurrent analysis for multiple parameters in several conditions. As presented by the previous results, the conduction losses are the major portion of the power losses as compared to the switching losses. However, for overload conditions, the switching losses can still impact the performance of the system, since the switching energy and the phase current have a linear dependency. Therefore, it is expected a better performance when the DPWM strategies are adopted (up to 50 % of reduction). By leveraging this behavior for the overcurrent capability, Si-IGBT power modules have demonstrated more robustness than SiC-MOSFET modules to withstand the overload conditions. For instance, in comparison to the SiC MOSFET power module, the overcurrent capability of the Si IGBT can be enlarged at least 40 % when the heat sink Type I is employed. Therefore, for the performed analysis, the IGBT power module presents more robustness than the SiC MOSFET one.

Appendix

Table V: Conduction Losses in a 2L-Inverter for the CPWM and DPWM strategies - IGBT power device.

	IGBT	Diode
SPWM	$P_{cond} = \dfrac{\alpha V_{CE}\hat{I}_g}{8\pi} + \dfrac{R_{on}\hat{I}_g^2}{24\pi}\left[3\pi + 8M\cos(\varphi)\right]$ $\alpha = 4 + \pi M\cos(\varphi)$	$P_{cond} = \dfrac{\beta V_F\hat{I}_g}{8\pi} + \dfrac{R_{on}\hat{I}_g^2}{24\pi}\left[3\pi - 8M\cos(\varphi)\right]$ $\beta = 4 - \pi M\cos(\varphi)$
THIPWM	$P_{cond} = \dfrac{\alpha V_{CE}\hat{I}_g}{8\pi} + \dfrac{R_{on}\hat{I}_g^2}{8\pi}\left[\pi + \dfrac{8M}{3}\cos(\varphi) - \dfrac{4M}{45}\cos(3\varphi)\right]$	$P_{cond} = \dfrac{\beta V_F\hat{I}_g}{8\pi} + \dfrac{R_D\hat{I}_g^2}{8\pi}\left[\pi - \dfrac{8M}{3}\cos(\varphi) + \dfrac{4M}{45}\cos(3\varphi)\right]$
SVPWM	$P_{cond} = \dfrac{V_{CE}\hat{I}_g}{2\pi}\left[1 + \dfrac{k_1 M}{32}\cos(\varphi) + \dfrac{\pi\sqrt{3}M}{96}\sin(\varphi)\right] +$ $+ \dfrac{R_{on}\hat{I}_g^2}{8\pi}\left[\pi - \dfrac{k_2 M}{4}\cos(\varphi) + \dfrac{k_3 M}{6}\sin(\varphi)\right]$ $k_1 = 7\pi + 2\sqrt{3}; \quad k_2 = \sqrt{3} - 12; \quad k_3 = 2\sqrt{3} - 3$	$P_{cond} = \dfrac{V_F\hat{I}_g}{2\pi}\left[1 - \dfrac{k_1 M}{32}\cos(\varphi) - \dfrac{\pi\sqrt{3}M}{96}\sin(\varphi)\right] +$ $+ \dfrac{R_D\hat{I}_g^2}{8\pi}\left[\pi + \dfrac{k_2 M}{4}\cos(\varphi) - \dfrac{k_3 M}{6}\sin(\varphi)\right]$ $k_1 = 7\pi + 2\sqrt{3}; \quad k_2 = \sqrt{3} - 12; \quad k_3 = 2\sqrt{3} - 3$
DPWM0	$P_{cond} = \dfrac{\alpha V_{CE}\hat{I}_g}{8\pi} + \dfrac{R_{on}\hat{I}_g^2}{8\pi}\left[\pi + \dfrac{k_4 M}{6}\cos(\varphi) + \dfrac{k_5 M}{6}\sin(\varphi)\right]$ $k_4 = 24 - 5\sqrt{3}; \quad k_5 = 15 - 8\sqrt{3}$	$P_{cond} = \dfrac{\beta V_F\hat{I}_g}{8\pi} + \dfrac{R_D\hat{I}_g^2}{8\pi}\left[\pi - \dfrac{k_4 M}{6}\cos(\varphi) - \dfrac{k_5 M}{6}\sin(\varphi)\right]$ $k_4 = 24 - 5\sqrt{3}; \quad k_5 = 15 - 8\sqrt{3}$
DPWM1	$P_{cond} = \dfrac{\alpha V_{CE}\hat{I}_g}{8\pi} + \dfrac{R_{on}\hat{I}_g^2}{12\pi}\left[\pi + \dfrac{3\sqrt{3}}{2} + 3M\cos(\varphi)\right]$	$P_{cond} = \dfrac{\beta V_F\hat{I}_g}{8\pi} + \dfrac{R_D\hat{I}_g^2}{6\pi}\left[\pi - \dfrac{3\sqrt{3}}{4} - \dfrac{3}{2}\cos(\varphi)\right]$
DPWM2	$P_{cond} = \dfrac{\alpha V_{CE}\hat{I}_g}{8\pi} + \dfrac{R_{on}\hat{I}_g^2}{8\pi}\left[\pi - \dfrac{k_5 M}{6}\sin(\varphi) + \dfrac{k_4 M}{6}\cos(\varphi)\right]$	$P_{cond} = \dfrac{\beta V_F\hat{I}_g}{8\pi} + \dfrac{R_D\hat{I}_g^2}{8\pi}\left[\pi + \dfrac{k_5 M}{6}\sin(\varphi) - \dfrac{k_4 M}{6}\cos(\varphi)\right]$
DPWM3	$P_{cond} = \dfrac{\alpha V_{CE}\hat{I}_g}{8\pi} + \dfrac{R_{on}\hat{I}_g^2}{6\pi}\left[\pi - \dfrac{3\sqrt{3}}{4} - \dfrac{k_6 M}{4}\cos(\varphi)\right]$ $k_6 = 5\sqrt{3} - 18$	$P_{cond} = \dfrac{V_F\hat{I}_g}{2\pi}\left[1 + \dfrac{k_7 M}{48}\cos(\varphi) + \dfrac{k_8 M}{48}\sin(\varphi)\right] +$ $+ \dfrac{R_D\hat{I}_g^2}{12\pi}\left[\pi + \dfrac{3\sqrt{3}}{2} + \dfrac{k_9 M}{4}\cos(\varphi) + \dfrac{7\sqrt{3}M}{4}\sin(\varphi)\right]$ $k_7 = \sqrt{3}(6 - \pi) - 9\pi; \quad k_8 = \sqrt{3}(6 + \pi) + 3\pi; \quad k_9 = 7\sqrt{3} - 24$
DPWMMIN	$P_{cond} = \dfrac{V_{CE}\hat{I}_g}{8\pi}\left[\dfrac{3}{2}\pi - \sqrt{3}M\cos(\varphi)\right] + \dfrac{R_{on}\hat{I}_g^2}{12\pi}\left[6 - \sqrt{3}\cos(\varphi)\right]$	$P_{cond} = \dfrac{V_F\hat{I}_g}{16\pi}\left[-\pi + 2\sqrt{3}M\cos(\varphi)\right] + \dfrac{R_D\hat{I}_g^2}{24\pi}\left[-12 + 7\sqrt{3}\cos(\varphi)\right]$
DPWMMAX	$P_{cond} = \dfrac{V_{CE}\hat{I}_g}{\pi}\left[1 - \dfrac{k_{10}M}{16}\cos(\varphi)\right] + \dfrac{R_{on}\hat{I}_g^2}{4\pi}\left[\pi - \dfrac{k_{11}M}{6}\cos(\varphi)\right]$ $k_{10} = 2\sqrt{3} - \pi; \quad k_{11} = 7\sqrt{3} - 12$	$P_{cond} = \dfrac{V_F\hat{I}_g}{\pi}\left[1 - \dfrac{k_{12}M}{16}\cos(\varphi)\right] + \dfrac{R_D\hat{I}_g^2}{4\pi}\left[\pi - \dfrac{k_{13}M}{3}\cos(\varphi)\right]$ $k_{12} = 2\sqrt{3} - 3\pi; \quad k_{13} = \sqrt{3} + 6$

Table VI: Conduction Losses in a 2L-Inverter for the CPWM and DPWM strategies - MOSFET power device.

Mod.	MOSFET	Mod.	MOSFET
SPWM	$P_{cond} = \dfrac{R_{on}\hat{I}_g^2}{8\pi}\left[\pi + \dfrac{8M}{3}\cos(\varphi)\right]$	**THIPWM**	$P_{cond} = \dfrac{R_{on}\hat{I}_g^2}{8\pi}\left[\pi + \dfrac{8M}{3}\cos(\varphi) - \dfrac{4M}{45}\cos(3\varphi)\right]$
SVPWM	$P_{cond} = \dfrac{R_{on}\hat{I}_g^2}{8\pi}\left[\pi - \dfrac{k_2 M}{4}\cos(\varphi) + \dfrac{k_3 M}{6}\sin(\varphi)\right]$	**DPWM0/2**	$P_{cond} = \dfrac{R_{on}\hat{I}_g^2}{8\pi}\left[\pi + \dfrac{k_4 M}{6}\cos(\varphi) + \dfrac{k_5 M}{6}\sin(\varphi)\right]$
DPWM1	$P_{cond} = \dfrac{R_{on}\hat{I}_g^2}{12\pi}\left[\pi + \dfrac{3\sqrt{3}}{2} + 3M\cos(\varphi)\right]$	**DPWM3**	$P_{cond} = \dfrac{R_{on}\hat{I}_g^2}{6\pi}\left[\pi - \dfrac{3\sqrt{3}}{4} + \dfrac{k_{14}M}{4}\sin(\varphi) + \dfrac{k_{15}M}{8}\cos(\varphi)\right]$ $k_{14} = 4\sqrt{3} - 7; \quad k_{15} = 5\sqrt{3} + 10$
DPWMMIN	$P_{cond} = \dfrac{R_{on}\hat{I}_g^2}{8\pi}\left[3\sqrt{3}M\cos(\varphi)\right]$	**DPWMMAX**	$P_{cond} = \dfrac{R_{on}\hat{I}_g^2}{8\pi}\left[4\pi - 3\sqrt{3}M\cos(\varphi)\right]$

$$T_{4n+2}[k+1] = \left(\frac{T_{4n+1}[k]\Delta T_s}{R_{th,4n+1}C_{th,4n+2}}\right) + \left(\frac{T_{4n+3}[k]\Delta T_s}{R_{th,4n+2}C_{th,4n+2}}\right) + \left[1 - \frac{\Delta T_s\left(R_{th,4n+1} + R_{th,4n+2}\right)}{R_{th,4n+1}R_{th,4n+2}C_{th,4n+2}}\right]T_{4n+2}[k]$$

$$T_{4n+3}[k+1] = \left(\frac{T_{4n+2}[k]\Delta T_s}{R_{th,4n+1}C_{th,4n+2}}\right) + \left(\frac{T_{4n+4}[k]\Delta T_s}{R_{th,4n+3}C_{th,4n+3}}\right) + \left[1 - \frac{\Delta T_s\left(R_{th,4n+2} + R_{th,4n+3}\right)}{R_{th,4n+2}R_{th,4n+3}C_{th,4n+3}}\right]T_{4n+3}[k] \quad (4)$$

$$T_{4n+4}[k+1] = \left(\frac{T_{4n+3}[k]\Delta T_s}{R_{th,4n+1}C_{th,4n+2}}\right) + \left(\frac{T_{TIM}[k]\Delta T_s}{R_{th,4n+4}C_{th,4n+4}}\right) + \left[1 - \frac{\Delta T_s\left(R_{th,4n+3} + R_{th,4n+4}\right)}{R_{th,4n+3}R_{th,4n+4}C_{th,4n+4}}\right]T_{4n+4}[k]$$

References

[1] T. Guillod, F. Krismer, and J. W. Kolar, "Protection of mv converters in the grid: The case of mv/lv solid-state transformers," *IEEE Journal of Emerg. and Selec. Topics in Power Electr.*, vol. 5, no. 1, pp. 393–408, 2017.

[2] R. Zhu, M. Liserre, M. Langwasser, and C. Kumar, "Operation and control of the smart transformer in meshed and hybrid grids: Choosing the appropriate smart transformer control and operation scheme," *IEEE Ind. Electr. Magazine*, vol. 15, no. 1, pp. 43–57, 2021.

[3] J. E. Huber and J. W. Kolar, "Applicability of solid-state transformers in today's and future distribution grids," *IEEE Trans. on Smart Grid*, vol. 10, no. 1, pp. 317–326, 2019.

[4] R. Zhu, V. Raveendran, and M. Liserre, "Overload operation of lv-side inverter in smart transformer," in *2019 IEEE Energy Conversion Congress and Exposition (ECCE)*, 2019, pp. 5997–6004.

[5] G. Hao, L. Zhou, H. Ren, M. Mao, B. Xie, Q. Zhang, and H. Li, "Study on improving the short-time over-current capability of press- pack igbts using phase change materials," in *2020 IEEE 9th International Power Electronics and Motion Control Conference (IPEMC2020-ECCE Asia)*, 2020, pp. 2173–2177.

[6] M. Bishop, S. Mendis, J. Witte, and K. Leix, "Selecting overcurrent protection for three-phase transformers," *IEEE Industry Applications Magazine*, vol. 2, no. 2, pp. 35–41, 1996.

[7] C. Chandraratne, W. L. Woo, T. Logenthiran, and R. T. Naayagi, "Adaptive overcurrent protection for power systems with distributed generators," in *2018 8th International Conference on Power and Energy Systems (ICPES)*, 2018, pp. 98–103.

[8] M. Liserre, G. Buticchi, M. Andresen, G. De Carne, L. F. Costa, and Z.-X. Zou, "The smart transformer: Impact on the electric grid and technology challenges," *IEEE Ind. Electr. Magazine*, vol. 10, no. 2, pp. 46–58, 2016.

[9] W. Liu, J. Yu, G. Li, J. Liang, C. E. Ugalde-Loo, and A. Moon, "Analysis and protection of converter-side ac faults in a cascaded converter-based mvdc link: Angle-dc project," *IEEE Transactions on Smart Grid*, pp. 1–1, 2021.

[10] A. M. Hava, R. J. Kerkman, and T. A. Lipo, "Simple Analytical and Graphical Methods for Carrier-Based PWM-VSI Drives," vol. 14, no. 1, p. 49, 1999.

[11] H. C. Skudelny and G. V. Stanke, "Analysis and Realization of a Pulsewidth Modulator Based on Voltage Space Vectors," *IEEE Transactions on Industry Applications*, vol. 24, no. 1, pp. 142–150, 1988.

[12] A. Hava, R. Kerkman, and T. Lipo, "A high-performance generalized discontinuous pwm algorithm," *IEEE Transactions on Industry Applications*, vol. 34, no. 5, pp. 1059–1071, 1998.

[13] J. W. Kolar, H. Ertl, and F. C. Zach, "Influence of the Modulation Method on the Conduction and Switching Losses of a PWM Converter System," vol. 21, no. 6, p. 1063, 1991.

[14] K. Zhou and D. Wang, "Relationship between space-vector modulation and three-phase carrier-based PWM: A comprehensive analysis," *IEEE Transactions on Industrial Electronics*, vol. 49, no. 1, pp. 186–196, feb 2002.

[15] "Influence of PWM methods in semiconductor losses of 15kVA three-phase SiC inverter for aircraft applications," in *2017 19th European Conference on Power Electronics and Applications, EPE 2017 ECCE Europe*, vol. 2017-January.   Institute of Electrical and Electronics Engineers Inc., nov 2017.

[16] A. Kwasinski, P. T. Krein, and P. L. Chapman, "Time domain comparison of pulse-width modulation schemes," *IEEE Power Electronics Letters*, vol. 1, no. 3, pp. 64–68, 2003.

[17] R. Mandrioli, A. Viatkin, M. Hammami, M. Ricco, and G. Grandi, "A comprehensive AC current ripple analysis and performance enhancement via discontinuous PWM in three-phase four-leg grid-connected inverters," *Energies*, vol. 13, no. 17, sep 2020.

[18] J. W. Kolar, "Calculation of the passive and active component stress of three-phase pwm converter systems with high pulse rate," in *Proc. of 3rd European Power Electronics and Applications Conf.(EPE'89), Aachen (Germany), 10*, 1989.

[19] P. Perruchoud and P. Pinewski, "Power losses for space vector modulation techniques," in *Power Electronics in Transportation*, 1996, pp. 167–173.

[20] K. Berringer, J. Marvin, and P. Perruchoud, "Semiconductor power losses in AC inverters," in *Conference Record - IAS Annual Meeting (IEEE Industry Applications Society)*, vol. 1.   IEEE, 1995, pp. 882–888.

[21] F. Blaabjerg, U. Jaeger, and S. Munk-Nielsen, "Power losses in pwm-vsi inverter using npt or pt igbt devices," *IEEE Transactions on Power Electronics*, vol. 10, no. 3, pp. 358–367, 1995.

[22] L. Dalessandro, S. D. Round, U. Drofenik, and J. W. Kolar, "Discontinuous space-vector modulation for three-level PWM rectifiers," *IEEE Transactions on Power Electronics*, vol. 23, no. 2, pp. 530–542, mar 2008.

[23] K. Ma, N. He, M. Liserre, and F. Blaabjerg, "Frequency-domain thermal modeling and characterization of power semiconductor devices," *IEEE Trans. on Power Electronics*, vol. 31, no. 10, pp. 7183–7193, 2016.

[24] S. Narumanchi, M. Mihalic, K. Kelly, and G. Eesley, "Thermal Interface Materials for Power Electronics Applications: Preprint," Tech. Rep., 2008. [Online]. Available: http://www.osti.gov/bridge

Adaptive Resonant-Valley Switching for a GaN HEMT Direct AC-AC Auxiliary Resonant Commutated Pole Converter

Kyle Steyn[*], Johan Beukes[†]
Department of Electrical and Electronic Engineering
University of Stellenbosch
Private Bag X1, Matieland, 7602
Stellenbosch, South Africa
[*] Email: 18171869@sun.ac.za, [†] Email: jbeukes@sun.ac.za

Keywords

≪AC-AC converter≫, ≪Adaptive control≫, ≪HEMT≫, ≪Resonant converter≫, ≪ZCZVS converters≫.

Abstract

Zero-voltage zero-current switching is not guaranteed with auxiliary resonant commutated pole (ARCP) based topologies. The optimal dead-time is dependant on various parameters and cannot easily be calculated in real-time. A control technique that ensures all phase-arm switches turn on during the first resonant valley is presented. The voltage across one phase-arm switch is sampled during a turn-on transition and stored. The samples are processed by an adaptive zero-crossing algorithm and the optimal dead-time for the next switching cycle is predicted. The proposed technique was validated during a load-transient and at steady-state in simulation and experimentally on a 5-kW direct ac-ac ARCP prototype.

Introduction

The auxiliary resonant commutated pole (ARCP) [1] is a controllable resonant circuit that is added to the phase-arm of a converter. The ARCP is activated during the phase-arm deadband via an auxiliary switch, causing the load current to transfer from the phase-arm to the auxiliary circuit. Once fully transferred, the auxiliary inductor and capacitors begin to resonate, resulting in the oscillation of the phase-arm pole voltage. If timed precisely, the phase-arm switch can turn on during a resonant valley and achieve zero-voltage zero-current switching (ZVZCS). If the resonant valley is missed, ZVZCS will not occur, and the energy stored in the auxiliary capacitors will be dissipated in the phase-arm switch.

The ARCP is controlled by adjusting the phase-arm dead time. Control techniques include fixed [2], load-current-based lookup table [1], equation-based estimation [3], input bus midpoint-voltage monitoring [4], and analog comparator-based zero-voltage detection (ZVD) of the phase-arm switches [5, 6, 7].

The dead time for ZVZCS is predominantly dependant on the input voltage, load current and auxiliary component values; therefore, a fixed dead-time would rarely achieve ZVZCS. Lookup table and estimation techniques require exact knowledge of the auxiliary component values and precisely timed measurements. Propagation delay in analog ZVD circuits results in the late detection of a zero-voltage state. This becomes problematic when the resonant frequency is high and timing-error tolerance low.

An adaptive zero-crossing (AZC) technique that ensures ZVZCS regardless of the input voltage, load current, propagation delay and resonant component values is presented. The voltage across one phase-arm switch is oversampled uniformly by a high-speed analog-to-digital converter (ADC) during each turn-on transition. The samples are used by an adaptive controller on a field-programmable gate array (FPGA) to calculate the dead time for the subsequent switching transition. The technique results in a dead time that rapidly converges to the optimal value at the first resonant valley.

The Direct AC-AC ARCP Converter

Topology Overview

The operation of the ARCP will be examined using the full-bridge direct ac-ac topology seen in Fig. 1.

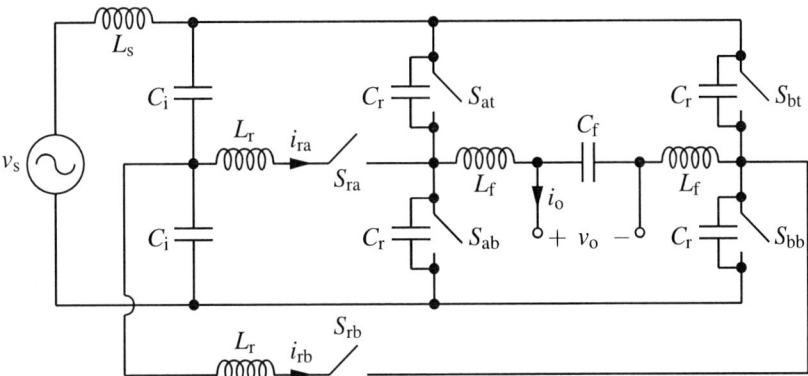

Fig. 1: Full-bridge direct ac-ac ARCP converter topology.

v_s and L_s represent the ac source, C_i the input capacitors, L_r and C_r the resonant inductors and capacitors, L_f and C_f the output filter, and v_o the converter output. Phase-arm switches are labelled S_{xy} and auxiliary switches S_{rx}, where x represents the connected phase-arm and y the bus. Each switch consist of two gallium nitride (GaN) high-electron-mobility transistors (HEMTs) in a common-source configuration.

Operating Principle

Using bipolar pulse-width modulation for the phase-arm switches, $i_{ra} = -i_{rb} = i_r$ and $S_{ra} = S_{rb} = S_r$. The operating states and waveforms during the turn-on of S_{at} & S_{bb} while in quadrant 1 can be seen in Fig. 2.

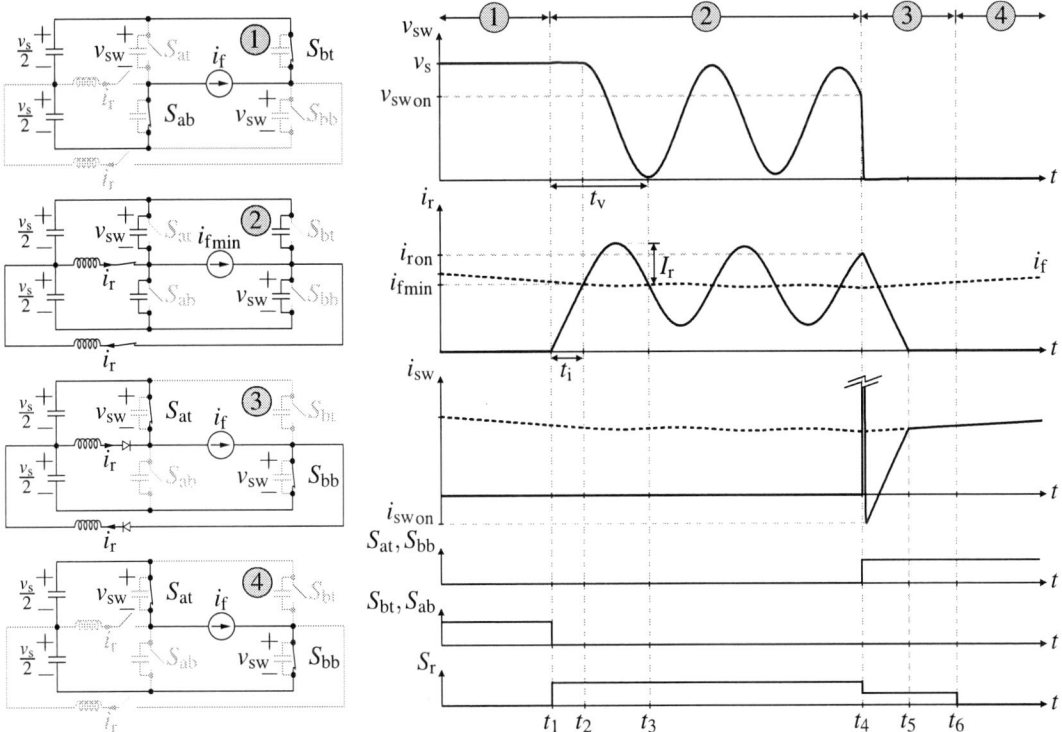

Fig. 2: Operating states and waveforms during turn-on of S_{at} & S_{bb}.

During ① the converter is near the end of the previous inverted-output state. The converter transitions to the deadband state ② where the filter current (i_f) reaches a minimum (i_{fmin}). The auxiliary switches (S_r) are activated at t_1 and the resonant inductor current (i_r) rises linearly until L_r fully conducts i_{fmin} at t_2. L_r and C_r resonate and the voltage at both phase-arm poles oscillate at the natural resonant frequency (f_r). To achieve ZVZCS, S_{at} & S_{bb} should turn on during the resonant valley at t_3. If the resonant valley is missed and S_{at} & S_{bb} turn on at t_4 for example, the charge in C_r would cause the transistors to turn on in the saturation region with high stress and power loss. The large peak current is combined with $i_{swon} = i_{fmin} - i_{ron}$ due to excess energy stored in L_r. S_r operates in diode mode during ③ to allow L_r to demagnetize and can be deactivated at t_6 (any time after t_5) without affecting the converter output ④.

Control Limitations

The dead time would have to be equal to t_v for the phase-arm switches to achieve ZVZCS. An estimation of t_v can be obtained by combining the current rise time (t_i) with half the resonant period as in (1).

$$t_v = t_i + \frac{1}{2f_r} = 2\,L_r \frac{i_{fmin}}{v_s} + \pi \sqrt{2\,L_r C_r} \tag{1}$$

Since v_s is sinusoidal and i_{fmin} is dependant on the duty cycle (D) and value of L_f, a fixed dead-time could only achieve ZVZCS with a resistive load at a constant duty cycle. This is provided that the exact values of L_r and C_r are known and remain constant under all operating conditions. Estimation techniques sample v_s and i_{fmin} and calculate the dead time every switching cycle using (1). Precise timing is required to effectively sample i_f at i_{fmin}. Alternatively, the load current (i_o) or the average value of i_f could be sampled and the steady-state dead-time estimated using additional converter parameters as in (2).

$$t_v = 2\,L_r \left[\frac{i_o}{v_s} - \frac{D(1-D)}{2\,L_f\,f_{sw}} \right] + \pi \sqrt{2\,L_r C_r} \tag{2}$$

The exact values of L_r and C_r are still required and drift in f_r cannot be detected. Analog techniques, such as comparator window-detection, solve the problem by detecting when the phase-arm switch voltage (v_{sw}) is within a set threshold (V_{th}). This can compensate for f_r drift but is sensitive to propagation delay.

Adaptive Zero-Crossing

The AZC technique utilizes one voltage sensor, consisting of a high-speed ADC, to sample v_{sw} and an FPGA to execute the control algorithms. The AZC controller consists of two algorithms, one for late turn-on and one for early turn-on conditions. A late turn-on occurs when the load current is decreased from steady-state, conversely, early turn-on occurs when the load current is increased from steady-state. The algorithms can be described using an illustration of the samples of v_{sw} during each switching cycle. The samples of v_{sw} during the late turn-on of S_{at} & S_{bb} for three switching cycles can be seen in Fig. 3.

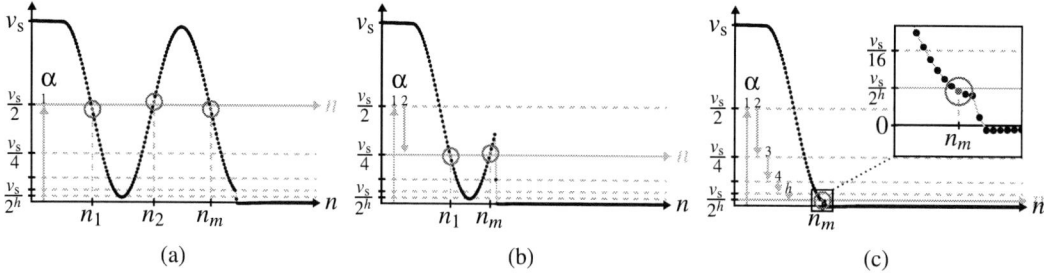

Fig. 3: Samples of v_{sw} during a late turn-on condition. (a) 1st, (b) 2nd and (c) 3rd switching cycle.

A set number of samples (N_{sample}) of v_{sw} are taken during the turn-on transition of S_{at}. The first sample is used to determine the value of v_s, then $\frac{v_s}{2}$ is subtracted from all subsequent samples to center the waveform at zero (Fig. 3a). The subtraction is illustrated as a shift in the x-axis for simplicity (e.g. the x-axis shifts to $\frac{v_s}{2}$ in Fig. 3a). The samples are iteratively searched for a sign change, indicating a zero-crossing. The zero-crossing sample index (n_i) and total count of zero-crossings (m) are stored. If $m > 1$ then the switch did not achieve ZVZCS and the dead time (t_d) for the next cycle is calculated using (3).

$$t_d(k+1) = t_d(k) - \left[\sum_{i=2}^{m} (n_i - n_{i-1}) - \frac{n_2 - n_1}{2} \right] T_{sample} \tag{3}$$

If $m \leq 1$, it is possible that the resonant oscillation was undetected. The x-axis can be shifted down by a further $\frac{v_s}{4}$ and the zero-crossings re-evaluated (Fig. 3b). To generalize, the x-axis can be shifted by $\frac{v_s}{2^\alpha}$, where α represents a halving parameter $\alpha \in \{1, \ldots, h\}$ with halving limit h. At the start of each cycle, α is reset to 1 and is only incremented each time $m \leq 1$. For $\alpha = 1$ the x-axis is shifted up and for $1 < \alpha \leq h$ the x-axis is shifted down. If $m > 1$, then (3) is repeated and the converter proceeds to the next cycle. If the halving limit is reached and $m \leq 1$ (Fig. 3c), it is assumed that either ZVZCS occurred, or S_{at} turned on too early. An illustration of the algorithm during an early turn-on condition can be seen in Fig. 4.

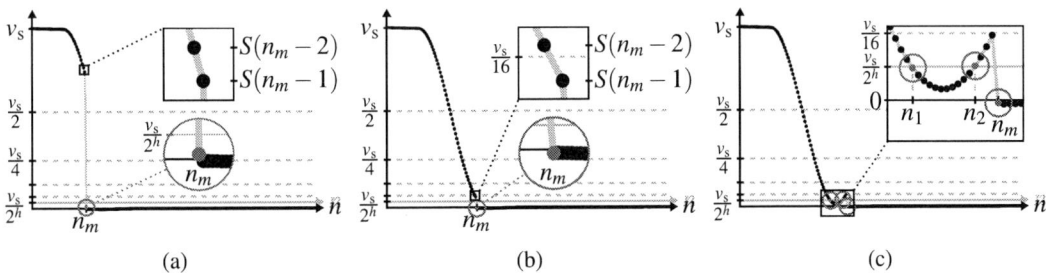

(a) (b) (c)

Fig. 4: Samples of v_{sw} during an early turn-on. (a) 1st and (b) 2nd switching transition. (c) Overshoot.

$S(n_m - 1)$ represents the value of the sample at $n = n_m - 1$. To differentiate between ZVZCS (Fig. 3c) and an early turn-on (Fig. 4a), the value of the samples before n_m can be used to determine how many T_{sample} should be added to t_d. Since the switch is on at $n = n_m$, the previous sample $S(n_m - 1)$ would be an estimate of the turn-on voltage, and the difference between $S(n_m - 2)$ and $S(n_m - 1)$ would indicate the voltage rate-of-change before turn on. The ratio of these values can be used to update t_d as in (4).

$$t_d(k+1) = t_d(k) + \left[\frac{S(n_m - 1)}{S(n_m - 2) - S(n_m - 1)} \right] T_{sample} \tag{4}$$

The ratio is an estimation of the amount of T_{sample} required to move $S(n_m - 1)$ to zero. The ratio decreases as $S(n_m - 1)$ approaches the resonant valley (Fig. 4b), thereby preventing over-compensation. If t_d is increased by too much (Fig. 4c), the resonant oscillation will be detected and (3) will ensure stability.

The controller effectively assumes that S_{at} turned on at the last zero-crossing (n_m). For (3), the amount of samples between n_m and the center of the first two zero-crossings (the valley) is calculated and multiplied by the sample period (T_{sample}). In (4), the ratio of the samples determine how many T_{sample} to increase t_d by. Since the load dynamics are slower than the switching frequency (f_{sw}), t_d is able to converge to t_v.

Simulation

The AZC controller was simulated in MATLAB Simscape with ac-ac converter parameters in Table I.

Table I: Converter and Simulation Parameters

Parameter	Value	Parameter	Value	Parameter	Value	Parameter	Value
P_s	5 kW	C_i	20 μF	h	5	L_f	110 μH
V_s	230 V	D	75%	T_{pd}	100 ns	C_f	1.5 μF
I_s	21.74 A	f_{sw}	100 kHz	N_{sample}	250	L_r	3.6 μH ±10%
L_s	230 μH	V_{th}	15 V	T_{sample}	10 ns	C_r	4.7 nF ±10%

Load-Transient Performance

The dynamic response of the AZC is evaluated using a load step-up and step-down from steady-state by examining v_{swon}. The results are compared with existing control techniques and can be seen in Fig. 5.

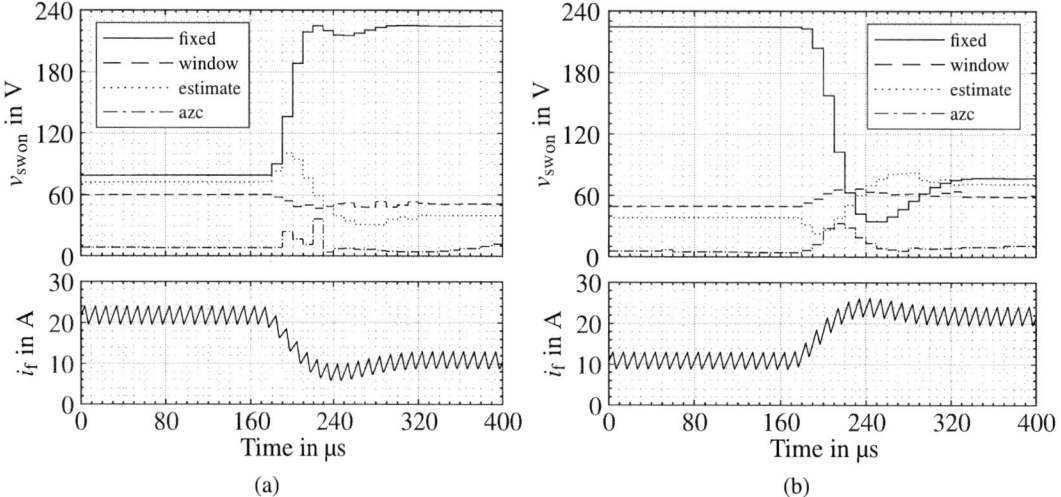

Fig. 5: Load step-response comparing control techniques. Load (a) decreased and (b) increased.

v_s is substituted with a dc source of value V_s and the converter operates with a constant duty-cycle (D). A tolerance of 10% is applied to L_r and C_r, and propagation delay (T_{pd}) is added to all sensors. For fixed control, t_d is chosen at rated load using (1). For estimated control, t_d is calculated each cycle using (2).

Fixed control performed the worst, taking thirteen cycles during step-down (Fig. 5a) and sixteen cycles during step-up (Fig. 5b) to settle. This is mainly due to the dependence of t_v on the load current. Estimated control performed the second worst, taking nine cycles during step-down/up. This is mainly due to the reliance of t_i on the value of L_r. The AZC performed the second-best, taking one cycle during step-down and four cycles during step-up to settle. The comparator window-detector performed the best, taking zero cycles during step-down/up to settle. This is due to V_{th} not being affected by the load current.

Voltage Sensor

The effect of sensor bandwidth and sampling rate on AZC performance was investigated by examining the rms of the turn-on voltage (v_{swon}) over one 50 Hz ac cycle for various ratios of sensor bandwidth (Fig. 6a) or sampling frequency (Fig. 6b) to resonant frequency (f_r). The results can be seen in Fig. 6.

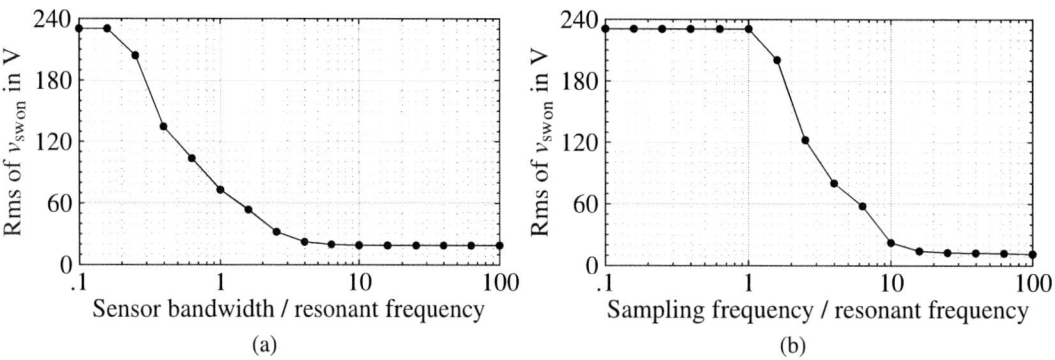

Fig. 6: Effect of (a) sensor bandwidth and (b) ADC sampling frequency on v_{swon} for AZC control.

From Fig. 6, it can be seen that for optimal performance of the AZC, a sensor bandwidth of $\geq 4f_r$ with sampling rate $T_{sample} \leq \frac{1}{20f_r}$ is required. With values from Table I, $f_c > 3.46$ MHz and $T_{sample} \leq 57.8$ ns.

Steady-State AC Performance

The steady-state performance of the AZC is evaluated by examining the rms of v_{swon} over one ac cycle vs. load current. The results are compared with existing control techniques and can be seen in Fig. 7.

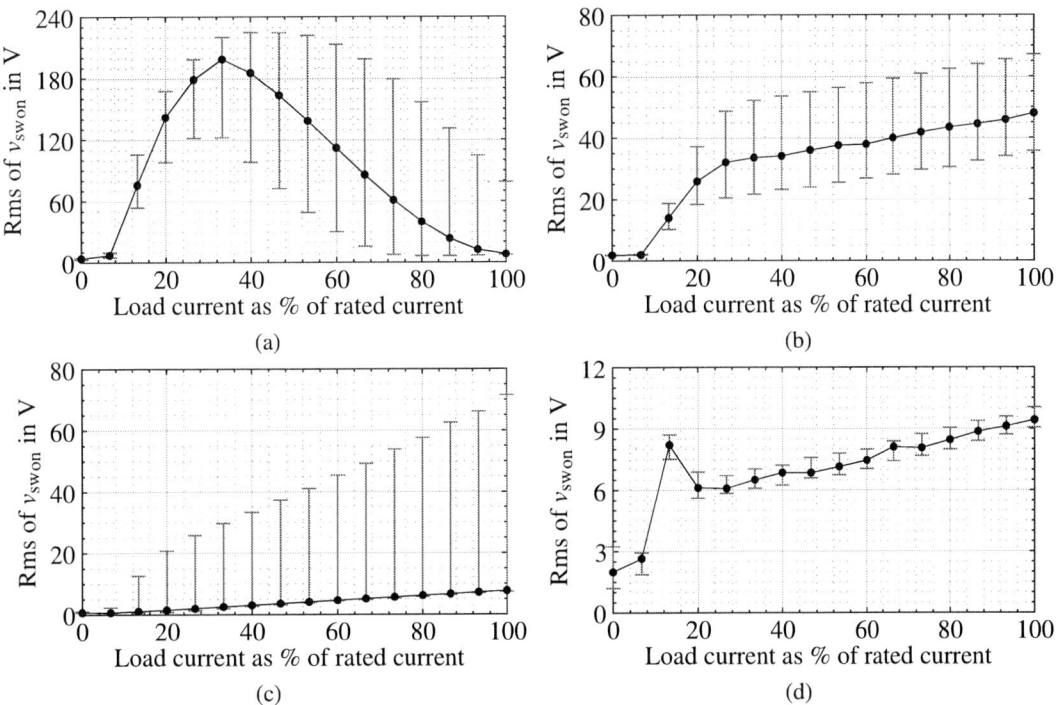

Fig. 7: Simulation results showing the rms of the switch turn-on voltage (with variance caused by L_r and C_r) vs. load current. (a) Fixed control. (b) Comparator window-detector. (c) Estimated control. (d) AZC.

v_s is a 50 Hz ac source with rms value V_s and the converter operates with a constant duty-cycle (D). A tolerance of 10% is applied to L_r and C_r, and propagation delay (T_{pd}) is added to all sensors. For each point (–●–), 128 simulations are performed, each having a random selection of L_r and C_r within tolerance. The points indicate the nominal values for L_r and C_r and the bars indicate the variance caused by the tolerance. Each simulation is run for one ac cycle and the rms is calculated of the resultant v_{swon}.

Fixed control (Fig. 7a) only performed well during low and rated current with nominal values of L_r and C_r. The variance of v_{swon} was high and propagation delay had no effect. This is due to (1) only being valid at rated load. The comparator window-detector (Fig. 7b) was impacted the most by propagation delay ($v_{swon} > V_{th}$) and had a large variance. Estimated control (Fig. 7c) was not affected by propagation delay, however, was impacted the most by tolerance. AZC control (Fig. 7d) was not affected by propagation delay and showed minimal variance due to tolerance, with the load slope being due to conduction losses.

Experimental Prototype

The prototype in Fig. 8 was designed with parameters as in Table I and notable components in Table II.

Table II: Notable Prototype Components

Component	Manufacturer	Product Number	Cost per Unit
FPGA	Xilinx	XC6SLX9	20 EUR
GaN HEMT	Infineon	IGT60R070D1	19 EUR
11-Bit Parallel ADC	Texas Instruments	ADC11C125	21 EUR
Fully Differential Op-Amp	Analog Devices	ADA4927	9 EUR

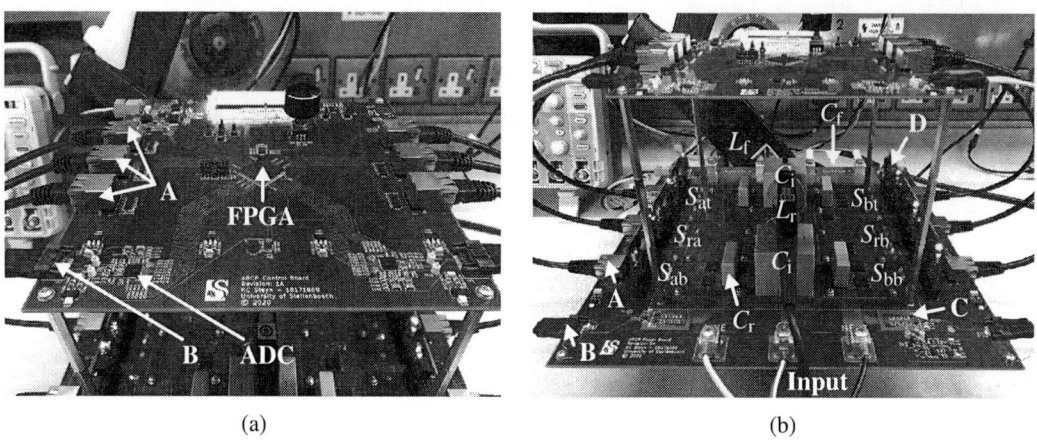

(a)   (b)

Fig. 8: 5-kW direct ac-ac ARCP prototype. (a) Control board. (b) Power board.

All switches consist of common-source GaN enhancement-mode HEMTs. One 11-bit parallel ADC is actively used with a differential op-amp for the voltage sensor, which interfaces with an FPGA. Notable indicators A: digital signals, B: analog differential signals, C: voltage sensor and D: gate driver boards.

A transient load test was performed to compare fixed control with the AZC. The results are seen in Fig. 9.

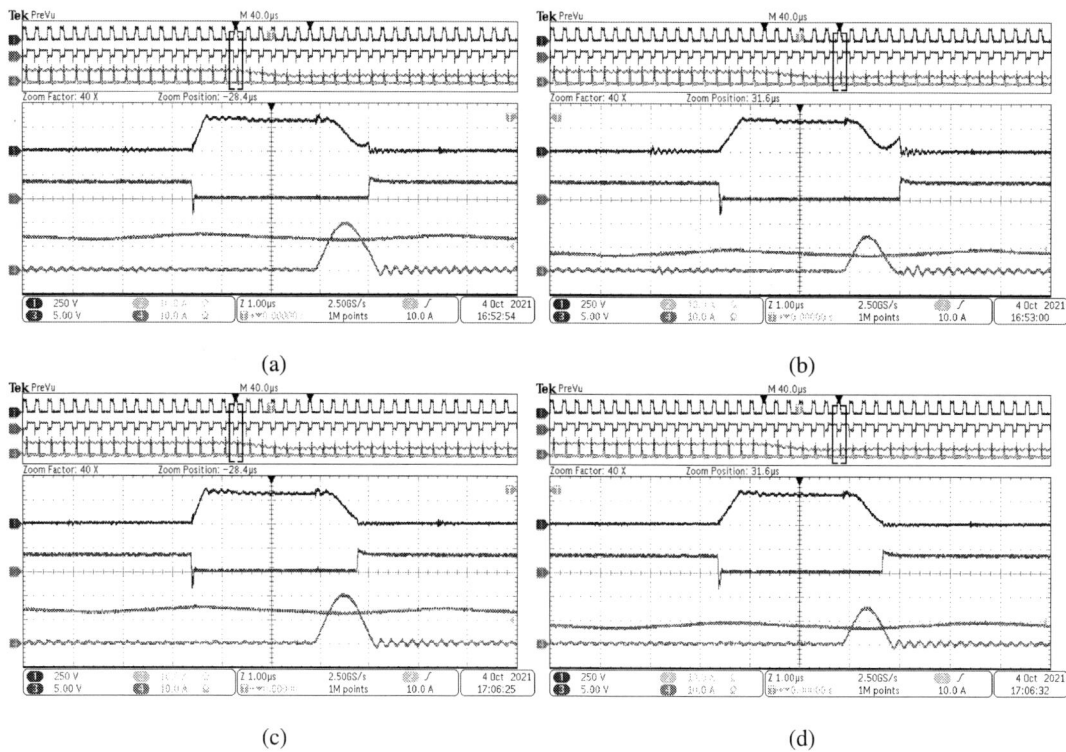

Fig. 9: Measurements during load-step. (a) - (b) Fixed control and (c) - (d) AZC. CH1: switch voltage (v_{sw}), CH2: filter inductor current (i_f), CH3: gate-source voltage and CH4: resonant inductor current (i_r).

With fixed control, there is a large difference in $v_{sw_{on}}$ before (Fig. 9a) and after (Fig. 9b) the load step. The AZC is able to track the resonant valley and achieve ZVZCS before (Fig. 9c) and after (Fig. 9d) the load step. The measurements also validate the transient simulation results in Fig. 5a and the steady-state values of $v_{sw_{on}}$ before and after the transient for fixed control (Fig. 7a) and AZC control (Fig. 7d).

The configuration of the voltage sensor utilized in the prototype for measuring v_{sw} can be seen in Fig. 10.

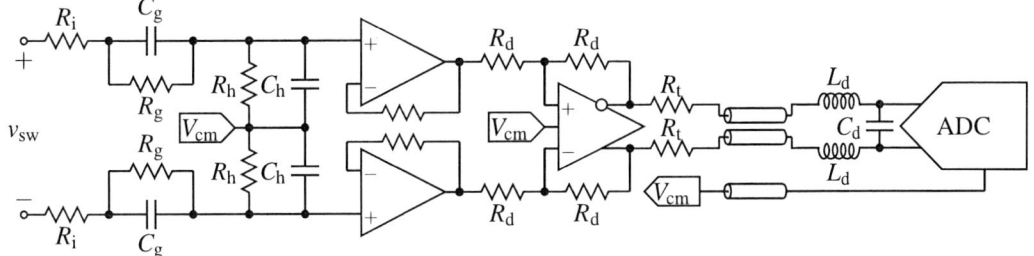

Fig. 10: Sensor for measuring phase-arm switch voltage (v_{sw}).

Resistors R_g and R_h with capacitors C_g and C_h form a voltage divider. R_i is added to limit the bandwidth and impact of the parasitic inductance of C_h. R_g consists of three series $1\,M\Omega$ resistors, R_h one $7.5\,k\Omega$ resistor, C_g four series $6.8\,pF$ capacitors, C_h one $680\,pF$ capacitor and R_i one $1\,k\Omega$ resistor. Buffer op-amps are added to drive a differential op-amp at unity gain ($R_d = 500\,\Omega$). Series termination resistors R_t of $50\,\Omega$ are added for the $100\,\Omega$ SATA transmission line. A filter ($L_d = 30\,nH$ and $C_d = 47\,pF$) is provided to decouple the line from sampling. The bandwidth of the sensor was measured at $32\,MHz$ in LTspice.

Conclusion

The AZC control technique was presented as a method to ensure ZVZCS for the ARCP topology. Uniform oversampling of the voltage across one phase-arm switch, with a sliding base-2 zero-crossing detector, could eliminate the effect that input voltage, load current, propagation delay and resonant frequency have on achieving ZVZCS with the ARCP. Simulation and experimental results indicated that AZC would reduce the turn-on voltage of the phase-arm switches during steady-state ac operation when compared to existing control techniques. The turn-on voltage exhibited low variance over a wide load range when tolerance was introduced on the resonant components and the sensor was subject to propagation delay.

References

[1] R. W. De Doncker and J. P. Lyons, "The auxiliary resonant commutated pole converter," Conference Record of the 1990 IEEE Industry Applications Society Annual Meeting, 1990, vol. 2, pp. 1228-1235.

[2] H. J. Beukes, J. H. R. Enslin and R. Spee, "Performance of the auxiliary resonant commutated pole converter in converter based utility devices," PESC Record. 27th Annual IEEE Power Electronics Specialists Conference, 1996, vol. 2, pp. 1033-1039.

[3] T. D. Batzel and K. Adams, "Variable timing control for ARCP voltage source inverters operating at low DC voltage," Proc. 2012 IEEE Transportation Electrification Conference and Expo (ITEC), 2012, pp. 1-8.

[4] F. R. Salberta, J. S. Mayer and R. T. Cooley, "An improved control strategy for a 50-kHz auxiliary resonant commutated pole converter," IECEC-97 Proceedings of the Thirty-Second Intersociety Energy Conversion Engineering Conference (Cat. No.97CH6203), 1997, vol. 1, pp. 332-336.

[5] K. Iida, H. Matsuo, T. Sakuma and H. Ochiai, "A novel firing control and overcurrent protection of the main power switches in the ARCP three phase inverter," PESC 98 Record. 29th Annual IEEE Power Electronics Specialists Conference (Cat. No.98CH36196), 1998, vol. 1, pp. 594-599.

[6] J. Voss, J. Henn and R. W. De Doncker, "Control techniques of the auxiliary-resonant commutated pole in the dual-active bridge," Proc. 2017 IEEE 12th International Conference on Power Electronics and Drive Systems (PEDS), 2017, pp. 972-978.

[7] K. Steyn and J. Beukes, "A Novel Alternating Current Auxiliary Resonant Commutated Pole Converter," Proc. IECON 2019 - 45th Annual Conference of the IEEE Industrial Electronics Society, 2019, vol. 1, pp. 6609-6614.

The Variation of Core Loss in High-Frequency Transformers Under Different Load Conditions

Navid Rasekh, Jun Wang, Xibo Yuan
Department of Electrical and Electronic Engineering, University of Bristol
Bristol, United Kingdom
Email: Navid.Rasekh@Bristol.ac.uk; Jun.Wang@Bristol.ac.uk; Xibo.Yuan@Bristol.ac.uk

Keywords

«High-frequency power converter», «Transformer», «Core loss», «Iron losses», «Eddy current loss»

Abstract

Typically, the core loss of the high-frequency (HF) transformers is considered constant regardless of the load condition on the secondary side. This paper indicates that the various load conditions can affect the core loss in HF transformers due to the variation of winding parasitics elements and subsequently the transformer's excitation voltage and current waveforms, and the field interactions between the primary and secondary windings and between the windings and core. The contribution of this work is demonstrating the core loss variation depending on different load conditions in HF transformers, which is not commonly assessed and considered in previous studies. The captured experimental results are compared with the three-dimensional (3D) finite element analysis (FEA) results conducted in Ansys Maxwell, to provide a better comprehension.

I. Introduction

Power converters are largely composed of different types of high-frequency (HF) transformers which are a key component in these such systems. The core loss of HF transformers is an essential parameter to characterize in converter design and enables accurately optimising the system specifications. In order to determine transformer losses, empirically measurement methods with potentially higher accuracy are widely used instead of the analytical methods. There is a great deal of literature on the core loss measurement approaches under sinusoidal excitations at the no-load test [1]. In contrast, not much has been done on measuring transformer core loss in high-frequency PWM excitation under different load conditions, or on practical issues and errors analysis related to these measurements.

In recent years, pulse-width modulation (PWM) converters have evolved toward higher switching frequencies along with the development of wide-bandgap power devices. A challenge for PWM converters has been to estimate precisely the core loss of magnetic components at high frequencies. Instead of sinusoidal excitation, in PWM converters, an HF transformer is exposed to rectangular voltage and different shapes of excitation current such as triangular, depending on various applications, which can affect the core loss. Since the transformers' core datasheets and analytical equations are based on sinusoidal excitation, calculating the core loss value in these methods is not accurate enough in the case of PWM converters [2-5]. Another challenge regarding obtaining the core loss is the dc bias factor, which is also called pre-magnetization and indicates the position of the B–H loop in relation to the origin, where $H_0 = 0$. The core loss of the magnetic components is reported to be impacted significantly by different dc biased currents [2-3], [5]. To sum it up, the analytical and direct schemes methods based on impedance or network analyzer (small-signal measurement) are not entirely suitable to obtain the transformer core loss [1], [6]. For that reason, experimentally measuring the core loss in a large signal is preferred compared to the other methods for achieving higher accuracy in the loss evaluation.

For the conventional power transformers with sinusoidal excitation, the core loss is carried out empirically with an open circuit or no-load test [7]. In this test, the transformer is excited from the low voltage (LV) side with the rated voltage/frequency, and the high voltage (HV) side is left open. As a result, power does not transfer from primary to secondary, and no current passes through the secondary winding. The generated flux in the core becomes the rated value, and the no-load current is quite small about 2 % to 10 % of the rated current during the test [8]. Accordingly, the wattmeter on the LV side records both the core and the LV winding losses. In spite of this, the winding loss is negligible compared with the loss in the core; because firstly, the frequency is low and the parasitic elements of the winding are negligible, and secondly, the core flux is in the rated value and the magnetizing current is insignificant.

In contrast, the advancement of power converter technologies enables intelligent and feasible operation at higher frequencies. At the same time, various modulation controls and distortions in current and voltage waveforms compared to the traditional sinusoidal excitation transformers may occur and impact the core loss. For instance, the dual active bridge (DAB) is an ideal choice at high frequencies for applications requiring high power density. The required HF transformer act as a magnetic link stage to transfer power between the two bridges, while the modulation control of DAB affects the waveform of flux distribution. As a consequence, the core loss estimation by traditional approaches becomes inaccurate if the flux varies due to the involvement of different types of switching patterns and excitation waveforms. For an example of core loss variation, the core loss of the HF transformers decreases by increasing the inner phase shift ratio of the excitation voltage in the dual-phase shift (DPS) modulation [9].

Additionally, as reported in previous high-frequency non-sinusoidal studies such as [9], with different load conditions, the voltage and current waveforms' shapes of the HF transformer are altered under light or heavy loads, which may, in turn, affect the core loss similar to the previous case. When a load presents in an HF transformer, the secondary winding conducts a current, and the primary current, as well as the field interactions between the windings or the mutual resistances, are changed compared to the no-load condition. Consequently, the winding loss will be changed accordingly due to the proximity effect [10-12]. On top of that, changing the AC resistance can influence the end/edge-winding leakage flux, and causes a local flux density variation in the edge of the core region, which leads to a change in the core loss as reported in [13]. Furthermore, depending on the application, the winding loss can also vary owing to phase shift between primary and secondary windings currents caused by the proximity effect on the different load conditions [14]. Similarly, this phase shift changes the winding resistances and alters the magnetic field distribution between the primary and secondary windings and also field interaction between the core and windings, which can subsequently change the core loss. Therefore, a precise empirical method with the same excitation waveforms and presence of the load used in the real test is needed to find these additional impacts on the HF transformer core loss, rather than a simplified no-load and small-signal tests.

Typically, the core loss of the HF transformers is considered constant regardless of the load conditions on the secondary side. This paper indicates that the presence of the load in the circuit can change the core loss value compared to the no-load condition. The core loss of the HF transformer is measured with the well-known two-winding method in the two conditions, i.e. the no-load and loaded. The load value is also varied to better investigate its impact on the core loss. Hence, The contribution of this work is demonstrating the core loss variation depending on different load conditions in an HF transformer, which is not commonly assessed and considered in previous studies. The captured experimental results are compared with the three-dimensional (3D) finite element analysis (FEA) results conducted in Ansys Maxwell, to provide a better comprehension.

II. Empirical HF Transformer Core Loss Measurement

A. No-Load Condition

The two-winding method is commonly used to measure the core loss of the HF transformers empirically as depicted in Fig. 1 [1-5], [15-17].

In this method, the secondary of the transformer is kept open and the primary winding is excited with the desired voltage and frequency. When the secondary winding is open-circuited and disconnected from the load, it does not conduct a current ($i_{Sec} = 0$). However, there is a nonzero current flowing through the primary winding as excitation current (i_{Exc}) and flows all the time during the operation of the transformer, which is assumed to be independent of the secondary load. The excitation current, which is equal to the primary current (i_{Pri}) for the no-load condition, is the summation of the core loss resistance (R_C) current, i_C, and the magnetizing inductance (L_M) current, i_M. Therefore, a part of the excitation current is used to create flux in the transformer core as i_M, and another part is required to overcome hysteresis and eddy current losses of the transformer core as i_C. The i_C is largely in phase with the supply voltage (V_{Pri}) while the i_M lags voltage by 90° since it is a purely reactive current and does not contribute directly to no-load losses [7-8]. In Fig. 1, R_{W1}, L_{l1}, and C_{P1} are the winding resistance, leakage inductance, and parasitic capacitance for the primary side, respectively.

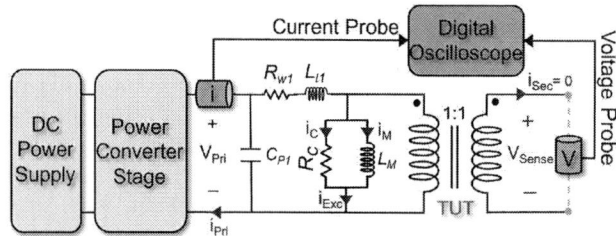

Fig. 1: Two-winding method for measuring the HF transformer core loss under the no-load conditions.

The core loss of the HF Transformer under test (TUT) is acquired by measuring the area of a closed *B–H* loop or integrating the product of the open-circuit voltage drop across the secondary winding (sensing winding, V_{Sense}) and the excitation current as defined in (1)

$$P_{Core} = \frac{N_1}{N_2} \frac{1}{T} \int_0^T i_{Exc}(t) \cdot v_{Sense}(t) \, dt \tag{1}$$

where T is the period of one cycle; $N1$ and $N2$ are the numbers of the primary and secondary winding turns, respectively. For simplicity of the loss calculation, by putting the equal turn ratios (1:1), the transformer has the same number of turns on each coil ($n = N_1/N_2 = 1$). The waveform signals are captured by the probes and digital oscilloscope over time as shown in Fig. 1.

In this paper for measuring the core loss, the TUT is excited with a bidirectional half-bridge structure proposed before in the former studies as [2-3]. By adjusting the output voltages of the two DC power supplies utilized in the half-bridge circuit, the asymmetric rectangular voltage generated by the device voltage drops can be compensated. Hence, this modification contributes to having a closed *B–H* loop, due to the consuming equal energies by the positive half cycle and the negative half cycle of the symmetric voltage excitation. If the magnetization or demagnetization process can not be completely finalized, it causes the unclosed *B–H* trajectory, which has the consequences of incorrect loss computation [17]. To excite the TUT, a refined discontinuous test procedure as a triple pulse test (TPT) is applied, presented before in [2-3]. To avoid unnecessary full-time operations and temperature rise, the TPT concept is formed, which operates only the necessary cycles and accelerates the estimation process. With the utilization of high-bandwidth voltage and current probes, a test rig has been built in the form of a two-winding method to experience TPT for measuring core loss. The test rig is depicted in Fig. 2, and the components and instruments in the test rig are listed in Table I.

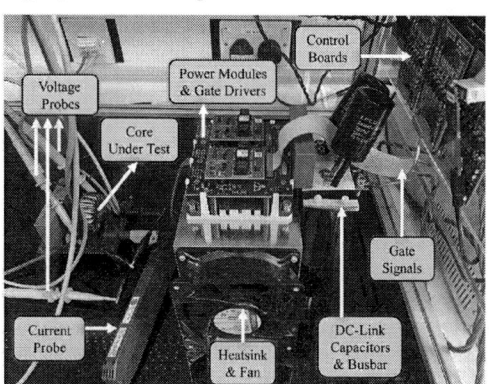

Fig. 2: The utilized test rig of the TUT core loss measurement.

TABLE I: COMPONENTS AND INSTRUMENTS IN THE TEST RIG

Power Supplies	Elektro-Automatik TS 8000 T
Voltage Probe	Keysight N2862B (150 MHz)
Differential Probe	Pico TA041 (25MHz)
Current Probe	Keysight N2783B (100 MHz)
Power Module	Semikron SKiM301TMLI12E4B
Gate Driver	Semikron SKYPER 42 J
Digital Oscilloscope	MSO-X 3054A (500 MHz, 4 GSa/s)
DC-Link Capacitance	$C1 = C2 = 2670\ \mu F$
Tested component	92 µH, T184-26, Micrometals©, $N1{:}N2{=}24{:}24$

As a result of square voltage excitation, the TUT observes the actual waveform that is emulated by TPT in the test setup. The experimental waveforms shown in Fig. 3 is illustrated a TPT procedure for one test point at the frequency of 10 kHz for the TUT presented in Table I in the no-load condition. The amplitude of the testing (V_{Pri}) is stabilized to half the dc-link voltage, which is 50 V at this test point. The selected cycle is chosen for calculating the core loss when the system has reached to steady-state and the waveforms are fixed. As a result, the closed B-H loop can be expected. A digital oscilloscope and software used for post-processing such as MATLAB are used to calculate the core loss through the selected cycle. In order to reduce phase discrepancy error, the voltage/current probe is aligned using Keysight U1880A deskew tool and calibrated on the oscilloscope with the deskew function. The core loss calculated for the target cycle of the testing point in Fig. 3 is 4.35 mJ, which can be obtained by (1) in the no-load condition. A short testing transition prevents any temperature rise in the TPT which makes it easy to control. Hence, all experiments presented in this article are performed at room temperature (T = 25°C).

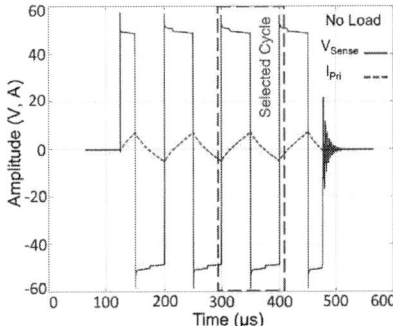

Fig. 3: Experimental current/voltage waveforms of one TPT test point for the no-load condition at 10 kHz and V_{Pri} = 50 V.

B. Loaded Condition

In this section, the load is added to the secondary winding of the HF transformer as shown in Fig. 4, to examine the differences between the loaded and no-load conditions of the transformer's core losses. Two various film resistors, i.e. 4 Ω and 8 Ω with low intrinsic parasitic inductance ($L_{Load} \approx 20$ nH), are selected to further demonstrate the differences between the loaded core loss results. Instead of using the secondary winding, auxiliary winding with the same number of turns equal to the primary and secondary sides is used to measure the voltage across the magnetizing inductance as V_{Sense}, to cancel out the voltage drops caused by the secondary side load resistor (R_{Load}), secondary winding resistance (R_{W2}) and leakage inductance (L_{l2}).

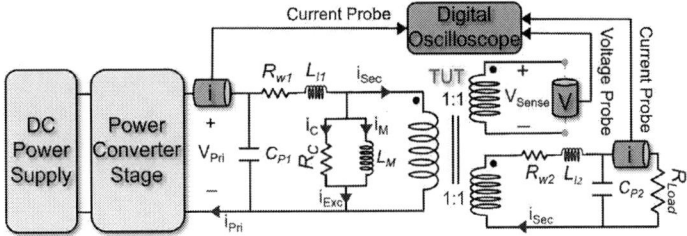

Fig. 4: Two-winding method for measuring the HF transformer core loss under the loaded conditions.

Fig 5 displays the TPT voltage and currents waveforms of the TUT presented in Table I for the 4 Ω and 8 Ω load resistors at the frequency of 10 kHz.

To measure the core loss in loaded conditions, the excitation current is needed, while it is not equal to the primary current to effortlessly attained similar to the no-load condition. The i_{Exc} can be achieved indirectly by subtracting the primary current from the secondary current when the turn ratio is equal to one (n = 1) [7].

$$i_{Exc} = i_{Pri} - \frac{i_{Sec}}{n} \tag{2}$$

As a result, the core losses can be calculated for the selected cycles in Fig. 5 by (1), which are 3.85 mJ and 3.92 mJ for 4 Ω and 8 Ω load resistors, respectively. The obtained loaded core loss values are not equal to the core loss value measured from the no-load condition in the previous section. Nevertheless, it was assumed that the presence of load would not change the core loss value considerably in different load conditions. To better investigate and clarify why the three attained core losses are different from each other, the obtained V_{Sense} and i_{Exc} waveforms for three different load conditions are shown in Fig 6.

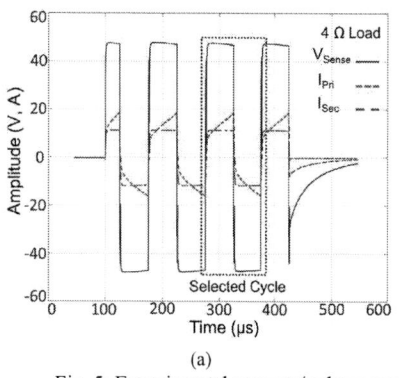
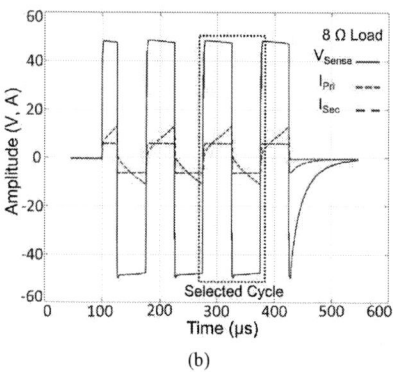

(a)   (b)

Fig. 5: Experimental currents/voltage waveforms of one TPT test point at 10 kHz and V_{Pri} = 50 V for the (a) 4 Ω and (b) 8 Ω load resistors.

Fig. 6 illustrates that the amplitude of the sensed voltage is changed with different load conditions. By reducing the load resistance from an open-circuit condition (assuming the lightest load resistance), or increasing the primary and secondary currents, it appears that the sensed voltage has decreased. However, for the excitation current cases in Fig. 6 (b), it is difficult to accurately investigate the current waveforms distinctions from the figure under varied load values compared to the no-load condition, due to the small values. By calculating the differences between the measured voltages and currents for different load conditions in Fig. 6, Fig. 7 depicts the results in more detail.

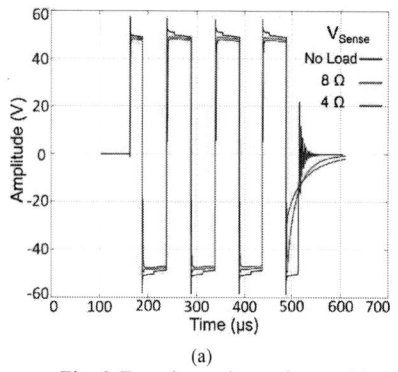
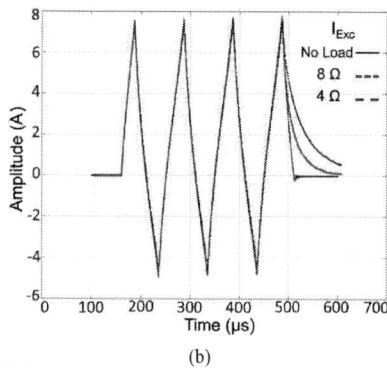

(a)   (b)

Fig. 6: Experimental waveforms of the various load conditions at 10 kHz and V_{Pri} = 50 V for the (a) sense voltages and (b) excitation currents.

According to Fig. 6 (a) and Fig. 7 (a), the sensing voltage amplitudes of loaded conditions are lower than the no-load test. In addition, the sense voltage amplitude of the 4 Ω load resistor is around 1 V lower compared to the 8 Ω load case. For the excitation current case, in Fig. 7 (b), there is a small amplitude difference between the loaded and no-load conditions. Similar to the sensed voltages, the higher load resistance values have resulted in more excitation current. Also, between the two cases of 8 Ω and 4 Ω excitation currents, there is around a 0.1 A amplitude dissimilarity. A higher sensed voltage amplitude suggests the presence of additional electromagnetic fields within and around the windings and core in higher load resistance conditions. Also, these results are indicated that there are more voltage drops across the winding of the loaded conditions. Furthermore, a higher excitation current proposes that one or both of the i_C and i_M has changed according to Fig. 4, which can lead to a more core loss. Consequently, as it is shown, the measured waveforms for computing the core loss may differ between the no-load and loaded conditions for the HF transformers. Variation in these parameters can be a key reason and reveal why the core loss is varied with various load conditions.

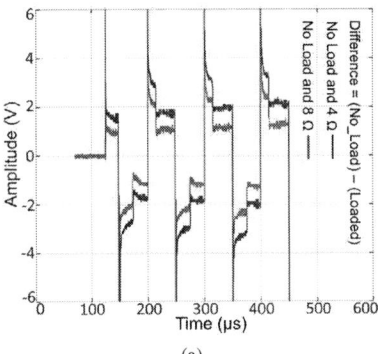 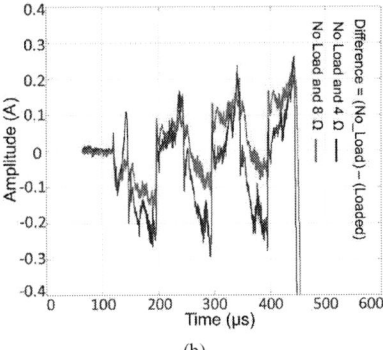

(a)   (b)

Fig. 7: Differences between the experimental waveforms of various loaded conditions compared to the no-load condition for the (a) sense voltages and (b) excitation currents.

The simplified equivalent circuit of the HF transformer referred to the primary side is shown in Fig. 8. The parasitic capacitances for the primary and secondary sides can be ignored due to their negligible values [10]. Since n=1, the impedance values of the transferred parameters to the primary side are kept constant. If the V_{Pri} is fixed to a certain voltage, by assuming constant primary and secondary winding impedances, R_C, and L_M, and increasing the load resistance value, both the primary and secondary currents decrease. Equations (3) and (4) are the results of the KVL over the closed-loop between the primary and secondary sides of the TUT and the magnetizing inductance in Fig. 8, respectively. According to (3), when the I_{Pri} increases at heavier loads, the voltage drop across the primary winding resistance and leakage inductance are increased which leads to a decrement of the V_{Sense}. Additionally, equation (4) demonstrates that by falling the secondary current at lighter loads, the V_{Sense} is also reduced.

$$V_{Sense} = V_{Pri} - I_{Pri}(R_{W1} + L_{l1}) \tag{3}$$

$$V_{Sense} = I'_{Sec}(R'_{W2} + L'_{l2} + R'_{Load}) \tag{4}$$

Therefore, one of the main reasons that change the sensed voltage is the voltage drops across the parasitic elements such as winding resistance and leakage inductance of the HF transformer at different load conditions owing to the variation of the windings' current. To sum up, the voltage that the magnetizing inductance needs to be excited will be reduced at heavier loads due to the voltage drops caused by the parasitic elements. Since the L_M had assumed to be constant, when the V_{Sense} is changed, the i_{Exc} is also varied as depicted in Fig. 7.

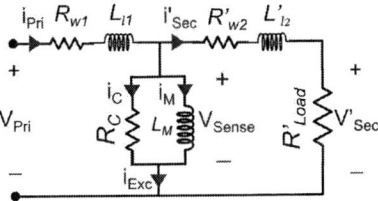

Fig. 8: Equivalent circuit of HF transformer for the loaded conditions with secondary parameters referred to the primary.

Enlargement of the V_{Sense} can result in an increment of the flux density (B) as well as flux density swing (ΔB) and subsequently change the core loss, as the core loss is generated by the varying magnetic flux field within a material. Fig. 9 (a) shows the obtained experimental ΔB of the TUT in various load conditions and excitation frequencies. In this figure, the applied TUT voltage, V_{Pri}, is held constant to 50 V regardless of the frequency increment, which is resulted in the reduction of flux density swings at higher frequencies. The no-load condition has a higher ΔB compared to the loaded conditions, and by reducing the load value, the ΔB is also decreased. By assuming the no-load ΔB result as a reference value, the differences between the loaded conditions and the no-load case are displayed in Fig. 9 (b). It can be seen that the mismatches are slightly increased when the frequency grows, which is more considerable for the heavier load value. The increase of the mismatches at higher frequencies is owing to the increment of the winding resistance and inductive leakage reactance. Also, since the 4 Ω load condition has higher windings currents, its voltage drops and mismatches from the no-load condition are more compared to the 8 Ω load condition. Variation in flux density at various load conditions confirms one of the main reasons that lead to a change in the core loss which can be caused by the voltage drops of the parasitics elements.

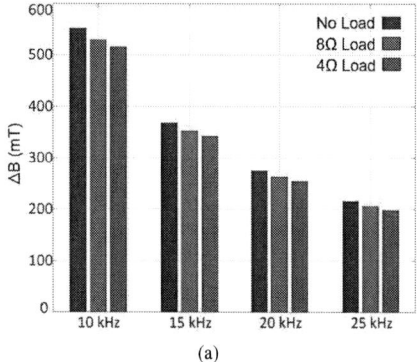
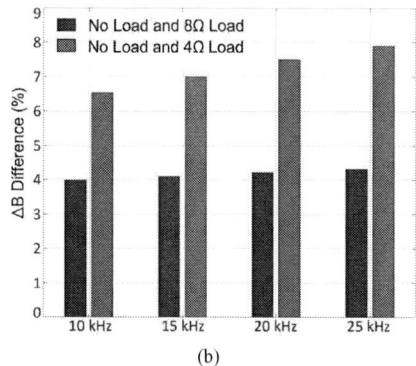

Fig. 9: (a) Experimental flux density swing of the TUT for various load conditions and excitation frequencies at V_{Pri}=50 V, (b) the differences between the loaded conditions and no-load when the no-load result considers as a reference value.

Fig. 10 (a) depicts the obtained TUT core losses at various frequencies for different load conditions. Similar to the flux density swings in Fig. 9, the core loss is decreased at heavier loads and these differences are intense at higher frequencies. For instance, as depicted in Fig. 10 (b), the difference between the no-load and 4 Ω load conditions at 10 kHz when the no-load result considers as a reference value is around 11.5 %; however, this difference is increased to about 16 % at 25 kHz.

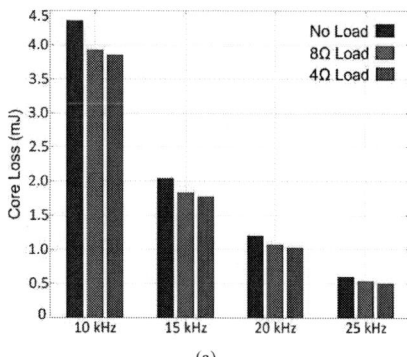
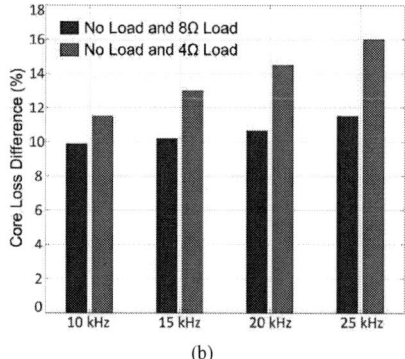

Fig. 10: (a) TUT core losses at various frequencies for different load conditions when V_{Pri} = 50 V, (b) the core loss differences between the loaded conditions and no-load when the no-load result considers as a reference value.

III. Finite Element Analysis (FEA) Evaluation

To entirely realize the causes of the core loss variation under different load conditions, the FEA analysis is employed. The HF transformer used in the experiment (Table I) is simulated through 3D ANSYS Maxwell and depicted in Fig. 11 (a). Fig. 11 (b) and (c) illustrate the magnetic flux density of the core for the no-load and 0.1 Ω load conditions at 10 kHz when the voltage and the magnetic flux density are at peak value (V_{Pri} = 50 V). The 0.1 load resistor is chosen to better compare the two conditions of lightest load, i.e. open circuit condition, and heaviest load condition which is similar and close to the short circuit test. The 0.1 Ω load condition has a small magnetic flux density and core loss as depicted in Fig. 11 (c) since the magnetizing inductance is almost shorted out and mainly the leakage inductance remained in the circuit. Hence, from Figs. 11 (b) and (c) can be concluded that the load condition plays an important role in the magnetic flux density of the core and core loss as well.

The core loss is computed through the FEA analysis and shown in Fig. 12 (a). There is around a 5% difference between the simulation and experimental method when the empirical result is considered as a reference value. For several reasons, the FEA captures a higher core loss than the experiment results. The main one is that Ansys Maxwell [18] uses the cores datasheet to compute the core loss which is based on the sinusoidal excitation. With the same magnitude of the flux density, normally the sinusoidal excitation has higher core loss compared to the square waveform excitation when the duty cycle is equal to 50% [19]. In the other duty cycles, the core loss starts to increase and more deviate from sinusoidal excitation as there are more intense high order harmonics included compared to the 50% case. In addition, it is not possible to completely and precisely simulate and replicate the TUT similar to the actual geometrical structure of the component considering the shape and winding configuration, especially when TUT is randomly wound.

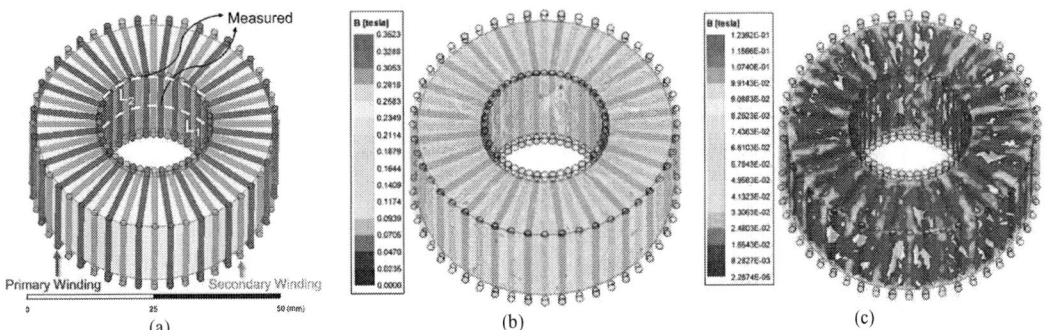

Fig. 11: (a) The FEA simulation model of the utilized HF transformer, (b) the magnetic flux density for the no-load condition at 10 kHz when V_{Pri} = 50 V, (c) the magnetic flux density for the 0.1 Ω load condition at 10 kHz when V_{Pri} = 50 V.

Fig. 12 (b) shows the FEA core loss results for different load conditions and the difference between the loaded and no-load conditions when the no-load core loss consider as a reference value. The difference between the 100 Ω load and 1 Ω load resistors compared to the no-load condition are around 6% and 16%, respectively. Fig. 12 confirms that similar to the results obtained in the previous empirical section, the core loss is changed with different load conditions, which is reduced from no-load to heaviest load conditions.

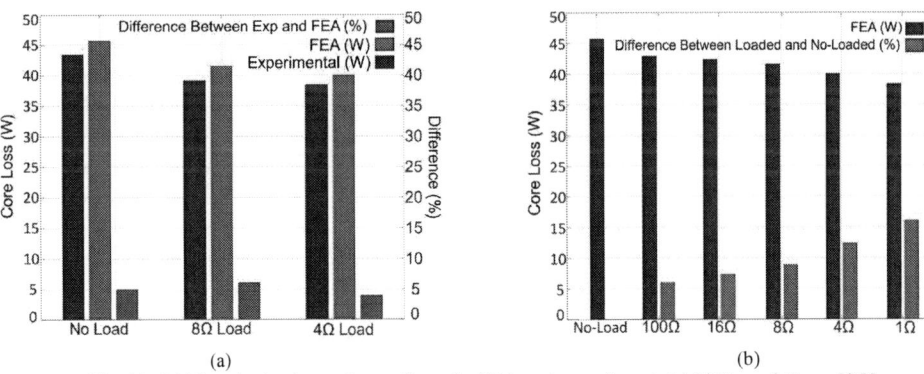

Fig. 12: (a) The obtained core losses from the FEA and experiment at 10 kHz and V_{Pri} = 50 V,
(b) the FEA core loss and the difference between the loaded and no-load conditions
when the no-load core loss is considered as a reference value at 10 kHz when V_{Pri} = 50 V.

Fig. 13 (a) demonstrates the magnetic flux density of the core for the two cases of no-load and 0.1 Ω load resistor for the middle line (L_1) depicted in Fig. 11 (a). The L_1 is at the middle of the toroidal core surface (throughout the inner circle of the core) and its length is equal to about 75 mm regarding the core specifications. Fig. 13 (a) shows the magnetic flux density at 10 kHz when the voltage is at peak value (V_{Pri} = 50 V), for each point of the L_1 length, which clarifies a large gap between the two different very high and low load conditions.

By calculating the average value of magnetic flux densities for the whole length of the two L_1 and L_2 lines shown in Fig. 13 (a), Fig. 13 (b) illustrates the average magnetic flux density (B_{Avg}) during the time. The L_2 line is selected to properly investigate the influence of the end/edge-winding leakage flux on the edge of the core region, where the presence of the windings is more concentrated compared to the other parts of the core. For the two load cases as no-load and 1 Ω load conditions and for both the L_1 and L_2 lines, the B_{Avg} of the no-load condition has higher peak values in contrast to the 1 Ω load resistor. Additionally, the B_{Avg} of the L_2 for both load conditions are higher than the L_1, which is confirmed the excessive amount of local flux density variation in the edge of the core region.

Fig. 13 (c) shows the difference between the B_{Avg} of the no-load and 1 Ω load cases for the L_1 and L_2 ($[B_{Avg}$ at No-load$] - [B_{Avg}$ at 1 $\Omega]$). At some points, this difference is less than zero because the no-load magnetic flux density is lower than the 1 Ω load case. However, the magnetic flux density of the no-load case is overall higher than the 1 Ω load resistor condition since the mean value of the L_1 case is 9.5 mT and for L_2 is 10.8 mT in Fig. 13 (c), which are both higher than zero. Also, it can be observed that the difference for the L_2 is higher than the L_1 for about 1.3 mT.

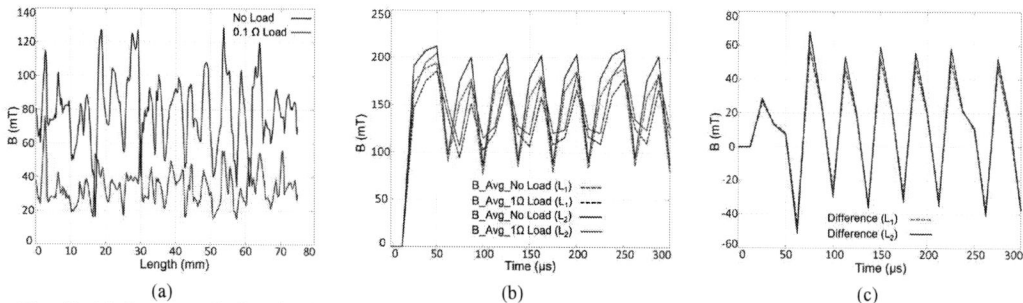

(a)   (b)   (c)

Fig. 13: (a) the magnetic flux density of the core for no-load and 0.1 Ω load resistors for the L_1 at voltage peak value,
(b) the B_{Avg} for the two L_1 and L_2 lines at different load conditions,
(c) the difference between the B_{Avg} of the no-load and 1 Ω load cases for the L_1 and L_2.

Fig. 13 clarifies that for the heavier loads, the B_{Avg} is lower compared to the light loads. Besides, the magnetic flux density in the edge of the core region is higher compared to the middle part. As explained, changing the AC resistance can influence the end/edge-winding leakage flux, and causes a local flux density variation in the edge of the core region, which leads to a change in the core loss value. Fig. 14 shows the AC resistance of the HF transformer winding, which is obtained by FEA analysis for different load conditions. Both the primary and secondary windings of the studied transformer have the same self-resistance value as they have the same number of turns and specifications. When the load impedance value is reduced (or the windings currents are increased), the HF transformer's windings resistance value is also decreased. The winding resistance value reduction is owing to the proximity effect and increasing the mutual resistance between the primary and secondary windings of the HF transformer [11-12]. Consequently, the higher resistance value at higher load conditions can be one of the factors of higher core loss, since the local magnetic flux density in the edge of the core region increases and causes more core loss. This also implies that the core loss value can alter considerably during the loaded conditions owing to the winding resistance and current amplitude variations.

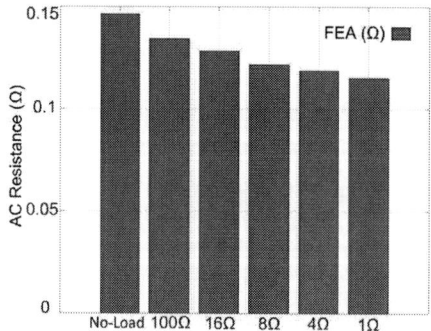

Fig. 14: The winding AC resistance of the HF transformer at different load conditions.

IV. Conclusion

The core loss of the HF transformers is previously assumed to be constant at different load conditions and measured experimentally with a no-load test. This paper demonstrates that the various load conditions can affect the core loss in HF transformers due to the variation of winding parasitics elements and subsequently the transformer's excitation voltage and current waveforms, and the field interactions between the primary and secondary windings and between the windings and core. It is concluded that a heavier load results in a lower core loss, due to the lesser value of the magnetic flux density in the core, and more voltage drops across the parasitic elements of the winding. The captured experimental results are compared with the three-dimensional (3D) finite element analysis (FEA) results conducted in Ansys Maxwell, to provide a comprehensive evaluation.

References

[1] Y. Han and Y. Liu, "A practical transformer core loss measurement scheme for high-frequency power converter," in *IEEE Transactions on Industrial Electronics*, vol. 55, no. 2, pp. 941-948, Feb. 2008.

[2] J. Wang, K. J. Dagan, X. Yuan, W. Wang and P. H. Mellor, "A Practical Approach for Core Loss Estimation of a High-Current Gapped Inductor in PWM Converters With a User-Friendly Loss Map," in *IEEE Transactions on Power Electronics*, vol. 34, no. 6, pp. 5697-5710, June 2019.

[3] J. Wang, X. Yuan and N. Rasekh, "Triple Pulse Test (TPT) for Characterizing Power Loss in Magnetic Components in Analogous to Double Pulse Test (DPT) for Power Electronics Devices," *IECON 2020 The 46th Annual Conference of the IEEE Industrial Electronics Society*, 2020, pp. 4717-4724.

[4] J. Wang, N. Rasekh, X. Yuan and K. J. Dagan, "An Analytical Method for Fast Calculation of Inductor Operating Space for High-Frequency Core Loss Estimation in Two-Level and Three-Level PWM Converters," in *IEEE Transactions on Industry Applications*, vol. 57, no. 1, pp. 650-663, Jan.-Feb. 2021.

[5] J. Muhlethaler, J. Biela, J. W. Kolar and A. Ecklebe, "Core Losses Under the DC Bias Condition Based on Steinmetz Parameters," in IEEE Transactions on Power Electronics, vol. 27, no. 2, pp. 953-963, Feb. 2012.

[6] B. X. Foo, A. L. F. Stein and C. R. Sullivan, "A step-by-step guide to extracting winding resistance from an impedance measurement," *2017 IEEE Applied Power Electronics Conference and Exposition (APEC)*, 2017, pp. 861-867.

[7] K. K. Marian, High-frequency magnetic components, 2nd ed. Chichester: WILEY, 2014.

[8] Mehta, V. K., Rohit Mehta, Objective Electrical Technology: For the Students of U.P.S.C. (Engg. Services), I.A.S. (Engg. Group), B. Sc. (Engg.), Diploma and Other Competitive Courses: Over 2800 Objective Questions with Hints. Ram Nagar, New Delhi: S. Chand, 2010.

[9] V. Karthikeyan, S. Rajasekar, S. Pragaspathy and F. Blaabjerg, "Core Loss Estimation of Magnetic Links in DAB Converter Operated in High-Frequency Non-Sinusoidal Flux Waveforms," *2018 IEEE International Conference on Power Electronics, Drives and Energy Systems (PEDES)*, 2018, pp. 1-5

[10] N. Rasekh, J. Wang and X. Yuan, "A Novel In-Situ Measurement Method of High-Frequency Winding Loss in Cored Inductors With Immunity Against Phase Discrepancy Error," in *IEEE Open Journal of the Industrial Electronics Society*, vol. 2, pp. 545-555, 2021.

[11] J. H. Spreen, "Electrical terminal representation of conductor loss in transformers," in *IEEE Transactions on Power Electronics*, vol. 5, no. 4, pp. 424-429, Oct. 1990.

[12] E. L. Barrios, D. Elizondo, A. Ursúa and P. Sanchis, "Winding Resistance Measurement in Power Inductors - Understanding the Impact of the Winding Mutual Resistance," in *IEEE Access*, vol. 9, pp. 92224-92238, 2021.

[13] A. Al-Timimy, P. Giangrande, M. Degano, M. Galea and C. Gerada, "Investigation of AC Copper and Iron Losses in High-Speed High-Power Density PMSM," *2018 XIII International Conference on Electrical Machines (ICEM)*, 2018, pp. 263-269

[14] C. Feeney, J. Zhang and M. Duffy, "AC Winding Loss of Phase-Shifted Coupled Windings," in *IEEE Transactions on Power Electronics*, vol. 31, no. 2, pp. 1472-1478, Feb. 2016.

[15] V. J. Thottuvelil, T. G. Wilson, and H. A. Owen, "High-frequency measurement techniques for magnetic cores," *IEEE Trans. Power Electron.*, vol. 5, no. 1, pp. 41–53, Jan. 1990.

[16] N. Rasekh, J. Wang and X. Yuan, "A New Method for Offline Compensation of Phase Discrepancy in Measuring the Core Loss With Rectangular Voltage," in *IEEE Open Journal of the Industrial Electronics Society*, vol. 2, pp. 302-314, 2021.

[17] T. Shimizu and S. Iyasu, "A practical iron loss calculation for AC filter inductors used in PWM inverters," *IEEE Trans. Ind. Electron.*, vol. 56, no. 7, pp. 2600–2609, Aug. 2009.

[18] ANSYS Maxwell, Oct. 2021. [Online]. Available: https://www.ansys.com/products/electronics/ansys-maxwell

[19] M. Mu and F. C. Lee, "A new core loss model for rectangular AC voltages," 2014 IEEE Energy Conversion Congress and Exposition (ECCE), 2014, pp. 5214-5220.

A COMPLETE PFC INDUCTOR DESIGN FOR LIGHTING EQUIPMENT APPLICATIONS

Wai Keung Mo[1] , Kasper M.Paasch[2], Thomas Ebel[3]
Centre for Industrial Electronics
Department of Mechanical and Electrical Engineering
University of Southern Denmark
Alsion 2, 6400
Sønderborg, Denmark
Tel.: +45 /65507627[1], +45/65501695[2], +45/65501288[3],
E-mail: wkmo@sdu.dk[1], Paasch@sdu.dk[2], ebel@sdu.dk[3]
URL: https://www.sdu.dk/da

Acknowledgements

This work was supported by Syddansk Vækstforum, the Bitten and Mads Clausen Foundation, the European Regional Development Fund as well as Interreg Deutschland-Danmark with funds from the European Regional Development Fund via the PE: Region and the PE-Region Platform projects. Find further information on Interreg Deutschland-Danmark on www.interreg5a.eu.

Keywords

«Active filter », «Boost», «Harmonics», «Magnetic device», «Simulation» «Winding topology»

Abstract

This paper presents the comparison of two novel air-gap inductor designs: two step air-gap solution (TSAG) and curved air gap solution (CAG). The effective mathematical inductance model of the proposal CAG design is derived and verified experimentally in terms of harmonic and EMI emissions. A complete optimal PFC inductor design procedure is given.

Introduction

Table 1: Definition of symbols used

Symbol	unit	definition	Symbol	unit	definition
$L_{req(min)}$	H	Minimum inductance for PFC inductor solution	ΔI	A	Ripple current
η		Converter efficiency	I_{av}	A	DC average current
$V_{in(min)}$	V	RMS minimum line voltage	P_o	W	Output power
k		$k = \dfrac{\Delta I}{I_{av}}$	L_{eff}	H	Effective inductance
f_s	Hz	Operation frequency	TSAG		Two step air gap inductor
CCM		Continuous conduction mode	I_{pk}	A	Peak excitation current
DCM		Discontinuous conduction mode	CAG		Curved-air gap inductor
D		Duty cycle	PFC		Power factor correction

Although LED lighting equipment with dimming function is used widely in many nations because of high reliability and high energy efficiency, it is a great challenge to satisfy Class C, published in IEC 61000-3-2 [1], regarding harmonic current regulation. Furthermore, the worst harmonic situation might not be in the maximum loading conditions or full dimming. Hence, the PFC solution [2],[3], boost pre-regulator, must be operated in CCM not only to improve the harmonic content but also to reduce the significant conduction loss because of I_{pk}. However, there are several drawbacks. Firstly, it is difficult to sustain a high-power factor when the output power has gradually decreased because of DCM operation in light loading. Secondly, the boost inductor requires an air gap to prevent core saturation, which results in extra-winding loss.

Practical PFC inductor considerations

The minimum inductance for CCM PFC booster converter, $L_{req(min)}$, is given by

$$L_{req(min)} = \frac{\sqrt{2}V_{in(min)}D}{\Delta I f_s}[4] \tag{1}$$

The ΔI at the peak of the $V_{in(min)}$ is stated by

$$\Delta I = \frac{k\sqrt{2}P_o}{\eta V_{in(min)(1+\frac{k}{2})}} \tag{2}$$

Substituting (2) into (1), it can be rewritten

$$L_{req(min)} \geq \frac{\eta D(1+\frac{k}{2})V_{in(min)}^2}{k f_s P_o} \tag{3}$$

The desired inductance for the CCM PFC converter must be larger than $L_{req(min)}$ given by equation (3). The $L_{req(min)}$ is inversely proportional to P_o implying two practical solutions to satisfy the harmonic requirements: 1. Large P_o (no dimming) demands on small $L_{req(min)}$ which means small inductor size. 2. Small P_o (100% dimming) demands on the large $L_{req(min)}$ which means large inductor volume. Although many PFC inductor solutions can be reduced size by an increase of fs and ΔI, the drawbacks result in large winding loss and significant EMI issues. Hence, a variable inductor concept based on a magnetic structure with novel air gap design is introduced.

In contrast to the recent publications, only few research papers focus on development of the inductance mathematical model with air-gap parameters to design an appropriate PFC inductor for satisfying harmonic and conducted emission limits. The TSAG and CAG analysis are given in section II. The inductance mathematical model of CAG is presented in section IV. Implementation of CAG in the CCM PFC boost converter and design procedure are provided in section IV, the L_{eff} equation of the CAG is examined to be applicable in the economical PFC boost inductor design in terms of D and f_s as well as different air gap dimension. In section VI, experimental verification is provided to verify the harmonic current and EMI emission.

Variable air gap inductor

TSAG

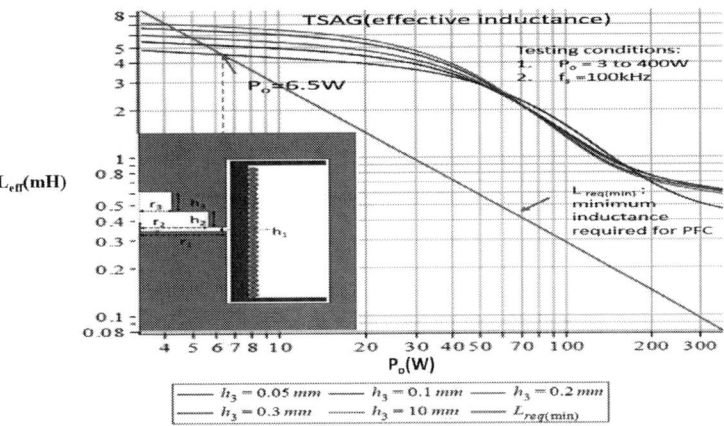

Fig 1: Choosing L_{eff} for PFC requirements by TSAG

Fig.1 illustrates the L_{eff} as a function of P_o with different h_3. Inspection of these curves, representing the

L_{eff}, shows that there is a region where L_{eff} is inversely proportional to P_o and the initial L_{eff} is proportional to h_3. At (4W $\leq P_o \leq$ 60W), this region commences, but, for (60W $< P_o <$ 200W), the green curve is lower than other curves. Finally, the purple curve is larger than other curves when 200W$< P_o <$ 300W. The red line represents the $L_{req(min)}$ for PFC solution, all the curves above this red line satisfy $L_{req(min)}$ with given air gap parameter (h_3). Hence, the minimum P_o is 6.5W for given TSAG solution [5] to fulfill $L_{req(min)}$ illustrated in fig.1.

CAG

Fig.2 illustrates L_{eff} as a function of P_o with different air gap parameter c. Inspection of these curves, representing L_{eff}, shows that there is a region where L_{eff} is inversely proportional to P_o and the initial L_{eff} is proportional to c.

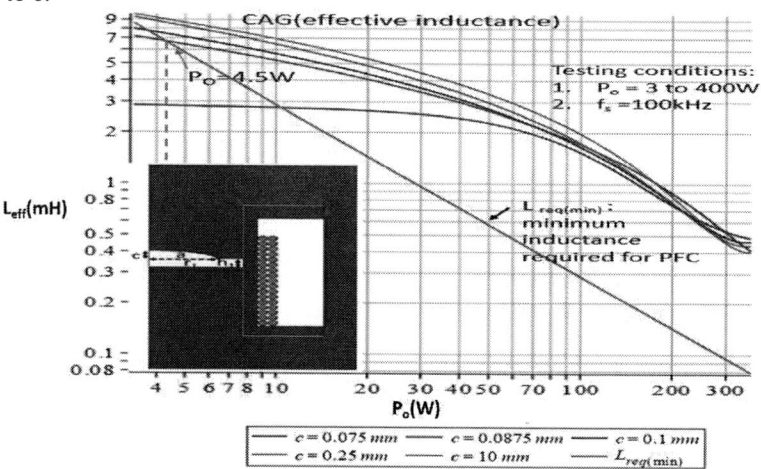

Fig.2: Choosing L_{eff} for PFC requirements by CAG

At (4W $\leq P_o \leq$ 200W), this region commences, but, for (200W $< P_o <$ 300W), the green curve is above than the other curves, it has gradually decreased to 0.4mH and is lower than other at P_o=340W. The red line represents the $L_{req(min)}$ for PFC solution, all the curves above this line satisfy $L_{req(min)}$ with given air gap parameter c. Hence, the minimum P_o is 4.5W except the blue curve for given CAG solution [6] illustrated in fig.2.

Result comparison between TSAG and CAG

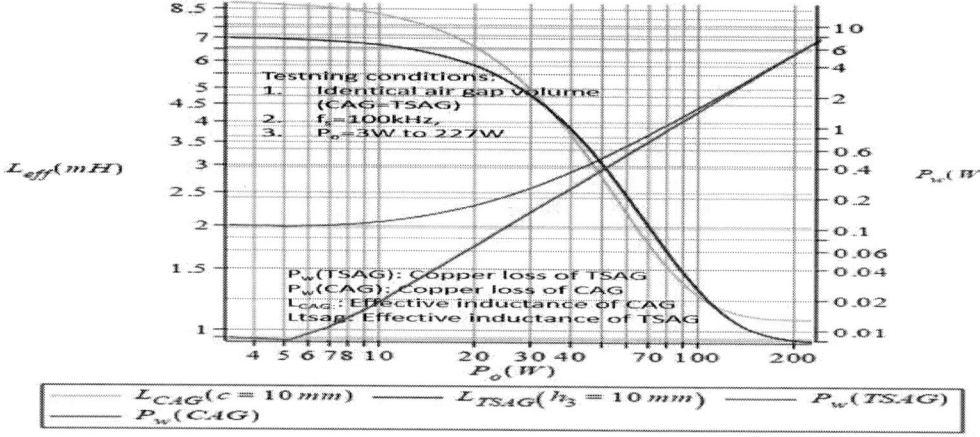

Fig.3: L_{eff} and copper loss (P_w) comparison between TSAG and CAG

Fig.3 illustrates the L_{eff} comparison between TSAG and CAG under identical test conditions, initial L_{eff} (CAG) is 28.5% larger than L_{eff} (TSAG), the L_{eff} difference has dramatically reduced at ($4W < P_o < 30W$) and lower than 5% at ($30W < P_o < 120W$), however, L_{eff} (CAG) is 10% larger than L_{eff} (TSAG) at ($120W<P_o<227W$). Apart from this, initial P_w (CAG), copper loss of CAG, is 90% smaller than P_w (TSAG) (copper loss of TSAG) and the copper loss difference has significantly reduced at $P_o > 100W$. Fig.4 shows the core loss comparison between TSAG and CAG, the loss difference has gradually increased for $4W < Po < 227W$.

Finally, the P_c (CAG) is 5% lower than P_c (TSAG) at Po=227W. Hence, the technical issues of TSAG are increased losses (P_c+P_w) and its initial L_{eff} is 30% lower than the CAG at $P_o \leq 4W$ with identical air gap parameters and test conditions.

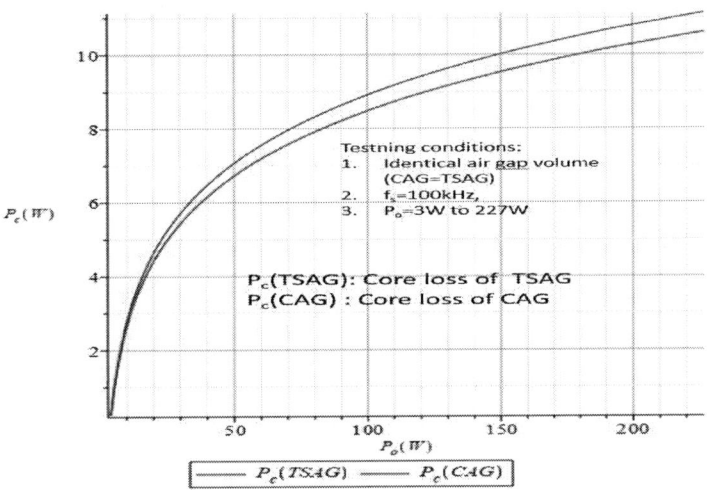

Fig.4: Core loss (P_c) comparison between TSAG and CAG

Consequently, the CAG is an optimal air gap solution to minimize the inductor volume and increase the overall PFC efficiency as well as lower harmonic regardless of loading conditions.

Analytical CAG inductance model

An innovative approach of CAG consists of the curved air gap (CAG1) and a small portion of standard air gap (minor gap) illustrated in fig.5, because this minor gap can extend its L_{eff} and maximize L_{eff} with high P_o capability shown in fig.13. The CAG1 works as a progressively accumulated air reluctance to provide large inductance at small P_o because of less core saturation and small inductance at large P_o because of large core saturation.

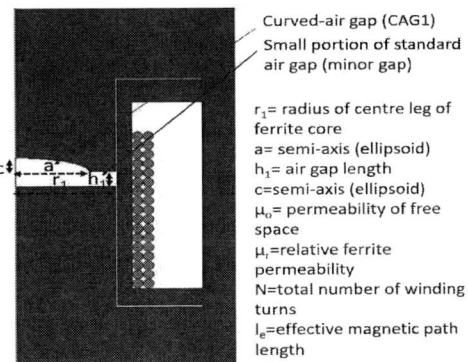

Fig. 5: Curved air gap inductor with air gap dimensions

The equivalent magnetic circuit of the CAG and its reluctance model is shown in fig.6 with n-individual reluctances ϑ_{fi}, which consists of the curved air gap reluctance R_g in series with the curved magnetic core reluctance R_m, to be connected in parallel with each other. Hence, the initial CAG inductance is

$$L_o \approx \frac{N^2}{R_c+\frac{1}{\sum_{i=1}^{n}\frac{1}{\vartheta_{fi}}}} \approx \frac{N^2}{(R_c+\frac{1}{\sum_{i=1}^{n}\left(\frac{1}{(R_{gi}+R_{mi})}\right)})} \approx \frac{N^2\mu_r\mu_o\pi r_1^2}{h_1\mu_r+l_e-h_1-c+\sum_{i=2}^{n-2}\frac{\frac{c}{(n-1)}}{1-(\frac{1}{r_1}(a-\frac{c(i-2)}{(n-1)})^2}} \tag{4}$$

where $R_c = \frac{l_{eff}-(c-h_1)}{\mu_r\mu_o\pi r_1^2}$, $R_{gc} = \frac{h_1}{\mu_o\pi r_1^2}$, $R_{gi} = \frac{h_i}{\mu_o\pi r_i^2}$, $R_{mi} = \frac{h_i}{\mu_o\pi(r_1^2-r_i^2)}$ and $a = \sum_{i=1}^{n}r_i$

All the equation parameters are given in fig.5 and 6.

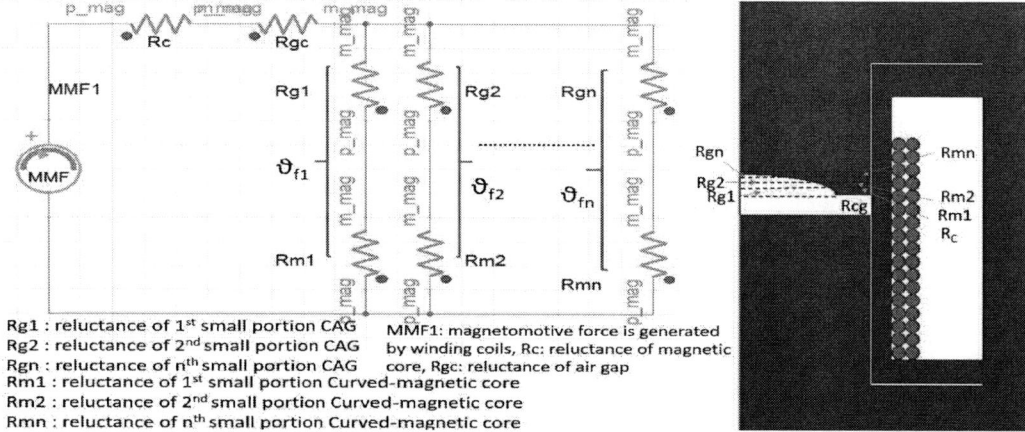

Rg1 : reluctance of 1st small portion CAG
Rg2 : reluctance of 2nd small portion CAG
Rgn : reluctance of nth small portion CAG
Rm1 : reluctance of 1st small portion Curved-magnetic core
Rm2 : reluctance of 2nd small portion Curved-magnetic core
Rmn : reluctance of nth small portion Curved-magnetic core

MMF1: magnetomotive force is generated by winding coils, Rc: reluctance of magnetic core, Rgc: reluctance of air gap

Fig.6: Equivalent magnetic circuit of the CAG inductor and reluctance model

Effective inductance mathematical model

The voltage across an inductor is related to its flux linkage and this in turn is related to the current, the dependence of the inductance on its current must be considered.

$$V = \frac{d(Li)}{dt} = \left(L_o + \frac{dL}{dI}I\right)\frac{di}{dt} = L_{eff}\frac{dI}{dt} \tag{5}$$

$$L_{eff} = L_o + \frac{dL}{dI}I_{pk} \tag{6}$$

where $I_{pk} = \frac{\sqrt{2}P_o}{\eta V_{in(min)}}$

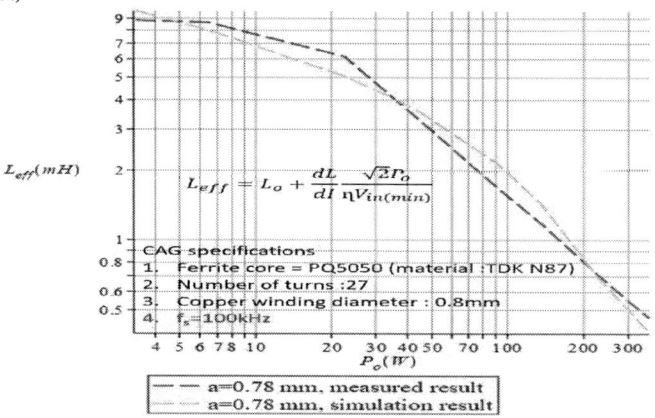

Fig. 7: L_{eff} (CAG) comparison between measured and mathematical simulation results (a=0.78mm)

Substituting of (4) and $\gamma = \dfrac{dL}{dI} = -\dfrac{(b\alpha\cosh\frac{\alpha x}{\delta}+\rho\varphi\sinh\frac{\varphi x}{\delta}\sinh\alpha x\delta)}{\delta\sinh^2\alpha\delta x\cosh^2\frac{\varphi x}{\delta}}$ into (6), it can be written as

$$L_{eff} = \dfrac{N^2\mu_r\mu_o\pi r^2}{h_1\mu_r+l_e-h_1-c+\sum_{i=2}^{n-2}\frac{\frac{c}{(n-1)}}{1-(\frac{1}{r_1}(a-\frac{c(i-2)}{(n-1)})^2}} - \dfrac{(b\alpha\cosh\frac{\alpha x}{\delta}+\rho\varphi\sinh\frac{\varphi x}{\delta}\sinh\alpha x\delta)}{\delta\sinh^2\alpha\delta x\cosh^2\frac{\varphi x}{\delta}}x \qquad (7)$$

where $x=\dfrac{\sqrt{2}P_o}{\eta V_{in(min)}}$, α, β, δ, φ are air-gap geometrical parameters.

Figs 7 and 8 illustrate the L_{eff} comparison against P_o by the air gap parameters (a & c). The differences between measured and simulation results, given in fig.7, are 13 %@a=0.78mm at (3W < P_o < 35W) and 10% @a=0.78mm at (35W < P_o < 200W). Comparing the simulation and measured L_{eff} against P_o, shown in fig.8, less 5% @c=0.075mm at (4W < P_o < 250W) and 10% @c=0.0875mm at (20W < P_o < 60W) as well as 15%@c=0.0875mm at (60W < P_o < 200W), is also promising and validates the L_{eff} mathematical model and its analysis. Although there are minor errors between them, the L_{eff} mathematical models [7] are sufficient to predict L_{eff} behavior against P_o.

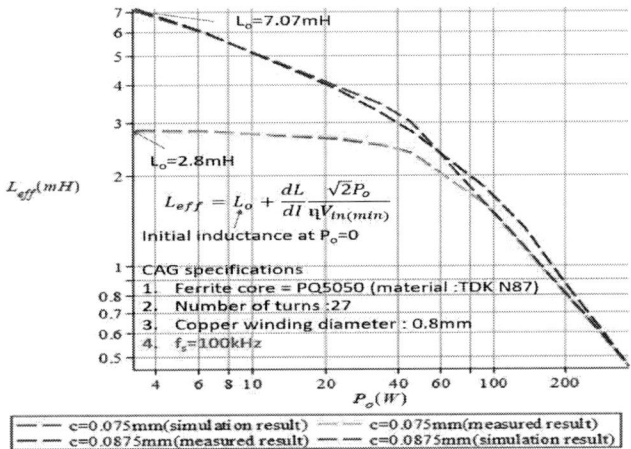

Fig.8: L_{eff}(CAG) comparison between measured and mathematical simulation results (c=0.075mm & 0.0875mm)

Implementation of CAG in the CCM PFC Boost converter and design procedure

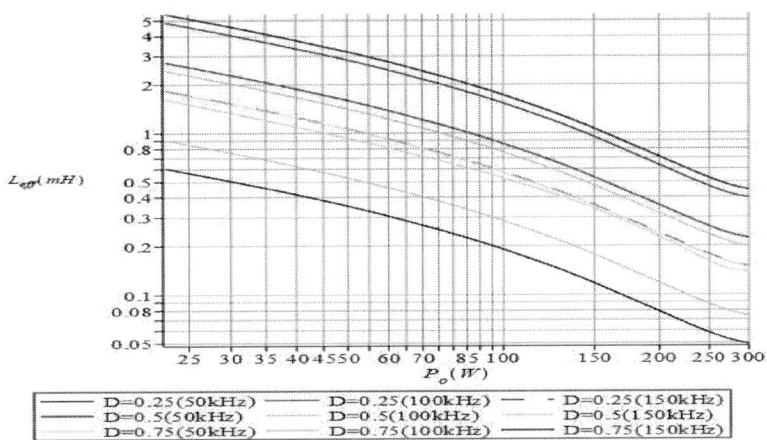

Fig. 9: L_{eff} with 3 duty cycles (D=0.25, 0.5 and 0.75) and 3 operation frequency (fs=50kHz,100kHz, 150kHz) against P_o

Fig.9 examines the L_{eff} of CAG with given conditions against the P_o and its design procedure is shown as follows.

1. Calculate the minimum inductance for sustaining boost converter in CCM operation by equation (3)
2. Determine the CAG air gap parameters by equation (4)
3. Determine the appropriately effective inductance from fig.9 with given operation conditions. E.g., DT=0.25, f=100kHz, P_o =300W, L_{eff}= 0.2mH

Fig.10 presents a complete optimal CAG inductance design algorithm [8] to optimize the winding loss by k and the effective inductance for satisfying harmonic requirements.

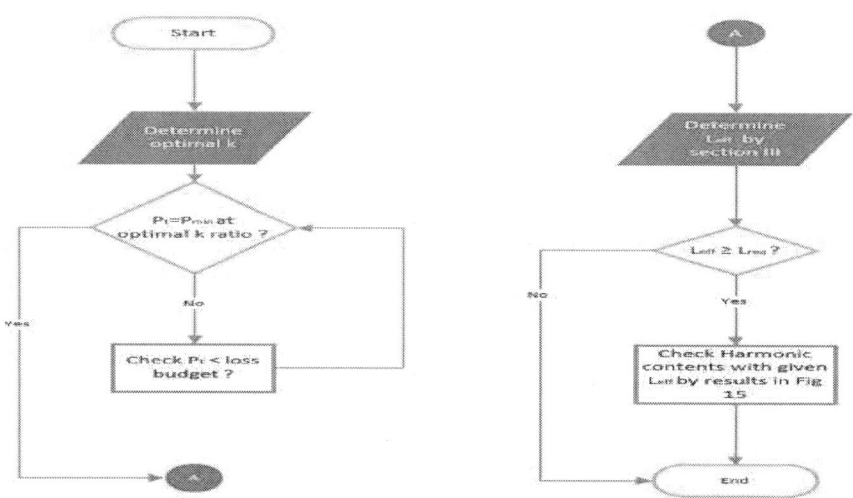

Fig. 10 Optimal CAG inductance design algorithm

Experimental results and measurement set-up

Experimental measurement set-up

Fig,11: Harmonic current measurement set-up

To verify the proposed CAG solution, several prototypes were built and test with PFC converter (UC3854) as well as full bridge converter (UCC22980) for operation of 200 LED lighting equipment shown in fig.11.

The specifications of the prototypes are shown in Table 2.

Table 2 : The specification of the entire system prototype

The specifications of the PFC boost converter	The specifications of the full bridge converter
1.Vin: 230Vrms /50Hz 2. Vo 385 V(dc) 3. Po: 300W 4. f_s:100 kHz 5.CAG: 6mH (initial inductance) 6. PFC controller: UC3854	1.Vin: 385V(dc) 2. V_o: 12V(dc) 3. P_o: 270W 4. f_s:200kHz 5. PWM controller: UCC22980
The specification of LED	The specifications of the CAG
CREE J series 2016 (V_f= 2.9V to 3.1V@ I_{fmax}=150mA)	1.Ferrite core = PQ 5050 (material: TDK N87), 2. Number of turns: 27 3. Copper winding diameter: 0.8mm

Experimental results

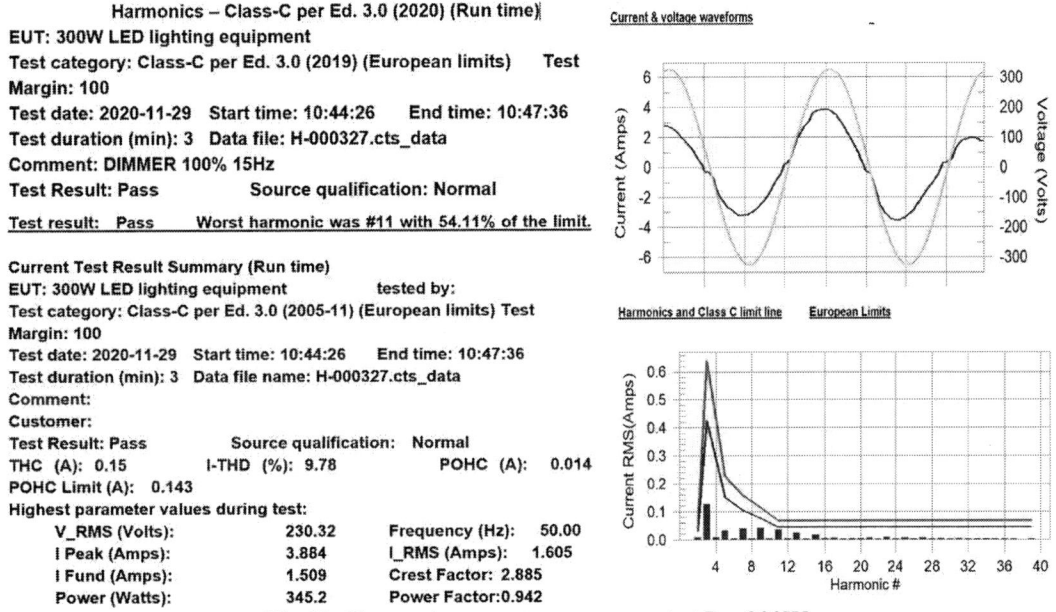

Fig.12: Harmonic current measurement at Po=300W

The harmonic result at P_o=300W with LED lighting application is shown in Fig.12. The blue and red curve represent 100% and 150% over the harmonic limit, the worst harmonic is 11th (550Hz) and 55.4% of the Class C limit.

Result comparison between CAG with and without minor gap solution

The result comparison between CAG with and without minor gap solution against P_o addresses the L_{eff} and EMI emission shown in fig.13, 14 and 15. Under identical testing conditions, the initial L_{eff} of CAG (no minor gap) is 75% larger than CAG (minor gap); however, this L_{eff} has been significantly reduced to 0.2mH ,which is one sixth of CAG (minor gap), at 35W < P_o < 90W and is below $L_{req(min)}$ at 90W < Po < 340W illustrated in fig.13, hence, this solution is inappropriate for PFC inductor design. The comparison reveals that the L_{eff} of CAG with minor gap is a better solution to sustain sufficient L_{eff} over $L_{req(min)}$ to satisfy the harmonic requirement.

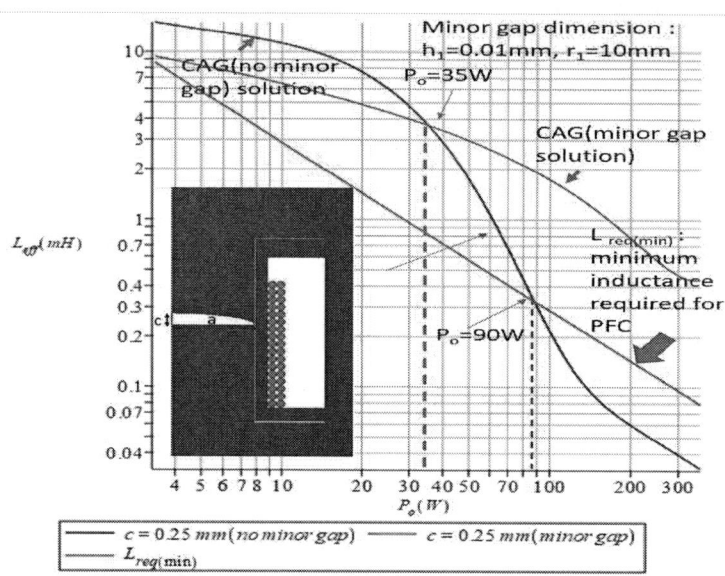

Fig.13: L_{eff} comparison between CAG (with and without minor gap solution given identical testing conditions) against P_o

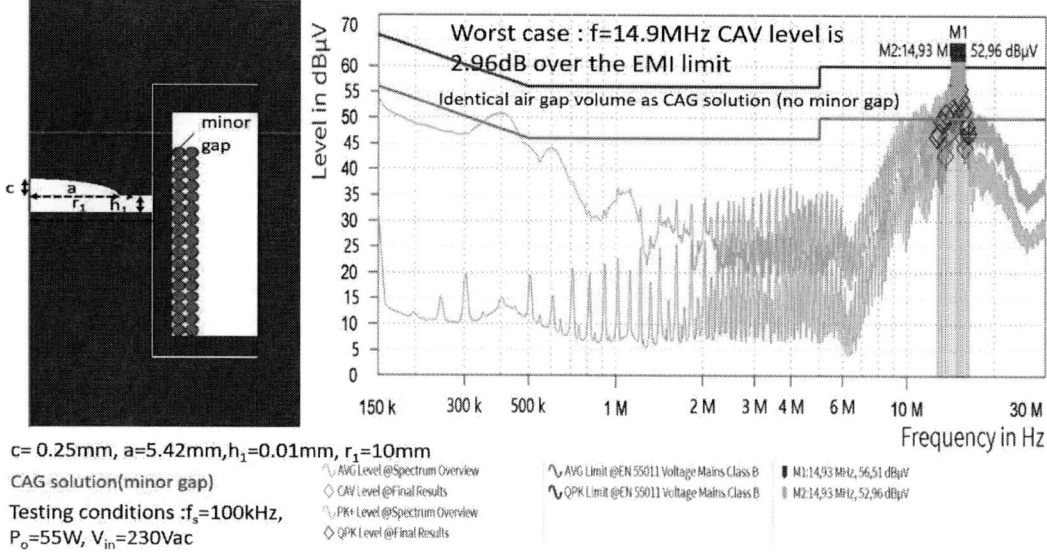

Fig.14: Conducted emission with CAG minor gap solution (air gap dimension: c=0.25mm, a=3.86 mm, h_1=0.01mm, r_1=10mm)

The prototypes of the CAG with and without minor gap solution together with the whole LED lighting system are measured on a R&S ESR EMI receiver under identical testing conditions (P_o=55W, fs=100kHz, V_{in}=230V). The test results indicate the highest peak of conducted emission level (53 dBμV shown in fig.14 (CAG with minor gap)) and (74.9dBμV shown in fig.15 (CAG without minor gap)) at 14.9MHz, which should be considered for the PFC inductor design. Hence, the result comparisons reveal that the CAG with optimal minor gap can improve the harmonic content and have compared EMI performance.

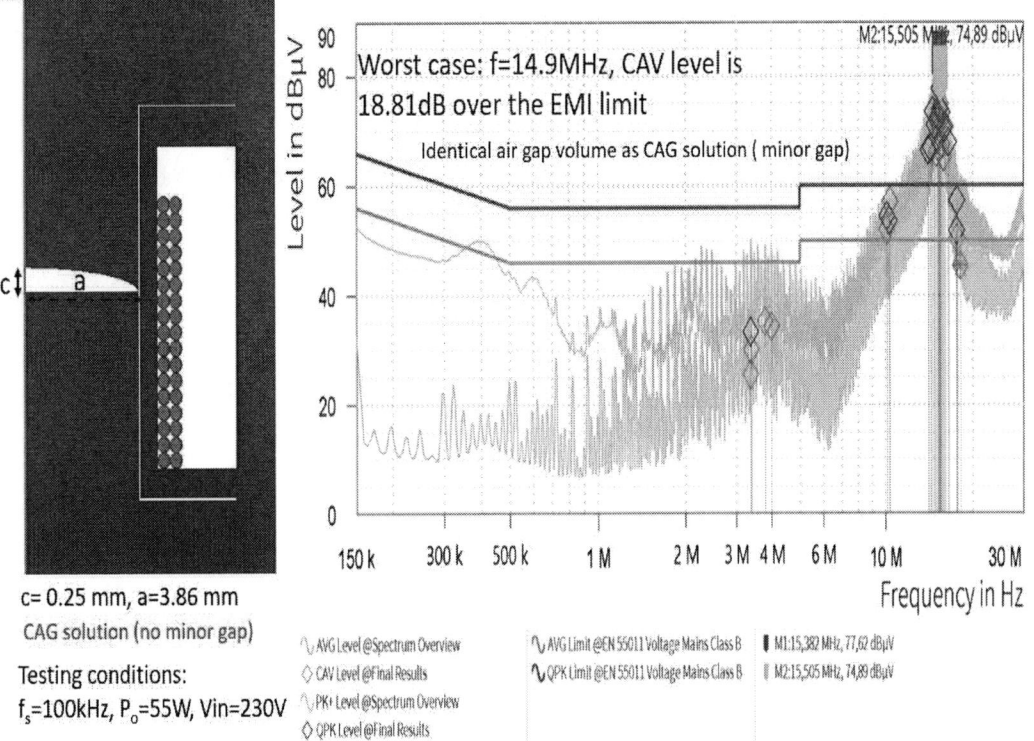

c= 0.25 mm, a=3.86 mm

CAG solution (no minor gap)

Testing conditions:

f_s=100kHz, P_o=55W, Vin=230V

Fig.15: Conducted emission with CAG no minor gap solution (air gap dimension: c=0.25mm, a=5.47mm)

Conclusion

The presented new concept of the curved air-gap inductor is a key solution to improve the harmonic content regardless of loading conditions and minimizing inductor volume. Given L_{eff}-P_o curves of the CAG solution against duty cycle and operation frequencies shown in fig.9 together with equation (7), the optimal CAG dimension can be determined effectively. Comparison of EMI measurement shows that the CAG with optimal minor gap structure is a preferrable solution to minimize conducted emission.

References

[1]. IEC, EN61000-3-2:2019, EMC-Part3-2: Limits for harmonic current emission (equipment input current ≤ 16A per phase)

[2]. Jee-Woo Lim and Bong Hwart-Kwon, "A power factor controller for single-stage PWM rectifiers," IEEE Trans. Ind.Electron., vol.46, pp.1035-1037, Oct.1999

[3]. R.Redl,L.Balogh,and N.O.Sokal, " A new family of single-stage isolated power-factor correctors with fast regulation of the output voltage," Rec. IEEE PESC, pp 1137-1144, 1994.

[4]. Philip C. Too, "UC3854 controlled Power Factor," Unitrode application note U-134

[5]. W.G. Hurley, "Power factor correction for ac/dc converters with cost effective inductive filtering," in Proc. IEEE 31st Power Electron. Spec. Conf. (PESC'00), vol.1, Galway, Ireland, June 2000 pp332-337.

[6]. Wai Keung Mo, Kasper M.Paasch and Thomas Ebel, " Optimal gapped boost inductor design for power factor correction applications," EPE 2021 23rd European Conference on Power Electronics and Applications , p.1-11.

[7]. E. Stenglein and M. Albach, "Analytical calculation method for the non-linear characteristic of ferrite-cored inductors with stepped air gap," Springer, Electr Eng (2017) 99: 421-429.

[8]. W.K.Mo, Kasper M Paasch and Thomas Ebel, " Improved method of power inductor design with DC current impact," IEEE 14th international conference on compatibility , power electronics and power engineering, pp120-126, 2020.

Automatic Generation Control-based Charging/Discharging Strategy for EV fleets to Enhance the Stability of a Vehicle-To-Weak Grid System

Majid Mehrasa[1], Mehrdad Gholami[1,2], Reza Razi[3], Khaled Hajar[3], Antoine Labonne[1], Ahmad Hably[3], and Seddik Bacha[1]

[1]Univ. Grenoble Alpes, CNRS, Grenoble INP, G2ELAB, 38000 Grenoble, France
[2]Faculty of engineering, University of Kurdistan, Sanandaj, Iran
[3]Univ. Grenoble Alpes, CNRS, Grenoble INP, GIPSA-Lab, 38000 Grenoble, France
majid.mehrasa@g2elab.grenoble-inp.fr

Acknowledgements

This work has been supported by the aVEnir project of the PIA operated by ADEME.

Keywords

«EVs Fleet», «Automatic Generation Control», «Weak Power Grid», «Balancing Power», «Frequency Control».

Abstract

In this paper, an automatic generation control (AGC)-based charging/discharging strategy is proposed for EV fleets to augment the stability of a weak power grid (WPG). The Vehicle-to-Weak Grid (V2WG) system under study is indeed a two-area microgrid including WPG, two EVs stations, PV units and loads. Firstly, the balancing power relation for three operating modes including no change mode, load increment and PV increment are comprehensively investigated. In doing so, a power sharing algorithm for each active EV is proposed to contribute to the WPG frequency deviation compensation. Then, AGC is developed for each area to attain the related angular frequency errors which are aimed to provide the tie-line power error needed for the dynamical frequency variation compensation using the appropriate EVs action. Moreover, the proposed strategy coefficients are analyzed to understand how much they can impact the stability margin of the system angular frequency errors. The simulation results in MATLAB/SIMULINK environment validate the ability of the proposed strategy-based EV fleets at providing a stable WPG under the increment of both load and PV units.

Introduction

Electric Vehicles (EVs) have continuously attracted the researcher's attention to be developed for Vehicle-to-Grid (V2G) applications using the bidirectional power transfer achieved from huge value of the battery energy of the Aggregated EVs [1-4]. Aggregated EVs can take the role of power grid frequency stabilizer while a high penetration of renewable energy resources (RERs) integrates into power grid or a large amount of load variations occurs [5-7]. Different challenges regarding EVs charging strategies have been discussed in recent years such as parking fleet scenario with modular converters [8], revenue assessment with wear cost model [9], PVs at the electric mobility's service [10-11], and different PEV parking scenarios [12]. But, the frequency vulnerability is an interesting challenge that is boosted in the presence of a weak power grid wherein much more accurate smart charging /discharging strategies must be designed for EVs by taking into account the required power for the appropriate frequency deviation compensation [13]. The appropriate compensation means both fast and accurate reactions achieved through i) the identification of EV rated power, ii) coordination between the EVs power sharing and the frequency deviation, and iii) maximum utilization of EV charging stations, and so on.

The secondary frequency control loop has been regarded as a target for researchers to be fortified by enhancing the EVs charging participation against power grid instabilities. To this end, the references [14], [15] and [16] have dealt with the grid secondary frequency control using respectively optimized fuzzy technique, mixed integer linear programming (MILP) and aggregator-based hierarchical strategy.

The secondary frequency control loop may be exposed to a stability uncertainty made by the aggregation dynamics and its delays. For this case, Ref [17] proposed an integration for the effects of heterogeneous delay and ascertained the sequential impact procedure for aggregation delays. On the other hand, some criteria in Ref [18] were taken into account such as various charging profiles, the SOC of EVs, and the numbers of EVs to enable a primary frequency loop aiming to attain flexible bidirectional power flow. The cooperation between wind system and EVs for completing the primary frequency loop of a microgrid was made by using small-signal analysis to establish frequency regulation strategy and droop or virtual inertia [19]. In the frequency regulation process, it has challenged a coordination between the power sharing control system and the SOC of EV batteries. This challenge was investigated through Ref [20] based on assigning the uncertain dispatch inside the control loop with no information regarding detailed EV charging/discharging in which the adjustment task was completed within the frequency control capacity of EVs. Using an aggregator-based coordination strategy, Ref [21] focused on assessment of a large number of EVs to fulfill the centralized supplementary frequency adjustment of the interlinked power systems. Also, the time delay of power systems was taken into account in [22] based on static output feedback frequency stabilization to provide the coordinate charge control for EVs. Among various control strategies incorporated by the charging schemes of EVs for the frequency control process, automatic generation control (AGC) has been also paid attention in recent literatures [23-24]. Several myopic/non-myopic-based real-time scheduling schemes were designed in [25] to provide suitable charging/discharging method of EVs fleet based on following out the signals generated through AGC. A wide-area virtual power plant (WVPP) was made in [26] including wind farms (WFs), photovoltaic units (PVs), and EVs wherein AGC was employed to noticeably guarantee the system response acceleration and the regulation cost mitigation due to unbalanced power mismatches between generation side and demand side. In Ref [27], the optimization factors including the performance-based compensation schedule, the AGC signals, and the EVs arrival and departure times were taken into account for solving a frequency regulation capacity scheduling problem to appoint the EV frequency adjustment service. In order to achieve the optimal solutions for a multi-area energy system, the DQ algorithm and double learning were utilized for the AGC-based frequency control under contributory obligation of battery energy storages, EVs and traditional power plants [28].

In this paper, a charging/discharging strategy is proposed for EV fleets to provide stable voltage for WPG while the PV units and interconnected loads are abruptly increased. To this end, a two-area microgrid is considered which consists of a WPG, two EV stations, PV units and loads. The proposed method is made through a pervasive balancing power and an AGC assigned for each area. The paper is organized as follows. The first section is allotted to the introduction. In section II, the considered two-area mcirogrid is described. The pervasive balancing power is discussed in section III. Section IV concentrates on the proposed AGC-based strategy. Simulation results and conclusion are respectively presented in Sections V and VI.

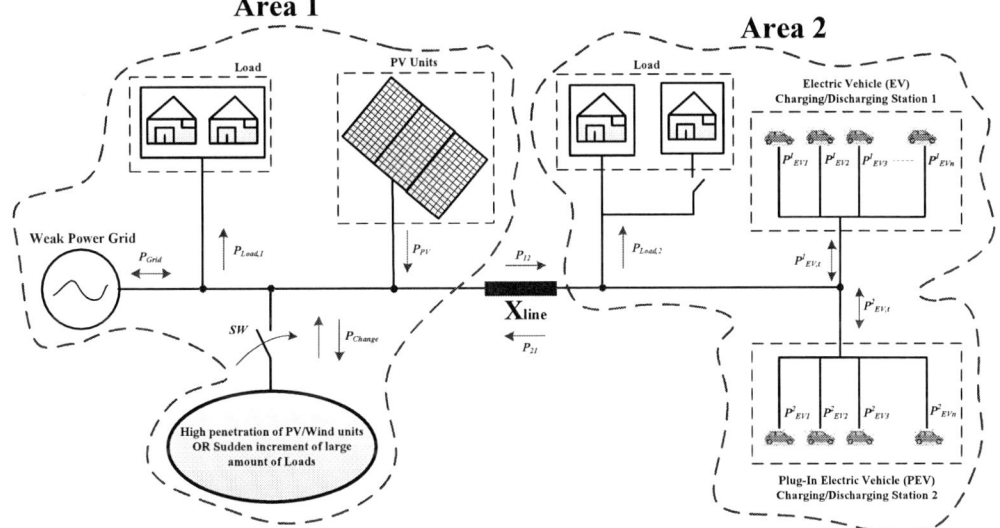

Fig. 1: The proposed V2WG-based two-area System.

Description of the considered two-area system

The two-area microgrid under study in this paper is displayed in Fig. 1. The WPG is placed within the first area and also the high-penetration renewable energies or a large amount of abrupt load variation will happen in this area. In addition, the PV units and load are included in the first area as well. As it can be seen from Fig. 1, two EVs charging/discharging stations are appointed in the second area that should be under a control process for an appropriate adjustment of WPG frequency. These EV fleets are also controlled to supply the loads in the second area. Two areas are interconnected through the inductance X_{line}.

Pervasive balancing power

The power in the microgrid shown in Fig. 1 must be controlled to enable a fast and precise frequency response for the WPG against high penetration of PV units or sudden increment of a large amount of loads. The power balance is the mandatory relation that should be taken into account by the proposed strategy. Based on Fig. 1, the balancing relation is acquired through the equation (1),

$$P_{Load} = P_{Grid} + P_{PV} + P_{EV,t} + (SW)(-1)^k P_{Change} + P_{X_{line}} \tag{1}$$

In the relation (1), when there is a new connection of PV units or Load, the SW (that acts as a switch) gets 1 and otherwise $SW=0$. Also, k is equalized to 1 and 2 for load connection and PV unit connection, respectively. The load power in (1) is achieved through $P_{Load}=P_{Load1}+P_{Load2}$. The P_{Xline} is the power due to the inductance X_{line} that has a key role at regulating the WPG frequency. It is realized from Fig. 1 that the total power of both EV charging/discharging stations is achieved through $P_{EV,t}=P^1{}_{EV,t}+P^2{}_{EV,t}$. The total power of each station is dependent on the number of the existing EVs as well as the instantaneous power of each EV. It is assumed that the first and second station respectively has "m" and "n" EVs. By noticing Fig. 1, the total power of EVs charging/discharging stations can meet,

$$P_{EV,t} = \sum_{j=1}^{n} P_{EV,j}^1 + \sum_{j=1}^{m} P_{EV,j}^2 \tag{2}$$

Using the relations (1) and (2), the total EVs power can be stated based on the power of load, the grid, PV, the inductance power and the additional load/PV,

$$\sum_{j=1}^{n} P_{EV,j}^1 + \sum_{j=1}^{m} P_{EV,j}^2 = P_{Load} - P_{Grid} - P_{PV} - (SW)(-1)^k P_{Change} - P_{X_{line}} \tag{3}$$

Three operating conditions are considered for the balancing power relation (3). These conditions are explained as the following to reach the most effective Frequency Regulation Strategy (FRS).

The microgrid with no PV/load increment

The first operating condition is fulfilled by $SW=0$. It means that there is no the PV/load increment. In this condition, while accurate prediction is accomplished for PV and load, the WPG encounters relatively a zero frequency alteration. Thus, the first and second stations tend to meet the following total power,

$$\sum_{j=1}^{n} P_{EV,j}^1 = \alpha \left(P_{Load} - P_{Grid} - P_{PV} - P_{X_{line}} \right) \tag{4}$$

$$\sum_{j=1}^{m} P_{EV,j}^2 = (1-\alpha)\left(P_{Load} - P_{Grid} - P_{PV} - P_{X_{line}} \right) \tag{5}$$

The coefficient α is determined with respect to the total existing rated power of each station wherein the higher power share belongs to the station with higher instantaneous power. As a consequence, the proposed charging/discharging method should enforce the EV of each station to be charged/discharged according to the relations (6) and (7),

$$P_{EV,x}^1 = \beta_x^1 \left[\alpha \left(P_{Load} - P_{Grid} - P_{PV} - P_{X_{line}} \right) \right] \tag{6}$$

$$P_{EV,y}^2 = \beta_y^2 \left[(1-\alpha)\left(P_{Load} - P_{Grid} - P_{PV} - P_{X_{line}} \right) \right] \tag{7}$$

Where $x=1, \ldots, n$ and $y=1, \ldots, m$. The coefficients β^1_x and β^1_y are adjusted by taking the EV specifications into account such as the rated power and the difference between the arrival and departure time. In addition, the relations $\sum_{x=1}^{n} \beta^1_x = 1$ and $\sum_{y=1}^{m} \beta^1_y = 1$ must be assured.

The microgrid with PV/load increment

High PV penetration and load increment are other challenges of WPG frequency that can be dealt with by the EV charging/discharging stations. Based on the relation (3), high PV penetration occurs when $SW=1$ and $k=2$ leading to,

$$\sum_{j=1}^{n} P^1_{EV,j} + \sum_{j=1}^{m} P^2_{EV,j} = P_{Load} - P_{Grid} - \left(P_{PV} + P^{PV}_{Change} \right) - P_{X_{line}} \tag{8}$$

Where P^{PV}_{Change} is the power injected from the PV units which are abruptly integrated into the WPG at a specified time. Since the PV units inject power into the WPG, it is highly probable that most EVs in both stations proceed at charging mode. To this end, the following inequality must be satisfied,

$$P_{PV} + P^{PV}_{Change} > \left[P_{Load} - P_{Grid} - P_{X_{line}} \right] \tag{9}$$

The inequality (9) depends on the load power and subsequently the power grid can be chosen as a constant value or within limited constraints because of the sensibility of WPG frequency. The load connection is taken into account through $SW=1$ and $k=1$ wherein the relation (3) can meet,

$$\sum_{j=1}^{n} P^1_{EV,j} + \sum_{j=1}^{m} P^2_{EV,j} = \left(P_{Load} + P^{Load}_{Change} \right) - P_{Grid} - P_{PV} - P_{X_{line}} \tag{10}$$

In equation (10), the P^{Load}_{Change} is the additional load interconnecting with the WPG at a specified time. For general discussion, the load variation in the second area also is involved with the P^{Load}_{Change} having approximately the same impact on the WPG. When the additional load is appeared, most EVs can be presumed to be discharged. This condition is made through (11),

$$P_{Load} + P^{Load}_{Change} > P_{Grid} + P_{PV} + P_{X_{line}} \tag{11}$$

In fact, the inequalities (9) and (11) can be employed to identify the charging or discharging states of two EVs fleet in various operating conditions.

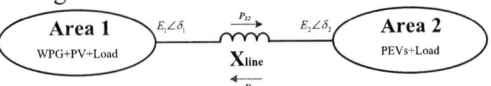

Fig. 2: The net tie-line power representation for a two-area system.

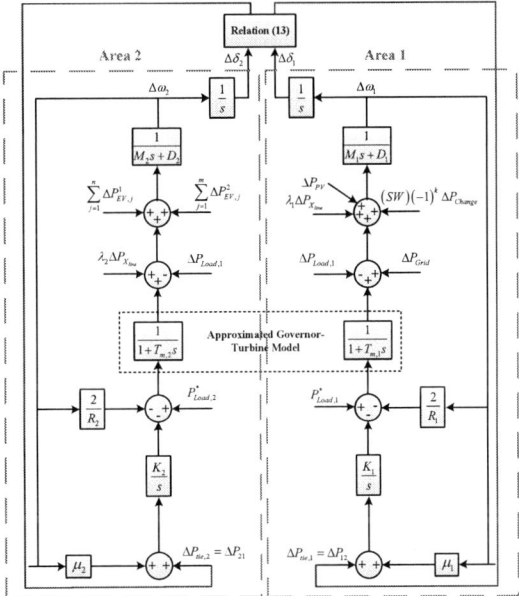

Fig. 3: The proposed AGC-based control.

Proposed AGC-based strategy

Each area in the proposed V2WG-based two-area microgrid shown in Fig. 1 is regarded as the independent generation unit that is responsible for reaching desired value for the frequency in its own area. In doing so, Automatic Generation Control (AGC) is allotted to each area commensurate with compensating the WPG frequency deviations and fulfilling effective charging/discharging procedures for EV stations. Before starting the AGC design for each area, it is worth investigating on the net tie-line power error $\Delta P_{12} = -\Delta P_{21}$ and the relation between the tie-line power and the area frequency. The tie-line power of a two-area system with the inductance interface that has the specifications according to Fig. 2 can be written,

$$P_{12} = \frac{E_1 E_2}{X_{line}} \sin(\delta_1 - \delta_2) \tag{12}$$

In relation (12), $E_{1(2)}$ and $\delta_{1(2)}$ are the voltage characteristics of the control areas named the voltage magnitude and the voltage angle, respectively. The small-signal linearization is exerted to (12) around the equilibrium points of P^*_{12} and $\delta^*_{1(2)}$ with the approximation $\Delta E_{1(2)} = 0$. This linearization results in the relation between the frequency and the tie-line power error according to (13),

$$\Delta P_{12} = \left(X_{line} \right)^{-1} \left[E_1^* E_2^* \cos\left(\delta_1^* - \delta_2^* \right) \times \int \left(\Delta \omega_1 - \Delta \omega_2 \right) dt \right] \tag{13}$$

The relation (13) determines that the tie-line power error has a key role at the instantaneous calculation of the area frequency. The following sub-sections concentrate on the frequency AGC-based control design for the areas.

The control design details for the area i^{th}

Each area is composed of various parts that can contribute to the frequency adjustment process as much as possible. To deeply understand this issue, it is mandatory to possess the angular frequency error formulation in the areas i^{th} based on the AGC that is just as,

$$\Delta \omega_i = \left(M_i s + D_i \right)^{-1} \left(\Delta P_{m,i} - \Delta P_{L,i} - \Delta P_{tie,i} \right) \tag{14}$$

Where M_i and D_i are the equivalent inertia and load-damping constant, respectively. The tie-line power error $\Delta P_{tie,i}$ is another variable that can impact on the frequency error of the area i^{th}. On the other hand, the variable $\Delta P_{m,i}$ in (14) is the input mechanical power that can be detailed as follows,

$$\Delta P_{m,i} = \left(1 + T_{m,i} s \right)^{-1} \left(P^*_{Load,i} - 2 \left(R_i \right)^{-1} \Delta \omega_i - \frac{K_i}{s} ACE_i \right) \tag{15}$$

It should be noticed that R_i and K_i are the droop and integral controller coefficients, respectively. In addition, ACE_i is named as the Area Control Error (ACE) dedicated to the area i^{th}. The ACE_i is defined as stated in (16) to keep the frequency and power interchanges at the scheduled values in which μ_i is the ACE i^{th} frequency response characteristic,

$$ACE_i = \mu_i \Delta \omega_i + \Delta P_{tie,i} \tag{16}$$

As it can be realized from Fig. 1, the first area consists of WPG, PV units, and load that is also exposed to both high penetration of PV units and abrupt increment of load. It can be concluded that the angular frequency error of the first area is equiponderated with,

$$\Delta \omega_1 = \left(M_1 s + D_1 \right)^{-1} \begin{pmatrix} \Delta P_{m,1} - \Delta P_{Load,1} + \Delta P_{Grid} \\ + \Delta P_{PV} + \left(SW \right) \left(-1 \right)^k \Delta P_{Change} + \lambda_1 \Delta P_{X_{line}} \end{pmatrix} \tag{17}$$

The sharing factor of the inductance power for the first area is the coefficient λ_1. On the other hand, the two EVs charging/discharging stations are situated in the second area and have a vital role at the frequency regulation process. The angular frequency error in (18) is presented for the second area,

$$\Delta \omega_2 = \left(M_2 s + D_2 \right)^{-1} \begin{pmatrix} \Delta P_{m,2} - \Delta P_{Load,2} \\ + \sum_{j=1}^{n} \Delta P_{EV,j}^1 + \sum_{j=1}^{m} \Delta P_{EV,j}^2 + \lambda_2 \Delta P_{X_{line}} \end{pmatrix} \tag{18}$$

As it was mentioned, λ_2 denotes the sharing factor of the inductance power for the second area. The sharing factors of the inductance power must be proportionate to the frequency deviation of each area.

The proposed AGC-based control for completing the EVs charging/discharging proceeding in the system under study is illustrated in Fig. 3.

The assessment of the AGC-based controller coefficients

The proposed AGC-based charging/discharging strategy consists of different coefficients including $\{M_i, D_i, T_{m,i}, R_i, K_i, \mu_i\}$ that all need to be appropriately selected. To this end, the relations (15) and (16) are substituted into (14) resulting in,

$$\Delta\omega_i = \left(M_i s + D_i\right)^{-1}\left(\left(1+T_{m,i}s\right)^{-1}\left(\begin{array}{c} P^*_{Load,i} - 2\left(R_i\right)^{-1}\Delta\omega_i \\ -\dfrac{K_i}{s}\left(\mu_i\Delta\omega_i + \Delta P_{tie,i}\right)\end{array}\right) - \Delta P_{L,i} - \Delta P_{tie,i}\right) \tag{19}$$

And finally the angular frequency error of the area i^{th} is attained after some mathematical simplifications according to (20),

$$\Delta\omega_i = H_{Load,\mathrm{Re}f}(s)P^*_{Load,i} + H_{tie,i}(s)\Delta P_{tie,i} + H_{L,i}(s)\Delta P_{L,i} \tag{20}$$

The angular frequency error consists of the transfer functions that are a consequence of the reference load power, tie-line power error, and load power error. The transfer functions in (20) are achieved using (21),

$$H_{Load,\mathrm{Re}f}(s) = s/\Delta_i, H_{tie,i}(s) = \left(T_{m,i}s^2 + s - K_i\right)/\Delta_i$$

$$H_{L,i}(s) = -\left(T_{m,i}s^2 + s\right)/\Delta_i \tag{21}$$

$$\Delta_i = \left(M_i T_{m,i}s^3 + \left(D_i T_{m,i} + M_i\right)s^2 + \left(2\left(R_i\right)^{-1} + D_i\right)s + K_i\mu_i\right)$$

The steady-state response of the angular frequency error is dependent on the transfer function $H_{tie,i}(s)$ while $t \rightarrow \infty$ leading to,

$$\Delta\omega_i(s=0) = \left(\mu_i\right)^{-1}\Delta P_{tie,i} \tag{22}$$

Based on (22), the frequency response characteristic and tie-line power error can impact the steady-state response of WPG frequency at the same time. The Nyquist diagrams are plotted in Fig. 4 to determine the contributions of AGC-based controller coefficients at the stability margin expansion. In Fig. 4(a), the Nyquist diagrams of the transfer functions (21) are separately depicted assuming the step response and exerting the appropriate values to the coefficients. By noticing this figure, it can be realized that the step response of load power error-based transfer function makes the angular frequency error unstable, however, the error stability is not disturbed through two other transfer functions. It is worth mentioning that the step response of transfer function $H_{tie,i}$ is able to aggrandize the stability margin of the angular frequency error accordingly. The step response-based Nyquist diagram of sum transfer functions in (20) is illustrated in Fig. 4(b). It should be paid attention that however the angular frequency error response is placed within a stable area, but the value of this error must be decreased by selecting more suitable coefficients.

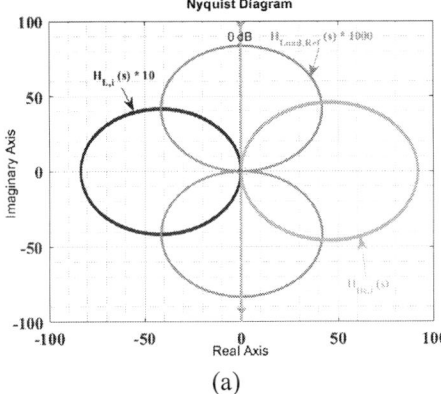
(a)

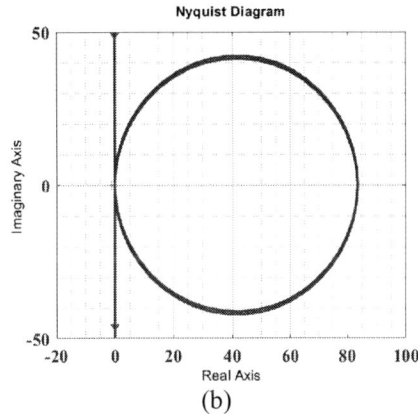
(b)

Fig. 4: The step response-based Nyquist diagrams of (a) each transfer function in (21), and (b) transfer function (20).

Simulation results

In this section, MATLAB/SIMULINK software is used to assay the proposed strategy performance for making the WPG voltage stable in presence of load/PV increment at t=0.8*60 s.

The EV rated power in the related EV charging stations are within the interval [80 kW, 95 kW]. Also, the simulation is executed in 1.6 min equalizing to 96 s. The inductance X_{line} is equalized to 0.002π. As it can be seen from Fig. 5, the arrival time of all EV is zero (the beginning of the simulation). According to this figure, the departure time of most EVs is the end of simulation excepted for some of them in which the departure times are 0.7 min, 1.1 min, 0.3 min, 0.55 min, 0.75 min, 0.4 min, 1.2 min, 0.75 min, and 1.3 min.

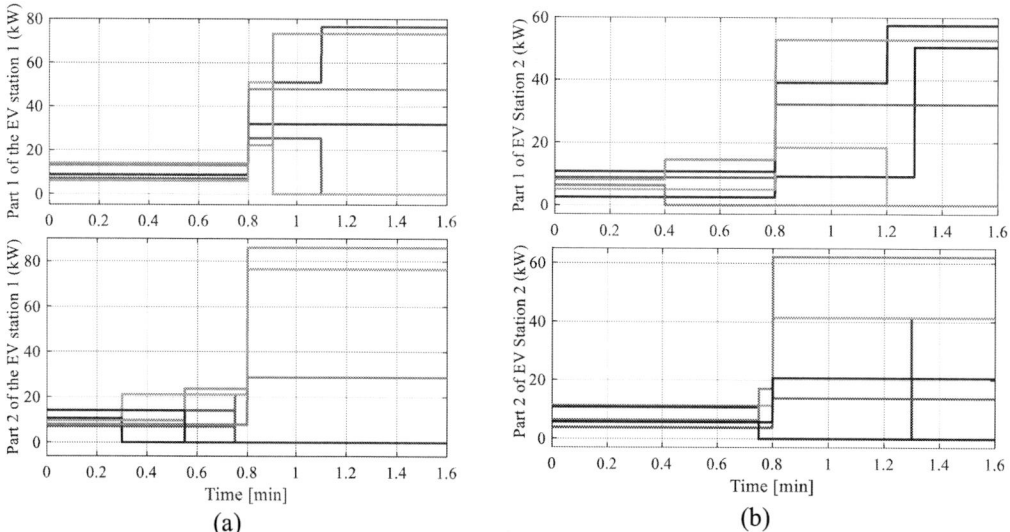

(a)   (b)

Fig. 5: The instantaneous EV power injected into the WPG from, (a) the first EV charging station, and (b) the second EV charging station.

The instantaneous EV power injected into the WPG from the first and second EV charging stations are presented in Fig. 5(a) and Fig.5(b), respectively. In fact, the appropriate operation of proposed strategy leads to the EV power of the EV charging stations illustrated in these figures wherein they also result in the WPG frequency and voltage magnitude responses given in the first part of Fig. 6(a) and (b). Fig. 6 exhibits the desirable response and inappropriate responses due to mistake at the proposed strategy coefficients selection for WPG frequency, and WPG voltage magnitude. When the proposed strategy coefficients are wrongly chosen including the coefficients $\{\alpha, \beta_x, \beta_y\}$ for the power balancing and the coefficients $\{M_i, D_i, T_{m,i}, R_i, K_i, \mu_i\}$ for AGC, two unacceptable responses for both WPG frequency and WPG voltage magnitude are obtained, as depicted in the second part of Fig. 6(a) and Fig. 6(b).

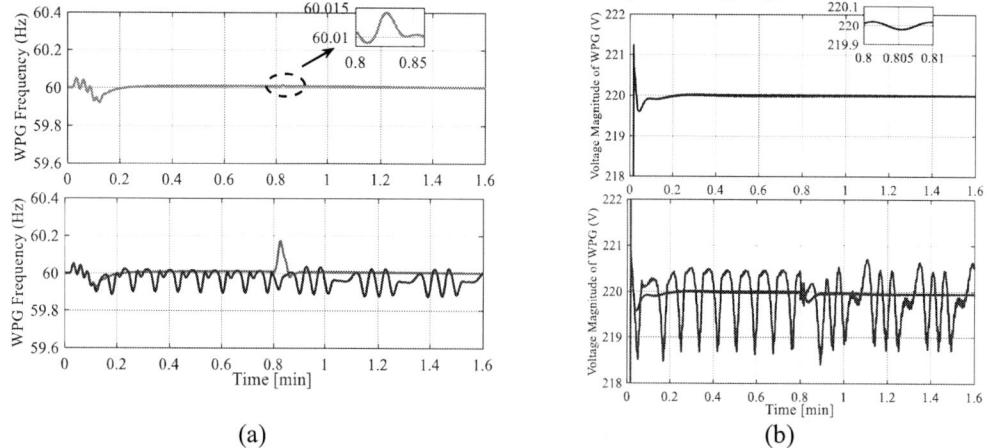

(a)   (b)

Fig. 6: The desirable response and inappropriate responses due to mistake at the proposed strategy coefficients selection for (a) WPG frequency, and (b) WPG voltage magnitude.

Conclusion

This paper has presented a charging/discharging strategy for EV fleets based on automatic generation control (AGC) to provide stable voltage for WPG in presence of load/PV increment. Using the EVs specifications, the EV station power has been specified for allotting to a proposed balancing power aimed at damping the steady-state frequency error of WPG. The proposed AGC method has been designed using the angular frequency error calculation for each area and the tie-line power error along with the suitable EV action based on the proposed balancing power to significantly mitigate the dynamic change of the WPG frequency. By assessing the variations of the proposed control coefficients, the efficient coefficients have been traced to see the stability margin trend of the angular frequency error. The proposed charging/discharging strategy correctness has been approved using simulation results in MATLAB/SIMULINK environment wherein the WPG frequency and voltage magnitude could be kept stable under the PV/load increment.

References

[1] R. Razi, K. Hajar, M. Mehrasa, A. Labonne, A. Hably and S. Bacha, "Limiting discharge cycles numbers for plug-in electric vehicles in bidirectional smart charging algorithm," IECON 2021 – 47th Annual Conference of the IEEE Industrial Electronics Society, 2021, pp. 1-6.

[2] M. Mehrasa, R. Razi, K. Hajar, A. Labonne, A. Hably and S. Bacha, "Power Management of a Smart Vehicle-to-Grid (V2G) System Using Fuzzy Logic Approach," IECON 2021 – 47th Annual Conference of the IEEE Industrial Electronics Society, 2021, pp. 1-6.

[3] A. Ovalle, A. Hably, S. Bacha, "Grid Optimal Integration of Electric Vehicles: Examples with Matlab Implementation," Springer International Publishing, 2018, pp. 1-213.

[4] H. Turker and S. Bacha, "Optimal Minimization of Plug-In Electric Vehicle Charging Cost with Vehicle-to-Home and Vehicle-to-Grid Concepts," IEEE Trans. Vehicular Technology, vol. 67, no. 11, pp. 10281-10292, Nov. 2018.

[5] M. Gholami, M. Mehrasa, R. Razi, A. Hably, S. Bacha and A. Labbone, "An Efficient Control Strategy for the Hybrid Wind-Battery System to Improve Battery Performance and Lifetime," IECON2021 – 47th Annual Conference of the IEEE Industrial Electronics Society, 2021, pp. 1-6.

[6] M. Mehrasa, E. Pouresmaeil, A. Sepehr, B. Pournazarian, and J. P. S. Catalão, "Control of power electronics-based synchronous generator for the integration of renewable energies into the power grid," International Journal of Electrical Power & Energy Systems, vol. 111, pp. 300-314, Oct 2019.

[7] Y. T. Holari, S. A. Taher, and M. Mehrasa, "Distributed energy storage system-based nonlinear control strategy for hybrid microgrid power management included wind/PV units in grid-connected operation," International Transactions on Electrical Energy Systems, vol. 30, no. 2, e12237, Feb 2020.

[8] R Razi, M Mehrasa, K Hajar, M Gholami, A Hably, S Bacha, and A Labonne, "Predictive smart charging of plug-in electric vehicles for parking fleet scenario with modular converters," CIRED Porto Workshop 2022 E-mobility and power distribution systems.

[9] M. Mehrasa, R. Razi, M. Gholami, K.Hajar, A. Labonne, A. Hably, and S. Bacha, "A Dynamic Real-Time Optimization Algorithm for the Revenue Assessment of a Vehicle-To-Grid System in Presence of Wear Cost Model," Electrimacs 2022: 14th International Conference of TC-Electrimacs Committee, 16-19 May 2022 Nancy (France).

[10] K.Hajar, R. Razi, M. Mehrasa, A. Labonne, A. Hably, and S. Bacha, "Photovoltaics at the electric mobility's service: French case study," Electrimacs 2022: 14th International Conference of TC-Electrimacs Committee, 16-19 May 2022 Nancy (France).

[11] K. Hajar, B. Guo, A. Hably and S. Bacha, "Smart charging impact on electric vehicles in presence of photovoltaics," 2021 22nd IEEE International Conference on Industrial Technology (ICIT), 2021, pp. 643-648.

[12] R. Razi, K. Hajar, A. Hably and S. Bacha, "A user-friendly smart charging algorithm based on energy-awareness for different PEV parking scenarios," 2021 29th Mediterranean Conference on Control and Automation (MED), 2021, pp. 392-397.

[13] M. F. M. Arani, Y. A. I. Mohamed and E. F. El-Saadany, "Analysis and Mitigation of the Impacts of Asymmetrical Virtual Inertia," IEEE Transactions on Power Systems, vol. 29, no. 6, pp. 2862-2874, Nov. 2014.

[14] S. Flahati, S. A. Taher and M. Shahidehpour, "Grid Secondary Frequency Control by Optimized Fuzzy Control of Electric Vehicles," IEEE Transactions on Smart Grid, vol. 9, no. 6, pp. 5613-5621, Nov. 2018.

[15] K. Kaur, N. Kumar and M. Singh, "Coordinated Power Control of Electric Vehicles for Grid Frequency Support: MILP-Based Hierarchical Control Design," IEEE Transactions on Smart Grid, vol. 10, no. 3, pp. 3364-3373, May 2019.

[16] K. Kaur, M. Singh and N. Kumar, "Multiobjective Optimization for Frequency Support Using Electric Vehicles: An Aggregator-Based Hierarchical Control Mechanism," IEEE Systems Journal, vol. 13, no. 1, pp. 771-782, March 2019.

[17] C. Dong, Q. Xiao, M. Wang, T. Morstyn, M. D. McCulloch and H. Jia, "Distorted Stability Space and Instability Triggering Mechanism of EV Aggregation Delays in the Secondary Frequency Regulation of Electrical Grid-Electric Vehicle System," IEEE Transactions on Smart Grid, vol. 11, no. 6, pp. 5084-5098, Nov. 2020.

[18] S. Iqbal et al., "Aggregation of EVs for Primary Frequency Control of an Industrial Microgrid by Implementing Grid Regulation & Charger Controller," IEEE Access, vol. 8, pp. 141977-141989, 2020.

[19] M. Fakhari Moghaddam Arani and Y. A. I. Mohamed, "Cooperative Control of Wind Power Generator and Electric Vehicles for Microgrid Primary Frequency Regulation," IEEE Transactions on Smart Grid, vol. 9, no. 6, pp. 5677-5686, Nov. 2018.

[20] H. Liu, J. Qi, J. Wang, P. Li, C. Li and H. Wei, "EV Dispatch Control for Supplementary Frequency Regulation Considering the Expectation of EV Owners," IEEE Transactions on Smart Grid, vol. 9, no. 4, pp. 3763-3772, July 2018.

[21] H. Liu, Z. Hu, Y. Song, J. Wang and X. Xie, "Vehicle-to-Grid Control for Supplementary Frequency Regulation Considering Charging Demands," IEEE Transactions on Power Systems, vol. 30, no. 6, pp. 3110-3119, Nov. 2015.

[22] T. N. Pham, S. Nahavandi, L. V. Hien, H. Trinh and K. P. Wong, "Static Output Feedback Frequency Stabilization of Time-Delay Power Systems with Coordinated Electric Vehicles State of Charge Control," IEEE Transactions on Power Systems, vol. 32, no. 5, pp. 3862-3874, Sept. 2017.

[23] P. M. Rocha Almeida, J. P. Iria, F. Soares and J. A. P. Lopes, "Electric vehicles in automatic generation control for systems with large integration of renewables," 2017 IEEE Power & Energy Society General Meeting, 2017.

[24] C. Battistelli, and A. J. Conejo, "Optimal management of the automatic generation control service in smart user grids including electric vehicles and distributed resources," Electric Power Systems Research, vol. 111, pp. 22-31, June 2014.

[25] G. Wenzel, M. Negrete-Pincetic, D. E. Olivares, J. MacDonald and D. S. Callaway, "Real-Time Charging Strategies for an Electric Vehicle Aggregator to Provide Ancillary Services," IEEE Transactions on Smart Grid, vol. 9, no. 5, pp. 5141-5151, Sept. 2018.

[26] X. S. Zhang, T. Yu, Z. N. Pan, B. Yang and T. Bao, "Lifelong Learning for Complementary Generation Control of Interconnected Power Grids with High-Penetration Renewables and EVs," IEEE Transactions on Power Systems, vol. 33, no. 4, pp. 4097-4110, July 2018.

[27] E. Yao, V. W. S. Wong and R. Schober, "Robust Frequency Regulation Capacity Scheduling Algorithm for Electric Vehicles," IEEE Transactions on Smart Grid, vol. 8, no. 2, pp. 984-997, March 2017.

[28] L. Xi, L. Zhou, Y. Xu and X. Chen, "A Multi-Step Unified Reinforcement Learning Method for Automatic Generation Control in Multi-Area Interconnected Power Grid," IEEE Transactions on Sustainable Energy, vol. 12, no. 2, pp. 1406-1415, April 2021.

Model-based Converter Control for the Emulation of a Wind Turbine Drive Train

Alexander Ernst, Wilfried Holzke, Dawid Koczy, Nando Kaminski, Bernd Orlik
University of Bremen, Institute for Electrical Drives, Power Electronics and Devices IALB
Otto-Hahn-Allee 1
Bremen, Germany
Tel.: +49 421 218-62691
Fax: +49 421 218 98-62691
E-Mail: aernst@ialb.uni-bremen.de
URL: https://www.uni-bremen.de/ialb

Acknowledgements

This work was funded by the German Federal Ministry for Economic Affairs and Climate Action (BMWK) as part of the project "HiPE-WiND - Multi-Dimensional Stresses in High Power Electronics of Wind Energy Plants" under grant no. 0324219A.

Keywords

Converter Control, EESM, Virtual Synchronous Generator (VSG), Wind Energy

Abstract

Failures in wind turbines are often attribute to faults in the power electronics. To investigate the fault mechanisms, a test bench was set up in which entire converters up to a power of 10 MW can be tested under changing climatic conditions and electrical loads. In order to generate realistic loads, one of the converters has to behave like a drive train of a wind turbine. Except from a few differences, the test bench which is used here and has a total power of 300 kW, is a scaled copy of the 10 MW test bench. This small test bench is used to perform preliminary developments of control concepts and test scenarios that can later be transferred to the large test bench. It consists mainly of two back-to-back converters and three transformers. The goal of this work is to force the behaviour of a drive train of a wind turbine on one of the converters. The second converter should be able to perform a generator current control, as is common in wind turbines. This paper shows the implementation of a wind rotor model, the connection with the existing generator model and the implementation of a speed-dependent generator load curve on the side of the device under test (DUT). The results demonstrate the functionality of the overall test bench. For this purpose, different curves of calculated and measured values such as wind speed, pitch angle, rotor speed, phase current and voltage under different conditions are shown. In summary, it can be shown that one of the converters behaves indeed like a drive train of a wind turbine and that realistic scenarios can be created based on measured wind speed curves.

Introduction

To gain a better understanding of the causes of malfunctions in the power electronics of wind turbine converters [1], a test bench was set up in which entire converters can be tested under electrical load and climatic conditions. In addition to the 10 MW test bench, another test bench with a rated power of 300 kW has been built. The basic structure of the two test benches is identical, however there are also some differences. The 300 kW test bench is used to develop and test different scenarios before they are transferred to the large-scale test bench. The focus of these scenarios is to simulate the electrical loads as realistically as possible. Therefore, this test bench does not have its own climate room. Furthermore, the smaller test bench does not have any converters working in parallel, as it is the case in the 10 MW test bench due to its high power. Fig. 1 shows the entire test bench, which consists of a total of three transformers and two full converters. The transformer T1 has an apparent power of only 40 kVA and is

Fig. 1: Photo of the test bench

responsible for compensating the losses of the converters and transformers during the tests. The transformers T2 and T3 with a rated apparent power of 270 kVA have adjustable output voltages, so that different devices under test (DUTs) can be connected. At this time, the two converters (load unit and DUT) are of the same type. The DUT should perform a generator current control as used in wind turbines and the load unit should emulate the drive train of a wind turbine. A full converter is used as load unit, so that the power is driven in a loop keeping the losses low.

System topology

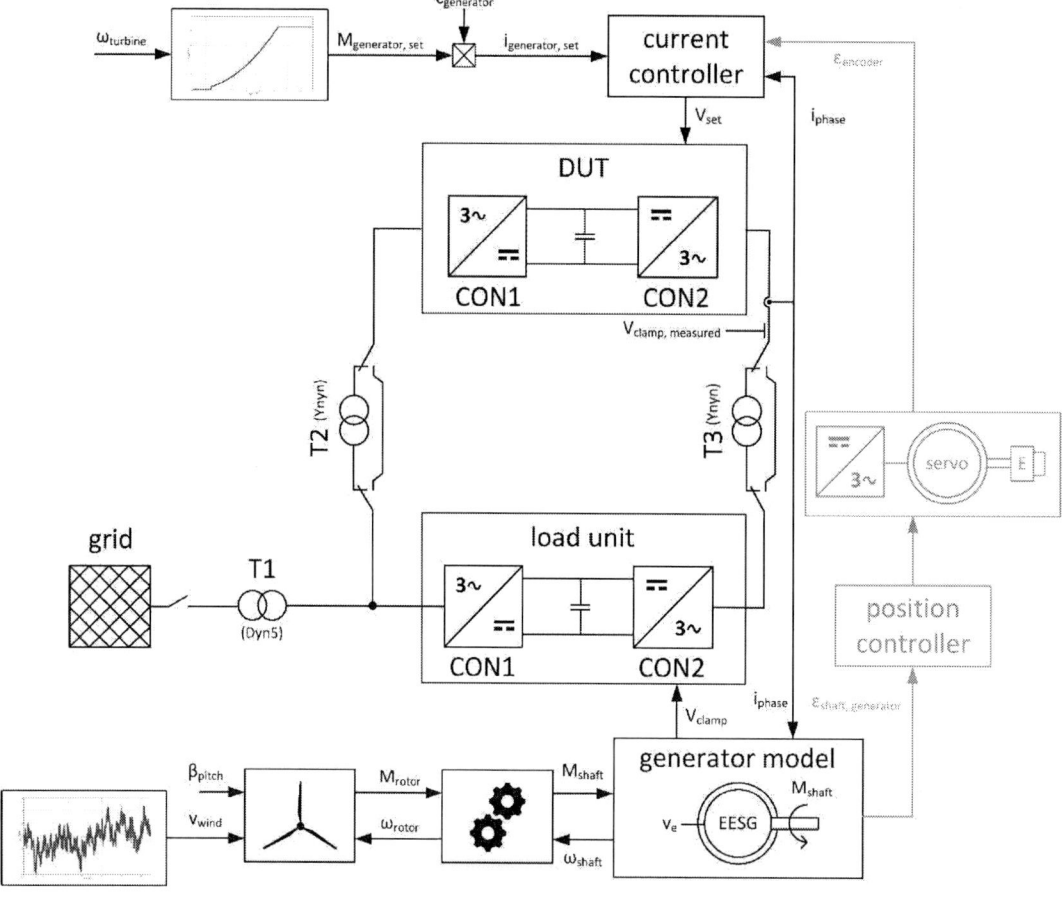

Fig. 2: System topology

Fig. 2 shows the topology of the entire system. The grid-side converters (CON1) are responsible for controlling the DC-link voltages of the load unit and the DUT. The power flow direction is arbitrary, so that the power can be driven in a loop. For this purpose, a current control is implemented on the DUT side, which receives a negative setpoint from a speed-dependent characteristic curve. The current ensures a load on the load unit and, thus, also on the generator model. The load unit is model-based controlled and gets its voltage setpoints from a synchronous generator model. The clamping voltage is influenced by the speed, the load and the excitation of the generator. The generator is part of a drive train, so that the speed depends on the current operating state of the system. The input variable is a wind profile, which generates a driving torque on the wind rotor depending on the current speed and pitch angle.

Generator model

For the model-based control of the converter, the model of an electrically excited synchronous machine according to [2] is used. This model receives three input variables, the currents i_{phase} transferred into the rotating coordinate system, the excitation voltage V_e and a mechanical torque M_{shaft}, which is applied to the shaft. This is used to calculate the clamping voltages of the generator V_{clamp}, also as d/q values, as well as the rotational speed. For the transformation of the currents and voltages into the rotating reference frame and the back transformation, the shaft angle is used. This model works with normalised values, so that the measured phase currents and the voltage setpoints must be normalised or denormalised. For the calculation on a PLC, the model has also been discretised. In order to simplify the discretisation, the reactance operators $x_d(s)$ and $x_q(s)$ were previously transformed into the controllable canonical form, resulting in a structure of simple integrators and coefficients that are easier to discretise. The parameters used for this model are taken from a data sheet of an electrically excited synchronous machine with damper windings and the parameters from Tab. 1.

Parameter	Value
Apparent power	9.375 MVA
Rated speed	500 min^{-1}
Rated torque	143.24 kNm

Tab. 1: Parameters of the synchronous generator

Wind rotor model

The emulation of the wind rotor is based on a characteristic set of curves of aerodynamic power coefficients C_P. This characteristic set of curves is described by the following equations with the coefficients from Tab 2. [3]:

$$\frac{1}{\lambda(i)} = \frac{1}{\lambda + 0.08 \cdot \beta} + \frac{0.035}{\beta^3 + 1} \tag{1}$$

$$C_P(\beta, \lambda) = C1 \cdot (C2 - C3 \cdot \beta(i) - C4 \cdot \beta(i)^x - C5) \cdot e^{-C6} \tag{2}$$

With the parameters:

C1	C2	C3	C4	C5	C6	x
0.6	$116/\lambda(i)$	0.4	0.001	5	$20/\lambda(i)$	2

Tab. 2: Parameters for analytic C_P curves

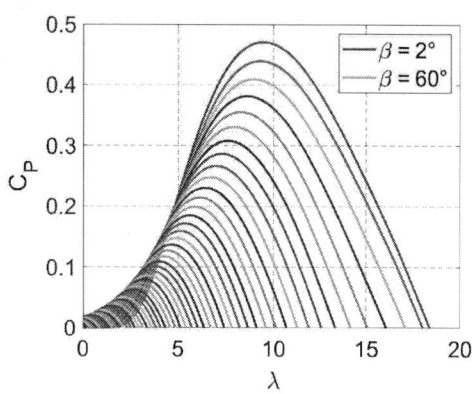

Fig. 3: Generated C_P curves

For a pitch angle β in each case, equations (1) and (2) are used to generate a characteristic curve that is dependent on the tip speed ratio $\lambda = \omega_{rotor} \cdot r_{rotor}/v_{wind}$. The working range in which the pitch angle β can be set is fixed here at 2° to 60°. Thus, at the smallest pitch angle, the highest C_P value of 0.4706 is obtained. Fig. 3 shows the generated C_P-curves. To calculate the torque that the wind generates at the rotor, it is necessary to have the current power that can theoretically be taken from the wind (Eq. 3) and the rotor speed ω_{rotor}.

$$P_{wind} = \frac{1}{2} \cdot \rho \cdot A_{rotor} \cdot v_{wind}^3 \tag{3}$$

The torque of the wind rotor is calculated as follows:

$$M_{rotor} = P_{wind} \cdot C_P(\beta, \lambda)/\omega_{rotor} \tag{4}$$

The rotor area A_{rotor} is chosen so that the power at the nominal point ($\lambda_{opt}, \beta_{min}, v_{wind\,nom}$) is equal to the active power of the generator model of 7.5 MW. To ensure that the calculated torque does not approach infinity, the speed in equation 4 must not become too small. A gear ratio factor of 0.046 ensures that the speed is increased towards the generator and that at the same time the torque is reduced. Since the generator model works with normalised quantities, the mechanical torque must be divided by the rated torque of the generator of 143.24 kNm between the gearbox and the connection. The moment of inertia of the entire drive train in relation to the generator shaft was set to $8.207 \cdot 10^5 \, kgm^2$.

Wind profile

The wind speed can be specified either as a changeable constant or from a lookup table. The wind speed must also not fall below a minimum of 3 m/s so that the turbine does not come to a standstill. The wind field in Fig. 4 was generated with the software "TurbSim". It is possible to set different requirements for the wind speed profile. These include the norm-dependent turbulence intensity, the step size of the data points or the mean wind speed. Here the mean wind speed has been set to 11.64 m/s, which is below the rated wind speed of the turbine of 14 m/s. Since this wind field is to be used cyclically, the first data point and the last data point must have approximately the same values. Due to the changeable constant wind speed, the entire wind field can be shifted up or down. This enables changing the operating point of the system between partial load and full load.

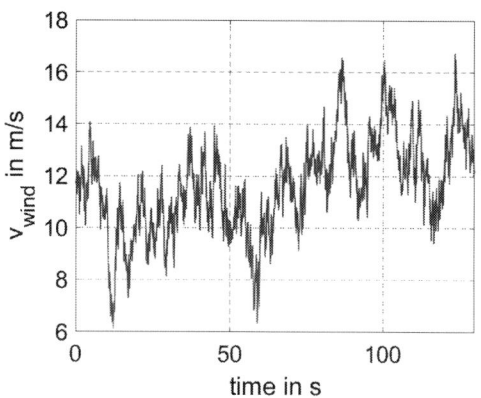

Fig. 4: Windfield generated with TurbSim (Configuration: IEC 61400-1 Ed. 3 (2010), step size: 0.05 s)

Generator load curve

The electrical load on the generator of a wind turbine in the partial load range is usually dependent on the speed. This means that the generator current increases with increasing shaft speed or wind speed. With a constant pitch angle, the speed of the wind rotor should be controlled in such a way that it

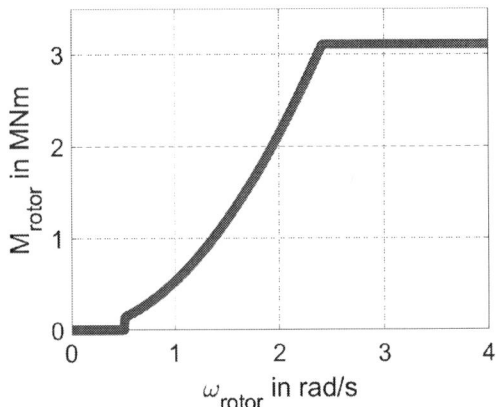

increases proportionally with the wind speed in order to always remain at the optimal tip speed ratio [4]. However, since the power that can be extracted from the wind increases with v_{wind}^3 (Eq. 3), the electromagnetic torque of the generator and thus the current must increase with ω^2 (Eq. 4). Therefore, the generator curve for the partial load range is quadratic. If the speed falls below the cut-in speed, the target torque is set to zero. Once the rated speed is reached, the wind turbine is in the full load range; here the target torque remains constant at its nominal value. Fig. 5 shows the generator curve generated.

Fig. 5: Generator load curve

Control strategies

Speed control

In this case, the wind rotor is started by a PI controller, which acts on the mechanical torque M_{shaft} via a first-order delay element. This has the advantage that the turbine is already at nominal speed when the rotor model is switched on. Furthermore, this speed control can be used, for example, to test the current control without the electromagnetic torque becoming greater than the driving torque and thereby slowing down the generator too much. The PI controller was designed according to the symmetrical optimum. Once the rated speed is reached, the torque of the wind rotor can be connected as M_{shaft} and the pitch control can be activated. The speed is stabilised indirectly in the partial load range via the electromagnetic torque of the generator. In the full load range, the pitch control must ensure that the turbine rotates on average at the rated speed.

Clamp voltage control

With decreasing speed and increasing electrical load, the clamping voltage of the generator decreases if the excitation voltage is kept constant [5]. The excitation voltage must be controlled so that the electrical power does not additionally decrease due to a lower clamp voltage. For this purpose, the magnitude of the voltage pointer of the clamping voltage is formed from the d and q components by $|V| = \sqrt{\underline{V}_d^2 + \underline{V}_q^2}$

and regulated to one. By regulating the clamping voltage, the generator can be loaded electrically higher, as this causes an increase of the excitation current and thus a shift of the operating point into the stable range of the V-curves of the synchronous generator. If the generator is temporarily accelerated, e.g. by an increasing wind speed, the clamp voltage can also be regulated to its nominal value by reducing the excitation voltage.

Simulation model

Compared to the concept shown in Fig. 2, the tasks of the load unit and the DUT are swapped. This means that the primary and secondary sides of the transformers T2 and T3 are reversed and the model-based control is implemented on the side of the DUT. The hardware of the interesting part of the test bench was modelled using the PLECS blockset, with all the control or actuation built in

Matlab/Simulink. Fig. 6 shows the section from the DC-link of the load unit via the transformer T3 to the DC-link of the DUT.

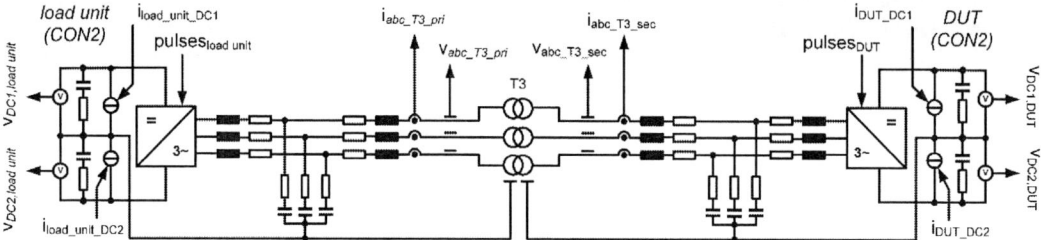

Fig. 6: Simplified PLECS model of the test bench hardware

To simplify the simulation model, the DC-link voltages are controlled by PI controllers acting on controlled current sources. This produces a larger voltage ripple of the DC-link voltages under load, as in the real system. The parameters of the LCL filters and the transformer T3 are partly read from the datasheet, measured or calculated.

Operation of the test bench

Before the actual model-based control and generator current control can be operated, the following steps must be completed:

- Pre-charge of T1 and T2
- Passive charging of the DC-links (load unit and DUT)
- Active boost control of the DC-link voltages

Once this state has been reached, the start-up of the drive train on the DUT side can begin. For this purpose, the drive train is accelerated up to the rated speed by a speed controller, as described above. Meanwhile, the excitation voltage of the generator is intentionally set low (0.02 p.u.) to avoid a high inrush current of the transformer T3 when activating the PWM. After the PWM has been activated, the excitation voltage can be increased to its no-load value (1 p.u.). The control of the excitation voltage and the torque calculation from the rotor model are started as soon as the clamping voltage has reached its nominal value. The torque applied by the wind rotor model now drives the generator. On the load unit side, the current control can now be activated and, shortly afterwards, the current setpoint from the generator curve. The current control is automatically shut down as soon as the measured frequency falls below 40 Hz, as the transformers would heat up too much at these low frequencies. Contrary to the system topology in Fig. 2, the generator angle is not yet transmitted to the DUT with the servo and the encoder. Instead, the pole wheel angle is estimated as a function of the current and the target current is divided accordingly in the d- and q-axes.

Results

Wind speed as step function

In order to illustrate the operation of the overall system, a step change in wind speed is first assumed. This case is not realistic, but it shows that the calculation of the rotor torque and the control of the pitch angle are working correctly. The falling wind speed causes a decreasing torque, which the wind rotor can deliver to the generator. This causes the speed to drop significantly at first. Fig. 7 shows how the speed controller reduces the pitch angle to the smallest value so that the rotor speed in Fig. 8 is stabilised.

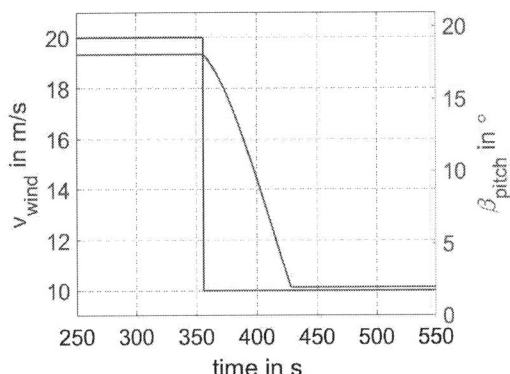
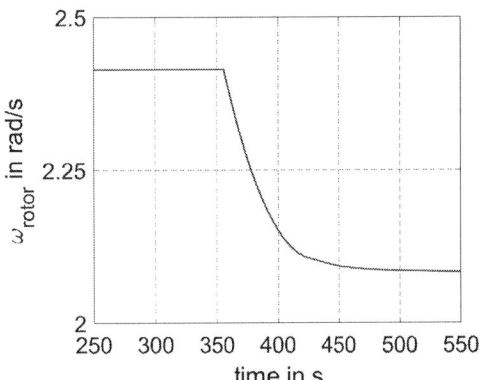

Fig. 7: Reaction of the pitch angle to a wind speed drop

Fig. 8: Reaction of the rotor speed to a wind speed drop

Of course, the electromagnetic torque in this case must not be greater than the available driving rotor torque. Fig. 10 shows that the calculated clamp voltage drops briefly due to the speed drop, but rises again to its nominal value by regulating the excitation voltage. However, there are deviations in the measured voltage, which can be justified by the fact that the frequency of the controlled voltage is further away from the resonance point of the LCL filters. As a result, the damping by the LCL filters is higher. Fig. 9 shows the curve of the set current and the measured current. The characteristic curve-based current control reduces the current setpoint depending on the speed so that the generator speed stabilises at an operating point below the rated speed.

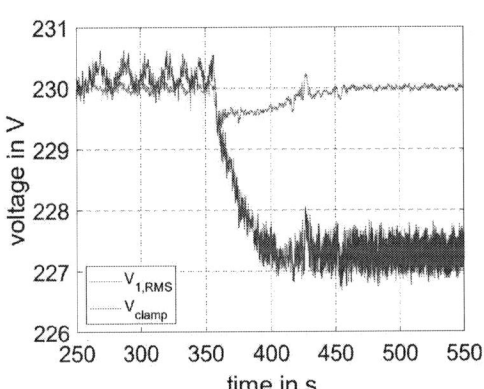

Fig. 9: Changing clamp voltage due to wind speed drop

In order to show the functioning of the frequency-dependent current control in this form, an offset of 5 Hz is subtracted from the input of the generator curve calculated by the PLL. This is necessary because the frequency must not fall below 40 Hz in total, but a reduction of the setpoint current is to be shown. In a real system without a transformer between the generator and the inverter, the frequency may be significantly lower.

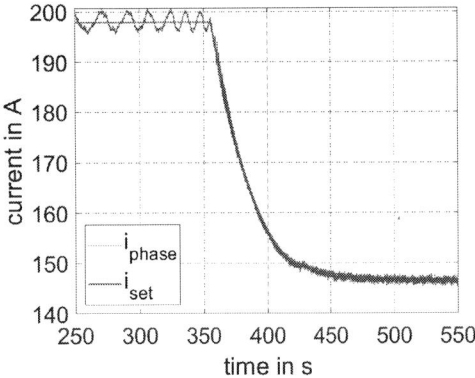

Fig. 10: Reduction of the current set value due to decreasing speed

Windfield

To investigate how the entire test bench reacts to a continuously changing wind speed, the wind field shown in Fig. 4 is now used instead of a step change in wind speed.

Simulation

The starting speed of the generator is below the rated speed (Fig. 11 and 12), so that the pitch angle is 0° at the beginning. The generator is accelerated by the average wind speed in the range between 600 s and 650 s, which is above the rated wind speed of 14 m/s. The pitch angle increases with increasing speed. As the speed increases, the pitch angle increases more and more to stop the acceleration. This is repeated cyclically as the wind profile is run through. Due to the very large inertia of the wind rotor, the speed in Fig. 12 is out of phase with the wind speed.

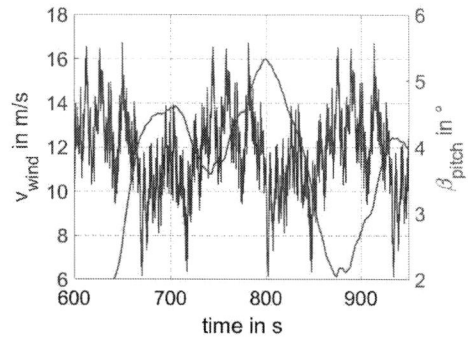

Fig. 11: Simulated wind speed and pitch angle

Fig. 12: Simulated mechanical generator speed

The phase voltage measured in the PLECS model is shown in Fig. 13. It can be noticed that this increases accordingly to the other curves with increasing speed. On the one hand, this means that the increase in clamping voltage due to the speed is higher than the reduction due to the higher current and, on the other hand, that the regulation of the excitation voltage is not fast enough to keep the voltage constant. The phase current curve in Fig. 14 also follows the generator speed.

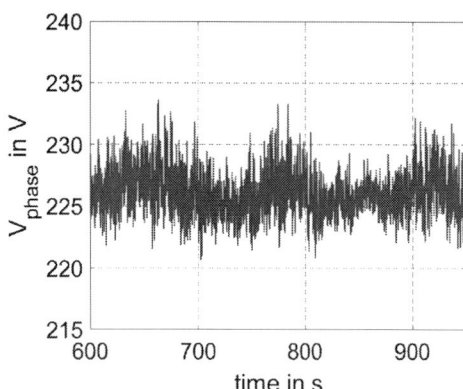

Fig. 13: Simulated phase voltage

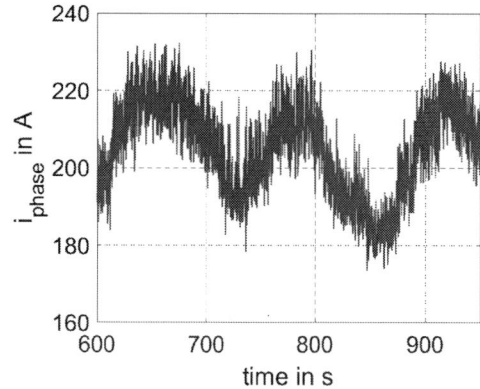

Fig. 14: Simulated phase current (RMS)

Test bench

The same measurement is now performed on the test bench. The wind field is accordingly stored on the PLC, where the rotor and generator model are also calculated. The frequency-dependent generator curve is stored in the PLC of the load unit.

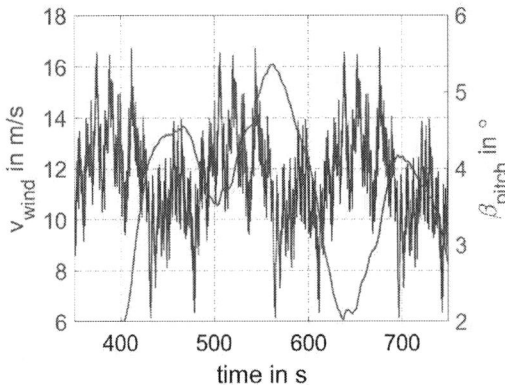

Fig. 15: Wind speed and pitch angle course

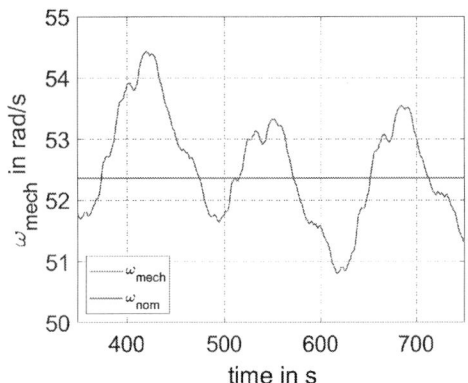

Fig. 16: Generator speed

Essentially, the curves correspond to the results from the simulation. There are differences especially in the measured currents and voltages in Fig. 17 and 18. These do not look as noisy on the test bench and are slightly offset, which is due to the damping properties of real components such as line inductance or deviating component parameters.

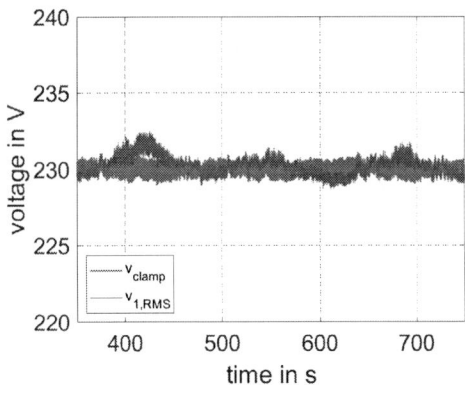

Fig. 17: Calculated and measured generator clamp voltage

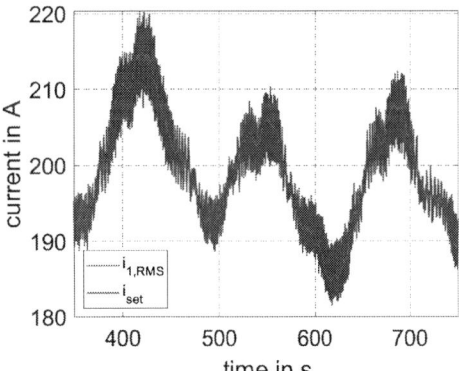

Fig. 18: Measured phase current and its set value

Conclusion

In this work, the model of a generator was extended by a rotor model with gearbox so that it represents a simple drive train of a wind turbine. First of all, it could be shown in simulation how a created wind profile affects the drive train. In the process, controllers were designed that enable simulating the partial-load and full load operation of a wind turbine. Parts of the simulation model could be used directly to generate a programme code for the PLCs of the test bench, which greatly simplifies commissioning and modifications. The results of the simulation and the measurements on the test bench show almost identical curves. The implementation of a frequency-dependent current control and operation in the partial load range were also successfully tested, whereby the setpoint current was separated into an active and a reactive component depending on the magnitude in order to emulate the increase in the pole wheel angle. For a real field-oriented control, the transmission of the shaft angle with the servo from Fig. 2 is still missing. The transfer of the shaft angle from the generator model to the DUT is part of current work.

References

[1] K. Fischer *et al.*, 'Reliability of Power Converters in Wind Turbines: Exploratory Analysis of Failure and Operating Data From a Worldwide Turbine Fleet', *IEEE Trans. Power Electron.*, vol. 34, no. 7, pp. 6332–6344, Jul. 2019, doi: 10.1109/TPEL.2018.2875005.

[2] H. Bühler, *Einführung in die Theorie Geregelter Drehstromantriebe: Band 1: Grundlagen*. 1977. Accessed: Feb. 24, 2021. [Online]. Available: https://link.springer.com/book/10.1007/978-3-0348-5941-7

[3] A. Schaffarczyk and Kompetenzzentrum Windenergie Schleswig-Holstein, Eds., *Einführung in die Windenergietechnik: mit 27 Tabellen sowie zahlreichen Beispielen und Übungen*. München: Fachbuchverl. Leipzig im Hanser-Verl, 2012.

[4] E. Hau, *Windkraftanlagen: Grundlagen, Technik, Einsatz, Wirtschaftlichkeit*, 5., neu Bearb. Aufl. Berlin Heidelberg: Springer-Vieweg, 2014. doi: 10.1007/978-3-642-28877-7.

[5] A. Ernst, D. Koczy, W. Holzke, and B. Orlik, 'Validation of a Converter Control based on a Generator Model as Voltage Source', May 2022. doi: DOI:10.30420/565822187.

A Novel Grid-demanded Power Point Tracking (GPPT) Control Method for Wind Turbines to Preserve Grid Stability with High Wind Energy Penetration

David Matthies, Alexander Ernst, Henning Sauerland, René Reimann, Wilfried Holzke, Bernd Orlik
IALB, University of Bremen
Otto-Hahn-Allee 1, NW1
D-28359 Bremen, Germany
Tel.: +49 421 218-98 626 92
E-Mail: dmatthies@ialb.uni-bremen.de
URL: https://www.uni-bremen.de/ialb

Keywords

«Wind energy», «Wind-generator systems», «Virtual Synchronous Generator (VSG)», «Test bench», «Smart power»

Abstract

To keep a stable grid operation despite shut-down power plants, wind turbines need to provide equal ancillary services. Therefore, a sophisticated grid-demanded power point tracking (GPPT) control method is proposed. This control strategy was successfully validated on a test bench, so that it may probably run on real wind turbines.

Introduction

In order to finally realize the implementation of the energy transition, large CO_2-emitting power plants need to be adequately replaced by renewable energy sources to a greater extent. Hereby, wind turbines are of central significance [1]. Figure 1 shows a commonly installed wind turbine with a full-scale converter, a permanent magnet synchronous generator (PMSG) and an active pitch control. Conventional control strategies aim at a maximum wind power point tracking (MPPT) without regarding the electrical consumer power demand, whereas a balance between the produced and consumed power actually is one mandatory prerequisite for a stable grid operation. Therefore, the highly grid-stabilizing

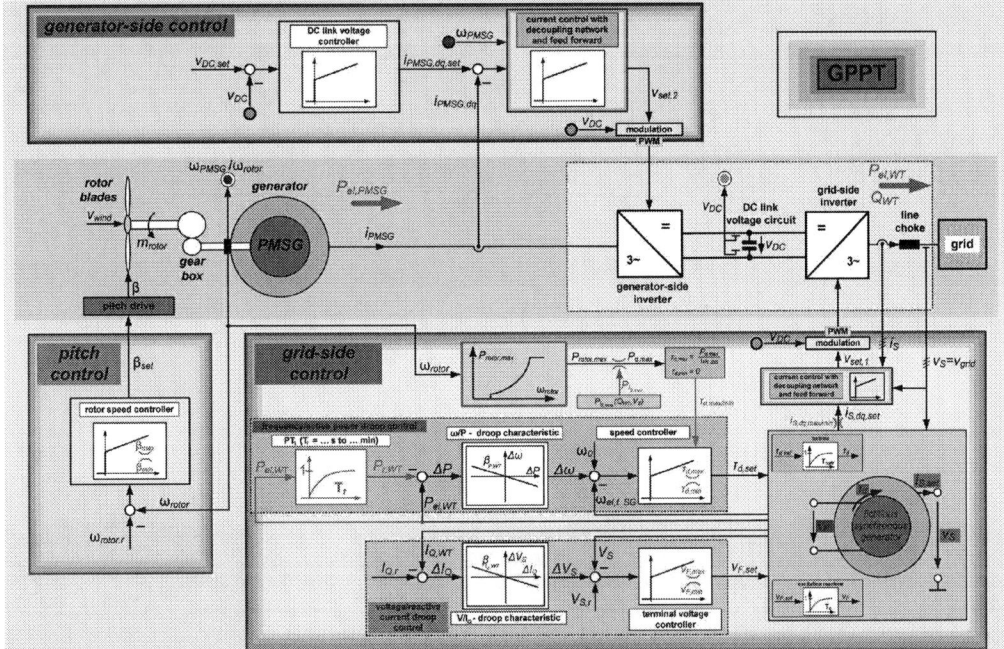

Fig. 1: Developed highly grid-stabilizing GPPT control strategy

wind turbine control strategy of Fig. 1 was developed to make the still indispensable provision of ancillary services by power plants, e. g. in terms of voltage and frequency stability, redundant. Here it must be considered, that current power plant dominated grid processes have to be appropriately adapted to the specific properties of wind turbines such as, inter alia, fast changing unpredictable wind speeds or their huge number with a relatively low power class in each case [2]. Keeping the given wind turbine structure unchanged, the three available control interventions were modified. Hereby, the resulting grid-demanded power point tracking (GPPT) control strategy has to consider all these complicating factors and additionally combine them with power plant behaviour as analogue as possible. Only thus, a reasonable interference in long-proven methods within the complex electricity grid is feasible, so that future grid stability can be sustainably preserved.

Grid-demanded Power Point Tracking (GPPT)

By analogy with conventional power plants and to keep proven physical grid characteristics, the grid-side inverter should behave like a directly grid-connected synchronous generator based on the principle of current synchronverter approaches [3]. Therefore, the used generator model [4] receives the measured terminal voltage v_S as an input value and the resulting set currents $i_{S,set}$ are set by a current control for a generator equivalent power input into the grid, validated in [2]. The fictitious mechanical drive

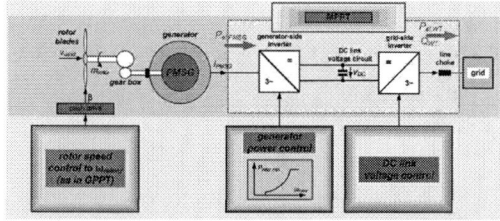

Fig. 2: Overview of an MPPT control structure

torque τ_d and the excitation voltage v_F serve for the generator control and are set by a turbine and an excitation machine. These are approximated by a respective PT1-element with the time constants $T_{turb} = 50\ ms$ and $T_e = 10\ ms$ [5]. Over this, the implemented frequency/active power droop control with an appropriate drive torque limitation to $\tau_{d,max/min}$ and the voltage/reactive current droop control intervene in the grid power flow. Opposed to the conventional MPPT control method of Fig. 2 [1], it is therefore now up to the generator-side inverter to control the DC link voltage v_{DC} cascaded with a subordinate current control [4]. The combination of the fed active power $P_{el,WT} = P_{el,PMSG}$ with the rotor speed control by pitch angle adjustment specifies the resulting operating points of the wind rotor. For the practical validation, the illustrated test bench with a central control in Fig. 3 was set up [6] to reproduce a wind turbine with real rotor behaviour feeding into an emulated electricity grid. The grid consists of a power plant and a common load to represent a grid state with high wind energy penetration [2] according to Fig. 4. Herein, the chosen wind turbine parametrization is listed as well.

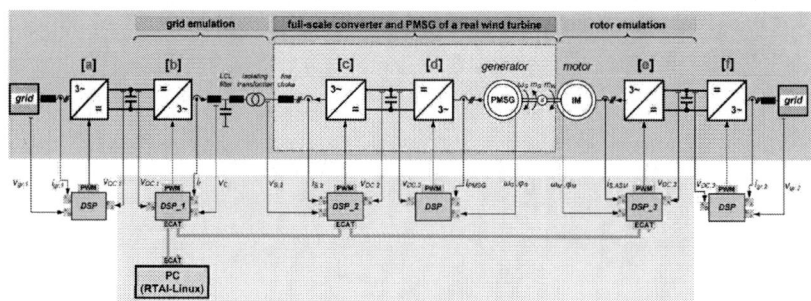

Fig. 3: Wind turbine test bench to validate the GPPT control method

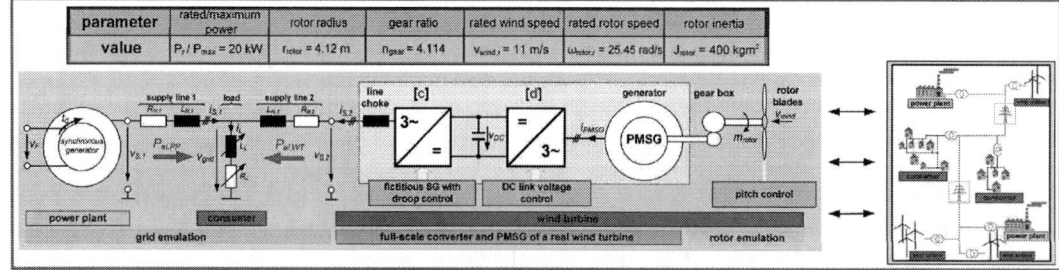

parameter	rated/maximum power	rotor radius	gear ratio	rated wind speed	rated rotor speed	rotor inertia
value	P_r / P_{max} = 20 kW	r_{rotor} = 4.12 m	n_{gear} = 4.114	$v_{wind,r}$ = 11 m/s	$\omega_{rotor,r}$ = 25.45 rad/s	J_{rotor} = 400 kgm²

Fig. 4: Chosen wind turbine parametrization and desired test bench functionality

Basic Element: Fictitious Synchronous Generator

Figure 5 depicts a single-phase equivalent circuit of the used separately excited synchronous generator (SESG) in Fig. 1 divided up into the rotor and stator circuit with their respective passive components. Here, the excitation voltage v_F adjusts the magnitude and the drive torque τ_d the phase angle of the internal generated voltage v_i [5]. Thereby, the generator active and reactive power is determined.

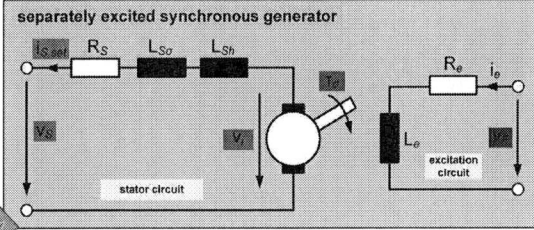

Fig. 5: Equivalent circuit of an SESG

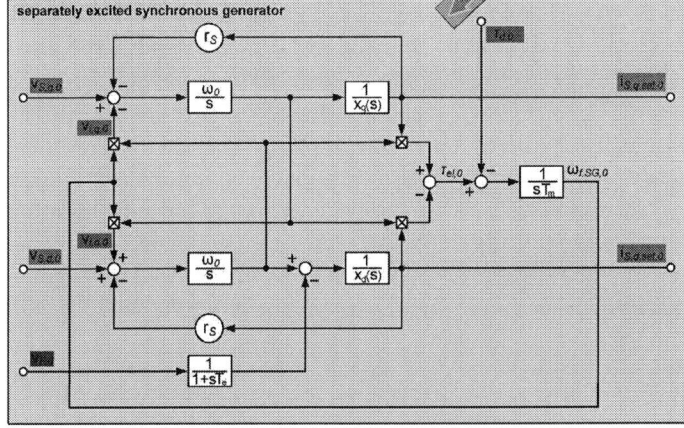

Fig. 6: Applied synchronous generator model [6]

Taking an additional damper winding into consideration, the corresponding applied generator model was derived in [4] and [7] and is shown in Fig. 6 in the d/q reference frame with redefined variable and parameter names.

The developed model requires, that the input and output values are normalized to their related rated values of Table 1, such as e. g. $v_{S,q,0} = \frac{v_{S,q}}{V_{S,r}}$. In accordance to the desired power class, the model variables in Fig. 6 $(x_d(s), x_q(s), \dots)$ can be parametrized by a fitting data sheet. The relevant parameters of it are listed in Table 1 and specify the properties of the synchronous machine. Hereby, equal values for the direct-axis and quadrature-axis synchronous reactances ($x_d = x_q$) indicate a non-salient pole synchronous generator as mostly applied in power plants. With the chosen inertia time constant $T_m = 100\ ms$ and the resulting electrical torque $\tau_{el,0}$ the dynamic behaviour is described by

$$\omega_{f.SG,0} = \frac{1}{T_m} \cdot \int \tau_{el,0} - \tau_{d,0}\, dt\ . \quad (1)$$

The damper winding and, therefore, the generator damping behaviour is adjusted by unequal values of the direct-axis transient and sub-transient reactances ($x'_d \neq x''_d$).

Table 1: Parametrization of the fictitious synchronous generator

general data (rated values):			
apparent power	$S_r = 37.5\ kVA$	power factor	$\cos\varphi_r = 0.8$
phase voltage	$V_{S,r} = 230\ V$	number of pole pairs	$p = 2$
phase current	$I_r = \dfrac{S_r}{3 \cdot V_{S,r}} = 54.1\ A$	angular frequency	$\omega_0 = 314.15\ \frac{rad}{s}$

resistances:	
stator winding resistance	$R_S = 0.1713\ \Omega$

reactances (per unit):		time constants:	
direct/quadrature-axis synchronous reactance	$x_d = 0.967$ / $x_q = 0.967$	direct-axis transient open-circuit time constant	$T'_{d0} = 0.57\ s$
direct/quadrature-axis transient reactance	$x'_d = 0.152$	direct-axis transient short-circuit time constant	$T'_d = 0.024\ s$
direct/quadrature-axis subtransient reactance	$x''_d = 0.083$ / $x''_q = 0.169$	direct/quadrature-axis subtransient open-circuit time constant	$T''_d = 0.015\ s$ / $T''_q = 0.015\ s$

The left side of Fig. 7 shows the actual values $i_{S,2,dq}$ and the set values $i_{S,2,dq,set}$ (test bench in Fig. 3 referred variable names) of the generator stator currents in the d/q reference frame at a constant terminal voltage $v_{S,2}$.

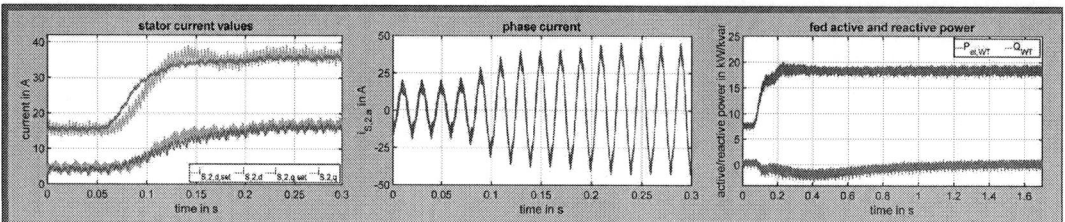

Fig. 7: Set and actual stator current values, phase current and active and reactive power on a torque jump $\Delta \tau_{d,set}$

At about $60\ ms$ a torque jump $\Delta \tau_{d,set,0}$ of 30 percent of the nominal torque value is performed. The control error reaches zero after several milliseconds, which is sufficiently fast so that the regarded system acts like a real generator. It appeared, that the existing harmonics in the measured stator voltage $v_{S,2}$ highly affect the set currents $i_{S,2,dq,set}$ caused by the transmission behaviour of the used generator model (Fig. 6). Therefore, the set currents are additionally filtered with the time constant $T_f = 0.005\ s$. During the measurement in Fig. 7, the fed reactive power Q_{WT} is controlled to zero by adjusting the excitation voltage v_F. Moreover, the resulting fed active power $P_{el,WT}$ and the phase current $i_{S,2,a}$ are depicted.

Before feeding into the grid it is mandatory, that the generator model synchronizes itself to the grid voltage v_{grid}, given out by inverter [b] (Fig. 3). The short-circuited damper winding behaves similar to the squirrel cage of an induction machine and, therefore, provides a starting torque at standstill. As soon as the measured grid voltage is applied ($v_{S,2} = v_{grid}$), the fictitious generator runs up to about the grid frequency f_{grid} as shown in Fig. 8 (asynchronous start-up). Then, the synchronous persistence completes the synchronization to $\omega_{el,f.SG} = \omega_{grid}$ and the damper winding has no more influence for the moment.

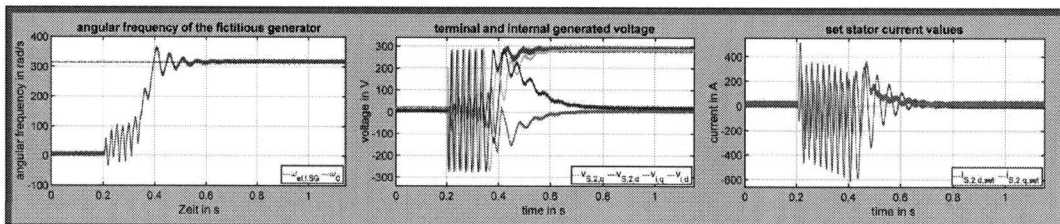

Fig. 8: Synchronization process of the fictitious generator

Because it is not a real machine, high set current values $i_{S,set}$ are absolutely permitted during the synchronization process. But at the time when the wind turbine starts to feed into the grid, these need to be around zero. According to the equivalent circuit in Fig. 5, the internal generated voltage v_i needs to get close to the terminal voltage v_S in order to minimize the voltage drop across the passive components. Therefore, the excitation voltage $v_{F,0}$ is fed forward by a value of 0.95 and the drive torque is set to $\tau_{d,0} = 0$. This leads to the courses of the terminal and internal generated voltages of Fig. 8 in the d/q reference frame. After the start-up, the respective d- and q-components are running towards the equal value, which also excludes a possible angular offset. In addition, the resulting set current values $i_{S,2,dq,set}$ approximate the desired value of zero after synchronization.

Basic Element: Generator-Side Boost Converter

At first, the conventional grid-side boost converter as applied on wind turbines with MPPT control and also on inverter [a] and [f] on the test bench (Fig. 3) is regarded on the left side of Fig. 9.

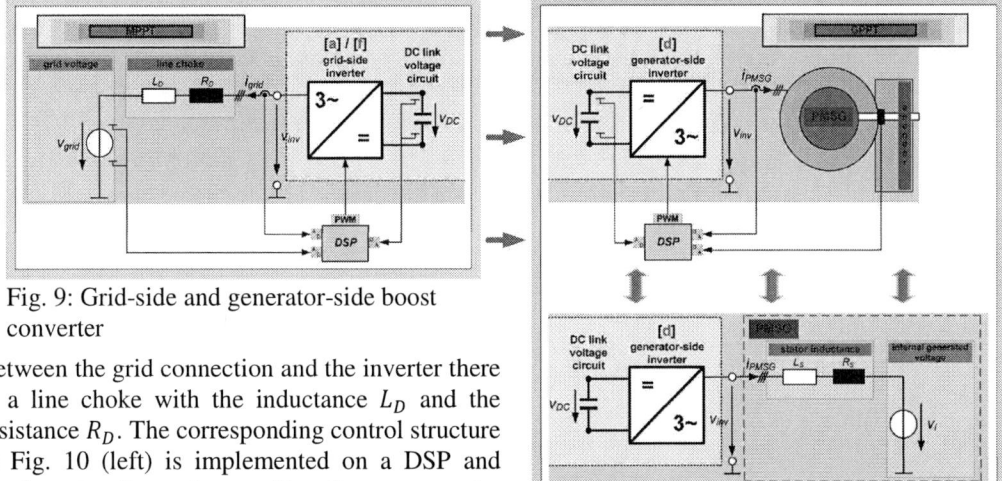

Fig. 9: Grid-side and generator-side boost converter

Between the grid connection and the inverter there is a line choke with the inductance L_D and the resistance R_D. The corresponding control structure in Fig. 10 (left) is implemented on a DSP and requires the three-phase grid voltage v_{grid}, the current i_{grid} and the DC link voltage v_{DC} as the measured input values. To determine the mains angle φ_{grid} needed for the d/q transformation, a phase-locked loop (PLL) is used. The DC link voltage v_{DC} is controlled with a subordinate current control of the active current component $i_{grid,d}$ in the d/q reference frame. The respective current controller output is fed forward by a decoupling network and the grid voltage $v_{grid,d}$ ($v_{grid,q} = 0$ for transformation angle φ_{grid}). Eventually, the PWM signal for the inverter control is generated out of the resulting set voltages $v_{set,dq}$.

Starting from this, the detailed control structure of the generator-side boost converter at GPPT on the right side of Fig. 10 is accordingly adapted for the connected PMSG in Fig. 9 (right).

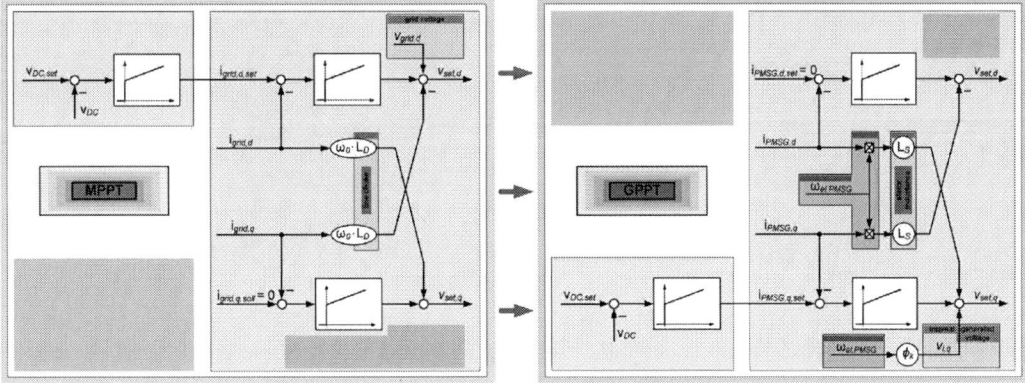

Fig. 10: Comparison between the control structure of a grid-side and generator-side boost converter

The relevant parameters of the PMSG are listed in Table 2. The related depicted equivalent circuit in Fig. 9 (right) directly emphasizes the analogy between both concepts. Hereby, the stator impedance (L_S and R_S) is comparable to the line choke (L_D and R_D) and the internal generated voltage v_i represents the grid voltage v_{grid}. In contrast, due to the applied electrical machine, $i_{PMSG,q}$ now represents the active current component subordinated to the DC link voltage controller. Moreover, the fed forward internal generated voltage $v_{i,q}$ hence affects the q-component. Opposed to the

Table 2: Parameters of the applied PMSG

parameter	value
rated power	$P_{r,PMSG} = 30\,kW$
rated speed	$\omega_{PMSG,r} = 104.72\,\frac{rad}{s}$
rated terminal voltage	$V_{S,PMSG,r} = 230\,V$
number of pole pairs	$Z_p = 3$
stator inductance	$L_S = 20\,mH$

measurable grid voltage v_{grid}, $v_{i,q} = \phi_k \cdot \omega_{el,PMSG}$ is determined by the known constant magnetic flux $\phi_k = \frac{\sqrt{2} \cdot V_{S,PMSG,r}}{Z_p \cdot \omega_{PMSG,r}} = 1.035\ Wb$ and the generator speed $\omega_{el,PMSG} = Z_p \cdot \omega_{PMSG}$. The speed is calculated out of the generator rotor angle φ_{gen} output by the encoder which is now also used for the d/q transformation. Contrary to the nearly constant grid frequency $\omega_0 = 2\pi \cdot 50\ Hz$, a variable generator speed range is resulting and, therefore, needs to be considered in the decoupling network as an input value.

Hereto, Fig. 11 shows a measurement for a linearly increasing generator speed ω_{PMSG} from $\frac{\omega_{PMSG,r}}{2}$ to $\omega_{PMSG,r}$ with an applied load of $\Delta P_{el,WT} = 7.5\ kW$ or rather $\Delta \tau_{d,set,0} = 0.2$ at the time $t = 4\ s$ by the fictitious generator. The DC link voltage v_{DC} is kept within an acceptable value range during the whole speed change and the phase current $i_{PMSG,a}$ is adapted to the increased power demand.

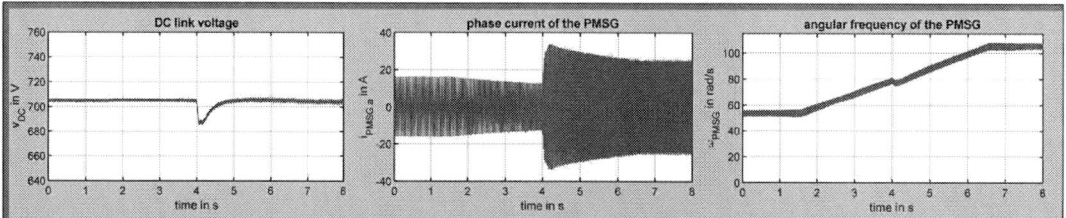

Fig. 11: Related measurement results of the DC link loaded during a speed change

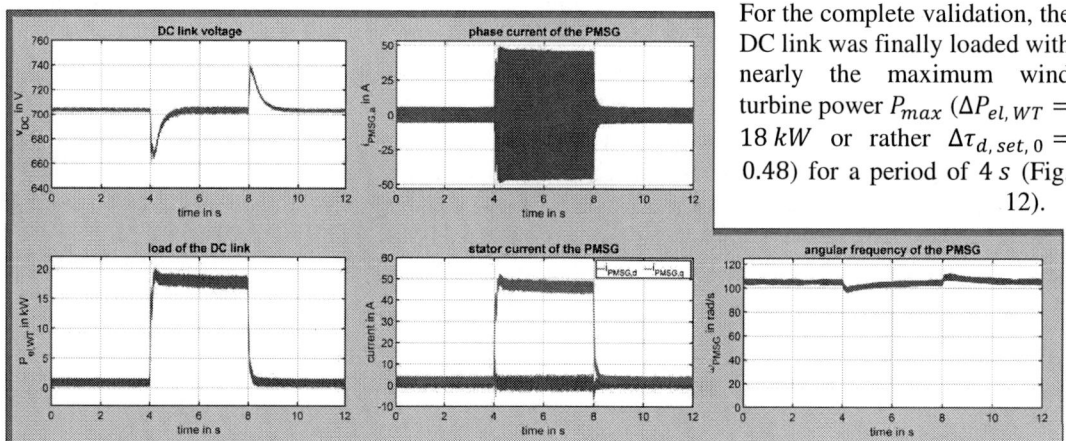

For the complete validation, the DC link was finally loaded with nearly the maximum wind turbine power P_{max} ($\Delta P_{el,WT} = 18\ kW$ or rather $\Delta \tau_{d,set,0} = 0.48$) for a period of $4\ s$ (Fig. 12).

Fig. 12: Measurement results of the DC link loaded with the nearly maximum wind turbine power

Active Power Flow

Whereas the always ensured availability of fossil fuels allows a power plant design with a constant rated power $P_{r,PP}$, wind turbines hereby encounter problems with fast unpredictable wind speeds. Therefore, the rated power $P_{r,WT}$ of the wind turbine is now directly shifted from the resource wind to the demanded power of the consumer by assuming it as the delayed fed-in power $P_{el,WT}$ (Fig. 1) [2], [8]. When a load change ΔP_{load} occurs (Fig. 4), the additional demanded power is extracted from the respective kinetic energy of the rotating generator inertia at first (instantaneous reserve). Then, the frequency/active power droop control provides a coordinated and defined load distribution between the power plant (PP) and the wind turbine (WT) according to Fig. 13 (primary reserve). The time constant T_r defines the time it takes for the frequency set value $\omega_{set,WT} = \omega_0 + \beta_{P,WT} \cdot (P_{el,WT} - P_{r,WT})$ to reach the nominal grid frequency

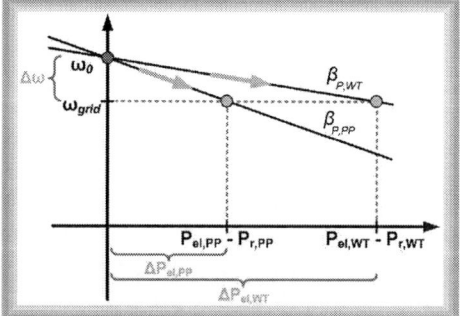

Fig. 13: Defined load distribution

ω_0. Meanwhile, the consistent grid frequency ω_{grid} forces the wind turbine to additionally take over the whole reserve power of the power plant so that it returns to its rated power $P_{r,PP} = P_{el,PP} = 40\ kW$ as well (Fig. 14, time range from 0 to 110 s and 195 s to 360 s for $T_r = 6\ s$).

Hence, the wind turbine even behaves like a regulating power plant and restores ω_0 as long as there is enough extractable wind power $P_{rotor,max}$ available. Otherwise, the settable fictitious mechanical drive torque $\tau_{d,set}$ is appropriately limited making use of the known wind turbine power curve $P_{rotor,max}(\omega_{rotor})$ in Fig. 1 originally developed for MPPT control:

$$\tau_{d,min} = 0 \quad \text{and} \quad \tau_{d,max} = \frac{P_{d,max}}{\omega_{f,SG}} \tag{2}$$

$$\text{for } P_{d,max} = P_{rotor,max}.$$

Without any wind speed measurement, the wind turbine can now pass through its whole power range and directly aim towards the value leading to the best load-demanded power flow:

$$0 \le P_{el,WT} \le P_{rotor,max}. \tag{3}$$

The grid feed-in of Fig. 14 requires, that the DC link voltage v_{DC} is kept constant to a value of $700\ V$ by the generator-side control, validated in Fig. 15.

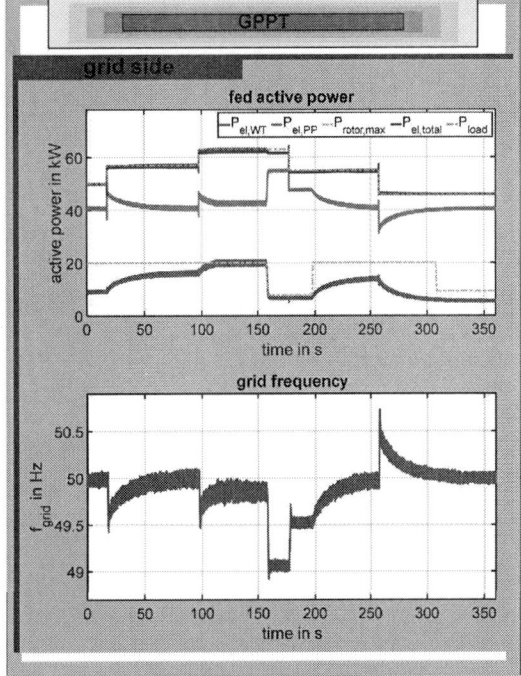

Fig. 14: Load distribution between PP and WT and resulting grid frequency

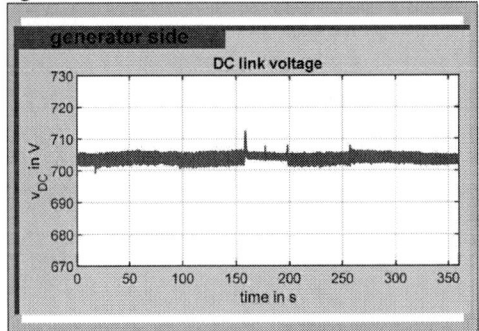

Fig. 15: DC link voltage course for the grid feed-in of Fig. 14

According to Eq. (3), the load-dependent generator power $P_{el,PMSG} = P_{el,WT}$ now also leads to a changed value range of the resulting rotor-side referred electrical generator torque $m'_{el,PMSG}$ (Eq. (4)) compared to a conventional MPPT control method ($P_{el,PMSG} = P_{rotor,max}$). Equation (4) determines, that the rotor speed ω_{rotor} varies until the wind-induced rotor torque m_{rotor} equals $m'_{el,PMSG}$.

$$J_{rotor} \cdot \dot{\omega}_{rotor} = m_{rotor} - m'_{el,PMSG}$$

$$\text{with } m'_{el,PMSG} = \frac{P_{el,PMSG}}{\omega_{rotor}} \cdot n_{gear} \quad \text{and} \quad m_{rotor} = \frac{P_{rotor}}{\omega_{rotor}} \tag{4}$$

Then, a stationary state is achieved and the resulting wind rotor operating points rotor speed ω_{rotor}, pitch angle β and power coefficient c_p are defined. Hereby, c_p describes the ratio between the extracted amount of rotor power P_{rotor} and the total available wind power P_{wind} with the rotor radius r_{rotor} and wind speed v_{wind}:

$$c_p = \frac{P_{rotor}}{P_{wind}} \quad \text{with} \quad P_{wind} = \frac{1}{2} \cdot \rho_0 \cdot \pi \cdot r_{rotor}^2 \cdot v_{wind}^3. \tag{5}$$

On the rotor specifying wind power coefficient characteristic of Fig. 16, the coefficient c_p is plotted over the tip speed ratio $\lambda = \frac{\omega_{rotor} \cdot r_{rotor}}{v_{wind}}$ for different pitch angles β.

Depending on the demanded power P_{load} of the consumer, a possible wind rotor operating range occurs to accordingly extract the current fed power $P_{el,WT} = P_{rotor}$ from the wind [8]. At full load operation, the maximum available rotor power $P_{rotor,max}$ is limited to P_{max} and specified by

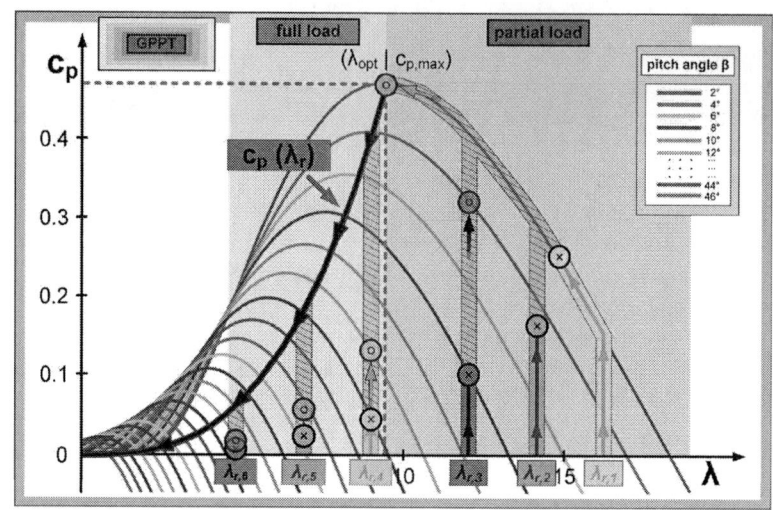

Fig. 16: Wind power coefficient characteristic with plotted operating points at full and partial load operation

$c_p(\lambda_r)$ at rated rotor speed $\omega_{rotor,r}$, whereas at partial load the wind rotor maximum operates at $(\lambda_{opt}|c_{p,max})$. Here, the rotor power P_{rotor} can only be increased by pitch angle adjustment till the minimum pitch angle $\beta_{min} = 2°$ is set. Therefore, the manipulated variable β of the rotor speed controller is limited to this value. Afterwards, the rotor is braked until at most $(\lambda_{opt}|c_{p,max})$ making use of the favourable aerodynamic effect, that the braking process itself directly implies an increasing wind power extraction [8]. Some resulting power coefficients c_p are exemplarily shown in Fig. 16 for two different assumed load changes $\Delta P_{load,1}$ (marked with "x") and $\Delta P_{load,2}$ (marked with "o") at six wind speeds $v_{wind,i}$ with the marked tip speeds ratios $\lambda_{r,i} = \frac{\omega_{rotor,r} \cdot r_{rotor}}{v_{wind,i}}$ at rated rotor speed $\omega_{rotor,r}$. These operating points are separately illustrated for β, ω_{rotor} and P_{rotor} over wind speed v_{wind} in Fig. 17 to underline the resulting operating area unlike in MPPT control with only maximum wind power extraction.

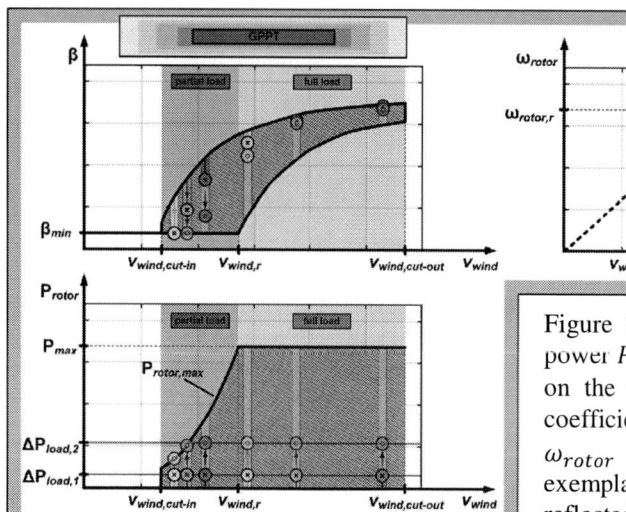

Fig. 17: Separated operating points of Fig. 16

Figure 18 illustrates, that the provided active power $P_{el,WT}$ of Fig. 14 is correctly readjusted on the rotor side with the demanded power coefficient c_p and the resulting rotor speed ω_{rotor} and pitch angle β. Hereby, the just exemplarily described rotor behaviour is reflected for the given load changes ΔP_{load}. For instance, the rated rotor speed $\omega_{rotor,r}$ is kept during partial load operation with $v_{wind} = 8.5 \frac{m}{s}$ from 310 to 360 s.

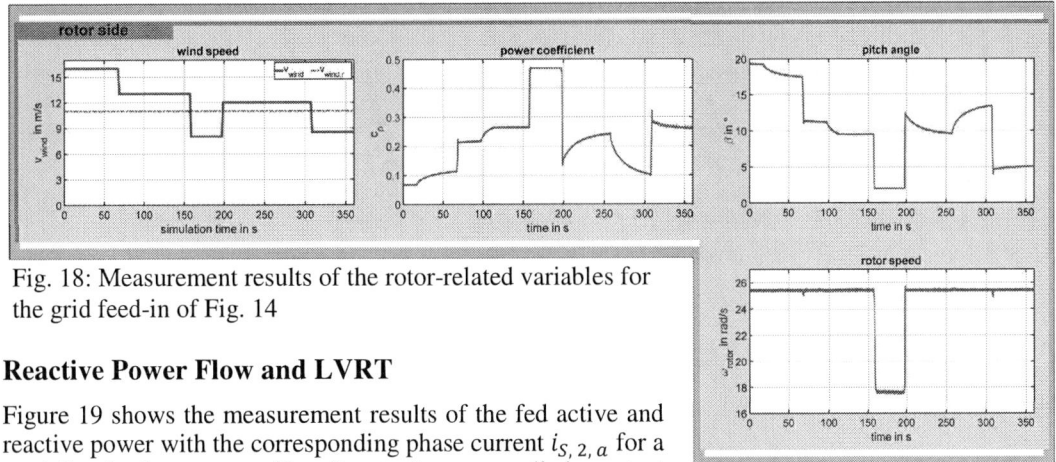

Fig. 18: Measurement results of the rotor-related variables for the grid feed-in of Fig. 14

Reactive Power Flow and LVRT

Figure 19 shows the measurement results of the fed active and reactive power with the corresponding phase current $i_{S,2,a}$ for a predefined voltage course $V_{S,2}$. Due to the now applied constant fixed grid frequency $f_{grid} < f_0$, the frequency/active power droop control then ensures, that the maximum drive power is fed into the grid ($P_{el,WT} = P_{d,max}$). According to the voltage/reactive current droop control [9], the fed reactive power Q_{WT} stays above the rated reactive power $Q_{r,WT} = 4048\ var$ for $V_{S,2} < (V_{S,r} = 230\ V)$ and also vice versa. If necessary due to the decreased terminal voltage, the implemented active power reduction with $P_{d,max} = P_{I_{S,max}}$ intervenes to keep $i_{S,2,a}$ at the permitted inverter maximum current of $\hat{I}_{S,max} = 47\ A$ unless $P_{d,max} = P_{rotor,max}$ (Eq. (2)) becomes effective in case of an occurring wind power drop (time range from 270 to 300 s).

Figure 20 shows the wind turbine riding through a voltage course $V_{S,2}$ according to the technical requirements in [9] ($V_{S,VLC}$ with an extended time range to fully set Q_{WT} defined by the droop control in the temporal size of power plants) without disconnecting from the grid (LVRT). Due to the now higher voltage gradients, an additional limitation of the fictitious generator set current values $i_{S,dq,set}$ to the constant values $i_{S,dq,max/min}$ is applied (Fig. 1) to avoid an otherwise resulting phase over current $i_{S,2,a} \geq \hat{I}_{S,max}$.

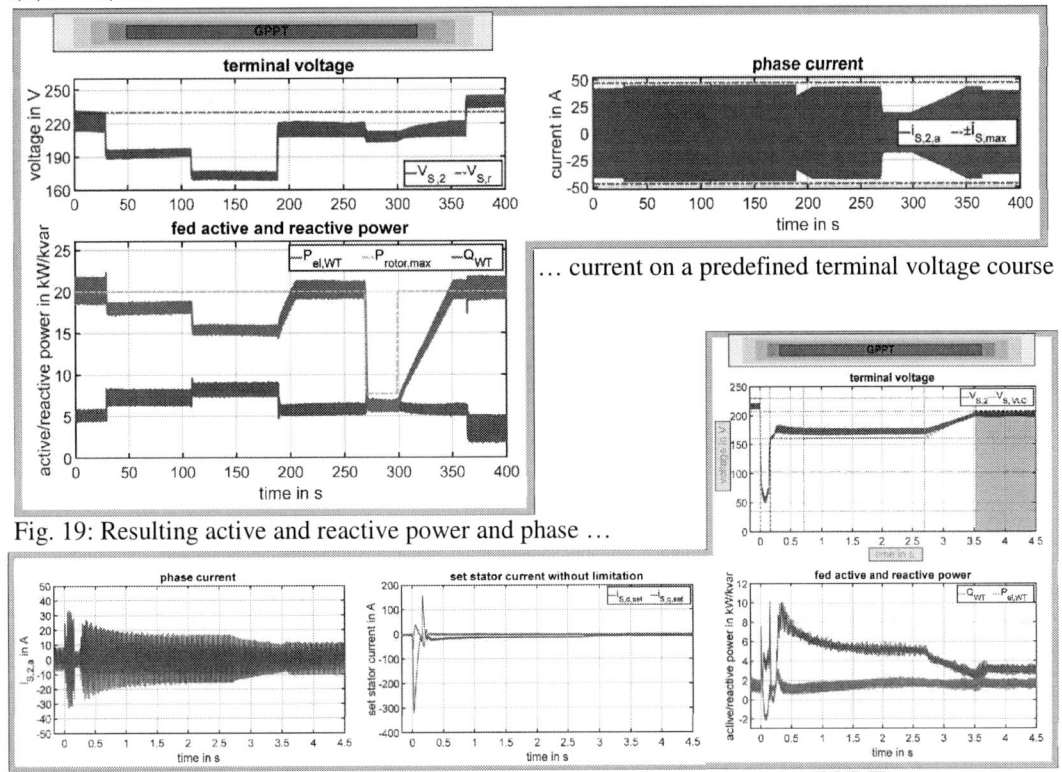

Fig. 19: Resulting active and reactive power and phase ...

... current on a predefined terminal voltage course

Fig. 20: Effectiveness of set current limitation on a predefined voltage course (LVRT)

Conclusion

Assuming wind energy as the most promising alternative to conventional power plants, a modified grid-stabilizing wind turbine control strategy called grid-demanded power point tracking (GPPT) was developed. Hereby, its basic elements as the applied fictitious synchronous generator and the generator-side boost converter and furthermore the control method as a whole were successfully validated on a test bench built up for this purpose. It has been proven, that equally to conventional power plants, the wind turbine provides the relevant operating reserves needed for a stable grid operation. Reconciling the two independent values wind and load profile, the wind turbine adapts its output power to the consumer and appropriately readjusts it on the rotor side. Moreover, the wind turbine fulfils grid supporting tasks and even performs LVRT. For this reason, the application of GPPT control could be suitable to become independent of the still indispensable ancillary services of power plants.

References

[1] J. Thongama and M. Ouhrouche, „*MPPT Control Methods in Wind Energy Conversion Systems*", chapter in book: Fundamental and Advanced Topics in Wind Power, 2011

[2] D. Matthies, A. Ernst and B. Orlik, „*Wind Energy Powered Electricity Grids*", in PCIM Europe digital days, Germany, 2020

[3] Q. Zhong and G. Weiss, „*Synchronverters: Inverters That Mimic Synchronous Generators*", in IEEE Transactions on Industrial Electronics, 2010

[4] A. Ernst, D. Matthies, W. Holzke and B. Orlik, „*Validation of a Generator-Side Boost Converter with Load by a Fictitious Synchronous Machine*", in PCIM Europe digital days, Germany, 2021

[5] W. Leonhard, „*Regelung in der elektrischen Energieversorgung*", Stuttgart: Teubner Studienbücher, 1980

[6] D. Matthies, W. Holzke, R. Reimann and B.Orlik, „*Practical Validation of a New Control Strategy for Wind Turbines by the Use of a Central PC for Model Calculation*", in PCIM Europe, Nuremberg, Germany, 2019

[7] H. Bühler, „*Einführung in die Theorie geregelter Drehstromantriebe*", Springer Base AG, 1977

[8] D. Matthies, A. Ernst, S. Sauerland, R. Reimann, W. Holzke and B. Orlik, „*Provision of Power Plant Equal Ancillary Services by Wind Turbines: From Maximum to Grid-demanded Power Point Tracking*", in PCIM Europe, Nuremberg, Germany, 2022

[9] TenneT TSO GmbH „*Netzanschlussregeln – Hoch- und Höchstspannung –*", 2015

Extension and Implementation of a Model-based Lifetime Monitoring System with Parallel Calculation of Multiple Power Semiconductors

Steffen Menzel, Wilfried Holzke, Michael Hanf, Holger Groke, Bernd Orlik, Nando Kaminski
University of Bremen, Institute for Electrical Drives, Power Electronics and Devices (IALB)
Otto-Hahn-Allee NW1
Bremen, Germany
Tel.: +49 421 218-62695
Fax: +49 421 218-98-62695
E-Mail: menzel@ialb.uni-bremen.de
URL: http://www.uni-bremen.de/ialb

Acknowledgements

This work has been funded by the BMWK and is a part of the project "DFWind Phase 2" under grant no. 0325936J.

Supported by:

Federal Ministry for Economic Affairs and Climate Action

on the basis of a decision by the German Bundestag

Keywords

«Power semiconductor device», «Lifetime», «Condition Monitoring», «IGBT», «Real-time processing»

Abstract

The importance of power electronics for future energy supply is steadily increasing. This implies a lot of semiconductor-based power converters and thus, a huge number of power semiconductors. Their operational availability becomes a critical feature of the power converters. Therefore, predictable maintenance is a key element for a stable and reliable energy supply. A new approach for a FPGA based implementation of a lifetime model for parallel monitoring of semiconductors will be presented.

Introduction

Whether in regenerative power generation such as wind and photovoltaics or in storage systems or in substations, frequency converters have become a key element of the electrical energy supply. Due to the fluctuation of regenerative energy sources such as wind and sun, there are always periods during which the systems do not feed in energy. Nevertheless, operational availability must be guaranteed at all times in order to feed energy into the electrical grid when sun and wind become available again. Thus, unplanned down times must be avoided as much as possible. However, wind turbines do fail unexpectedly and power converters contribute their fair share [1]. For this reason, a condition monitoring system for their power electronics would be of utmost value.

There are different approaches for the condition monitoring [2], [3], [4]. For this work a model-based system had been selected and implemented in software. Compared to other approaches, it does not use special hardware circuits. In [5] it was shown that the model can be implemented on a central processing unit (CPU) of an Industrial-PC for the execution in real-time. The measured average execution time for the implemented model was 1.52 μs on an Intel Core i7 CPU running at 2.3 GHz. Compared to a cycle time of 20 μs, this leaves enough time for data acquisition and transmission. The software already implemented, can be parameterised and stores the state variables for each IGBT in a data structure [6][7]. Only an additional data structure needs to be created for each additional IGBT, while the calculation procedures can all be reused. A disadvantage arises when many IGBTs have to be monitored, e.g. in case of parallel operation of power semiconductors to reach the megawatt range. Then, capturing the needed measurements and the data transfer to the CPU can take a long time. Furthermore, the calculations must be done one by one. A further drawback is, that a CPU has only data types with a fixed width. It was shown in [5], that at least a 64-bit double precision floating point data type has to be used to accumulate the extremely small values, which describe the damage of individual cycles or time steps. Depending on the environmental condition these values can be as low as 10^{-30} and to overcome these disadvantages, an FPGA should be used.

In the following sections, the model for calculating the remaining lifetime is described. Subsections will follow for preparing a thermal model and advantages of additional sensors are discussed. Finally, the advantages and challenges of the new implementation approach will be described.

The life time model

To estimate the remaining life time of a power semiconductor, for example an IGBT, two predominant degradation mechanisms have to be considered. During the blocking phase, i.e. when the switch is open, the electrochemical state is updated with respect to the main acceleration factors: temperature, humidity and bias voltage. The remaining lifetime is estimated using an extended Peck model [8].

Thermomechanical stress is the second, main degradation mechanism covered in this lifetime model. Switching as well as conduction lead to power losses and subsequently to a temperature increase of the chip. Due to volatile operation conditions, the collector current and the duty cycle are continuously changing, resulting in temperature swings. These recurring temperature cycles lead to a thermomechanical stress. The determining parameters are the junction temperature, the applied collector-emitter voltage, the current flowing through the component, and the switching times of the power semiconductors. All variables must be transferred to the model as a "process image", i.e. as a time-consistent, synchronised data set. For the calculation of the damage, an extension of the Coffin-Manson model is used.

The overall model is shown in Fig. 1. The input variables are the voltages applied to the semiconductor, if available the switching times, the currents flowing through the IGBT, the temperature of the heat sink and the relative humidity of the environment. The output value is the remaining lifetime of the extended models according to Coffin-Manson and Peck, respectively.

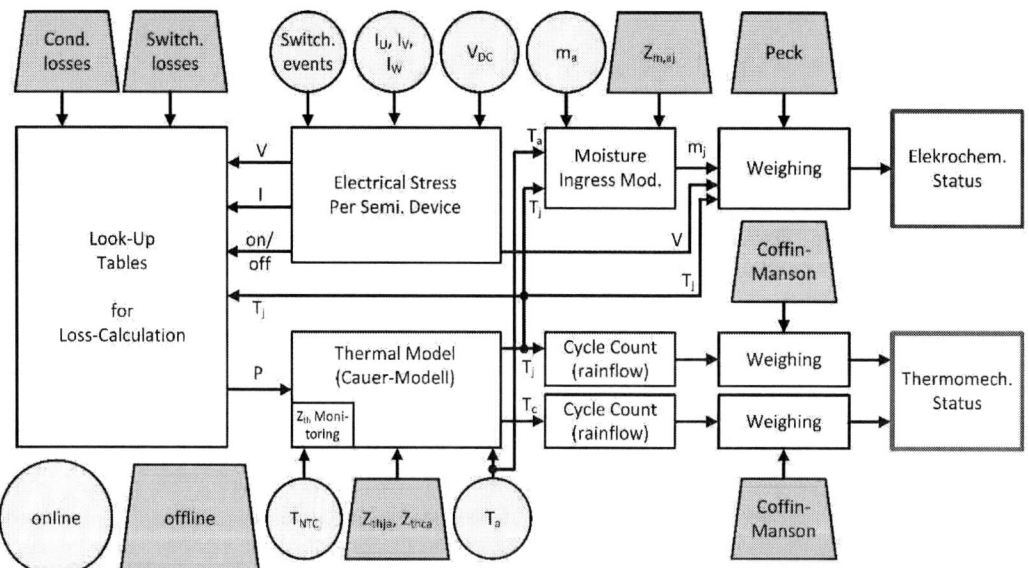

Fig. 1: Lifetime model according to [6] with yellow circles representing online measurements, while green trapezia represent offline data measured in the lab before.

From the input variables obtained by measurements the electrical load per semiconductor is determined. For this purpose, the phase currents I_U, I_V and I_W, the DC-link voltage V_{DC} and the switching events are required. Conduction losses and switching losses occurring during the operation of the semiconductors are taken from look-up tables. The parameters of these look-up tables have to be determined from laboratory measurements, because data sheet information usually includes a safety margin with respect to the typical values.

As previously described, the junction and case temperatures are required for the calculations. Since not all semiconductors are equipped with a temperature sensor and such sensors would anyway yield a package temperature, the required junction temperature T_j is determined using a thermal model.

For the Coffin-Manson model, the temperature swing ΔT_j and the mean temperature $T_{j,m}$ are needed as input values. With a cycle counter, here a rain-flow algorithm [9], completed temperature cycles can be extracted. The parameters α and β will be determined with a power cycle test (PCT) in the laboratory. E_A is the activation energy and k_B Boltzmann's constant. The result of the equation (1) is the life time consumption LC_{CF} for the temperature swing ΔT_j.

$$LC_{CF} = \frac{1}{\alpha} \cdot \Delta T_j^{\beta} \cdot e^{-\frac{E_A}{k_B \cdot T_{j,m}}} \tag{1}$$

The life time consumption LC_{CF} for the corresponding temperature swing ΔT_j is accumulated to $LC_{CF,total}$ with equation (2). If this $LC_{CF,total}$ reaches a value of one, the overall life time limit regarding the thermomechanical state is reached.

$$LC_{CF,total\,n} = LC_{CF,total,n-1} + LC_{CF,n} \tag{2}$$

To calculate the life time consumption for electrochemical state a modified model of Peck is used [8]. The input variables are the moisture at the junction m_j, the collector-emitter voltage V_{CE} and the temperature of the junction T_j. In a Temperature Humidity Bias test (THB) the reference lifetime is identified. This reference lifetime is determined at a reference relative humidity RH_{REF}, the reference Temperature $T_{j,REF}$ and the blocking voltage $V_{CE,REF}$. The parameters x and y are adjustment constants determined by the THB test. While the device is operated in the field, the relative humidity RH, ambient temperature T_a and the collector-emitter voltage V_{CE} are measured. With all these input values and equation (3) the acceleration factor a_f can be calculated. E_A is the activation energy and k_B Boltzmann's constant.

$$a_f = \frac{L_{REF}}{L} = \left(\frac{RH}{RH_{REF}}\right)^x \cdot e^{\frac{E_A}{k_B} \cdot \left(\frac{1}{T_{j,REF}} - \frac{1}{T_j}\right)} \cdot \left(\frac{V_{CE}}{V_{CE,REF}}\right)^y \tag{3}$$

For a time interval Δt_i the acceleration factor is constant if all inputs values are constant. The lifetime consumption $LC_{Peck,total}$ (4) is calculated as the sum of the products of the acceleration factor $a_{f,i}$ and the corresponding time interval Δt_i referred to the reference lifetime L_{REF}.

$$LC_{Peck,total} = \frac{1}{L_{REF}} \sum_{i=0}^{n} \Delta t_i \cdot a_{f,i} \tag{4}$$

If $LC_{Peck,total}$ reaches a value of one, the overall life time limit regarding the electrochemical state is reached.

While most of the described submodels are generic for power semiconductor devices with only the parameters varying, the thermal model and the modelling of the moisture ingress depend highly on the monitored device. The look-up tables have a fixed structure, but their contents depend on the IGBT module as well. How to implement such a thermal model will be described in the following subsection.

Thermal model

In the first approach, a passive one-dimensional equivalent thermal network is implemented for each IGBT module based on the parameters given by the data sheet of the module and the data of the cooling plate and system. The overall goal is to determine the junction temperature $T_{j,n}$ of each chip inside the module. In order to do so, the Foster network specified in the IGBT module's data sheet is converted to an equivalent Cauer network for the diodes as well as the IGBT switches. Thus, a physical and thermal description of the power device can be obtained. Finally, the determined parameters of the Cauer model are fitted in such way, that the time response of both models match exactly in case of subdividing the semiconductor construction into the same number of thermal layers. By knowing the material stack of the power module, the number of layers can be adopted to the number of physically existing or relevant material layers. The physical parameters like the thermal conductivity and thermal capacity describing the thermal behaviour of the material for each layer. Therefore, the mean value and the span of variation for the corresponding parameter per material (copper, ceramics, die attach, etc.) is determined from the literature [9,10,11,12,13] and used as initial and boundary value for the fitting algorithm, respectively. Finally, a Cauer network that matches the data sheet values is established, but has to be verified by Z_{th} measurements. A temperature online monitoring can be further tuned if an integrated NTC-resistor, connected to the base plate inside the power module, is utilised. In this case, the network needs to be extended with a transfer function, considering the position of the NTC with respect to the chip positions. This approach will observe long-term changes of the thermal behaviour of the module itself and offers the possibility to compare calculated values to an actual temperature measurement. In the future, a reduction of calculation time and hardware requirements is pursued with the implementation of a three-layer variant of the matched Cauer model for an online Z_{th} condition monitoring measurement.

Additional environmental sensors

The measurement campaign, carried out in [6] was done with a minimally invasive system on an existing wind turbine converter. The measurements were used in [7]. Within the new approach, the sensors are included in the development process at the manufacturer to ensure an improved climatic mapping inside the cabinet. Every cabinet contains three RHT-, three PT100-, and one flow-sensor together with NTC-resistors inside the power modules. Estimating the micro climate at specific components requires a proper dataset of critical environmental conditions at well-known positions. The remaining thermomechanical lifetime is determined via PT100 on the base plate and NTCs inside the power module, while the electrochemical status is estimated via the same temperature sensors and an RHT-sensor close to the devices. Furthermore, the climate around the DC-link is measured by an RHT-sensor at the bottom and a PT100-sensor at the top to improve the data set. By the integration of the model into the control of the converter and by the accurate climate mapping, the resolution of the lifetime modelling is improved significantly. Moreover, it is possible to obtain critical environmental conditions like condensation within the operational life in real-time.

Implementation of the model

Figure 2 shows the minimum required measurements and the sequence in which the individual sub models have to be processed. For calculation of the damage due to thermomechanical stress switched on, the Coffin-Manson model is used. Therefore, the switching event S_{Evt}, the ambient temperature T_a and the load current I have to be assessed, i.e. must be measured. In the blocking phase of the IGBT the Peck model is used to calculate damage due to electrochemical stress. The switching event S_{Evt} is also used. Additionally, the ambient moisture m_a and the DC-link voltage V_{DC} are required.

The order of execution is as follows: The procedure of accessing the look-up table has to be executed each time a switching event occurs to calculate the switching losses and while the IGBT is in on-state to calculate the conduction losses. The result is fed into the thermal model. This model must be executed with a fixed sampling rate. The sampling rate depends on the desired accuracy, ideally it is in the range of the PWM frequency and must not be higher than the sampling rate of the current. Additionally, the time for processing the sub models has to be considered. The Peck model has to be executed each time an IGBT is switched off and must be executed after a (significant) change in the input values occurred. The cycle counter has to be executed each time when a new output value is available from the thermal model. The Coffin-Manson model has to be executed after each half or full temperature cycle of the cycle counter.

As previously discussed, implementation on an FPGA offers some advantages, if many IGBTs need to be monitored. Nevertheless, some challenges also arise. While the implementation of the Coffin-Manson and Peck models on a CPU, e.g. in the "C" programming language, represents one line of program code, it is more complex if implemented on an FPGA. For the execution of the whole mathematical description of the models, a state machine has to be designed, which coordinates the correct sequence of mathematical operations. The advantage is parallel execution of several parts of the equations.

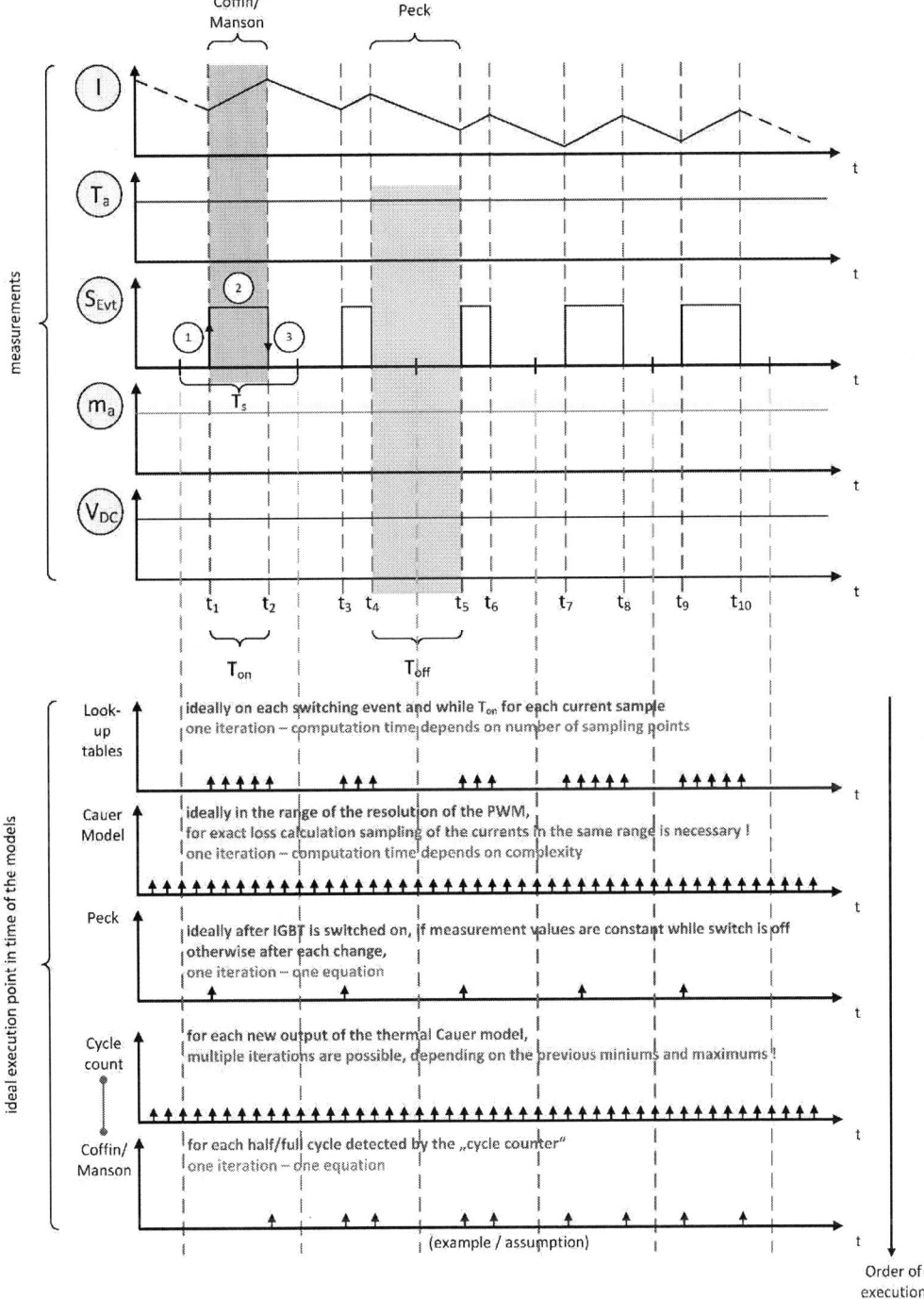

Fig. 2: Switching events and sequence of execution of the submodels

Tools such as Matlab/Simulink can be used to support the developer. However, they do not necessarily generate the hardware description with the necessary degree of optimisation. The limited resources of an FPGA must be considered. Especially the DSP slices, which are used for multiplications and other complex operations such as the exponential function, are limited. Therefore, a hybrid approach should be used. Tools for the automatic code generation for the hardware description shall be used, while other parts will be optimised by hand. That means that some DSP slices must be reused. Nevertheless, sufficient DSP slices must be available that a parallel calculation of the submodels is still possible, otherwise a CPU with only one thread could be used. This has to be observed when the FPGA-type is selected. To estimate the consumption of the FPGA resources, such as look-up tables (LUT), flip-flops (FF) as well as DSP slices, the loss calculation part of the lifetime model was set up in Matlab/Simulink in a first step. The resulting VHDL code for a Xilinx FPGA target was generated using the HDL Coder and integrated in Xilinx Vivado.

Since the switching events – respectively the gate signals – are provided by the overlaying power converter controller and are transmitted to the FPGA, the current path of each phase leg has to be determined by these signals as well as the measured current values. This functionality has been implemented by an analysis of the states of the components in one phase leg (see Fig. 3). The fundamental table to determine the conducting elements is shown in Table 1.

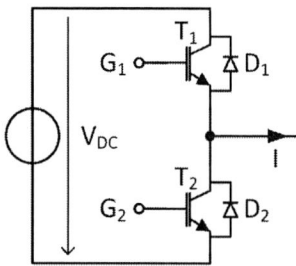

Fig. 3: Power components of one phase leg

Table 1: States of the components

gate signals		I	conducting
G_1	G_2		component
0	0	>0	D_2
0	0	<0	D_1
1	0	>0	T_1
0	1	>0	D_2
1	0	<0	D_1
0	1	<0	T_2

For the conduction states of T_1 and T_2 in one phase leg it was assumed, that the dominant current is always flowing through the transistors and not through the anti-parallel diodes D_2 and D_1. This holds for any allowed switching action according to the direction of the current, which relates to the potential at the output of the phase leg. This detection of the conducting components was implemented together with an edge detection of the switching signals by using logic blocks in the loss calculations based on look-up tables, which is shown in Fig. 4.

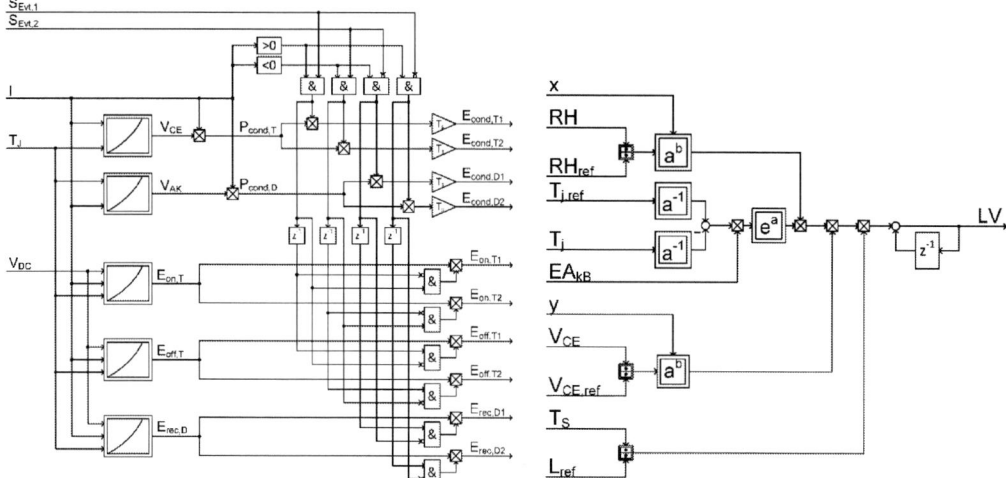

Fig. 4: Implementation of the look-up tables   Fig. 5: Implementation of the Peck-model

For the validation of the model, basic input and output periphery was mapped to avoid the erroneous deletion of code segments, which seem to be unused, by the code generation toolchain. After running the synthesis and implementation, the usage of the resources for different implementation methods of the exponential function was compared. Furthermore, the impact of different data types on the resources for the same calculations was analysed by choosing different fixed-point formats as well as running the complex parts in floating-point blocks and additionally, using a floating-point core in Vivado. The model that was used for the implementation of the Peck model in that case, is shown in Fig. 5.

Data processing

The input values must be fetched from the analog-to-digital converters at equidistant times and buffered in order to make them available to the sub-modules. To determine the amount of memory for the buffers and the data transfer rates, the number of power semiconductors, which should be monitored have to be evaluated. The number of power semiconductors depends on the topology of the power converter. In case of two-level power converter [14] twelve switches (IGBT) and diodes must be considered. If a chopper is applied to the DC-link an additional switch has to be monitored. If a power converter in three-level neutral-point-clamped (NPC) topology [14] should be monitored, the number of switches is double and six additional diodes must be taken into consideration. Power converters in the range of several megawatt are often a parallel connection of multiple modules, which are implemented in a two-level topology. For the following investigations it is assumed that all measurements are collected in one FPGA.

Initially the sampling rates for the different measurements must be defined. While the ambient temperature and relative humidity changes slowly, the voltages and currents change faster. The measurements have to be taken depending on the switching algorithm and its frequency. If the switching frequency is fixed, the measurements are commonly taken with the same frequency but not necessarily at the moment when the IGBTs are switched. If the switching of the IGBTs is done based on the measured current (current hysteresis method) then the sampling frequency is variable. Thus, the minimum frequency for sampling the currents is given by the change of the signal frequency. The maximum required sampling frequency to fulfil the requirements of the lifetime model is limited by the calculation time of the model itself. The calculation times of the previously described submodels must be determined. Thus, depending on the needed calculation time, the upper limit of the sampling frequency is defined.

For the following example, it will be assumed, that the feedback control of the DC-link keeps the voltage constant, so that the same sampling frequency as for the currents can be applied. For a system consisting of one two-level converter there are 6 IGBTs per inverter and 14 IGBTs for a full converter including a chopper when two IGBTs are in one module. If four of these systems are operated in parallel, the total amount of IGBTs is 56, in case of 8 systems it results in 112 IGBTs.

If a switching frequency of 4 kHz is assumed and only one instance of the lifetime model exists, i.e. updating the lifetime can only be processed for one IGBT at a time, that means: The whole model must be calculated in at most 4.4 µs to calculate updates for 56 IGBTs and 2.2 µs to calculate updates of the model for 112 IGBTs. If the switching frequency is increased, the calculation time must be lower. However, if the calculation time is too high, more instances of the lifetime modules have to be calculated in parallel, which is the reason to use an FPGA. In turn, the number of instances that can be implemented in parallel depends on the necessary resources for the implementation of the lifetime model in the FPGA. This will be dealt with in the next chapter.

Simulation and Result

After implementing the parts of the lifetime model in Matlab/Simulink, the following results were achieved by generating VHDL code using the HDL coder and implementing the design in Xilinx Vivado on the target hardware. A comparison of the resources' consumption of the Peck model is shown in the following table. In this case a Zynq-7000 SoC including an Artix-7 FPGA core from Xilinx was used for the reference design. The calculation of the exponential function can be implemented in two different ways. First, the calculation can be done by using the Vivado floating point library and second, the exponential function can be implemented by using a fifth order Taylor series to do the calculation by using fixed point variables to avoid type casting between floating and fixed-point numbers, which requires a large amount of FPGA resources and can lead to inaccurate results of the calculations.

Table 2: Usage of FPGA resources of the Peck model

size of variables	32bit	
calculation of exp()	floating point	taylor series
LUT	13871 (26.07 %)	12133 (22.81 %)
FF	9399 (8.83 %)	8479 (7.97 %)
DSP	44 (20.00 %)	72 (32.73 %)

The relative count refers to the overall resources of the used Zynq module. The results of these comparisons lead to a compromise between maximum parallel execution of the model and minimum resources. Another advantage of a FPGA is the use of data types with an adapted bit width to be able to calculate and add up very small damage values. Afterward, a timing analysis of the implementation has been performed for two clock rates of the FPGA, which are shown in the following table.

Table 3: Timing analysis of the Peck model implementation

size of variables	32bit			
calculation of exp()	floating point		taylor series	
FPGA clock [MHz]	100	250	100	250
Calculation time [ns]	780	312	830	332
No. of clock cycles per calculation	78	78	83	83

As expected, the overall calculation time decreases for higher clock rates, but the numbers of the necessary clock cycles to perform the calculations are equal. During the calculation of the Peck model several multiplications need to be performed simultaneously, while their results depend on each other. Therefore, many DSP slices as well as LUTs are required for this part.

The results of the lookup table implementation are shown in the following Table 4. Due to restrictions of the array size for 2D-lookup tables in the HDL coder, the calculation of the lookup tables has only been implemented for 32-bit variables. But, in this case the resolution of 32-bit fixed-point datatypes is also sufficient for calculating the losses of the transistor and the diode.

Table 4: Resource consumption and timing analysis of the lookup table calculation

size of variables	32bit	
FPGA clock [MHz]	100	250
LUT usage	1706 (3.21 %)	1706 (3.21 %)
FF usage	870 (0.38 %)	870 (0.38 %)
DSP slice usage	8 (3.64 %)	8 (3.64 %)
Calculation time [ns]	80	32
No. of clock cycles per calculation	8	8

The results underline one key advantage of the FPGA at performing lookup table operations, since the overall resource consumption is quite low compared to the Peck model calculation before. In this case the lookup tables were stored in the LUTRAM, which significantly decreases the usage of LUTs.

For the calculation of the junction temperature, a simple Cauer model consisting of four layers has been implemented. Each layer represents a physical layer of the thermal network of the power semiconductor module mounted to the heat sink. Therefore, each layer can be modelled with a first-order lag according to the thermal network. The results are shown in the following Table 5. This model requires a high number of DSP slices for the calculation of the temperatures at the different layers and the heat transfer between the layers. But since the DSP slice usage is at only 18.18 % in total, even more complex thermal networks can be implemented.

Table 5: Resource consumption and timing analysis of the Cauer model

size of variables	32bit	
FPGA clock [MHz]	100	250
LUT usage	3626 (6.82 %)	3626 (6.82 %)
FF usage	2380 (2.24 %)	2380 (2.24 %)
DSP slice usage	40 (18.18 %)	40 (18.18 %)
Calculation time [ns]	400	160
No. of clock cycles per calculation	40	40

According to figure 1, the evaluation of the thermomechanical status of the semiconductor is done by a rain flow cycle counter for the thermal cycles and the impact of the life time consumption is weighed by using the Coffin-Manson model. This model has also been implemented on the FPGA and the results are shown in the following table.

Table 6: Resource consumption and timing analysis of the Coffin-Manson model

size of variables	32bit	
FPGA clock [MHz]	100	250
LUT usage	7357 (13.83 %)	7357 (13.83 %)
FF usage	5876 (5.52 %)	5876 (5.52 %)
DSP slice usage	32 (14.55 %)	32 (14.55 %)
Calculation time [ns]	780	312
No. of clock cycles per calculation	78	78

For the overall investigation of one calculation of the lifetime model, the usage of the FPGA resources of all sub modules have been added. This overview is presented in the following table.

Table 7: Resource consumption of the lifetime model for one phase leg

	32bit (100 MHz)					
	available	Peck	Lookup	Cauer	Coffin-Manson	sum
LUT	53200	13871	1706	3626	7357	26560 (49.92 %)
FF	106400	9399	870	2380	5876	18525 (17.41 %)
DSP	220	44	8	40	32	124 (56.36 %)
Calculation time	-	780	80	400	780	2040
Clock cycles	-	78	8	40	78	204

The results in Table 7 show, that the main part of the resources of the FPGA core is consumed solely for the calculation of one phase leg. But, in this investigation the shared usage of DSP slices of the different sub modules has not been considered, especially regarding pipelining of different calculations. Therefore, the values represent a worst-case implementation for analysing the feasibility of the FPGA. But according to the number of necessary clock cycles, this FPGA is theoretically capable of performing the calculation of over 490,000 phase legs per second. Considering a switching frequency of 4 kHz of the power converter this results in the calculation of 123 phase legs in only one switching cycle. Assuming a 2-level power converter topology, the online calculation of 20 full-scale converters is possible in this period.

In this work, some resources of the FPGA needed to be used for interfacing the model by physical ports to avoid undesired deletion of parts of the generated program code. So, there is the opportunity of further optimisations of the proposed implementation method. For the calculation of the complete lifetime model, high bit sizes need to be achieved to correctly represent the smallest fractions of lifetime consumption. But regarding the overall calculation capacity per switching cycle, an increase of the bit size of the variables is even possible on this device. For larger power converter systems, the calculation model can easily be ported to a larger FPGA.

Conclusion

This article describes the extension of a model for estimating the remaining lifetime of power semiconductors and the implementation of this system. The model takes the damage caused by electrochemical and thermomechanical degradation into account. An implementation has been drawn up and implemented for an FPGA. The determined timing measurement based on the FPGA based implementation showed an increase in performance of the new approach. Nevertheless, while mathematical operation can be implemented using Matlab/Simulink, for loops which are needed for the cycle counting algorithm this does not hold. Furthermore, Matlab/Simulink currently only supports data types of 64 bits in this toolchain, data types with e. g. 128 bits must be implemented directly in VHDL. That means, a hybrid approach is the best solution, using the FPGA for parts of the model, which can highly be parallelised, and e. g. a CPU for the cycle counting. For this combination the Xilinx Zynq is an example for an optimal choice. Due to the extensible implementation based on basic mathematical operations, the model can easily be adapted to further degradation mechanisms.

References

[1] K. Fischer; K. Pelka; A. Bartschat; B. Tegtmeier; D. Coronado; C. Broer; J. Wenske: Reliability of Power Converters in Wind Turbines: Exploratory Analysis of Failure and Operating Data From a Worldwide Turbine Fleet, IEEE Transactions on Power Electronics, Volume: 34, Issue: 7, July 2019

[2] C. Bhargava, P. K. Sharma, M. Senthilkumar, S. Padmanaban, V. K. Ramachandaramurthy, Z. Leonowicz, F. Blaabjerg, M. Mitolo: Review of Health Prognostics and Condition Monitoring of Electronic Components, in IEEE Access, vol. 8, pp. 75163-75183, 2020

[3] S. Mollov and F. Blaabjerg, "Condition and Health Monitoring in Power Electronics," CIPS 2018; 10th International Conference on Integrated Power Electronics Systems, 2018

[4] P. Chauhan, M. Osterman, M. Pecht and Q. Yu, "Use of temperature as a health monitoring tool for solder interconnect degradation in electronics," Proceedings of the IEEE 2012 Prognostics and System Health Management Conference (PHM-2012 Beijing), 2012

[5] W. Holzke: Entwicklung eines Messsystems für Feldmessungen in Windenergieanlagen und echtzeitfähige Implementierung eines Lebensdauermodells zur Zustandsüberwachung von Leistungshalbleitern, Dissertation,VDI Verlag, 2020

[6] W. Holzke, A. Brunko, H. Groke, N. Kaminski, B. Orlik: A Condition Monitoring System for Power Semiconductors in Wind Energy Plants, PCIM Europe 2018

[7] A. Brunko, W. Holzke, H. Groke, B. Orlik, N. Kaminski: Model-Based Condition Monitoring of Power Semiconductor Devices in Wind Turbines, EPE 2019 ECCE Europe

[8] C. Zorn; N. Kaminski: Temperature Humidity Bias (THB) Testing on IGBT Modules at High Bias Levels, CIPS 2014; 8th International Conference on Integrated Power Electronics Systems, 2014

[9] M. Musallam, C. M. Johnson: An Efficient Implementation of the Rainflow Counting Algorithm for Life Consumption Estimation. In: IEEE Transactions on Reliability Bd. 61 (2012), Nr. 4

[9] F. Qin, X. Bie, T. An, J. Dai, Y. Dai, und P. Chen: A Lifetime Prediction Method for IGBT Modules Considering the Self-Accelerating Effect of Bond Wire Damage, IEEE Journal of Emerging and Selected Topics in Power Electronics, Vol. 9, Nr. 2, S. 2271–2284, April 2021

[10] A. Wintrich, U. Nicolai, W. Tursky, und T. Reimann: Applikationshandbuch Leistungshalbleiter, 2. Auflage Ilmenau: ISLE Verlag, 2015

[11] M. Bartram: IGBT-Umrichtersysteme für Windkraftanlagen: Analyse der Zyklenbelastung, Modellbildung, Optimierung und Lebensdauervorhersage, Serie Aachener Beiträge des ISEA, Aachen: Shaker, 2006, Nr. Bd. 40

[12] F. Masana: A new approach to the dynamic thermal modelling of semiconductor packages, Microelectronics Reliability, Vol. 41, Nr. 6, S. 901–912, June 2001

[13] W. Zhihong, S. Xiezu, und Z. Yuan: IGBT junction and coolant temperature estimation by thermal model, Microelectronics Reliability, Vol. 87, S. 168–182, August 2018

[14] B. Wu: High-Power Converters and AC Drives. Hoboken, New Jersey: John Wiley & Sons, Inc., 2006

Smart Charging Strategy for Electric Vehicles Using an Optimized Fuzzy Logic System

M. Gholami (1,2), M. Mehrasa (1), R. Razi (3), K. Hajar (3), A. Hably (3), S. Bacha (1,4), A. Labonne (1)
1 Univ. Grenoble Alpes, CNRS, Grenoble INP*, G2Elab
2 Faculty of engineering, University of Kurdistan, Sanandaj, Iran
3 Univ. Grenoble Alpes, CNRS, Grenoble INP*, GIPSA-Lab
4 SuperGrid Institute
*Institute of Engineering Univ. Grenoble Alpes

Acknowledgments

This work has been supported by the aVEnir project of the PIA operated by ADEME.

Keywords

≪Electric vehicle≫, ≪Energy management≫, ≪Optimized Fuzzy system≫, ≪Genetic algorithm≫, ≪Smart charging≫.

Abstract

The increasing growth of electric vehicles (EVs) may arise as a challenge of increasing the load. Therefore, energy management in microgrids, including renewable energy resources such as PV systems, would be essential. Moreover, providing a smart charging pattern can optimize the overall cost of energy in a microgrid. In this paper, a genetic algorithm-based optimized fuzzy technique is developed, which has simple implementation such as rule-based methods and provides the optimal operation. The proposed scheme is simulated in MATLAB/Simulink environment for a case study. Results show the effectiveness of the proposed approach in comparison to conventional models.

Introduction

Concerning environmental issues and the tendency to reduce fossil fuel consumption, the number of electric vehicles (EVs) is growing significantly [1, 2]. Despite their advantages, it can lead to a challenge by increasing the grid load, especially during peak hours [3, 4, 5]. Therefore, managing and shifting them to the light-load times would be not only effective but also essential. Since EVs are stayed at parking lots for most of the time, a flexible bidirectional operation including vehicle to grid (V2G) and grid to vehicle (G2V) modes can be considered to minimize the microgrid cost [6, 7, 8]. Moreover, concerning the variable market, charging planning has become very important and has been conducted in many studies as an interesting topic called smart charging [9, 10, 11]. In addition, with the increasing penetration of renewable energy resources, and the emergence of microgrids, the tendency to manage energy at the micro-grid level has increased [12, 13]. One of the solutions proposed in previous studies is the formation of microgrids including charging stations, local loads, and renewable resources such as wind and PV [14, 15, 16].

There are a variety of strategies have been proposed for smart charging implementation. The use of rule-based algorithms such as fuzzy systems [17, 18, 19], the use of optimization programming methods [20, 21, 22, 23], and also predictive optimization methods [11, 24, 25, 26] to implement the smart charging pattern have been presented in various studies. The efficiency of these methods can be compared based on two criteria, including model complexity (cost of calculations) and model flexibility in the presence of system uncertainties such as load, solar system power, and energy price. Rule-based methods, although simple to implement and the rules can be used in different uncertainty conditions do not necessarily create the optimal situation. While methods based on programming and predictive optimization, despite providing optimal solutions, to manage system uncertainties, model calculations must be repeated over consecutive periods to ensure an optimal solution. Therefore, these models have a high computational load, and their implementation is problematic.

In this paper, we proposed a smart strategy based on an optimized fuzzy system that has both the benefits of fuzzy systems simplicity and providing an optimal solution. In this method, Fuzzy system parameters are obtained using genetic algorithm-based optimization. This method does not require repetitive calculations. Once in the beginning, the optimization is performed to find a suitable fuzzy system, and then the control is performed in a rule-based manner. Furthermore, to ensure the requested state of charge (SOC) at the time of departure, a supporting controller is provided to decide according to the time remaining and the maximum rated power of the EV. The rest of the paper is organized as follows. In Section II, the system model is described. Then, in Section III, the proposed model is presented. In Section IV, the simulation results for a case study are shown. Finally, Section V is devoted to the conclusions.

System model

The studied system is shown in Fig 1. This grid-connected the system includes local loads, a charge station, and a PV system. In this study, we followed the cost minimization of the microgrid by participating EVs in the energy management process. Since EVs usually stay for a long time at the station, they can experience not only the charging mode but also the discharging mode in a smart strategy to achieve the maximum benefit.

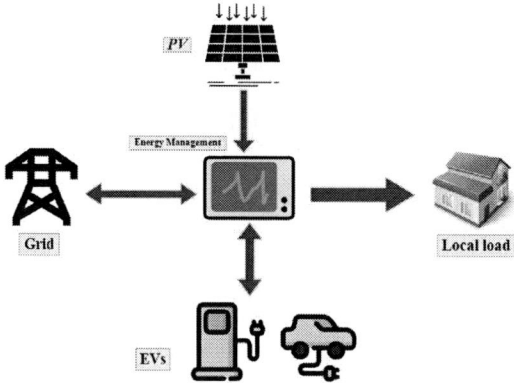

Fig. 1: The studied microgrid.

The objective function is defined based on the total cost for exchanged power with the grid, which is expressed in (1).

$$Cost = \sum_{k=1}^{N} [\Delta T * (\frac{P_{grid(k)} + |P_{grid(k)}|}{2}) * Price_{pos}(k) \\ - (\frac{P_{grid(k)} - |P_{grid(k)}|}{2}) * Price_{neg}(k)]$$

(1)

In which, $P_{grid}(k)$ is the exchanged power with the grid at the instant k, ΔT is the duration of each interval, N is the number of intervals during the stopping in a parking lot. Also, the prices of positive and negative exchanged powers with the grid ($Price_{pos}, Price_{neg}$) are considered different. The constraints of the problem includes the rated power of EVs, the state of charge (SOC) of EV's batteries, and the microgrid power balance as bellow:

$$subject\ to$$
$$P_{load} + P_{PV} + P_{EV} = P_{grid}$$
$$P_{EV} \leq P_{EV,n}$$
$$SOC_{min} \leq SOC \leq SOC_{max}$$

(2)

The uncertainty in the load and PV power profiles and using the forecasted profiles is the main challenge for this problem where the optimization result will be affected by the prediction error. Although using some methods such as the predictive method can be useful to reduce the effect of prediction error, they need to be performed during subsequent intervals which leads to a high computational burden. The fuzzy inference system (FIS) method can deal well with the uncertainty so that it makes a decision based on the current information rather than forecasting data. A FIS system including membership functions and rules is designed based on the knowledge of the system. In addition to the knowledge, parameters of member functions and rules can affect the results of a FIS. Therefore, it is needed to optimize the FIS system based on the objective function. In this paper, we proposed an optimized FIS which is discussed in the next section.

Proposed smart charging strategy

The system given in Fig.1 is studied. In this paper, we consider one EV which can be extended to several EVs. An optimized fuzzy system is developed to perform the smart charging algorithm in this paper. First, the fuzzy system design is described and then the optimized fuzzy approach is presented.

Fuzzy system

The fuzzy system is used to obtain the power of EVs based on the status of the microgrid in terms of the energy price, the SOC of EVs, and the power of PV. Therefore a fuzzy system including three input variables and one output variable is designed. A Mamdani fuzzy inference system with triangular membership functions is used. Furthermore, rules are defined based on an overall view of the desired operation to achieve more benefit. According to these rules, the EV is charged during times with low energy price and high PV power and it is discharged vice versa. The intensity of charge/discharge power is determined through the fuzzy system. The membership functions for input and outputs are shown in Fig.2, and rules are given in Table I.

Optimized FIS

As mentioned the fuzzy system is designed based on an overall view and selection of parameters for membership functions are intuitive. In this paper, an optimized fuzzy system is presented and parameters are obtained based on minimizing the microgrid cost function as below:

$$min\{Cost(x,u)\}$$
$$S.t.\quad g(x,u) = 0$$
$$f(x,u) \leq 0$$

(3)

In which, u is the set of decision variables including three parameters for input membership functions and three ones for the output membership function, x is the set of independent variables, g and f are equality and inequality constraints mentioned in eq(2).

Due to the nonlinear and complex relationship between the cost function and the parameters of fuzzy variables, we use the Genetic algorithm (GA) optimization technique.

Obtaining the final SOC

To ensure the required SOC at the departure time, a support controller is provided which does not allow the SOC to be lower than the allowable level at any time. The allowable level at any time is determined based on the time remaining until the exit and the maximum rated power of the EV. It is defined in Eq(4).

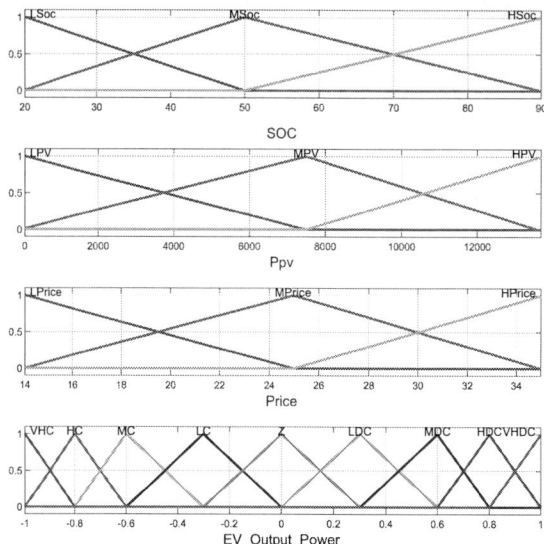

Fig. 2: Membership functions for normal-fuzzy system.

Table I: Fuzzy system rules.

SOC	P_{PV}	Price	EV	SOC	P_{PV}	Price	EV
L	L	L	HC	M	L	L	HC
L	L	M	LC	M	L	M	Z
L	L	H	Z	M	L	H	HDC
L	M	L	VHC	M	M	L	VHC
L	L	M	HC	M	M	M	LC
L	M	H	MC	M	M	H	MDC
L	H	L	VHC	M	H	L	VHC
L	H	M	HC	M	H	M	MCH
L	H	H	MC	M	H	H	MDC
H	L	L	LC	L:Low, M:Medium, H:High, Z:Zero			
H	L	M	LDC	LC:Low Charging			
H	L	H	VHDC	MC:Medium Charging			
H	M	L	LC	HC:High Charging			
H	M	M	LDC	VHC:Very-High Charging			
H	M	H	VHDC	LDC:Low Discharging			
H	H	L	LC	MDC:Medium Discharging			
H	H	M	MDC	HDC:High Discharging			
H	H	H	VHDC	VHDC:Very-High Discharging			

$$SOC - Line = \begin{cases} SOC_{min} + m*(t-t_0), & \text{if } t \geq t_0 \\ SOC_{min}, & \text{otherwise.} \end{cases}$$

$$m = \frac{(SOC_f - SOC_{min})}{\Delta T} \tag{4}$$

$$\Delta T = ((SOC_f - SOC_{min}) * E_r/P_{max}) * 3600$$

$$t_0 = t_{out} - \Delta T$$

where, SOC_f is the final SOC, E_r is the rated energy capacity, P_{max} is the maximum (rated) power of EV, and t_{out} is the departure time. Whenever the SOC level reaches the support line, the EV power is set to the maximum value.

Case study

The simulation result for one EV is given in this section to verify the proposed method. The characteristics of the microgrid are shown in Table II.

Table II: Microgrid characteristics.

Rated power of PV Array (P_{PV})	10kW
Rated power of load (P_{PV})	6kW
Rated power of EV (P_{EV})	15kW
Rated energy capacity of EC (E_{EV})	50kWh
SOC_{min} and SOC_{max}	0.2 - 0.95

The load, PV power ,and energy price profiles are shown in Fig. 3 and Fig. 4 for 24 hours. The same profiles are also considered for the next day in the simulation.

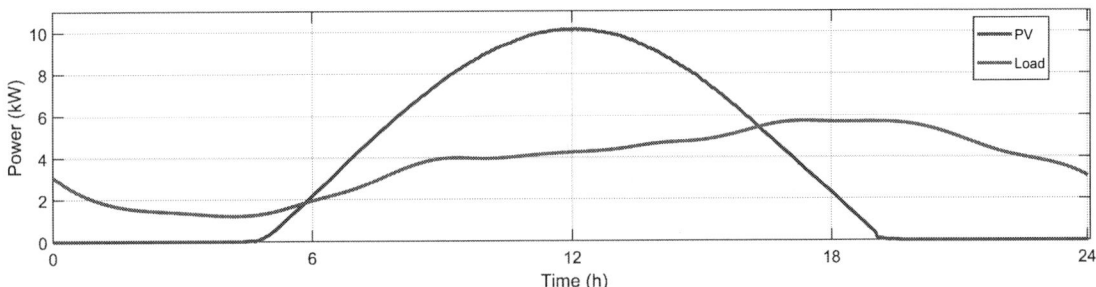

Fig. 3: Load and PV power profiles for 24 hours.

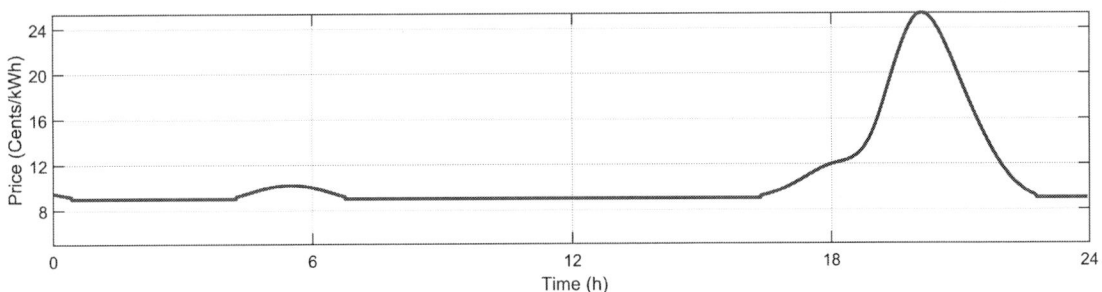

Fig. 4: Energy price profile for24 hours.

The studied scenario is defined as follows:

"The Ev is assumed to arrive at the parking lot at 4 p.m with the 60% of initial SOC and stays there until 8 a.m next day. The requested SOC at the time of departure is 90%."

The support line for this scenario is shown in Fig. 5.

The simulation has been done in three cases. In the first case, there is no smart charging, and the EV just is charged to get the final SOC. In the two other cases, the smart charging is followed with normal fuzzy system and optimized fuzzy system. The fuzzy system is optimized using the GA technique in MATLAB, the new membership functions for the optimized fuzzy system are shown in Fig. 6. As can be seen, they are different than the normal system (Fig. 1).

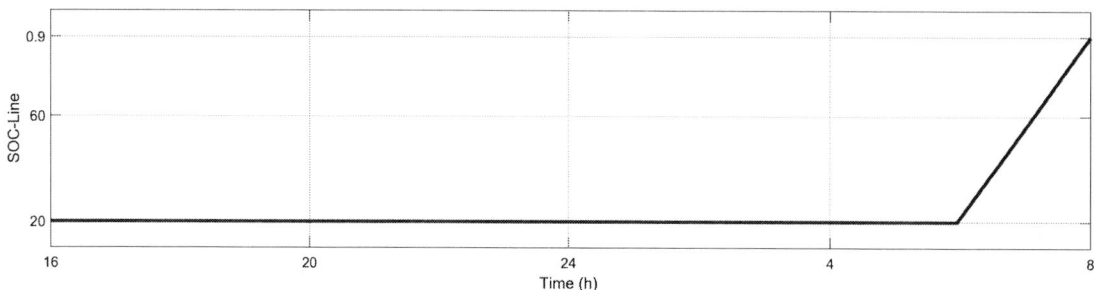

Fig. 5: Protective SOC Line.

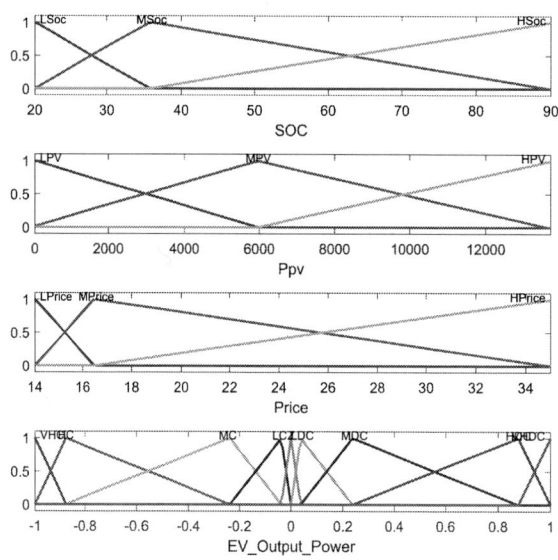

Fig. 6: Membership functions for optimized fuzzy system.

The simulation results including the power and SOC of EV and the cost of microgrid for three cases are shown in Figures 7, 8, 9.

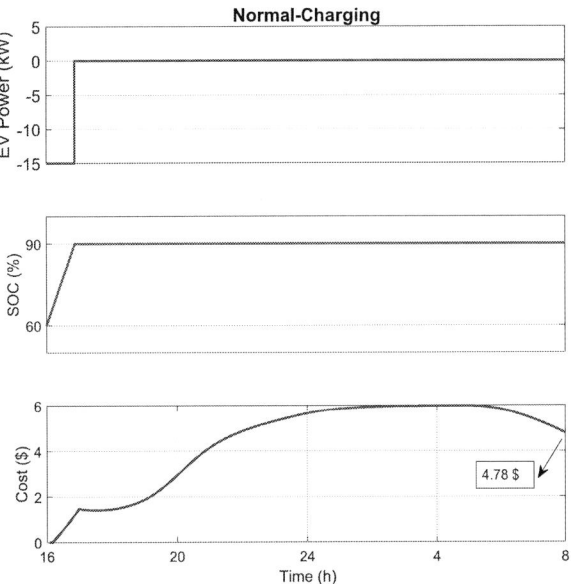

Fig. 7: Results for Normal-Charging mode; EV power, SOC, and microgrid cost.

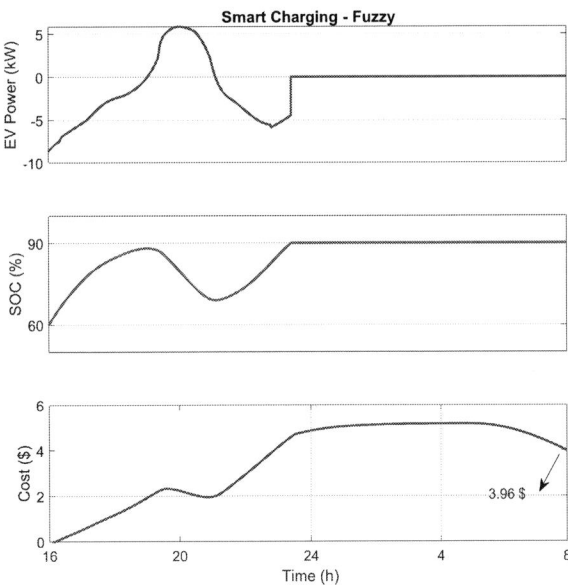

Fig. 8: Results for Smart-Charging mode, Normal-Fuzzy; EV power, SOC, and microgrid cost.

In the first case, the EV does not participate in the energy management and is charged with the nominal power to get to the final SOC. The final cost, in this case is 4.78$ which is much more than other cases. While the fuzzy controller has a significant effect in case II. The final cost is significantly decreased to 3.96$, which is 18% less than that of case I. Furthermore, the optimized fuzzy controller has much more effect in case III. In this case, the final cost is just 2.55$ which the final cost has been reduced to 50%. A comparison is given in Table III. As mentioned earlier, optimization in this method is done only once at the beginning of the EV entering the parking lot, so the implementation of this algorithm is much faster than other methods that require online optimization.

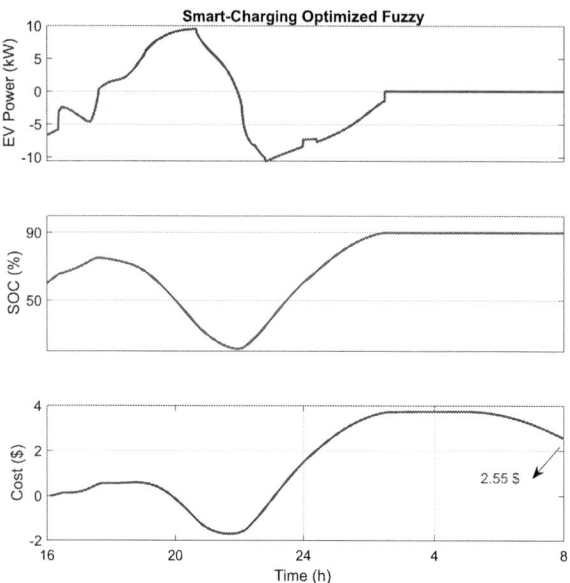

Fig. 9: Results for Smart-Charging mode, Optimized-Fuzzy; EV power, SOC, and microgrid cost.

Table III: Comparing the results.

	Case I Normal-Charging	Case II Fuzzy	Case III Optimized Fuzzy
Energy Cost ($)	4.78	3.96	2.25
Reduction percent (%)	0	18	53

Conclusions

The involvement of electric vehicles in the energy management of microgrids, including renewable and intermittent energy resources, can significantly reduce energy costs. In this process, charging of EVs is postponed to times when we have renewable production and energy price is low. In addition, when the energy price is high, they can discharge to provide some of the energy of local loads. Therefore, an algorithm is needed that can determine the power profile of EVs by considering the stopping time in the parking lot and their requested SOC at the departure. This paper presents an intelligent charging algorithm using an optimized fuzzy system that is both easy to implement and ensures an optimal response. In this method, the parameters of the fuzzy system membership functions are obtained based on the optimization of the cost function and using a genetic algorithm. The advantage of this method is that the optimization operation is performed only once when the EV enters the parking lot, and then, like the fuzzy method, it operates on a rule-based basis. Therefore, there is no need for optimization calculations during successive intervals. The next step in this strategy is to implement the algorithm in the presence of other vehicles, which will be examined in future studies.

References

[1] K. Hajar, B. Guo, A. Hably, S. Bacha, Smart charging impact on electric vehicles in presence of photo-voltaics, in: 2021 22nd IEEE International Conference on Industrial Technology (ICIT), Vol. 1, IEEE, 2021, pp. 643–648.

[2] R Razi, M Mehrasa, K Hajar, M Gholami, A Hably, S Bacha, A Labonne: Predictive smart charging of plug-in electric vehicles for parking fleet scenario with modular converters. CIRED Porto Workshop 2022 E-mobility and power distribution systems.

[3] A. Gerossier, R. Girard, G. Kariniotakis, Modeling and forecasting electric vehicle consumption profiles, Energies 12 (7) (2019) 1341.

[4] H. Turker, S. Bacha, D. Chatroux, A. Hably, Modelling of system components for vehicle-to-grid (v2g) and vehicle-to-home (v2h) applications with plug-in hybrid electric vehicles (phevs), in: 2012 IEEE PES Innovative Smart Grid Technologies (ISGT), IEEE, 2012, pp. 1–8.

[5] M. Mehrasa, R. Razi, M. Gholami, K. Hajar, A. Labonne, A. Hably, and S. Bacha: A Dynamic Real-Time Optimization Algorithm for the Revenue Assessment of a Vehicle-To-Grid System in Presence of Wear Cost Model. Electrimacs 2022: 14th International Conference of TC-Electrimacs Committee, 16-19 May 2022 Nancy (France).

[6] Q. Hoarau, Y. Perez, Interactions between electric mobility and photovoltaic generation: A review, Renewable and Sustainable Energy Reviews 94 (2018) 510–522.

[7] V.-L. Nguyen, T. Tran-Quoc, S. Bacha, B. Nguyen, Charging strategies to minimize the peak load for an electric vehicle fleet, in: IECON 2014- 40th Annual Conference of the IEEE Industrial Electronics Society, IEEE, 2014, pp. 3522–3528.

[8] H. Turker, S. Bacha, Optimal minimization of plug-in electric vehicle charging cost with vehicle-to-home and vehicle-to-grid concepts, IEEE Transactions on Vehicular Technology 67 (11) (2018) 10281–10292.

[9] Y. Cao, H. Wang, D. Li, G. Zhang, Smart online charging algorithm for electric vehicles via customized actor-critic learning, IEEE Internet of Things Journal (2021).

[10] Y. Deng, Y. Mu, X. Dong, H. Jia, J. Wu, J. Zhang, S. Li, Hierarchical operation management of electric vehicles for depots with pv on-site generation, IEEE Transactions on Smart Grid (2021).

[11] R. Razi, K. Hajar, A. Hably, S. Bacha, A user-friendly smart charging algorithm based on energy-awareness for different pev parking scenarios, in: 2021 29th Mediterranean Conference on Control and Automation (MED), IEEE, 2021, pp. 392–397.

[12] M. Gholami, M. Mehrasa, R. Razi, A. Hably, S. Bacha and A. Labbone, "An Efficient Control Strategy for the Hybrid Wind-Battery System to Improve Battery Performance and Lifetime," IECON2021 – 47th Annual Conference of the IEEE Industrial Electronics Society, 2021, pp. 1-6.

[13] K. Hajar, R. Razi, M. Mehrasa, A. Labonne, A. Hably, and S. Bacha: Photovoltaics at the electric mobility's service: French case study. Electrimacs 2022: 14th International Conference of TC-Electrimacs Committee, 16-19 May 2022 Nancy (France).

[14] Y. Yang, M. Wang, G. Geng, Q. Jiang, Energy management of microgrid considering ev integration based on charging reservation information, in: 2021 IEEE/IAS Industrial and Commercial Power System Asia (I & CPS Asia), IEEE, 2021, pp. 609–614.

[15] A. Alsharif, C. W. Tan, R. Ayop, K. Y. Lau, C. L. Toh, Sizing of photovoltaic wind battery system integrated with vehicle-to-grid using cuckoo search algorithm, in: 2021 IEEE Conference on Energy Conversion (CENCON), IEEE, 2021, pp. 22–27.

[16] P. K. Ray, A. Bharatee, S. Panda, I. N. W. Satiawan, Modeling and power management of electric vehicle charging system, in: 2021 International Conference on Smart-Green Technology in Electrical and Information Systems (ICSGTEIS), IEEE, 2021, pp. 100–105.

[17] Y. Lu, S. Liu, M. Wu, D. Kong, Energy management of dual energy sources pure electric vehicle based on fuzzy control, in: 2021 IEEE International Conference on Internet of Things and Intelligence Systems (IoTaIS), IEEE, 2021, pp. 228–233.

[18] M. Mehrasa, R. Razi, K. Hajar, A. Labbone, A. Hably, S. Bacha, Power management of a smart vehicle-to-grid (v2g) system using fuzzy logic approach, in: IECON 2021–47th Annual Conference of the IEEE Industrial Electronics Society, IEEE, 2021, pp. 1–6.

[19] D. Sharma, B. Faujdar, B. S. Surjan, Fuzzy logic based power management of electric vehicle with battery and ultracapacitor energy storage system, in: 2021 4th International Conference on Recent Developments in Control, Automation & Power Engineering (RDCAPE), IEEE, 2021, pp. 246–250.

[20] Y. Lin, K. Wang, M. Quan, Z. Zhang, X. Dong, Optimal allocation method for microgrid system capacity with electrical vehicles, in: 2021 6th International Conference on Power and Renewable Energy (ICPRE), IEEE, 2021, pp. 1181–1185.

[21] N. Korolko, Z. Sahinoglu, Robust optimization of ev charging schedules in unregulated electricity markets, IEEE Transactions on Smart Grid 8 (1) (2015) 149–157.

[22] A. Pal, A. Bhattacharya, A. K. Chakraborty, Allocation of ev fast charging station with v2g facility in distribution network, in: 2019 8th International Conference on Power Systems (ICPS), IEEE, 2019, pp. 1–6.

[23] S. Bacha, A. Hably, A. Ovalle, Grid Optimal Integration of Electric Vehicles: Examples with Matlab Implementation, Springer, 2018.

[24] C. Diaz, F. Ruiz, D. Patino, Smart charge of an electric vehicles station: A model predictive control approach, in: 2018 IEEE Conference on Control Technology and Applications (CCTA), IEEE, 2018, pp. 54–59.

[25] M. Yousefi, A. Hajizadeh, M. N. Soltani, B. Hredzak, Predictive home energy management system with photovoltaic array, heat pump, and plug-in electric vehicle, IEEE Transactions on Industrial Informatics 17 (1) (2020) 430–440.

[26] F. Luo, G. Ranzi, C. Wan, Z. Xu, Z. Y. Dong, A multistage home energy management system with residential photovoltaic penetration, IEEE Transactions on Industrial Informatics 15 (1) (2018) 116–126

Analysis and Discussion of a Concept for an Adjustable Inductance Based on an Impact of an Orthogonal Magnetic Field

Guido Schierle, Michael Meissner, Klaus F. Hoffmann
HELMUT SCHMIDT UNIVERSITY
UNIVERSITY OF THE FEDERAL ARMED FORCES HAMBURG
Holstenhofweg 85
22043 Hamburg, Germany
Tel.: +49 / (0)40 6541 2768
Fax: +49 / (0)40 6541 2018
Email: guido.schierle@hsu-hh.de
URL: https://www.hsu-hh.de/lek/en/

Keywords

≪Coupled inductor≫, ≪Passivity≫, ≪Passive component≫, ≪Passive component integration≫, ≪Magnetic coupling≫, ≪Magnetic device≫, ≪Winding topology≫.

Abstract

In this paper the realisation of a concept for an adjustable inductance by the use of an auxiliary control current is analysed and discussed. Using a construction of two magnetic toroid cores the magnetic field of a load core, and by this its effective inductance, is influenced by an orthogonal magnetic field induced by an additional control core. It will be shown that with an increase of the magnetic field generated by the control circuit the inductance of the load circuit decreases.

Introduction

Inductors are essential components for power electronic applications. Their effective inductivity mainly depends on the number of turns of a winding and, when using a core, the geometry and the characteristics of the used material. Therefore, an influence on the effective inductance of an electric circuit apart from an adjustment of the number of turns or the core itself needs to be achieved otherwise. In this paper one of various concepts for adjustable inductances is presented and analysed [1, 4, 5, 6]. In the past, current-controlled inductors were used for many applications, e.g. for AC power control or special gate drivers [2, 4, 5]. The motivation for this extended investigation is a possible integration of the described principle in e.g. resonant converters with high switching frequencies and a power range of 2-50 kW. An adjustable inductor would allow to control both the resonant frequency and the characteristic impedance of the resonant tank, in addition to the widely used parameter of switching frequency.

Inductive core setup

The objective of the below described setup (Fig.1) is to achieve a controllability on the effective inductance during operation. Therefore, a magnetic field orthogonal to the magnetic field of the load inductance will be used. The inductive core setup is comprised on the one hand by a winding around a magnetic core, the load core. On the other hand, another magnetic toroid core, in the following named control core, is implemented perpendicularly. Fig. 1 depicts the theoretical setup of the orthogonal circuit. It also shows that the control circuit provides a DC current. Following from this the control core is permeated with a constant magnetic flux. It can be seen that the resulting magnetic field of the control core in an idealized construction should be orthogonal to the magnetic field of the load core. In the area of the shared volume there will be a resulting magnetic flux, with an orientation depending on the strength of both magnetic fields.

For construction, a segment of the control core was removed according to the dimensions of the load core. This ensures the area of the control core which is implemented onto the load core fits as tightly

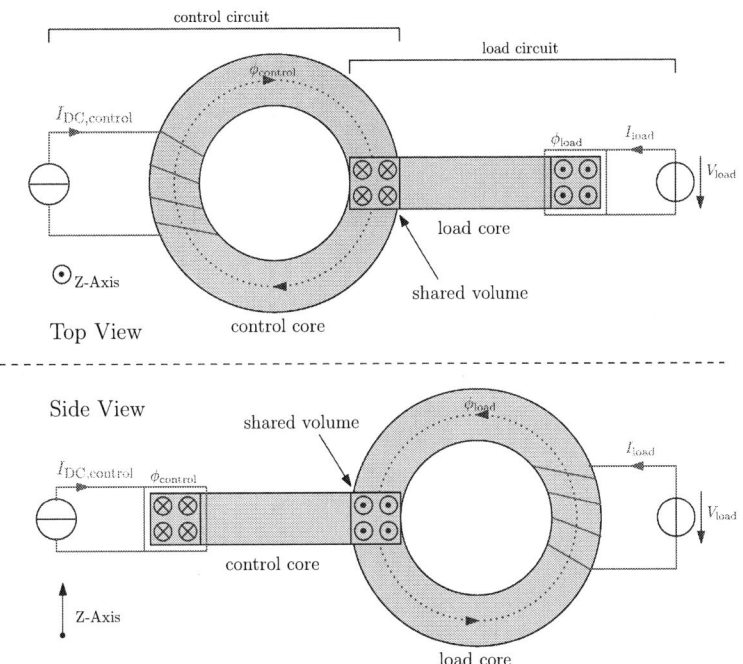

Fig. 1: Design of the orthogonal connection of load and control core [8]

as possible to minimize the resulting parasitic air gap. Fig. 2a presents a prototype of the theoretical setup. For better visualisation the load and the control core of the first setup are marked. The core in the background will be used for investigations regarding the influence of the size of the shared volume and is made from the same material as the bigger core. From the construction of the inductive core setup a simplified magnetic circuit was derived and is presented in Fig. 2b. The magnetic resistance $R_{m,shv}$ describes the reluctance of the shared volume of both cores. The magnetic resistances $R_{m,air}$ depicts the reluctance of the parasitic air gap.

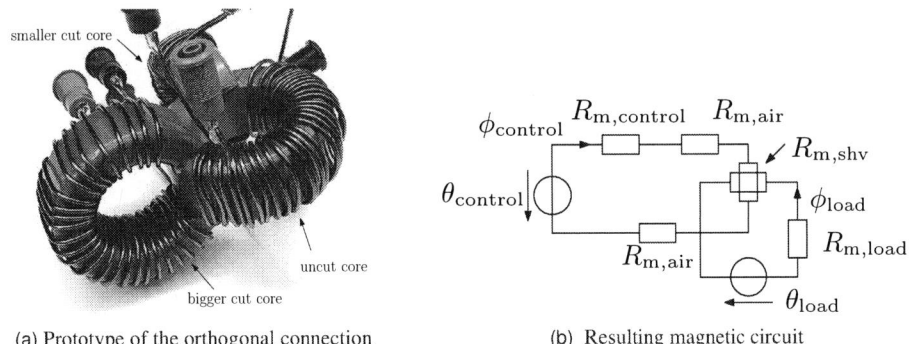

(a) Prototype of the orthogonal connection   (b) Resulting magnetic circuit

Fig. 2: Prototype of the orthogonal connection of load and control core with resulting magnetic circuit

The impact of the air gap will be examined and discussed in this paper. Therefore, both cores will be swapped in their function, which means that the cut core will be used as the load and the uncut core will operate as the control core.

This concept for a current-controlled inductance was chosen because of two reasons. First, Fig. 1 shows that the influence of the control core can be limited to a specific area of the load core, depending on the size of the control core. Second, following the assumption that both magnetic fields are orthogonal to each other the magnetic field of the load core should not affect the magnetic flux of the control core beyond the shared volume. According to Faraday's Law no voltage is induced in the control circuit so that the used power source is protected, which could differ from other concepts of current-controlled inductors [3, 4, 9].

The control core

The magnetic field of the control core shall influence the effective inductance of the load core. Therefore, magnetic fundamentals will be described in the following.

Simplified, a constant and homogeneous magnetic flux density in the magnetic cross section is assumed and stray fields are neglected. Considering the described simplifications and Ampère's circuital law (displacement currents are not taken into account) the magnetic field strength and the current in the winding are related as equation (1) shows [7].

$$\theta = \oint \vec{H}(t)\, d\vec{l} = \sum_n H_n(t) \cdot l_{e,n} = N \cdot i(t) \tag{1}$$

For an impact of the control core on the load core an according magnetic field strength is needed. One additional requirement for the analysis is the saturation of the control core and therefore the saturation of the shared volume of both cores during fixed operating points. Here, it should be considered that in the parasitic air gap, resulting from construction, magnetic energy is stored [3]. Hence, a higher magnetic field strength, and for this reason a higher number of turns of the winding N and/or a higher current I is needed to saturate the core as equation (1) clarifies. Using a defined core the effective magnetic path length l_e remains constant. Regarding the used steps of control current the number of turns of the control winding were defined and remained constant.

Measurement method

The load voltage of Fig. 1 was generated by the Power Choke Tester DPG 20. To determine the effective inductance the Choketester applies a pulsed measurement method. At the beginning of the measurement a capacitor bank is charged and afterwards a square wave DC voltage is applied to the DUT. The current in the load circuit starts to rise with a slew rate, dependent on the inductance L, which is also a function of the current. As soon as a pre-defined maximum current is reached, the measurement ends. Since this measurement device has a maximum of 100 sampling points the simplified equation (2) is used to calculate the inductance. According to the equation it can be calculated by the time interval for the rise of current with an almost constant voltage. Further information to the measurement method are given in [10].

$$v_L(t) = L \cdot \frac{\triangle i(t)}{\triangle t} \tag{2}$$

A distinction is made between incremental inductance L_{inc} and secant inductance L_{sec}. The incremental inductance describes the tangent at an operating point for a small modulation. However, L_{sec} is defined by a secant between an operating point and the origin. Especially during saturation of the core material, the values show significant differences. For power electronic applications often the currently effective inductance L_{inc} is of interest. Thus, only the incremental inductance will be analysed [7].

Additionally, the Choketester was used to examine the magnetic flux density. Because of the saturation of the shared volume it is expected that it behaves like an air gap or more specifically the magnetic reluctance should increase for the load core. For this reason, a higher magnetic field strength is needed to reach the same magnetic flux density in the load core as without the impact of the control core. Hence, the B-H characteristic of the inductance of the load core should decrease with increasing control current.

The flux linkage $\Psi(i)$ is calculated as the product of the inductance, given by equation (2), and the current as equation (3) shows, which is equivalent to the the product of the number of turns N and the magnetic flux Φ.

$$\Psi(i) = L(i) \cdot i = N \cdot \Phi \tag{3}$$

With information about the number of turns and the effective magnetic cross section A_e the Choketester calculates the magnetic flux density B using the equation (4).

$$B = \frac{\Psi}{A_e \cdot N} \tag{4}$$

The calculation used is sufficiently accurate for a general analysis of the concept. The results of the inductance of the load core and the B-H characteristics are presented and discussed hereinafter.

Measurement results

Based on the rising control current the different characteristics of the incremental inductance L_{inc} and the magnetic flux density of the load core are depicted and interpreted. For the load circuit the Choketester was configured to stop the measurement once the load current reaches 11 A, because it was observed in earlier measurements that the load core was almost entirely saturated at this value.

Impact of the control current on differential inductance

The steps of the control current were chosen to follow the E-12 series. After a first analysis of the measurement results some current steps were added in order to fill occurring gaps, because a relatively significant decrease in the effective inductance was detected. The maximum value was defined to 100 A. The results of the measurement with different control currents are shown in Fig. 3.

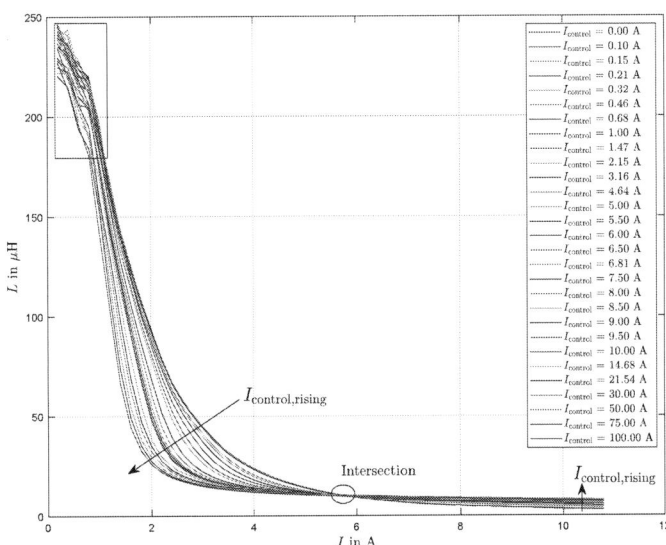

Fig. 3: Dependency of the effective inductance to the applied control current

Fig. 3 presents that the load core saturates increasingly with a rising load current right from the start of the measurement even when there is no control current applied. Because of measurement inaccuracies at low currents, indicated by a rectangle in the top left corner, the results of the beginning of the curves are not taken into account for analysis.

Clearly, it can be seen that the incremental inductance decreases from the beginning of the analysed measurement with a rising control current. With no control current the effective inductance has a value of about 90 μH at a load current of 2 A. With rising control current the effective inductance decreases down to a value of around 35 μH at a control current of 30 A and 25 μH for 100 A. The drop of effective inductance between the step of control current from 21.54 A to 30 A or even from 50 A to 100 A is significantly smaller than between 4.64 A and 6.81 A although the absolute increase in control current is obviously higher, justifying the chosen E-12 series for the values. It is noticeable, that all measurement curves intersect in the area of a load current around 5.7 A. In the following this intersection will be examined more closely. The area of intersection is indicated by an ellipse in Fig. 3. From the point of intersection, it can be observed that the effective inductance decreases more slowly with increasing control current. The magnetic domain of a ferrite core could provide an explanation for this behaviour. The rising control current generates a magnetic field in the shared volume of both cores. Due to an increase in the load current the resulting magnetic field in the shared volume aligns more and more in the direction of the magnetic field of the load core. To verify this assumption, the impact of the size of the shared volume and therefore the area in which both magnetic fields are superpositioned will be investigated further. Additionally, the ratio of the magnetic field of the load core and the magnetic field of the control core will be varied by increasing the number of turns of the load core.

Investigation of the size of the shared volume

In order to investigate the influence of the size of the shared volume a smaller control core, made of the same material, was implemented similar to the bigger core. The final setup is shown in Fig. 2a with the smaller control core in the background. To obtain equally effective magnetic fields for both cores the same magnetic field strength needs to be achieved. For this reason and in due consideration of equation 1, assuming the same length of air gaps, the number of turns for the smaller core was chosen accordingly.

Subsequently, the setup was operated with the smaller core as control core. For better visualization, the used control current was limited to 30 A and only chosen values are presented. Fig. 4a shows the influence on the differential inductance of both, the bigger and the smaller control core.

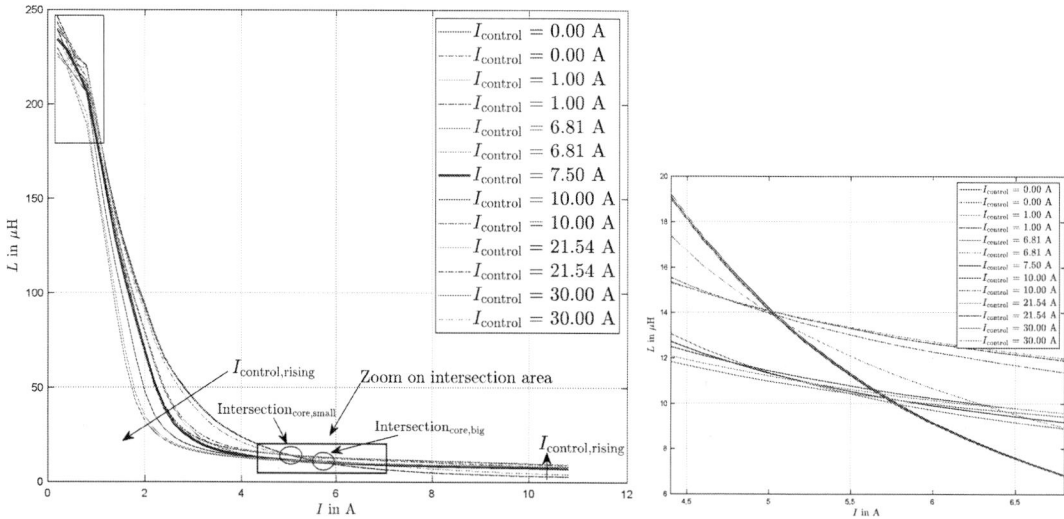

(a) Comparison of impact of size of shared volume. Dotted lines represent the smaller control core, solid lines the bigger control core

(b) Zoom on both intersection areas from Fig. 4a

Fig. 4: Impact of the size of shared volume with focus on both intersection areas

The comparison of the measurement curves shows that a larger shared volume has a greater impact on the differential inductance, especially at the beginning of the measurement around 2 A. While the smaller core starts to have a slight influence on the differential inductance (first dotted light blue curve) at a control current of 6.81 A the bigger core already lowered the effective inductance from around 90 μH to about 75 μH with the same current (first solid green curve). For 30 A, the maximum of the applied control current, the difference is even more significant. The inductance decreases from the initial 90 μH to around 70 μH for the smaller control core but only has a value of around 30 μH for the bigger core. The curves converge with rising load current and therefore the differences between the cores drop. It is conspicuous that both cores have an intersection area but the curves of the smaller core intersect with a lower load current and for this reason a higher effective inductance. After the intersection point the curves only drop slightly with increasing load current. For a control current of 21.54 A and 30 A regarding the smaller core the effective inductance still has a greater value in comparison to the bigger core. However, the curves show a larger negative slope which is hard to see in the figure but could be observed for all measurements with higher load currents. The bigger the core and the higher the control current the lower was the negative slope of the respective differential inductance after the intersection area. This behaviour could be addressed in further investigations.

To outline the opportunity of a possible trade-off between the size of the shared volume and the applied control current an additional measurement with 7.5 A, using the bigger core as control core, was added to fig. 4a and is represented by the dark red solid line. It is noticeable that right from the start until a load current of 2.5 A is reached, the effective inductance of the load core shows similar behaviour. It is influenced by the bigger core with a control current of 7.5 A and by the smaller core with a control current of 30 A. With rising load currents both graphs split but the curve for 30 A of the smaller control core intersects exemplarily with the curve for 6.81 A of the bigger control core. This illustrates that nearly every desired operating point of the effective inductance could be reached, varying the size of the shared volume or adapting the control current correspondingly. Future examinations should focus on the influence of possible differences of the length of the airgap resulting from construction.

Fig. 4b shows a selected area around the intersection from Fig. 4a. The dotted lines represent the smaller core and the solid lines the bigger core. The figure depicts that the measurement curves of the smaller control core intersect with a lower load current of about 5 A. Furthermore, the effective inductance dropped to a value of 14 µH. For the bigger core the intersection of the curves with an applied control current and the one with no control current occur in an area from 5.6 A - 5.75 A. The effective inductance for the chosen curves varies only slightly from nearly 11 µH to 10 µH at the respective intersection point. It can be derived that the larger the magnetic field of the load core in ratio to the magnetic field of the control core the less load current is needed to reach the intersection area.

Impact of the ratio of the magnetic fields on differential inductance

The influence of the magnetic field generated by the load current should be examined more closely to test its influence on the intersection area. To obtain a higher magnetic field strength and therefore a higher magnetic flux density, the number of turns on the load side was raised, according to equation 1. The range of the load current and the magnetic path length remained constant. Based on the previous results, it can be presumed that following from the higher magnetic field of the load side the measurement curves of the differential inductance should intersect with a smaller load current.

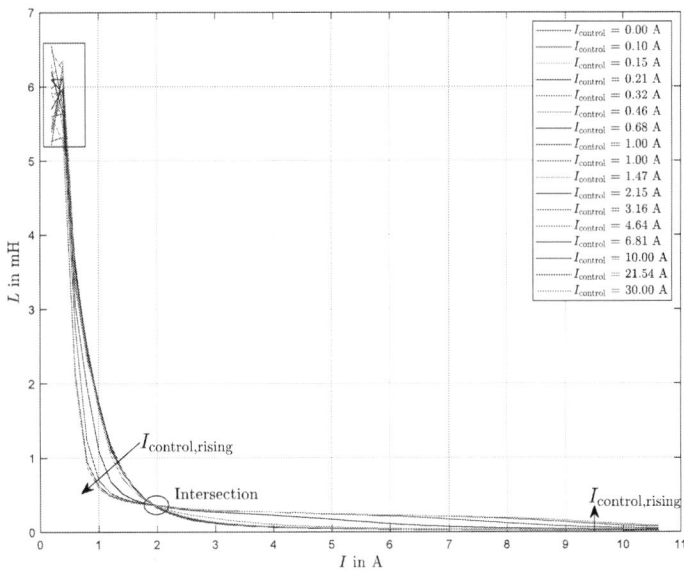

Fig. 5: Influence of the magnetic field of the load core on intersection area

Fig. 5 shows the results of the measurement with control currents up to 30 A. It can clearly be seen that with a rising control current the effective inductance decreases, as in case of the investigation with a lower number of turns. The change of the intersection area is visible distinctly, changing from 5.8 A with a lower number of turns to about 2 A with a higher number of turns.

It can be derived that the intersection area depends strongly on the ratio of the magnetic field of the load core and the magnetic field of the control core, because both magnetic fields are superpositioned in the area of the shared volume. The stronger the magnetic field of the load core the earlier the measurement curves of the differential inductance will intersect. An explanation could be found in the magnetic domains of the core material. Due to the magnetic field of the control side the Weiss domains of the shared volume are magnetized corresponding to the direction of the magnetic flux. For this reason, the magnetic resistance of the shared volume rises and the total reluctance of the magnetic circuit of the load core increases. With the same magnetomotive force of the load core a higher magnetic resistance leads to a lower magnetic flux and therefore the effective inductance decreases, which is observable in the previous measurement curves [3]. The premagnetized Weiss domains need to be aligned in the direction of the magnetic flux of the load core. With rising load currents the magnetic field of the load core increases and will dominate the superposition of the magnetic fields in the shared volume, because the Weiss domains will align more and more in the direction of the magnetic flux of the load core.

Trade-off number of turns vs control current

As already explained, there is a trade-off between the size of the shared volume and the control current. Regarding the efficiency for a possible implementation in an electrical circuit another degree of freedom is examined. Because of the dependency of the magnetic field on the magnetic field strength and therefore on the number of turns and the current of the control core, it is investigated if the number of turns and the current can be balanced equally.

Equation 5 presents the possible trade-off between the number of turns and the current. The index describes the number of turns. Even if the factors vary, the magnetic field strength still has to have the same value to generate the same magnetic field. Hence, the influence of the control core on the effective inductance of the load core should only be dependent on the ratio of the number of turns multiplied with the current. The equation clarifies that a needed current can be substituted with a higher number of turns.

$$H_{N27} = H_{N22} = \frac{N_{27} \cdot I_{27}}{l_e} = \frac{N_{22} \cdot I_{22}}{l_e}$$

$$I_{27} = \frac{N_{22}}{N_{27}} \cdot I_{22} \tag{5}$$

For exemplary values the results of this verification by measurement are observable in fig. 6.

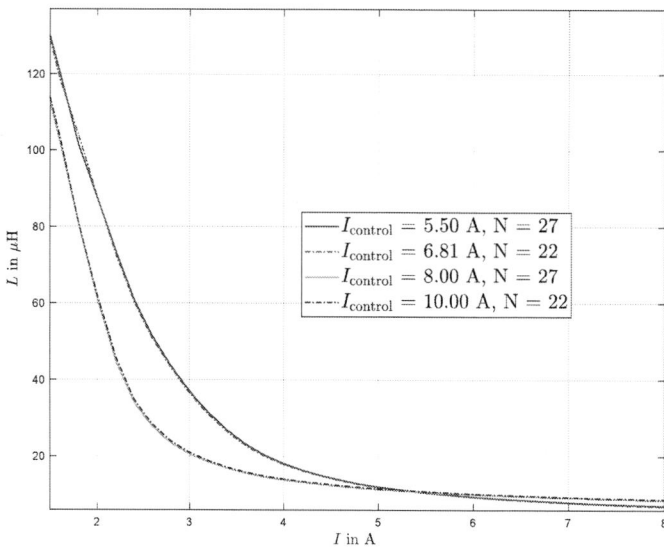

Fig. 6: Trade-off between control current and number of turns for the load core

The outcomes illustrate that the number of turns and the control current can be exchanged using equation 5. Considering that for the electrical losses the current is squared and the ohmic resistance, which increases with a higher number of turns, is taken into account only linearly a sensible trade-off can be found. Together with the size of the shared volume there are three degrees of freedom which can influence the efficiency of a system using this setup. However, according to Faraday's Law the higher number of turns on the control core could lead to a disadvantage regarding the magnetic back coupling of the system on the control circuit. Though, as described in the inductive core setup, this construction was chosen because both magnetic fields are orthogonal and therefore should not affect each other beyond the shared volume. This needs to be investigated further in the future to clarify its influence on a possible realization of the setup for application in an electrical circuit and to find potential limits of the trade-off.

Swap of control and load core in their function

As explained in the inductive core setup there will be a parasitic air gap resulting from construction. To examine its influence the setup will be reversed, so that the uncut core will work as the control core and the bigger cut core as the load core. The resulting magnetic circuit is shown in fig. 7b. This means that independent of the applied control current an air gap already exists in the load circuit. For this reason the load current was increased to the devices technical maximum value of 100 A to ensure a saturation of the load core [11]. For additional comparison the effective inductance of the cut core with pure air gap and thus without the implementation of the uncut core was measured for reference. The results are depicted in fig. 7a. The horizontal blue curve which has a value of a around 4.5 µH represents the measurement with no core implemented, so with a pure and because of this a relatively "big" air gap. It displays that there is no saturation at all without the implementation of the uncut core for the chosen parameters.

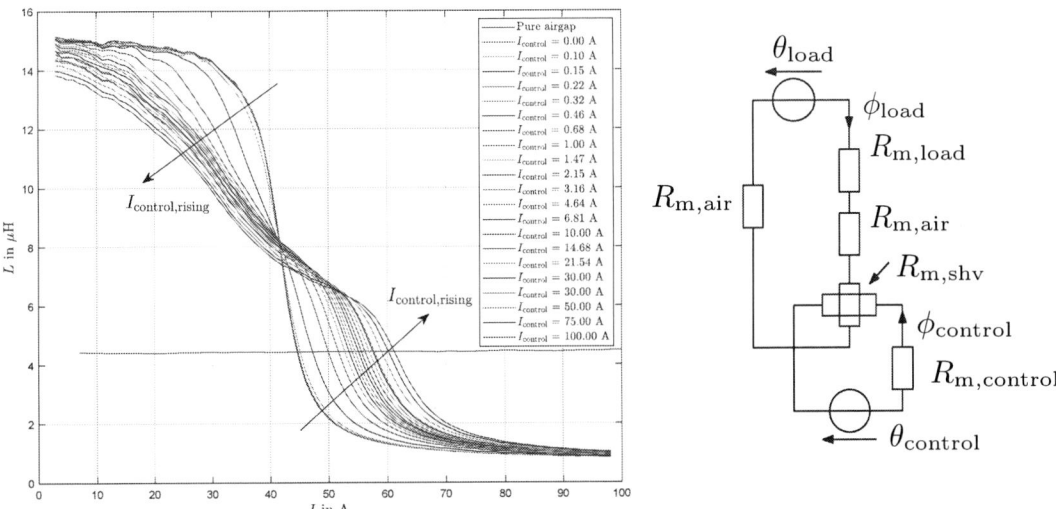

(a) Dependency of inductance to applied control current for swapped cores   (b) Resulting magnetic circuit for swapped cores

Fig. 7: Dependency of the effective inductance with swapped cores with corresponding magnetic circuit

The green and purple curve representing 30 A are introduced for comparison, because for the higher currents another current source was used. Both validate that the effect on the differential inductance is sufficiently enough the same. With no control current applied the effective inductance shows the characteristic behaviour of an inductance with a small air gap compared to the state when no core is implemented. The inductance remains constant until a load current of 30 A is reached. At higher values the core starts to saturate, the permeability decreases and therefore the value of the inductance drops rapidly to an almost constant value for which the core is saturated completely [3]. It is visible that the reached constant value is smaller than for measurement with pure air gap. This behaviour should be adressed further, an explanation could be the widening of the air gap. This widening leads to a larger magnetic cross section A_e, therefore to a lower magnetic flux density as equation 4 shows and because of this to a larger value of the effective inductance as equation 3 clarifies [3].

Beyond that, fig. 7a depicts that with rising control current the effective inductance decreases. Complementary to the previous setup a comparatively small control current of 0.46 A (first dark red line) already has an noticeable impact on the effective inductance. Furthermore, the effective inductance decreases right from the start and does not remain constant with increasing control current. The decline of the measurement curves kind of linearize until the point of sudden saturation is reached. Following from this, the setup should be taken into consideration for realizations of the concept, especially concerning a required closed loop control, which would be needed to adjust the inductance based on the requirements of the load side. Additionally, the typical sharp decline because of the saturation of the core happens with a higher load current and with a smaller negative slope. The absolute drop of the effective inductance from the point of sudden saturation lowers with rising control current as well. Finally, there is an intersection area visible, too. It can been seen that with rising control currents the curves intersect at a lower effective inductance and with a slightly higher load current, which supports the previous assumptions.

In order to verify the earlier described effect of the ratio of both magnetic fields and to show that larger, reasonable values of effective inductance can be reached the number of turns of the load core was increased. The results are presented in fig. 8a.

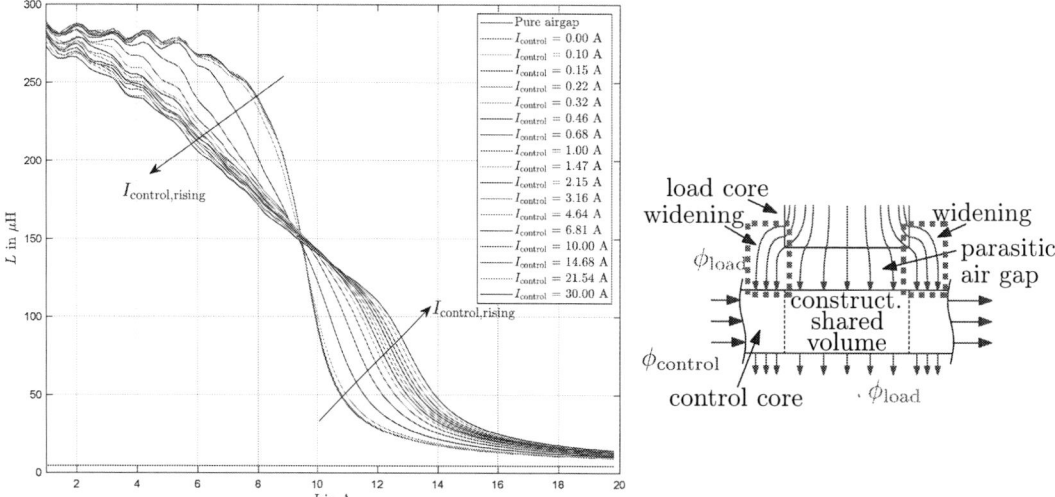

(a) Dependency of the effective inductance on ratio of magnetic fields   (b) Widening of the air gap

Fig. 8: Dependency of inductance on higher ratio of magnetic fields and resulting widening of air gap

According to the results from previous examination the curves intersect at a significantly lower load current. As observed in the examination with a lower number of turns, the higher the control current the higher is the needed load current to intersect the respective curve with the initial one and the lower is the effective inductance. The effective inductance of the pure air gap was added to the figure as reference like in the examination before. It is observable that the saturated inductances does not have the same values as the one with pure airgap or with a lower number of turns. An explanation could be found in the widening of the air gap of the load core following from the increase of number of turns. In contrast to the pure air gap the magnetic flux can permeate more permeable material of the control core, because of the widening, which is symbolized in fig. 8b. This leads to larger values of the effective inductance.

Corresponding to the slope of the decrease of the effective inductance this setup seems to be promising showing less variation in the inductance-value with small changes of the load current. Nevertheless, the effective inductance could be varied for this example at a load current of 6 A from 270 µH to 210 µH, giving a large range of different values realizable.

Impact of the control current on $B - H$ characteristic

It was anticipated that the shared volume would behave like a variable air gap. To verify that, the magnetic flux density is plotted against the load current. To show the basic effect on the $B - H$ characteristic, the setup and the according results of the measurement presented in fig. 3 were used. The magnetic field strength could be used instead of load current but regarding equation (1) the number of turns and the effective magnetic path length are constant and therefore only a proportionality factor would scale the measurement curves shown in fig. 9.

This shows that with an increase in control current the $B - H$ characteristics starts to flatten. The slope of the curves decreases with rising control currents for lower load currents. Thus, a higher magnetic field strength is aquired to reach the same magnetic flux density. Moreover, it can be seen for further rising load currents, that the slope of the curves start to behave contrary, because of the saturation of the core. From this point on, the slope approximates to the vacuum permeability. Furthermore, with increasing control current the curvature of the $B - H$ characteristics decreases. Consistent to the results presented in fig. 3 the first noticeable influence can be observed for a control current of 4.64 A.

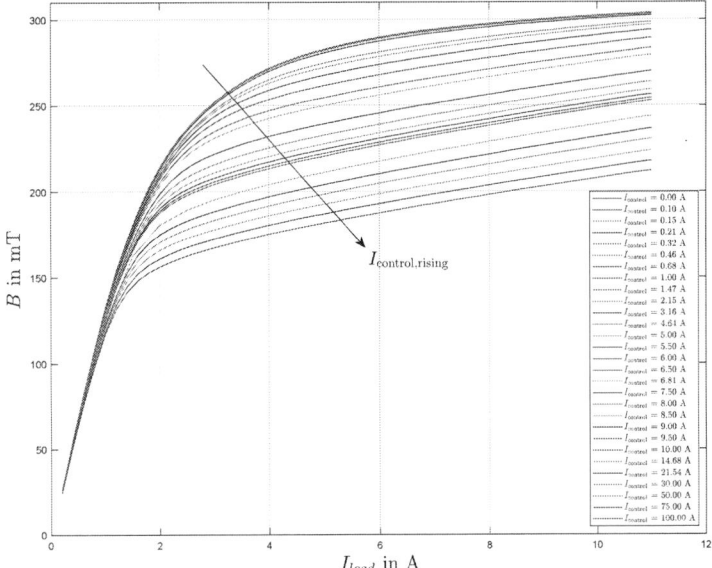

Fig. 9: *B-H* characteristics for an increase in control current

Conclusion and prospective

This paper presents a concept for an adjustable inductance by the use of a control current. A decrease of the effective inductance with an increase in control current was clearly observable in the measurement curves. As expected, the corresponding $B - H$ characteristics flatten. It was outlined that the ratio of the magnet fields of load and control core are essential for the intersection area. Furthermore, degrees of freedom were explained and therefore possibilities of optimization were derived concerning the efficiency of the system.

In future investigations the magnetic back coupling on the control circuit, the usage of several control cores at once, AC current on the load side and different materials will be analysed. Generally, a closed loop control needs to be implemented for realization, which adjusts the control current. A mathematical approach to calculate the influence on the effective inductance should be a target to derive control parameters. Otherwise, the measured change of the current in the inductivity and the measured voltage must be used as feedback. Additionally, the results of the measurements should be verified with another method to examine systematical and statistical errors using the Choketester measurement.

References

[1] E. S. Tez : "The Parametric Transformer". PhD Thesis, Loughborough University of Technology, 1977

[2] F. Kümmel: "Regel-Transduktoren. Theorie und Anwendungen in der Regelungstechnik". Springer, 1961

[3] M. Albach: "Induktivitäten in der Leistungselektronik. Spulen, Trafos und ihre parasitären Eigenschaften". Springer Vieweg, 2017

[4] J. P. Karst, K. Hoffmann, "Design Rules for Transductor based Gate Drives". Power Conversion and Intelligent Motion Conference-PCIM , 2005

[5] J. P. Karst, K. Hoffmann, "Transductor based high speed gate drive". 2004 IEEE 35th Annual Power Electronics Specialists Conference, 2004

[6] J. P. Karst, K. Hoffmann, "High Speed Complementary Gate Drives Utilising a Single Planar Transformer". Power Conversion and Intelligent Motion Conference-PCIM, 2004

[7] S. Fahlbusch: "Messsystem zur Charakterisierung von Leistungsspulen auf Basis eines 15-Level-Siliziumkarbid-Wechselrichters". PhD Thesis, Helmut Schmidt University, 2020

[8] D. Magel: "Implementation, Analysis and Comparison of Different Prototypes of Current Controlled Inductors for Power Electronic Applications". Bachelor Thesis, Helmut Schmidt University, 2022

[9] S. Brandt, M. Meissner, N. Polap, G. Schierle, K. F. Hoffmann: "A Survey on Adjustable Inductances for Power Electronic Circuits". PCIM Europe digital days 2022, 2022

[10] H. Kreis: "Pulsed Inductance Measurement on Magnetic Components from 0.1A to 10kA.", in Bodos Power Systems, Nov 2021

[11] ed-k: "Power Choke Tester DPG10 B - Serie & DPG 20. Manual". 01/2020

A Field Programmable and Dynamic Configurable Power Electronic Converter Concept

Bjarte Hoff
UiT The Arctic University of Norway
Lodve Langes gt. 2
Narvik, Norway
Phone: +47 76966318
Email: bjarte.hoff@uit.no
URL: http://www.uit.no

Keywords

≪Converter circuit≫, ≪Emerging topology≫, ≪Emerging technology≫, ≪System integration≫, ≪Synthesis≫.

Abstract

This paper proposes the concept of a field programmable and configurable power electronic converter together with a new synthesis method. By considering the converter as a pack of resources, the converter topology and configuration of inputs and outputs can be programmed in the field through, utilising the available device resources.

Introduction

Power electronic converters have been widely adopted for most areas requiring efficient control of power, including renewable energy, electric transport, industrial drives, and numerous household appliances and power supplies. So far, these power converters have mainly been designed as a fixed topology where only the firmware or control software can be upgraded. There has been a trend to create more application specific converters, for instance can converters can be integrated inside the electrical machine housing [1], but it is also possible to go in the opposite direction and make the converter more universal.

In many cases, several converters are wired together in the same system for form a complete solution, where electric transport is a good example. For an electric ship, numerous converters are used to handle loads like propulsion, charging, hydraulic pumps and on-board power supply. Not all converters are used simultaneously, such as propulsion and charging. It is an obvious possibility to allow the same power electronic devices to serve multiple purposes. Although a converter could be built to have several predefined configurations, such customised solution could limit the application if the requirements are changing.

An upgrade of the solution, in this example the ship, could result in replacing the converter to include new functionality. If a power electronic converter could be seen as a generic field programmable and configurable pack of resources, new functionality can be programmed within the limitations of available resources. By making it dynamic configurable, the utilisation of those resources can be increased.

For a converter to be field programmable and dynamic configurable, one can imagine a power converter along the same idea as the field programmable gate array (FPGA). An FPGA uses an synthesis procedure to realise programmable digital electronics. While an FPGA is less efficient than an application-specific integrated circuit (ASIC), its programmability and off-the-shelf versatility makes it the preferred choice in an increasing number of applications.

Methods for synthesis of converter in the literature are used as design tools, mainly to explore different topologies before a final topology is selected for the physical implementation. It needs to be evaluated if existing methods can be expanded in to a programming tool, where the structure in the synthesis process can be realised in hardware. Input requirements to the synthesis process is also important, as it will decide what type of converter structures that can be synthesised.

In [2], Erickson describes the synthesis process based on the DC conversion ratio, absence of pulsating input or output currents, the number of reactive elements, and the number of switching elements used. The method is based on a general state-space description of RLC networks, where there is a finite number of configuration options once the number of reactive elements are known. The advantage of Erickson's method in [2] is that all possible topologies is considered, but the method is based on manually inspecting all feasible configurations, where the value of the reactive elements and ripple current requirements are not taken into account. A state-space approach is also used in [3] to realise arbitrary converter configurations and [4] expands with a synthesis method, but relies in intuition to select the right topology among the feasible alternatives.

Synthesis through combination of basic converter blocks are explored in [5], although design objectives are limited to voltage and current conversion ratio and single input single output configurations. The concept of using basic converter cells is also applied to three-port DC-DC converters in [6], where an iterative computer program reduces the time spent on manual examination. Flux balance equations are used to synthesise converter topologies in [7, 8] based on a chosen voltage conversion equation. Instead of examining a pool of topologies, [9] proposes a method based on the voltage conversion equation that is decomposed before circuits are synthesised.

Common for the synthesis methods reported in literature, is the ambition to use the synthesis process to realise new converter topologies that is unlikely to be discovered by manual circuit manipulation and combination of known basic converter types.

Power converter with configurable structures have been presented to increase voltage range of a series resonant DC-DC converter in [10], where a DC-DC converter has four predefined configurable operation states to increase its allowed voltage range. The voltage range for a LLC resonant converter is increased with a configurable structure in [11], while a non-isolated configurable bidirectional DC-DC converter is used to increase voltage range for interfacing retired batteries from electric vehicles in [12]. Common for [10, 11, 12] is that a configurable structure is used to increasing the voltage range, not fundamentally change the converter topology.

This paper proposes the idea of a field programmable and dynamic configurable converter (FPCC) as an universal power control device by combining synthesis with a configurable converter design, allowing the converter to be seen as a programmable pack of active and passive resources.

A new synthesis method is proposed for the FPCC together with a configurable structure, where the objective is to realise a converter based on a known physical structure and performance requirements. The ambition is not to synthesise new converter topologies, but to realise a specified functionality by mapping circuit configurations into known predefined structures.

Concept of Configurable Converter

The FPCC is considered to be a pack or collection of active switches and passive components, such as a selection of transistors, diodes, inductors and capacitors. A set of power IO's (inputs and outputs) will interface the converter with electrical sources and loads, while control IO's will be the interface to controlling the switches for whatever topology that is realised. The converter topology and internal connections can be realised based on a synthesis process, where configuration switches are used to realise the desired topology through a set of configuration inputs. Power switches can also be used as part of the configuration in addition to the switching action to increase the flexibility in configurations. An overview of how such converter can be arranged is shown in Fig. 1.

The internal construction of the converter and the way different active and passive elements are connected will act as constraints in the synthesis process, together with the user requirements. To design such

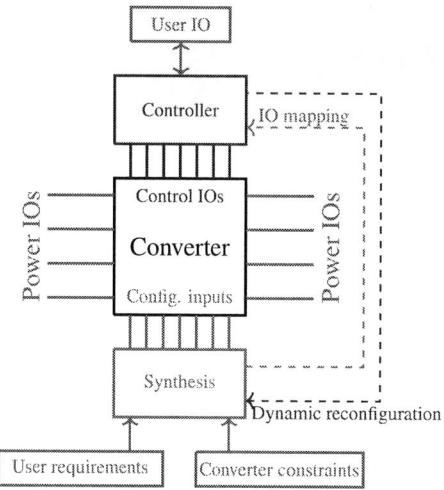

Fig. 1: Configurable converter

converter structure and include the right number of active and passive elements, their parameter values, and interconnection possibilities is an optimisation and synthesis process in its own. There is also the question if each sub-component in the converter can be configured individually, or if there should be basic cells with predefined groups of components. This internal structure is critical for the converter to achieve the versatility, where the greatest (or required) number of topologies can be realised given the available resources. As with FPGAs, one can imagine the converters to come in different standard sizes with a given number of internal resources and power levels.

Applications

While an example of configurable converter has been presented in [10] to increase voltage range, the concept can be taken further. This section presents two examples where such converter could be applied.

Electric Vessels

Electric vessels uses numerous power converters for propulsion, charging, internal power distribution and other vessel specific loads. Not all converters are used simultaneously, where several functions could be combined into one configurable converter as shown in Fig. 2. This is especially relevant for smaller vessels, where space is limited. The concept has been applied to electric vehicles by combining charging with the electric drive in [13], but reconfiguration was made outside the converter.

If spare resources are available in the converter, redundancy for component failure could also be implemented on small single motor vessels by temporary reconfigure to another optimal or a sub-optimal topology until repairs can be made.

For retrofit application converting from diesel to batteries or fuel cells, it would be an advantage if the converter was one integrated box, where all sources and loads were connected and functionality of the power electronics could be programmed during commissioning. This will not only reduce the number of individual components, but also reduce the wiring on the vessel, reduce the need for a complex communication network between converters and the integration of those in energy management systems.

Mobile Fuel Cell based Electricity Supply

If hydrogen is used as energy carrier on an off-grid construction site, it might not be known if the machines would require AC or DC, and at what voltage level. Such mobile generation units would be moved from site to site and would benefit from a versatile design in terms of electricity output. In that case, a configurable converter can change its topology according to the required voltage, current and frequency, making a versatile electricity source for the site. One could also imagine supplying different

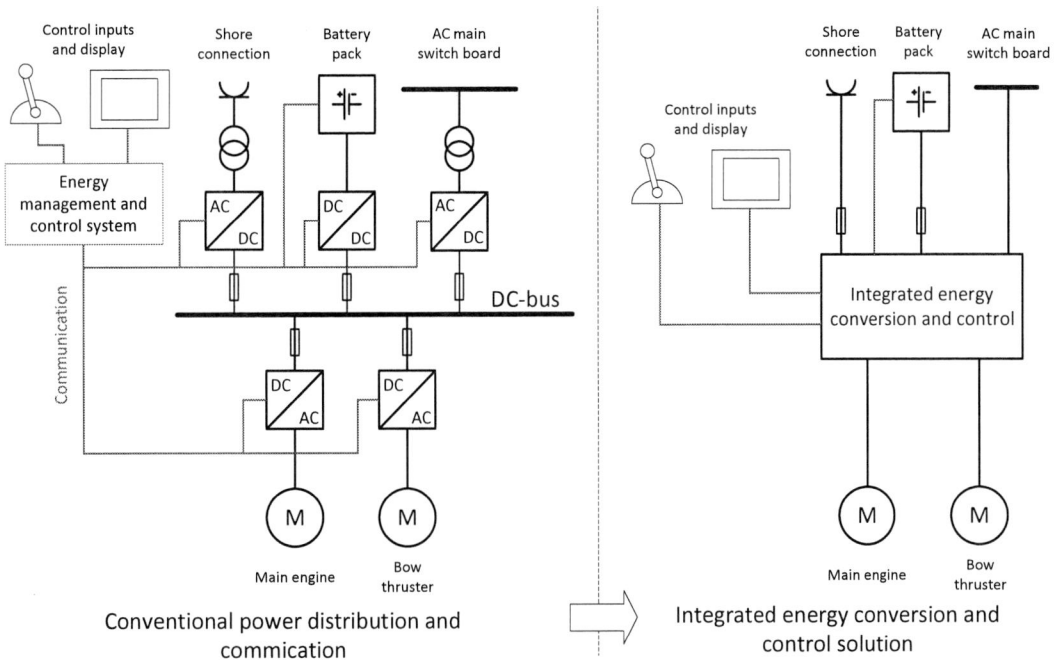

Fig. 2: Integrated power conversion in electric vessels

types of electricity supply simultaneous if required, or switches could be paralleled to increase current in high demand periods.

Work towards realisation

To realise this field programmable and dynamically re-configurable converter, several advanced in research is required.

Some key challenges are considered to be:

1. Optimal internal design of the converter and configuration options

2. Optimal choice of parameters for the internal components

3. Synthesis method taking constraints regarding available components and configuration options into account

4. Practical realisation of the configurable converter

5. Develop control algorithms to support dynamic configurable converters

While the realisation of the proposed converter would initially be by wire or printed circuit board tracks, one could imagine a future of integration on a semiconductor level inspired by the potential in additive manufacture process [14]. At this stage, it is not known if the realisation and the advantage of field programmable converter will make a this cost competitive solution, although the author's opinion is that the concept is worth exploring. Each of these five key challenges require a significant research effort and it is out of the scope of this paper to address all these in detail. To illustrate the feasibility of the concept, a simplified configurable structure and new synthesis methods is proposed in the following two subsections.

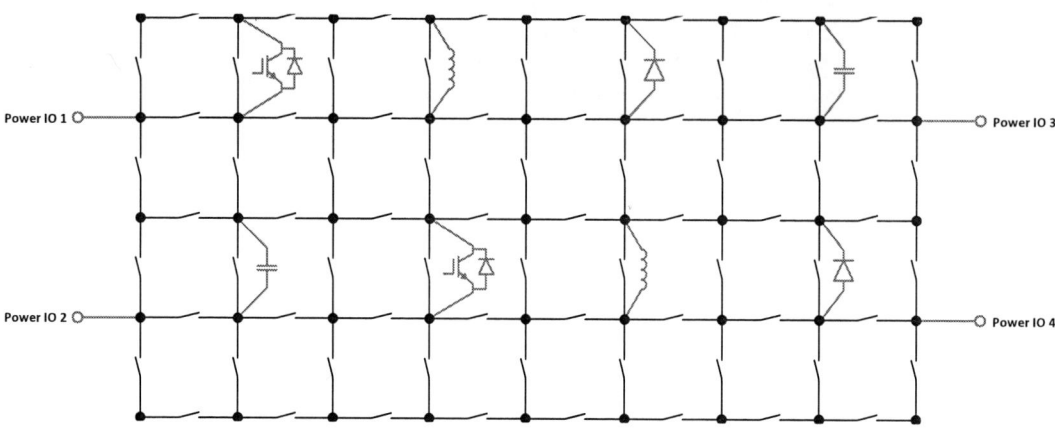

Fig. 3: Configurable structure with power electronic components

Configurable structure

A configurable structure will consist of a given number of active switching elements and passive components, connected by a grid of configuration switches. Nodes in this grid is used as connection points for active and passive elements and power IO's. Note that the structure may contain components of different sizes for increased flexibility.

To demonstrate the concept, a small 2 dimensional structure with a few selected components are shown in Fig. 3, although a three dimensional structure is also possible. How to optimally arrange this structure, the number of configuration switches, and where to connect active and passive element is an area where more research is required and out of the scope in this paper.

Synthesis

The proposed synthesis procedure consist of five steps. If the intended application has different operating modes, each mode can be synthesised such that the structure can alternate between then as required.

Step 1 - Set requirements

The requirements for the converter has to be defined with the functionality and performance parameters relevant for the solution. As a minimum one should specify which inputs and outputs sources and loads are connected to, including the properties of those. That will include if there is a voltage or current source, conversion ratio, voltage or current limits, and how many quadrants the converter should work within. The structure can include several independent converter functions.

In addition, parameters like voltage and current ripple, efficiency and harmonic distortion can be added to later choose between several feasible topologies.

Step 2 - Identify all feasible configurations

By cycle through all configurations options, those that does not connect to the assigned power IO's, provide no path from input to output, or violate basic electric constraints such a short circuit is ruled out in this step.

Step 3 - Mapping of known structures

The synthesis process has a predefined list of known converter topologies, or basic cells that is identified and mapped to all the feasible configurations in Step 2. Each of those predefined topologies have a known set of parameters like voltage conversion range, ripple current and voltages, and efficiency, taking into account known parameters for the individual components. Depending on the requirements and available resources, several converters can be mapped within the configurable structure as indicated in Fig. 4.

A simple example of synthesis of a step-down converter based on the structure in Fig. 3 is shown in Fig. 5 and Fig. 6.

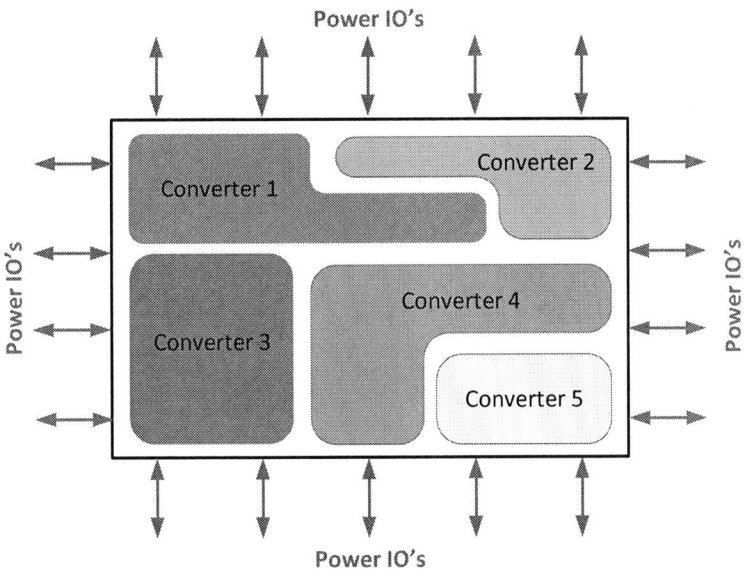

Fig. 4: Multiple converters mapped in one FPCC structure

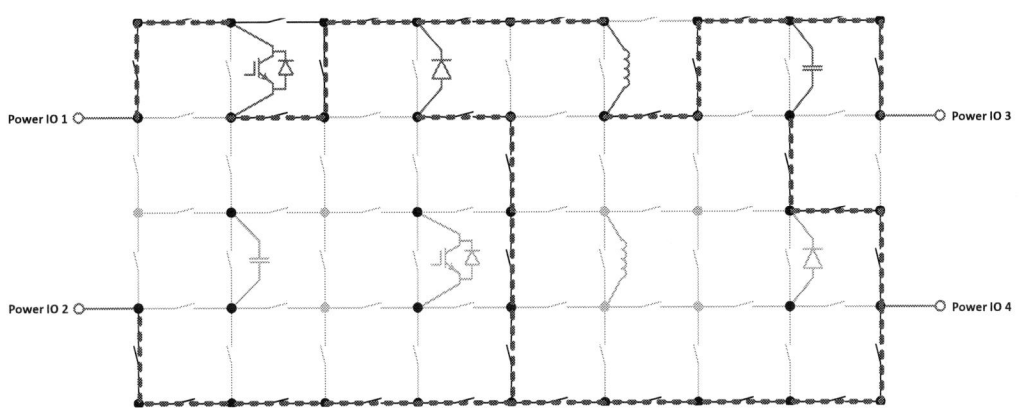

Fig. 5: Structure configured as an asynchronous Buck converter

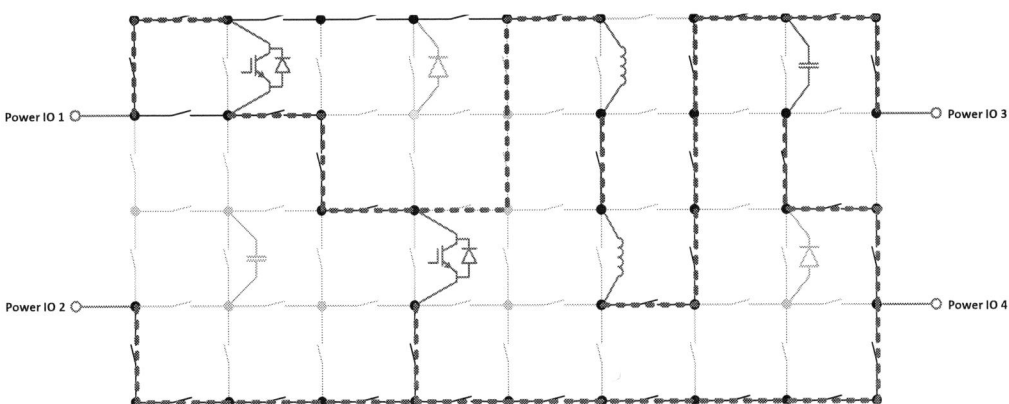

Fig. 6: Structure mapped and configured as s synchronous Buck converter

From all configurations that connects the four Power IO's, two of them was mapped as asynchronous in Fig. 5 and synchronous Buck converter in Fig. 6, the latter one with both inductors in series. Both will perform as a step-down converter, but will have different performance in terms of efficiency and ripple. Note that this example is simplified for illustrative purposes and multiple converter structures are expected to be mapped within the same configuration, together with more Power IO's.

Step 4 - Performance evaluation

In this step, all mapped converter structures from the previous step is evaluated and compared to the requirements set in step 1. Based on this, the best configuration that achieves the desired converter functions can be selected. That might include several simultaneous converters in the same configuration, and several configuration options if the application is specified to have several operation modes.

Step 5 - Implementation

The final step is to implement the chosen configuration from step 5 by programming the configurable structure. This includes mapping the control algorithm to the active switches in use from step 5.

Practical limitations

With an ideal setup as shown in Fig. 3, a full matrix of ideal switches can configure the converter in any feasible configuration. In a practical realisation of the concept, those configuration switches will result in losses, depending on the type of switch. Semiconductor switches will introduce a power loss due to their on-state resistance, in addition to an additional constraint if they are not bidirectional devices. Both losses on the configurable structure itself and limitations in the configuration switches should be part of the synthesis process, making it more complex than the ideal case. The use of transistors as configuration switches will also require isolated power supplies to drive the gate-source voltage for MOSFET's or gate-emitter for IGBT's, complicates the control circuitry for the converter.

If reconfiguration is expected to occur seldom with low timing requirements in the reconfiguration event, one could realise a structure using mechanical relays. The on-state resistance will be lower, but there will be some power loss in the relay coils used to activate the switch. Efficiency can be improved if bi-stable relays is used, which does not require the coil to be kept energised for the relay to stay closed. Depending on the power level, mechanical relays can result in a large unit with complex wiring, unless it can be integrated on a circuit board together with the power electronic components. On the other hand, relay coils are galvanic isolated from the relay switch, removing the need for isolated power supplies in the control circuitry.

In any practical realisation, there will be advantageous to reduce the number of configuration switches as much as possible while maintaining necessary configurability of the structure. This is highlighted in the key challenge number 1 in the beginning of this section and something that as to be addressed for this to be a commercial viable solution.

Conclusion

The concept of a field programmable and dynamic configurable converter (FPCC) has been suggested, together with two application examples and a new synthesis method to realise the potential. Such converters could never compete with a single application specific converter, but fundamentally change how power electronics is implemented in a system perspective that normally would require multiple power converters. Its programmability can also provide component redundancy and versatility with changing requirements.

There is fundamental research that remains to realise such converter and this paper does not attempt to solve all of these, but merely introduce the concept to inspire future research.

References

[1] T. M. Jahns and B. Sarlioglu, "The incredible shrinking motor drive: Accelerating the transition to integrated motor drives," *IEEE Power Electronics Magazine*, vol. 7, no. 3, pp. 18–27, 2020.

[2] R. W. Erickson, "Synthesis of switched-mode converters," in *1983 IEEE Power Electronics Specialists Conference*, 1983, pp. 9–22.

[3] J. Burdio and A. Martinez, "A unified discrete-time state-space model for switching converters," *IEEE Transactions on Power Electronics*, vol. 10, no. 6, pp. 694–707, 1995.

[4] J. Burdio, A. Martinez, and J. Garcia, "A synthesis method for generating switched electronic converters," *IEEE Transactions on Power Electronics*, vol. 13, no. 6, pp. 1056–1068, 1998.

[5] A. Purwadi, K. A. Nugroho, A. Rizqiawan, and P. A. Dahono, "A new approach to synthesis of static power converters," in *2009 International Conference on Electrical Engineering and Informatics*, vol. 02, 2009, pp. 627–633.

[6] G. Chen, Z. Jin, Y. Liu, Y. Hu, J. Zhang, and X. Qing, "Programmable topology derivation and analysis of integrated three-port dc–dc converters with reduced switches for low-cost applications," *IEEE Transactions on Industrial Electronics*, vol. 66, no. 9, pp. 6649–6660, 2019.

[7] S. S. Nag, R. Panigrahi, S. K. Mishra, A. Joshi, K. D. T. Ngo, and S. Mandal, "A theory to synthesize nonisolated dc–dc converters using flux balance principle," *IEEE Transactions on Power Electronics*, vol. 34, no. 11, pp. 10 910–10 924, 2019.

[8] R. Panigrahi, S. K. Mishra, and A. Joshi, "Synthesizing a family of converters for a specified conversion ratio using flux balance principle," *IEEE Transactions on Industrial Electronics*, vol. 68, no. 5, pp. 3854–3864, 2021.

[9] T. S. Ambagahawaththa, D. Nayanasiri, and A. Pasqual, "A four-step method to synthesize a dc–dc converter for multi-inductor realizable arbitrary voltage conversion ratio," *IEEE Transactions on Industrial Electronics*, vol. 69, no. 6, pp. 5594–5603, 2022.

[10] Y. Shen, H. Wang, A. Al-Durra, Z. Qin, and F. Blaabjerg, "A structure-reconfigurable series resonant dc–dc converter with wide-input and configurable-output voltages," *IEEE Transactions on Industry Applications*, vol. 55, no. 2, pp. 1752–1764, 2019.

[11] Y. Zuo, X. Pan, J. Zhu, J. Ye, and Y. Wang, "A bidirectional isolated llc resonant converter with configurable structure for wide output voltage range applications," in *2020 IEEE 9th International Power Electronics and Motion Control Conference (IPEMC2020-ECCE Asia)*, 2020, pp. 1716–1721.

[12] J. Teng, P. Shen, B. Liu, and S. Chen, "Circuit configurable bidirectional dc-dc converter for retired batteries," *IEEE Access*, vol. 9, pp. 156 187–156 199, 2021.

[13] I. Subotic, N. Bodo, and E. Levi, "An ev drive-train with integrated fast charging capability," *IEEE Transactions on Power Electronics*, vol. 31, no. 2, pp. 1461–1471, 2016.

[14] L. Lopera, R. Rodriguez, M. Yakout, M. Elbestawi, and A. Emadi, "Current and potential applications of additive manufacturing for power electronics," *IEEE Open Journal of Power Electronics*, vol. 2, pp. 33–42, 2021.

DAB converter discrete ADRC control into real-time CHIL simulation of a MVDC/LVDC power grid

Alessio Clerici, Riccardo Chiumeo, Diego Raggini, Alessandro Veroni
Ricerca Sul Sistema Energetico - RSE S.p.A.
Via Rubattino 54, 20134
Milan, Italy
Tel.: +39 02 399 21
E-Mail: alessio.clerici@rse-web.it
URL: https://www.rse-web.it

Acknowledgements

This work has been financed by the Research Fund for the Italian Electrical System in compliance with the Decree of April 16, 2018.

Keywords

«Real-time simulation», «Distribution of electrical energy», «Microcontrollers», «Converter control», «Dual Active Bridge (DAB) DC-DC converter»

Abstract

Active Disturbance Rejection Control (ADRC) is implemented into commercial microcontroller to drive a Dual Active Bridge converter fed by a Medium Voltage Direct Current network. A real-time Control Hardware In the Loop system is successfully implemented and performance meets conventional off-line simulations results. ADRC proves to be a robust control strategy for a distribution network converters and Hardware In the Loop a valid technique for power electronics tests.

Introduction

Active Disturbance Rejection Control (ADRC) is currently developed for high-performance and high-dynamics demanding applications, like Permanent Magnet Synchronous Motors (PMSM) drivers and speed controllers [1] [2]; in such cases, ADRC stability margin is approximately the same as traditional PI controller, however, ADRC is superior in rapidity and overshoot performance [3].

According to these encouraging results, this work aims to implement ADRC in a Dual Active Bridge (DAB) power converter by means of real-time Control Hardware In the Loop (CHIL) simulations.

CHIL testing is widely used in automotive, aerospace, and robotics, while usage in power electronics is still in early stage [4].

The growing need for advanced grid support functions and the study of innovative distribution networks and power converters can be a first step to increase the development of such test technique in this field. ADRC is based on optimal control strategy: the regulator, through an Extended State Observer (ESO), can estimate system variables, allowing disturbance rejection.

Professor Han theorized ADRC [5] starting from the PID controller analysis; in the last decade, ADRC has seen growing development both theoretical [6] and practical [7] [8] [9] [10].

The control is intrinsically versatile since it combines simple feasibility of PID with modern approach based on model analysis and state observers.

Compared to PID control, ADRC can achieve better setpoint tracking capability and superior disturbance rejection.

ESO operates in parallel to the controlled process; it is mathematically modeled on the process itself, keeping the characteristic matrices and the system order.

The observer acquires the input and output values of the controlled process and calculates a disturbance estimate.

Equation (1) describes first order state observer algorithm in Laplace domain.

$$\begin{cases} \hat{x}_1 = \dfrac{l_1 \cdot \left(y - \hat{x}_1\right) + \hat{f} + b_0 \cdot u}{s} \\[3mm] \hat{f} = \dfrac{l_2 \cdot \left(y - \hat{x}_1\right)}{s} \end{cases} \qquad (1)$$

Equation (2) briefly models the concept of ADRC technique; $\dot{y}(t)$ represents the time derivative of process output. By definition of time derivative, it is possible to rewrite $\dot{y}(t)$ as the difference between reference setpoint $q(t)$ and system output $y(t)$ in time domain; an arbitrary constant K_A can thus modulate the control speed response.

$$\dot{y}(t) = u_0(t) = K_A \cdot \left[q(t) - y(t)\right] \qquad (2)$$

Coefficients b_0, K_A, l_1 and l_2 have been derived from the theoretical analysis of [6].
Fig. 1 shows whole ADRC control structure to be implemented in digital simulations, with a detail of ESO transfer function.

(a)   (b)

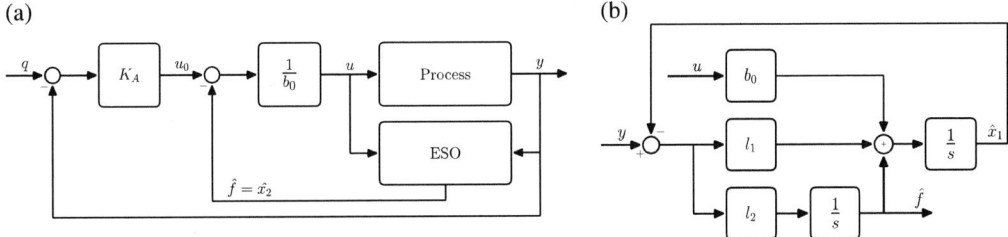

Fig. 1: ADRC block scheme. (a) Whole plant structure. (b) ESO transfer function.

ADRC implementation into a digital, general-purpose microcontroller is the aim of the present article. Assuming T_s as sampling period, ESO transfer function in Laplace domain has been discretized by using the bilinear transformation (or Tustin's method) [11], whose general formulation is shown in (3):

$$y(n) = y(n-1) + \frac{T_s}{2} \cdot \left[x(n) + x(n-1)\right] \qquad (3)$$

Test bench for ADRC analysis is on voltage regulation of a Dual Active Bridge (DAB) converter. DAB converter operates into a Control Hardware in the Loop (CHIL) simulation, where the Medium Voltage Direct Current (MVDC) grid of Fig. 2 is simulated.

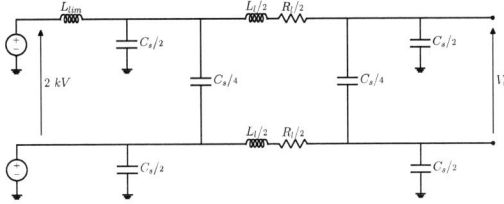

Fig. 2: MVDC grid model used in CHIL simulations.

MVDC grid parameters are shown in Table I:

Table I: MVDC grid parameters

Parameter	Quantity
Line length (l)	1 km
Line resistance (R_l)	0.0283 Ohm
Line inductance (L_l)	0.4 mH
Upstream inductance (L_{lim})	0.4 mH
Service capacitance (C_s)	0.92 µF

ADRC validation in CHIL context is then composed by two steps:

1) respect of hard real-time constraints and synchronization;

2) system response comparison with off-line implementation of the same ADRC control.

In next paragraph, discretized ADRC algorithm implementation strategy will be presented, with great focus on time constraints imposed by the hardware. The target is to maximize the system performances to guarantee an effective exchange of data with the CHIL real-time simulator.

Later on, a summary of non-idealities introduced by the CHIL simulation will serve to define a correct off-line reference model to validate CHIL real-time simulation results.

Both off-line and CHIL simulations results comparison will be shown in the third paragraph: ADRC characteristics will be investigated through the response of a DAB converter.

Paper nomenclature is reported in Table II:

Table II: **Nomenclature**

Symbol	Description
ADC	Analog to Digital Converter
$ADRC$	Active Disturbance Rejection Control
b_0	ADRC rated value of input variable disturbance
$CHIL$	Control Hardware in the Loop
CPU	Central Processing Unit
d	Quantized signal
DAB	Dual Active bridge (converter)
$FPGA$	Field Programmable Gate Array
$IGBT$	Insulated Gate Bipolar Transistor
ISR	Interrupt Service Routine
K_A	ADRC static gain
l_1	Luenberger matrix first coefficient
l_2	Luenberger matrix second coefficient
$LVDC$	Low Voltage Direct Current
$MVDC$	Medium Voltage Direct Current
N	ADC n° of bits
OC	Output compare
PID	Proportional Integral Derivative (control)
$PMSM$	Permanent Magnet Synchronous Motors
PWM	Pulse-Width Modulation
q	System setpoint
$round(\ )$	Round-to-nearest mathematical function
SPS	Single Phase Shift
T_c	Control period
$T_{s\mu c}$	Microcontroller discretization period
T_{sCPU}	Real-time simulator CPU module discretization period
T_{sFPGA}	Real-time simulator FPGA module discretization period
T_φ	Time-delay between DAB H-bridges voltages
V_{in}	DAB input voltage
V_{out}	DAB output voltage
x	System state
y	System output
ΔV_{MAX}	Sampled voltage range

Timing and discretization of real-time algorithm

ADRC implementation in a physical microcontroller for real-time simulation implies a discretization of the control strategy. Facing the inevitable computational limitations, the chosen time-step must not interfere with system dynamics.

In a real-time system, control computation time, here called control cycle (or control period) T_c, cannot be greater than the microcontroller discretization period $T_{s\mu c}$. For the entire control cycle execution, this trivial condition must be always true:

$$T_c \leq T_{s\mu c} \tag{4}$$

Control is fed with inputs sampled at the beginning of the control cycle. In this way, $T_{s\mu c}$ also becomes the input sampling period and the output update rate.
Implementation strategy chosen in this work allows for an easy and effective structure, guaranteeing a constant $T_{s\mu c}$ period.
It makes use of the hardware resources of the available microcontroller (ST NUCLEO-F767ZI) and of the abstraction layer provided by the MicroPython programming language, already employed in previous CHIL setups [12], to keep the code simple.
The discretization period is tied to a high-precision hardware timer rising an interrupt to periodically update the control variables (the result of the computation of the ADRC algorithm).
The same timer also triggers ADRC code execution, allowing for a strict scheduling of the computation times. A high-level description of the process is shown in Fig. 3.

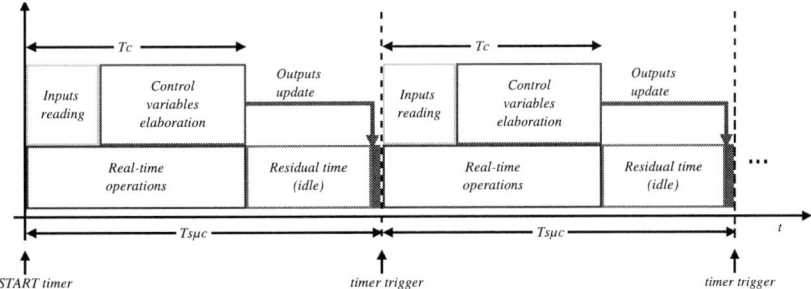

Fig. 3: Time diagram of the control cycle execution.

Notwithstanding a residual (idle) time, output update is realized at the end of $T_{s\mu c}$ to achieve a coherently discretized system. The new value of the control variable is then stored in the relevant output compare (OC) registers of the output channels. Those are natively managed by the hardware interrupts and thus run independently from the main execution cycle, effectively decoupling the two main tasks assigned to the microcontroller. This strategy allows for a stable and constant generation of all the control signals, minimizing the latency and jitter typical of software-based gate generation, leveraging the advanced ISR (Interrupt Service Routine) capabilities of the microcontroller. IGBT pulses are generated at the DAB switching frequency (5 kHz) and are therefore independent from $T_{s\mu c}$.
The selection of $T_{s\mu c}$ must then be related to the discretization periods of the various components of the CHIL simulation. In this work, an OPAL-RT simulator is used. The simulated model is divided between a CPU and an FPGA module, working according to the scheme of Fig. 4.

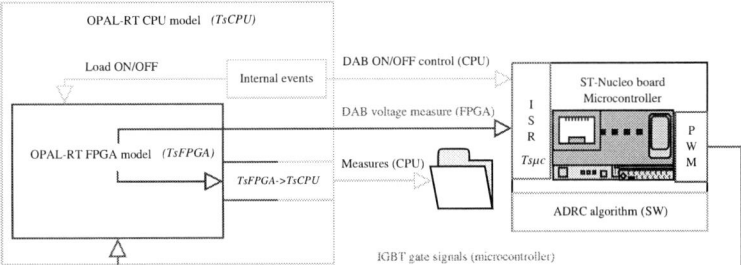

Fig. 4: CHIL simulation structure, composed by systems having different discretization periods: OPAL-RT CPU module (green), OPAL-RT FPGA module (blue) and ST-Nucleo microcontroller (orange). Code execution is scheduled by an ISR system (red), the PWM signals generation (brown) is managed by microcontroller hardware resources.

Besides ADRC sampling rate, OPAL-RT simulation of the network is then discretized as follows:

$$T_{sCPU} = 20\,\mu s$$
$$T_{sFPGA} = 1\,\mu s \qquad (5)$$

The ADRC regulator is used to control the DAB secondary voltage. This converter is then represented in the FPGA module, allowing for the real-time exchange of the gate signals and feedback measures between the machine and the implemented control. At each $T_{s\mu c}$ step, the microcontroller must sample the voltage value and compute a new reference for DAB control variable (time-delay between H-bridges voltages T_{φ}[1]). Considering the control characteristics of the DAB converter [13] and the typical execution time of the digital implementation of the ADRC algorithm, a 300 µs discretization period was selected.

$$T_{s\mu c} = 300\,\mu s \qquad (6)$$

The result is the coexistence of three distinct discrete systems, with different discretization intervals, together with a "low level hardware" layer, responsible for the scheduling of microcontroller operations and the generation of pulses to the gates of the IGBTs.

The signals exchange between the different sections, where necessary, implies a "conversion" in terms of discretization period; for example, in order to store key measures, some signals coming from the FPGA module are transmitted to the CPU module and therefore must be properly decimated; this is necessary because the system allows the saving of measurements through CPU module only.

Signal quantization aspects

Off-line simulations have been developed by means of ATPDraw graphical interface software, thus using ATP (Alternative Transient Program) as model solver.

The same model has been then implemented in MATLAB/Simulink environment, by using SimPower System Toolbox blocks. OPAL-RT also uses SimPower System Toolbox blocks to program FPGA module, thus, once the model is set, translation between off-line and real time simulation is quite straightforward.

Effective comparison between ADRC CHIL results and off-line reference simulation is possible by slightly tweaking the off-line implementation: this is necessary to reduce the effects of non-idealities arising when implementing the control strategy in a physical microcontroller.

In this specific case, a key effect is the quantization of the ADC (Analog-to-Digital Converter), performed by the microcontroller to sample the feedback measure.

The simplest model to uniform quantization introduced by the ADC is reported in (7). As mentioned above, it is then straightforward to implement the same quantizer block in the reference off-line simulations and in CHIL real time environments:

$$d = \frac{\Delta V_{MAX}}{2^N - 1} \cdot round\left(\frac{2^N - 1}{\Delta V_{MAX}} y \right) \qquad (7)$$

with: d is the quantized signal, y is the input signal from the sampling stage, N is the bit resolution of the hardware selected (in this case, 12-bit [14]).

Off-line vs. CHIL simulation comparison

The complexity of the implemented circuital model is not primarily relevant for the present work, as the focus is related to preliminary analysis of the ADRC on a microcontroller.

The model (Fig. 5), implemented both in the off-line and CHIL real-time simulation, consists of a 1 MW DAB converter which interconnects a MVDC grid to a LVDC one.

DAB is fed by MVDC at nominal 2000 V and must regulate LVDC voltage at nominal 750 V.

---

[1] In this work, for simplicity, the DAB converter is controlled in Single Phase Shift (SPS) mode [15].

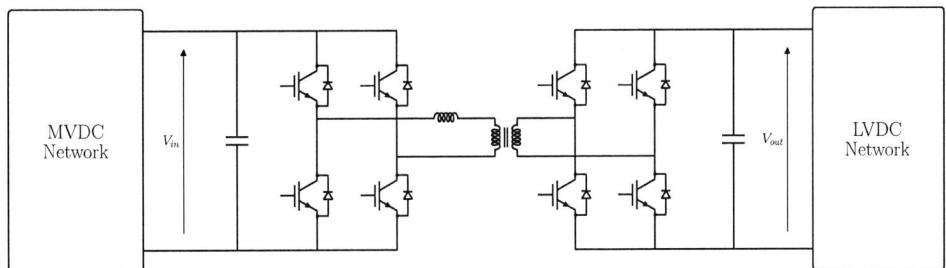

Fig. 5: Implemented model.

The simulation test setup has been arranged to verify the response of the ADRC regulator facing grid disturbances.

Events are organized as follows:

- t<0.1 s: the DAB regulates, at no-load, the voltage (750 V), no power transfer from MVDC to LVDC side;

- t=0.1s: a heavy resistive load is connected to DAB LVDC side (resistance R_{load} = 0.5625 Ω has been calculated to absorb 1 MW at nominal 750 V); DAB must keep LVDC voltage stable;

- t=0.5s: the resistive load is disconnected; DAB must handle such heavy load rejection keeping LVDC voltage stable.

Comparison between off-line reference simulations (ATPDraw and Simulink) and CHIL test (OPAL-RT) is shown in Fig. 6 and Fig. 7.

Following signals have been plot:

- DAB power profile;

- DAB output voltage V_{out};

- DAB control variable[2] T_{φ}.

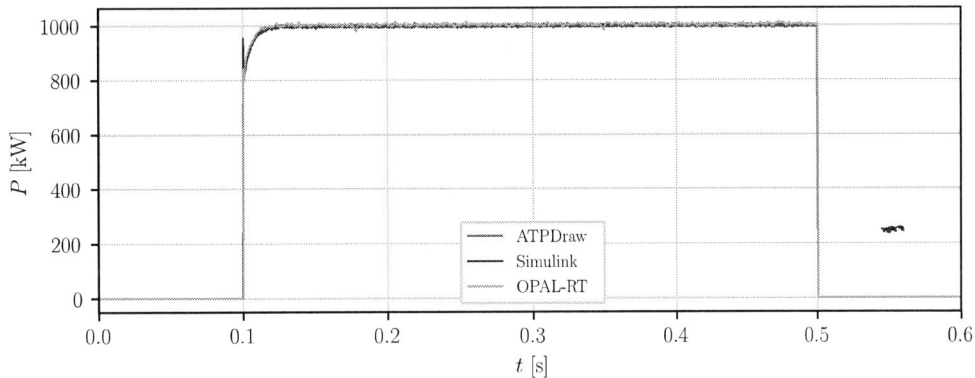

Fig. 6: DAB output power. Off-line reference simulations (ATPDraw and Simulink) and CHIL test (OPAL-RT).

---

[2] T_{φ} computed value is not readily available for storage in the real-time simulator as other variables, since the microcontroller outputs directly the gate signals. The value used here is the oscilloscope time-delay measure between PWM pulses applied to the converter.

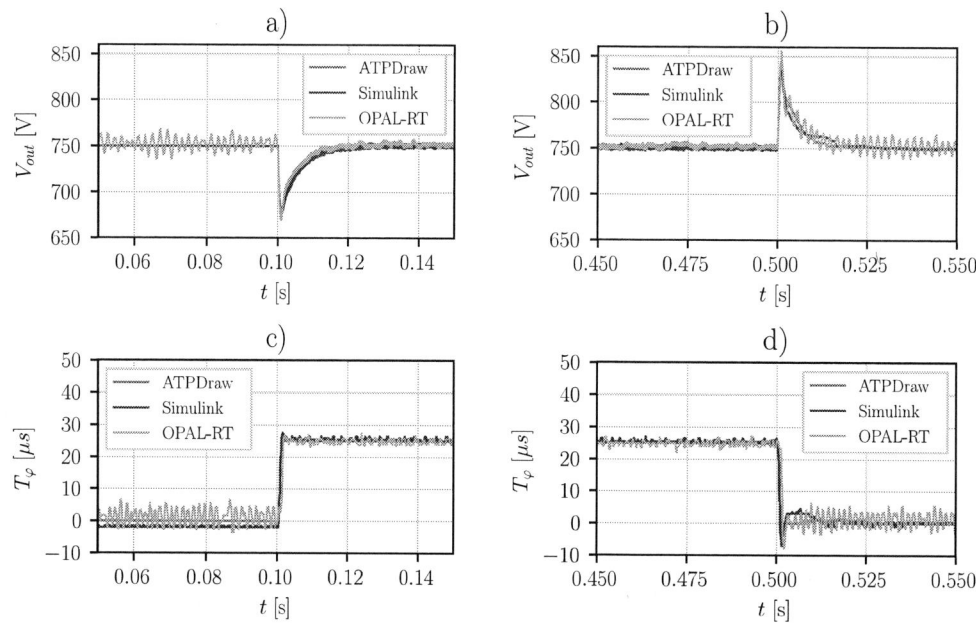

Fig. 7: DAB output voltage V_{out} and control variable T_φ at loading (a and c) and unloading (b and d). Off-line reference simulations (ATPDraw and Simulink) and CHIL test (OPAL-RT).

Both figures show a generally good level of correspondence between the results obtained off-line and real-time. The system response following each transient is correctly represented and shows no deviation across the various ADRC control algorithm implementations. V_{out} regulated voltage (Fig. 7 – a and b) shows the expected step variation at t = 0.10 s and t = 0.50 s, after which the control imposes an almost first order exponential dynamics to recover the nominal voltage.

The same can be said about T_φ converter control variable (Fig. 7 – c and d): it reaches the same steady state value corresponding to the nominal phase-shift of the converter. Transient deviations in rising and falling fronts are compatible with slight inaccuracies in measurement synchronization.

In Fig. 7 – c it can be observed a small steady-state error between ATPDraw and Simulink. This minor difference is given essentially by the behavior the DAB converter manifests at no-load: very small variations in control variable (T_φ) cause a consistent variation in controlled variable (V_{out}); as a result, in both cases output voltage is set properly at 750 V, with two slightly different values of T_φ in the two simulations. Such aspects can't be completely compensated, mainly depending in simulation environments asymmetries.

In CHIL results there is also a high-frequency component superimposed to control variable profile at no-load. Once again, despite OPAL-RT uses the same SimPower System blocks to implement the circuit, FPGA code translation and then compilation produces slight differences in the final model setup. Such effect is observable also in the regulated voltage profile, but it is greatly reduced at given load. Anyway, no significant impact of such oscillation can be seen in the general voltage regulation performance, the value being on average the same of reference off-line simulations.

The key phenomenon giving rise to this oscillation can be:

- the control variable higher instability in CHIL simulations can be due to the way T_φ is acquired. Since microcontroller computed value cannot be outputted, it has been acquired by means of an external oscilloscope capture setup. No-load values are then more difficult to measure and are therefore subject to lower accuracy;

- DAB constructive characteristic is such that at no-load small variations in control variable produce a very amplified effect over controlled variable. Although this characteristic is common to all simulations (off-line and CHIL), in real-time this effect is further amplified by the non-ideality of the generated gate pulses.

Besides minor differences between off-line simulations and real-time CHIL test, it was possible to:

- verify the correct discretization of the off-line simulations in order to make a direct and consistent comparison with the CHIL test;

- confirm with CHIL simulation the proper implementation of discrete ADRC algorithm into the microcontroller board.

In the last analysis, real-time timing constraints have been assessed during multiple runs of simulation cycles. According to Table III, timings of different control cycle computational phases are well below maximum allowed time. The detail of typical microcontroller cycle execution is reported here below:

Table III: Measured control cycle execution timing

	Input readings	**Control**	**Output conditioning**	**Idle**
Measure(μs)	*105*	*75*	*35*	*85*

Conclusion

This work addressed the feasible implementation on a general-purpose microcontroller board of an ADRC control for use in real-time CHIL simulations. ADRC has been used into an OPAL-RT simulator environment to regulate output voltage of a DAB converter and compensate heavy load variations.

Performances obtained from real-time CHIL simulations have been compared with the same configuration in off-line simulation environments (ATPDraw and Simulink).

The comparison is positive: the proposed control is sufficiently robust through the simulation in CHIL environment, and discretization process didn't introduce substantial changes to the algorithm studied in the ideal scheme of continuous time control.

This result represents a first step for the development of future real-time CHIL implementations of ADRC control for more complex circuit structures interconnecting different power electronics systems.

References

[1] Y. Gao, H. Zhao, K. Ma and X. Huo, "Control Design for Current Loop of PMSM Using A Modified Model-Compensation ADRC Controller," *2020 39th Chinese Control Conference (CCC)*, p. 2545–2550, 2020.

[2] F. Deng and Y. Guan, "PMSM Vector Control Based on Improved ADRC," *2018 IEEE International Conference of Intelligent Robotic and Control Engineering (IRCE)*, p. 154–158, 2018.

[3] L. E. I. Yang, X. U. Jing, M. Shuang, H. A. O. Qiang and S. Yue, "A Comparative Study of The First Order Linear ADRC and PI Controller in The Speed Control System of Permanent Magnet Synchronous Motor," *2020 IEEE 9th Data Driven Control and Learning Systems Conference (DDCLS)*, p. 498–503, 2020.

[4] H. Magnago, H. Figueira, O. Gagrica and D. Majstorovic, "HIL-based certification for converter controllers: Advantages, challenges and outlooks (Invited Paper)," *2021 21st International Symposium on Power Electronics (Ee)*, p. 1–6, 2021.

[5] J. Han, "From PID to Active Disturbance Rejection Control," *IEEE Transactions on Industrial Electronics,* vol. 56, no. 3, pp. 900-906, 2009.

[6] G. Herbst, "A Simulative Study on Active Distrubance Rejection Control (ADRC) a control tool for Practitioners," *Electronics,* vol. 2, no. 3, pp. 246-279, Aug. 2013.

[7] B. Sun; Z. Gao, "A DSP-Based Active Disturbance Rejection Control Design for a 1-kW H-Bridge DC–DC Power Converter," *IEEE Transactions on Industrial Electronics,* vol. 52, no. 5, pp. 1271-1277, October 2005. doi:10.1109/TIE.2005.855679..

[8] Y. X. Su; C. H. Zheng; B. Y. Duan, "Automatic disturbances rejection controller for precise motion control," *IEEE Transactions on Industrial Electronics,* vol. 52, no. 3, p. 814–823, 2005. doi:10.1109/TIE.2005.847583.

[9] Q. Zheng; Z. Gao, "On practical applications of active disturbance rejection control," in *In Proceedings of the 29th Chinese Control Conference*, 2010.

[10] R. Mandonski; M. Nowicki; P. Herman, "Application of active distrubance rejection controller to water supply system.," in *In Proceedings of the 33rd Chinese Control Conference*, Nanjing, China, 2014.

[11] A. Oppenheim, Discrete Time Signal Processing Third Edition, Pearson, Ed., 2010.

[12] A. Clerici, R. Chiumeo and C. Gandolfi, "Real Time Control Hardware in The Loop test of a novel MVDC solid-state breaker," in *Proc. 22nd European Conf. Power Electronics and Applications (EPE'20 ECCE Europe)*, 2020.

[13] A. Clerici, R. Chiumeo, D. Raggini and A. Veroni, "Digital strategy to compensate offset currents into DC-DC Dual Active Bridge converter simulations," *2021 IEEE Fourth International Conference on DC Microgrids (ICDCM)*, p. 1–6, July 2021.

[14] ST Microelectronics, "STM32F765xx - STM32F767xx - STM32F768Ax - STM32F769xx Datasheet - production data," DS11532 Rev 7, 2021.

[15] O. Yade, J.-Y. Gauthier, X. Lin-Shi, M. Gendrin and A. Zaoui, "Modulation strategy for a Dual Active Bridge converter using Model Predictive Control," in *2015 IEEE International Symposium on Predictive Control of Electrical Drives and Power Electronics (PRECEDE)*, Valparaiso, 2015.

SNNFT: Sequential Neural Network-Fuzzy Thermal Early Warning System for Lithium-ion Batteries

Marui Li[1], Chaoyu Dong[2,*], Yunfei Mu[1], Qian Xiao[1], Jingming Cao[1], Hongjie Jia[1]
[1]School of Electrical and Information Engineering, Tianjin University, Tianjin, China
[2]Nanyang Technological University, Singapore
Tel.:+86 / 18322694116
E-Mail: dong0120@e.ntu.edu.sg

Acknowledgments

This work was supported by the National Natural Science Foundation of China (No. 52107121), and the joint project of NSFC of China and EPSRC of UK (No. 52061635103 and EP/T021969/1).

Keywords

≪Battery≫, ≪Artificial intelligence ≫, ≪Deep Learning≫, ≪Optimization method≫, ≪Safety≫.

Abstract

Due to the promotion of electric vehicles and new energy sources, lithium-ion batteries have been widely used. However, temperature has a great influence on the performance and safety of lithium-ion batteries during operation. Therefore, it is very important to predict the temperature of lithium-ion batteries and implement thermal early warning. In order to solve this problem, this paper designed a Sequential neural network-fuzzy thermal early warning system (SNNFT). First, the SNNFT uses a denoising autoencoder to eliminate the noise in real-time measurement. Then it combines the long short-term memory network and the temporal convolutional network that can handle the time series problem well to realize the accurate prediction of the lithium-ion battery temperature. And the SNNFT applies interpretable adaptive network-based fuzzy inference system model to build thermal early warning system. Complete experiments are conducted to verify the reliability advantages.

Introduction

Lithium-ion batteries have been widely used as the important energy storage supplies. However, the status of the lithium-ion batteries need to be monitored during operation to ensure its safe operation. Temperature is critical to the safe use of lithium-ion batteries. High temperatures will accelerate the side reactions of lithium-ion batteries, so lithium-ion batteries cannot continue to operate at high temperatures. Therefore, it is very important to predict the temperature of the lithium-ion battery and make a correct thermal diagnosis, which can ensure that there is enough time to adjust the heat generation and heat dissipation of the lithium-ion battery to avoid battery damage.

Part of the thermal research is devoted to the development of a physical or chemical model that can accurately estimate the temperature changes of lithium-ion batteries [1, 2, 3]. However, model-based methods usually require complex parameter measurement experiments in advance to obtain some modeling parameters of lithium-ion batteries. As a black-box model, methods based on neural networks do not need to know any physical or chemical characteristics of lithium-ion batteries in advance, so they are gradually being applied in the temperature prediction of lithium-ion batteries [4, 5, 6]. The traditional artificial neural network method has been proved to be effective, but for the time series problem of temperature prediction, the long short-term memory network is currently applied more [7]. However, the applications of these models in temperature prediction are relatively basic and single. Different from the traditional single simple network model, this paper combines multiple advanced network models for processing time series.

In addition, adaptive network-based fuzzy inference system (ANFIS) is also a nonlinear system modeling tool that can reflect the complex relationship between input and output [8]. ANFIS can construct an input and output mapping in the form of fuzzy if-then rules [9]. Fuzzy inference system based on fuzzy if-then rules can simulate human knowledge and reasoning process, so it is more explanatory and reliable than neural networks. However, ANFIS has not been tested and applied in lithium-ion battery state estimation and thermal diagnosis. The application of ANFIS in the thermal diagnosis of lithium-ion batteries not only avoids the complexity of model establishment and the need for a large number of expert experience, but also is more interpretative and more acceptable than the neural network model.

This paper attempts to make some contributions and improvements to the current technology of temperature prediction and thermal diagnosis of lithium-ion batteries. The temperature prediction part and the thermal diagnosis part constitute the complete thermal early warning system established in this paper. The four main contributions are as follows:

1. Due to various reasons such as measurement errors and signal transmission errors in the measurement of lithium-ion battery status, noise often exists in the data obtained from real-time measurement. Therefore, an LSTM denoising autoencoder is designed to reduce the noise of the measured data.

2. A temporal convolution-recurrent network (TCRN) is then proposed to accurately predict the surface temperature changes of lithium-ion batteries. TCRN not only integrates long short-term memory network (LSTM) and temporal convolutional network (TCN) to combine the advantages of those two models, but also adds a third input to enhance important features. During the training process of TCRN, Bayesian optimization is applied to optimize and adjust the model parameters to improve its prediction ability.

3. An interpretable adaptive network-based fuzzy inference system model (ANFIS) is established to realize the thermal early warning of lithium-ion battery. ANFIS model based on fuzzy system and neural network has interpretable rules, which makes it more credible and easier to be used in engineering field. In the training process, unlike traditional ANFIS, this paper uses a swarm intelligence technology—particle swarm algorithm to optimize the premise parameters and consequence parameters of ANFIS, which further enhances the inference capability.

4. This paper combines TCN, LSTM, and ANFIS for the first time to form a complete sequential neural network-fuzzy thermal early warning system (SNNFT). SNNFT realizes multi-step ahead prediction of the surface temperature of lithium-ion batteries and conducts thermal diagnosis of lithium-ion batteries according to the prediction results to determine whether there is a danger of thermal instability.

Methods

Prediction of surface temperature of lithium-ion battery

Denoising autoencoder

Autoencoders (AE) usually consist of an encoder and a decoder. Firstly, the input vector is mapped into feature vector by weighting, which is represented as the process from input layer to hidden layer in neural network. This process is usually completed by the encoder. Then the decoder reconstructs the feature vector into the original input vector by reverse weighting, which is the process from the hidden layer to the output layer in the neural network [10]. Ideally, the input and output should be the same [11].

The denoising autoencoder (DAE) is based on the autoencoder. In the training process, add noise to the input vector, and then make the autoencoder learn to obtain the real input that has not been polluted by noise, that is, the DAE has the function of removing noise [12]. Its architecture is shown in Fig. 1. Encoders and decoders can usually be composed of convolutional networks, recurrent networks and other ways. Since the task here is to reduce the noise of the measurement data of lithium-ion batteries, the input is a sequence. Therefore, an LSTM DAE is constructed, which is implemented by the sequence data autoencoder of encoder-decoder LSTM architecture. Fig. 2 shows the LSTM DAE established by Keras in this paper.

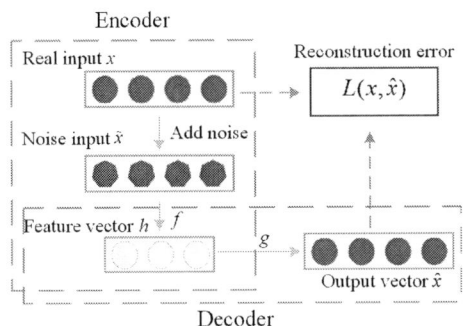

Fig. 1: The architecture of denoising autoencoder.

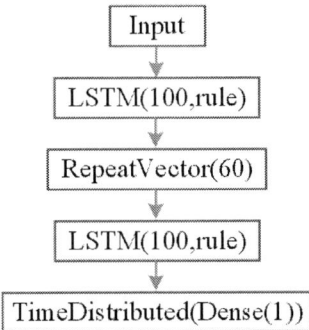

Fig. 2: LSTM denoising autoencoder.

Temporal convolution-recurrent network

Long short-term memory network (LSTM) is a special recurrent neural network, which not only solves the problem of gradient explosion and disappearance during the back-propagation process of RNN, but also overcomes the long-term dependencies of RNN [13]. The key to overcoming the long-term dependence of LSTM lies in the addition of cell state. Each LSTM cell is made of input, forget, and output gates. The calculation process is shown in (1).

$$
\begin{cases}
f_t = \sigma(W_f \cdot [h_{t-1}, x_t] + b_f) \\
i_t = \sigma(W_i \cdot [h_{t-1}, x_t] + b_i) \\
\bar{C}_t = \tanh(W_c \cdot [h_{t-1}, x_t] + b_C) \\
C_t = f_t \cdot C_{t-1} + i_t \cdot \bar{C}_t \\
\tanh(x) = \frac{e^x - e^{-x}}{e^x + e^{-x}} \\
o_t = \sigma(W_o[h_{t-1}, x_t] + b_0) \\
h_t = o_t \cdot \tanh(C_t)
\end{cases}
\tag{1}
$$

Temporal convolutional network (TCN) is a special convolutional network for processing sequential tasks with causal constraints. Based on 1D convolution, causal convolution and dilated convolution, it can process and predict different length sequences according to the causal relationship between the latter and the former. LSTM will forget the useless information, but TCN based on causal convolution will not miss the past information, nor will it reveal the future information. Causal convolution conforms to a strict time constraint. Dilated convolution is proposed to ensure enough receiving fields and reduce the amount of computation. In TCN, the simple convolution layer is replaced by a residual block. The composition of a residual block is shown in Fig.3 .

Based on the unique advantages of LSTM and TCN networks, the paper designed a temporal convolution-recurrent network (TCRN). TCRN is a network with three inputs and a single output and its stucture is shown in Fig. 4. Input 1 and input 2 are fed into LSTM and TCN respectively, and then connected to the fully connected layer to ensure that the output dimension is consistent with input 3. Then the three parts are connected in series through a concatenation layer, and the final output is obtained through the fully connected layer. Input 3 is added to enhance important features, thereby enhancing the predictive ability of the network.

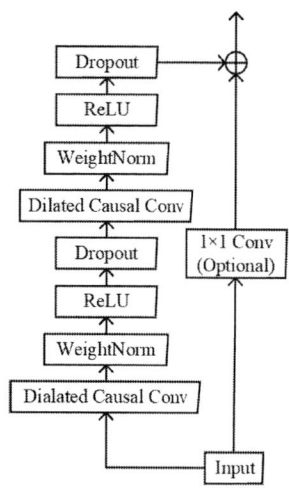

Fig. 3: Residual Block

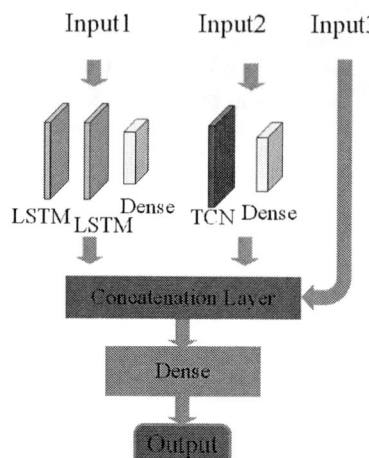

Fig. 4: Temporal convolution-recurrent network

The thermal early warning system of lithium-ion battery

Based on the adaptive network-based fuzzy inference system optimized by the particle swarm optimization algorithm, this paper designs a classifier to determine whether the surface temperature of lithium-ion batteries is out of bounds, and decides whether to issue an early warning signal.

Adaptive network-based fuzzy inference system

Adaptive network-based fuzzy inference system (ANFIS) integrates the fuzzy system and the artificial neural network to combine the advantages of the two methods [14]. Fuzzy system is a promotion of deterministic system. It captures the fuzzy characteristics of human brain thinking and imitates the way people make decisions on inaccurate and non-digital information [15]. Fuzzy system uses If-Then rules to realize the conversion from input to output, so as to deal with the difficult problem of fuzzy information processing. As a fuzzy inference system based on the Takagi-Sugeno model, ANFIS has rules as shown in 1 and 2 [16].

1. Rule1: if x is A_1 and y is B_1 then $z = p_1x + q_1y + r_1$;

2. Rule2: if x is A_2 and y is B_2 then $z = p_2x + q_2y + r_2$;

where x and y are two inputs, A and B are the linguistics labels and z is the output. Therefore, compared with artificial neural network, ANFIS is interpretable and more meaningful.

However, the traditional fuzzy inference system is generally formed based on expert experience, which lacks an effective learning mechanism. ANFIS realizes the three basic processes of fuzzy control through the method of neural network: fuzzification, fuzzy inference and defuzzification. Thus the process of establishing fuzzy inference system is simplified. It can automatically extract the dependent rules from the training data set. The structure of the first-order TSK-type ANFIS is shown in Fig. 5 [9]. Where the square nodes are the node with variable parameters and the circular nodes represent the node with immutable parameters.

The functions of each layer of ANFIS are shown in (2):

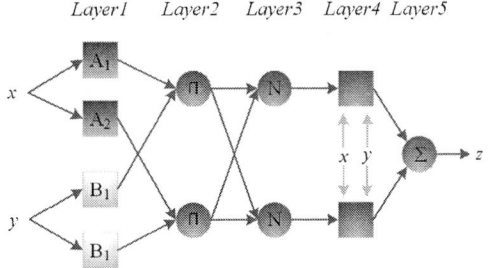

Fig. 5: The structure of the first-order TSK-type ANFIS.

$$\begin{cases} O_{1,ij} = \mu_j(x_i) = 1/(1 + |(x_i - c_{ij})/a_{ij}|^{2b_{ij}}); \\ i = 1,2,...,N; j = 1,2,...,M \\ O_{2,r} = \prod_{i=1}^{N} \mu_j(x_i) = w_r; r = 1,2,...,R \\ O_{3,r} = w_r/(\sum_{r=1}^{R} w_r) = \bar{w}_r \\ O_{4,r} = \bar{w}_r(\sum_{i=0}^{N} p_{ri}x_i), x_0 = 1 \\ O_5 = \sum_{r=1}^{R} O_{4,r} \end{cases} \qquad (2)$$

where N is the number of input features, M is the number of membership functions corresponding to the ith input and R is the total number of rules.

Particle swarm optimization

There are two parts of the parameters in the ANFIS architecture that can be adjusted during the training process: the premise parameters (a_{ij}, b_{ij}, c_{ij}) in layer 1 and the consequence parameters (p_{ri}) in layer 4. The traditional ANFSI uses the least square method to update the consequence parameters in the forward propagation process, and then uses the gradient descent method to update the premise parameters in the backpropagation process. In the paper, the particle swarm algorithm (PSO) is used to adjust ANFIS parameters.

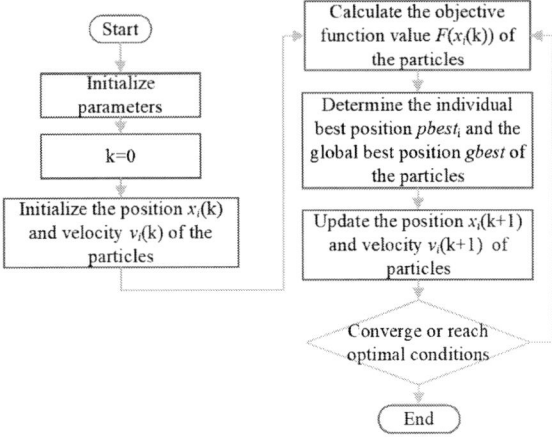

Fig. 6: The particle swarm optimization algorithm.

Originating from the study of the behavior of bird flocking, Eberhart and Kennedy proposed a new swarm intelligence technology—particle swarm optimization(PSO) [17]. As a population-based search method, PSO designed a massless particle to simulate birds in a flock of birds, in which each particle with the two attributes of speed and position is a candidate solution [18]. By simulating the cooperation and competition mode of individuals in the group, the particles of particle swarm will adjust their speed and position according to the current individual extreme value and the global extreme value shared by the whole particle swarm until they reach equilibrium or optimal state. The PSO algorithm is shown in Fig. 6. The update equations for the position and velocity of the particles are as follows:

$$v_i(k+1) = wv_i(k) + \rho_1(pbest_i - x_i(k)) + \\ \rho_2(gbest - x_i(k)) \tag{3}$$

$$\begin{cases} \hat{v}_i(k+1) = \min(\max(v_{\min}, v_i(k+1)), v_{\max}), \\ \quad LB \le x_i(k) \le UB \\ \hat{v}_i(k+1) = -r * \min(\max(v_{\min}, v_i(k+1)), v_{\max}), \\ \quad x_i(k) \le LB \text{ or } UB \le x_i(k) \end{cases} \tag{4}$$

$$x_i(k+1) = x_i(k) + \hat{v}_i(k+1) \tag{5}$$

where v_{max} and v_{min} are the maximum and minimum particle velocity. LB and UB are the lower and upper bounds of the particle position.

Sequential neural network-fuzzy thermal early warning system

This paper established a sequential neural network-fuzzy thermal early warning system (SNNFT), and its architecture is shown in Fig. ??. First, the LSTM denoising autoencoder is used to reduce the noise of the lithium battery data measured in real-time. Then, the denoising data are used as the input of TCRN to predict the surface temperature of lithium battery in the next 4 seconds. Finally, the predicted surface temperature of 4 seconds is fed into the ANFIS-based classifier to determine whether a thermal early warning signal is required. In the training process, we use Bayesian optimization to optimize the training parameters of network and use PSO to adjust the premise and consequence parameters of the ANFIS model.

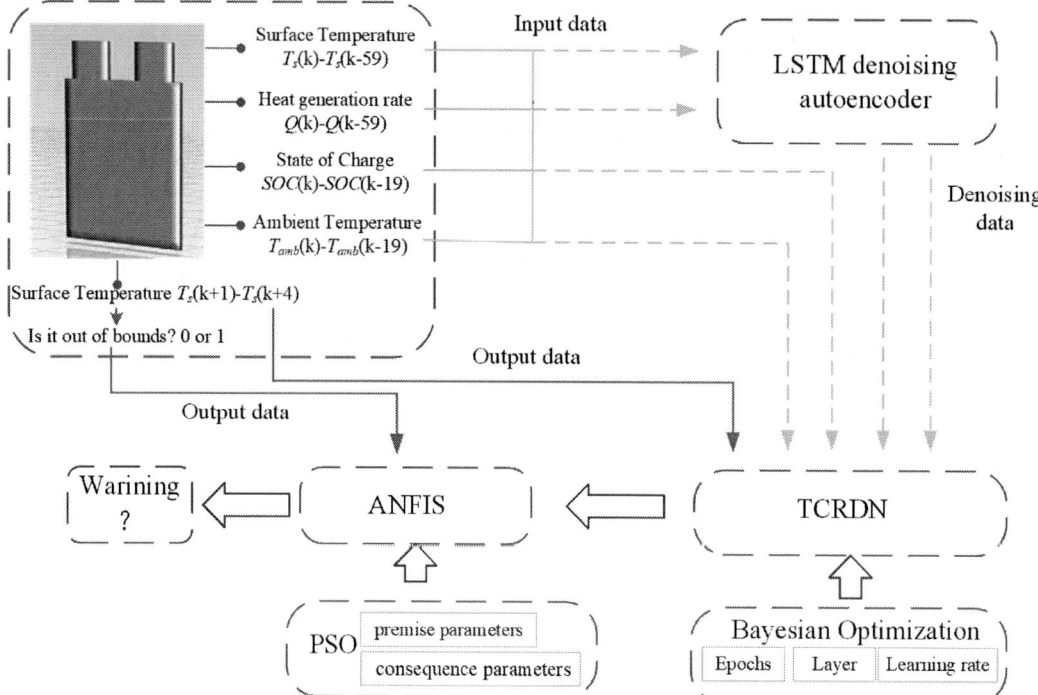

Fig. 7: Sequential neural network-fuzzy thermal early warning system.

Prediction results and discussion

Data preparation

We used the commercial software FLUENT to perform finite element simulation of a 10Ah pouch LiFePO4-Graphite battery to obtain the battery data. First, a finite element model of the lithium-ion battery is established, as shown in Fig. 4. The parameters of the battery are set as follows: Nominal Cell Capacity = 10Ah; Min Stop Voltage = 3V; Max Stop Voltage = 4.3V; Solution Method = MSMD; E-Chemistry Models = NTGK Empirical Model; Initial Soc = 0.7. The profile of the load current of the battery is shown in Fig.8(a), which is mainly for the thermal performance of the lithium-ion battery under high rate current, and the maximum load current can reach 4C. Because the SOC of the battery is usually cycled within a limited range of 20%-80% or 30%-70% which the SOC has little effect on the heat generation of the battery, this paper sets a cycle of SOC in a small range [19]. Then use the commercial software FLUENT to calculate and solve the lithium-ion battery terminal voltage, terminal current, SOC, internal resistance and temperature under certain working conditions. The result obtained by finite element simulation is shown in Fig. 5. In order to simulate the noise signal under real conditions, this paper used the randn function to add 1.5 times Gaussian noise to the temperature and heat generation rate signals obtained by the finite element simulation. The randn function returns a random value obtained from the standard normal distribution, thereby generating Gaussian random noise.

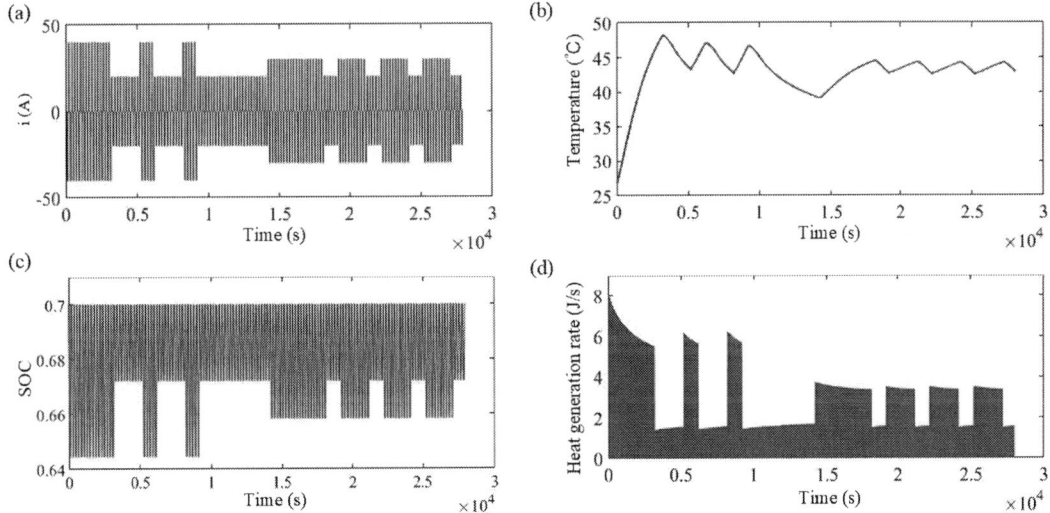

Fig. 8: Results of finite element simulation. (a) is the terminal current of the battery; (b) is the surface temperature of the battery; (c) is the state of charge; (d) is the heat generation rate of the battery.

Prediction results of surface temperature of lithium-ion battery

This paper obtains the measurement data of finite element simulation in real-time, and feeds the noise-added data into the LSTM denoising autoencoder through the sliding window method. The window length is 60 seconds. Take the first second test of each window as an example to show the noise reduction capability of the LSTM denoising autoencoder, as shown in Fig. 9.

It can be seen from Fig. 9 that the LSTM denoising autoencoder has a very good effect in reducing noise. After noise reduction, the data fluctuates less, the temperature change trend is clearer, and big error messages are removed.

The denoising data and measurement data are then fed into TCRN to predict the surface temperature for a period of time in the future. This paper used Keras to build the TCRN architecture. The input 1 and input 2 of TCRN are the surface temperature, heat generation rate, ambient temperature and SOC for 20 consecutive seconds. Because the surface temperature is a very important feature, the input 3 of

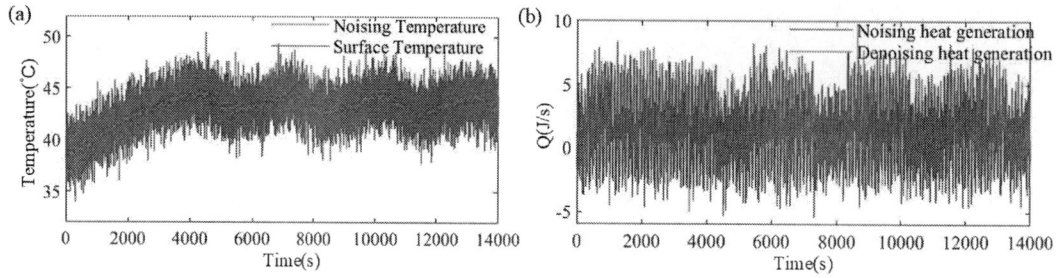

Fig. 9: The test results of LSTM denosing autoencoder. (a) surface temperature. (b) heat generation rate.

TCRN is the denoising measured value of the surface temperature for 60 seconds. In order to get a better prediction model, Bayesian optimization algorithm is used to optimize the layer, learning rate, epochs and other important parameters of the model during the training process. Set the prediction length to 4 seconds, Table 1 compares the mean square error of the prediction results when the input is noising data, the input is denoising data and Bayesian optimization is added.

Table I: The mse of prediction with the prediction length of 4s.

MSE	Noising Data	Denoising Data	Bayesian optimization
Train	0.0541	0.0305	0.0296
Test	0.0597	0.0475	0.0378

It can be seen from Table 1 that the prediction of TCRN that uses the denoising data as the input is more accurate. Under the same learning rate, the number of training epochs required with the noisy input is much more than the denoising input. This shows that the advantages of the denoising data obtained by the LSTM denoising autoencoder are more obvious, which can enhance the effect of prediction. On the other hand, Bayesian optimization is a very effective hyperparameter optimization method. Using Bayesian optimization to adjust hyperparameters not only saves a lot of manual tuning time, but also is more efficient and performs better than grid search and random search. After Bayesian optimization, a better TCRN model was obtained, which improved the accuracy of prediction.

Thermal diagnosis results of lithium-ion battery

Finally, the output of TCRN is fed into the classifier based on ANFIS to judge whether it is necessary to send an early warning signal, where 0 is not required and 1 is required. The ANFIS model established in this paper adopts the generalized bell membership function, and its equation is shown in (2). In the training process, PSO algorithm is used to adjust the premise membership function parameters and consequence parameters of ANFIS. When the prediction length is 4 seconds, the main PSO parameters used in ANFIS model training are as follows: the number of agents is 42; and the number of iterations is 250. The membership function finally obtained after PSO algorithm training is shown in Fig 10.

Each input corresponds to three generalized bell membership functions, and the linguistic labels are high, medium and low respectively. As the number of membership functions increases, the rules will increase exponentially. Therefore, when the prediction length is 5 seconds, in order to speed up the calculation time and save calculation resources, the membership function corresponding to each input is reset to 2.

In this paper, several experiments have been done under the scenarios of different prediction lengths, and compared with other advanced methods. The paper uses four indicators of accuracy, precision, recall and F1 score to evaluate these methods. The results are shown in Table 2. Where ANFIS-PSO: the method proposed in this paper; ANFIS: the traditional ANFIS method trained using least square method and gradient descent method; ANFIS-ANFIS: Train two traditional ANFIS models for 0 and 1 signals respectively, and finally perform MAX operation [20]; Skmoefs-ngsa3: Multi-Objective Evolutionary

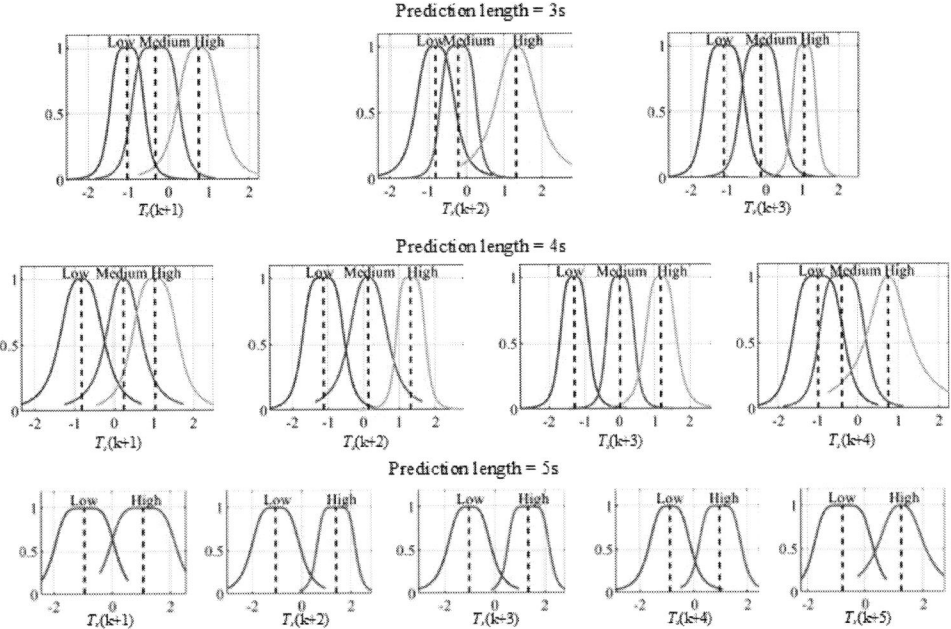

Fig. 10: The membership function.

Fuzzy Systems- Non-dominated Sorting Genetic Algorithm3 [21]; NN: neural network with 16 neurons in the hidden layer; MLP: Multilayer perceptron with three hidden layers and the number of neurons is 8, 16, 4, respectively. It can be seen from Table 2 that the ANFIS-PSO model established in this paper has achieved the highest prediction accuracy in scenarios with various prediction lengths. And when the prediction length is 4 seconds and 5 seconds, the ANFIS-PSO model outperformed other models in the four indicators of accuracy, precision, recall and F1 score. The ANFIS-PSO model maintains the performance not inferior to traditional neural network methods while possessing stronger interpretability.

Table II: Comparison of diagnosis ability of different methods

Prediction Length = 3s (Accuracy, precision, recall and F1 score)								
ANFIS-PSO	**0.8941**	0.9005	**0.8941**	0.8965	ANFIS	0.8799	0.8933	0.8799	0.8865
ANFIS-ANFIS	0.8905	0.8916	0.8905	0.8910	NN	0.8913	**0.9008**	0.8913	0.8944
Skmoefs-ngsa3	0.8950	0.9007	0.8950	**0.8971**	MLP	0.8919	0.9007	0.8919	0.8949
Prediction Length = 4s (Accuracy, precision, recall and F1 score)								
ANFIS-PSO	**0.8943**	**0.9007**	**0.8943**	**0.8966**	ANFIS	0.8658	0.8881	0.8658	0.8763
ANFIS-ANFIS	0.8773	0.8825	0.8773	0.8794	NN	0.8756	0.9912	0.8756	0.8819
Skmoefs-ngsa3	0.8824	0.8887	0.8824	0.8848	MLP	0.8763	0.8828	0.8763	0.8789
Prediction Length = 5s (Accuracy, precision, recall and F1 score)								
ANFIS-PSO	**0.8924**	**0.9005**	**0.8924**	**0.8952**	ANFIS	0.8641	0.8956	0.8641	0.8732
ANFIS-ANFIS	0.8687	0.8958	0.8687	0.8758	NN	0.8911	0.8994	0.8911	0.8940
Skmoefs-ngsa3	0.8878	0.8943	0.8878	0.8902	MLP	0.8904	0.8991	0.8904	0.8934

Conclusion

This paper established a sequential neural network-fuzzy thermal early warning system (SNNFT), which includes two parts: lithium-ion battery temperature prediction and thermal early warning. Firstly, aiming at the problem of noise in real-time measurement, an LSTM denoising autoencoder is applied. Then a temporal convolution-recurrent network (TCRN) integrating LSTM and TCN is designed to predict the surface temperature change of lithium-ion batteries in the future. The experiments show that when the input is denoising data, the MSE of the predicted result of the TCRN model after Bayesian optimization is reduced by 45%. Finally, the ANFIS model optimized by the particle swarm algorithm constitutes the thermal early warning part. By comparing the four indicators of accuracy, precision, recall and F1 score, it can be seen that the SNNFT established in the paper is notably the best compared to other methods.

References

[1] Gümüşsu E., Özgür E., Köksal M.: 3-D CFD modeling and experimental testing of thermal behavior of a li-Ion battery, Applied Thermal Engineering Vol 120, pp. 484- 495, 2017

[2] Chen L., Hu M., Cao K., Li S., Su Z. Jin G. et al: Core Temperature estimation based on electro-thermal model of lithium-ion batteries,International Journal of Energy Research Vol. 44 no 7, pp. 5320-5333, 2020

[3] Pan Y., Hua Y., Zhou S., He R., Zhang Y., Yang S., et al: A computational multi-node electro-thermal model for large prismatic lithium-ion batteries, Journal of Power Sources Vol. 459, pp. 2228070, 2020

[4] Fang K., Mu D., Chen S., Wu B., Wu F.: A prediction model based on artificial neural network for surface temperature simulation of nickel–metal hydride battery during charging, Journal of Power Sources Vol. 208, pp. 378-382, 2012

[5] Hong J., Wang Z., Chen W., Yao Y.: Synchronous multi-parameter prediction of battery systems on electric vehicles using long short-term memory networks, Applied Energy Vol. 254, pp. 113648, 2019

[6] Hussein A.A., Chehade A.A.: Robust artificial neural network-based models for accurate surface temperature estimation of batteries, IEEE Transactions on Industry Applications Vol.56 no 5, pp. 5269-5278, 2020

[7] Ojo O., Lang H., Kim Y., Hu X., Mu B., Lin X.: A Neural Network Based Method for Thermal Fault Detection in Lithium-Ion Batteries, IEEE Transactions on Industrial Electronics Vol. 68 no 5, pp. 4068-4078, 2021

[8] Lan G., Wang Y., Zeng G., Zhang J.: Compressive strength of earth block masonry: estimation based on neural networks and adaptive network-based fuzzy inference system, Composite Structures Vol. 235, pp.111731, 2019

[9] Jang J.-S. R.: ANFIS: adaptive-network-based fuzzy inference system, IEEE Transactions on Systems, Man, and Cybernetics Vol.23 no 3, pp. 665-685, 1993

[10] Sun W., Shao S., Zhao R., Yan R, Zhang X, Chen X: A sparse auto-encoder-based deep neural network approach for induction motor faults classification, Measurement Vol. 89, pp. 171-178, 2016

[11] Majumdar A.: Blind denoising autoencoder, IEEE Transactions on Neural Networks and Learning Systems Vol. 30 no 1, pp. 312-317, 2019

[12] Ashfahani A., Pratama M., Lughofer, E., Ong, Y.-S.: DEVDAN: Deep evolving denoising autoencoder, Neurocomputing Vol.390, pp. 297-314, 2020

[13] Hochreiter, S., Schmidhuber, J.: Long short-term memory, Neural Computation Vol.9 no 8, pp.1735-1780, 1997

[14] Turabieh H., Mafarja M., Mirjalili S.: Dynamic adaptive network-based fuzzy inference system (D-ANFIS) for the imputation of missing data for internet of medical things applications, IEEE Internet of Things Journal Vol. 6 no 6, pp. 9316-9325, 2019

[15] Chen H.-Y., Lee C.-H.: Electricity consumption prediction for buildings using multiple adaptive network-based fuzzy inference system models and gray relational analysis, Energy Reports Vol. 5, pp. 1509-1524, 2019

[16] Yang J., Shang C., Li Y., Li F., Shen Q.: ANFIS construction with sparse data via group rule interpolation, IEEE Transactions on Cybernetics Vol. 51 no 5, pp. 2773-2786, 2021

[17] Eberhart R., Kennedy J.: A new optimizer using particle swarm theory, MHS'95 Proceedings of the Sixth International Symposium on Micro Machine and Human Science , pp. 39-43, 1995

[18] Aliyari Shoorehdeli M., Teshnehlab M., Sedigh A.K.: Identification using ANFIS with intelligent hybrid stable learning algorithm approaches, Neural Computing and Applications Vol. 18 no 2, pp. 157-174, 2009

[19] Zhang C., Li K., Deng J.: Real-time estimation of battery internal temperature based on a simplified thermo-electric model, Journal of Power Sources Vol. 302, pp. 146-154, 2016

[20] Rostaghi M., Khatibi M.M., Ashory M.R., Azami H.: Bearing fault diagnosis using refined composite generalized multiscale dispersion entropy-based skewness and variance and multiclass FCM-ANFIS, Entropy Vol.23 no 11, pp. 1510, 2021

[21] Antonelli M., Ducange P., Marcelloni F.: A fast and efficient multi-objective evolutionary learning scheme for fuzzy rule-based classifiers, Information Sciences Vol. 283, pp. 36-54, 2014

Fine-grained Dynamics Representation and Stability Analysis for MMC-based Hybrid AC/DC Power Systems

Jingming Cao[1], Chaoyu Dong[2*], Qian Xiao[1], Marui Li[1], Xiaodan Yu[1], Hongjie Jia[1]

[1]School of Electrical and Information Engineering, Tianjin University, Tianjin, China

[2]Nanyang Technological University, Singapore

Tel.: +86 / 18322694146.

E-Mail: dong0120@e.ntu.edu.sg

Acknowledgements

This work was supported by the National Natural Science Foundation of China (No. 52107121), and the joint project of NSFC of China and EPSRC of UK (No. 52061635103 and EP/T021969/1).

Keywords

« Modelling», «Modular Multilevel Converters (MMC)», « State-space model», « Stability analysis ».

Abstract

The hybrid AC/DC power system is favored because of its huge energy transmission capacity and excellent steerability. However, the real system is large in scale and the inner dynamics interaction and coupling are intricate, which introduces a series of stability issues for the system operation. To investigate AC/DC interaction dynamics, an MMC-based hybrid power system model is established in the state space. The model considers the detailed dynamics of both AC and DC, and their interaction, thus it can be used for AC/DC interaction study, and offer a more precise stability assessment result than either AC power system model or MMC converter model. After the verification in MATLAB/Simulink, The interaction is studied thoroughly by the proposed model, which clearly identifies the harmonic coupling between MMC and traditional power systems. The small signal stability gaps of different systems are also compared. Case results prove that the reduced-order model can be deployed to assess stability efficiently under specific circumstances. The interaction causes extra harmonics in AC affecting the hybrid system stability.

Introduction

In order to consume sustainable energy more properly and deploy large-scale transmission more productively, the interconnection between AC and DC is important even necessary. In the future, the MTDC (Multi-Terminal DC) grids will interconnect and operate in parallel with ac systems [1]. Nevertheless, AC and DC differ in the physical behaviours, control design, and time scale. Both the AC system and the DC system need complicated control to guarantee normal work and they might also need coordination control for resource sharing. Many corresponding issues like inner dynamics interaction and control design, remain to be developed. In most cases, the AC/DC interaction is considered weak. However, some articles have declared that the mutual effect could be fierce in certain circumstances [1, 2]. Therefore, it is of interest to build the model of the entire hybrid AC/DC power system and study its dynamics, especially considering the participation of power electronic devices.

A few researches have attached the detailed hybrid AC/DC power system model issue so far. Reference [3] took the converter high-order harmonics into account, established the elaborate MTDC state-space model based on the insert index theory and modular modelling design. Reference [4] considered the equations of the MMC circuit, the control system and phase-locked loop (PLL), derived the MMC-HVDC impedance model and investigated the mechanism and damping control of high frequency resonance (HFR) occurred in HVDC project. However, both [3] and [4] simplified the AC model too much to retain its dynamics. Reference [5] treated MMC as the equivalent generator or equivalent load and built the hybrid AC/DC power system model whereas the assumptions about MMC are too strict and the DC dynamics are ignored. Reference [1] combined the AC network model and the simple MMC model together, built the multi-terminal hybrid system model and pointed out that there existed weak coupling between asynchronous grids linked via MTDC. Reference [2] applied modal coupling theory to examine the AC/DC dynamic interactions and claimed that when a complex pole of the open-loop MTDC subsystem is close to an electromechanical oscillation mode (EOM) of concern in the open-loop AC subsystem on the complex plane, the strong interaction might be caused and the stability would be degraded.

It can be found that existing researches simplify either the AC system or DC system during the analysis, which barely reflect the accurate system dynamics. To investigate the interaction between AC and DC in the hybrid system and analyse the stability more precisely, the hybrid system model considering the fine-grained dynamics of both AC and DC is established. the physical dynamics are first summarised. Given that the real hybrid system includes many controllers to ensure the normal operation, LFC (local frequency control) and several controls in MMC are considered, and the detailed hybrid system model is derived by integrating both the physical dynamics and control dynamics. After the verification in virtue of Simulink, the model is used to study the transient interaction and stability. The results indicate that the common power and voltage might be the key factor of interaction and the hybrid system has better stability compared to the AC system in specific scenarios.

Physical dynamics in hybrid AC/DC power system

The hybrid AC/DC power system here is assumed as the combination of the AC network and the converters linked to the DC lines. The AC network is connected to the converter through the AC line and the AC network can be equivalent to one single machine. The equivalent circuit diagram is shown in Fig. 1.

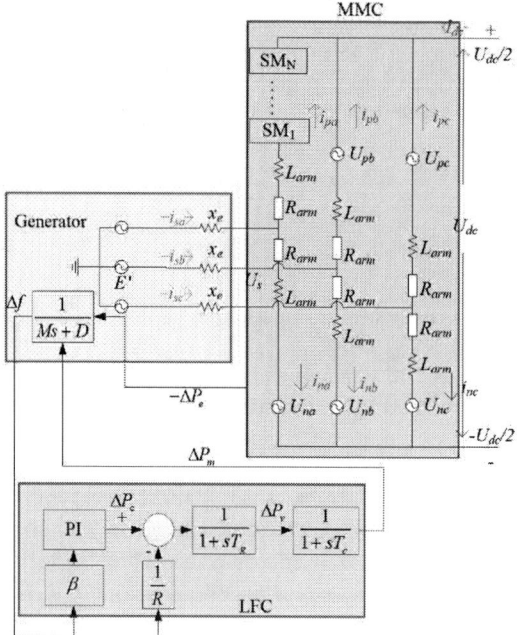

Fig. 1: The equivalent circuit diagram of the hybrid AC/DC power system.

Considering the excitation system dynamics, A fourth-order generator model [6] is deployed.

$$\Delta\dot{\delta} = 100\pi\Delta f \tag{1}$$

$$M\Delta\dot{f} = -D\Delta f + \Delta P_m - \Delta P_e \tag{2}$$

$$T'_{d0}\dot{E}' = -E' + \frac{x_d - x'_d}{x_e + x'_d}[E' - U_s \sin(\delta)] + E_{fd} \qquad (3)$$

$$T_A \dot{E}_{fd} = -K_A(U_s - U_{ref}) - (E_{fd} - E_{fd0}) \qquad (4)$$

where $\Delta\delta$, Δf, ΔP_m and ΔP_e are the deviation from the power angle, frequency, mechanical power and electrical power in the equilibrium respectively. M, D, E', x_d, x'_d and T'_{fd0} are the moment of inertia, damping coefficient, transient voltage, synchronous reactance, transient reactance and transient time constant of the generator. x_e is the reactance of the AC line, U_s is the voltage of the PCC (point of common coupling) and E_{fd} is field excitation voltage. T_A and K_A are the time constant and controller gain of the excitation system. U_{ref} and E_{fd0} are the references of U_s and E_{fd} respectively.

When SM (submodule) number is big enough (for example, 250), the unbalancing effect in the different modules of a stack can be negligible and the insert index theory is considered appliable (some results in [7, 8] validated it). Based on the insert index theory, considering the harmonic dynamics in MMC, the 12th-order state-space model is presented in [7]. The AC network is connected to the converter through the AC line, thus they share the same U_s and P_e, which are the connections between them. The generator's and the converter's dynamics make up the total physical dynamics in a hybrid AC/DC power system.

However, the dynamics mentioned above are far from enough to assess the system. In practical engineering, the hybrid system is faced with the LFC need, coordination between AC and DC, and coordination among different areas, which bring the extra control dynamics to the system.

MMC-based hybrid AC/DC power system detailed state-space model

Control dynamics in hybrid AC/DC power system

The physical dynamics fail to describe the whole hybrid system, thus it's obligatory to establish a model containing both the physical dynamics and control dynamics, to study the small signal stability and transient. Frequency is a critical index of the system's safety and stable operation. According to (2), frequency is decided by the trade-off between P_m and P_e. P_m is influenced by the LFC loop and P_e equals to the power transmitted to the DC side through the converter.

The LFC loop is shown in Fig. 1, from which we derive the equations below:

$$T_{ch}\Delta\dot{P}_m = \Delta P_v - \Delta P_m \qquad (5)$$

$$T_g \Delta\dot{P}_v = \Delta P_c - \Delta P_v - \Delta f / R \qquad (6)$$

$$\dot{x}_{lf} = ACE = \beta\Delta f \qquad (7)$$

$$\Delta P_c = -k_{plf} ACE - k_{ilf} x_{lf} = -k_{plf}\beta\Delta f - k_{ilf} x_{lf} \qquad (8)$$

where ΔP_c and ΔP_v are the deviation of the PI controller output and valve position, T_g, T_{ch}, R, β and ACE are the time constant of the governor, time constant of the turbine, speed drop, frequency bias factor and the area control error. k_{plf}, k_{ilf} and x_{lf} are the proportional gain, integral gain and integrator output respectively.

PQ control, CCSC (Circulating Current Suppressing Controller) and PLL in [7] are considered in the paper. Owing to the phase adjustment of PLL, U_s becomes interrupted by the MMC and PLL dynamics. So equation (3) is rewritten as:

$$T'_{d0}\dot{E}' = -E' + \frac{x_d - x'_d}{x_e + x'_d}[E' - U_s \sin(\delta - x_{pll})] + E_{fd} \qquad (9)$$

All these MMC control dynamics may cause fluctuation of U_s and P_e, whose expressions are:

$$\begin{cases} U_{sd} = E'\sin(\delta - x_{pll}) - Ldi_{sd}/dt - \omega Li_{sq} \\ U_{sq} = E'\cos(\delta - x_{pll}) - Ldi_{sq}/dt + \omega Li_{sd} \end{cases} \tag{10}$$

$$P_e = 3(U_{sd}i_{sd} + U_{sq}i_{sq})/2 \tag{11}$$

MMC-based hybrid AC/DC power system detailed state-space model

With the physical model and the control dynamics presented above, we know that U_s and P_e in the generator model are affected by MMC. In return, MMC dynamics are relevant to δ and E' in AC. The AC system and the DC system depend on each other and can't be solved by decoupling. Thus, the hybrid system model (12) is derived by combining all the equations together:

$$\begin{bmatrix} \dot{x}_{AC}^T & \dot{x}_{MMC}^T & \dot{x}_c^T \end{bmatrix}^T = f(x_{AC}, x_{MMC}, x_c, u) \tag{12}$$

where $x_{AC} = [E', E_{fd}, \delta, f, P_m, P_v, x_{lf}]^T$, $x_{MMC} = [\bar{u}_c, u_c^{1d}, u_c^{1q}, u_c^{2d}, u_c^{2q}, u_c^{3x}, u_c^{3y}, I_{dc}, i_{sd}, i_{sq}, i_{cird}, i_{cirq}]^T$, $x_c = [f_1,$
$f_2, x_1, x_2, x_3, x_4, x_5, x_{pll}]^T$, $u = [U_{dc}, P_{ref}, Q_{ref}, U_{ref}, E_{fd0}]^T$.

According to model (12), the AC sub-model and DC sub-model are coupled by P_e and U_s because they share the same electrical power and PCC voltage, which indicates AC and DC might have transient interaction through P_e and U_s. Considering that MMC is a typical multiple harmonic response system [9, 10], the intricate DC harmonics might affect AC and make AC's harmonic distribution complicated. After linearisation, 3 simple state-space sub-models can be obtained directly:

$$\Delta \dot{x}_{MMC} = A\Delta x_{MMC} + B\Delta u_{MMC} \tag{13}$$

$$\Delta \dot{x}_c = A_c \Delta x_c + B_c \Delta u_c \tag{14}$$

$$\Delta \dot{x}_{AC} = A_{AC}\Delta x_{AC} + B_{AC}\Delta u_{sdq} + C_{AC}\Delta x_c \tag{15}$$

where $\Delta u_{sdq} = [\Delta U_{sd}, \Delta U_{sq}]^T$ is linearised from (10). Δu_{MMC} and Δu_c also contain ΔU_{sd} and ΔU_{sq} terms.

Therefore, even though the several sub-models can't be combined together directly, solving (16) and (17) below simultaneously to eliminate the Δu_{sdq} term can be an accessible approach to derive the linearised hybrid system state-space model.

$$\begin{bmatrix} \Delta \dot{x}_{MMC} \\ \Delta \dot{x}_c \\ \Delta \dot{x}_{AC} \end{bmatrix} = \begin{bmatrix} A & O & O \\ O & A_c & O \\ O & C_{AC} & A_{AC} \end{bmatrix} \begin{bmatrix} \Delta x_{MMC} \\ \Delta x_c \\ \Delta x_{AC} \end{bmatrix} + \begin{bmatrix} B_1 & B_2 & O & O \\ O & B_{c1} & B_{c2} & O \\ O & O & O & B_{AC} \end{bmatrix} \begin{bmatrix} \left. \begin{matrix} \Delta u_{MMC1} \\ \Delta u_{sdq} \end{matrix} \right\} \Delta u_{MMC} \\ \left. \begin{matrix} \Delta u_{c1} \\ \Delta u_{sdq} \end{matrix} \right\} \Delta u_c \\ \Delta u_{sdq} \end{bmatrix} \tag{16}$$

$$\Delta u_{sdq} = A'\Delta x_{MMC} + B'\Delta x_c + C'\Delta \dot{x}_{MMC} + D'\Delta x_{AC} \tag{17}$$

After daedal algebraic deduction, equation (16) becomes (18) below, denoting $\Delta x = [\Delta x_{MMC}^T, \Delta x_c^T, \Delta x_{AC}^T]^T$, $\Delta u = [\Delta U_{dc}, \Delta P_{ref}, \Delta Q_{ref}, \Delta U_{ref}, \Delta E_{fd0}]^T$, the entire hybrid system detailed state-space model (19) is established.

$$\begin{bmatrix} \Delta \dot{\boldsymbol{x}}_{MMC} \\ \Delta \dot{\boldsymbol{x}}_c \\ \Delta \dot{\boldsymbol{x}}_{AC} \end{bmatrix} = \begin{bmatrix} A_{hy}^{11} & A_{hy}^{12} & A_{hy}^{13} \\ A_{hy}^{21} & A_{hy}^{22} & A_{hy}^{23} \\ A_{hy}^{31} & A_{hy}^{32} & A_{hy}^{33} \end{bmatrix} \begin{bmatrix} \Delta \boldsymbol{x}_{MMC} \\ \Delta \boldsymbol{x}_c \\ \Delta \boldsymbol{x}_{AC} \end{bmatrix} + \boldsymbol{B}_{hy} [\Delta U_{dc}, \Delta P_{ref}, \Delta Q_{ref}, \Delta U_{ref}, \Delta E_{fd0}]^T \tag{18}$$

$$\Delta \dot{\boldsymbol{x}} = \boldsymbol{A}_{hy} \Delta \boldsymbol{x} + \boldsymbol{B}_{hy} \Delta \boldsymbol{u} \tag{19}$$

The differences between sub-models and the hybrid model are shown in Table I. It can be seen that the AC sub-model (15)'s dimension is much smaller than the dimension of the DC sub-model (13)+(14). DC dynamics coupling is tighter thus it has more harmonics and dimension consequently. The sub-models are integrated together through Δu_{sdq}, forming the hybrid model (18). Nonetheless, A_{hy} doesn't equal to the block combination of A_{AC} and A_{DC}. It's denser and has no zero block matrix because AC and DC have a close link due to the same input Δu_{sdq}. It means that the model considers the physical coupling between AC and DC, and the control dynamics effect. Both AC and DC disturbances can incur the system transient process. With model (12) and model (19), we're able to assess the small signal stability and transient.

Table I: Different models comparison

Model	State variables	Input	Jacobian matrix
Sub-model (15)	$\Delta \boldsymbol{x}_{AC\ 7*1}$	$\Delta \boldsymbol{u}_{sdq}$, $\Delta \boldsymbol{x}_c$, $\Delta \boldsymbol{u}$	$A_{AC\ 7*7}$
Sub-model (13)+(14)	$[\Delta \boldsymbol{x}_{MMC}^T, \Delta \boldsymbol{x}_c^T]_{20*1}^T$	$\Delta \boldsymbol{u}_{sdq}$, $\Delta \boldsymbol{u}$	$A_{DC\ 20*20}$
Hybrid model (18)	$\Delta \boldsymbol{x}_{27*1}$	$\Delta \boldsymbol{u}$	$A_{hy\ 27*27}$

Model verification and modal identification

Part of the parameters of the Zhoushan multi-terminal VSC-HVDC transmission project [6, 11] are selected to verify the established model and study the hybrid system behaviours. The generator parameters are $[M, D, x_d, x_d', x_e, T_{d0}', T_A, U_{ref}, E_{fd0}, K_A]$=[10,1,1,0.4,0.5,10,10,1,1,190], base values $[U_B, f_B]$=[$220/\sqrt{3}$ kV, 50Hz]. The LFC parameters are presented as $[T_{ch}, T_g, R, \beta, k_{plf}, k_{ilf}]$=[0.3,0.1,0.05, 21,0.2,0.2]. As for the MMC, SM number N=250, arm inductance and resistance $[L_{arm}, R_{arm}]$=[0.09H, 1Ω] and SM capacitance C=6000μF. The parameters of the MMC control are: PQ control $[k_{p1}, k_{i1}, k_{p2}, k_{i2}, P_{ref}, Q_{ref}]$=[0.001,0.1,0.1,1,400MW,0]; CCSC $[k_{p3}, k_{i3}]$=[10,1000]; PLL $[k_{ppll}, k_{ipll}]$=[5,100].

To test the capability of the proposed model during the dynamic representation, we give P_{ref} a step change from 400MW to 420MW at t=1.8s and the time-domain responses of AC (f) and DC (i_{sd}) are given both in Simulink and the Runge-Kutta simulation (RK simulation) based on model (12). The comparison results are shown in Fig. 2. It shows that the waveforms of Simulink and the RK

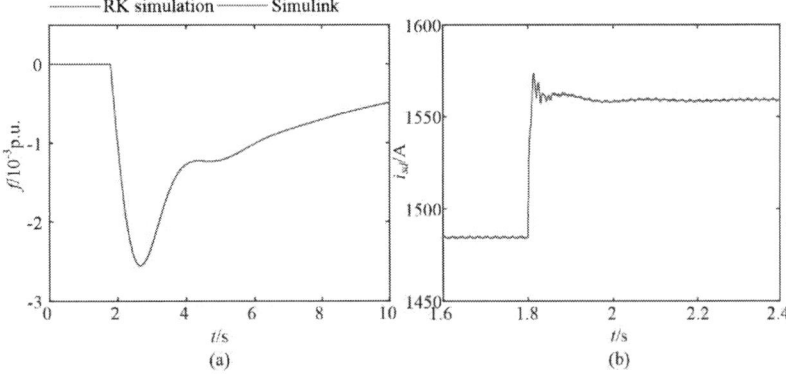

Fig. 2: The time-domain response of: (a) f, (b) i_{sd}.

simulation almost coincide, which indicates the model is able to represent system dynamics correctly.

To verify the small signal stability model (19), the eigenvalues with certain parameters varying are plotted and observed. The corresponding parameters' stability margins are compared among different approaches, which are listed in Table II. The results demonstrate the accuracy of the model (19).

Table II: Parameters margins comparison

	Simulink	RK simulation	Eigenvalues
(UM,CMF) of k_{plf}	(5.67,5.883)	(5.66,5.883)	(5.66,5.886)
(UM,CMF) of k_{ilf}	(2.54,2.501)	(2.54,2.497)	(2.54,2.498)
(LM,CMF) of M	(1.17,7.005)	(1.19,6.950)	(1.19,6.946)
UM means Upper Margin, LM means Lower Margin, CMF means Critical Mode Frequency.			

To investigate the small signal characteristics, the hybrid system eigenvalues (HSE) should be identified and classified. The eigenvalues of the hybrid system, the pure MMC system (MMC and MMC control) and the pure AC system (generator and LFC loop) are plotted and compared in Fig. 3. We can divide HSE into 3 types: b1) the ones almost coinciding with pure MMC roots; b2) the ones almost coinciding with pure AC roots; b3) the conjugate roots E1 and E2 marked in Fig. 3(c).

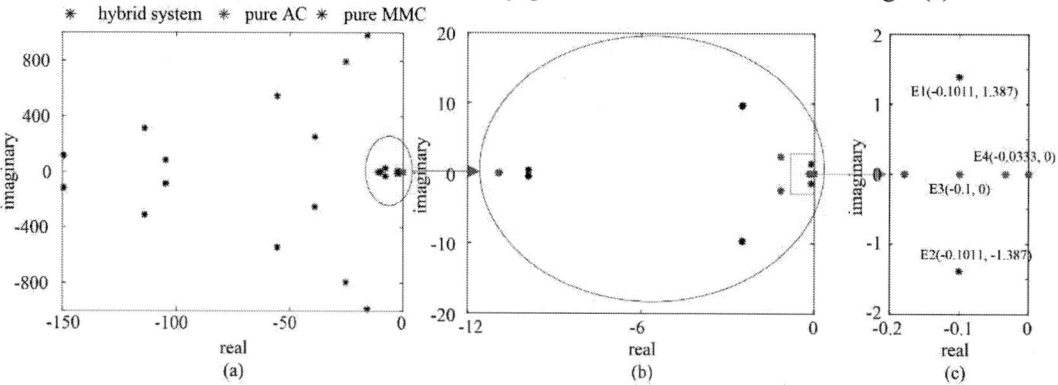

Fig. 3: The eigenvalues of the hybrid system, the pure MMC system and the pure AC system.

Among all the roots of pure AC and pure MMC, only 2 roots don't coincide with HSE, which are E3 and E4. We analysed the participate factor (PF) [3] of E3 and E4, found that they were dominated by E_{fd0} and E' respectively. In the pure AC system, E_{fd0} and E' are mainly affected by themselves, and their dynamics are not involved with oscillation. While in the hybrid system, the oscillatory U_s dynamic is considered, U_s is affected by E', E' is affected by E_{fd0} and E_{fd0} is affected by U_s. The closed-loop is formed and U_s brings the loop oscillation mode, so the real roots E3 and E4 become the conjugate roots E1 and E2 in the hybrid system. PF of E1 and E2 shown in Fig. 4 demonstrates the

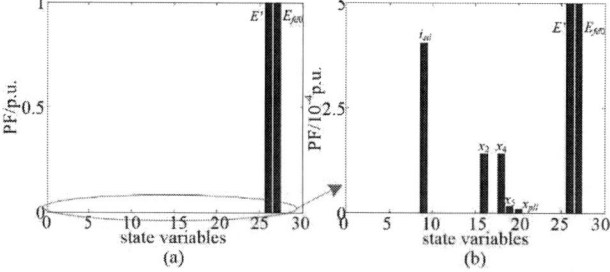

Fig. 4: The PF of E1 and E2 in the hybrid system.

mutual effect of E_{fd0} and E', and their dynamic's participation. We name b1) as DC-related roots, name b2) and b3) collectively as AC-related roots.

Hybrid system interaction investigations

We give an additional power P_{ad}=-0.2p.u. at t=1.8s to simulate part of the generator set lose connection, the DC side is affected by the AC transient and its MMC circulating current response is shown in Fig. 5(a). Once P_{ad} is exerted, f starts to deviate according to (2), then power angle will change and the power balance decided by the angle difference is broken, causing the converter couldn't transmit the wanted P_{ref}. In the meantime, PLL starts to trace the changed δ, the original dq rotating frame steady state is also broken. Thus the current increases initially. With the LFC regulation and PQ control adjustment, the power balance can be rebuilt. PLL also tends to be stable once f remains constant. Then the circulating current returns to zero under the CCSC control.

Similarly, the sub-module capacitor voltage decrement Δu_c=-150V is added at t=1.8s to simulate part of the sub-modules getting bypassed out of the fault. According to the MMC equations, It's easy to see that di_{sd}/dt increases and U_s decreases with the u_c decreases, thus P_e increases as the potential difference between the generator and the converter is reduced. So f declines at the very start. The excess power charges the capacitor to the original state and f returns to 50Hz as the power balance is reconstructed under the influence of PQ control and LFC. The frequency waveform in Fig. 5(b) coincides with the analysis above.

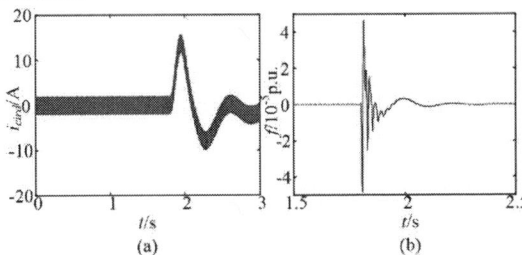

Fig. 5: The Simulink time-domain response of: (a) i_{cird}, (b) f.

To distinguish the differences between the traditional power system and the hybrid system, assume 2 cases: a1) traditional case in which the generator connects with AC load directly; a2) hybrid case that the generator is linked to the equal-sized DC load through the MMC converter. We sample and observe the harmonic components of the transmitted current i_{sd} after the system startup. The sampled harmonics are shown in Fig. 6 which is plotted with the aid of FFT Analysis. Apparently, harmonics of a2) are far bigger than harmonics of a1). And a2) has distinct components that a1) does not have, in f=150Hz and f=300Hz. Other than the fundamental, the biggest current component is of 5 orders (the 3-order component is eliminated). After the Park transformation, it will turn to the peak whose f=300Hz (and a small f=200Hz peak). As for another peak, MMC introduces the even-order circulating current inevitably. The biggest component is of 4 orders since the 2-order current is suppressed by CCSC. Likewise, it becomes an f=150Hz peak and a small f=250Hz peak. With the corresponding analysis and histograms, we know that the hybrid system has more abundant harmonics than the traditional system owing to its DC part.

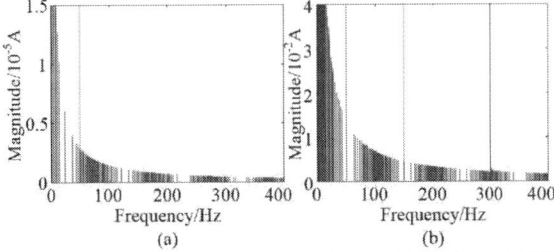

Fig. 6: Harmonic histogram: (a) traditional AC, (b) hybrid AC/DC system.

The researches above testify the indication, that AC and DC have transient interaction through P_e and U_s mutually indeed. Also the results show that the established model is able to reflect the transient interaction between AC and DC.

The root locus with M varying and the root locus with k_{i1} varying are plotted in Fig. 7 and Fig. 8. It can be seen that DC-related roots seldom move when M varies and AC-related roots seldom move when k_{i1} varies. Unlike notable transient change, the small signal disturbance might be insufficient to affect the other side through power or voltage directly.

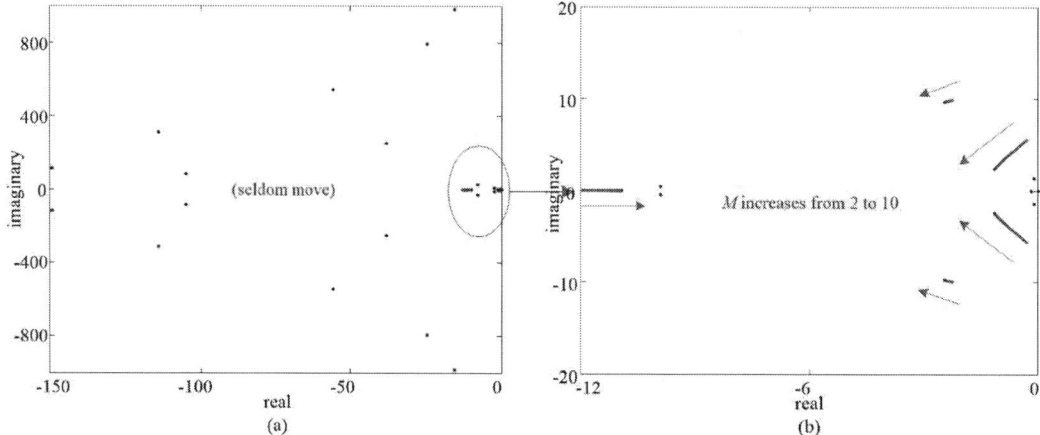

Fig. 7: The root locus when M increases from 2 to 10.

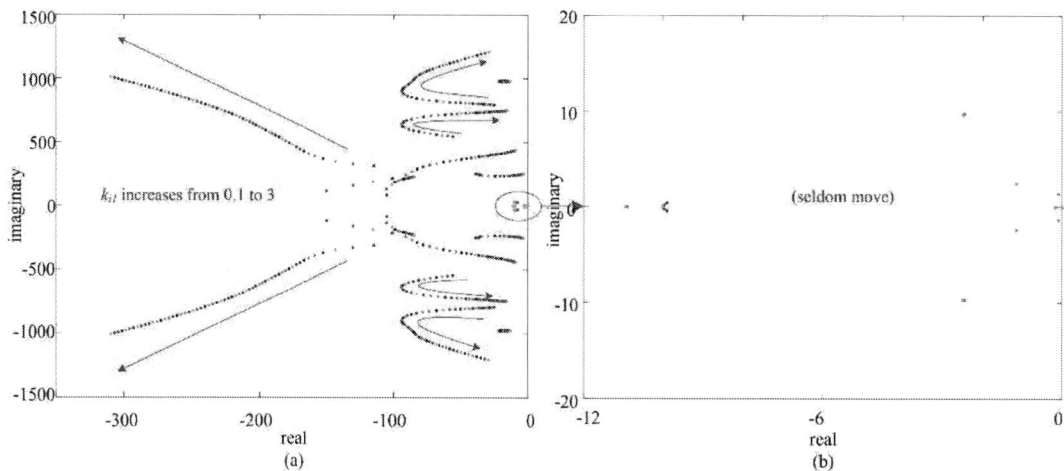

Fig. 8: The root locus when k_{i1} increases from 0.1 to 3.

As for the stability margins comparison among the hybrid system, the pure MMC system and the pure AC system, the obtained results reflect that the hybrid system has almost the same margins as the pure MMC system. As the AC time scale is much larger than the DC time scale, from DC's view, the AC side changes so slow that it can be regarded as fixed. AC dynamics do no contribution to the DC small signal stability, leading to the same margins with pure MMC system. With this conclusion, deploying the reduced-order MMC model to study the DC-side stability might be a more efficient idea.

On the other side, however, we found that compared to the pure AC system, the hybrid system has larger AC parameters margins. The comparison whose data are calculated and checked by the RK simulation and eigenvalues is listed in Table III. Contrast to the pure AC system's constant PCC voltage situation, the hybrid system's PCC voltage is affected by the converter's fast response. Under the PQ control effect, the converter response helps to maintain power within a certain range in most cases, thus the hybrid system has better AC-side stability due to the DC's fast regulation.

Table III: AC and hybrid system margins comparison

	(UM,CMF) of k_{plf}	(UM,CMF) of k_{ilf}	(LM,CMF) of M
AC system	(5.57,5.851)	(2.53,2.499)	(1.51,6.303)
MMC-based Hybrid system	(5.66,5.883)	(2.54,2.497)	(1.19,6.950)

Conclusion

The physical dynamics in hybrid AC/DC power systems are modelled with a generator model and MMC state-space model based on the insert index theory. Taking load frequency control and MMC control into consideration, the detailed hybrid system model is derived by integrating both the physical dynamics and control dynamics. The established model is verified with Simulink and parameter stability margins. After the elaborate investigation, 5 key conclusions are drawn:

c1) The established model is able to reflect the system transient and small signal dynamics accurately, which is competent for the quantitative study on hybrid AC/DC power system stability.

c2) In the hybrid system, AC and DC interact through P_e and U_s. The interaction might be more intense once one part breaks down, so applying the built model to study the hybrid system transient issue is of great significance.

c3) The change brought by AC/DC interaction strengthens the coupling between the roots dominated by E_{fd0} and E' and turns them into a pair of conjugate roots.

c4) AC dynamics barely affect the DC small signal stability due to the time scale difference. Hence, the reduced-order MMC model instead of the whole hybrid system model is more efficient to study DC-side stability.

c5) The hybrid system has better AC-side stability due to the DC's fast regulation, which confirms that the paper's meticulous hybrid system modelling is worthwhile.

References

[1] A. G. Endegnanew, G. Bergna-Diaz and K. Uhlen, "Avoiding AC/DC grid interaction in MMC based MTDC systems," in 2017 Twelfth International Conference on Ecological Vehicles and Renewable Energies (EVER), 2017, pp. 1-8.

[2] W. Du, Q. Fu and H. Wang, "Strong dynamic interactions between multi‑terminal DC network and AC power systems caused by open-loop modal coupling," IET Generation, Transmission & Distribution, vol. 11, pp. 2362-2374, 2017.

[3] J. Cao, C. Dong, Q. Xiao, X. Yu, and H. Jia, "State Modeling and Stability Analysis of the Multi-terminal MMC-HVDC Cyber-physical System Considering the Control and Communication Delay," Proceedings of the CSEE, vol. 41, pp. 3547-3560, 2021(in Chinese).

[4] Y. Li, T. An, D. Zhang, X. Pei, K. Ji, and G. Tang, "Analysis and Suppression Control of High Frequency Resonance for MMC-HVDC System," IEEE Transactions on Power Delivery, vol. 36, pp. 3867-3881, 2021.

[5] Y. Liu, Z. Lin, K. Xiahou, Y. Lin, and Q. H. Wu, "Equivalent hamiltonian equations modelling and energy function construction for MMC-HVDC in hybrid AC/DC power systems," CSEE Journal of Power and Energy Systems, vol. 7, pp. 821-831, 2021.

[6] H. Jia, J. Chen, and X. Yu, "Impact of Delay on Power System Small Signal Stability," Automation of Electrical Power System, vol. 30, pp. 5-8, 2006(in Chinese).

[7] T. Li, A. M. Gole and C. Zhao, "Small-Signal Model of the Modular Multilevel Converter Considering the Internal Dynamics," Proceedings of the CSEE, vol. 36, pp. 2890-2899, 2016(in Chinese).

[8] T. Li, A. M. Gole and C. Zhao, "Harmonic Instability in MMC-HVDC Converters Resulting From Internal Dynamics," IEEE Transactions on Power Delivery, vol. 31, pp. 1738-1747, 2016.

[9] K. Ji, G. Tang, J. Yang, Y. Li, and D. Liu, "Harmonic Stability Analysis of MMC-Based DC System Using DC Impedance Model," IEEE Journal of Emerging and Selected Topics in Power Electronics, vol. 8, pp. 1152-1163, 2020.

[10] J. Cao, C. Dong, X. Yu, R. Wang, Q. Xiao, and H. Jia, "Modelling and stability assessment of the MMC‐HVDC energy interconnected system with the cyber delay of communication network," IET Energy Systems Integration, vol. 3, pp. 86-98, 2021.

[11] L. Liu, X. Cai, E. Yu, and X. Yu, "Zhoushan Multi-Terminal VSC-HVDC Transmission Demonstration Project and Its Evaluation," Southern Power System Technology, vol. 13, pp. 79-88, 2019(in Chinese).

Adaptive Pontryagin's Minimum Principle-Inspired Supervised-Learning-based Energy Management for Hybrid Trains Powered by Fuel Cells and Batteries

Hujun Peng[1], Feifei Li[2], Zhu Chen[1], Kai Deng[1], Sebina Jeschke[2], Kay Hameyer[1]

1. Institute of Electrical Machines (IEM)
2. Institute for Information Management in Mechanical Engineering (IMA)
RWTH Aachen University
Aachen, Germany
Email: hujun.peng@iem.rwth-aachen.de
URL: https://www.iem.rwth-aachen.de/go/id/poti/?lidx=1

Acknowledgments

This study is funded by the German Federal Ministry of Transport and Digital Infrastructure (BMVi) under the National Innovation Program Hydrogen and Fuel Cell Technology (NIP). The funding numbers are 03B10502B and 03B10502B2. The authors gratefully thank the support of Siemens AG, Ballard, and NIP.

Keywords

≪Fuel Cell Electric Vehicle (FCEV)≫, ≪Energy System Management≫, ≪Neural Network≫

Abstract

This work develops a supervised learning-based strategy using Long Short-Term Memory Networks (LSTMN) to distribute power between fuel cells and batteries for fuel cell trains. The learning-based strategy exceeds the adaptive Pontryagin's minimum principle (PMP)-based strategy in all driving conditions, which is state-of-the-art. In the case with the most significant difference between the learning-based and the adaptive PMP-based strategy, more consumption of 1.84 % than the off-line PMP strategy, which determines the minimal hydrogen consumption for the same load profiles, is observed for the learning-based strategy. In comparison, 2.54 % more consumption is required by the adaptive PMP-based strategy compared to the off-line strategy.

Introduction

The world's first commercialized hybrid passenger train powered by fuel cells is developed by Alstom and began its operation in Germany in 2018, whose energy source system consists of batteries and proton-exchange membrane (PEM) fuel cells. Since the power change rate of fuel cell systems in hybrid vehicles is limited due to lifetime consideration, another battery system is required to cover the frequently varying load power in various driving conditions [1, 13]. Benefited from this kind of hybrid power source structure, the hydrogen usage efficiency of the hydrogen vehicles can be maximized by possibly optimal power distribution between fuel cells and batteries [14]. Many research works about energy management for hydrogen-powered vehicles are published in recent years, and some comprehensive reviews can be found in [3, 15]. The energy management strategies can be classified into three primary types: the rule-based method, the optimization-based method, and the learning-based method [4]. Developing the rule-based strategies and tuning their parameters are time-consuming, and the rule-based strategy is not optimal at the same time due to insufficient expertise.

The optimization-based methods are divided into global and local optimization-based methods. Furthermore, the global optimization-based energy management strategy is further classified into Pontryagin's Minimum Principle (PMP) as well as Dynamic Programming (DP). It has to be mentioned that the global optimization-based methods are off-line strategies as references to on-line strategies. For the local optimization-based power distribution methods, the most famous ones are the adaptive Pontryagin's Minimum Principle-based strategy (adaptive PMP) as well as the Equivalent Consumption Minimization Strategy (ECMS). Mathematically, ECMS can be obtained from the adaptive PMP because the equivalent factor defined in ECMS is linked to the co-state in the adaptive PMP [6]. The most crucial challenge faced with ECMS and the adaptive PMP is estimating the equivalent factor or the co-state [16, 17]. To our best knowledge, an analytical formula is proposed based on estimated mean values of fuel cell system power as well as the actual battery SoC values in [6] to estimate the co-state accurately and physically. However, the formula in [6] is based on average fuel cell power, and does not consider a so-called time effect. After checking the off-line PMP results, the co-state amplitude at the beginning of driving cycles is larger than that at the end. However, the formula in [6] calculates the same co-state values for the initial and end time point of the driving cycles. Therefore, this kind of sequence or time effect is not considered by the adaptive PMP method.

With machine learning technology developed, the learning-based strategies for hybrid vehicles attract more and more attention. The kind of sequence effect mentioned before can be taken into account by machine learning [8]. However, an enormous amount of data is required to train the learning-based method's control policy or prediction mechanism. Among the learning-based methods, the reinforcement learning is widely applied to develop energy management for hybrid vehicles in recent years [10]. Unlike the supervised learning, the reinforcement learning is a learning process of the policy principle by utilizing the interaction between agents and the environment under the pre-definition of a reward function. The advantage of reinforcement learning is that it can learn long-term accumulated rewards. However, for hybrid vehicles, its most significant drawback lies in the fact that the limited transferability of the trained control models. In this work, the long short-term memory network (LSTMN), as the most useful variant of RNN, is utilized to distribute load power for hybrid trains powered by fuel cells and batteries. Following contributions are included:

- The input variables are chosen based on the physical correlation between them and the output fuel cell power under the optimal control theory. Therefore, a small amount of data is required to train the learning-based strategy.

- After comparing the supervised-learning-based strategy to the off-line PMP strategy and the most energy efficient adaptive PMP-based strategies to our knowledge, a higher hydrogen usage efficiency than the adaptive PMP-based strategy is observed for the developed supervised-learning-based strategy.

Modeling of the serial hybrid powertrain

The entire drivetrain's configuration is displayed in Fig. 1. The fuel cell system *HD8* is supplied by the company *Ballard*, and it has a maximum net output power of 200 kW. The battery system has an energy capacity of about 200 kWh and a nominal voltage of 850 V, which can provide a maximal charge and discharge current of 900 A. The DC/DC converter is supplied by *Siemens*, which can provide a peak power of more than 1 MW and has a DC-link voltage of 1650 V.

The SoC of the entire battery system and the train speed are chosen as the state variables, while lookup tables are applied for modeling other subsystems' power losses. The exact models of each subsystem of the fuel cell train, as well as their related parameters, can be found in [6]. The altitude and speed profiles of three train lines are considered. The driving cycle 1 in Fig. 2a will be applied for training the neural network in the next section. The driving cycle 2 in Fig. 2b will be tested to validate the trained model, which comes from Regional Express 1 running between Aachen and Cologne. The driving cycle 3 in Fig. 2c corresponds to the Regional Express 27 running between Brandenburg and Berlin. This driving cycle is an entire-day velocity trajectory of 19 hours and will be used for training or validation in an on-line simulation environment.

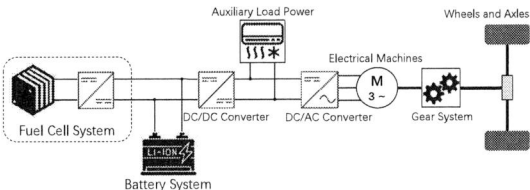

Fig. 1: Configuration of the serial hybrid drivetrain.

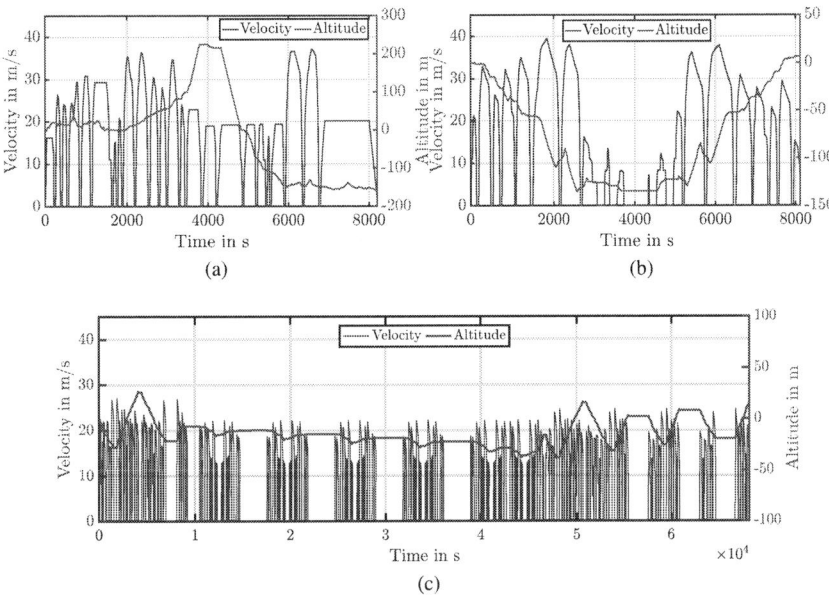

Fig. 2: Three driving cycles used in training and test: (a) Driving cycle 1 for training, (b) Driving cycle 2 for test, (c) Driving cycle 3 between Berlin and Brandenburg for training and test.

Machine learning-based strategies

This section consists of three parts: in the first part, the process of choosing input variables is explained, which has not been found in other works related to machine learning-based strategies for hybrid vehicles. In the second part, the LSTMN method is briefly introduced. Finally, in the third part, the training and test of models are displayed, and the most promising control block is prepared for validation in an on-line simulation environment in the next chapter. The whole process of developing the machine learning-based strategy is presented in Fig. 3.

Selection of input variables

The selection of input variables of the neural network is inspired by the adaptive PMP-based algorithm developed by the authors before. In [6], the adaptive PMP-based algorithm is described in detail. Therefore, a full review is not done here, and merely the process of choosing the input variables is explained. In the adaptive PMP-based strategy, the control variable is determined by minimizing the so-called Hamiltonian in every time instant as follows:

$$H(SoC(t), P_{fc}(t), \lambda(t), t) = \dot{m}_{H_2}(P_{fc}(t)) + \lambda \cdot \dot{SoC}(t), \tag{1}$$

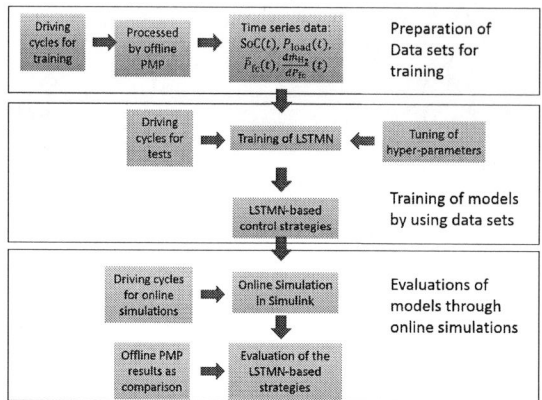

Fig. 3: Flow chart of developing the machine learning-based strategy, whereby the LSTMN method is used for the machine learning.

whereby $\dot{m}_{H_2}$ is the hydrogen mass flow mentioned before, and λ represents the co-state which is eventually a Lagrange factor. Then, by solving the following minimization problem, the optimal fuel cell power is determined in every time instant:

$$P_{fc}^*(t) = \underset{P_{fc}(t)}{\arg\min}(H(SoC(t), P_{fc}(t), \lambda(t), t)). \tag{2}$$

The estimation of the co-state is crucial for the performance of the adaptive PMP-based strategy. In [6], an analytic formula for estimating the co-state is derived from the energy conservation principle as follows:

$$\lambda = -V_{oc} \cdot Q \cdot \left.\frac{d\dot{m}_{H_2}}{dP_{fc}}\right|_{P_{fc} = \overline{P}_{fc}}. \tag{3}$$

The battery capacity Q is assumed constant in this work. Then, according to (3), The mean value of fuel cell power and the battery open-circuit voltage V_{oc}, which is dependent on the SoC, influence mainly the co-state. After inserting (1) and (3) into (2), the control variable, namely the fuel cell power, depends on various inputs, including SoC, battery voltage, battery capacity, hydrogen consumption curves, average values of fuel cell power, demand load power as well as time.

The influence of the time or sequence effect on the output control variable is not apparent. In order to show the time effect, the co-state trajectories resulting from off-line PMP strategies for various driving cycles are demonstrated in Fig. 4. Thereby, the dependency of the co-state on SoC is presented. Furthermore, the estimated co-state by using the formula in (3) is also added. After comparison between the estimated co-state values and those resulting from off-line PMP, the average values of the co-state are well approximated by using the analytical formula (3). However, in the off-line PMP results, there are different co-state values at the same SoC. More precisely, the co-state amplitude near the end of the driving cycles is less than the initial co-state although they have the same SoC values and fuel cell system conditions. So far, the influence of the so-called time effect on the control variable is described. In order to have a control mechanism considering this kind of time effect, the LSTMN method will be used and explained in the next part.

Performance study of the LSTMN under different combinations of input variables

The training of the LSTMN requires the selection of features. The possible candidates of features include the open-circuit voltage V_{oc}, SoC, power demand P_{load}, average fuel cell power $\overline{P}_{fc}$, and the derivative of hydrogen mass flow with respective to the fuel cell power $\frac{d\dot{m}_{H_2}}{dP_{fc}}$ as well as the time. The control variable is the fuel cell power P_{fc}, which is also called the label in LSTMN. The training data sets correspond to the driving cycle 1, and the test of the trained model is applied to driving cycle 2. In the case of training

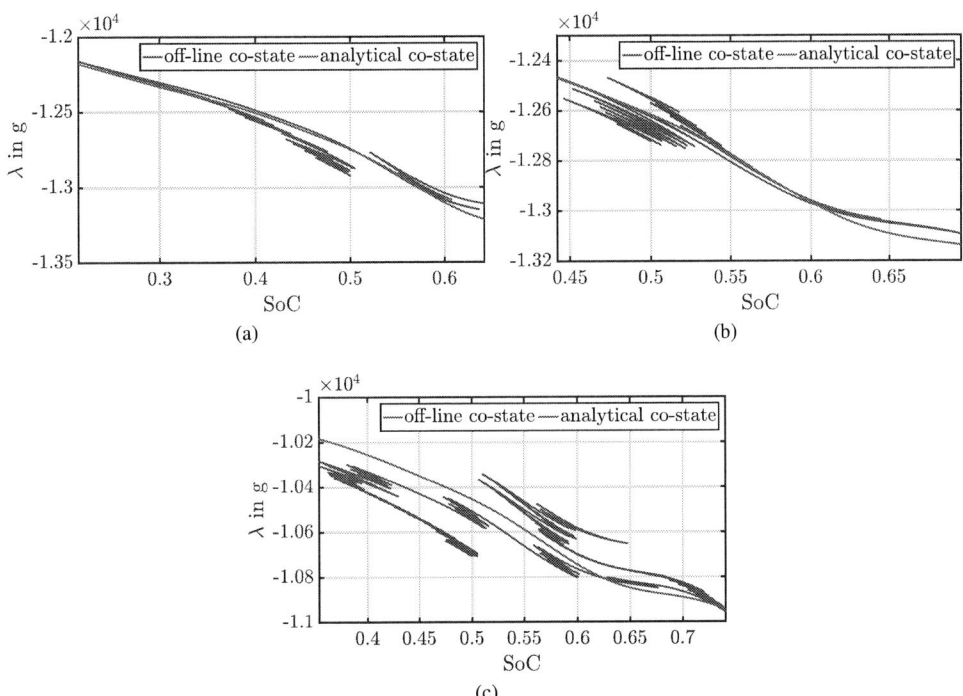

Fig. 4: Comparison between the analytically calculated co-state values and those resulting from off-line PMP algorithm under three different driving cycles: (a) Driving cycle 1 with average fuel cell power of 106.5 kW, (b) Driving cycle 2 with average fuel cell power of 114.1 kW, (c) Driving cycle 3 with average fuel cell power of 78.5 kW.

with more data, the data corresponding to driving cycle 3 is also used besides the driving cycle 1. In all experiments, the training data of the hybrid trains in summer and winter conditions are concatenated together. Before using the LSTMN, some parameter study for choosing best inputs combination will be implemented.

The first experiment aims to analyze which input combination performs best in predicting the label variable under test data sets. Numerical results are presented in Table I. The rMSE stands for root Mean Square Error in the field of machine learning, and MAE represents Mean Absolute Error. In above study

Table I: Performance study of the LSTMN under different inputs combinations.

Training inputs	rMSE (kW)	Training loss in MAE (kW)	Test loss in MAE (kW)
$SoC + P_{load}$	14.820	0.0094	0.0880
$SoC + P_{load} + V_{oc}$	14.729	0.0093	0.0878
$SoC + P_{load} + \overline{P}_{fc} + V_{oc}$	8.310	0.0066	0.0497
$SoC + P_{load} + \overline{P}_{fc} + V_{oc} + \frac{d\dot{m}_{H_2}}{dP_{fc}}$	5.117	0.0062	0.0286
$SoC + P_{load} + \overline{P}_{fc} + V_{oc} + \frac{d\dot{m}_{H_2}}{dP_{fc}} + Time$	5.482	0.0065	0.0302

results, some conclusion can be made: P_{load} is the necessary input variable, $\overline{P}_{fc}$ is useful to set a baseline of season trend, and $\frac{d\dot{m}_{H_2}}{dP_{fc}}$ has positive influence on the prediction results. The time leads to a negative change in the trained model. As for adding the V_{oc}, the performance is not improved significantly, as the similar nature of SoC and V_{oc}, whereby the open-circuit voltage strongly depends on the SoC. In the following parts, more experiments are operated to explore their influences on the model. The numerical results of the comparison are listed in Table II. Through the comparison study, a pretty close influence is generated between SoC and V_{oc} since their nature is similarly related in batteries. Given the rMSE results, the SoC has a slight advantage compared to V_{oc} in all three comparison experiments.

Table II: Comparative influence study between V_{oc} and SoC.

Training inputs	rMSE (kW)	Training loss in MAE (kW)	Test loss in MAE (kW)
$V_{oc} + P_{load} + \overline{P}_{fc}$	7.579	0.0071	0.0459
$SoC + P_{load} + \overline{P}_{fc}$	7.009	0.0068	0.0417
$V_{oc} + P_{load} + \overline{P}_{fc} + \frac{d\dot{m}_{H_2}}{dP_{fc}}$	4.255	0.0062	0.0227
$SoC + P_{load} + \overline{P}_{fc} + \frac{d\dot{m}_{H_2}}{dP_{fc}}$	4.175	0.0057	0.0225
$V_{oc} + P_{load} + \overline{P}_{fc} + \frac{d\dot{m}_{H_2}}{dP_{fc}} + $ Time	5.384	0.0066	0.0306
$SoC + P_{load} + \overline{P}_{fc} + \frac{d\dot{m}_{H_2}}{dP_{fc}} + $ Time	4.721	0.0061	0.0259

Due to the advantage of choosing SoC over the open-circuit voltage, the standard inputs are assigned as SoC, P_{load}, $\overline{P}_{fc}$ and $\frac{d\dot{m}_{H_2}}{dP_{fc}}$. The input of time is not explicitly further adopted because the performance of the trained model without an explicit input of time is better than that with time as the input.

After choosing the input variables, the fine-tuning experiments of hyper-parameters are implemented. This step tests if using more training data by utilizing driving cycle 3, changing the learning rate, and adding an early-stop policy influence the training and testing results. The fine-tuning results are expressed in Table III. From the results, the change of learning rate is illustrated as a strong influence factor, which is capable of improving the model's performance twofold.

While monitoring the training and test loss, as shown in Fig. 5, the over-fitting phenomenon are much more significant with more training data from driving cycle 3. Hence, the early stop policy is adopted when involving more training data by using driving cycle 3. The computational time takes about ten seconds to train one epoch by using a computer with *2.6 GHz 6Core Intel Core i7*. Since 50 epochs are used to train the neural network, it takes about five hundred seconds for a combination of input variables per a setup of hyper-parameters. The low computational time results from that merely one driving cycle is applied to train the neural network, and this driving cycle contains about 16000 sampled time points. Furthermore, the test results of the best two models in Table III, corresponding to the last two rows, are shown in Fig. 6, whereby negligible difference among models is observed.

Table III: Results of hyper-parameters fine tuning. Thereby, *lr* stands for the learning rate and more training data comes from driving cycle 3.

Methods	rMSE (kW)	Training loss in MAE (kW)	Test loss in MAE (kW)
Standard inputs / $lr = 0.001$	4.175	0.0057	0.0225
Standard inputs / $lr=0.0001$	2.213	0.0075	0.0114
Standard inputs / More training data / $lr=0.0001$	3.607	0.0051	0.0179
Standard inputs / $lr=0.0001$ / early stop	2.012	0.0078	0.0097
Standard inputs / More training data / $lr=0.0001$ / early stop	1.718	0.0047	0.0072

In the next section, the machine learning-based strategy's fuel economy will be validated by applying simulation. It is worth mentioning that the trained model includes four input signals: load power, SoC, fuel cell average power, and the derivative of the mass flow regarding the fuel cell output power. Furthermore, the training data set merely uses the driving cycle 1 to keep the training effort low, and the learning-based strategy will be tested under driving cycles 2 and 3 for different weather conditions in the on-line simulation environment.

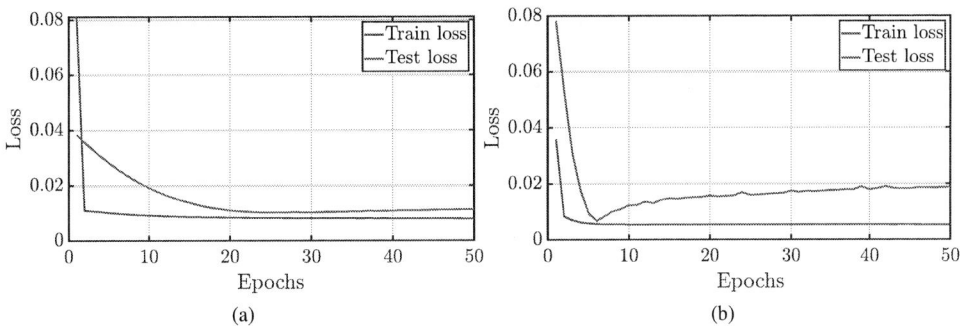

(a)   (b)

Fig. 5: Training loss and test loss monitoring: (a) Training process of standard inputs with learning rate equals 0.0001 and training data from driving cycle 1, (b) Training process of standard inputs with more training data from driving cycle 3 besides the data from driving cycle 1 at learning rate equals 0.0001.

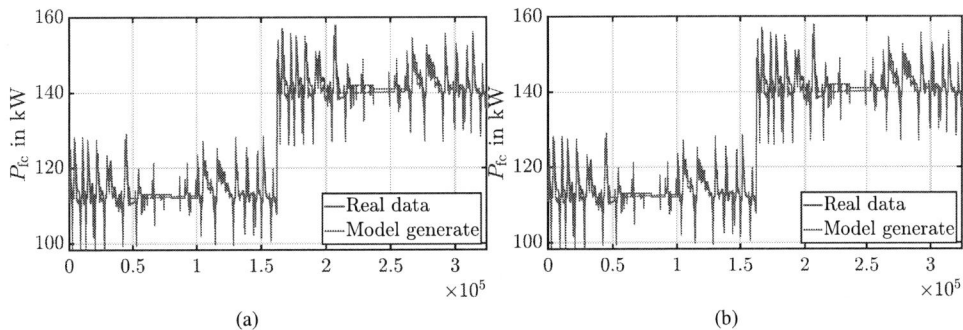

(a)   (b)

Fig. 6: Test results of the best two trained models, whereby driving cycle 2 is used for testing. In the legend, the *Real data* represents the control sequences under the off-line PMP strategy, and the *Model generate* is the output of the trained model. (a) The trained model with standard inputs and training data from driving cycle 1, (b) The trained model with standard inputs with training data from driving cycles 1 and 3.

On-line simulation results with comparisons to off-line and adaptive PMP strategies

In the on-line simulation environment, the learning-based strategy will be compared to the off-line PMP strategy to show how close the machine learning-based strategy is to the optimal strategy. Besides that, the learning-based strategy will be compared to the adaptive PMP-based strategy developed by authors before, which is the most fuel-efficient strategy so far, to our knowledge. The high fuel economy and adaptivity of the developed learning-based strategies will be emphasized through comparisons to either the off-line or on-line strategies.

It is worth mentioning that in the on-line simulation environment, the control signal from the energy management system, which is the fuel cell system power, suffers from a rate limit in order to reduce fuel cell aging, which is about 20 kW/s for a system with a maximum power of 200 kW. Furthermore, due to simplicity, this rate limit is set to be the same for increasing and decreasing fuel cell power. The charge and discharge current of the battery system is determined by the driving cycle because the battery system assists the train acceleration and regenerative braking. Due to the battery system's high power dynamic capability, there are no other rate limiters in the on-line simulation.

In Fig. 7a, the fuel cell power trajectories under the learning-based strategy are displayed, together with the off-line PMP results. The fuel power trajectories are close to their mean values to a large degree and above the mean values during acceleration while below the average values in the regenerative braking phase. Therefore, the battery system provides and absorbs peak power during acceleration and regenerative braking phases, either under the machine learning-based strategy or the off-line PMP-based strategy, as shown in Fig. 7c. As the result of the battery power trajectories, the corresponding SoC trajectories can be found in Fig. 7b. It is worth mentioning that the change rate in the fuel cell power trajectories, corresponding to the acceleration and regenerative braking phases, adheres to the dynamic limits of the fuel cell power mentioned before. These acceleration and braking phases take more than one minute, and the variation in the fuel cell power is less than 40 kW. Regarding the fuel cell trajectories, the closeness of the trajectories to their mean values means utilizing the convexity of the hydrogen consumption curve, which enables working points of the fuel cell system with high efficiency. The increase of fuel cell power during acceleration and the decrease during regenerative braking help to reduce battery losses because the batteries cover the difference between load demand and supplied fuel cell power. Thereby, an oscillation of fuel cell power, to some degree, helps reduce battery current, ohmic losses, and total hydrogen consumption. Therefore, a compromise between maintaining the fuel cell system in operational points with high efficiency and reducing inner battery losses has to be met to reduce the total hydrogen consumption. After comparing the off-line PMP strategy, which realizes the best fuel economy, it is evident that the machine learning-based strategy can work close to the best compromise and leads to good fuel economy. In Fig. 7b, the SoC trajectory has an end value close to its initial value, which shows that the charge-sustaining mode is maintained for the fuel cell trains without an extra charger. Regarding hydrogen consumption, the learning-based strategy consumes 1.84 % more compared to the off-line PMP strategy in summer for driving cycle 3.

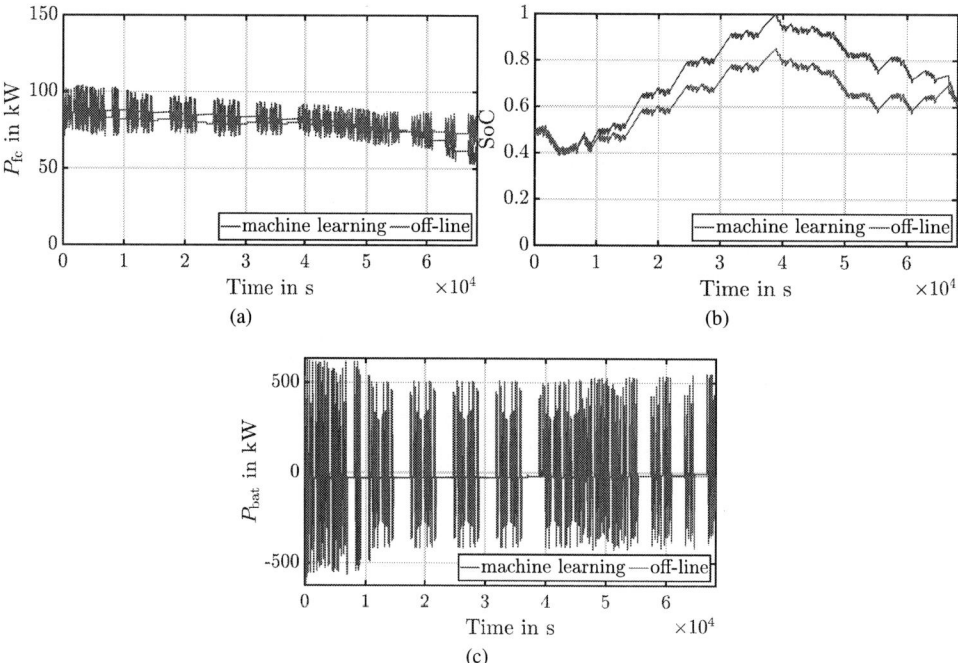

Fig. 7: On-line simulation results under the learning-based strategy for driving cycle 3 in summer with the off-line optimal results as references: (a) Fuel cell power trajectories, (b) SoC trajectories, (c) Battery power trajectories.

The validation is also performed for the winter weather conditions. Moreover, the difference between the learning-based strategy and the off-line PMP strategy decreases in winter conditions compared to the difference in summer because a higher average load demand power is estimated with lower relative errors. The fuel cell power trajectories are also close to their average values, which are about 30 kW larger than in summer, and the SoC trajectories are also similar to the cases in summer since the higher auxiliary consumption due to air conditioning will be covered by the fuel cell system. Regarding hydrogen consumption, the learning-based strategy consumes 1.58 % more than the off-line strategy in winter for driving cycle 3.

The validation of the fuel economy of the learning-based strategy, with comparison to the off-line strategies, is also done for driving cycle 2 and different weather conditions. Some related parameters to fuel economy can be found Table IV.

Besides comparing the learning-based strategy and the off-line PMP, another comparison will be given between the learning-based strategy and the adaptive PMP-based strategy since this work aims to develop a strategy better than the best adaptive PMP-based strategy known so far. The comparison between them is carried under driving cycle 3 in summer weather conditions.

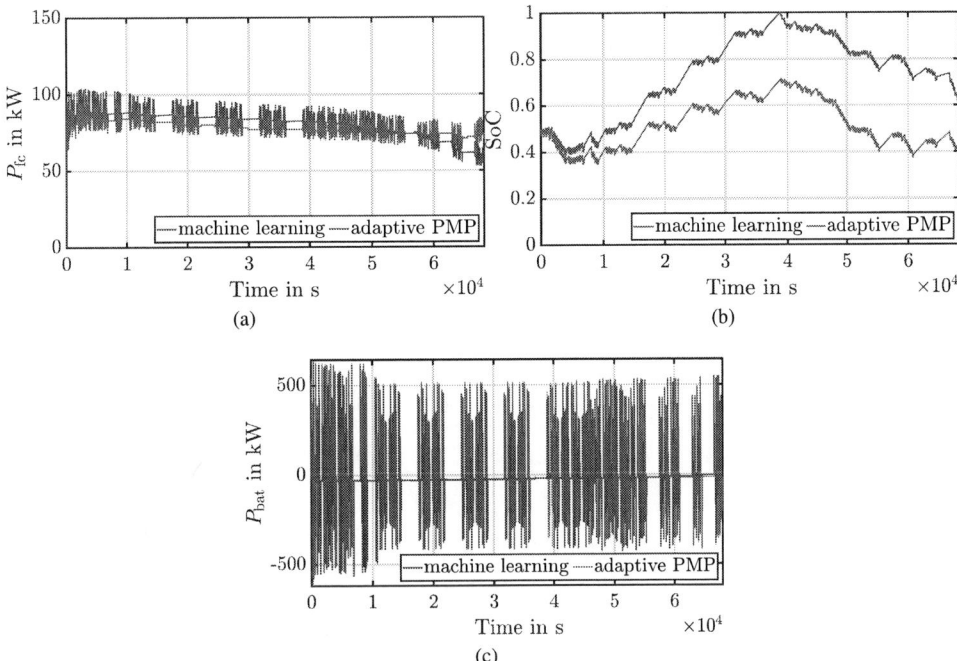

Fig. 8: Comparison of the learning-based strategy and the adaptive PMP-based strategy under driving cycle 3 in summer weather conditions: (a) Fuel cell trajectories, (b) SoC, (c) Battery power trajectories.

The resulting trajectories, which make the learning-based strategy different from the adaptive PMP-based strategy, can be found under the driving cycle 3 with a longer running time. The battery power trajectories of the learning-based strategy and the adaptive PMP-based strategy, as shown in Fig. 8c, are slightly different. In Fig. 8a and Fig. 8b, more differences regarding the fuel cell power and SoC trajectories can be identified. The fuel cell power under the learning-based strategy has a higher average value than that under the adaptive PMP-based strategy in the beginning and midterm of the driving cycle, while near the end of driving cycle 3, the fuel cell power under the learning-based strategy is less than that of the adaptive PMP-based strategy. In other words, under the learning-based strategy, more fuel cell power than the estimated average power will be used at the beginning of the driving cycle and less fuel cell power than the estimated average power used near the end of the driving cycle. This phenomenon is similar to the time or sequence effect mentioned before, whereby in the off-line PMP, the same SoC values correspond to different co-state values due to different time points within driving cycles. More detailed to speak, the amplitude of the co-state at the beginning of the driving cycle is larger than that near the end of the driving cycle, even though the same SoC values exist. Due to the larger co-state amplitude at the beginning of the driving cycle, the off-line PMP strategy outputs fuel cell power more than its average first and decreases the power later less than its average. Therefore, it is evident that the learning-based strategy can consider the time or sequence effect to minimize hydrogen consumption. A simple explanation of the benefits of considering this kind of time or sequence effect follows. Due to the difference in fuel cell power, the SoC trajectory under the learning-based strategy lies in higher values than the adaptive PMP-based strategy, as shown in Fig. 8b. According to the battery characteristics, a higher SoC results in a higher open-circuit voltage and lower battery resistance. Then, for the same battery power, a lower battery current and inner ohmic losses result. As mentioned before, the basic principle of minimizing hydrogen consumption for fuel cell trains lies in the compromise between maintaining the fuel cell system working at higher efficiency and reducing inner battery losses. By considering this kind of time or sequence effect in the learning-based strategy, the battery losses will be further decreased.

The comparison between the learning-based strategy and the adaptive PMP-based strategy is also done for driving cycle 2 and other weather conditions. The results can be found in Table IV. Summarily, the

Table IV: Comparison of the learning-based strategy and the adaptive PMP-based strategy with off-line PMP as references.

		Driving cycle 2		Driving cycle 3	
		Summer	Winter	Summer	Winter
Learning-based	m_{H_2} in g	15599	20843	87589	120379
	m_{H_2} in g under PMP	15200	20339	86007	118504
	Compared to PMP	2.63 %	2.48 %	1.84 %	1.58 %
Adaptive PMP	m_{H_2} in g	15617	20862	84954	120390
	m_{H_2} in g under PMP	15195	20349	82847	118486
	Compared to PMP	2.78 %	2.52 %	2.54 %	1.61 %

most important results of the supervised-learning-based strategy, with comparisons to the off-line PMP and the on-line adaptive PMP-based strategies, are included in Table IV. The learning-based strategy has a higher fuel economy than the adaptive PMP-based strategy in all driving conditions, independent of the length of driving cycles and weather conditions. The reason is that the learning-based strategy considers the time or sequence effect, besides all the merits included in the adaptive PMP-based strategy.

Conclusions

In this work, a learning-based strategy is developed using LSTMN, a variant of the recurrent neural network, to distribute power between fuel cells and batteries for fuel cell trains. The inputs of the neutral network are chosen based on inspiration from the adaptive PMP-based strategy. Due to their physical relation to the output fuel cell power, much less data is required to train the learning-based strategy than those in the literature about learning-based strategies. Based on simulation results, the learning-based strategy outperforms the best adaptive PMP strategy, which is the best strategy to our knowledge, in all driving conditions. In the case with the most significant difference between the learning-based and the adaptive PMP-based strategy, more consumption of 1.84 % compared to the off-line PMP results for the learning-based strategy for a typical driving cycle of regional trains in summer, while 2.54 % more consumption is required by the adaptive PMP strategy than the off-line PMP strategy for the same driving cycle.

References

[1] Fletcher, T. and Ebrahimi, K., 2020. The effect of fuel cell and battery size on efficiency and cell lifetime for an L7e fuel cell hybrid vehicle. Energies, 13(22), p.5889

[2] Lü, X., Wu, Y., Lian, J., Zhang, Y., Chen, C., Wang, P. and Meng, L., 2020. Energy management of hybrid electric vehicles: A review of energy optimization of fuel cell hybrid power system based on genetic algorithm. Energy Conversion and Management, 205, p.112474

[3] Sulaiman, N., Hannan, M.A., Mohamed, A., Ker, P.J., Majlan, E.H. and Daud, W.W., 2018. Optimization of energy management system for fuel-cell hybrid electric vehicles: Issues and recommendations. Applied energy, 228, pp.2061-2079

[4] Peng, H., Chen, Z., Deng, K., Dirkes, S., Ünlübayir, C., Thul, A., Löwenstein, L., Sauer, D.U., Pischinger, S. and Hameyer, K., 2021. A comparison of various universally applicable power distribution strategies for fuel cell hybrid trains utilizing component modeling at different levels of detail: From simulation to test bench measurement. eTransportation, 9, p.100120

[5] Peng, H., Chen, Z., Li, J., Deng, K., Dirkes, S., Gottschalk, J., Ünlübayir, C., Thul, A., Löwenstein, L., Pischinger, S. and Hameyer, K., 2021. Offline optimal energy management strategies considering high dynamics in batteries and constraints on fuel cell system power rate: From analytical derivation to validation on test bench. Applied Energy, 282, p.116152

[6] Peng, H., Li, J., Löwenstein, L. and Hameyer, K., 2020. A scalable, causal, adaptive energy management strategy based on optimal control theory for a fuel cell hybrid railway vehicle. Applied Energy, 267, p.114987

[7] Chen, Z., Liu, Y., Ye, M., Zhang, Y. and Li, G., 2021. A survey on key techniques and development perspectives of equivalent consumption minimisation strategy for hybrid electric vehicles. Renewable and Sustainable Energy Reviews, 151, p.111607

[8] Li, W., Cui, H., Nemeth, T., Jansen, J., Ünlübayir, C., Wei, Z., Zhang, L., Wang, Z., Ruan, J., Dai, H. and Wei, X., 2021. Deep reinforcement learning-based energy management of hybrid battery systems in electric vehicles. Journal of Energy Storage, 36, p.102355

[9] Zhou, D., Ravey, A., Al-Durra, A. and Gao, F., 2017. A comparative study of extremum seeking methods applied to online energy management strategy of fuel cell hybrid electric vehicles. Energy conversion and management, 151, pp.778-790

[10] Tan, H., Zhang, H., Peng, J., Jiang, Z. and Wu, Y., 2019. Energy management of hybrid electric bus based on deep reinforcement learning in continuous state and action space. Energy Conversion and Management, 195, pp.548-560

[11] Han, L., Jiao, X. and Zhang, Z., 2020. Recurrent neural network-based adaptive energy management control strategy of plug-in hybrid electric vehicles considering battery aging. Energies, 13(1), p.202

[12] Munoz, P.M., Correa, G., Gaudiano, M.E. and Fernández, D., 2017. Energy management control design for fuel cell hybrid electric vehicles using neural networks. International Journal of Hydrogen Energy, 42(48), pp.28932-28944

[13] Kandidayeni, M., Trovo, J.P., Soleymani, M. and Boulon, L., 2022. Towards health-aware energy management strategies in fuel cell hybrid electric vehicles: A review. International Journal of Hydrogen Energy.

[14] Manoharan, Y., Hosseini, S.E., Butler, B., Alzhahrani, H., Senior, B.T.F., Ashuri, T. and Krohn, J., 2019. Hydrogen fuel cell vehicles; current status and future prospect. Applied Sciences, 9(11), p.2296.

[15] Ahmadi, S., Bathaee, S.M.T. and Hosseinpour, A.H., 2018. Improving fuel economy and performance of a fuel-cell hybrid electric vehicle (fuel-cell, battery, and ultra-capacitor) using optimized energy management strategy. Energy Conversion and Management, 160, pp.74-84.

[16] Li, H., Ravey, A., NDiaye, A. and Djerdir, A., 2019. Online adaptive equivalent consumption minimization strategy for fuel cell hybrid electric vehicle considering power sources degradation. Energy Conversion and Management, 192, pp.133-149.

[17] Musardo, C., Rizzoni, G., Guezennec, Y. and Staccia, B., 2005. A-ECMS: An adaptive algorithm for hybrid electric vehicle energy management. European journal of control, 11(4-5), pp.509-524.

A Case Study of Pole-Phase Changing Induction Machine Performance

Konstantina Bitsi[1], and Sjoerd G. Bosga[1,2]

[1]KTH Royal Institute of Technology, Division of Electric Power
and Energy Systems, Stockholm, Sweden
[2]ABB Corporate Research, Västerås, Sweden
E-mails: bitsi@kth.se, sjoerd.bosga@se.abb.com

Acknowledgments

This research project is supported in part by the Swedish Energy Agency (Energimyndigheten).

Keywords

≪FEM modeling≫, ≪independently-controlled stator coils≫, ≪induction machine≫, ≪maximum efficiency operation≫, ≪maximum torque per ampere operation≫, ≪phase-changing≫, ≪pole-changing≫.

Abstract

Pole-phase changing (PPC) induction machines (IMs) can achieve improved efficiency as well as wider torque-speed range compared to their fixed pole-phase counterparts. In this paper, a 4-pole IM is designed and evaluated in terms of its pole-phase changing performance.

Introduction

The exponential growth of the global electric vehicle market in recent years represents a significant on-going environmental sustainable development. In order to provide viable solutions in the competitive automotive industry, the produced electric motors and drivetrain designs need to meet demanding requirements, including high efficiency, increased torque and power density capabilities and a wide high-speed range. Induction machines (IMs) constitute a robust, reliable and rare-earth-free alternative to widely adopted interior permanent magnet synchronous machines [1–3]. However, due to their high values of leakage inductance, they inherently showcase poor flux weakening capability.

To improve the high-speed torque and power output of IMs, leakage-minimization strategies [4, 5], dual drive topologies [6, 7] as well as torque-speed range manipulation methods have been proposed. The latter include also the electronic pole-changing, which involves the utilization of the converter in order to dynamically change the pole number of the machine [8–11]. In accordance with this technique, the wound independently-controlled stator coils (WICSC) machine is introduced in [12]. This topology comprises an asynchronous rotor and individually excited toroidal coils in each stator slot. The adopted stator-winding configuration allows the independent current control of each stator-slot coil and, thus, WICSC topologies exhibit the ability of real-time pole-phase changing (PPC) during operation [13]. To independently control each stator toroidal coil, it should be connected to an individual converter leg, as shown in Fig. 1.

A widely selected pole number for machines used in automotive applications is four, as 4-pole designs offer a good compromise between speed range and torque capability. Therefore, the goal of this paper is to demonstrate the benefits that PPC can offer to the performance of a 4-pole IM, originally intended for fixed pole-phase operation. Specifically, the study will compare the produced torque capability, efficiency and flux-weakening range while operating with a PPC strategy versus typical fixed pole-phase operation.

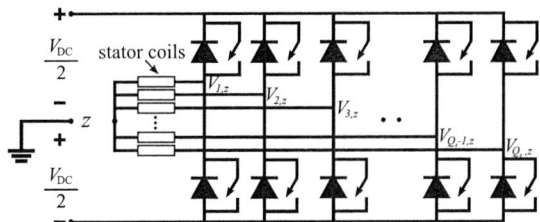

Fig. 1: Schematic of an m-phase bridge-type voltage source inverter.

This paper is organized as follows: in Section II, the design and the electromagnetic model of the examined 4-pole IM are discussed. Section III presents the identification processes of the optimal pole-phase modes based on the selected criteria. In Section IV, the performance of the studied topology as a PPCIM is evaluated in terms of torque-speed capability and efficiency using different operating strategies. The main conclusions are summarized in the final section.

WICSC Electromagnetic Model

The geometry of the investigated WICSC IM is shown in Fig. 2. This geometry is designed within the scope of this analysis for fixed 4-pole operation, following the design algorithm described in [14], with the specifications of Table I as input parameters.

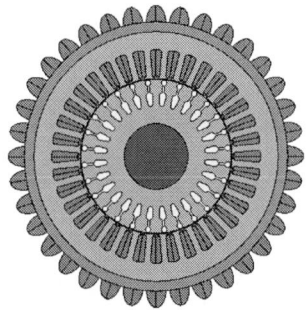

Fig. 2: Geometry of the investigated WICSC IM, originally designed for 4-pole fixed pole-phase operation.

Table I: Geometrical and nominal operating parameters of the original 4-pole IM

Stator outer diameter (mm)	256
Axial length (mm)	200
Torque (Nm)	64
Speed (rpm)	3000
Mechanical power (kW)	20
Number of poles	4

The PPC analysis is performed numerically using 2D FEM models, with sinusoidal stator current supply in each independent coil. The current in each phase n can be defined as [12]

$$i_{s,n} = \hat{i}_s \cos\left(\frac{p}{2}\omega_s t - (n-1)\frac{\pi}{m}\right) \tag{1}$$

where $\hat{i}_s$ is the stator current amplitude, ω_s the angular electrical frequency, m the number of phases and p the number of poles. The number of stator slots per pole per phase is expressed as

$$q_s = \frac{Q_s}{mp} \tag{2}$$

where Q_s is the number of stator slots. In this study, the parameter q_s is set to unity, to maximize the winding factor and minimize the harmonic content. Therefore, during pole-phase modulation, the highest permissible phase number is selected for a given pole number of the stator current excitation according to (2). It should be noted that with the term "phase", we do not mean coils that are connected physically between them, but rather share the current excitation pattern of a typical "phase" in a conventional IM.

In the FEM implementation, the stator-coil end-winding resistance and inductance are present in the 2D

FEM model, using external electric circuits for all stator toroidal coils. The end-winding inductances are approximated by manipulating the expression for an air-cored solenoid inductance to take into account the presence of iron parts in the vicinity [15]. On the rotor side, the resistances and the inductances of the short-circuit ring elements are included in the FEM model, by implementing an external electric circuit to connect the rotor bars. The end-winding squirrel-cage inductance is estimated based on the method of calculating the self-inductance of an identical circular ring, as in [16].

Optimal Pole-Phase Selection Methods

The first step towards determining the optimal pole-phase selection of a PPC machine at each operating point is to establish the goal of the operating strategy. In this paper, two strategies will be investigated: (a) the maximization of the torque per ampere capability, i.e. maximum extension of the torque-speed envelope, as well as (b) the maximization of the machine efficiency. For these strategies, the appropriate current-contol schemes need to be determined.

Maximum Torque Per Ampere Operation

To accurately specify the pole-phase selection that outputs the highest torque per ampere at each specific operating point, all maximum torque capabilities of the possible pole-phase selections should be determined. For this purpose, the operation of each of these selections should be controlled based on the following schemes:

1. Maximum torque per ampere (MTPA) control: At low speeds, below the maximum voltage imposed by the DC bus voltage limit, the maximum permissible produced torque is only dependent on the current excitation. When i_s is low and the behavior of the iron is linear, the angular slip frequency ω_{slip} in the rotor-flux reference frame can be analytically estimated using the following expression

$$\omega_{\text{slip}} = s\omega_s = \frac{R'_r}{L_m}\frac{i_{sq}}{i_{sd}} \tag{3}$$

   where R'_r is the rotor resistance referred to the stator side in the steady-state equivalent circuit of the IM model [17]. As the current level is increased, saturation becomes more prominent and the machine parameters R'_r and L_m are no more constant [18]. The maximum current limit is imposed by the converter current rating.

2. Flux weakening (FW) control: When the DC bus voltage limit is reached at nominal speed, the voltage of the machine cannot be further increased. Therefore, the stator flux λ_{sd} should be decreased in order for the machine to reach higher speeds, resulting inevitably in a lower maximum torque capability.

3. Maximum Torque Per Volt (MTPV) Control: In the case that both the current and voltage limits must be violated to reach a higher-speed region, λ_{sd} should be further reduced. To achieve this, the operating current level of the machine should be decreased, while remaining at the maximum voltage, and the slip frequency should be reduced accordingly to output the maximum permissible torque.

Maximum Efficiency Operation

Based on the strategy of the the previous section, the most extended torque-speed envelope of the machine can be obtained. Within that envelope, the optimal efficiency map can be determined, by estimating and comparing the total losses of all possible pole-phase selections. For this purpose, the following loss components are considered:

1. Resistive losses: The stator copper losses $P_{\text{Cu},s}$ can be calculated using the following expression

$$P_{\text{Cu},s} = \frac{Q_s}{2}\hat{i}_s^2 R_{s,\text{coil}} \tag{4}$$

where $R_{s,\text{coil}}$ is the resistance of each stator-slot coil.

The losses of the aluminum squirrel-cage rotor $P_{\text{Al},r}$ can be estimated as follows

$$P_{\text{Al},r} = \frac{Q_r}{2}\hat{\imath}_r^2 R_r \tag{5}$$

subject to

$$R_r = R_b + \frac{2R_{\text{scr}}}{\sin^2\left(\dfrac{\pi p}{Q_r}\right)} \tag{6}$$

where $\hat{\imath}_r$ is the rotor-bar current amplitude, R_r the equivalent rotor-bar resistance, corresponding to the resistance of one bar and two short-circuit ring elements, R_b the rotor-bar resistance and R_{scr} the resistance of one short-circuit ring element [19].

2. Iron losses: The magnetic flux density in each mesh element κ in the stator and rotor core can be numerically evaluated and approximated as a Fourier series expansion

$$B_\kappa(t) = \sum_{\nu=1}^{\infty} |B_{\kappa,\nu}| \cos\left[\omega_\nu t + \arg\left(B_{\kappa,\nu}\right)\right] \tag{7}$$

where $B_{\kappa,\nu}$ is the complex Fourier coefficient of the magnetic flux density in the mesh element κ and for frequency ν [20]. The stator and rotor iron losses P_{Fe} of the machine can be estimated then as

$$P_{\text{Fe}}L_a \sum_{\kappa=1}^{\kappa_{\max}} \sum_{\nu=1}^{\nu_{\max}} \left\{ k_{\text{e}}\left(|B_{\kappa,\nu}|\right) f_\nu^2 |B_{\kappa,\nu}|^2 + k_{\text{h}}\left(|B_{\kappa,\nu}|\right) f_\nu |B_{\kappa,\nu}|^2 \right\} \Delta_\kappa \tag{8}$$

where Δ_κ is the area of the mesh element κ and k_{e} and k_{h} the eddy-current and hysteresis loss coefficients respectively. These coefficients are assumed to be a function of the magnetic flux density and are expressed as third-order polynomials [21].

3. Mechanical losses: The windage loss component P_w is approximated using the following empirical formula

$$P_w = C_D \pi \rho r_r L_\alpha v_r^3 \tag{9}$$

subject to

$$\sqrt{C_D} = 2.04 + 1.768 \ln\left(\text{Re}\sqrt{C_D}\right)$$
$$\text{Re} = \frac{\rho}{\mu}\omega_m r_r \delta \tag{10}$$

where C_D is the skin friction coefficient, ρ the air density, v_r the rotor surface speed, μ the kinematic viscosity of air, Re the Reynolds number and ω_m the angular mechanical frequency [22]. Friction losses are considered negligible.

The efficiency η of each examined pole-phase selection can be determined numerically at each operating point as follows:

$$\eta = \frac{P_m}{P_m + P_{\text{Cu},s} + P_{\text{Al},r} + P_{\text{Fe},s} + P_{\text{Fe},r} + P_w}. \tag{11}$$

where P_m is the mechanical power.

Results and Discussion

In this section, the PPC performance of the designed original 4-pole IM is investigated. For this study, the DC bus voltage limit is set to 550 V. Moreover, the machine is excited up to 2 times the original rated current, as the aim is to focus on the possible benefits of PPC operation at higher-load operation. This high torque capability is feasible for short-time operation, as the saturation on the stator and rotor iron remain on a low level.

Maximum Torque Per Ampere Operation

With maximum torque per ampere operating mode, the torque as well as the torque per ampere capability of all pole-phase selections at $n = 500$ rpm are shown in Fig. 3. For the examined geometry, the torque per ampere capacity of the 2-pole/18-phase configuration starts to saturate already at $\hat{\imath}_s = 5.4$ A, rendering the 4-pole/9-phase operation superior. However, saturation seems to impact the torque output of the 4-pole/9-phase only at the highest examined excitation levels. As a result, the configurations with pole number of 6 or higher do not manage to offer better torque outputs within the examined range, although the 6-pole is very close to overcome the 4-pole selection.

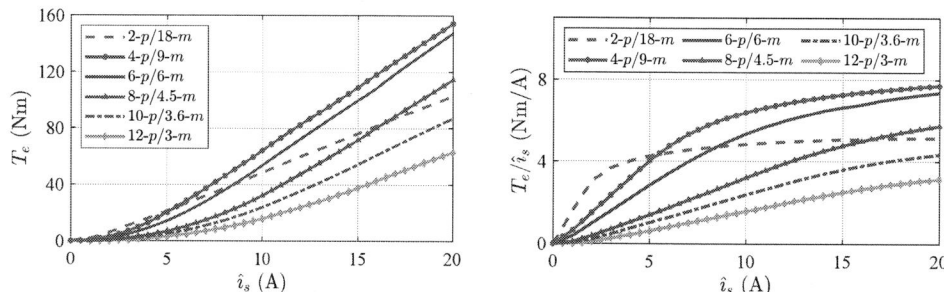

Fig. 3: Plots of $\hat{\imath}_s$ vs (a) torque T_e and (h) $T_e/\hat{\imath}_s$ at $n = 500$ rpm for different pole-phase selections of an original 4-pole machine at maximum torque per ampere operation.

The optimal torque-speed map for maximum torque per ampere operation is shown in Fig. 4. The lowest pole-number operation (i.e. 2-pole/18-phase) significantly widens the flux weakening range compared to a fixed 4-pole operation above 3600 rpm. Specifically, it can be observed in Fig. 5 that pole-phase changing helps to extend the torque-speed envelope significantly at higher speeds, offering an increase equal to 142% in the produced torque per ampere capability at the maximum speed.

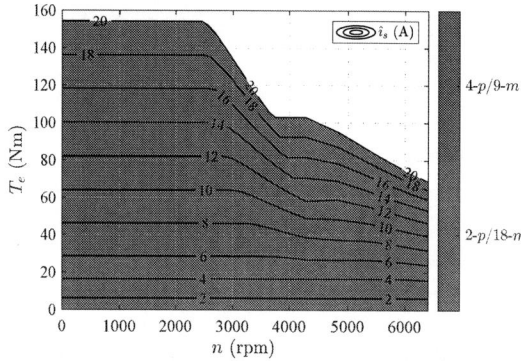

Fig. 4: Optimal torque-speed map of an original 4-pole IM operated as PPC machine at maximum torque per ampere operation for chosen peak current level of $\hat{\imath}_s = 20$ A.

Maximum Efficiency Operation

In this section, the WICSC IM is operated using a maximum efficiency strategy. The total losses P_l, the efficiency η as well as the percentages of stator and rotor losses for the different pole-phase operations

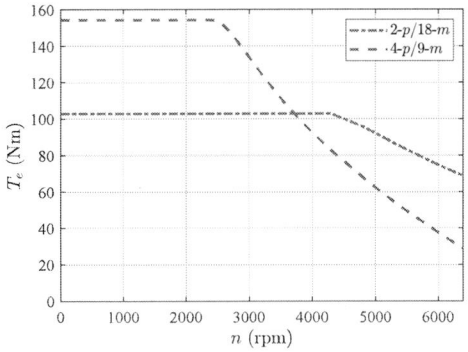

Fig. 5: Torque-speed envelopes of 4-pole/9-phase and 2-pole/18-phase operations of the studied IM with maximum torque per ampere strategy for chosen peak current level of $\hat{\imath}_s = 20$ A.

at $n = 1500$ rpm are shown in Fig. 6 and Fig. 7 respectively. At the examined speed, the total losses of the 4-pole/9-phase operation are the lowest for torque levels higher than $T_e = 45$ Nm, and therefore this pole-phase selection is the most beneficial at the higher range of current excitation levels. Seeing the comparison of stator vs rotor losses shown in Fig. 7, it can be observed that when changing from 2 to 4 poles, the losses shift from the rotor to the stator side. A similar observation is true for the 6-pole/6-phase operation, as it demonstrates higher stator losses and lower rotor losses compared to the 4-pole/9-phase case.

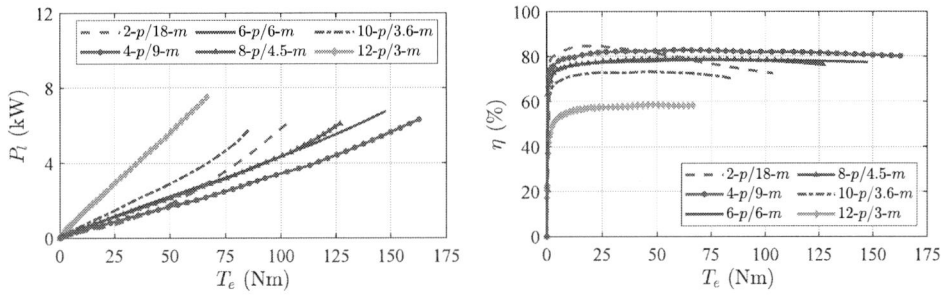

Fig. 6: Plots of T_e vs total losses P_l and η at $n = 1500$ rpm for different pole-phase selections of an original 4-pole machine at maximum efficiency operation.

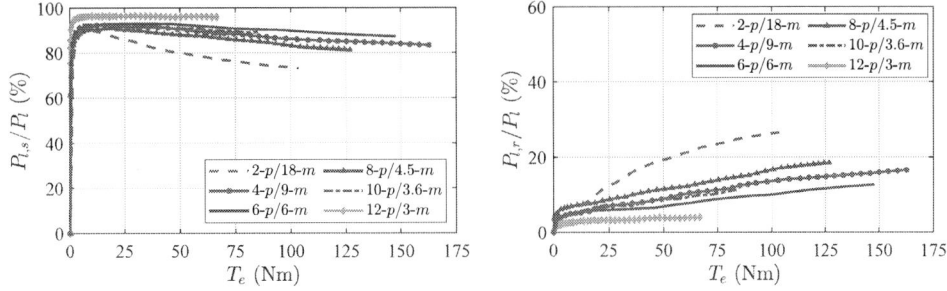

Fig. 7: Plots of T_e vs percentages of stator and rotor losses over total losses, $P_{l,s}/P_l$ and $P_{l,r}/P_l$ respectively, at $n = 1500$ rpm for different pole-phase selections of an original 4-pole IM at maximum efficiency operation.

In Fig. 8, the optimal efficiency map for the whole speed range is depicted. The 2-pole/18-phase operation accomplishes wider flux-weakening range to a noteworthy extent, as wells as improved efficiency at a big part of the lower torque levels. The 4-pole/9-phase operation offers the best efficiency for the highest

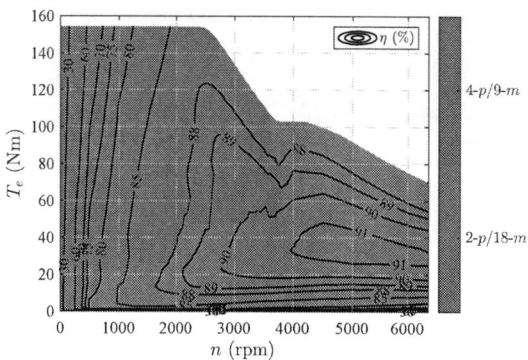

Fig. 8: Optimal efficiency map of an original 4-pole IM operated as PPC machine at maximum efficiency operation for chosen peak current level of $\hat{\imath}_s = 20$ A..

torque capabilities at lower speeds.

Conclusion

The goal of this paper is to examine the impact of pole-phase changing in the performance of an IM topology, originally designed for fixed 4-pole operation. The assessment is carried out focusing on the efficiency and the maximum torque output over the entire speed range. It is shown that by switching to the lowest pole number configuration (i.e. 2-pole/18-phase), a considerable extension of the available torque capability is achieved at higher speeds compared to the 4-pole/9-phase operation, while operation with lower stator current is possible at the lowest load levels. In terms of efficiency, the 2-pole operation is superior at high speed as well as low load conditions, while the 4-pole/9-phase offers the best efficiency for higher torque capabilities at low speeds. The pole-phase combinations with higher pole number than 4 demonstrate poorer performance for the examined excitation range. Further analysis is required in order to assess the influence of the various design variables towards the full utilization of the pole-phase changing capability of IMs.

References

[1] M. V. Terzic, D. S. Mihic and S. N. Vukosavić: Design of High-Speed, Low-Inertia Induction Machines With Drag-Cup Rotor, IEEE Transactions on Energy Conversion, vol. 29, no. 1, pp. 169-177, March 2014.

[2] M. Zeraouila, M. E. H. Benbouzid and D. Diallo: Electric motor drive selection issues for HEV propulsion systems: a comparative study, 2005 IEEE Vehicle Power and Propulsion Conference, 2005.

[3] K. Bitsi, O. Wallmark and S. Bosga, "Many-objective Optimization of IPM and Induction Motors for Automotive Application," 2019 21st European Conference on Power Electronics and Applications (EPE '19 ECCE Europe), 2019, pp. P.1-P.10.

[4] Z. M. Zhao, S. Meng, C. C. Chan and E. W. C. Lo: A novel induction machine design suitable for inverter-driven variable speed systems, IEEE Transactions on Energy Conversion, vol. 15, no. 4, pp. 413-420, Dec. 2000.

[5] J. L. Oldenkamp and S. C. Peak: Selection and Design of an Inverter-Driven Induction Motor for a Traction Drive System, IEEE Transactions on Industry Applications, vol. IA-21, no. 1, pp. 259-265, Jan. 1985.

[6] U. R. Muduli, A. R. Beig, R. K. Behera, K. A. Jaafari and J. Y. Alsawalhi: Predictive Control With Battery Power Sharing Scheme for Dual Open-End-Winding Induction Motor Based Four-Wheel Drive Electric Vehicle, IEEE Transactions on Industrial Electronics, vol. 69, no. 6, pp. 5557-5568, June 2022.

[7] M. Oh and I. Husain: Optimal Torque Distribution of Dual-Motor All-Wheel Drive Electric Vehicles for Maximizing Motor Energy Efficiency, 2021 IEEE Transportation Electrification Conference & Expo (ITEC), 2021, pp. 188-193.

[8] M. Osama and T. A. Lipo: A new inverter control scheme for induction motor drives requiring wide speed range, IAS '95. Conference Record of the 1995 IEEE Industry Applications Conference Thirtieth IAS Annual Meeting, 1995.

[9] S. Mallampalli, Z. Q. Zhu, J. C. Mipo and S. Personnaz: Six-Phase Pole-Changing Winding Induction Machines With Improved Performance, IEEE Transactions on Energy Conversion, vol. 36, no. 1, pp. 534-546, March 2021.

[10] M. P. Magill and P. T. Krein: A dynamic pole-phase modulation induction machine model, 2015 IEEE International Electric Machines & Drives Conference (IEMDC), 2015, pp. 13-19.

[11] H. Yano and K. Sakai: Integrated Motor-Controlled Independently by Multi-Inverters with Pole and Phase Changes, 2019 21st European Conference on Power Electronics and Applications (EPE '19 ECCE Europe), 2019, pp. P.1-P.10.

[12] K. Bitsi, O. Wallmark and S. Bosga: An Induction Machine with Wound Independently-Controlled Stator Coils, 2019 22nd International Conference on Electrical Machines and Systems (ICEMS), 2019, pp. 1-5.

[13] O. Wallmark, K. Bitsi and S. G. Bosga, "A Transient Model of WICSC and ISCAD Machines Based on Permeance Networks," 2020 International Conference on Electrical Machines (ICEM), 2020, pp. 2048-2054.

[14] A. Boglietti, A. Cavagnino, M. Lazzari, and S. Vaschetto: Preliminary induction motor electromagnetic sizing based on a geometrical approach, IET Electric Power Applications, vol. 6, no. 9, pp. 583–592, November 2012.

[15] P. Ponomarev, Y. Alexandrova, I. Petrov, P. Lindh, E. Lomonova, and J. Pyrhönen: Inductance calculation of tooth-coil permanent-magnet synchronous machines, IEEE Transactions on Industrial Electronics, vol. 61, no. 11, pp. 5966–5973, 2014.

[16] S. Williamson and M. A. Muller: Calculation of the impedance of rotor cage end rings, IEE proceedings.-B, vol. 140, no. 1, p. 51, 1993.

[17] T. Murata, T. Tsuchiya and I. Takeda: Vector control for induction machine on the application of optimal control theory, IEEE Transactions on Industrial Electronics, vol. 37, no. 4, pp. 283-290, Aug. 1990.

[18] H. Cai, L. Gao and L. Xu: Calculation of Maximum Torque Operating Conditions for Inverter-Fed Induction Machine Using Finite-Element Analysis, IEEE Transactions on Industrial Electronics, vol. 66, no. 4, pp. 2649-2658, April 2019.

[19] T. Lipo: Introduction to AC Machine Design, ser. IEEE Press Series on Power Engineering. Wiley, 2017.

[20] O. Wallmark and K. Bitsi: Iron-loss computation using matlab and comsol multiphysics, 2020 International Conference on Electrical Machines (ICEM), vol. 1, 2020, pp. 916–920.

[21] A. Boglietti, A. Cavagnino, D. M. Ionel, M. Popescu, D. A. Staton, and S. Vaschetto: A general model to predict the iron losses in pwm inverterfed induction motors, IEEE Transactions on Industry Applications, vol. 46, no. 5, pp. 1882–1890, 2010.

[22] C. T. Krasopoulos, M. E. Beniakar, and A. G. Kladas: Multicriteria pm motor design based on anfis evaluation of ev driving cycle efficiency, IEEE Transactions on Transportation Electrification, vol. 4, no. 2, pp. 525–535, 2018.

New Topology of Superconducting Fault Current Limiter with Bypass Resistor

D. Baimel[1], Eli Barbi[1], S. Bronstein[1], N. Baimel[2] and A. Kuperman[3]

[1]Shamoon College of Engineering, Beer-Sheva, Israel

[2]Sapir Academic College, Hof Ashkelon, Israel

[3]Ben-Gurion University of Negev, Beer-Sheva, Israel

Email: dmitrba@sce.ac.il

Abstract

This paper presents a new type of bridge type SFCL with a bypass resistor. The proposed topology overcomes the drawbacks of the conventional diode bridge SFCL that can limit the fault current only during the first cycles. The proposed SFCL can limit the fault current also in the following subsequent cycles. Extensive simulations validate the proposed topology and demonstrate its advantages. The simulation results are discussed and compared to the conventional diode bridge topology. The conclusions are presented at the end of the paper.

Keywords

<<Fault handling strategy>>,<<Fault ride-through>>.

Introduction

Different types of faults in the power systems can result in destructive short circuit currents that may cause significant damage to the system's components such as generators, transformers, transmission lines, bus bars, and loads. One of the popular approaches for their limiting is to use superconducting fault current limiters (SFCL), which change their state from low impedance to high impedance during the fault currents. One of the advantages of SFCL is a very fast response that allows limiting the first peak of the fault current until the corresponding circuit breaker responds and opens the faulted line. Furthermore, more advanced SFCLs can also limit the current in the following cycles of the short circuit current.

The most common types of SFCLs that are mentioned in the literature are Inductive type SFCL which increases it's inductive impedance during high fault currents [1, 2]; Resistive type SFCL [3-7], Flux-lock type SFCL [8-12]; SMES type SFCL [13]; Shield type SFCL [14]; flux-coupling type [15-17]; Transformer type SFCL [18]–[21]; Matrix-Type SFCL [22-23]; Non-inductively wound solenoid type [24]; Dual reactor type SFCL [25].

This paper aims to improve the conventional diode bridge type SFCL. The bridge type SFCL has several advantages over other types of SFCLs: it has a fast recovery speed and can be used in situations where the protected power circuit requires auto-reclose; the superconducting coil is at the DC side of the bridge, so there is no ac power loss; in a normal situation, the voltage across the FCL is very small, and there is almost no harmonics; without magnetic core, the cost and the weight of the SFCL is lower [26-34]. The configuration of the basic diode bridge type SFCL integrated into the power line fed by voltage V_{in} is shown in Fig. 1. The superconducting coil in the center of the bridge has zero resistance and high inductance. Therefore, the voltage drops are present only on the diodes. During the fault, the reactor has high impedance and limits the first cycle of the short currents.

The major advantage of the diode bridge topology is its simple structure and low price. However, the main drawback associated with the diode bridge topology is the absence of any control, which results in a fault current limiting effect only for the first cycles, until the reactor is fully charged to the maximal value of the fault current. If the fault current continues beyond these first cycles and the corresponding circuit breaker does not respond, then the fault current will cause damage to the power system. The purpose of this paper is to solve the previously mentioned drawback of the conventional diode bridge SFCL. The paper proposes a more advanced bridge-type topology with a bypass resistor that allows limiting fault current not only during the first cycles but also for the next coming cycles.

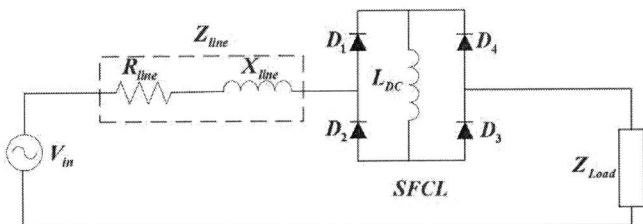

Fig. 1. Conventional diode bridge type SFCL.

The rest of the paper is organized as follows: Section II presents the proposed topology, section III presents simulation results and their analysis, and section IV presents the conclusions.

The proposed topology and its control

The proposed SFCL is shown in Fig. 2. It is comprised of the bridge and parallel-connected bypass resistor R_{Bypass}. The resistor can be connected or disconnected from the bridge by the bypass switch S_{Bypass} comprised of two back-to-back connected thyristors. The bypass switch is used only as an on/off switch without using turn on and off angles. This fact significantly simplifies the control of the SFCL.

The proposed SFCL bridge has two gate turn-off thyristors (GTOs) T_1 and T_4 that allow connection or disconnection of the reactor L_{DC} from the line. It is important to note that the state of the thyristors T_1 and T_4 is always opposite to the state of the bypass switch.

The pair of GTOs is turned on only during the first cycles of the fault current, during which the fault current is limited by the impedance of the reactor L_{DC}. The number of these cycles is defined by the controller. If during these first cycles the corresponding circuit breaker did not trip the power line, then the GTOs T_1 and T_4 are turned off disconnecting the reactor from the line and the bypass switch S_{Bypass} is turned on. In this case, the fault current will be limited by the bypass resistor. When GTOs T_1 and T_4 are turned off, the reactor parallel diode prevents high voltage spikes on the reactor by allowing the reactor to discharge through it.

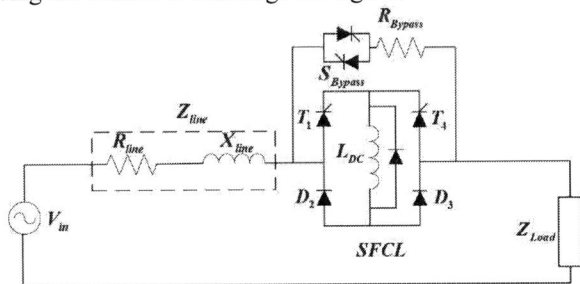

Fig. 2. The proposed SFCL.

The proposed topology can be designed for different values of the fault current by connecting several bypass resistors as shown in fig. 3. In this case, seven combinations of the bypass resistance values can be obtained by simultaneous switching of two or three switches: R_{Bypass_1}, R_{Bypass_2}, R_{Bypass_3}, $R_{Bypass\_1}//R_{Bypass\_2}$, $R_{Bypass\_1}//R_{Bypass\_3}$, $R_{Bypass\_2}//R_{Bypass\_3}$ and $R_{Bypass\_1}//R_{Bypass\_2}//R_{Bypass\_3}$. The value of the bypass resistor can be calculated by:

$$R_{Bypass} = \frac{V_{in}}{I_k} - (Z_{line} + Z_{fault} \| Z_{load}), \tag{9}$$

where I_k is the value of the limited fault current, Z_{line} is the impedance of the line, Z_{fault} is the impedance of the fault to the ground and Z_{load} is the impedance of the load.

The number of bypass resistors is defined by the designer of the SFCL according to the desired resolution of the current limiting and the system's price. The energy rating of the bypass resistors depends on the value of the fault current at the time point of switching off GTOs and turning on bypass switches.

When the fault current is limited by the reactor (bypass resistor is switched off), the limited fault current has two operation modes. The first mode takes place twice in each period during source voltage positive and negative half-cycles. In the charging mode, during which the current through the reactor equals the absolute value of the load current, the KVL of the charging circuit can be described by:

$$V_{in} = (R_{line} + R_{Load})i_{load} + (L_{line} + L_{Load} + L_{DC})\frac{di_{load}}{dt} + 2V_{TF}.$$ (1)

The fault current during the charging mode is described by:

$$i_{load\_ch}(t) = e^{\frac{-(R_{line}+R_{Load})}{L_{line}+L_{Load}+L_{DC}}(t-t_0)}[i_{load\_ch(I.C.)} - \frac{\sqrt{2}V}{|Z_0|}\sin(\omega t_0 - \theta_0) + \frac{2V_{TF}}{R_{line}+R_{Load}}] + \frac{\sqrt{2}V}{|Z_0|}\sin(\omega t - \theta_0) - \frac{V_{TF}+V_{DF}}{R_{line}+R_{Load}},$$ (2)

where V_{TF} is the voltage drop across thyristors (its value is typically 0.8 to 1.0 V), V_{DF} is the voltage drop across diodes, and $i_{load\_ch(I.C.)}$ is the initial condition of the reactor charging current. The charging mode impedance modulus is given by

$$|Z_0| = \sqrt{(R_{line} + R_{Load})^2 + (X_{line} + \omega L_{DC} + X_{Load})^2},$$ (3)

and the angle of the charging circuit impedance, which includes the line, reactor, and load impedances, is given by

$$\theta_0 = tg^{-1}\frac{\omega(L_{line} + L_{Load} + L_{DC})}{R_{line} + R_{Load}}.$$ (4)

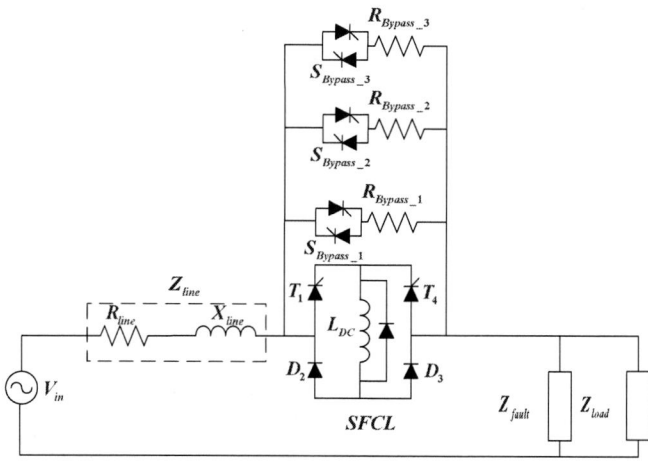

Fig. 3. The proposed SFCL with three bypass resistors.

During the freewheeling mode that takes place when the reactor is charged to its maximum value, the reactor current begins to decrease and the polarity of the voltage drop across the reactor reverses. At this moment, the pair of previously non-conducting thyristors and the diode is turned ON additionally to the thyristors and diode that were already conducting, i.e., all four semiconductors conduct and the freewheeling mode starts. The reactor L_{DC} freewheels, discharging through all four thyristors. The KVL of the freewheeling circuit can be described by:

$$V_{in} = (R_{line} + R_{Load})i_{load} + (X_{line} + X_{Load})\frac{di_{load}}{dt}$$ (5)

The fault current during the freewheeling mode is described by:

$$i_{load\_fw}(t) = e^{\frac{-(R_{line}+R_{Load})}{L_{line}+L_{Load}}(t-t_1)}[i_{load\_fw(I.C.)} - \frac{\sqrt{2}V}{|Z_1|}\sin(\omega t_1 - \theta_1)] + \frac{\sqrt{2}V}{|Z_1|}\sin(\omega t - \theta_1),$$ (6)

where and $i_{load\_fw(i.c)}$ is the initial condition of the freewheeling current.

The freewheeling mode impedance modulus is given by $|Z_1| = \sqrt{(R_{line} + R_{Load})^2 + (X_{line} + X_{Load})^2}$, (7)

while the angle of the freewheeling mode impedance, which includes line and load impedances, is given by

$$\theta_1 = tg^{-1} \frac{\omega(L_{line} + L_{Load})}{R_{line} + R_{Load}}. \tag{8}$$

The reactor current is given by

$$i_{LDC}(t) \cong \frac{\sqrt{2}V}{|Z_1|} - \frac{V_{TF} + V_{DF}}{L_{DC}} t.$$

The control unit of the proposed SFCL is shown in Fig. 4. The "fault current detection" block receives at its input the measured line current and checks if its value exceeds the defined normal threshold value. While the value of the line current is below this threshold, the current is defined as "normal" and GTOs are turned on while all bypass switches are turned off. When the value of the line current becomes higher than the threshold, the control will turn off the GTOs after the delay that defines how many cycles will be limited by the bridge. The "Bypass resistance selection block" receives at its input the gating pulse of the GTOs and line current. It reverts the GTOs gating pulse by using the "Not" logical operator. Afterward, it calculates the corresponding bypass resistance that will limit the fault current to the desired value in the following cycles. As was explained before, the number of bypass resistor combinations is limited by the number of resistors. Therefore, this block can calculate and choose the limited fault current from several possible options that are defined by the number of the bypass resistors combinations.

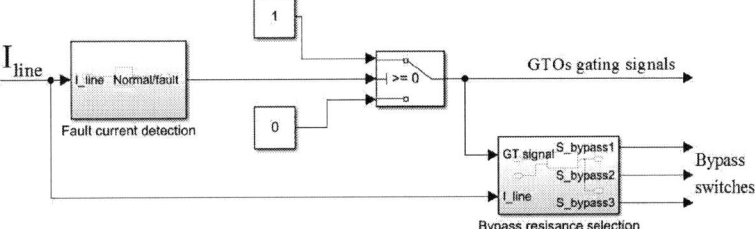

Fig. 4. The control unit of the proposed SFCL with three bypass resistors.

Validation and analysis

This section presents the simulations of the proposed SFCL for different delay times and resistors combinations. All simulations were performed by Simulink program. They are compared to the simulation of the conventional diode bridge that is shown in Fig. 5. The simulation parameters are shown in Table I. In all simulations, the short circuit to the ground starts at the time point of 0.04sec.

TABLE I. SYSTEM PARAMETER VALUES USED IN ACCOMPANYING SIMULATIONS

Parameter	Value	Units
Supply voltage, V_{in}	22	kV
SFCL reactance, L_{DC}	40	mH
Line impedance, Z_{line}	0	Ω
Base frequency, ω	100π	rad/s
Load impedance, Z_{Load}	$100 + j\pi$	Ω
Thyristors forward voltage drop, V_{TF}	1	V
R_{Bypass_1}	20	Ω
R_{Bypass_2}	30	Ω
R_{Bypass_3}	40	Ω

The simulations presented in Fig. 6 show that the proposed topology is working and capable of limiting the fault current to the desired values as long as needed and not only during the first cycles as a conventional diode bridge does. The timing of the transition from the bridge to the bypass path is defined by the user according to the expected fault currents. As shown in the simulations, the value of the limited current can change according to the combination of the bypass resistances. When the fault current is limited by the reactor, its waveform is not sinusoidal and behaves according to the charging and freewheeling modes. When the fault current is limited by the bypass resistors, its waveform is purely sinusoidal. As a result, harmonic pollution is eliminated.

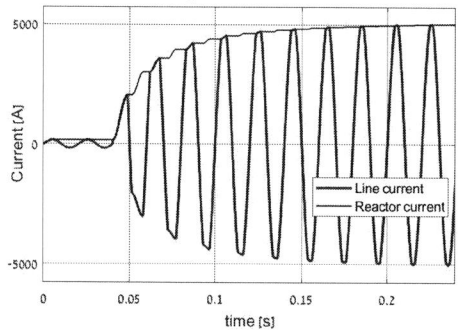

Fig. 5. The simulated line current and reactor current of the conventional diode bridge. Normal operation mode is during the time interval of 0<t<0.04s while 0.04<t- fault state. It can be seen from Fig. 5 that the diode bridge SFCL limits the fault current only during the first six cycles after which the fault current is maximal again.

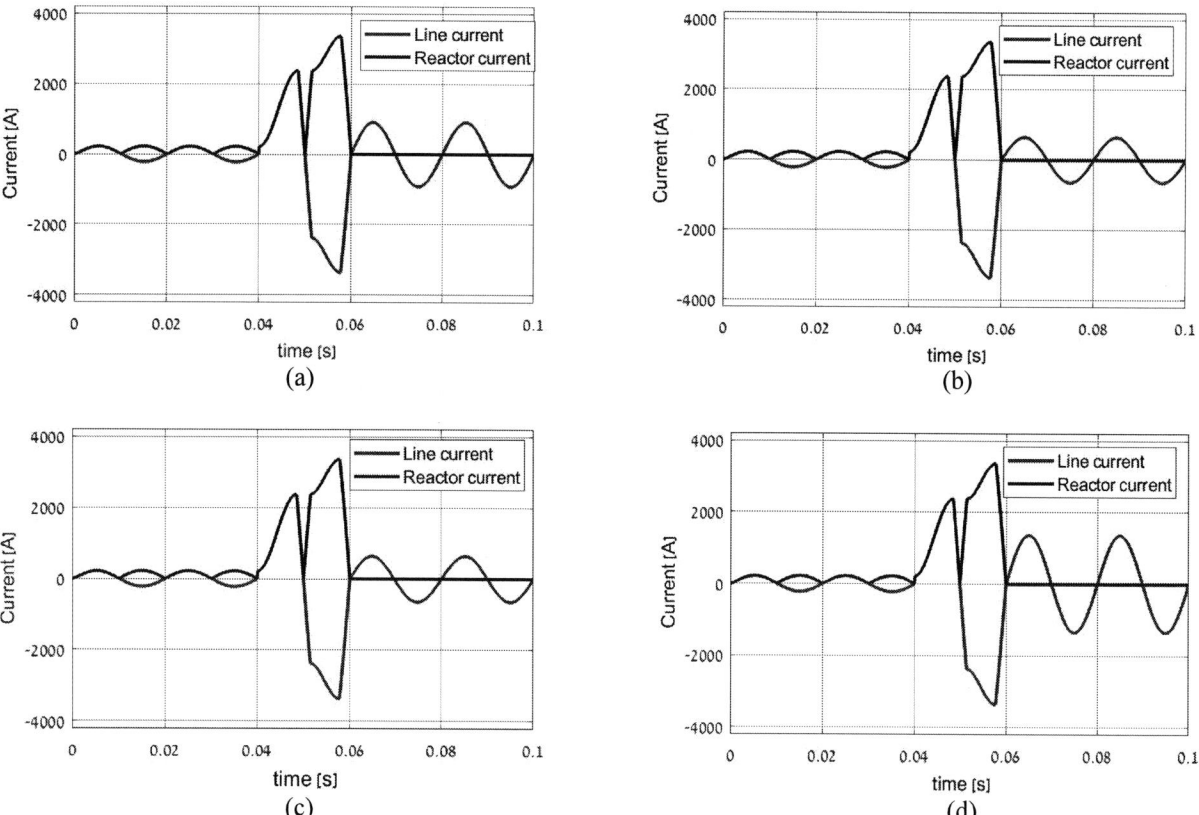

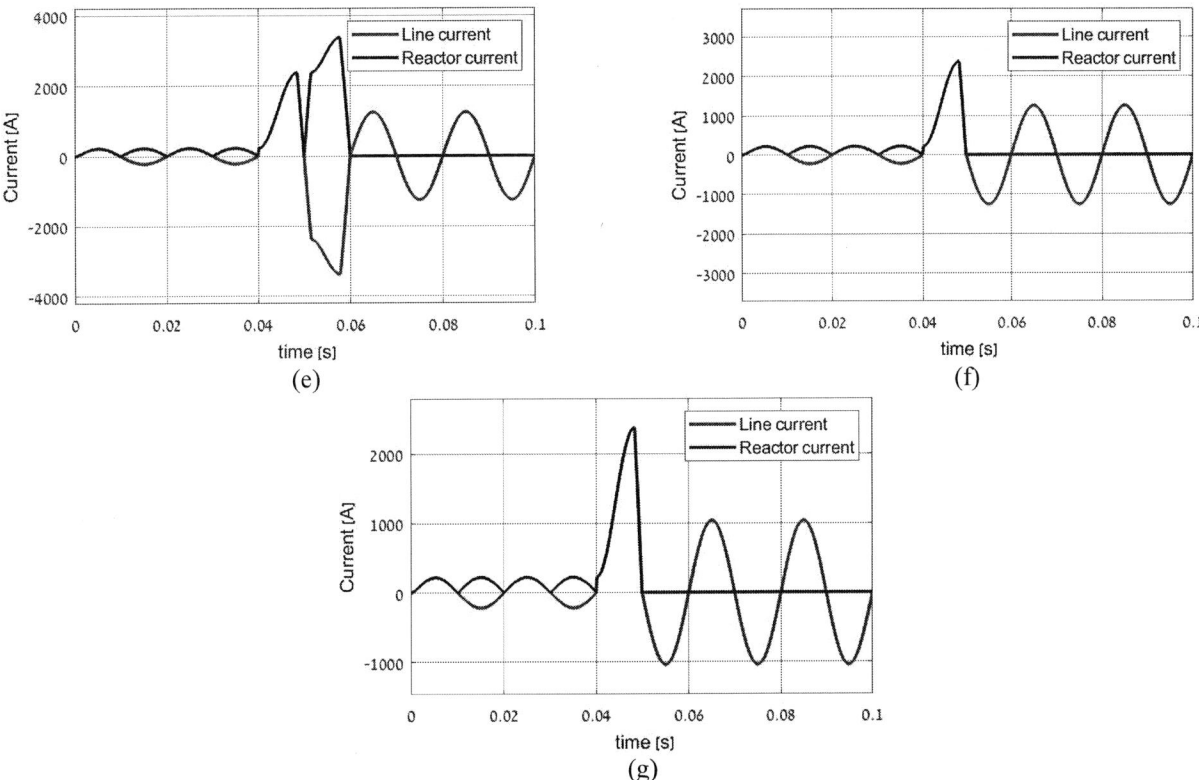

Fig. 6. The simulated line current and reactor current of the proposed SFCL. (a) The switch $S_{Bypass\_}1$ is switched on at a time point of 0.06sec. The bypass resistance is 20Ω; (b) The simulated line current and reactor current of the proposed SFCL. The switch $S_{Bypass\_}2$ is switched on at a time point of 0.06sec. The bypass resistance is 30Ω; (c) The switch $S_{Bypass\_}3$ is switched on at time point of 0.06sec. The bypass resistance is 40Ω; (d) The switches $S_{Bypass\_}1$ and $S_{Bypass\_}2$ are switched on at a time point of 0.06sec. The total bypass resistance is 12Ω; (e) The switches $S_{Bypass\_}1$ and $S_{Bypass\_}3$ are switched on at a time point of 0.06sec. The total bypass resistance is 13.33Ω; (f) The simulated line current and reactor current of the proposed SFCL. The switches $S_{Bypass\_}1$ and $S_{Bypass\_}3$ are switched on at a time point of 0.05sec. The total bypass resistance is 13.33Ω; (g) The simulated line current and reactor current of the proposed SFCL. The switches $S_{Bypass\_}2$ and $S_{Bypass\_}3$ are switched on at a time point of 0.05sec. The total bypass resistance is 17Ω.

Conclusions

This paper proposed a new topology of the bridge type SFCL with parallel bypass resistors. The operation of the SFCL was validated by extensive simulations. The proposed topology has several advantages. The first advantage is simple control and structure. The second advantage is its ability to limit the fault current to the desired value as long as needed and not only during the first cycles. The third advantage is the ability to change limited fault current according to the possible combinations of the bypass resistors. However, the increase in the number of possible combinations results in more parallel paths that complicate the structure of the SFCL and results in a bigger size and higher price.

References

[1] T. Janowski, G. Wojtasiewicz, B. Kondratowicz-Kucewicz, S. Kozak, J. Kozak, and M. Majka, "Superconducting winding for inductive type SFCL made of HTS tape with increased resistivity," *IEEE Trans. Appl. Supercond.*, vol. 19, no. 3, pp. 1884–1887, 2009.

[2] S. Kozak, T. Janowski, G. Wojtasiewicz, J. Kozak, B. Kondratowicz- Kucewicz, and M. Majka, "The 15 kv class inductive SFCL," *IEEE Trans. Appl. Supercond.*, vol. 20, no. 3, pp. 1203–1206, 2010.

[3] S. Imparato, A. Morandi, L. Martini, M. Bocchi, G. Grasso, M. Fabbri, F. Negrini, and P. Ribani, "Experimental evaluation of ac losses of a DC restive SFCL prototype," *IEEE Trans. Appl. Supercond.*, vol. 20, no. 3, pp. 1199–1202, 2010.

[4] A. Morandi, S. Imparato, G. Grasso, S. Berta, L. Martini, M. Bocchi, M. Fabbri, F. Negrini, and P. Ribani, "Design of a DC resistive SFCL for application to the 20 kv distribution system," *IEEE Trans. Appl. Supercond.*, vol. 20, no. 3, pp. 1122–1126, 2010.

[5] Z. He and S. Wang, "Design of the electromagnetic repulsion mechanism and the low-inductive coil used in the resistive-type superconducting fault current limiter," *IEEE Trans. Appl. Supercond.*, vol. 24, no. 5, pp. 1–4, 2014.

[6] H. S. Ruiz, X. Zhang, and T. Coombs, "Resistive-type superconducting fault current limiters: concepts, materials, and numerical modeling," *IEEE Trans. Appl. Supercond.*, vol. 25, no. 3, pp. 1–5, 2015.

[7] S. H. Lim and J. C. Kim, "Quench and recovery characteristics of series connected resistive type SFCL s with magnetically coupled shunt-reactors," IEEE Trans. Appl. Supercond., vol. 18, no. 2, pp. 729–732, 2008.

[8] S. H. Lim, J. C. Kim, and B.-W. Lee, "Improvement of recovery characteristics of a flux-lock type SFCL using a superconductor's trigger," *IEEE Trans. Appl. Supercond.*, vol. 20, no. 3, pp. 1182–1185, 2010.

[9] H. S. Choi and S.-H. Lim, "Operating performance of the flux-lock and the transformer type superconducting fault current limiter using the YBCO thin films," *IEEE Trans. Appl. Supercond.*, vol. 17, no. 2, pp. 1823–1826, 2007.

[10] S. C. Ko and S. H. Lim, "Analysis on magnetizing characteristics due to peak fault current limiting operation of a modified flux-lock-type SFCL with two magnetic paths," *IEEE Trans. Appl. Supercond.*, vol. 26, no. 4, pp. 1–5, 2016.

[11] S. H. Lim, S. W. Lee, Y. H. Moon, J. C. Kim, and T. K. Ko, "Power quality improving operations of a flux-lock type SFCL integrated with voltage or current controlled-voltage source inverter," *IEEE Trans. Appl. Supercond.*, vol. 18, no. 2, pp. 644–647, 2008.

[12] J. S. Kim, S. H. Lim, and J. Kim, "Study on protection coordination of a flux-lock type SFCL with over-current relay," *IEEE Trans. Appl. Supercond.*, vol. 20, no. 3, pp. 1159–1163, 2010.

[13] G. Zhu, Z. Wang, X. Liu, G. Zhang, and X. Jiang, "Transient behavior research on the combined equipment of SMES-SFCL," *IEEE Trans. Appl. Supercond.*, vol. 14, no. 2, pp. 778–781, 2004.

[14] A. Hekmati, "Proposed design for a tunable inductive shield-type SFCL," *IEEE Trans. Appl. Supercond.*, vol. 24, no. 4, pp. 1–7, 2014.

[15] L. Ren, Y. Tang, Z. Li, L. Chen, J. Shi, F. Jiao, and J. Li, "Techno- economic evaluation of a novel flux-coupling type superconducting fault current limiter," *IEEE Trans. Appl. Supercond.*, vol. 20, no. 3, pp. 1242–1245, 2010.

[16] S. Yan, L. Ren, Z. Wang, Z. Yang, Y. Xu, L. Chen, B. Yu, Y. Tang, and S. Liang, "Simulation analysis and experimental tests of a small-scale flux-coupling type superconducting fault current limiter," *IEEE Trans. Appl. Supercond.*, vol. 27, no. 4, pp. 1–5, 2016.

[17] H. S. Choi, B. I. Jung, and Y. S. Cho, "Transient characteristics of a flux- coupling type superconducting fault current limiter according to winding direction," *IEEE Trans. Appl. Supercond.*, vol. 19, no. 3, pp. 1827–1830, 2009.

[18] S. G. Choi, H. S. Choi, and K. H. Ha, "Analysis of recovery characteristics of three-phase transformer type vol. 20, no. 3, pp. 1242–1245, per types of faults according to reclosing system," *IEEE Trans. Appl. Supercond.*, vol. 22, no. 3, pp. 5 601 204–5 601 204, 2012.

[19] H.-S. Choi and Y.-S. Cho, "Critical current equalization via neutral lines in a transformer-type vol. 20, no. 3, pp. 1242–1245," *IEEE Trans. Appl. Supercond.*, vol. 18, no. 2, pp. 733–736, 2008.

[20] K. Fushiki, T. Nitta, J. Baba, and K. Suzuki, "Design and basic test of SFCL of transformer type by use of Ag sheathed BSSCO wire," *IEEE Trans. Appl. Supercond.*, vol. 17, no. 2, pp. 1815–1818, 2007.

[21] H. Yamaguchi, K. Yoshikawa, M. Nakamura, T. Kataoka, and K. Kaiho, "Current limiting characteristics of transformer type superconducting fault current limiter," *IEEE Trans Appl. Supercond.*, vol. 15, no. 2, pp. 2106–2109, 2005.

[22] D. C. Chung, B. H. Pak, Y. S. Cho, B. I. Jung, H. S. Han, M. H. Kwak, S. B. Kang, H. W. Oh, T. H. Sung, T. H. Han et al., "Increase of current limiting capacity of SFCLs by using matrix-type SFCL

module," *IEEE Trans. Appl. Supercond.*, vol. 21, no. 3, pp. 1280–1283, 2011.

[23] B. I. Jung, Y. S. Cho, H. S. Choi, and D.-C. Chung, "Recovery characteristics of three-phase matrix-type SFCL in ground fault," *IEEE Trans. Appl. Supercond.*, vol. 20, no. 3, pp. 1229–1232, 2010.

[24] J. B. Na, H. Kang, and T. K. Ko, "Numerical analysis and electrical insulation design of a single-phase 154kv class non-inductively wound solenoid type superconducting fault current limiter," *IEEE Trans. Appl. Supercond.*, vol. 22, no. 3, pp. 5 602 104–5 602 104, 2012.

[25] S.-H. Lim, H.-S. Choi, and B.-S. Han, "Fault current limiting characteristics of DC dual reactor type SFCL using switching operation of HTSC elements," *IEEE Trans. Appl. Supercond.*, vol. 16, no. 2, pp. 723–726, 2006.

[26] Fei, Wanmin, and Bin Wu. "A novel topology of bridge-type superconducting fault current limiter" *Canadian Conference on Electrical and Computer Engineering, CCECE'09*, vol. 21, no. 7, pp. 2201 – 2204, 2009.

[27] L. Li, L. Gong, X. Xu, J. Lu, Z. Fang, and H. Zhang, "Field test and demonstrated operation of 10.5 kv/1.5 kA HTS fault current limiter," *IEEE Trans. Appl. Supercond.*, vol. 17, no. 2, pp. 2055–2058, 2007.

[28] H. J. Boenig, C. H. Mielke, B. L. Burley, H. Chen, J. A. Waynert, and J. O. Willis, "The bridge-type fault current controller-a new facts controller," *Power Engineering Society Summer Meeting,* vol. 1. pp. 455–460, 2002.

[29] T. Hoshino, K. M. Salim, M. Nishikawa, I. Muta, and T. Nakamura, "DC reactor effect on bridge type superconducting fault current limiter during load increasing," *IEEE Trans Appl. Supercond.*, vol. 11, no. 1, pp. 1944–1947, 2001.

[30] K. M. Salim, T. Hoshino, A. Kawasaki, I. Muta, and T. Nakamura, "Waveform analysis of the bridge type SFCL during load changing and fault time," *IEEE Trans Appl. Supercond.*, vol. 13, no. 2, pp. 1992–1995, 2003.

[31] L. Jiang, J. X. Jin, and X. Y. Chen, "Fully controlled hybrid bridge type superconducting fault current limiter," *IEEE Trans. Appl. Supercond.*, vol. 24, no. 5, pp. 1–5, 2014.

[32] H. You and J. Jin, "Characteristic analysis of a fully controlled bridge type superconducting fault current limiter," *IEEE Trans. Appl. Supercond.*, vol. 26, no. 7, pp. 1–6, 2016.

[33] M. C. Ahn and T. K. Ko, "Proof-of-concept of a smart fault current controller with a superconducting coil for the smart grid," *IEEE Trans. Appl. Supercond.*, vol. 21, no. 3, pp. 2201–2204, 2011.

[34] M. Firouzi, G. Gharehpetian, and B. Mozafari, "Bridge-type superconducting fault current limiter effect on distance relay characteristics," *Int. J. Electr. Power Energy Syst.*, vol. 68, pp. 115–122, 2015.

A Pre- and Discharge Unit for Capacitive DC-Links Based on a Dual-Switch Bidirectional Flyback Converter

Madlen Hoffmann and Martin März
Institute of Power Electronics, Friedrich-Alexander-Universität Erlangen-Nürnberg
Fürther Straße 248
90429 Nürnberg, Germany
Phone: +49 (0) 911 56854 99296
Email: madlen.hoffmann@fau.de, martin.maerz@fau.de
URL: https://www.lee.tf.fau.de

Keywords

≪Bi-directional converters≫, ≪DC-DC converter≫, ≪Flyback Converter≫,
≪Grid-connected converter≫, ≪DC grid component≫

Abstract

This paper proposes a novel DC grid component that supplements conventional DC circuit breakers with a pre- and discharging functionality, thereby replacing conventional electromechanical solutions including lossy, bulky, and fault-prone pre- and discharging resistors. To reduce operational costs and energy consumption, this article proposes a bidirectional DC-DC converter for pre- and discharging purposes, which is based on an extended flyback topology with a dual-switch configuration on both primary and secondary side. This reduces the high voltage stress on the power semiconductors and provides both a constant current behavior over a wide voltage range and an inherent overload protection for all capacitors on the load side. The operating principle, characteristic waveforms, and the converter design are presented. The proposed concept was verified on the basis of a realized prototype setup. With this, we are able to charge and discharge a 500 μF DC-link capacitor within less than 500 ms to and from a voltage of 700 V. Furthermore, the pre- and discharging unit fulfills increased functional safety requirements as the converter is self-sufficient after start-up and does not require an auxiliary power supply. Due to a very low quiescent current in the stand-by mode of merely 8.5 μA, it is particularly suited for battery-powered applications.

Introduction

With the aim of an increased sustainability and the transition to a clean power generation combined with an improvement in power electronic technologies, renewable energy sources are facing an ever increasing demand [1, 2, 3]. Among the most prominent examples are photovoltaic systems, hydroelectric generating stations, and solar thermal as well as wind power plants [4]. Since many renewable energy sources, stationary storage devices such as batteries and fuel cells as well as most modern consumers (e.g. LEDs, computers, mobile devices, ...) generate or use direct current (DC), grids based on DC, especially decentralized DC microgrids, avoid the conversion from DC to alternating current (AC) and vice versa. Therefore, these grids represent a more efficient, sustainable, and economical solution. Nowadays, DC microgrids are already applied in data and telecommunication centers, ships, airplanes as well as in electric, hybrid, and hydrogen vehicles. [5, 6, 7]

Fig. 1a presents a typical DC grid for applications where an increased security of supply is required, e.g. in industrial production plants. It deploys multiple feeds and exhibits a ring-shaped, zonally grouped structure. Therefore, a large number of DC contactors are required to isolate faulty network sections

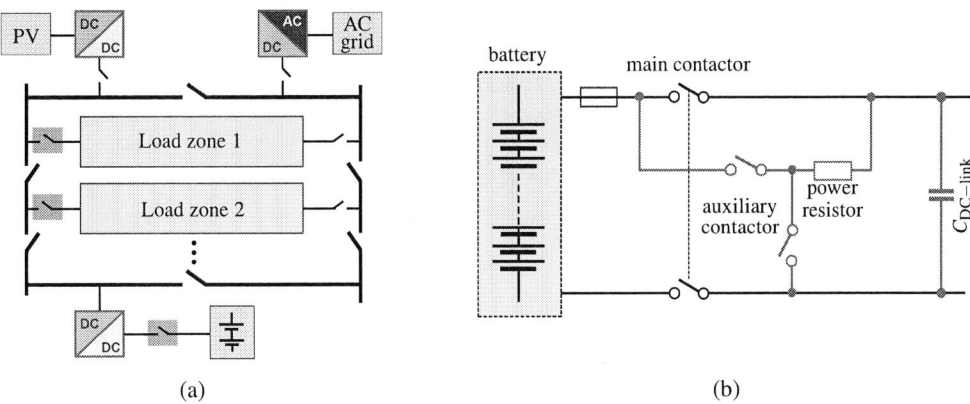

(a)   (b)

Fig. 1: a) Exemplary presentation of a ring-shaped, zonally grouped DC grid structure with load zones allowing a local enclosing of fault impacts, b) Conventional pre- and discharging circuit, which is at least necessary for each red highlighted contactor in a).

as locally as possible. The capacitive storages in each grid section must be precharged before the main contactors are closed and, when a grid section is disconnected, discharged to a safety extra-low voltage below 60 V within a few seconds. In conventional systems, this functionality is still typically implemented by a pre- and discharging unit combining an electromechanical auxiliary contactor with a series power resistor as illustrated in Fig. 1b. For DC applications, the required auxiliary contactors have to be quite bulky, in order to avoid inextinguishable arcs during disconnection. In addition, the precharging resistor is an interference-prone component, as it can be overloaded very quickly if a residual mains load is present during precharging. Furthermore, during pre- and discharging the power resistor must be capable of absorbing the energy stored in the sum of all connected capacitors on the load side $C_{\text{DC}-\text{link}}$, which increases quadratically with the DC-link voltage V_c and can easily reach the range of several kilojoules. Equivalently to the auxiliary contactors, only voluminous and cost-intensive power resistors can cope with this requirement. For the large number of pre- and discharging units required for the ring-shaped and zonally grouped grid structure shown in Fig. 1a, this results in significant costs and a problematically high installation space demand.

As a first mitigation, the electromechanical contactors can by substituted by semiconductor switches [8, 9] to ensure an active control of the pre- and discharge current and to enhance switching times. On this basis, the authors in [10] and [11] examine a current-controlled precharging circuit featuring semiconductor switches with a series inductance. In [12], a DC-link discharge device using a switched resistor is proposed, thus reducing the peak power by shifting the discharge curve to a constant power characteristic.

To further improve and to merge the pre- and discharging functionality, we propose to combine both within a single bidirectional, isolated DC-DC converter. While maintaining a small and compact design, we consequently avoid the use of power resistors and electromechanical auxiliary contactors. Therefore, the energy efficiency is increased greatly. The novel setup is depicted schematically in Fig. 2a. The proposed pre- and discharging device will be explained using the example of a DC microgrid within an electrical car. At this, the DC-link capacitor is connected to a traction battery with a maximum DC voltage of 850 V. This is based on rather new 800 V systems, which are gaining importance next to typical 400 V systems in passenger vehicles [13, 14]. Regardless of the system voltage, the proposed device can equally substitute any of the above mentioned pre- and discharging circuits in other DC microgrids. Merely the voltage constraints of the semiconductor switches deployed have to be matched according to the application.

This paper discusses the basic requirements and proposes a topology for a converter-based pre- and discharge unit. Furthermore, its operating principle during pre- and discharging is explained in detail and the prototype design and specification are described. Finally, the paper presents the experimental evaluation of the prototype and concludes with a discussion of the measurement results.

Proposed Converter-based Topology

The proposed pre- and discharging converter in Fig. 2a replaces the conventional disadvantageous electromechanical solution and other above-mentioned approaches featuring a series resistor. It therefore covers the urgent demand for a smart electronic solution. However, without voluminous electromechanical auxiliary contactors and because the proposed converter bypasses the main contactor, several additional, especially safety-related requirements have to be met and a high functional safety must be ensured.

Because of the bypassing, a safe galvanic isolation must be guaranteed. To avoid damaging of the pre- and discharge unit by excessive loads, an inherent overload protection must be considered. Additionally, an overcharging of the DC-link must be prevented. To be compatible with battery storage applications, a very low quiescent current at the supply side must also be realized. Furthermore, the unit must work independently of an external power supply and must safely activate the DC-link discharge mode in the event of a control signal failure. Also, a bidirectional energy transfer and a current source behavior including a steady short-circuit current capability is required. While the voltage range is usually quite narrow on the source side, the proposed converter must be able to handle a very wide voltage range on the load or DC-link side. During discharging, it has to be fully functional down to a safety extra-low voltage of 60 V DC, delivering a constant discharge power at the same time. In the example considered here, this corresponds to a wide voltage range ($V_{c,max}$ / $V_{c,min}$) of almost 12 (700 V / 60 V). Taking the bidirectional operation into account, only the flyback converter has been considered as a possible topology, although there is a wide range of galvanically isolated DC-DC converters, including forward converters as well as resonant converters. This is due to the fact that a flyback topology concurrently combines all requirements at a simple structure, low costs with a wide voltage range capability, and an intrinsic current source behavior. Bidirectional dual active bridges (DABs) require a full-bridge topology with eight semiconductor switches. The higher device count not only increases the costs, it also reduces the reliability of the circuit and increases the control complexity. Apart from that, DABs and resonant converters in particular reach their limit and lose efficiency for the wide voltage range required.

Inherent disadvantages of a regular single-switch flyback converter, such as high voltage stress and spikes on the semiconductor switches due to the energy stored in the leakage inductance, can be avoided by extending the basic topology to a dual-switch bidirectional flyback converter. The topology of the proposed dual-switch flyback converter for pre- and discharging of capacitive DC-links is shown in Fig. 2b. The left circuit part represents the source or battery side, e.g. being directly connected to a high-voltage (HV) battery. Using an asymmetric half bridge on the battery side, dissipative snubber circuits can be avoided, because the energy stored in the leakage inductance L_s is fed back directly into the battery via the two clamp diodes D_2 and D_4. Therefore, the maximum voltage stress the two MOSFETs T_1 and T_3 are exposed to is limited to the battery voltage [15, 16]. Furthermore, together with a transformer ratio ü of 1, D_2 and D_4 provide an intrinsic overcharge protection, because they clamp the DC-link voltage to the value of the battery voltage. This will be further described in section "Charging Process". A detailed

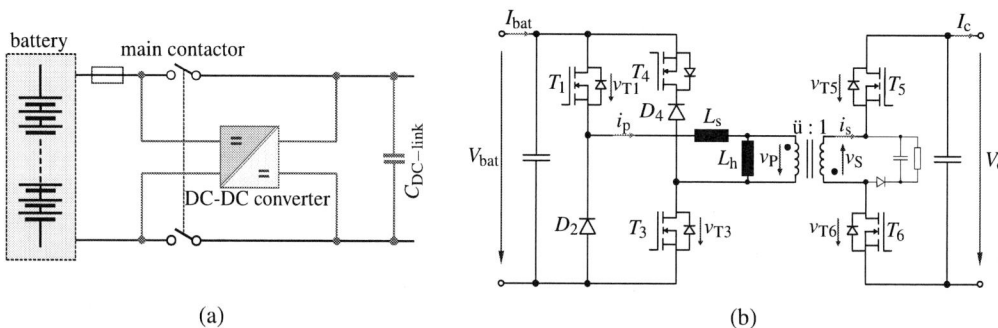

(a)   (b)

Fig. 2: a) Simplified diagram of the pre- and discharging converter replacing conventional components, b) Simplified schematic of the proposed dual-switch flyback converter.

explanation of the added switch T_4, which is conductive while charging and non-conductive while discharging, follows.

The circuit part on the right, representing the DC-link side, also corresponds to a typical dual-switch configuration except for the two missing clamp diodes. They are left out because they would prevent discharging under the assumption $V_c \leq V_{bat}$.

Operating Principle

Charging Process

While charging, the DC-link voltage V_c is increased from $0\,V$ to V_{bat} and thereby transferring energy from the battery to the DC-link capacitor. In this process, only T_1 and T_3 are actively switched. This is performed synchronously and without a phase shift. The body diodes of T_5 and T_6 ensure the required freewheel path. In the following, the steady-state charging process in the discontinuous-conduction mode (DCM) is described under ideal conditions and with the leakage inductance L_s neglected. For an extensive description of an unidirectional dual-switch flyback in DCM and in CCM (continuous-conduction mode) refer to [16] and [17], respectively. The three resulting time intervals are:

1. Interval $t_0 \leq t < t_1$:
   The switches T_1 and T_3 are conducting, the battery-sided transformer winding is clamped by the battery voltage V_{bat}, and the current i_p increases linearly, as depicted in Fig. 3. Because the clamp diodes D_2 and D_4 as well as the freewheeling body diodes of T_5 and T_6 are reverse-biased, the flyback transformer is magnetized.

2. Interval $t_1 \leq t < t_2$:
   At time t_1, the switches T_1 and T_3 are turned off simultaneously. The magnetized transformer now requires a freewheeling path. For this, there are two options available:

   a) As long as the DC-link voltage V_c is lower than the battery voltage V_{bat} ($V_c < V_{bat}$), the conduction condition for the body diodes of T_5 and T_6 is fulfilled and the energy stored in the flyback transformer can be transferred to the DC-link capacitor.

   b) When the DC-link voltage V_c reaches or even exceeds the battery voltage V_{bat} ($V_c \geq V_{bat}$), the conduction condition for both clamp diodes D_2 and D_4 is fulfilled. Assuming T_4 is conducting, D_2 and D_4 enable a freewheeling path and the energy stored in the transformer is fed back to the battery. Due to this operating principle, an intrinsic overcharge protection is realized. As soon as the DC-link voltage V_c reaches V_{bat}, D_2 and D_4 prevent an energy transfer to the DC-link, thereby clamping the maximum DC-link voltage $V_{c,max}$ to V_{bat}. Therefore and to protect the switches T_1 and T_3 against over voltage, T_4 must be conducting during the whole charging process. This can be realized via a charge pump that is merely connected to the gate driver of T_1 and T_3, enabling a continuous on-state of the MOSFET T_4, the meaning of which will be described later.

3. Interval $t_2 \leq t \leq t_3$:
   At the end of the second and the beginning of the third time interval, the flyback transformer is completely demagnetized. Neither the switches nor the diodes are conducting. Time t_3 marks the end of the third time interval and the beginning of a new switching period.

Discharging Process

During the discharging process, energy is transferred from the DC-link capacitor to the battery. Both DC-link-sided switches T_5 and T_6 are controlled phase-synchronously, while T_1 and T_3 are turned off. Similar to the charging process, the steady-state analysis of the DCM under ideal conditions can be subdivided into the following three time intervals:

1. Interval $t_4 \leq t < t_5$:
   The switches T_5 and T_6 are conducting, the DC-link-sided transformer winding is clamped to the DC-link voltage V_c, and the magnitude of the current i_s is linearly increasing. For $V_{bat} \geq V_c$, the

body diodes of T_1 and T_3 are reverse-biased and the flyback transformer is magnetized. Under normal operating conditions, V_{bat} will always be greater or equal to V_c. However, should $V_{bat} < V_c$ apply, the forward-converter operation caused by the forward-biased clamp diodes D_2 and D_4 has to be prevented. Therefore, the switch T_4 is turned off during the whole discharging process to ensure a flyback operation.

2. Interval $t_5 \leq t < t_6$:

At time t_5, the switches T_5 and T_6 are turned off simultaneously. Once the voltage v_p across the battery-sided transformer winding reaches V_{bat}, the forward-biased body diodes of T_1 and T_3 ensure the necessary freewheeling path for the magnetized flyback transformer.

3. Interval $t_6 \leq t \leq t_7$:

At time t_6, the transformer is completely demagnetized and all switches and diodes are non-conducting. As in the charging process, all currents become zero. Time t_7 marks the end of the third time interval and the beginning of a new switching period.

As depicted in Fig. 3, the maximum voltage stress on the DC-link-sided switches can be halved due to the additional high-side switch T_5. Under real conditions, the leakage inductance L_s of the transformer, leading to voltage spikes at the turn-off, cannot be neglected. To reduce the switch voltage stress, an RCD-clamp is added over the DC-link-sided transformer winding (refer to Fig. 2b).

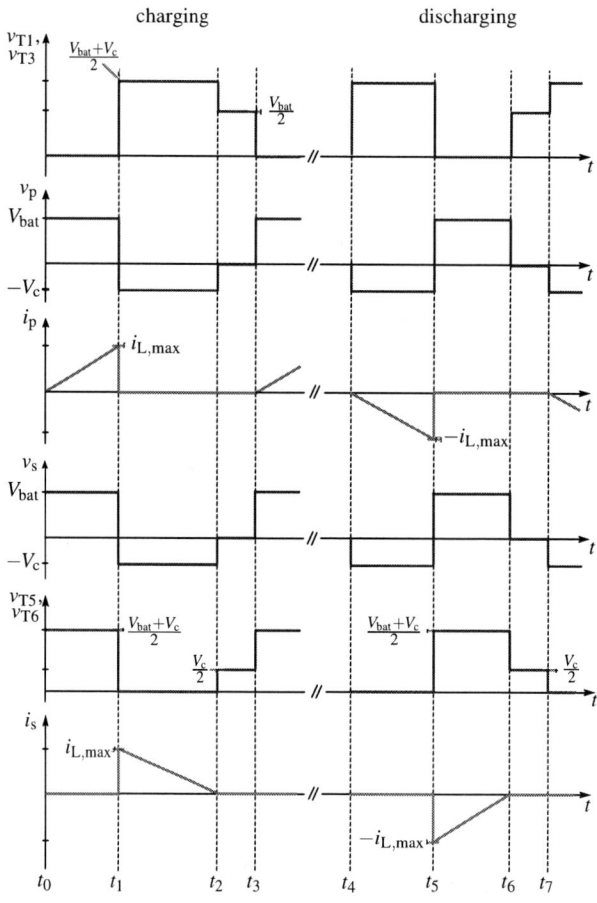

Fig. 3: Steady state waveforms for the pre- and discharging process in the DCM (ideal conditions, neglecting all parasitics) with the passive sign convention used in Fig. 2b. Time interval $[t_0, t_3]$ represents a switching period of the charging process, whereas time interval $[t_4, t_7]$ shows a switching period of the discharging process.

Operating Mode

The proposed topology can be operated either at fixed or at variable switching frequency f_s. Due to the wide voltage range on the DC-link side, a fixed switching frequency, e.g of $100\,\mathrm{kHz}$, would be accompanied by several drawbacks. Assuming a fixed switching frequency, a mixed CCM and DCM operation would occur during the pre- and discharging. At the beginning of the precharging process, i.e. at a low DC-link voltage V_c, the required demagnetization time t_{demag} (equivalent to time interval t_1 to t_2, cf. Fig. 3), which is according to (1) inversely proportional to V_c, is greater than the maximum available demagnetization time t^*_{demag} (cf. Eq. (2)).

$$t_{\mathrm{demag}} = \frac{i_{\mathrm{L,max}} \cdot L_h}{V_c} \tag{1}$$

$$t^*_{\mathrm{demag}} = t_3 - t_1 \overset{t_0=0}{=} T_p - t_1 \tag{2}$$

Therefore, the current i_s on the secondary side cannot reach $0\,\mathrm{A}$ within the fixed period time, thus forcing a CCM operation. In this operation, the resulting initial current combined with a low DC-link voltage and a chosen peak current mode control lead to switch-on times of the power semiconductors in the range of nanoseconds – which can hardly be realized by conventional control ICs. Additionally, the switching of the power semiconductors cannot occur currentless, thus increasing the switching losses. With an increasing DC-link voltage V_c, the required demagnetization time t_{demag} decreases, which corresponds to a transition to the DCM. Alternatively, by choosing a significant lower fixed switching frequency, the DCM can be forced almost immediately after first start-up, but especially for increased DC-link voltages, this would result in long pause times in the third time interval (cf. section "Charging Process") and decreased power transfer. However, in order to complete charging and, more importantly, discharging within a few hundred milliseconds, the transfer of energy packages has to occur as fast as possible. Therefore, the DCM is also considered unsuitable for this application.

To reduce the charging and discharging time as well as the switching losses, the proposed topology is operated in the boundary conduction mode (BCM) or in transition mode (TM) with a variable frequency f_s. In this mode, the period time T_p is composed of the switch-on time t_{on} and the demagnetization time t_{demag} of the magnetization inductance L_h. Due to the constant battery voltage V_{bat} and because the BCM is implemented with a constant maximal transformer current $i_{\mathrm{L,max}}$, t_{on} is constant while charging:

$$t_{\mathrm{on}} = \frac{i_{\mathrm{L,max}} \cdot L_h}{V_{\mathrm{bat}}} \tag{3}$$

Equivalent to (3), t_{demag} can be expressed by substituting V_{bat} with the DC-link voltage V_c to which it is inversely proportional (refer to Eq. (1)). For the discharging process, t_{on} and t_{demag} are interpreted vice versa. In either case, the switching frequency f_s can be expressed according to:

$$f_s = \frac{1}{T_p} = \frac{1}{t_{\mathrm{on}} + t_{\mathrm{demag}}} = \frac{V_{\mathrm{bat}} \cdot V_c}{i_{\mathrm{L,max}} \cdot L_h \cdot (V_{\mathrm{bat}} + V_c)} \tag{4}$$

Consequently, the switching frequency f_s increases with the DC-link voltage V_c.

Prototype Design

To prove the feasibility of the proposed converter, a prototype of the dual-switch flyback topology was designed according to the topology in Fig. 2b and the specification summarized in Table I. For the operation in transition mode with variable switching frequency, a control IC with primary side regulation was chosen. Because of the maximum battery- and DC-link voltage of $850\,\mathrm{V}$ and the maximum control IC switching frequency f_s of $130\,\mathrm{kHz}$, the prototype is designed with $1000\,\mathrm{V}$ silicon carbide MOSFETs (SiC-MOSFETs) for all switches. In this voltage range, SiC-MOSFETs offer the best performances of switching times, drain-source on-state resistance $R_{\mathrm{DS,on}}$, and parasitic capacities. To improve efficiency in rectifier mode SiC-Schottky Barrier Diodes (SiC-SBDs) are used antiparallel to the switches.

Table I: Specification of the evaluated prototype

	Description	Min.	Typ.	Max.
V_{bat}	battery voltage	550 V	700 V	850 V
V_c	DC-link voltage	0 V	700 V	850 V
$C_{DC-link}$	DC-link capacitance		500 µF	
f_s	switching frequency	1 kHz		130 kHz
t_c	charge/discharge time			500 ms
I_c	charge current			1 A
I_q	quiescent current			10 µA

Due to the intrinsic current source behavior of the proposed topology, the DC-link capacitor can be charged with a constant current. This way, the realized linear pre- and discharging process allows a much more precise defined charging time, since the final value is reached faster as with conventional RC elements and their exponential characteristics.

The operating mode described in section "Operating Mode" can be realized using a primary side regulation. For charging and discharging, the flyback converter has to be extended by a battery- and a DC-link-sided auxiliary winding, respectively. With the first, the battery-sided control IC can monitor the DC-link-sided transformer voltage v_s during the switch-off time of T_1 and T_3 and can therefore detect the current zero-crossing (cf. $N_{aux,bat}$ in Fig. 4). The latter is used vice versa for the discharging process. With this approach, the required safety standards of an isolated DC-DC converter are maintained without the necessity of additional isolated control or measurement signals.

A further advantage of the proposed design is that due to the additional auxiliary winding on the flyback transformer no external auxiliary power supply is needed. Apart from the auxiliary winding, the prototype only requires a specially designed start-up circuit on both battery and DC-link side. To prevent an excessive discharging of the battery in the stand-by mode, the quiescent current I_q on the battery side has to be limited to a few µA. With a deactivatable start-up circuit consisting of two stacked small-signal MOSFETs, a quiescent current of 8.5 µA is achieved with this prototype. This start-up circuit is deactivated after the start-up. The converter can then supply itself via the auxiliary winding and the voltage doubler circuit consisting of C_1, D_{10}, and D_{11} as depicted in Fig. 4. Here, the auxiliary winding voltage $v_{aux,bat}$ equals the transformed battery voltage during the switch-on time of T_1 and T_3 and the negative transformed DC-link voltage during their switch-off time. Due to the diode D_{10}, the voltage across C_1 is clamped to the maximum auxiliary voltage which equals the transformed battery voltage. Therefore, a constant auxiliary supply voltage $V_{DD,bat}$ is provided despite the wide voltage range of V_c. For the discharging process, the DC-link-sided auxiliary power supply is implemented similarly and under consideration of an opposite winding direction of the auxiliary winding. Here, a quiescent current limitation is less critical, thus simplifying the start-up circuit design. With this configuration, the converter is practically self-sufficient after start-up. Without external power supply, the circuitry complexity is reduced and the efficiency is further improved compared to conventional state-of-the-art approaches.

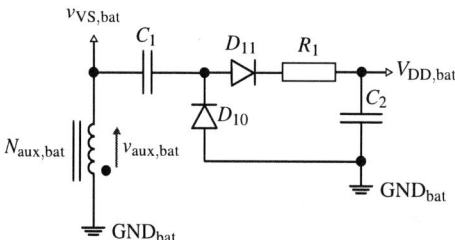

Fig. 4: Circuit diagram of the battery-sided auxiliary power supply.

Experimental Evaluation and Measurement Results

Measurement Results of the Charging Process

The charging process of the prototype demonstrator was experimentally examined using a battery voltage V_{bat} of 700 V. Fig. 5a depicts that a DC-link capacitance of 500 µF can be charged from an initial 0 V to 700 V within approximately 400 ms after a 40 ms delay that is caused by a delay in the battery-sided start-up circuit. The associated steady-state currents on the battery and the DC-link side are depicted in Fig. 5b. It can be verified that the average charging current I_c is approximately constant over the increasing DC-link voltage V_c. As expected, the average battery current I_{bat} is increasing with V_c. This is due to the fact that the duty cycle increases for a constant switch-on time t_{on} and an increasing switching frequency f_s.

Furthermore, the implemented overcharge protection, which is realized by the clamp diodes D_2 and D_4, was proven in Fig. 5a. Here it is shown that as soon as V_c reaches V_{bat}, the energy transfer to the DC-link stops and both charging and battery current decrease rapidly (cf. Fig. 5b). This inherent overcharge protection is independent of controllers or any type of supplementary software and ensures a high functional safety.

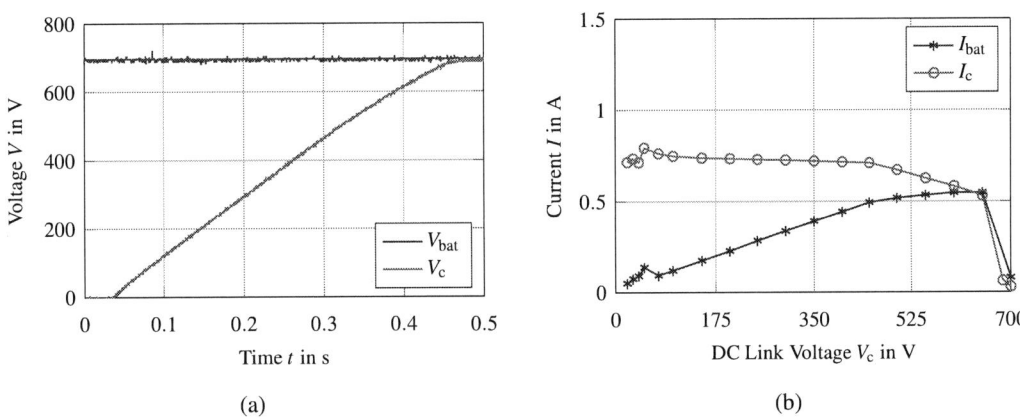

(a) (b)

Fig. 5: Measurement results while charging a DC-link capacitor with a constant battery voltage V_{bat} of 700 V: a) DC-link voltage V_c over charging time t, b) average steady-state currents I_{bat} and I_c.

Measurement Results of the Discharging Process

For the discharging process, the measurement results shown in Fig. 6a confirm that the DC-link capacitor is discharged to a safety extra-low voltage below 60 V within less than 500 ms. As this is accompanied by an inverse energy transfer compared to the charging process, the average steady-state currents involved can be interpreted reversely. Therefore, the battery current I_{bat} can be approximated as being almost constant, while the magnitude of the discharging current I_c increases with a decreasing DC-link voltage. The associated measurement results are depicted in Fig. 6b.

Efficiency

In the intended application, the efficiency of the converter is not of great importance, since the converter typically operates only a few hundred milliseconds. However, it is inherently much more efficient than the conventional solutions with power resistors. As shown in Fig. 7, the efficiency exceeds 92 % for charging and 90 % for discharging over a wide voltage range. The discrepancy of 2 % is caused by the additional snubber losses of the RCD-clamp while discharging. For a bidirectional isolated DC-DC converter, the achieved efficiency is very satisfactorily, especially considering the wide voltage-range on the DC-link side.

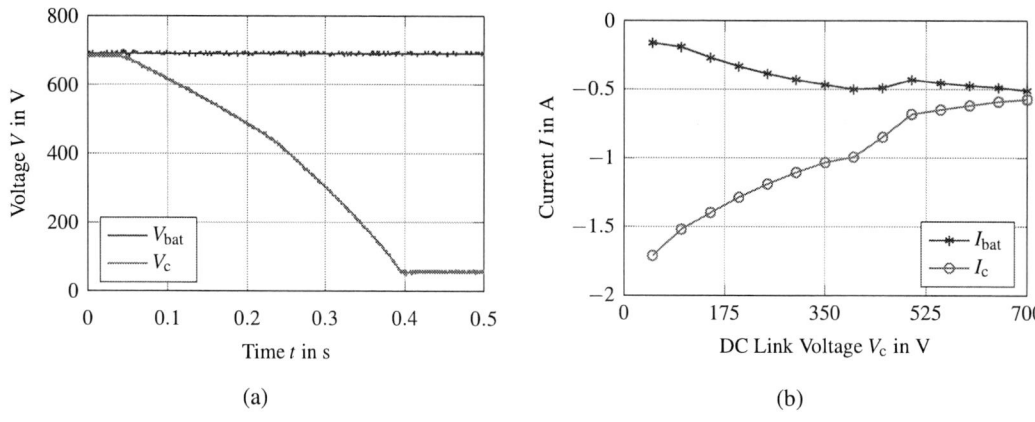

(a)   (b)

Fig. 6: Measurement results while discharging a DC-link capacitor from 700 V: a) DC-link voltage V_c over discharging time t, b) average steady-state currents I_c and I_{bat}.

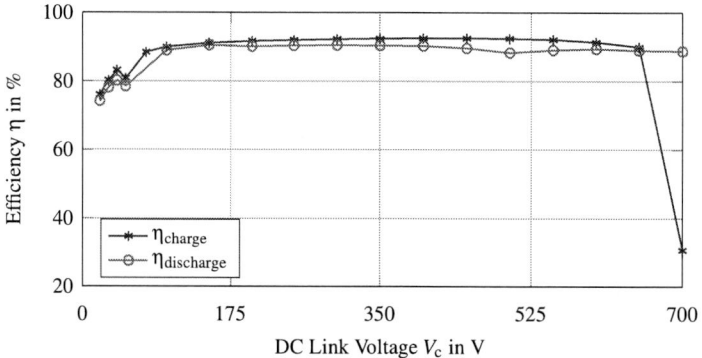

Fig. 7: Pre- and discharging efficiency η_{charge} and $\eta_{discharge}$ over the DC-link voltage V_c with a battery voltage V_{bat} of approximately 700 V.

Conclusion

In this paper, a new pre- and discharging unit bypassing the main contactor and replacing the conventional solution with bulky and lossy electromechanical auxiliary contactors and power resistors is presented. The proposed topology is based on a specially designed flyback converter that concurrently ensures the combination of a bidirectional galvanically isolated power flow and a high efficiency with a constant pre- and discharging current over a wide voltage range. With the dual-switch configuration on the source or battery side, an inherent overcharge protection is automatically provided and a dissipative snubber is no longer required. Furthermore, the switch voltage stress can be halved due to the dual-switch configuration. The operating principle of the charging and discharging process are explained in detail and measurement results are provided and analyzed. With a specially built prototype with a quiescent current of only 8.5 µA, the functionality could be confirmed by pre- and discharging a 500 µF capacitor by a 700 V voltage source within less than 500 ms and with a maximum efficiency of 92.52 % and 90.52 %, respectively.

Consequently, the proposed device represents a smart electric solution for pre- and discharging purposes of capacitive DC-links or DC microgrids. It is only limited by the voltage constraints of the semiconductor switches deployed, and can therefore readily substitute the conventional pre- and discharging circuits in any other DC microgrid, including data and telecommunication centers, aircraft electrical systems, DC grids for commercial and building infrastructure as well as electric, hybrid, and hydrogen automobiles.

References

[1] J. M. Carrasco et al., "Power-Electronic Systems for the Grid Integration of Renewable Energy Sources: A Survey," *in IEEE Transactions on Industrial Electronics*, vol. 53, no. 4, pp. 1002-1016, June 2006, doi: 10.1109/TIE.2006.878356.

[2] T. Chun, H. Kim and E. Nho, "Charging and discharging strategies of grid-connected super-capacitor energy storage systems," *2018 IEEE International Conference on Industrial Technology (ICIT)*, 2018, pp. 1743-1747, doi: 10.1109/ICIT.2018.8352446.

[3] M. Jebali Ben Ghorbal, S. Moussa, J.A. Ziani and I. Slama-Belkhodja, "A comparison study of two DC microgrid controls for a fast and stable DC bus voltage," *in Mathematics and Computers in Simulation*, Volume 184, 2021, Pages 210-224, https://doi.org/10.1016/j.matcom.2020.02.008.

[4] R. Singh and K. Shenai, "DC Microgrids and the Virtues of Local Electricity," *IEEE Spectrum*, 2014, [Online]. Available: https://spectrum.ieee.org/green-tech/buildings/dc-microgrids-and-the-virtues-of-local-electricity

[5] L. E. Zubieta, "Are Microgrids the Future of Energy?: DC Microgrids from Concept to Demonstration to Deployment," *in IEEE Electrification Magazine*, vol. 4, no. 2, pp. 37-44, June 2016, doi: 10.1109/MELE.2016.2544238.

[6] S. Backhaus et. al., "DC Microgrids Scoping Study-Estimate of Technical and Economic Benefits," *Los Alamos National Laboratory*, 2015, [Online]. Available: https://www.energy.gov/oe/downloads/dc-microgrids-scoping-study-estimate-technical-and-economic-benefits-march-2015

[7] L. E. Zubieta, "Power management and optimization concept for DC microgrids," *2015 IEEE First International Conference on DC Microgrids (ICDCM)*, 2015, pp. 81-85, doi: 10.1109/ICDCM.2015.7152014.

[8] T. Schwinn, "Frequency inverter with precharge series resistor and method for operating same: European patent," Patent EP2 523 332A1, 2012.

[9] M. Stadler, "Discharging device for actively discharging high-voltage intermediate circuit in high-voltage power supply of electrically driven vehicle, has discharge circuit arranged outside electronics module and integrated in high-voltage component," Patent DE102012020019A1, 2014.

[10] A. K. Mensah-Brown et al., "Capactitor precharging and capacitance/resistance measurement in electric vehicle drive systems: US patent," Patent US 9,573.474 B2, Mar. 6, 2014.

[11] R. Dillig et al., "Frequency converter with dc link capacitor and method for pre-charging the dc link capacitor: US patent," Patent 13/930,561, 2016.

[12] O. Kreutzer, B. Eckardt and M. März, "A simple, reliable, cost and volume saving DC-link discharge device for electric vehicles," *2015 IEEE Transportation Electrification Conference and Expo (ITEC)*, 2015, pp. 1-6, doi: 10.1109/ITEC.2015.7165800.

[13] C. Jung, "Power Up with 800-V Systems: The benefits of upgrading voltage power for battery-electric passenger vehicles," *in IEEE Electrification Magazine*, vol. 5, no. 1, pp. 53-58, March 2017, doi: 10.1109/MELE.2016.2644560.

[14] D. Meyer and J. Wang, "Integrating ultra-fast charging stations within the power grids of smart cities: A review," *in IET Smart Grid*, vol. 1, pp.3–10, 2018, https://doi.org/10.1049/iet-stg.2018.0006

[15] K. Marn-Go and J. Young-Seok, "A novel soft-switching two-switch flyback converter with a wide operating range and regenerative clamping," *in Journal of Power Electronics*, vol. 9, no. 5, pp. 772–780, 2009.

[16] D. Murthy-Bellur and M. K. Kazimierczuk, "Two-switch flyback pwm dc-dc converter in continuous-conduction mode," *in International Journal of Circuit Theory and Applications*, vol. 39, no. 1, pp. 1145–1160, 2010, DOI 10.1002/cta.690

[17] D. Murthy-Bellur and M. K. Kazimierczuk, "Two-switch flyback pwm dc-dc converter in discontinuous-conduction mode," *in International Journal of Circuit Theory and Applications*, vol. 39, no. 8, pp. 849–864, 2010, DOI 10.1002/cta.672,

Control and Integration of a multiphase Brushless Wounded Synchronous Motor Drive

Rémi Perrin, Guilherme Bueno-Mariani
Mitsubishi Electric R&D Centre Europe
1 Allee de beaulieu
Rennes, France

Email: r.perrin@fr.merce.mee.com

Keywords

« Brushless drive », « Control of drive », « Multiphase drive », « Packaging »

Abstract

This paper presents a fully integrated seven-phase Brushless Wounded Rotor Synchronous Motor BWRSM machine drive and its control. A direct integration strategy of the power electronics on the motor case is performed. Advanced PCB integration with embedded die are used for low profile rotary transformer. For the control side, the machine was modelled by Finite Element Method (FEM) in order to extract the flux tables. These tables were then used to improve the precision of analytic simulation model. By using so obtained analytic simulation with integrated flux tables designed field-oriented control was verified. Details on design and the model are presented in the paper. Finally, the controller's efficiency and the power electronics of the motor drive are confirmed with experiments on described prototype machine.

Introduction

HVAC (Heating, ventilation and air-conditioning) is one of the main applications for electrical machines. The machines implemented in these systems are dominantly PMSM (Permanent Magnet Synchronous Machine) because of their high torque density and efficiency. The main disadvantage with this type of machines remains unstable price of rare-earth magnets [23]. Nowadays, other machines appear as potential alternatives to the PMSM [1], [2]. Among the potential candidates, the BWRSM, was identified due to a torque density close to the PMSM [3], [4]. BWRSM has also the advantage of having a wider field weakening region at high speeds in conjunction with a lower cost.

The main advantages of multi-phase machines are known to be reliability, higher torque density, fault tolerance capability and reduced torque pulsation [5]. The motivation of this study is to develop a high integrated BWRSM that could reach the same performances of a PMSM with the advantage of reliability and fault tolerance capability for harsh environments.

Simplified machine modelling is used on many different studies ever since Park transformation theory was introduced. This transformation simplified the studies of transient behavior of the electrical machines, making controlling the machine much easier. Typically, Park transformation allows to turn a 3-phases machine into 2-components (d-q) simplifying the modelling of the machine. However, Park transformation can be applied for any kind of multiphase machine [6] into 2-phases modelling.

Recent improvements in the power density, packaging of power electronics technologies and new semiconductor improvement are opening new opportunities for the integration of more power electronics within the motor case [7], [8]. Motor volume reduction has clear advantages, especially for application with constrained volume, dust sensitive industry [9] and cost reduction due to less interconnection.

This volume reduction is due to the elimination of the need of two separated housing for power electronic and motor as well as the cable connection in between the drive and the motor [10]. The absence of cable connection also goes along a reduction of the total weight of the assembly. In terms of EMI, the use of short connection reduces the propagated noise while issues related to long connection with control signal are mitigated. This results in lowering surge voltages for long cables and slowing down the progressive insulation aging [11], [12] due to non-homogeneous distribution over the windings

Moreover, the opportunity of power electronic integration in the motor comes along new outlooks for optimizing both cost and overall complexity of the system. Placement of power electronics inside the motor case also offers new opportunities for self-diagnosis and simplifies application of additional functions such as determining rotor position and lifetime prediction [13], [14].

The paper will address first different aspects of the motor drive for this 7 phases BWRSM. In the second section, the model of the machine to be used on the simulations is presented. The main objective is to obtain a simple model of the machine that considers the saturation effect and can be therefore used for control design purposes. The third section presents the control design and design verification by simulation. The fourth section is focused on the power electronics integration on the machine. Some preliminary experimental results are given. Finally, the last section concludes the paper and outlines the future research

Machine modeling

One of the main characteristics of the Park transformation is that the flux linkages are independent of the rotor position. Beyond that the voltage and current are constant when the steady state is achieved. This is the main reason why Park transformation is widely used on the analysis of steady state of electrical machines [4].

The analytical BWRSM model in the Park reference is described by the following equations. This model considers the machine without any damper windings [4].

$$v_d = R_s i_d - \omega \varphi_q(I_d, I_q, i_f) + \frac{d\varphi_d(I_d, I_q, i_f)}{dt} \ (1)$$

$$v_q = R_s i_q \mp \omega \varphi_d(I_d, I_q, i_f) + \frac{d\varphi_q(I_d, I_q, i_f)}{dt} \ (2)$$

$$v_f = R_s i_f + \frac{d\varphi_f(I_d, I_q, i_f)}{dt} \ (3)$$

$$T_{em} = \frac{7p}{2}(\varphi_d I_q - \varphi_q I_d) \ (4)$$

This model can consider the saturation of the inductances since the Park transformation is merely a geometric transformation. On the next paragraph, the FEM modelling of the machine used to obtain the flux tables with respect to the current is detailed.

A. FEM Modelling

A FEM model was built using FEMM software [15], which provides the low-frequency electromagnetic analysis on two-dimensional planar and axisymmetric problems. From [3], [15]–[17] a 2D FEM model is built for the BWRSM with 7-phase/7-slots/6-poles and concentrated-windings. The results can be found in Fig. 1

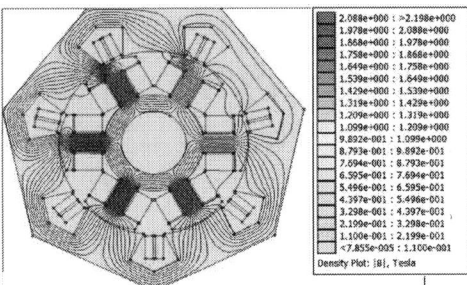

Fig. 1: Simulation model in FEMM

Fig. 1 also shows the simulation results of the machine in FEMM making it is possible to observe the flux lines and the saturation of the machine in different positions.

One weakness of FEMM software is that transient simulation is not available. Making it impossible to co-simulate with the control. Even if this would be achievable, this kind of simulation would be time consuming. In order to overcome this difficulty and obtain a controller simulation model that would consider the saturation of the machine flux tables in function of the current were extracted from the FEMM simulation.

These tables will be used in the analytical machine model. The model considers the fundamental space harmonics component of the air-gap, the flux density and the BEMF (Back Electromotive force). The flux and the current values are measured on the FEMM model and Park transformation is applied in order to obtain a d-q model. This allows the flux to be independent of the rotor position.

Since the position of the rotor does not have any influence on the flux linkages, the simulation procedure to design the flux tables is as follows:
- The position where d axis is aligned with the flux in phase A.
- The current on the rotor is maintained at 5.1A, since constant current on the rotor is predicted.

- I_d and I_q changes from 0 A 20 A
- The corresponding flux linkage φ_d and φ_q are recorded on each point of operation.

As previously mentioned, the Park transformation can be applied for any multiphase machine [6]. For a seven-phase machine Park transformation of the flux can be found on (5), while the Inverse Park transformation for the currents can be found on (6).

$$
\begin{bmatrix} \varphi_d \\ \varphi_q \end{bmatrix} = \frac{2}{7} \begin{bmatrix} cos\theta_m & \cos(\theta_m - \frac{2\pi}{7}) & \dots & \cos(\theta_m - \frac{6\cdot2\pi}{7}) \\ -sin\theta_m & -\sin(\theta_m - \frac{2\pi}{7}) & \dots & -\sin(\theta_m - \frac{6\cdot2\pi}{7}) \end{bmatrix} \begin{bmatrix} \varphi_a \\ \varphi_b \\ \vdots \\ \varphi_g \end{bmatrix} \quad (5)
$$

$$
\begin{bmatrix} I_a \\ I_b \\ I_c \\ \vdots \\ i_g \end{bmatrix} = \begin{bmatrix} cos\theta_m & -sin\theta_m \\ \cos(\theta_m - \frac{2\pi}{7}) & -\sin(\theta_m - \frac{2\pi}{7}) \\ cos(\theta_m - \frac{2\cdot2\pi}{7}) & -\sin(\theta_m - \frac{2\cdot2\pi}{7}) \\ \vdots & \vdots \\ \cos(\theta_m - \frac{6\cdot2\pi}{7}) & -\sin(\theta_m - \frac{6\cdot2\pi}{7}) \end{bmatrix} \begin{bmatrix} I_d \\ I_q \end{bmatrix} \quad (6)
$$

The proposed simulation provides the tables from the relationships $\varphi_d(i_d, i_q)$ and $\varphi_q(i_d, i_q)$. Using a technique presented on [4], the inversed relationship $i_d(\varphi_d, \varphi_q)$ and $i_q(\varphi_d, \varphi_q)$ is obtained and used directly on the control simulation. The two flux tables can be understood as a set of data points $(\varphi_d, \varphi_q, i_d, i_q)$. The independent variables and the dependent variables can be swapped. The flux linkage is considered as the independent variables, the new inversed flux tables of $i_d(\varphi_d, \varphi_q)$ and $i_q(\varphi_d, \varphi_q)$ can be seen in Fig. 2.

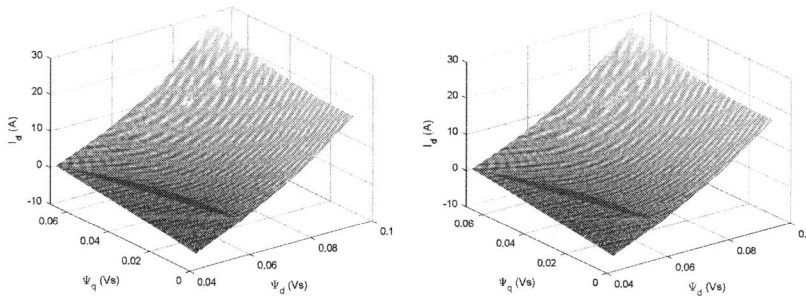

Fig. 2: Inversed flux tables a) $I_d(\varphi_d, \varphi_q)$ and b) $I_q(\varphi_d, \varphi_q)$

The generated look-up table (LUT) $i_d(\varphi_d, \varphi_q)$ and $i_q(\varphi_d, \varphi_q)$ can be used in the machine model for the control simulation, taking into account the saturation of the machine. The proposed block scheme for the model used on the simulation can be found in Fig. 3.

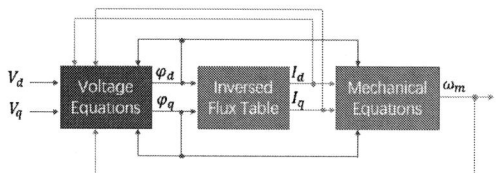

Fig. 3: Motor model with the inverse flux table

II. Open Loop Simulation

In this section, the simulation of the open loop control will be detailed, and the results will be presented. The model considers the saturation of the machine by means of the LUT previously built. It also takes into consideration the design of the stator board with the hysteresis controller for the PWM generation.

The controller scheme can be found in Fig. 4.

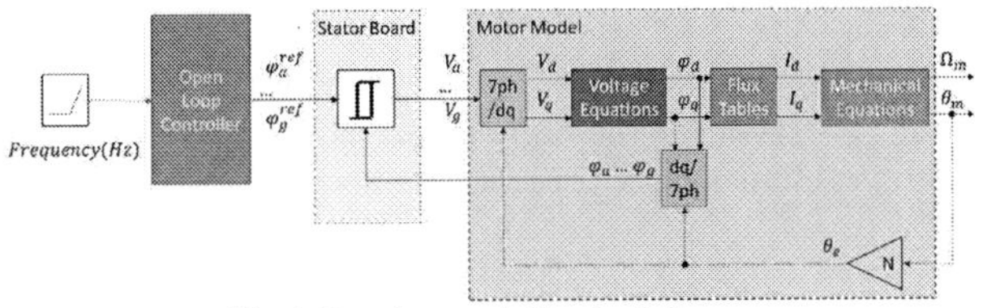

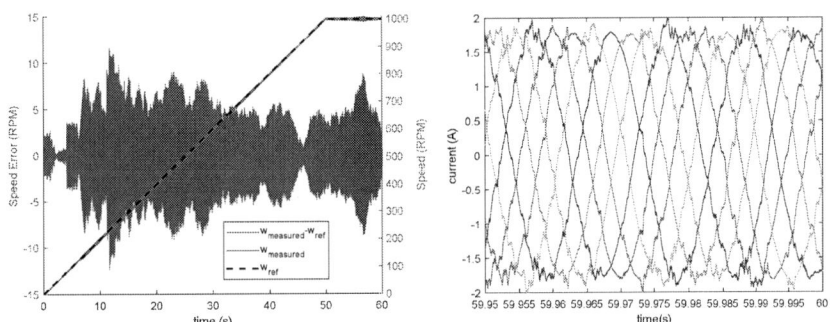

Fig. 4: Open loop controller simulation

Normally open loop control of synchronous machines considers V/f controllers in order to obtain constant φ_d and φ_q. The problem in the case of this integrated machine is that the stator board has already the feedback from the flux in each tooth of the motor from a search coil. This feedback is used on a hysteresis manner in order to generate the needed PWM. In order to be able to control the machine under open loop we need to generate the flux references in the seven phases rather than voltage reference. Instead of V/f controller is a φ/f controller. The amplitude of the flux is maintained at a constant level while the frequency is raised over the time.

Fig. 5: Speed response to ramp (left) and Current response on the 7 phases (right)

Fig. 5 shows the speed response to slow ramp. This ramp needs to be slow since there is no feed-back on the speed and the machine do not have any kind of starter squirrel cage or damper winding on the rotor. The output speed follows the reference even on open-loop control with minimum error.

The current output when the speed 1000 rpm can be seen in Fig. 5. There are some deformations on the current that are linked to the switching of the 7 inverters and the saturation phenomenon of the machine.

These results will be later compared with the experimental results on the next section

Power electronics integration for compact motor drive

Wireless transmission for rotary application is designed to avoid electrical contact for reliability and lifetime improvement. The rotor winding of the BWRSM needs to be energized from the static part. The most common way to supply is through brushes and slip rings [24], [25]. Nevertheless, the slip ring produces carbon dust that restricts the application range excluding clean room environment (medical, food industry). In addition, its lifetime is impeded and a dedicated space is required on the motor what increases the overall volume [26].

Inductive (IPT) and Capacitive (CPT) transfer are two different ways to wirelessly transfer the energy and replace direct electrical contact such as slip ring.

The CPT can be achieved by using a modulation of the electrical field to transfer the energy between two electrodes [27]. This system can be very cost effective compared to an IPT due to the absence of magnetic material and copper windings. The CPT has also the advantage of being more tolerant to misalignment between primary and secondary sides. However, most applications are using the capacitive coupling in a resonant tank that requests a nano-farad range capacitance value. The application is therefore limited to short air-gap leading to a limited capability of withstanding high voltage.

The IPT can be achieved by the magnetic field to transfer the energy in between two coupled coils [28]. IPT is easily implemented with different resonant topologies due to the large leakage inductance coming from a large air-gap. The IPT system can be classified in function of the type of topology, series-series, series-parallel, parallel-series, and parallel-parallel [29]. In the state of the art, the IPT efficiency can be up to 97 % for 7 kW [30]. However, the magnetic core and the copper needed to design the coils increase the cost of the system [31]. In this application, IPT is selected to feed the rotor due to its capability to work with larger air-gap.

A. Rotary Transformer

Rotary transformer design is different than a classical transformer due to its large air-gap to accommodate with the inherent distance between stator and rotor (primary-secondary). Consequently, high values of leakage inductance are usually obtained, along with high magnetizing currents in the case of an inductive transfer.

Parameters	Value
Input Voltage	100 V
Input Current	3 A
Output Voltage	60 V
Output Current	5.1 A
Switching Frequency	100 kHz
Max. Temperature	125°C
Air-gap	1 mm
ZVS	T1 & T2 On and Off
ZCS	Synchronous rectifier MOSFET at Turn-Off

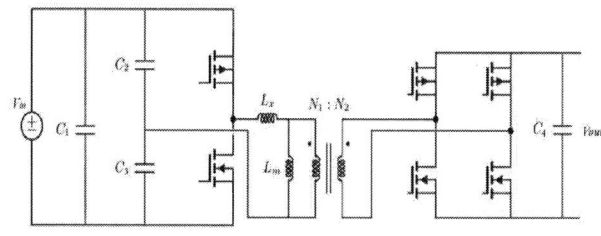

Fig. 6 Series resonant converter input parameters and schematic.

The IPT system includes, a DC/AC source that excites the primary of the rotary transformer with a periodic AC current, the secondary of the transformer that needs an AC/DC rectifier to feed the rotor coils and the respective control electronics. The input parameters for the converter selected are detailed in Fig. 6.

The series resonant topology is able to achieve ZVS (Zero Voltage Switching) on both primary transistors and ZCS (Zero Current Switching) on the synchronous rectifier at turn-off [32], [33].The resonant tank is made with the leakage inductance of the transformer Lx and the two capacitors C2 and C3. In this case, the tank does not need any additional discrete components which helps keep a low volume and complexity. The half-bridge power stage operates at fixed duty-cycle (50%) and switching frequency (100 kHz), set according to the tank resonant frequency. The control is made through a phase-leg gate driver with a dead-time circuitry and external resistors.

The secondary side of the transformer embeds the synchronous rectifier MOSFETs in the PCB using a stacking method and a process detailed in [34]. This method allows a better optimization of the PCB footprint as well as a lower resistivity of the connection, resulting in lower losses and less thermal stress on the power dies. This point will be fully explained in the following section.

B. Stator Inverters

An H-bridge circuit is used to supply each phase following the reference current. Four MOSFETs with two associate bridge gate drivers and local control are chosen for fitting the parameters in Table I.

Table I: Input parameters

Parameters	Value
Bus Voltage	150 V
Input Current	3.1 A
Maximum ripple bus voltage	5%
Switching Frequency Max	40 kHz
Max Temperature	125°C
Mode	Hard switching
Size Max	55 X 55 mm

The stator control is made in two different parts. The first part is the master control board made with the FPGA. It receives the position and speed information from a rotary encoder and sends a signal equivalent to the current shape expected in the stator coil to the stator board a reference. The reference signal mentioned above is injected in the second control loop, here called as local control loop, based on a hysteresis control type.

As shown in Fig.4, the control principle is based on a hysteresis comparator. The flux reference from the FPGA is compared with the flux measured by the search coil. The low-pass filter is taking the differential input signal from the FPGA board and linearize it to generate the reference signal for the hysteresis comparator. The comparator has two complementary outputs sending those PWM signals directly to the gate driver. The whole control loop circuit is supplied by a symmetrical ±5V power supply in order be compliant with the sinusoidal signal from reference and search coil while keeping a symmetric amplitude on the input signal.

C. Motor Drive Integration

1) Rotary transformer and Smart stator tooth

As shown in [17], the motor design, is the result of a maximum power density optimization. In a similar manner, the volume allocated to the rotary converter and the stator tooth also needs to be reduced. Consequently, the mechanical design of those converters is driven by the compactness. In low power converters, interconnections can be a significant part of the overall volume due to the reduce power conversion area. In order to reduce the impact both sides - primary and secondary side of the converter are made on a single PCB. The windings, the power stage and the control system are in the PCB. On the rotor side, the available volume is even more constrained. In particular, the thickness of the rotating part is limited. One trend in power electronics is to embed the power dies in the PCB. This technology drastically reduces the parasitic inductance in the switching cell, enables double side cooling and improves reliability. One additional advantage of this technology, from which this converter will benefit, is the low-profile shape of the PCB. Both sides of the converter, stator or rotor side, are built symmetrically as shown in Fig. 7. Each PCB, which embeds one side of the transformer, will be surrounded by a magnetic core concentrating the field towards the other PCB. The used material is a mix of epoxy and ferrite powder that can be molded, in any shape then cured to get its final mechanical strength. This assembly is further strengthened by an aluminum yoke. One side of the PCB is flushed with the yoke surface in order to maximize the coupling without any mechanical interference.

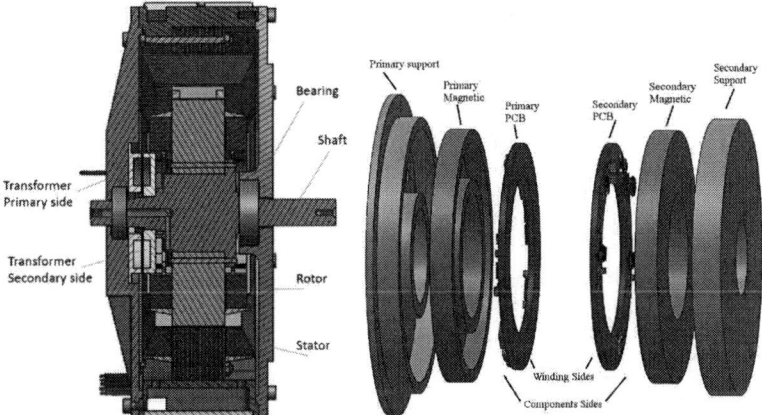

Fig. 7: Cross view of the motor (left) and exploded view of the rotary transformer (right).

Since the transformer is split between the stator and the rotor side, a functional air gap is mandatory. The thickness of this air gap is driven by mechanical constraints and will affect the magnetizing inductance of the transformer. Based on the assembly tolerances and unbalance of parts, the air gap is fixed at 1 mm. As demonstrated in [32], the resonant frequency is fixed by the leakage inductance and capacitors C2 and C3. The switching frequency is mostly limited by the conducting losses in the switches since their intrinsic capacitances are negligible. To set this frequency, the value of the resonant capacitor needs to be selected based on the leakage inductor value.

As shown in 8, the rotary transformer is placed on the rear of the motor. The secondary side of the transformer is placed in between the shaft and the stator windings. This assembly allows to limit the increase of the length of the back case of the motor. The primary side is fixed on the back case in a cavity. The bearing is on the outside of the transformer in order to reduce the connection complexity between the rotor and the transformer.

As shown in Fig. 8, for the stator tooth integration specific cavity all around the stator within the motor case. Two holes in the bottom of the cavity allow the connection with the search coil, used for the motor control, and the stator coil on the other side. A cable path is drilled all around the motor case to make the connection easier in between the 7 phases. The PCB has an interface to connect the thermal pad against the lid of the stator slot in order to extract heat from the MOSFETs bridge. The interface is a bit higher than the decoupling capacitor and is fixed with 4 screws in between the lid and the PCB board.

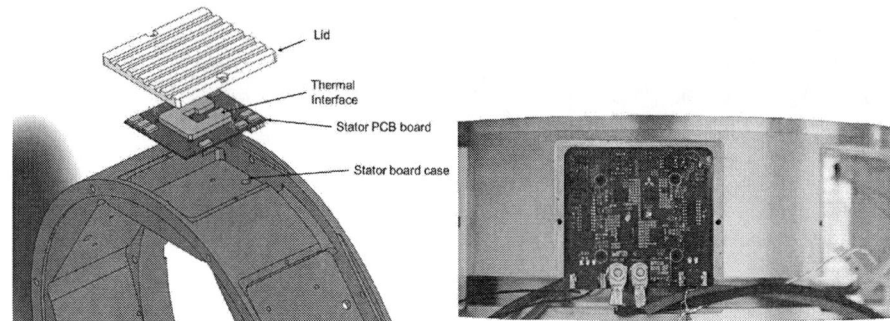

Fig. 8: Assembly view (top) and picture (bottom) of the PCB board the lid on the motor case

2) Manufacturing of the rotary transformer

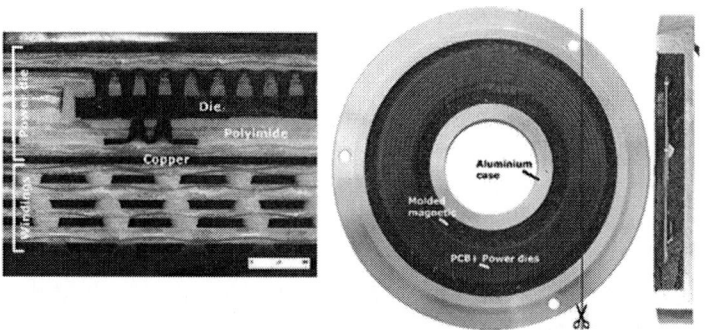

Fig. 9: Cross section of the PCB and of the transformer after potting

On the rotary transformer side, four silicon N-MOSFETs from Infineon were embedded in a polyamide substrate (Arlon 35N) in order to withstand the high temperature (Tj=160°C) with 70 μm copper layers. The top circuit components are soldered following a standard process in order to realize the synchronous rectifier function associated with the four transistors embedded. The PCB populated is then embed in an epoxy molded magnetic material within an aluminum container. A picture of the final assembly is shown in Fig. 9.

The transformer is characterized in order to adapt the resonant tank to the right leakage inductance value. The magnetizing inductance at primary is measured with an impedancemeter at the primary and secondary are 695 μH, 449 μH for respectively, and 110 μH for the leakage. Those values confirm the poor coupling factor and prove the necessity of a resonant topology.

3) Motor drive experimental test

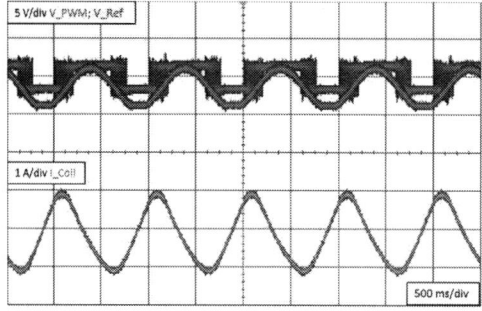

Fig. 10: Sinus reference and the PWM generated accordingly and the stator coil current

The built test bench has a torque meter and the assembled machine was coupled with a load machine. It was driven by a FPGA controller developed for the test, applying speed control in open loop. As described in Fig. , each stator coil is equipped with a search coil for the local hysteresis controller. The flux value is compared to the reference sent by the FPGA to operate the local hysteresis controller. The search coil value is converted into a flux by the mean of an integrator directly on each stator inverter. This control architecture gives redundancy to the machine and helps with operating under fault. A ramp from 0 to 1000 rpm is applied on the control and the rotary transformer supplies the rotor coils with 5 A showed in Fig. . The rotary transformer and the machine showed the predicted behavior at the operating point. The rotary transformer was successful in its purpose to be a "transparent" replacement for slip ring system.

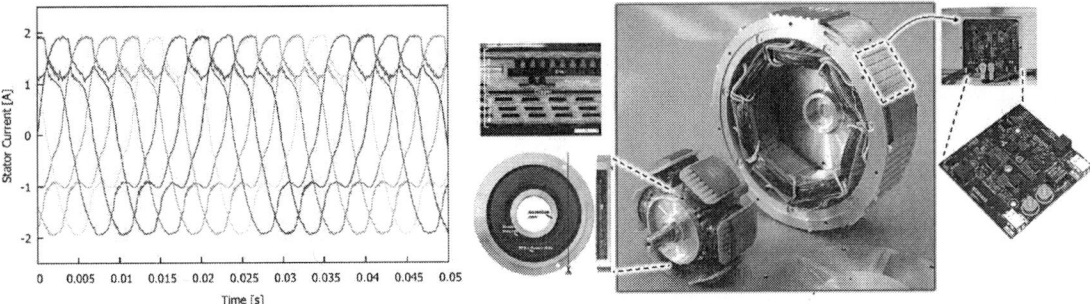

Fig. 11: Seven stator currents and final assembly of the BWRSM motor

The current on the motor at 1000 rpm operating with the open loop control is shown in Fig. 11 as well as the final assembly process. The simulation presents the same amplitude for the currents but there a phenomenon of deformation on the seven phase currents that is not predicted from the simulation.

Conclusion

In this paper the FEM modelling of a BWRSM was discussed with the goal of control simulation. The model turned out to be effective for the modelling of nonlinear saturation of the machine. It also made possible to simulate the control with less computational effort than a co-simulation between the FEM simulation and control simulation. The open-loop control is then simulated in MATLAB/SIMULINK using the previous cited BWRSM model. In a second step a 350 W rotary transformer with PCB embedded dies is designed and implemented in a 3.3 kW BWRSM. A 100 kHz series resonant topology circuit design for the converter is given, as well as the rotary transformer and the smart stator tooth assembly structure. The result is an optimized structure for a brushless motor drive in a volume constrained motor context. Final experimentation shows the motor drive with the rotary transformer, its converter and the smart stator tooth within the motor case both control in open loop up to 1000 rpm.

References

[1] R. Parsons, "Things in Motion: How to model a BLDC (PMSM) motors Kv (velocity constant) and Kt (torque constant) in FEMM," *Online*, 2019. https://things-in-motion.blogspot.com/2019/02/how-to-model-bldc-pmsm-motors-kv.html (accessed Dec. 15, 2020).

[2] D. C. Meeker, "Rotating Losses in a Surface Mount Permanent Magnet Motor:Finite Element Method Magnetics," *Online*, 2017. http://www.femm.info/wiki/SPMLoss (accessed Dec. 15, 2020).

[3] H. T. LE LUONG, "Optimal Design of Modular High Performance Brushless Wound RotorSynchronous Machine for embedded systems," *PhD Thesis, Univ. Toulouse*, no. Umr 5215, pp. 1–190, 2010.

[4] T. Wisniewski, J. C. Vannier, B. Lorcet, J. Saint-Michel, and X. Jannot, "Wound-rotor synchronous machine dq modeling with saturation for transient analysis," *2017 IEEE Int. Electr. Mach. Drives Conf. IEMDC 2017*, pp. 11–16, 2017, doi: 10.1109/IEMDC.2017.8002063.

[5] L. Parsa, "On advantages of multi-phase machines," in *IECON Proceedings (Industrial Electronics Conference)*, 2005, vol. 2005, pp. 1574–1579, doi: 10.1109/IECON.2005.1569139.

[6] A. A. Rockhill and T. A. Lipo, "A generalized transformation methodology for polyphase electric machines and networks," *Proc. - 2015 IEEE Int. Electr. Mach. Drives Conf. IEMDC 2015*, pp. 27–34, 2016, doi: 10.1109/IEMDC.2015.7409032.

[7] N. R. Brown, T. M. Jahns, and R. D. Lorenz, "Power Converter Design for an Integrated Modular Motor Drive," *Ind. Appl. Conf. 2007. 42nd IAS Annu. Meet. Conf. Rec. 2007 IEEE*, pp. 1322–1328, 2007, doi: 10.1109/07IAS.2007.205.

[8] P. Brockerhoff, Y. Burkhardt, K. Egger, and H. Rauh, "Highly integrated drivetrain solution: Integration of motor, inverter and gearing," *2014 4th Int. Electr. Drives Prod. Conf. EDPC 2014 - Proc.*, 2014, doi: 10.1109/EDPC.2014.6984412.

[9] "Motors/generators for traction/propulsion applications: A review - IEEE Journals & Magazine." https://ieeexplore.ieee.org/document/6470756 (accessed Feb. 04, 2021).

[10] J. Wang, Y. Li, and Y. Han, "Integrated Modular Motor Drive Design With GaN Power FETs," *IEEE Trans. Ind. Appl.*, vol. 51, no. 4, pp. 3198–3207, 2015, doi: 10.1109/TIA.2015.2413380.

[11]   M. Kaufhold, H. Auinger, M. Berth, J. Speck, and M. Eberhardt, "Electrical stress and failure mechanism of the winding insulation in PWM-inverter-fed low-voltage induction motors," *IEEE Trans. Ind. Electron.*, vol. 47, no. 2, pp. 396–402, 2000, doi: 10.1109/41.836355.

[12]   P. Wang, G. C. Montanari, and A. Cavallini, "Partial discharge phenomenology and induced aging behavior in rotating machines controlled by power electronics," *IEEE Trans. Ind. Electron.*, vol. 61, no. 12, pp. 7105–7112, Dec. 2014, doi: 10.1109/TIE.2014.2320226.

[13]   I. Vernica, H. Wang, and F. Blaabjerg, "Uncertainties in the Lifetime Prediction of IGBTs for a Motor Drive Application," in *Proceedings - 2018 IEEE International Power Electronics and Application Conference and Exposition, PEAC 2018*, Dec. 2018, doi: 10.1109/PEAC.2018.8590417.

[14]   F. Barbieri, J. W. Hines, M. Sharp, and M. Venturini, "Sensor-Based Degradation Prediction and Prognostics for Remaining Useful Life Estimation: Validation on Experimental Data of Electric Motors," 2015.

[15]   H. T. Le Luong, F. Messine, C. Henaux, G. Bueno Mariani, N. Voyer, and S. Mollov, "Comparison between fmincon and NOMAD optimization codes to design wound rotor synchronous machines," in *International Journal of Applied Electromagnetics and Mechanics*, 2019, doi: 10.3233/JAE-191108.

[16]   H. T. Le Luong, C. Hénaux, F. Messine, G. Bueno-Mariani, N. Voyer, and S. Mollov, "Finite element analysis of a modular brushless wound rotor synchronous machine," *J. Eng.*, vol. 2019, no. 17, pp. 3521–3526, 2019, doi: 10.1049/joe.2018.8206.

[17]   H. T. Le Luong, F. Messine, C. Henaux, G. B. Mariani, N. Voyer, and S. Mollov, "3D Electromagnetic and thermal analysis for an optimized wound rotor synchronous machine," *Proc. - 2018 23rd Int. Conf. Electr. Mach. ICEM 2018*, pp. 455–460, 2018, doi: 10.1109/ICELMACH.2018.8507020.

[18]   C. H. v. d. Broeck, "Lecture 9 Modeling And Current Control." RWTH Aachen University, 2019.

[19]   J. Espina, A. Arias, J. Balcells, and C. Ortega, "Speed anti-windup PI strategies review for field oriented control of permanent magnet synchronous machines," *CPE 2009 - 6th Int. Conf. - Compatability Power Electron.*, no. June 2009, pp. 279–285, 2009, doi: 10.1109/CPE.2009.5156047.

[20]   J. Espina, A. Arias, J. Balcells, and C. Ortega, "Speed anti-windup PI strategies review for field oriented control of permanent magnet synchronous machines," in *CPE 2009 - 6th International Conference-Workshop - Compatability and Power Electronics*, 2009, pp. 279–285, doi: 10.1109/CPE.2009.5156047.

[21]   S. Huang, Z. Chen, K. Huang, and J. Gao, "Maximum torque per ampere and flux-weakening control for PMSM based on curve fitting," in *2010 IEEE Vehicle Power and Propulsion Conference, VPPC 2010*, 2010, doi: 10.1109/VPPC.2010.5729024.

[22]   W. Ahmed Khan, "Torque Maximizing and Flux Weakening Control of Synchronous Machines Title: Torque Maximizing and Flux Weakening Control of Synchronous Machines," *Master's Thesis, Aalto Univ.*, p. 63, 2016, [Online]. Available: https://core.ac.uk/download/pdf/80719727.pdf.

[23]   A. Chiba, K. Kiyota, N. Hoshi, M. Takemoto, and S. Ogasawara, "Development of a rare-earth-free SR motor with high torque density for hybrid vehicles," *IEEE Trans. Energy Convers.*, vol. 30, no. 1, pp. 175–182, 2015, doi: 10.1109/TEC.2014.2343962.

[24]   R. D. Hall and R. P. Roberge, "Carbon brush performance on slip rings," *IEEE Conf. Rec. Annu. Pulp Pap. Ind. Tech. Conf.*, 2010, doi: 10.1109/PAPCON.2010.5556522.

[25]   Morgan Advanced Materials, "Carbon Brush & Holder Technical Handbook," [Online]. Available: http://www.morganelectricalmaterials.com/media/1996/technicalhandbookglobalproof_0.pdf.

[26]   M. A. Badr, A. I. Alolah, and A. F. Almarshood, "Transient performance of series connected three phase slip-ring induction motors," *IEEE Trans. Energy Convers.*, vol. 13, no. 4, pp. 305–310, 1998, doi: 10.1109/60.736314.

[27]   D. Shmilovitz, S. Ozeri, and M. M. Ehsani, "A resonant LED driver with capacitive power transfer," in *Conference Proceedings - IEEE Applied Power Electronics Conference and Exposition - APEC*, 2014, pp. 1384–1387, doi: 10.1109/APEC.2014.6803487.

[28]   Q. Li and Y. C. Liang, "An Inductive Power Transfer System with a High-Q Resonant Tank for Mobile Device Charging," *IEEE Trans. Power Electron.*, vol. 30, no. 11, pp. 6203–6212, Nov. 2015, doi: 10.1109/TPEL.2015.2424678.

[29]   O. Knecht and J. W. Kolar, "Comparative evaluation of IPT resonant circuit topologies for wireless power supplies of implantable mechanical circulatory support systems," in *Conference Proceedings - IEEE Applied Power Electronics Conference and Exposition - APEC*, May 2017, pp. 3271–3278, doi: 10.1109/APEC.2017.7931166.

[30]   J. Deng, F. Lu, S. Li, T. D. Nguyen, and C. Mi, "Development of a high efficiency primary side controlled 7kW wireless power charger," in *2014 IEEE International Electric Vehicle Conference, IEVC 2014*, 2014, doi: 10.1109/IEVC.2014.7056204.

[31]   J. Deng, F. Lu, W. Li, R. Ma, and C. Mi, "ZVS double-side LCC compensated resonant inverter with magnetic integration for electric vehicle wireless charger," in *Conference Proceedings - IEEE Applied Power Electronics Conference and Exposition - APEC*, May 2015, vol. 2015-May, no. May, pp. 1131–

1136, doi: 10.1109/APEC.2015.7104490.

[32] B. Lu, W. Liu, Y. Liang, F. C. Lee, and J. D. Van Wyk, "Optimal design methodology for LLC resonant converter," in *Conference Proceedings - IEEE Applied Power Electronics Conference and Exposition - APEC*, 2006, vol. 2006, pp. 533–538, doi: 10.1109/apec.2006.1620590.

[33] "A 90-w, high-efficiency, llc series-resonant converter withsecondary-side synchronous rectification - Recherche Google." https://www.google.com/search?client=firefox-b-d&q=A+90-w%2C+high-efficiency%2C+llc+series-resonant+converter+withsecondary-side+synchronous+rectification (accessed Feb. 04, 2021).

[34] G. Regnat, P. O. Jeannin, J. Ewanchuk, D. Frey, S. Mollov, and J. P. Ferrieux, "Optimized power modules for silicon carbide MOSFET," in *ECCE 2016 - IEEE Energy Conversion Congress and Exposition, Proceedings*, 2016, doi: 10.1109/ECCE.2016.7855324.

A Way Forward to Achieve Interoperability in Multi-Vendor HVDC Systems

Adil Abdalrahman[*1], Ying-Jiang Häfner[1], Philippe Maibach[2] and
Christoph Haederli[2]
[1]Hitachi Energy - HVDC, Ludvika, Sweden
[2]Hitachi Energy - FACTS, Turgi, Switzerland
[*]Email: adil.abdalrahman@hitachienergy.com

Keywords

≪Multi-terminal HVDC≫, ≪Railway power supply≫, ≪Interoperability≫, ≪Multi-vendor≫

Abstract

Interoperability for multi-vendor voltage source converter-based high voltage direct current (VSC-based HVDC) systems has been recognized as a new challenge. However, interoperability for multi-vendor power-converter-based islanded systems is not new. For example, well established practices and standards can be found in the areas of railway power supply and rolling stock. This paper provides a brief review of the challenges, and research completed in this area. Guidelines and standards developed to address these challenges will be summarized. A comparison between islanded systems with power converters for railway power supply and multi-terminal HVDC (MTDC) systems is presented. Finally, a way forward to achieve interoperability for multi-vendor VSC-based HVDC systems is described.

Introduction

The power system is undergoing a rapid transition, where power electronics interfaced devices are becoming the norm. This is driven by the ongoing transition towards a low-carbon society, where conventional synchronous generators are phased out and replaced by renewable energy sources that are connected via power converters. Power converters based on semiconductors with both turn-on/turn-off capability, e.g., insulated-gate bipolar transistors (IGBTs), have the advantages of flexible control and a fast response. However, interaction between power converters connected to the same network, especially where there are no other stronger sources, is a challenge. Interoperability is a well-known challenge in power converter domain AC and DC grids. The dynamic behavior and steady-state characteristics of power converters are very much dependent on the design of the controls. The converter control has been the core intellectual property (IP) of power converter manufactures from low voltage (e.g., photovoltaics and wind inverters), to medium voltage (e.g., traction applications), and up to high voltage (e.g., HVDC). It is expected that the industry will continue in this manner to maintain a competitive market and promote constant innovation, making it difficult to perform a thorough analysis of potential interactions. Nonetheless, there are not only analytic approaches to address the stability issues due to multiple converters coupling/interaction [1]-[3], but also mitigation methods [4].

An offshore windfarm is an islanded AC grid with "source" and "load" connected to it via power converters. Often the power converters may be provided by different manufacturers, thus it is a typical multi-vendor multi-converter system. References [2] and [3] present an overview, status, and outline of the CIGRE working group C4.49 on converter stability in power systems. Together with converter modelling assumptions, the following stability analysis methods were presented and evaluated using the proposed converter-based benchmark system:

- impedance-based stability analysis,

- passivity-based stability analysis,
- eigenvalue-based stability analysis,
- and time domain stability analysis.

In particular, a tutorial [3] was held to introduce procedures and guidelines to industry and academia on how to perform small-signal stability studies in modern power electronic based power systems. At the time of writing this paper, there are more than 10 offshore windfarms integrated by point-to-point VSC-based HVDC (from multiple vendors) that are operating with good operating experience indicating that converter interoperability can be realized in practice. Furthermore, there is rich experience in traction applications where guidelines and criteria have been established based on experiences and many years of research work [5]-[7]. This paves a path for realizing interoperability in multi-vendor systems without compromising the manufacturer's IP.

So far, there is no existing multi-vendor HVDC grid system. However, "Guidelines and Parameter Lists for Functional Specifications" featuring planning, specification, and execution of multi-vendor HVDC grid systems are published in [8]. This is a result of common efforts by a team of experts from leading vendors of HVDC technology, Transmission System Operators (TSOs), academia, and institutions. In general, [8] provides a common agreement for an open market of compatible equipment and solutions for HVDC grid systems. In relation to interoperability, the recommendation of [8] is that an assessment through electro-magnetic transient (EMT) simulations is necessary for ensuring HVDC grid system stability. Upon applying this recommendation, the research project, Best Path, identified potential instability due to interactions of converter control from different vendors [9]. As a result, challenges related to interoperability in multi-vendor HVDC grids have attracted a lot of attention from TSOs, academia, and industry. A multi-level sensitivity analysis approach has been developed in [1] to analyze the stability of multi-vendor MTDC systems. Although the approach is readily applicable for black-box systems, the analysis requires effort and it may only be necessary when the system is unstable, or close to instability with poor damping. DC impedances used for assessing DC dynamic interaction of a MTDC system is proposed in [10]. There are likely other ideas under investigation by researchers.

The purpose of this paper is to give a brief introduction to the experience in traction applications. The characteristics of a DC grid in comparison with traction electric systems is discussed. The established process, methodology and criteria in [7] can be readily applied in multi-vendor HVDC systems. This leads to the conclusion that interoperability in multi-vendor HVDC systems can be achieved.

Interoperability in Railway Systems

Historical Experience

Single-phase AC railway systems were introduced more than 100 years ago and have been in successful operation since then. In countries that electrified their railway system very early, a different frequency than the public grid frequency was introduced due to the technical capabilities at that time. For example, Switzerland, Germany, Austria, Sweden and Norway are using 16.7 Hz and parts of the USA are using 25 Hz. An impressive variety of technologies, systems and products are co-operating in such grids, such as dedicated single-phase generators, rotary frequency converters and many different kinds of static frequency converters. Examples of these static converters include:

- thyristor-based direct converters,
- VSC from several technology generations and/or vendors,
- locomotive and other traction vehicles ranging from old vehicles which are controlled by mechanically switched passive components (such as transformers, motors, resistors, capacitors and reactors) to thyristor-based converter locomotives,
- and VSC based vehicles of all technology generations and many vendors.

Traction vehicles represent highly varying loads to the system, both with respect to time and location. Furthermore, wire-based track safety systems of different functional principles, generations, and vendors are in operation. They communicate at a certain frequency or within a frequency band. They are part of the safety back-bone of the system since they report if a track section is occupied by a train or not.

Safe operation of the railway system has always been a key requirement and therefore, analysis of interoperability has played an important role from the very beginning. During the first couple of decades, when railway infrastructure and traction vehicle equipment were based on passive components, the system was sufficiently damped, and stability issues were not often observed.

Introducing a substantially large fleet of power-electronic-converter-based traction vehicles whilst decommissioning more and more old passive vehicles, increased interoperability challenges. An early and prominent example is from the Zurich main station area in 1995. After the introduction of a large fleet of VSC based locomotives, many legacy locomotives with traditional tap-changer-operated transformers were out of operation during a low-traffic period. In this situation, an instable situation tripped the complete railway system in the area. The root cause of this instability was not immediately understood, but a detailed analysis revealed that the modern VSC based locomotives were active in a certain frequency range and excited existing resonances. The resonances were sufficiently damped when old locomotives were in operation [11]. This experience made it clear that new processes were required to ensure interoperability among traction vehicles (both same and different types) and between traction vehicles and infrastructure.

Not only within the traction vehicles, but also on the infrastructure side, there has been an ongoing technology shift. For example, rotary frequency converters have increasingly been replaced by static frequency converters, to a point where expansion of the railway power systems has been almost exclusively done with static frequency converters.

Establishing EN 50388 standard

Following the experience in Zurich, the European research project ESCARV (Electrical System Compatibility for Advanced Rail Vehicles) was executed between 1998 and 2000. The experience gained [12] substantially influenced the elaboration of the EN 50388 standard by CENELEC (European Committee for Electromechanical Standardization) [5]-[7] to achieve interoperability. This standard has been available in a first release since 2012. It is under revision and will be split into two parts, EN 50388-1 [6] and EN 50388-2 [7].

EN 50388-1 will cover the general requirements, including a process of how to achieve interoperability. Apart from railway system specific topics, this standard covers important aspects relevant to other electrical systems. It can therefore be seen as a reference document describing how challenges derived from interoperability between very different assets in an electric grid can be managed. Such topics are discussed in Chapter 10 of the standard:

- Overvoltages caused by instability due to interaction between the controls of one or several active elements (converters) and passive elements. Normally, such instabilities occur in the frequency range of up to about 1 kHz, which is the bandwidth of the relevant controllers.
- Low frequency oscillations below and around supply frequency, which are the result of the nonlinear characteristics of modern converter systems.
- Overvoltages caused by harmonics and existing resonances.

The standard requires a compatibility study to be run when introducing a new element into the system. A process to check compatibility is laid out in the standard, defining roles and responsibilities of the involved stakeholders:

1. Characterization of the existing system and setting of acceptance criteria for the new element.
   Concerned parties: Organization in charge of the compatibility study, infrastructure manager, and operator/owner of rolling stock.
2. Design, characterization and testing of the new element.
   Concerned party: Supplier of the new element.
3. Perform compatibility tests.
   Concerned parties: Organization in charge of the compatibility study and infrastructure manager.

The "Organization in charge of the compatibility study" is not further specified, it could be a third-party consultant company or a dedicated expert team on the customer side. In any case, this organization collects all relevant data needed to perform the compatibility study.

EN 50388-2 establishes the acceptance criteria according to EN 50388-1 in relation to:

- co-ordination between controlled elements, and between controlled elements and resonances in the electrical infrastructure to achieve network system stability,
- and co-ordination of harmonic behavior with respect to excitation of electrical resonances.

Main phenomena identified and treated in this standard are:

- electrical resonance stability,
- low frequency stability,
- and overvoltages caused by harmonics.

Electrical resonance stability requirements

Electrical resonance stability concerns the excitation of electrical resonances in the power systems caused by feedback loop effects in the line converter controllers of rolling stock or static converters for power supply systems.

The following requirements shall be fulfilled:

- The lowest resonance in the power system shall not fall below the frequency limit (f_L). This limit is 87 Hz for a 16.7 Hz railway system or 300 Hz for a 50 Hz railway system.
- All components of the power system shall be included and considered. If resonances below f_L are unavoidable, sufficient damping shall be provided.
- All controlled elements shall be passive for all frequencies higher than f_L, which means that the phase for its frequency dependent input admittance lies between $\pm$ 90 degrees.

Low frequency stability requirements

Low frequency stability concerns oscillations at a frequency below the line frequency. These oscillations appear between rolling stock and power supply containing inductive and capacitive elements. They are initiated by feedback loops as well as limitations and protection functions within the system. Since low-frequency stability is multi-dimensional (coupled feedback loops for both magnitude and phase of voltage and current), no simple interface requirements for single components are defined so far. The system to be analyzed is a simplified case of a railway system which consists of:

- a constant or controlled voltage source (power supply system),
- a linear network (power supply system),
- and one or several vehicles or other new elements at a single location.

The investigation for small-signal stability shall be done:

- by time domain simulation,
- or according to the dq method (dq frequency response and multiple-input multiple-output (MIMO) Nyquist method) as detailed in Annex A.2 of the standard.

Interestingly, in the context of applications other than for railway, the standard includes the study case 'E' in which converters/compensators are to be tested against other converters/compensators.

Requirements related to overvoltages caused by harmonics

Specific overvoltage limits due to harmonics are given. Additionally, requirements specifically applicable to the railway application are provided, specifying how to handle the effect of multiple sources for harmonic current or voltage components. Other than that, the calculation of harmonics is straight forward and comparable to other standards.

Discussion

By far, not all aspects described in EN 50388 are applicable for other applications. However, even if the acceptance criteria will presumably have to be adapted, the described process on how to assess interoperability of a new asset in an existing grid seems to be applicable in general.

Comparison between electric traction system and multi-terminal HVDC system

There are three types of electric traction systems: DC, AC and composite electrification systems. A composite system includes both an AC and DC system, see Fig. 1. The most common is the AC electrification system, typically 16.7 Hz or 50-60 Hz single-phase. The AC system is among the most challenging systems from a stability and interaction point of view.

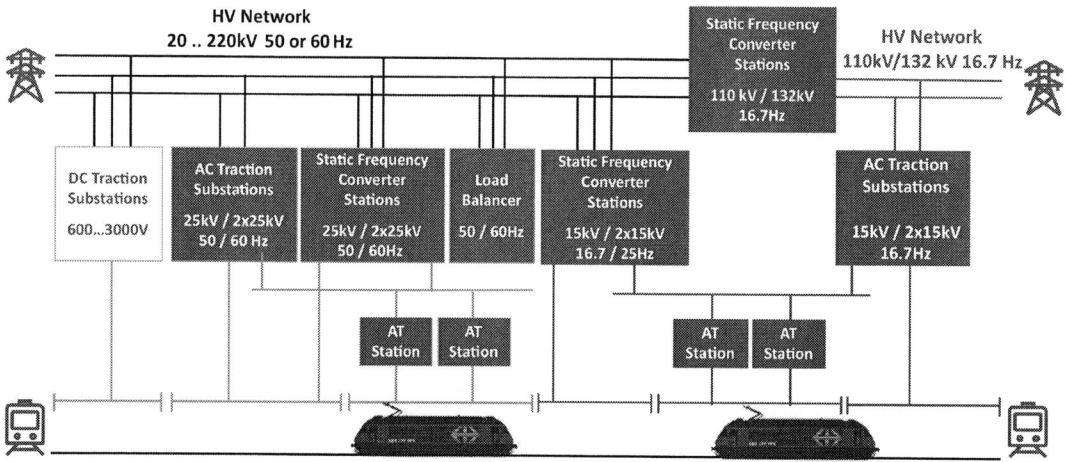

Fig. 1: A simplified overview of an electric traction system.

An example of an MTDC system is presented in Fig. 2. The terminals in grey are planned to be added to the existing Caithness–Moray HVDC link (between Spittal and Blackhillock).

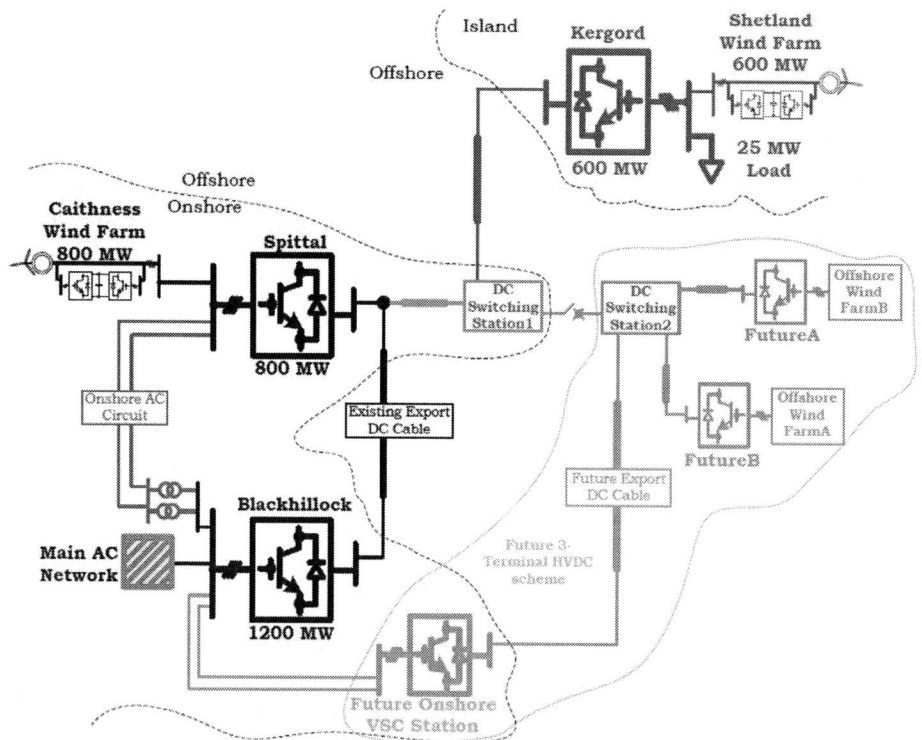

Fig. 2: A simplified overview of an MTDC system [13].

In the following comparison, the 16.7 Hz AC traction system will be used. The comparison will mainly cover three aspects: basic elements in the system, system configuration, and the potential stability aspects.

Basic elements

The basic building blocks may be categorized as "active elements" and "passive elements". An active element may be defined as a control loop which is based on direct or indirect signals/variables sensed in the system of concern, i.e., 16.7 Hz single-phase for electric traction systems and DC network for MTDC. The active power sensed on the main grid side is an indirect variable since the power on either side of the converter is balanced (disregarding the minor converter losses). In a traction system, the line converter of rolling stock or a static converter for power supply systems are active elements. In an MTDC system, any converter which controls the DC voltage or active power is an active element. As shown in Fig. 3, the converter, including its control system, may be represented as a medium (blue box) which interconnects the concerned system and other systems. Other systems may be an AC main grid Fig. 3(**a**)-1/Fig. 3(**b**)-1, locomotive motor Fig. 3(**a**)-2 or an islanded grid Fig. 3(**b**)-2.

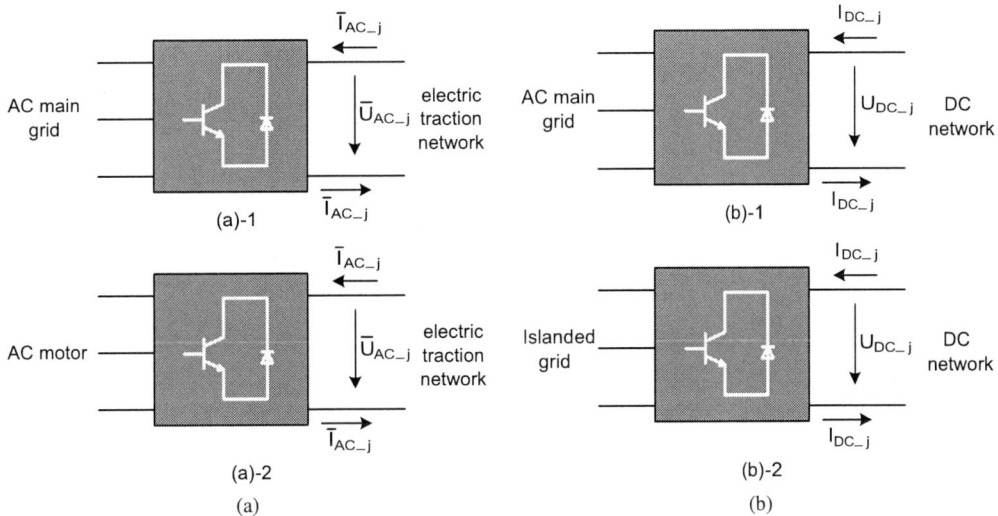

Fig. 3: Active element in (**a**) electric traction systems; (**b**) in MTDC.

In general, the active elements in both systems look similar. However, there may be some differences by looking at the details. As seen from the electric traction system, the voltage $\overline{U}_{AC\text{-}j}$ and the current $\overline{I}_{AC\text{-}j}$ are phasors having both a magnitude and an angle. This means that the coupling impedance between the voltage and current is a matrix related MIMO system. According to [7], it can be generalized as (1).

$$\begin{bmatrix} \Delta u_{d_j} \\ \Delta u_{q_j} \end{bmatrix} = \begin{bmatrix} Z_{dd_j} & Z_{dq_j} \\ Z_{qd_j} & Z_{qq_j} \end{bmatrix} \begin{bmatrix} \Delta i_{d_j} \\ \Delta i_{q_j} \end{bmatrix} \tag{1}$$

On the other hand, as seen from the DC network, the coupling impedance between voltage $U_{DC\text{-}j}$ and current $I_{DC\text{-}j}$ is a transfer function of a single-input single-output (SISO) system $\Delta u_{dc_j} = Z_{d_j} \Delta i_{dc_j}$ for one converter. The DC network may often be bipolar with a common neutral point connected to the ground to provide redundancy and to avoid significantly high voltages when there is a DC line fault. This means that at each terminal there are two converters: one connecting to the positive pole and the other connecting to the negative pole. Therefore, as seen from DC network at node j, the converter station may be described by (2).

$$\begin{bmatrix} \Delta u_{p_j} \\ \Delta u_{n_j} \end{bmatrix} = \begin{bmatrix} Z_{d_j} & Z_{pn_j} \\ Z_{pn_j} & Z_{d_j} \end{bmatrix} \begin{bmatrix} \Delta i_{p_j} \\ \Delta i_{n_j} \end{bmatrix} \tag{2}$$

where Z_{pn_j} irepresents the interaction between the positive and negative poles.

Furthermore, the active element is also a source of harmonics. In a traction system, the pulse width modulation (PWM) converter may generate harmonics in a wide range, from very low frequency (near fundamental frequency) to high frequency. In an MTDC system, the modular multilevel converter (MMC) may generate harmonics typically higher than N_c*50 Hz with a magnitude limited to below $1/N_c$, where N_c is the number of modules which is linearly related to the DC voltage level. For example, $N_c > 170$ for 500 kV DC. Thus, the active elements from a feedback control point of view are similar in both an electric traction system and an MTDC, but there is a difference from a harmonic source point of view.

The passive elements are similar in both electric traction and MTDC systems, which basically consist of distributed inductors and capacitors (in lines and cables) or inductance and capacitance from dedicated equipment such as auto-transformers, filters or current limiting reactors.

System configuration

In both electric traction and MTDC systems, the active elements are interconnected via the passive elements. The active elements are connected in parallel as seen from the concerned system point of view. Thus, the same equivalent can be used from system point of view as shown in Fig. 4.

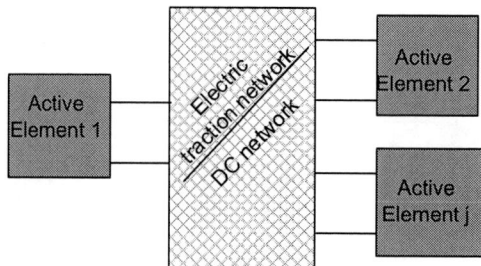

Fig. 4: Equivalent system configuration for electric traction and MTDC system.

Potential stability issues

As has been described earlier, the potential stability issues in electric traction systems for two types of instability (namely Electrical Resonance and Low Frequency Stability) are related to the control of a certain active element interacting with electric resonance frequency in the network. The interaction between the control of active elements is at low frequency.

Similarly, these two types of instability could occur in MTDC systems. This is firstly because the DC network could also have some resonances due to capacitive and inductive elements, and the resonance frequencies could be in the range of control bandwidth of active elements. Secondly, the converters are nonlinear devices, as voltage and current must be limited within the converters rating. Furthermore, the short circuit power ratio (SCR), which is defined as the short circuit power at the AC common coupling point divided by the converter rated active power, may be much lower for the HVDC converter. When the grid SCR is low i.e., the AC grid is weak, the HVDC function of supporting the AC grid has to be prioritized. In case of some severe events, it may even be necessary to transfer DC voltage control mode to other stations [14]. Overvoltages due to harmonics are unlikely in the MTDC system, mainly because the frequencies of generated harmonics are very high and losses in a DC network at high frequencies are often high due to the skin effect.

Proposal to achieve interoperability for a multi-vendor VSC-based HVDC system

It is the responsibility of each HVDC vendor to ensure a stable and robust performance when their converters are integrated into a multi-vendor MTDC system, but on the other hand vendors must also protect their IP. Accordingly, there should be a practical and simple approach to assess interoperability with sufficient accuracy whilst respecting IP rights.

Proven practice and guidelines [7] in electric traction systems may be used to achieve interoperability for a multi-vendor VSC-based HVDC system, whilst considering possible differences in pre-conditions related to VSC-based HVDC systems. Like electric traction systems, the challenges related to interoperability between HVDC terminals supplied from different vendors can be addressed in two aspects:

1. Stability related to interaction between HVDC control and resonance frequencies in the DC network.
2. Stability related to interaction between the control of different HVDC converter stations.

Both aspects could be assessed primarily by using the scanned impedance at the DC coupling point of the respective HVDC converter stations. Limited time domain simulations [8, 13] of defined cases may be used as a complementary assessment or verification.

The scanned impedance can be either provided by the vendor using the relevant pre-conditions provided by the DC network owner/TSO (such as the maximum and minimum short circuit power ratio at the point of common coupling at the converter station AC side). Alternatively, the impedance can be scanned by the DC network owner/TSO based on the black-box model of the HVDC link (with identical control as in the real system) provided by the vendor using the relevant AC network models. Note that for the DC network it is important to consider all elements and all possible configurations, from minimum three terminals to future expanding multi-terminals.

Interaction between HVDC control and resonance frequencies in the DC network

As for the interaction between HVDC control and resonance frequencies in the DC network, it can be avoided as long as the active control of HVDC converter remains passive at the resonance frequencies. This may be translated into a requirement on the impedance observed at the point of DC connection.

The following requirement shall be fulfilled:

- The phase of the scanned impedance (in case of matrix, element at the diagonal of matrix, refer to (2)) at the point of DC connection at resonance frequencies of a DC network shall lie between ± 90 degrees.

Interaction between the control of different HVDC converter stations

The interaction between the converter control at different terminals can also be examined by using the scanned impedance. The theory behind the detailed stability criterion [7] is based on the generalized Nyquist criteria. Using the equivalent system configuration for an MTDC system as shown in Fig. 4, the DC network may be described with an admittance matrix $\mathbf{Y}$, arranged according to terminals, for example:

$$\begin{bmatrix} \Delta i_{p1} \\ \Delta i_{n1} \\ \Delta i_{p2} \\ \Delta i_{n2} \\ \Delta i_{p3} \\ \Delta i_{n3} \end{bmatrix} = \begin{bmatrix} \mathbf{Y} \end{bmatrix} \begin{bmatrix} \Delta u_{p1} \\ \Delta u_{n1} \\ \Delta u_{p2} \\ \Delta u_{n2} \\ \Delta u_{p3} \\ \Delta u_{n3} \end{bmatrix} \tag{3}$$

A $\mathbf{Z}$ matrix collects the impedance of each terminal as seen from the respective connected DC point:

$$
\begin{bmatrix} \Delta u_{p1} \\ \Delta u_{n1} \\ \Delta u_{p2} \\ \Delta u_{n2} \\ \Delta u_{p3} \\ \Delta u_{n3} \end{bmatrix} = \underbrace{\begin{bmatrix} \boxed{1} & & \\ & \boxed{2} & \\ & & \boxed{3} \end{bmatrix}}_{\mathbf{Z}} \begin{bmatrix} \Delta i_{p1} \\ \Delta i_{n1} \\ \Delta i_{p2} \\ \Delta i_{n2} \\ \Delta i_{p3} \\ \Delta i_{n3} \end{bmatrix}
\tag{4}
$$

where blocks 1, 2, and 3 are 2×2 matrices as shown in (2) for a bipolar HVDC system for terminal 1, 2, and 3. The other blocks are 2×2 matrices with 0 as elements.

The stability of the system is derived from the local minimum of the absolute value of:

$$
N(j\omega) = \det(1 + \mathbf{Y}(j\omega)\,\mathbf{Z}(j\omega))
\tag{5}
$$

As described in Section A.2.7 in [7], the system will be stable if the origin is on the left-hand side as seen from the trace of $N(j\omega)$ for all local minima of the absolute value. If the origin is on the right-hand side for at least one local minimum, the system will be unstable and oscillate at the corresponding frequencies as illustrated in Fig. 5.

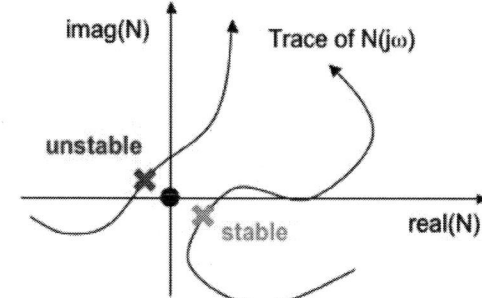

Fig. 5: Nyquist diagram for a stable and unstable system [7].

The control interaction may be also assessed via the impedance directly, as presented in [10], instead of plotting the trace of $N(j\omega)$.

Time domain simulation

Time domain simulations using EMT tools (PSCAD/EMTDC or alike) can be used for further verification. Black-box models would be sufficient for that purpose as suggested in [8]. The EMT verification can be done by TSOs, independent parties or HVDC vendors. Sometimes black-box models cannot be exchanged among different vendors due to proprietary reasons. In this case, the vendor can perform a verification using EMT software with co-simulation capabilities using separate computers (at the same or different physical locations) linked via fast telecommunication schemes [15]. In this case, simulations are run over a network where each vendor only gets access to a transmission line interface and the rest of model is running elsewhere (invisible to the user). This also requires an independent third-party that is organizing the setup. An example of such a verification can be found in [13] where a Dynamic Performance Study (DPS) verification was performed for the Caithness–Moray future MTDC. To speed up the simulation, parallel and high-performance computing, as proposed by PSCAD/EMTDC [16], can be utilized. Alternatively, a hardware-in-the-loop (or real time) simulation can be used [9, 17].

Conclusion

Interoperability for multi-vendor VSC-based HVDC systems has been recognized as a new challenge. However, interoperability for multi-vendor power-converter-based islanded systems is not new. A brief review of the challenges, and research completed in the areas of electric railway system have been presented. Guidelines and standards for addressing these challenges in this area have been presented. The comparison between the islanded system with power converters in a railway electric system with DC grids or multi-terminal HVDC systems shows that interoperability for multi-vendor VSC-based HVDC systems is not a new challenge. A way forward to achieve interoperability for multi-vendor VSC-based HVDC systems has been discussed. In summary, the interoperability in multi-vendor HVDC systems can be managed, through the process disclosed in the established standard [7] and the performance of time domain simulations [13], [15]-[17] without compromising the IP of vendors.

References

[1] Y. Liao et al., "Stability and Sensitivity Analysis of Multi-Vendor, Multi-Terminal HVDC Systems", 22nd Power Systems Computation Conference (PSCC 2022).

[2] L. Kocewiak et al., "Overview, Status and Outline of Stability Analysis in Converter-based Power Systems" 19th Int'l Wind Integration Workshop, 11-12 November 2020.

[3] M. Larsson, "Part 2: Harmonic stability and converter interoperability", SC B4 Tutorial / Workshop for CIGRE session 2021, August 26th, HVDC harmonics – topical and emerging issues for AC and DC sides.

[4] Ł. Kocewiak et al., "Instability Mitigation Methods in Modern Converter-based Power Systems". 20th Int'l Wind Integration Workshop, 29-30 September 2021.

[5] DIN EN 50388: "Railway Applications – Power supply and rolling stock – Technical criteria for the coordination between power supply (substation)," VDE, December 2012.

[6] prEN 50388-1:2017: "Railway Applications – Fixed installations and rolling stock – Technical criteria for the coordination between traction power supply and rolling stock to achieve interoperability – Part 1: general".

[7] prEN 50388-2:2017: "Railway Applications – Fixed installations and rolling stock – Technical criteria for the coordination between traction power supply and rolling stock to achieve interoperability – Part 2: stability and harmonics".

[8] CLC/FprTS 50654-1: "HVDC Grid Systems and connected Converter Stations - Guideline and Parameter Lists for Functional Specifications - Part 1: Guidelines," CENELC, December 2019.

[9] BestPaths, "Deliverable D9.3: BEST PATHS DEMO#2 Final Recommendations for Interoperability of Multivendor HVDC Systems.". Available at: http://www.bestpaths-project.eu/contents/publications/d93–final-demo2-recommendations–vfinal.pdf.

[10] F. Loku et al., "Equivalent Impedance Calculation Method for Control Stability Assessment in HVDC Grids," MDPI Energies 2021, 14, 6899. Available at: https://www.mdpi.com/1996-1073/14/21/6899/pdf.

[11] M. Aeberhard et al., "Resonanzproblematik im SBB Energienetz.", available at: http://www.news.admin.ch /NSBSubscriber/message/attachments/34330.pdf%202012#: :text=Mitte%20der%201990er%2DJahre%20s ind,Abschaltung%20der%20Loks%20Re%20450.

[12] M. Lörtscher et al., "Kompatibilitätsuntersuchungen am schweizerischen 16,7-Hz Bahnstromnetz," Elektrische Bahnen, vol. 6–7, pp. 292–300, 2001.

[13] O. D. ADEUYI et. al., "Multi-terminal Extension of Embedded Point-to-Point VSC-HVDC Schemes," Cigre B4-120, Paris, 2020.

[14] Y. Hafner et al., "Stability Enhancement and Blackout Prevention by VSC Based HVDC" Paper 14-02, CIGRE 2011, BOLOGNA.

[15] Manitoba Hydro International. Electric Network Interface. Online. Available at: https://hvdc.ca/webhelp /PSCAD/Features_and_Operations/Electric_Netwrok_Interface_(ENI).htm.

[16] https://www.pscad.com/webhelp/PSCAD/Features_and_Operations/Parallel_and_High_Performance_Compu ting.htm.

[17] http://www.bestpaths-project.eu/contents/publications/d93–final-demo2-recommendations–vfinal.pdf.

Model Predicitve Position Control of Electrical Drives on an Industrial PC

Fabian Karau, Michael Leuer
Bielefeld University of Applied Science
Schulstraße 10
Gütersloh, Germany
Email: fabian.karau1@fh-bielefeld.de, michael.leuer@fh-bielefeld.de

Keywords

≪MPC (Model-based Predictive Control)≫, ≪Motion control≫, ≪Dead-time≫, ≪Servo-drive≫, ≪State-space model≫.

Abstract

Dynamic position control of electric drives is of high importance in many industrial applications. In addition to the classic cascade control based on P and PI controllers for this task, advanced controls such as model predictive control (MPC) are receiving increasing attention. However, one drawback of MPC is the high computational power required, which makes it difficult to implement, especially for older servo drives. In this paper, we show how an industrial PC (IPC) that is available at many plants anyway can be used to implement MPC for position control of a simple drive system. Occurring dead times can be considered directly by the MPC.

Introduction

Electric drives are used for highly dynamic, precise positioning tasks in many industrial applications. Examples include machine tools, robotics applications and printed circuit board assembly. The quality of the drive control has a significant effect on the quality of the manufactured workpieces and also influences the wear, energy consumption and productivity of the system. The control concept is therefore highly important and has been subject of research for years.

The foundation for many of today's industrially used drive control structures is a cascade control based on linear P and PI controllers. This consists of the current or torque control loop with the superimposed velocity and position control loops. It is characterized by simple implementation and parameterization. In addition, it can be easily extended with feedforward controls and setpoint as well as actual value filters. However, complex drive systems with non-linear influences such as gear backlash, friction or elasticities in the drive train and manipulated variable or state variable constraints can only be handled to a limited extent by PI-based cascade control [1, 2].

Model predictive control (MPC) is a universal control method capable of solving the above problems. With the help of a mathematical model of the controlled system, the reaction to certain manipulated variables can be predicted. By solving an optimization problem, the optimal manipulated variables in terms of a freely formulable cost function are determined. The model and the optimization problem can include nonlinearities and constraints. This potentially enables MPC to achieve a superior control performance compared to PI-based cascade control.

The high computational power required by MPC to solve the optimization problem is a major drawback compared to P and PI controllers. Thanks to the generally increasing computational power, the concept of model-based predictive control has also become more of a focus in drive technology. Numerous MPC algorithms have already been published for current and torque control [3, 4, 5] as well as for velocity and position control [6, 7, 8, 9, 10, 11]. In many publications, powerful dSPACE rapid control prototyping systems are used for this purpose [6, 9, 8]. But also the implementation on modern SoC FPGAs [10] and state of the art digital signal processors (DSP)[11] has already been realized. Industrial PCs (IPCs) represent another exciting target platform. Newer IPCs have standard extensive memory and computing power. In addition, almost every industrial plant has a programmable logic controller (PLC), which is increasingly being implemented as an IPC. The goal of this paper is therefore to implement a first approach of a model predictive position control on an IPC for a simple drive system. In this way, older servo drives without sufficient computing power can benefit from advanced control methods as well.

The upper part of Fig. 1 displays the classical, widespread cascade control in a simplified way. Here, control takes place entirely within the servo drive and setpoint generation runs on the IPC. The lower part shows the desired control structure, in which both the position/velocity MPC and the setpoint generation are implemented on the IPC. The servo drive receives a torque reference from the IPC and is used only for torque control and setting the motor voltages. Strictly speaking, current control takes place, but with the permanent magnet synchronous motors used, the current can be converted into a torque.

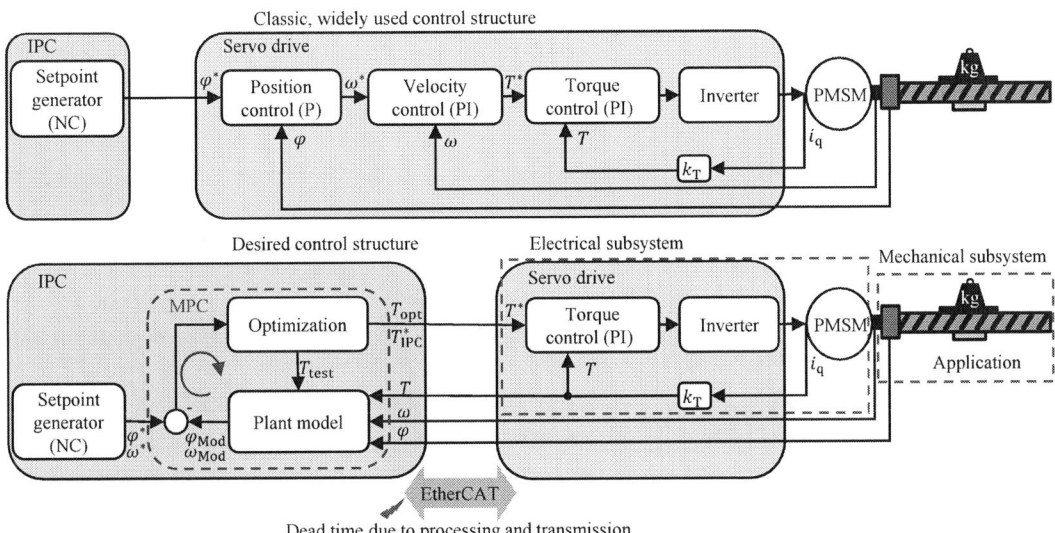

Fig. 1: Simplified overview of classic cascade control and the desired MPC control structure

Model Predictive Control

This section describes the applied MPC approach. The basis of the control is the model of the controlled system, which is explained in the following.

Prediction Model

The controlled system of the position control of a drive system can be divided into an electrical and a mechanical subsystem. The electrical subsystem comprises the closed torque control loop including the inverter and motor windings. Since there is no direct access to the inverter from the IPC in the form of a switching state or voltage reference, the torque control structure is not changed and the preset PI controller is used. Furthermore, the exchange of reference and actual values between the IPC and servo drive via the EtherCAT bus system and the processing time in the PLC introduces additional dead time into the system. The closed torque control loop and the total dead time are modeled by the transfer function (1)

$$G_{\text{Trq}}(s) = \frac{T_{\text{M}}(s)}{T_{\text{MIPC}}^*(s)} = e^{-\tau_{\text{Delay}} \cdot s} \cdot \frac{1}{\tau_{\text{Trq}}^2 \cdot s^2 + 2D_{\text{Trq}}\tau_{\text{Trq}} \cdot s + 1} \, . \tag{1}$$

Fig. 2 shows the measured frequency response with a sampling time $T_S = 250\,\mu\text{s}$ and the model fitted according to (1). The parameters determined are $\tau_{Trq} = 154.64\,\mu s$, $\tau_{Delay} = 843.75\,\mu s$ and $D_{Trq} = 0.7071$.

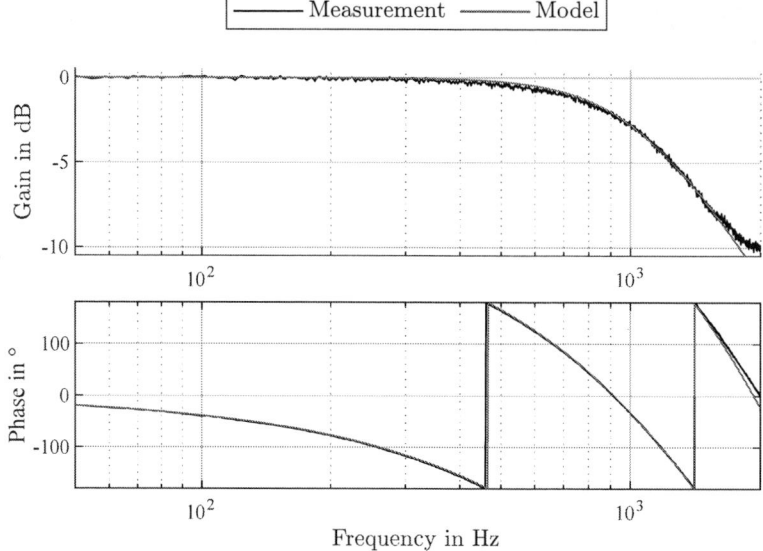

Fig. 2: Measured frequency response and estimated model

Here, the PT2 element mainly describes the torque control loop and τ_{Delay} the accumulated dead time in the system as well as the remaining model deviations of the torque control loop. For the hardware and configuration used, the total system dead time is thus $\tau_{Delay} = 3.375 \cdot T_S$. By optimizing the configuration, a further reduction of the dead time is probably possible, but a dead time will always remain in the described experimental setup.

The mechanical subsystem is composed of the motor moment of inertia and the driven application. For rigid mechanics, modeling as a single-mass system with the total moment of inertia Θ_T and the viscous friction μ_{FV} is permissible. The disturbance torques caused by the application and the coloumb friction are summarized as load torque T_L. This load torque is usually not measurable and must be reconstructed by an observer [7], Kalman filter [11] or moving horizon estimator [8, 9]. Here a Luenberger disturbance observer is used, whose poles are placed according to Butterworth with a cutoff frequency of 400 Hz. The single mass system is described by (2)

$$\dot{\omega}_M = \frac{T_M - \mu_{FV} \cdot \omega_M - T_L}{\Theta_T}, \tag{2}$$

where ω_M is the motor angular velocity. The dead-time free part of (1) and (2) are combined and brought into the continuous-time state space representation according to (3) and (4). The integration of the angular velocity leads to the angular position

$$
\underbrace{\begin{pmatrix} \ddot{T}_M \\ \dot{T}_M \\ \dot{\omega}_M \\ \dot{\varphi}_M \\ \dot{T}_L \end{pmatrix}}_{\dot{x}(t)} = \underbrace{\begin{pmatrix} \frac{-2D_{Trq}}{\tau_{Trq}} & \frac{-1}{\tau_{Trq}^2} & 0 & 0 & 0 \\ 1 & 0 & 0 & 0 & 0 \\ 0 & \frac{1}{\Theta_G} & \frac{-\mu_{RV}}{\Theta_G} & 0 & \frac{-1}{\Theta_G} \\ 0 & 0 & 1 & 0 & 0 \\ 0 & 0 & 0 & 0 & 0 \end{pmatrix}}_{\mathbf{A}_c} \cdot \underbrace{\begin{pmatrix} \dot{T}_M \\ T_M \\ \omega_M \\ \varphi_M \\ T_L \end{pmatrix}}_{x(t)} + \underbrace{\begin{pmatrix} \frac{1}{\tau_{Trq}^2} \\ 0 \\ 0 \\ 0 \\ 0 \end{pmatrix}}_{b_c} \cdot \underbrace{T_M^*}_{u(t)}, \tag{3}
$$

$$
\underbrace{\varphi_M}_{y(t)} = \underbrace{\begin{pmatrix} 0 & 0 & 0 & 1 & 0 \end{pmatrix}}_{c_c^T} \cdot \underbrace{\begin{pmatrix} \dot{T}_M \\ T_M \\ \omega_M \\ \varphi_M \\ T_L \end{pmatrix}}_{x(t)}. \tag{4}
$$

For the implementation in a PLC program the model must be discretized. This is done with the known methods as described for example in [12]. To represent the dead time, an augmentation of the state space model is done. The discrete-time state space model finally has the form of (5)

$$
x(k+1) = \mathbf{A}x(k) + bu(k), \qquad y(k) = c^T x(k). \tag{5}
$$

Control Approach

Over the years, numerous MPC algorithms have been developed. Nevertheless, there are some common features that essentially all algorithms share. Two important variables of MPC are the prediction horizon n_p and the control horizon n_c, where $n_p \geq n_c$. Starting from the current state $x(k)$ and a certain manipulated variable sequence $u(k+i)$, the course of the controlled variable $y(k+i)$ is predicted for n_p sampling steps, with $i = 1, 2, ..., n_p$. The manipulated variable can be varied for n_c times by the optimization and remains constant afterwards. Similarly, the cost function is a central element of all algorithms. Often the predicted quadratic control deviation $(y^*(k+i) - y(k+i))^2$ and the manipulated variable changes $\Delta u(k+i-1)^2$ are evaluated, resulting in the cost function for the SISO case according to (6)

$$
J(u,x) = \sum_{i=1}^{n_p} (y^*(k+i) - y(k+i))^2 q + \Delta u(k+i-1)^2 r. \tag{6}
$$

The aggressiveness/dynamics of the control can be set via the weights q and r. Instead of the sum equation, a compact representation in the matrix-vector notation according to (7) is also possible

$$
J(\Delta \bar{u}) = (\bar{y}^* - \bar{y})^T \cdot \bar{\mathbf{Q}} \cdot (\bar{y}^* - \bar{y}) + \Delta \bar{u}^T \cdot \bar{\mathbf{R}} \cdot \Delta \bar{u} \tag{7}
$$

with the vectors and matrices

$$
\bar{y}^* = \begin{pmatrix} y^*(k+1) \\ y^*(k+2) \\ \vdots \\ y^*(k+n_p) \end{pmatrix}, \Delta \bar{u} = \begin{pmatrix} \Delta u(k) \\ \Delta u(k+1) \\ \vdots \\ \Delta u(k+n_c-1) \end{pmatrix}, \bar{y} = \begin{pmatrix} y(k+1) \\ y(k+2) \\ \vdots \\ y(k+n_p) \end{pmatrix}, \tag{8}
$$

$$
\bar{\mathbf{Q}} = \mathrm{diag}(q), \bar{\mathbf{R}} = \mathrm{diag}(r).
$$

In many industrial motion control applications, the reference trajectory is known a priori, since it is given by the workpiece contour, for example. The MPC can take advantage of this by previewing the future position references and using them in the reference vector $\bar{y}^*$. The optimal sequence of manipulated variables $\Delta \bar{u}_{opt}$ is the one that minimizes the cost function while complying with the constraints

$$\Delta \bar{u}_{opt} = \underset{\Delta \bar{u}}{\operatorname{argmin}} J(\Delta \bar{u}). \tag{9}$$

Only the first value $u_{opt}(k) = \Delta u_{opt}(k) + u(k-1)$ is set. In the next sampling step the process starts again.

This principle underlies all MPC algorithms. In addition, a fundamental distinction can be made between whether the optimization problem is solved at runtime (online) or before commissioning (offline). The offline variant is also called explicit MPC. Here, the state space is decomposed into regions and the optimal control law is determined for each region. At runtime, only the current state has to be assigned to a region and the corresponding control law has to be applied. This approach avoids the computationally intensive online solution of the optimization problem, but requires a lot of memory to store the control laws. This is where the advantages of a modern IPC come into play, for which memory space is not an obstacle, making explicit MPC an attractive approach. Nevertheless, this paper is limited to the online solution. Investigations on explicit MPC are planned for further work.

The Finite Control Set MPC (FCS-MPC) and the Continuous Control Set (CCS-MPC) are the two main categories of MPC algorithms. With the FCS-MPC, the manipulated variable can only assume a limited number of discrete values, such as the switching states of the inverter. Therefore, the FCS-MPC is often used for current or torque control [3, 4, 5]. However, the computational complexity of the FCS-MPC increases exponentially to the prediction horizon. Since the time constants of the position control loop are usually much larger than the electrical time constants in the torque control loop, a long prediction horizon is required making FCS-MPC uncommon for direct position control.

For this purpose, the concept of CCS-MPC is usually used, where the manipulated variables can assume any value within the constraints. To limit the numerical effort and to avoid a non-convex optimization problem, simplified linear models are often aimed at. As long as no constraints have to be considered, a computationally efficient analytical solution is possible. If this solution exceeds a manipulated variable limit, it is simply limited to this value. This procedure has been successfully implemented in [9, 10], for example. If constraints have to be considered in the optimization problem, methods of quadratic programming (QP) can be used as in [6, 8].

Since the IPC is also to be used for further control tasks, the analytical solution is selected as the solution approach despite its high computing power. This is described briefly below; a detailed description can be found in [13], for example.

The state space model (5) can also be formulated as $x(k+1) = \mathbf{A}x(k) + bu(k-1) + b\Delta u(k)$. Using the output equations, all predicted control variables can now be calculated up to the prediction horizon. Thereby predicted states $x(k+1), x(k+2), ..., x(k+n_p)$ have to be formulated depending on the current state $x(k)$. This is shown exemplarily for $y(k+1)$ and $y(k+2)$ in (10) and (11)

$$\begin{aligned} y(k+1) &= c^T x(k+1) \\ &= c^T \mathbf{A}x(k) + c^T bu(k-1) + c^T b\Delta u(k) \end{aligned} \tag{10}$$

$$\begin{aligned} y(k+2) &= c^T x(k+2) \\ &= c^T \mathbf{A}^2 x(k) + c^T(\mathbf{A}+\mathbf{I})bu(k-1) + c^T(\mathbf{A}+\mathbf{I})b\Delta u(k) + c^T b\Delta u(k+1) \end{aligned} \tag{11}$$

By continuing this scheme up to the prediction horizon, the predicted controlled variables can be computed in compact form according to (12)

$$\bar{y} = \mathbf{F}x(k) + \mathbf{G}u(k-1) + \mathbf{H}\Delta \bar{u}, \tag{12}$$

with the vectors and the matrices already introduced in (8) and the following

$$
\mathbf{F} = \begin{pmatrix} c^{\mathrm{T}}\mathbf{A} \\ c^{\mathrm{T}}\mathbf{A}^2 \\ c^{\mathrm{T}}\mathbf{A}^3 \\ \vdots \\ c^{\mathrm{T}}\mathbf{A}^{n_{\mathrm{p}}} \end{pmatrix}, \ \mathbf{G} = \begin{pmatrix} c^{\mathrm{T}}b \\ c^{\mathrm{T}}(\mathbf{A}+\mathbf{I})b \\ c^{\mathrm{T}}(\mathbf{A}^2+\mathbf{A}+\mathbf{I})b \\ \vdots \\ c^{\mathrm{T}}(\mathbf{A}^{n_{\mathrm{p}}-1}+\ldots+\mathbf{I})b \end{pmatrix},
$$

$$
\mathbf{H} = \begin{pmatrix} c^{\mathrm{T}}b & 0 & \cdots & 0 \\ c^{\mathrm{T}}(\mathbf{A}+\mathbf{I})b & c^{\mathrm{T}}b & \cdots & 0 \\ c^{\mathrm{T}}(\mathbf{A}^2+\mathbf{A}+\mathbf{I})b & c^{\mathrm{T}}(\mathbf{A}+\mathbf{I})b & \cdots & 0 \\ \vdots & \vdots & \ddots & \vdots \\ c^{\mathrm{T}}(\mathbf{A}^{n_{\mathrm{c}}-1}+\ldots+\mathbf{I})b & c^{\mathrm{T}}(\mathbf{A}^{n_{\mathrm{c}}-2}+\ldots+\mathbf{I})b & \cdots & c^{\mathrm{T}}b \\ c^{\mathrm{T}}(\mathbf{A}^{n_{\mathrm{c}}}+\ldots+\mathbf{I})b & c^{\mathrm{T}}(\mathbf{A}^{n_{\mathrm{c}}-1}+\ldots+\mathbf{I})b & \cdots & c^{\mathrm{T}}(\mathbf{A}+\mathbf{I})b \\ \vdots & \vdots & \ddots & \vdots \\ c^{\mathrm{T}}(\mathbf{A}^{n_{\mathrm{p}}-1}+\ldots+\mathbf{I})b & c^{\mathrm{T}}(\mathbf{A}^{n_{\mathrm{p}}-2}+\ldots+\mathbf{I})b & \cdots & c^{\mathrm{T}}(\mathbf{A}^{n_{\mathrm{p}}-n_{\mathrm{c}}}+\ldots+\mathbf{I})b \end{pmatrix} \tag{13}
$$

The predicted control error that would occur without a change in the manipulated variable is now denoted by $\tilde{e}$

$$
\tilde{e} = \bar{y}^* - \mathbf{F}x(k) - \mathbf{G}u(k-1). \tag{14}
$$

By substituting (12) and (14) into the cost function (7) it follows

$$
J(\Delta\bar{u}) = (\tilde{e} - \mathbf{H}\Delta\bar{u})^{\mathrm{T}} \cdot \bar{\mathbf{Q}} \cdot (\tilde{e} - \mathbf{H}\Delta\bar{u}) + \Delta\bar{u}^{\mathrm{T}} \cdot \bar{\mathbf{R}} \cdot \Delta\bar{u}. \tag{15}
$$

Setting the gradient of the cost function to zero corresponds to the necessary condition for a minimum, from which the analytical solution is determined

$$
\frac{\partial J(\Delta\bar{u})}{\partial \Delta\bar{u}} = 2(\mathbf{H}^{\mathrm{T}}\bar{\mathbf{Q}}\mathbf{H} + \bar{\mathbf{R}})\Delta\bar{u} - 2\mathbf{H}^{\mathrm{T}}\bar{\mathbf{Q}}\tilde{e} = 0, \tag{16}
$$

$$
\Delta\bar{u} = (\mathbf{H}^{\mathrm{T}}\bar{\mathbf{Q}}\mathbf{H} + \bar{\mathbf{R}})^{-1}\mathbf{H}^{\mathrm{T}}\bar{\mathbf{Q}}\tilde{e}. \tag{17}
$$

Experimental results

The described algorithm is implemented on a Beckhoff IPC CP6930-0050 with an i7-4700EQ 2.4GHz processor and 2 times 4096 MB DDR3L-RAM SO memory and tested with the laboratory setup shown in Fig. 3. Two PMSM of type Beckhoff AM3021 are connected via a rigid coupling. One motor acts as the test motor, the other serves as the load motor. Torque control is performed on a Beckhoff AX5203 servo drive. The configured maximum torque is $T_{\mathrm{L}} = 0.9\,\mathrm{Nm}$.

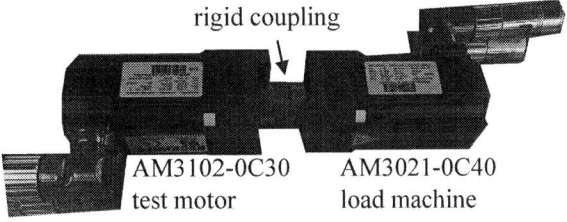

Fig. 3: Laboratory setup

The MPC parameters were set experimentally. Here $n_{\mathrm{p}} = 20$, $n_{\mathrm{c}} = 10$, $q = 1$ and $r = 0.2$ are selected. Moreover, the MPC controller is implemented with reference previewing. The cascade control was also parameterized manually and operates with active velocity feedforward. The sampling time of the

MPC and the position controller is $T_S = 250\,\mu s$, that of the subordinate velocity controller is $125\,\mu s$. It is emphasized again that for the comparative representation the cascade control is completely closed within the servo drive and is therefore not influenced by the dead time.

Two reference trajectories are compared. On the left side in Fig. 4, a typical trajectory with limited jerk is shown. Due to the dead time that acts between the numerical control (NC on the IPC) and the servo drive with the classic cascade control, the actual position follows the NC setpoint position delayed with this control, as can be seen in the detailed view in the upper left figure. However, in order to evaluate the path accuracy when tracking contours, the tracking error measured in the servo drive is relevant. Therefore this tracking error is shown. For the MPC, the tracking error is calculated in the IPC. Without disturbances, the MPC has a slightly higher maximum tracking error, which is, however, compensated for faster than with classic cascade control. This is especially the case for velocity transitions. At the time 100 ms a load step is switched on. Both controllers are able to compensate for the disturbance, although a higher tracking error is observed with the MPC due to the delayed response caused by the dead time. Furthermore, small oscillations in the tracking error are to be noted with the MPC, which can be explained by the high noise sensitivity of the MPC.

On the right side in Fig. 4, the results of a step-shaped excitation are illustrated. This excitation leads to an operation at the manipulated variable limit resulting in anti-windup procedures becoming active. Therefore, the tracking error is compensated faster with the MPC, albeit with an overshoot. In addition, a load step is switched on again. The influence of the dead time can be clearly seen in the detailed picture, which is why the MPC reacts later and a larger tracking error is formed. This is still compensated faster than with the cascade control. To improve the disturbance behavior, the dead time must be reduced by decreasing the sampling time.

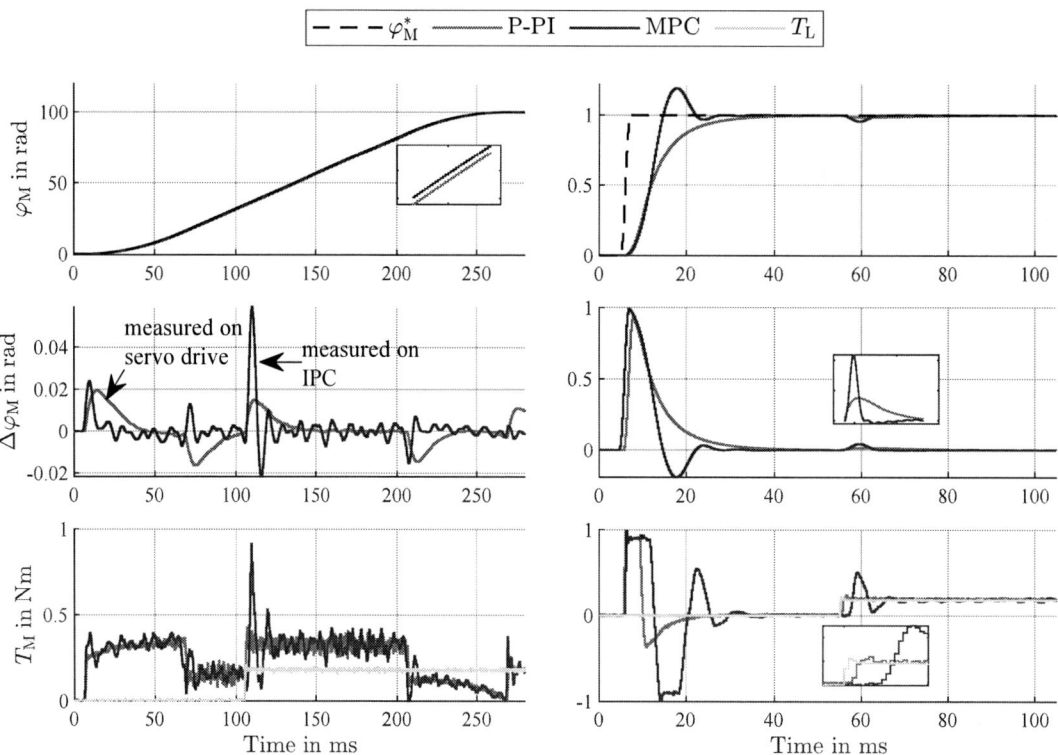

Fig. 4: Comparison of MPC and P-PI cascade control: Left typical position trajectory, Right step-like excitation. Disturbance $T_L = 0.18\,\mathrm{Nm}$

Furthermore, during the test the CPU utilization of the IPC was measured with the function block TC_CpuUsage. The resulting CPU usage or required computing power strongly depends on the particular implementation. The code of the MPC algorithm was generated with the standard Simulink PLC coder as well as with a special Beckhoff TE1400 + Simulink coder [14]. The results are shown in Table I.

Table I: CPU usage of the IPC

	only PLC programm	PLC programm with MPC (Simulink Coder with TE1400 (C++))	PLC programm with MPC (Simulink PLC Coder)
CPU usage	$\approx 8\%$	$\approx 12\%$	$\approx 20\%$

About 8 % are required as base load for the TwinCAT system and the test program. The MPC increases the utilization to 12 % or 20 %, depending on the implementation. It should be noted that in the extended state space it is already a ninth-order system and quite long horizons were used. Model simplifications and optimization of the MPC parameters can potentially save further computation time. Nevertheless, it can be stated that the computational load on the CPU is moderate and further control tasks can definitely be executed. With the IPC used, it is even possible to increase the complexity of the MPC and thus achieve further improvements.

Finally, the significance of the results on the path accuracy in the interaction of several axes is to be investigated simulatively, since no real plant was available. For this purpose, a diamond-shaped trajectory in the XY-plane is given. The results are illustrated in Fig. 5. In the cascade control considered, the velocity feedforward is not balanced by a position setpoint filter. The result is a strong overshoot at contour corners. With the MPC, the path accuracy is significantly improved. A similar result is achieved if the position setpoint filter is implemented correctly. For this purpose, the filter time constant must be set to the equivalent time constant of the velocity control loop, which therefore means additional tuning effort.

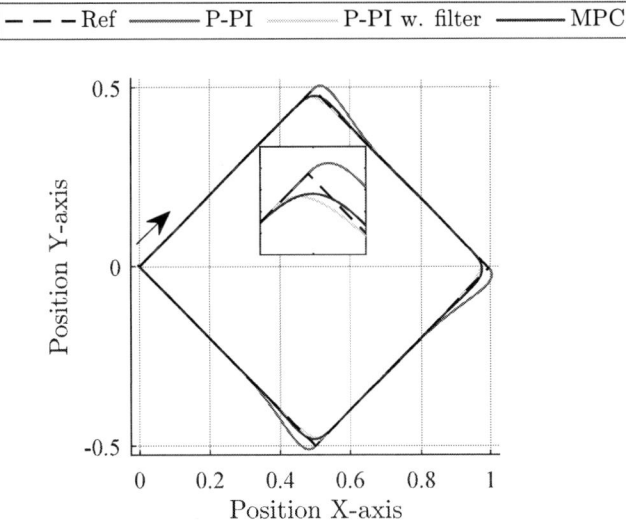

Fig. 5: Simulation: Tracking performance for a diamond reference trajectory

Conclusion

Model predictive control is well known for its superior control performance compared to PI control. However, to take advantage of this in practice, a target platform with sufficient computing power is required. Industrial PCs, in contrast to standard servo drives, have sufficient computing power and are available in many industrial plants. This paper gives evidence that they are suitable for the implementation of high dynamic model predictive position control. A main problem with the implementation on the IPC is the occurring dead time due to the communication between the IPC and the servo drive. This leads to a delayed response to disturbances compared to the internal cascaded control implemented on the servo drive. Since the IPC has sufficient computing capacity for further improvements subsequent work will investigate how to improve the disturbance behavior. Another exciting research topic is explicit MPC, for which the IPC is an ideal target platform thanks to its large storage capacity.

References

[1] S. Thomsen and F. W. Fuchs, "Speed control of torsional drive systems with backlash," 2009 13th European Conference on Power Electronics and Applications, 2009, pp. 1-10

[2] S. Thomsen, N. Hoffmann and F. W. Fuchs: PI Control, PI-Based State Space Control, and Model-Based Predictive Control for Drive Systems With Elastically Coupled Loads—A Comparative Study, in IEEE Transactions on Industrial Electronics, vol. 58, no. 8, pp. 3647-3657, Aug. 2011

[3] T. Geyer, G. Papafotiou and M. Morari, "Model Predictive Direct Torque Control—Part I: Concept, Algorithm, and Analysis," in IEEE Transactions on Industrial Electronics, vol. 56, no. 6, pp. 1894-1905, 2009

[4] M. Leuer and J. Böcker: Fast online model predictive control of IPMSM using parallel computing on FPGA, 2013 International Electric Machines Drives Conference, 2013, pp. 1017-1022

[5] H. A. Young, M. A. Perez, J. Rodriguez and H. Abu-Rub, "Assessing Finite-Control-Set Model Predictive Control: A Comparison with a Linear Current Controller in Two-Level Voltage Source Inverters," in IEEE Industrial Electronics Magazine, vol. 8, no. 1, pp. 44-52, March 2014

[6] P. Serkies and K. Szabat: Predictive Control of the Two-Mass Drive with an Induction Motor for a Wide Speed Range, 2018 IEEE 18th International Power Electronics and Motion Control Conference (PEMC), 2018, pp. 750-755

[7] P. Serkies and K. Szabat: Predictive position control of the induction two-mass system drive, 2014 IEEE 23rd International Symposium on Industrial Electronics (ISIE), 2014, pp. 871-876

[8] O. Wallscheid, E. F. Bouna Ngoumtsa and J. Böcker: Hierarchical Model Predictive Speed and Current Control of an Induction Machine Drive with Moving-Horizon Load Torque Estimator, 2019 IEEE International Electric Machines Drives Conference (IEMDC), 2019, pp. 2188-2195

[9] P. G. Carlet, F. Toso, A. Favato and S. Bolognani: A speed and current cascade Continuous Control Set Model Predictive Control architecture for synchronous motor drives, 2019 IEEE Energy Conversion Congress and Exposition (ECCE), 2019, pp. 5682-5688

[10] S. Wendel, B. Haucke-Korber, A. Dietz and R. Kennel: Cascaded Continuous and Finite Model Predictive Speed Control for Electrical Drives, 2018 20th European Conference on Power Electronics and Applications (EPE'18 ECCE Europe), 2018, pp. P.1-P.10

[11] S. Wendel, P. Löhdefink, M. Hoerner, A. Dietz and R. Kennel: Dynamic model predictive position control for linear actuators in automotive applications, 2018 Thirteenth International Conference on Ecological Vehicles and Renewable Energies (EVER), 2018, pp. 1-6

[12] J. Lunze: Regelungstechnik 2, 12th ed. Berlin, Heidelberg: Springer Berlin Heidelberg, 2020

[13] J. Adamy: Nichtlineare Systeme und Regelungen. Berlin, Heidelberg: Springer Berlin Heidelberg, 2018

[14] Beckhoff Automation GmbH & Co. KG: Manual Target for Simulink, 2022

Bidirectional Active EMC Filter for Industrial Power Converters

Bernhard Wunsch[1], Stanislav Skibin[1], Ville Forsstrom[2]
[1]ABB Corporate Research, 5405 Baden, Switzerland
[2]ABB Oy Drives, 00381 Helsinki, Finland
E-Mail: bernhard.wunsch@ch.abb.com, stanislav.skibin@ch.abb.com,
ville.forsstrom@fi.abb.com
URL: https://abb.com

Keywords

«EMC/EMI», «active filter», «power converter», «Adjustable speed drive», «Filtering», «Parasitic inductance»

Abstract

Size and cost of EMC filters of power converters can be reduced using active circuits. In this work we demonstrate the filter performance of a bidirectional active EMC filter in an industrial 75kW power converter driving a motor. The active filter is bidirectional since it suppresses EMC noise on either side of the filter. The voltage noise source on the grid side stems from magnetic coupling between input and output of the converter. For a purely passive EMC filter, grid side noise needs to be filtered by large inductors. We show that the active filter which is based on current sensing and current injection strongly reduces the magnetic filter components required.

Introduction

Switch mode power supplies emit high frequency noise towards the grid. The emitted EMC noise is limited by international EMC norms like CISPR 11 [1] or product norms like EN61800-3 [2]. EMC filters are placed at the input of the converters in order to comply with these norms. These filters contribute significantly to the cost and size of the converter. One way to reduce the size of the passive filter elements is to improve their filter performance using HF-active circuits. However, despite of many publications on active and hybrid EMC filters [3]-[11] their use in high power converters especially with AC-input is still challenging. Depending on the characteristic impedances on power converter or grid side, different topologies of the active circuit are preferred [7].

In this work we discuss an additional important selection criterion for the active circuit topology, namely bidirectionality, as explained in the following. While in a simplified EMC analysis of power converters it is often assumed that the EMC noise source is placed on the converter side of the EMC filter, parasitic couplings can lead to induced voltages at the grid side of the EMC filter, see Fig. 2. Such a mechanism is analyzed in detail in [12] for the example of an industrial motor drive connected between the 3-phase AC-grid and a three-phase motor, a similar setup is used here. Fig. 1 shows a photo of the setup consisting of a 75 kW power converter driving a motor. As shown in the subfigure (b) input and output cables are located next to each other thus causing a magnetic coupling between the noisy currents on the motor cable and current on the supply cable on the grid side, which is required to have low EMC noise.

This input-output coupling can be represented by an induced voltage source that drives current from the grid side towards the EMC filter see Fig. 2. Since the EMC filter needs to suppress noise on either side of the filter, EMC noise is filtered much more efficiently by inductive filter elements than by shunt capacitors.

(a)   (b)

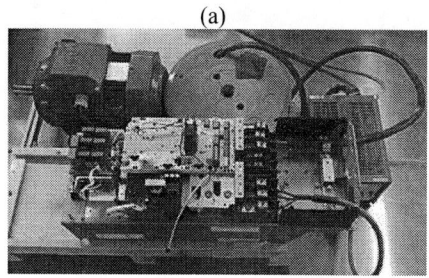 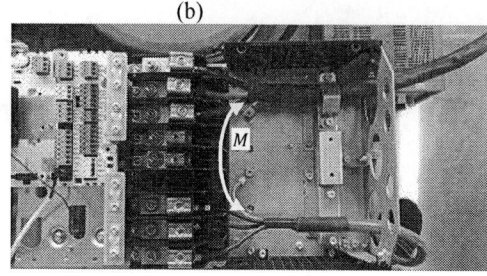

Fig. 1 (a) Photo of the EMC setup consisting of LISN, AC-AC power converter, motor cable and motor. (b) The grid side cable is connected close to and in parallel to the motor side cable which introduces a large input-output coupling bypassing the passive EMC filter.

Based on the requirement on bidirectionality of the EMC filter, we select the topology of the active circuit to be based on a current-controlled current-source. In this publication we demonstrate the active EMC filter in a 75 kW power converter driving a motor [12]. The active filter allows strong reduction of magnetic components of a purely passive filter, and it is also immune to the main parasitic bypass degrading the passive filter. Compared to previous work on active EMC filter, the novelty here is the high power of the drive (75 kW) and the focus on bidirectionality of the active filter. In the next section, we discuss the operation principle of the active circuit. Then we introduce the active board and show the performance of the active EMC filter (AEF) in the power converter. In the final section we conclude this work.

Operation principle of active circuit

In this section we summarize key features of active circuits used for EMC filtering based on a simple single line equivalent representation of noise source and victim. For the power converter studied in the next chapter, such a single line equivalent circuit has been derived for the common mode (CM) noise in [12].

Active circuits used for filtering realize controlled sources and can be classified by the quantity that is sensed and the quantity that is controlled. Both sensed and output quantity can be either current or voltage which results in four possible combinations, like current-controlled current-source etc. Additionally, the circuits can be distinguished as feed backward or feed forward depending on the relative location of the sensing part and actuation part of the circuit. If the sensing occurs closer to the noise source than the actuation, then the circuit is called feed forward type and if the sensing occurs closer to the victim than the actuation, then the circuit is of backward type [6], [7].

In active circuits based on voltage actuation the injection mechanism is realized by an inductor on the power line between source and victim. For large power converters the power lines are realized by thick cables or large busbars so that the inductors are bulky and costly. Therefore, an active circuit based on current actuation was chosen in this work, where the controlled source is placed on a shunt connection between noise source and victim.

A circuit based on voltage-sensing current-actuation behaves as a controlled shunt impedance. This topology can effectively filter noise from the converter, if the controlled shunt impedance is smaller than the converter output impedance and the victim impedance. It can be argued that this topology is the easiest to implement since there is no need of inductors for sensing or actuation [13]. However, the low shunt impedance will be ineffective and in fact even harmful for a voltage noise source on the victim/grid side. Therefore, such an active filter is not bidirectional.

Therefore, the selected topology uses current-sensing current-actuation in either feedback or feedforward topology. Both topologies enable a bidirectional filtering of voltage noise sources and can be understood as an effective multiplication of the source impedance by the current gain [3]. The operation principle is visualized in Fig. 2 for ideal controlled current sources. The LISN current is driven by the noise source U_{src} of the converter and the induced noise source U_{ind} on the grid side. Both sources are suppressed by the same total impedance Z_{tot} consisting of the series connection of the LISN impedance Z_{LISN} and the input impedance Z_{in} of the converter. The active circuit increases the input impedance of the converter by a factor depending on the current gain and topology.

$$I_{LISN} = \frac{U_{ind}+U_{src}}{Z_{LISN}+Z_{in}}; \; Z_{in} = \begin{cases} Z_{src}\,(1+F) \text{ feedback} \\ Z_{src}/(1-F) \text{ feed-forward} \end{cases} \tag{1}$$

In principle both topologies are suited as bidirectional filter. For good filter performance a high gain $|F|\gg 1$ is required for the feedback configuration whereas an accurate gain of $F\rightarrow 1$ is required for the feedforward configuration. After first tests with both configurations, we selected the feedback configuration since we found the accurate control of both magnitude and phase required for the feedforward topology more challenging. The same topology has been assessed in previous works [3]-[10].

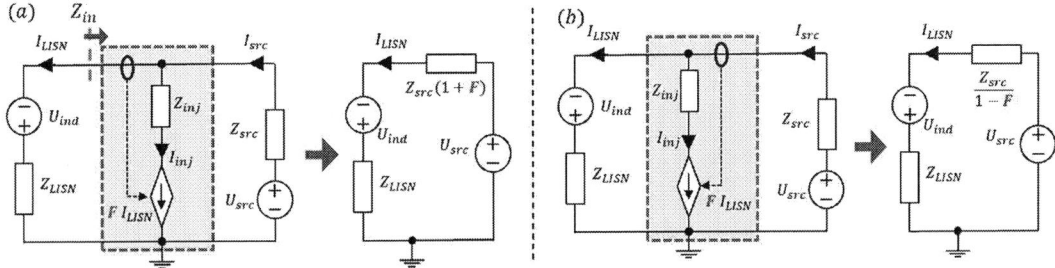

Fig. 2 The active circuit based on current-sensing current-actuation leads to a multiplication of the source impedance [3], see (1). (a) Feed-backward configuration has multiplication factor $Z_{in} = Z_{src}(1+F)$. (b) Feed-forward configuration has multiplication factor $Z_{in} = Z_{src}/(1-F)$.

Active EMC Filter Board and the test setup

The AEF board is used to reduce the CM filter of conducted EMC noise in a high-power motor drive. The active circuit board realizes a current-controlled current-source $I_{AEF} = F\,I_{LISN}$ where F labels the unitless current amplification. This topology has been selected since it enables bidirectional noise filtering as discussed above.

A photo of the board together with its schematic circuit diagram is shown in Fig. 3. The controlled current-source is realized by a series combination of a current transformer and a Howland current source, which also includes a push-pull stage to increase the output current capabilities, see [4] for a similar design. The current amplification F of the active filter is set by the current transformer load resistor R_{CT}, its turn number N_2, as well as by the three impedances of the Howland current source, namely the input resistance R_{in}, the feedback resistance R_f, and the sense resistance R_{sense}, see Fig. 3:

$$I_{AEF} = \frac{R_{CT}}{N_2}\frac{R_f}{R_{in}\,R_{sense}}\,I_{LISN}. \tag{2}$$

Beside of the Howland current source, the AEF board contains a power supply unit (PSU) and the injection circuit. The PSU is powered from the local supply (+24 V) and provides the active circuit with a bipolar voltage (±15 V). The injection circuit consists of a Y2 rated capacitor, C_{inj}, and three X2 rated capacitors. The Howland current source is built up around a TI LM7171 voltage feedback OpAmp with Unity-Gain Bandwidth of 200 MHz and specified for ±15 V operation. The push-pull stage uses Diodes Inc. ZDT6753 complimentary medium power transistors which are rated for 100 V collector-emitter voltage and 2 A continuous current. The current transformer showed in Fig. 4 (a) is built up using 10 turns of the secondary winding around a Vacuumschmelze T60004-L2050-W626 core. The detailed analysis of the active filter board and its hardware realization is beyond the scope of the current article and will be published elsewhere.

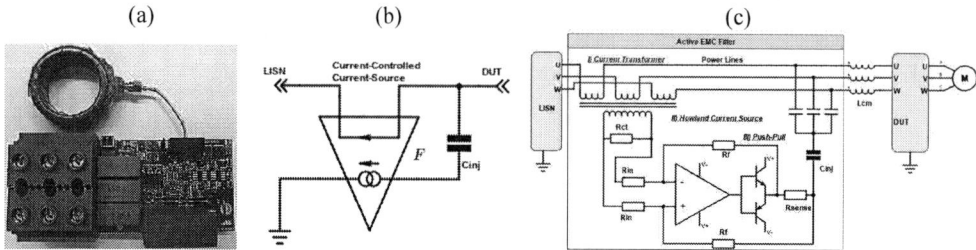

Fig. 3: AEF board prototype (a), its operational principle with respect to the filtering of CM noise (b) and a simplified circuit diagram (c). Due to the active filtering the common mode choke can be strongly reduced. A schematic of the power converter is shown in Fig. 5 (a).

Fig. 4 AEF board installation inside the test setup on Fig.1 (a)

The developed AEF board has been tested in a setup shown in Fig. 1 (a). The footprint of the designed AEF board allows its installation inside the motor drive for the normal operation (see Fig. 4). However, in the tests below, the board is mounted externally to ensure equal conditions in the benchmark of the active filter versus passive one.

The schematic of the test setup is shown in Fig. 5 and has been discussed in detail in [12]. The 75 kW motor drive is connected on the grid side to a 32 A, 3-phase LISN and the load consists of 30 m long 3-phase shielded motor cable with 16 mm^2 wire cross-section and 15 kW motor. The coupling between motor and grid cables is captured by coupled loop inductances L_{p1}, L_{p3} with a coupling parameter k_{13}. This parasitic coupling introduces the induced common mode voltage U_{ind} on the grid side of the filter (see Fig. 2). The induced voltage $U_{ind} = j \, \omega \, M \, i_3$ with $M = k_{13}\sqrt{L_{p1}L_{p2}}$ denoting the mutual inductance and i_3 the motor side current, see Fig. 5 (b).

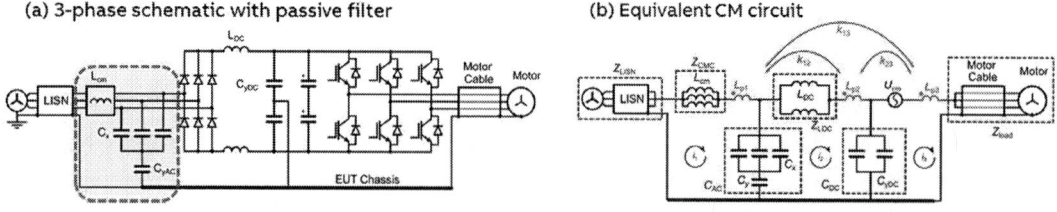

Fig. 5 (a) Simplified schematic of the EMC setup shown in Fig. 4. The EMC input filter is highlighted. Objective of the active board is to reduce the size of the common mode inductance L_{CM} using the idea of impedance multiplication discussed in section 2. (b) Corresponding common mode circuit. Details can be found in [12].

Test setup and measurement results

Benchmark 1: Influence of cross coupling between motor and grid cables on emitted EMC noise

(a)   (b)

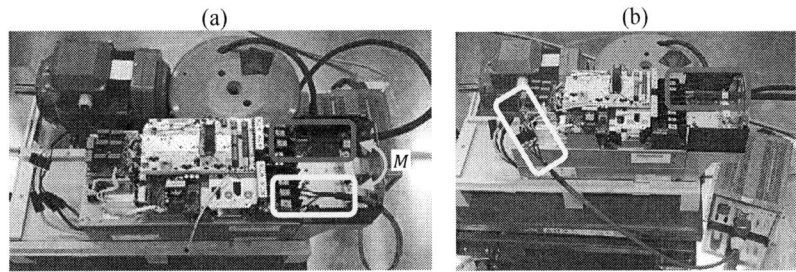

Fig. 6: Test setup for the evaluation of the cross-coupling effect. (a) – original cable connection where motor and grid cable are attached on same side and in parallel, corresponding to large mutual inductance M, (b) – attachment of the grid cable at the back of the drive opposite to the motor cable, thus minimizing the coupling.

In the first test, we quantify the importance of the magnetic input-output coupling which introduces the grid side noise source thus requiring a bi-directional EMC filter. The coupling can be strongly reduced by modifying the connection of the AC grid cable to the backside of the converter opposite to the motor cable connection as shown in Fig. 6 (b). Note that the same components are present but only the parasitic mutual coupling k_{13} is changed. Fig. 6 (a) shows the original cable connection where motor and grid cables are attached to the same side of the motor drive and run in parallel which maximizes the magnetic coupling. Fig. 6 (b) shows a modified setup where the grid cable is attached at the back of the drive, avoiding cable cross-coupling. In this test, the motor drive is attached to a 15 kW asynchronous motor via a shielded 30 m long motor cable with 16 mm^2 wire cross-section. The measured CM-noise without CM-choke on the grid side for the setups shown in Fig. 6 are shown by the red and green curve in Fig. 7. The blue curve in Fig. 7 shows the EMC noise in presence of input-output coupling but with the CM-chokes added see Fig. 8 (a).

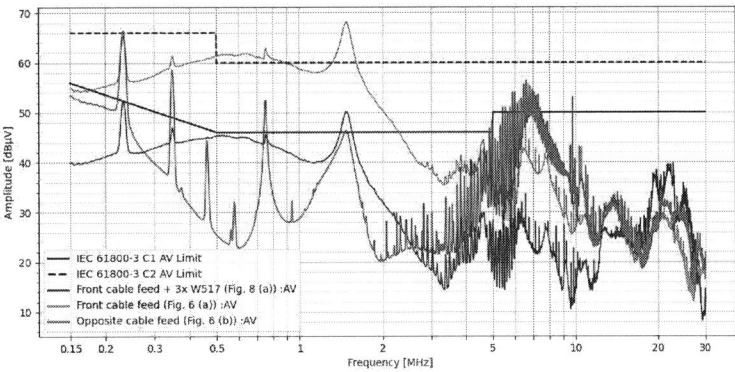

Fig. 7: Various measurements of EMC noise (CM noise using average detector). Red and green curves illustrate the impact of input-output coupling (two different connections of grid side cable see Fig. 6) without CM-choke. Blue curve shows EMC filtering of blue curve when CM-inductance is added as in Fig. 8 (a).

The following conclusions can be drawn from the results shown in Fig. 7:

- The EMC noise in the range 300 kHz – 4 MHz is dominated by the cross-coupling effect between motor and grid cables. The difference between green and red curves shown in Fig. 7 is caused by the noise source U_{ind} caused by the magnetic coupling of input and output cables.
- To comply with the EMC in presence of cross-coupling corresponding to the original cable connection shown in Fig. 6 (a), a large common mode inductor consisting of 3 × Vacuumschmelze (VAC) W517 cores is placed on the grid side of the inverter as shown in Fig. 8 (a). EMC measurements of this setup are shown by blue curve in Fig. 7.

Benchmark 2: Application of the bi-directional active filter within motor drive

As discussed in previous sections, the bi-directional AEF efficiently filters HF noise stemming from noise sources located on DUT and/or grid sides (U_{src} and U_{ind} from the equivalent circuit on Fig. 2). In this benchmark, we evaluate the performance of the AEF using a 30 m motor cable with 16 mm^2 wire cross-section and an asynchronous 15 kW motor and compare it to the conventional passive filter, consisting of three VAC W517 cores (see Fig. 7 (a)). The designed AEF board has also been successfully tested in our lab using different combinations of motors and cables which will not be shown here because of the lack of space

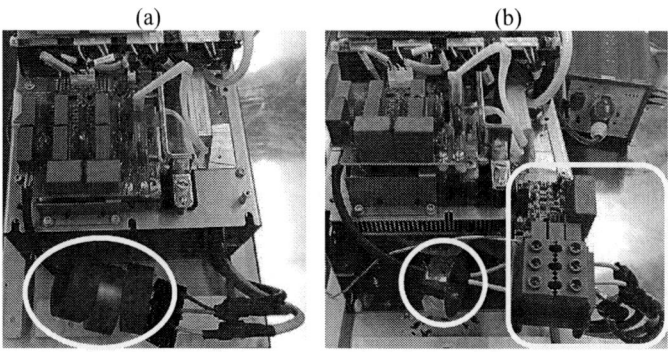

Fig. 8: Test setup for the evaluation of the AEF performance: using passive filter (a), using active filter (b). The common choke is strongly reduced when using the active circuit board.

The AEF board is connected as shown in Fig. 8 (b) to ensure similar cross-coupling effects for passive and active EMC filtering. Motor and grid cables are connected on the same side as shown in Fig. 7 (a). In presence of the active filter the common mode choke can be reduced to a single VAC W625 core on the DUT side due to the impedance multiplication effect from the active circuit, see (1).

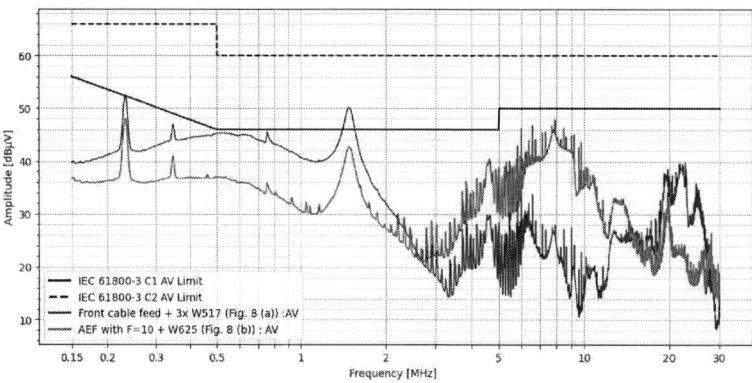

Fig. 9: Measured AV EMC noise of the passive and active (F=10) filters in a setup with 30 m long motor cable

In the Fig. 9, measurement results of the AEF board with a single VAC W625 on the DUT side vs. pure passive filter with 3 × VAC W517 cores are shown. Motor and grid cables are connected on the same side as shown in Fig. 7 (a). The designed AEF performs significantly better than the pure passive filter in the range up to few MHz with a much smaller CM choke. The CM noise even complies with the C1 limit of EN 61800-3. The filter performance of the currently designed board degrades in the range above few MHz .

Conclusion

In this paper, a successful application of the Active EMC Filter (AEF) in a power inverter driving a motor has been presented, leading to a strong reduction of magnetic filter components. It has been shown that the coupling between input and output cables introduces a voltage noise source on the AC-gird side of the EMC filter so that bi-directionality of the active filter is essential. Therefore, a current-sensing current-actuation active filter topology (both feed backward and feed forward) has been selected. In the scope of this paper, we showed an implementation of the above-mentioned topology and its successful application in an industrial motor drive for realistic operation conditions. The designed Active EMC Filter showed very good performance up to few MHz which is the typical "problem" range for the high-power motor drive applications, providing significant cost cut (70% reduction of expensive Nanocrystalline cores) in the EMC filter design.

References

[1] CISPR 11-Industrial Scientific and Medical Equipment-Radio-Frequency Disturbance Characteristics-Limits and Methods of Measurement. CISPR 11:2015|IEC Webstore|Electromagnetic Compatibility, EMC, Smart City. Available online: https://webstore.iec.ch/publication/22643 (accessed on 11 March 2021).

[2] EN/IEC 61800-3, Ed. 3.0, Adjustable Speed Electrical Power Drive Systems–Part 3: EMC Requirements and Specific Test Methods, 2017. IEC 61800-3:2017|IEC Webstore|Electromagnetic Compatibility, EMC, Smart City, Pump, Motor, Water Management. Available online: https://webstore.iec.ch/publication/31003 (accessed on 11 March 2021).

[3] W. Chen, X. Yang, and Z. Wang, "A novel hybrid common-mode EMI filter with active impedance multiplication," *IEEE Trans. Ind. Electron.*, vol. 58, no. 5, pp. 1826–1834, 2011.

[4] S. Wang, Y. Y. Maillet, F. Wang, D. Boroyevich, and R. Burgos, "Investigation of hybrid EMI filters for common-mode EMI suppression in a motor drive system," *IEEE Trans. Power Electron.*, vol. 25, no. 4, pp. 1034–1045, 2010.

[5] Y. Chu, S. Wang, and Q. Wang, "Modeling and Stability Analysis of Active/Hybrid Common-Mode EMI Filters for DC/DC Power Converters," *IEEE Trans. Power Electron.*, vol. 31, no. 9, pp. 6254–6263, 2016.

[6] N. K. Poon, J. C. P. Liu, C. K. Tse, and M. H. Pong, "Techniques for input ripple current cancellation: Classification and implementation," *PESC Rec. - IEEE Annu. Power Electron. Spec. Conf.*, vol. 2, no. 6, pp. 940–945, 2000.

[7] Y. C. Son and S. K. Sul, "Generalization of active filters for EMI reduction and harmonics compensation," *IEEE Trans. Ind. Appl.*, vol. 42, no. 2, pp. 545–551, 2006.

[8] M. L. Heldwein, H. Ertl, J. Biela, and J. W. Kolar, "Implementation of a transformerless common-mode active filter for offline converter systems," IEEE Trans. Ind. Electron., vol. 57, no. 5, pp. 1772–1786, 2010.

[9] Amaducci, "Design of a wide bandwidth active filter for common mode EMI suppression in automotive systems," IEEE Int. Symp. Electromagn. Compat., pp. 612–618, 2017.

[10] E. Mazzola, F. Grassi, and A. Amaducci, "Enhanced Circuit Model for Insertion Loss Prediction of Active EMI Filters Considering Non-ideal Parameters," in *2020 International Symposium on Electromagnetic Compatibility - EMC EUROPE*, Sep. 2020, pp. 1–5.

[11] S. Jeong, D. Shin, and J. Kim, "A Transformer-Isolated Common-Mode Active EMI Filter Without Additional Components on Power Lines," *IEEE Trans. Power Electron.*, vol. 34, no. 3, pp. 2244–2257, 2019.

[12] B. Wunsch, S. Skibin, V. Forsström, and I. Stevanovic, "EMC Component Modeling and System-Level Simulations of Power Converters: AC Motor Drives", *Energies*, vol. 14, no. 6, p. 1568, 2021.

[13] M. L. Heldwein, J. Biela, H. Ertl, T. Nussbaumer, and J. W. Kolar, "Novel Three-Phase CM/DM Conducted Emission Separator," *IEEE Trans. Ind. Electron.*, vol. 56, no. 9, pp. 3693–3703, Sep. 2009.

A General Method to Measure Parasitic Capacitance of Transformer Using Guarding Technique

Shaokang Luan[1], Stig Munk-Nielsen[1], Bruce Wakelin[2], Magnus Hortans[2], Jan Schupp[2],
Hongbo Zhao[1]
[1]DEPARTMENT OF ENERGY, AALBORG UNIVERSITY
Pontoppidanstræde 111, Aalborg, Denmark
[2]DANFOSS A/S
Nordborgvej 81, Nordborg, Denmark
slu@energy.aau.dk

Acknowledgements

The authors would like to acknowledge the financial support from Innovation Fund Denmark and thank the editors and reviewers for providing valuable comments and suggestions.

Keywords

«Parasitic elements», «Measurement», «Transformer»

Abstract

Precise, simple, and general measurement methods are essential for parasitic capacitance of transformer. Conventional two-terminal method has some limitations on the measurement of multi-winding transformer, like complex measurement processes. This paper introduces guarding technique into a new method to measure parasitic capacitance of multi-winding transformer. The principle and measurement processes of proposed three-terminal guarding method are introduced in detail. Comprehensive comparisons between two-terminal method and guarding method are made based on circuit verification and three case studies. Guarding method has high accuracy, but simpler measurement processes, and higher feasibility for multi-winding transformer compared to the conventional two-terminal method.

I. Introduction

Due to the rapid development of wide-bandgap (WBG) semiconductors with higher switching speed and blocking voltage, the frequency and voltage level of DC/DC converters are on the increase in a wide range of applications. But much higher dv/dt during switching transient in WBG semiconductors is a potential source of electromagnetic interference (EMI) [1-4]. In the meanwhile, the sizes of passive components, e.g., transformers, can be significantly reduced at higher operation frequency [5]. More compact structure of transformer can result in larger parasitic capacitances because of smaller distances between turn to turn and turn to core [6, 7]. Large parasitic capacitances will offer paths for common mode (CM) noise and, together with high dv/dt, lead to severe EMI issues in converters [8]. Accordingly, parasitic capacitances are drawing more and more attention in transformer design [9, 10].

Proper measurement methods are crucial to evaluate the parasitic capacitance of transformer, which can provide verifications and guidelines for the transformer design. Six-lumped-capacitance equivalent circuit shown in Fig. 1 is precise and widely used in measurement and modelling of parasitic capacitance in two-winding transformer, which is derived based on the conservation of energy [11]. [12] introduced a conventional two-terminal measurement method, which is normally used in the measurement of two-terminal circuit, like inductor [13], to two-winding transformer based on six-capacitance equivalent circuit. In this method, at least six different measurement setups shorting the transformer into six different two-terminal circuits are needed. Then six two-terminal measurements are carried out by impedance analyzer to obtain equivalent capacitances of six two-terminal circuits. The values of six parasitic capacitances were calculated by matrix operations based on measurement results. Proper selection of measurement setups is essential to simplify the measurement processes and matrix operations. Pre-analyses and data post-processing complicated the measurements and, together with the propagation of errors, became the obstacles to generalizing this method to the measurements for parasitic capacitances of multi-winding transformer.

Guarding technique is a three-terminal measurement function provided by impedance analyzer E4990A together with adaptor 16047E from Keysight [14], which is initially used in in-circuit-test (ICT) to detect the component and manufacture flaws of printed circuit board assemblies (PCBA) [15]. This method has been used to measure the parasitic capacitances of three-terminal inductors with grounded cores in [16, 17]. In this paper, guarding technique will be applied in a more general measurement method for parasitic capacitance of multi-winding transformer. Through proper uniform measurement setup, each individual parasitic capacitance of multi-winding transformer can be accurately measured in simple processes.

This paper is organized as follows: In Section II, conventional two-terminal method is reviewed together with its limitations. In Section III, the principle of proposed guarding method is introduced in detail, taking two-winding transformer as an example. In Section IV, the proposed guarding method is generalized to multi-winding transformer. In Section V, circuit verification and three case studies are conducted to compare guarding method with conventional two-terminal method on the accuracy, measurement simplicity, and feasibility for multi-winding transformer.

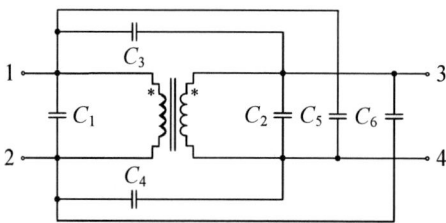

Fig. 1 Six-lumped-capacitance equivalent circuit for two-winding transformer

II. Review of conventional two-terminal method and its limitations

Conventional two-terminal method introduced in [11] only focused on two-winding transformer with four terminals, which is based on a six-lumped-capacitance equivalent circuit, shown in Fig. 1. C_1 and C_2 are intra-winding parasitic capacitances of primary and secondary sides. C_3, C_4, C_5 and C_6 are inter-winding parasitic capacitances between primary and secondary sides. 1 ~ 4 are four terminals of two-winding transformer. The component under test (CUT), taking C_1 as an example, cannot be directly measured, because there are some shunt paths around CUT, like $C_2 \sim C_6$ connected to the two terminals of C_1. So, there was an essential precondition in conventional two-terminal method: two-winding transformer can be seen as a linear system under small test signal from impedance analyzer, whether it be air-core transformers, ferromagnetic-core transformers, or ferrimagnetic-core transformers. In this condition, six parasitic capacitances in Fig. 1 can be treated as six variables, which can be deduced by six independent linear equations.

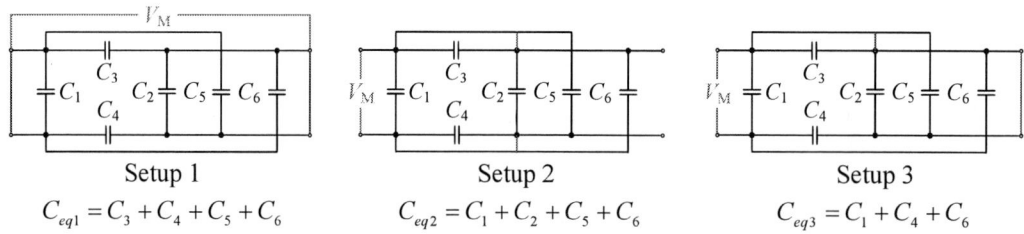

$$C_{eq1} = C_3 + C_4 + C_5 + C_6 \qquad C_{eq2} = C_1 + C_2 + C_5 + C_6 \qquad C_{eq3} = C_1 + C_4 + C_6$$

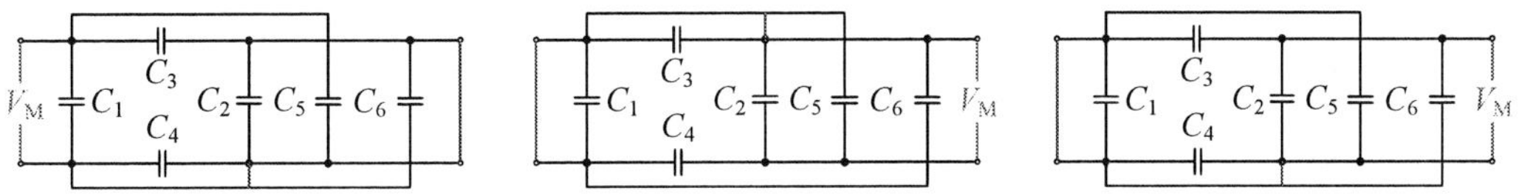

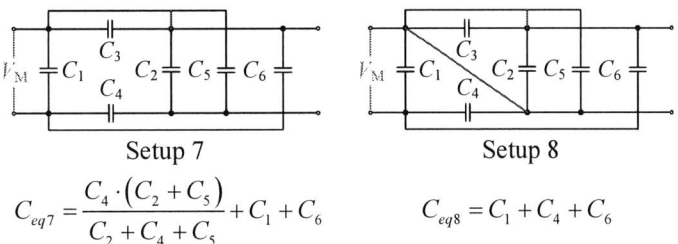

Setup 4

$$C_{eq4} = C_1 + C_3 + C_5$$

Setup 5

$$C_{eq5} = C_2 + C_4 + C_5$$

Setup 6

$$C_{eq6} = C_2 + C_3 + C_6$$

Fig. 2 Six independent measurement setups

Fig. 2 shows six independent measurement setups and six independent linear equations generated by two-terminal method. Red lines represent the short-circuit lines simplifying the four-terminal circuits of two-winding transformer to two-terminal circuits, V_M is the two-terminal source of impedance analyzer, $C_{eq1} \sim C_{eq6}$ are equivalent capacitances of six two-terminal circuits. The results obtained by impedance analyzer are frequency responses of six two-terminal circuits, from which $C_{eq1} \sim C_{eq6}$ are derived. All parasitic capacitance $C_1 \sim C_6$ can be deduced through matrix operations as (1).

$$
\begin{bmatrix} C_1 \\ C_2 \\ C_3 \\ C_4 \\ C_5 \\ C_6 \end{bmatrix} = \frac{1}{2} \cdot \begin{bmatrix} -1 & 0 & 1 & 1 & 0 & 0 \\ -1 & 0 & 0 & 0 & 1 & 1 \\ 0 & -1 & 0 & 1 & 0 & 1 \\ 0 & -1 & 1 & 0 & 1 & 0 \\ 1 & 1 & -1 & 0 & 0 & -1 \\ 1 & 1 & 0 & -1 & -1 & 0 \end{bmatrix} \cdot \begin{bmatrix} C_{eq1} \\ C_{eq2} \\ C_{eq3} \\ C_{eq4} \\ C_{eq5} \\ C_{eq6} \end{bmatrix} \tag{1}
$$

It should be noticed that proper selections of measurement setup are crucial for two-terminal method, which need thorough analyses before setting up the measurements. The reasons are as follows:

1. The measurement results of some setups, like Setup 7 shown in Fig. 3, can significantly increase the complexity of data post-processing. To simplify data post-processing, it requires that two-winding transformers need to be simplified by short-circuiting lines into two-terminal circuits with all parasitic capacitances paralleled to each other in all measurement setups. In other words, there should be only two nodes left in every simplified two-terminal circuits.
2. The measurement results of different setups can be same to each other, like Setup 8 shown in Fig. 3 and Setup 3 shown in Fig. 2, which are called dependent setups. There will be invalid measurements if dependent setups are chosen.

Setup 7

$$C_{eq7} = \frac{C_4 \cdot (C_2 + C_5)}{C_2 + C_4 + C_5} + C_1 + C_6$$

Setup 8

$$C_{eq8} = C_1 + C_4 + C_6$$

Fig. 3 Other two measurement setups

Through analysis, the limitations of conventional two-terminal method can be summarized as follows:

1. Pre-analyses for the proper selection of measurement setups is needed. Otherwise, large amounts of invalid measurements and unnecessary data post-processing will significantly complicate the measurement processes.

2. Data post-processing is always needed. All measurement results in this method are equivalent capacitances of different simplified two-terminal circuits. Additional matrix operations are needed to obtain the parasitic capacitances.

3. Low feasibility for muti-winding transformers. In the n-terminal equivalent circuit for multi-winding transformer, there will be $n(n-1)/2$ parasitic capacitances to be measured. At least $n(n-1)/2$ independent linear equations generated by $n(n-1)/2$ different measurement setups are needed. Limitation 1) and 2) will be more severe in parasitic capacitance measurement of multi-winding transformer if n is much larger.

III. Proposed guarding method for two-winding transformer

Guarding technique can eliminate the mentioned influences from shunt paths around CUT in Section II, which means that CUT can be directly measured. The schematic diagram of guarding technique, provided by impedance analyzer E4990A together with adaptor 16047E from Keysight, is shown in Fig. 4. There are three terminals in adaptor 16047E: S for source terminal, M for measurement terminal, G for guard terminal. V_s for source voltage, V_m for measurement voltage. I_1, the current of Z_1, which flows between S and G, does not flow to the auto-balancing bridge. Thus, I_1 has nearly no effect on the measurements of CUT. Because the equivalent input impedance of auto-balancing bridge is negligible, the voltage potential of M is nearly to zero (virtual ground due to auto-balancing), which basically shorts Z_2. I_2, which flows between M and G, is much smaller and negligible compared to I_x, current of CUT. Hence $I_x \approx I_m$, current of auto-balancing resistance R_m. CUT can be derived as (2). The measurement result of guarding technique is the frequency response of CUT. Therefore, direct measurement of CUT is achieved.

$$I_x \approx I_m \qquad \frac{V_s}{\text{CUT}} \approx \frac{V_m}{R_m} \qquad \text{CUT} \approx \frac{V_s \cdot R_m}{V_m} \tag{2}$$

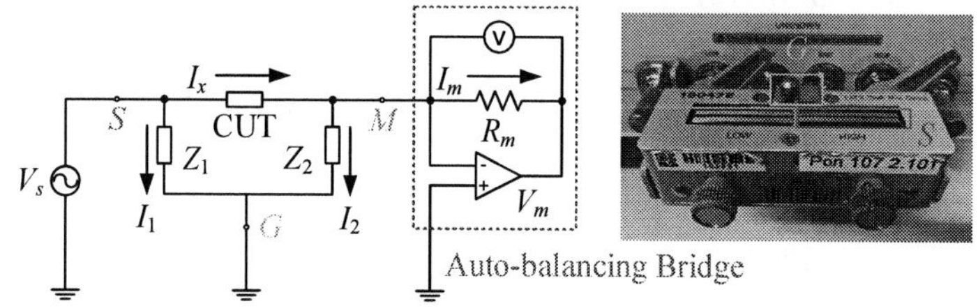

Fig. 4 Schematic diagram of guarding technique

Since guarding technique is provided for three-terminal circuit measurements, proper connections between two-winding transformers, which have four terminals, and adaptor 16047E are crucial to the feasibility of guarding method. The proposed guarding method to measure parasitic capacitance of two-winding transformer is shown in Fig. 5 (also based on the six-lumped-capacitance equivalent circuit in Fig. 1). Taking the measurement of C_1 as an example, S and G should be connected to the two terminals directly connected to C_1, 1 and 2. The rest of two terminals, 3 and 4, should be connected to G. As shown in Fig. 5 (a), C_2 is shorted. Four-terminal circuit of two-winding transformer can be reduced to three-terminal circuit suitable for guarding technique. Based on former analyses in Fig. 4, C_1 can be directly measured through impedance analyzer E4990A together with adaptor 16047E. The other parasitic capacitances $C_2 \sim C_6$ can be measured in similar way with guarding technique shown in Fig. 5 (b) – (f).

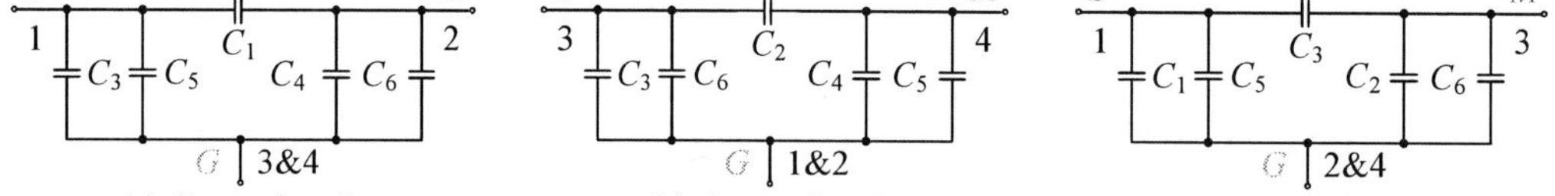

(a) Setup for C_1 (b) Setup for C_2 (c) Setup for C_3

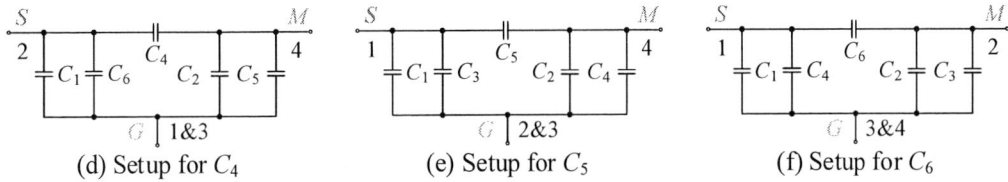

(d) Setup for C_4 (e) Setup for C_5 (f) Setup for C_6

Fig. 5 Proposed guarding method to measure parasitic capacitance of two-winding transformer

IV. Proposed guarding method for multi-winding transformer

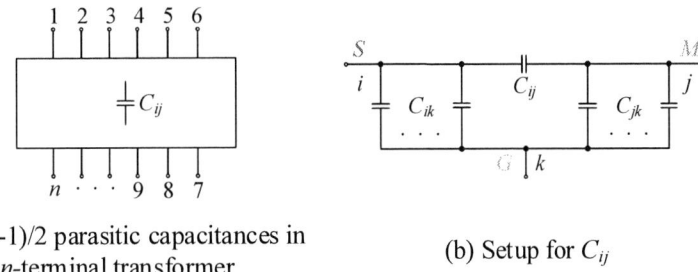

(a) $n(n\text{-}1)/2$ parasitic capacitances in n-terminal transformer

(b) Setup for C_{ij}

Fig. 6 Proposed guarding method to measure parasitic capacitance of multi-winding transformer

Guarding method can be extended to general multi-winding transformers (number of windings ≥ 2), there will be n terminals ($n \geq 4$) and $n(n\text{-}1)/2$ parasitic capacitances to be measured, shown in Fig.6 (a). Randomly choose one parasitic capacitance C_{ij} ($i, j \in \{1, 2, ..., n\}$) as CUT, S and G should be connected to two terminals directly connected to C_{ij}, i and j. The rest of (n-2) terminals k ($k \in \{1, 2, ..., n\} - \{i, j\}$) should be connected to G. As shown in Fig. 6 (b), all parasitic capacitances which are not directly connected to i or j are shorted and therefore only C_{ik} and C_{jk} are left. N-terminal circuit of multi-winding transformer is reduced to three-terminal circuit suitable for guarding technique. Based on former analyses in Fig. 4, C_{ij} can be directly measured through impedance analyzer E4990A together with adaptor 16047E.

Through analysis, guarding method shows several advantages compared to conventional two-terminal method, shown as follows:

1. Streamlined measurement processes. There is only one measurement setup for any parasitic capacitance. And all parasitic capacitances can be directly measured based on guarding technique. There are no needs for pre-analyses and data post-processing.
2. Higher feasibility for general muti-winding transformers. Uniform measurement setups and direct measurements will always valid no matter how many parasitic capacitances in all for general multi-winding transformers.

V. Experimental verifications

The experimental verifications will include two parts:

1. Circuit verification. The reason why circuit verification is needed is that standard values are always necessary no matter what comparison will be conducted. The comparison between two-terminal method and guarding method cannot be only based on transformer samples, in which all components are unknown and not satisfied for comparison standards. Inspired by ICT, circuit verification can be set as standards for measurements, which is based on a PCB assembled by passive components with known values.
2. Case studies based on three different two-winding transformers.

Through the circuit verification and three case studies, advantages, and limitations of both two measurement methods can be shown. All these verifications are conducted by impedance analyzer 4990A and adaptor 16047E, which are well calibrated. The values of capacitance are calculated based on measured frequency response.

For circuit verification, Fig. 7 (a) and (b) show the PCB assembled by passive components with known values and its schematic diagram. All values in schematic diagram are obtained by measurement results of six different branches shown in Fig. 7 (c) to eliminate the errors caused by tracks and components themselves. And all values of capacitance in every single branch will be set as standard values. Fig. 8 shows the measurement setups of two-terminal method and guarding method. Two-terminal method is based on six setups in Fig. 2, in which sixth order matrix operation shown in (1) is needed to obtain all values of capacitance in every single branch. Guarding method are based on six setups in Fig. 5, in which every single capacitance is directly measured. Table I shows the comparison of standard values, results of two-terminal method, and results of guarding method for all capacitances.

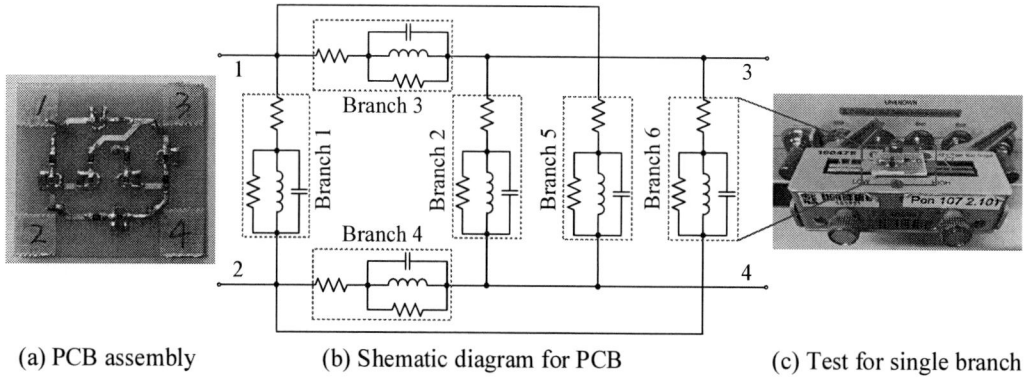

(a) PCB assembly   (b) Shematic diagram for PCB   (c) Test for single branch

Fig. 7 Circuit verification

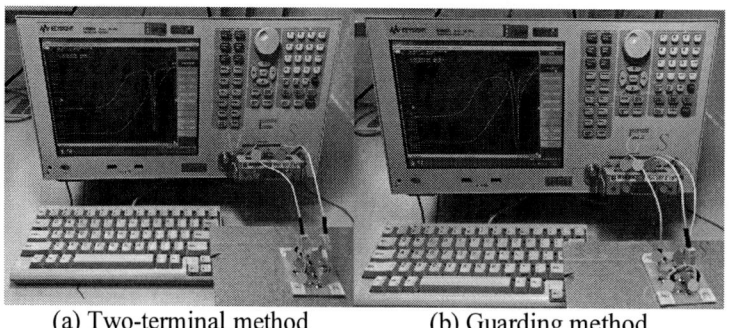

(a) Two-terminal method   (b) Guarding method

Fig. 8 Measurement setups of circuit verification

Table I: Comparison of standard values, two-terminal method results, and guarding method results in circuit verification

(pF)	Standard	Two-terminal	|Errors of Two-terminal|	Guarding	|Errors of Guarding|
C_1	1060	1134	6.98%	982	7.36%
C_2	1190	1046	12.10%	1088	8.57%
C_3	326	308	5.52%	312	4.29%
C_4	326	324	0.61%	330	1.22%
C_5	450	408	9.33%	424	5.78%
C_6	450	440	2.22%	436	3.11%

For two-terminal method, the calculated capacitances have 0.61% ~ 12.10% absolute values of error compared to standard values. For guarding method, the directly measured capacitances have 1.22% ~ 8.57% absolute values of error compared to standard values. Both two-terminal method and guarding method are capable to obtain all values of capacitance in every single branch with acceptable errors. Errors are caused by:

1. Extra cables connecting PCB to adaptor.
2. Discrepancies between selected frequencies and real frequencies of first resonant peak.
3. Couplings between different branches.
4. Propagation of errors through matrix operations. This will only occur in two-terminal method, which might have the potential to enlarge errors.

For case studies, Table II shows the comparisons between two-terminal method results and guarding method results for all parasitic capacitances. The test setups for case studies are same to circuit verification shown in Fig. 8. Because there are no accurate values of the parasitic capacitances as the references for the errors, no accuracy comparisons are shown in these samples. The mean values of results obtained by two different methods are set as the references for absolute values of discrepancy between two methods $|\Delta|$. Maximum discrepancy is 14.81%, which is still acceptable.

Table II: Comparison between two-terminal method results and guarding method results in case studies

	Case 1			Case 2			Case 3		
(pF)	Two-terminal	Guarding	$|\Delta|$	Two-terminal	Guarding	$|\Delta|$	Two-terminal	Guarding	$|\Delta|$
C_1	684	752	9.47%	61	60	1.65%	238	247	3.71%
C_2	1100	1210	9.52%	62	62	0.00%	210	209	0.48%
C_3	354	343	3.16%	22	21	4.65%	1390	1260	9.81%
C_4	321	344	6.92%	29	25	14.81%	1440	1260	13.33%
C_5	647	678	4.68%	43	48	10.99%	962	1010	4.87%
C_6	698	693	0.72%	41	40	2.47%	998	1030	3.16%

After circuit verification and case studies, guarding method can be selected as the better measurement method based on followed advantages and limitations shown in Table III.

Table III: Advantages and limitations of two-terminal method and guarding method

	Two-terminal method	Guarding method
1. Accuracy	High	High
2. Measurement processes	Complex	Simple
3. Applicability	Narrow	Broad
4. Instrument Requirement	Low	High

1. Both two-terminal method and guarding method are feasible to obtain parasitic capacitances in two-winding transformers with similar and acceptable errors. Propagation of errors through matrix operations should be noticed in two-terminal method.
2. To obtain all parasitic capacitances in two-winding transformers, both methods need six tests. But there are pre-analyses for the proper selection of measurement setups and data post-

processing in two-terminal method. Guarding method can directly measure each individual parasitic capacitance.

3. Measurement processes of two-terminal method will be much more complex for multi-winding transformer with more terminals. Meanwhile, it will keep simple in guarding method due to uniform measurement setups and direct measurements.

4. Any impedance analyzer with any two-terminal adaptor is sufficient for two-terminal method. But guarding method can only be conducted using impedance analyzers and adaptors with guarding function.

Error analyses of both two methods are complex but important for more accurate results, especially the propagation of errors in two-terminal method. Therefore, error analyses will be discussed in future work.

VI. Conclusion

A more general method for parasitic capacitance measurement of transformer is proposed in this paper, which is based on three-terminal guarding technique. Without pre-analyses, data post-processing, and propagation of errors, guarding method can directly measure each individual parasitic capacitance, which is crucial for multi-winding transformer. Circuit verification based on a PCB with known components and three case studies of two-winding transformers was conducted for the comparisons between conventional two-terminal method and guarding method. With high accuracy, simpler measurement processes, guarding method is a better measurement method for parasitic capacitance in multi-winding transformer. Error analyses, especially propagation of errors in two-terminal method, will be priorities in future research.

References

[1] Zhang, B. and Wang, S.: A survey of EMI research in power electronics systems with wide-bandgap semiconductor devices, IEEE Journal of Emerging and Selected Topics in Power Electronics 8.1 (2019), pp.626-643.

[2] Dalal, D.N., Zhao, H., Jørgensen, J.K., Christensen, N., Jørgensen, A.B., Bęczkowski, S., Uhrenfeldt, C. and Munk-Nielsen, S.: Demonstration of a 10 kV SiC MOSFET based medium voltage power stack, 2020 IEEE Applied Power Electronics Conference and Exposition (APEC), pp. 2751-2757.

[3] Thummala, P., Schneider, H., Zhang, Z. and Andersen, M.A.: Investigation of transformer winding architectures for high-voltage (2.5 kV) capacitor charging and discharging applications, IEEE Transactions on Power Electronics 31.8 (2015), pp.5786-5796.

[4] Nguyen-Duy, K., Ouyang, Z., Petersen, L.P., Knott, A., Thomsen, O.C. and Andersen, M.A.: Design of a 300-W isolated power supply for ultrafast tracking converters, IEEE Transactions on Power Electronics 30.6 (2014), pp.3319-3333.

[5] Bahmani, M.A., Thiringer, T., Rabiei, A. and Abdulahovic, T.: Comparative study of a multi-MW high-power density DC transformer with an optimized high-frequency magnetics in all-DC offshore wind farm, IEEE Transactions on power delivery 31.2 (2015), pp.857-866.

[6] Biela, J. and Kolar, J.W.: Using transformer parasitics for resonant converters-a review of the calculation of the stray capacitance of transformers, IEEE Transactions on Industry Applications, Vol. 44, No. 1 (2008), pp. 223-233.

[7] Liu, X., Wang, Y., Zhu, J., Guo, Y., Lei, G. and Liu, C.: Calculation of capacitance in high-frequency transformer windings, IEEE Transactions on Magnetics 52.7 (2016), pp.1-4.

[8] Zhang, H., Wang, S., Li, Y., Wang, Q. and Fu, D.: Two-capacitor transformer winding capacitance models for common-mode EMI noise analysis in isolated DC–DC converters, IEEE Transactions on Power Electronics 32.11 (2017), pp.8458-8469.

[9] Dalessandro, L., da Silveira Cavalcante, F. and Kolar, J.W.: Self-capacitance of high-voltage transformers, IEEE Transactions on power electronics 22.5 (2007), pp.2081-2092.

[10] Fei, C., Yang, Y., Li, Q. and Lee, F.C.: Shielding technique for planar matrix transformers to suppress common-mode EMI noise and improve efficiency, IEEE Transactions on Industrial Electronics 65.2 (2018), pp.1263-1272.

[11] Blache, F., Keradec, J.P. and Cogitore, B.: Stray capacitances of two winding transformers: equivalent circuit, measurements, calculation and lowering, 1994 IEEE Industry Applications Society Annual Meeting Vol. 2, pp. 1211-1217.

[12] Biela, J., Bortis, D. and Kolar, J.W.: Modeling of pulse transformers with parallel-and non-parallel-plate windings for power modulators, IEEE Transactions on Dielectrics and Electrical Insulation 14.4 (2007), pp.1016-1024.

[13] Zhao, H., Luan, S., J Hanson, A., Gao, Y., Dalal, D.N., Wang, R., Zhou, S. and Munk-Nielsen, S.: Rethinking Basic Assumptions for Modeling Parasitic Capacitance in Inductors, IEEE Transactions on Power Electronics, Early Access Article.

[14] Impedance Measurement Handbook, Keysight Technologies.

[15] Hults, C., Schwedner, F. and Grossman, S.: In-Circuit Test Systems-An Evolution, IEEE Transactions on Manufacturing Technology 4.2 (1975), pp.42-48.

[16] Zhao, H., Dalal, D.N., Jørgensen, A.B., Jørgensen, J.K., Wang, X., Bęczkowski, S., Munk-Nielsen, S. and Uhrenfeldt, C.: Physics-based modeling of parasitic capacitance in medium-voltage filter inductors, IEEE Transactions on Power Electronics 36.1(2021), pp.829-843.

[17] Zhao, H., Shen, Z., Dalal, D.N., Jørgensen, A.B., Wang, X., Munk-Nielsen, S. and Uhrenfeldt, C.: Parasitic Capacitance Modeling of Inductors Without Using the Floating Voltage Potential of Core. IEEE Transactions on Industrial Electronics 69.3 (2021), pp.3214-3222.

Inductance Analysis of Electric Machines by Classical and Numerical Methods

J.J. Germishuizen
Siemens Mobility GmbH
Nürnberg, Germany
Email: johannes.germishuizen@siemens.com

T.J.E. Miller
Emeritus Professor
University of Glasgow
United Kingdom
Email: tjem@retrospeed.co.uk

Keywords

≪AC Machine≫, ≪Finite-element analysis≫, ≪Impedance analysis≫.

Abstract

In electric machine design calculations, strong compatibility is essential between the methods used to design windings, the equivalent circuit model, and the overall electromagnetic performance. This paper presents a structured classification of complementary methods to achieve a unified approach, with particular focus on inductance calculations using classical and finite-element methods.

Introduction

Computer-aided design is now by far the preferred tool for the development of electric machines. Together with advances in materials and manufacturing methods, numerical analysis has helped to increase efficiencies and power densities, to advance the art of motion control, and even to introduce new types of machines such as superconducting machines, permanent-magnet machines of unprecedented size and power density, axial-flux machines, nano-scale machines, and others.

The fundamental theory of the electric machine is rooted in certain definite precepts related to its equivalent-circuit impedances, including generated EMF in some cases. The equivalent circuit is necessarily a *circuit* representation of the machine, since the machine is connected to an external supply circuit and controlled as a circuit element by other circuit elements such as switches and inverters. Internally, however, it is represented by different kinds of model: for example, an equivalent-circuit model and a field model, or a set of field models comprising thermal and electromagnetic fields.

The classical equivalent-circuit model and its parameters are not automatically compatible with the finite-element model. Indeed the dichotomy between "circuit models" and "field models" is very old, and pre-dates modern numerical analysis by many years. But the question of compatibility and the "working together" of these methods is now acute, because both models are essential to the modern designer: numerical analysis for its accuracy and its ability to solve nonlinear problems with complex geometry (including time-dependent problems), and classical theory for its structure and its role in the synthesis of windings, machine configurations, terminal behaviour, and the general understanding of performance.

This paper presents a structure of inductance calculations by different methods, intended to exploit the strengths of both sets of models with the highest possible degree of compatibility, avoiding conflicts, duplication, and wasted effort on inappropriate methods. The structure follows the classical partition of inductances into components, while taking the further step of capturing the effect of magnetic saturation by means of the finite-element method. An important distinction is made between "whole machine" parameters, which belong to the "terminal" equivalent-circuit model, and "partitioned components".

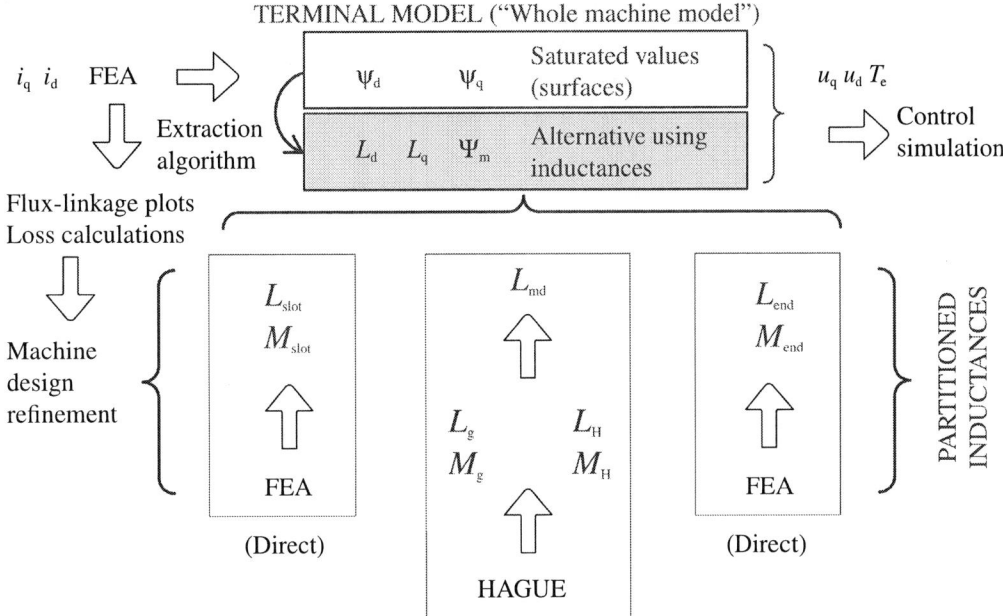

Fig. 1: "Road-map" or structure of inductance calculations

These two classes of parameters are generally used for different purposes and by different engineers, so it is helpful – almost essential – to have them sorted in a definite structure or scheme.

Although the methods described are not in themselves entirely new, the comprehensive treatment of the different methods is rare. The most original aspect is the emergence of quite categorical conclusions about the accuracy, compatibility, and convenience of the different methods and the relationships between them. The presentation of the air-gap magnetizing inductance and the components of harmonic leakage directly from Hague's solution for a "large-airgap" configuration is also rare, and has not been presented before in a comparison with traditional narrow-gap formulations and finite-element methods. Further, the prediction of induced saliency in a surface-magnet motor is not known to have been published before, while the "extraction algorithm" for the synchronous inductances is shown to be equally suitable for the nonsalient-pole machine and the salient-pole machine.

This paper is essentially a theoretical comparison of calculation methods, and we would make the unusual argument that physical test results would not help to clarify the methods. All the calculation methods are known from wide experience to provide accurate results when used in appropriate circumstances, while the differences between calculation methods are in many cases less than the errors that arise in ordinary physical measurements. Actual physical measurements are possible only in the case of the terminal inductances and not in the case of the components, and even then, inductance measurements are performed relatively less often than for other parameters (such as torque, efficiency, temperature rise, or even resistance) [11, 12]. And while we could not advance the finite-element method as a substitute for physical test, we certainly would hold it up as a standard of accuracy in calculation.

Structure of inductance calculations

The proposed structure is shown in Fig. 1, and it can be viewed as a "road-map" in the sense that it suggests or prescribes "best practice" methods for the calculation of the various inductances and their components.

A simple but general concept of the equivalent circuit is shown in Fig. 2, in which EMFs and impedances are combined in a circuit of greater or lesser complexity. Fig. 2 shows the beginnings of the partition of EMFs and inductances into "fundamental" and "harmonic" components, which are generally attributable

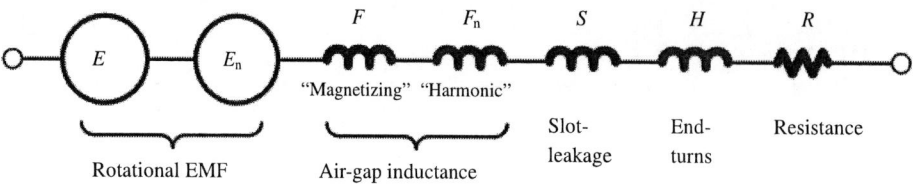

Fig. 2: General concept of equivalent circuit

to the space-harmonic properties of the winding distribution. A single circuit is usually sufficient for non- salient-pole machines, while a double circuit is needed for salient-pole machines such as the interior permanent-magnet motor (IPM), the wound-field synchronous motor, or the synchronous reluctance motor. In AC machines where the focus is on steady-state performance, the EMFs and impedances in Fig. 2 are usually invariant in time, but liable to saturation as a function of current and/or the strength of the excitation (magnets or field current).

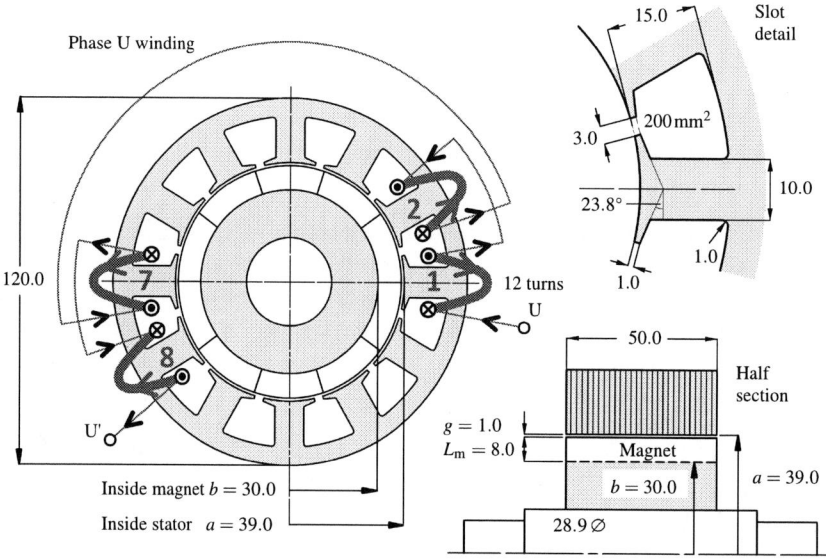

Fig. 3: Example surface-magnet PM brushless AC motor [14]

Example – a surface-magnet PM motor

At this point it may be helpful to introduce an example motor, Fig. 3. While this class of motor can be viewed as simpler than the IPM or the induction motor, it does in fact have *all* the inductance components necessary to illustrate *all* the methods of calculation, including the effects of saturation and even a small degree of saliency.

Table I shows the results of the inductance calculations.

Components and Methods

A key distinction in Fig. 1 is to separate the inductance parameters into "terminal" inductances and "partitioned" inductances. These are used for different purposes, and even by different engineers, as is suggested in the right-hand column of Table I. From an external viewpoint, the terminal inductances are

Table I: Inductance analysis

Component	Partitioned			Description	Main function
	FEA	Hague	MEC		
L_d	0.2555	0.2815	0.2366	Synchronous d	CSP
L_q	0.2368	0.2815	0.2366	Synchronous q	CSP
L_{ph}	0.2489	0.2554	0.2160	Phase self	TM
M_{ph}	−0.0263	−0.0261	−0.0206	Phase mutual	TM
L_{md}		0.05532	0.04356	Magnetizing	DMD
L_g		0.1354	0.0857	Air-gap self	W
M_g		−0.0106	0	Air-gap mutual	W
$L_{g\ \&\ slot}$	0.2424	0.02489		2D self	W
$M_{g\ \&\ slot}$	−0.0263	0.0261		2D mutual	W
L_{g5}		0.03688		5th harmonic self	W
M_{g5}		−0.01844		5th harmonic mutual	W
L_H		0.09071		Harmonic	W
L_{slot}	0.1238		0.1238	Slot self	DMD
M_{slot}	−0.0206		−0.0206	Slot mutual	DMD
L_{end}			0.0065	End-winding	DMD

W = winding design; **TM** = test & measurement; **DMD** = detailed machine design; **CSP** = control, simulation, performance

obviously of key importance, and this is reflected in Table I where the synchronous inductances L_d and L_q are highlighted all the way across the table – in other words, regardless of the method used to calculate them. L_{ph} and M_{ph} fall into the same category.

The pre-eminence given to L_d and L_q presupposes a terminal model in dq axes, because this is by far the most common circumstance for this type of machine, indeed for any inverter-fed machine. On the other hand, if direct phase variables are used, the phase inductances L_{ph} and M_{ph} (self and mutual) would be the natural ones to use. In salient-pole machines, of course these are rotor-position-dependent, and (as is well known) the basic reason for transforming to dq axes is to eliminate this dependence.

These important inductances L_d and L_q can be calculated directly and accurately using the finite-element method, no matter how much they are affected by saturation. Fig. 4 shows the result for the example motor; the flux-linkages ψ_d and ψ_q are presented as functions of the currents i_d and i_q [13, 12].

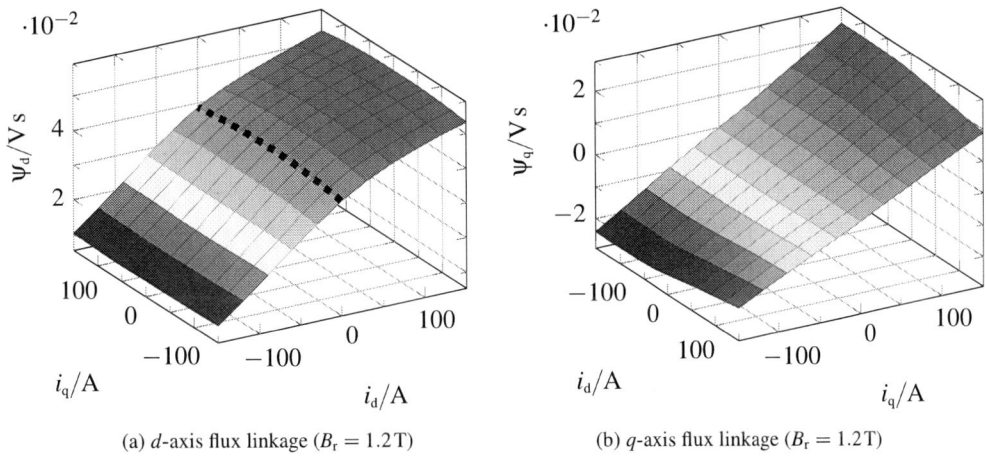

(a) d-axis flux linkage ($B_r = 1.2\,\text{T}$) (b) q-axis flux linkage ($B_r = 1.2\,\text{T}$)

Fig. 4: Flux-linkage surfaces of ψ_d and ψ_q calculated by the finite-element method

Although L_d and L_q can be calculated from the assembly of components by the other methods in the "partitioned" category, this method is generally less accurate.

However, the magnetizing inductance L_{md} should be singled out for consideration because it forms an important link between the terminal model and the partitioned model. It is also one of the central parameters in the theory of winding configurations. It can be defined as that component of L_d that is attributable to the air-gap flux produced by the fundamental space-harmonic of the winding distribution (with all three phases acting together), and it can be calculated for the example motor by (1) derived from the original analysis of Hague [9, 10]. (See also [7, 8]).

$$L_{md} = \frac{3}{2} k_w T_{ph} \Phi_{a1} = \frac{6 \mu_0 L_{stk} (k_w T_{ph})^2}{\pi p} \left[\frac{(a/b)^{2p} + 1}{(a/b)^{2p} - 1} \right] \quad \text{mH.} \tag{1}$$

The form of this equation shows its usefulness in machine design, containing as it does the fundamental winding factor k_w, the turns in series per phase T_{ph}, the number of pole-pairs p, the radial dimensions a, b and the stack length L_{stk}. Yet L_{md} cannot be calculated directly by the finite-element method, because it requires a harmonic analysis of the winding distribution *and* the separation of the air-gap flux from all other flux components such as slot-leakage, end-winding leakage, and harmonics.

Hague's method is the pure harmonic analysis of an ampere-conductor distribution on the surface of a smooth cylinder. This method and the finite-element method are complementary. In other words, what one method does, the other does not. When the slot-leakage and end-winding leakage are added to the air-gap component, the two methods do everything needed, and agree closely in the unsaturated case.

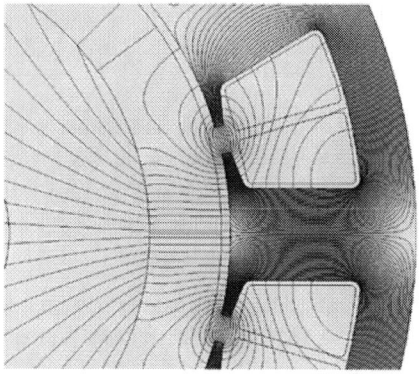

Fig. 5: Slot leakage [15]   Fig. 6: 3D eddy-currents [15]

Fig. 5 shows the calculation of inductance by the finite-element method. The classical magnetic-equivalent-circuit (MEC) method assumes often that the flux excited by current in the slot goes across the slot, with flux-lines everywhere perpendicular to the slot centre-line. That this is far from reality is immediately clear in Fig. 5. The MEC method often underestimates the slot-leakage inductance by a significant amount (although to be fair, this has been recognized and allowed for in many cases, [1, 2, 3, 4, 5]).

Fig. 6 shows a 3D eddy-current calculation in the end-region of an AC motor stator – one of the most sophisticated instances of modern numerical analysis. In simple terms the end-winding inductance would be obtained from a magnetostatic version of this calculation. This figure is included to underline the power of the FEA method in 3D calculations with time-dependent fields, while magnetic saturation is included in the formulation in both Figs. 5 and 6. This type of calculation is not possible with any serious level of accuracy or robustness by means of traditional methods (although of course the classical approximations are quick and still widely used).

The terminal or "whole machine" model

We return now to the "whole machine" or "terminal" model to consider how it may be efficiently calculated (without any partitioning); how it is used; and the effect of saturation [14]. The model is expressed in the classical Park's equations

$$u_d = R_1 \cdot i_d - \omega \cdot \psi_q$$
$$u_q = R_1 \cdot i_q + \omega \cdot \psi_d \qquad (2)$$

from which the phasor or r.m.s. terminal voltage is obtained as

$$U_{1,\text{phase}} = \sqrt{\frac{u_d^2 + u_q^2}{2}} \quad V_{\text{r.m.s.}} \qquad (3)$$

The torque equation is used at the same time in the form

$$T_e = \frac{m}{2} \cdot p \cdot (\psi_d \cdot i_q - \psi_q \cdot i_d). \qquad (4)$$

When the motor is fed by a current-regulated PWM inverter, the current and phase angle are controlled to have specific values i_d and i_q. For "sinewave" machines an efficient finite-element solution can be obtained at a fixed rotor position, for example when the d-axis is aligned with phase U, $\theta = 0$. By Park's inverse transformation the instantaneous phase currents to be used in the finite-element formulation then have the fixed values

$$i_u = i_d \cos(\theta) - i_q \sin(\theta) = i_d;$$
$$i_v = i_d \cos(\theta - 120°) - i_q \sin(\theta - 120°) = -\frac{1}{2} i_d + \frac{\sqrt{3}}{2} i_q;$$
$$i_w = i_d \cos(\theta + 120°) - i_q \sin(\theta + 120°) = -\frac{1}{2} i_d - \frac{\sqrt{3}}{2} i_q; \qquad (5)$$

The finite-element calculation produces the phase flux-linkages ψ_u, ψ_v, ψ_w from which ψ_d and ψ_q are obtained by Park's transformation. This completes the description of a "whole machine" terminal model in which the circuit equations are perfectly integrated with the finite-element method. The necessary finite-element data is shown in Fig. 4 in the form of flux-linkage surfaces computed for ψ_d and ψ_q over a range of currents i_d and i_q for the example motor in Fig 3. The data in these surfaces completely characterizes the motor at the terminals, making it possible to calculate the torque and the required fundamental terminal voltage at any load or speed; this data is thus in a form suitable for immediate integration with a field-oriented control scheme.

Fig. 4 shows the calculations in which the magnet remanent flux-density has its normal value (1.2 T). The surfaces show the variation in the saturation level caused by the variation in *current*; of course they include the effect the magnets as well. The actual flux-linkage/current relationships are of the form $\psi_d = \psi_d(i_d, i_q)$ and $\psi_d = \psi_d(i_d, i_q)$, described by the surfaces in Fig. 4. Because of the magnetic nonlinearity, it is not possible to resolve or segregate the saturation effects *uniquely* into separate components attributable to the current and the magnet. However, a common form used for segregation is

$$\psi_d = \Psi_m(i_q) + L_d \cdot i_d;$$
$$\psi_q = L_q \cdot i_q \qquad (6)$$

where L_d and L_q are the synchronous inductances in the d- and q-axes, and Ψ_m is the flux-linkage per phase produced by the magnet. It is shown here as a function of i_q, to recognize the effect of cross-saturation (which is particularly strong in the IPM). In general, Ψ_m is a function of both i_d and i_q, although it is common to take it as constant equal to the open-circuit value. Furthermore, in the most general case L_d and L_q are functions of both i_d and i_q to a greater or lesser extent. Only in the case of a magnetically linear machine is it possible to define *unique* constant values of Ψ_m, L_d and L_q.

Extraction of inductance and identification of saturation effects – In routine design calculations it is necessary to have an 'extraction' algorithm to obtain working values of Ψ_m, L_d and L_q, especially when these values are required in the design of field-oriented controllers. A review of the literature shows that many such algorithms or procedures are in use, but it is not essential to standardize them as long as consistent results are obtained within one company. (See, for example, [13, 12]).

The surface-magnet motor in Fig. 3 is often considered to be "simple" and free from saturation effects, in contrast with the IPM which has very strong and variable saturation effects. However, finite-element analysis shows that saturation due to the *magnet* can be significant. This result appears immediately from the comparison of L_d and L_q values in Table I. L_d falls to 91 % and L_q to 84 % of the unsaturated values which can be attributed, at least in part, to the small saturated regions in the rotor hub just inside the magnets. We can identify this as "induced saliency".

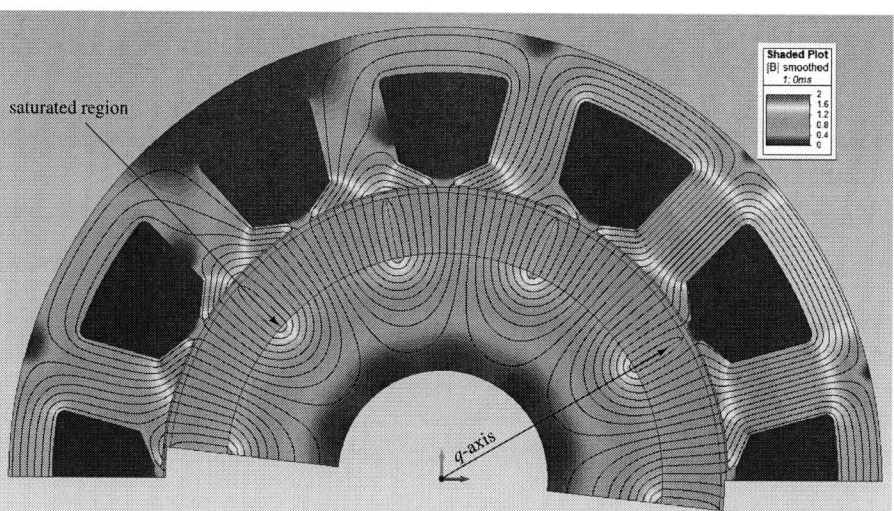

Fig. 7: Finite-element flux-plot showing "induced saliency"

It is natural to ask where the saturation effects are located – in which parts of the magnetic circuit? For the purpose of creating a simulation model to be used with control-system simulations, this question could be seen as irrelevant, on the grounds that the saturated terminal model of (2)-(4), together with the surfaces in Fig. 4, are completely adequate.

However, the machine *designer* is faced with the challenges of alleviating saturation effects as far as possible, and the immediate resource to be used for this is the flux-plot, an example of which appears in Fig. 5. It is often possible to tell at a glance, whether a tooth-width or a yoke section is too narrow, and remedial action can be taken very quickly in the CAD environment. The finite-element method provides many other tools for the assessment of design dimensions, for example the calculation of local loss densities.

But the flux-plot and related intensity maps do not answer all the designer's questions. For example, the slot-leakage inductance is a significant component (not greatly affected by saturation), and its value should be worked out in the process of optimizing both the slot shape and the winding pattern. Likewise the air-gap inductance needs to be separated into its space-harmonic components, again in relation to the winding pattern. Both these inductance components have self- and mutual components between phases, which are relevant to the design of the winding layout. The value of the separate inductance components gradually becomes clear, whether they are saturated or not.

The total terminal inductances can be expressed as a sum [14],

$$L_d \cong L_q = L_{md} + \overbrace{\left(L_{diff} - M_{diff}\right)}^{L_H} + \left(L_{slot} - M_{slot}\right) + \left(L_{end} - M_{end}\right) \tag{7}$$

where L_{md} is that part of the synchronous inductance attributable to the space-harmonic of the winding at the working harmonic; L_H is the harmonic or differential leakage inductance; and L_{slot}, M_{slot}, L_{end} and M_{end} are the self- and mutual slot-leakage and end-winding leakage inductances. Not only the individual values, but also the balance between these components is important.

If, then, we imagine starting with the "whole machine" terminal inductances in Table I and moving into the "partitioned" part of the table, we can begin to seek the most appropriate methods for determining the components. It is not a question of splitting up the finite-element results for the terminal inductances, so much as generating a set of component values from the three main sources in the Table:

- Direct magnetic equivalent-circuit methods [1, 2, 3, 4, 5, 6]

- Hague-type analysis of air-gap fields and their harmonics [9, 10, 7, 8]

- Finite-element calculations on isolated parts of the machine [1, 2]

This structure is expressed in a more generalised way in Fig. 1.

Conclusion

A summary of findings can now be given, based on the comprehensive set of inductance components calculated by the three classes of method and summarized in Table I.

(a) Direct magnetic equivalent-circuit methods are the quickest but also the least accurate, although it is dangerous to make sweeping statements because there is a huge variety of these methods going back over 120 years, often supported or verified by test data.

(b) The Hague-type analysis of the air-gap field and its harmonics is very accurate within the confines of its idealised model. It is especially valuable for the surface-magnet motor in dealing with the air-gap fields (from both the magnet and the winding), while for all machines it is intimately connected with the theory of winding factors. This raises its value in support of the design of windings. It is also very quick, being algebraic in formulation.

(c) Finite-element calculations on isolated parts of the machine are useful only where such isolation is clearly valid: for example, in calculating slot-leakage and end-winding leakage (which is not seriously possible by any other method).

When the results of these methods are put together, we find in particular that the Hague-type methods are generally the best for unsaturated air-gap and harmonic-leakage components, while the finite-element method is the most accurate (but also the slowest) for all the other components.

It therefore becomes possible to treat Fig. 1 as a road-map or classification of methods that helps to choose the most appropriate method for each task; and most importantly, to avoid wasting time on inappropriate methods. The finite-element method is also completely sufficient for the terminal model or "whole machine"model.

We have described a structured "road-map" of design calculation methods for the inductances of inverter-fed electric machines, in which the most appropriate tools are used for the respective components. It is shown that the methods can be applied in a "non-overlapping" manner that maximizes their compatibility, especially in regard to the combination of classical theory and finite-element analysis. Two objectives are achieved without the need to reconcile conflicts in calculated values: (i) providing accurate simulation models for control-system simulation and (ii) producing detailed component data for the refinement of machine designs. The structure includes both the "whole machine" terminal models and the detailed partition of inductance components that is necessary for detailed design and the synthesis and analysis of windings. Saturation of parameters due to currents and/or magnets is included in the "whole machine" model through finite-element methods without the need for artificial simplifying assumptions.

References

[1] A. Boglietti, A. Cavagnino and M. Lazzari: Modelling of the closed rotor slot effects in the induction motor equivalent circuit, International Conference on Electric Machines, ICEM, paper ID 781, 2008, pp. 1-4.

[2] A. Tessarolo: Analytical Determination of Slot Leakage Field and Inductances of Electric Machines With Double-Layer Windings and Semiclosed Slots, IEEE Transactions on Energy Conversion, Vol. 30, No. 4, December 2015, pp. 1528-1536.

[3] M. Bortolozzi, L. Branzo, A. Tessarolo and C. Bruzzese: An Improved Analytical Expression for Computing the Leakage Inductance of a Circular Bar in a Semi-Closed Slot, International Conference on Sustainable Mobility Applications, Renewables and Technology (SMART), 2015, pp. 1-5.

[4] M. Bortolozzi, L. Branzo, A. Tessarolo and C. Bruzzese: Improved Analytical Computation of Rotor Rectangular Slot Leakage Inductance in Squirrel-Cage Induction Motors, International Conference on Sustainable Mobility Applications, Renewables and Technology (SMART), 2015, pp. 1-5.

[5] A.F. Puchstein: Calculation of Slot Constants, AIEE Transactions, Vol. 66, pp. 1315-1323, 1947.

[6] M. Caruso, A.O. Di Tommaso, F. Genduso, R. Miceli and G.R. Galluzzo: A General Mathematical Formulation for the Determination of Differential Leakage Factors in Electrical Machines With Symmetrical and Asymmetrical Full or Dead-Coil Multiphase Windings, IEEE Transactions on Industry Applications, Vol. 54, No. 6, November/December 2018, pp. 5930-5940

[7] A. Hughes and T.J.E. Miller: Analysis of fields and inductances in air-cored and iron-cored synchronous machines, IEE Proceedings Vol. 124, pp. 121-128, 1977.

[8] T.J.E. Miller and A. Hughes: Comparative design and performance analysis of air-cored and iron-cored synchronous machines, IEE Proceedings Vol. 124, pp. 127-132, 1977.

[9] B. Hague: Electromagnetic Problems in Electrical Engineering, Oxford University Press, 1929.

[10] Z.P. Xia: Analytical Magnetic Field Analysis of Halbach Magnetized Permanent-Magnet Machines, IEEE Transactions on Magnetics, Vol. 40, No. 4, July 2004, pp. 1864-1872.

[11] T.J.E. Miller: Methods for testing permanent-magnet AC motors, IEEE Industry Applications Society Annual Meeting, Toronto, pp. 494-499, 1981

[12] T.J.E. Miller, M. Popescu, C. Cossar, M.I. McGilp, M. Olaru, A. Davies, J. Sturgess and A. Sitzia: Embedded Finite-Element Solver for Computation of Brushless Permanent-Magnet Motors, IEEE Transactions on Industry Applications, Vol. 44, No. 4, July/August 2008, pp. 1124-1133.

[13] J. Germishuizen, S. Stanton and V. Delafosse: Integrating FEM in an Everyday Design Environment to Accurately Calculate the Performance of IPM Motors, Studies in Applied Electromagnetics and Mechanics, Vol. 34, 2010, pp. 235-243.

[14] J.R. Hendershot and T.J.E. Miller: Design Studies in Electric Machines, MotorDesignBooks.com, ISBN 978-0-9840687-4-6 (e-Book 978-0-9840687-3-9).

[15] Acknowledgement: Powersys / JSOL Corporation and H. Sano, Figs. 5 and 6; Figs. 1, 2, 3, 5 and 6 and eqns. (1)–(7) are reproduced by kind permission of the authors of Ref. [14].

Dynamic Wireless Power Transfer DWPT Time Domain model: xyz position and speed coupling effect

Iosu Aizpuru[1], Eneko Agirrezabala[1], Mikel Mazuela[1], Unai Iraola[1], Estanis Oyarbide[2], Carlos Bernal[2]
[1]Mondragon Unibertsitatea, Mondragon, Spain
[2]University of Zaragoza, Zaragoza, Spain
E-Mail: iaizpuru@mondragon.edu
URL: https://www.mondragon.edu/en/research-transfer/engineering-technology/research-transfer-group/energy-storage

Acknowledgements

This research project is financially supported by CDTI program MISIONES. The name of the project is CARDHIN. The project is also supported by Ministerio de Ciencia, Innovación y Universidades.

Keywords

« Wireless power transmission », « Electric vehicle », « Time-domain analysis », « Battery charger», « Contactless Energy Transfer»

Abstract

The paper presents a DWPT system time domain model which considers speed $(\dot{x}, \dot{y}, \dot{z})$ and position (x, y, z) coupling effects. The speed effect compared to static WPT, presents an active behavior that should be considered during the coil design stage. The model is generalized and validated for resonant WPT systems and dynamic speed influenced DWPT systems.

1. Introduction

Greenhouse Gas Emissions GHG are increasing and are the main contributors to the global warming effect, breaking new temperature records since data calculation started in 1880 [1]. 72% of the GHG emissions generated are due to the energy sector, where road transportation (light-mid-heavy vehicles) are responsible of the 11.9% of the emissions [2]. For this reason, the electric vehicle is a great asset or candidate to help in the mitigation of the GHG emissions in the XXI century [3].

For the massive insertion in society of the electric vehicle, two major challenges must be solved: improving battery technology and designing a sustainable charging infrastructure. Referring to the charging infrastructure, this infrastructure must offer a great diversity of charging methods, be sustainable and at the same time have the capacity to supply the power of electric vehicles. There are different charging methods. Some differ by charging power (slow charging - fast charging). Others differ by the charging method (wired charging, wireless etc). Wireless charging systems are mainly static systems but dynamic wireless power transfer systems DWPT [4] are emerging as a new alternative within the charging infrastructure. DWPTs **Fig. 1 a)** allow the reduction of stops for battery charging, minimize the DOD of the batteries, increasing the useful life and even allow the optimization of the size of the battery [5]. In turn, this enables the use of smaller batteries, reducing the costs and facilitating the insertion in society of the electric vehicle [6].

Although DWPT technology is incipient and novel, there are studies that try to analyze and understand the problem of wireless charging systems. DWPT systems suffer mainly from misalignments, with some studies analyzing the effect of misalignments in the longitudinal road direction (x axis in this article) in steady state [7]–[9]. Low number of Publications make reference to misalignments in other axis as lateral misalignment, and they are classically limited to steady state analysis [10]. Limited analysis has been

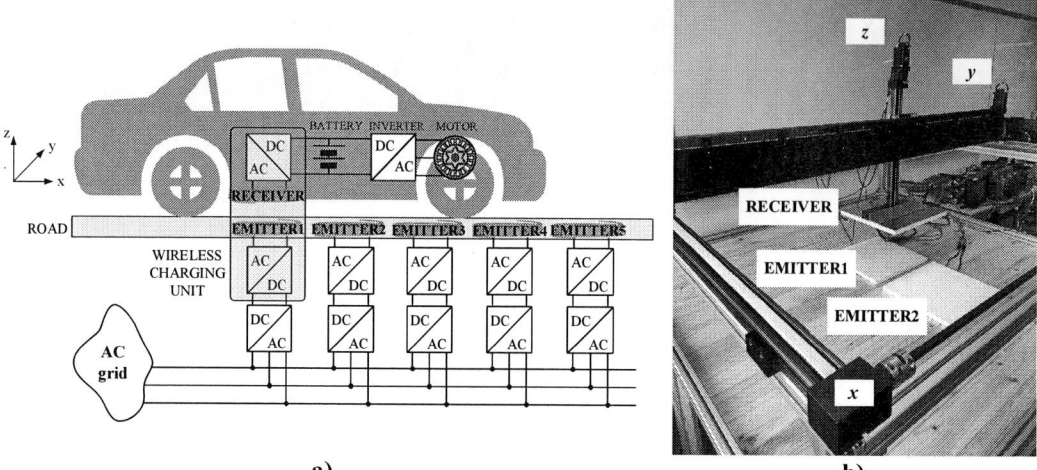

Fig. 1: a) Classical Dynamic Wireless Power Transfer DWPT schematic diagram. EMITTER: The energy comes from the AC grid, rectified and transformed to the emitter coil through a high frequency wireless charging unit. RECEIVER: The receiver is the Electric Vehicle where the energy is obtained through the wireless charging unit receiver coil, inserted to the battery through a high frequency converter and after transformed to mechanical energy by a power train composed of an inverter and a motor. WIRELESS CHARGING UNIT: Wireless charging unit composed of 2 high frequency AC-DC converters in emitter and receiver coupled by 2 air coupled coils. **b)** Developed 3D experimental platform for speed $(\dot{x}, \dot{y}, \dot{z})$ and position (x, y, z) misalignments analysis in DWPT systems.

done related to the influence of speed in DWPT systems, mainly concluding that the speed is not relevant in DWPT systems [11], [12]. Different control structures have been analyzed in order to maximize the transferred energy, but they are limited to steady state analysis [11], [13]–[16].

After the analysis of DWPT state of the art it is clear that there is a lack of modelling effort regarding different topics.

- DWPT modelling is mainly focused on steady state phasor modelling and there is a lack of investigation regarding dynamic time domain models.
- DWPT is clearly influenced by misalignments. However, the state of the art presents reduced number of analysis regarding misalignment effects. Longitudinal (x), lateral (y) and vertical (z) misalignments should be taken into account.
- DWPT systems permit to charge EVs during movement. The moving effect inserts another uncertainty to the modelling issue, the coupling of this speed $(\dot{x}, \dot{y}, \dot{z})$ in the DWPT system. This coupling is not deeply analyzed in the state of the art.

The contributions of this research paper will try to mitigate the lack of investigation in DWPT modelling, by developing a dynamic time domain model for DWPT systems where longitudinal (x, $\dot{x}$), lateral (y, $\dot{y}$) and vertical (z, $\dot{z}$) position and speed coupling are considered.

The research paper starts with an introduction regarding the state art and the main contributions of the research paper. Section 2 presents the basics regarding wireless power transfer systems modelling. Section 3 develops the 3-axis speed and position coupled model. Section 4 analyses how to augment the model order to implement a static resonant WPT system. Section 5 validates the model under a dynamic behavior analyzing the model sensitivity to speed variations. Finally, the main conclusions of the paper are presented.

2. Wireless charging basics modelling

Wireless charging units are composed by two power converters working in the emitter (infrastructure side) and the receiver (EV side) and a wireless coupling system composed of coils. The power converters

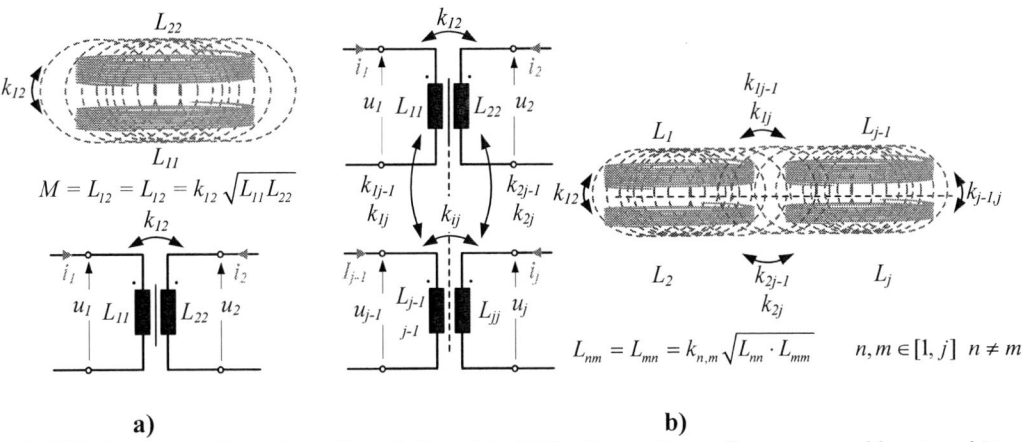

Fig. 2: Wireless power Transfer coil modelling. Model for 2 coupling coils represented by **a)** and **b)** generalized model for *j* coupled coils, where coils could be part of the transmitter or receiver circuit.

could be represented as 2 voltage sources for the sake of simplicity. The coupling between the emitter and the receiver is developed by 2 coils. Instead of a high coupling magnetic core, due to the inherit need of Wireless Power Transfer systems, the coupling media of these 2 coils is air.

The basic WPT system is modelled by the 2 auto inductances L_{11} and L_{22} of the emitter and transmitter coils, and the coupling inductor $L_{12} = L_{21}$ which models the coupling flux between both coils **Fig. 2 a)**. This coupling inductance is related with the auto inductances by the coupling factor k which could be identified as k_{12} for the basic coupling between 2 coils. The voltage induced in each coil u is directly the time derivative of the flux linkage Ψ of each of the coils. This magnetic analysis can be written by equations (1) and (2) and wrapped in a matrixial way by (3).

$$M = L_{12} = L_{21} = k_{12}\sqrt{L_{11} \cdot L_{22}} = k\sqrt{L_{11} \cdot L_{22}} \qquad (1)$$

$$u_1 = \frac{\partial \Psi_1}{\partial t} = L_{11}\frac{\partial i_1}{\partial t} + M\frac{\partial i_2}{\partial t} \;\; ; \;\; u_2 = \frac{\partial \Psi_2}{\partial t} = L_{22}\frac{\partial i_2}{\partial t} + M\frac{\partial i_1}{\partial t} \qquad (2)$$

$$\begin{bmatrix} u_1 \\ u_2 \end{bmatrix} = \begin{bmatrix} \dfrac{\partial \Psi_1}{\partial t} \\ \dfrac{\partial \Psi_2}{\partial t} \end{bmatrix} = \begin{bmatrix} L_{11} & L_{12} \\ L_{21} & L_{22} \end{bmatrix} \cdot \begin{bmatrix} \dfrac{\partial i_1}{\partial t} \\ \dfrac{\partial i_2}{\partial t} \end{bmatrix} \qquad (3)$$

The problem analyzed for 2 coils WPT could be generalized to *j* number of coils as presented in **Fig. 2 b)**. The generalized form composed by *j* coils could implement different configurations (1 receiver, j-1 transmitters; j-1 receivers, 1 transmitter, etc.) which leads to a general model that could model any DWPT or WPT configuration. The general model presented in **Fig. 2 b)** could be expressed by equations (4) and (5) for any coil *n,m* which belongs to the *j* coils. Equations (4) and (5) could be merged in (6). The voltage of each inductor could be merged by equation (7) where the auto inductances and mutual inductances are represented by the general inductance L_{nm}.

$$L_{nm} = L_{mn} = k_{n,m}\sqrt{L_{nn} \cdot L_{mm}} \qquad n,m \in [1,j] \;\; n \neq m \qquad (4)$$

$$u_n = \frac{\partial \Psi_n}{\partial t} = L_{nn}\frac{\partial i_n}{\partial t} + \ldots + \sum_{\substack{m=1 \\ n \neq m}}^{j} L_{nm}\frac{\partial i_m}{\partial t} \;\; ; \;\; u_m = \frac{\partial \Psi_m}{\partial t} = L_{mm}\frac{\partial i_m}{\partial t} + \ldots + \sum_{\substack{m=n \\ n \neq m}}^{j} L_{mn}\frac{\partial i_n}{\partial t} \qquad (5)$$

$$\begin{bmatrix} u_1 \\ \vdots \\ u_j \end{bmatrix} = \begin{bmatrix} \dfrac{\partial \Psi_1}{\partial t} \\ \vdots \\ \dfrac{\partial \Psi_j}{\partial t} \end{bmatrix} = \begin{bmatrix} L_{11} & \cdots & L_{1j} \\ \vdots & \ddots & \vdots \\ L_{j1} & \cdots & L_{jj} \end{bmatrix} \cdot \begin{bmatrix} \dfrac{\partial i_1}{\partial t} \\ \vdots \\ \dfrac{\partial i_j}{\partial t} \end{bmatrix} \qquad (6)$$

$$u_n = \frac{\partial \Psi_n}{\partial t} = \sum_{m=1}^{j} L_{nm} \frac{\partial i_m}{\partial t} \qquad (7)$$

However, this generalized model does not include the dynamic (speed $(\dot{x}, \dot{y}, \dot{z})$ and position (x, y, z) misalignments) effects of DPWT systems, so this model should be modified in order to undercome DPWT system modelling and simulations.

3. DWPT modelling: *(x,y,z)* speed and position coupling

Compared to the general static model presented in Section 2 which represents constant values for the auto inductance and mutual inductances of different transmission coils, a DWPT system has variable parameter values that could be modelled due to the changing value of the coupling factor *k.*

According to equation (7) any *n* inductor could be defined by the sum of all the coupling voltages. However in a DPWT system the inductance value is variable as a function of the position (x, y, z) as defined by equation (8).

$$L_{nm}(t) = f_{nm}(x(t), y(t), z(t)) \qquad (8)$$

Equations (2),(5) and (7) are a simplified version of the derivative of flux due to the assumption of constant inductors in static WPT systems; as it is presented in equation (9). However, DWPT systems do not have constant inductance values as presented in equation (8), so the derivative of flux is expressed by equation (10).

$$u_n = \frac{\partial \Psi_n}{\partial t} = \sum_{m=1}^{j} \frac{\partial \{ L_{nm} \cdot i_m(t) \}}{\partial t} = \sum_{m=1}^{j} L_{nm} \frac{\partial i_m}{\partial t} \; ; \text{static WPT equation (7)} \qquad (9)$$

$$u_n = \frac{\partial \Psi_n}{\partial t} = \sum_{m=1}^{j} \frac{\partial \{ L_{nm}(t) \cdot i_m(t) \}}{\partial t} = \sum_{\substack{p=x,y,z \\ m=1}}^{j} \frac{\partial L_{nm}}{\partial p} \frac{\partial p}{\partial t} i_m(t) + \sum_{m=1}^{j} L_{nm}(t) \frac{\partial i_m(t)}{\partial t} \; ; \text{dynamic DWPT} \qquad (10)$$

The general DWPT expression presented in equation (10) presents a speed term *δp/δt*. If this speed term is expanded as *δp/δt* $= (\dot{x}, \dot{y}, \dot{z})$ equation (10) could be expressed as equation (11).

$$u_n = \frac{\partial \Psi_n}{\partial t} = \underbrace{\sum_{m=1}^{j} \left(\frac{\partial L_{nm}}{\partial x} \dot{x} + \frac{\partial L_{nm}}{\partial y} \dot{y} + \frac{\partial L_{nm}}{\partial z} \dot{z} \right) i_m(t)}_{\textit{ACTIVE BEHAVIOUR}} + \underbrace{\sum_{m=1}^{j} L_{nm}(t) \frac{\partial i_m(t)}{\partial t}}_{\textit{REACTIVE BEHAVIOUR}} \qquad (11)$$

Equation (11) presents the main difference of a DWPT system vs a static WPT: The influence of speed to the system. Analyzing equation (11) 2 different components could be analyzed:

- REACTIVE BEHAVIOUR ($v_L = L \cdot di/dt$): Same behavior as in a static WPT system presented in equation (7) but with variable and position dependent inductances *$L_{nm}(t)$* defined in (8).
- ACTIVE BEHAVIOUR ($v = R \cdot i$): The voltage induced in the inductor is influenced by a speed dependent parameter multiplied by the current flowing in the inductor, modelled by an active behavior.

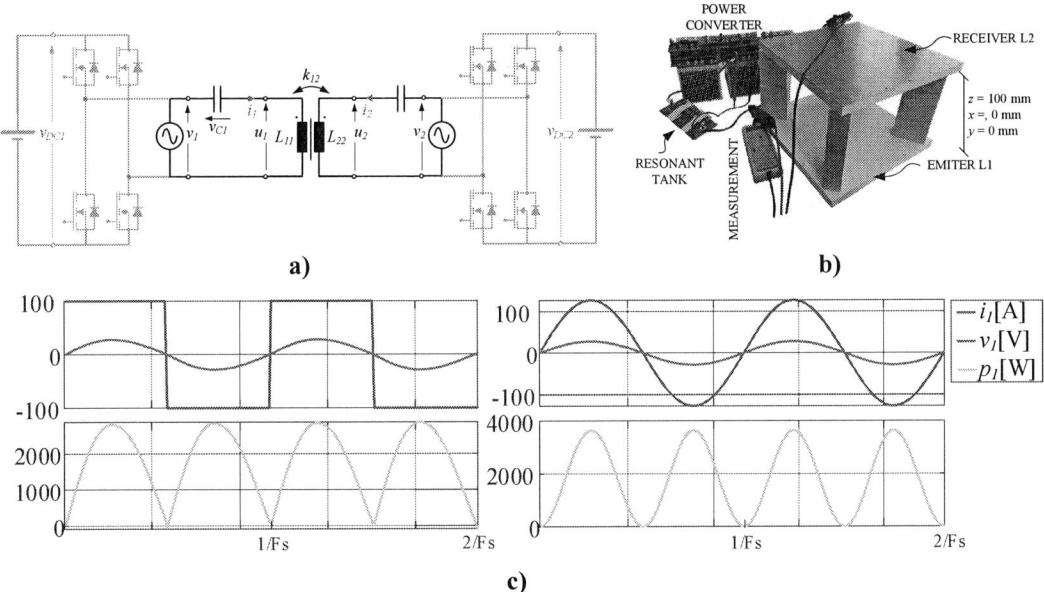

Fig. 3: Analysis of a series compensated resonant system. **a)** Modelled circuit with equivalent phasor/sinusoidal voltage generation or real converter operation. **b)** Developed experimental WPT system for model validation with 350 mm x 350 mm square WPT coils. **c)** Model results WPT resonant converter: LEFT: Real converter square excitation RIGHT: Phasor/Sinusoidal excitation. Results for primary side instantaneous voltage, current and power (i_1, v_1, p_1).

So as a conclusion, for modelling a DWPT system is mandatory to obtain the variable magnetic coupling components $L_{nm}(t)$ and the model will suffer some active behavior due to the speed of the DWPT system.

4. Model Use Case 1: Augmented static model for resonant coupling.

Models developed in Section 2 and 3 are representative for static and dynamic wireless transfer systems. The developed models only focus on the magnetic coupling between the different coils of a wireless power transfer system. However, wireless power transfer systems need a compensation network in order to maximize the transferred energy between the emitter and the receiver coils. This compensation networks are resonant networks that are tuned in order to maximize the coupling effect of the WPT coils. The model presented in Section 2 and 3 should be easily augmented to a resonant circuit in order to represent a real WPT system with the compensation resonant network.

Fig. 3 represents a classical series compensated resonant WPT system architecture. In order to increase the model order to a series resonant configuration, input voltages v_1, v_2 and resonant capacitor voltages v_{C1}, v_{C2} should be inserted to the model.

Input voltages v_1 and v_2 could be modelled by the simplified sinusoidal approach used in resonant converters or could be modelled by the square wave generated by a classical front-end full bridge converter.

Converter voltages v_1 and v_2 could be defined by equations (12) and (13) depending on the adopted modelling option (sinusoidal approach or square wave approach)

$$v_1 = \frac{4 \cdot V_{DC1}}{\pi} \sin(\omega t + \varphi_1); \quad v_2 = \frac{4 \cdot V_{DC2}}{\pi} \sin(\omega t + \varphi_2) \text{ ; Sinusoidal fundamental approach} \qquad (12)$$

$$v_1 = V_{DC1} \cdot \text{square}(t, \varphi_1); \quad v_2 = V_{DC2} \cdot \text{square}(t, \varphi_2); \text{ Power converter square wave approach} \qquad (13)$$

The square function (square(t, φ)) represents the power converter modulation technique. φ_1 and φ_2 represent the phase shift difference between the voltage introduced to the WPT system and is classically used to control the power transferred between V_{DC1} and V_{DC2}.

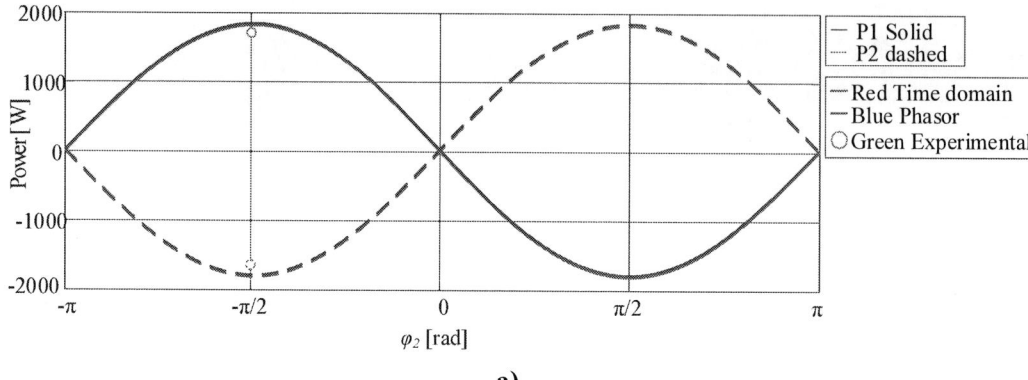

a)

Fig. 4: Simulation results of the developed validation environment presented in **Fig. 3** and parametrized in **Table I**. Compared variables: average power $\left(P = \frac{1}{T}\int_0^T p(t)dt\right)$ P1(solid) and P2 (dashed), in the proposed time domain model (red), steady state phasor model (blue) and experimental platform (green circles)

The voltage u_1 and u_2 in the coupling WPT system is defined by equations.

$$u_1 = v_1 - v_{C1}; \quad u_2 = v_2 - v_{C2}; \tag{14}$$

$$i_1 = i_{C1} = C_1\frac{dv_{C1}}{dt}; \quad i_2 = i_{C2} = C_2\frac{dv_{C2}}{dt}; \tag{15}$$

The combination of equations (14) and (15) with equation (3) permits to model the dynamic time domain behavior of a series resonant WPT system.

In order to validate the time-domain modelling approach, the time domain model is compared to a classical phasor steady state model and an experimental platform. The modelled system is described in **Table I**.

Table I: Parameter and variable definition for the simulation analysis of a series resonant WPT system.

Common parameters		Primary parameters / <u>variables</u>					Secondary parameters / <u>variables</u>				
$\omega = 2\pi f$	k	C_1	L_{11}	φ_1	v_1	V_{DC1}	C_2	L_{22}	φ_2	v_2	V_{DC2}
$2\pi85kHz$	0.22	94 nF	38 µH	0	sinusoidal	100 V	94 nF	38 µH	$-\pi...\pi$	sinusoidal	100 V

Regarding **Table I** parameters, L_{11} and L_{22} self inductances presented in **Fig. 3 b)** are experimentally measured by a LCR equipment. C_1 and C_2 are calculated to be in full resonance with the self inductances L_{11} and L_{22} at 85 kHz working frequency. The coupling factor k is experimentally measured for a constant z distance of 100 mm.

The obtained results are presented in **Fig. 4** for a secondary phase variation φ_2 from $-\pi$ to π. The compared parameters are the active power of the input ports P1 and P2. The active power is calculated from the time-domain simulation as the average of the instantaneous power during a whole period. The active power from the phasor analysis is the real part of the apparent power which is defined by the multiplication between the voltage and the conjugate of the current. The active power of the experimental platform is calculated for a single φ_2 value equal to $-\pi/2$ in order to check the maximum power transfer point.

The presented results validate the developed WPT series resonant time-domain model. The results are nearly identical to the steady state phasor model. The experimental results have similar values that the simulation results. The experimental power is slightly lower due to losses in the power converter and circuit parasitic resistance.

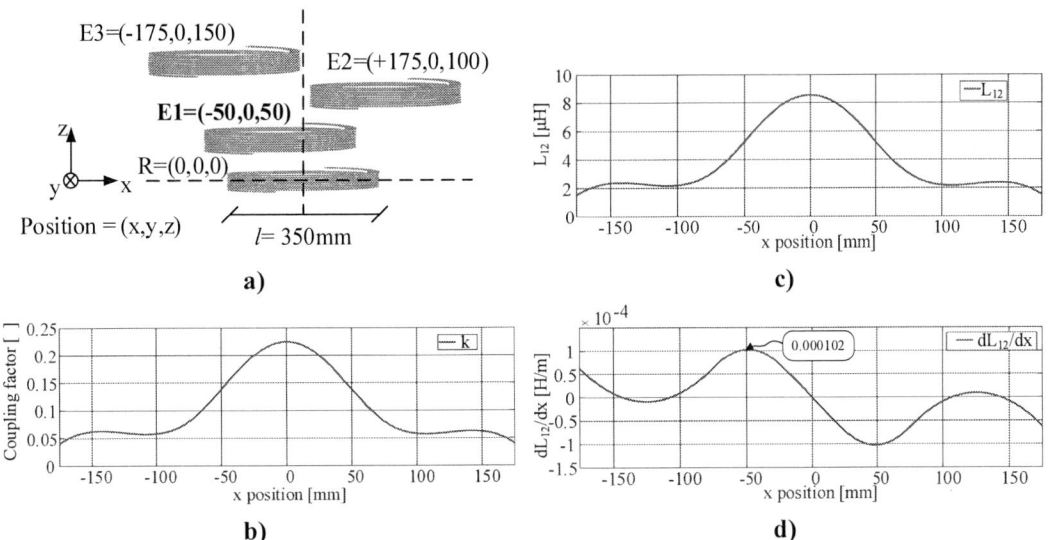

Fig. 5: Simulated DWPT system. **a)** Example of coil alignment and position reference definition in a DWPT system. **b)** Coupling factor k as a function of longitudinal misalignment x. **c)** Mutual inductance L_{12} as function of longitudinal misalignment x. **d)** Derivative of the mutual inductance respect to longitudinal position with a maximum value in $x=$ -50mm of 0.102 μH/mm

5. Model Use Case 2: Speed influence in power loses.

The DWPT model proposed in Section 3 will be parametrized in order to analyze the speed influence in the "active behaviour" component presented in equation (11). The speed influenced model will be analyzed for the coupling effect between 2 single coils (Emitter E and Receiver R) in the longitudinal position $x =$ -50mm **E1 Fig. 5 a)**, where the change in the mutual inductance due to the position is high. In order to simplify the problem, only the longitudinal speed ($\dot{x}$) effect will be analyzed. The misalignment in the lateral position y and height z will be kept constant. The auto inductance of the receiver and the emitter L_{11} and L_{22} will be assumed as constant. This assumption is valid for high z distances, where the self inductance of the receiver is not affected by the emitter and vice versa. Thus, only the mutual inductance $M(x,t) =L_{21}(x,t) =L_{12}(x,t)$ will vary in function of longitudinal position x. In order to obtain the active behavior of the speed influenced model presented in (11), the derivative of the mutual inductance with respect to the longitudinal position x. $(\delta L_{12}/\delta x)$ is required.

First, the mutual inductance variation effect between 2 DWPT coils is defined. The $L_{21}(x,t)$ inductance variation is totally dependent on the geometry, type and construction of the DWPT coils. The maximum coupling factor k is defined as 0.22 and is obtained when the longitudinal position is equal to 0 mm **Fig. 5 b)**. The mutual inductance is plotted for a ± 175 mm deviation which is represented in **Fig. 5 c)** for a defined coil of 350 mm total length.

After defining $L_{21}(x,t)$ the derivative $\delta L_{12}/\delta x$ is obtained. The derivative will be analyzed in $x =$-50 mm, where the influence is maximum. As presented in **Fig. 5 d)** the $\delta L_{12}/\delta x$ value for $x=$ -50 mm is 0.102 μH/mm. (0.000102 H/m).

The $\delta L_{12}/\delta x$ value 0.102 μH/mm for $x=$ -50 mm will be analyzed in order to check the voltage drop derived from the active behavior analyzed in equation (11). The analysis will be performed for RMS currents from 0 to 150 A (Ranges in the order of 0 to 50 kW DWPT systems) and a speed range in the longitudinal axis x from 0 km/h to 120 km/h.

The results presented in **Fig. 6** show that voltage drops of 0,5 V and power losses of 80 W are possible in DWPT system. These drop/losses could also increase depending on the coil geometry/sizing and if $(\dot{y}, \dot{z})$ misalignments are considered, so should be analyzed during the design process.

The developed model will be validated in future research works by the 3D experimental platform presented in **Fig. 1 b)** where speed $(\dot{x}, \dot{y}, \dot{z})$ and position (x, y, z) misalignments will be analyzed.

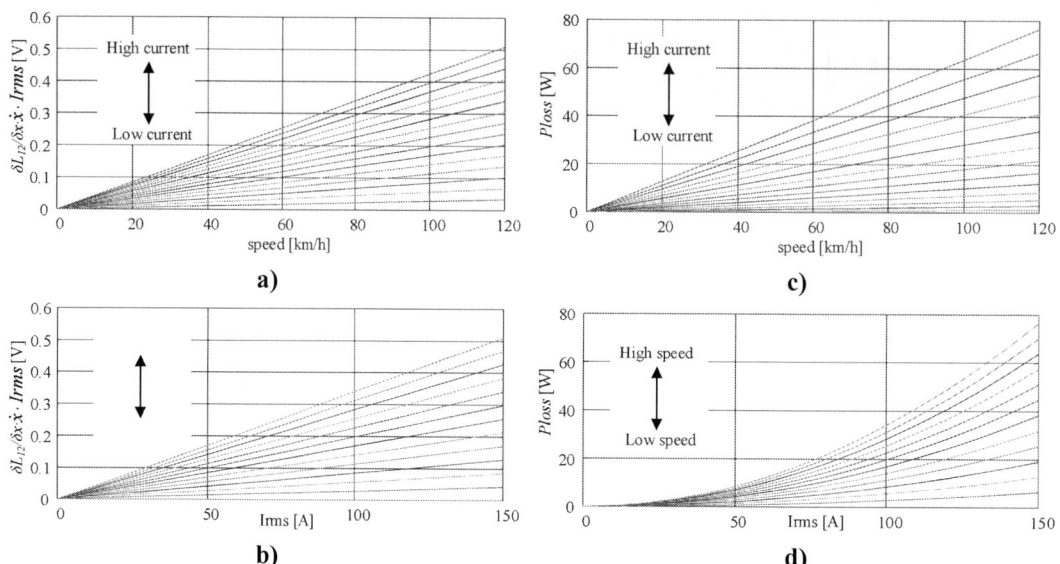

Fig. 6: Parametric analysis of primary RMS current Irms and speed $\dot{x}$ influence (Irms =0…150A, $\dot{x}$ = 0…120 km/h) .**a)** Voltage drop as a function of speed for different Irms. **b)** Voltage drop as a function of Irms for different speed. **c)** *Ploss* as a function of speed for different Irms. **d)** *Ploss* as a function of Irms for different speed

Conclusions

The develop research work presents a full Dynamic Wireless Power Transfer DWPT time-domain model that simulates the behavior of any generalized DWPT system. The developed model considers misalignments and coupling effects in the space/position domain (x, y, z) and speed $(\dot{x}, \dot{y}, \dot{z})$ domain. It is clearly presented that in DWPT systems there is a classical REACTIVE BEHAVIOR of magnetic coupling systems, but there is also an ACTIVE BEHAVIOR due to the influence of speed that should be considered.

The developed time domain model is easily augmented to reproduce a classical series resonant WPT system. The time-domain model presents same average power results as a classical phasor steady state model, and it is also validated by an experimental platform with 350mm x 350mm square WPT coils.

In order to analyze the DWPT ACTIVE BEHAVIOR under speed and current variations, the model is parametrized for a varying coupling factor k in function of the position x. The model presents a maximum voltage drop of 0.5V and power losses of 80 W for a 120 km/h and 150 A RMS values. The voltage drop and the power loss is directly dependent on the coupling factor shape and the derivative with respect to position, so it should be taken into account in the manufacturing process of DWPT coils.

The developed model will be fully validated under (x, y, z) - $(\dot{x}, \dot{y}, \dot{z})$ variations in future research works. A 3D experimental platform presented in **Fig. 1 b)** will be used in order to validate the developed time domain model.

References

[1]   NOAA National Centers for Environmental Information, "State of the Climate: Global Climate Report for Annual 2019," 2019. [Online]. Available: https://www.ncdc.noaa.gov/sotc/global/201913.

[2]   World resources Institute, "Climate analysis indicators tool," 2017, [Online]. Available:

https://www.wri.org/blog/2020/02/greenhouse-gas-emissions-by-country-sector.

[3] G. Hill, O. Heidrich, F. Creutzig, and P. Blythe, "The role of electric vehicles in near-term mitigation pathways and achieving the UK ' s carbon budget," *Appl. Energy*, vol. 251, no. July 2018, p. 113111, 2019, doi: 10.1016/j.apenergy.2019.04.107.

[4] R. Zeng, V. P. Galigekere, O. C. Onar, and B. Ozpineci, "Grid Integration and Impact Analysis of High-Power Dynamic Wireless Charging System in Distribution Network," *IEEE Access*, vol. 9, pp. 6746–6755, 2021, doi: 10.1109/ACCESS.2021.3049186.

[5] X. Mou, Y. Zhang, J. Jiang, and H. Sun, "Achieving Low Carbon Emission for Dynamically Charging Electric Vehicles through Renewable Energy Integration," *IEEE Access*, vol. 7, pp. 118876–118888, 2019, doi: 10.1109/ACCESS.2019.2936935.

[6] S. Jeong, Y. J. Jang, and D. Kum, "Economic Analysis of the Dynamic Charging Electric Vehicle," *IEEE Trans. Power Electron.*, vol. 30, no. 11, pp. 6368–6377, 2015, doi: 10.1109/TPEL.2015.2424712.

[7] F. Lu, H. Zhang, H. Hofmann, and C. C. Mi, "A Dynamic Charging System with Reduced Output Power Pulsation for Electric Vehicles," *IEEE Trans. Ind. Electron.*, vol. 63, no. 10, pp. 6580–6590, 2016, doi: 10.1109/TIE.2016.2563380.

[8] N. Teerakawanich, "Dynamic Modeling of Wireless Power Transfer Systems with a Moving Coil Receiver," *ITEC Asia-Pacific 2018 - 2018 IEEE Transp. Electrif. Conf. Expo, Asia-Pacific E-Mobility A Journey from Now Beyond*, pp. 1–5, 2018, doi: 10.1109/ITEC-AP.2018.8433269.

[9] A. Rakhymbay, A. Khamitov, M. Bagheri, B. Alimkhanuly, M. Lu, and T. Phung, "Precise analysis on mutual inductance variation in dynamic wireless charging of electric vehicle," *Energies*, vol. 11, no. 3, 2018, doi: 10.3390/en11030624.

[10] R. Tavakoli and Z. Pantic, "Analysis, Design, and Demonstration of a 25-kW Dynamic Wireless Charging System for Roadway Electric Vehicles," *IEEE J. Emerg. Sel. Top. Power Electron.*, vol. 6, no. 3, pp. 1379–1393, 2018, doi: 10.1109/JESTPE.2017.2761763.

[11] T. Fujita, T. Yasuda, and H. Akagi, "A Dynamic Wireless Power Transfer System Applicable to a Stationary System," *IEEE Trans. Ind. Appl.*, vol. 53, no. 4, pp. 3748–3757, 2017, doi: 10.1109/TIA.2017.2680400.

[12] Y. Guo, L. Wang, Q. Zhu, C. Liao, and F. Li, "Switch-On Modeling and Analysis of Dynamic Wireless Charging System Used for Electric Vehicles," *IEEE Trans. Ind. Electron.*, vol. 63, no. 10, pp. 6568–6579, 2016, doi: 10.1109/TIE.2016.2557302.

[13] A. Babaki, S. Vaez-Zadeh, and A. Zakerian, "Performance Optimization of Dynamic Wireless EV Charger under Varying Driving Conditions without Resonant Information," *IEEE Trans. Veh. Technol.*, vol. 68, no. 11, pp. 10429–10438, 2019, doi: 10.1109/TVT.2019.2944153.

[14] Z. Zhou, L. Zhang, Z. Liu, Q. Chen, R. Long, and H. Su, "Model Predictive Control for the Receiving-Side DC-DC Converter of Dynamic Wireless Power Transfer," *IEEE Trans. Power Electron.*, vol. 35, no. 9, pp. 8985–8997, 2020, doi: 10.1109/TPEL.2020.2969996.

[15] F. Liu, Y. Yang, Z. DIng, X. Chen, and R. M. Kennel, "A Multifrequency Superposition Methodology to Achieve High Efficiency and Targeted Power Distribution for a Multiload MCR WPT System," *IEEE Trans. Power Electron.*, vol. 33, no. 10, pp. 9005–9016, 2018, doi: 10.1109/TPEL.2017.2784566.

[16] H. He *et al.*, "Maximum Efficiency Tracking for Dynamic WPT System Based on Optimal Input Voltage Matching," *IEEE Access*, vol. 8, pp. 215224–215234, 2020, doi: 10.1109/ACCESS.2020.3041769.

Dynamic average small signal model of the SAB converter

Alexis A. Gómez[1], Alberto Rodríguez[1], Marta M. Hernando[1], Diego G. Lamar[1], Javier
Sebastián[1], Ibán Ayarzaguena[2], Jose Manuel Bermejo[2], Igor Larrazabal[2], David Ortega[2],
Francisco Vázquez[3]
[1] University of Oviedo, [2] Ingeteam Power Technology S.A., [3] Ingeteam R&D Europe.
[1] Gijón, Spain; [2] Zamudio, Spain; [3] Zamudio, Spain
E-Mail: gomezalexis@uniovi.es
URL: https://sea.grupos.uniovi.es/

Acknowledgements

This work was financed by the European project UE-18-POWER2POWER-826417, by the Principado
de Asturias through project SV-PA-21-AYUD/2021/51931, and by the Spanish Ministry of Science,
Innovation and Universities through projects MCI-21-PDC2021-121242-I00 and MCI-20-PID2019-
110483RB-I00.

Keywords

«Small signal», «Modelling», «Single Active Bridge», «Isolated converter», «DC-DC power converter»

Abstract

In this article the average small signal model of the Single Active Bridge (SAB) converter is obtained.
This converter can operate in two different conduction modes, Discontinuous Conduction Mode (DCM)
and Continuous Conduction Mode (CCM). The SAB and the Phase-Shifted Controlled Full Bridge
(PSFB) converter when operating in DCM present the same static behavior if the value of the inductor
of the SAB is the same as the value of the output inductor of the PSFB referred to the primary side of
the transformer. This conclusion can be extrapolated from the static analysis to the dynamic average
small signal model, as such a first order model is obtained for this operating mode. However, when the
SAB converter operates in CCM, the current through the inductor does not start at zero at the beginning
of the switching period. Both small signal models are analyzed and confirmed by means of simulations.

Introduction

One of the most common DC-DC bidirectional converters is the Dual Active Bridge (DAB) [1]–[4]. It
is an attractive converter for applications in which the power flow is reversible. However, in applications
where the power flow is always in the same direction, one port is the input and the other the output, it is
possible to substitute the secondary active bridge for a diode H bridge. This converter is referred as the
Single Active Bridge (SAB) and although the bidirectional power flow capability is lost due to this
change; the power density, reliability and cost are potentially improved. In Fig. 1 a schematic of the
SAB converter is shown, and a detailed static analysis is carried out in [5], [6]. Dynamic average small
signal modelling of the DAB converter can be found in several publications [3], [4]. The objective of
this article is precisely to perform a similar analysis for the SAB converter.

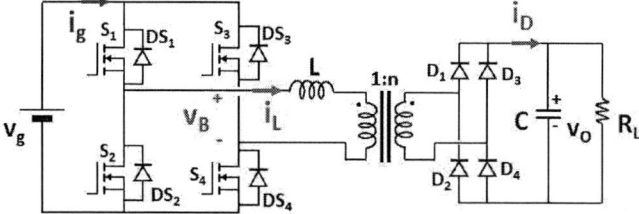

Fig. 1: Single Active Bridge (SAB) converter descriptive schematic

As exposed in [5], [6] the SAB converter can operate in two distinct conduction modes according to the inductor current (i_L) waveform. If the current only crosses and does not remain at zero, the converter operates in Continuous Conduction Mode (CCM), on the contrary if the current level remains at zero as seen in Fig. 2 the converter operates in Discontinuous Conduction Mode (DCM). In Fig. 1, i_D is the injected current into the R_LC net and corresponds to a rectified and scaled version of the current i_L. From this and Fig. 3 it is possible to understand that when the converter operates in DCM the current at the beginning of the switching period is zero. However, when the converter operates in CCM the value at the beginning of the switching period is non zero. This means that in DCM the electrical charge transferred to R_LC is not dependent on the previous switching period as in this mode always ends with no current flowing through the inductor. This implies a cycle-by-cycle update on the transferred power, therefore the resulting dynamic model corresponds to a first order model; similar to other converters, such as the PSFB when operating in DCM. In CCM the current i_L is not always the same at the start of the switching period and depends on the value at the end of the previous switching period. This fact makes for a model of higher order.

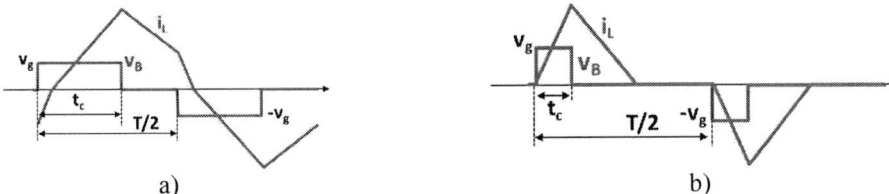

a)   b)

Fig. 2: Current through the inductor i_L and voltage applied to the inductor transformer set. a) in CCM. b) in DCM.

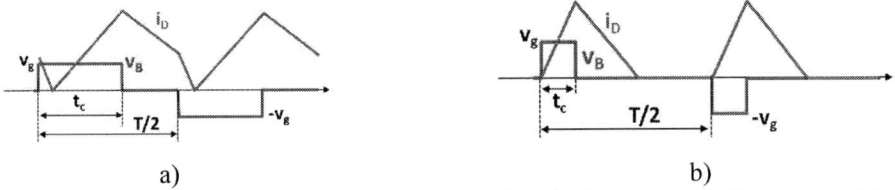

a)   b)

Fig. 3: Current injected to the R_LC net and voltage applied to the inductor transformer set. a) in CCM. b) in DCM.

The method used to obtain the dynamic average small signal models is the Current Injected Equivalent Circuit Approach (CIECA), explained in [7], [8].

Dynamic average small signal model in DCM

The basic idea of CIECA is represented in Fig. 4. The converter shown in Fig. 1 can be divided into three parts: The input voltage source, the set of switches and magnetic elements, and the R_LC output net. The basic idea of CIECA is to substitute the converter for a quadripole composed by two non-linear current sources. These current sources must behave in the same way of the average converter currents they replace. The input current source replaces the average input current, i_{g_avg}. The value of the second current source represents the average current through the secondary diodes bridge, and it is called i_{D_avg}. Therefore, the obtained model possesses the sampling delay characteristic of the averaging process. This delay will only be noticeable when trying to visualize variations close to the switching frequency of the converter, so this effect will be neglected for the model, as it is in most of the other proposed averaged models [7]–[9].

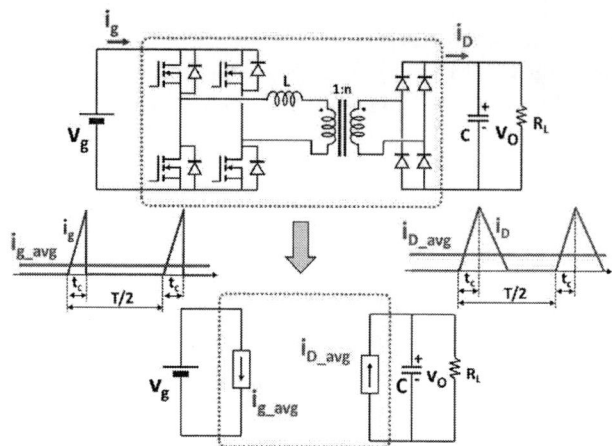

Fig. 4: Process for obtaining an average model using the Current Injected Equivalent Circuit Approach (CIECA).

Considering the waveform of i_D, on [5], [6], and depicted in Fig. 3, (1) was obtained:

$$i_{D\_avg} = \frac{v_g}{LTv_o}\left[v_g - \frac{v_o}{n}\right]t_c^2 \tag{1}$$

Where T is the switching period, L is the inductor, n the transformer relation, v_g is the input voltage, v_o is the output voltage and t_c corresponds to the simultaneous conduction interval of S1 and S4 or S2 and S3, in other words, when + v_g or - v_g is applied to the inductor-transformer arrangement (see Fig. 2 and Fig. 3) in one semi cycle. By conducting a power assessment during a switching period and taking into account that the energy stored on the inductor at the beginning and the end of a switching period is zero, we obtain:

$$I_{g\_avg} = \frac{v_o}{v_g}i_{D\_avg} = \frac{1}{LT}\left[v_g - \frac{v_o}{n}\right]t_c^2 \tag{2}$$

Transforming equations (1) and (2) as functions of the duty cycle, (4) and (5) are obtained. The duty cycle is defined in (3). From (4) and (5) it can be seen that the duty cycle is related to the electrical variables by means of multiplications and divisions, which implies a non-linear model, valid for small and big signal analysis. However, in order to use classic control theory for the feedback loop it is necessary to obtain transfer functions that relate the current sources with the electrical and control variables. To do this a linearization of the current sources around an operating point is carried out using capital letters (D, V_g…). The linearization makes the new model only valid for small signal analysis.

$$d = \frac{t_c}{T} \tag{3}$$

$$i_{D\_avg} = \frac{Tv_g}{Lv_o}\left[v_g - \frac{v_o}{n}\right]d^2 \tag{4}$$

$$i_{g\_avg} = \frac{T}{L}\left[v_g - \frac{v_o}{n}\right]d^2 \tag{5}$$

The linearization process transforms the quadripole from Fig. 4 into the one shown in Fig. 5. In this figure the variables represented by capital letters are associated with the operating point, whereas the ones with a circumflex accent correspond to the small signal perturbations.

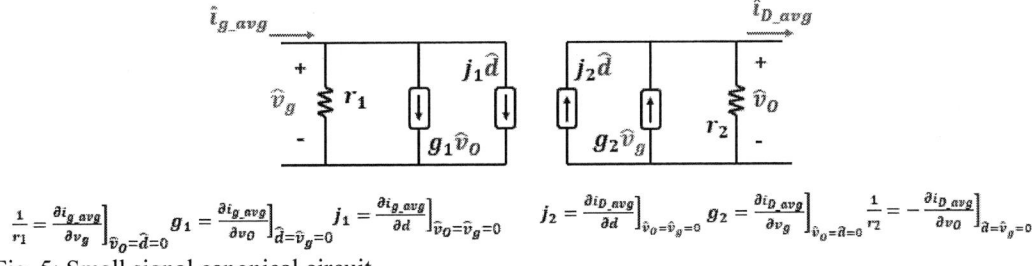

$$\frac{1}{r_1} = \frac{\partial i_{g\_avg}}{\partial v_g}\bigg]_{\hat{v}_o=\hat{d}=0} \quad g_1 = \frac{\partial i_{g\_avg}}{\partial v_o}\bigg]_{\hat{d}=\hat{v}_g=0} \quad j_1 = \frac{\partial i_{g\_avg}}{\partial d}\bigg]_{\hat{v}_o=\hat{v}_g=0} \quad j_2 = \frac{\partial i_{D\_avg}}{\partial d}\bigg]_{\hat{v}_o=\hat{v}_g=0} \quad g_2 = \frac{\partial i_{D\_avg}}{\partial v_g}\bigg]_{\hat{v}_o=\hat{d}=0} \quad \frac{1}{r_2} = -\frac{\partial i_{D\_avg}}{\partial v_O}\bigg]_{\hat{d}=\hat{v}_g=0}$$

Fig. 5: Small signal canonical circuit.

If the control variable is the duty cycle, the different transfer functions can be calculated as:

$$G_{od} = \frac{\widehat{v_o}}{\hat{d}}\bigg|_{\widehat{v_g}=0} = \frac{j_2 R_{eq}}{1 + R_{eq}Cs} \tag{6}$$

$$G_{og} = \frac{\widehat{v_o}}{\widehat{v_g}}\bigg|_{\hat{d}=0} = \frac{g_2 R_{eq}}{1 + R_{eq}Cs} \tag{7}$$

Where:

$$R_{eq} = \frac{R_L r_2}{R_L + r_2} = \frac{V_o(1-N)}{I_o(2-N)} \tag{8}$$

$$j_2 R_{eq} = \frac{2}{nN(2-N)}\sqrt{\frac{[V_o(1-N)]^3}{I_o L f}} \tag{9}$$

$$g_2 R_{eq} = nN \tag{10}$$

$$N = \frac{V_o}{nV_g} \tag{11}$$

As anticipated, G_{od} and G_{og} are first order transfer functions with a pole in the negative semi-plane. A comparative analysis with the transfer functions obtained by means of an analogue analysis of the PSFB converter when operating in DCM would reveal the same equations with the value of L transferred to the other side of the transformer. The parameters for both transfer functions are summarized in Table . Two equivalent equation formats are provided in Table I, the first uses the duty cycle, whereas the second uses external magnitudes.

Table I: Parameter values of the averaged small signal canonical circuit in DCM.

Parameter	j_1	g_1	r_1	j_2	g_2	r_2
Form 1	$\frac{2TD}{L}\left[V_G - \frac{V_o}{n}\right]$	$-\frac{TD^2}{nL}$	$\frac{L}{TD^2}$	$\frac{2TV_gD}{LV_o}\left[V_g - \frac{V_o}{n}\right]$	$\frac{TD^2}{L}\left[\frac{2V_g}{V_o} - \frac{1}{n}\right]$	$\frac{LV_o^2}{TD^2V_g^2}$
Form 2	$2\sqrt{I_o}\sqrt{\frac{V_o(1-N)}{Lf}}$	$-I_o\frac{nN^2}{V_o(1-N)}$	$\frac{V_o(1-N)}{n^2N^2I_o}$	$\frac{2\sqrt{I_o}}{nN}\sqrt{\frac{V_o(1-N)}{Lf}}$	$\frac{NnI_o(2-N)}{V_o(1-N)}$	$\frac{V_o(1-N)}{I_o}$

Dynamic average small signal model in CCM

As stated before, when the SAB converter operates in CCM the transferred electrical charge from the secondary diode bridge to the R_LC network during a switching period it is not independent from the previous switching periods. This is because the current through the inductor does not start at zero, as it did for DCM. Therefore, the area under the current waveform, the transferred charge, will depend on the starting current value. It is clear then that a change of the control variable does not determine the final transferred charge value during the next switching period, and it is logical to assume that several periods are necessary for the transferred charge on each switching period to settle on the stationary value. The number of cycles necessary to reach steady state will be obtained through simulations and if this effect can or cannot be deemed negligible.

Assuming that the change of the control variable results on a change of the transferred charge value during the next switching period equal to the steady state value, the same process applied to DCM can be used for CCM. The equations for the average input and secondary side diodes currents in CCM are (12) and (13).

$$i_{D\_avg} = \frac{T}{2Ln}\left[v_g d - v_g d^2 - \frac{v_o^2}{4n^2 v_g}\right] \tag{12}$$

$$i_{g\_avg} = \frac{T}{2Ln}\left[v_o d - v_o d^2 - \frac{v_o^3}{4n^2 v_g^2}\right] \tag{13}$$

The linearization process results in the same canonical circuit of Fig. 5. However, on this occasion, the parameters are those of Table II and Table III. The transfer functions can be expressed as before in (6) and (7) with the new parameters from Table and Table . As r_2 changed, the value of R_{eq} also changed. The value of j_2 R_{eq} does not coincide with (9), but g_2 R_{eq} does correspond with (10). A comparison between the different parameters of both conduction modes when operating at the border from each mode is carried out in Table IV, the equations obtained consider border mode operating point, therefore $2nD = V_o/V_g$. From this table, parameters of the canonical circuit differ from one conduction mode to another, creating an abrupt change at the border between conduction modes.

Table II: Parameter values of the averaged small signal canonical circuit in CCM. Primary side.

Parameter	j_1	g_1	r_1
Form 1	$\frac{TV_o}{2Ln}(1-2D)$	$\frac{T}{2nL}\left[D(1-D) - \frac{3}{4n^2}\left(\frac{V_g}{V_o}\right)^2\right]$	$\frac{4n^3 L}{T}\left(\frac{V_g}{V_o}\right)^3$
Form 2	$\frac{V_o}{2Lfn}\sqrt{1 - N^2 - \frac{8NLn^2 f I_o}{V_o}}$	$\frac{I_o Nn}{V_o} - \frac{N^2}{4Lfn}$	$\frac{4Lf}{N^3}$

Table III: Parameter values of the averaged small signal canonical circuit in CCM. Secondary side.

Parameter	j_2	g_2	r_2
Form 1	$\frac{TV_g}{2nL}(1-2D)$	$\frac{T}{2nL}\left[D(1-D) + \frac{1}{4n^2}\left(\frac{V_o}{V_g}\right)^2\right]$	$\frac{4n^3 L}{T}\left(\frac{V_g}{V_o}\right)$
Form 2	$\frac{V_o}{2Lfn^2 N}\sqrt{1 - N^2 - \frac{8NLn^2 f I_o}{V_o}}$	$\frac{I_o nN}{Vo} + \frac{N^2}{4Lfn}$	$\frac{4Lfn^2}{N}$

Table IV: Parameter values of the canonical circuits at both sides of the operation modes border.

Parameter	j_1	g_1	r_1	j_2	g_2	r_2
DCM	$\frac{TV_o}{nL}(1-N)$	$-\frac{TN^2}{4nL}$	$\frac{4L}{TN^2}$	$\frac{TV_o}{n^2 NL}(1-N)$	$\frac{TN}{4nL}(2-N)$	$\frac{4n^2 L}{T}$
CCM	$\frac{TV_o}{2nL}(1-N)$	$\frac{TN}{4nL}(1-2N)$	$\frac{4L}{TN^3}$	$\frac{TV_o}{2n^2 NL}(1-N)$	$\frac{TN}{4nL}$	$\frac{4n^2 L}{TN}$

Simulation results

In order to check the theoretical model, a PSIM simulation has been done of a converter designed using the design guide presented on [5], [6]. The resulting converter designed in [5] can be defined by f=100 kHz, n=0.55, L=78.96 µH. All the semiconductors are considered ideal, the magnetizing inductance is 789.6 mH and no transformer leakage inductance is considered.

With this converter, the parameter values are evaluated for two operating points, one in DCM and other in CCM, but near the border as one of the most important characteristics is the change in the parameters. The nominal input and output voltages are V_g= 400 V and V_o= 44 V. Using (11) this means a value of N=0.2. Knowing that the normalized voltage conversion ratio at the border between CCM and DCM is N=2D_{crit}, the critical duty cycle is D_{crit}=0.1. The chosen duty cycles are slightly under 0.1 for DCM and slightly above for CCM.

To calculate the canonical circuit parameters from the simulations, denoted by the suffix sim, the following expressions are used.

$$r_{1\,sim} = \frac{V_{g2} - V_{g1}}{I_{g\,avg\,sim}(V_{g2}) - I_{g\,avg\,sim}(V_{g1})} \tag{14}$$

$$g_{1\,sim} = \frac{I_{g\,avg\,sim}(V_{o2}) - I_{g\,avg\,sim}(V_{o1})}{V_{o2} - V_{o1}} \tag{15}$$

$$j_{1\,sim} = \frac{I_{g\,avg\,sim}(D_2) - I_{g\,avg\,sim}(D_1)}{D_2 - D_1} \tag{16}$$

To calculate r_1 the variation of average input current is observed when the input voltage changes, while maintaining the output voltage and duty cycle constant. For g_2 is the output voltage that suffers a variation, while maintaining the input voltage and duty cycle fixed. Finally, j_1 is calculated by observing the change in input average current when the duty cycle varies. To calculate the second side parameters, the same equations can be used by substituting the first side parameters for the second side and vice versa.

For this the simulated circuit corresponds to that of Fig. 1. To calculate j_1 and j_2 the input and output voltages are kept at nominal values, whereas the duty cycle changes from 0.090 to 0.100 for DCM and from 0.105 to 0.115 for CCM. For r_1 and g_2 the input voltage varies from 390 V to 400 V. And for g_1 and r_2 the output voltage varies from 44 V to 46 V. The nominal duty cycle values are 0.10 for DCM and 0.11 for CCM.

Theoretical and simulations results for all the parameters in both conduction modes are compared in Table IV. As shown, theoretical results exhibit a good correlation with simulation values.

Table V: Comparison of theoretical and simulation of canonical circuit parameters.

Parameters in CCM	j_1 (A)	g_1 (Ω^{-1})	r_1 (Ω)	j_2 (A)	g_2 (Ω^{-1})	r_2 (Ω)
Theoretical	4.05	0.0069	3953.6	36.8	0.0115	47.8
Simulated	4.15	0.0076	5000.0	37.0	0.0120	50.0

Parameters in DCM	j_1 (A)	g_1 (Ω^{-1})	r_1 (Ω)	j_2 (A)	g_2 (Ω^{-1})	r_2 (Ω)
Theoretical	8.11	-0.0023	789.9	73.7	0.0207	9.6
Simulated	7.40	-0.0021	909.1	66.00	0.0180	10.5

In order to confirm the validity of the first order model approximation when the converter operates in CCM, Bode diagrams of the G_{od} transfer functions for both conduction modes are represented in Fig. 6, these graphs were obtained by simulating the full switching circuit with a sinusoidal perturbation. These graphs show an almost perfect correlation between theoretical and simulation results for DCM, and a very good approximation for CCM. This confirms the initial assumption, the dynamic CCM model can be approximated as a first order model. A theoretical (G_{od}) and simulation step response of the output voltage when the duty cycle changes is depicted in Fig. 7. Fig. 7 a) corresponds when the converter operates in CCM, the duty cycle changes from 0.11 to 0.13. Fig. 7 b) does the same for DCM, the duty cycle changes from 0.077 to 0.097.

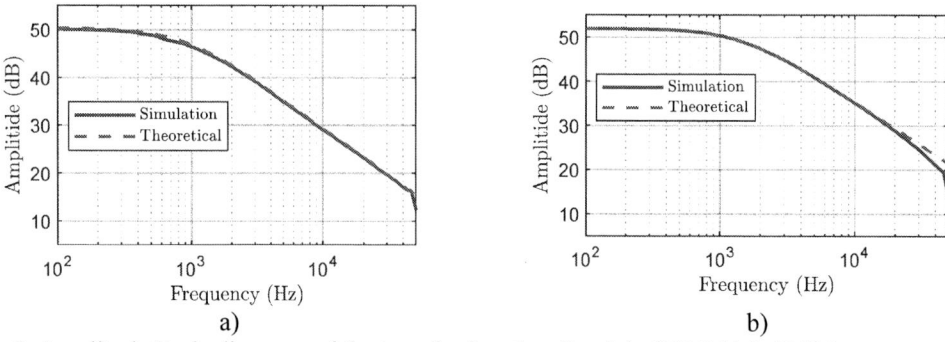

a)   b)

Fig. 6: Amplitude Bode diagrams of the transfer function G_{od} a) in CCM. b) in DCM.

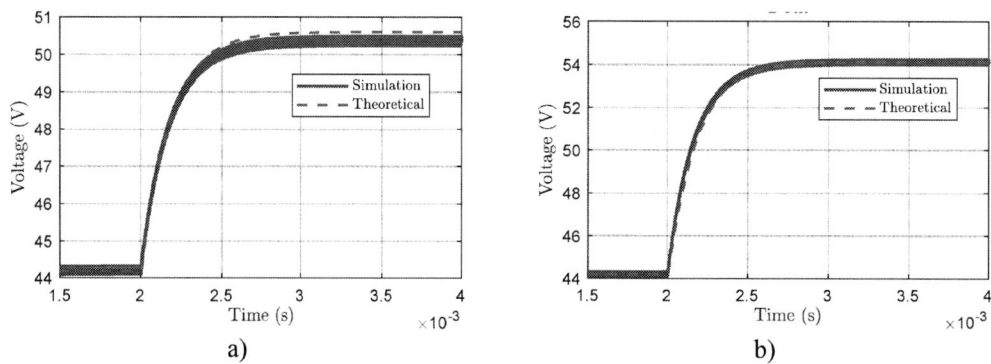

a)   b)

Fig. 7: Step response of the output voltage to a change in the duty cycle. a) in CCM. b) in DCM.

If the converter is subject to the same duty cycles steps as before but maintaining the input and output voltages constant, at 400 V and 44 V respectively; the current through the inductor and the diodes bridge changes instantly, when the converter operates in DCM. However, if the converter is working on CCM the change does not happen instantly and the current oscillates for a few cycles. In Fig. 8, the current through the secondary diode bridge is represented when the duty cycle changes from 0.77 to 0.97 in DCM and 0.11 to 0.13 in CCM while maintaining the input and output voltages.

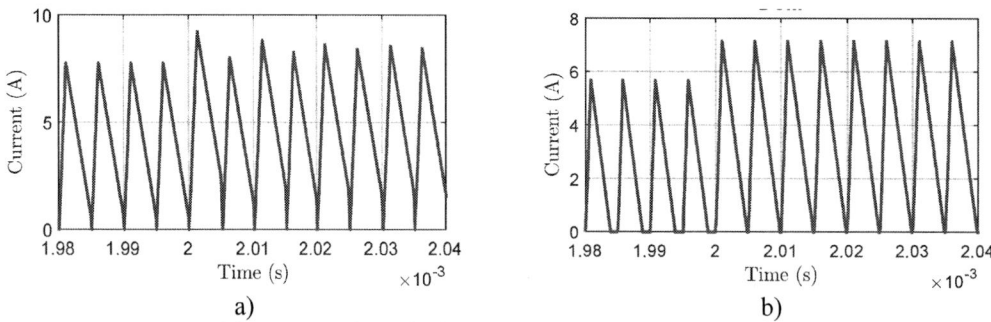

a)   b)

Fig. 8: Step response of the current through the secondary diode bridge. a) in CCM. b) in DCM.

Conclusion

This article provides a dynamic average small signal model of the SAB converter. With the obtained results it is possible to assume a first order model for either operation mode DCM or CCM. Linearized small signal models are presented for DCM and CCM, as well as transfer functions between duty cycle and output voltage perturbations, and input and output voltages perturbations; both necessary to design a control loop. Parameters of these models are found to be different at different sides of the border between CCM and DCM. This is not a surprising fact, as it happens in the case of other converters.

References

[1] R. W. A. A. De Doncker, D. M. Divan, and M. H. Kheraluwala, "A three-phase soft-switched high-power-density DC/DC converter for high-power applications," *IEEE Transactions on Industry Applications*, vol. 27, no. 1, pp. 63–73, Jan. 1991, doi: 10.1109/28.67533.

[2] M. N. Kheraluwala, R. W. Gascoigne, D. M. Divan, and E. D. Baumann, "Performance characterization of a high-power dual active bridge DC-to-DC converter," *IEEE Transactions on Industry Applications*, vol. 28, no. 6, pp. 1294–1301, Nov. 1992, doi: 10.1109/28.175280.

[3] A. R. Rodríguez Alonso, J. Sebastian, D. G. Lamar, M. M. Hernando, and A. Vazquez, "An overall study of a Dual Active Bridge for bidirectional DC/DC conversion," in *2010 IEEE Energy Conversion Congress and Exposition*, Sep. 2010, pp. 1129–1135. doi: 10.1109/ECCE.2010.5617847.

[4] A. Rodríguez, A. Vázquez, D. G. Lamar, M. M. Hernando, and J. Sebastián, "Different Purpose Design Strategies and Techniques to Improve the Performance of a Dual Active Bridge With Phase-Shift Control," *IEEE Transactions on Power Electronics*, vol. 30, no. 2, pp. 790–804, Feb. 2015, doi: 10.1109/TPEL.2014.2309853.

[5] F. J. Sebastián *et al.*, "Estudio estático completo del convertidor Single Active Bridge," *Seminario Anual de Automática, Electrónica industrial e Instrumentación (SAAEI)*, 2021, Accessed: Feb. 16, 2022. [Online]. Available: https://digibuo.uniovi.es/dspace/handle/10651/60273

[6] A. Rodriguez *et al.*, "An Overall Analysis of the Static Characteristics of the Single Active Bridge Converter," *Electronics*, vol. 11, no. 4, Art. no. 4, Jan. 2022, doi: 10.3390/electronics11040601.

[7] P. R. K. Chetty, "Current Injected Equivalent Circuit Approach to Modeling of Switching DC-DC Converters in Discontinuous Inductor Conduction Mode," *IEEE Transactions on Industrial Electronics*, vol. IE-29, no. 3, pp. 230–234, Aug. 1982, doi: 10.1109/TIE.1982.356670.

[8] P. R. K. Chetty, "CIECA: Application to Current Programmed Switching Dc-Dc Converters," *IEEE Transactions on Aerospace and Electronic Systems*, vol. AES-18, no. 5, pp. 538–544, Sep. 1982, doi: 10.1109/TAES.1982.309266.

[9] S. Cuk and R. D. Middlebrook, "A general unified approach to modelling switching DC-tO-DC converters in discontinuous conduction mode," in *1977 IEEE Power Electronics Specialists Conference*, Jun. 1977, pp. 36–57. doi: 10.1109/PESC.1977.7070802.

Algorithm for optimal selection of drive motor transmission combination

Santiago Ramos Garces[*1], Dries Jacques[*1], Stijn Derammelaere[1], Simon Houwen[2],
Nick Van Oosterwyck[1], Bart Vanwalleghem[2]
[1] Department of Electromechanics, CoSysLab, University of Antwerp
Groenenborgerlaan 171, Antwerp 2020, Belgium
[2] Department of Electromechanical, Systems and Metal Engineering, Ghent University
Graaf Karel de Goedelaan 5, Kortrijk 8500, Belgium
Phone: +32 (0) 32653088
[*] Email: Santiago.RamosGarces@uantwerpen.be, Dries.Jacques@uantwerpen.be

Keywords

≪Computation Cost≫, ≪Drive≫, ≪Load torque≫, ≪Optimization≫, ≪PMSM≫

Abstract

For machine builders, it is essential to have an optimal balance between economic benefit and fulfillment of the technical requirements when selecting an appropriate driveline. This paper proposes a practical tool for finding the optimal drive-motor-transmission combination from a database of available devices. The algorithm denoted as the optimal selection tool consists of four main steps. The three first steps are dedicated to efficiently finding the triplets that meet the technical requirements and feasibilities (such as current-, torque-, speed- and ratio-limitations). The last step implements the branch and bound algorithm [1] to find the optimal solution in terms of price. The algorithm avoids unnecessary checks of not technically feasible combinations. This is possible because the motors and drives are sorted with ascending peak current and the transmissions with an ascending ratio in the database. The algorithm's utility is demonstrated in a validation example of a transfer robot (pick-and-place unit), where a drive-motor-transmission combination with the lowest possible price must be selected. This combination must fulfill the technical requirements. Because of the optimization, the possible 3276 combinations were reduced to 302 technically feasible combinations, which means a significant reduction of 90%. Furthermore, this reduction implies a decrease in computational load. After that, the fourth step selects the combination with the lowest price, which is 1668 units. This price is five times lower than the most expensive combination (8408 units).

Nomenclature

Parameter	Drive	Motor	Transmission
Symbol	d	m	g
Population size	r	h	k
Index	j	w	q
Cost	C_d	C_m	C_g
Peak current	I_{dp}	I_{mp}	-
Standstill current	I_{d0}	I_{m0}	-
Inertia	-	J_m	J_g
Peak torque	-	T_{mp}	-
Nominal torque	-	$T_{m_{nom}}$	-
Peak speed	-	n_{mp}	-
Ratio	-	-	i
Nominal speed	-	$n_{m_{nom}}$	-

- T_{lp} Peak load torque
- n_{lp} Peak load speed
- $J_{l_{max}}$ Maximum load inertia
- J_l Total load inertia
- i_{min} Minimum gear ratio
- i_{max} Maximum gear ratio
- T_{lv} Load torque reflected to the rotor, including the transmission
- n_{lv} Load speed reflected to the rotor, including the transmission
- $T_{lv_{RMS}}$ Root Means Square load torque reflected to the rotor, including the transmission
- N Number of samples
- b Best combination index

Introduction

One complex task in designing electric drivelines is selecting a proper drive-motor-transmission combination that allows driving the load with a specific motion profile, demanded torque, and speed. In addition, machine builders are also interested in solutions that contribute to economic benefits like price, weight, or energy consumption. Indeed, the authors in [2] proposed a method to find the best motor-transmission combination minimizing the power losses of the motor. In [3], the optimal combination for an exoskeleton is found by minimizing the power losses after eliminating non-feasible combinations based on motor and load characteristics. For machines where the component's weight plays an essential role, like an exoskeleton, an optimization can be performed under constraints in the weight of the motor-transmission set as proposed in [4].

Another interesting optimization is the determination of the optimal gear ratio for achieving maximum acceleration, as suggested in [5], [6], and [7]. Authors in [5] based the selection on price, where the best solution out of two motors for an industrial application is found. In [6], a normalized technique to compare different types of motors is presented. Then, the drive's current is minimized to get the combination with the lowest cost. The method presented in [7] minimizes the RMS torque by estimating the effect of the transmission on the other parameters like inertia and motor torque.

On the other hand, this paper aims to provide a tool for machine designers that finds the optimal combination of drive-motor-transmission from a database of commercially available devices that allows driving specific load inertia with a particular motion profile considering all elements simultaneously. Computational efficiency is essential when the available devices in the database increase. On that account, a procedure consisting of four steps is proposed. The three first steps are dedicated to find feasible triplets efficiently. Then, the last step consists of an optimization routine that minimizes or maximizes a specific cost function based on the so-called branch and bound algorithm [1].

Optimal selection tool

The optimal selection tool presented in this paper aims to find a suitable combination of drive, motor, and transmission. Therefore, the peak load torque (T_{lp}) and peak load speed (n_{lp}) are needed. In addition, a database with commercial drives, motors, and transmissions is used. The selection of a proper combination of drive-motor-transmission is challenging because it relies on finding an economical solution that can meet the application motion profile's load torque/speed requirements. Moreover, let r, h, and k be the drives', motor's, and transmission's population size. Then, the number of possible combinations is $(r \cdot h \cdot k)$, leading to a high computational effort to consider all these combinations. Therefore, the optimal selection tool consists of four main steps that reduce the possible combinations to a set of practical triplets and select the one that minimizes a specific objective function. These four steps are performed after sorting the motors and drives with ascending peak current while the transmissions are ordered by the ascending ratio (i). A summary of the optimal selection tool is depicted in the flow diagram in Fig. 1. Furthermore, the MATLAB® code is available upon request to the authors.

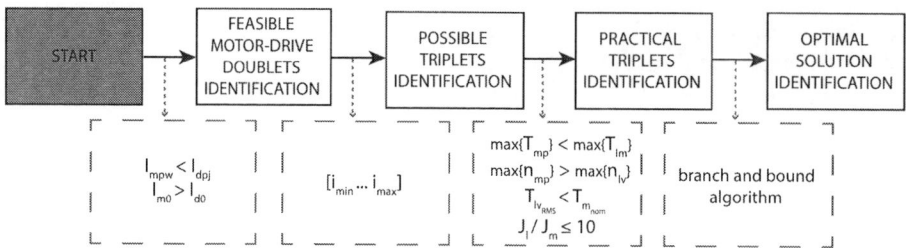

Fig. 1: Flow diagram of the optimal selection tool

Feasible drive-motor doublets identification

Not all drive-motor doublets are feasible from a technical point of view. One of the main reasons is that a driver has a specific current range delivered to the motor. As a consequence, the drive is not able to power all motors available in the database. To check which drive-motor doublets are feasible, the first step of the procedure takes advantage of the fact that the drives and motors are sorted from the lowest to the highest peak current (I_{dp}, I_{mp}).

This step is labeled as the feasible drive-motor doublets identification. The drive and motor indexes are represented as j and w. Hence, the algorithm starts selecting the motor with the highest peak current on the list ($I_{mpw}, w = h$) and compares it with all the drives starting from the one with the highest peak current ($I_{dpj}, j = r$), and decrements the drive selection j until the condition $I_{mpw} > I_{dpj}$ is met, as depicted in Fig. 2. The latter means neither the drive (j) can power the motor (w) nor can the remaining drives ($j-1, j-2, \cdots, 1$) in the list because they have a lower peak current. In that way, the number of checks is minimal. The algorithm does the same for all list's motors ($w = h$ to 1), and the output is a list of feasible drive-motor doublets. Furthermore, the standstill current of the motor (I_{m0}) must be higher than the standstill permissible current of the drive (I_{d0}). This condition is checked for the identified doublets.

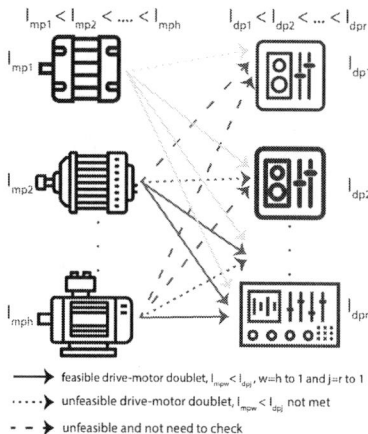

Fig. 2: Feasible drive-motor doublets identification

A reduction of combinations is achieved through the doublets identification step that only considers the drives and motors. However, including transmissions in the tool requires checking each doublet and database's transmission, leading to a list of triplets.

Possible triplets identification

The remaining steps of the optimal selection algorithm require the explanation of some functions and variables that are employed. First, the load torque and speed are defined by a trajectory with multiple samples (N) taken at a specific sample instant (t). They are defined as ($T_{lp}[t]$), and ($n_{lp}[t]$), where $t = 1, 2, \cdots, N$. These quantities can be defined graphically through the (N) samples ($n_{lp}[t], T_{lp}[t]$) represented by the red curve depicted in Fig. 3 (left). Furthermore, the torque-speed curves for commercial motors are available in the datasheet or provided by the manufacturer. In this paper, the peak and nominal torque-speed curves are relevant. The former characterizes the motor's peak capacity, and the latter is related to the nominal motor's operation. There are also many samples taken at time instant (u) for these curves, which size is not necessarily equal to (N). Thus, both curves are formed by the samples ($n_{mp}[u], T_{mp}[u]$) and ($n_{m_{nom}}[u], T_{m_{nom}}[u]$). The last function to define is the load torque after including the transmission, represented in (1):

$$T_{lv}[t] = \frac{T_{lp}[t]}{i} + (J_m + J_g) \cdot \ddot{\theta}[t] \cdot i, \tag{1}$$

where $\ddot{\theta}$ refers to the load acceleration. It can be seen from (1) that the expression is computed from $T_{lp}[t]$; therefore, it contains the same number of samples (N). In addition, the load torque after considering the transmission can be represented graphically in the same way as the load torque through the pairs ($n_{lv}[t], T_{lv}[t]$), where ($n_{lv}[t] = i \cdot n_{lp}[t]$) is the reflected load speed.

On the other hand, the second step of the optimal selection tool, the so-called possible triplets identification, consists of selecting the viable range of gear ratio for the feasible drive-motor doublets based on the load torque ($T_{lp}[t]$) and speed ($n_{lp}[t]$)). Then, only the transmissions inside the correct range are considered. The minimum gear ratio (i_{min}) and maximum ratio (i_{max}) are computed for every feasible drive-motor doublet. The minimal gear ratio (i_{min}) is obtained by dividing the maximum value of the peak load torque (T_{lp}) by the maximum motor peak torque (T_{mp}) as represented by (2)

$$i_{min} = \frac{max\{T_{lp}[t]\}}{max\{T_{mp}[u]\}} \tag{2}$$

The above means that the necessary ratio to reach the maximum peak load torque is given by i_{min}. Likewise, the maximum peak motor speed is divided by the maximum peak load speed to determine the maximum ratio (3), which can be interpreted as the highest transmission ratio that allows fulfilling the load speed.

$$i_{max} = \frac{max\{n_{mp}[u]\}}{max\{n_{lp}[t]\}} \tag{3}$$

The graphical description of the procedure is depicted in Fig. 3 (left). Furthermore, the algorithm searches into the database for the transmissions with the minimum and maximum ratio inside the range $[i_{min} \cdots i_{max}]$. Then, it assigns a starting and ending transmission index (q) for each motor belonging to the doublets. Consequently, the possible triplets are obtained by paring the doublets with the transmissions between the starting and ending index (q).

Practical triplets identification

The third step of the optimal selection tool is denoted as practical triplets identification. In addition, four conditions can be performed to reduce the number of possible triplets. The first condition consists of checking that the maximum value of the motor's peak torque ($max\{T_{mp}[u]\}$) is higher than the maximum value of the peak load torque reflected to the rotor after including the transmission ($max\{T_{lv}[t]\}$). The second condition states that after including the transmission inertia, the maximum value of the peak load speed ($max\{n_{lv}[t]\}$) must be lower than the maximum value of the peak motor's speed ($max\{n_{mp}[u]\}$). In the third condition, the Root Mean Square (RMS) reflected load torque ($T_{lv_{RMS}}$) must be lower than the nominal torque of the motor ($T_{m_{nom}}[u]$) at the mean load speed[1] ($\bar{n}_l$). Finally, the last condition verifies that the inertia ratio between the total load and the motor's rotor must be lower or equal to ten, as suggested

---

[1]Statistical mean computed as $\bar{n}_l = \frac{1}{N} \sum_{t=0}^{N} n_{lv}[t]$

in [8]. In this case, the total load inertia is composed of the maximum load's inertia reflected to the motor shaft and the transmission's inertia, as indicated in (4).

$$J_l = \frac{J_{l_{max}}}{i^2} + J_g \tag{4}$$

After checking the previous conditions, the practical triplets identification is performed for the resulting possible triplets. As shown in Fig. 3 (right), the algorithm checks that the load torque after including the transmission ($T_{lv}[t]$) is lower than the motor peak torque[2] ($T_{mp}[t]$) for all operation points or samples (N). If the condition is not met, the algorithm stops, and the triplet is discarded. In that way, the output is the practical triplets.

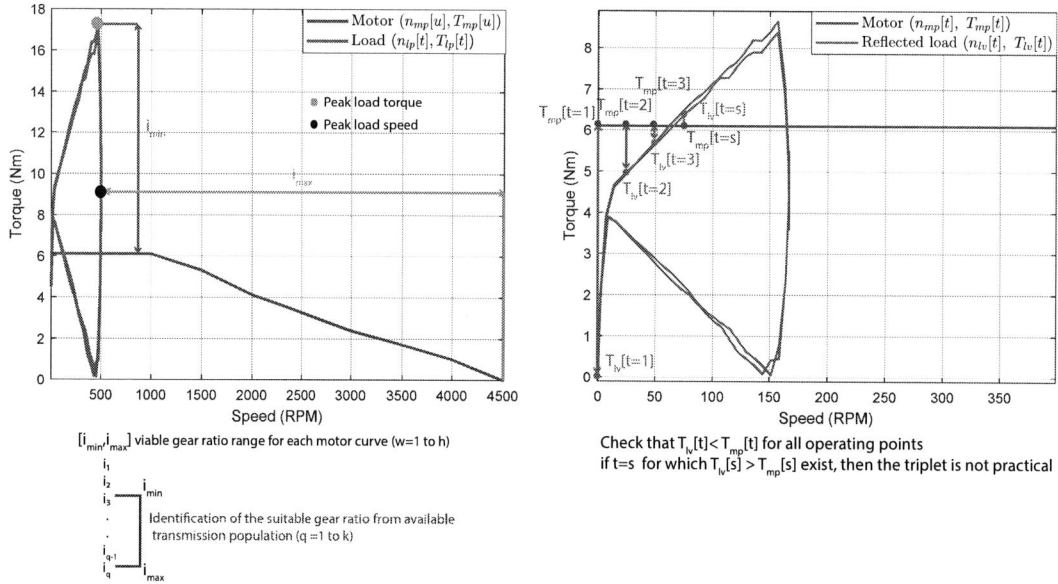

Fig. 3: Viable gear ratio for available motors (left). Practical triplets identification based on load torque (right)

Optimal solution identification

The optimal solution identification step aims to find the optimal triplet from the available practical triplets. For that, the objective function C_b must be minimized. This objective function can be computed in three different ways; only the drive's price, the combination of drive-motor prices, or even the addition of all combination's elements prices (drive-motor-transmission)[3]. On that account, from this point in this manuscript, the objective function C_b will be referred to as the actual best price. The so-called branch and bound algorithm [1] is used for solving this problem.

The mentioned algorithm searches efficiently for the triplet with the lowest price. The strategy is to discard as many unnecessary checks as possible. The way the algorithm achieves discarding combinations is to check progressively. The latter can be illustrated through some steps of the algorithm. One of the steps checks that the ongoing drives' price is lower than the price of the actual best combination. If the condition is not met, all triplets containing the current drive can be discarded, contributing to substantially reducing operations. On the contrary, the algorithm goes to the next step if the condition is met. Here, the price of the drive-motor combination is compared with the actual best combination. If

---

[2]$T_{mp}[t]$ is obtained from the linear interpolation of $T_{mp}[u]$

[3]The units of the objective function are arbitrary and not related to any currency, so it is only illustrative

the current drive-motor price exceeds the price of the actual best combination, all triplets containing the drive-motor can be discarded, also reducing operations. The complete flow diagram of the algorithm is depicted in Fig. 4.

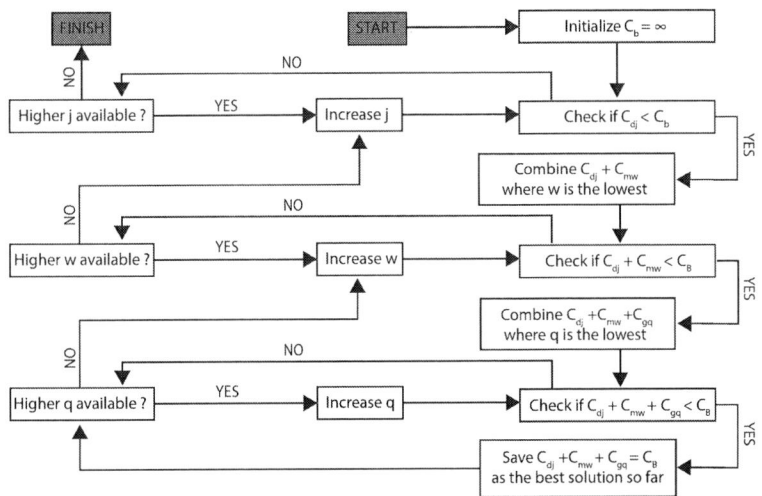

Fig. 4: Branch and bound algorithm for finding the optimal drive-motor-transmission triplet

Finally, it is relevant to mention that the objective function C_b can be extended to consider other relevant factors that machine builders are interested in optimizing, e.g., total losses, weight, and size, among others. For instance, the necessary modification to the optimal solution identification step is twofold. First, the relevant factors have to be normalized using their highest value. Second, the objective function C_b is built from the sum of those weighted factors, where the wights allow for prioritizing some factors over others.

Validation Case

This section aims to validate the functionality of the optimal selection tool. Therefore, a machine builder provides the load torque, speed, and acceleration of a transfer robot installed in a manufacturing cell performing repetitive movements to apply the proposed algorithm. The robot's task is to take an object from one position and place it on a conveyor belt on the other side of the production line. The manufacturer is interested in the drive-motor-transmission with the lowest possible price to drive a payload with the specific torque, speed, and acceleration profiles from Fig. 5. The algorithm is implemented in MATLAB®. In this case, a database with $r = 6$ drives, $h = 21$ motors, and $k = 26$ transmissions is used resulting in $(6 \cdot 21 \cdot 26 = 3276)$ possible combinations. The number of combinations at each step for the transfer robot is shown in Fig. 6. It is relevant to mention that the only motor technology considered in the validation case is the permanent magnet synchronous machine (PMSM).

For this example, the number of allowable drive-motor combinations is the product $(r \cdot h = 126)$. However, after applying the feasible drive-motor doublets identification step, feasible doubles are the 36 shown in Table I. Furthermore, the total number of possible triplets is obtained by multiplying feasible drive-motor doublets and the total number of transmissions, resulting in $(36 \cdot 26 = 936)$. On that account, after applying the first step of the algorithm, the total number of possible triplets $(6 \cdot 21 \cdot 26 = 3276)$ is reduced to around one-third. In addition, the possible triplets identification step is applied. As a result, the starting and ending index for the transmissions for each feasible drive-motor doublet can be observed in the fourth, fifth, ninth, tenth, fourteenth, and fifteenth columns from Table I.

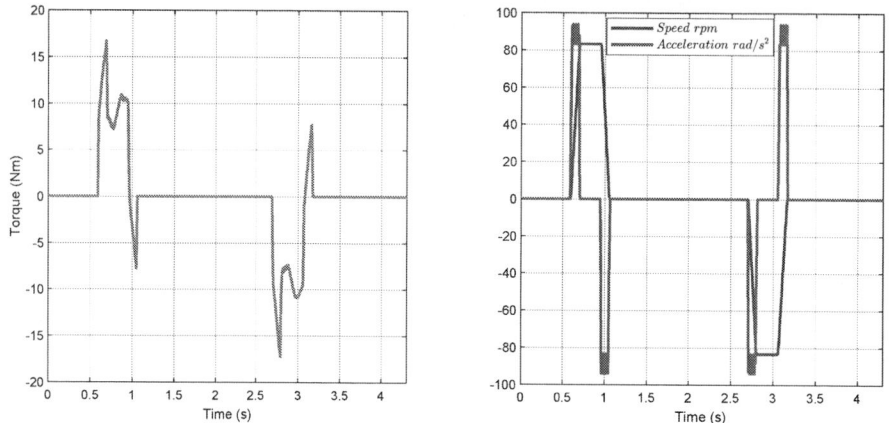

Fig. 5: Torque (left), speed and acceleration (right) required for the transfer robot

The interpretation of the latter can be explained with an example. Considering the left part of the first row from Table I, which corresponds to the first feasible drive-motor doublet, the triplets formed by considering only the transmission with index one until the transmission with index eleven are the possible triplets for the specific feasible doublet. The advantage of this step is clear at this point because for this feasible doublet, only 11 transmissions out of the 26 are worth checking, and the number of operations is reduced. Finally, the possible triplets for the transfer robot (650) are formed by adding all the feasible doublets from Table I.

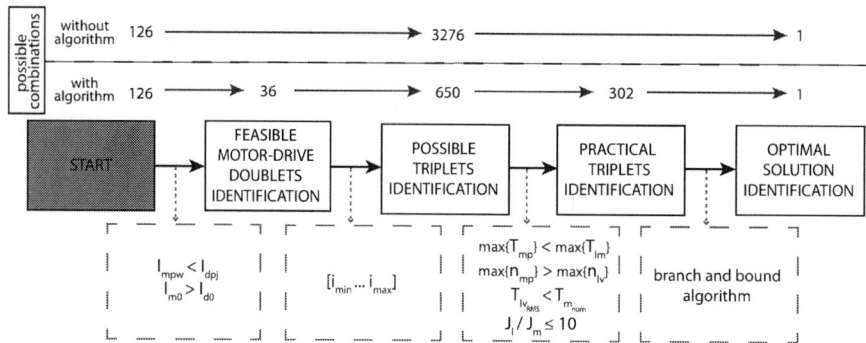

Fig. 6: Flow diagram of the optimal selection tool for the transfer robot including the number of combinations with and without algorithm

To continue with the practical triplets identification step, first the condition ($max\{T_{mp}[u]\} > max\{T_{lv}[t]\}$) is checked for the possible triplets, leading to (617) triplets. Furthermore, conditions ($max\{n_{mp}[u]\} > max\{n_{lv}[t]\}$) and (($\bar{n}_l, T_{m_{non}}) > (\bar{n}_l, T_{lv_{RMS}})$) are applied, resulting in 564 triplets fulfilling both requirements. Finally, the condition ($J_l/J_m \leq 10$) is verified. As a result, the number of possible triplets is reduced to 328. The above intermediate steps are justified after considering the number of operations that the optimal selection tool requires in the practical triplets identification step. In this case, the speed profile contains ($N = 1001$) samples, which in the case of the possible initial triplets takes ($1001 \cdot 650 = 650650$) checks as in Fig. 3 (right). Instead, if the possible triplets after applying the intermediate conditions are used, the number of necessary checks is $(1001 \cdot 328 = 328328)$, which saves (322322) operations. In this example, (26) triplets from the (328) possible drive-motor-transmission triplets do not fulfill the condition $T_{lv}[t] < T_{mp}[t]$ for all (N) samples, which implies that (302) are the practical triplets.

Table I: Doublets for the transfer robot

Index	j	w	S q [4]	E q [5]	Index	j	w	S q	E q	Index	j	w	S q	E q
1	6	4	1	11	13	5	5	12	23	25	5	1	11	24
2	6	7	1	23	14	1	13	2	26	26	1	14	2	21
3	1	6	1	26	15	5	13	2	26	27	5	14	2	21
4	1	2	2	24	16	1	20	2	26	28	1	11	8	23
5	1	17	1	21	17	5	20	2	26	29	5	11	8	23
6	1	16	2	26	18	1	19	2	26	30	1	10	7	26
7	5	16	2	26	19	5	19	2	26	31	5	10	7	26
8	1	3	2	23	20	1	9	2	26	32	4	10	7	26
9	5	3	2	23	21	5	9	2	26	33	1	21	10	24
10	1	12	4	23	22	1	18	2	24	34	4	21	10	24
11	5	12	4	23	23	5	18	2	24	35	1	15	12	26
12	1	5	12	23	24	1	1	11	24	36	4	15	12	26

Table II: Specification optimal solution (left). Specification second-best solution (right)

Optimal solution				Second-best solution			
Motor		**Drive**		**Motor**		**Drive**	
T_{mp}	2.67 Nm	I_{dp}	13 A	T_{mp}	6.36 Nm	I_{dp}	13 A
$T_{m_{nom}}$	0.5 Nm	I_{d0}	1 A	$T_{m_{nom}}$	1 Nm	I_{d0}	1 A
$n_{m_{nom}}$	9000 rpm	C_d	840 units	$n_{m_{nom}}$	8000 rpm	C_d	840 units
J_m	0.134 kgm^2	**Transmission**		J_m	0.373 kgm^2	**Transmission**	
I_{mp}	8.6 A	Ratio	40	I_{mp}	11.4 A	Ratio	40
I_{m0}	1.6 A	J_g	0.35 $kgcm^2$	I_{m0}	2.2 A	J_g	0.35 $kgcm^2$
C_m	320 units	C_g	710 units	C_m	382 units	C_g	710 units

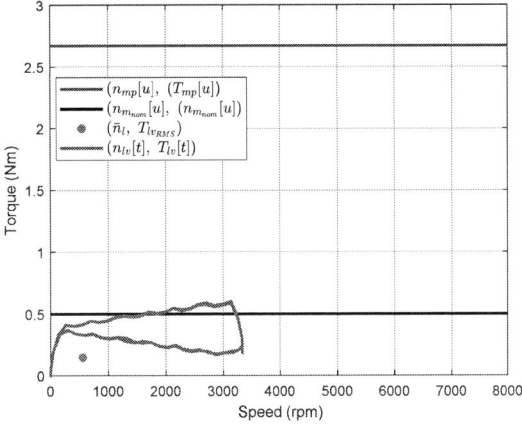

Fig. 7: Speed-torque curves for the optimal solution reflected load and motor

---

[4] Starting q index

[5] Ending q index

Finally, the optimal solution identification step allows finding the triplet with the lowest price of the 302 practical triplets. Instead of checking the 302 triplets, the necessary number of checked triplets is reduced to 191 based on the branch and bound algorithm described in the optimal solution identification step. The optimal solution parameters are indicated in Table II (left), and the comparison between the motor curves and the application curve for the optimal solution is depicted in Fig. 7. The magenta point represents $(\bar{n}_l, T_{lv_{RMS}})$ for the optimal solution. The second-best solution information is presented in Table II (right) to compare both solutions. It can be observed that the drive and transmission are the same, but the motor is different. Therefore, the motor is the critical element because the second-best triplet uses a motor with higher torque which is more expensive. Consequently, the price leads to a difference of around 60 units related to the best solution.

Conclusion

The optimization tool presented in this paper identifies the most cost-efficient drive-motor-transmission triplet for machine builders when an extensive database of devices is available. The first three steps of the algorithm reduce the initial number combinations $(r \cdot h \cdot k)$ to a limited number of practical triplets, as demonstrated with the validation case where the initial 3276 combinations were reduced to only 302 practical triplets. In addition, the optimization algorithm found the optimal solution efficiently instead of checking all the 302 practical triplets. Finally, to outline the scope of the optimal selection tool, it is relevant to compare the price of the best and the most expensive practical triplet. The drive, motor and transmission prices of the most expensive triplet are 3328, 2190, and 2890 units, respectively. The total price is 8408 units compared to the optimal triplet total price of 1668 units. Therefore, it is clear that even though both solutions are technically feasible, the optimal selection tool identifies a solution with a saving factor of five.

References

[1] Narendra P.: A branch and bound algorithm for feature subset selection, 1997 IEEE Transactions on Computers Vol 26 no 9, pp. 917- 922

[2] Dresscher D.: Motor-gearbox selection for energy efficiency, 2016 IEEE International Conference on Advance Intelligent Mechatronic AIM, pp. 669- 675

[3] Aftab Z.: Systematic method for selection of motor-reducer units to power a lower-body robotic exoskeleton, 2021 Journal of Applied Science and Engineering Vol 24, pp. 457- 465

[4] Barjuei E. P.: Optimal selection of motors and transmissions in back-support exoskeleton applications, 2020 IEEE Transactions on Medical Robotics and Bionics Vol 2 no 3, pp. 320- 330

[5] Giberti H.: A practical approach to the selection of the motor-reducer unit in electric drive systems, 2011 Mechanics Based Design of Structures and Machines Vol 39 no 3, pp. 303- 319

[6] Van de Straete H.: An efficient procedure for checking performance limits in servo drive selection and optimization, 1999 IEEE/ASME Transactions on Mechatronics Vol 4 no 4, pp. 378- 386

[7] Richiedei D.: Integrated selection of gearbox, gear ratio, and motor through scaling rules, 2018 Mechanics Based Design of Structures and Machines Vol 46 no 6, pp. 712- 729

[8] Voss, Wilfried.: A comprehensible guide to servo motor sizing, Copperhill Media, 2007.

Evaluation of Drain-Source Voltage in Switch Transient Time Intervals as Gate Oxide Degradation Precursor of SiC Power MOSFETs

Javad Naghibi[1], Sadegh Mohsenzade[2], Kamyar Mehran[1], and Martin P. Foster[3]

1 School of Electronics Engineering and Computer Science, Queen Mary University of London, London, UK.
2 Electrical Engineering Department, K. N. Toosi University of Technology, Tehran, Iran.
3 Department of Electronic and Electrical Engineering, The University of Sheffield, Sheffield, UK

Emails: *s.naghibinasab@qmul.ac.uk; s.mohsenzade@kntu.ac.ir; k.mehran@qmul.ac.uk;*
m.p.foster@sheffield.ac.uk

Keywords

«Degradation», «MOSFET», «Reliability», «Silicon Carbide (SiC)».

Abstract

Gate oxide degradation is a major chip-related reliability issue in Silicon Carbide power MOSFETs. Being focused on turn-on/-off transient behavior of the switch, drain-source voltage waveform is employed as a gate oxide degradation precursor in this paper. Precursor evaluation is carried out in various operating conditions of the switch.

Introduction

Silicon Carbide (SiC) metal–oxide–semiconductor field-effect transistors (MOSFETs) present low conduction power loss, high temperature endurability, and high frequency operation capabilities. These features pave the way for high-density power electronic applications, including electric vehicle charger, electrified aircraft systems, and DC/DC converters [1–4]. Assuring reliable performance is a major requirement in all of the above-mentioned application areas [5], [6]. SiC MOSFET reliability issues have been categorised into two main groups: Package-level (wire bond and solder layer failure modes) and chip-level (gate oxide layer and body diode failure modes) [1]. The focus of this paper is on gate oxide degradation modes.

Tunneling current into the gate oxide layer leads to gate oxide degradation [7]. Due to thin gate-oxide layer in SiC MOSFETs, a Fowler–Nordheim tunneling can inject additional electrons into oxide layer in high electric conditions [8]. Comparing SiC- and Silicon (Si)-MOSFETs, interface trapped charge at SiC-SiO$_2$ interface is much higher than Si counterpart. As a consequence, the device mobility is decreased by high coulombic scattering and few numbers of free carriers [9]. A total positive charge created by oxide and interface trapped charges result in a negative shift of the threshold voltage (V_{th}) and vice versa. The effect of trapped charges on V_{th} depends on the distance between trapped charges in the oxide and SiC conduction channel [8].

To detect the gate oxide degradation level in SiC MOSFETs, V_{th} [10], drain leakage current [11], gate leakage current [5], gate Miller plateau voltage (V_{GP}) [12], gate Miller plateau time (t_{GP}) [7], switch turn-on delay [13], switch junction capacitance [8], and on-state resistance (R_{DS-on}) [14] have been employed as condition monitoring (CM) precursors in the literature. The mentioned CM precursors are generally adopted from the conventional Si-based MOSFETs [7]. However, some of the above-mentioned precursors behave differently during the SiC MOSFET gate oxide degradation process in comparison to the ones of Si MOSFETs. For example, in transitioning from brand-new condition to degraded condition in Si MOSFETs, there is a rebound in t_{GP} value [15], while t_{GP} is a strictly increasing function of gate oxide degradation level in SiC MOSFETs [7]. One other problem in the process of adaption of Si MOSFET CM precursors for SiC MOSFETs is that SiC MOSFET operational parameters, such as R_{DS-on}, on-state voltage drop (V_{DS-on}), gate input capacitor (C_{rss}), total gate charge (Q_g), diode reverse recovery time, t_{GP}, etc., are considerably smaller in value in comparison to those of Si-based MOSFETs and IGBTs. Therefore, specialized characterization of the proposed gate oxide CM precursor is required for SiC MOSFETs.

In the process of developing a precursor for CM purposes, it is important to examine the proposed CM precursor in various circuit configurations and conditions. Because the CM precursor should be able to monitor the gate oxide health status in real circuit conditions and without interrupting the switch normal performance [16].

The presented CM precursors in the literature are usually affected by the operational conditions of the switch. For example, increasing switch junction temperature (T_j) leads to a negative shift in V_{th} value [3]. Besides, the transient behavior of the SiC MOSFET in inductive loading condition is different in comparison to the transient behavior in resistive loading condition. Considering this, examining the proposed CM precursor in various switch conditions and circuit configurations is a fundamental requirement in the process of developing a CM precursor for gate oxide degradation.

In this paper, the effect of gate oxide degradation on switch turn-on and -off transient behavior is studied and characterized using drain-source voltage (V_{DS}) waveform. Using V_{DS} waveform, V_{GP} and t_{GP}, as the two main parameters of Miller plateau, are obtained for brand-new and degraded switch conditions. Besides, the changes in switch transient behavior under different circuit and switch conditions are studied. To achieve more comprehensiveness:

- Using the developed degradation set-up based on high electric field stress (HEFS) mechanism, various rates of gate oxide degradation is applied to the switch, and the transient behavior is studied in different rates of degradation.

- Switch transient behavior is evaluated for both resistive and inductive loading conditions on the switch. This basically enables the studies of this paper to be used for resistive loads (such as pulsed power applications) and inductive loads (such as DC/DC converters and resonant converters).

- The evaluations are carried out for normal and high T_j levels of the switch.

Gate Oxide Degradation in SiC MOSFETs

V_{GP} of the switch in both turn-on and -off transient time intervals of the switch is depended on V_{th} and can be written as (1) [17].

$$V_{GP} = V_{th} + \sqrt{\frac{I_D L_{CH}}{\mu C_{ox} Z}} \tag{1}$$

, where I_D is the drain channel current, L_{CH} is the channel length, C_{ox} is the specific gate oxide capacitance, Z is the channel width, and μ is the channel carrier mobility. Since μ is a decreasing function with respect to the applied gate oxide stress time (t_{stress}) [18], V_{GP} is an increasing function of t_{stress}.

In the turn-on process, V_{DS} falls from the off-state blocking voltage (V_{Bus}) to V_{DS-on}. The main part of the V_{DS} decrement process occurs when V_{GS} is constant, and the output current of the driver discharges the gate-drain capacitance (C_{GD}). This time interval is known as Miller plateau. Assuming V_{DS} falls linearly from V_{Bus} to V_{DS-on} during switch turn-on process, Miller time in turn-on transient time of the switch (t_{GP-on}) can be written as (2).

$$t_{GP-on} = R_G C_{GD,avg} \frac{V_{Bus} - V_{DS-on}}{V_{Dr} - V_{GP}} \tag{2}$$

, where R_G and V_{Dr} are the total gate resistance and the applied voltage of the gate driver respectively. Since V_{GP} increases due to gate oxide degradation (according to (1) and Figure.1(a)), it can be concluded that t_{GP-on} also increases over the degradation process. As a result, $d(V_{DS})/dt$ in turn-on process of the switch is decreased during the degradation, as shown in Figure.1(b).

In the turn-off process, V_{DS} increases from V_{DS-on} to V_{Bus}. The same as the turn-on process, the major part of the changes in V_{DS} occurs in the Miller plateau time interval. In this time interval, V_{GS} is approximately constant, and the driver current charges C_{DG}. On this basis, Miller time in turn-on transient time of the switch (t_{GP-off}) is described as (3).

$$t_{GP-off} = R_G C_{GD,avg} \frac{V_{Bus}}{V_{GP}} \tag{3}$$

Regarding (1), V_{GP} increases over the gate oxide degradation (see Figure.1(c)), and t_{GP-off} decreases in the degradation process as the result.

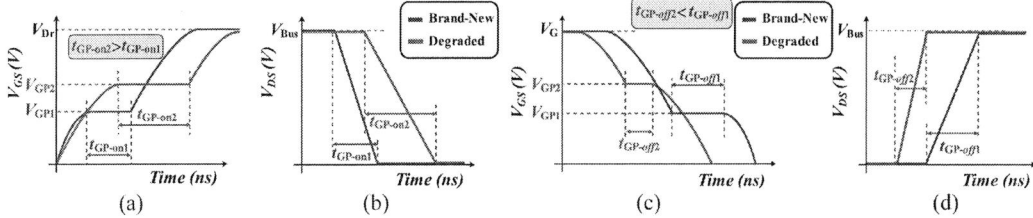

Figure 1: (a) Increment in V_{GP} and t_{GP}, and (b) decrement in $d(V_{DS})/dt$ in turn-on process during degradation; (c) Increment in V_{GP} and decrement in t_{GP-off}, and (d) increment in $d(V_{DS})/dt$ in turn-off process during degradation

Experimental Set-up

Commercially available SiC discrete MOSFETs 650V/22A are used to examine transient changes over the gate oxide degradation process. The switch is employed in a power circuit as shown in Figure.2(a). In Table.1, the circuit specifications and parameters of the power circuit are listed. $R_L = 10\ \Omega$ is considered for the resistive load tests. $R_L = 3.4\ \Omega$ and $L_L = 220\ \mu H$ is considered for the inductive load tests. $R_{G_{ext}}$ is the external gate resistance (see Figure.2(a)). In Figure.2(b), the laboratory set-up, consisting of the direct switch structure, resistive load and inductive load, is shown.

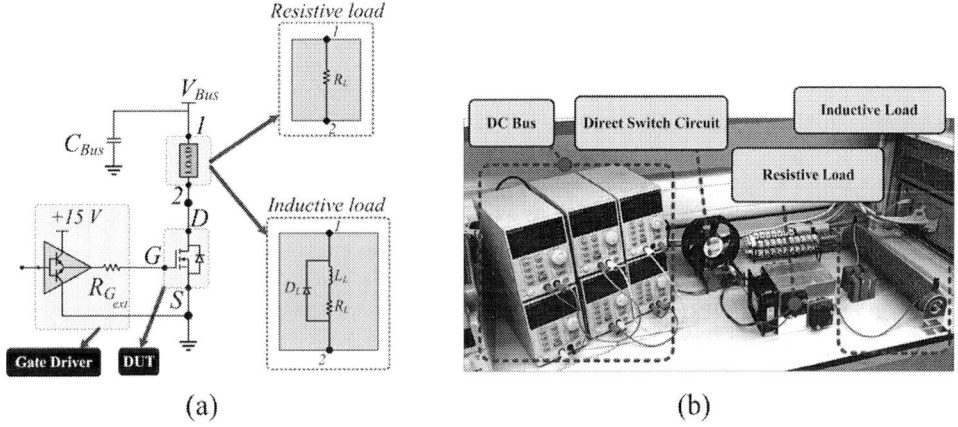

Figure 2: (a) Detailed view of the employed power circuit in the experimental set-up, and (b) overall view of the implemented laboratory set-up

Table 1: The overall specifications and ratings of the implemented power circuit

V_{Bus}	93 V to 232V
C_{Bus}	34 μF
f_s	200 kHz
D	10%
$R_{G_{ext}}$	10 Ω

Gate Oxide Degradation Rate Effect

Two stressors are introduced for the accelerated gate oxide degradation testings, i.e., high electric field stress (HEFS) and high temperature stress [1], [7]. In Figure.3(a), the employed degradation circuit based on HEFS mechanism is shown. Using different values of gate stressor voltage (V_{stress}), Miller plateau changes due to the gate oxide degradation is examined. The gate oxide breakdown voltage of the SiC MOSFET case study switch is found as 39 V. In [19], it is shown that for a Si MOSFET with similar gate structure and similar ratings, the breakdown voltage is 65 V which clearly demonstrates the vulnerability of SiC MOSFET gate oxide layer against HEFS.

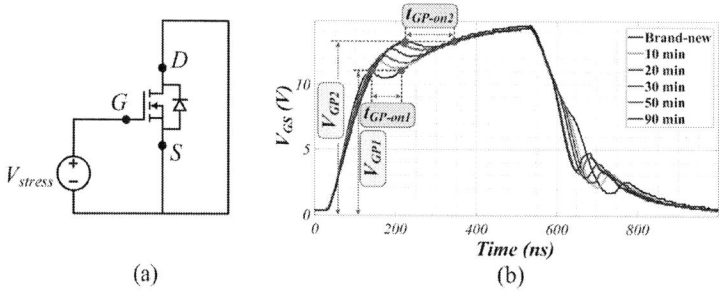

(a) (b)

Figure 3: (a) HEFS circuit for gate oxide degradation, and (b)V_{GP} and t_{GP-on} changes during the gate oxide degradation tests with V_{stress} = 37V

In Figure.3(b), the effect of degradation on V_{GP} and t_{GP-on} of the switch is shown in V_{stress} = 37V and stress time of 90min. It can be seen that V_{GP} value is changed from V_{GP1}=11.9V to V_{GP2}=13.4V due to gate oxide degradation. Moreover, in the turn-on transient, t_{GP-on} has experienced a value change from 68ns to 114ns.

Based on the gate breakdown voltage of the switch, V_{stress} is chosen as 31V and 37V for the experimental evaluation of the proposed CM technique. In Figure.4 and Figure.5, the overall changes in the transient behavior of the switch are shown for V_{stress} = 31V and V_{stress} = 37V respectively. The gate oxide complete failure occurs at t_{stress} = 93min with V_{stress} = 37V, while it occurs at t_{stress} = 566min with V_{stress} = 31V . It can be inferred that although the rate of gate oxide degradation is different, V_{DS} rise and fall time intervals are changed similarly in both the experiments.

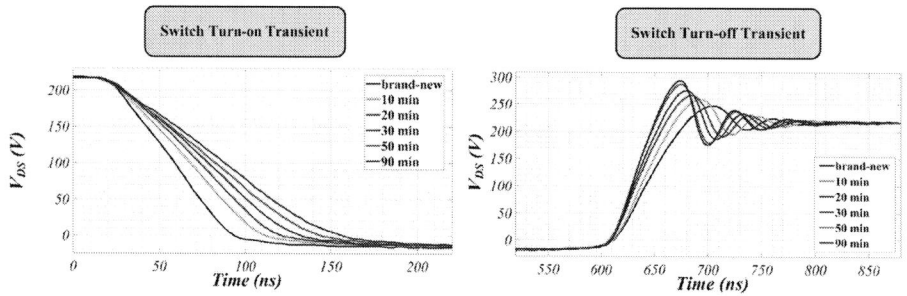

Figure 4: V_{DS} rise time and fall time changes during the gate oxide degradation tests with V_{stress} =37V and V_{Bus}= 217V

Load Type Effect on Switch Transient Behavior

The circuit diagrams of generic resistive and inductive loads are depicted in Figure.2(a). The existence of the free-wheeling diode (D_L) and uninterrupted current of the inductor in the turn-on and turn-off intervals cause I_D to be different in the mentioned transient times.

In the turn-on process for inductive load, D_L conducts the whole of the load current. When V_{DS} decreases from V_{Bus} to V_{DS-on}, D_L is reversed-biased, and the load current is commutated to the switch drain-source terminals. Therefore, in the turn-on process, I_D is constant while V_{DS} decays from V_{Bus} to V_{DS-on} (see Figure.6(a)). Accordingly, (1), (2), and (3) are correct for the inductive load without any approximation. It is because V_{GP} (see

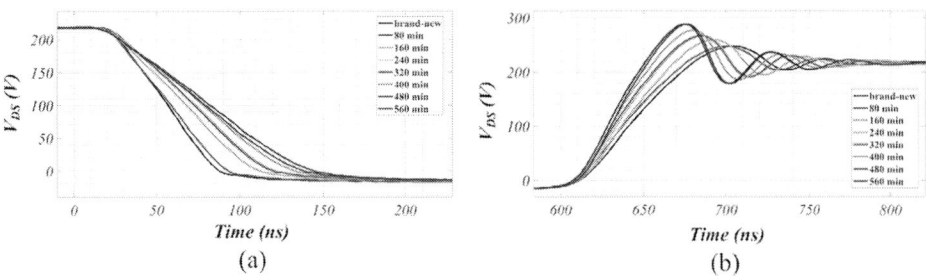

Figure 5: V_{DS} rise time and fall time changes during the gate oxide degradation tests with V_{stress} =31V and V_{Bus}= 217V

(1)) is constant during the turn-on process. For the resistive load case, changes in I_D occur simultaneously with V_{DS} changes, and the sudden change in the load current is not detectable as presented in Figure.6(a). On this basis, according to (1), V_{GP} is not constant, and it changes dynamically during the turn-on process. Based on (1), at the beginning of the switch turn-on process, V_{GP} approximately equals to V_{th}, and when the turn-on transient time is terminated, it has its maximum value. Therefore, for the resistive load cases, (1), (2), and (3) can be used with some approximations.

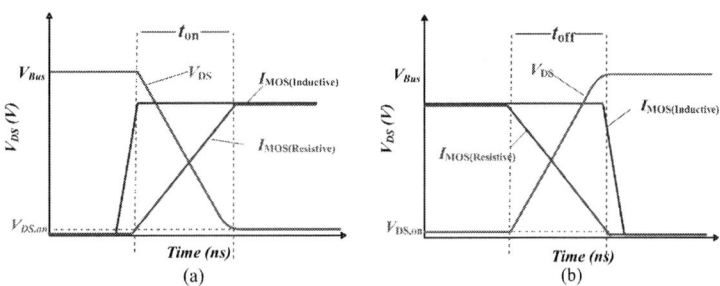

Figure 6: V_{DS} and I_D waveforms of the switch in the resistive and inductive loads; (a) turn-on and (b) turn-off transient behavior

In the turn-off process, V_{DS} increases from V_{DS-on} to V_{Bus}. I_D is constant, since D_L is reverse-biased during this time interval. When V_{DS} reaches to V_{Bus}, D_L starts conducting. Hence, at the end of the turn-off interval, the inductor current is commutated to the D_L path. Accordingly, it can be concluded that during turn-off time interval, V_{GP} is constant, thus (1), (2), and (3) can be served without approximation.

The developed gate oxide degradation set-up is employed to degrade the gate oxide layer with V_{stress} = 37V. The variations in turn-on and turn-off transients of the switch due to the gate oxide degradation in V_{Bus} = 217 V are shown in Figure.4 in power circuit with resistive load. The variations in turn-on and turn-off transients of the switch due to the gate oxide degradation in V_{Bus} = 232 V are shown in Figure.7 in power circuit with inductive load. At the beginning of the turn-on transient time in inductive load tests, V_{DS} fall time shows a slowed-down behavior, which is due to the power path parasitic inductance out of the free-whiling diode path. This inductance acts similar to an on-state snubber and does not allow I_D of the switch to have an instant incremental change. Accordingly, at the beginning of the turn-on process, the current of the parasitic inductance starts increasing. Its voltage is proportional to the integration of its current and approximately has a second order profile. Thus, V_{DS} will have two slopes at the beginning, if parasitic inductance values of the path are considerable. In Figure.7, during the transitioning from brand-new condition to fully degraded gate oxide, the rise time value of the switch has changed from 73 ns to 108 ns and fall time has changed from 54 ns to 45 ns.

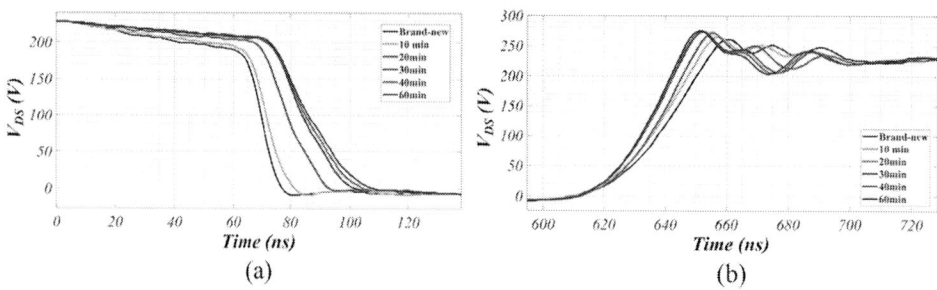

(a)   (b)

Figure 7: V_{DS} changes during gate oxide degradation test with V_{stress} = 37 V and V_{Bus} = 232V in inductive load; (a) turn-on transient and (b) turn-off transient

Junction temperature effect on Miller plateau

To investigate the effect of T_j on V_{DS} transient time intervals, two different heat dissipation topologies are implemented. The case study switch is examined both in the brand-new and degraded conditions in the two heat dissipation topologies. Using the thermal equivalent circuit model for the structure of switch and heatsink, T_j, switch case temperature (T_c), heatsink temperature (T_{HS}), and ambient temperature (T_{Am}) can be modeled as shown in Figure.8. On this basis, T_c and T_j can be obtained using (4) and (5).

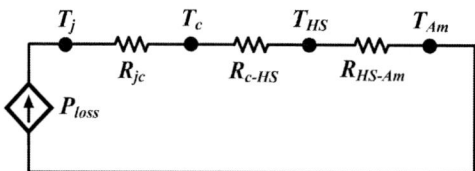

Figure 8: Thermal equivalent circuit of the switch and heatsink configuration

$$P_{loss} = \frac{T_c - T_{Am}}{R_{c-HS} + R_{HS-Am}} \tag{4}$$

$$P_{loss} = \frac{T_j - T_c}{R_{jc}} \tag{5}$$

, where P_{loss} is switch total power loss. Moreover, R_{c-HS}, R_{HS-Am}, and R_{jc} are thermal resistance from case to heatsink, thermal resistance from heatsink to ambient, and thermal resistance from junction to case of the switch respectively. Based on the power circuit specification, which is listed in Table.1, P_{loss} = 7.41 W.

A) Normal junction temperature configuration: In this configuration, a forced air-cooled heatsink with R_{HS-Am} = 0.07 K/W is employed. A thermal pad with thermal conductivity of 8.2 W/m.K is used as the thermal interface between the switch and the heatsink. In Figure.9(a), the mounted case study switch on the heatsink is shown. Using (4) and (5), it is calculated that T_c = 34°C and T_j = 45°C. T_c measurement is carried out using thermal camera (model: FLIR ONE PRO LT), and the captured thermal image is shown in Figure.9(c).

B) High junction temperature configuration: In this configuration, a heatsink with relatively large value thermal resistance parameter is used to achieve a high temperature profile of the switch. No air cooling is considered, and the heatsink thermal resistance is R_{HS-Am} = 15.2 K/W. Similar thermal pad is employed as the thermal interface between the switch and the heatsink. In Figure.9(b), the switch and the small-size heatsink are shown. In this mode, using (4) and (5), it is calculated that T_c = 146°C and T_j = 157°C. The thermal image of the switch in this thermal configuration is shown in Figure.9(d).

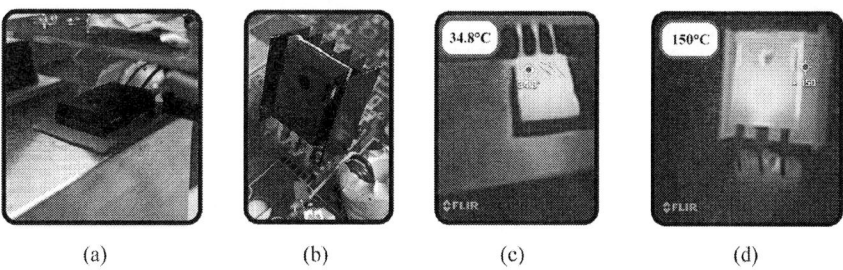

(a)   (b)   (c)   (d)

Figure 9: Heatsink and switch configurations of (a) normal temperature design and (b) high temperature design; Thermal images of the switch during operation in (c) normal temperature configuration and (d) high temperature configuration

For the brand-new switch, the results are shown in Figure.10. As can be seen, the rise time and fall time changes are 10ns and 2ns respectively. For the degraded switch, the rise time and fall time changes are 4ns and 1ns respectively, as shown in Figure.11. It can be inferred that changes in V_{DS} rise and fall time values due to temperature rise are considerably smaller than the changes in rise and fall time values due to gate oxide degradation.

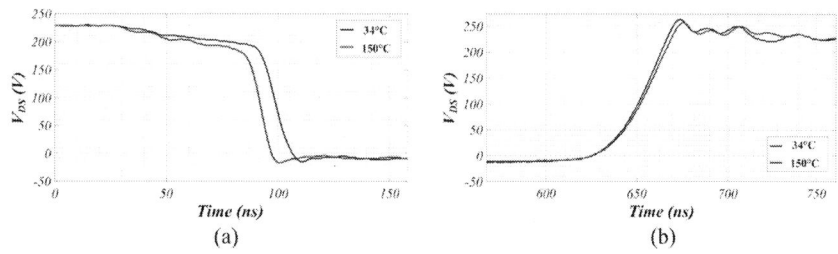

(a)   (b)

Figure 10: (a) Rise time and (b) fall time changes of V_{DS} due to temperature increment for brand-new switch

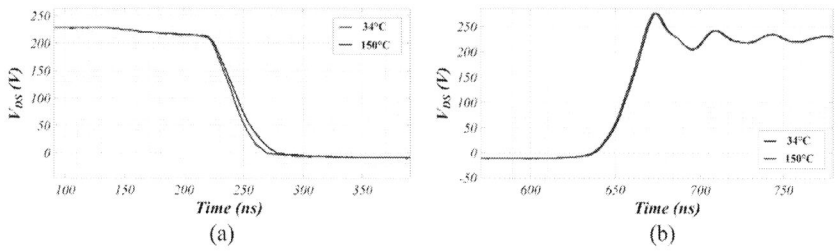

(a)   (b)

Figure 11: (a) Rise time and (b) fall time changes of V_{DS} due to temperature increment for degraded switch

Conclusion

Being focused on gate oxide degradation failure of SiC MOSFETs, the transient behavior of the switch was studied in different circuit configurations and switch operational conditions to develop an understanding of the effect of gate oxide degradation on drain-source voltage waveform of the switch. Studies on switch drain-source voltage waveform showed that although the drain-source voltage waveform is partially affected by the load type, junction temperature, and rate of degradation, it undergoes major changes in rise time and fall time values during the gate oxide degradation process. Therefore, it can be concluded that the turn-on and -off transient time intervals of the drain-source voltage waveform of the SiC MOSFET switches can be used as a robust indicator of gate oxide degradation in a wide range of applications.

References

[1] Z. Ni, X. Lyu, O. P. Yadav, B. N. Singh, S. Zheng, and D. Cao, "Overview of real-time lifetime prediction and extension for sic power converters," *IEEE Transactions on Power Electronics*, vol. 35, no. 8, pp. 7765–7794, 2020.

[2] J. Naghibi, K. Mehran, and M. P. Foster, "An online failure assessment approach for sic-based mosfet power modules using iterative condition monitoring technique," in *2020 IEEE 21st Workshop on Control and Modeling for Power Electronics (COMPEL)*, pp. 1–5, 2020.

[3] A. Griffo, J. Wang, K. Colombage, and T. Kamel, "Real-time measurement of temperature sensitive electrical parameters in sic power mosfets," *IEEE Transactions on Industrial Electronics*, vol. 65, no. 3, pp. 2663–2671, 2018.

[4] J. Naghibi, K. Mehran, and M. P. Foster, "High-frequency non-invasive magnetic field-based condition monitoring of sic power mosfet modules," *Energies*, vol. 14, no. 20, p. 6720, 2021.

[5] R. Ouaida, M. Berthou, J. León, X. Perpiñà, S. Oge, P. Brosselard, and C. Joubert, "Gate oxide degradation of sic mosfet in switching conditions," *IEEE Electron Device Letters*, vol. 35, no. 12, pp. 1284–1286, 2014.

[6] J. Naghibi and K. Mehran, "Multiphysics condition monitoring technique for reliability assessment of wide bandgap-based power modules in electric vehicle application," in *2020 Fifteenth International Conference on Ecological Vehicles and Renewable Energies (EVER)*, pp. 1–6, 2020.

[7] U. Karki, N. S. Gonzalez-Santini, and F. Z. Peng, "Effect of gate-oxide degradation on electrical parameters of silicon carbide mosfets," *IEEE Transactions on Electron Devices*, vol. 67, no. 6, pp. 2544–2552, 2020.

[8] M. Farhadi, F. Yang, S. Pu, B. T. Vankayalapati, and B. Akin, "Temperature-independent gate-oxide degradation monitoring of sic mosfets based on junction capacitances," *IEEE Transactions on Power Electronics*, vol. 36, no. 7, pp. 8308–8324, 2021.

[9] S. Potbhare, N. Goldsman, G. Pennington, A. Lelis, and J. M. McGarrity, "Numerical and experimental characterization of 4 h-silicon carbide lateral metal-oxide-semiconductor field-effect transistor," *Journal of Applied Physics*, vol. 100, no. 4, p. 044515, 2006.

[10] A. J. Lelis, R. Green, D. B. Habersat, and M. El, "Basic mechanisms of threshold-voltage instability and implications for reliability testing of sic mosfets," *IEEE Transactions on Electron Devices*, vol. 62, no. 2, pp. 316–323, 2014.

[11] R. Green, A. Lelis, and D. Habersat, "Application of reliability test standards to sic power mosfets," in *2011 International Reliability Physics Symposium*, pp. EX.2.1–EX.2.9, 2011.

[12] S. Pu, F. Yang, E. Ugur, C. Xu, and B. Akin, "Sic mosfet aging detection based on miller plateau voltage sensing," in *2019 IEEE Transportation Electrification Conference and Expo (ITEC)*, pp. 1–6, 2019.

[13] S. Pu, E. Ugur, F. Yang, and B. Akin, "In situ degradation monitoring of sic mosfet based on switching transient measurement," *IEEE Transactions on Industrial Electronics*, vol. 67, no. 6, pp. 5092–5100, 2019.

[14] Y. Zhang and Y. C. Liang, "A simple approach on junction temperature estimation for sic mosfet dynamic operation within safe operating area," in *2015 IEEE Energy Conversion Congress and Exposition (ECCE)*, pp. 5704–5707, 2015.

[15] U. Karki and F. Z. Peng, "Effect of gate-oxide degradation on electrical parameters of power mosfets," *IEEE Transactions on Power Electronics*, vol. 33, no. 12, pp. 10764–10773, 2018.

[16] S. Mohsenzade, J. Naghibi, and K. Mehran, "Reliability enhancement of power igbts under short-circuit fault condition using short-circuit current limiting-based technique," *Energies*, vol. 14, no. 21, p. 7397, 2021.

[17] J. Liu, G. Zhang, Q. Chen, L. Qi, Y. Geng, and J. Wang, "In situ condition monitoring of igbts based on the miller plateau duration," *IEEE Transactions on Power Electronics*, vol. 34, no. 1, pp. 769–782, 2018.

[18] A. J. Lelis, D. Habersat, R. Green, A. Ogunniyi, M. Gurfinkel, J. Suehle, and N. Goldsman, "Time dependence of bias-stress-induced sic mosfet threshold-voltage instability measurements," *IEEE transactions on Electron Devices*, vol. 55, no. 8, pp. 1835–1840, 2008.

[19] X. Ye, C. Chen, Y. Wang, G. Zhai, and G. J. Vachtsevanos, "Online condition monitoring of power mosfet gate oxide degradation based on miller platform voltage," *IEEE Transactions on Power Electronics*, vol. 32, no. 6, pp. 4776–4784, 2017.

AUTHOR INDEX

Abdalrahman, Adil 2241, 3282, 3757
Abdullah, Ahmed...554
Abedini, Hossein...865
Aceña, Javier Cañas..484
Adabi, Jafar..2537
Addin, Ali Sharaf...1824
Afonso, Luciana C. 4018
Aganza-Torres, Alejandro......................................1328
Agarwal, Ritika..3615
Agirrezabala, Eneko 3327
Aguglia, D. 1955
Ahmed, Emad M...1015
Aiello, Giuseppe...2628
Aillerie, Michel..315
Aizpuru, I..2903
Aizpuru, Iosu 325, 3327, 3574, 3750
Akuru, Udochukwu B. 2958
Al-Haddad, Kamal...1025
Alaluss, Mohamed 1424
Alatise, Olayiwola 1497, 2477
Albrecht, Fabian 2726
Aldarmon, Mohamed...2574
Ali, Mohammad 2392, 3022
Ali, Ramy 390
Ali, Rana Asad...698
Allard, Bruno 169, 3862
Allioua, Abdelmoumin..2835
Alvarez, Asier...279
Alvarez-Herault, Marie-Cecile 1147
Alves, Wendell Da Cunha......................................1046
Alvi, Muhammad H..1692
Aly, Mokhtar...1015
Andersen, Michael A. E. 1561
Ando, Y. 1785
Andresen, Jan..1684
Ansari, Sajad A. 3440
Antonopoulos, Antonios 297, 432
Anzola, J. 2903, 2967
Anzola, Jon 3574
Apostolidou, Nena...1796
Appel, Tobias...1121
Apte, Pramod..2773
Arabsalmanabadi, Bita...1025
Arias, Manuel...152
Arrizabalaga, Antxon...325
Arrozy, Juris 681
Arruti, Asier.............................. 3574, 3750
Artal-Sevil, J. S................................ 2903, 2967

Arza, Joseba................................484, 2011
Asllani, Besar..2515
Asoodar, Mohsen 2843
Atzler, Frank...3391
Aunsborg, Thore Stig 825
Aviñó, Oriol..2715
Ayarzaguena, Ibán..............................1765, 3336
Aztiria, Jon 325
Baars, Nico 2788
Babin, Anthony 3696
Baburske, Roman 1424
Bacha, S. 2422, 3179
Bacha, Seddik...............................3140, 3928
Bacheti, Gabriel Gaburro 421
Bachmann, Matthias..3501
Badenhop, Niklas.............................145, 1939
Baek, Seung-Hyuk 2877
Bagaber, Bakr...............................3037, 3711
Baimel, D. 3254
Baimel, N. 3254
Bak, Claus Leth 2504
Bakhos, Gianni 3928
Bakran, Mark-M...............................805, 1036, 2744
Bakri, Reda 1046
Balachandran, Arvind 1456
Balasubramanian, Sridhar 2030
Ballestín-Bernad, V.............................2903, 2967
Banana, Shady 1064
Banavath, Satish Naik 730
Banda, Joseph...............................187, 289
Barba, V. 1975
Barbi, Eli 3254
Barg, Sobhi 361
Barman, Subhranil...2462
Barón, Kevin Muñoz...2698
Bashar, Erfan...2477
Basic, Duro 125
Basler, Michael 242
Basler, Thomas....................1424, 1713, 1733, 3373
Bauer, Luca 971
Bauer, Pavol..............................1319, 3607, 3729
Baumann, Michael 1167
Baumann, Timm Felix 2355
Bäumler, Christian 1733
Bayer, Markus.. 115
Bayhan, Sertac 3518
Bayram, Islam Safak 3518
Beck, Simon..............................1434, 2038

Beckemeier, Christian 2327
Beczkowski, Szymon Michal 2661
Beineke, Stephan 3501
Beiranvand, Hamzeh 833, 3092, 3846, 3966
Belhaouane, Mohamed Moez 582
Benchaib, Abdelkrim 3928
Bendfeld, Christian 1620
Benech, Philippe 169
Bensetti, Mohamed 3883
Bergmann, Lukas 1036
Bergveld, Henk Jan 3796
Bermejo, Jose Manuel 1765, 3336
Bernal, Carlos 3327
Bernal-Agustín, J. L. 2967
Bernal-Ruiz, Carlos 3750
Bernichon, Thomas 3920
Bertilsson, Kent 361
Bertin, Matthieu 534
Beukes, Johan 3112
Beye, Mamadou Lamine 2736
Beza, Mebtu 1187
Bezerra, Vinicius Freire 2689
Bhatnagar, Pallavee 3804
Bhattacharya, Arghyadip 178
Bhoi, Sachin Kumar 3031
Biadene, Davide 865
Biela, Juergen 1402
Biela, Jürgen 651, 662, 933, 1391, 1434, 2038, 2544
Bieler, Arne 1121
Bier, Anthony 922, 2736
Billa, Laxma R. 2301
Bimmel, Luc 2736
Binder, Andreas 2316
Bitsi, Konstantina 3246
Blaabjerg, Frede 2110, 2182, 2496, 2504, 2939
Blanes, J. M. 3382, 3401
Blank, Thomas 232
Blanquez, Francisco R. 2189, 2451
Blasco-Gimenez, Ramon 2189, 2451
Blasuttigh, Nicola 3846
Blatsi, Zoe 2824, 3813
Blömeke, Alexander 4025
Böcker, Joachim 2276, 2432, 2754, 3625, 3686
Bockholt, Jan 1286
Boettcher, Norman 1128
Bohllaender, Marco 4016
Bohne, David 514
Boige, Francois 944
Boisson, Guillaume Piquet 960
Bolzoni, A. 1371
Bongiorno, Massimo 1187
Bonten, Remco 634

Böorngen, Hannes 1754
Borcherding, Holger 2852
Börngen, Hannes 3362
Boroyevich, Dushan 2806
Bosch, Swen 2219
Bosga, Sjoerd G. 3246
Bouscayrol, Alain 2175
Boutleux, Emmanuel 251
Boutry, Arthur 2515
Brabetz, Ludwig 2383
Branco, Cesar Augusto Santana Castelo 2948
Braun, Gerrit 2205
Braz, Cesar 1445
Briff, Pablo 451
Bringezu, Thilo 662
Brinker, Tobias 2977
Brogioli, Doriano Constantino 833
Brommer, Volker 2726
Bronstein, S. 3254
Brooks, Michael 279
Brückner, Thomas 1824
Brulin, Pierre-Yves 3831
Brunner, Andreas 593
Brunner, Frank 3775
Brüns, Michael 474
Bruyere, Antoine 1046
Bruyere, Paul 960
Bucarey, Victor 1074
Budo, Kohei 213, 351
Bueno, Emilio José 421
Bueno-Mariani, Guilherme 3272
Bugarski, Stevan 2334
Bünte, Andreas 380
Burgos, Rolando 1692, 2806
Burgos-Mellado, Claudio 1074, 3429
Burkart, Ralph M. 203
Burke, Richard 3696
Bushra, Rehnuma 2392, 3022
Busquets-Monge, Sergio 2715
Buticchi, Giampaolo 3014
Buttay, Cyril 2049, 2515
Byen, Byengjoo 1207
Caarls, Esin Ilhan 681
Cabrera, Michel 169
Cacciato, Mario 2628
Caillierez, Antoine 3883
Cajander, D. 1955
Cakal, Gokhan 3947
Caldognetto, Tommaso 865
Camargo, Renner Sartório 421
Camurca, Luis 3101
Can, Görkem 3092

Cano, Tania C. 335
Cao, Jingming 3215, 3225
Cao, Yongtao 2003
Cappelle, Jan 1300
Cárcamo, Alberto 1083
Carcouet, S. 843
Carpita, Mauro 1543
Carrasco, Miguel 370
Casado, P. 3382, 3401
Castellazzi, Alberto 689, 2156, 2285, 2402, 2893, 3084
Castelli-Dezza, Francesco 1476
Castro, Ignacio 335
Catalán, Pedro 2011
Catellani, Stéphane 922, 990
Ceccarelli, Lorenzo 681
Chakraborty, Sajib 2101, 3031
Chang, Che-Wei 1692
Charkaoui, Abdelmouneim 442
Chatterjee, Kishore 178, 2462
Chen, Zhe 2011
Chen, Zhu 3235
Chevalier, Florian 3582
Chida, Makoto 1580
Chinthavali, Madhu Sudhan 344
Chiumeo, Riccardo 3206
Choksi, Kushan 344
Choudhury, Soham 1966
Chub, Andrii 730
Cimetiere, Xavier 1046
Clerc, Guy 251
Clerici, Alessio 3206
Cobaleda, Diego Bernal 2581
Cogitore, Bruno 1216
Colmenero, Manuel 2189, 2451
Cosso, Simone 2919
Coumont, Martin 1966
Crovetti, Paolo 554
Cui, Yi 3986
Czerwenka, Philipp 593
Dahmen, Christopher 1824, 1855
Damian, Ioan Catalin 2266
Damm, Gilney 3590
Danielsson, Christer 2843
Dargahi, Vahid 2073
Davari, Pooya 2496
Davidson, Jonathan N. 3440
De Bernardinis, Alexandre 315
De Carne, Giovanni 3014
De Cesaris, Ivan 223
De Donato, Giulio 1569
De Doncker, Rik W. .... 709, 1266, 2119, 3599, 3676, ... 3740, 3766, 3893

De Lillo, Liliana 3450
De Matos, Jose Gomes 2948
De Oliveira, Eduardo Facanha 2441
De, Dipankar 689
Deb, Arkadeep 1497
Deblecker, Olivier 504
Deboy, Gerald 3984
Deck, Patrick 514
Deckers, Martijn 2795
Degaa, Laid 3696
Delette, Gérard 922
Deng, Kai 3235
Dennetiere, Sébastien 582
Derammelaere, Stijn 3344
Despouys, Olivier 2486
Dick, Christian P. 514
Dickmann, Stefan 758
Dieckerhoff, Sibylle 1466, 2596, 2607, 2644, 3775
Dieng, A. 2092, 2930
Dierks, Rebecca 1533
Dietrich, Tim-Hendrik 1094
Disselkamp, Simon 2912
Domae, Shinichi 3084
Domes, Daniel 2744
Domes, Konrad 1137
Dong, Chaoyu 3215, 3225
Dong, Dong 1692, 2515
Dong, Jianning 1319
Dong, Tenghui 3084
Dorner, Oscar 1177
Dos Santos, Pedro Leal 604
Dragicevic, Tomislav 2496, 2939, 3429
Drexler, Christoph 411, 1167
Driesen, J. 3655
Driesen, Johan 2795
Drimizi, Youssef 2869
Drissi, Khalil El Khamlichi 3786
Duarte, Jorge L. 681, 798
Duarte, Jorge 2788
Duchamp, Jean-Marc 169
Dujic, Drazen 2049
Dumtzlaff, Jacob 1865
Duquesne, Thierry 3582
Dürbaum, Thomas 88, 307
Duun, Sune Bro 825
Dworakowski, P. 2422
Dworakowski, Piotr 2049
Ebel, Thomas 3130
Ebner, Kathrin 4015
Eckart, Martin 3646
Eckel, Hans-Guenter 3460

Eckel, Hans-Günter 11, 59, 70, 980, 1294, 1703,
.... 1744, 1885, 1895, 2308, 4003
Eckstein, Mattea 1277
Effenberger, Thomas 1754
Eggers, Malte... 1466
Ehlich, Martin 2852
El Baghdadi, Mohamed 2101, 2293, 3031
El Sherif, Alaa 3796
El-Refaie, Ayman 719, 1692
Ellinger, Thomas... 2885
Emmers, G. 3655
Emmers, Glenn 2795
Empringham, Lee 3450
Encarnação, Lucas Frizera... 421
Endo, Yusuke.. 2285
Epping, Daniel 749
Erckrath, Tobias.. 1350, 1620
Eremia, Mircea 2266
Eriksson, Lars 1456
Erlbacher, Tobias.. 1128
Ernst, Alexander 3149, 3159
Es-Seghier, Hajar.. 922
Escoffier, René 990
Etoz, Burhan 1497
Faber, Samuel 307
Falchi, Daniele.. 2486
Faramehr, Soroush... 3822
Farhangi, Shahrokh.. 787
Fauth, Leon 2003, 2638, 3838
Fayolle-Lecocq, Murielle.. 990
Fazli, Nastaran 11
Fehr, Hendrik.. 49, 3391
Felgemacher, Christian 442, 4004
Fernández, Arturo 152
Ferreyra, Fabio.. 554
Festerling, Tobias 1237
Finney, Stephen 80, 3470, 3813
Fischer, Katharina.. 1674, 1804
Fischer, Manuel 749
Fischer-Baeumer, Rico 1137
Fölkel, Lorandt 279
Formentini, Andrea.. 2919, 3975
Forouzesh, Mojtaba 1590, 1601
Forsstrom, Ville 3301
Förster, Nikolas 2432
Foster, Martin P. 3353, 3440
Foteinopoulos, Georgios.. 1985
Fräger, Lukas...................... 145, 641, 1939, 2588, 2773
Frank, S. R. 3411
Franzki, Jonas... 261
Frey, David 1147
Fricke, Tobias.. 1247

Fricke, Torben 1381
Friebe, Jens 1914, 2003, 2327, 2392, 2588, 2638,
.... 2655, 2689, 2773, 2977, 3022, 3059, 3545, 3838
Fritze, Eric.. 758
Fröhling, Sören.. 1674
Fuchs, Simon... 1434, 2038
Fuhrmann, Jan... 980
Fukunaga, Shuhei.. 108
Ganeshpure, Dhanashree Ashok................................. 3729
Gao, Xiang 3014
Gaona, Daniel.. 2441
Garces, Santiago Ramos... 3344
Garcia, Raul Murillo 2355
Garrigós, A..3382, 3401
Gaubert, Jean-Paul 1525
Gauthier, Jean-Yves... 3862
Gavelle, Mathieu... 2618
Gehl, Adrian 2912
Geiss, Michael... 2554
Gemma, Filippo 3975
Geng, Weiwei.. 3722
Geng, Xiaomeng 2596, 2644, 3775
Gennaro, Francesco.. 2628
Gensior, Albrecht..49, 370, 3391
Gerges, Tony 169
German, Ronan 2175
Germishuizen, J. J. 3318
Geury, Thomas 2101
Gholami, M. 3179
Gholami, Mehrdad 3140
Ghumman, Sukhjit S 2763
Gieraths, Antje 767
Gierschner, Magdalena... 1294
Gierschner, Sidney 11
Gillon, Frédéric 1046
Girona-Badia, Jaume.. 3704
Glaser, Martin 4020
Gleissner, Michael..805, 1036
Gnärig, Lasse 370
Goetz, Stefan............................ 1025, 1064, 1197, 3636, 3665
Gohler, Katherina 1804
Gohrmann, Kai.. 1137
Golev, Victor... 1286
Goller, Maximilian 1733
Gomes, Lucas Vinícius De Araújo............................... 3059
Gomes, Zariff Meira... 3590
Gómez, Alexis A. 1765, 3336
Gomis-Bellmunt, Oriol 2486, 3704
Gonzalez, Jose Ortiz.. 1497
Gonzalez-Hernando, Fernando.................................... 3938
Gonzalez-Torres, Juan-Carlos..................................... 3928
Götz, Georg Tobias 709

Gräber, Hendrik 2977
Grabs, Volker 97
Gradinger, Thomas B. 203
Grant, Thomas 2301
Grass, Norbert 2366
Grau, Vivien 854
Gremme, Florian 4021
Griepentrog, Gerd 160, 2780, 2835
Grodnichev, Anton 624
Groke, Holger 3169
Groon, Fabian 3092
Groten, Jonas 279
Gruson, François 582
Guerrero, Bruno 944
Gui, Qiuye 49
Guillaud, Xavier 582
Günes, Ece Olcay 1361
Gupta, Kirti 2110
Gupta, Krishna Kumar 3615, 3804
Gutierrez, Alonso 2618
Haag, Felix 2726
Haake, Daniel 624
Haarer, Jörg 971, 1237, 1277
Habersetzer, Antoine 4015
Hably, A. 3179
Hably, Ahmad 3140
Hackl, Philipp 39
Haederli, Christoph 3282
Häfner, Ying-Jiang 2241, 3282, 3757
Hagedorn, Maximilian 1875
Hajar, K. 3179
Hajar, Khaled 3140
Hajian, Masood 468
Hakkila, Akseli 297
Hald, Alex 380
Hameyer, Kay 3005, 3235
Hammes, David 11
Handt, Karsten 2607
Hanf, Michael 3169
Hanisch, Lucas Vincent 261
Hanisch, Lucas 1094
Hänsel, Stefan 572
Hansen, Sandra 3966
Hanson, Alex J. 1722
Hanson, Jutta 1966
Hao, Chuantong 80, 3470
Hardan, Faysal 468
Harmand, Souad 2996
Hasan, Md. Mahamudul 3031
Hasler, J. P. 1371
Hassan, Tayssir 1466
Hatori, K. 1785

Hatori, Kenji 777
Hattori, Takato 739
Hauenschild, Philipp 1506
Haug, Martin 279, 698
Hayes, John G. 2470
Hegazy, Omar 2101, 2293, 3031
Heide, Daniel 3711
Heien, Christian 1294
Heimler, Patrick 1713
Hein, Yves 1294
Helmholdt-Zhu, Ting 97, 854
Hembel, Ahmed 3947
Henke, Markus 261, 1094, 2030
Henkenjohann, Jonas 1684
Henn, Jochen 3599
Henneberg, Dustin 2885, 3491
Herbold, Johannes 749
Hernando, Marta M. 1083, 1765, 3336
Herzog, Hans-Georg 952
Heydari, Rasool 2682
Hikihara, Takashi 108
Hiller, M. 3411
Hiller, Marc 115, 999
Hillmer, Hartmut 2383
Hilt, Oliver 2596, 2644, 3775
Himker, Niklas 1631
Himmelmann, Patrick 999
Hiraki, Eiji 2164
Hirning, David 971, 1237, 1277, 3536
Hissel, Daniel 315
Hjerrild, Jesper 2504
Hoerner, Michael 1754
Hofer, Heimo 1445
Hofer, Matthias 2251
Hoff, Bjarte 3198
Hoffmann, Klaus F. 758, 2726, 3188
Hoffmann, Madlen 3262
Hoffstadt, Thorben 1157
Hofmann, Viktor 195, 400
Hofmann, Wilfried 3957
Hofstetter, Patrick 195, 400
Hölscher, Jonas 2432
Holtje, Pauline 1665
Holzke, Wilfried 3149, 3159, 3169
Horn, Markus 2383
Hortans, Magnus 3309
Hoshi, Nobukazu 1776, 1844
Hosseinabadi, Farzad 3031
Hosseini, Elham 1025
Hou, Jingning 3722
Houwen, Simon 3344
Hridya, I 187

Hu, Anliang 651
Hu, Bin 2182
Hu, Xiaowei 3722
Huang, Jiasheng 1561
Huerta, Gabriel Ramos 1226
Huesgen, Till 2230
Huisman, Henk 634, 673, 681
Hutzler, Michael 1445
Idir, Nadir 2996, 3582, 3822
Igic, Petar 3822
Iida, Masaki 2164
Iman-Eini, Hossein 787
Imgart, Paul 1187
Incurvati, Maurizio 223, 268
Inoue, Michiko 3420
Iraola, Unai 3327
Ishihara, Mastaka 2164
Itoh, Jun-Ichi 902, 1104, 2127
Ittamveettil, Hridya 289
Izurza, Pedro 484
Jaber, Hamzeh J. 2156, 2285, 3084
Jacques, Dries 3344
Jagannath, Sriram 3362
Jahdi, Saeed 1497, 2477
Jain, Anekant 3615
Jain, Sanjay K. 3615, 3804
Jamal, Adeel 2780
Jaman, Shahid 3031
Jankovic, Marija 442
Jayathurathnage, Prasad 1947
Jena, Kasinath 3804
Jenhani, Firas 1343
Jeong, Byunghwang 1207
Jeschke, Sebina 3235
Jha, Kapil 187, 289
Jia, Hongjie 3215, 3225
Jia, Ming 1266
Joebges, Philipp 1266
Johansson, N. 1371
Johnson, C. Mark 3450
Jonsson, Tomas 1456
Jordà, Xavier 2715
Jørgensen, Asger Bjørn 825, 1641, 2661
Jöst, Dominik 4025
Jovanovic, Raka 3518
Juchem, Ralf 4023
Judge, Paul 80
Junemann, Lennart 1665
Jung, Marco 624, 1515, 1611, 1620
Junghans, Christoph 3460
Junyent-Ferre, Adria 2574
Kabbara, Wassim 3883

Kacetl, Jan 1197, 3636, 3665
Kacetl, Tomáš 1197, 3636, 3665
Kacki, Marcin 2470
Kadem, Karim 3590
Kaerst, Jens Peter 544
Kaiser, Jeremias 307
Kallfass, Ingmar 2698, 3565
Kamel, Tamer 468
Kaminski, Nando 2230, 3149, 3169
Kamm, Simon 2698
Kampen, Dennis 145, 1939, 2588
Kamper, Maarten J. 2958
Karakasli, Vefa 2835
Karamanakos, Petros 297, 1476, 1754
Karau, Fabian 3292
Karnehm, Dominic 767
Karwatzki, Dennis 195
Kasten, Henning 3501
Kayser, Felix 59, 4003
Keilmann, Robert 891
Kempchen, Malte 2912
Kemper, Philipp 749
Kennel, Ralph 1754, 2366, 3362
Kerekes, Tamas 1933
Keshavarzi, Davood 1064
Khader, Meriem 2655
Khan, Basit Ali 2537
Khan, Mohammed Ali 135
Khan, Nameer 3796
Khan, Siam Hasan 484
Khanzadeh, Babak 2344
Khenfri, Fouad 3831
Kiehnle, Philip 999
Kiffe, Axel 1157
Kikuchi, Naoto 1104
Kim, Dong-Uk 1207
Kim, Sungmin 1207, 2877
Kinzer, Dan 3987
Kirsch, Andreas 380
Kitagawa, Wataru 739
Kjærsgaard, Benjamin Futtrup 825
Klee, Matthias 1515
Klever, Severin 3676
Klötzer, Sebastian 4011
Knebusch, Benjamin 1665, 3048
Ko, Youngjong 3014
Kobayashi, Hiroyasu 1580
Kocewiak, Lukasz 2504
Koch, Jan-Niklas 2852
Koczy, Dawid 3149
Kohlhepp, Benedikt 88, 307
Kojima, Tetsuya 3740

Kondo, Keiichiro 1580
Kondratenko, Dmytro 1906
Kopp, Tobias 912
Kormska, Tomáš 1114
Körner, Patrick 2021
Korthauer, Bastian 3625
Kosesoy, Yusuf 634
Kostka, Benedikt 1649
Kostynski, Daniel 3855
Koteich, Mohamad 534
Kouro, Samir 1015
Koutroulis, Eftychios 1985
Kowal, Julia 4014
Kragl, Robert 2554
Krick, Alexander 3989
Krigar, Tim 2375
Krishnamoorthy, Harish Sarma 730
Krüger, Helge 3966
Krümpelmann, Marcel 1631
Kubulus, Pawel Piotr 2661
Kuder, Manuel 767
Kumar, Amit 451
Kumar, Kaushik Naresh 1486
Kumar, Manish 3511
Kuperman, A. 3254
Kuprat, Johannes 3067
Kuring, Carsten 2596, 2644, 3775
Kurrat, Michael 912
Kurukuru, V S Bharath 135
Kusaka, Keisuke 1104, 2127
Kusche, Stephan 3704
Kusebauch, Manuel 3491
Küster, Pierre 411
Kwak, Jaedon 2893
Kyyrä, Jorma 1947
La Mantia, Fabio 833
Labonne, A. 3179
Labonne, Antoine 3140
Labrousse, D. 843
Lacerda, Vinícius Albernaz 3704
Laclaverie, Julien 944
Laforet, David 1445
Lamar, Diego G. 335, 1083, 1765, 3336
Lange, Jarren 2276
Lange, Yannic 2644
Langfermann, Sascha 1939, 2588
Lanzarotto, D. 2564
Larrañaga, Uxue 3938
Larrazabal, Igor 1765, 3336
Larsson, Anders 1456
Lataire, Philippe 2293
Laumen, Michael 3766

Lauri, Andrea 865
Laza, Saioa Burutxaga 370
Lazkano, Markel Zubiaga 484
Le Leslé, Johan 2526
Le Métayer, Pierre 2049
Lee, Jaehong 2877
Lee, Seung-Hwan 2877
Lee, Yonghwa 2402
Lefebvre, Bruno 2515
Lefevre, Guillaume 2526
Legay, Florian 3529
Lehn, Peter W. 1995, 2084, 2145, 2763
Leifert, Torsten 4013
Lemaire-Semail, Betty 2175, 2996
Lembeye, Yves 1216
Lenz, Kevin 442
Lenzen, Patrick 2413
Leuer, Michael 3292
Leuzzi, Riccardo 3975
Lévy, PE 843
Lewicki, Arkadiusz 1906
Lexow, Daniel 1744
Li, Feifei 3235
Li, Ke 3822
Li, Marui 3215, 3225
Li, Qiang 3722
Li, Weihan 4025
Li, Xiang 2301
Li, Xupeng 3373
Li, Zheming 2744
Liang, Mincui 3786
Lichtenstein, Timo 1674
Liebfried, Oliver 2726
Liegmann, Eyke 1754, 3362
Lievre, Aurelien 2175
Lin, Siqi 1914, 2638
Lin-Shi, Xuefang 3862
Lindemann, Georg 3555
Linder, Stefan 3992
Lippold, Florian 1506
Liserre, Marco 421, 833, 3014, 3067, 3092, 3101,
.... 3846, 3966
Liu, Chao 1561
Liu, Steven 604
Liu, Xing 1733, 3373
Liu, Yan-Fei 1590, 1601
Liu, Yining 1947
Llanos, Jacqueline 3429
Löfgren, Jonas 3920
Lombard, Philippe 169
López, Abraham 152
Lorenz, Andreas 814

Lorenz, Erwin 1167
Lorenz, Malte 1875
Lorenz, Oscar 873
Loudot, Serge 3883
Lu, Xuyang 3822
Lu, Yizhou 883
Luan, Shaokang 3309
Luckert, Franz 2706
Luecke, Stefan 3075
Luh, Matthias 232
Luo, Fang 344, 2860
Lusardi, Federico 3975
Lutsch, Michael 88
Lutz, Josef 1713
Lutzen, Hauke 2230
Ma, Wenhao 80
Maamri, Nezha 1525
Maibach, Philippe 3282
Maier, Robert W. 2744
Maitra, Abhishek 1424
Mallwitz, Regine 891, 912, 1094, 1247, 1506
Mambetow, Arthur 145
Manthey, Tobias 2655, 2689, 3059
Marca, Ygor Pereira 798
Marcaide, Inko 3920
Marcault, Emmanuel 2618
Marchesoni, Mario 2919
Margreiter, Thomas 223
Margueron, Xavier 1046
Marks, Hendrik 2030
Marquardt, Rainer 1855
Marroquí, D. 3382, 3401
Martin, Jérémy 990, 2736
Martinez, Wilmar 1914, 2197, 2581
Martinez-Garcia, Herminio 1056
Martinez-Padron, Daniel S. 1256
Martnez, Wilmar 2638
Marx, Philipp 1237, 1277, 3536
März, Martin 493, 3262
Mashaly, Aly 442
Mashayekh, Ali 767
Mathúna, Cian Ó 4006
Mattavelli, Paolo 865
Matthies, David 3159
Maussion, Pascal 2869
Maynard, X. 843
Mazuela, Mikel 325, 3327, 3574
Meddour, Aissam Riad 3696
Mehran, Kamyar 614, 3353
Mehrasa, M. 3179
Mehrasa, Majid 3140
Meier, Hans 2021

Meinert, Janus Dybdahl 825
Meissner, Michael 758, 3188
Mellor, Phil 2477
Mendoza-Araya, Patricio 1177, 1226
Meng, Qingchao 933
Menzel, Steffen 3169
Merlin, Michael M. C. 2824, 3813
Merlin, Michael 80, 3470
Mersche, Stefan 115
Mertens, Axel 641, 1350, 1533, 1631, 1649, 1665,
..1684, 1865, 1875, 2003, 2066, 2392, 2706, 3022, 3037, 3048,
3075, 3555, 3711
Miaja, Pablo F. 152
Mijatovic, Nenad 2496, 2939
Miller, T. J. E. 3318
Minami, Masataka 2285
Mir, Tabish Nazir 468
Mirza, Abdul Basit 344
Mirzadeh, Mina 1350
Mirzaeva, Galina 3903
Miskiewicz, Rafal 1486
Mistretta, C. 1975
Mita, Salvatore 2628
Mo, Wai Keung 3130
Möckel, Andreas 3391
Moench, Stefan 242
Mogorovic, Marko 203
Mohanta, MK Kharabela 689
Möhlenkamp, Georg 3993
Mohsenzade, Sadegh 614, 3353
Moldenhauer, Deniz-Heinz 2205
Mondal, Gopal 572
Mondzik, Andrzej 3804
Monmasson, Eric 1256
Mönninghoff, Sebastian 3005
Montero, E. Rodriguez 1834
Morales-Paredes, Helmo K. 1074
Morand, Julien 2526
Morel, F. 2422, 2564
Morey, Philippe 1543
Morshed, Muhammad 2301
Motte-Michellon, Denis 1216
Mouselinos, Theodoros P. 1551
Moussa, Hassan 3590
Movagharnejad, Hedieh 3048
Mu, Yunfei 3215
Müller, Jonas 2230
Müller, Tankred 474
Munk-Nielsen, Stig 825, 1641, 2661, 3309
Muñoz-Carpintero, Diego 1074, 3429
Muruaga, Endika Bilbao 3529
Musolino, Francesco 554

Mustafeez-Ul-Hassan 2860
Musumeci, S. 1975
Muyllaert, Koenraad 2383
Mysore, Madhu Lakshman 1424
Naeve, Tomasz 1445
Nagayasu, Kiwa 2164
Naghibi, Javad 614, 3353
Nahalparvari, Mehrdad 2843
Najjar, Mohammad 2682
Nakamura, Keiichi 777
Nakamura, Taketsune 3084
Nami, Ashkan 2241, 3757
Nannen, Hauke 160
Nassurdine, B. Mohamed 843
Nayak, Khirod Kumar 2241, 3757
Nayampalli, Vishwas Acharya 1703
Nazeri, Ahmad Ali 1309, 1336, 1343, 2670, 3871
Neal, Harley 2301
Nee, Hans-Peter 2843
Nehmer, Dominik 1036
Neira, Sebastian 2824, 3813
Neuland, Tanja 3991
Neumann, Christian 1895
Neumann, Ingmar 1445
Neumeister, Matthias 572
Nguyen, Allen 1722
Nguyen, Khanh-Hung 562, 1309
Nguyen, Van-Sang 922, 990
Nguyen, Xuan Viet Linh 169
Nian, Heng 2182
Niasar, Mohamad Ghaffarian 3729
Nie, Shuang 2145
Niedernostheide, Franz-J. 2744
Niedernostheide, Franz-Josef 1424
Nielebock, Sebastian 493, 2607
Niemetz, Michael 2021
Niggemann, Oliver 3545
Nikowitz, Mario 2251
Nishio, Atsushi 351
Nishitani, Yota 3420
Nishizawa, Shin-Ichi 1128
Noboru, Wakana 777
Noisette, Philippe 3910
Nooshabadi, Morteza Tadbiri 787
Nordström, Lars 883, 1006
Nymand, Morten 2682
O'Donnell, Terence 390
O'Driscoll, Seamus 4006
Obernolte, Urs 854
Odeh, Charles 1906
Okada, Ryohei 1776, 1844
Olbrich, Markus 2912

Oliveira, Hercules Araujo 2948
Orbay, Raik 3920
Orchard, Marcos 3429
Orfanoudakis, Georgios I. 1985
Örgüt, Osman 1361
Orlik, Bernd 3149, 3159, 3169
Ortega, David 1765, 3336
Ortiz-Gonzalez, Jose 2477
Orts, C. 3382, 3401
Oshnoei, Arman 2939
Ota, Ryosuke 1776, 1844
Ouyang, Ziwei 1413, 1561
Owzareck, Michael 1939, 2588
Oyarbide, Estanis 3327
Paasch, Kasper M. 3130
Pace, Loris 3582
Páez, J. D. 2422
Pagnani, Daniela 2504
Panigrahi, Bijaya Ketan 2110, 3511
Papadopoulos, Georgios 1391
Papadopoulos, Theofilos 432
Papafotiou, George 2788
Papanikolaou, Nick 1796, 2257
Papastergiou, Konstantinos 2355
Pascal, Yoann 3067
Pasquier, Christophe 3786
Passalacqua, Massimiliano 2919
Passmore, Brandon 4005
Pathmanathan, Mehanathan 1995, 2084, 2145, 2763
Patin, Nicolas 1256
Patti, Dario 2628
Patzelt, Nikolaus 1923
Paul, Arup Ratan 178
Pauls, Denis 2441
Pavone, Mario 554
Pedroso, Douglas 335
Peftitsis, Dimosthenis 1486, 2355
Pelletier, Sebastien 223
Penczek, Adam 3804
Peng, Hujun 3235
Péra, Marie-Cécile 315
Pereda, Javier 2824
Pereira, Thiago 3014, 3092, 3101, 3846
Perez, Gaëtan 960
Perez-Cebolla, Francisco Jose 3574, 3750
Peroutka, Zdenek 1114
Perpiñá, Xavier 2715
Perrin, Rémi 2526
Perrin, Remi 3272
Petritz, Andreas 279
Petzoldt, Jürgen 2885, 3491
Peyghami, Saeed 2939

Pfeiffer, Jonas 411, 1167
Pfost, Martin 2375, 2413
Phanse, Ajinkya 1722
Phulpin, Tanguy 3883
Pichon, Pierre-Yves 2526
Pickert, Phil Leon 1381
Piepenbrock, Till 2432
Pietrzak-David, Maria 2869
Pigott, John 3796
Pinheiro, José Renes 3590
Piqué, Gerard Villar 3796
Piróg, Stanislaw 3804
Placzek, Julius M. 833
Plat, Arnaud 3862
Plötz, Till-Mathis 980
Pogulaguntla, Aditya 730
Pohlmann, Sebastian 767
Polezhaev, Vladimir 2230
Ponick, Bernd 1381, 1665, 3048, 3711
Poormohammadi, Fereshteh 2795
Pöschke, Florian 3704
Pouresmaeil, Edris 2537
Pouresmaeil, Kaveh 2788
Pouresmaeil, Mobina 2537
Pramanick, Sumit 1658, 3511
Pree, Elias 1445
Prenleloup, Pierre 3529
Prieto-Araujo, Eduardo 2486, 3704
Puls, Simon 2852
Puschmann, Frank 749
Qin, Zian 3607
Quabeck, Stefan 3893
Quade, Katharina Lilith 4025
Quay, Rüdiger 242
Rabkowski, Jacek 1486, 3938
Rädel, Uwe 2885, 3491
Radha, Krishna Moorthy 344
Rafiq, Aamir 1658
Raggini, Diego 3206
Raghavendra, I Venkata 730
Rahmani, Mehdi 2496
Raison, Bertrand 1147
Rajabian, Amir Azam 614
Ramdane, Brahim 1216
Ramirez, Fernando 289
Rasekh, Navid 3120
Rasool, Haaris 2101, 2293
Raßmann, Rando 1286
Rathjen, Kai-Uwe 758
Rault, Pierre 582
Ravyts, Simon 1300
Raya, Mariana 2715

Razi, R. 3179
Razi, Reza 3140
Regnat, Guillaume 2526
Rehlaender, Philipp 2432, 2754, 3625
Reimann, René 3159
Reincke-Collon, Carsten 370, 3391
Reindl, Andrea 2021
Reiner, Richard 242
Reißenweber, Lukas 525
Reitmeier, Dominik 2211
Remón, Daniel 1083
Rettner, Cornelius 4019
Reyes-Chamorro, Lorenzo 3429
Reynaud, Jean-François 3529
Ribeiro, Luiz Antonio De Souza 2948
Richard, Lucas 1147
Rickert, Kai 115
Rigbers, Klaus 4023
Rigogiannis, Nick 2257
Ringbeck, Florian 4025
Risch, Raffael 651
Rizoug, Nassim 3696, 3831
Robinson, Jonathan 572
Rocha, Gabriel Silva 2948
Roche, Jan-Philipp 3545
Rodríguez, Alberto 335, 1083, 1765, 3336
Rodriguez, Daniel C. 3893
Rodriguez, Joan Marc 2574
Rodriguez, José 1015
Roes, Maurice G. L. 798
Roes, Maurice 2788
Roß, Tilo 3391
Rossi, Mattia 1476
Rothenburger, Max 2383
Roth-Stielow, Jörg 971, 1237, 1277, 3536
Rouphael, Rosalie 1525
Rudolph, Christian 474
Rueß, Manuel 3565
Rufer, Alfred 30
Ruppert, Lukas A. 3766
Ruthardt, Johannes 971
Rylko, Marek S. 2470
Sadarnac, Daniel 3883
Saeidi, Mahmoud 1336, 1343, 3871
Safdarzadeh, Omid 2316
Sah, Gyanendra Kumar 1885
Sahan, Benjamin 1137
Sahin, Ilker 1361
Sahoo, Subham 2110, 2182
Sahu, Malaya Kumar 2241, 3757
Sahu, Silpashree 689
Said, Nasri 2618

Saito, Wataru 1128
Sakai, J. 1785
Salehi, Navid 1056
Samples, Ben 4005
Sanchez, Juan 873
Sanchez-Ruiz, Alain 484
Santos, Francisco 3101
Sanusi, Bima Nugraha 1413
Sanz-Alcaine, José Miguel...................................3750
Sarlioglu, Bulent 3947
Sato, Kota 1580
Sato, Takashi 3420
Sauer, Dirk Uwe 4012, 4025
Sauerland, Henning...3159
Sawicki, Jean–paul 315
Scarcella, Giuseppe 1569
Scelba, Giacomo 1569, 2628
Schäffner, Philipp 279
Schafmeister, Frank 2432, 2754, 3625, 3686
Schanen, Jean-Luc 787
Schanen, JL 843
Schefer, Hendrik 891, 912, 1094
Schellekens, Jan 634
Schierle, Guido 3188
Schiestl, Martin 223, 268
Schillinger, Tobias 3646
Schillingmann, Henning 2030
Schlegel, Christian 1923
Schlegel, Ludwig 3957
Schmid, Markus 268
Schmidhuber, Michael 411, 1167
Schmies, Dominik...2276
Schmitz, Laurids 3599
Schnabel, Fabian 624, 1515
Scholjegerdes, Moritz 3005
Schön, André 814
Schrödl, Manfred 2251
Schueltzke, Jens 1167
Schuerhuber, Robert 39
Schuhmann, Thomas 3646
Schullerus, Gernot 593, 2334
Schulte, Horst 3704
Schulz, D. 3411
Schulze, Gerold..2383
Schulze, Hans-Joachim 1424
Schumann, Christian 2058
Schumann, Sven 4022
Schümann, Ulf 1286
Schupp, Jan...3309
Schütt, Michael 1885, 2308
Schwarz, Babette 1381
Schwendemann, R. 3411

Scohier, Martin..504
Scrimizzi, F. 1975
Sebastián, Javier 1765, 3336
Seibel, Axel 1515, 1620
Seitz, Arne 4015
Seliger, Norbert 22
Semail, Eric 2996
Sen, Paresh C. 1590, 1601
Sepehr, Amir 2537
Serdyuk, Yuriy 2344
Sergentanis, Grigorios 3450
Serra, Amiron Wolff Dos Santos 2948
Seybold, Felix 3536
Shahparasti, Mahdi 2682
Sharma, Kanuj 2698
Shawky, Ahmed 1015
Shen, Chengjun 2477
Shen, Xiaobing 1914, 2197
Shinoda, Kosei 3928
Shintani, Michihiro 3420
Shousha, Mahmoud....................................279, 698
Shuqin, Wang 1815
Siala, Sami 125
Siemaszko, Daniel 3910
Siemieniec, Ralf 1445
Sievers, Markus 3855
Singh, Rupam 135
Singh, Shashank Shekhawat..................................279
Singh, Sukhjit 2084
Skala, Aleksander...3804
Skibin, Stanislav..3301
Soeiro, Thiago Batista 1319, 3729
Solomentsev, Michael 1722
Solovyov, Vyacheslav 2860
Soltau, N. 1785
Soltau, Nils 777
Sönmez, Ertugrul 593, 2334
Soundararajan, Ajeeth Phrassanna 3729
Soupremanien, Ulrich 922
Spieler, Matthias 1692
Sprunck, Sebastian 1611
Sreekanth, T 730
Stadler, Alexander...525
Stadlober, Barbara 279
Staiger, Jochen 2219
Stala, Robert..3804
Stalleicken, Frederik 2607
Stallmann, Frederik 641
Stärz, Ronald...223, 268
Stathis, Spyridon 1402
Staubach, Christian 1137
Steckler, P. B...2564

Stefanski, L.	3411
Steffen, Jonas	1515
Steinhart, Heinrich	2219
Štengl, Josef	1114
Stevic, Marija	2985
Stewart, Joshua	2806
Steyn, Kyle	3112
Stille, Karl Stephan	2276
Stock, Alexander	1
Stöckl, Thomas	952
Stone, David A.	3440
Strunk, Robin	1350
Stul, Koen	1300
Stutz, Christian	493
Suberski, Martin	2885, 3491
Sujeeth, Arjun	2628
Sullivan, Charles R.	2470
Svensson, Jan R.	1187
Tabrizi, Gholamreza	1611
Takamori, Taro	1128
Takayama, Hajime	108
Takeshita, Takaharu	213, 351, 739
Talla, Jakub	1114
Tang, Chengjun	2813
Tang, Zhongting	1933
Tashakor, Nima	1025, 1064, 1197, 3636, 3665
Tatakis, Emmanuel C.	1551
Tegtmeier, Bernd	1674
Teske, Peter	1466
Thiringer, Torbjörn	2344, 2813, 3920
Thoma, Jürgen	2554
Thönelt, Nick	1713
Thönnessen, André	3676
Tian, Fanghao	2581
Tillmann, Philipp	3740
Tiwari, Arvind Kumar	289
Tiwari, Arvind	187
To, Pham Ha Trieu	59, 70, 4003
Tornello, Luigi Danilo	1569
Torres, C.	3382, 3401
Torrico, Grover	361, 1815
Tournez, Florian	2175
Tran, Dai Duong	2293
Tran, Manh Tuan	2101, 2293
Tresca, Giulia	3975
Trescases, Olivier	3796
Tricoli, Pietro	468
Trochimiuk, Przemyslaw	1486
Tschepp, Andreas	279
Turrisi, Gaetano	1569
Tzanakis, Athanasios	3920
Uicich, Simon	3862

Ulbing, Alexander	3855
Ulmer, Sabrina	593, 2334
Ulrich, Burkhard	459
Umetani, Kazuhiro	2164
Unruh, Peter	1620
Unruh, Roland	3686
Urkizu, June	325
Vaccaro, Luis	2919
Vaessen, Peter	3729
Vagg, Christopher	3696
Vagnon, Eric	2515
Vahid, Sina	719
Vala, Sama Salehi	344
Valderrama, Carlos	504
Valenzuela, Rodrigo Alonso Alvarez	814
Van Cappellen, Leander	2795
Van Mierlo, Joeri	2101
Van Oosterwyck, Nick	3344
Van Tuan, Mai	351
Vandenbussche, Thomas	1300
Vanfretti, Luigi	3928
Vanwalleghem, Bart	3344
Vasiladiotis, Michail	1923
Vatamanu, Lucian	1046
Vázquez, Aitor	1083
Vázquez, Francisco	1765, 3336
Velasco-Quesada, Guillermo	1056
Velazco, Diego	251
Vellvehi, Miquel	2715
Venkataramanan, Giri	3480
Venugopal, Ravinder	2985
Verdier, Jacques	169
Vermeerch, Pierre	582
Veroni, Alessandro	3206
Vershinin, K.	2564
Viana, Caniggia	1995, 2084
Viarouge, I.	1955
Viarouge, P.	1955
Vidal-Albalate, Ricardo	2189
Videau, Nicolas	944
Videt, Arnaud	3822
Villar, Irma	3529, 3938
Vitorino, Montiê Alves	2689, 3059
Vogelsberger, M.	1834
Volzer, Benjamin	2554
Von Hoegen, Anne	3740
Wada, Keiji	1128
Wagner, Valentin	514
Wakelin, Bruce	3309
Wallart, Francois	251
Wallscheid, Oliver	2276, 2432
Waltereit, Patrick	242

Wang, Chu	3722	Yadav, Sachin	3607
Wang, Jun	2136, 3120	Yamaguchi, Masamichi	2127
Wang, Kangan	3014	Yamashita, Shota	213
Wang, Rui	673, 1641	Yamauchi, Kohei	2119
Wang, Xiaoya	3722	Yang, Huoming	1466
Wang, Xin	315	Yang, Jiajun	3014
Wang, Yanbo	2011	Yang, Juefei	2477
Wang, Yangang	2301	Yang, Yinghui	3993
Waradzyn, Zbigniew	3804	Yang, Yongheng	2257
Watanabe, Hiroki	1104	Yaqoob, M.	1815
Wattenberg, Martin	873	Yasuda, Takumi	902
Weicker, Martin	2316	Yeganeh, Mohammad Sadegh Orfi	2496, 2939
Weires, Jonas	604	Yu, Guangyao	1319
Weiser, Mathias C. J.	3565	Yu, Xiao	562, 1309, 2383
Weiss, Xavier	1006	Yu, Xiaodan	3225
Wenzel, Johannes C.	2066	Yuan, Xibo	2136, 3120
Werlig, Christian	3966	Zacharias, Peter	411, 562, 1309, 1328, 1336, 1343,
Weyh, Thomas	767		2383, 2670, 3871
Wicht, Bernhard	2912	Zacher, Benjamin H.	2058
Wieczorek, Nick	3775	Zampardi, Giorgia	833
Wiemer, Adrian	2544	Zanchetta, Pericle	3975
Wiesemann, Julius	1865	Zatocil, Heiko	160
Wiesner, E.	1785	Zdanowski, Mariusz	3938
Wiesner, Eugen	777	Zhang, Bo	1733
Wijnands, Korneel	673, 798, 2788	Zhang, Shimin	709
Wilkowski, Matt	4008	Zhang, Yaqian	2182
Willer, Felix	3838	Zhang, Zhe	1561
Willich, Viktor	3555	Zhang, Zhuoqi	1776
Wohlrath, Fritz	525	Zhang, Ziqian	39
Wolbank, T.	1834	Zhao, Hongbo	1641, 3309
Wolf, Mihaela	2596, 2644, 3775	Zheng, Zhixue	315
Wolfstädter, Simon	4017	Zhetessov, Aidar	3480
Wölk, Alexander	279	Zhu, Zi-Qiang	2958
Wouters, Hans	2197	Ziani, Adel	944
Woywode, Oliver	758	Ziegler, Philipp	971, 1237, 1277, 3536
Wu, Weimin	1985	Zilic, Rufad	1336
Wu, Xiangqiang	1933	Zocher, Markus	2366
Wu, Yuxuan	2860	Zolfi, Pouya	719
Wunsch, Bernhard	3301	Zou, Zhixiang	3014
Würfl, Joachim	2596, 2644, 3775	Zsurzsan, Tiberiu Gabriel	1561
Würsig, Andreas	3966		
Xia, Peizhou	3470		
Xiao, Qian	3215, 3225		
Xiao, Xiong	1966		
Xie, Jun	2885, 3491		
Xie, Lihong	2136		
Xu, Huihui	709		
Xu, James	3796		
Xu, Qianwen	883, 1006		
Xu, Wei	2136		
Xu, Zhongqing	912		
Xu, Zixiao	2182		

IEEE
445 Hoes Lane
Piscataway, NJ 08854-4141

ISBN 978-1-6654-8700-9